T0155686

# ENCYCLOPEDIA OF MATHEMATICS AND ITS APPLICATIONS

ENCYCLOPEDIA OF MATHEMATICS AND ITS APPLICATIONS

# Quantum Field Theory for Mathematicians

### R. TICCIATI

*Maharishi University of Management*

**CAMBRIDGE**
UNIVERSITY PRESS

CAMBRIDGE UNIVERSITY PRESS
Cambridge, New York, Melbourne, Madrid, Cape Town, Singapore, São Paulo

Cambridge University Press
The Edinburgh Building, Cambridge CB2 8RU, UK

Published in the United States of America by Cambridge University Press, New York

www.cambridge.org
Information on this title: www.cambridge.org/9780521632652

First published 1999
This digitally printed version (with corrections) 2008

*A catalogue record for this publication is available from the British Library*

*Library of Congress Cataloguing in Publication data*
Ticciati, Robin.
Quantum field theory for mathematicians / Robin Ticciati.
p.    cm. – (Encyclopedia of mathematics and its applications; v. 72)
Includes bibliographical references.
ISBN 0 521 63265 X (hc.)
1. Quantum field theory.   I. Title.   II. Series.
QC174.45.T53   1999
530.14′3–dc21   98-39473 CIP

ISBN 978-0-521-63265-2 hardback
ISBN 978-0-521-06025-7 paperback

# Contents

# Preface

After completing my dissertation in differential geometry, I returned to Maharishi University of Management to join the faculty there. The greatest need for my services was in the physics department, and the chairman, the well-known John Hagelin, pointed the finger of authority and said 'quantum field theory!' The class was to start in a few weeks. I laughed, but John was serious.

Fortunately, I had audited Sidney Coleman's outstanding Harvard lectures and had taken very good notes. Equally fortunate, I had Robert Brandenburger's official solutions to all the homework sets. I rolled up my sleeves and waded in.

As we battled through the material, the beautiful architecture of Coleman's course became apparent. It introduced the primary concepts – canonical quantization, renormalization, spin, functional integral quantization – one at a time and made each one practical before advancing to the next abstraction. It started with simple models and provided motivation for each elaboration.

The students, however, pinned me to the board with questions about every step in the logic. Could I produce some mathematics to fill the gap? Was there a physical principle which would justify the proposed step? The standard references failed to meet the need, and for the most part I was stumped. It was a couple of years later, when the next group of graduate students was ripening, that I found time to think out some answers. The result was a draft of the first nine chapters of this book.

Student feedback was most helpful for refining this draft, and as I began to take on the teaching of more advanced material and continued to work with successive classes, the notes developed into a pretty good book.

As this history indicates, this book is unique. The common approach in other texts is generally either that of a model builder who aims at efficient presentation of what works, or that of an applied mathematician who is trying to find a foundation for the techniques of the practicing field theorist. The model builder tends to treat technical mountains as molehills and conserves energy by omitting the details of every derivation. The applied mathematician gets absorbed in the technical mountains and does not get close to presenting anything practical. My work is half way between these two. If I may be permitted a Fullerian sentence, my text aims to provide the usual practical knowledge of QFT in comfortable stages naturally structured by the systematic introduction of fundamental principles, pleasantly enriched with overviews, summaries, and delightful mathematical details.

It has been a pleasure to work with Coleman's notes, with the many students who took my classes, and with Blue Sky Research's implementation of TeX. I would particularly like to thank Kurt Kleinschnitz for excellent suggestions, Sunil Rawal for help with the closing chapters, Geoff Golner for explaining Wilson's exact renormalization group, John Hagelin for pushing me off the deep end, and Maharishi University of Management for the financial support that made this work possible.

This book is a great satisfaction to me, and I would like to thank Cambridge University Press for making it possible to share this satisfaction with others. I hope it will be a useful reference for physics students with questions in this area and an intelligible introduction for mathematicians who are trying to find out what the physicists are up to.

Robin Ticciati
Maharishi University of Management

# Introduction

This text provides a mathematical account of the principles and applied techniques of quantum field theory. It starts with the need to combine special relativity and quantum mechanics, and culminates in a basic understanding of the Standard Model of electroweak and strong interactions. The exposition is enriched with as much mathematical detail as is useful for developing a practical knowledge of quantum field theory. All the details can be brought out in a three semester course, and the main ideas can comfortably occupy two semesters. The prerequisites for this presentation are (1) familiarity with the Hilbert-space formalism of quantum mechanics, (2) assimilation of the basic principles of special relativity, and (3) a goodly measure of mathematical maturity.

The book is divided into five parts. These cover (1) canonical quantization of scalar fields, (2) Weyl, Dirac, and vector fields, (3) functional integral quantization, (4) the Standard Model of the electroweak and strong interactions, and (5) renormalization. Each of the five parts introduces one primary new principle and, having developed that principle to a goodly degree of practicality, naturally motivates its own extension in the next part.

Part 1, Chapters 1–5, presents the principles that transform quantum mechanics and special relativity into canonical perturbative quantum field theory. Its goal is to arrive at cross sections with minimal technical machinery. Part 2, Chapters 6–9, provides fields for realistic models like QED. It covers representation theory of the Lorentz group and extends the principles of Part 1 to Weyl spinors, Dirac spinors, and massive vector fields. At the end of Part 2, three weaknesses of perturbative canonical quantization are apparent, namely, its application to unstable particles, derivative interactions, and gauge theories. This motivates Part 3, Chapters 10–13, in which Haag–Ruelle scattering theory, renormalization, and functional integral quantization are developed. Using this new formalism, Part 4, Chapters 14–17, presents the structure of the Standard Model, introduces hadrons, and provides tree-level applications of the Standard Model. Finally, having taken tree-level theory about as far as is useful, Part 5 goes more deeply into regularization, renormalization, renormalizability, and the renormalization group.

The principles, examples, and techniques presented in this text are, for the most part, quite standard. However, the detailed treatment of the representation theory of the Lorentz group and $SU(3)$ in Chapters 6 and 14 respectively is unusual and helps to make this text self-sufficient. The use in Chapter 7 of Weyl spinor fields as an aid to deriving Feynman rules for Fermi fields is also unusual. The two-component spinor notation built up in that chapter is more commonly used in superfield theory.

Internal references are written without parentheses in the form '12.3.6' rather than '(12.3.6)'. To save the reader from looking up references unnecessarily, references are qualified by descriptive phrases rather than by an unhelpful noun. For example, we write 'the winding-number formula 13.7.11' rather then 'Theorem 13.7.11'.

# Chapter 1

# Relativistic Quantum Mechanics

Uniting the operator and state-space formalism of quantum mechanics with special relativity through a unitary representation of the Poincaré group.

## Introduction

Chapters 1 to 5 constitute the first part of this book. They develop the theory of the scalar field from its roots in special relativity and quantum mechanics to its fruits in cross sections and decay rates. The technique of quantization developed here will be applied in the second part – Chapters 6 to 9 – to spinor and vector fields.

Taking quantum mechanics, with its formalism of state space, Hamiltonian, and observables, together with relativity, with its emphasis on invariance under Lorentz transformations, as the two major pillars or principles in our understanding of particle physics, the purpose of this chapter is to introduce a framework in which both principles coexist.

Section 1.1 clarifies the concept of a state space, putting physical states, position eigenstates, and momentum eigenstates into proper relationship. Section 1.2 takes the first step towards a relativistic quantum theory by promoting the energy and momentum observables into a Lorentz vector. Section 1.3 uses this vector to construct a unitary representation of translations on state space, Section 1.4 uses an independent construction to build a unitary representation of the Lorentz group, and Section 1.5 shows that these two representations determine a unitary representation of the Poincaré group. At this point, the principle of special relativity is effectively built into the formalism of quantum mechanics. Only the position operator has yet to be introduced. Section 1.6 shows that, if we adjoin eigenstates of the position operator to our state space, then there is a probability for signals to travel faster than light.

Although this probability is too small to be ruled out by experiments to date, we decide to uphold relativistic causality and drop the position operator from the theory. To model localization of events, we localize measurement using fields of observables.

For simplicity of notation, we shall choose units such that the speed of light $c$ and the reduced Planck constant $\hbar$ are both unity:

$$c = 1 \quad \text{and} \quad \hbar = 1. \tag{1.0.1}$$

This choice of units reduces the three units time, length, and mass, to a single unit which we will generally call mass:

$$T = L = M^{-1}. \tag{1.0.2}$$

When it is necessary to compare with experiment, it will be easy to insert the factors of $c$ and $\hbar$ which convert the units of a prediction into the units of the data. Indeed, if our result has units $M^a$ and the physical units for the result should be $M^b L^c T^d$, then consistency requires $a = b - c - d$ and the conversion is accomplished by the equation

$$(\text{Answer in units } M^b L^c T^d) = (\text{Answer in units } M^a) \times \hbar^{b-a} \times c^{a-b-d}. \quad (1.0.3)$$

Throughout this text, we shall take the metric on Minkowski space to be

$$g^{\mu\nu} = \begin{pmatrix} 1 & 0 & 0 & 0 \\ 0 & -1 & 0 & 0 \\ 0 & 0 & -1 & 0 \\ 0 & 0 & 0 & -1 \end{pmatrix}, \quad (1.0.4)$$

and we use the dummy suffix conventions

$$a^\mu = g^{\mu\nu} a_\nu,$$

$$a \cdot b = a^\mu b_\mu = \sum_{\mu=0}^{3} a^\mu b_\mu = g^{\mu\nu} a_\mu b_\nu, \quad (1.0.5)$$

$$a^{2k} = (a \cdot a)^k.$$

# 1.1 One-Particle State Space – Mathematics

We want our one-particle state space to include states with definite energy and momentum. In fact we want more. We want a basis for state space in which the energy and momentum operators are simultaneously diagonal. To fulfill this desire, an excursion into mathematics is necessary. However, a little explanation will be sufficient for practical purposes. In this section, we present the little explanation and indicate the relevant mathematics in two remarks.

The function space used to model quantum-mechanical states is a Hilbert space $H$ of square-integrable functions on physical space. The functions in $H$ can be written either as functions of position $\bar{x}$ or as functions of momentum $\bar{p}$. The position-space representation is related to the momentum-space representation by the Fourier transform $\mathcal{F}$:

$$\mathcal{F}(f)(\bar{p}) \stackrel{\text{def}}{=} \int dx \, e^{i\bar{x} \cdot \bar{p}} f(\bar{x}). \quad (1.1.1)$$

Despite the change of independent variables from $\bar{x}$ to $\bar{p}$, the function $\mathcal{F}(f)$ is in $H$, and $\mathcal{F}$ maps $H$ to itself.

**Remark 1.1.2.** Actually, if $f$ is square-integrable, then the function $e^{i\bar{x} \cdot \bar{p}} f(\bar{x})$ is certainly square-integrable but it may not be integrable. To define the Fourier transform on $H$, it is necessary to define it first on some good subspace of $H$ (such as the space of smooth functions with compact support) and then to use the fact that the Fourier transform is an isometry (up to a scale factor) on this subspace to extend the Fourier transform to an isometry of $H$ itself.

The position operators $X_r$ and momentum operators $P_r$ are represented respectively by multiplication by coordinate functions $x_r$ and the partial derivative operators $-i\partial_r$. The Fourier transform converts multiplication by $x_r$ on functions of $\bar{x}$ into the differential operator $-i\partial_r$ on functions of $\bar{p}$:

$$\mathcal{F}(x_r f)(\bar{p}) = -i\partial_r \mathcal{F}(f)(\bar{p}). \quad (1.1.3)$$

The eigenstates of the position operator are the delta functions:

$$(X_r \delta_{\bar{q}})(\bar{x}) = x_r \delta(\bar{x} - \bar{q}) = q_r \delta(\bar{x} - \bar{q}) = (q_r \delta_{\bar{q}})(\bar{x}), \qquad (1.1.4)$$

and the eigenstates for the momentum operator $P_r$ are the exponential functions $e_{\bar{p}}(\bar{x}) = \exp(i\bar{p}\cdot\bar{x})$:

$$(P_r e_{\bar{p}})(\bar{x}) = -i\partial_r \exp(i\bar{p}\cdot\bar{x}) = p_r \exp(i\bar{p}\cdot\bar{x}) = (p_r e_{\bar{p}})(\bar{x}). \qquad (1.1.5)$$

A technical problem now arises: neither $X_r$ nor $P_r$ act on the whole of $H$, and $H$ does not contain the eigenstates of either operator. The first part of the problem is solved by defining a subspace $S$ of physical states in $H$ which will be mapped into itself by $X_r$ and $P_r$. The second part is solved by defining the kets (ideal states) as elements of the space $S^*$ of continuous anti-linear functionals on $S$. Indeed, since $P_r$ acts on the functions in $S$, these functions must be infinitely differentiable, and so $S^*$ will contain the $\delta$ functions and all their derivatives. Similarly, taking the Fourier transform of this point, since $X_r$ acts on $S$, $S^*$ will contain the exponential functions $\exp(i\bar{p}\cdot\bar{x})$.

Thus, instead of one Hilbert space $H$, quantum theory uses a triple of normed vector spaces:

$$S \subset H \subset S^*. \qquad (1.1.6)$$

The physical states live in $S$ and the operator eigenstates live in $S^*$. With appropriate conditions on $S$, such a triple is called a *rigged Hilbert space*.

**Remark** 1.1.7. In fact, the triple $S \subset H \subset S^*$ is a rigged Hilbert space if $S$ is a nuclear subspace of $H$. One condition for this is that (a) there exist a countable family $\| \ \|_k$ of norms on $S$ with respect to which convergence is defined by

$$f_n \to f \quad \Longleftrightarrow \quad \|f_n - f\|_k \to 0 \quad \text{for all } k \geq 0, \qquad (1.1.8)$$

(b) $S$ is complete with respect to this notion of convergence, and (c) there exists a Hilbert–Schmidt operator on $S$ with a continuous inverse.

To link our example above with this structure, let $H_k$ be the Hilbert space of functions whose $k$th derivatives are square-integrable, let $S_k$ be the intersection of $H_k$ with its Fourier transform $\mathcal{F}(H_k)$, and let $\| \ \|_k$ $(k \geq 0)$ be the norm which makes $S_k$ a Hilbert space. Then we can take $S$ to be the intersection of the $S_k$. If $N$ is the number operator of the quantum oscillator, then $1 + N$ is a Hilbert–Schmidt operator on $S$ which has a continuous inverse.

This area of mathematics is well developed. However, its impact on field theory has been small. So we shall make no further remark on these technicalities.

In a rigged Hilbert space, the physical states have eigenstate expansions. Translating into the notation of quantum mechanics, let $|f\rangle$ be the state represented by the function $f$, and let $|\bar{x}\rangle$ be the position eigenstate represented by the distribution $\delta_{\bar{x}}$. We assume that the relationship between functions and kets is such that

$$f(\bar{x}) = \langle \bar{x}|f\rangle. \qquad (1.1.9)$$

Then the expansion of the state $|f\rangle$ in terms of the position eigenstates $|\bar{x}\rangle$ should have the form

$$|f\rangle = \int d^3\bar{x} \, |\bar{x}\rangle\langle \bar{x}|f\rangle = \int d^3\bar{x} \, f(\bar{x})|\bar{x}\rangle. \qquad (1.1.10)$$

The conditions on $S$ in a rigged Hilbert space ensure that $f(\bar{x})|\bar{x}\rangle$ is integrable for all $f \in S$.

We conclude from this section that we can achieve our desire of finding a basis in which the energy-momentum operators are diagonal, but that we should be aware that the space of physical states, the space of states that can be expanded in terms of this basis, is smaller than a Hilbert space, while the space of states that contains the energy-momentum eigenstates is larger than a Hilbert space. In particular, the Hilbert space norm cannot be extended to the eigenstates. With this outline of the mathematics of a one-particle state space, we have enough insight into the technicalities to successfully manage operator eigenstates.

## 1.2 One-Particle State Space – Physics

A model for relativistic quantum mechanics must include a state space and a Hamiltonian. Being a relativistic theory, the Hamiltonian $H$ must be the time component of a four-vector energy-momentum operator $P$. Choose a basis of momentum eigenvectors for the state space:

$$\bar{P}|\bar{k}\rangle = \bar{k}|\bar{k}\rangle. \tag{1.2.1}$$

Assume that these states are normalized so that:

$$\langle\bar{k}|\bar{k}'\rangle = \delta^{(3)}(\bar{k} - \bar{k}'). \tag{1.2.2}$$

Notice that this normalization leaves the length of a ket undefined, infinite. It is a normalization suitable for integration over momentum (refer to 1.1.10 for the analogous integral in position-space). This shows that these eigenstates are not physical states, that they are not even in a Hilbert space, but that they could be distributions like the delta function, dual states to the physical states. This normalization condition leads to a resolution of the identity

$$\mathbf{1} = \int d^3\bar{k} \, |\bar{k}\rangle\langle\bar{k}|. \tag{1.2.3}$$

Since energy-momentum is a four-vector, we must make $H^2 - |\bar{P}|^2$ constant on orbits of the Poincaré group. Furthermore, given two states $|\bar{k}\rangle$ and $|\bar{k}'\rangle$ of a single particle, there exists a Lorentz transformation which (up to scale) will convert one state into the other. Hence we assume the existence of a constant $\mu$, the particle mass, which satisfies the operator equation

$$H^2 - |\bar{P}|^2 = \mu^2. \tag{1.2.4}$$

This equation implies that the Hamiltonian $H$ is diagonal on the basis of momentum eigenstates:

$$H|\bar{k}\rangle = \left(|\bar{k}|^2 + \mu^2\right)^{\frac{1}{2}}|\bar{k}\rangle. \tag{1.2.5}$$

The eigenvalues of $H$ crop up sufficiently often that there is a definition

$$\omega(\bar{k}) \stackrel{\text{def}}{=} \left(|\bar{k}|^2 + \mu^2\right)^{\frac{1}{2}}. \tag{1.2.6}$$

**Remark** 1.2.7. This setup is the limit of the state space for a cube of side $L$ under periodic boundary conditions. In such a cube the spectrum is discrete and the normalization is given by the Kronecker delta:

$$\bar{k} = \frac{2\pi}{L}(n_x, n_y, n_z) \quad \text{and} \quad \langle \bar{k} | \bar{k}' \rangle = \delta_{\bar{k}, \bar{k}'}. \tag{1.2.8}$$

In Chapter 5, we use this discrete approximation in the derivation of the differential transition probability per unit time for particle scattering.

## 1.3 The Action of Translations on States

The translations, rotations, and boosts of the Poincaré group must act on the space of states. A Poincaré group element $g$ acts as a unitary operator $U(g)$ on the state space. The action must satisfy a multiplication condition:

$$U(gh) = U(g)U(h) \qquad \text{for all } g, h \text{ in the Poincaré group.} \tag{1.3.1}$$

Translation of space-time by a four-vector $a^\mu$ is defined by:

$$\Delta_a(x) = x + a. \tag{1.3.2}$$

Translation of a state $\psi$, however, should mean moving the graph by $a$. Hence $\Delta_a \psi(x) = \psi(x - a)$. The unitary representation $U(\Delta_a)$ of $\Delta_a$ must therefore be defined by

$$U(\Delta_a)|\psi\rangle \stackrel{\text{def}}{=} |\Delta_a \psi\rangle. \tag{1.3.3}$$

Now we need to find an expression for $U(\Delta_a)$ in terms of the energy-momentum four-vector $P$.

Evolution in time is translation of the observer forwards in time, or, equivalently, of the state backwards in time:

$$\exp(-iHt)|\psi(x)\rangle = |\psi(x_0 + t, \bar{x})\rangle. \tag{1.3.4}$$

Writing $\tau$ for the four-vector $(-t, \bar{0})$, we see that this equation can be written

$$\exp(i\tau \cdot P)|\psi\rangle = |\Delta_\tau \psi\rangle. \tag{1.3.5}$$

Lorentz invariance implies that this equation is true whenever $\tau$ is time-like, and the additivity of translations then shows that it is true for all four-vectors $\tau$. From the definition of $U(\Delta_a)$ above, we therefore conclude that

$$U(\Delta_a) = \exp(ia \cdot P). \tag{1.3.6}$$

The unitary representation 1.3.6, though derived in the position-space formulation of quantum mechanics, will work equally well in the momentum-space formulation. We deduce that the unitary representation of translations on momentum eigenstates is given by

$$U(\Delta_a)|\bar{k}\rangle = e^{ia \cdot P}|\bar{k}\rangle = e^{ia \cdot k}|\bar{k}\rangle \tag{1.3.7}$$

where $k_0 = \omega(\bar{k})$.

**Remark 1.3.8.** If we drop the distinction between $|\psi\rangle$ and the function $\psi$, then the definition 1.3.3 of the unitary representation of translations implies that $\exp(ia\cdot P)\psi = \Delta_a\psi$. If we now replace the momentum operator $P_\mu$ by the differential operator $i\partial/\partial x^\mu$, then the resulting equation is simply Taylor's theorem: $e^{-a\cdot\partial}\psi(x) = \psi(x - a)$.

# 1.4 The Action of the Lorentz Group on States

The space of particle states is three-dimensional; the energy $k_0$ of a particle with momentum $\bar{k}$ is constrained by $k_0 \geq 0$ and $k^2 = k\cdot k = \mu^2$. Therefore the possible energy-momentum vectors lie on a hyperbolic sheet in $k$-space, the mass hyperboloid. In the interests of Lorentz-invariant computation, we need an integration measure on this hyperboloid.

Let $\theta(t)$ be the Heaviside function

$$\theta(t) = \begin{cases} 0, & \text{if } t < 0; \\ 1, & \text{if } t > 0. \end{cases} \tag{1.4.1}$$

Define an integration measure $d\lambda(k)$ on the positive hyperboloid as follows:

$$d\lambda(k) \stackrel{\text{def}}{=} d^4k\, \delta(k^2 - \mu^2)\theta(k_0). \tag{1.4.2}$$

The Lebesgue measure $d^4k$ is Lorentz invariant because Lorentz transformations have unit determinant. Since the $\delta$ and $\theta$ factors are obviously Lorentz invariant, $d\lambda(k)$ is also Lorentz invariant.

Recalling the delta function equality

$$\delta\big(f(k)\big) = \sum_{\{k\,|\,f(k)=0\}} \frac{1}{|f'(k)|}\delta(k), \tag{1.4.3}$$

we see that

$$\begin{aligned}
\delta(k^2 - \mu^2)\theta(k_0) &= \frac{1}{2\omega(\bar{k})}\Big(\delta\big(k_0 - \omega(\bar{k})\big)\theta(k_0) + \delta\big(k_0 + \omega(\bar{k})\big)\theta(k_0)\Big) \\
&= \frac{1}{2\omega(\bar{k})}\delta\big(k_0 - \omega(\bar{k})\big)\theta(k_0).
\end{aligned} \tag{1.4.4}$$

Hence we can eliminate $k_0$ from any integral with respect to $d\lambda(k)$ as follows:

$$\begin{aligned}
\int d\lambda(k)\, f(k) &= \int d^3\bar{k}\, dk_0\, \frac{1}{2\omega(\bar{k})}\delta\big(k_0 - \omega(\bar{k})\big)\theta(k_0)f(k) \\
&= \int \frac{d^3\bar{k}}{2\omega(\bar{k})} f\big(\omega(\bar{k}), \bar{k}\big).
\end{aligned} \tag{1.4.5}$$

The integral and the arbitrary function $f$ are commonly eliminated from this result, leaving an equality of measures:

$$d\lambda(k) = \frac{d^3\bar{k}}{2\omega(\bar{k})} \quad \text{with} \quad k_0 = \omega(\bar{k}). \tag{1.4.6}$$

**Homework 1.4.7.** Verify that the measure $d^3\bar{k}/\omega(\bar{k})$ is invariant under a Lorentz transformation $\Lambda$ by direct computation of the Jacobian for the change of variables $l^r = \Lambda^r{}_\mu k^\mu$. (Factorize $\Lambda$ intelligently.)

If we define Lorentz-normalized kets $|k\rangle$ by

$$|k\rangle = \left(2\omega(\bar{k})\right)^{\frac{1}{2}}(2\pi)^{\frac{3}{2}}|\bar{k}\rangle \qquad (1.4.8)$$

with $k_0 = \omega(\bar{k})$, then the new normalization equation is

$$\langle k|k'\rangle = 2\omega(\bar{k})(2\pi)^3\delta^{(3)}(\bar{k}-\bar{k}'), \qquad (1.4.9)$$

and the resolution of the identity is based on the Lorentz invariant measure:

$$1 = \int \frac{d^3\bar{k}}{(2\pi)^3\,2\omega(\bar{k})}\,|k\rangle\langle k|. \qquad (1.4.10)$$

With these Lorentz-normalized states, we can define the unitary representation of the Lorentz group simply:

**Theorem 1.4.11.** *If we define $U(\Lambda)$ by $U(\Lambda)|k\rangle = |\Lambda k\rangle$, then $U$ is a unitary representation of the Lorentz group.*

**Proof.** The multiplicative property $U(\Lambda\Lambda') = U(\Lambda)U(\Lambda')$ is immediate from the definition. Now we must show that $U(\Lambda)$ is unitary; for this, we use the resolution of the identity:

$$U(\Lambda)U(\Lambda)^\dagger = \int \frac{d^3\bar{k}}{(2\pi)^3\,2\omega(\bar{k})}\,U(\Lambda)|k\rangle\langle k|U(\Lambda)^\dagger$$

$$= \int \frac{d^3\bar{k}}{(2\pi)^3\,2\omega(\bar{k})}\,|\Lambda k\rangle\langle\Lambda k| \qquad (1.4.12)$$

$$= 1,$$

since the measure is Lorentz invariant. $\qquad\qquad\square$

It is surprising that the $U(\Lambda)$ defined in Theorem 1.4.11 is a unitary operator since $|k\rangle$ and $|\Lambda k\rangle$ appear to have different lengths when $\Lambda$ is a boost. However, $\delta^{(3)}(0)$ is undefined, so the normalization of the kets does not actually determine a length. We regard the uniformly unlocalized state described by $|k\rangle$ as unphysical. The physical states have the form

$$|\psi\rangle \stackrel{\text{def}}{=} \int \frac{d^3\bar{k}}{(2\pi)^3\,2\omega(\bar{k})}\,\psi(\bar{k})|k\rangle, \qquad (1.4.13)$$

where the measure is so structured that $\langle k|\psi\rangle = \psi(\bar{k})$. It is easy to check that the length of $|\psi\rangle$ is well defined whenever $\psi(\bar{k})$ is square integrable, and that our definition of $U(\Lambda)$ makes $U(\Lambda)$ unitary on the space of physical states.

## 1.5 Representing the Poincaré Group

It remains to check that the map of the Poincaré group determined by our definition of $U$ on translations (equation 1.3.7) and on the Lorentz group (Theorem 1.4.11) is a representation. The representation condition, $U(gh) = U(g)U(h)$, must hold for all $g$ and $h$ in the Poincaré group. This condition is clearly satisfied when $g$ and $h$ are either both translations or both Lorentz group elements. It remains to check the case where one is a translation and the other an element of the Lorentz group.

Any element in the Poincaré group can be factored uniquely as a product of a translation and a Lorentz group element:

$$g = \Delta_a \Lambda. \qquad (1.5.1)$$

Multiplication in the Poincaré group depends on multiplication in the Lorentz group and addition of translations through an interchange in the order of the two factors:

$$\begin{aligned} gh &= \Delta_a \Lambda \Delta_b M \\ &= \Delta_a (\Lambda \Delta_b \Lambda^{-1}) \Lambda M \\ &= \Delta_a \Delta_{\Lambda b} \Lambda M, \end{aligned} \qquad (1.5.2)$$

where we have used the identity

$$\Lambda \Delta_b \Lambda^{-1} = \Delta_{\Lambda b}, \qquad (1.5.3)$$

a relation easily verified by application to a vector $x$.

Our definition of $U$ so far covers translations and Lorentz group elements only; when we extend to the Poincaré group, we do so through the definition

$$U(\Delta_a \Lambda) \stackrel{\text{def}}{=} U(\Delta_a)U(\Lambda). \qquad (1.5.4)$$

We can now see that $U$ is a representation of the Poincaré group if and only if $U$ preserves the action 1.5.3 of Lorentz group elements on translations:

$$\begin{aligned} & U(\Delta_a \Lambda)U(\Delta_b M) = U(\Delta_a \Delta_{\Lambda b} \Lambda M) \\ \Longleftrightarrow \quad & U(\Delta_a)U(\Lambda)U(\Delta_b)U(M) = U(\Delta_a)U(\Delta_{\Lambda b})U(\Lambda)U(M) \\ \Longleftrightarrow \quad & U(\Lambda)U(\Delta_b) = U(\Delta_{\Lambda b})U(\Lambda) \\ \Longleftrightarrow \quad & U(\Lambda)U(\Delta_b)U^{\dagger}(\Lambda) = U(\Delta_{\Lambda b}). \end{aligned} \qquad (1.5.5)$$

We verify this final condition by evaluating both sides on $|k\rangle$. From the right-hand side, we have:

$$U(\Delta_{\Lambda b})|k\rangle = \exp(i\Lambda b \cdot k)|k\rangle, \qquad (1.5.6)$$

and from the left:

$$\begin{aligned} U(\Lambda)U(\Delta_b)U^{\dagger}(\Lambda)|k\rangle &= U(\Lambda)U(\Delta_b)|\Lambda^{-1}k\rangle \\ &= U(\Lambda)\exp(ib \cdot \Lambda^{-1}k)|\Lambda^{-1}k\rangle \\ &= \exp(ib \cdot \Lambda^{-1}k)|k\rangle. \end{aligned} \qquad (1.5.7)$$

The equality of the two sides now follows from the Lorentz invariance of the inner product.

We summarize the discussion of $U$ with a theorem:

**Theorem 1.5.8.** *The map $U$ from the Poincaré group to operators on state space defined by*
1. $U(\Delta_a)|k\rangle = e^{ia\cdot k}|k\rangle$;
2. $U(\Lambda)|k\rangle = |\Lambda k\rangle$;
3. $U(\Delta_a \Lambda) = U(\Delta_a)U(\Lambda)$,

*is a unitary representation of the Poincaré group.*

**Homework 1.5.9.** Show that the Heisenberg action of $U(\Lambda)$ on the energy-momentum vector $P$ simply multiplies $P$ by $\Lambda$:

$$U(\Lambda)^\dagger P^\mu U(\Lambda) = \Lambda^\mu{}_\nu P^\nu.$$

The unitary representation $U$ successfully combines the principle of relativity as represented by the Poincaré group with the principle of quantum mechanics as represented by the operator and state-space formalism. This combined structure of a one-particle state space provides the foundation for the many-particle state space used in all quantum field theories.

The following section shows that we cannot localize measurement in this space by adding a position operator to the theory without violating relativistic causality. This obstruction to localization provides the motivation for quantum fields.

## 1.6 The Position Operator

The purpose of this section is to show that including a position operator in our relativistic quantum mechanics will lead to inconsistencies. This is simply understood in terms of the uncertainty principle. If, for example, we try to squeeze a pion into a box which has sides that are small compared to its Compton wavelength $\lambda$, then the uncertainty in position satisfies $\Delta x \ll \lambda$, and the uncertainty in momentum satisfies $\Delta p \gg m_\pi$. But this makes the range of energies so large that pair production becomes possible. Hence, from first principles, the position of a one-particle system is not well defined. The following argument shows how Lorentz causality is also violated by measurement of position.

The axioms for operators $\bar{X}$ to be position operators are:

**Axiom 1.** $\bar{X} = \bar{X}^\dagger$.
**Axiom 2.** If $\Delta_{\bar{a}}$ is a space translation, then $U(\Delta_{\bar{a}})^\dagger \bar{X} U(\Delta_{\bar{a}}) = \bar{X} + \bar{a}$.
**Axiom 3.** If $R$ is a space rotation, then $U(R)^\dagger \bar{X} U(R) = R\bar{X}$.

From Axiom 2 and the formula 1.3.6 for $U(\Delta_a)$ we see that:

$$e^{i\bar{a}\cdot\bar{P}}\bar{X}e^{-i\bar{a}\cdot\bar{P}} = \bar{X} + \bar{a}. \tag{1.6.1}$$

The sign in the exponential reflects the relationship between the Lorentz dot product and the Euclidean dot product of 3-vectors. Differentiating this equation with respect to $a^r$ and setting $\bar{a} = \bar{0}$, we find the usual canonical commutation relations

$$[iP_r, X^s] = \delta_r^s. \tag{1.6.2}$$

**Homework 1.6.3.** Show that, although $\bar{P}$ is undefined on $|\bar{x}\rangle$, the axioms for the position operator imply that the exponential $\exp(-i\bar{a}\cdot\bar{P})$ must be defined on these states:

$$e^{-i\bar{a}\cdot\bar{P}}|\bar{x}\rangle = |\bar{x} + \bar{a}\rangle.$$

Show also that, if $\langle\bar{k}|\bar{x} = \bar{0}\rangle = 1$, then $\langle\bar{k}|\bar{a}\rangle = e^{-i\bar{a}\cdot\bar{k}}$.

**Remark 1.6.4.** In this remark, we demonstrate that the position operator is essentially unique. Indeed, suppose that $\bar{X}$ and $\bar{Y}$ both satisfy the position operator axioms. We will show that there exists a unitary operator $U$ such that $\bar{Y} = U^{\dagger}\bar{X}U$.

Assume that $\bar{Y}$ is the position operator with respect to the basis $|\bar{k}\rangle$. The canonical commutation relation 1.6.2 shows that $P_r$ commutes with $X_s - Y_s$. Therefore, assuming that any operator can be expressed as a function of $X_r$ and $P_r$, $Y_r$ must be of the form $X_r + f_r(\bar{P})$. Axiom 3 implies that $f_r(\bar{P})$ is of the form $g(|\bar{P}|^2)P_r$. This vector-valued function of a vector has zero curl and may therefore be written as the gradient of a scalar function $\phi(|\bar{P}|^2)$, where

$$\phi(\xi) \stackrel{\text{def}}{=} \int_0^{\xi} d\eta\, g(\eta). \tag{1.6.5}$$

If we define new kets by using a unitary operator $U$ to change phases,

$$|\bar{k}\rangle_{\text{new}} \stackrel{\text{def}}{=} U|\bar{k}\rangle \stackrel{\text{def}}{=} \exp\left(-i\phi(|\bar{k}|^2)\right)|\bar{k}\rangle, \tag{1.6.6}$$

then, since

$$\langle\psi'|\bar{Y}|\psi\rangle = {}_{\text{new}}\langle\psi'|U\bar{Y}U^{\dagger}|\psi\rangle_{\text{new}}, \tag{1.6.7}$$

the new position operators are $U\bar{Y}U^{\dagger}$. Writing $U^{\dagger} = e^A$, we find that

$$\begin{aligned}
UY_rU^{\dagger} &= U\left(U^{\dagger}Y_r + [Y_r, U^{\dagger}]\right) \\
&= Y_r + U[Y_r, 1 + A + \tfrac{1}{2}A^2 + \cdots] \\
&= Y_r + U(1 + A + \tfrac{1}{2}A^2 + \cdots)[Y_r, A] \\
&= Y_r + \left[Y_r, i\phi(|\bar{P}|^2)\right] \\
&= Y_r - g(|\bar{P}|^2)P_r \\
&= X_r.
\end{aligned} \tag{1.6.8}$$

We conclude that any two sets of position operators $\bar{X}$ and $\bar{Y}$ are related by a change of basis. We also note that, since the change of basis operator $U$ is a function of momenta alone, the new momentum operators $U\bar{P}U^{\dagger}$ are the old ones $\bar{P}$.

This shows that the axioms determine the position operator uniquely up to a choice of phase in the momentum eigenstate basis.

The simplest inconsistency emerges when we consider a state initially localized at the origin and see whether it can be detected outside the forward lightcone of the origin.

Suppose that we have a position operator $\bar{X}$. Let $|\bar{x}\rangle$ be a basis of position eigenstates. Then, from non-relativistic quantum mechanics, we can choose the normalization of these kets so that

$$\langle\bar{x}|\bar{k}\rangle = e^{i\bar{x}\cdot\bar{k}}. \tag{1.6.9}$$

Now consider the evolution $|\psi\rangle$ of a state $|\psi_0\rangle$ initially localized at the origin:

$$\psi_0(\bar{x}) \stackrel{\text{def}}{=} (2\pi)^3\delta^{(3)}(\bar{x}) \quad\Longrightarrow\quad \hat{\psi}_0(\bar{k}) = 1 \quad\Longrightarrow\quad |\psi_0\rangle = \int d^3\bar{k}\,|\bar{k}\rangle, \tag{1.6.10}$$

where $\hat{\psi}_0$ is the Fourier transform of $\psi_0$. The evolution of this state is given by

$$\psi(t, \bar{x}) = \langle \bar{x} | e^{-iHt} | \psi_0 \rangle$$

$$= \int d^3\bar{k} \, \langle \bar{x} | e^{-iHt} | \bar{k} \rangle$$

$$= \int d^3\bar{k} \, \langle \bar{x} | e^{-i\omega(\bar{k})t} | \bar{k} \rangle \qquad (1.6.11)$$

$$= \int d^3\bar{k} \, e^{-i\omega(\bar{k})t} e^{i\bar{x} \cdot \bar{k}}.$$

If the theory is to be relativistic, then a state initially localized at the origin should have zero amplitude outside the lightcone. We therefore proceed to estimate $\psi(t, \bar{x})$ outside the lightcone. Using spherical polars with $k = |\bar{k}|$ and $r = |\bar{x}|$, we find that

$$\psi(t, \bar{x}) = \int_{-1}^{1} d(\cos\theta) \int_0^{2\pi} d\phi \int_0^{\infty} dk \, k^2 e^{-it\sqrt{k^2+\mu^2}} e^{ikr\cos\theta}$$

$$= \frac{2\pi}{ir} \int_0^{\infty} dk \, k e^{-it\sqrt{k^2+\mu^2}} (e^{ikr} - e^{-ikr}) \qquad (1.6.12)$$

$$= \frac{2\pi}{ir} \int_{-\infty}^{\infty} dk \, k e^{-it\sqrt{k^2+\mu^2}} e^{ikr}.$$

When $r > t$, this integral may be evaluated by treating $k$ as a complex variable and deforming the contour of integration from the real axis to the branch cut from $i\mu$ to $i\infty$. Substituting $k = iz$ gives

$$\psi(t, \bar{x}) = \frac{4\pi i}{r} \int_{\mu}^{\infty} dz \, z \sinh\left(t\sqrt{z^2 - \mu^2}\right) e^{-zr}, \qquad (1.6.13)$$

which is clearly non-zero.

**Remark 1.6.14.** The integral that we have been manipulating is actually divergent. This is a consequence of the extreme nature of the initial state $|\psi_0\rangle$. If we had started with a physical state instead of a position eigenstate, then there would be no convergence problem. The moral, then, is to treat divergent integrals which arise in such situations as defining distributions.

The outcome, then, is that a position operator is inconsistent with relativity. This compels us to find another way of modeling localization of events. In field theory, we model localization by making the observables dependent on position in space-time.

# 1.7 Summary

This chapter brings together the two great foundational viewpoints of physics, relativity and quantum mechanics. Quantum mechanics is represented by the state space and operator formalism and relativity by the unitary representation of the Poincaré group on the state space. We have also seen that a quantum-mechanical position operator leads to a violation of relativistic causality. Though this violation may not be large enough to conflict with experiment, its very existence is sufficient to motivate a change in approach to localization.

In fact, observation of position can never be absolutely precise, so there is no need to model the classical concept of localization at a point. Observation of position involves structuring a sensitive environment, and this process can be modeled by a field. In the next chapter, we shall continue to use the state space, the energy-momentum observables, and the unitary representation presented in this chapter, but we shall model localization of events by giving the dynamic observables space-time dependence, that is, promoting them to quantum fields.

# Chapter 2

# Fock Space, the Scalar Field, and Canonical Quantization

Providing complete knowledge of the free scalar field and the space of many-particle states – the foundation for perturbative canonical quantization.

## Introduction

In this chapter, we shall develop a relativistic quantum theory of a single scalar field using the one-particle state space and the unitary representation of the Poincaré group on that space described in the last chapter. There are two spurs that give direction to the construction of a scalar field theory. The first is that at relativistic energies, pair production contributes significantly to scattering processes. This motivates us to build an extended state space (a Fock space) in which states can have any number of particles. The second is that, if we localize measurement by making observables $\phi$ functions of position $x$ in space-time, then we must construct $\phi$ in such a way that this field of operators has Lorentz transformation properties appropriate to its status as a field of observables.

In Section 2.1, we construct the Fock space for a non-interacting assemblage of particles of a single type. Section 2.2 reviews the construction of ladder operators for the simple harmonic oscillator, and Section 2.3 transforms them into the creation and annihilation operators which connect the whole of Fock space to its foundation, the vacuum state. In Section 2.4, we use these operators to construct a free relativistic field and, by investigating this field, motivate the principle of canonical quantization of field theories. Section 2.5 reviews canonical quantization of particle mechanics, and Section 2.6 shows that though this quantization principle applies to the free field well enough, there is a little twist in the application which makes it impractical if not impossible to apply to the interacting field. This provides the springboard for Chapter 4 in which we see how to describe the interacting field in terms of the free field. Section 2.7 is a luxury; it indicates the mathematical character of quantum field theory and, perhaps more importantly, introduces the highly non-trivial structure of the vacuum state of a free scalar field.

## 2.1 Bosonic Fock Space

Relativity becomes significant when the energy $T$ in a process satisfies $T = E - mc^2 \geq mc^2$. The threshold energy for pair production is $T = 2mc^2$. Hence particle number need not be conserved in a process in which relativity is important.

The effects of relativistic kinematics as a perturbation of Newtonian physics are of order $T/mc^2$. But, from quantum mechanics,

$$\delta E = \langle \psi | V | \psi \rangle + \sum_n \frac{|\langle \psi | V | n \rangle|^2}{E_0 - E_n} + \cdots, \qquad (2.1.1)$$

so the effect of pair-states is of order (typical energy)$/2mc^2$. Hence the pair-states contribution is generally of the same order of magnitude as the relativistic correction. (Quantum mechanics with relativistic corrections but no pair-state corrections gives a good approximation to the hydrogen atom because the relativistic energy of motion $T$ is much larger than the typical energies of transitions in the atom.) Therefore, if we are to have a relativistic quantum theory, we need a state space that contains states with arbitrarily many particles. Such a state space is called a Fock space.

The one-particle state space $S$ is the foundation for Fock space, the many-particle state space. The tensor product of $S$ with itself $n$ times contains all $n$-particle states. Bosonic $n$-particle states correspond to symmetric tensors. Bosonic Fock space is the direct sum of all symmetric tensors in all the tensor products of $S$ with itself. As indicated in Section 1.1, the energy-momentum eigenstates are dual to these spaces of physical states. We build a basis of these eigenstates as follows.

The no-particle state is written $|0\rangle$ and called the vacuum state. This state is normalized by $\langle 0|0\rangle = 1$. We write $|\bar{k}_1, \ldots, \bar{k}_n\rangle$ for the symmetric tensor product of one-particle states $|\bar{k}_r\rangle$. Since this $n$-particle state is symmetric, the order of the momenta in the ket is not significant. The energy or momentum of the $n$-particle state is the sum of the energies or momenta of the constituent one-particle states:

$$P^\mu|0\rangle = 0,$$

$$P^\mu|\bar{k}_1, \ldots, \bar{k}_n\rangle = \left(\sum_{r=1}^n k_n^\mu\right)|\bar{k}_1, \ldots, \bar{k}_n\rangle. \tag{2.1.2}$$

The normalization of the states makes the Bose symmetry a manifest part of the formalism, for example:

$$\langle \bar{k}_1', \bar{k}_2'|\bar{k}_1, \bar{k}_2\rangle = \delta^{(3)}(\bar{k}_1' - \bar{k}_1)\delta^{(3)}(\bar{k}_2' - \bar{k}_2) + \delta^{(3)}(\bar{k}_1' - \bar{k}_2)\delta^{(3)}(\bar{k}_2' - \bar{k}_1). \tag{2.1.3}$$

Of course, states with different numbers of particles are orthogonal.

These states form a basis for Fock space. A general state would have the form

$$|\psi\rangle = \psi_0|0\rangle + \int d^3\bar{k}\,\psi_1(\bar{k})|\bar{k}\rangle + \frac{1}{2!}\int d^3\bar{k}_1\,d^3\bar{k}_2\,\psi_2(\bar{k}_1, \bar{k}_2)|\bar{k}_1, \bar{k}_2\rangle + \cdots \tag{2.1.4}$$

We may assume that $\psi_k$ is a symmetric function. The norm of $|\psi\rangle$ is given by

$$\langle\psi|\psi\rangle = |\psi_0|^2 + \int d^3\bar{k}\,|\psi_1(\bar{k})|^2 + \frac{1}{2!}\int d^3\bar{k}_1\,d^3\bar{k}_2\,|\psi_2(\bar{k}_1, \bar{k}_2)|^2 + \cdots \tag{2.1.5}$$

For each momentum $k$, we can interpret the sequence of states

$$|0\rangle, \quad |\bar{k}\rangle, \quad |\bar{k}, \bar{k}\rangle, \quad \ldots \tag{2.1.6}$$

as states of a harmonic oscillator. We could identify these states simply by occupation numbers 0, 1, 2, etc. and write $|n\rangle$ for the general state in this sequence. A different momentum $\bar{k}'$ would have its own sequence of excitations represented by $|n'\rangle$. For different momenta, these sequences of states coexist without interfering. A basis element is therefore determined by a finite occupation number for each of a finite number of momenta. Such a state can be also written $|n\rangle$, where $n$ now gives the occupation number as a function of momentum.

Thus bosonic Fock space, with its basis described by occupation-number functions, is a state space for an infinite set of independent harmonic oscillators, one oscillator for each momentum. In the next section, we shall therefore review the raising and lowering operator formalism for a single harmonic oscillator in preparation for the annihilation and creation operators which act on Fock space.

## 2.2 Harmonic Oscillator Review

The Hamiltonian for the oscillator is $H = \frac{1}{2}(p^2 + q^2 - 1)$, where the operators $p$ and $q$ satisfy $[q, p] = i$. We will assume that $p$ and $q$ are complete in the set of operators in the sense that, if an operator $r$ satisfies $[q, r] = [p, r] = 0$, then $r$ is a complex number.

The raising and lowering operators $a^\dagger$ and $a$ are defined by

$$a = \frac{1}{\sqrt{2}}(q + ip) \qquad \text{and} \qquad a^\dagger = \frac{1}{\sqrt{2}}(q - ip). \qquad (2.2.1)$$

Then the pair $\{a, a^\dagger\}$ is complete, $[a, a^\dagger] = 1$ and $H = a^\dagger a$. These equations imply that $[H, a] = -a$ and $[H, a^\dagger] = a^\dagger$, so that $a$ and $a^\dagger$ are eigenvectors of the action of $H$ on the Hilbert space operators. If we rewrite these equations in the form

$$Ha = aH - a = a(H - 1) \qquad \text{and} \qquad Ha^\dagger = a^\dagger H + a^\dagger = a^\dagger(H + 1), \qquad (2.2.2)$$

then we can compute the effect of $a$ and $a^\dagger$ on energy levels:

$$H|n\rangle = E_n|n\rangle \quad \Longrightarrow \quad Ha|n\rangle = a(H-1)|n\rangle = a(E_n-1)|n\rangle = (E_n-1)a|n\rangle, \qquad (2.2.3)$$

and similarly

$$Ha^\dagger|n\rangle = (E_n + 1)a^\dagger|n\rangle. \qquad (2.2.4)$$

These two equations show that $a|n\rangle$ is an eigenstate of $H$ with energy $E_n - 1$, while $a^\dagger|n\rangle$ is an eigenstate of $H$ with energy $E_n + 1$. Thus $a$ is an energy lowering operator and $a^\dagger$ is an energy raising operator.

Energy is bounded below:

$$E_n = \langle n|H|n\rangle = \langle n|a^\dagger a|n\rangle = \big|a|n\rangle\big|^2 \geq 0. \qquad (2.2.5)$$

Hence, if we continue to lower the energy with $a$, at some point we will find a ground state $|n_0\rangle$ which is annihilated by $a$. Since $H = a^\dagger a$, the energy of this state is zero. We therefore relabel states by writing $|0\rangle$ for the ground state and applying $a^\dagger$ to generate successively all the excited states $|1\rangle$, $|2\rangle$, and so on. Since $a^\dagger$ raises energy by one unit, the energy of $|n\rangle$ is $n$.

Normalize $|n\rangle$ by $\langle n|n\rangle = 1$. Define $c_n$ by $c_n|n\rangle = a^\dagger|n - 1\rangle$, choosing the phase of $|n\rangle$ to make $c_n$ real and positive. Then

$$|c_n|^2 = \langle n - 1|aa^\dagger|n - 1\rangle = \langle n - 1|H + 1|n - 1\rangle = n. \qquad (2.2.6)$$

Hence $c_n = \sqrt{n}$ and $(a^\dagger)^n|0\rangle = \sqrt{n!}\,|n\rangle$.

**Homework 2.2.7.** Show that $a^n|n\rangle = \sqrt{n!}\,|0\rangle$.

This structure of states $|n\rangle$ and raising and lowering operators $a^\dagger$ and $a$ is called a ladder. The completeness assumption on $p$ and $q$ implies that all states of the quantum harmonic oscillator are linear combinations of the $|n\rangle$, and that there is a unique ground state for this system.

## 2.3 Application to Fock Space

As we saw above, an assemblage of non-interacting particles has the energy spectrum of an assemblage of uncoupled oscillators. The interpretation of the operators $a$ and $a^\dagger$ is different, however. Acting on Fock space, they are creation and annihilation operators. There is one pair for each momentum, and so we have an operator $a(\bar{k})$ that destroys a particle of momentum $\bar{k}$, and an operator $a(\bar{k})^\dagger$ that creates such a particle. The canonical commutation relations on these operators are:

$$
\begin{aligned}
\left[a(\bar{k}), a(\bar{k}')^\dagger\right] &= \delta^{(3)}(\bar{k} - \bar{k}'), \\
\left[a(\bar{k}), a(\bar{k}')\right] &= \left[a(\bar{k})^\dagger, a(\bar{k}')^\dagger\right] = 0.
\end{aligned}
\tag{2.3.1}
$$

The vacuum state (as the ground state is called in this context) is determined by the conditions $\langle 0|0\rangle = 1$ and $a(\bar{k})|0\rangle = 0$ for all momenta $\bar{k}$. Using the commutation properties of the creation and annihilation operators, we can reconstruct bosonic Fock space by applying creation operators to the vacuum. Thus we can define $|\bar{k}\rangle$ and $\langle \bar{k}|$ by

$$
|\bar{k}\rangle = a(\bar{k})^\dagger|0\rangle \quad \text{and} \quad \langle \bar{k}| = \langle 0|a(\bar{k}),
\tag{2.3.2}
$$

and use the properties of $|0\rangle$ and the commutation relations to deduce

$$
\langle \bar{k}|0\rangle = 0 \quad \text{and} \quad \langle \bar{k}|\bar{k}'\rangle = \delta^{(3)}(\bar{k} - \bar{k}').
\tag{2.3.3}
$$

Similarly we can define two-particle states $|\bar{k}_1, \bar{k}_2\rangle$ by

$$
|\bar{k}_1, \bar{k}_2\rangle = a(\bar{k}_1)^\dagger a(\bar{k}_2)^\dagger|0\rangle.
\tag{2.3.4}
$$

Since the creation operators commute with each other, this two-particle state is naturally bosonic. Making similar definitions for many particle states, all of Fock space can be recreated from the vacuum.

**Homework** 2.3.5. Use the commutation properties 2.3.1 of the creation and annihilation operators to show that the two-particle states defined above satisfy the normalization condition 2.1.3.

The number operator $N$ and energy-momentum operators $P^\mu$ can be expressed in terms of these creation and annihilation operators:

$$
N = \int d^3\bar{k}\, a(\bar{k})^\dagger a(\bar{k}) \quad \text{and} \quad P^\mu = \int d^3\bar{k}\, k^\mu a(\bar{k})^\dagger a(\bar{k}).
\tag{2.3.6}
$$

**Homework** 2.3.7. Use the properties and definitions of this section to show that $H|\bar{k}\rangle = k^0|\bar{k}\rangle$. (Remember, $H = P^0$.)

We could change the normalization of the creation operators so that they create Lorentz-covariant states:

$$
\alpha(k)^\dagger = (2\pi)^{3/2}\left(2\omega(\bar{k})\right)^{1/2} a(\bar{k})^\dagger \quad \Longrightarrow \quad \alpha(k)^\dagger|0\rangle = |k\rangle.
\tag{2.3.8}
$$

As in the definition of $|k\rangle$, the notation $\alpha(k)$ implies that $k^0$ is equal to $\omega(\bar{k})$. The Lorentz action on these creation operators is

$$U(\Lambda)\alpha(k)^\dagger U(\Lambda)^\dagger = \alpha(\Lambda k)^\dagger. \qquad (2.3.9)$$

**Homework 2.3.10.** Show that $U(\Delta_a)\alpha(k)U(\Delta_a)^\dagger = e^{-ia\cdot k}\alpha(k)$.

The many-particle state space for a fixed particle type at fixed energy and momentum is simply a ladder like that of the quantum harmonic oscillator. This suggests defining a basis for many-particle states using symmetric tensor products of finite sets of ladder states. The Fock space or full many-particle state space can be understood as the linear combinations of these many-particle basis states or more compactly as a tensor product of the oscillator state spaces over all momenta.

For particles of a single type, the canonical commutation properties of the ladder operators determine the orthogonality and normalization of the Fock-space basis. The Fock space for a variety of particle types is a tensor product of the various Fock spaces, one for each particle type. States whose particle types do not match are by definition orthogonal.

The annihilation and creation operators of the ladders dynamically connect the basis states in Fock space to the vacuum state. The free scalar quantum field will be built out of these operators.

## 2.4 The Free Scalar Field

Now we have a Fock space describing an assemblage of non-interacting bosons of a single type. We have seen that it is not possible to include a position operator in the theory. Hence we allow the observables to be functions of position, thereby localizing measurement, not particles. The result is a quantum field $\phi$, an operator-valued function of space-time. In this section, we shall show that there is a non-interacting or linear scalar field $\phi$ which satisfies the natural axioms for a field specified below and that these axioms can be reformulated in a way which in principle makes it easy to quantize a classical field theory.

Writing $|i\rangle$ for an initial state and $\langle f|$ for a final state, a matrix element $\mathcal{M}$ of a scalar field,

$$\mathcal{M}(x) \overset{\text{def}}{=} \langle f|\phi(x)|i\rangle, \qquad (2.4.1)$$

is a function of position in space-time. Let $\Lambda \cdot \mathcal{M}$ be the matrix element obtained by transforming the initial and final states by the Lorentz group element $\Lambda$. (Here the use of a dot in $\Lambda \cdot M$ represents some means by which the group element $\Lambda$ transforms the function $M$. The dot is used to alert one that matrix multiplication is not implied.) Then

$$\Lambda \cdot \mathcal{M}(x) \overset{\text{def}}{=} \langle f|U(\Lambda)^\dagger \phi(x)U(\Lambda)|i\rangle. \qquad (2.4.2)$$

However, we expect matrix elements to obey the transformation law

$$\Lambda \cdot \mathcal{M}(x) = \mathcal{M}(\Lambda^{-1}x), \qquad (2.4.3)$$

meaning that shifting the momenta in the states by $\Lambda$ is equivalent to shifting the position of a phenomenon by $\Lambda^{-1}$. Hence

$$\langle f|U(\Lambda)^\dagger \phi(x)U(\Lambda)|i\rangle = \langle f|\phi(\Lambda^{-1}x)|i\rangle. \qquad (2.4.4)$$

We conclude that the Heisenberg action of the Lorentz group on the fields satisfies

$$U(\Lambda)^\dagger \phi(x) U(\Lambda) = \phi(\Lambda^{-1}x). \tag{2.4.5}$$

Similarly, since a shift in position of the states by a vector $a$ corresponds to a shift of observed phenomenon by $a$,

$$U(\Delta_a)^\dagger \phi(x) U(\Delta_a) = \phi(x - a). \tag{2.4.6}$$

**Remark 2.4.7.** The transformation laws 2.4.5 and 2.4.6 look odd to the mathematical eye because $U^\dagger$ on the left appears to break the natural multiplicativity. Indeed, the equation

$$U(L)^\dagger \phi U(L) = L \cdot \phi \tag{2.4.8}$$

leads to

$$\begin{aligned}
U(LM)^\dagger \phi U(LM) &= U(M)^\dagger U(L)^\dagger \phi U(L) U(M) \\
&= U(M)^\dagger (L \cdot \phi) U(M).
\end{aligned} \tag{2.4.9}$$

In general, this last expression will not equal $(LM) \cdot \phi$. In field theory, however, $L\cdot$ is a matrix of complex numbers which act on the column vector of fields by ordinary matrix multiplication. Hence $L\cdot$ commutes with the action of quantum operators, and the argument concludes

$$\begin{aligned}
U(M)^\dagger (L \cdot \phi) U(M) &= L \cdot \left( U(M)^\dagger \phi U(M) \right) \\
&= L \cdot (M \cdot \phi),
\end{aligned} \tag{2.4.10}$$

which will equal $(LM) \cdot \phi$ by the nature of the operation represented by the dot.

**Remark 2.4.11.** The previous remark established the consistency of the transformation laws 2.4.5 and 2.4.6, but the equality

$$U^\dagger(M) U^\dagger(L)^\dagger \phi U(L) U(M) = L \cdot M \cdot \phi \tag{2.4.12}$$

still looks weird because we are equating a right group action with a left group action.

To appreciate the validity of this, consider the action of the permutation group $S_3$ on a triangle $\triangle$. How should the permutation $R = (abc)$ act? We can either think of $a$, $b$, and $c$ as fixed locations in the plane and move the vertex at $a$ to $b$ and so on, or we can think of $a$, $b$, and $c$ as labels for the vertices of the triangle and move vertex $a$ to the position of vertex $b$ and so on. At this stage the two actions are indistinguishable, but if we follow the $R$ with an $F = (bc)$, then the difference will show up. Assuming the usual multiplication rule for permutations, the first action is a left action $R \cdot \triangle$ and the second action is a right action $\triangle \circ R$. We therefore have

$$\triangle \circ R \circ F = R \cdot F \cdot \triangle. \tag{2.4.13}$$

In light of these transformation properties, we propose the following four axioms for the field $\phi$ as a relativistic, operator-valued function of space-time:

**Axiom 1.** The field is a field of observables: $\phi(x)^\dagger = \phi(x)$.

**Axiom 2.** The unitary representation of translation acts on the field:

$$U(\Delta_a)^\dagger \phi(x) U(\Delta_a) = \phi(x - a).$$

**Axiom 3.** The unitary representation of the Lorentz group acts on the field:

$$U(\Lambda)^\dagger \phi(x) U(\Lambda) = \phi(\Lambda^{-1}x).$$

**Axiom 4.** The interaction between two observations is constrained by causality:

$$|x - y|^2 < 0 \quad \Longrightarrow \quad [\phi(x), \phi(y)] = 0.$$

**Remark 2.4.14.**    As the following theorem of Wightman indicates, these axioms are almost inconsistent:

**Theorem.** *Let $U(\Delta_a)$ be a unitary representation of the translations on a Hilbert space $H$. Let $\phi(x)$ is a field of bounded operators on a Hilbert space $H$ which satisfies Axioms 2 and 4 above. Assume that the spectrum of $P^\mu$ satisfies $p^2 > 0$ and that the vacuum is the unique translation-invariant state in $H$. Then $\phi(x)|0\rangle = c|0\rangle$ for some complex constant $c$.*

This result indicates that, if our quantum field $\phi$ is to be non-trivial, then it cannot be an unbounded operator on Hilbert space. Indeed, knowing the structure of the free field from the following theorem, it is easy to see that the norm of $\phi(x)|\psi\rangle$ does not converge for any physical particle state $|\psi\rangle$. Knowing this, we shall not be surprised when powers of $\phi$ generate divergent quantities.

We now prove that there is a unique linear solution to the first three axioms, and that this solution, the free scalar field, satisfies Axiom 4.

**Theorem 2.4.15.** *If a field $\phi$ is linear in creation and annihilation operators and satisfies Axioms 1–3, then, up to a real factor,*

$$\phi(x) = \int \frac{d^3\bar{k}}{(2\pi)^3 \, 2\omega(\bar{k})} \left( e^{ix\cdot k}\alpha(k)^\dagger + e^{-ix\cdot k}\alpha(k) \right)$$

$$= \int \frac{d^3\bar{k}}{(2\pi)^{3/2} \left(2\omega(\bar{k})\right)^{1/2}} \left( e^{ix\cdot k}a(\bar{k})^\dagger + e^{-ix\cdot k}a(\bar{k}) \right).$$

**Proof.** The assumption of linearity in the operators $\alpha(k)^\dagger$ and $\alpha(k)$ implies that $\phi$ must have the form

$$\phi(x) = \int \frac{d^3\bar{k}}{(2\pi)^3 \, 2\omega(\bar{k})} \left( f_+(x, k)\alpha(k)^\dagger + f_-(x, k)\alpha(k) \right). \tag{2.4.16}$$

Using Axiom 2, we eliminate the $x$ dependence of the coefficients:

$$\phi(x) = e^{ix\cdot P}\phi(0)e^{-ix\cdot P}$$
$$= \int \frac{d^3\bar{k}}{(2\pi)^3 \, 2\omega(\bar{k})} \left( f_+(0, k)e^{ix\cdot k}\alpha(k)^\dagger + f_-(0, k)e^{-ix\cdot k}\alpha(k) \right). \tag{2.4.17}$$

Using Axiom 3, we determine the $k$ dependence of the coefficients:

$$\phi(0) = U(\Lambda)^\dagger\phi(0)U(\Lambda)$$
$$= \int \frac{d^3\bar{k}}{(2\pi)^3 \, 2\omega(\bar{k})} \left( f_+(0, k)\alpha(\Lambda^{-1}k)^\dagger + f_-(0, k)\alpha(\Lambda^{-1}k) \right) \tag{2.4.18}$$
$$= \int \frac{d^3\bar{k}}{(2\pi)^3 \, 2\omega(\bar{k})} \left( f_+(0, \Lambda k)\alpha(k)^\dagger + f_-(0, \Lambda k)\alpha(k) \right).$$

Equating coefficients of the annihilation and creation operators on both sides of this equation we find:

$$f_+(0, k) = f_+(0, \Lambda k) \qquad \text{and} \qquad f_-(0, k) = f_-(0, \Lambda k). \tag{2.4.19}$$

Hence $f_+$ and $f_-$ are constants determined by their values on the rest frame of the particle, $k = (\mu, 0, 0, 0)$. Now Axiom 1 shows that $f_+^* = f_-$. The phases can be absorbed by redefining $\alpha(k)$, so we may assume that $f_+$ is real and positive. In this case $f_+ = f_-$, and, up to a constant factor, the field has the form given in the theorem. $\qquad\square$

Being linear, the field $\phi$ of Theorem 2.4.15 is a free field. As it has no Lorentz index, it is known as the free scalar field. Now we must check that the free scalar field satisfies the causality axiom:

**Corollary 2.4.20.** *The free scalar field determined by Axioms 1–3 satisfies the causality axiom, Axiom 4.*

**Proof.** Writing $d\lambda(p)$ for the Lorentz-invariant measure,

$$d\lambda(p) \overset{\text{def}}{=} \frac{d^3\bar{p}}{(2\pi)^3 \, 2\omega(\bar{p})}, \tag{2.4.21}$$

we find that:

$$
\begin{aligned}
& [\phi(x), \phi(y)] && (2.4.22) \\
&= \left[ \int d\lambda(p) \left( e^{ix\cdot p}\alpha(p)^\dagger + e^{-ix\cdot p}\alpha(p) \right), \int d\lambda(q) \left( e^{iy\cdot q}\alpha(q)^\dagger + e^{-iy\cdot q}\alpha(q) \right) \right] \\
&= \int d\lambda(p)\, d\lambda(q) \left[ e^{ix\cdot p}\alpha(p)^\dagger + e^{-ix\cdot p}\alpha(p), e^{iy\cdot q}\alpha(q)^\dagger + e^{-iy\cdot q}\alpha(q) \right] \\
&= \int d\lambda(p)\, d\lambda(q) \left( e^{ix\cdot p}e^{iy\cdot q}[\alpha(p)^\dagger, \alpha(q)^\dagger] + e^{ix\cdot p}e^{-iy\cdot q}[\alpha(p)^\dagger, \alpha(q)] \right. \\
&\qquad\qquad \left. + e^{-ix\cdot p}e^{iy\cdot q}[\alpha(p), \alpha(q)^\dagger] + e^{-ix\cdot p}e^{-iy\cdot q}[\alpha(p), \alpha(q)] \right) \\
&= \int d\lambda(p)\, d\lambda(q) \left( -e^{ix\cdot p}e^{-iy\cdot q} + e^{-ix\cdot p}e^{iy\cdot q} \right) (2\pi)^3 \left( 2\omega(\bar{q}) \right) \delta^{(3)}(\bar{p} - \bar{q}) \\
&= \int d\lambda(p) \left( e^{-i(x-y)\cdot p} - e^{-i(y-x)\cdot p} \right) \\
&\overset{\text{def}}{=} \Delta_+(x - y) - \Delta_+(y - x) \overset{\text{def}}{=} i\Delta(x - y).
\end{aligned}
$$

Now $\Delta_+(z)$ is Lorentz invariant. If $|z|^2 < 0$, then there is a Lorentz transformation $\Lambda$ such that $\Lambda z = -z$. Hence $|z|^2 < 0$ implies $\Delta_+(z) = \Delta_+(-z)$, and so $\Delta(z) = 0$. This shows that the effectively unique field that satisfies Axioms 1–3 also satisfies Axiom 4. $\qquad\square$

**Homework 2.4.23.** The field $\phi$ is roughly a Fourier transform of the creation and annihilation operators. Using the inverse transform formula

$$f(\bar{x}) = \int \frac{d^3\bar{k}}{(2\pi)^3} e^{i\bar{x}\cdot\bar{k}} \hat{f}(\bar{k}) \quad \Longleftrightarrow \quad \hat{f}(\bar{k}) = \int d^3\bar{x}\, e^{-i\bar{x}\cdot\bar{k}} f(\bar{x}),$$

find expressions for $\alpha(k)$ and $\alpha^\dagger(k)$ in terms of $\phi$ and $\partial_0\phi$.

**Remark 2.4.24.** One approach to Homework 2.4.23 leads to a formula which we shall need later.

Define a time-dependent sesquilinear form $\langle,\rangle$ on complex-valued functions of space-time by

$$\langle f,g\rangle(t) \stackrel{\text{def}}{=} i \int d^3\bar{x}\, f^*(t,\bar{x})\partial_0 g(t,\bar{x}) - g(t,\bar{x})\partial_0 f^*(t,\bar{x}). \qquad (2.4.25)$$

The functions $e_p^{\pm}(x) \stackrel{\text{def}}{=} \exp(\pm ix\cdot p)$, where $p_0 = \omega(\bar{p})$, are orthogonal with respect to this sesquilinear form:

$$\langle e_p^r, e_q^s\rangle = (2\pi)^3\delta^{(3)}(\bar{p}-\bar{q})(rp_0+sq_0)e^{-it(rp_0-sq_0)}. \qquad (2.4.26)$$

From the form of the free field in Theorem 2.4.15, we see at once that

$$\alpha^\dagger(p) = \langle e_p^+,\phi\rangle \quad \text{and} \quad \alpha(p) = -\langle e_p^-,\phi\rangle. \qquad (2.4.27)$$

Taking $\phi$ to be an interacting field, these equations could be used to define time-dependent creation and annihilation operators.

From Homework 2.4.23, we can see that the set of operators $\phi(x)$ is complete. Any operator that commutes with $\phi(x)$ for all $x$ must also commute with all the creation and annihilation operators and therefore be a complex number.

We could replace the causality axiom by

**Axiom 4′.** $[\phi(x),\phi(y)] = i\triangle(x-y)$,

but this axiom is specific to the free field; it will not apply to interacting fields. However, if we knew the equation of evolution for the field, then we could reconstruct this commutator from Cauchy initial data. We shall see that the Cauchy data could apply to interacting fields.

By direct computation using Theorem 2.4.15, we find the equation of motion $\partial_\mu\partial^\mu\phi + \mu^2\phi = 0$ for the free field, where $\mu$ is the mass of the associated particles (also called the mass of the field). This is the Klein–Gordon equation, generally written in terms of the box operator $\square = \partial_\mu\partial^\mu$:

$$(\square + \mu^2)\phi = 0. \qquad (2.4.28)$$

In the presence of the equation of motion, Axiom 4′ contains redundant information: since the field satisfies a second order wave equation, so does the commutator. Hence we do not need to know the commutator everywhere, we only need to know its value and time derivative on a spacelike surface. Thus Axiom 4′ may be weakened to the following:

**Axiom 4′a.** $[\phi(x),\phi(y)]_{x^0=y^0} = 0$,

**Axiom 4′b.** $[\phi(x),\partial_0\phi(y)]_{x^0=y^0} = i\delta^{(3)}(\bar{x}-\bar{y})$,

**Axiom 5′.** Equation of motion for the field.

Axioms 4′a and 4′b are equal-time commutation relations. They are sufficiently weak as assumptions that they can be taken as axioms in the canonical quantization of interacting fields. Indeed, causality makes it impossible for interactions of a field to reveal themselves in the equal-time commutators.

**Homework 2.4.29.** Define the time-ordered product of operators $A(x)$ and $B(y)$ by

$$T\big(A(x)B(y)\big) = \begin{cases} A(x)B(y), & \text{if } x^0 > y^0; \\ B(y)A(x), & \text{if } y^0 > x^0. \end{cases}$$

Show, using only the Klein–Gordon equation and the equal-time commutation relations Axioms 4′a and 4′b, that the time-ordered product of a free scalar field $\phi$ satisfies

$$(\Box_x + \mu^2)T\big(\phi(x)\phi(y)\big) = -i\delta^{(4)}(x - y).$$

**Homework 2.4.30.** Using the definition of time ordering in Homework 2.4.29, show that the vacuum expectation of the time-ordered product of free scalar fields can be expressed as an integral:

$$\langle 0|T\big(\phi(x)\phi(y)\big)|0\rangle = \lim_{\epsilon \to 0^+} \int \frac{d^4k}{(2\pi)^4} \frac{-ie^{i(x-y)\cdot k}}{k^2 - \mu^2 + i\epsilon}.$$

The number $\langle 0|T\big(\phi(x)\phi(y)\big)|0\rangle$ is the amplitude for $\phi$ to create a state out of the vacuum at the earlier of the times $x^0$ and $y^0$, for that state to propagate for a time $|x^0 - y^0|$, and then to be destroyed by $\phi$ at the later time. (Hint: it may be useful to establish the formula

$$\lim_{\epsilon \to 0^+} \int \frac{dE}{2\pi} \frac{ie^{itE}}{E^2 - \omega^2 + i\epsilon} = \frac{e^{-i|t|\omega}}{2\omega}$$

as a lemma.)

At this point we have shown that there is a unique, free, relativistic scalar field which is linear in creation and annihilation operators, and we have identified its essential properties in Axioms 1–3, 4′a and 4′b, and the Klein–Gordon equation of motion. This characterization provides the basis for the canonical quantization procedure presented in the next two sections.

# 2.5 Canonical Quantization of Classical Mechanics

The previous section introduced the notion of canonical quantization as the imposition of equal-time commutation relations on classical position and momentum variables. In this section, in preparation for application to field theory, we review canonical quantization in particle mechanics.

The structure of classical mechanics is that the natural observables are generalized coordinates $q = (q^1, \ldots, q^n)$ and their rates of change $\dot{q} = (\dot{q}^1, \ldots, \dot{q}^n)$ and the laws of motion are encapsulated in a scalar function of these observables, the Lagrangian $L(q, \dot{q})$. Application of Hamilton's least-action principle to the action

$$S(q) = \int dt\, L(q, \dot{q}) \tag{2.5.1}$$

brings the Euler–Lagrange equations of motion out of the Lagrangian:

$$\frac{d}{dt}\frac{\partial L}{\partial \dot{q}^r} = \frac{\partial L}{\partial q^r} \quad \text{for } r = 1, \ldots, n. \tag{2.5.2}$$

The Euler–Lagrange equations are generally second-order differential equations. If we define generalized or canonical momenta $p_r$ by

$$p_r \stackrel{\text{def}}{=} \frac{\partial L}{\partial \dot{q}^r}, \tag{2.5.3}$$

and if the map

$$(q^1, \ldots, q^n, \dot{q}^1, \ldots, \dot{q}^n) \mapsto (q^1, \ldots, q^n, p_1, \ldots, p_n) \qquad (2.5.4)$$

is invertible, then we can change coordinates from $q$'s and $\dot{q}$'s to $q$'s and $p$'s. Let $p = (p_1, \ldots, p_n)$ and $\dot{p} = (\dot{p}_1, \ldots, \dot{p}_n)$. Define the Hamiltonian $H$ by

$$H(q, p) \stackrel{\text{def}}{=} p \cdot \dot{q} - L(q, \dot{q}) \quad \text{where } \dot{q} = \dot{q}(q, p). \qquad (2.5.5)$$

By direct computation, using the Euler–Lagrange equations, we find Hamilton's equations of motion:

$$\frac{\partial H}{\partial p_r} = \dot{q}^r \quad \text{and} \quad \frac{\partial H}{\partial q^r} = -\dot{p}_r. \qquad (2.5.6)$$

These equations may be derived from Hamilton's principle by writing the action $S$ in Hamilton's form

$$S(q, p) = \int dt\, p \cdot \dot{q} - H(q, p), \qquad (2.5.7)$$

and finding the functions $q(t)$ and $p(t)$ which minimize this action.

The rate of change of an observable $A(q, p)$ is given by

$$\begin{aligned}
\frac{d}{dt} A(q, p) &= \frac{\partial A}{\partial q^r} \dot{q}^r + \frac{\partial A}{\partial p_r} \dot{p}_r \\
&= \frac{\partial A}{\partial q} \frac{\partial H}{\partial p} - \frac{\partial A}{\partial p} \frac{\partial H}{\partial q} \stackrel{\text{def}}{=} \{A, H\},
\end{aligned} \qquad (2.5.8)$$

where $\{\,,\,\}$ is the Poisson bracket. In general the Poisson bracket $\{A, B\}$ gives the rate of change of $A$ when we take $B$ as the Hamiltonian, or, since $\{A, B\} = -\{B, A\}$, as the rate of change of $-B$ when we take $A$ as the Hamiltonian. The canonical Poisson bracket equation $\{q^r, p_s\} = \delta^r_s$ indicates that $q$ and $p$, as classical Hamiltonian operators, do not commute.

In quantum mechanics we interpret $q$ and $p$ as operators on a Hilbert space of states, and the defining property of these operators is the bracket equation $[q^r, p_s] = i\delta^r_s$. The factor of $i$ enters because of the requirement that quantum observables be hermitian:

$$A = A^\dagger \quad \text{and} \quad B = B^\dagger \quad \Longrightarrow \quad [A, B]^\dagger = -[A, B], \qquad (2.5.9)$$

so naturally we map the bracket structures by

$$\{A, B\} \to -i[A, B]. \qquad (2.5.10)$$

Similarly, the Heisenberg equations of motion in quantum mechanics are $[A, H] = i\dot{A}$. Thus the Poisson bracket in the algebra of classical observables becomes a Lie bracket or commutator in the algebra of quantum observables.

Canonical quantization is simply the process of (a) promoting the classical coordinates $q$ and $p$ into operators, (b) assuming canonical commutation relations,

$$[q^r, p_s] = i\delta^r_s, \quad [q^r, q^s] = 0, \quad [p_r, p_s] = 0, \qquad (2.5.11)$$

and (c) defining the quantum Hamiltonian as the classical Hamiltonian evaluated on the quantum $p$ and $q$. Of course, since the quantum $p$ and $q$ do not commute, the quantum Hamiltonian will not in general be unique.

**Remark 2.5.12.** Actually, the correspondence between quantum and classical algebras of observables breaks down on the level of cubic polynomials.

On the one hand we have a commutative algebra with Poisson bracket generated by $p$ and $q$. On the other hand we have a non-commutative algebra with commutator bracket generated by $P$ and $Q$, the quantum equivalents of $p$ and $q$. Now we want a map between these two which extends the map $p \to P$, $q \to Q$ and preserves brackets.

Even at this stage, the map cannot be an algebra map, so let us try to construct an extension of $p \to P$, $q \to Q$ without regard to multiplication or linearity which maps $\{\,,\,\}$ to $-i[\,,\,]$.

By considering the brackets of $p^2$ with $p$ and $q$, we see that $p^2$ must map to $P^2 + c$ for some constant $c$. Similarly $q^2$ maps to $Q^2 + d$. From the equations

$$\{q^2, p^2\} = 4qp \quad \text{and} \quad -i[Q^2 + d, P^2 + c] = 2(QP + PQ) \qquad (2.5.13)$$

we see that

$$qp \to \frac{1}{2}(QP + PQ). \qquad (2.5.14)$$

Now, by considering brackets with $qp$, we find that $p^k \to P^k$ and $q^k \to Q^k$. Finally, by considering $\{p^{m+1}, q^{n+1}\}$, we see that

$$p^m q^n \to \frac{i}{(m+1)(n+1)}[P^{m+1}, Q^{n+1}]. \qquad (2.5.15)$$

Thus the map is determined on all monomials.

The map 2.5.15 breaks down on the relation $\{pq^2, p^2q\} = 3p^2q^2$, for

$$-i\left[\frac{i}{6}[P^2, Q^3], \frac{i}{6}[P^3, Q^2]\right] = 3 \times \frac{i}{9}[P^3, Q^3] + 1. \qquad (2.5.16)$$

We conclude that there is no map of any sort from the Poisson algebra of classical functions to the commutator algebra of quantum operators which extends $p \to P$, $q \to Q$ and preserves the brackets.

The significance of canonical quantization is that, with any one of the possible quantum Hamiltonians, the classical limit is correct. To see this, start from the Heisenberg equations of motion for $q$ and $p$. If $H$ is a polynomial in $q$ and $p$, compute $[q^r, H]$ and $[p_r, H]$ using the commutation relation for $q$ and $p$, and, after computing these brackets, take the classical limit to obtain:

$$[q^r, p_s] = i\delta_s^r \hbar \quad \Longrightarrow \quad \begin{cases} i\dot{q}^r = \dfrac{1}{\hbar}\left[q^r, H(q,p)\right] \xrightarrow{\hbar \to 0} i\dfrac{\partial H}{\partial p_r}, \\[2mm] i\dot{p}_r = \dfrac{1}{\hbar}\left[p_r, H(q,p)\right] \xrightarrow{\hbar \to 0} -i\dfrac{\partial H}{\partial q^r}. \end{cases} \qquad (2.5.17)$$

Canceling the $i$'s yields Hamilton's equations.

**Homework 2.5.18.** Verify the details of this argument by finding a formula for $\left[p_r, (q^r)^k\right]$ and proving that Heisenberg's equation has the correct classical limit when $H(q,p) = \frac{1}{2}p^2 + V(q)$ for some polynomial $V$. How about $H(q,p) = \frac{1}{2}(p^2 + 2qpq + q^2)$?

Canonical quantization appears simple: just promote the classical $p$'s and $q$'s to operators which satisfy the canonical commutation condition, and define their evolution by commutation with the resulting Hamiltonian operator. We should just note two subtleties. First, due to the canonical commutation condition itself, the quantum $p$'s and $q$'s do not commute, and so there are in general many quantum Hamiltonians associated with a given classical one. Second, sensitivity to the order of $p$'s and $q$'s will in general distinguish Heisenberg's equations of motion, $\dot{p} = [iH, p]$ and $\dot{q} = [iH, q]$, from Hamilton's equations, $\dot{p} = -H_q$ and $\dot{q} = H_p$.

In field theory, as we shall see in the following sections, because Lorentz invariance constrains the canonical momentum to occur only in the kinetic term, there are no products of fields and their canonical momenta in the Lagrangian densities of interest, and so these subtleties do not arise.

## 2.6 Canonical Quantization of Classical Fields

The principles presented in the previous section apply to fields with little more than a change in notation. The classical coordinates $q^r(t)$ with their index $r$ correspond to classical fields $\phi^a(t, \bar{x})$ with their indices $a$ and $\bar{x}$. Let $\phi$ be the column vector of the fields $\phi^1, \ldots, \phi^n$. Corresponding to the particle Lagrangian we have a field Lagrangian

$$L(\phi) = \int d^3\bar{x}\, \mathcal{L}(\phi, \partial_\mu \phi), \qquad (2.6.1)$$

where $\mathcal{L}$ is a Lagrangian density function. Hamilton's principle applies to the action

$$S(\phi) = \int dt\, L(\phi, \partial_\mu \phi), \qquad (2.6.2)$$

and we find Euler–Lagrange equations

$$\frac{d}{dx^\mu}\frac{\partial \mathcal{L}}{\partial(\partial_\mu \phi^a)} = \frac{\partial \mathcal{L}}{\partial \phi^a}. \qquad (2.6.3)$$

These equations are generally second-order partial differential equations.

The Euler–Lagrange equations can be obtained directly from the action $S$ by functional differentiation with respect to the field. Taking $\phi$ to be a single scalar field for simplicity, functional differentiation is defined by

$$\frac{\delta S}{\delta \phi(x)} \overset{\text{def}}{=} \lim_{\epsilon \to 0} \frac{1}{\epsilon}\big(S(\phi + \epsilon \delta_x^{(4)}) - S(\phi)\big), \qquad (2.6.4)$$

where $\delta_x^{(4)}$ is the delta function at $x$: $\delta_x^{(4)}(y) \overset{\text{def}}{=} \delta^{(4)}(x-y)$. Application to the action above yields:

$$\begin{aligned}
\frac{\delta S}{\delta \phi(x)} &= \lim_{\epsilon \to 0} \frac{1}{\epsilon} \int d^4y\, \mathcal{L}(\phi + \epsilon \delta_x^{(4)}, \partial_\mu \phi + \epsilon \partial_\mu \delta_x^{(4)}) - \mathcal{L}(\phi, \partial_\mu \phi) \\
&= \int d^4y\, \frac{\partial \mathcal{L}}{\partial \phi}\delta_x^{(4)} + \frac{\partial \mathcal{L}}{\partial(\partial_\mu \phi)}\partial_\mu \delta_x^{(4)} \\
&= \int d^4y\, \frac{\partial \mathcal{L}}{\partial \phi}\delta_x^{(4)} - \frac{d}{dx^\mu}\frac{\partial \mathcal{L}}{\partial(\partial_\mu \phi)}\delta_x^{(4)} \\
&= \frac{\partial \mathcal{L}}{\partial \phi}\Big|_x - \frac{d}{dx^\mu}\frac{\partial \mathcal{L}}{\partial(\partial_\mu \phi)}\Big|_x.
\end{aligned} \qquad (2.6.5)$$

Thus the condition that the action has a stationary value at the actual motion of the field yields the Euler–Lagrange equations:

$$\frac{\delta S}{\delta \phi(x)} = 0 \quad \Longleftrightarrow \quad \frac{d}{dx^\mu} \frac{\partial \mathcal{L}}{\partial(\partial_\mu \phi)} = \frac{\partial \mathcal{L}}{\partial \phi}. \tag{2.6.6}$$

Taking $\phi$ to be a vector of scalar fields again, the conjugate or canonical momentum of $\phi^a$ is

$$\Pi_a = \frac{\partial \mathcal{L}}{\partial(\partial_0 \phi^a)}. \tag{2.6.7}$$

Write $\Pi$ for the row vector $(\Pi_1, \ldots, \Pi_n)$. If the map

$$(\phi, \partial_0\phi, \partial_1\phi, \partial_2\phi, \partial_3\phi) \mapsto (\Pi, \phi, \partial_1\phi, \partial_2\phi, \partial_3\phi) \tag{2.6.8}$$

is invertible, then we can define the Hamiltonian density function by

$$\mathcal{H}(\Pi, \phi, \partial_1\phi, \partial_2\phi, \partial_3\phi) = \Pi{\cdot}\partial_0\phi - \mathcal{L}. \tag{2.6.9}$$

As in the particle case, by direct computation using the Euler–Lagrange equations, we find Hamilton's equations:

$$\dot{\phi}^a = \frac{\partial \mathcal{H}}{\partial \Pi_a} \quad \text{and} \quad \dot{\Pi}_a = -\frac{\partial \mathcal{H}}{\partial \phi^a} + \frac{d}{dx^r} \frac{\partial \mathcal{H}}{\partial(\partial_r \phi^a)}. \tag{2.6.10}$$

These equations also follow from an application of Hamilton's least-action principle to Hamiltonian form of the action:

$$S(\Pi, \phi) = \int d^4x \, \Pi{\cdot}\partial_0\phi - \mathcal{H}. \tag{2.6.11}$$

**Homework 2.6.12.** Take

$$\mathcal{L} = \frac{1}{2}\partial_\mu\phi \, \partial^\mu\phi - \frac{\mu^2}{2}\phi^2,$$

and show that the Euler–Lagrange equation is the Klein–Gordon equation. Find the Hamiltonian density and Hamilton's equations.

The Hamiltonian is the integral of the Hamiltonian density over space,

$$H(\Pi, \phi) = \int d^3\bar{x} \, \mathcal{H}(\Pi, \phi, \partial_1\phi, \partial_2\phi, \partial_3\phi). \tag{2.6.13}$$

Since the integral here is over $d^3\bar{x}$, the functional derivative should be defined here as in the definition 2.6.4 but with by $\delta_{\bar{x}}^{(3)}$ replacing $\delta_x^{(4)}$. To define the three-dimensional functional derivative in this four-dimensional context, we make $t$ a parameter and let $\phi_t(\bar{x})$ stand for $\phi(x)$:

$$\frac{\delta}{\delta \phi_t(\bar{x})} A(\phi_t) \overset{\text{def}}{=} \lim_{\epsilon \to 0} \frac{1}{\epsilon} \left( A(\phi_t + \epsilon\delta_{\bar{x}}^{(3)}) - A(\phi_t) \right). \tag{2.6.14}$$

With this definition, we see that

$$\begin{aligned}
\frac{\delta H}{\delta \phi_t(\bar{x})} &= \int d^3\bar{y} \, \frac{\partial \mathcal{H}}{\partial \phi} \delta_{\bar{x}}^{(3)} + \frac{\partial \mathcal{H}}{\partial(\partial_r \phi)} \partial_r \delta_{\bar{x}}^{(3)} \\
&= \frac{\partial \mathcal{H}}{\partial \Pi}\Big|_x - \frac{d}{dx^r} \frac{\partial \mathcal{H}}{\partial(\partial_r \phi)}\Big|_x
\end{aligned} \tag{2.6.15}$$

and similarly

$$\frac{\delta H}{\delta \Pi_t(\bar{x})} = \frac{\partial \mathcal{H}}{\partial \Pi}\bigg|_x. \tag{2.6.16}$$

Consequently, Hamilton's equations for fields can be written in the same form as the corresponding equations for particles:

$$\dot{\phi}(x) = \frac{\delta H}{\delta \Pi_t(\bar{x})} \quad \text{and} \quad \dot{\Pi}(x) = -\frac{\delta H}{\delta \phi_t(\bar{x})}. \tag{2.6.17}$$

Seeing Hamilton's equations thus expressed in terms of the functional derivatives of the Hamiltonian suggests defining a Poisson bracket for functions of the field, its space derivatives and its momentum as follows:

$$\{A, B\} \stackrel{\text{def}}{=} \int d^3\bar{x} \left( \frac{\delta A}{\delta \phi_t(\bar{x})} \frac{\delta B}{\delta \Pi_t(\bar{x})} - \frac{\delta A}{\delta \Pi_t(\bar{x})} \frac{\delta B}{\delta \phi_t(\bar{x})} \right). \tag{2.6.18}$$

Straight away from this definition and Hamilton's equations 2.6.17, we see that the Poisson bracket with the Hamiltonian determines time evolution:

$$\frac{dA}{dt} = \int d^3\bar{x} \left( \frac{\delta A}{\delta \phi_t(\bar{x})} \dot{\phi}(x) + \frac{\delta A}{\delta \Pi_t(\bar{x})} \dot{\Pi}(x) \right) = \{A, H\}. \tag{2.6.19}$$

Also, the Poisson bracket of the field with its momentum has the usual canonical value:

$$\{\phi^a(t, \bar{y}), \Pi_b(t, \bar{z})\} = \delta_b^a \delta^{(3)}(\bar{y} - \bar{z}). \tag{2.6.20}$$

Canonical quantization follows the principles of the previous section. First assume that $\phi$ and $\Pi$ are operator-valued functions of space-time, second impose the equal-time commutation relations which correspond to the Poisson bracket relations above:

$$\begin{aligned} \left[\phi^a(t, \bar{x}), \Pi_b(t, \bar{y})\right] &= i\delta_b^a \delta^{(3)}(\bar{x} - \bar{y}), \\ \left[\phi^a(t, \bar{x}), \phi^b(t, \bar{y})\right] &= \left[\Pi_a(t, \bar{x}), \Pi_b(t, \bar{y})\right] = 0, \end{aligned} \tag{2.6.21}$$

and third, let the quantum field Hamiltonian be the same function of the quantum fields as the classical one is of the classical fields (up to ordering ambiguities).

The Hamiltonian for a single free field is

$$H(\phi, \Pi) = \int d^3\bar{x} \, \frac{1}{2}(\Pi^2 + \delta_{rs}\partial^r \phi \, \partial^s \phi + \mu^2 \phi^2). \tag{2.6.22}$$

If we now substitute the integral form of the free field from Theorem 2.4.15 into this expression for the Hamiltonian, we find that

$$\begin{aligned} H &= \frac{1}{2} \int d^3\bar{k} \, \omega(\bar{k}) \big(a(\bar{k})^\dagger a(\bar{k}) + a(\bar{k})a(\bar{k})^\dagger\big) \\ &= \int d^3\bar{k} \, \omega(\bar{k})a(\bar{k})^\dagger a(\bar{k}) + \frac{1}{2} \int d^3\bar{k} \, \omega(\bar{k})\delta^{(3)}(0). \end{aligned} \tag{2.6.23}$$

So, in order to obtain the desired result, we have to drop an undefined constant.

**Homework 2.6.24.** Do this computation: obtain 2.6.23 from 2.6.22.

Since the order of quantum theoretic operators cannot be deduced from the classical formalism, this is not a problem. We can simply define the Hamiltonian as the first integral in the sum above and never mention the undefined constant. The proper way to do this is to introduce the concept of *normal order*. If we write out a product of operators $\phi^1(x_1) \cdots \phi^k(x_k)$ as an integral of sums of products of complex functions, creation operators, and annihilation operators, then the normal-ordered product $:\phi^1(x_1) \cdots \phi^k(x_k):$ is that same integral of sums of products, but with all the creation operators on the left and all the annihilation operators on the right in each term.

Using this definition of normal ordering, we can write the formula for the free-field Hamiltonian as

$$H(\phi, \Pi) \stackrel{\text{def}}{=} \int d^3\bar{x} \, \frac{1}{2} :(\Pi^2 - \partial_r \phi \, \partial^r \phi + \mu^2 \phi^2): = \int d^3\bar{k} \, \omega(\bar{k}) a(\bar{k})^\dagger a(\bar{k}). \quad (2.6.25)$$

To conclude the example, we should verify that the Heisenberg equations of motion for the free field have the same form as Hamilton's classical equations of motion:

**Homework 2.6.26.** Using the commutation relations, show that Heisenberg's equations of evolution for $\phi$ and $\Pi$ with the Hamiltonian 2.6.22 have the same form as Hamilton's equations of evolution as worked out in Homework 2.6.12.

**Remark 2.6.27.** For convenience in application, we must assume that normal ordering is a linear operation:

$$:\lambda A + \mu B: = \lambda:A: + \mu:B: \quad (2.6.28)$$

However, by intention, normal ordering does not preserve equalities between operators on state space:

$$A = B \quad \not\Longleftrightarrow \quad :A: = :B: \quad (2.6.29)$$

(For example, take $A = aa^\dagger$ and $B = a^\dagger a + 1$.) Hence normal ordering must be defined on an algebraic structure in which the commutator is not evaluated, for example, on the formal non-commutative algebra generated by $a$ and $a^\dagger$ over the complex numbers.

To apply normal ordering to an interacting field, we start by defining creation and annihilation operators in terms of the field. First, we define time-dependent operators $q_{\bar{k}}$ and $p_{\bar{l}}$ as the Fourier coefficients of $\phi(x)$ and $\Pi(x)$:

$$q_{\bar{k}}(t) \stackrel{\text{def}}{=} \int d^3\bar{x} \, e^{-i\bar{k}\cdot\bar{x}} \phi(\bar{x}, t) \quad \text{and} \quad p_{\bar{l}}(t) \stackrel{\text{def}}{=} \int d^3\bar{y} \, e^{-i\bar{l}\cdot\bar{y}} \Pi(\bar{y}, t). \quad (2.6.30)$$

The canonical commutation relations for $\phi$ and $\Pi$ imply

$$\left[ q_{\bar{k}}(t), p_{\bar{l}}(t) \right] = i(2\pi)^3 \delta^{(3)}(\bar{x} - \bar{y}). \quad (2.6.31)$$

Then we absorb the factor of $i$ by defining operators $\alpha_{\bar{k}}(t)$ by

$$\alpha_{\bar{k}}(t) \stackrel{\text{def}}{=} \omega_{\bar{k}} q_{\bar{k}}(t) + ip_{-\bar{k}}(t). \quad (2.6.32)$$

Using the fact that $\phi$ and $\Pi$ are hermitian, we can verify the canonical commutation relations for $\alpha$ and $\alpha^\dagger$:

$$\left[ \alpha_{\bar{k}}(t), \alpha_{\bar{l}}^\dagger(t) \right] = 2\omega_{\bar{k}}(2\pi)^3 \delta^{(3)}(\bar{k} - \bar{l}). \quad (2.6.33)$$

Recreating $\phi$ and $\Pi$, we find the expansions

$$\phi(x) = \int \frac{d^3\bar{k}}{(2\pi)^3\, 2\omega(\bar{k})}\, e^{-i\bar{k}\cdot\bar{x}}\alpha^\dagger_{\bar{k}}(t) + e^{i\bar{k}\cdot\bar{x}}\alpha_{\bar{k}}(t),$$

$$\Pi(x) = \int \frac{d^3\bar{k}}{(2\pi)^3\, 2\omega(\bar{k})}\, i\omega_{\bar{k}} e^{-i\bar{k}\cdot\bar{x}}\alpha^\dagger_{\bar{k}}(t) - i\omega_{\bar{k}} e^{i\bar{k}\cdot\bar{x}}\alpha_{\bar{k}}(t). \tag{2.6.34}$$

Now let us apply this to the concrete example obtained by adding a $\phi^4$ interaction to the free-field Lagrangian density. In this case, $\Pi$ is simply $\partial_0\phi$. In light of the expansions 2.6.34, we conclude that $\partial_0\alpha_{\bar{k}} = -i\omega_{\bar{k}}\alpha_{\bar{k}}$. Hence the time dependence of the $\alpha$'s is trivial and 2.6.34 reduces to the free-field expansion. To avoid this disaster, we must assume that $\partial_0\alpha_{\bar{k}}$ does not exist. Reservations about the validity of the formalism inevitably follow.

Continuing bravely, the Hamiltonian density is

$$\mathcal{H} = \frac{1}{2}\Pi^2 + \frac{1}{2}(\nabla\phi)^2 + \frac{\mu^2}{2}\phi^2 + \frac{\lambda}{4!}\phi^4. \tag{2.6.35}$$

The next step in normal ordering is to substitute the expansions 2.6.34 into $\mathcal{H}$ and integrate over space. The integration over space generates an overall momentum-conserving delta function out of the exponentials. We see at once that the $\phi^4$ term will yield a sum of quartic monomials in $\alpha$ and $\alpha^\dagger$ which cannot be canceled by any other term. Using the commutation relations and moving the $\alpha^\dagger$ to the right will leave some terms proportional to $\int d^3\bar{p}\, d^3\bar{q}\, \alpha^\dagger_{\bar{p}}\alpha_{\bar{p}}$ – four momentum integrals up against two delta functions leaves one integral too many.

This divergent operator does not contribute to the vacuum expectation of the Hamiltonian, but to its one-particle expectations. Normal ordering, of course, discards this operator and its divergent coefficient, and so modifies the Hamiltonian by an operator which is not even a polynomial in $\phi$ and $\Pi$. Hence, the normal-ordered Hamiltonian is not a polynomial in the field and its momentum, and the quantum field will not only fail to satisfy the classical equation of motion, but will not evolve according to any polynomial-type differential equation.

**Homework 2.6.36.** Verify the assertion that normal ordering in $\phi^4$ theory leads to making a modification of the Hamiltonian by an operator which is not a polynomial in $\phi$ and $\Pi$.

At this point it is clear that the brute force approach to canonical quantization is not working. One practical remedy is to apply canonical quantization perturbatively as set out in Chapter 4. The functional integral quantization introduced in Chapter 11 provides a more informative perspective. Putting this to work in Chapter 13, we find that quantum fields do evolve according to the corresponding classical equations. Since $\phi(x)$ and $\Pi(x)$ form a complete set of operators, the fact that Heisenberg's equations have the same polynomial form as Hamilton's indicates that normal ordering only changes the Hamiltonian by a constant after all.

The simplicity of the principle of canonical quantization – just impose a commutation relation on the field and its momentum – makes it easy to apply to any classical theory. In field theory, however, the free-field Hamiltonian turns out to be divergent even when evaluated on the vacuum. This introduces the need to normal order this Hamiltonian. It is not so clear how to extend this analysis to interacting fields. To proceed, we assume that an interacting Hamiltonian can be made finite on all particle states by merely subtracting a constant. (This assumption is justified in Chapter 13.) Then the evolution of all operators is known from Heisenberg's equations and the quantum theory is completely specified. On this basis, Chapter 3 develops an operator approach to quantum conservation laws. Chapter 4, however, avoids the issue by developing a perturbative implementation of canonical quantization.

## 2.7 The Structure of the Vacuum State

In this section, we shall describe the structure of the vacuum state of a free scalar quantum field $\phi$ from the viewpoint of classical field theory. When going from particle mechanics to quantum mechanics, we represent the classical particle by a delta function and develop the mathematical description of a quantum particle, the wave function $\psi$, by considering a weighted coexistence of all possible classical particles:

$$\psi(\bar{x}) = \int d^3\bar{y}\, \psi(\bar{y})\delta^{(3)}(\bar{x} - \bar{y}), \tag{2.7.1}$$

or, in terms of kets labeled with quantum numbers,

$$|\psi\rangle = \int d^3\bar{y}\, \psi(\bar{y})|\bar{y}\rangle. \tag{2.7.2}$$

Now we shall extend this idea to fields in three steps.

First, the analogue of a wave function with a definite position is a field state with definite, classical shape. The test for localization in quantum mechanics is the application of the position operator:

$$\bar{X}|\bar{x}\rangle = \bar{x}|\bar{x}\rangle. \tag{2.7.3}$$

The analogous test in quantum field theory would be:

$$\phi(0, \bar{x})|f\rangle = f(\bar{x})|f\rangle. \tag{2.7.4}$$

Since $\phi$ is hermitian, $f$ is a real-valued function. Note that, just as we expect the localized quantum particle $|\bar{x}\rangle$ to spread out in time, so we expect the *shape state* $|f\rangle$ of the quantum field $\phi$ to spread out, and we must therefore make the time $t = 0$ of the measurement explicit.

Second, by analogy with the position eigenstates in quantum mechanics, we can represent $|f\rangle$ by a delta functional on the function space $H$ of classical field shapes with respect to some measure $[df]$ on $H$. Then the shape states are orthogonal:

$$\begin{aligned}\langle g|f\rangle &= \int [dh]\, \delta(g - h)\delta(f - h) \\ &= \int [dh]\, \delta(g - f)\delta(f - h) \\ &= \delta(g - f).\end{aligned} \tag{2.7.5}$$

Third, we assume that the states $|\Omega\rangle$ of the quantum field $\phi$ can be expanded in terms of the shape states using integration over $H$:

$$|\Omega\rangle = \int [df]\, \Omega(f)|f\rangle, \tag{2.7.6}$$

where $\Omega$ is given by

$$\langle g|\Omega\rangle = \int [df]\,\Omega(f)\langle g|f\rangle$$

$$= \int [df]\,\Omega(f)\delta(g-f) = \Omega(g). \tag{2.7.7}$$

At this point we see that the mathematical model for the states of a free scalar field resembles the model presented in Section 1.1 for the free quantum-mechanical particle. The space $S$ of states $|\Omega\rangle$ for the quantum field is represented by a space of 'good' functions $\Omega$ on the space $H$ of classical field configurations, and the basis of shape eigenstates $|f\rangle$ are elements in the space $S^*$ of distributions on $H$.

To avoid problems with delta functions, it is convenient to smooth out the field $\phi$ by local averaging. Thus, for $g$ in $H$, define $\phi_g$ by

$$\phi_g = \int d^3\bar{x}\,g(\bar{x})\phi(0,\bar{x}). \tag{2.7.8}$$

(Notice that we are simplifying the whole discussion in this section by focusing on $t = 0$.) Define an inner product on $H$ by

$$(g,h) \overset{\text{def}}{=} \int d^3\bar{x}\,g(\bar{x})h(\bar{x}). \tag{2.7.9}$$

Then the action of $\phi_g$ on a state $|\Omega\rangle$ takes the form

$$\phi_g|\Omega\rangle = \int d^3\bar{x}\,g(\bar{x})\phi(0,\bar{x})\int [df]\Omega(f)|f\rangle$$

$$= \int [df]\,d^3\bar{x}\,g(\bar{x})\Omega(f)f(\bar{x})|f\rangle \tag{2.7.10}$$

$$= \int [df]\,(f,g)\Omega(f)|f\rangle.$$

If we define $c_g$ to be the coordinate function on $H$ corresponding to projection onto $g$, that is, $c_g(f) = (f,g)$, then the action of $\phi_g$ is represented by multiplication by the coordinate $c_g$:

$$\phi_g|\Omega\rangle = |c_g\Omega\rangle. \tag{2.7.11}$$

Motivated by the canonical representation of $q$ and $ip$ as multiplication by a coordinate $x$ and $\partial_x$, clearly we should now represent the action of $i\dot{\phi}_g$ by differentiation with respect to the variable $f$ in the direction $g$, that is, the operator $g\cdot\partial_f$ defined by

$$g\cdot\partial_f \overset{\text{def}}{=} \int d^4x\,g(x)\frac{\delta}{\delta f(x)}. \tag{2.7.12}$$

From the definition of functional differentiation 2.6.4, we find that

$$g\cdot\partial_f\Omega(f) = \lim_{\epsilon\to 0}\frac{1}{\epsilon}\left(\Omega(f+\epsilon g) - \Omega(f)\right). \tag{2.7.13}$$

In summary, we propose that the representation $\phi_g \to c_g$ and $\dot{\phi}_g \to -ig\cdot\partial_f$ preserves the canonical commutation relations.

To verify this proposition, we first derive the averaged equal-time commutation relations (ETCR) from the ordinary ones:

$$\begin{aligned}
[\phi_g, \dot{\phi}_h] &= \int d^3\bar{x}\, d^3\bar{y}\; g(\bar{x})h(\bar{y})[\phi(0,\bar{x}), \dot{\phi}(0,\bar{y})] \\
&= \int d^3\bar{x}\, d^3\bar{y}\; g(\bar{x})h(\bar{y})i\delta^{(3)}(\bar{x}-\bar{y}) \\
&= i(g,h),
\end{aligned} \tag{2.7.14}$$

and then check that the representation satisfies this ETCR by applying the representation to a state:

$$\begin{aligned}
[c_g, -ih\cdot\partial_f]\Omega(f) &= -ic_g(f)h\cdot\partial_f\Omega(f) + ih\cdot\partial_f\big(c_g(f)\Omega(f)\big) \\
&= i\big(h\cdot\partial_f c_g(f)\big)\Omega(f) \\
&= i(h,g)\Omega(f).
\end{aligned} \tag{2.7.15}$$

We conclude that the ETCR is satisfied by the representation.

To find the structure of the quantum-field vacuum state, we use the defining property of the vacuum:

$$\alpha(k)|0\rangle = 0. \tag{2.7.16}$$

We shall therefore define an averaged annihilation operator from our averaged field operators. Let $B$ be the operator on functions $g$ in $H$ defined in terms of the Fourier transform by

$$\widehat{Bg}(\bar{k}) \overset{\text{def}}{=} \omega(\bar{k})\hat{g}(\bar{k}). \tag{2.7.17}$$

With this definition, and writing $d\lambda(p)$ for the measure

$$d\lambda(p) = \frac{d^3\bar{p}}{(2\pi)^3\, 2\omega(\bar{p})}, \tag{2.7.18}$$

we find that

$$\begin{aligned}
\phi_{Bg} &= \int d^3\bar{x}\; Bg(\bar{x})\phi(0,\bar{x}) \\
&= \int d^3\bar{x}\, \frac{d^3\bar{k}}{(2\pi)^3}\, d\lambda(p)\; e^{-i\bar{x}\cdot\bar{k}}\omega(\bar{k})\hat{g}(\bar{k})\big(e^{-i\bar{x}\cdot\bar{p}}\alpha^\dagger(p) + e^{i\bar{x}\cdot\bar{p}}\alpha(p)\big) \\
&= \int \frac{d^3\bar{k}}{(2\pi)^3}\, d\lambda(p)\; \omega(\bar{k})\hat{g}(\bar{k})\big((2\pi)^3\delta^{(3)}(\bar{k}+\bar{p})\alpha^\dagger(p) + (2\pi)^3\delta^{(3)}(\bar{k}-\bar{p})\alpha(p)\big) \\
&= \int d\lambda(p)\, \omega(\bar{p})\hat{g}(\bar{p})\big(\alpha^\dagger(\tilde{p}) + \alpha(p)\big),
\end{aligned} \tag{2.7.19}$$

where $\tilde{p} = (\omega(\bar{p}), -\bar{p})$, and

$$\begin{aligned}
i\dot{\phi}_g &= i\int d^3\bar{x}\; g(\bar{x})\dot{\phi}(0,\bar{x}) \\
&= i\int d^3\bar{x}\, \frac{d^3\bar{k}}{(2\pi)^3}\, d\lambda(p)\; e^{-i\bar{x}\cdot\bar{k}}\hat{g}(\bar{k})\big((ip_0)e^{-i\bar{x}\cdot\bar{p}}\alpha^\dagger(p) + (-ip_0)e^{i\bar{x}\cdot\bar{p}}\alpha(p)\big) \\
&= \int \frac{d^3\bar{k}}{(2\pi)^3}\, d\lambda(p)\; \hat{g}(\bar{k})\big(-(2\pi)^3\delta^{(3)}(\bar{k}+\bar{p})p_0\alpha^\dagger(p) + (2\pi)^3\delta^{(3)}(\bar{k}-\bar{p})p_0\alpha(p)\big) \\
&= \int d\lambda(p)\, \omega(\bar{p})\hat{g}(\bar{p})\big(-\alpha^\dagger(\tilde{p}) + \alpha(p)\big),
\end{aligned} \tag{2.7.20}$$

which together imply

$$\phi_{Bg} + i\dot{\phi}_g = \int \frac{d^3\bar{k}}{(2\pi)^3}\,\hat{g}(\bar{k})\alpha(k). \tag{2.7.21}$$

From this result we may define the vacuum $|0\rangle$ by

$$(\phi_{Bg} + i\dot{\phi}_g)|0\rangle = 0 \tag{2.7.22}$$

or define the vacuum functional $\Omega_0$ that represents $|0\rangle$ by

$$g\cdot\partial_f\Omega_0(f) + c_{Bg}(f)\Omega_0(f) = 0. \tag{2.7.23}$$

Since $(Bg, f) = (g, Bf)$, the solution of this functional differential equation is simply

$$\Omega_0(f) = Ce^{-\frac{1}{2}(Bf,f)}. \tag{2.7.24}$$

The function $B$ can be written as follows:

$$(Bf, f) = \int \frac{d^3\bar{k}}{(2\pi)^3}\,\omega(\bar{k})\hat{f}(\bar{k})\hat{f}^*(\bar{k}), \tag{2.7.25}$$

which shows that both high frequency and large amplitude functions are damped in the vacuum functional.

The argument above correctly indicates the relationship between classical shape states and the vacuum state. We should note, however, that the 'Lebesgue' measure $[df]$ is not well defined on the function space $H$. If we choose an orthonormal basis of functions $f_r$ such that

$$(Bf_r, f_s) = \delta_{rs}, \tag{2.7.26}$$

then, using the coordinates $x = (x_1, x_2, \ldots)$ associated with the basis $f_r$, we can define a Gaussian measure $d\mu(x)$ on $H$ by:

$$d\mu(x) \stackrel{\text{def}}{=} \lim_{n\to\infty} \exp\left(-\frac{1}{2}\sum_{r=1}^{n} x_r^2\right) \prod_{r=1}^{n} \frac{dx_r}{(2\pi)^{1/2}}. \tag{2.7.27}$$

Formally, this measure is related to $[df]$ by

$$d\mu(f) = Ne^{-\frac{1}{2}(Bf,f)}[df], \tag{2.7.28}$$

where $N$ is an infinite normalization factor coming from the $\sqrt{2\pi}$'s.

The results of the formal argument above can be translated into results for $d\mu$ on the basis of one observation: where we had an inner product

$$(\phi, \phi') = \int [df]\,\phi(f)\phi'(f) \tag{2.7.29}$$

we now need an inner product

$$(\psi, \psi') = \int d\mu(f)\,\psi(f)\psi'(f) = N\int [df]\,e^{-\frac{1}{2}(Bf,f)}\psi(f)\psi'(f). \tag{2.7.30}$$

Thus the map transforming 'Lebesgue' square-integrable functions into Gaussian ones is

$$\psi(f) = N^{-1/2} e^{\frac{1}{4}(Bf,f)} \phi(f) \tag{2.7.31}$$

and so, with respect to the Gaussian measure $d\mu$, the vacuum state is described by the function

$$\Omega_0^\mu(f) = c e^{-\frac{1}{4}(Bf,f)}. \tag{2.7.32}$$

This analysis of the vacuum state of the free scalar quantum field is intended simply to give insight into the structure of the quantum vacuum from the classical viewpoint. It is possible to pursue the analysis. For example, from the formulae

$$\bar{P} = \int d^3\bar{x}\, \dot{\phi}(0,\bar{x}) \bar{\partial}\phi(0,\bar{x}),$$

$$H = \frac{1}{2} \int d^3\bar{x} \left( \dot{\phi}(0,\bar{x})^2 + \left(\nabla\phi(0,\bar{x})\right)^2 + \mu^2 \phi(0,\bar{x})^2 \right), \tag{2.7.33}$$

one can show that

$$\bar{P}|0\rangle = \bar{0},$$

$$H|0\rangle = \left( \int \frac{d^3\bar{k}}{(2\pi)^3} \frac{\omega(\bar{k})}{2} \right) |0\rangle, \tag{2.7.34}$$

which indicates that the vacuum state is translation invariant and has a half unit of energy for every wavelength; and one may easily show that the vacuum is rotation invariant, and with a little more difficulty, that it is also invariant under boosts.

Further, from the formula

$$g = e^{i\bar{x}\cdot\bar{p}} \implies \left(2\omega(\bar{k})\right)^{1/2} \phi_g = a^\dagger(\bar{k}) + a(-\bar{k}), \tag{2.7.35}$$

one can obtain the expression

$$|\bar{k}_1,\ldots,\bar{k}_n\rangle = \int [df] \left( \prod_{r=1}^n \left(2\omega(\bar{k}_r)\right)^{1/2} \hat{f}(\bar{k}_r) \right) \Omega_0(f)|f\rangle \tag{2.7.36}$$

for the $n$-particle states with all momenta different and see how the vacuum as a balanced coexistence of classical shapes is distorted in the particle states.

The insights presented in this section are worthy of contemplation as they bring out the relationship between classical and quantum world views. The formalism used here, however, does not extend to the interacting quantum field, and so we shall not develop it any further. The idea of the functional integral $\int [df]$ will be introduced afresh when we need it for functional integral quantization.

# 2.8 Summary

In the previous chapter, we established a formalism to combine state space and observables with a unitary representation of the Poincaré group, but we found that including a position operator to model localization of observation leads to a violation of relativistic causality. In this chapter, we pointed out a deeper, more intrinsic problem of localization: the tendency of that process to destroy particle number through pair production. Hence, in order to model localization, we constructed a state space (Fock space) which can accommodate particle creation and annihilation. This Fock space has the structure of a state space for an uncountable assembly of harmonic oscillators indexed by momentum.

On the foundation of Fock space, we built a free relativistic scalar field out of the particle creation and annihilation operators. Investigating the properties of this field led to a procedure, canonical quantization, which should be sufficient to promote classical field theories into quantum field theories. The essence of canonical quantization is the imposition of the equal-time canonical commutation relations on classical fields. The apparent simplicity of this quantization procedure is marred by the need to normal order the Hamiltonian, an operation which cannot readily be applied to Hamiltonians with interactions.

Chapter 3, assuming that normal ordering is not a significant obstacle, investigates the consequences of classical symmetries for quantum field theories. Then, in Chapter 4, we see how to circumvent the normal-ordering problem by developing a perturbation expansion for the evolution of interacting quantum fields, an expansion in which, miraculously, every term can be computed using only our current knowledge of the free quantum field.

# Chapter 3

# Symmetries and Conservation Laws

Providing the understanding of the algebra of conserved quantities for application to quantum numbers, gauge theories, and flavor symmetry of hadrons.

## Introduction

Through canonical quantization we have established a link between classical field theory and quantum field theory. Just as a particle in quantum mechanics may be thought of as a smeared version of a classical particle, just so a particle in quantum field theory may be considered a smeared form of a classical field shape. Consequently, the notion of a propagating state is far more subtle in quantum field theory than in classical field theory. For example, a $\pi^+$ particle will gradually change into a sum of many-particle states such as $\mu^+\nu_\mu$, $\mu^+\nu_\mu\gamma$, $\pi^0 e^+\nu_e$, and $e^+\nu_e e^+e^-$, since the pion can decay into these combinations of particles. To identify a state and to limit its protean propensities, it is useful to know what aspects of that state do not change as it propagates. This chapter presents a theory of symmetries, conserved quantities, and quantum numbers which provides such knowledge.

Section 3.1 explains the concept of symmetry, bringing out and connecting its different manifestations on the different levels of a particle theory. Sections 3.2 proves Noether's famous theorem on the existence of a conserved quantity in a classical particle theory with a symmetry, and Section 3.3 comments on the effects of canonical quantization on this theorem. Having understood what kind of conservation laws may be expected to survive quantization, we turn to Noether's Theorem in classical field theory in Section 3.4, and make a detailed study of the simplest example in Section 3.5. This example establishes the complex scalar field which we shall use throughout the text.

To provide a foundation for more elaborate examples, Section 3.6 summarizes the structure of the most common symmetry groups and describes their infinitesimal structure. Then Section 3.7 shows how to construct field theories with internal symmetries, and Section 3.8 constructs algebras of conserved quantities in such theories. Sections 3.9 and 3.10 show respectively how translations and Lorentz transformations give rise to the energy-momentum and angular-momentum operators. Section 3.11 uses the results of the previous sections to show how maximal sets of commuting conserved quantities can be chosen and used as quantum numbers.

Having completed the discussion of differentiable symmetries, it remains to discuss the separate topic of discrete symmetries and associated multiplicative conservation laws. Sections 3.12, 3.13, and 3.14 cover the most common examples of discrete symmetries, focusing respectively on charge conjugation, parity, and time reversal.

# 3.1 Four Levels of Symmetry in an Example

The purpose of this section is to introduce the concept of a differentiable symmetry and the relationship between symmetries and conservation laws in the context of a concrete example, the spherical pendulum and its rotation symmetry.

There are four levels of symmetry: trajectory, equation of motion, Lagrangian, and infinitesimal. The trajectory of a system is the path it takes in configuration space as a function of time. Symmetry on the level of trajectories means that a rotated trajectory is itself a trajectory, or more precisely, that time evolution commutes with rotation. This level of symmetry may be obvious to the intuition, but it is generally hard to verify explicitly. However, if each trajectory is uniquely determined by some initial conditions and an equation of motion, then implicit verification is available through the uniqueness of solutions to initial value problems. If we can verify symmetry of the equation of motion, then we can deduce symmetry of the space of trajectories.

Symmetry of an equation of motion may be tricky to verify, but if the equation of motion is derived from a Lagrangian, then all we need to do is verify symmetry of the Lagrangian, and that is simple. Finally, to make the connection to conserved quantities, we go one step further and describe infinitesimal symmetries of the Lagrangian.

To illustrate these four levels of symmetry, we introduce an example. Consider a pendulum with bob mass $m$ suspended from a fixed point by the traditional massless string of length $l$, the whole acted upon by a constant force of gravity $g$. Let $\theta$ be the angle between the string and the vertical defined by gravity, and let $\phi$ be the angle of rotation about this vertical. Then the Lagrangian for this spherical pendulum is:

$$L = \tfrac{1}{2}l^2\dot{\theta}^2 + \tfrac{1}{2}l^2\sin^2\theta\,\dot{\phi}^2 + mgl\cos\theta. \tag{3.1.1}$$

From the Lagrangian we obtain the Euler–Lagrange equations

$$\frac{d}{dt}\frac{\partial L}{\partial \dot{\theta}} = \frac{\partial L}{\partial \theta} \quad \text{and} \quad \frac{d}{dt}\frac{\partial L}{\partial \dot{\phi}} = \frac{\partial L}{\partial \phi}, \tag{3.1.2}$$

which expand to

$$l^2\ddot{\theta} = l^2\sin\theta\,\cos\theta\,\dot{\phi}^2 - mgl\sin\theta,$$
$$\frac{d}{dt}(l^2\sin^2\theta\,\dot{\phi}) = 0. \tag{3.1.3}$$

Physical intuition indicates that two spherical pendulums related initially by a rotation about the vertical axis will move in such a way that this relation is always preserved. In other words, for the spherical pendulum, rotation about the vertical axis commutes with time evolution.

First, to develop the mathematics of the symmetry, we express the rotation of the pendulum through an angle $\tau$ by a one-parameter family of functions:

$$(\theta, \phi) \to \tau \cdot (\theta, \phi) \overset{\text{def}}{=} (\theta, \phi + \tau). \tag{3.1.4}$$

The dot here indicates action of the real number $\tau$ on the configuration $(\theta, \phi)$ of the pendulum. This family of rotations is defined for all $\tau \in \mathbf{R}$. It is continuously differentiable in all its parameters and has the properties

$$\sigma \cdot \big(\tau \cdot (\theta, \phi)\big) = (\sigma + \tau) \cdot (\theta, \phi) \quad \text{and} \quad 0 \cdot (\theta, \phi) = (\theta, \phi), \tag{3.1.5}$$

which reveal a harmony between the action of the real numbers as rotations and the additive structure of the set of real numbers. These properties make the one-parameter family of rotations a *real action*.

In general, since real actions mix coordinates, the action of $\tau$ on a configuration can only be computed if we know all the coordinates of the configuration. However, the action factorizes in the case of the pendulum, and an expression like $\tau \cdot \theta$, which is not in general well defined, is meaningful in this case.

When the pendulum is rotated, its velocity is also rotated by the same amount, so that, since $\tau$ is constant,

$$(\dot{\theta}, \dot{\phi}) \rightarrow \frac{d}{dt} \tau \cdot (\theta, \phi) = \frac{d}{dt}(\theta, \phi + \tau) = (\dot{\theta}, \dot{\phi}). \tag{3.1.6}$$

Second, we introduce a notation for the time evolution of the pendulum. Let

$$\big(\theta(t), \phi(t)\big) = \big(\theta(\theta_0, \phi_0, \dot{\theta}_0, \dot{\phi}_0; t), \phi(\theta_0, \phi_0, \dot{\theta}_0, \dot{\phi}_0; t)\big) \tag{3.1.7}$$

be the solution of the Euler–Lagrange equations 3.1.3 with initial conditions

$$(\theta, \phi, \dot{\theta}, \dot{\phi})|_{t=0} = (\theta_0, \phi_0, \dot{\theta}_0, \dot{\phi}_0). \tag{3.1.8}$$

Then the mathematical expressions for the rotated solution (RS) and for the solution with the rotated initial condition (SR) may be written:

$$\begin{aligned}
\text{RS} &= \big(\theta(t), \phi(t) + \tau\big) \\
&= \big(\theta(\theta_0, \phi_0, \dot{\theta}_0, \dot{\phi}_0; t), \phi(\theta_0, \phi_0, \dot{\theta}_0, \dot{\phi}_0; t) + \tau\big) \\
\text{SR} &= \big(\theta(\theta_0, \phi_0 + \tau, \dot{\theta}_0, \dot{\phi}_0; t), \phi(\theta_0, \phi_0 + \tau, \dot{\theta}_0, \dot{\phi}_0; t)\big).
\end{aligned} \tag{3.1.9}$$

Third, we express the property of commuting with time evolution by equating these two:

$$\begin{aligned}
&\big(\theta(\theta_0, \phi_0, \dot{\theta}_0, \dot{\phi}_0; t), \phi(\theta_0, \phi_0, \dot{\theta}_0, \dot{\phi}_0; t) + \tau\big) \\
&= \big(\theta(\theta_0, \phi_0 + \tau, \dot{\theta}_0, \dot{\phi}_0; t), \phi(\theta_0, \phi_0 + \tau, \dot{\theta}_0, \dot{\phi}_0; t)\big).
\end{aligned} \tag{3.1.10}$$

This equation asserts that, if the pendulum starts from a fixed initial state and evolves for a time $t$ before it is rotated by $\tau$, its final position is the same as it would have been if it had first been rotated by $\tau$ and then allowed to evolve freely for time $t$.

As it is impossible to obtain explicit general solutions even in this simple example, the trajectory-space symmetry condition 3.1.10 cannot be verified directly. The symmetry condition can be verified indirectly by using the uniqueness of solutions to initial value problems: clearly the two sides are equal at $t = 0$, and the governing equations 3.1.3 are invariant under the change of variables from $(\theta, \phi)$ to $(\theta, \phi + \tau)$; hence the commutation condition 3.1.10 is true for the spherical pendulum.

Since, however, the governing equations are derived from the Lagrangian 3.1.1, the invariance of these equations follows at once from the invariance of the Lagrangian. In this example, since

$$\tau \cdot (\theta, \phi) = (\theta, \phi + \tau) \quad \text{and} \quad \frac{d}{dt} \tau \cdot (\theta, \phi) = (\dot{\theta}, \dot{\phi}), \tag{3.1.11}$$

and since $L$ does not depend on $\phi$ but only on $\dot{\phi}$, the following computation is sufficient:

$$\tau \cdot L(\theta, \phi, \dot{\theta}, \dot{\phi}) \overset{\text{def}}{=} L\left(\tau \cdot (\theta, \phi), \frac{d}{dt}\tau \cdot (\theta, \phi)\right) \tag{3.1.12}$$

$$= L(\theta, \phi, \dot{\theta}, \dot{\phi}).$$

The fact that our real action is differentiable makes it possible to reduce the study of this action to the study of its velocity and, in fact, to evaluate the derivative at parameter-value zero without losing any information. Indeed,

$$\frac{\partial}{\partial \tau}\tau \cdot (\theta, \phi) = \frac{\partial}{\partial \sigma}(\sigma + \tau) \cdot (\theta, \phi) \Big|_{\sigma=0} \qquad \text{by definition of derivative,}$$

$$= \frac{\partial}{\partial \sigma}\sigma \cdot (\tau \cdot (\theta, \phi)) \Big|_{\sigma=0} \qquad \text{by the group action property.} \tag{3.1.13}$$

Thus knowledge of $\frac{\partial}{\partial \sigma}\sigma \cdot (\theta, \phi)$ for all $(\theta, \phi)$ at $\sigma = 0$ gives complete knowledge of $\frac{\partial}{\partial \tau}\tau \cdot (\theta, \phi)$. This differential equation, together with the initial condition $(\theta_0, \phi_0) = (\theta, \phi)$, determines the real action on all points $(\theta, \phi)$.

The velocity of the real action,

$$(\theta', \phi') \overset{\text{def}}{=} \frac{\partial}{\partial \sigma}\sigma \cdot (\theta, \phi)|_{\sigma=0} = (0, 1), \tag{3.1.14}$$

is said to be the *infinitesimal real action*. It gives the action of the *generator* of the real numbers, that is, the positive unit velocity vector at zero in **R**.

Using this principle of reduction from real action to infinitesimal real action, we finally conclude that a sufficient condition for rotational symmetry of the spherical pendulum can be expressed in the equation

$$\frac{\partial}{\partial \tau}\tau \cdot L \Big|_{\tau=0} = 0. \tag{3.1.15}$$

Since, by the Lagrangian symmetry condition 3.1.12, $\tau \cdot L$ is independent of $\tau$, this condition is obviously satisfied for the spherical pendulum. Note, however, that because rotation is a real action, this infinitesimal symmetry condition 3.1.15 is actually equivalent to the manifest symmetry condition 3.1.12.

**Homework** 3.1.16. From the real action 3.1.6 on velocities, find the infinitesimal real action on velocities. Using this result and the infinitesimal action 3.1.14 on configurations, verify the infinitesimal symmetry condition 3.1.15 directly by differentiating the Lagrangian.

In summary, we have at this point identified four levels of symmetry: trajectory 3.1.10, equation of motion 3.1.3, Lagrangian 3.1.12, and infinitesimal 3.1.15; and we have understood that if these four levels of structure are available in a model, then symmetry on any level implies symmetry on all levels. We conclude this section by connecting the infinitesimal symmetry to a conservation law.

The rotation symmetry is apparent on the level of the Lagrangian 3.1.1 in the absence of explicit dependence on the coordinate $\phi$. The Euler–Lagrange equation for $\phi$ itself expresses the conservation of angular momentum:

$$\frac{d}{dt}\frac{\partial L}{\partial \dot{\phi}} = \frac{\partial L}{\partial \phi} = 0 \quad \Rightarrow \quad l^2 \sin^2 \theta \, \dot{\phi} = \text{const.} \tag{3.1.17}$$

The underlying principle that connects the symmetry to this conservation law is that the infinitesimal symmetry shows which combination of generalized momenta will be conserved:

$$\textbf{Conserved Quantity} = \left(\frac{\partial L}{\partial \dot{\theta}}, \frac{\partial L}{\partial \dot{\phi}}\right)\left(\begin{array}{c} \theta' \\ \phi' \end{array}\right) = \frac{\partial L}{\partial \dot{\phi}}, \qquad (3.1.18)$$

where $(\theta', \phi')$ are given by 3.1.14.

The purpose of going through this long derivation of a near obvious result is to establish the stages of transformation that connect a symmetry of path space to an infinitesimal symmetry and a conserved quantity. Symmetry is most apparent on the level of trajectories, it is most easily verified on the level of the Lagrangian, and it is most readily connected to the conserved quantity on the infinitesimal level. This knowledge will be useful whenever the symmetry is more readily apparent than the conserved quantity. In particular, it is essential to quantum field theory because many theories in particle physics are built about specific symmetries.

## 3.2 Symmetry and Conserved Quantities in Classical Mechanics

The previous section has brought out the concepts essential to a discussion of symmetry. In this section, we shall take the conclusion that all four levels of symmetry – symmetry of trajectories, of equations of motion, of the Lagrangian, and infinitesimal symmetry of the Lagrangian – are equivalent, and develop the concepts and notation necessary for applications of symmetry to field theory.

We saw in 3.1.5 that rotation of the spherical pendulum is an action of the real numbers. In general, a real action has the following structure. Let $M$ be the set of configurations $m$ that a system of particles can take, and let $q = (q^1, \ldots, q^n)^\top$ be generalized coordinates for the system. In general, a system of generalized coordinates $q$ will identify only a part of $M$ with part of $\mathbf{R}^n$. To avoid unnecessary differential geometry, however, we shall treat $M$ as a part of $\mathbf{R}^n$, and $q$ as the coordinate functions on $\mathbf{R}^n$. A real action on $M$ is given by a map

$$\begin{aligned} e_R \colon \mathbf{R} \times M &\to M \\ (\tau, m) &\mapsto \tau \cdot m \end{aligned} \qquad (3.2.1)$$

which satisfies

$$\begin{aligned} 0 \cdot m &= m, \\ (\sigma + \tau) \cdot m &= \sigma \cdot (\tau \cdot m). \end{aligned} \qquad (3.2.2)$$

The notation '$e_R$' indicates 'evolution under real action'. For example, if $M$ were the unit sphere in $\mathbf{R}^3$, then there would be $n = 2$ generalized coordinates, $\theta$ and $\phi$, and if $R(\bar{a}, \tau)$ is the matrix for right-handed rotation of $\mathbf{R}^3$ through the angle $\tau$ about the axis $\bar{a}$, then matrix multiplication $(\tau, m) \mapsto R(\bar{a}, \tau)m$ would be a real action on $M$.

Since $\mathbf{R}$ acts on the whole space $M$, it also acts on the velocity vectors at each point in $M$. We need a notation for this map in order to discuss the symmetry of

Lagrangians, as in 3.1.12. For example, consider the map $F$ from $M = \mathbf{R}^2$ to itself given by

$$F \equiv \begin{pmatrix} X \\ Y \end{pmatrix} \stackrel{\text{def}}{=} \begin{pmatrix} x^2 - y^2 \\ xy \end{pmatrix}. \tag{3.2.3}$$

The velocities $\dot{x}$ and $\dot{y}$ map to velocities $\dot{X}$ and $\dot{Y}$ under the chain rule

$$\begin{pmatrix} \dot{X} \\ \dot{Y} \end{pmatrix} = \begin{pmatrix} 2x & -2y \\ y & x \end{pmatrix} \begin{pmatrix} \dot{x} \\ \dot{y} \end{pmatrix}. \tag{3.2.4}$$

The map on velocities is a linear map $TF$ determined by the matrix of partial derivatives

$$TF \stackrel{\text{def}}{=} \begin{pmatrix} X_x & X_y \\ Y_x & Y_y \end{pmatrix}, \tag{3.2.5}$$

where the notation $A_b$ stands for the partial derivative of $A$ with respect to $b$ and $TF$ stands for the tangent map derived from $F$.

Clearly, in general, the linear map of velocities associated with a map $F\colon M \to M$ will be a function of position, and we may define the matrix of partial derivatives $TF$ at the point $q(t)$ by

$$TF\big(q(t)\big)\dot{q}(t) \stackrel{\text{def}}{=} \frac{d}{dt} F\big(q(t)\big). \tag{3.2.6}$$

At this point, treating the Lagrangian as a function of position and velocity variables, it would be proper to introduce a phase space for the domain of the Lagrangian and extend the real action to this space. It is, however, far simpler to treat the Lagrangian as a function of path and time. For a velocity vector $\dot{q}$, $\tau \cdot \dot{q}$ cannot be evaluated without reference to the location of the basepoint of this vector in $M$, but for a path $q\colon \mathbf{R} \to M$, as the example above demonstrates, it is entirely proper to write

$$\tau \cdot \frac{dq(t)}{dt} \stackrel{\text{def}}{=} \frac{d}{dt} \tau \cdot q(t), \tag{3.2.7}$$

and, for a function $F$ of position and velocity evaluated on a path $q$,

$$\tau \cdot F\big(q(t), \dot{q}(t)\big) \stackrel{\text{def}}{=} F\big(\tau \cdot q(t), \tau \cdot \dot{q}(t)\big). \tag{3.2.8}$$

**Remark 3.2.9.** The action of the real numbers is a special case of a group action. If $G$ is a group and $S$ is a set, then a *left group action* of $G$ on $S$ is a map from $G \times S$ to $S$ which satisfies

$$\mathbf{1} \cdot s = s \quad \text{and} \quad (gh) \cdot s = g \cdot (h \cdot s) \tag{3.2.10}$$

where $\mathbf{1}$ is the identity element of $G$, $g, h \in G$, and $s \in S$. Naturally, if $G$ acts on $S$, then $G$ also acts on the functions $F$ defined on $S$. There are two ways of framing a definition. The first introduces a right group action:

$$(F \cdot g)(s) \stackrel{\text{def}}{=} F(g \cdot s), \tag{3.2.11}$$

and the second introduces an inverse:

$$(g \cdot F)(s) \stackrel{\text{def}}{=} F(g^{-1} \cdot s). \tag{3.2.12}$$

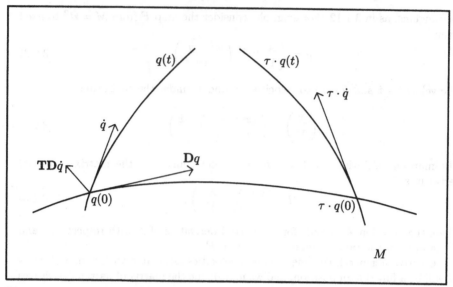

**Fig. 3.2.** This figure displays a portion of the state space $M$, a path $q(t)$, and a real action. The velocity of the path at $q(0)$ is denoted $\dot{q}$, that of the real action $\mathbf{D}q$. As the real action transforms $\dot{q}$ into $\tau \cdot \dot{q}$, it gives rise to an instantaneous rate of change of the velocity, $\mathbf{TD}\dot{q}$. We need $\mathbf{D}q$ and $\mathbf{TD}\dot{q}$ in order to write out the rate of change of a function $F(q, \dot{q})$ under the real action.

Either definition is consistent with the appropriate multiplicative condition. In the case of the real action above, we are being lax; we should either use a right action or an inverse. However, as $\mathbf{R}$ is commutative, our usage is in fact free from inconsistency.

As we intend to focus on infinitesimal symmetry, we will use the definitions of $\tau \cdot \dot{q}$ and $\tau \cdot F$ above to define differential operators $\mathbf{D}$ and $\mathbf{TD}$ which will act on points $q$, path velocities $\dot{q}(t)$, and functions $F(q(t), \dot{q}(t))$ as follows:

$$\mathbf{D}q \overset{\text{def}}{=} \frac{\partial}{\partial \tau} \tau \cdot q \Big|_{\tau=0},$$

$$\mathbf{TD}\dot{q}(t) \overset{\text{def}}{=} \frac{\partial}{\partial \tau} \tau \cdot \dot{q}(t) \Big|_{\tau=0} = \frac{d}{dt}\mathbf{D}q(t), \tag{3.2.13}$$

$$\mathbf{D}F(q(t), \dot{q}(t)) \overset{\text{def}}{=} \frac{\partial}{\partial \tau} \tau \cdot F(q(t), \dot{q}(t)) \Big|_{\tau=0} = F_q \mathbf{D}q(t) + F_{\dot{q}}\mathbf{TD}\dot{q}(t).$$

It is helpful to bear in mind that $\mathbf{D}$ maps configurations to velocity vectors and $\mathbf{TD}$ is simply the matrix of partial derivatives of $\mathbf{D}$. Figure 3.2 above depicts the connection between these notations and the geometry they represent.

We can now apply these definitions to the Lagrangian in order to expand the infinitesimal symmetry condition 3.1.15:

$$\mathbf{D}L = L_q \mathbf{D}q(t) + L_{\dot{q}}\mathbf{TD}\dot{q}(t) = 0. \tag{3.2.14}$$

This expression shows in detail what we need to compute in order to verify symmetry. Clearly, if a Lagrangian density obeys this infinitesimal symmetry condition, then the Lagrangian density and the action are invariant under the real action:

$$\mathbf{D}L = 0 \quad \Longrightarrow \quad \tau \cdot L = L \quad \text{and} \quad \tau \cdot S = S. \tag{3.2.15}$$

From this the invariance of the Euler–Lagrange equations is obvious.

If we apply the Euler–Lagrange equations to the infinitesimal action on the Lagrangian density, we find that $\mathbf{D}L$ evaluated on an orbit is always a total derivative:

$$\begin{aligned}
\mathbf{D}L &= L_q \mathbf{D}q(t) + L_{\dot{q}} \mathbf{T}\mathbf{D}\dot{q}(t) \\
&= \left(\frac{d}{dt}\frac{\partial L}{\partial \dot{q}}\right)\mathbf{D}q(t) + L_{\dot{q}}\frac{d}{dt}\mathbf{D}q(t) \\
&= \frac{d}{dt}\left(\frac{\partial L}{\partial \dot{q}}\mathbf{D}q(t)\right).
\end{aligned} \qquad (3.2.16)$$

If the real action generates an infinitesimal symmetry of the Lagrangian density, then of course we have

$$\mathbf{D}L = 0 \quad \Longrightarrow \quad \frac{d}{dt}Q = 0, \quad \text{where} \quad Q = \frac{\partial L}{\partial \dot{q}}\mathbf{D}q(t). \qquad (3.2.17)$$

Consequently, the function $Q(q, \dot{q})$ is a *conserved quantity* – it is constant along any orbit.

## 3.3 Symmetry and Conserved Quantities in Quantum Mechanics

In quantum mechanics, the Lagrangian formalism is unpopular because, when $q$ and $\dot{q}$ are operators which do not commute, $\dot{q}$ does not in general have a simple relationship to $q$, the Hamiltonian defined by Legendre transformation is not uniquely determined, and it is not easy to write equations of motion using $\partial/\partial q$ and $\partial/\partial \dot{q}$ when $q$ and $\dot{q}$ are operators. The Hamiltonian operator, however, gives the time evolution of a system directly:

$$[F, H] = i\dot{F}, \qquad (3.3.1)$$

where $F$ is any function of $p$ and $q$. The three deficiencies of the Lagrangian approach to quantum mechanics are avoided in the Hamiltonian approach. Both approaches, however, suffer from the problem of ordering monomials involving $p$'s and $q$'s.

As explained in Section 2.5, canonical quantization starts from the classical theory, converts Poisson brackets of classical observables into Lie brackets of quantum observables and hopefully has the same symmetries as the classical limit. The formula for the conserved quantity $Q$ is as above in 3.2.17:

$$Q(p, q) = p\mathbf{D}q. \qquad (3.3.2)$$

It should follow that $i\dot{Q} = [Q, H] = 0$, and that the action of $Q$ on observables gives the infinitesimal real action:

$$[q, Q] = i\mathbf{D}q \quad \text{and} \quad [p, Q] = i\mathbf{D}p. \qquad (3.3.3)$$

In the case of time translation invariance with the Hamiltonian as conserved quantity, these are Heisenberg's equations of motion.

**Homework 3.3.4.** Assuming that $\mathbf{D}q$ is independent of the momenta $p$, verify 3.3.3.

Actually, since classical functions only determine their quantum counterparts up to ordering ambiguities, it is possible for classical conservation laws to break down on the quantum level. To gain an idea of the problem, consider the Hamiltonian for two uncoupled harmonic oscillators:

$$H = \frac{1}{2m}(p_1^2 + p_2^2) + \frac{k}{2}(q_1^2 + q_2^2). \tag{3.3.5}$$

The rotation symmetry for this system generates the angular-momentum conserved quantity

$$Q = q_1 p_2 - q_2 p_1. \tag{3.3.6}$$

It is easy to check that the canonical commutation relations $[q, p] = i$ imply that $[Q, H] = 0$. Now classically $Q^3$ is also a conserved quantity, but if we start from the classical expression it is not obvious how to order the operators. In fact, if we write $p_1$'s and $q_1$'s to the left of $p_2$'s and $q_2$'s and eliminate $p_r q_r p_r$ and $q_r p_r q_r$ products using the generic formulae

$$pqp = \frac{1}{2}(qp^2 + p^2 q) \quad \text{and} \quad qpq = \frac{1}{2}(q^2 p + pq^2), \tag{3.3.7}$$

then there is a unique order for the operators in $Q^3$ for which $Q^3$ is hermitian and satisfies $[Q^3, H] = 0$. Similarly, we may take a classical Hamiltonian with rotational symmetry and re-order its $p$'s and $q$'s to break the symmetry.

**Homework 3.3.8.** Show that any arrangement of the operators in $Q^2$ which makes $Q^2$ hermitian also makes it a conserved quantity.

We do not, then, expect classical conservation laws to survive quantization. That they generally do survive (making those that fail 'anomalous') reflects the simplicity of the common conservation laws. Thus, for example, if the Hamiltonian is of the form $H = K(p) + V(q)$, and if $A$ is a square matrix and $b$ a column vector (such matrices would come from a linear symmetry of the system), then the quantum evolution of $Q = p(Aq + b)$ is given by:

$$\begin{aligned}
\dot{Q} &= [iH, Q] \\
&= [iK(p) + iV(q), p(Aq + b)] \\
&= [iV, p](Aq + b) + pA[iK, q] \\
&= -V_q(Aq + b) + pAK_p.
\end{aligned} \tag{3.3.9}$$

The last line above is free from ordering ambiguities; making $p$ and $q$ classical variables, it gives the classical evolution of the classical quantity $Q$. Hence, in this example, the quantity $Q = p(Aq + b)$ is conserved in the quantum system if and only if it is conserved in the classical system. Note that, though this result is rather trivial, it still covers many examples in which linear or angular momentum is conserved.

We conclude that a classical conserved quantity $Q$ is conserved in the corresponding quantized theory if $H(p, q)$ has the form $K(p) + V(q)$ and if $Q$ has the form $p(Aq + b)$ for some constant matrix $A$ and constant vector $b$. In other cases, ordering ambiguities inherent in quantization may cause the breakdown of classical conservation laws; so quantum conservation laws must generally be checked by hand.

# 3.4 Symmetry and Conservation Laws in Classical Field Theory

In this section, we shall proceed quickly to the desired formulae following the outline presented above for classical mechanics. The main difference between the field and particle cases is that the field Lagrangian density contains space derivatives. This gives rise to the analogue of canonical momenta in the spatial directions. Our particle conserved quantity formula, therefore, carries an extra index when applied to fields. This leads at first to a conserved current rather than a conserved quantity.

Let $\mathcal{L}(\phi, \partial_\mu\phi, x)$ be a Lagrangian density, a function of fields $\phi = (\phi^1, \dots, \phi^n)^\top$, their derivatives $\partial_\mu\phi$, and perhaps $x$. The Lagrangian $L$ is defined as the integral of the Lagrangian density over space:

$$L(\phi; t) \stackrel{\text{def}}{=} \int d^3\bar{x}\, \mathcal{L}\big(\phi(x), \partial_\mu\phi(x), x\big), \tag{3.4.1}$$

and the action $S$ is defined as usual by the integral of the Lagrangian:

$$\begin{aligned} S(\phi; t_0, t_1) &\stackrel{\text{def}}{=} \int_{t_0}^{t_1} dt\, L(\phi; t) \\ &= \int_{t_0}^{t_1} \int dt\, d^3x\, \mathcal{L}\big(\phi(x), \partial_\mu\phi(x), x\big). \end{aligned} \tag{3.4.2}$$

The principle of stationary action, $\delta S = 0$ when $\phi$ is the classical path of motion, implies the Euler–Lagrange equations for the field:

$$\partial_\mu\left(\frac{\partial\mathcal{L}}{\partial(\partial_\mu\phi^a)}\right) = \frac{\partial\mathcal{L}}{\partial\phi^a}. \tag{3.4.3}$$

It is convenient to simplify the appearance of this and other equations by introducing a notation

$$\Pi_a^\mu \stackrel{\text{def}}{=} \frac{\partial\mathcal{L}}{\partial(\partial_\mu\phi^a)}. \tag{3.4.4}$$

The function $\Pi_a^0$ is the canonical momentum corresponding to the field $\phi^a$. It is also generally best to suppress the $a$ index and think of $\Pi^\mu$ as a row vector. With these conventions, the equations of motion can be written:

$$\partial_\mu\Pi^\mu = \frac{\partial\mathcal{L}}{\partial\phi}. \tag{3.4.5}$$

Let $\phi'(\tau, \phi)$ be a real action on the vector of fields $\phi$. The velocities $\mathbf{D}$ and $\mathbf{TD}$ of the real action on $\phi$ and $\partial_\mu\phi$ are given by:

$$\mathbf{D}\phi = \frac{\partial}{\partial\tau}\phi'(\tau, \phi)\bigg|_{\tau=0}, \tag{3.4.6}$$

$$\mathbf{TD}\partial_\mu\phi = \partial_\mu\mathbf{D}\phi.$$

**Remark 3.4.7.** The notation $\phi'(\tau, \phi)$ is helpful here because the value of $\phi'$ at $x$ may depend on the value of $\phi$ somewhere else. To be entirely correct, we should specify the analogue of the definition 3.1.5 of a real action:

$$\phi'(s + t, \phi) = \phi'\big(s, \phi'(t, \phi)\big) \quad \text{and} \quad \phi'(0, \phi) = \phi. \tag{3.4.8}$$

As we saw in 3.2.16, if we use the equations of motion, we can convert the infinitesimal action on the Lagrangian density into a total derivative:

$$\mathbf{D}\mathcal{L} = \frac{\partial \mathcal{L}}{\partial \phi}\mathbf{D}\phi + \frac{\partial \mathcal{L}}{\partial(\partial_\mu \phi)}\mathbf{T}\mathbf{D}\partial_\mu \phi$$
$$= (\partial_\mu \Pi^\mu)\mathbf{D}\phi + \Pi^\mu \partial_\mu \mathbf{D}\phi \qquad (3.4.9)$$
$$= \partial_\mu(\Pi^\mu \mathbf{D}\phi).$$

Then, following 3.2.17, there is a simple condition on the infinitesimal action from which a conservation law can be derived:

$$\mathbf{D}\mathcal{L} = \partial_\mu f^\mu \quad \Longrightarrow \quad \partial_\mu j^\mu = 0, \quad \text{where} \quad j^\mu = \frac{\partial \mathcal{L}}{\partial(\partial_\mu \phi)}\mathbf{D}\phi - f^\mu. \qquad (3.4.10)$$

Of course, to avoid triviality of the conserved current $j^\mu$, we must satisfy the symmetry condition $\mathbf{D}\mathcal{L} = \partial_\mu f^\mu$ without using the equations of motion.

The significance of a conserved current is that we can interpret $j^0$ as a density of stuff and $\bar{j}$ as the velocity of that stuff. If the current is conserved, then the amount of stuff is also conserved. In detail, define the amount of stuff $Q$ by

$$Q(t) \stackrel{\text{def}}{=} \int d^3\bar{x}\, j^0(t,\bar{x})$$
$$= \int d^3\bar{x}\, \Pi_0(\mathbf{D}\phi). \qquad (3.4.11)$$

Then $Q$ is actually constant:

$$\dot{Q} = \partial_0 \int d^3\bar{x}\, j^0(x)$$
$$= -\int d^3\bar{x}\, \partial_r j^r(x) = 0, \qquad (3.4.12)$$

assuming that the velocity $\bar{j}(x)$ is zero at spatial infinity. Summarizing, we have:

**Theorem 3.4.13.** *Let $\phi'(\tau,\phi)$ be a real action on a vector $\phi$ of fields. If the infinitesimal action on a Lagrangian density $\mathcal{L}$ satisfies $\mathbf{D}\mathcal{L} = \partial_\mu f^\mu$, then $j^\mu = \Pi^\mu \mathbf{D}\phi - f^\mu$ is a conserved current, and $Q = \int d^3\bar{x}\, j^0(x)$ is a conserved quantity.*

**Remark** 3.4.14. Note that $f^\mu$ in the symmetry condition 3.4.10 is not unique since only its divergence is determined. Indeed, if $a^{\mu\nu}$ is any anti-symmetric tensor which vanishes at spatial infinity, then

$$\partial_\mu f^\mu = \partial_\mu(f^\mu + \partial_\nu a^{\mu\nu}). \qquad (3.4.15)$$

Adding $\partial_\nu a^{\mu\nu}$ to $f^\mu$ changes the conserved current to $j^\mu - \partial_\nu a^{\mu\nu}$. However, since $a^{00} = 0$, $Q$ is not affected:

$$Q' = \int d^3\bar{x}\, j^0 - \partial_\nu a^{0\nu} = \int d^3\bar{x}\, j^0 = Q. \qquad (3.4.16)$$

This freedom to adjust $f^\mu$ can be used to simplify the form of the current $j^\mu$.

As noted in Section 3.2 on symmetries in classical mechanics, the problem now is to show that a real action which satisfies the condition in 3.4.10 is a symmetry;

it is not obvious that it commutes with time evolution. Certainly the conserved quantity $Q$ derived from the real action generates a symmetry, but it is not easy to show that the Poisson action of $Q$ generates the original real action.

For simplicity, we shall split the symmetries of field theory into *internal* symmetries, which leave the point of evaluation unchanged, and *external* symmetries, which in this case means Poincaré symmetries.

To cover internal symmetries, we can reasonably write

$$\phi'(\tau, \phi) = \tau \cdot \phi, \tag{3.4.17}$$

and, since the infinitesimal action of an internal symmetry cannot generate extra spatial derivatives, assume the symmetry condition $\mathbf{D}\mathcal{L} = 0$. With this assumption, it is clear that the Lagrangian density and action are invariant, and so the real action maps trajectories into trajectories. To summarize, we state a theorem:

**Theorem 3.4.18.** *If a real action $\phi' = \tau \cdot \phi$ on a vector $\phi$ of fields satisfies $\mathbf{D}\mathcal{L} = 0$, then $\phi$ is a solution of the Euler–Lagrange equations if and only if $\phi'$ is a solution.*

**Homework 3.4.19.** Assuming that $\mathbf{D}\mathcal{L} = 0$, verify that Poisson action of the conserved quantity is the infinitesimal real action.

For Poincaré or *external* symmetries, a Lagrangian density $\mathcal{L}$ is said to be invariant if the action of the Poincaré group element simply shifts the point of evaluation of $\mathcal{L}$. This implies that $\mathcal{L}$ is translation invariant if there is no explicit dependence on $x$, and is Lorentz invariant if all Lorentz indices are contracted. These conditions on the Lagrangian density lead to the same conditions on the Euler–Lagrange equations. Obviously, if the equations of motion are translation invariant, then a translated solution will also be a solution, and similarly for real actions derived from Lorentz generators.

**Theorem 3.4.20.** *If a Poincaré generator is used to define a real action $\phi'(\tau, \phi)$ on a vector $\phi$ of fields, and if a Lagrangian density $\mathcal{L}$ satisfies*

$$\tau\mathcal{L}\big(x, \phi(x), \partial_\mu\phi(x)\big) = \mathcal{L}\big(\tau \cdot x, \phi(\tau \cdot x), \partial_\mu\phi(\tau \cdot x)\big), \tag{3.4.21}$$

*then $\phi$ is a solution of the Euler–Lagrange equations if and only if $\phi'$ is a solution.*

**Remark 3.4.22.** In classical mechanics, a conserved quantity acting through the Poisson bracket generates a symmetry, and a symmetry through Noether's formula gives rise to a conserved quantity. In classical field theory, the situation is similar: a symmetry gives rise to a conserved current from which we obtain a conserved quantity which in turn generates the symmetry again through the Poisson bracket. The only difficulty here is that the Poisson bracket in field theory uses functional differentiation and so will only apply to conserved quantities which are integrals over space. It is therefore possible to find conserved quantities in field theories which are not associated with symmetries.

**Homework 3.4.23.** Using the definition in Section 2.6 of the Poisson bracket in field theory, show that $\{\phi, Q\} = \mathbf{D}\phi$.

**Remark 3.4.24.** In the case where the Lagrangian density $\mathcal{L}$ has a standard kinetic term $\partial_\mu\phi \, \partial^\mu\phi$ and no other occurrences of derivatives, we can show that the Poisson action of the conserved quantity $Q$ is the original infinitesimal action.

The Poisson bracket which gives the action of $Q$ on configuration space works out as follows:

$$\{\phi, Q\} = \int d^3\bar{x} \, \{\phi, \Pi\} \mathbf{D}\phi + \Pi\{\phi, \mathbf{D}\phi\} - \{\phi, f^0\}$$

$$= \mathbf{D}\phi + \Pi \frac{\partial \mathbf{D}\phi}{\partial \Pi} - \frac{\partial f^0}{\partial \Pi}.$$

(3.4.25)

Note that the assumptions on $\mathcal{L}$ imply that $\Pi = \partial_0 \phi$.

To evaluate further we need information about $f^0$. From the defining property of $f^\mu$ we find

$$\mathbf{D}\mathcal{L} = \frac{\partial \mathcal{L}}{\partial \phi} \mathbf{D}\phi + \frac{\partial \mathcal{L}}{\partial(\partial_\mu \phi)} \partial_\mu \mathbf{D}\phi + \frac{\partial \mathcal{L}}{\partial x^\mu} \mathbf{D}x^\mu_\cdot$$

$$= \frac{\partial f^\mu}{\partial \phi} \partial_\mu \phi + \frac{\partial f^\mu}{\partial(\partial_\nu \phi)} \partial_\mu (\partial_\nu \phi) + \frac{\partial f^\mu}{\partial x^\mu}.$$

(3.4.26)

In this equation, we see a unique occurrence of $\partial f^0 / \partial \Pi$; it has coefficient $\partial_0^2 \phi$. Again using our assumptions on the specific form of $\mathcal{L}$, this coefficient could only occur in the first line as a linear contribution to the second term. Hence, differentiating the equation with respect to $\partial_0 \Pi$,

$$\frac{\partial f^0}{\partial \Pi} = \frac{\partial \mathcal{L}}{\partial(\partial_0 \phi)} \frac{\partial}{\partial(\partial_0 \Pi)} \partial_0 \mathbf{D}\phi = \Pi \frac{\partial}{\partial \Pi} \mathbf{D}\phi.$$

(3.4.27)

Substituting in the equation above for $\{\phi, Q\}$, we conclude that

$$\{\phi, Q\} = \mathbf{D}\phi.$$

(3.4.28)

Thus the method of generating a conservation law from a symmetry generalizes from classical mechanics to classical field theory with but one significant difference: in field theory, a symmetry generates a conserved current rather than a conserved quantity. One consequence of this difference is that in Hamiltonian particle mechanics every conserved quantity generates a symmetry, but in field theory there may be conserved quantities that do not come from conserved currents and have no relation to symmetries.

## 3.5  Application: Complex Quantum Fields

The simplest example of symmetry in field theory is the $SO(2)$ symmetry on a Lagrangian density containing two hermitian fields. In this section, we shall analyze this example for two purposes. First, the example is important in its own right as it describes fields with electric charge. Second, it introduces in a concrete fashion the general idea of matrix groups acting as symmetries, thereby preparing the ground for the following sections.

Let $\phi = (\phi^1, \phi^2)^\top$, where $\phi^a$ are hermitian fields. Let $X$ be the generator of $SO(2)$:

$$X \overset{\text{def}}{=} \begin{pmatrix} 0 & -1 \\ 1 & 0 \end{pmatrix},$$

(3.5.1)

so that by direct computation

$$e^{\tau X} \overset{\text{def}}{=} 1 + \tau X + \frac{\tau^2}{2!} X^2 + \cdots$$

$$= \cos \tau \, 1 + \sin \tau \, X$$

$$= \begin{pmatrix} \cos \tau & -\sin \tau \\ \sin \tau & \cos \tau \end{pmatrix}.$$

(3.5.2)

Then the Lagrangian density

$$\mathcal{L} = \frac{1}{2}\partial_\mu\phi_a\partial^\mu\phi^a - \frac{1}{2}\mu^2(\phi_a\phi^a) \tag{3.5.3}$$

has a symmetry

$$\phi'(\tau,\phi) = \begin{pmatrix} \cos\tau & -\sin\tau \\ \sin\tau & \cos\tau \end{pmatrix}\begin{pmatrix} \phi^1 \\ \phi^2 \end{pmatrix}. \tag{3.5.4}$$

The velocity $\mathbf{D}\phi$ of the real action is given by

$$\mathbf{D}\phi = \begin{pmatrix} 0 & -1 \\ 1 & 0 \end{pmatrix}\begin{pmatrix} \phi^1 \\ \phi^2 \end{pmatrix} = \begin{pmatrix} -\phi^2 \\ \phi^1 \end{pmatrix}. \tag{3.5.5}$$

Notice the explicit appearance of the symmetry generator $X$ in this equation.

Clearly $\mathbf{D}\mathcal{L} = 0$. Following the classical formula 3.4.11 for a conserved quantity, the quantum conserved quantity $Q$ is defined by:

$$
\begin{aligned}
Q &\overset{\text{def}}{=} \int d^3\bar{x}\, \Pi^0 X\phi \\
&= \int d^3\bar{x}\, (\Pi^0_2\phi^1 - \Pi^0_1\phi^2) \\
&= \int d^3\bar{x}\, (\partial^0\phi^2)\phi^1 - (\partial^0\phi^1)\phi^2 \\
&= i\int d^3\bar{k}\, (a^2(\bar{k})^\dagger a^1(\bar{k}) - a^1(\bar{k})^\dagger a^2(\bar{k})),
\end{aligned}
\tag{3.5.6}
$$

where we have substituted the creation/annihilation operator expansion of the free field into the integral. Since $[a^{1\dagger}, a^2] = 0$ and $[a^{2\dagger}, a^1] = 0$, the operator $Q$ can be put into normal order without introducing any extra constants, undefined or otherwise.

**Remark** 3.5.7. Normal ordering will generally be necessary for conserved quantities which involve Fermi fields. For a free Fermi field, normal ordering of a conserved quantity will involve throwing away an undefined constant. This will not affect commutators or the property of conservation. For interacting theories, however, if normal ordering could be defined, it might cause an anomaly by destroying the conservation property of the 'conserved quantity'.

Using the formula 3.5.6 and the invariance of the vacuum $Q|0\rangle = 0$, we can evaluate the action of $Q$ on the one-particle states:

$$
\begin{aligned}
Q|\bar{p}^{(r)}\rangle &= Qa^r(\bar{p})^\dagger|0\rangle \\
&= (Qa^r(\bar{p})^\dagger - a^r(\bar{p})^\dagger Q)\,|0\rangle \\
&= i\int d^3\bar{k}\, [a^2(\bar{k})^\dagger a^1(\bar{k}) - a^1(\bar{k})^\dagger a^2(\bar{k}), a^r(\bar{p})^\dagger]\,|0\rangle \\
&= i\int d^3\bar{k}\, \delta^{(3)}(\bar{k} - \bar{p})(\delta^{r1}a^2(\bar{k})^\dagger - \delta^{r2}a^1(\bar{k})^\dagger)\,|0\rangle \\
&= i\delta^{r1}|\bar{p}^{(2)}\rangle - i\delta^{r2}|\bar{p}^{(1)}\rangle \\
&= -iX^r{}_s|\bar{p}^{(s)}\rangle.
\end{aligned}
\tag{3.5.8}
$$

Thus $Q$ appears to act like the matrix $-iX$, where $X$ is the symmetry generator defined in 3.5.1. The factor of $i$ in this relationship arises because, to be a real generator of an orthogonal rotation, $X$ must be skew-symmetric, while to be an observable, $Q$ must be hermitian. The factor of $-1$ can be absorbed in the last line by substituting $X^\top$ for $-X$. The matrix representing an operator in coordinates arises transposed in the action of the operator on a basis of vectors. It is therefore natural to substitute $X^\top$ for $-X$. Hence $Q$ actually corresponds to the matrix $iX$.

**Homework 3.5.9.** Use 3.5.6 to show that the Heisenberg action of the conserved quantity on the fields $\phi$ is given by

$$[\phi^a, Q] = iX^a{}_b\phi^b.$$

Associating $Q$ with $iX$ leads to associating $-iQ$ with $X$. Thus, when $Q$ is used to generate the unitary representation $U(\exp(\tau X))$ of the symmetry, the usual convention defines $U$ by

$$U(\exp(\tau X)) \overset{\text{def}}{=} e^{-i\tau Q}. \tag{3.5.10}$$

Notice that $U$ is defined on elements of $SO(2)$. Since $\exp(\tau X) = \exp(\tau_0 X)$ has solutions $\tau_0 + 2k\pi$, this means that we must check that the right-hand side is constant on this set. This condition is clearly equivalent to $\exp(-2\pi iQ) = 1$. But the action of $Q$ on particle states and on fields is basically the action of $X$ with no scale factor, so this condition is satisfied. Things do not always work out so well; as we shall see in Chapter 6, unitary representations can be many-valued.

**Homework 3.5.11.** Let $U_\tau = \exp(-i\tau Q)$, $R = \exp(\tau X)$, and $\phi = (\phi_1, \phi_2)^\top$. Prove that $U_\tau^\dagger \phi U_\tau = R\phi$. (Hint: differentiate and use Homework 3.5.9 to show that both sides satisfy the same initial-value problem.)

The following display summarizes the applications of the conserved quantity in field theory. It shows the relationship between the action of the conserved quantity and the action of the unitary representation on states and fields:

$$\textbf{Schrödinger}: \begin{pmatrix} e^{-i\tau Q}|p\rangle \\ {\scriptstyle\frac{\partial}{\partial\tau}}\Big\downarrow \quad \Big\uparrow{\scriptstyle\exp} \\ -iQ|p\rangle \end{pmatrix}, \quad \textbf{Heisenberg}: \begin{pmatrix} e^{i\tau Q}\phi e^{-i\tau Q} \\ {\scriptstyle\frac{\partial}{\partial\tau}}\Big\downarrow \quad \Big\uparrow{\scriptstyle\exp} \\ [\phi, -iQ] \end{pmatrix}. \tag{3.5.12}$$

Since $Q$ mixes particle types by destroying type-one particles and creating type-two particles and vice versa, in order to find the particles which carry $Q$-quantum numbers it is necessary to change basis and diagonalize $Q$. Since the action 3.5.8 of $Q$ on one-particle states is the action of $iX$, we can use the change of basis that diagonalizes $iX$ to diagonalize $Q$:

$$\frac{1}{\sqrt{2}}\begin{pmatrix} 1 & -i \\ 1 & i \end{pmatrix}\begin{pmatrix} 0 & -i \\ i & 0 \end{pmatrix}\frac{1}{\sqrt{2}}\begin{pmatrix} 1 & 1 \\ i & -i \end{pmatrix} = \begin{pmatrix} 1 & 0 \\ 0 & -1 \end{pmatrix}. \tag{3.5.13}$$

(Note that the change of basis must be unitary if we are to preserve the canonical form of the kinetic energy term.)

Define a new basis $\psi$, $\psi^\dagger$ for the fields by

$$\begin{pmatrix} \psi \\ \psi^\dagger \end{pmatrix} \overset{\text{def}}{=} \frac{1}{\sqrt{2}}\begin{pmatrix} 1 & -i \\ 1 & i \end{pmatrix}\begin{pmatrix} \phi_1 \\ \phi_2 \end{pmatrix}. \tag{3.5.14}$$

Substituting the integral forms of the free fields $\phi_r$ into this definition, we find that

$$\psi(x) = \int \frac{d^3\bar{k}}{(2\pi)^{3/2}(2\omega(\bar{k}))^{1/2}}(e^{ix\cdot k}c(\bar{k})^\dagger + e^{-ix\cdot k}b(\bar{k})),$$

$$\psi^\dagger(x) = \int \frac{d^3\bar{k}}{(2\pi)^{3/2}(2\omega(\bar{k}))^{1/2}}(e^{ix\cdot k}b(\bar{k})^\dagger + e^{-ix\cdot k}c(\bar{k})),$$

(3.5.15)

where the new creation and annihilation operators are given by:

$$\begin{pmatrix} c^\dagger & b \\ b^\dagger & c \end{pmatrix} = \frac{1}{\sqrt{2}}\begin{pmatrix} 1 & -i \\ 1 & i \end{pmatrix}\begin{pmatrix} a^{1\dagger} & a^1 \\ a^{2\dagger} & a^2 \end{pmatrix}.$$

(3.5.16)

From the commutation relations of the $a^r(\bar{k})$'s and $a^r(\bar{k})^\dagger$'s, we find that the only non-zero commutators among the new set of operators are

$$[b(\bar{p}), b(\bar{q})^\dagger] = \delta^{(3)}(\bar{p} - \bar{q}) \quad \text{and} \quad [c(\bar{p}), c(\bar{q})^\dagger] = \delta^{(3)}(\bar{p} - \bar{q}).$$

(3.5.17)

From these commutation relations or directly from the equal-time commutation relations for the $\phi_r$, we find the equal-time commutation relations for $\psi$ and $\psi^\dagger$ to be

$$[\psi(t,\bar{x}), \psi(t,\bar{y})] = [\psi(t,\bar{x}), \partial_0\psi(t,\bar{y})] = [\partial_0\psi(t,\bar{x}), \partial_0\psi(t,\bar{y})] = 0,$$

$$[\psi(t,\bar{x}), \partial_0\psi(t,\bar{y})^\dagger] = [\psi(t,\bar{x})^\dagger, \partial_0\psi(t,\bar{y})] = i\delta^{(3)}(\bar{x} - \bar{y}),$$

(3.5.18)

$$[\psi(t,\bar{x})^\dagger, \psi(t,\bar{y})^\dagger] = [\psi(t,\bar{x})^\dagger, \partial_0\psi(t,\bar{y})^\dagger] = [\partial_0\psi(t,\bar{x})^\dagger, \partial_0\psi(t,\bar{y})^\dagger] = 0.$$

With respect to the new basis of operators, $Q$ is diagonal:

$$Q = \int d^3\bar{k}\,(b(\bar{k})^\dagger b(\bar{k}) - c(\bar{k})^\dagger c(\bar{k})).$$

(3.5.19)

An eigenvalue of $Q$ acting on the Fock space is called the *charge* of the associated eigenstate.

Either from the new formula for $Q$ or directly from the diagonalization 3.5.13 of $iX$, we see that the new operators are eigenoperators of $Q$:

$$[Q, b(\bar{k})^\dagger] = b(\bar{k})^\dagger, \qquad [Q, b(\bar{k})] = -b(\bar{k});$$

$$[Q, c(\bar{k})^\dagger] = -c(\bar{k})^\dagger, \qquad [Q, c(\bar{k})] = c(\bar{k}).$$

(3.5.20)

Just as a ladder operator, by virtue of being an eigenoperator of the Hamiltonian, was seen in Section 2.2 to raise energy by the amount of its eigenvalue, so these new creation and annihilation operators, by virtue of being eigenoperators of $Q$, raise charge by their eigenvalues. For example,

$$Q|\bar{k}^{(b)}\rangle = Qb(\bar{k})^\dagger|0\rangle$$

$$= (b(\bar{k})^\dagger Q + b(\bar{k})^\dagger)|0\rangle$$

(3.5.21)

$$= |\bar{k}^{(b)}\rangle.$$

Hence $b(\bar{k})^\dagger$ creates particles of positive $Q$-charge at momentum $\bar{k}$. Similarly, $c(\bar{k})^\dagger$ creates particles of negative $Q$-charge with momentum $\bar{k}$. By convention, if we call the $b$-type particles 'particles,' then we call the $c$-type particles 'anti-particles.' Overall then, we see $Q$ as a conserved quantity analogous to electric charge. Indeed, $Q$ could be electric charge.

Because of the form of the definition of $\psi$, the original hermitian fields are known as *real fields* and $\psi$ is called a *complex field*. The expansion of $\psi$ in terms of the pure charge creation and annihilation operators shows that $\psi$ either destroys a positively charged particle or creates a negatively charged one. In either case, $\psi$ lowers charge by one unit.

Similarly, $\psi^\dagger$ is a complex quantum field which creates positively charged particles and destroys negatively charged ones. Note, however, that $\psi^\dagger$ does not contain any degrees of freedom not already present in $\psi$. The two real degrees of freedom with which we started are both present in both $\psi$ and $\psi^\dagger$.

By convention, the charge of a complex field is the charge of the particles it annihilates. The following homework uses the Heisenberg action of $Q$ on the complex fields to justify this convention.

**Homework 3.5.22.** What is the formula for the conserved current $j^\mu$ and quantity $Q$ in terms of $\psi$ and $\psi^\dagger$? Use the commutation relations 3.5.18 to show that $[\psi, Q] = \psi$.

From the change of basis 3.5.13, we find that the real action on the complex fields is

$$\tau \cdot \begin{pmatrix} \psi \\ \psi^\dagger \end{pmatrix} = \begin{pmatrix} e^{-i\tau} & 0 \\ 0 & e^{i\tau} \end{pmatrix} \begin{pmatrix} \psi \\ \psi^\dagger \end{pmatrix} = \begin{pmatrix} e^{-i\tau}\psi \\ e^{i\tau}\psi^\dagger \end{pmatrix}. \tag{3.5.23}$$

This result supports the general principle that a complex field $\psi'$ with charge $q$ is transformed by $\tau \cdot \psi' = \exp(-iq\tau)\psi'$.

As a final technical point on complex fields, note that $\psi$ and $\psi^\dagger$ cannot be varied separately in the action $S(\psi, \psi^\dagger)$ since the value of one field determines the value of the other. However, if we treat $\psi$ and $\psi^\dagger$ as formally independent, so that

$$\delta S = \int d^4 x \left( F\,\delta\psi + G\,\delta\psi^\dagger \right) = 0$$
$$\implies F = G = 0, \tag{3.5.24}$$

and then do the computation properly by varying $\phi^1$ and $\phi^2$,

$$\delta S = \frac{1}{\sqrt{2}} \int d^4 x \left( (F+G)\,\delta\phi^1 + i(F-G)\,\delta\phi^2 \right) = 0$$
$$\implies F + G = F - G = 0, \tag{3.5.25}$$

we actually arrive at the same conclusion. Hence, when a Lagrangian is written in terms of complex fields, one may compute variations of functionals in the complex fields with the assumption that $\delta\psi$ and $\delta\psi^\dagger$ are independent.

The free field Lagrangian density for the hermitian fields $\phi_1$ and $\phi_2$ may be written in terms of the complex field $\psi$:

$$\mathcal{L} = \partial_\mu \psi^\dagger \partial^\mu \psi - \mu^2 \psi^\dagger \psi. \tag{3.5.26}$$

From the above remarks on variation with respect to a complex field and its conjugate, we see that the usual Euler–Lagrange formula determines the equation of motion for $\psi$:

$$\partial_\mu \frac{\partial \mathcal{L}}{\partial(\partial_\mu \psi^\dagger)} = \frac{\partial \mathcal{L}}{\partial \psi^\dagger} \quad \Longrightarrow \quad \partial_\mu \partial^\mu \psi = -\mu^2 \psi. \tag{3.5.27}$$

The conserved quantity and the change of basis to complex fields $\psi$ and $\psi^\dagger$ depend only on the existence of the symmetry, not on the particular form of the Lagrangian. Thus the commutation relations for $\psi$ and $\psi^\dagger$ follow from the commutation relations for $\phi$ and do not depend on the fact that we were working with free fields above.

It is easy to write Lagrangians representing the dynamics of charged scalar particles. For example, if we write $\psi$ for a charged scalar and $\phi$ for a real scalar, then

$$\mathcal{L}^{(1)} = \partial_\mu \psi^\dagger \partial^\mu \psi + \partial_\mu \phi\, \partial^\mu \phi - m^2 \psi^\dagger \psi - \frac{\mu^2}{2}\phi^2 - \lambda \psi^\dagger \psi \phi \tag{3.5.28}$$

roughly represents a charged nucleon interacting with a neutral meson, and

$$\mathcal{L}^{(2)} = \partial_\mu \psi^\dagger \partial^\mu \psi - \frac{\alpha}{2}(\psi^\dagger \psi - v^2)^2 \tag{3.5.29}$$

illustrates the concept of spontaneous symmetry breaking – see the following homework. Note that the masses of all fields are determined from the quadratic part of the Lagrangian density.

**Homework 3.5.30.** The potential energy in the Lagrangian density 3.5.29 has a classical minimum at $\psi^\dagger \psi = v^2$. Expand the Lagrangian density around a classical minimum $\psi_{\text{cl}} = v e^{i\alpha}$ by substituting

$$\psi(x) = \big(v + r(x)\big)e^{i(\alpha + \theta(x))}$$

where $r$ and $\theta$ are two real quantum fields. Show that $\theta$ is massless, while $r$ is massive.

This homework illustrates a general result due to Goldstone, that when a global symmetry breaks, the theory must contain a corresponding massless boson. Note that, though $\alpha$ could be a function of $x$, we expect that in the volume of space occupied by laboratory apparatus $\alpha$ will effectively be constant.

When a Lagrangian density, expressed in terms of real fields, has an $SO(2)$ symmetry, then the conserved quantity may be diagonalized by a linear change of basis to complex charged fields $\psi_r$ and real charge-zero fields $\phi_s$. The charge $q_r$ of the field $\psi_r$ is defined as the charge of the particles that $\psi_r$ annihilates. Thus

$$[\psi_r, Q] = q_r \psi_r. \tag{3.5.31}$$

In this basis, the symmetry is given by

$$\tau \cdot \psi_r = e^{-iq_r \tau}\psi_r = e^{i\tau Q}\psi_r e^{-i\tau Q} \quad \text{and} \quad \tau \cdot \phi_s = \phi_s. \tag{3.5.32}$$

Finally, for the purpose of variation we may treat the fields $\psi_r$ and $\psi_r^\dagger$ as independent.

## 3.6 Groups of Matrices and their Lie Algebras

In classical mechanics, symmetries may be non-linear transformations of the canonical coordinates, but in quantum field theory symmetries are almost always linear in the fields. (Supersymmetries, however, are often non-linear.) Linear symmetries are represented in coordinates by groups of matrices. Therefore, before introducing the general discussion of symmetry in quantum field theory, some preliminary remarks on the common matrix groups will be appropriate.

There are two types of symmetry in field theory, internal and external. An internal symmetry acts on the fields but not on space-time, while external symmetry means Poincaré or Lorentz symmetry, which acts on space-time and also on some types of field.

The characteristic property of a Lorentz group element $\Lambda$ is that it leaves the relativistic interval between events invariant:

$$g_{\mu\nu}(\Lambda x)^{\mu}(\Lambda y)^{\nu} = g_{\alpha\beta}x^{\alpha}y^{\beta} \quad \text{for all } x \text{ and } y, \tag{3.6.1}$$

Writing $(\Lambda x)^{\mu} = \Lambda^{\mu}{}_{\alpha}x^{\alpha}$ implies that

$$g_{\mu\nu}\Lambda^{\mu}{}_{\alpha}\Lambda^{\nu}{}_{\beta} = g_{\alpha\beta}. \tag{3.6.2}$$

This is a set of sixteen quadratic equations in the entries in $\Lambda$. The set of solutions is the *Lorentz group*, $O(1,3)$. There are four distinct components of $O(1,3)$: any two points in any one component can be connected by a path in $O(1,3)$, while any two elements in different components cannot be so connected. The component of $O(1,3)$ that contains the identity matrix is called the *connected Lorentz group*, $SO(1,3)$. Two useful Lorentz transformations not in the connected Lorentz group are parity $P$ and time reversal $T$:

$$P = \begin{pmatrix} 1 & 0 & 0 & 0 \\ 0 & -1 & 0 & 0 \\ 0 & 0 & -1 & 0 \\ 0 & 0 & 0 & -1 \end{pmatrix} \quad \text{and} \quad T = \begin{pmatrix} -1 & 0 & 0 & 0 \\ 0 & 1 & 0 & 0 \\ 0 & 0 & 1 & 0 \\ 0 & 0 & 0 & 1 \end{pmatrix}. \tag{3.6.3}$$

Each component of the Lorentz group contains exactly one of the four matrices $\mathbf{1}$, $P$, $T$, and $PT$.

The characteristic of the internal symmetries is that, with a correct choice of basis for the fields, they leave Euclidean distance invariant:

$$\delta_{ab}(R\phi)^{a}(R\phi)^{b} = \delta_{rs}\phi^{r}\phi^{s} \quad \text{for all } x \text{ and } y. \tag{3.6.4}$$

Writing $(R\phi)^{a} = R^{a}{}_{r}\phi^{r}$ implies that

$$\delta_{ab}R^{a}{}_{r}R^{b}{}_{s} = \delta_{rs}. \tag{3.6.5}$$

This is a set of $n^2$ quadratic equations in the entries in the matrix $R$. The set of solutions to these equations is the *orthogonal group*, $O(n)$. The orthogonal group has two components, one with determinant $+1$, the other with determinant $-1$. The component with determinant $+1$ is called the *special orthogonal group*, $SO(n)$.

If we choose a basis of complex fields, then we must amend the foregoing by including complex conjugates:

$$\delta_{ab}\big((U\psi)^a\big)^*(U\psi)^b = \delta_{rs}(\psi^r)^*\psi^s \quad \text{for all } x \text{ and } y, \tag{3.6.6}$$

and so

$$\delta_{ab}(U^a{}_r)^*U^b{}_s = \delta_{rs}. \tag{3.6.7}$$

Again we find a set of $n^2$ quadratic equations, this time specifying the *unitary group*, $U(n)$. This group has only one component.

The defining equations of these groups can be written in terms of matrices if we represent the metrics $g_{\mu\nu}$ and $\delta_{ab}$ by the following matrices:

$$g_{\mu\nu} \mapsto G = \text{diag}(1, -1, -1, -1) \quad \text{and} \quad \delta_{ab} \mapsto \mathbf{1}. \tag{3.6.8}$$

Now the groups are defined by the equations:

$$\begin{aligned} O(1,3): & \quad \Lambda^\mathsf{T} G \Lambda = G; \\ O(n): & \quad R^\mathsf{T} R = \mathbf{1}; \\ U(n): & \quad U^\dagger U = \mathbf{1}. \end{aligned} \tag{3.6.9}$$

In general, it is hard to handle a set of simultaneous quadratic equations. To reduce our work to linear equations, we differentiate the defining equations of the groups. To be more precise, let $\Lambda(t)$ be a path in $SO(1,3)$ that passes through $\mathbf{1}$ at time $t = 0$, differentiate the defining equation 3.6.2 with respect to $t$ and evaluate at $t = 0$:

$$g_{\mu\nu}\dot\Lambda^\mu{}_\alpha\delta^\nu{}_\beta + g_{\mu\nu}\delta^\mu{}_\alpha\dot\Lambda^\nu{}_\beta = 0. \tag{3.6.10}$$

Applying the same process to the defining equations for $SO(n)$ and $U(n)$, and contracting the $\delta$'s we find:

$$\begin{aligned} SO(1,3): & \quad g_{\mu\beta}\dot\Lambda^\mu{}_\alpha + g_{\alpha\nu}\dot\Lambda^\nu{}_\beta = 0; \\ SO(n): & \quad \delta_{as}\dot R^a{}_r + \delta_{br}\dot R^b{}_s = 0; \\ U(n): & \quad \delta_{as}(\dot U^a{}_r)^* + \delta_{br}\dot U^b{}_s = 0. \end{aligned} \tag{3.6.11}$$

Write $X$ for the derived matrices and use the matrix forms $G$ and $\mathbf{1}$ of the metrics to obtain the matrix form of these equations:

$$\begin{aligned} SO(1,3): & \quad X^\mathsf{T} G + GX = 0; \\ SO(n): & \quad X^\mathsf{T} + X = 0; \\ U(n): & \quad X^\dagger + X = 0. \end{aligned} \tag{3.6.12}$$

These are the linear equations which constrain infinitesimal movements in the group at $\mathbf{1}$. The solutions are the spaces of tangent vectors at $\mathbf{1}$ to the group considered as the solution set of the quadratic equations 3.6.9 on the appropriate Euclidean vector space of matrices. These tangent spaces are known as the *Lie algebras* of the groups, and called $so(1,3)$, $so(n)$, and $u(n)$ respectively.

If we use the metric $g_{\mu\nu}$ to raise and lower indices, then the component linear equation for $so(1,3)$ becomes:

$$X_{\beta\alpha} + X_{\alpha\beta} = 0. \tag{3.6.13}$$

Therefore $X_{\alpha\beta}$ is an anti-symmetric $4 \times 4$ matrix. This set of matrices can be parameterized by the entries above the diagonal and therefore form a six-dimensional real vector space. This vector space is the space of tangent vectors to the group $SO(1,3)$ at $\mathbf{1}$ regarded as a surface in the Euclidean space $M_4(\mathbf{R})$ of real, $4 \times 4$ matrices.

Similarly $so(n)$ consists of real anti-symmetric $n \times n$ matrices, so $so(n)$ has dimension $\frac{1}{2}n(n-1)$, and $u(n)$ is the vector space of anti-hermitian $n \times n$ matrices, so $u(n)$ has dimension $n^2$.

The beauty of this particular type of tangent vector is that we can reconstruct group elements from these infinitesimal motions by exponentiating. Let $X_{\alpha\beta}$ be an anti-symmetric matrix in $M_4(\mathbf{R})$. Use the inverse metric $g^{\mu\alpha}$ to raise the first index of $X_{\alpha\beta}$

$$X^{\mu}{}_{\alpha} \stackrel{\text{def}}{=} g^{\mu\alpha} X_{\alpha\beta}. \tag{3.6.14}$$

With one upper and one lower index, $X = X^{\mu}{}_{\alpha}$ can multiply itself, so that the powers $X^k$ may be defined by the dummy suffix convention. Define $\Lambda(t)$ by

$$\Lambda(t) = 1 + tX + \frac{t^2}{2!}X^2 + \cdots = e^{tX}. \tag{3.6.15}$$

To show that $\Lambda(t)$ satisfies the defining equation 3.6.2 for all $t$, we work with the matrix form of this equation, 3.6.9, and the derived Lie algebra condition, 3.6.12. It is clear that $\Lambda(t)$ satisfies 3.6.9 at $t = 0$. Now differentiate the left-hand side of 3.6.9 evaluated at $\Lambda = \Lambda(t)$ to find that it never changes:

$$\begin{aligned}
\frac{d}{dt}\left(\Lambda(t)^{\top} G \Lambda(t)\right) &= \frac{d}{dt}\left(e^{tX^{\top}} G e^{tX}\right) \\
&= e^{tX^{\top}} X^{\top} G e^{tX} + e^{tX^{\top}} G X e^{tX} \\
&= e^{tX^{\top}}(X^{\top}G + GX)e^{tX} = 0.
\end{aligned} \tag{3.6.16}$$

Consequently, $\Lambda(t)$ satisfies 3.6.9 for all $t$.

**Homework 3.6.17.** Show that $\Lambda(t)$ is uniquely determined by the properties

$$\Lambda(0) = \mathbf{1}, \quad \dot{\Lambda}(0) = X, \quad \Lambda(s+t) = \Lambda(s)\Lambda(t).$$

Similarly, by exponentiating elements in $so(n)$ and $u(n)$ we obtain matrices in the groups $SO(n)$ and $U(n)$. Clearly, any element near $\mathbf{1}$ is the exponential of some matrix in the Lie algebra of the group, but, while this is true for every element of $SO(n)$ and $U(n)$, there are elements in $SO(1,3)$ that are not the exponential of any element in $so(1,3)$.

Exponentiation as a map from small Lie algebra elements to the group is actually a parameterization of the group near the identity element. Thus the dimension of the Lie algebra (which is the dimension of the tangent space to the group) is also the dimension of the group.

If we have a group action on a system, such as translation of the solar system in a fixed direction or rotation of an experimental set-up about some fixed axis, it is clear that, given the initial velocity of movement, we can continue moving the system in exactly the same way at exactly the same rate and generate from that initial velocity a continuing manifest motion of the system. This is the meaning of Homework 3.6.17. Therefore we need not consider group actions, but only infinitesimal group actions, that is, actions of Lie algebras.

**Homework 3.6.18.** Show that $\det(e^X) = e^{\operatorname{tr}(X)}$.

In addition to $SO(1,3)$, $O(n)$, $SO(n)$ and $U(n)$, the other common symmetry group is the *special unitary group*, $SU(n)$, the set of unimodular ($\det U = 1$) unitary matrices. The constraint of unimodularity in the group becomes, using Homework 3.6.18, a constraint of tracelessness on the Lie algebra. Thus $su(n)$ is the set of traceless, $n \times n$ anti-hermitian matrices. The dimension of $su(n)$ is therefore $n^2 - 1$.

Note that, though $u(n) = su(n) \oplus u(1)$, $U(n)$ is not a product of $SU(n)$ and $U(1)$. Indeed, the only multiplicative maps from $U(1) \times SU(n)$ to $U(n)$ are the $f^k$ defined by

$$\begin{aligned} f^k : U(1) \times SU(n) &\rightarrow U(n) \\ (e^{i\tau} \ , \quad U) &\mapsto e^{ik\tau}U, \end{aligned} \tag{3.6.19}$$

and $f^k$ maps $kn$ points in $U(1) \times SU(n)$ map to each point in $U(n)$. Thus, at best, $U(1) \times SU(n)$ must be regarded as $n$ times as big as $U(n)$.

**Homework 3.6.20.** $SL(2, \mathbf{R})$ is the group of real, $2 \times 2$ matrices with unit determinant. Find the Lie algebra $sl(2, \mathbf{R})$ of $SL(2, \mathbf{R})$. Describe the geometry of $SL(2, \mathbf{R})$ and find the image of the exponential map from $sl(2, \mathbf{R})$ to $SL(2, \mathbf{R})$.

In this section, we have described the groups $O(1,3)$, $SO(1,3)$, $O(n)$, $SO(n)$, $U(n)$, and $SU(n)$, and their Lie algebras. The defining properties of these groups and of their Lie algebras, and the exponentiation of generators should be intimately understood, as these groups are the most common symmetry groups in physics.

# 3.7 Internal Symmetries in Quantum Field Theory

In quantum field theory (before supersymmetry) we only need linear symmetries. There are two types of linear symmetries, namely the Poincaré or external symmetry and the internal symmetries that commute with the Poincaré symmetry. As the internal symmetries commute with all space-time differentiation operators they are simpler to analyze than the external symmetries. In this section, we shall show how a symmetry group can act on the fields of a theory, how to build symmetric Lagrangian densities, and present the formulae for the associated conserved currents and quantities.

In general, if $\phi = (\phi^1, \dots, \phi^n)^\top$ is the vector of fields in a theory, we can define an action of a group $G$ on the theory by choosing $n \times n$ matrices $M(g)$, one for each element $g$ of $G$, subject to the multiplicative condition

$$M(gh) = M(g)M(h), \tag{3.7.1}$$

and letting the group element $g$ act on the fields through the matrix $M(g)$ as follows:

$$g \cdot \phi = M(g)\phi. \tag{3.7.2}$$

Such a matrix-valued function of a group is called a *group representation*.

Now if $\phi$ are real fields and $M$ represents $G$ in $O(n)$, then $\phi^\top\phi$ will be invariant under the group action:

$$
\begin{aligned}
g \cdot (\phi^\top\phi) &= (g \cdot \phi)^\top (g \cdot \phi) \\
&= \left(M(g)\phi\right)^\top \left(M(g)\phi\right) \\
&= \phi^\top M(g)^\top M(g)\phi \\
&= \phi^\top\phi.
\end{aligned} \tag{3.7.3}
$$

Similarly, if $\phi$ are complex fields – let us write $\psi$ for these – and $N$ represents $G$ in $U(n)$, then $\psi^\dagger\psi$ will be invariant. As $g$ is independent of space-time, the scalars $\partial_\mu\phi^\top\partial^\mu\phi$ and $\partial_\mu\psi^\dagger\partial^\mu\psi$ will also be $O(n)$ and $U(n)$ invariant respectively. This permits us to form $G$-symmetric Lagrangian densities out of vectors of fields:

$$
\begin{aligned}
\mathcal{L}_v = {}&\frac{1}{2}\partial_\mu\phi^\top\partial^\mu\phi + \partial_\mu\psi^\dagger\partial^\mu\psi && \text{(kinetic terms)} \\
&-\frac{\mu^2}{2}\phi^\top\phi - m^2\psi^\dagger\psi && \text{(mass terms)} \\
&-a(\phi^\top\phi)^2 - b\phi^\top\phi\psi^\dagger\psi - c(\psi^\dagger\psi)^2 && \text{(interaction terms)}.
\end{aligned} \tag{3.7.4}
$$

The groups $O(n)$ and $U(n)$ also act respectively on real and complex $n \times n$ matrices $M_n(\mathbf{R})$ and $M_n(\mathbf{C})$ as follows. Let $R \in O(n)$, $U \in U(n)$, and let $\Phi$ and $\Psi$ be matrices of real and complex fields respectively. Define the actions as follows:

$$
\begin{aligned}
R \cdot \Phi &\stackrel{\text{def}}{=} R\Phi R^\top; \\
U \cdot \Psi &\stackrel{\text{def}}{=} U\Psi U^\dagger.
\end{aligned} \tag{3.7.5}
$$

Note that these actions preserve symmetry properties of the matrices – symmetric or anti-symmetric, hermitian or anti-hermitian – and their traces.

The invariants associated with this action are the traces of products of powers of $\Phi$ and $\Phi^\top$ or $\Psi$ and $\Psi^\dagger$ and their derivatives. Thus, if we have a group $G$ and representations $M$ and $N$ of $G$ in $O(m)$ and $U(n)$ respectively, then we can form $G$-invariant Lagrangian densities like:

$$
\mathcal{L}_m = \frac{1}{2}\operatorname{tr}(\partial_\mu\Phi^\top\partial^\mu\Phi) + \operatorname{tr}(\partial_\mu\Psi^\dagger\partial^\mu\Psi) \tag{3.7.6}
$$

$$
-\frac{\mu^2}{2}\operatorname{tr}(\Phi^\top\Phi) - m^2\operatorname{tr}(\Psi^\dagger\Psi) - A\operatorname{tr}(\Phi^4) - B\operatorname{tr}(\Phi^2)\operatorname{tr}(\Psi^2) - C\operatorname{tr}(\Psi^4).
$$

(Other quartic invariant interactions terms are possible here since we have not made any symmetry assumptions on the matrices of fields.)

With matrices and vectors of fields, we can structure more invariants. For example, if $O(k)$ acts on a vector $\phi$ and a matrix $\Phi$, we have invariants like

$$
\phi^\top\Phi\phi \quad \text{and} \quad \phi^\top\phi\operatorname{tr}(\Phi^2). \tag{3.7.7}
$$

Constructions of invariants such as those presented here and above are central to field theories.

The action of the matrix groups on matrices is really just a special case of their actions on vectors. For example, if we have a traceless, symmetric matrix $\Phi$ of $n = \frac{1}{2}k(k-1)$ real fields, then we can organize these fields into a vector $\phi$. Hence the action of $O(k)$ on $\Phi$ can be expressed as an action of $O(n)$ on $\phi$ through a quadratic map $M$ from $O(k)$ to $O(n)$. This shows that the matrix group action on matrices is a special case of a representation of that matrix group. Thus the invariant Lagrangian density 3.7.6 constructed out of matrices is actually a special case of that constructed out of vectors 3.7.4. However, it is much more convenient to work with $k \times k$ matrices than vectors of length $\sim k^2$.

**Homework 3.7.8.** Let $SU(2)$ act on 2×2 hermitian matrices $\Psi$ by $g \cdot \Psi = g \Psi g^\dagger$. Let $\sigma_0 = 1$ and $\bar{\sigma}$ be the Pauli matrices. Show that $g$ can be written in the form $g = a^0 \sigma_0 - i a^r \sigma_r$ where $\delta_{\mu\nu} a^\mu a^\nu = 1$. Show that $\Psi$ has a unique representation $\Psi = \psi^\mu \sigma_\mu$. Work out the detailed structure of the map $M$ from $SU(2)$ to $SU(4)$ defined by

$$\Psi' = g \Psi g^\dagger \quad \Longleftrightarrow \quad \psi'^\mu = M(g)^\mu{}_\nu \psi^\nu.$$

(Actually, $M(g)$ is in $SO(4)$.)

As any field theory can be described in a basis of real fields, we shall develop the theory of $SO(n)$ symmetry. It is generally convenient to express a group action in terms of the generators of the group. Thus a group action of $SO(n)$ on a theory of fields $\phi$ may be written:

$$\phi'(\tau, \phi) = e^{\tau X} \phi, \tag{3.7.9}$$

where $X$ is in the Lie algebra $so(n)$. For fixed $X$ the group action becomes a real action. The velocity operator $\mathbf{D}$ of this real action acts as follows:

$$\mathbf{D}\phi^a = X^a{}_b \phi^b,$$
$$\mathbf{TD}\partial_\mu \phi^a = X^a{}_b \partial_\mu \phi^b, \tag{3.7.10}$$
$$\mathbf{D}\mathcal{L} = \Pi^\mu_a X^a{}_b \partial_\mu \phi^b + \frac{\partial \mathcal{L}}{\partial \phi^a} X^a{}_b \phi^b.$$

The condition for the real action of $\phi \mapsto \exp(\tau X)\phi$ to be a symmetry of the Lagrangian is $\mathbf{D}\mathcal{L} = 0$. If this condition is satisfied, $X$ is called a *symmetry generator*. The set of symmetry generators, the Lie algebra of the symmetry group $G$, forms a vector space $\mathcal{G}$ which generates some subgroup of $SO(n)$.

Letting $\Pi^\mu$ be the row vector $(\Pi^\mu{}_1, \dots, \Pi^\mu{}_n)$, for each generator $X$ there is a conserved current $j^\mu$ and a conserved quantity $Q$ defined by

$$j^\mu(X) \overset{\text{def}}{=} \Pi^\mu X \phi \quad \text{and} \quad Q(X) \overset{\text{def}}{=} \int d^3\bar{x}\, \Pi^0 X \phi. \tag{3.7.11}$$

Note that $j^\mu$ and $Q$ are linear functions of $X$.

In this section, we have applied our understanding of the matrix groups to field theory through group representations. We have seen some simple constructions for symmetric Lagrangian densities and presented Noether's formula for the associated conserved current and quantity. In the next section, we deepen our understanding of the conserved quantity.

# 3.8 The Actions and Algebra of Conserved Quantities

Using the formula for the conserved current and quantity derived in the previous section, we shall now see how the conserved quantity acts on fields, states, other conserved quantities, and currents, and how exponentiation of the conserved quantities gives rise to a unitary representation of the original symmetry group.

If we have a theory of interacting hermitian fields with $SO(n)$ symmetry, then we can use the equal-time commutation relations for hermitian fields,

$$\left[\phi^a(t,\bar{x}), \Pi_b(t,\bar{y})\right] = i\delta_b^a \delta^{(3)}(\bar{x} - \bar{y}),$$
$$\left[\phi^a(t,\bar{x}), \phi^b(t,\bar{y})\right] = \left[\Pi_a(t,\bar{x}), \Pi_b(t,\bar{y})\right] = 0, \tag{3.8.1}$$

to compute the effect of the conserved quantity on fields:

$$\begin{aligned}
\left[\phi^c(x), Q(X)\right] &= \int d^3\bar{y}\, \left[\phi^c(x), \Pi_a^0(y) X^a{}_b \phi^b(y)\right] \\
&= \int d^3\bar{y}\, \left[\phi^c(x), \Pi_a^0(y)\right] X^a{}_b \phi^b(y) \\
&= \int d^3\bar{y}\, i\delta_a^c \delta^{(3)}(\bar{x} - \bar{y}) X^a{}_b \phi^b(y) \\
&= i X^c{}_b \phi^b(x).
\end{aligned} \tag{3.8.2}$$

Thus the bracket action of $Q$ reproduces the action of the symmetry generator on the hermitian fields. Hence, as in 3.5.10 for the special case of $SO(2)$, we associate the matrix $iX$ with the operator $Q(X)$.

From this action on the fields, we can deduce that when $\phi$ is a family of free hermitian fields, then the conserved quantity acts in the same way on the creation and annihilation operators as it acts on the fields:

$$\begin{aligned}
\left[Q(X), a^a(k)\right] &= -iX^a{}_b a^b(k), \\
\left[Q(X), a^a(k)^\dagger\right] &= -iX^a{}_b a^b(k)^\dagger.
\end{aligned} \tag{3.8.3}$$

This enables us to deduce the action of $Q(X)$ on Fock space. Assume that the vacuum state $|0\rangle$ is symmetric, so that $Q(X)|0\rangle = 0$. Then:

$$\begin{aligned}
Q(X)|k^{(a)}\rangle &= Q(X) a^a(k)^\dagger |0\rangle \\
&= \left(\left[Q(X), a^a(k)^\dagger\right] + a^a(k)^\dagger Q(X)\right)|0\rangle \\
&= -iX^a{}_b a^b(k)^\dagger |0\rangle = -iX^a{}_b |k^{(b)}\rangle.
\end{aligned} \tag{3.8.4}$$

This is the generalization of the $SO(2)$ result 3.5.8; as explained following that result, $X$ here should really act on the right as $X^\top$.

The tangent spaces to groups have extra structure, the *Lie bracket*. For matrix groups we can make a simple definition of this operation:

$$[X, Y] \overset{\text{def}}{=} XY - YX. \tag{3.8.5}$$

The Lie algebras are closed under this operation. The conserved quantity operators reflect this structure:

**Theorem 3.8.6.** *The conserved quantity operators defined by 3.7.11 satisfy*

$$[-iQ(X), -iQ(Y)] = -iQ([X,Y]).$$

This theorem enables us to select commuting conserved quantities to define quantum numbers just by selecting commuting symmetry generators. The property of the linear function $Q$ established in the theorem makes $-iQ$ a *representation* and $Q$ a *hermitian representation* of the Lie algebra $\mathcal{G}$ of symmetry generators.

**Homework 3.8.7.** Verify that if $X$ and $Y$ are symmetry generators in any one of the Lie algebras $so(3)$, $u(n)$, $su(n)$, or $so(1,3)$, then the Lie bracket $[X,Y]$ of $X$ and $Y$ is in the same Lie algebra.

**Homework 3.8.8.** Use the definition 3.7.11 of the conserved quantity function $Q(X)$ to prove Theorem 3.8.6.

**Homework 3.8.9.** From the definitions 3.7.11 and the canonical commutation relations 3.8.1, show that $[Q(X), j^\mu(Y)] = ij^\mu([X,Y])$. Deduce Theorem 3.8.6.

**Homework 3.8.10.** Use the formula 3.8.4 to directly compute the action of $Q([X,Y])$, $Q(X)Q(Y)$, and $Q(Y)Q(X)$ on $|k^{(a)}\rangle$. The results should confirm the commutation formula in Theorem 3.8.6.

From Theorem 3.8.6 we can see that the map $X \mapsto -iQ(X)$ respects the bracket operation. This implies (by some non-trivial mathematics – see Remark 3.8.21) that the unitary operators defined by

$$U(\exp(X)) \stackrel{\text{def}}{=} \exp(-iQ(X)) \tag{3.8.11}$$

form a unitary representation of the symmetry group, that is, the function $U$ satisfies

$$U(gh) = U(g)U(h). \tag{3.8.12}$$

Note that, as observed in Section 3.5 for the $SO(2)$ example, the function $U$ is single valued on the group because the action of $Q(X)$ on fields and particles is given by $X$ itself with no scale factor.

**Remark 3.8.13.** The minus sign in the exponent of 3.8.11 is found in the evolution operator $\exp(-itH)$, which translates the time coordinate from $t_0$ to $t_0 + t$. The unitary representation of translations, $U(\Delta_a) = \exp(ia \cdot P)$, has the opposite sign. However, since the time-reflection matrix $T$ generates an automorphism $M \to TMT^\top$ of the Poincaré Lie algebra which reverses the signs of the boost generators, reversing the sign of the momenta does not break the multiplicativity condition 3.8.12.

**Homework 3.8.14.** Generalize the $SO(2)$ result in Homework 3.5.11 to show that the action 3.8.2 of $Q(X)$ on fields implies

$$e^{i\tau Q(X)} \phi e^{-i\tau Q(X)} = e^{\tau X} \phi.$$

This equation may be written $U^\dagger(g)\phi U(g) = g\phi$. Show that this last equation is multiplicative, that is, consistent with the representation property 3.8.12 of $U$.

The following diagram summarizes the relationship between the representation $-iQ$ of the Lie algebra $\mathcal{G}$ and the unitary representation $U = e^{-iQ}$ of the Lie group $G$:

$$
\begin{array}{cccc}
\textbf{Lie Algebra} & & \textbf{Lie Group} & \\
\text{Structure} & \text{Element} & \text{Element} & \text{Structure} \\
[X,Y] = Z & X \xrightarrow{\;exp\;} e^{\tau X} & & gh = k \\
\Big\downarrow & \Big\downarrow & \Big\downarrow & \Big\downarrow \\
[-iQ_X, -iQ_Y] = -iQ_Z & -iQ_X \xrightarrow{\;exp\;} e^{-i\tau Q_X} & & U(g)U(h) = U(k)
\end{array}
\tag{3.8.15}
$$

The analysis above generalizes readily to complex fields and unitary symmetries. Writing $\psi^*$ for the column vector of complex fields $\psi^{\dagger\mathsf{T}}$, assuming that $\mathbf{D}\mathcal{L} = 0$, the conserved current and quantity are defined by

$$
\begin{aligned}
X \in u(n) \implies j_\mu(X) &= \Pi_\mu \mathbf{D}\psi + \Pi_\mu^* \mathbf{D}\psi^* \\
&= \Pi_\mu X\psi + \Pi_\mu^* X^* \psi^* \\
\implies Q(X) &= \int d^3\bar{x}\, \Pi_0 X\psi + \Pi_0^* X^* \psi^*.
\end{aligned}
\tag{3.8.16}
$$

From the definition of the conserved quantity, the canonical commutation relations imply that

$$
\left.\begin{aligned}
{}[\psi, Q(X)] &= iX\psi \\
[\psi^*, Q(X)] &= iX^*\psi^*
\end{aligned}\right\}
\tag{3.8.17}
$$

Notice that we have kept the momenta on the left of the fields in order to define $j_\mu$ and $Q$ without normal ordering. If we transpose the last term in formula 3.8.16, then a commutator term will be generated:

$$
\Pi_0^* X^* \psi^* = \psi^\dagger X^\dagger \Pi_\mu^\dagger - i\delta^{(3)}(0)\,\mathrm{tr}\,X^*,
\tag{3.8.18}
$$

and so normal ordering would again be required:

$$
j_\mu = {:}\Pi_0 X\psi - \psi^\dagger X \Pi_\mu^\dagger{:}
\tag{3.8.19}
$$

**Homework 3.8.20.** Show that if we write the complex field $\psi$ in terms of a pair of real fields $\phi_1$ and $\phi_2$, the conserved current 3.8.16 for $\psi$ is equivalent to the real-field formula 3.7.11.

**Remark 3.8.21.** The multiplicativity 3.8.12 of $U$ is not at all obvious. However, we can base a proof on the Campbell–Hausdorff formula for the solution $Z$ of $e^Z = e^X e^Y$:

$$
Z(X,Y) \overset{\text{def}}{=} \ln(e^X e^Y) = X + \int_0^1 dt\, F(e^{\mathrm{ad}\,X} e^{t\,\mathrm{ad}\,Y})Y.
\tag{3.8.22}
$$

In this formula, $\mathrm{ad}\,X$ is the linear transformation of the vector space $\mathcal{G}$ of symmetry generators determined by the Lie bracket with $X$:

$$
\mathrm{ad}(X)Y \overset{\text{def}}{=} [X,Y],
\tag{3.8.23}
$$

and the function $F$ is defined by

$$F(x) \overset{\text{def}}{=} \frac{x \ln x}{x - 1}. \tag{3.8.24}$$

Expanding $F$ around $x = 1$ and integrating yields the first few terms in the formula:

$$Z(X, Y) = X + Y + \frac{1}{2}[X, Y] + \frac{1}{12}[X, [X, Y]] + \frac{1}{12}[Y, [Y, X]] + \cdots \tag{3.8.25}$$

The formula shows that, for sufficiently small $X$ and $Y$, $Z(X, Y)$ is defined by a convergent series whose terms are determined entirely by the Lie algebra structure; one never needs to compute a product like $X^2$ or $XY$.

The application to the unitary representation is as follows. If $X$ and $Y$ are small, then $Q(X)$ and $Q(Y)$ are small as operators on the one-particle state space. Restricting attention to the one-particle state space, since $-iQ$ is a representation of $\mathcal{G}$, and since $Z(X, Y)$ can be computed from Lie brackets, we find that

$$-iQ\big(Z(X, Y)\big) = Z\big(-iQ(X), -iQ(Y)\big). \tag{3.8.26}$$

Now the Campbell–Hausdorff formula is a power-series identity which applies to small operators as well as small matrices, so we can exponentiate the last equation to find

$$\begin{aligned}
\exp\big(-iQ(X)\big) \exp\big(-iQ(Y)\big) &= \exp\Big(Z\big(-iQ(X), -iQ(Y)\big)\Big) \\
&= \exp\Big(-iQ\big(Z(X, Y)\big)\Big).
\end{aligned} \tag{3.8.27}$$

In terms of $U$, this is the desired multiplicative property:

$$U(e^X) U(e^Y) = U(e^{Z(X,Y)}) = U(e^X e^Y). \tag{3.8.28}$$

Since the action of $Q$ on one-particle states determines its action on Fock space, this equality extends from the one-particle subspace to the whole of Fock space.

Note that we have demonstrated multiplicativity in a small neighborhood of the identity element in the group $G$. Any group element in $G$ (assuming $G$ is connected) is a product of small group elements, and so this property extends readily to $G$. The lack of uniqueness in expressing a group element as a product of small group elements introduces the possibility of multi-valuedness for $U$. The fact that our $U$'s are single valued depends on the particular structure of $Q$, as already noted.

This section presents the structure of quantum symmetries which we would hope to find as a consequence of symmetries of the classical limit. Indeed, our proof that the Noether currents are conserved used the classical equations of motion. In Section 2.6 on canonical quantization, however, we noted that a naïve approach to normal ordering an interacting Hamiltonian raises doubts about the validity of these equations for quantum fields. At this point, we will continue with the assumption that normal ordering merely modifies the Hamiltonian by a constant, in which case Heisenberg's equations have the same polynomial form as Hamilton's. After we have introduced functional integral quantization in Chapter 11, we shall investigate quantum currents directly in Chapter 13 and conclude that the present analysis is correct.

# 3.9 Translation Symmetry

The homogeneity of space-time on the scales at which quantum field theory is a useful description of nature is represented in the theory by invariance under the action of the Poincaré group. In this section, we shall show how the translation invariance gives rise to the stress-energy tensor as a set of four currents which have the energy-momentum operators as conserved quantities.

Translations in the direction $e$ are represented by the real action

$$\phi'(\tau, \phi)(x) = \phi(x - \tau e). \tag{3.9.1}$$

The infinitesimal form of the real action is

$$\begin{aligned} \mathbf{D}\phi &= -e^{\nu} \partial_{\nu} \phi, \\ \mathbf{TD}\partial_{\mu}\phi &= \partial_{\mu} \mathbf{D}\phi = -e^{\nu} \partial_{\nu} \partial_{\mu} \phi, \end{aligned} \tag{3.9.2}$$

and so the action on the Lagrangian density $\mathcal{L}(\phi, \partial_{\mu}\phi)$ is

$$\begin{aligned} \mathbf{D}\mathcal{L} &= \Pi_a^{\mu} \mathbf{TD}\partial_{\mu}\phi^a + \frac{\partial \mathcal{L}}{\partial \phi^a} \mathbf{D}\phi \\ &= -\frac{\partial \mathcal{L}}{\partial(\partial_{\mu}\phi^a)} e^{\nu} \partial_{\nu} \partial_{\mu} \phi^a - \frac{\partial \mathcal{L}}{\partial \phi^a} e^{\nu} \partial_{\nu} \phi^a \\ &= -e^{\nu} \partial_{\nu} \mathcal{L} = -\partial_{\mu}(\mathcal{L}e^{\mu}). \end{aligned} \tag{3.9.3}$$

Theorem 3.4.19 (or Remark 3.4.24 on the symmetry condition $\mathbf{D}\mathcal{L} = \partial_{\mu} f^{\mu}$) allows us to deduce from this equation that translation is a symmetry of the dynamics.

Since the unitary representation of translations is

$$U(\Delta_a) = e^{ia \cdot P}, \tag{3.9.4}$$

whereas the usual convention for the unitary representation of internal symmetries contains a $-i$ in the exponential, we will uphold the convention for the sign of $P$ by inserting a minus sign in the definition of the conserved current associated with translations. Thus the translation conserved currents $j^{\mu}(e)$ are defined by:

$$\begin{aligned} j^{\mu} &= -\Pi_a^{\mu} \mathbf{D}\phi^a + f^{\mu} \\ &= \Pi_a^{\mu} e^{\nu} \partial_{\nu} \phi^a - \mathcal{L}e^{\mu} \\ &= e_{\nu}(\Pi_a^{\mu} \partial^{\nu} \phi^a - g^{\mu\nu}\mathcal{L}). \end{aligned} \tag{3.9.5}$$

From this equation we identify the stress-energy tensor $T$:

$$T^{\mu\nu} \stackrel{\text{def}}{=} \Pi_a^{\mu} \partial^{\nu} \phi^a - g^{\mu\nu}\mathcal{L}. \tag{3.9.6}$$

Then the conservation law for the current is $\partial_{\mu} T^{\mu\nu} = 0$.

**Homework 3.9.7.** Verify by direct computation that $\partial_{\mu} j^{\mu} = 0$ despite the dependence of $f^{\mu}$ on $\partial_0 \phi$.

The conserved stuff, energy-momentum, is given by integrating $T^{0\,\nu}$ over space:

$$H = \int d^3\bar{x}\,(\Pi_a^0 \partial^0 \phi^a - \mathcal{L}),$$
$$P^r = \int d^3\bar{x}\,\Pi_a^0 \partial^r \phi^a. \tag{3.9.8}$$

For the free field, this gives the previous formula for energy-momentum up to normal ordering. For the free field, this is no problem since the effect of normal ordering, namely modification of $P^\mu$ by a constant, does not affect commutators. Normal ordering in an interacting field theory, if it could be defined, must however change commutation relations, so $[Q, H] = 0$ may not imply $[Q, :H:]$ vanishes; this is a possible source of anomalies. Chapter 13 resolves the question of the breaking of classical conservation laws by quantum effects in the context of functional integral quantization.

**Homework 3.9.9.** Prove the Poisson bracket relations $\{\phi, P^r\} = \Pi^r$ and $\{\Pi, P^r\} = \partial^r \Pi$.

**Homework 3.9.10.** Use the canonical commutation relations to show that $[P^\mu, P^\nu] = 0$. (It helps to write the Lagrangian density in the form $\mathcal{L} = \frac{1}{2}\Pi^2 + \frac{1}{2}\partial_r \phi \partial^r \phi - V(\phi)$. Note that surface terms may be dropped when integrating by parts because they will evaluate to zero on wave packets.)

**Remark 3.9.11.** Since, as Homework 3.9.9 shows, $P^\mu$ defined by the Noether formula generates translation in the direction $x^\mu$ on the complete set of observables $\phi(x)$ and $\Pi(x)$, it appears that normal ordering of the $P^\mu$ can only change $P^\mu$ by a constant. However, the Homework 2.6.36 on normal ordering in $\phi^4$-theory rules out this possibility. To resolve this paradox, normal ordering must change the canonical momentum $\Pi(x)$.

This completes the application of Noether's formalism to translations. Next we cover the Lorentz group and find the angular momentum conserved quantities.

## 3.10 Lorentz Symmetry

The action of the Lorentz group on scalar fields is given by

$$\phi'(\tau, \phi)(x) = \phi\big(\exp(-\tau X)x\big), \tag{3.10.1}$$

where $X$ is in $so(1, 3)$. The significance of the restriction to scalar fields is that these fields are not mixed among themselves by the Lorentz transformations. The infinitesimal action is

$$\mathbf{D}\phi = \frac{\partial}{\partial \tau} \phi\big(\exp(-\tau X)x\big)\,\Big|_{\tau=0}$$
$$= \partial_\alpha \phi(x)\,\frac{\partial}{\partial \tau}\big(\exp(-\tau X)x\big)^\alpha\,\Big|_{\tau=0} \tag{3.10.2}$$
$$= -X^\alpha{}_\beta x^\beta \partial_\alpha \phi,$$

and

$$\mathbf{T}\mathbf{D}\partial_\mu \phi = \partial_\mu \mathbf{D}\phi$$
$$= \partial_\mu(-X^\alpha{}_\beta x^\beta \partial_\alpha \phi) \tag{3.10.3}$$
$$= -X^\alpha{}_\beta \delta_\mu^\beta \partial_\alpha \phi - X^\alpha{}_\beta x^\beta \partial_\alpha \partial_\mu \phi.$$

Using these formulae we compute the infinitesimal action on the Lagrangian density:

$$\mathbf{D}\mathcal{L} = \Pi_a^\mu \mathbf{T} D \partial_\mu \phi^a + \frac{\partial L}{\partial \phi^a} \mathbf{D}\phi^a$$

$$= -\Pi_a^\mu \left( X^\alpha{}_\mu \partial_\alpha \phi^a + X^\alpha{}_\beta x^\beta \partial_\alpha \partial_\mu \phi^a \right) - \frac{\partial L}{\partial \phi^a} X^\alpha{}_\beta x^\beta \partial_\alpha \phi^a \qquad (3.10.4)$$

$$= -\Pi_a^\mu X^\alpha{}_\mu \partial_\alpha \phi^a - X^\alpha{}_\beta x^\beta \partial_\alpha \mathcal{L}.$$

Lorentz invariance restricts derivatives to occur in combinations $\partial_\mu \phi^a \partial^\mu \phi^b$. Renormalizability, as we shall see in Chapter 19, prohibits square or higher powers of such terms. If we have a term $m_{ab} \partial_\mu \phi^a \partial^\mu \phi^b$, then for the theory to be meaningful, $m_{ab}$ must diagonalize so that the derivatives of the fields only occur in $\mathcal{L}$ in the standard kinetic terms $\partial_\mu \phi \partial^\mu \phi$. In this case, $\Pi_a^\mu = \partial^\mu \phi_a$, and so, since $X_{\alpha\mu}$ is anti-symmetric, the first term in $\mathbf{D}\mathcal{L}$ vanishes. Then, since $X^\alpha{}_\alpha = 0$,

$$\mathbf{D}\mathcal{L} = -X^\alpha{}_\beta x^\beta \partial_\alpha \mathcal{L} = \partial_\mu (-X^\mu{}_\beta x^\beta \mathcal{L}). \qquad (3.10.5)$$

This shows that we should take

$$f^\mu \overset{\text{def}}{=} -X^\mu{}_\beta x^\beta \mathcal{L}. \qquad (3.10.6)$$

As for translations, Theorem 3.4.19 (or Remark 3.4.24) allows us to deduce from this equation that Lorentz transformations are symmetries of the dynamics.

The established tradition for the sign of angular momentum agrees with our sign convention for the conserved current. Thus the conserved current associated with the generator $X$ is

$$j^\mu = \Pi_a^\mu \mathbf{D}\phi^a + X^\mu{}_\beta x^\beta \mathcal{L}$$

$$= -\Pi_a^\mu X^\alpha{}_\beta x^\beta \partial_\alpha \phi^a + X^\mu{}_\beta x^\beta \mathcal{L}$$

$$= -X_{\alpha\beta} x^\beta (\Pi_a^\mu \partial^\alpha \phi^a - g^{\alpha\mu} \mathcal{L}) \qquad (3.10.7)$$

$$= -X_{\alpha\beta} x^\beta T^{\mu\alpha}$$

$$= \tfrac{1}{2} X_{\alpha\beta} (x^\alpha T^{\mu\beta} - x^\beta T^{\mu\alpha}).$$

Let $M^{\mu\alpha\beta} = x^\alpha T^{\mu\beta} - x^\beta T^{\mu\alpha}$. Then the conservation law for the current is

$$\partial_\mu M^{\mu\alpha\beta} = 0.$$

The conserved quantities are

$$J^{\alpha\beta} = \int d^3\bar{x}\, M^{0\alpha\beta}. \qquad (3.10.8)$$

The operators $J^{rs}$ are the angular-momentum operators:

$$J^{rs} = \int d^3\bar{x}\, (x^r T^{0s} - x^s T^{0r}). \qquad (3.10.9)$$

The conservation of $J^{0s}$ leads to the center of energy law:

$$J^{0s} = \int d^3\bar{x}\, (x^0 T^{0s} - x^s T^{00}) = tP^s - \int d^3\bar{x}\, x^s T^{00}, \qquad (3.10.10)$$

which implies that

$$\frac{\partial}{\partial t} \int d^3\bar{x} \, x^s T^{00} = P^s. \qquad (3.10.11)$$

**Homework 3.10.12.** From the formulae of this and the previous section, we have linear and angular momentum conserved quantities

$$P^\sigma = \int d^3\bar{x} \, \Pi\partial^\sigma\phi - \delta^{0\sigma}\mathcal{L},$$

$$J^{\mu\nu} = \int d^3\bar{x} \, x^\mu\Pi\partial^\nu\phi - x^\nu\Pi\partial^\mu\phi + (\delta^{0\mu}x^\nu - \delta^{0\nu}x^\mu)\mathcal{L}.$$

The operator $J^{\mu\nu}$ is derived from the anti-symmetric generator $X^{\mu\nu}$ which has a $+1$ in row $1+\mu$ and column $1+\nu$, a $-1$ in row $1+\nu$ and column $1+\mu$, and 0's elsewhere. From the generator relation

$$[X^{\mu\nu}, T^\sigma] = (\delta^{\sigma\nu}T^\mu - \delta^{\sigma\mu}T^\nu),$$

we deduce that the operators should satisfy

$$[J^{\mu\nu}, P^\sigma] = i(\delta^{\sigma\nu}P^\mu - \delta^{\sigma\mu}P^\nu).$$

Show that this last equation is true. (You will have to drop surface terms when integrating by parts; this is acceptable because they will vanish on wave packets.)

**Homework 3.10.13.** Following on from the previous homework, use the equal-time canonical commutation relations to show that the angular momentum operators $J^{\mu\nu}$ form a hermitian representation of the Lorentz Lie algebra.

The conserved currents and quantities associated with Poincaré symmetry are derived from the formula for the Noether current. As with internal symmetry, however, our proof of the conservation law uses the classical equations of motion, and the justification of this usage must wait till Chapter 13.

## 3.11 Application: Quantum Numbers

The original goal in discussing symmetries was to find quantities that would identify a state even as it propagates. The way that conserved quantities are put to use is in labeling initial and final states. Then the sum of the conserved quantities of each type in the final state must be equal to the corresponding sums in the initial state if the initial state is to have any probability of being detected in that final state. Therefore, it is natural to use a basis of eigenstates for the conserved quantity operators, so that each basis vector has well-defined quantum numbers. But if the basis states are eigenstates, the conserved quantity operators must be diagonal, and hence must also commute. Since, by Theorem 3.8.6, the algebra of the conserved quantity operators is the same as the algebra of the symmetry generators, clearly the conserved quantity operators do not in general commute with each other. The best we can do is to find a linear basis for a maximal set of commuting generators and define the quantum numbers by the associated conserved quantity operators. Since internal symmetries commute with the Poincaré symmetry, we can choose such bases for the two symmetries separately. Note that the term 'charge' is often used to denote a quantum number associated with any internal symmetry.

If we choose the translation generators as a commuting set in the Poincaré Lie algebra, then it is easy to see that it is maximal. Thus the energy-momentum operators $P^\mu$ are a maximal set of commuting conserved quantities associated with Poincaré symmetry that we can use to label states.

If the internal symmetry is $U(n)$, then any maximal commuting set of generators will be similar under change of basis to the set of diagonal generators. The set of diagonal generators is a real vector space of dimension $n$, so there are $n$ commuting conserved charges that we can use to label states. If the internal symmetry is $SU(n)$, then the standard maximal commuting set of generators is the set of traceless diagonal generators. This set has dimension $n-1$, and so we can choose $n-1$ commuting conserved charges to label the states.

Suppose that the internal symmetry is $SO(n)$. A generator $X$ in $so(n)$ is an anti-symmetric matrix. The standard form for $X$ is found by diagonalizing $X$ over the complex numbers and separating the real and imaginary parts of the eigenvectors. Let $v = v_1 + iv_2$ be an eigenvector of $X$ with eigenvalue $i\lambda$. Since $X$ is anti-symmetric, $\lambda$ is a real number. Therefore $Xv = i\lambda v$ implies $Xv_1 = -\lambda v_2$ and $Xv_2 = \lambda v_1$, and so the action of $X$ on the subspace spanned by $v_1$ and $v_2$ is represented by the matrix

$$\begin{pmatrix} 0 & \lambda \\ -\lambda & 0 \end{pmatrix}. \tag{3.11.1}$$

There could also be null vectors of $X$, that is, real vectors $v$ that satisfy $Xv = 0$. Therefore, if $X$ has $k$ complex eigenvectors and $n - 2k$ real null ones, the normal form of $X$ will be

$$X \sim \begin{pmatrix} 0 & \lambda_1 & & & & & & \\ -\lambda_1 & 0 & & & & & & \\ & & \ddots & & & & & \\ & & & 0 & \lambda_k & & & \\ & & & -\lambda_k & 0 & & & \\ & & & & & 0 & & \\ & & & & & & \ddots & \\ & & & & & & & 0 \end{pmatrix}. \tag{3.11.2}$$

The maximum number of parameters in a matrix of this type is $[n/2]$, the largest integer less than $n/2$. No other type of anti-symmetric matrix commutes with all the matrices of this type. Hence the maximal number of commuting generators in $so(n)$ is $[n/2]$, and this is the maximal number of commuting conserved quantities available to us for labeling states.

There are conserved quantities that are not linear in the operators $Q(X)$ which commute with the set of diagonal charges above, for example the total spin operator. These operators are called Casimir operators, and correspond to homogeneous polynomials in Lie algebra elements. For each of the groups above, the number of algebraically independent Casimir operators is equal to the dimension of its maximal commuting sub-algebras.

**Remark** 3.11.3. For these Lie algebras, every maximal commuting sub-algebra has the same dimension. This dimension is called the *rank* of the Lie algebra.

Instead of computing Casimir numbers, however, it is simpler to use the machinery of group representation theory to guarantee their conservation. We shall introduce the representation theory of $SU(2)$ and $SO(1,3)$ in Chapter 6 in preparation for the description of fields that are not Lorentz scalars and that of $SU(3)$ in Chapter 14 for the analysis of color and flavor symmetries.

At all energies there is the $SU(3)$ color symmetry of the strong interaction. But since there appears to be no chance of observing a colored particle, there are no names for the two associated conserved quantities. At the energies of daily life, the internal symmetry is $SU(3) \times U(1)$ of color and electromagnetism, and so states are identified by electric charge and energy-momentum. At higher energies, more symmetry is manifest. When the energy is high enough for the ready creation of W and Z bosons, then electroweak $SU(2) \times U(1)$ symmetry is effectively restored and the states may be labeled by two internal quantum numbers, weak isospin and hypercharge. At higher energies, there may be a grand unified symmetry group that unites the electroweak force with the strong force. Such a group must contain the $SU(3) \times SU(2) \times U(1)$ of the standard electroweak theory. The best candidate for a grand unified symmetry is Hagelin's $SU(5) \times U(1)$ theory, which, at the appropriate energy level, has five conserved quantities that we can use to label states.

To round off this introduction to continuous symmetries in quantum field theory, we observe that states with indefinite electric charge or baryon number are simply not observed. The general principle proposed to cover this is that two states $|A\rangle$ and $|B\rangle$ form a physically meaningful linear combination if and only if there is some observable $\mathcal{O}$ which converts one into the other: $\mathcal{O}|A\rangle = |B\rangle$. It follows that the eigenspaces of any hermitian operator $Q$ which commutes with all observables will contain all physically meaningful states; any linear combination of states with different $Q$ eigenvalues will not be meaningful. Note that such an operator $Q$ must commute with the Hamiltonian and hence be a conserved quantity operator. Any principle which like this identifies sections of Fock space as unphysical is called a *superselection rule*.

With the concept of spontaneous symmetry breaking comes a possibility that conservation laws (and therefore superselection rules) might be temperature dependent. Thus, for example, at high temperature the Standard Model of electroweak interactions has an $SU(2) \times U(1)$ symmetry. The associated conserved quantities mix the three $W$ vector bosons, so there is no superselection rule on their one-particle state space. At low temperatures, however, the symmetry breaks to the $U(1)$ of quantum electrodynamics, and so there is a superselection rule which prohibits mixing of the $W^+$, $W^-$, and $W_3$.

In general, the primary value of symmetry is in visibly constraining the evolution of states. In this section, we have discussed how to use symmetry to generate quantum numbers. Later, when we have developed scattering theory and investigated group actions in more detail, we will see how to use symmetry to rotate one scattering amplitude into another, thereby generating predictions for the results of related experiments.

# 3.12 Charge Conjugation

In this and the two following sections, we shall deal with charge conjugation $C$, parity $P$, and time reversal $T$. These three are not separately symmetries of the Standard Model of strong and electroweak interactions, but their product $CPT$ is. Indeed, any theory which computes scattering amplitudes from Feynman rules will be $CPT$ invariant. Separately, $C$, $P$, and $T$ are useful symmetries of parts of the Standard Model.

Since the discrete symmetries are not one-parameter symmetries, we cannot apply Noether's Theorem to find conserved currents and additive conserved quantities. Instead, as we shall see, they give rise to multiplicative conservation laws.

Charge conjugation, like the other discrete symmetries, has three levels of action: on the particle states, on the creation and annihilation operators, and on the fields. The action on any one level determines the action on the other two.

Charge conjugation on the states reverses the quantum numbers of particles that are associated with internal symmetries. The charge conjugate of a particle is another particle with the same energy and momentum but opposite charges.

Charge conjugation on the fields converts a field $\psi(x)$ into a field $\psi_c(x)$ with opposite internal quantum numbers. This transformation is a symmetry of a Lagrangian density $\mathcal{L}$ if the application of the transformation to every field in $\mathcal{L}$ changes $\mathcal{L}$ by a total divergence. It is symmetry of the quantum field theory defined by $\mathcal{L}$ if in addition it preserves the canonical commutation relations.

**Remark** 3.12.1.   There is a theorem of quantum mechanics which states that any canonical change of coordinates $(q, p) \mapsto (q', p')$ can be implemented by a unitary operator $U$:

$$q' = U^\dagger q U, \quad p' = U^\dagger p U. \tag{3.12.2}$$

The field theory extension of this theorem asserts that if $\phi'$ and $\Pi'$ are a change of coordinates – that is, they act on the same state space as $\phi$ and $\Pi$ – that preserves the canonical commutation relations, then there exists a unitary operator $U$ such that

$$\phi' = U^\dagger \phi U, \quad \Pi' = U^\dagger \Pi U. \tag{3.12.3}$$

Hence, if charge conjugation is a symmetry of a quantum field theory, there must exist a unitary operator $U_C$ which represents $C$.

Charge conjugation may easily be understood from an example. Let

$$\mathcal{L} = \frac{1}{2} \partial_\mu \phi^\top \partial^\mu \phi - \frac{\mu^2}{2} \phi^\top \phi - g(\phi^\top \phi)^2, \tag{3.12.4}$$

where $\phi = (\phi_1, \phi_2)^\top$. Then $\mathcal{L}$ is $O(2)$ invariant, and we can choose a charge conjugation symmetry from the determinant $-1$ part of $O(2)$:

$$C = \begin{pmatrix} 1 & 0 \\ 0 & -1 \end{pmatrix}. \tag{3.12.5}$$

The charge conjugation $\phi \to C\phi$ may be represented by a unitary transformation $U_C$ as follows:

$$U_C^\dagger \phi U_C \stackrel{\text{def}}{=} C\phi. \tag{3.12.6}$$

Since in the interaction picture (which we present in the next chapter) the interacting fields evolve as free fields, and since the creation and annihilation operators are linear functions of the free fields and their momenta, the charge conjugation of the fields induces a charge conjugation of the creation and annihilation operators:

$$U_C^\dagger a^1(\bar{k}) U_C = a^1(\bar{k}) \quad \text{and} \quad U_C^\dagger a^2(\bar{k}) U_C = -a^2(\bar{k}). \tag{3.12.7}$$

These equations for $U_C$ are solved by

$$U_C = (-1)^{N_2}, \quad \text{where} \quad N_2 = \int d^3\bar{k} \, a^2(\bar{k})^\dagger a^2(\bar{k}). \tag{3.12.8}$$

We can expand $U_C$ as an exponential by substituting $-1 = \exp(i\pi)$. Then we can evaluate $U_C$ on states by first computing $N_2$ and then exponentiating. This unitary operator is normalized to satisfy $U_C|0\rangle = |0\rangle$. Note that $U_C^2 = 1$.

The charge conjugate of a particle is called its *anti-particle*. In the example above, we see that $c$-particles are anti-particles of $b$-particles and *vice versa*:

$$U_C|k^{(b)}\rangle = |k^{(c)}\rangle \quad \text{and} \quad U_C|k^{(c)}\rangle = |k^{(b)}\rangle. \tag{3.12.9}$$

With our previous definition $\psi = (\phi^1 - i\phi^2)/\sqrt{2}$ of a complex field, we see that the action of $U_C$ on $\psi$ is

$$U_C^\dagger \begin{pmatrix} \psi \\ \psi^\dagger \end{pmatrix} U_C = \begin{pmatrix} \psi^\dagger \\ \psi \end{pmatrix}, \tag{3.12.10}$$

and its action on the creation and annihilation operators is

$$U_C^\dagger \begin{pmatrix} c^\dagger(\bar{k}) & b(\bar{k}) \\ b^\dagger(\bar{k}) & c(\bar{k}) \end{pmatrix} U_C = \begin{pmatrix} b^\dagger(\bar{k}) & c(\bar{k}) \\ c^\dagger(\bar{k}) & b(\bar{k}) \end{pmatrix}. \tag{3.12.11}$$

In general, we could insert a phase $\eta$ here so that

$$U_C^\dagger \psi U_C \stackrel{\text{def}}{=} \eta \psi^\dagger \quad \text{and} \quad U_C^\dagger \psi^\dagger U_C = \eta^* \psi. \tag{3.12.12}$$

If we write $\psi_c$ for $\eta \psi^\dagger$, then we find

$$U_C^\dagger \psi U_C = \psi_c \quad \text{and} \quad U_C^\dagger \psi_c U_C = \psi. \tag{3.12.13}$$

Note that a definition of charge conjugation must include a distinction between fields and their coefficients; we must know what to conjugate.

**Homework 3.12.14.** Show that if $\psi' = \zeta\psi$, then there is a value of the complex number $\zeta$ for which $U_C^\dagger \psi' U_C = \psi'^\dagger$ and $U_C^\dagger \psi'^\dagger U_C = \psi'$.

**Remark 3.12.15.** We would expect any hermitian field that transforms non-trivially under charge conjugation to be (like the real components of $\psi$ here) part of a complex field that carries charge. However, this is not a theoretical necessity.

**Homework 3.12.16.** Verify that charge conjugation preserves the canonical commutation relations 3.5.18.

**Homework 3.12.17.** Show that charge conjugation changes the sign of the conserved current and quantity 3.8.16 in a theory of complex scalar fields.

In general, charge conjugation can mix the fields and their particle states. The superselection rules indicate that it is sensible to change basis and diagonalize this action. Since $U_C^2 = 1$, the eigenvalues of charge conjugation are $\pm 1$. After diagonalization, fields and particles can therefore be associated with their eigenvalues under charge conjugation. In particular, any $n$-particle state $|k_1, \ldots, k_n\rangle$ will be an eigenstate of charge conjugation, and its eigenvalue will be the product of the eigenvalues of the individual particles in the state.

The application of charge conjugation is to eliminate final states for scattering and decay processes and to provide a link between different processes involving charged particles. For example, charge conjugation reverses the sign of an electromagnetic current and so must also reverse the sign of the electromagnetic field. Hence it reverses the sign of photons in an initial or final state. Now quantum electrodynamics is charge-conjugation invariant. We can conclude then that in quantum electrodynamics there is no possibility of an odd number of photons scattering into an even number or *vice versa*. For a second example, the neutral pion $\pi^0$ has $C = 1$. Hence the decay $\pi^0 \to 2\gamma$ into two photons is permitted but the decay $\pi^0 \to 3\gamma$ into three photons is not. Experiment confirms these points.

This section shows how charge conjugation acts on fields, creation and annihilation operators, and states, and how it gives rise to a multiplicative conservation law and thereby constrains the evolution of an initial state. The following section on parity is similar.

# 3.13 Parity

Classical parity is any element in the component of the Lorentz group that contains the matrix $P = \text{diag}(1, -1, -1, -1)$. We shall use $P$ itself as our classical parity. The action of $P$ on fields is

$$P: \phi^a(t, \bar{x}) \mapsto M^a{}_b \phi^b(t, -\bar{x}). \tag{3.13.1}$$

Since $P^2 = 1$, the matrix $M$ satisfies $M^2 = 1$. This may be seen as the minimal polynomial for $M$, showing that there is a basis of fields in which $M$ is diagonal with eigenvalues $\pm 1$. Thus the action of $P$ on fields is of two types:

$$
\begin{aligned}
P^+ \phi^a(t, \bar{x}) &\stackrel{\text{def}}{=} \phi^a(t, -\bar{x}), \\
P^- \phi^a(t, \bar{x}) &\stackrel{\text{def}}{=} -\phi^a(t, -\bar{x}).
\end{aligned}
\tag{3.13.2}
$$

Fields that transform according to $P^+$ are called *scalar fields*; those that transform according to $P^-$ are called *pseudoscalar fields*. The eigenvalue of $P$ associated with an eigenfield is its *intrinsic parity*.

As for charge conjugation, so for parity we deduce the action of $P$ on creation and annihilation operators and on states from its action on fields. Also there exists a unitary operator $U_P$ which represents $P$:

$$
\begin{aligned}
U_P^\dagger a(k) U_P &= \pm a(\tilde{k}) \quad \text{and} \quad U_P^\dagger a(k) U_P = \pm a(\tilde{k}), \\
U_P |k_1, \ldots, k_n\rangle &= (-1)^{n_-} |\tilde{k}_1, \ldots, \tilde{k}_n\rangle,
\end{aligned}
\tag{3.13.3}
$$

where $\tilde{k}_\mu = k^\mu$ and $n_-$ is the number of pseudoscalar particles in the $n$-particle state.

To illustrate the varieties of parity, we present two examples. Let $\epsilon^{\mu\nu\rho\sigma}$ be the totally antisymmetric tensor normalized by $\epsilon^{0123} = 1$. This tensor is Lorentz invariant. Let $\phi_a$ be four hermitian fields with four different masses $\mu_a$. Define a Lagrangian density by

$$\mathcal{L} = \mathcal{L}_{\text{Free}} - g\epsilon^{\mu\nu\rho\sigma}\partial_\mu\phi_1\,\partial_\nu\phi_2\,\partial_\rho\phi_3\,\partial_\sigma\phi_4. \qquad (3.13.4)$$

Since each interaction term involves three space derivatives and one time derivative, $\mathcal{L}$ is not invariant under $P^+$. Suppose that one field, say $\phi_4$, is a pseudoscalar and the other three are scalars. Then $\mathcal{L}$ is $P$-invariant. If we add a cubic interaction term $h(\phi_1^3 + \phi_2^3 + \phi_3^3)$, then this assignment of scalar/pseudoscalar is unique. If we also add $h\phi_4^3$, then the resulting Lagrangian density does not admit any parity invariance.

For the second example, let $\phi_a$ be as before but add a complex field $\psi$ with mass $\mu$. Define an interaction Lagrangian density by

$$\mathcal{L}' = -g\epsilon^{\mu\nu\rho\sigma}\partial_\mu\phi_1\,\partial_\nu\phi_2\,\partial_\rho\phi_3\,\partial_\sigma\phi_4\big(\psi^2 + (\psi^\dagger)^2\big) - h(\phi_1^3 + \phi_2^3 + \phi_3^3 + \phi_4^3). \qquad (3.13.5)$$

Then we may define a discrete symmetry $P$ by

$$P : \phi_r(t, \bar{x}) \mapsto \phi_r(t, -\bar{x}), \quad P : \psi(t, \bar{x}) \mapsto i\psi(t, -\bar{x}). \qquad (3.13.6)$$

This $P$ is actually a mixture of parity and the broken charge $U(1)$ symmetry. It satisfies $P^4 = 1$ and may be regarded as a parity operator insofar as it acts as parity on space-time.

A Lagrangian density $\mathcal{L}$ is parity invariant if, when parity is applied to all its fields, it merely changes from $\mathcal{L}(x)$ to $\mathcal{L}(\bar{x})$ plus a total divergence. Since the action involves integration over space, invariance of $\mathcal{L}$ implies invariance of the Euler–Lagrange equations.

Invariance of the quantum theory derived from a parity-invariant Lagrangian density follows from invariance of the canonical commutation relations. This invariance is a trivial consequence of the parity invariance of the delta function.

Parity, like charge conjugation, gives rise to a multiplicative conservation law. For example, the $\eta$ meson and the pions are pseudoscalars, and so the decay $\eta \to \pi^+\pi^-$ is forbidden by conservation of parity. However, since parity transforms space, the eigenvalues of parity depend on the orbital angular momentum of a state and the intrinsic parity of a state is not in general conserved. We shall cover this point when we return to parity in Section 8.10.

From this section we have learned that particles and fields have intrinsic parity, and that, when we include the parity associated with orbital angular momentum, parity gives rise to a multiplicative conservation law in parity-invariant theories. We now conclude our treatment of symmetries of scalar field theories with a section on time reversal.

# 3.14 Time Reversal

In the previous sections on charge conjugation and parity, we have defined the operations on fields and then derived the actions on operators and states. In this section, we build up time reversal from a definition on states.

The idea of time reversal is to take a film of the evolution of some system and run the film in reverse. To separate the effects of charge conjugation from those of time reversal, it is customary to assume that time reversal preserves the internal quantum numbers of all particles. The law governing the evolution is symmetric under time reversal if it governs the time-reversed film.

In classical mechanics, time reversal can be implemented by changing the sign of the Hamiltonian. If we suppose that this effect is achieved in quantum theory by a unitary transformation $U_T$, then

$$U_T^\dagger e^{-iHt} U_T = e^{iHt} \quad \Rightarrow \quad U_T^\dagger H U_T = -H$$

$$\Rightarrow \quad H U_T |n\rangle = -E_n U_T |n\rangle, \tag{3.14.1}$$

for any energy eigenstate $|n\rangle$. This contradicts the principle that energy be bounded below. We conclude that time reversal is not represented by a unitary operator on state space.

The conventional resolution of this paradox is to represent time reversal by an *anti-unitary* operator $\Omega_T$, that is, an operator which preserves lengths but is anti-linear. Thus in the argument above, $\Omega_T$ commutes with the Hamiltonian $H$ but anti-commutes with the $i$, and we escape the contradiction.

**Remark 3.14.2.** This resolution is justified by a theorem (due to Wigner) which states that any continuous invertible map $F$ from a Hilbert space $\mathbf{H}$ to itself which satisfies

$$\left| \big(F(a), F(b)\big) \right| = \left| (a, b) \right| \quad \text{for } a, b \text{ in } \mathbf{H} \tag{3.14.3}$$

has a factorization $Fa = \exp\big(i\theta(a)\big) Ga$, where $G$ is either unitary or anti-unitary. This theorem gives the form of the most general symmetries used in quantum theory, and enables us to conclude that, modulo a redefinition of the phases of the kets to absorb the $\exp\big(i\theta(a)\big)$, there exists an anti-unitary operator $\Omega_T$ for time reversal.

In more detail, letting $a$, $b$, and $(a, b)$ respectively stand for the vectors $|a\rangle$, $|b\rangle$, and their inner product $\langle a|b\rangle$, we may define an anti-unitary operator as an invertible operator $\Omega$ which satisfies

$$(\Omega a, \Omega b) = (b, a), \quad \text{and} \quad \Omega(\alpha a + \beta b) = \alpha^* \Omega(a) + \beta^* \Omega(b). \tag{3.14.4}$$

With respect to a *real* orthonormal basis, this implies that $\Omega^T \Omega = 1$. However, since $(a, \Omega^\dagger \Omega b)$ must be antilinear in $a$, it cannot equal $(b, a)$. Consequently, $\Omega^\dagger$ does not exist. Nevertheless, we can get the effect of $\Omega^\dagger$ by using $\Omega^{-1}$ and the hermitian conjugate as follows:

$$(a, \Omega b) = (\Omega \Omega^{-1} a, \Omega b) = (b, \Omega^{-1} a). \tag{3.14.5}$$

In the Dirac notation, the non-existence of $\Omega^\dagger$ makes the use of anti-unitary operators delicate. The key point is that $\Omega$ and $\Omega^{-1}$ always act on kets, never on bras. The effect of $\Omega$ on a bra is only well defined in the context of a matrix element, in which case the expression 3.14.5 above can be used to turn the bra into a ket:

$$\langle a|\Omega|b\rangle = \langle b|\Omega^{-1}|a\rangle. \tag{3.14.6}$$

With a small abuse of notation, this enables us to express the effect of an anti-unitary operator on a matrix element as follows:

$$\begin{aligned}
\langle a|A|b\rangle &= \langle a|\Omega^{-1}\Omega A\Omega^{-1}\Omega|b\rangle \\
&= \langle a|\Omega^{-1}A_\Omega|\Omega b\rangle \\
&= \langle \Omega b|A_\Omega{}^\dagger\Omega|a\rangle \\
&= \langle \Omega b|A_\Omega{}^\dagger|\Omega a\rangle,
\end{aligned} \tag{3.14.7}$$

where $A_\Omega = \Omega A\Omega^{-1}$. Note that both sides of this equation are linear in $A$.

The action of time reversal on a state leaves positions unchanged but reverses momenta because of the implicit time derivative. We can therefore define an anti-unitary operator $\Omega_T$ representing time reversal by the following action on kets:

$$\Omega_T|k\rangle \stackrel{\text{def}}{=} |\tilde{k}\rangle, \tag{3.14.8}$$

where $\tilde{k}_\mu = g_{\mu\mu}k_\mu$. As in charge conjugation, we could insert a phase in this definition, but because $\Omega_T$ is anti-unitary we will in any case find $\Omega_T^2 = 1$.

The action of $\Omega_T$ on kets determines its action on operators:

$$\begin{aligned}
A(x)|k\rangle &\xrightarrow{\;\;T\;\;} \Omega A(x)|k\rangle = \Omega A(x)\Omega^{-1}\Omega|k\rangle \\
&\Longrightarrow \quad A(x) \to A(x)_T \stackrel{\text{def}}{=} \Omega_T A(x)\Omega_T^{-1}.
\end{aligned} \tag{3.14.9}$$

Assuming invariance of the vacuum, from the action of time reversal on states we deduce its action on creation operators:

$$\Omega_T \alpha^\dagger(k)\Omega_T^{-1} = \alpha^\dagger(\tilde{k}). \tag{3.14.10}$$

From its action on $\alpha(k)|l\rangle$, we deduce its action on annihilation operators:

$$\Omega_T \alpha(k)\Omega_T^{-1} = \alpha(\tilde{k}). \tag{3.14.11}$$

With these results in hand, it is easy to compute the effect of time reversal on a real field $\phi$:

$$\begin{aligned}
\Omega_T\phi(t,\bar{x})\Omega_T^{-1} &= \int \frac{d^3\bar{k}}{(2\pi)^3\,2\omega(\bar{k})}\; \Omega_T\left(e^{ix\cdot k}\alpha^\dagger(k) + e^{-ix\cdot k}\alpha(k)\right)\Omega_T^{-1} \\
&= \int \frac{d^3\bar{k}}{(2\pi)^3\,2\omega(\bar{k})}\; \left(e^{-ix\cdot k}\alpha^\dagger(\tilde{k}) + e^{ix\cdot k}\alpha(\tilde{k})\right) \\
&= \phi(-t,\bar{x}).
\end{aligned} \tag{3.14.12}$$

This result suggests that a complex field $\psi$ should be mapped to $\psi^\dagger$. Actually, however, since time reversal does not change particles into anti-particles, $\psi$ must be mapped into itself:

$$\Omega_T \psi(t, \bar{x}) \Omega_T^{-1} = \psi(-t, \bar{x}). \tag{3.14.13}$$

For this to be true, it is necessary that the imaginary component of $\psi$ has time-reversal phase $-1$. Following the logic above from definition on particle states to action on field confirms the necessity and consistency of this transformation.

Applying the formula 3.14.7 for the action of an anti-unitary operator on a matrix element, we find that

$$\langle p | A | q \rangle = \langle \tilde{q} | A_T^\dagger | \tilde{p} \rangle, \quad \text{where } A_T = \Omega_T A \Omega_T^{-1}. \tag{3.14.14}$$

We may define $\langle \tilde{q} | A_T^\dagger | \tilde{p} \rangle$ as the time-reversed matrix element and $A_T^\dagger$ as the time-reversed operator – the extra adjoint coming from switching initial and final states.

It remains to see whether time reversal preserves the dynamics. For this, it will be convenient to allow the Hamiltonian to be time dependent. Then the evolution operator $U(t'', t')$ is given by

$$U(t'', t') = T \int dt e^{-iH(t)}$$
$$\stackrel{\text{def}}{=} \lim_{n \to \infty} \prod_{r=0}^{n} (1 - i\Delta t\, H(t' + r\Delta t)), \tag{3.14.15}$$

where $n\Delta t = t'' - t'$ and multiplication law is 'later on the left'. From the expansion it is clear that

$$U(t'', t')_T^\dagger(t) = \lim_{n \to \infty} \prod_{r=0}^{n} (1 - i\Delta t\, H_T^\dagger(t' + r\Delta t)), \tag{3.14.16}$$

where now the multiplication law is 'later on the right'.

Time-reversal invariance of the dynamics means

$$U(t'', t')_T^\dagger(t) = U(-t', -t''). \tag{3.14.17}$$

From the expressions above, it is clear that this is equivalent to

**Condition for Time-Reversal Invariance:**    $H_T^\dagger(t) = H(-t). \tag{3.14.18}$

When $H$ is hermitian, this reduces to $H_T = H$.

The Lagrangian density is related to the Hamiltonian density by a Legendre transform:

$$\mathcal{L} = \frac{\partial \mathcal{H}}{\partial \Pi} \Pi - \mathcal{H}. \tag{3.14.19}$$

Thus the condition $H_T^\dagger(t) = H(-t)$ on the Hamiltonian density is equivalent a Lagrangian density condition

$$\mathcal{L}_T^\dagger(t, \bar{x}) = \mathcal{L}(-t, \bar{x}). \tag{3.14.20}$$

**Homework 3.14.21.** Let $\mathcal{L}(x, \phi, \partial\phi)$ be a Lagrangian density which is time-reversal invariant. Show that, if $\phi(x)$ is a solution of the Euler–Lagrange equations derived from $\mathcal{L}$, then $\phi_T{}^\dagger(x)$ is also.

Now that we have defined time reversal on fields, we can easily check time-reversal invariance of Lagrangian densities. For example, consider the following interactions between charged scalars $\psi$ and a neutral scalar $\phi$:

$$\mathcal{L}_1' = \lambda \psi^\dagger \psi \phi,$$
$$\mathcal{L}_2' = \lambda \psi_1^\dagger \psi_2 \phi + \lambda^* \psi_2^\dagger \psi_1 \phi, \qquad (3.14.22)$$
$$\mathcal{L}_3' = \lambda_1 \psi_2^\dagger \psi_3 \phi + \lambda_2 \psi_3^\dagger \psi_1 \phi + \lambda_3 \psi_1^\dagger \psi_2 \phi + \text{h.c.}$$

If $\lambda$ is real, then $\mathcal{L}_1'$ is hermitian and obviously time-reversal invariant. $\mathcal{L}_2'$ will become time-reversal invariant after absorbing the phase of $\lambda$ into $\psi_2$ and its particle states. However, we cannot in general choose phases for the fields in $\mathcal{L}_3'$ in order to make $\mathcal{L}_3'$ invariant under time reversal.

To demonstrate invariance of the quantum field theory associated with an invariant Lagrangian density, it remains to verify invariance of the canonical commutation relations. On account of the time derivative in the canonical momentum $\Pi$, we have

$$\Omega_T \Pi(t, \bar{x}) \Omega_T^{-1} = -\Pi(-t, \bar{x}). \qquad (3.14.23)$$

Consequently time reversal generates a minus on each side of the canonical commutation relation:

$$\Omega_T \left[\phi(t, \bar{x}), \Pi(t, \bar{y})\right] \Omega_T^{-1} = \left[\phi(-t, \bar{x}), -\Pi(-t, \bar{y})\right]$$
$$\text{and} \quad \Omega_T i \delta^{(3)}(\bar{x} - \bar{y}) \Omega_T^{-1} = -i\delta^{(3)}(\bar{x} - \bar{y}). \qquad (3.14.24)$$

In scalar field theories, the only essential phase factors in the three discrete symmetries are $\pm 1$. Consequently, $U_C$, $U_P$, and $\Omega_T$ in their simplest manifestations commute.

In practice, it is generally more convenient to work with $P$ and $PT$ rather than with $P$ and $T$ because $PT$ acts simply on vectors and commutes with the Lorentz group. If we write $\Omega_{PT}$ for $U_P \Omega_T$, then for scalar particles,

$$\Omega_{PT} | k_1, \ldots, k_n \rangle = | k_1, \ldots, k_n \rangle. \qquad (3.14.25)$$

For pseudoscalar particles there would be factors of $(-1)$, of course.

The action of $PT$ on a typical interaction of scalars leaves it invariant but conjugates its coefficient:

$$\Omega_{PT} (\lambda \psi_1^\dagger \psi_2 \phi) \Big|_x \Omega_{PT}^{-1} = (\lambda^* \psi_1^\dagger \psi_2 \phi) \Big|_{-x}, \qquad (3.14.26)$$

but adding the action of $C$ produces the hermitian conjugate of the original term:

$$\Omega_{CPT} (\lambda \psi_1^\dagger \psi_2 \phi) \Big|_x \Omega_{CPT}^{-1} = (\lambda^* \psi_2^\dagger \psi_1 \phi) \Big|_{-x}. \qquad (3.14.27)$$

In general, from the actions of $C$ and $PT$ on operators, it is apparent that the action of $CPT$ will transform any Lagrangian density $\mathcal{L}(x)$ into $\mathcal{L}^\dagger(-x)$.

This implies that $\Omega_{CPT}$ tranforms $\mathcal{H}(t)$ into $\mathcal{H}^\dagger(-t)$. But the full time-reversal transformation involves an extra adjoint coming from interchange of initial and final states. Therefore the $CPT$ transform of $\mathcal{H}(t)$ is in fact $\mathcal{H}(-t)$. Consequently the Hamiltonian is $CPT$ invariant whenever it is a conserved quantity. We conclude that $CPT$ is a symmetry of all quantum field theories.

**Homework 3.14.28.** Show that the canonical commutation relations are invariant under $C$, $P$, and $T$.

Since charge conjugation, parity, and time reversal are not symmetries of the universe, there is generally no point in labeling states with $C$, $P$, or $T$ indices. Partial theories of natural law, however, may be $C$, $P$, or $T$ invariant. For example, electromagnetic and strong interactions appear to be parity invariant. In processes where such aspects of natural law dominate, it is useful to use the discrete symmetries to reduce the range of the transformations a propagating state may undergo. Note that discrete symmetries give rise to multiplicative conserved quantities while continuous symmetries give rise to additive ones. This is because, in the case of continuous symmetries, we effectively took a logarithm when we chose to work with symmetry generators rather than group elements.

## 3.15 Summary

Assuming that canonical quantization provides a simple means for transforming classical field theories into quantum field theories, we have naively assumed that the validity of Noether's Theorem in classical field theory is sufficient to justify application of the associated conserved current formula in quantum field theory. The proof that the Noether current is conserved depends on the classical equations of motion, and we note with a touch of apprehension that, after quantization in an interacting field theory, these evolution laws may not be valid. The analysis of quantum operator evolution in Chapter 13, however, justifies the use of Noether's formula in the quantum context.

The conserved quantities resulting from the Noether current in quantum field theory make it possible to identify one-particle states and generally support a deeper understanding of states and their evolution. Indeed, a state will evolve into a linear combination of all other states not made inaccessible by conservation laws. In the next chapter, we develop the Feynman technique for computing the coefficients in such a linear combination, so that we can predict the probability of observing any particular piece of it.

# Chapter 4

# From Dyson's Formula to Feynman Rules

On the basis of canonical quantization of free fields, by means of the
interaction picture which expresses the evolution of interacting fields
in terms of the free field, the scattering matrix is first expressed in
Dyson's formula as a perturbation series in a free-field Hamiltonian,
then converted to Wick's operator expansion, and finally reduced
to matrix elements and the Feynman perturbation series.

## Introduction

Now that, through symmetries and conserved quantities, we have some constraints
on the possibilities for the evolution of a state, it is time to develop the machinery
for computing the structure of an evolving state. At the end of Chapter 2, we
observed that canonical quantization of an interacting field theory could not be
carried out on account of the normal-ordering problem. This chapter describes
Dyson's technique for viewing an interacting field in terms of a free field, thereby
quantizing an interacting theory on the basis of the canonical quantization of a free
theory. This technique is intrinsically perturbative, being in effect equivalent to
the asymptotic series defined by Feynman diagrams and rules, and could be called
*perturbative canonical quantization.*

Section 4.1 presents the fundamental principle of perturbative canonical quantization, namely the interaction picture, in which the states evolve in a modified
Schrödinger fashion, and the operators evolve like Heisenberg operators in a free-
field theory. Working in the interaction picture, we easily find Dyson's formula for
the $S$ or scattering matrix as a function of free fields. Section 4.2 takes the first
step towards Feynman rules, namely Wick's operator expansion of Dyson's formula.
Sections 4.3 to 4.5 develop the theory of the Wick expansion, and Sections 4.6 and
4.7 compute the $S$ matrix for two examples using this theory. On the basis of these
computations, Section 4.8 points out some similarities and fundamental differences
between classical and quantum fields.

Introducing a third example, Section 4.9 derives Feynman's perturbation series
for $S$-matrix elements by evaluating the Wick expansion between specific initial
and final states. In detail, Feynman diagrams and Feynman rules are obtained
from Wick operators. Section 4.10 analyzes the simplest or tree-level Feynman
diagrams which contribute to particle scattering and compares the results with
analogous computations in quantum mechanics. Sections 4.11 and 4.12 cover a few
basic points in simplifying, manipulating, and representing scattering amplitudes.
Finally, in Section 4.13, we comment on the more complicated or loop-level Feynman
diagrams, introducing the concept of renormalization, the system for bringing finite
values out of divergent Feynman integrals.

Overall, we shall find that it is not a practical possibility to compute the time
dependence of even a simple initial state or to compute in one stroke the corresponding final state, but that, thanks to Feynman, it is a practical possibility to
compute $S$-matrix elements.

# 4.1 The Interaction Picture and Dyson's Formula

In this section, we shall prepare for a perturbative viewpoint on the canonical quantization of interacting quantum fields by explaining in and out states, the interaction picture, and Dyson's formula for the time evolution operator in non-relativistic quantum mechanics.

In quantum mechanics, when an incoming wave scatters off a potential $V$ which vanishes for large $|\bar{x}|$, then the propagation of incoming waves, $t \ll 0$, and of outgoing waves, $t \gg 0$, is simple. Scattering amplitudes relate the incoming wave to the outgoing wave. (In one spatial dimension, for example, all the information about scattering is contained in the reflection and transmission coefficients.)

In order to identify the particle content of the incoming and outgoing waves, it is necessary to compare them with free-particle states. If the Hamiltonian $H$ is a sum $H = H_0 + V$, where $H_0$ contains no interaction terms, then we associate a free-particle state $|\psi^{\text{in}}\rangle$ (a solution of $i\frac{\partial}{\partial t}|\psi\rangle = H_0|\psi\rangle$) with an incoming wave $|\psi\rangle$ (a solution of $i\frac{\partial}{\partial t}|\psi\rangle = H|\psi\rangle$) if $|\psi^{\text{in}}\rangle$ is equal to $|\psi\rangle$ for $t \sim -\infty$. This association is a unitary map from ideal initial states $|\psi^{\text{in}}\rangle$ to real incoming waves $|\psi\rangle$. Similarly one can define a map from ideal final states to real outgoing waves by associating $|\psi^{\text{out}}\rangle$ with $|\psi\rangle$ if $|\psi^{\text{out}}\rangle$ is equal to $|\psi\rangle$ for $t \sim \infty$.

Scattering is the relationship between in and out states that reflects the real process of passing through the support of the potential $V$. Scattering is summarized in the $S$ matrix, an operator on ideal initial states defined in terms of the real physical scattering process:

$$\langle \psi_2^{\text{out}}|S|\psi_1^{\text{in}}\rangle \overset{\text{def}}{=} \langle \psi_2|\psi_1\rangle. \tag{4.1.1}$$

Having defined the $S$ matrix on wave packets, we can extend the definition to plane waves. Note that, since the operators which assign in states to kets and out states to bras are both unitary, $S$ is also unitary.

This definition will not apply as it stands to an interacting quantum field theory because interaction terms form in effect a potential which has the same magnitude everywhere in space-time. To circumvent this obstruction, we shall for the moment assume that we can switch the interactions on and off in such a way that:

1. the process of switching on and off is so slow that the effect on the scattering process is small;
2. the interactions are left on for long enough that the scattering process is near enough complete when we switch the interactions off;
3. the period of time taken for switching on and off is vanishingly small compared to the interaction period so that the scattering is dominated by the interaction period.

A turning on and off function $f(t)$ which satisfies these conditions is called adiabatic.

Switching on the interactions would, of course, affect decay rates of unstable particles; switching them off would cause the decay of stable bound states. For the time being, we shall therefore focus on theories with no unstable particles or stable bound states. The gauge invariance of a gauge theory like quantum electrodynamics depends on the interactions, so switching the interactions off also breaks gauge invariance. Clearly, we shall need a better scattering theory in the end, but

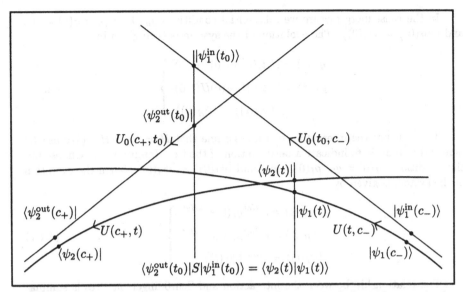

**Fig. 4.1.** This figure depicts the link between the $S$ matrix and physical states in the Schrödinger picture. The curved lines represent the time evolution of the physical in and out states and the slanted straight lines represent the free evolution of the conceptual in and out states. The vertical lines represent arbitrary times at which the inner products between in and out states of either type may be taken.

this approach to quantum field theory enables us to introduce some fundamental elements of the general formalism in a simple theoretical environment.

There are three frames for viewing scattering in quantum mechanics. The Schrödinger picture regards the states as evolving and the observables as constant. The Heisenberg picture regards the states as constant and the observables as evolving, and the interaction picture regards the states as evolving in such a way that the evolution of the observables can be computed in terms of a Hamiltonian for free particles.

In the Schrödinger picture we have observables $q_S(t) \equiv q_S(0)$, $p_S(t) \equiv p_S(0)$, and states $|\psi_S(t)\rangle$ which satisfy

$$i\frac{\partial}{\partial t}|\psi_S(t)\rangle = H(p_S, q_S, t)|\psi_S(t)\rangle. \tag{4.1.2}$$

Evolution of states from time $t'$ to time $t$ can be written in terms of a unitary operator $U(t, t')$:

$$|\psi_S(t)\rangle = U(t, t')|\psi_S(t')\rangle. \tag{4.1.3}$$

$U(t, t')$ is determined by the initial value problem

$$i\frac{\partial}{\partial t}U(t, t') = H(t)U(t, t'), \quad U(t', t') = 1. \tag{4.1.4}$$

If $H$ does not depend on $t$ then we can write

$$U(t, t') = e^{-iH(t-t')}. \tag{4.1.5}$$

Figure 4.1 shows the relationship between the $S$ matrix and the physical in and out states in the Schrödinger picture.

In the Heisenberg picture we take initial conditions $q_H(0) = q_S$, $p_H(0) = p_S$, and $|\psi_H(0)\rangle = |\psi_S(0)\rangle$. The evolution of the system is then given by

$$\left.\begin{array}{r} q_H(t) = U(t,0)^\dagger q_H(0) U(t,0) \\ p_H(t) = U(t,0)^\dagger p_H(0) U(t,0) \\ |\psi_H(t)\rangle = |\psi_H(0)\rangle \end{array}\right\} \qquad (4.1.6)$$

To construct the interaction picture, assume that $H = H_0 + H'$. (We use $H'$ here rather than $V$ to indicate a perturbation of the free Hamiltonian.) Choose initial conditions $q_I(0) = q_S$, $p_I(0) = p_S$, and $|\psi_I(0)\rangle = |\psi_S(0)\rangle$. Then the evolution of the system is given by

$$\left.\begin{array}{r} q_I(t) = e^{iH_0 t} q_I(0) e^{-iH_0 t} \\ p_I(t) = e^{iH_0 t} p_I(0) e^{-iH_0 t} \\ |\psi_I(t)\rangle = e^{iH_0 t} |\psi_S(t)\rangle \end{array}\right\} \qquad (4.1.7)$$

The relationship between the interaction and Schrödinger pictures is summarized in the evolution of a matrix element $M(t)$ of an observable $A$. The following algebra starts from the definition of $M$ in the Schrödinger picture and concludes with the interaction picture expression for $M$:

$$\begin{aligned} M(t) &= \langle \phi_S(t)|A_S|\psi_S(t)\rangle \\ &= \langle \phi_I(t)|e^{iH_0 t} A_I(0) e^{-iH_0 t}|\psi_I(t)\rangle \\ &= \langle \phi_I(t)|A_I(t)|\psi_I(t)\rangle. \end{aligned} \qquad (4.1.8)$$

Since $H_0$ does not depend on time, we find the following differential equation for $|\psi_I(t)\rangle$ by direct computation:

$$\begin{aligned} i\frac{\partial}{\partial t}|\psi_I(t)\rangle &= -e^{iH_0 t} H_0(p_S, q_S)|\psi_S(t)\rangle + e^{iH_0 t} H(p_S, q_S, t)|\psi_S(t)\rangle \\ &= e^{iH_0 t} H'\big(p_I(0), q_I(0), t\big) e^{-iH_0 t}|\psi_I(t)\rangle \\ &= H'\big(p_I(t), q_I(t), t\big)|\psi_I(t)\rangle. \end{aligned} \qquad (4.1.9)$$

$H'\big(p_I(t), q_I(t), t\big)$ is called the interaction Hamiltonian $H_I(t)$. Define the interaction unitary operator $U_I(t, t')$ by

$$U_I(t, t')|\psi_I(t')\rangle = |\psi_I(t)\rangle. \qquad (4.1.10)$$

Then $U_I(t, t')$ is determined by the initial value problem

$$i\frac{\partial}{\partial t} U_I(t, t') = H_I(t) U_I(t, t'), \quad U_I(t', t') = \mathbf{1}. \qquad (4.1.11)$$

**Homework 4.1.12.** By comparison of the Schrödinger picture with the interaction picture, show that the interaction unitary operator is related to the Schrödinger unitary operator by

$$U_I(t, t') = e^{iH_0 t} U_S(t, t') e^{-iH_0 t'}.$$

Since the interaction-picture states do not evolve when $H_I(t) = 0$ and since we have an adiabatic turning on and off function $f(t)$ which guarantees that $H_I(t) = 0$ for large $|t|$, the physical states are the incoming and outgoing waves. Hence, in the interaction picture, the $S$ matrix is given by

$$S = \lim_{t \to \infty} \lim_{t' \to -\infty} U_I(t, t'). \tag{4.1.13}$$

**Homework 4.1.14.** Assume that $f(t) = 0$ for $|t| > T$. Show that in the Schrödinger picture the relationship between real and conceptual states is

$$\left|\psi(t)\right\rangle = U(t, c_-) e^{-iH_0(c_- - t_0)} \left|\psi^{\text{in}}(t_0)\right\rangle, \quad c_- < -T;$$

$$\left|\phi(t)\right\rangle = U(t, c_+) e^{-iH_0(c_+ - t_0)} \left|\phi^{\text{out}}(t_0)\right\rangle, \quad c_+ > T.$$

Show that, if $|t| > T$ and $|t + \Delta t| > T$, then

$$U(t + \Delta t, t) = e^{-iH_0 \Delta t},$$

and deduce that $S = U_I(c_+, c_-)$ when $c_- < -T$ and $c_+ > T$.

In a theory with no particles which decay and no bound states, the turning on and off of the interactions merely serves to limit the effective range of forces. In this case, turning the interactions on and off adiabatically will not significantly affect the evolution of any state, so the initial energy and the final energy as determined by the full Hamiltonian $H$ will be equal to that determined by the free Hamiltonian $H_0$, so we find that scattering preserves the $H_0$-energy of the state:

$$[H_0, S] = 0. \tag{4.1.15}$$

**Homework 4.1.16.** Show that, if there are no bound states or particle decays in the theory, then $S$ is independent of the parameter $t_0$ in Figure 4.1.

Since $H_I(t)$ is a function of time, it is not possible to solve the evolution equation 4.1.11 for $U_I$ by simply exponentiating the interaction Hamiltonian. We can, however, generalize the exponential function to the time-ordered exponential function and write a formal solution using this idea. A time-ordered product $T\big(A(t_1) \cdots A(t_k)\big)$ of operators $A(t_r)$ is the product of the operators with the later on the left. Define the time-ordered exponential by

$$T \exp\left(-i \int_{t_1}^{t_2} dt\, H_I(t)\right) \tag{4.1.17}$$

$$\stackrel{\text{def}}{=} 1 - i \int_{t_1}^{t_2} dt\, H_I(t) - \frac{1}{2!} \int_{t_1}^{t_2} \int_{t_1}^{t_2} dt\, dt'\, T\big(H_I(t) H_I(t')\big) + \cdots$$

$$= 1 - i \int_{t_1}^{t_2} dt\, H_I(t) - \int_{t=t_1}^{t_2} \int_{t'=t_1}^{t} dt\, dt'\, H_I(t) H_I(t') + \cdots$$

It is clear that this formula implies

$$i\frac{\partial}{\partial t_2} T \exp\left(-i \int_{t_1}^{t_2} dt\, H_I(t)\right) = H_I(t_2) T \exp\left(-i \int_{t_1}^{t_2} dt\, H_I(t)\right),$$

$$\tag{4.1.18}$$

$$T \exp\left(-i \int_{t_1}^{t_1} dt\, H_I(t)\right) = 1.$$

But these equations have the same form as the initial value problem 4.1.11 for $U_I$. Hence we have found a formula for the evolution operator in the interaction picture,

$$U_I(t_2, t_1) = T \exp\left(-i \int_{t_1}^{t_2} dt\, H_I(t)\right). \qquad (4.1.19)$$

We at once deduce the main result of this section, namely

**Dyson's Formula** :   $S = T \exp\left(-i \int_{-\infty}^{\infty} dt\, H_I(t)\right). \qquad (4.1.20)$

**Homework** 4.1.21. Solve the initial value problem for $U_I(t, t')$ by iterating the discrete-time approximation to the defining differential equation 4.1.11

$$U_I(t + \triangle t, t) \simeq 1 - i H_I(t) \triangle t,$$

and taking the limit as $\triangle t \to 0$ to obtain 4.1.17.

**Remark** 4.1.22. A serious formal weakness in applying the interaction picture to field theory is pinpointed in Haag's Theorem, a part of which follows:

**Theorem.** *Let $\phi_0$ be a free, hermitian quantum field of mass $m$ which acts on a space $H_0$ with an invariant vacuum $|0\rangle_0$ and let $\phi$ be a hermitian quantum field which acts on a space $H$ with invariant vacuum $|0\rangle$. Assume that the sets of operators $\phi_0(t, \bar{x})$ and $\phi(t, \bar{x})$ are complete at some fixed time $t$ and that there exists a unitary transformation $U$ from $H_0$ to $H$ for which $\phi_0 = U^\dagger \phi U$ at time $t$. Then $\phi$ is also a free field of mass $m$.*

This theorem shows that when we apply the interaction picture to field theory, if $H$ contains interactions, then the transformation of kets used to define the interaction picture cannot be unitary.

Conceivably, one could allow the transformation not to be unitary. (The following remark gives an example of non-unitary canonical transformation.) This is related to rendering Haag's Theorem inapplicable by using state spaces without a free vacuum. First, a fact – the representations of a free field as operators on state spaces are of two types: one type has a vacuum state and is therefore the standard Fock space representation, the other type does not have a vacuum state and is known as 'strange.' In a strange representation, every state has infinitely many particles. Second, a suggestion: from the perspective of the free field, it might appear that evolution under $H$ takes a state through a sequence of inequivalent strange representations. This suggestion, however, falls foul of another mathematical result, that there is no operator relating inequivalent representations of a free field.

Then again, in quantum electrodynamics, because the photon is massless, it is possible for a moving electron to emit a countable infinity of photons. Techniques have been devised to manage this 'infrared divergence' and track the evolution of such states. Even so, the mathematical content of the interaction picture is in doubt.

In practice, as we shall see in the next sections, the result of applying the interaction picture to field theory is a perturbation series which does not converge but whose first few terms yield wonderfully accurate predictions. It appears then that the interaction picture provides a sound approach to perturbation theory but may have no non-perturbative validity.

**Remark** 4.1.23. We expect switching the interactions on and off to change the normalization of the field with respect to the one-particle states. To understand such a transformation, consider first rescaling of canonical variables in quantum mechanics:

$$Q' = e^{-\alpha} Q \quad \text{and} \quad P' = e^{\alpha} P. \qquad (4.1.24)$$

Clearly, this rescaling preserves the canonical commutation relation, and is therefore a canonical transformation. Furthermore, it can be implemented by a unitary transformation $V$:

$$V = e^{i\alpha(PQ+QP)} \quad \implies \quad X' = VXV^\dagger. \tag{4.1.25}$$

If we now return to field theory, rescaling is still a canonical transformation, but the sum in the definition of $V$ becomes a divergent integral. Thus, we have a canonical transformation which cannot be implemented by a unitary operator.

The great significance of the interaction picture for field theory is that it enables us to quantize an interacting field theory on the basis of the quantum theory of free fields. This phenomenon, that there is a viewpoint in which the evolution of interacting fields appears to be the evolution of free fields, is so remarkable that it is not surprising to find that Dyson's formula has a restricted range of validity and cannot properly be applied to field theory. Nevertheless, we shall take it as a guiding principle for the derivation of the usual Feynman rules. The Feynman rules will provide the perturbative definition of the canonically quantized field theory. The theoretical weakness of the argument will not be apparent on the simplest (tree) level of the resulting theory, but higher order (loop) effects will often be undefined, forcing us to include another process, perturbative renormalization, in the definition of perturbative canonical quantization.

## 4.2 The Wick Expansion of the Scattering Matrix

In this section, we shall begin to explain the consequences of applying the interaction picture and Dyson's formula for the $S$ matrix to field theory. Even though such application is not strictly valid, the concepts and principles that result from it provide an effective foundation for perturbative quantum field theory.

Dyson's formula expresses the interaction dynamics in terms of complicated behavior of free fields. Computing the effect of a time-ordered product on a ket is difficult because creation and annihilation can happen at any time, so particles come and go. The effect of a normal-ordered product on a ket is easy to compute because we do all the annihilating first and all the creating second. The next thing to do then is to simplify the computation of the time-ordered exponential by converting time-ordered products to normal-ordered products. We shall call the resulting sum of operators the *Wick expansion* of the $S$ matrix.

Let $A_r = A_r^+(x_r) + A_r^-(x_r)$ be operators which are linear in creation and annihilation operators. The *contraction* $\acute{A}_1\grave{A}_2$ of $A_1$ and $A_2$ is by definition the difference between the time-ordered and normal-ordered products:

$$\acute{A}_1\grave{A}_2 \overset{\text{def}}{=} T(A_1 A_2) - {:}A_1 A_2{:} \tag{4.2.1}$$

If we assume that $t_1 > t_2$, then

$$T(A_1 A_2) = (A_1^+ + A_1^-)(A_2^+ + A_2^-) = {:}A_1 A_2{:} + [A_1^-, A_2^+], \tag{4.2.2}$$

which implies that

$$\acute{A}_1\grave{A}_2 = \begin{cases} [A_1^-, A_2^+], & \text{when } t_1 > t_2; \\ [A_2^-, A_1^+], & \text{when } t_2 > t_1. \end{cases} \tag{4.2.3}$$

Since normal-ordered products have zero vacuum expectation, and since the contraction of $A_1$ and $A_2$ is a complex number, the contraction is equal to the vacuum expectation of the time-ordered product:

$$\acute{A}_1\grave{A}_2 = \langle 0|T(A_1 A_2)|0\rangle. \tag{4.2.4}$$

Contraction takes precedence over normal ordering so that, for example,

$$:A_1\acute{A}_2 A_3\grave{A}_4: = \acute{A}_2\grave{A}_4:A_1 A_3: \tag{4.2.5}$$

From Homework 2.4.30 on the time-ordered product of free scalar fields, we have an integral formula for the contraction of such fields:

$$\acute{\phi}(x)\grave{\phi}(y) = \lim_{\epsilon\to 0^+} \int \frac{d^4 k}{(2\pi)^4} \frac{ie^{-i(x-y)\cdot k}}{k^2 - \mu^2 + i\epsilon};$$

$$\acute{\psi}(x)\grave{\psi}(y)^\dagger = \acute{\psi}(x)^\dagger\grave{\psi}(y) = \lim_{\epsilon\to 0^+} \int \frac{d^4 k}{(2\pi)^4} \frac{ie^{-i(x-y)\cdot k}}{k^2 - m^2 + i\epsilon}. \tag{4.2.6}$$

This shows that the contraction of free fields is a complex number as expected, and so commutes with all operators. Note that the integrals here are invariant under the change of parameter $k \to -k$, but that the sign convention used here is significant as it represents momentum $k$ flowing from $y$ to $x$.

Wick's Theorem relates the time-ordered product to the normal-ordered product. Before stating the theorem, we will give an example of it, so that the proposition will be intelligible. Write $C_{rs}$ for the operator that contracts $A_r$ and $A_s$. Then, for a product of three or four fields the results are

$$
\begin{aligned}
T(A_1 A_2 A_3) &= :A_1 A_2 A_3: + :(C_{12} + C_{13} + C_{23})A_1 A_2 A_3: \\
&= :A_1 A_2 A_3: + (\acute{A}_1\grave{A}_2)A_3 + (\acute{A}_1\grave{A}_3)A_2 + (\acute{A}_2\grave{A}_3)A_1,
\end{aligned} \tag{4.2.7}
$$

$$
\begin{aligned}
T(A_1 A_2 A_3 A_4) &= :A_1 A_2 A_3 A_4: \\
&\quad + :(C_{12} + C_{13} + C_{14} + C_{23} + C_{24} + C_{34})A_1 A_2 A_3 A_4: \\
&\quad + :(C_{12}C_{34} + C_{13}C_{24})A_1 A_2 A_3 A_4: \\
&= :A_1 A_2 A_3 A_4: + \acute{A}_1\grave{A}_2:A_3 A_4: + \acute{A}_1\grave{A}_3:A_2 A_4: \\
&\quad + \acute{A}_1\grave{A}_4:A_2 A_3: + \acute{A}_2\grave{A}_3:A_1 A_4: + \acute{A}_2\grave{A}_4:A_1 A_3: \\
&\quad + \acute{A}_3\grave{A}_4:A_1 A_2: + (\acute{A}_1\grave{A}_2)(\acute{A}_3\grave{A}_4) + (\acute{A}_1\grave{A}_3)(\acute{A}_2\grave{A}_4).
\end{aligned}
$$

**Theorem 4.2.8.** *The time-ordered product of $n$ free fields $A_r(x_r)$ is equal to the sum of the normal-ordered products of all possible partial and complete contractions of $A_1 \cdots A_n$.*

**Proof.** The theorem is true for $n \leq 2$ by the definitions of time and normal ordering. Proceed by induction, assuming that $t_1 \geq \cdots \geq t_n$. Let $W(A_1 \cdots A_n)$ be the sum of all contractions of $A_1 \cdots A_n$ normally ordered. Assume the theorem is true for $n - 1$ operators. Then:

$$
\begin{aligned}
T(A_1 \cdots A_n) &= A_1 T(A_2 \cdots A_n) = A_1 W(A_2 \cdots A_n) \\
&= A_1^+ W + W A_1^- + [A_1^-, W] \\
&= W(A_1 \cdots A_n).
\end{aligned} \tag{4.2.9}
$$

We note that $[A_1^-, W]$ can be reduced to a sum of brackets of the type used to define contraction in 4.2.3. The last step hides all the combinatorics, but the argument is sufficient to indicate the logic behind the proposition. □

In order to rearrange Dyson's expansion of the $S$ matrix into a sum of more convenient normal-ordered terms, we begin by working on second order terms and find it necessary to introduce the contraction of pairs of fields. Wick's Theorem shows that this small additional concept is sufficient to systematically reorganize Dyson's expansion into a sum of normal-ordered terms, the Wick expansion of the $S$ matrix.

# 4.3 Wick Diagrams

Dyson's time-ordered exponential $T \exp(-i \int dt\, H_I)$ is a sum of time-ordered operators, graded by power of $H_I$. Wick's Theorem enables us to convert this sum of time-ordered monomials into a sum (which we will call the Wick expansion of $S$) of normal-ordered monomials. Where the Dyson expansion is a simple sum of complicated terms, the Wick expansion is a complicated sum of simple terms. The variety of terms in the Wick expansion is so great that it is sensible to devise a systematic labeling for them. We shall now introduce a labeling system that leads naturally to Feynman diagrams. We will call the labels *Wick diagrams*. Note that a Wick diagram labels an operator in the Wick expansion, whereas a Feynman diagram, as we shall see, labels a number, a matrix element of an operator.

A Wick diagram consists of points and lines. A line that joins two vertices is called an internal line of the diagram, and a line with a free end is called an external line. Each point represents an interaction, a monomial in the interaction Hamiltonian, each internal line represents a contraction of fields, an influence propagating from one point to another through a field, and each external line represents an uncontracted field.

The basic element in a Wick diagram is not a point or a line but a vertex, that is, a point with line segments emanating from it. A Wick diagram is constructed from a finite set of vertices by joining any number of line ends in pairs. The line segments at the vertices may be of distinguishable types; for example, they may be dashed, or they may carry arrows. The purpose of distinguishing line types is to make it clear which line pairs can be joined; the common convention is that we may join lines only if they are of the same type (solid, dashed, dotted, etc.) and, if there are arrows on the lines, only if one line has an outgoing arrow and the other an incoming one.

For a fixed interaction Hamiltonian density $\mathcal{H}_I$, if a Wick diagram is to represent a unique operator in the Wick expansion of the $S$ matrix, we must be able to identify the interaction monomials and contractions from the diagram alone. Hence we introduce line types, one for each scalar field in $\mathcal{H}_I$, and, using these typed lines, we introduce a set of vertices, one for each monomial in $\mathcal{H}_I$. Now we can associate an operator with a Wick diagram constructed from these vertices: start from the product of the monomials corresponding to the vertices in the diagram, contract the pairs of fields corresponding to the internal lines, normal order the remaining fields, and integrate over space-time variables. Since a contraction between different scalar fields is zero, if we join a pair of lines of different types, the resulting operator will be zero. Hence, to eliminate zero operators, we impose the constraint that only pairs of lines of the same type may be joined.

We can accommodate a complex field $\psi$ as follows. As the only non-zero contraction involving $\psi$ is $\psi\psi^\dagger$, if we label a $\psi$ line with an outgoing arrow and a $\psi^\dagger$ line with an incoming arrow, then the appropriate joining constraint permits matching of outgoing arrows only with incoming arrows and *vice versa*.

Clearly, with this procedure we have established a one-to-one correspondence between terms in the Wick expansion and Wick diagrams constructed from the appropriate set of vertices with the appropriate joining constraints. We shall refer to an operator associated with a Wick diagram as a *Wick operator*.

To make these notions concrete, we introduce an example, a scalar model for nucleon-meson dynamics:

$$\mathcal{L} = \frac{1}{2}(\partial_\mu\phi\,\partial^\mu\phi - \mu^2\phi^2) + (\partial_\mu\psi^\dagger\partial^\mu\psi - m^2\psi^\dagger\psi) - \lambda\psi^\dagger\psi\phi. \tag{4.3.1}$$

This Lagrangian density represents a charged scalar interacting with itself through the medium of a neutral scalar. If we interpret the charged field $\psi$ as protons and the neutral field as the neutral pion, then we have a model for understanding roughly how pion exchange holds protons together in an atomic nucleus. For this Lagrangian density, the interaction Hamiltonian density $\mathcal{H}_I$ is given by

$$\mathcal{H}_I = \lambda\psi_I^\dagger\psi_I\phi_I \overset{\text{relabel}}{=} \lambda\psi^\dagger\psi\phi. \tag{4.3.2}$$

Note that we have taken the adiabatic limit here, setting $f(t) = 1$.

Represent $\phi$ propagation by a line, and $\psi$ or $\psi^\dagger$ propagation by a line with an arrow on it indicating the flow of positive charge. Represent each occurrence of the interaction monomial $\psi^\dagger\psi\phi$ by a vertex of a $\phi$-line, a $\psi$-line, and a $\psi^\dagger$-line:

**Wick Vertex:** (4.3.3)

The $-i\lambda$ at the vertex represents the coefficient of $\psi^\dagger\psi\phi$ – $\lambda$ from $\mathcal{H}_I$ and $-i$ from Dyson's formula.

To build a Wick diagram of order $n$, just take $n$ vertices and create as many internal lines as desired subject to the appropriate joining constraints. Note that charge conservation is reflected in every Wick diagram by 'conservation of arrows': because at each vertex one arrow comes in and one goes out, the arrow lines either form closed loops or have both ends on external lines; hence, in the diagram as a whole, the number of external arrows going in is equal to the number coming out.

The rules above enable us to associate an operator $\mathcal{O}(D)$ in the $n^{\text{th}}$ order term in Dyson's formula with an $n$-vertex Wick diagram $D$. Due to the variety of possible contraction schemes, the operator $\mathcal{O}(D)$ may arise from the same term in Dyson's formula in many different ways. We need therefore to compute the proper coefficient for it. Rather than count contraction schemes in a product of fields we can count joining schemes in Wick diagrams. Given $n$ disconnected vertices, let $c(D)$ be the number of different ways of joining lines to form the diagram $D$. The coefficient of the $n^{\text{th}}$ order integral is $(-i)^n/n!$ and there is a factor of $\lambda^n$ from the interaction

Hamiltonian. Divide this up into $(-i\lambda)^n$ times $1/n!$, and include the $(-i\lambda)^n$ in $\mathcal{O}(D)$. Then the total contribution of operators with diagrams $D$ to the $S$ matrix is

$$S(D) \stackrel{\text{def}}{=} \frac{c(D)}{n!} \mathcal{O}(D). \tag{4.3.4}$$

With $\mathcal{H}_I = \lambda \psi^\dagger \psi \phi$, it is easy to compute $c(D)/n!$ directly, as we now show. Since there is only one monomial in $\mathcal{H}_I$, a product of $n$ copies of this monomial can be permuted by the permutation group $S_n$ of $n$ symbols. Since there are no quadratic factors in the monomial, the set of fields in the monomial has no symmetry. Hence $S_n$ is the symmetry group of the uncontracted product of monomials, or equivalently, of the set of $n$ Wick vertices.

Now impose the desired contractions in some specific way. All $c(D)$ contraction patterns that are represented by the same Wick diagram can be obtained from the chosen one by applying a permutation from $S_n$. Let $Z$ be the subgroup of $S_n$ which leaves the specified contraction pattern invariant; $Z$ is the symmetry group of $D$, moving lines and vertices around in such a way that the appearance of $D$ is unaltered. Let $s(D)$, the symmetry number of $D$, be the number of elements in $Z$. Since the number of elements in $S_n$ is $n!$, from a standard theorem of finite group theory we know that $s(D)c(D) = n!$. Rearranging this equality, we obtain

$$\frac{1}{s(D)} = \frac{c(D)}{n!}. \tag{4.3.5}$$

The benefit of this shift in emphasis from $c(D)$ to $s(D)$ is that $s(D)$ can often be more easily computed from the specified contraction pattern, or by inspection of the Wick diagram. For the given interaction, we simply regard the Wick diagram as a network of strings, marked to indicate flow of charge, and count the number of permutations of the vertices which, after moving the strings around or through each other, leave the network looking unchanged.

For example, we have the following Wick diagrams and operators:

$$\mathcal{O}\left(\text{><}\right) = (-i\lambda)^2 \int d^4x\, d^4y\ \psi^\dagger(x)\dot\psi(y) : \psi(x)\phi(x)\psi^\dagger(y)\phi(y):$$

$$\mathcal{O}\left(\text{-O-}\right) = (-i\lambda)^2 \int d^4x\, d^4y\ \psi^\dagger(x)\dot\psi(y)\,\dot\psi(x)\dot\psi^\dagger(y) :\phi(x)\phi(y): \tag{4.3.6}$$

$$\mathcal{O}\left(\text{O}\right) = (-i\lambda)^2 \int d^4x\, d^4y\ \psi^\dagger(x)\dot\psi(y)\,\dot\psi(x)\dot\psi^\dagger(y)\,\dot\phi(x)\dot\phi(y)$$

The first diagram has contraction number 2 and symmetry number 1, so that $S(D) = \mathcal{O}(D)$. The last two diagrams have contraction number 1 and symmetry number 2, implying that $S(D) = \frac{1}{2}\mathcal{O}(D)$ in these cases.

For a final example, consider the fourth order vacuum bubble

$$D = \text{OO} \tag{4.3.7}$$

Given four vertices, there are three ways of joining them up to form this fourth order diagram. Hence $c(D) = 3$ and $s(D) = 8$. Observe that this diagram does indeed have three independent symmetries of order two.

**Homework 4.3.8.** Draw all the diagrams $D$ up to third order for the interaction Hamiltonian density $\mathcal{H}_I = \lambda\psi^\dagger\psi\phi$ and compute their contraction and symmetry numbers, $c(D)$ and $s(D)$. (As a check, note that the number of ways of contracting $\phi$ and $\psi$ lines is:

First Order:    $(1)\big(1 + 1(\dot\psi^\dagger\dot\psi)\big)$;

Second Order:   $\big(1 + 1(\dot\phi\dot\phi)\big)\big(1 + 4(\dot\psi^\dagger\dot\psi) + 2(\dot\psi^\dagger\dot\psi)^2\big)$;

Third Order:    $\big(1 + 3(\dot\phi\dot\phi)\big)\big(1 + 9(\dot\psi^\dagger\dot\psi) + 18(\dot\psi^\dagger\dot\psi)^2 + 6(\dot\psi^\dagger\dot\psi)^3\big)$.

So, for example, at third order there are three ways of joining one pair of $\phi$ lines and 18 ways of joining two $\psi^\dagger$ lines to two $\psi$ lines. Thus the sum of the contraction numbers for diagrams with one $\dot\phi\dot\phi$ line and two $\psi^\dagger\psi$ lines should be $3 \times 18$.)

In order to compute Wick operators and their effects, it is convenient to use the free-field expansions to split the fields up into creation and annihilation terms:

$$\phi_+(x) \overset{\text{def}}{=} \int d\lambda(\bar k)\, e^{ix\cdot k}\alpha^\dagger(k), \qquad \phi_-(x) \overset{\text{def}}{=} \int d\lambda(\bar k)\, e^{-ix\cdot k}\alpha(k),$$

$$\psi_+(x) \overset{\text{def}}{=} \int d\lambda(\bar k)\, e^{ix\cdot k}\gamma^\dagger(k), \qquad \psi_-(x) \overset{\text{def}}{=} \int d\lambda(\bar k)\, e^{-ix\cdot k}\beta(k), \qquad (4.3.9)$$

$$\psi^\dagger_+(x) \overset{\text{def}}{=} \int d\lambda(\bar k)\, e^{ix\cdot k}\beta^\dagger(k), \qquad \psi^\dagger_-(x) \overset{\text{def}}{=} \int d\lambda(\bar k)\, e^{-ix\cdot k}\gamma(k),$$

where $d\lambda(k)$ is the usual Lorentz invariant measure:

$$d\lambda(\bar k) \overset{\text{def}}{=} \frac{d^3\bar k}{(2\pi)^3\, 2\omega(\bar k)}. \qquad (4.3.10)$$

The action of these operators on bras and kets is simple. For example, if $|p\rangle$ is a nucleon – a particle created by $\beta^\dagger(p)$ – then

$$\psi_-(x)|p\rangle = \int d\lambda(k)\, e^{-ix\cdot k}\beta(k)\beta^\dagger(p)|0\rangle$$

$$= \int d^3\bar k\, e^{-ix\cdot k}\delta^{(3)}(\bar k - \bar p)|0\rangle \qquad (4.3.11)$$

$$= e^{-ix\cdot p}|0\rangle.$$

**Homework 4.3.12.** In the theory defined by the Lagrangian density 4.3.1, consider the second-order diagram $D$ formed by joining the meson lines in the two vertices. Find $c(D)$, $s(D)$, $\mathcal{O}(D)$, and $S(D)$. Write $S(D)$ in terms of creation and annihilation operators using 4.3.9 and expand to obtain $2^4$ monomial, normal-ordered operators. Describe what these operators do, and compute $\langle p', q'|S(D)|p, q\rangle$, the contribution of $S(D)$ to elastic nucleon-nucleon scattering. (In particle theory, scattering is 'elastic' when the initial and final states have the same particle content.)

This section establishes the idea of using diagrams as labels for the terms in the Wick expansion of the $S$ matrix. Note that every Wick diagram contains at least one vertex, so the initial $\mathbf{1}$ in Dyson's formula is not represented by a diagram. (Mathematicians would represent it by the empty diagram, of course.) The sum of the operators in the Wick expansion is therefore $S - \mathbf{1}$. As the example is too simple to reveal the delicacy of the combinatorics, and as these factors are seldom needed, we complete this part of the theory separately in the next section.

## 4.4 The Combinatorics of Wick Diagrams

In this section, we show that the formula for the symmetry factor deduced from 4.3.4 and 4.3.5 is actually universal as long as we normalize each coupling constant with the symmetry factors appropriate to its interaction. Figure 4.4 indicates some possibilities. To state the result formally, we need some multi-index notation. So we begin with the proof.

We consider theories with a vector of distinct neutral fields $\phi = (\phi_1, \ldots, \phi_n)$ and $k$ monomial interactions $V = (V_1, \ldots, V_k)$. For simplicity we shall use multi-index notation conventions, so that, if $A$ is an $m$-component vector of operators and $I$ is an $m$-component vector of integers, then

$$A^I \stackrel{\text{def}}{=} (A_1^{I_1}, \ldots, A_m^{I_m}), \quad I! \stackrel{\text{def}}{=} (I_1!, \ldots, I_m!);$$
$$\sum A \stackrel{\text{def}}{=} A_1 + \cdots + A_m, \quad \prod A \stackrel{\text{def}}{=} A_1 \times \cdots \times A_m. \tag{4.4.1}$$

Let $I_r$ be the multi-index for which $V_r = \prod \phi^{I_r}$. Write the coefficient of $V_r$ in $\mathcal{H}_I$ as $\lambda_r / \prod I_r!$, so that

$$\mathcal{H}_I = \frac{\lambda_1}{\prod I_1!} \prod \phi^{I_1} + \cdots + \frac{\lambda_k}{\prod I_k!} \prod \phi^{I_k}. \tag{4.4.2}$$

Suppose we are working on the term $\prod V^J$ in Dyson's formula and have specified a contraction pattern on it through a $\sum J$-vertex Wick diagram $D$. Write $\mathcal{O}(D)$ for a generic operator constructed from $:\prod V^J:$ through the contraction pattern prescribed by the diagram $D$ together with the coupling constant factor $\prod(-i\lambda)^J$ and integration over space-time variables; write $S(D)$ for the total contribution of operators associated with the diagram $D$ to the $S$ matrix. Note finally that the multinomial coefficient for $\prod V^J$ in $\mathcal{H}_I^n$ is $\binom{n}{J}$. Then the generalization of relationship 4.3.4 between $S(D)$ and $\mathcal{O}(D)$ is simply

$$S(D) = \binom{n}{J} \frac{c(D)}{n!} \frac{1}{\prod(\prod I!)^J} \mathcal{O}(D), \tag{4.4.3}$$

where $\prod I!$ is $(\prod I_1!, \ldots, \prod I_k!)$ – the operator $\prod$ acts on the innermost level of a tensor.

Continuing the generalization, the monomial $V_r = \prod \phi^{I_r}$ has symmetry group

$$G_r = S_{I_{r,1}} \times \cdots \times S_{I_{r,n}}. \tag{4.4.4}$$

The number of elements in $G_r$ is $\prod I_r!$, which matches the combinatorial factors in the coupling constants in the interaction Hamiltonian density 4.4.2. The group of symmetries that interchanges monomials in a product $\prod V^J$ is

$$G_J = S_{J_1} \times \cdots \times S_{J_k}. \tag{4.4.5}$$

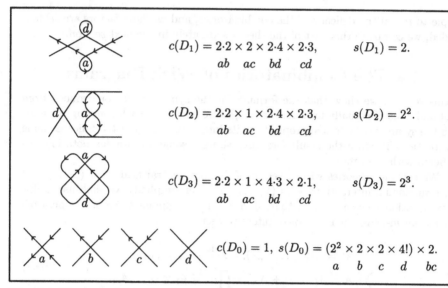

$$c(D_1) = 2 \cdot 2 \times 2 \times 2 \cdot 4 \times 2 \cdot 3, \qquad s(D_1) = 2.$$
$$\qquad\qquad ab \quad ac \quad bd \quad cd$$

$$c(D_2) = 2 \cdot 2 \times 1 \times 2 \cdot 4 \times 2 \cdot 3, \qquad s(D_2) = 2^2.$$
$$\qquad\qquad ab \quad ac \quad bd \quad cd$$

$$c(D_3) = 2 \cdot 2 \times 1 \times 4 \cdot 3 \times 2 \cdot 1, \qquad s(D_3) = 2^3.$$
$$\qquad\qquad ab \quad ac \quad bd \quad cd$$

$$c(D_0) = 1, \; s(D_0) = (2^2 \times 2 \times 2 \times 4!) \times 2.$$
$$\qquad\qquad\qquad a \quad b \quad c \quad d \quad bc$$

**Fig. 4.4.** The examples in this figure show how to compute contraction and symmetry numbers for Wick diagrams. At the bottom of the figure we see four vertices associated with interaction Hamiltonian terms $\frac{1}{4}(\psi^\dagger \psi)^2$, $\frac{1}{2}\psi^\dagger \psi \phi^2$, and $\frac{1}{4!}\phi^4$. Above are three Wick diagrams constructed from this set of four vertices. The order $\#G$ of the symmetry group of the vertices is $s(D_0)$. All the products $c(D_r)s(D_r)$ are equal to $\#G$.

The full group of symmetries $G$ for $\prod V^J$ combines the shuffling of the monomials by $G_J$ with the shuffling of their fields described by the groups $G_r$. The number of elements $\#G$ in the combined symmetry group is

$$\#G = J_1!(I_1!)^{J_1} \times \cdots \times J_k!(I_k!)^{J_k} = \prod J! \prod \left( \prod I! \right)^J. \qquad (4.4.6)$$

**Remark** 4.4.7. The total symmetry group $G$ is actually a product of wreath products:

$$G = \left( S_{J_1} \wr G_1 \right) \times \cdots \times \left( S_{J_k} \wr G_k \right).$$

As before, if we specify a contraction pattern for $\prod V^J$ and label it with a Wick diagram $D$, then all the contraction patterns that give rise to $D$ are connected by the action of $G$. Letting $Z$ be the subgroup of $G$ which leaves the specified contraction pattern invariant, and letting $s(D)$ be the number of elements in $Z$, we find as before that $s(D)c(D) = \#G$. (This is the group-theory formula $\#\text{Centralizer} \times \#\text{Orbit} = \#\text{Group}$.) Hence

$$S(D) = \binom{n}{J} \frac{c(D)}{n!} \frac{\prod J!}{\#G} \mathcal{O}(D)$$
$$= \frac{1}{s(D)} \mathcal{O}(D). \qquad (4.4.8)$$

The approach to computing $s(D)$ described in the previous section also generalizes: think of $D$ as a network of colored (and possibly directed) strings, and count the number of ways of moving the network which leave the appearance of the whole unchanged. Figure 4.4 provides examples in which symmetries move strings alone and examples in which they move vertices too.

**Homework 4.4.9.** One of simplest interacting field theories is the $\phi^4$-theory defined by the Lagrangian density

$$\mathcal{L} = \frac{1}{2}\partial_\mu\phi\partial^\mu\phi - \frac{\mu^2}{2}\phi^2 - \frac{g}{4!}\phi^4.$$

Write down all the second order diagrams together with their contraction numbers $c(D)$ and symmetry factors $s(D)$.

**Homework 4.4.10.** If an interaction Hamiltonian density is given by

$$\mathcal{H}_I = \lambda\phi\psi^\dagger\psi + \frac{g}{4!}\phi^4 + \frac{\kappa}{2^2}(\psi^\dagger\psi)^2,$$

what are the second order diagrams $D$, and what are the symmetry factors $s(D)$ associated with them?

The notation necessary for a general proof obscures the simplicity of the conclusion. All we need to know is the following theorem.

**Theorem 4.4.11.** *If $\mathcal{H}_I$ is an interaction Hamiltonian density in which, for each monomial $\phi^J$, the coupling constant is normalized by the symmetry factor $1/\prod J!$, then a Wick diagram $D$ corresponds to a contribution $S(D) = \mathcal{O}(D)/s(D)$ to the $S$ matrix, where $\mathcal{O}(D)$ is the operator constructed from the interaction monomials in $-i\mathcal{H}_I$ by multiplication, contraction, normal ordering, and integration and $s(D)$ is the number of symmetries of $D$.*

# 4.5 Reduction to Connected Wick Diagrams

Wick's Theorem applied to Dyson's formula produces diagrams that are disconnected. A disconnected Wick diagram corresponds to an integral that factorizes. In fact,

$$D = D_1 + D_2 \quad\Longrightarrow\quad \mathcal{O}(D) = :\mathcal{O}(D_1)\mathcal{O}(D_2): \qquad (4.5.1)$$

Maximal factorization corresponds to splitting a diagram into its connected components. Seeing the fundamental role played by connected Wick diagrams in computation of operators, it is natural to seek a formula for the $S$ matrix which is based on them. The following cluster expansion theorem presents the relevant result:

**Theorem 4.5.2.** *The $S$ matrix, which is the sum of operators associated with Wick diagrams, is equal to the normal-ordered exponential of the sum of the operators associated with connected Wick diagrams:*

$$S = \sum_i S(D_i) = :\exp\left(\sum_r S(C_r)\right):$$

*where $D_i$ is a list of all diagrams and $C_r$ is a list of all connected diagrams.*

**Proof.** Let $D_i$ be a list of all diagrams. Let $s(D_i)$ be the symmetry factor of $D_i$. Let $\mathcal{O}(D_i)$ be the operator associated with $D_i$, and let $S(D_i)$ be the corresponding contribution $\mathcal{O}(D_i)/s(D_i)$ to the $S$ matrix. Let $C_r$ be a list of all connected diagrams. Now every diagram is a sum of connected diagrams, so we can define matrices of numbers $k_{ir}$ by

$$D_i = \sum_{r=1}^{\infty} k_{ir}C_r. \qquad (4.5.3)$$

Now the symmetry of $D_i$ can be expressed in terms of the symmetry of its components. There are two contributions, the first from the symmetry of the components taken one at a time, the second from interchanging components:

$$s(D_i) = \Big(\prod_{r=1}^{\infty} s(C_r)^{k_{ir}}\Big)\Big(\prod_{r=1}^{\infty} k_{ir}!\Big). \qquad (4.5.4)$$

Finally, noting that the operator associated with a sum of diagrams is the normal-ordered product of the operators associated with the individual diagrams, we are ready for the algebra:

$$
\begin{aligned}
S &= \sum_i S(D_i) = \sum_i \frac{1}{s(D_i)} \mathcal{O}(D_i) \\
&= \sum_i \frac{1}{s(D_i)} :\prod_r \mathcal{O}(C_r)^{k_{ir}}: \\
&= :\sum_i \prod_r \frac{1}{k_{ir}!}\Big(\frac{\mathcal{O}(C_r)}{s(C_r)}\Big)^{k_{ir}}: \\
&= :\sum_i \prod_r \frac{1}{k_{ir}!} S(C_r)^{k_{ir}}:
\end{aligned}
\qquad (4.5.5)
$$

But, in order to give all diagrams $D_i$ in terms of the connected diagrams $C_i$, the set of vectors

$$\{(k_{i1}, k_{i2}, \dots) \,|\, i = 1, 2, \dots\} \qquad (4.5.6)$$

must contain all infinite vectors with non-negative integer components and a finite number of positive components. Hence we can interchange sum and product as follows:

$$:\prod_r \sum_n \frac{1}{n!} S(C_r)^n: = :\sum_i \prod_r \frac{1}{k_{ir}!} S(C_r)^{k_{ir}}: \qquad (4.5.7)$$

Therefore

$$S = :\prod_r \exp\big(S(C_r)\big): = :\exp\Big(\sum_r S(C_r)\Big): \qquad (4.5.8)$$

and the theorem is established. $\qquad\qquad\qquad\qquad\qquad\qquad\qquad\qquad\qquad\Box$

**Homework** 4.5.9. Use Wick diagrams and contraction or symmetry factors to verify Theorem 4.5.2 for $\mathcal{H}_I = g\psi^\dagger\psi\phi$ up to third order.

With this section, we have completed our presentation of the Wick expansion of the $S$ matrix and its diagram representation. The combinatorics worked out here is equally applicable to Feynman diagrams, as we shall see. Connected diagrams will continue to dominate in that context too. The cluster expansion theorem of this section is an operator theorem and will be used in the examples of the next two sections.

# 4.6 First Example: A Time-Dependent Classical Source

In this section, we apply the methods and results of the last section to an example obtained from the scalar nucleon-meson theory 4.3.1 by replacing the charged field by a classical source $-j$ which is zero for large $|t|$; the resulting interaction Hamiltonian density is

$$\mathcal{H}_I(x) = -\lambda j(x)\phi(x). \tag{4.6.1}$$

The fundamental vertex is simply a dot with a line emerging from it, and the only connected diagrams are the vertex itself $C_1$ and the diagram $C_2$ where a single line joins two vertices:

$$\textbf{Vertex} = C_1 = \bullet\!\!-\!\!- \quad \text{and} \quad C_2 = \bullet\!\!-\!\!\bullet \tag{4.6.2}$$

The contraction numbers for these diagrams are both one, so that $S(C_1) = \mathcal{O}(C_1)$ and $S(C_2) = \frac{1}{2}\mathcal{O}(C_2)$. The operators $S(C_r)$ are given by

$$S(C_1) = i\lambda \int d^4x \, j(x)\phi(x),$$
$$S(C_2) = \frac{(i\lambda)^2}{2!} \int d^4x_1 \, d^4x_2 \, j(x_1)\phi(x_1)j(x_2)\phi(x_2) \stackrel{\text{def}}{=} \ln(A), \tag{4.6.3}$$

since $S(C_2)$ is simply a number.

Let $\hat{\jmath}$ be the Fourier transform of $j$:

$$\hat{\jmath}(k) = \int d^4x \, e^{-ix\cdot k} j(x). \tag{4.6.4}$$

Then we can substitute the free-field integral into $S(C_1)$ and do the $x$-integration to obtain

$$S(C_1) = i\lambda \int \frac{d^3\bar{k}}{(2\pi)^{3/2}\big(2\omega(\bar{k})\big)^{1/2}} \Big(\hat{\jmath}\big(\omega(\bar{k}),\bar{k}\big)^* a(\bar{k})^\dagger + \hat{\jmath}\big(\omega(\bar{k}),\bar{k}\big)a(\bar{k})\Big). \tag{4.6.5}$$

Define a function $\sigma$ by

$$\sigma(\bar{k}) = \frac{i\lambda\hat{\jmath}\big(\omega(\bar{k}),\bar{k}\big)^*}{(2\pi)^{3/2}\big(2\omega(\bar{k})\big)^{1/2}} \tag{4.6.6}$$

so that

$$S(C_1) = \int d^3\bar{k} \left(\sigma(\bar{k})a(\bar{k})^\dagger - \sigma(\bar{k})^* a(\bar{k})\right). \tag{4.6.7}$$

To compute the $S$ matrix we exponentiate the sum of connected diagrams and normal order, which in this example gives

$$S = :\exp\big(S(C_1) + S(C_2)\big):$$
$$= A \exp\left(\int d^3\bar{k}\,\sigma(\bar{k})a(\bar{k})^\dagger\right) \exp\left(-\int d^3\bar{k}\,\sigma(\bar{k})^* a(\bar{k})\right) \tag{4.6.8}$$

Applying $S$ to the vacuum $|0\rangle$ gives rise to particle production:

$$S|0\rangle = A \exp\left(\int d^3\bar{k}\, \sigma(\bar{k}) a(\bar{k})^\dagger\right) |0\rangle$$

$$= A \sum_{n=0}^{\infty} \frac{1}{n!} \int d^3\bar{k}_1 \cdots d^3\bar{k}_n \prod_{r=1}^{n} \sigma(\bar{k}_r) \,|\bar{k}_1, \ldots, \bar{k}_n\rangle. \tag{4.6.9}$$

Therefore the amplitude for finding $n$ particles at prescribed momenta is

$$\mathcal{A}(\bar{k}_1, \ldots, \bar{k}_n) = \langle \bar{k}_1, \ldots, \bar{k}_n | S | 0 \rangle = A \sigma(\bar{k}_1) \cdots \sigma(\bar{k}_n), \tag{4.6.10}$$

and the final state $S|0\rangle$ is a sum of coherent states.

Notice that the only aspect of the source $j$ which is effective in creating particles is the Fourier component with the right frequency, $\hat{j}(\omega(\bar{k}), \bar{k})$. If $\hat{j}$ is zero on the mass hyperboloid of the mesons, then $j$ will not create any mesons.

Taking into account the Bose statistics of the final state, the probability $P(n)$ of finding $n$ mesons after $j$ has become zero is

$$P(n) = \frac{1}{n!} |A|^2 \int d^3\bar{k}_1 \cdots d^3\bar{k}_n \,\left|\sigma(\bar{k}_1) \cdots \sigma(\bar{k}_n)\right|^2$$

$$= \frac{|A|^2 \alpha^n}{n!} \quad \text{where} \quad \alpha = \int d^3\bar{k} \,\left|\sigma(\bar{k})\right|^2. \tag{4.6.11}$$

The sum of these probabilities must be one, and so we can find $|A|$ in terms of $\alpha$:

$$|A| = e^{-\alpha/2}. \tag{4.6.12}$$

This shows that $P(n)$ is a Poisson distribution:

$$P(n) = \frac{\alpha^n}{n!} e^{-\alpha}. \tag{4.6.13}$$

Using this distribution of final states we can compute the expected final energy, momentum and particle number:

$$\langle P \rangle_{\text{final}} = \sum_{n=0}^{\infty} \frac{e^{-\alpha}}{n!} \int d^3\bar{k}_1 \cdots d^3\bar{k}_n \,\left|\sigma(\bar{k}_1) \cdots \sigma(\bar{k}_n)\right|^2 (k_1 + \cdots + k_n)$$

$$= \sum_{n=0}^{\infty} \frac{e^{-\alpha}}{(n-1)!} \alpha^{n-1} \int d^3\bar{k} \,\left|\sigma(\bar{k})\right|^2 k$$

$$= \int d^3\bar{k} \,\left|\sigma(\bar{k})\right|^2 k \tag{4.6.14}$$

$$\langle N \rangle_{\text{final}} = \sum_{n=0}^{\infty} n \frac{e^{-\alpha}}{n!} = \alpha.$$

**Remark** 4.6.15. One can simplify $S(C_2)$ from 4.6.3 by substituting the value of $\dot{\phi}(x_1)\dot{\phi}(x_2)$ from 4.2.6 and doing the integrations over $x_1$ and $x_2$. The result is

$$S(C_2) = -\frac{\lambda^2}{2} \int \frac{d^4k}{(2\pi)^4} \frac{i\hat{j}(k)\hat{j}(-k)}{k^2 - \mu^2 + i\epsilon}. \tag{4.6.16}$$

One is now tempted to do the integral over $k_0$ by residue theory using either the upper half circle $C_+$ or the lower half circle $C_-$ to form a loop with the real axis. The resulting integral would make $S(C_2)$ a real number. In general, however, neither completion of the contour is valid, as the following paragraph explains.

For some fixed $\bar{x}$, let $f(t) = j(t, \bar{x})$. Assume that $f(t)$ is real, vanishes for large $|t|$, and is square-integrable. Then $\hat{f}(k_0)^* = \hat{f}(-k_0)$, $\hat{f}(k_0)$ is analytic, and $|\hat{f}(k_0)|^2$ is integrable. If we now suppose that we have a bound on $\hat{f}(k_0)\hat{f}(-k_0)$ so that the integral in $S(C_2)$ vanishes over one half circle, then this bound also holds on the other, implying that $\hat{f}(k_0)\hat{f}(-k_0)$ is a bounded analytic function and hence a constant. But $\hat{f}(k_0)\hat{f}(-k_0)$ is integrable over the real axis, and so its constant value must be zero.

To apply residue theory we must instead use the integrability of $\hat{j}(k)\hat{j}(-k)$ in $k_0$ to split it into a sum $F_+(k) + F_-(k)$ of positive and negative frequency $k_0$ plane waves. Then we can use $C_\pm$ on the $F_\pm$ summand to obtain

$$S(C_2) = -\frac{\lambda^2}{2} \int \frac{d^3\bar{k}}{(2\pi)^3 \, 2\omega(\bar{k})} \, F_+\big(-\omega(\bar{k}), -\bar{k}\big) + F_-\big(\omega(\bar{k}), \bar{k}\big). \tag{4.6.17}$$

The outcome is that $\operatorname{Im} S(C_2)$ is generally non-zero.

**Homework** 4.6.18. We found $|A|$ by a consistency argument. Use the formula

$$\lim_{\epsilon \to 0^+} \left( \frac{1}{x - i\epsilon} - \frac{1}{x + i\epsilon} \right) = 2\pi i \delta(x)$$

to compute the real part of $S(C_2)$ and thus to find $|A|$ directly. Justify this method in light of the analysis in Remark 4.6.15.

This example makes the first link between our developing machinery of quantum field theory and the sensory world, the world of experiment. It helps us to refine our intuition of the field-theoretic processes underlying the simplest phenomena.

# 4.7 Second Example: A Time-Independent Classical Source

In this section, we modify the example of the last section by taking $j$ to be a constant classical distribution of nucleons. In this case we must introduce an adiabatic turning on and off function $f(t)$ as explained in Section 4.1. The product $f(t)j(\bar{x})$ plays the role of $j(x)$ in the previous example. The interaction Hamiltonian density is therefore

$$\mathcal{H}_I(x) = -\lambda f(t) j(\bar{x}) \phi(x), \tag{4.7.1}$$

and the Fourier transform $\hat{j}(k)$ of $j(x)$ factorizes:

$$\hat{j}(k) = \int d^4x \, e^{ix \cdot k} f(x^0) j(\bar{x}) = \hat{f}(k_0) \hat{j}(\bar{k}). \tag{4.7.2}$$

In the adiabatic limit $\hat{f}(k_0) = 2\pi\delta(k_0)$, and so the static source has no chance of creating or absorbing mesons. The interest of this example is not in its dynamics but in the structure of the ground state. We shall see that the ground state energy corresponds to the potential energy of the nucleon distribution due to a Yukawa potential generated by unmanifest activity of the meson field.

**Homework** 4.7.3. Show that $\hat{f}(k_0) \to 2\pi\delta(k_0)$ as $T \to \infty$ by evaluating the integral

$$\int dk_0 \, \hat{f}(k_0) \hat{g}(k_0)$$

for a function $g(t)$ with bounded support.

Let $|0\rangle_0$ be the vacuum state of the non-interacting system. Let $|0\rangle_P$ be the vacuum state of the interacting system. Let $E_0$ be the energy of the vacuum state when $f = 1$. The energy of $|0\rangle_0$ is zero, so up to the time when $f(t)$ stops being zero, $|0\rangle_0$ does not evolve. As $f(t)$ grows from zero to one, $|0\rangle_0$ evolves into $|0\rangle_P$ multiplied by some phase. While $f(t) = 1$, $|0\rangle_P$ picks up a phase because of $E_0$. When $f(t)$ shrinks from one back to zero, $|0\rangle_P$ evolves back into $|0\rangle_0$ multiplied by some phase. After $f(t)$ arrives at zero there is no further evolution of the system. Therefore the Schrödinger picture timetable is

$$
\begin{array}{ll}
t \leq -(\Delta + \tfrac{1}{2}T) & |0\rangle_0 \\
t = -\tfrac{1}{2}T & e^{-i\alpha}|0\rangle_P \\
t = \tfrac{1}{2}T & e^{-i\alpha}e^{-iE_0 T}|0\rangle_P \\
t \geq \tfrac{1}{2}T + \Delta & e^{-i\alpha}e^{-iE_0 T}e^{-i\beta}|0\rangle_0
\end{array}
\tag{4.7.4}
$$

The vacuum to vacuum amplitude of $S$ is therefore

$$
{}_0\langle 0|S|0\rangle_0 = e^{-i(\alpha+\beta+E_0 T)}.
\tag{4.7.5}
$$

Translating this into the interaction picture and writing $|0\rangle$ for $|0\rangle_0$ we find that, for $t' \sim -\infty$ and $t \sim \infty$:

$$
\langle 0|U_I(t,t')|0\rangle = e^{-i(\alpha+\beta+E_0 T)}.
\tag{4.7.6}
$$

We can absorb this phase factor by adjusting the Hamiltonian with an *energy counterterm*:

$$
H_I(t) = -af(t) - \int d^3\bar{x}\,\lambda f(t) j(\bar{x})\phi(x),
\tag{4.7.7}
$$

choosing $a$ so that $\langle 0|S|0\rangle = 1$. Since the term $af(t)$ is a number, it commutes with everything and merely contributes to the ground state energy. Thus $a$ satisfies

$$
a \int dt\, f(t) = \alpha + \beta + E_0 T.
\tag{4.7.8}
$$

In the adiabatic limit this yields $a = E_0$. The term $-af(t)$ is introduced (like normal ordering in Section 2.6) simply to normalize the vacuum energy.

In this example there are three connected Wick diagrams:

1. The cross $C_0$ representing the counterterm;
2. The vertex $C_1$ representing the interaction;
3. And the line segment $C_2$ representing two vertices with one contraction.

All these diagrams have contraction number $c(D) = 1$, and so $S(C_0) = \mathcal{O}(C_0)$, $S(C_1) = \mathcal{O}(C_1)$, and $S(C_2) = \tfrac{1}{2}\mathcal{O}(C_2)$.

Applying the observations following 4.7.2 to $S(C_1)$, the operators are evaluated as follows:

$$
S(C_0) = -i \int dt\,(-af(t)) \simeq iaT,
$$

$$
S(C_1) = -i\lambda \int d^4x\, f(t) j(\bar{x})\phi(x) \simeq 0,
\tag{4.7.9}
$$

$$
S(C_2) = \frac{(-i\lambda)^2}{2!} \int d^4x_1\, d^4x_2\, f(t_1) j(\bar{x}_1)\acute{\phi}(x_1) f(t_2) j(\bar{x}_2)\grave{\phi}(x_2).
$$

$S(C_2)$ is a number, as in the previous example. Exponentiating we find that $S = \exp(S(C_2) + iaT)$. But in the adiabatic limit, $\langle 0|S|0\rangle = 1$, and so $S = 1$. This shows that

$$E_0 = a = \lim_{T \to \infty} \frac{iS(C_2)}{T}. \qquad (4.7.10)$$

We will now show that the integral form of $E_0$ is derived from a Yukawa potential on the distribution $j(\bar{x})$. To get started, substitute the formula 4.2.6 for the contraction of free scalar fields in 4.7.9, write $f(t)j(\bar{x})$ in terms of the Fourier transform 4.7.2, and perform the $x$-integrations. Then, using the fact that $|\hat{f}(k_0)|^2/T$ tends to $2\pi\delta(k_0)$ as $T \to \infty$, evaluate the $k_0$ integral:

$$
\begin{aligned}
E_0 &= \lim_{T \to \infty} \frac{iS(C_2)}{T} \\
&= \lim_{T \to \infty} \lim_{\epsilon \to 0^+} \frac{i}{T} \frac{(i\lambda)^2}{2} \int \frac{d^4k}{(2\pi)^4} \left|\hat{f}(k_0)\hat{j}(\bar{k})\right|^2 \frac{i}{k^2 - \mu^2 + i\epsilon} \\
&= \lim_{\epsilon \to 0^+} \lim_{T \to \infty} \frac{\lambda^2}{2} \int \frac{d^4k}{(2\pi)^4} \frac{1}{T} \left|\hat{f}(k_0)\hat{j}(\bar{k})\right|^2 \frac{1}{k^2 - \mu^2 + i\epsilon} \qquad (4.7.11) \\
&= \lim_{\epsilon \to 0^+} \frac{-\lambda^2}{2} \int \frac{d^3\bar{k}}{(2\pi)^3} \frac{\left|\hat{j}(\bar{k})\right|^2}{\bar{k}^2 + \mu^2 - i\epsilon} \\
&= \frac{-\lambda^2}{2} \int \frac{d^3\bar{k}}{(2\pi)^3} \frac{\left|\hat{j}(\bar{k})\right|^2}{\bar{k}^2 + \mu^2}.
\end{aligned}
$$

**Homework** 4.7.12. Show that $|\hat{f}(k_0)|^2/T$ tends to $2\pi\delta(k_0)$ as $T \to \infty$ by evaluating

$$\frac{1}{T} \int dk_0 \left|\hat{f}(k_0)\right|^2 \hat{g}(k_0) \qquad (4.7.13)$$

for a function $g(t)$ with bounded support.

Now substitute the integral definition of $\hat{j}$ into this expression to find

$$E_0 = -\frac{\lambda^2}{2} \int d^3\bar{x}\, d^3\bar{y}\, j(\bar{x}) G(\bar{x} - \bar{y}) j(\bar{y}), \qquad (4.7.14)$$

where

$$G(\bar{x} - \bar{y}) = \int \frac{d^3\bar{k}}{(2\pi)^3} \frac{e^{i(\bar{x} - \bar{y}) \cdot \bar{k}}}{\bar{k}^2 + \mu^2}. \qquad (4.7.15)$$

If we choose a distribution $j$ to represent two particles

$$j(\bar{x}) = \delta^{(3)}(\bar{x}_1 - \bar{x}) + \delta^{(3)}(\bar{x}_2 - \bar{x}), \qquad (4.7.16)$$

substitute this in the formula for $E_0$ and extract the interaction term, then the energy $E_{\text{int}}$ due to the interactions is

$$E_{\text{int}} = -\lambda^2 G(\bar{x}_1 - \bar{x}_2), \qquad (4.7.17)$$

which shows that $G$ may be interpreted as a pair potential.

To evaluate $G$, change variables to spherical polars, treat the radial variable as a complex variable, complete the contour of integration in the upper half plane, and take the residue at $k = i\mu$:

$$
\begin{aligned}
G(r) &= \int_{-1}^{1} d(\cos(\theta)) \int_{0}^{2\pi} d\phi \int_{0}^{\infty} dk \, \frac{1}{(2\pi)^3} \frac{k^2}{k^2 + \mu^2} e^{irk\cos(\theta)} \\
&= \frac{1}{(2\pi)^2} \int_{0}^{\infty} dk \, \frac{k^2}{k^2 + \mu^2} \frac{1}{irk} (e^{irk} - e^{-irk}) \\
&= \frac{1}{(2\pi)^2} \frac{1}{ir} \int_{-\infty}^{\infty} dk \, \frac{k}{k^2 + \mu^2} e^{irk} \\
&= \frac{1}{(2\pi)^2} \frac{1}{ir} (2\pi i) \frac{i\mu}{2i\mu} e^{-r\mu} \\
&= \frac{1}{4\pi r} e^{-\mu r},
\end{aligned}
\tag{4.7.18}
$$

which is indeed the Yukawa potential.

**Remark** 4.7.19. The formulae developed in the first example allow us to consider turning the interactions off at time $t = 0$. The idea is to create the physical vacuum at $t = -\epsilon$ and view it from the perspective of the free vacuum at $t = \epsilon$. As this switching off is indistinguishable from a sudden removal of the classical source, we simply take

$$
j(x) \stackrel{\text{def}}{=} j(\bar{x}) f(t)\theta(-t),
\tag{4.7.20}
$$

where $\theta$ is the Heaviside function, and compute the $\sigma$ function from its definition 4.6.6 in the adiabatic limit,

$$
\begin{aligned}
\sigma(k) &= \frac{i\lambda}{(2\pi)^{3/2} (2\omega(\bar{k}))^{1/2}} \lim_{T \to \infty} \int d^4 x \, e^{ix \cdot k} j(\bar{x}) f(t)\theta(-t) \\
&= \frac{2\lambda \hat{j}(\bar{k})^*}{(2\pi)^{3/2} (2\omega(\bar{k}))^{3/2}}.
\end{aligned}
\tag{4.7.21}
$$

Now the formulae of the first example yield the free-particle content of the physical vacuum,

$$
_0\langle \bar{k}_1, \ldots, \bar{k}_n | 0 \rangle_P = e^{-\alpha/2} \sigma(\bar{k}_1) \cdots \sigma(\bar{k}_n),
\tag{4.7.22}
$$

and the expectations of $H_0$ and the number operator $N$ in the physical vacuum,

$$
\begin{aligned}
_P\langle 0 | H_0 | 0 \rangle_P &= \frac{\lambda^2}{2} \int \frac{d^3 \bar{k}}{(2\pi)^3 \omega(\bar{k})^2} \left| \hat{j}(\bar{k}) \right|^2, \\
_P\langle 0 | N | 0 \rangle_P &= \frac{\lambda^2}{2} \int \frac{d^3 \bar{k}}{(2\pi)^3 \omega(\bar{k})^3} \left| \hat{j}(\bar{k}) \right|^2.
\end{aligned}
\tag{4.7.23}
$$

If we take $j(\bar{x}) \to \delta^{(3)}(\bar{x})$, then $\hat{j}(\bar{k}) \to 1$, $\langle H_0 \rangle_P \to \infty$, and $\langle N \rangle_P \to \infty$. These are ultra-violet or high-energy divergences. However, $\langle H_0 \rangle_P$ cannot be observed because it is the sum of the vacuum bubbles, and $\langle N \rangle_P$ cannot be observed because there is an upper bound on the number of particles any apparatus can detect.

If we take $\mu \to 0$, then $\omega(\bar{k}) \to |\bar{k}|$, $\langle H_0 \rangle_P \to E_0' < \infty$, and $\langle N \rangle_P \to \infty$. This is an infra-red or low-energy divergence. It arises because we can distribute a small amount of energy among infinitely many massless particles. However, the limited sensitivity of experimental apparatus provides a low-energy cut-off. Hence this divergence does not make the theory inconsistent.

Following Remark 4.1.22 on Haag's Theorem, this computation suggests that $H$ should act on a strange representation of the free field. The argument is inconclusive, however, because the violence of the turning off procedure could be seen as the source of the free particles.

**Remark 4.7.24.** From the example of the last section and Remark 4.7.19 above on the physical vacuum, we see that $\operatorname{Re} S(C_2)$ is associated with the expected number of particles into which the vacuum decays, while from this section we see that $\operatorname{Im} S(C_2)$ is associated with the physical vacuum energy. Remark 4.6.15 shows that in general $S(C_2)$ contributes to the decay and the energy of the physical vacuum. This observation will hold for all vacuum bubbles, that is, Wick diagrams with no external lines.

This example illustrates the important principle that the structure of the vacuum, which classically should have no structure, is intimately dependent on the Lagrangian. Furthermore, we see here an indication of how the creation and annihilation operators of a quantum field can give rise to a classical structure like the Yukawa potential.

## 4.8 Comparison with Classical Field Theory

In the preceding sections, we have introduced a scalar approximation to nucleon-meson dynamics based on the Lagrangian density 4.3.1 and analyzed its structure when the charged particles are replaced by a classical source. In this section, we comment on the classical theory associated with this Lagrangian density in order to clarify similarities and differences between classical and quantum analysis.

The Lagrangian density 4.3.1, an *a priori* classical concept, is

$$\mathcal{L} = \tfrac{1}{2}(\partial_\mu \phi \, \partial^\mu \phi - \mu^2 \phi^2) + (\partial_\mu \psi^\dagger \partial^\mu \psi - m^2 \psi^\dagger \psi) - \lambda \psi^\dagger \psi \phi, \tag{4.8.1}$$

where $\phi$ and $\psi$ represent the meson and nucleon fields respectively. The Euler–Lagrange equations derived from $\mathcal{L}$ are

$$\begin{aligned}
(\partial_\mu \partial^\mu + \mu^2)\phi &= -\lambda \psi^\dagger \psi, \\
(\partial_\mu \partial^\mu + m^2)\psi &= -\lambda \psi \phi.
\end{aligned} \tag{4.8.2}$$

To understand the classical dynamics, we start with a simple case. Suppose that the nucleon fields are replaced by a distribution $\Psi$ of nucleons:

$$\mathcal{L}(\phi) = \tfrac{1}{2}(\partial_\mu \phi \, \partial^\mu \phi - \mu^2 \phi^2) - \lambda \Psi \phi \quad \Rightarrow \quad (\partial_\mu \partial^\mu + \mu^2)\phi = -\lambda \Psi. \tag{4.8.3}$$

Taking the Fourier transform of the Euler–Lagrange equation we find

$$\hat{\phi}(k) = \frac{\lambda \hat{\Psi}(k)}{k^2 - \mu^2} + f(k)\delta(k^2 - \mu^2). \tag{4.8.4}$$

The $\delta$ term gives the plane waves, the solutions to the homogeneous equation. It is generated when we divide by $k^2 - \mu^2$ and vanishes when we multiply by $k^2 - \mu^2$. Taking the inverse Fourier transform we find the solution to the field equation

$$\begin{aligned}
\phi(x) &= \int \frac{d^4 k}{(2\pi)^4} \frac{\lambda \hat{\Psi}(k)}{k^2 - \mu^2} e^{ix \cdot k} + \int \frac{d^4 k}{(2\pi)^4} f(k)\delta(k^2 - \mu^2) e^{ix \cdot k} \\
&\overset{\text{def}}{=} \phi_P(x) + \phi_H(x),
\end{aligned} \tag{4.8.5}$$

where $\phi_P$ is a particular solution to the equation and $\phi_H$ is the general solution of the homogeneous equation.

Observe that $\phi$ responds to $\hat{\Psi}(k)$ for all $k$. In particular, if $\hat{\Psi} \neq 0$ on the mass hyperboloid $k^2 = \mu^2$, then the resonance between the movement of the nucleon distribution and the natural frequency of the meson field generates plane waves $\phi_P$ which look like those in $\phi_H$. When we quantized this theory, these plane waves became particles: the Wick diagram $C_1$ represented the operator responsible for the emission and absorption of these particles.

To simplify further, consider the case where $\Psi$ is a static nucleon distribution. Then $\phi_P$ will also be static. Take $\Psi(\bar{x}) = \delta^{(3)}(\bar{x})$ to represent a single, unmoving nucleon. Then we can solve for $\phi_P$:

$$(-\nabla^2 + \mu^2)\phi_P(\bar{x}) = \delta^{(3)}(\bar{x}) \quad \Longrightarrow \quad \hat{\phi}_P(\bar{k}) = \frac{1}{k^2 + \mu^2}$$

$$\Longrightarrow \quad \phi_P(\bar{x}) = \frac{1}{4\pi r}e^{-\mu r}. \tag{4.8.6}$$

The plane wave solutions are unaffected by either the static distribution or the corresponding soliton in the meson field. This is because, firstly, a static distribution can only emit and absorb momentum, not energy, and secondly, our mesons do not interact with themselves directly. The first point corresponds to the vanishing of $S(C_2)$ in this case. The second point is a consequence of our simple approximation to nucleon-meson dynamics; chiral Lagrangian theories introduce meson self-interactions.

Finally, consider taking a classical distribution $\Phi$ of mesons:

$$\mathcal{L} = \partial_\mu \psi^\dagger \partial^\mu \psi - m^2 \psi^\dagger \psi - \lambda \psi^\dagger \psi \Phi. \tag{4.8.7}$$

Now the field equations are

$$(\partial_\mu \partial^\mu + m^2)\psi = -\lambda \psi \Phi, \tag{4.8.8}$$

and so there are no plane wave solutions, only scattering solutions. The Wick diagram for the interaction is a directed line through a dot. This represents the simultaneous absorption and emission of energy-momentum by the meson distribution. In this case, scattering is still possible even when the distribution is static since energy can be conserved without asking the distribution to emit or absorb it.

**Homework** 4.8.9. List the connected Wick diagrams for the scalar nucleon-meson model when the meson field is replaced by a classical distribution. Write out the integrals for the first and second order diagrams. Show from the first order diagram that nucleon-antinucleon pair production is possible in general, but is not possible if the meson distribution is static.

This classical note serves to show which aspects of the previous two sections have direct analogues in the classical theory of fields and which are unique to quantum field theory. Thus both levels of field theory contain the phenomenon of resonance between a source and the field it is coupled to and the emergence of pair potentials from interactions; but where a classical field can be induced to carry a plane wave with any wave number $k$ by a source with that wave number, a quantum field can only carry ideal states with on-shell momenta and will not respond at all to off-shell stimulation.

# 4.9 Feynman Diagrams and Feynman Rules

In this section, we shall derive Feynman diagrams and Feynman rules for computing $S$-matrix elements from Wick diagrams and operators in the context of the Lagrangian density 4.3.1 for our ongoing example of a scalar field theory. This derivation transforms the Wick expansion of the $S$ matrix into the Feynman perturbation expansion of $S$-matrix elements.

Wick diagrams correspond to operators, and each such operator contains many different processes which contribute to many different scattering amplitudes. Even $\mathcal{O}(C_1)$ of our time-dependent source example contributes to both emission and absorption. In general for cubic interactions, if, following 4.3.9, we divide each field in the theory into its creation part and its annihilation part, then an $n^{\text{th}}$-order Wick diagram with $c$ contractions splits up into $2^{3n-2c}$ terms. This is too many for comfort, even in our simple scalar approximation to nucleon-meson dynamics.

Feynman diagrams and rules simplify the analysis by focusing on processes, not operators. Our goal is to compute matrix elements of the form

$$\mathcal{M} = \langle p'_1, \ldots, p'_{r'}; k'_1, \ldots, k'_{s'} | S - 1 | p_1, \ldots, p_r; k_1, \ldots, k_s \rangle, \qquad (4.9.1)$$

where the $p$'s and $k$'s are nucleon and meson momenta respectively. Hence we do not need those parts of Wick operators which, because of their annihilation-creation structure, cannot contribute. Since normal ordering makes it trivial to see which of the $2^{3n-2c}$ terms do contribute, focusing on matrix elements greatly simplifies our work.

Indeed, an external line in a Wick diagram represents a field factor in the operator, a factor capable either of annihilating an incoming particle or of creating an outgoing particle. To identify a process by which the Wick operator can contribute to the overall matrix element, all we need to do is separate the external lines of the Wick diagram into incoming lines (put on the right in our convention) and outgoing lines (put on the left according to the rule 'later on the left'), and label each external line with the momentum of the appropriate particle. The result is a Feynman diagram. It is of course possible that one Wick diagram may give rise to many Feynman diagrams, for example,

**Wick Diagram**     **Associated Feynman Diagrams**

$$(4.9.2)$$

but such proliferation of diagrams is a small price for the consequent simplification of computation.

Notice that it is not the Wick operator that has been broken into parts represented by Feynman diagrams, but the contribution of the operator to a matrix element. Thus the initial and final state are built into the Feynman diagram, which represents a complex number. At this point, to compute the complex number, we have to go from the Feynman diagram to a part of the associated Wick operator and then to the matrix element. The Feynman rules for the theory connect the diagram directly to the matrix element. They specify a factor for each internal line and each vertex in the Feynman diagram. These factors are then to be multiplied and integrated. To show how to derive Feynman rules we will return to our nucleon-meson Lagrangian density 4.3.1 and follow a specific calculation from Dyson's formula to a matrix element. The form of the final integral will enable us to read off the Feynman rules.

Let $D$ be the connected Wick diagram with two vertices and one nucleon line contracted:

$$D = \text{⟩−⟨} \tag{4.9.3}$$

As noted below the examples 4.3.6, there are two ways of joining the vertices that give this diagram, so $c(D) = 2$, $s(D) = 1$, and the operator contribution to the $S$ matrix associated with this diagram is

$$S(D) = (-i\lambda)^2 \int d^4x_1\, d^4x_2 \; :\!\psi_1^\dagger \dot\psi_1 \phi_1 \dot\psi_2^\dagger \psi_2 \phi_2\!: \tag{4.9.4}$$

According to the splitting of $\phi$ and $\psi$ in 4.3.9, this operator breaks up into $2^4$ monomials in creation and annihilation operators.

Suppose our interest is in the matrix element

$$\mathcal{M} = \langle p', k' | S - 1 | p, k \rangle. \tag{4.9.5}$$

Then, since

$$\langle 0|\phi|k\rangle = \langle 0|\phi_-|k\rangle = e^{-ix\cdot k}, \qquad \langle k'|\phi|0\rangle = \langle k'|\phi_+|0\rangle = e^{ix\cdot k'},$$
$$\langle 0|\psi|p\rangle = \langle 0|\psi_-|p\rangle = e^{-ix\cdot p}, \qquad \langle p'|\psi^\dagger|0\rangle = \langle p'|\psi_+^\dagger|0\rangle = e^{ix\cdot p'}, \tag{4.9.6}$$

we clearly need that part of the Wick operator which contains the term $\psi_+^\dagger \phi_+ \psi_- \phi_-$. Therefore, the contribution $\mathcal{M}(D)$ of the Wick operator $S(D)$ to the matrix element $\mathcal{M}$ may be represented as follows:

$$\mathcal{M}(D) \stackrel{\text{def}}{=} \langle p', k' | \begin{pmatrix} \psi_{1+}^\dagger + \psi_{1-}^\dagger \searrow \qquad \nearrow \phi_{2+} + \phi_{2-} \\ \phi_{1+} + \phi_{1-} \nearrow \qquad \searrow \psi_{2+} + \psi_{2-} \end{pmatrix} | p, k \rangle \tag{4.9.7}$$

$$= \langle p', k' | \left( \begin{matrix} \psi_{1+}^\dagger \searrow \quad \nearrow \phi_{2-} \\ \phi_{1+} \nearrow \quad \searrow \psi_{2-} \end{matrix} \quad + \quad \begin{matrix} \psi_{1+}^\dagger \longleftarrow \phi_{1-} \\ \phi_{2+} \longleftarrow \psi_{2-} \end{matrix} \right) | p, k \rangle.$$

Our next task is to find an integral form of $\mathcal{M}(D)$ from which we can deduce the Feynman rules. Deriving Feynman rules by inspection of an integral for an amplitude is rather loose. It is not so clear that the rules will apply to all diagrams. We are taking this route because it is a quick way of getting Feynman rules from Dyson's Formula. In Section 11.8, we prove the universality of our Feynman rules using functional integral formalism.

We can do the integrals over spatial momenta if we use the field normalization conditions 4.9.6. To keep the contributions of the different parts of the Feynman diagram separate, we do not eliminate the resulting delta functions:

$$\langle p', k'|S(D)|p, k\rangle \tag{4.9.8}$$

$$= \langle p', k'|(-i\lambda)^2 \int d^4x_1\, d^4x_2\, :\!\psi_1^\dagger\psi_1\phi_1\psi_2^\dagger\psi_2\phi_2\!:|p, k\rangle$$

$$= (-i\lambda)^2 \int d^4x_1\, d^4x_2\, \psi_1^\dagger\psi_2^\dagger$$

$$\times \langle p', k'|:(\psi_{1+}^\dagger + \psi_{1-}^\dagger)(\phi_{1+} + \phi_{1-})(\psi_{2+} + \psi_{2-})(\phi_{2+} + \phi_{2-}):|p, k\rangle$$

$$= (-i\lambda)^2 \int d^4x_1\, d^4x_2\, \psi_1^\dagger\psi_2^\dagger\langle p', k'|\psi_{1+}^\dagger\phi_{1+}\psi_{2-}\phi_{2-} + \psi_{1+}^\dagger\phi_{2+}\psi_{2-}\phi_{1-}|p, k\rangle$$

$$= (-i\lambda)^2 \int d^4x_1\, d^4x_2\, \frac{d^4q}{(2\pi)^4}\, \frac{i}{q^2 - m^2 + i\epsilon}e^{-i(x_1 - x_2)\cdot q}$$

$$\times \left(e^{ix_1\cdot p'}e^{ix_1\cdot k'}e^{-ix_2\cdot p}e^{-ix_2\cdot k} + e^{ix_1\cdot p'}e^{ix_2\cdot k'}e^{-ix_2\cdot p}e^{-ix_1\cdot k}\right)$$

$$= (-i\lambda)^2 \int \frac{d^4q}{(2\pi)^4}\, \frac{i}{q^2 - m^2 + i\epsilon}\left((2\pi)^4\delta^{(4)}(k + p - q)\right)\left((2\pi)^4\delta^{(4)}(k' + p' - q)\right)$$

$$+ (-i\lambda)^2 \int \frac{d^4q}{(2\pi)^4}\, \frac{i}{q^2 - m^2 + i\epsilon}\left((2\pi)^4\delta^{(4)}(k' - p + q)\right)\left((2\pi)^4\delta^{(4)}(k - p' + q)\right).$$

The last two integrals are the Feynman integrals to be analyzed below.

The variable $q$ is interpreted as the momentum on the internal line of the Wick diagram. (We have chosen to have $q$ flow from $x_2$ to $x_1$.) The activity of the $\psi$ field represented by this line is not a particle state since $q^2$ is not required to be $m^2$. It is called a virtual particle, though it really has nothing to do with particles. It represents unobservable, transient behavior of the field which gives rise to scattering and is the quantum field theory version of a force.

From the origin of the integrals in the creation-annihilation monomials and from the Wick diagram analysis 4.9.7, we see that the two Feynman integrals represent different processes: in the first, the incoming nucleon absorbs the incoming meson, propagates, and then emits the outgoing meson; in the second, the incoming nucleon emits the outgoing meson, propagates, and then absorbs the incoming meson. These two distinct processes are naturally represented by two distinct Feynman diagrams:

$$\psi_{2+}^\dagger\phi_{2+}\psi_{1-}\phi_{1-} \longleftrightarrow$$

$$\psi_{2+}^\dagger\phi_{1+}\psi_{1-}\phi_{2-} \longleftrightarrow \tag{4.9.9}$$

When we use the 'later on the left' rule with Feynman diagrams, we must remember that time is generally only defined for the incoming and outgoing particles; the virtual particles may be subject to an $\int d^4q$, or equivalently (under Fourier transform) an $\int d^4x$, and hence the left to right location of vertices in a Feynman diagram is generally of no significance.

By inspection of the final integral we see that we obtain the integral from the associated diagram by multiplying factors for vertices and virtual particles according to the following list, and then integrating.

> **Vertex, incoming momenta $p$, $q$, $k$:**    $-i\lambda(2\pi)^4\delta^{(4)}(k+p+q)$

> **Virtual nucleon, momentum $p$:**    $\dfrac{d^4p}{(2\pi)^4}\dfrac{i}{p^2-m^2+i\epsilon}$    (4.9.10)

> **Virtual meson, momentum $k$:**    $\dfrac{d^4k}{(2\pi)^4}\dfrac{i}{k^2-\mu^2+i\epsilon}$

These associations of integrand factors with diagram elements are the Feynman rules for the nucleon-meson theory. The factors associated with virtual particles are called *propagators*. For brevity in a statement of Feynman rules, the factors of $2\pi$, $\delta^{(4)}(P_{\rm in})$, and $d^4p$ are generally not written out explicitly.

**Homework 4.9.11.** Just as these Feynman rules were deduced from a consideration of Wick operators, deduce the following Feynman rule for a source term $\mathcal{L}' = j\pi$:

> **Source Term:**    $i\hat{j}(-k)$    $\overset{\longrightarrow}{k}$    (4.9.12)

To work out the coefficient of the Feynman integral, it is also necessary to compute a combinatoric factor. The Feynman integral inherits the symmetry factor $1/s(D)$ of the associated Wick diagram, but this factor must be multiplied by the number of ways in which the same Feynman diagram can be created from a given Wick diagram. Let $WD$ be a Wick diagram, and let $FD$ be a Feynman diagram derived from it. Then all possible ways of obtaining this Feynman diagram are related to any given way by the symmetry group of $WD$, a group of $s(D)$ elements. Let $i(WD)$ be the number of internal symmetries, that is, the number of group elements which leave all the external lines of $WD$ individually invariant. Then the number of ways of obtaining $FD$ from $WD$ is $n(FD) = s(WD)/i(WD)$. Hence the combinatoric factor for $FD$ is

$$\frac{1}{s(FD)} \overset{\text{def}}{=} \frac{n(FD)}{s(WD)} = \frac{1}{i(WD)}. \qquad (4.9.13)$$

We conclude then that the symmetry factor $s(FD)$ of the Feynman diagram is simply the number of permutations $i(WD)$ of its lines and vertices that leave the external lines individually fixed and leave the diagram as a whole invariant.

It will be convenient to write the matrix element $\mathcal{M}(D)$ corresponding to a Feynman diagram $D$ as a product

$$\mathcal{M}(D) = \frac{1}{s(D)}I(D), \qquad (4.9.14)$$

where $s(D)$ is the symmetry factor just defined, and $I(D)$ is the integral derived from the vertex and propagator rules.

In our nucleon-meson theory, the symmetry number of any connected Feynman diagram with external lines is actually one. This is because, starting from any external line (a fixed feature of the diagram under the allowed symmetries), one can systematically identify all the other lines and all the vertices by providing travel instructions intrinsic to the diagram: follow the meson line to the next vertex, then take the nucleon line out, and so on. Note that the rule for the symmetry factor is also a Feynman rule, a necessary part of specifying the Feynman integral associated with a Feynman diagram.

**Homework 4.9.15.** Show by examples that, for the interaction Hamiltonian density $\mathcal{H}_I = \lambda \psi^\dagger \psi \phi$, the internal symmetry factor for a vacuum diagram can be arbitrarily large.

**Homework 4.9.16.** A diagram is called one-particle irreducible (1PI) if, when any one line is removed, the diagram remains connected. For the $\phi^4$-theory, write down the three second-order 1PI Feynman diagrams for particle-particle scattering. Compute their symmetry factors.

Pick one of these diagrams. Work out the corresponding Feynman integral from the matrix element for the associated Wick operator. Propose Feynman rules for $\phi^4$-theory.

Since energy-momentum is conserved at each vertex, we can always rearrange the $\delta$ functions in such a way that an overall energy-momentum $\delta$ function factorizes out of the integral. It will be convenient to define the invariant scattering amplitude $\mathcal{A}$ by

$$(2\pi)^4 \delta^{(4)}(P_f - P_i) i \mathcal{A} \stackrel{\text{def}}{=} \langle f | S - 1 | i \rangle. \tag{4.9.17}$$

The $-1$ is present because, just as there is no Wick diagram for $1$ in Dyson's formula, so there is no Feynman diagram for it either. Note that, since we generally compute $i\mathcal{A}$ directly, we often refer to $i\mathcal{A}$ as the amplitude.

We will illustrate this by continuing the analysis of nucleon-meson scattering from the integral 4.9.8. First, in order to perform the integration over internal momenta, we need to manipulate the products of delta functions. The following lemma summarizes the mathematics; it may be proved by multiplying by a test function and integrating over all momenta.

**Lemma 4.9.18.** *If $f = (f^1, \ldots, f^m)$ is a differentiable vector-valued function of a vector variable $x = (x^1, \ldots, x^n)$, and if $f(x) = 0$ has a unique solution $x = x_0$ at which the vectors $\partial_r f$, $1 \leq r \leq m$, are linearly independent, then:*

$$\delta^{(m)}\big(f(x)\big) = \left| \frac{\partial(f^1, \ldots, f^m)}{\partial(x^1, \ldots, x^m)} \right|^{-1} \prod_{r=1}^m \delta(x^r - x_0^r).$$

In our case, $f$ will be a linear function of momenta. Thus, for example, if $f$ is given by

$$f(p, q, r) \stackrel{\text{def}}{=} \begin{pmatrix} 2p - q \\ p + 2q - 5r \end{pmatrix}, \tag{4.9.19}$$

then $m = 8$, $n = 12$, the zero of $f$ is located at $(p, q) = (r, 2r)$, and so the lemma implies

$$\delta^{(4)}(2p - q)\delta^{(4)}(p + 2q - 5r) = 5^{-4}\delta^{(4)}(p - r)\delta^{(4)}(q - 2r). \tag{4.9.20}$$

As this example demonstrates, we can form linear combinations of the arguments of the delta functions as long as we compensate with an appropriate Jacobian factor. The following corollary of the lemma above states this result formally:

**Corollary 4.9.21.** *Following the notation of Lemma 4.9.18, if $A$ is an invertible real $m \times m$ matrix, then*

$$\delta^{(m)}(f) = \big|\det(A)\big| \delta^{(m)}(Af).$$

Using Corollary 4.9.21 with

$$A \stackrel{\text{def}}{=} \begin{pmatrix} 1 & -1 \\ 1 & 0 \end{pmatrix}, \tag{4.9.22}$$

we can rewrite the products of delta functions in the final integrals of 4.9.8:

$$\begin{aligned}
\delta^{(4)}(k+p-q)\delta^{(4)}(k'+p'-q) &= \delta^{(4)}(k+p-k'-p')\delta^{(4)}(k+p-q), \\
\delta^{(4)}(k'-p-q)\delta^{(4)}(k-p'+q) &= \delta^{(4)}(k'+p'-k-p)\delta^{(4)}(k'-p+q),
\end{aligned} \tag{4.9.23}$$

to obtain:

$$\begin{aligned}
\langle p', k' | S(D) | p, k \rangle & \tag{4.9.24} \\
&= (-i\lambda)^2 (2\pi)^4 \delta^{(4)}(k+p-k'-p') \\
&\quad \times \int d^4q \, \frac{i}{q^2 - m^2 + i\epsilon} \left( \delta^{(4)}(k+p-q) + \delta^{(4)}(k'-p+q) \right) \\
&= (-i\lambda)^2 (2\pi)^4 \delta^{(4)}(k+p-k'-p') \left( \frac{i}{(k+p)^2 - m^2 + i\epsilon} + \frac{i}{(p-k')^2 - m^2 + i\epsilon} \right).
\end{aligned}$$

In this case we have

$$i\mathcal{A} = (-i\lambda)^2 \left( \frac{i}{(k+p)^2 - m^2 + i\epsilon} + \frac{i}{(p-k')^2 - m^2 + i\epsilon} \right). \tag{4.9.25}$$

The example illustrates a general principle, that the invariant amplitudes for Feynman diagrams with no loops can be written down by using the vertex energy-momentum conserving delta functions to find the momenta on internal lines and simply substituting these momenta into the appropriate propagators. Diagrams without loops are called tree diagrams.

A Feynman diagram is a label for a contribution of a Wick operator to a matrix element of $S - 1$. Feynman rules form a system for deriving an integral for this matrix element from the Feynman diagram; knowing the Feynman rules for internal lines, vertices, and symmetry factors, we can translate Feynman diagrams directly into Feynman integrals by multiplying all three types of factor. We therefore have no further use for Wick diagrams or Wick operators. Indeed, Feynman diagrams and rules can be taken as the perturbative definition of a quantum field theory. However, lacking the genius of Feynman, it is not possible for most of us to guess consistent Feynman rules without a derivation of the kind presented here.

## 4.10 Third Example: Tree-Level Scattering

In this section, we will compute scattering amplitudes in the scalar nucleon-meson model 4.3.1 and interpret our results in terms of quantum mechanics. At second order, there are fourteen connected Feynman diagrams which represent scattering processes. These diagrams are all tree diagrams. Since $PT$ changes incoming particles into outgoing particles, $CPT$ acts on these diagrams as reflection about the vertical, making the following exchanges:

$$CPT: \begin{cases} \textbf{Incoming:} & \Phi \quad N \quad \bar{N} \\ & \updownarrow \quad \updownarrow \quad \updownarrow \\ \textbf{Outgoing:} & \Phi \quad \bar{N} \quad N \end{cases} \tag{4.10.1}$$

The fourteen diagrams therefore give amplitudes for seven scattering processes, namely, $NN \rightarrow NN$, $N\bar{N} \rightarrow N\bar{N}$, $N\Phi \rightarrow N\Phi$, $N\bar{N} \rightarrow \Phi\Phi$, and their $CPT$ transforms. ($N\bar{N} \rightarrow N\bar{N}$ is $CPT$ invariant.)

Since $CPT$ is always a symmetry of a quantum field theory, the amplitudes for a process and its $CPT$ transform are always equal. So we will consider just four processes and the eight relevant Feynman diagrams, as set out in the following table:

$$NN \leftarrow NN \qquad N\bar{N} \leftarrow N\bar{N} \qquad N\Phi \leftarrow N\Phi \qquad \Phi\Phi \leftarrow N\bar{N}$$

$$(4.10.2)$$

The computation of an invariant amplitude is basically the same for all tree diagrams. We shall do it once and then generalize from the conclusion. Taking the first $NN$ scattering diagram at specific momenta,

$$(4.10.3)$$

we apply the Feynman rules by first, noting that the symmetry factor $s(D)$ is 1, second, writing down the propagator for the internal line and the two vertex factors, and third, using the product of delta functions trick 4.9.23 to do the integral:

$$
\begin{aligned}
I(D) &\stackrel{\text{def}}{=} \int \frac{d^4k}{(2\pi)^4} \frac{i}{k^2 - \mu^2 + i\epsilon} \\
&\quad \times (-i\lambda)(2\pi)^4\delta^{(4)}(p_1 - k - p_1')(-i\lambda)(2\pi)^4\delta^{(4)}(p_2 + k - p_2') \\
&= (2\pi)^4\delta^{(4)}(p_1 + p_2 - p_1' - p_2')(-i\lambda)^2 \\
&\quad \times \int \frac{d^4k}{(2\pi)^4} \frac{i}{k^2 - \mu^2 + i\epsilon}(-ig)(2\pi)^4\delta^{(4)}(p_1 - k - p_1') \\
&= (2\pi)^4\delta^{(4)}(p_1 + p_2 - p_1' - p_2')(-i\lambda)^2\frac{i}{(p_1 - p_1')^2 - \mu^2 + i\epsilon}.
\end{aligned}
$$

$$(4.10.4)$$

This yields the matrix element. The invariant amplitude is

$$i\mathcal{A}(D) = (-i\lambda)^2\frac{i}{(p_1 - p_1')^2 - \mu^2 + i\epsilon}. \qquad (4.10.5)$$

From this derivation of an invariant amplitude for a tree diagram, we see again the procedure for writing down such amplitudes directly: first, use the vertex delta functions to determine the momenta on internal lines in terms of external momenta, and second, multiply symmetry, vertex, and propagator factors omitting all $2\pi$'s, delta functions, and integrals to obtain the invariant amplitude. With this general principle in hand, we return to our analysis of tree-level scattering diagrams.

Consider nucleon-nucleon scattering, $NN \rightarrow NN$:

$$\langle p_1', p_2' | S - 1 | p_1, p_2 \rangle. \tag{4.10.6}$$

The two second-order Feynman diagrams give an invariant amplitude

$$i\mathcal{A} = (-i\lambda)^2 \left( \frac{i}{(p_1 - p_1')^2 - \mu^2 + i\epsilon} + \frac{i}{(p_1 - p_2')^2 - \mu^2 + i\epsilon} \right). \tag{4.10.7}$$

In the center of momentum frame $\bar{P} = \bar{0}$, and so

$$
\begin{aligned}
p_1 &= ((p^2 + m^2)^{1/2}, \bar{p}), & p_2 &= ((p^2 + m^2)^{1/2}, -\bar{p}), \\
p_1' &= ((p^2 + m^2)^{1/2}, \bar{p}'), & p_2' &= ((p^2 + m^2)^{1/2}, -\bar{p}'),
\end{aligned}
\tag{4.10.8}
$$

where $|\bar{p}| = |\bar{p}'| = p$. Hence, naming the momentum transfer in each process:

$$
\left.
\begin{aligned}
\bar{\Delta} &\overset{\text{def}}{=} \bar{p}_1 - \bar{p}_1' \\
\bar{\Delta}_c &\overset{\text{def}}{=} \bar{p}_1 - \bar{p}_2'
\end{aligned}
\right\}
\implies
\left.
\begin{aligned}
\Delta^2 &= (\bar{p}_1 - \bar{p}_1')^2 = -(p_1 - p_1')^2 \\
\Delta_c^2 &= (\bar{p}_1 - \bar{p}_2')^2 = -(p_1 - p_2')^2
\end{aligned}
\right\}
\tag{4.10.9}
$$

the scattering amplitude becomes

$$i\mathcal{A} = \lambda^2 \left( \frac{i}{\Delta^2 + \mu^2 + i\epsilon} + \frac{i}{\Delta_c^2 + \mu^2 + i\epsilon} \right). \tag{4.10.10}$$

In quantum mechanics, if

$$H = \frac{\bar{p}_1^2}{2m} + \frac{\bar{p}_2^2}{2m} + V(\bar{r}_1 - \bar{r}_2) \quad \text{where} \quad V(r) = g^2 \frac{e^{-\mu r}}{r}, \tag{4.10.11}$$

then the first-order Born approximation to this scattering amplitude is

$$\mathcal{A}_{\text{Born}} \propto \langle f | V | i \rangle \propto \int d^3\bar{r} \, e^{-i\bar{\Delta} \cdot \bar{r}} g^2 \frac{e^{-\mu r}}{r} \propto \frac{\lambda^2}{\Delta^2 + \mu^2}. \tag{4.10.12}$$

To obtain the complete amplitude we must add the exchange potential $V(r)E$ (where $E\psi(\bar{r}_1, \bar{r}_2) = \psi(\bar{r}_2, \bar{r}_1)$) onto the Hamiltonian.

Consider now $N\bar{N} \rightarrow N\bar{N}$ scattering. Again there are two second-order diagrams that contribute to the invariant amplitude, and we find

$$i\mathcal{A} = (-i\lambda)^2 \left( \frac{i}{(p_1 - p_1')^2 - \mu^2 + i\epsilon} + \frac{i}{(p_1 + p_2)^2 - \mu^2 + i\epsilon} \right). \tag{4.10.13}$$

Again take the center of momentum frame. In this case the first term in $\mathcal{A}$ is the first-order Born approximation, but the second term is the second-order Born approximation:

$$(p_1 + p_2)^2 - \mu^2 = \left( 2(p^2 + m^2)^{1/2} \right)^2 - \mu^2 = E_T^2 - \mu^2, \tag{4.10.14}$$

and

$$\mathcal{A}_{\text{Born}} \propto \langle f | V | i \rangle - \sum_n \frac{\langle f | V | n \rangle \langle n | V | i \rangle}{E_T - E_n + i\epsilon}. \tag{4.10.15}$$

If there is an isolated energy eigenstate $|n\rangle$ with $E_n \geq 2m$, then there is an energy eigenvalue pole in the scattering amplitude at $E_T = E_n$.

Next consider $N\Phi \to N\Phi$ scattering. In this case the invariant amplitude is

$$i\mathcal{A} = (-i\lambda)^2 \left( \frac{i}{(p+k)^2 - m^2 + i\epsilon} + \frac{i}{(p-k')^2 - m^2 + i\epsilon} \right). \qquad (4.10.16)$$

This loosely corresponds to a second-order Born approximation and an exchange potential in quantum mechanics. The exchange potential again comes from a Yukawa potential. To find the mass in the Yukawa potential, we need to put the denominator $(p - k')^2 - m^2$ into standard form. In the center of momentum frame:

$$\begin{aligned} p &= ((p^2 + m^2)^{1/2}, \bar{p}), & k &= ((p^2 + \mu^2)^{1/2}, -\bar{p}), \\ p' &= ((p^2 + m^2)^{1/2}, \bar{p}'), & k' &= ((p^2 + \mu^2)^{1/2}, -\bar{p}'). \end{aligned} \qquad (4.10.17)$$

Hence

$$(p - k')^2 - m^2 = -((\bar{p} + \bar{p}')^2 + m_{\text{eff}}^2), \qquad (4.10.18)$$

where

$$m_{\text{eff}}^2 = m^2 - \left((p^2 + m^2)^{1/2} - (p^2 + \mu^2)^{1/2}\right)^2. \qquad (4.10.19)$$

As $p^2 \to \infty$, $m_{\text{eff}} \to m$, while as $p^2 \to 0$, $m_{\text{eff}} \to \sqrt{2m\mu + \mu^2}$. This shows that the effective potential is energy dependent.

**Homework 4.10.20.** Derive all the second-order invariant amplitudes used in this section from their Feynman diagrams.

**Homework 4.10.21.** Find the invariant amplitude for $N\bar{N} \to \Phi\Phi$, and interpret it in terms of quantum mechanics.

From these examples of scattering amplitudes in field theory, we see that the Feynman formalism provides a marvelous and effective means for computing $S$-matrix elements. Furthermore, the Born approximation, Yukawa potentials, exchange potentials, and energy-dependent Yukawa potentials can all be discerned in Feynman's formalism, but this classification is an unnecessary complication in quantum field theory.

## 4.11 Lorentz Invariance and Elimination of Variables

Since an invariant amplitude $\mathcal{A}(p_1, \ldots, p_n)$ is indeed Lorentz invariant, it is always possible to simplify the dependence on external momenta. In this section, we shall briefly present some practical results on the structure of Lorentz-invariant functions.

There are just two isotropic tensors for the Lorentz group, the metric tensor $g^{\mu\nu}$ and the completely anti-symmetric tensor $\epsilon^{\alpha\beta\gamma\delta}$. It is convenient to introduce a shorthand for contraction with the $\epsilon$ tensor:

$$\epsilon(p_r p_s p_t p_u) \overset{\text{def}}{=} \epsilon_{\alpha\beta\gamma\delta} p_r^\alpha p_s^\beta p_t^\gamma p_u^\delta. \qquad (4.11.1)$$

We now take recourse to a helpful theorem from mathematics:

**Theorem 4.11.2.** *If $A$ is a Lorentz-invariant function of vectors $p_r$, then it is a function of the Lorentz-invariant quantities $p_r \cdot p_s$ and $\epsilon(p_r p_s p_t p_u)$.*

Hence we can reduce the dependence of Lorentz-invariant functions on momenta as follows:

$$\mathcal{A}_1(p) = \mathcal{A}_1'(p^2), \tag{4.11.3}$$
$$\mathcal{A}_2(p, q) = \mathcal{A}_2'(p^2, q^2, p \cdot q),$$
$$\mathcal{A}_3(p, q, k) = \mathcal{A}_3'(p^2, q^2, k^2, p \cdot q, p \cdot k, q \cdot k),$$
$$\mathcal{A}_4(p, q, k, l) = \mathcal{A}_4'(p^2, q^2, k^2, l^2, p \cdot q, p \cdot k, p \cdot l, q \cdot k, q \cdot l, k \cdot l, \epsilon(pqkl)).$$

Generally, we can use either on-shell or energy-momentum constraints or both to eliminate more variables. In practice, we are primarily concerned with the elimination of quadratic monomials in $\mathcal{A}_3$ and $\mathcal{A}_4$.

Taking $\mathcal{A}_3$ first and working in the center of momentum frame, it is clear that the mass-shell and energy-momentum constraints can be solved by three vectors $p_r$ if and only if the masses $m_r$ of the associated particles satisfy $m_1 \geq m_2 + m_3$ (or some permutation of this constraint). In this case, the solutions $p_r$ form a single orbit under the Lorentz group, and so a Lorentz-invariant function $\mathcal{A}_3$ of these three vectors is a constant. The following theorem summarizes the possibilities for a Lorentz-invariant function of three vectors.

**Theorem 4.11.4.** *A Lorentz-invariant function $\mathcal{A}_3$ of three vector $p_r$ can be expressed as a function of the six quadratic terms $p_r \cdot p_s$.*

1. *If the $p_r$ all satisfy shell constraints, then $\mathcal{A}_3$ is a function of $p_r \cdot p_s$ for $r < s$.*
2. *If the $p_r$ satisfy energy-momentum conservation, then $\mathcal{A}_3$ can be expressed as a function of any three quadratic monomials.*
3. *If the $p_r$ satisfy both the shell constraints and energy-momentum conservation, then $\mathcal{A}_3$ is a constant.*

In the case of $\mathcal{A}_4$, the energy-momentum constraint can be written in seven different ways, for example:

$$p_1 = -p_2 - p_3 - p_4 \quad \text{or} \quad p_1 + p_2 = -p_3 - p_4. \tag{4.11.5}$$

Squaring these equations yields seven constraints on the quadratic monomials $p_r \cdot p_s$. However, only the four of the first type are linearly independent. The coefficient matrix for these four follows:

|      | $p_1 \cdot p_2$ | $p_1 \cdot p_3$ | $p_1 \cdot p_4$ | $p_2 \cdot p_3$ | $p_2 \cdot p_4$ | $p_3 \cdot p_4$ | $p_1^2$ | $p_2^2$ | $p_3^2$ | $p_4^2$ |
|------|------|------|------|------|------|------|------|------|------|------|
| $C(1)$ | 0 | 0 | 0 | $-2$ | $-2$ | $-2$ | 1 | $-1$ | $-1$ | $-1$ |
| $C(2)$ | 0 | $-2$ | $-2$ | 0 | 0 | $-2$ | $-1$ | 1 | $-1$ | $-1$ |
| $C(3)$ | $-2$ | 0 | $-2$ | 0 | $-2$ | 0 | $-1$ | $-1$ | 1 | $-1$ |
| $C(4)$ | $-2$ | $-2$ | 0 | $-2$ | 0 | 0 | $-1$ | $-1$ | $-1$ | 1 |

$$\tag{4.11.6}$$

From this matrix, it is clear for example that we can simultaneously eliminate $p_1 \cdot p_4$, $p_2 \cdot p_4$, $p_3 \cdot p_4$, and $p_2 \cdot p_3$ from $\mathcal{A}_4$, but that we cannot simultaneously eliminate $p_1 \cdot p_3$, $p_1 \cdot p_4$, $p_2 \cdot p_3$, and $p_2 \cdot p_4$. Note further that the energy-momentum constraint makes $\epsilon(p_1 p_2 p_3 p_4)$ vanish. The following theorem summarizes the possibilities:

**Theorem 4.11.7.** *A Lorentz-invariant function $\mathcal{A}_4$ of four vector $p_r$ can be expressed as a function of the ten quadratic terms $p_r \cdot p_s$ and the anti-symmetric term $\epsilon(p_1 p_2 p_3 p_4)$.*

1. *If the $p_r$ all satisfy shell constraints, then $\mathcal{A}_4$ is a function of $p_r \cdot p_s$ for $r < s$ and $\epsilon(p_1 p_2 p_3 p_4)$.*
2. *If the $p_r$ satisfy energy-momentum conservation, then $\mathcal{A}_4$ can be expressed as a function of $p_r \cdot p_s$ for $r < s$.*
3. *If the $p_r$ satisfy both the shell constraints and energy-momentum conservation, then $\mathcal{A}_4$ can be expressed as a function of $p_1 \cdot p_2$ and $p_1 \cdot p_3$.*

**Remark 4.11.8.** The following theorem shows that the energy-momentum constraint on $n$ momenta can be used to eliminate up to $n$ terms of the form $p_r \cdot p_s$.

**Theorem.** *For $n \geq 3$, the linear energy-momentum constraint $\sum_1^n p_r = 0$ leads to $2^{n-1}$ quadratic constraints*

$$C(r_1, \ldots, r_k) \equiv (p_{r_1} + \cdots p_{r_k})^2 - (p_{r_{k+1}} + \cdots p_{r_n})^2 = 0,$$

*where the $p_{r_s}$ are the $p_r$ in some order. Taking these constraints as linear relations on the $p_r \cdot p_s$, the $n$ constraints $C(r) = 0$ form a basis for the linear space of constraints.*

**Proof.** The linear independence of the $C(r)$ is obvious. The linear dependence of the other constraints follows from the identity

$$(n-2)C(1, \ldots, r) = (n - r - 1)\big(C(1) + \cdots + C(r)\big) \qquad (4.11.9)$$
$$- (r-1)\big(C(r+1) + \cdots + C(n)\big). \qquad \square$$

Obviously one possibility is to eliminate all $p_r \cdot p_n$ for $1 \leq r \leq n$, a second is to eliminate $p_r^2$ for $1 \leq r \leq n$, and a third is to eliminate $p_r \cdot p_n$ for $1 \leq r \leq n - 1$ and any other quadratic monomial. To show that these last two options are valid, one must examine the determinant of coefficients in the constraints $C(r) = 0$.

The results above extend readily to Lorentz-tensor functions since the tensor indices must be provided by momentum variables or the isotropic tensors. For example, a Lorentz-vector function $F^\mu$ of three momenta will have the form

$$F^\mu(p_1, p_2, p_3) = F_1 p_1^\mu + F_2 p_2^\mu + F_3 p_3^\mu + F_\epsilon \epsilon^\mu{}_{\nu\rho\sigma} p_1^\nu p_2^\rho p_3^\sigma, \qquad (4.11.10)$$

where the coefficients are Lorentz-invariant functions of the same three momenta, and therefore subject to the analysis above.

The points presented above are very useful when we need to write down the form of a Lorentz-tensor function in the absence of detailed knowledge of its structure. The results on $\mathcal{A}_3$ and $\mathcal{A}_4$ find immediate application in parameterization of invariant amplitudes.

## 4.12 Crossing and Mandelstam Parameters

It is obvious from the results of Section 4.10 that there are really only two types of basic graph in the fourteen Feynman diagrams considered there: one with two arrows in, two arrows out, the other with one arrow in, one out, and two neutral lines. These basic graphs are, of course, the two underlying Wick diagrams. The four-arrow Wick diagram gives rise to six Feynman scattering diagrams, the two-arrow diagram gives rise to eight. The Feynman diagrams in each of these two sets of Feynman graphs are related to each other by moving external lines. The term for moving external lines in a Feynman diagram is *crossing*.

The foundation for applying crossing to amplitudes is the following property of Feynman amplitudes: Feynman amplitudes are analytic functions of the external momenta which only distinguish between incoming and outgoing particles by the sign of the energy term. Thus, if two diagrams $D_1$ and $D_2$ are connected by crossing, then their invariant amplitudes $\mathcal{A}_1$ and $\mathcal{A}_2$ will be identical functions of external momenta, but the regions of momentum space which correspond to physical particles will be different for the two diagrams.

In more detail, let $D$ be a Feynman diagram with all the momenta $p_1, \ldots, p_n$ oriented inwards. Then incoming particles will have positive energies and outgoing particles will have negative energies. The corresponding invariant amplitude $\mathcal{A}$ is defined in terms of the Feynman integral $I(D)$ by

$$I(D) = (2\pi)^4 \delta^{(4)}(p_1 + \cdots + p_n) i \mathcal{A}(p_1, \ldots, p_n). \tag{4.12.1}$$

$\mathcal{A}$ is an analytic function of the external momenta $p_r$. From this function we can derive $2^n$ scattering amplitudes $\mathcal{A}_P$ by choosing which momenta should be incoming particles and which ones should be outgoing. A physical amplitude $\mathcal{A}_P$ is the same function as $\mathcal{A}$, but the domain is restricted to the appropriate positive and negative mass hyperboloids.

If a diagram $D$ has four external lines, and if the momenta on these lines are on shell, then it is convenient to write the amplitude $\mathcal{A}_4$ in terms of the Mandelstam parameters

$$s = (p_1 + p_2)^2, \quad t = (p_1 + p_3)^2, \quad u = (p_1 + p_4)^2. \tag{4.12.2}$$

Clearly, under our hypotheses, $s$, $t$, and $u$ are equivalent as variables to $p_1 \cdot p_2$, $p_1 \cdot p_3$, and $p_1 \cdot p_4$ respectively. Since according to the previous section we can reduce dependence on three momenta to dependence on two quadratic scalars, it is no surprise that $s$, $t$, and $u$ are linearly dependent. Indeed, since the sum of the momenta is zero, we also have

$$s = (p_3 + p_4)^2, \quad t = (p_2 + p_4)^2, \quad u = (p_2 + p_3)^2, \tag{4.12.3}$$

and so

$$\begin{aligned} s + t + u &= \frac{3}{2}(p_1^2 + p_2^2 + p_3^2 + p_4^2) + \sum_{r<s} p_r p_s \\ &= \sum_r p_r^2 + \frac{1}{2}\left(\sum_r p_r\right)^2 \\ &= \sum_r m_r^2. \end{aligned} \tag{4.12.4}$$

We can choose any scattering process $AB \to CD$ as the *s-channel process*. With this choice, the *t-channel process*, $\bar{D}B \to C\bar{A}$, is obtained by exchanging the $A$ and $D$ external lines, and the *u-channel process*, $A\bar{D} \to C\bar{B}$, by exchanging the $B$ and $D$ external lines.

Since the three channels are related by crossing, they are all described by one amplitude $\mathcal{A}(p_1, p_2, p_3, p_4)$, where (following the momentum convention used above for crossing) all four momenta are incoming. The three channels are distinguished by the signs of the energies in the momenta:

$$
\begin{array}{lccccc}
\textbf{Process} & & p_1 & p_2 & p_3 & p_4 \\
s: AB \to CD & & + & + & - & - \\
t: \bar{D}B \to C\bar{A} & & - & + & - & + \\
u: A\bar{D} \to C\bar{B} & & + & - & - & +
\end{array} \tag{4.12.5}
$$

Thus, if $p_1^0 > 0$, we have an incoming $A$ particle, and if $p_1^0 < 0$, we have an outgoing anti-particle $\bar{A}$, and so on.

If we define the *s*-channel by the $N\bar{N} \to N\bar{N}$ diagram

$$
D_s \stackrel{\text{def}}{=} \quad \overset{-p_3}{\diagdown} \hspace{-0.3em} \underset{-p_4}{\diagup} \hspace{-0.5em} \Large{>\!\!\!-\!\!\!<} \hspace{-0.5em} \overset{p_1}{\diagup} \underset{p_2}{\diagdown} \tag{4.12.6}
$$

then the diagrams for the *t*- and *u*-channels are given by

$$
D_t \stackrel{\text{def}}{=} \begin{array}{c} -p_3 \longleftarrow \quad \longleftarrow p_4 \\ \big| \\ -p_1 \longrightarrow \quad \longrightarrow p_2 \end{array} \quad \text{and} \quad D_u \stackrel{\text{def}}{=} \begin{array}{c} -p_3 \longleftarrow \quad \longleftarrow p_1 \\ \big| \\ -p_2 \longrightarrow \quad \longrightarrow p_4 \end{array} \tag{4.12.7}
$$

The generic amplitude, the one with four incoming momenta, is

$$
i\mathcal{A}(p_1, p_2, p_3, p_4) \stackrel{\text{def}}{=} (-i\lambda)^2 \frac{i}{(p_1 + p_2)^2 - \mu^2 + i\epsilon}. \tag{4.12.8}
$$

The amplitudes associated with the three channels, $i\mathcal{A}_s = i\mathcal{A}(D_s)$ and so on, satisfy

$$
\begin{aligned}
i\mathcal{A}_s(p_3, p_4; p_1, p_2) &= i\mathcal{A}(p_1, p_2, -p_3, -p_4), \\
i\mathcal{A}_t(p_1, p_3; p_2, p_4) &= i\mathcal{A}(-p_1, p_2, -p_3, p_4), \\
i\mathcal{A}_u(p_2, p_3; p_1, p_4) &= i\mathcal{A}(p_1, -p_2, -p_3, p_4),
\end{aligned} \tag{4.12.9}
$$

where all the energies $p_n^0$ are now positive. Hence, for example, a mass-eigenstate pole in the *s*-channel implies a Yukawa potential in the other channels.

Since there are only two linearly independent Mandelstam parameters, it is possible to plot the values of $(s, t, u)$ on a plane. Indeed, $s + t + u = \sum_r m_r^2$ is the equation of a plane in three dimensions. Clearly the lines $s = 0$, $t = 0$, and $u = 0$ form a triangle in that plane. This triangular diagram of $(s, t, u)$ values is called a Mandelstam plot.

For particles of equal masses, the areas of the Mandelstam plot which correspond to physical processes (that is, are consistent with mass-shell conditions and conservation of energy and momentum) are as follows:

$$
\begin{array}{llll}
s\text{-channel:} & & t < 0 & u < 0 \\
t\text{-channel:} & s < 0 & & u < 0 \\
u\text{-channel:} & s < 0 & t < 0 &
\end{array} \tag{4.12.10}
$$

When the masses are unequal the physical regions are more complicated, but differences between masses are often not significant at large energies.

Of course, if an external line has an arrow on it, then we can regard it either as an incoming particle or as an outgoing anti-particle, depending on the sign of the energy in the associated momentum. Since Lorentz-invariant quantities must have even order in the external momenta, all invariant amplitudes satisfy

$$\mathcal{A}(p_1, \ldots, p_n) = \mathcal{A}(-p_1, \ldots, -p_n). \tag{4.12.11}$$

But $PT$ converts an incoming particle with momentum $p_r$ into an outgoing particle with the same momentum, and charge conjugation $C$ converts particles into anti-particles and *vice versa*. Hence this amplitude identity may be interpreted as $CPT$ invariance of Feynman amplitudes.

There are two practical ideas in this section: first, the use of the generic amplitude for a diagram which has all momenta oriented inwards to obtain amplitudes for all processes related to this one by crossing; second, the use of Mandelstam parameters to provide a standardized description of scattering amplitudes.

## 4.13 Preliminary Points on Renormalization

So far, Chapter 4 has shown us how to transform a classical field theory defined by a Lagrangian density into a quantum field theory defined by a set of Feynman rules and how to compute and interpret amplitudes derived from tree diagrams. The next logical step is to extend computation and interpretation to amplitudes derived from loop diagrams.

Since many loop diagrams give rise to divergent Feynman integrals, management of loop diagrams contains two parts: a technique for integrating convergent loop integrals and a renormalization procedure for obtaining finite values from divergent loop integrals. The philosophy of this text is to take the most naive viewpoint on quantization as far as possible towards application before introducing new levels of theory. Hence, in this section, we shall only introduce loop integrals and renormalization; a fuller treatment of these topics follows the improved scattering theory of Chapter 10 which gets rid of the adiabatic turning on and off function.

We have seen that tree diagrams lead to trivial Feynman integrals. In general, if a connected diagram $D$ has $V$ vertices and $I$ internal lines, then the Feynman integral will contain $V$ delta functions and $I$ internal momenta to integrate over. Using Corollary 4.9.21 to manipulate products of delta functions, we find that we can extract one overall energy-momentum conserving delta function and use the remaining $V - 1$ delta functions to perform $V - 1$ integrals over internal momenta. This leaves $I - V + 1$ integrals over internal momenta. The remaining momenta may be identified as *loop momenta* insofar as they effectively circulate around loops in $D$. We can therefore define the number of loops $L$ in $D$ by

$$L \stackrel{\text{def}}{=} I - V + 1. \tag{4.13.1}$$

**Remark** 4.13.2. The connection between loops identified by the Feynman integral and loops as topological entities may be understood as follows. Let $D$ be any connected Feynman diagram with all external lines removed. Consider the transformation of $D$ to a diagram $D'$ with one less edge and one less vertex obtained by shrinking any edge with two endpoints to a single point. Clearly $D$ and $D'$ have the same number of loops from either the perspective of the Feynman rules or the perspective of topology. Repeatedly applying this transformation to $D$, we obtain finally a diagram $D''$ in which every edge is a loop and, since $D$ is connected, there is only one vertex. Hence the number $L$ of loops in $D$ is equal to the number of edges in $D''$, that is, the number $I$ of edges in $D$ minus the number $V - 1$ of transformations connecting $D$ to $D''$.

For example, we have the following:

**Eliminating delta functions:** $\quad l \longrightarrow\!\bigcirc\!\longleftarrow k \quad \longrightarrow \quad l \longrightarrow\!\bigcirc\!\longleftarrow k \quad$ (4.13.3)

The arrows by the external momenta indicate the orientation of the momenta with respect to the diagram – flowing in or out; the internal momenta follow the charge arrows. (It is often necessary to indicate direction of momentum flow on internal lines with little arrows too.)

The Feynman amplitude $\mathcal{A}$ for the diagram above can now be written down:

$$i\mathcal{A}(k^2) = (-i\lambda)^2 \int \frac{d^4p}{(2\pi)^4} \frac{i}{p^2 - m^2 + i\epsilon} \frac{i}{(p+k)^2 - m^2 + i\epsilon}. \qquad (4.13.4)$$

We notice at once that this integral is logarithmically divergent.

**Homework** 4.13.5. Draw three loop diagrams in each of the theories defined by the interaction Hamiltonian densities $H_1 = \lambda\psi^\dagger\psi\phi$ and $H_2 = g\phi^4/4!$. Write down the corresponding invariant amplitudes.

A clear understanding of the symbols in the Lagrangian density provides the foundation for making finite predictions in the presence of divergent loop contributions. The Lagrangian density is a classical expression which does not directly govern quantum evolution. Since an experiment is modeled by an infinite sum of Feynman diagrams, any particle production process, measured mass, or coupling strength will generally depend on all the symbols in the Lagrangian density. Therefore neither the fields nor the mass and coupling parameters in the Lagrangian density have direct experimental interpretations.

The meaning of symbols in the Lagrangian density must therefore be fixed by equating predictions for measurements of particle production, mass, and coupling strength to experimental data and then solving these equations order by order in perturbation theory. Such equations are called *renormalization conditions*.

When we solve these equations, we generally find that every symbol in the Lagrangian density has a value infinitely far from that predicted by a naive experimental interpretation.

First, the free-field vacuum has zero energy, but the interacting-field vacuum energy is represented by a sum of vacuum bubble diagrams. To normalize the vacuum energy, we therefore need to modify the Lagrangian by an energy counterterm $a$. Second, the free field has zero vacuum expectation value (VEV), but the VEV $\langle\phi\rangle$ for the interacting field is represented by a sum of 'tadpole' diagrams, diagrams with a single external line. To normalize the VEV we change variables from the original bare field $\phi$ to

$$\phi_1(x) \stackrel{\text{def}}{=} \phi(x) - \langle\phi\rangle. \qquad (4.13.6)$$

Third, the free field has normalized particle production, $\langle k|\phi_{\rm F}(0)|0\rangle = 1$, but since the interacting field can produce many-particle states from the vacuum, the interacting field satisfies

$$\sqrt{z_\phi} \stackrel{\text{def}}{=} \langle k|\phi(0)|0\rangle < 1. \tag{4.13.7}$$

To normalize particle production, we change variables from $\phi_1$ to

$$\phi_2 \stackrel{\text{def}}{=} z_\phi^{-1/2}\phi_1, \tag{4.13.8}$$

and rewrite the kinetic term in the form

$$\frac{1}{2}\partial_\mu\phi_2\,\partial^\mu\phi_2 = \frac{1}{2}\partial_\mu\phi_2\,\partial^\mu\phi_2 - \frac{1}{2}(1-z_\phi)\partial_\mu\phi_2\,\partial^\mu\phi_2. \tag{4.13.9}$$

At this point, the quantum field theory of the renormalized field $\phi_2$ defined by the original Lagrangian has the normalization property that $\phi_2$ is the field most closely associated with $\phi$-type particles.

Fourth, as we shall see later, the mass of $\phi_2$ is identified from the sum of all diagrams which represent propagation of the associated particle, and gives rise to an expression of the form $\mu^2 = \mu_0^2 + \delta\mu^2$, where $\mu_0^2/2$ is the coefficient of $\phi_2^2$ in the Lagrangian density. Commonly, one defines $z_\mu$ by $\mu_0^2 = z_\mu\mu^2$, which implies that

$$\delta\mu^2 = (1-z_\mu)\mu^2. \tag{4.13.10}$$

Here, the bare mass $\mu_0$ of the field is masked by an amount $\delta\mu^2$ by the interactions of the field with itself and other fields, just as the mass of a rock is masked as it falls through the ocean. Since we can never turn the interactions off, measurement determines the renormalized mass $\mu$; $\mu_0$ is a theoretical fiction.

Fifth, the measured coupling between some number $n$ of fields depends on scattering $n$ particles. Again, the prediction for this will be represented by a sum of diagrams, and so the bare coupling $\lambda_0$, the coupling parameter in the Lagrangian density, will not be measured. In fact, the measured coupling strength $\lambda$ will not even be a constant; it will depend on the momenta of the incoming and outgoing particles. Assuming that we have specified a set of these momenta, then we can write the renormalized coupling $\lambda$ in the form $\lambda_0 + \delta\lambda$, or we can define a constant $z_\lambda$ by $z_\lambda\lambda = \lambda_0$ so that

$$\delta\lambda = (1-z_\lambda)\lambda. \tag{4.13.11}$$

This gap between the measured coupling strength $\lambda$ and the bare coupling strength $\lambda_0$ is a feature of classical electromagnetism. The response of the electromagnetic field to the charge of an electron within it effectively masks the bare charge of the electron with a cloud of virtual electron-positron pairs. (This response of the vacuum to an electron is known as *vacuum polarization*). The energy dependence of $\lambda$ is an experimental fact; higher energy scattering penetrates this cloud and discovers a larger effective charge.

It so happens that when we use Feynman rules to compute $z_\phi$, $z_\mu$ and $z_\lambda$, these quantities are generally infinite. Therefore, in order for predicted particle production rates, masses, and couplings to match experimental data, it is necessary for the Lagrangian parameters to be infinite also.

As it is clearly impossible to work directly with infinite parameters, if we are to make sense of loop diagrams, it is necessary to invoke a *regularization procedure*, a systematic way of introducing a parameter to express divergent integrals as limits of convergent ones. Then $z_\phi$, $z_\mu$, $z_\lambda$, and the Lagrangian density itself have finite regularized values but diverge as the regularization parameter approaches its limit.

It turns out that, if we are to make all matrix elements finite by modifying a finite number of parameters (which might originally have zero values) in the Lagrangian density, it is necessary that the operators in the Lagrangian density have mass dimension at most four.

Since we have set $\hbar = c = 1$, we have the dimension relationships $L = T = M^{-1}$. The analysis of mass dimension goes as follows: $Ht$ has dimension zero, so the Hamiltonian has dimension 1; since $dx$ has dimension $-1$, the Hamiltonian density and the Lagrangian density have dimension 4, the partial derivative operators $\partial_\mu$ have dimension 1, and so from the kinetic term we deduce that the dimension of a scalar field is 1. Hence the renormalizable interactions of scalar fields are either cubic or quartic.

The non-renormalizable nature of higher-order interactions is easily seen from an example. Consider $\phi^5$. By contracting two lines between two $\phi^5$ vertices we can generate a Feynman loop diagram with six external lines. The loop integral diverges, so we need a term $\lambda_6 \phi^6$ in the Lagrangian density to mop up the divergence. Now with a $\phi^5$ and a $\phi^6$ vertex we can create a loop diagram with seven external lines, and so we need to introduce a $\lambda_7 \phi^7$ interaction. Clearly this sequence continues, generating infinitely many parameters in the Lagrangian density.

Such a Lagrangian density may be consistent, but it not so useful for prediction since all of the parameters must be determined from experiment. The last section of the book, Section 21.13, describes the role and utility of non-renormalizable Lagrangian densities in field theory. Until then, we shall assume that our Lagrangian densities are renormalizable.

Note that the adiabatic turning on and off function $f(t)$ must eliminate the $1 - z$ terms 4.13.9, 4.13.10, and 4.13.11 from the Lagrangian density; we must treat these terms as interactions. It is tricky to find the Feynman rule for interactions with derivatives of fields in our formalism. When we return to renormalization in Chapter 10, we shall guess a Feynman rule, and when we develop functional integral quantization in Chapter 11, we shall prove that this rule is correct.

Fortunately, at tree level, all these renormalization effects are trivial: the VEV's are zero, $z_\phi$ and $z_\mu$ are both unity, and $z_\lambda$ is at worst finite. We shall therefore continue to develop tree-level perturbative canonical quantization without further reference to renormalization until we have reached the point where for further progress we must eliminate the adiabatic turning on and off function from our scattering theory.

**Homework 4.13.12.** Cook up a theory in which $z_\lambda$ is not unity even at tree level and then find its value.

First, we saw how application of the Feynman rules to loop diagrams leads to integrals for the corresponding invariant amplitudes. Second, we introduced the concept of renormalization, that is, the modification of parameters in the Lagrangian density in order to produce experimentally correct predictions for particle production rates, masses, and coupling strengths. Third, finding that loop integrals often diverge, we introduced regularization in order to express the divergence as a limit of finite quantities. Fourth, we concluded that renormalization yields a Lagrangian density which is a well-defined function of a regularization parameter.

Fifth, becoming aware that renormalization is technically non-trivial and observing that renormalization can be ignored at tree level, we decided to pursue tree-level quantum field theory as far as we can. This decision will make it possible to introduce the basic field types (scalar, Weyl, Dirac, and vector), the Feynman rules for the associated propagators and interactions, and the machinery for making predictions in the simplest possible technical framework.

# 4.14 Summary

In the previous chapter, we looked at the evolution of states from the outside, finding conservation laws that restrict the possibilities for evolution. In this chapter, we have looked at the evolution of states from the inside, finding an infinite collection of Wick operators in Dyson's formula for the $S$ matrix and many Feynman integrals in the contribution of each Wick operator to a particular scattering amplitude. Finally, we have seen how the Feynman rules for a theory contain all the information we need for computing matrix elements.

We are not justified in applying Dyson's formula to field theory, yet the resulting Feynman rules indicate the merit of this derivation. In principle, Feynman rules are so self-sufficient that specification of Feynman rules constitutes a procedure for perturbative canonical quantization. If one is not Feynman, however, one needs a guiding principle such as Dyson's formula to come up with consistent and meaningful Feynman rules.

Since Feynman loop integrals often diverge, it is necessary to choose a regularization scheme in order to define a renormalized Lagrangian density. This introduces derivative interactions. Since it is not easy to find the Feynman rules for derivative interactions within the present formulation of scattering theory and since renormalization can be ignored at tree level, we shall continue to develop quantum field theory at tree level. When the development of gauge theories and theories with unstable particles demands the removal of the adiabatic turning on and off function, then we shall revamp scattering theory, introduce functional integral quantization, find the missing Feynman rules, and present renormalization in detail.

Despite the problem posed by renormalization, we have found in perturbative canonical quantization a means to circumvent the normal-ordering dilemma of canonical quantization. Indeed, the motivation for normal ordering was to make the eigenvalues of the Hamiltonian finite on the vacuum and particle states. Now, with perturbative canonical quantization and its renormalization procedure, we can actually make all matrix elements finite.

Even with only the tree-level Feynman machinery for computing matrix elements, we are still in a position to compute scattering amplitudes and to make predictions for scattering and decay processes. The next chapter therefore introduces that essential link between the unobservable level of quantum evolution and the results of measurement, the differential cross section.

# Chapter 5

# Differential Transition Probabilities and Predictions

Providing the connection between theory and experiment by transforming the invariant amplitude into a prediction for probabilities of final states.

## Introduction

In the last chapter we derived the Feynman formalism for computing scattering amplitudes. In this chapter, we take the scattering amplitude and convert it into probability distributions for prediction and comparison with experiment.

Actually, the only example of interacting scalars in nature are the three pions, but their interactions are greatly complicated by non-perturbative effects. Even the magnitude of the effective coupling constant is too large for perturbation theory to make sense beyond the tree level. Hence, we can hardly make any realistic predictions at this point. The formulae of this chapter are introduced now to provide the final stage in the development of scalar field theory from foundations to predictions. Having covered this vertical range of scalar field theory, we will find it a simple matter to make a horizontal extension to include spinor and vector fields.

The central concept for making a prediction from an amplitude is the differential transition probability per unit time. The DTP/T is derived in Section 5.1. Sections 5.2 and 5.3 focus on one factor in the DTP/T, the invariant density of final states, providing convenient formulae for the two-particle and three-particle cases respectively. Section 5.4 covers the case of a one-particle initial state, giving formulae for the decay width for decays into two or three final-state particles. Section 5.5 similarly covers the case of a two-particle initial state, giving formulae for scattering cross sections. Finally, Section 5.6 proves the optical theorem, a perturbative formulation of unitarity.

## 5.1 Differential Transition Probabilities

The basic measurement in scattering experiments is the number of particles of a given type within a range of momenta that are detected in unit time and unit volume. To obtain a prediction for this measurement, we need a formula involving the square of the scattering amplitude. Since this square contains the square of a delta function, we shall work in the box approximation where a delta function can be squared and then take the limit.

In a box with volume $V$, the states $|\bar{k}_1, \ldots, \bar{k}_n\rangle$ are normalized with Kronecker deltas $\delta_{\bar{k}\bar{k}'}$. We restrict attention to the case where the initial state $|i\rangle$ is either the one-particle state $|\bar{k}\rangle$ for decays or the two-particle state $V^{1/2}|\bar{k}_1, \bar{k}_2\rangle$ for volume-uniform beams in a scattering experiment.

The free field in a box has the form

$$\phi(x) = \sum_{\bar{k}} \left( \frac{e^{ix\cdot k}}{\left(2\omega(\bar{k})V\right)^{1/2}} \, a(\bar{k})^\dagger + \frac{e^{-ix\cdot k}}{\left(2\omega(\bar{k})V\right)^{1/2}} \, a(\bar{k}) \right), \qquad (5.1.1)$$

where $k_0 = \omega(\bar{k})$. The scattering amplitude for an experiment in the box with time duration $T$ is

$$\langle f|S - 1|i\rangle = \left(\prod_{\text{out}} \frac{1}{(2E_b V)^{1/2}}\right)\left(\prod_{\text{in}} \frac{1}{(2E_a)^{1/2}}\right)\frac{1}{V^{1/2}} \tag{5.1.2}$$
$$\times i\mathcal{A}_{T,V}(2\pi)^4 \delta_{T,V}^{(4)}(P_{\text{in}} - P_{\text{out}}).$$

The Feynman rules give rise to amplitudes for matrix elements defined by relativistically normalized bras and kets, but the matrix element here is taken between non-relativistic bras and kets; the first line of the right-hand side comes from the ratio of normalization factors and the special nature of $|i\rangle$.

The transition probability is the square of the amplitude:

$$\text{TP} = \left(\prod_{\text{out}} \frac{1}{2E_b V}\right)\left(\prod_{\text{in}} \frac{1}{2E_a}\right)V^{-1}|\mathcal{A}_{T,V}|^2(2\pi)^8\left(\delta_{T,V}^{(4)}(P_{\text{in}} - P_{\text{out}})\right)^2, \tag{5.1.3}$$

and the differential transition probability is the transition probability multiplied by density of final state factors $V d^3\bar{p}/(2\pi)^3$, one for each particle in the final state:

$$\text{DTP} = \left(\prod_{\text{out}} \frac{d^3\bar{p}_b}{(2\pi)^3}V\right)\left(\prod_{\text{out}} \frac{1}{2E_b V}\right)\left(\prod_{\text{in}} \frac{1}{2E_a}\right)\frac{1}{V}$$
$$\times |\mathcal{A}_{T,V}|^2(2\pi)^8\left(\delta_{T,V}^{(4)}(P_{\text{in}} - P_{\text{out}})\right)^2, \tag{5.1.4}$$

and so the differential transition probability per unit time is

$$\frac{\text{DTP}}{\text{Time}} = \left(\prod_{\text{out}} \frac{d^3\bar{p}_b}{(2\pi)^3 2E_b}\right)\left(\prod_{\text{in}} \frac{1}{2E_a}\right)|\mathcal{A}_{T,V}|^2 \frac{1}{TV}(2\pi)^8\left(\delta_{T,V}^{(4)}(P_{\text{in}} - P_{\text{out}})\right)^2. \tag{5.1.5}$$

The differential transition probability per unit time can be split into two factors, the invariant amplitude squared normalized by the incoming energies and the invariant density of final states $\mathcal{D}$:

$$\frac{\text{DTP}}{\text{Time}} = |\mathcal{A}_{T,V}|^2\left(\prod_{\text{in}} \frac{1}{2E_a}\right)\mathcal{D}, \tag{5.1.6}$$

where

$$\mathcal{D} = \frac{1}{TV}(2\pi)^8\left(\delta_{T,V}^{(4)}(P_{\text{in}} - P_{\text{out}})\right)^2\prod_{\text{out}} \frac{d^3\bar{p}_b}{(2\pi)^3 2E_b}. \tag{5.1.7}$$

To take the continuum limit of these expressions, we need to convert summation into integration. Consider first one spatial dimension. If $L$ is the length of an interval, then the volume of continuous momentum space associated with the momentum of any periodic wave on the interval is $\epsilon = 2\pi/L$. Let $\delta P(k)$ be the interval with length $\epsilon$ centered on momentum $k$. Let $\delta_\epsilon(k)$ be the step function defined by

$$\delta_\epsilon(k - k_0) \stackrel{\text{def}}{=} \begin{cases} \epsilon^{-1}, & \text{if } k \text{ is in } \delta P(k_0); \\ 0, & \text{otherwise.} \end{cases} \tag{5.1.8}$$

If $f$ is a function on one-dimensional discrete momentum space, define a step function representation $f_\epsilon$ of $f$ by

$$f_\epsilon(k) = \epsilon \sum_n f(k_n)\delta_\epsilon(k - k_n). \tag{5.1.9}$$

Then $\delta_\epsilon$ works properly on the step function $f_\epsilon$:

$$\int_V dk\, f_\epsilon(k)\delta_\epsilon(k - k') = f_\epsilon(k'). \tag{5.1.10}$$

The energies $\omega(\bar{k}_{\bar{n}})$ are not rationally related. However, if $T$ is large enough, we can approximate them all as closely as we like. Assume then that we have periodic boundary conditions in time over the interval $T$. The interval in energy space associated with each discrete energy is $\epsilon' = 2\pi/T$. Let $\delta E(k_0)$ be the interval of length $\epsilon'$ centered on the energy $k_0$. Now we can define a step function representation $\delta_{\epsilon'}$ of the Kronecker delta for energy and step function representations $f_{\epsilon'}$ of functions $f$ of energy as we did for momentum above.

In four dimensions of energy-momentum, the step function representation $\delta^{(4)}_{\epsilon'\epsilon}$ of the energy-momentum Kronecker delta is therefore defined by

$$\delta^{(4)}_{\epsilon'\epsilon}(k - k') \overset{\text{def}}{=} \delta_{\epsilon'}(k_0 - k_0')\delta_\epsilon(k_1 - k_1')\delta_\epsilon(k_2 - k_2')\delta_\epsilon(k_3 - k_3'). \tag{5.1.11}$$

The square of this function satisfies

$$\left(\delta^{(4)}_{\epsilon'\epsilon}(k - k')\right)^2 = \frac{1}{\epsilon'\epsilon^3}\delta^{(4)}_{\epsilon'\epsilon}(k - k') = TV(2\pi)^4\delta^{(4)}_{\epsilon'\epsilon}(k - k'). \tag{5.1.12}$$

As $V$ and $T$ tend to infinity, we therefore have the following limits:

$$\mathcal{A}_{T,V} \to \mathcal{A}_{\epsilon'\epsilon} \to \mathcal{A};$$

$$(2\pi)^4\delta^{(4)}_{T,V}(P_{\text{tot}}) \to (2\pi)^4\delta^{(4)}_{\epsilon'\epsilon}(P_{\text{tot}}) \to (2\pi)^4\delta^{(4)}(P_{\text{tot}}); \tag{5.1.13}$$

$$\frac{1}{TV}\left((2\pi)^4\delta^{(4)}_{T,V}(P_{\text{tot}})\right)^2 \to \frac{1}{TV}\left((2\pi)^4\delta^{(4)}_{\epsilon'\epsilon}(P_{\text{tot}})\right)^2 \to (2\pi)^4\delta^{(4)}(P_{\text{tot}}).$$

The formula for the differential transition probability per unit time in the continuum limit follows at once:

**Theorem 5.1.14.** *The differential transition probability per unit time is given by*

$$\frac{\text{DTP}}{\text{Time}} = |\mathcal{A}|^2 \left(\prod_{\text{in}} \frac{1}{2E_a}\right)\mathcal{D},$$

*where*

$$\mathcal{D} = (2\pi)^4\delta^{(4)}(P_{\text{in}} - P_{\text{out}})\prod_{\text{out}} \frac{d^3\bar{p}_b}{(2\pi)^3\, 2E_b}.$$

For the purposes of this book, all we need to know are the formulae for the differential transition probability per unit time and the invariant density of final states presented in Theorem 5.1.14. But it is interesting to see from the derivation of these formulae that the square of a delta function can make sense!

## 5.2 The Invariant Density for Two-Particle Final States

The general principle for simplifying densities is that integration variables which occur in delta functions can be replaced by constraints. For example, if $f$ is a function of two real variables and if $f(x, y) = 0$ has a unique solution $x = h(y)$ for each $y$, then

$$\int dx\, dy\, \delta\big(f(x, y)\big) g(x, y) = \int dx \left| \frac{\partial f}{\partial y} \right|^{-1} g\big(x, h(x)\big). \qquad (5.2.1)$$

It is convenient to write this conclusion without reference to the arbitrary function $g$ or even the integral. The result is an equality of measures:

$$dx\, dy\, \delta\big(f(x, y)\big) = dx \left| \frac{\partial f}{\partial y} \right|^{-1} \quad \text{with } f(x, y) = 0. \qquad (5.2.2)$$

In the case where there are two particles in the final state, we can work in the center of mass frame determined by $\bar{P}_T = \bar{0}$ and simplify the invariant density $\mathcal{D}_2$ of two-particle final states as follows.

The first step is to replace the momentum delta function in Theorem 5.1.14 by a constraint:

$$
\begin{aligned}
\mathcal{D}_2 &= \frac{1}{4E_1 E_2} (2\pi)^4 \delta^{(3)}(\bar{p}_1 + \bar{p}_2) \delta(E_1 + E_2 - E_T) \frac{d^3\bar{p}_1\, d^3\bar{p}_2}{(2\pi)^6} \qquad (5.2.3)\\
&= \frac{1}{16\pi^2 E_1 E_2} \delta(E_1 + E_2 - E_T) d^3\bar{p}_1 \quad \text{(with } \bar{p}_2 \text{ set equal to } -\bar{p}_1\text{)}\\
&= \frac{1}{16\pi^2 E_1 E_2} \delta(E_1 + E_2 - E_T) p^2 dp\, d\Omega,
\end{aligned}
$$

where $p = |\bar{p}_1|$ is the magnitude of the momentum of a single outgoing particle.

The second step is to eliminate the energy delta function. Since

$$E_r^2 = \bar{p}_r^2 + m_r^2 = p^2 + m_r^2 \quad \Longrightarrow \quad \frac{\partial E_r}{\partial p} = \frac{p}{E_r}, \qquad (5.2.4)$$

we have an equality of measures

$$\delta(E_1 + E_2 - E_T)\, dp = \left( \frac{p}{E_1} + \frac{p}{E_2} \right)^{-1} = \frac{E_1 E_2}{E_T} \frac{1}{p}. \qquad (5.2.5)$$

Applying this to the last form of $\mathcal{D}_2$ in 5.2.3, we find that

$$
\begin{aligned}
\mathcal{D}_2 &= \frac{1}{16\pi^2 E_1 E_2} \left( \frac{p}{E_1} + \frac{p}{E_2} \right)^{-1} p^2 d\Omega \quad \text{(with } E_1 + E_2 = E_T\text{)}\\
&= \frac{1}{16\pi^2} \frac{p}{E_T}\, d\Omega.
\end{aligned}
\qquad (5.2.6)
$$

The following theorem summarizes this discussion:

**Theorem 5.2.7.** *In the center of momentum frame, the invariant density of final states for a two-particle final state $|p_1, p_2\rangle$ simplifies to*

$$D_2 = \frac{1}{16\pi^2} \frac{p_{\text{out}}}{E_T} \, d\Omega,$$

*where $p_{\text{out}} = |\bar{p}_1|$ is the magnitude of momentum for either of the outgoing particles, $E_T = p_1^0 + p_2^0$ is the total energy, and the dynamical variables $\bar{p}_1$ and $\bar{p}_2$ are subject to the constraints*

$$\bar{p}_1 + \bar{p}_2 = 0 \quad \text{and} \quad E_1 + E_2 = E_T,$$

*where $E_r^2 = m_r^2 + |\bar{p}_r|^2$.*

**Homework 5.2.8.** Let $\mathcal{A}(p_1 \cdot p_2, p_1 \cdot p_3)$ be an invariant amplitude simplified by use of energy-momentum conservation and mass-shell conditions as explained in Section 4.11. Being invariant, $\mathcal{A}$ is independent of $\phi$. Use the constraints in Theorem 5.2.7 to express $\mathcal{A}$ as a function of $\theta$.

**Homework 5.2.9.** For the scattering of light off charged particles, the lab frame in which the charged particle is at rest is often used. For this we take $q = (\omega, \bar{\omega})$, $q' = (m, \bar{0})$, and $p = (\omega', \bar{\omega}')$. Since we are describing elastic scattering, $p'^2 = m^2$. Define the scattering angle $\theta$ by

$$\omega \omega' \cos \theta = \bar{\omega} \cdot \bar{\omega}'.$$

Show that

$$D_2 = \frac{\omega'^2}{16\pi^2 m\omega} \, d\Omega, \quad \text{where} \quad \omega' = \frac{m\omega}{m + \omega(1 - \cos\theta)}.$$

**Homework 5.2.10.** Assume that the invariant density of final states $D_2$ for a two-particle state $|\bar{q}_1, \bar{q}_2\rangle$ scattering into a two-particle final state $|\bar{p}_1, \bar{p}_2\rangle$ is given by the first line in 5.2.3 above. Define the lab frame by setting the $z$-axis along $\bar{q}_1$ and boosting to $\bar{q}_2 = \bar{0}$. This frame is convenient when a light particle is accelerated into a heavy, stationary target: $q_1^2 = m^2$ and $q_2^2 = M^2$ with $m \ll M$.

Let $p = |\bar{p}_1|$, let $m_1$ be the mass associated with $\bar{p}_1$, and let $P_T = |\bar{q}_1|$. Use the delta function to eliminate variables in $D_2$ to arrive at first,

$$D_2 = \frac{1}{16\pi^2} \frac{p^2}{pE_T - P_T \sqrt{m_1^2 + p^2} \, \cos\theta} \, d\Omega,$$

and second,

$$D_2 = \frac{1}{16\pi^2 P_T} \frac{p}{\sqrt{m_1^2 + p^2}} \, d\phi \, dp.$$

In both cases, give the formulae for the variables which have been eliminated in terms of the variables appearing in the differentials. ($E_T$ and $P_T$ are parameters of the initial state and we observe either $\theta$ and $\phi$ or $p$ and $\phi$.)

## 5.3 The Invariant Density for Three-Particle Final States

For three-particle final states, the invariant density of states takes the form

$$
\begin{aligned}
\mathcal{D}_3 &= \frac{1}{(2\pi)^5 8 E_1 E_2 E_3} \delta^{(4)}(p_1 + p_2 + p_3 - P_T)\, d^3\bar{p}_1\, d^3\bar{p}_2\, d^3\bar{p}_3 \\
&= \frac{1}{256\pi^5 E_1 E_2 E_3} \delta(E_1 + E_2 + E_3 - E_T)\, p_1^2 dp_1\, d\Omega_1\, p_2^2 dp_2\, d\Omega_2,
\end{aligned}
\tag{5.3.1}
$$

with the substitution $\bar{p}_3 = \bar{P}_T - \bar{p}_1 - \bar{p}_2$.

Switch to the c.o.m. frame, so that $\bar{P}_T = 0$. Now, if we use Euler angles $\phi_{12}$ and $\theta_{12}$ with the direction of motion $\bar{p}_1$ of particle one as $z$-axis, then we can use this conservation of momentum condition to express $\theta_{12}$ in terms of $E_3$. First, replace $d\Omega_2$:

$$
d\Omega_2 = d\phi_{12}\, d\cos\theta_{12}.
\tag{5.3.2}
$$

Second, view momentum conservation as a constraint on $E_3$ to find a link between $E_3$ and $\cos\theta_{12}$:

$$
\begin{aligned}
E_3^2 &= p_3^2 + m_3^2 = p_1^2 + p_2^2 + 2p_1 p_2 \cos\theta_{12} + m_3^2 \\
&\implies \quad dE_3 = \frac{p_1 p_2}{E_3}\, d\cos\theta_{12}.
\end{aligned}
\tag{5.3.3}
$$

Third, eliminate $\cos\theta_{12}$ from $\mathcal{D}_3$ and perform the integral $dE_3$ with the help of the remaining delta function to find:

$$
\begin{aligned}
\mathcal{D}_3 &= \frac{1}{256\pi^5 E_1 E_2}\, p_1 dp_1\, p_2 dp_2\, d\Omega_1\, d\phi_{12} \\
&= \frac{1}{256\pi^5}\, dE_1\, dE_2\, d\Omega_1\, d\phi_{12},
\end{aligned}
\tag{5.3.4}
$$

with the additional condition $E_3 = E_T - E_1 - E_2$.

The integral of $\mathcal{D}_3$ over the angular variables $\Omega_1$ and $\phi_{12}$ is a uniform distribution on the allowed domain of the $E_1 E_2$ plane:

$$
\int_{\Omega_1, \phi_{12}} \mathcal{D}_3 = \frac{1}{32\pi^3} dE_1\, dE_2.
\tag{5.3.5}
$$

Hence, if we integrate the DTP per unit time over the angular variables $\Omega_1$ and $\phi_{12}$ and plot the result as a density function on the $E_1 E_2$ plane, we can compare this density plot directly with experimental data. An experimental or theoretical plot of this kind is known as a *Dalitz plot*.

Our analysis of Lorentz-invariant functions in Section 4.11 shows that an amplitude $\mathcal{A}_4$ which depends on four energy-momentum vectors can be reduced to a function $\mathcal{A}_4'$ of two scalar variables if energy-momentum conservation and mass-shell constraints are in force. If $\mathcal{A}_4'$ can be interpreted as one particle decaying into three, since an invariant amplitude will never depend on $\Omega_1$ or $\phi_{12}$, a contour plot of $|\mathcal{A}_4'|^2$ will be a Dalitz plot. (In this case, the Dalitz plot is to decay as the Mandelstam plot is to scattering.)

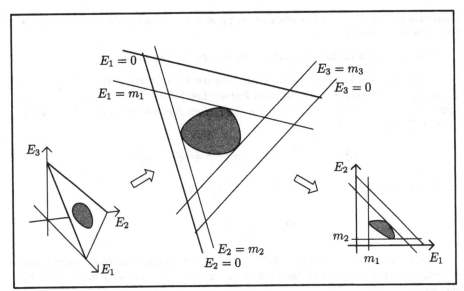

**Fig. 5.3.** The central diagram shows the Dalitz plot in triangular coordinates. Here the three energies $E_r$ are measured perpendicular to the lines $E_r = 0$. They are constrained by the obvious inequalities $E_r \geq m_r$ and also by triangle inequalities which determine the shaded oval. The diagram on the left gives the origin of the triangle in three dimensions, and that on the right gives the usual rectilinear form of the Dalitz plot.

The dynamical variables for a three-particle final state are $\bar{p}_1$, $\bar{p}_2$, and $\bar{p}_3$. Momentum conservation eliminates $\bar{p}_3$. The shell conditions determine the energies, $E_r^2 = m_r^2 + |\bar{p}_r|^2$, and so energy conservation eliminates one more, leaving five independent variables.

Choose the $z$-axis to lie along $\bar{p}_1$. This eliminates the angular variables $\Omega_1$. Choose the $y$-axis so that $\bar{p}_2$ lies in the $yz$-plane and has positive $y$-coordinate. This eliminates the variable $\phi_{12}$, leaving two independent variables. We shall choose $E_1$ and $E_2$ as the independent variables so that the density 5.3.5 is uniform.

Given $E_1$ and $E_2$, we determine the other variables as follows:

1. Use energy conservation to determine $E_3$;
2. Use shell conditions to determine $|\bar{p}_1|$, $|\bar{p}_2|$, and $|\bar{p}_3|$;
3. Use the choice of axes to determine $\bar{p}_1$, $\bar{p}_2$, and $\bar{p}_3$.

For the last step to be possible, $E_1$ and $E_2$ must be chosen in such a way that the $|\bar{p}_r|$ satisfy the triangle inequalities:

$$|\bar{p}_1| \leq |\bar{p}_2| + |\bar{p}_3| \quad \text{etc.} \tag{5.3.6}$$

In terms of $E_1$, $E_2$, and $E_3$, these become

$$(E_1^2 - m_1^2)^{1/2} \leq (E_2^2 - m_2^2)^{1/2} + (E_3^2 - m_3^2)^{1/2} \quad \text{etc.} \tag{5.3.7}$$

The intersection of these three constraints defines the boundary of the Dalitz plot in the $E_1$-$E_2$ plane. Figure 5.3 shows the Dalitz plot in $E_1E_2E_3$ space, in triangular coordinates, and in rectangular coordinates.

**Homework** 5.3.8. Find the boundary of the physical region in the Dalitz plot when the three masses $m_r$ are all zero.

The discussion in this section may be summarized as follows:

**Theorem 5.3.9.** *Consider the invariant density $\mathcal{D}_3$ of three-particle final states $|\bar{p}_1, \bar{p}_2, \bar{p}_3\rangle$ in the center of momentum frame. Let $E_r$ be the energy $\omega(\bar{p}_r)$. Then energy conservation $E_1 + E_2 + E_3 = E_T$ determines $E_3$ as a function of $E_1$ and $E_2$, $E_1$ and $E_2$ satisfy the constraints*

$$(E_r^2 - m_r^2)^{1/2} \leq (E_s^2 - m_s^2)^{1/2} + (E_t^2 - m_t^2)^{1/2}$$

*for all permutations $(r, s, t)$ of $(1, 2, 3)$, and $\mathcal{D}_3$ simplifies to*

$$\mathcal{D}_3 = \frac{1}{32\pi^3} \, dE_1 \, dE_2.$$

*Furthermore, using the mass-shell conditions and momentum conservation, any Lorentz-invariant function of the vectors $p_r$ can be expressed as a function of $E_1$ and $E_2$.*

# 5.4 Decays

Though our adiabatic turning on and off function prohibits introducing initial states which can decay, this is a good place to introduce decay rates. In our current formalism, decay of initial states is not a problem at tree level: it is an obstacle to the formulation of renormalization conditions. In due course, we shall change our viewpoint on renormalization and in the process remove this weakness from our theory.

In any case, the probability $d\Gamma$ of a single incoming particle decaying into a final state $|f\rangle$ per unit time is simply

$$d\Gamma_f(E_i) \stackrel{\text{def}}{=} \frac{\text{DTP}}{\text{Time}} = \frac{1}{2E_i} |\mathcal{A}_f|^2 \mathcal{D}, \tag{5.4.1}$$

where $E_i$ is the energy of the particle. The corresponding decay width $\Gamma$ is defined by

$$\Gamma(E_i) = \frac{1}{2E_i} \sum_{\substack{\text{decay} \\ \text{modes}}} \int_{\substack{\text{final} \\ \text{momenta}}} |\mathcal{A}_f|^2 \mathcal{D}. \tag{5.4.2}$$

Of course, we can apply the formulae of Theorems 5.2.7, 5.3.4, and 5.3.9 for $\mathcal{D}$ when computing decay rates and widths. In particular, we have:

**Theorem 5.4.3.** *The decay width for a particle of mass $M$ in its rest frame decaying into a two-particle final state $|\bar{p}_1, \bar{p}_2\rangle$ with specific particle types is given by*

$$\Gamma = \frac{p_{\text{out}}}{8\pi M^2} |\mathcal{A}|^2,$$

*where $p_{\text{out}} = |\bar{p}_1|$.*

In the absence of a background field, the decay rate for a particle of mass $M$ in its rest frame decaying into a three-particle final state $|\bar{p}_1, \bar{p}_2, \bar{p}_3\rangle$ with specific particle types is given by

$$d\Gamma = \frac{1}{64\pi^3 M} |\mathcal{A}|^2 \, dE_1 \, dE_2,$$

where $E_r$ is the energy of the particle associated with $\bar{p}_r$.

Note that, on account of the factor $1/2E_i$, neither $d\Gamma$ nor $\Gamma$ is Lorentz invariant. Indeed, since decay width $\Gamma$ and half-life $\tau$ are related by

$$\tau = \frac{\hbar}{\Gamma}, \tag{5.4.4}$$

the factor of $1/2E_i$ correctly gives a particle greater life expectancy at higher momentum.

**Homework 5.4.5.** The most common decay mode of the short-lived neutral kaon $K_S$ (mass 498 MeV) is into two charged pions (mass 145 MeV). For this process $\Gamma$ is $0.776 \times 10^{10}$ sec$^{-1}$. Use this data to find $|\mathcal{A}|$. Make the unrealistic assumption that the decay is modeled by $\mathcal{H}_I = \lambda \psi^\dagger \psi \phi$ with the pion and kaon represented by $\psi$ and $\phi$ respectively, and compute the value of the dimensionless quantity $\lambda/m_K$ to one significant figure. (Work in the center of mass frame and take $\hbar$ to be $6.58 \times 10^{-22}$ MeV sec.)

**Homework 5.4.6.** Let $A$, $B$, $C$ and $D$ be scalar fields with dynamics determined by the Lagrangian density

$$\mathcal{L} = \tfrac{1}{2}\left(\partial_\mu A \, \partial^\mu A + \partial_\mu B \, \partial^\mu B + \partial_\mu C \, \partial^\mu C + \partial_\mu D \, \partial^\mu D - m^2 D^2\right) + gABCD.$$

Compute the decay width of $D$ to lowest non-vanishing order in perturbation theory.

## 5.5 Cross Sections

The differential cross section gives the probability of a two-particle incoming state $|\bar{q}_1, \bar{q}_2\rangle$ scattering into a region of $n$-particle momentum space per unit time:

$$\begin{aligned} d\sigma &\overset{\text{def}}{=} \frac{\text{DTP}}{\text{Flux} \times \text{Time}} \\ &= \frac{1}{|\bar{v}_1 - \bar{v}_2|} \frac{1}{4E_1 E_2} |\mathcal{A}|^2 \mathcal{D} \\ &= \frac{1}{4|E_1 \bar{q}_2 - E_2 \bar{q}_1|} |\mathcal{A}|^2 \mathcal{D}, \end{aligned} \tag{5.5.1}$$

since $\bar{v}_r = \bar{q}_r / E_r$.

**Homework 5.5.2.** Use the results of Homework 5.2.10 to give two forms of $d\sigma$ appropriate to a two-particle final state in the lab frame.

In the c.o.m. frame, $\bar{q}_2 = -\bar{q}_1$. Write $p_{\text{in}}$ for $|\bar{q}_1|$. Then

$$|E_1 \bar{q}_2 - E_2 \bar{q}_1| = p_{\text{in}}(E_1 + E_2) = p_{\text{in}} E_T, \tag{5.5.3}$$

and so

$$d\sigma = \frac{1}{4E_T p_{\text{in}}} |\mathcal{A}|^2 \mathcal{D}. \tag{5.5.4}$$

Note that, given the mass-shell conditions, $E_T$ can be determined from $p_{\text{in}}$ and *vice versa*.

The differential cross section $d\sigma$ has units of area. Comparison with Rutherford's classical model for a beam of particles scattering off a particle target, we can interpret $d\sigma$ as the area of the beam that is scattered into the infinitesimal volume of apparatus represented by $\mathcal{D}$. This mixture of classical and quantum thinking is loose but helpful.

The cross section $\sigma$ is obtained by integrating $d\sigma$ over all outgoing momenta. The resulting number is the probability that incoming particles at a given momentum will scatter into a final state made up of specific particles at arbitrary momenta.

We can find the angular density of scattering into a two-particle state in the c.o.m. frame by substituting the formula of Theorem 5.2.7 for $\mathcal{D}$ in 5.5.4 above:

$$\frac{d\sigma}{d\Omega} = \frac{1}{64\pi^2 E_T^2} \frac{p_{\text{out}}}{p_{\text{in}}} |\mathcal{A}|^2. \tag{5.5.5}$$

This density blows up as $p_{\text{in}} \to 0$. The formula is then trying to describe two particles sitting together in space. The source of the divergence is the flux factor, which is obviously meaningless in this limit.

Note that, given the mass-shell conditions $p_r^2 = m_r^2$ and energy-momentum conservation, the value of any one of $E_T$, $p_{\text{in}}$, and $p_{\text{out}}$ determines the values of the other two:

$$(p_{\text{in}}^2 + m_1^2)^{1/2} + (p_{\text{in}}^2 + m_2^2)^{1/2} = E_T = (p_{\text{out}}^2 + m_3^2)^{1/2} + (p_{\text{out}}^2 + m_4^2)^{1/2}. \tag{5.5.6}$$

An invariant amplitude can be expressed as a function of $\theta$ and any one of these parameters.

**Homework 5.5.7.** Show that, in elastic scattering where the incoming particles are the same as the outgoing particles, $p_{\text{out}} = p_{\text{in}}$.

**Homework 5.5.8.** Using the interaction Hamiltonian density $\mathcal{H}_I = \lambda\psi^\dagger\psi\phi$ of Section 4.10 and working in the center of momentum frame at lowest non-trivial order, show that the differential cross section $d\sigma$ for $N\bar{N} \to N\bar{N}$ scattering is given by

$$\frac{d\sigma}{d\Omega} = \frac{\lambda^4}{256\pi^2(p^2 + m^2)} \left( -\frac{1}{2p^2(1 - c) + \mu^2} + \frac{1}{4(p^2 + m^2) - \mu^2} \right)^2,$$

where $p = p_{\text{in}}$, $c = \cos\theta$, and $\theta$ is the angle through which $N$ is scattered. Sketch the graph of $d\sigma/d\Omega$ as a function of $\theta$. Compute the cross section for this process.

**Remark 5.5.9.** In the preceding homework, the annihilation diagram contributes a $\theta$-independent piece to $d\sigma$, indicating that the direction of the incoming particles is forgotten in the annihilation process. This term becomes infinite when the total energy $2\sqrt{p^2 + m^2}$ is equal to the mass $\mu$, which is the very energy where the annihilation of the $N\bar{N}$ pair could produce a physical $\Phi$ particle. This event could only occur if the $\Phi$ could decay into an $N\bar{N}$ pair, and the condition for this decay to be possible is $\mu \leq 2m$. In general then, when decays are possible, tree-level annihilation diagrams can give meaningless amplitudes. The solution to this breakdown of our scattering theory is the inclusion of loop diagrams and renormalization. As we shall see, the $\mu^2$ representing an unstable particle in its propagator should have a negative imaginary part.

**Remark 5.5.10.** Also in the preceding homework, the $\Phi$-exchange diagram produces a contribution to $d\sigma$ which has the form of a Rutherford scattering prediction with a Yukawa potential. As the $\Phi$ mass $\mu$ goes to zero, $d\sigma$ becomes singular at $\theta = 0$ and $\sigma$ diverges, indicating that $\Phi$ exchange becomes a long-range force like electromagnetism. ($\Phi$ exchange is, however, a purely attractive force.)

It is generally convenient to express amplitudes as functions of $E_T$ and $\theta$. Writing $|\bar{p}_1, \bar{p}_2\rangle$ and $|\bar{p}_3, \bar{p}_4\rangle$ for the initial and final state respectively, towards this end we find repeated application of the formulae for $p_r \cdot p_s$ and $(p_r \pm p_s)^2$. In our applications of the differential cross section 5.5.5, we shall generally simplify these formulae by assuming the total energy is so high that masses are negligible.

In the center of momentum frame, we can then give specific form to the $p_r$ as follows:

$$p_1 \overset{\text{def}}{=} (p, \bar{p}), \qquad p_2 = (p, -\bar{p}), \qquad p = |\bar{p}|;$$
$$p_3 \overset{\text{def}}{=} (p, \bar{p}'), \qquad p_4 = (p, -\bar{p}'), \qquad p = |\bar{p}'|. \tag{5.5.11}$$

We define $\theta$ in the range $[0, \pi]$ by

$$\bar{p} \cdot \bar{p}' = p^2 \cos \theta \tag{5.5.12}$$

and compute the various scalars:

$$
\begin{aligned}
p_1 \cdot p_2 &= p_3 \cdot p_4 = 2p^2, & (p_1 + p_2)^2 &= (p_3 + p_4)^2 = 4p^2, \\
p_1 \cdot p_3 &= p_2 \cdot p_4 = p^2(1 - \cos \theta), & (p_1 - p_3)^2 &= (p_2 - p_4)^2 = -2p^2(1 - \cos \theta), \\
p_1 \cdot p_4 &= p_2 \cdot p_3 = p^2(1 + \cos \theta), & (p_1 - p_4)^2 &= (p_2 - p_3)^2 = -2p^2(1 + \cos \theta),
\end{aligned} \tag{5.5.13}
$$

where $p = p_{\text{in}} = p_{\text{out}} = \frac{1}{2} E_T$.

In non-relativistic elastic scattering, $p_{\text{out}} = p_{\text{in}}$, and so

$$\frac{d\sigma}{d\Omega} = |f(E_T, \cos \theta)|^2, \tag{5.5.14}$$

implying

$$f(E_T, \cos \theta) \propto \frac{1}{8\pi E_T} \mathcal{A}_{fi}. \tag{5.5.15}$$

The constant of proportionality, a phase factor, turns out to be unity.

When the final state is a two-particle state of one Bose field, it is symmetric under interchange of these particles, and the two orientations $(\theta, \phi)$ and $(\pi - \theta, \pi + \phi)$ represent the same final state. Hence, in this case, when we integrate to obtain the cross section we must compensate for counting each final state twice by dividing the integral by two. (Obviously, this point generalizes to symmetric many-particle final states.)

**Homework 5.5.16.** Using $\mathcal{H}_I = \lambda \psi^\dagger \psi \phi$ again, compute to lowest non-vanishing order in $\lambda$ the c.o.m. differential cross section for $N\bar{N} \to \Phi\Phi$. Show that the cross section is given by

$$\sigma = \frac{\lambda^4}{256\pi p^2(p^2 + m^2)(2p^2 + 2m^2 - \mu^2)} \left( \frac{2\alpha}{1 - \alpha^2} + \ln \frac{1 + \alpha}{1 - \alpha} \right),$$

where $p = p_{\text{in}}$ and $\alpha$ is a parameter

$$\alpha = \frac{2p(p^2 + m^2 - \mu^2)^{1/2}}{2p^2 + 2m^2 - \mu^2},$$

which satisfies $0 \leq \alpha < 1$.

For three-particle final states in the c.o.m. frame, when there is a background field, we use the formula 5.3.4 for $\mathcal{D}_3$ and substitute this in 5.5.4 to find the differential cross section:

$$d\sigma = \frac{1}{256\pi^5} \frac{1}{4E_T p_{\text{in}}} |\mathcal{A}|^2 \, dE_1 \, dE_2 \, d\Omega_1 \, d\phi_{12}, \qquad (5.5.17)$$

or, for a Lorentz-invariant experiment, use the value of $\mathcal{D}$ in Theorem 5.3.9 to find

$$d\sigma = \frac{1}{32\pi^3} \frac{1}{4E_T p_{\text{in}}} |\mathcal{A}|^2 \, dE_1 \, dE_2, \qquad (5.5.18)$$

where $p_{\text{in}}$ is the magnitude of momentum of one incoming particle, $E_1$ and $E_2$ are the energies of two of the outgoing particles, $\Omega_1$ is the orientation of the $E_1$ particle, and $\phi_{12}$ is the angle between the momenta of the $E_1$ and $E_2$ particles.

The main results of this section are:

**Theorem 5.5.19.** *In the center of momentum frame, the differential cross section for two particles scattering into two particles is given by*

$$d\sigma = \frac{1}{64\pi^2 E_T^2} \frac{p_{\text{out}}}{p_{\text{in}}} |\mathcal{A}|^2 \, d\Omega,$$

*where $p_{\text{in}}$ and $p_{\text{out}}$ are respectively the magnitude of momentum of a single incoming particle and a single outgoing particle.*

*In the center of momentum frame, in the absence of background fields, the differential cross for two particles scattering into three particles is given by*

$$d\sigma = \frac{1}{128\pi^3 E_T p_{\text{in}}} |\mathcal{A}|^2 \, dE_1 \, dE_2,$$

*where $p_{\text{in}}$ is the magnitude of momentum of a single incoming particle and $E_1$ and $E_2$, the energies of two of the final-state particles, are subject to the triangle inequalities as explained in the derivation of Theorem 5.3.9.*

## 5.6 The Optical Theorem

Clearly the total cross section can be deduced by measuring the unscattered wave and the forward scattered wave. In fact, the imaginary part of the forward wave is enough to determine the cross section, as we shall now show.

**Optical Theorem 5.6.1.** *The imaginary part of the forward wave determines the cross section for an initial two-particle state $|i\rangle$ in the c.o.m. frame through the equation*

$$2E_T p_{\text{in}} \sigma = \text{Im}\,\mathcal{A}_{ii},$$

*where $p_{\text{in}}$ is the magnitude of momentum for an incoming particle and $\sigma$ is the cross section for $|i\rangle$ to scatter into all possible final states.*

**Proof.** Let $|i\rangle = |p_1, p_2\rangle$ and $|f\rangle = |p_1', p_2'\rangle$. Working in the c.o.m. frame, the invariant amplitude $\mathcal{A}_{fi}$ is given by

$$\langle f|(S-1)|i\rangle = i\mathcal{A}_{fi}(2\pi)^4 \delta^{(4)}(P_{\text{out}} - P_{\text{in}}). \qquad (5.6.2)$$

The $S$ matrix is unitary, and so

$$(S-1)^\dagger(S-1) = S^\dagger S - S^\dagger - S + 1 = -\left((S-1)^\dagger + (S-1)\right). \tag{5.6.3}$$

Computing the $\langle f|\text{-}|i\rangle$ matrix element of this equation enables us to connect a sum of amplitudes to sums of products of amplitudes:

$$(i\mathcal{A}_{fi} - i\mathcal{A}_{if}^*)(2\pi)^4\delta^{(4)}(P_f - P_i) \tag{5.6.4}$$

$$= \langle f|(S-1)|i\rangle + \left(\langle f|(S-1)|i\rangle\right)^*$$

$$= -\langle f|(S-1)^\dagger(S-1)|i\rangle$$

$$= -\sum_n \frac{1}{n!}\int\prod\frac{d^3\bar{q}_r}{(2\pi)^3 2E_r}\,\langle f|(S-1)^\dagger|q_1,\ldots,q_n\rangle\langle q_1,\ldots,q_n|(S-1)|i\rangle$$

$$= -\sum_n \frac{1}{n!}\int\prod\frac{d^3\bar{q}_r}{(2\pi)^3 2E_r}\,\mathcal{A}_{nf}^*(2\pi)^4\delta^{(4)}(P_f - P_n)\mathcal{A}_{ni}(2\pi)^4\delta^{(4)}(P_n - P_i)$$

$$= -\sum_n \frac{1}{n!}\int\prod\frac{d^3\bar{q}_r}{(2\pi)^3 2E_r}\,\mathcal{A}_{nf}^*\mathcal{A}_{ni}(2\pi)^8\delta^{(4)}(P_f - P_i)\delta^{(4)}(P_n - P_i)$$

$$= -(2\pi)^4\delta^{(4)}(P_f - P_i)\sum_n \frac{1}{n!}\int\mathcal{D}_n\,\mathcal{A}_{nf}^*\mathcal{A}_{ni}.$$

Canceling the common factors leaves

$$\sum_n\frac{1}{n!}\int\mathcal{D}_n\,\mathcal{A}_{nf}^*\mathcal{A}_{ni} = -(i\mathcal{A}_{fi} - i\mathcal{A}_{if}^*), \tag{5.6.5}$$

from which we deduce the optical theorem by setting $f = i$ and recalling formula 5.5.4 for the differential cross section. $\qquad\square$

From this proof we see how the unitarity of the $S$ matrix is reflected in a property of invariant amplitudes. This property, order by order in perturbation theory, provides a test for the consistency of the Feynman rules.

**Homework 5.6.6.** A loop diagram has a non-zero imaginary part if it can be cut into two tree diagrams both of which could represent physical scattering processes. With $\mathcal{H}_I = \lambda\psi^\dagger\psi\phi$, write down the second-order diagrams $D_r$ that contribute to the cross section for an initial state $N\bar{N}$. Representing the amplitude derived from $D_r$ by the diagram $D_r$ itself, write down the cross section. See how squaring the amplitudes effectively combines the diagrams into six one-loop diagrams which contribute to $N\bar{N}$ forward scattering. Show that the other four connected fourth-order diagrams which contribute to this process have no imaginary part.

**Remark 5.6.7.** The principle of the previous homework follows directly from the Feynman rules by integration over loop energies. For example, the diagram

$$D = \tag{5.6.8}$$

can be cut into two copies of

$$D' = \tag{5.6.9}$$

The associated transformation of Feynman integrals follows.

Applying the Feynman rules to $D$, we obtain the forward amplitude

$$i A(k^2) = (-i\lambda)^2 \int \frac{d^4 p}{(2\pi)^4} \frac{d^4 q}{(2\pi)^4} (2\pi)^4 \delta^{(4)}(p+q-k) \frac{i}{p^2 - m^2 + i\epsilon} \frac{i}{q^2 - m^2 + i\epsilon}. \quad (5.6.10)$$

An expression for the imaginary part of this forward amplitude follows immediately:

$$\mathrm{Im}\, A = \frac{\lambda^2}{2} \int \frac{d^4 p}{(2\pi)^4} \frac{d^4 q}{(2\pi)^4} (2\pi)^4 \delta^{(4)}(p+q-k) \quad (5.6.11)$$

$$\times \left( \frac{i}{p^2 - m^2 + i\epsilon} \frac{i}{q^2 - m^2 + i\epsilon} + \frac{i}{p^2 - m^2 - i\epsilon} \frac{i}{q^2 - m^2 - i\epsilon} \right).$$

Making the substitution

$$(2\pi)^4 \delta^{(4)}(p+q-k) = \int d^4 x\, e^{ix \cdot (p+q-k)} \quad (5.6.12)$$

and using contour integration to evaluate the integrals over energies (as in Homework 2.4.30), we find that

$$\mathrm{Im}\, A = \frac{\lambda^2}{2} \int \frac{d^3 \bar{p}}{(2\pi)^3} \frac{d^3 \bar{q}}{(2\pi)^3} \frac{1}{2\omega(\bar{p})} \frac{1}{2\omega(\bar{q})} (2\pi)^4 \delta^{(3)}(\bar{p} + \bar{q} - \bar{k}) \delta\big(\omega(\bar{p}) + \omega(\bar{q}) - k_0\big). \quad (5.6.13)$$

The integrand here is simply the invariant density for a two-particle final state. As the final state is bosonic, there is a factor of one-half associated with the integral. Finally, the factor $\lambda^2$ is the square of the amplitude for the decay diagram $D'$.

The non-relativistic form of the optical theorem is

$$f(E_T, \cos\theta) = \frac{1}{8\pi E_T} A_{ii} \quad \Longrightarrow \quad \mathrm{Im}\, f(E_T, 1) = \frac{p_{\mathrm{in}}\sigma}{4\pi}. \quad (5.6.14)$$

Note that the forward amplitude $A_{ii}$ for a single particle state is proportional to $S - 1$, so that $A_{ii}$ is naturally associated with off-diagonal or decay activity. The following homework converts this observation into a formula.

**Homework 5.6.15.** Prove a version of the optical theorem for one-particle initial states $|i\rangle$: $E_i \Gamma = \mathrm{Im}\, A_{ii}$.

**Remark 5.6.16.** From Homework 5.6.15 we find that if a particle can decay, then its forward scattering amplitude has a positive imaginary part. As we shall see in Chapter 10, this aspect of the forward scattering amplitude generates a modification of the propagator for the scalar field:

$$\frac{i}{k^2 - \mu^2 + i\epsilon} \rightarrow \frac{i}{k^2 - \mu^2 + i\mu\Gamma}.$$

With such a modification, the meaningless pole problem described in Remark 5.5.9 will not arise.

**Remark 5.6.17.** We saw in Section 4.6 that the real part of the vacuum bubble contribution to the $S$ matrix exemplified by $S(C_2)$ was equal to $-\alpha/2$, where $\alpha$ was the expected particle number of the final state. Now $S(C_2)$ corresponds to $iA_{00}$, where $A_{00}$ is the vacuum to vacuum amplitude. We have seen that $A_{ii}$ has positive imaginary part when $|i\rangle$ is either a one-particle state which can decay or a two-particle state. It now appears that $A_{00}$ also has a positive imaginary part when the vacuum can decay. This suggests the following homework question.

**Homework** 5.6.18. In light of the foregoing remark, what is the analogue of the optical theorem for the vacuum to vacuum amplitude $\mathcal{A}_{00}$ when there is a background field which makes vacuum decay possible?

# 5.7 Summary

In this chapter, we have taken the abstract material of the previous chapters and converted it into numbers directly related to measurable phenomena. The central concept is the differential transition probability per unit time (DTP/T), which gives the rate at which particles with specific momenta should be detected. In fact the DTP/T is our only contact with experimental data. All decay rates and differential cross sections are deduced from it. This chapter effectively establishes the outpost of theory in prediction. Further material on prediction will simply be applications of the same ideas to more complicated theories.

This ends Part 1 of the text, concluding the development of scalar field theory from its foundations in special relativity and quantum mechanics to its applications in decay rates and cross sections. At this stage, however, we have no realistic models of elementary particles. In Part 2 – Chapters 6 to 9 – we apply the same quantization technique to spinor and vector fields and present QED with a massive photon, which is a realistic theory if the mass is small enough.

In the next chapter, we study representations of the Lorentz group as a preliminary to introducing fields with non-trivial Lorentz transformation properties. Among such fields we find the fermions and vector bosons with which to build not only QED but also the Standard Model of the electroweak and strong interactions.

# Chapter 6

# Representations of the Lorentz Group

A mathematical introduction to representation theory of Lie groups and Lie algebras providing details of the irreducible representations of the Lorentz groups in preparation for theories of spinor fields, vector fields, and their interactions.

## Introduction

This Chapter begins the second part of the text, which comprises Chapters 6 to 9. Here, we develop the representation theory of the Lorentz group, thereby providing the spinor and vector fields which the following chapters quantize. Part 2 introduces all the types of quantum field used in the Standard Model.

At this point, we have quite a well-developed view of scalar field theory. Scalar fields, however, are necessarily bosonic. If we tried to change their statistics, we would find that the Hamiltonian is no longer bounded below. Furthermore, scalar fields do not have a polarization and so cannot represent the electromagnetic field.

The problem is the Lorentz transformation law of Axiom 3, Section 2.4, which only permits scalar fields. Now we want to change this axiom and allow the Lorentz group to mix field components. We will see just how many possibilities there are and find candidates for fermion and photon fields among these possibilities.

This chapter is a fairly mathematical one. The reason for this is that representation theory is mainly linear algebra, and so it is convenient to clarify the aspects of representation theory used in physics by plunging into the appropriate mathematical terminology. Section 6.1 begins the dive into mathematics by using what we know of the unitary representation of the Lorentz group on Fock space to identify the representation property of the infinitesimal action of the Lorentz group on the vector space of fields. Section 6.2 presents the language of Lie algebra representation theory. Section 6.3 shows that the Lorentz Lie algebra comprises two copies of $su(2)$. This motivates the thorough analysis of the structure of the representations of $su(2)$ described in Sections 6.4, 6.5, and 6.6.

The next phase is the application of what we know of $su(2)$ to the Lorentz Lie algebra. Section 6.7 gives a complete description of the irreducible representations of the Lorentz Lie algebra, and Section 6.8 investigates the splitting of the tensor products of these representations. The last two sections, Sections 6.9 and 6.10, cover details of the theory useful for applications in later chapters, namely the effects of complex conjugation, parity, and restriction to rotations on representations of the Lorentz Lie algebra, and symmetric and anti-symmetric rank-two tensors. With all these facts, the building of Lorentz-invariant Lagrangian densities with fields that transform non-trivially under the action of the Lorentz Lie algebra will be easy.

The method of analysis that we use in this chapter is central to the understanding of symmetry groups in grand unified theories and string theories. It will be useful for any particle theorist. The conclusions for $su(2)$ and $so(1,3)$, however, may be understood on the basis of the definitions alone.

# 6.1 From Lie Group to Lie Algebra Representations

The action of the Lorentz group on states is given by the unitary representation

$$\Lambda \cdot |k\rangle = U(\Lambda)|k\rangle = |\Lambda k\rangle. \tag{6.1.1}$$

The Heisenberg view of this transformation is

$$\phi^a(x) \mapsto U(\Lambda)^\dagger \phi^a(x) U(\Lambda). \tag{6.1.2}$$

The hypothesis that the fields transform into each other is expressed in the statement

$$U(\Lambda)^\dagger \phi^a(x) U(\Lambda) = D^a{}_b(\Lambda) \phi^b(\Lambda^{-1}x). \tag{6.1.3}$$

To understand this better, we shall write out the implication of this hypothesis for the transformation of matrix elements. Define the matrix element $\mathcal{M}^a$ by

$$\mathcal{M}^a(f, x, i) \overset{\text{def}}{=} \langle f|\phi^a(x)|i\rangle. \tag{6.1.4}$$

If we rotate the incoming state, the outgoing state, and the point of evaluation all in the same way, then the polarization of the matrix element should change accordingly:

$$
\begin{aligned}
\mathcal{M}^a(\Lambda \cdot f, \Lambda x, \Lambda \cdot i) &= \langle \Lambda \cdot f|\phi^a(\Lambda x)|\Lambda \cdot i\rangle \\
&= \langle f|U(\Lambda)^\dagger \phi^a(\Lambda x) U(\Lambda)|i\rangle \\
&= \langle f|D^a{}_b(\Lambda) \phi^b(x)|i\rangle \\
&= D^a{}_b(\Lambda)\langle f|\phi^b(x)|i\rangle \\
&= D^a{}_b(\Lambda)\mathcal{M}^b(f, x, i).
\end{aligned}
\tag{6.1.5}
$$

Notice that the first line gives the Lorentz action on the matrix element $\mathcal{M}^a(f, x, i)$, and the subsequent lines simply transfer the action to the index $a$. The fact that this is possible makes $a$ a Lorentz index and indicates that all together the matrix elements form a covariant Lorentz tensor.

The question is, what matrices $D^a{}_b(\Lambda)$ can occur? The principle restriction on these matrices follows from the unitary representation condition 6.1.3:

$$
\begin{aligned}
D^a{}_c(\Lambda\Lambda')\phi^c(\Lambda'^{-1}\Lambda^{-1}x) &= U(\Lambda\Lambda')^\dagger \phi^a(x) U(\Lambda\Lambda') \\
&= U(\Lambda')^\dagger U(\Lambda)^\dagger \phi^a(x) U(\Lambda) U(\Lambda') \\
&= U(\Lambda')^\dagger D^a{}_b(\Lambda) \phi^b(\Lambda^{-1}x) U(\Lambda') \\
&= D^a{}_b(\Lambda) U(\Lambda')^\dagger \phi^b(\Lambda^{-1}x) U(\Lambda') \\
&= D^a{}_b(\Lambda) D^b{}_c(\Lambda') \phi^c(\Lambda'^{-1}\Lambda^{-1}x),
\end{aligned}
\tag{6.1.6}
$$

which implies that

$$D^a{}_c(\Lambda\Lambda') = D^a{}_b(\Lambda) D^b{}_c(\Lambda'). \tag{6.1.7}$$

Thus the map $D$ represents the elements of the group as matrices and the multiplication in the group as matrix multiplication. This multiplicative property of $D$ makes it a matrix representation of the Lorentz group.

Our problem then is to find all matrix representations of the Lorentz group. As we remarked in Section 3.4, it is simpler to work with linear properties than quadratic ones. We shall use differentiation to transform the quadratic representation condition 6.1.7 which applies on the group level into a linear condition which will apply on the Lie algebra level. The first step is to linearize the representation, the second, to linearize the representation condition.

To linearize a representation, we simply differentiate it and evaluate the derivative at the identity element of the group. Since the tangent space at the identity is the Lie algebra, the derivative is naturally a linear map from the Lie algebra to a space of matrices.

Recall that $X$ is in $so(1,3)$ if and only if it satisfies $X^\top G + GX = 0$, where $G = \text{diag}(1, -1, -1, -1)$. Any element of the Lorentz group $SO(1,3)$ that is close to $1$ can be written as the exponential of a Lie algebra element. Hence we can define the linear version $\bar{D}$ of $D$ by evaluating $D$ on $\Lambda = \exp(\tau X)$, differentiating with respect to $\tau$, and evaluating at $\tau = 0$:

$$\bar{D}^a{}_b(X) \stackrel{\text{def}}{=} \frac{\partial}{\partial \tau} D^a{}_b(e^{\tau X}) \Big|_{\tau=0}. \tag{6.1.8}$$

Note that, since the unitary representation condition 6.1.3 is linear in $D$, the matrices $\bar{D}^a{}_b(X)$ act on the fields by matrix multiplication.

In order to complete the transfer of attention from representations of the Lorentz group to maps of its Lie algebra, we need to find the condition on a Lie algebra map that characterizes maps like $\bar{D}$ which come from representations of the group. As it is the multiplicative structure of the group that underlies the representation condition 6.1.7, we need to find the corresponding structure in the Lie algebra. Then we can use this structure to define representations of the Lie algebra without reference to the group.

In the following two subsections, we shall find the structure of the Lorentz Lie algebra which represents multiplication in the Lorentz group. To do this, we define three levels of an action of the Lorentz group on itself. All three levels are known as adjoint maps, each level being simply a derivative of the one before. It may be helpful to refer to Figure 6.1 for an overview of the geometry and concepts presented in these subsections.

To generate a structure in the Lie algebra from group multiplication, we make the group act on itself in such a way that the identity element in the group (the origin of the Lie algebra) is not moved. Thus, for each $\Lambda$ in $SO(1,3)$, we define the adjoint action $\mathbf{Ad}_\Lambda$ of $SO(1,3)$ on itself by

$$\begin{aligned}\mathbf{Ad}_\Lambda &: SO(1,3) \to SO(1,3), \\ M &\mapsto \Lambda M \Lambda^{-1}.\end{aligned} \tag{6.1.9}$$

**Homework 6.1.10.** Show that $\mathbf{Ad}_\Lambda$ is multiplicative:

$$\mathbf{Ad}_\Lambda(MM') = \mathbf{Ad}_\Lambda(M)\,\mathbf{Ad}_\Lambda(M');$$

and that $\mathbf{Ad}$ is also multiplicative:

$$\mathbf{Ad}_{\Lambda\Lambda'} = \mathbf{Ad}_\Lambda\,\mathbf{Ad}_{\Lambda'}.$$

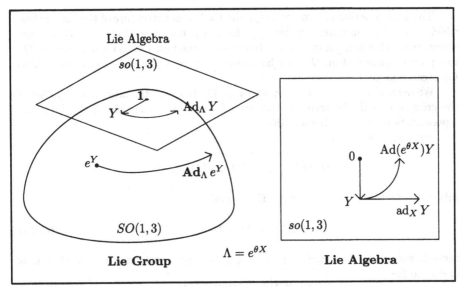

**Fig. 6.1.** The left half of the figure illustrates the relationship between the group $SO(1,3)$ and its Lie algebra $so(1,3)$, the tangent space to $SO(1,3)$ at the identity. The point $Y$ in $so(1,3)$ maps to $e^Y$ in $SO(1,3)$; these points are rotated by $\mathbf{Ad}_\Lambda$ and $\mathbf{Ad}_\Lambda$ respectively. The right half represents $so(1,3)$ and illustrates the relationship between Ad and ad.

Let $\mathbf{D}_\tau$ be the operation of differentiating with respect to $\tau$ and then evaluating at $\tau = 0$. Then we can find the infinitesimal version of **Ad** in the Lie algebra by applying $\mathbf{D}_\tau$ twice as follows.

First, make the adjoint act on $\exp(\tau Y)$ instead of $M$ in 6.1.9 and apply $\mathbf{D}_\tau$ to define the second level of the adjoint, the adjoint action of the $SO(1,3)$ on its Lie algebra $so(1,3)$:

$$\mathbf{Ad}_\Lambda(Y) \stackrel{\text{def}}{=} \mathbf{D}_\tau \, \mathbf{Ad}_\Lambda(e^{\tau Y})$$
$$= \Lambda \mathbf{D}_\tau e^{\tau Y} \Lambda^{-1} \qquad (6.1.11)$$
$$= \Lambda Y \Lambda^{-1}.$$

**Homework 6.1.12.** Show that Ad is multiplicative, i.e., that $\mathbf{Ad}_{\Lambda\Lambda'} = \mathbf{Ad}_\Lambda \mathbf{Ad}_{\Lambda'}$.

**Homework 6.1.13.** Show that $\mathbf{Ad}_\Lambda \exp(Y) = \exp(\mathbf{Ad}_\Lambda Y)$.

Second, make Ad act on $\exp(\tau X)$ instead of $\Lambda$ and apply $\mathbf{D}_\tau$ to bring out the third level of the adjoint, the adjoint action of $so(1,3)$ on itself:

$$\mathrm{ad}_X Y \stackrel{\text{def}}{=} \mathbf{D}_\tau \, \mathbf{Ad}_{\exp(\tau X)}(Y)$$
$$= \mathbf{D}_\tau e^{\tau X} Y e^{-\tau X} \qquad (6.1.14)$$
$$= XY - YX.$$

This combination of $X$ and $Y$ is also written as a Lie bracket:

$$[X, Y] \stackrel{\text{def}}{=} XY - YX. \qquad (6.1.15)$$

**Homework 6.1.16.** Show that $\mathbf{Ad}_{\exp(X)} Y = \exp(\mathrm{ad}_X Y)$. Hint: the traditional method is to substitute $\tau X$ for $X$ and then show that both sides of the resulting hypothesis satisfy the same initial value problem.

The adjoint action is summarized in the Lie bracket structure of the Lie algebra, which is itself an infinitesimal form of the group multiplication law. This is the structure which a map on the Lie algebra should preserve. Let us therefore apply $\mathbf{D}_\tau$ to a group representation $D$ to see how this structure is preserved by the associated Lie algebra map $\bar{D}$.

We derived the Lie bracket by applying $\mathbf{D}_\tau$ twice to the adjoint action $\mathbf{Ad}$ of the group on itself. To arrive at the desired equation, we simply apply the group representation $D$ to $\mathbf{Ad}$ before differentiating.

First, starting from

$$D(\Lambda M \Lambda^{-1}) = D(\Lambda)D(M)D(\Lambda)^{-1}, \tag{6.1.17}$$

substitute $M = \exp(\tau Y)$ and apply $\mathbf{D}_\tau$ to find:

$$\bar{D}(\Lambda Y \Lambda^{-1}) = D(\Lambda)\bar{D}(Y)D(\Lambda)^{-1}. \tag{6.1.18}$$

Second, substitute $\Lambda = \exp(\tau X)$ and apply $\mathbf{D}_\tau$ again (bearing in mind that $\bar{D}$ is linear) to find:

$$\bar{D}([X,Y]) = [D(X), D(Y)]. \tag{6.1.19}$$

We conclude that the Lie algebra map $\bar{D}$ satisfies

$$\bar{D}([X,Y]) = [\bar{D}(X), \bar{D}(Y)]. \tag{6.1.20}$$

This equation shows what it means for a map to preserve a Lie bracket structure. Thus, a linear matrix-valued map on a Lie algebra is called a *Lie algebra representation* if it has the property 6.1.20. The following homework shows that ad is a Lie algebra representation.

**Homework** 6.1.21. Show that the Jacobi identity

$$\big[X,[Y,Z]\big] = \big[[X,Y],Z\big] + \big[Y,[X,Z]\big]$$

holds whenever the Lie bracket is defined by a commutator as in 6.1.15.

Starting from the most general form 6.1.3 for the action of the Lorentz group on the fields in a quantum field theory, we derived the notion of a matrix representation of the group at 6.1.7 and, by differentiating to bring attention to the group generators, discovered the Lie algebra structure of $so(1,3)$ and the notion of a matrix representation of Lie algebras, 6.1.20. This shows that matrix representations of the Lorentz group generate matrix representations of its Lie algebra. We will find it simple to understand the matrix representations of the group when we have classified the representations of its Lie algebra.

## 6.2 Definitions for Representation Theory

In order to understand and analyze representations, it will be necessary to have a good grasp of the language of Lie algebra representation theory. In this section, we simply give the necessary definitions. These definitions are couched in the language of abstract linear algebra in order to draw attention to the essential structure and away from coordinate-dependent details. We begin with the definition of a Lie algebra:

**Definition 6.2.1.** A *Lie algebra* is a vector space $\mathcal{G}$ with scalars $\mathbf{F}$ together with a bracket operation $[\,,\,]$ which, for all $X$, $Y$, and $Z$ in $\mathcal{G}$, and scalars $a$ and $b$, satisfies:

1. Anti-Symmetry: $[X, Y] = -[Y, X]$;
2. Linearity: $[aX + bY, Z] = a[X, Z] + b[Y, Z]$;
3. Jacobi Identity: $\big[X, [Y, Z]\big] = \big[[X, Y], Z\big] + \big[Y, [X, Z]\big]$.

The basic example of a Lie algebra is the vector space $gl(V)$ of linear transformations from a vector space $V$ to itself. If $V$ has dimension $n$, and if we choose a basis for $V$, then this space $gl(V)$ becomes the space of $n \times n$ matrices $M_n(\mathbf{R})$ or $M_n(\mathbf{C})$ depending on whether $V$ is a real or complex vector space. Notice, however, that the definition of a Lie algebra simply prescribes algebraic structure and makes no reference to matrices.

**Definition 6.2.2.** The natural relationships between Lie algebras are like those between vector spaces, namely:

1. A *Lie subalgebra* of a Lie algebra $\mathcal{G}$ is a subspace of $\mathcal{G}$ which is closed under the bracket operation.
2. A *direct sum* of Lie algebras $\mathcal{G} \oplus \mathcal{G}'$ is the direct sum of the vector spaces equipped with the bracket operation

$$[(X, X'), (Y, Y')] = ([X, Y], [X', Y']). \tag{6.2.3}$$

3. A *homomorphism* of Lie algebras from $\mathcal{G}$ to $\mathcal{G}'$ is a linear map $D$ on the vector spaces which is compatible with the two bracket structures:

$$D([X, Y]) = [D(X), D(Y)]. \tag{6.2.4}$$

4. An *isomorphism* of Lie algebras is a Lie algebra homomorphism which is $1 - 1$ and onto.

The Lie algebras $so(n)$ and $su(n)$ are Lie subalgebras of $M_n(\mathbf{R})$ and $M_n(\mathbf{C})$ respectively. Also $M_n(\mathbf{R})$ is a Lie subalgebra of $M_n(\mathbf{C})$ considered as a real Lie algebra.

**Homework 6.2.5.** What is the natural bracket which makes the vector space of linear differential operators $X(x) = a^r(x)\partial_r$ on $\mathbf{R}^n$ a Lie algebra?

Note that there is no tensor product of Lie algebras. There is no way of defining a new bracket operation on the tensor product of the underlying vector spaces in terms of the original bracket operations.

**Definition 6.2.6.** A *representation* of a Lie algebra $\mathcal{G}$ on a vector space $V$ is a Lie algebra homomorphism $D$ from $\mathcal{G}$ to $gl(V)$. The Lie algebra is said to *act* on $V$ through $D$. The *degree* of the representation is the dimension of $V$.

The representation of a Lie algebra $\mathcal{G}$ on $V$ which maps every element of $\mathcal{G}$ to the zero transformation of $V$ is called the *trivial representation*.

A representation $D$ effectively transforms the abstract Lie algebra $\mathcal{G}$ into a subalgebra of $gl(V)$. In fact, through the Lie bracket, a Lie algebra can generate a representation of itself on itself as vector space:

**Definition 6.2.7.** The *adjoint* representation of a Lie algebra $\mathcal{G}$ is the map ad from $\mathcal{G}$ to $gl(\mathcal{G})$ given by $\mathrm{ad}_X(Y) = [X, Y]$.

**Homework 6.2.8.** Verify that the adjoint satisfies the representation condition

$$[\mathrm{ad}_X, \mathrm{ad}_Y] = \mathrm{ad}_{[X,Y]}.$$

When we think of a representation, we should not think of any particular set of matrices but only of the abstract pattern of the Lie algebra $\mathcal{G}$ acting on an $n$-dimensional vector space. More precisely, we are not interested in representations *per se* but in equivalence classes of representations, where equivalence is defined as follows:

**Definition 6.2.9.** Two representations $D$ and $F$ of a Lie algebra $\mathcal{G}$ on vector spaces $U$ and $V$ are *equivalent* or *isomorphic* if there exists a linear isomorphism $T$ from $U$ to $V$ such that

$$F(X)T = TD(X).$$

If we choose bases for $U$ and $V$, then this definition of equivalence of representations becomes similarity of matrices. Thus equivalent representations $F$ and $D$ are really just two superficially different forms of the same algebraic entity. We should be aware, then, that the essential algebraic structure of a Lie algebra representation is a structure in common to all similar representations.

To understand all representations, it is sufficient to understand how an arbitrary representation is a simple combination of simple parts. The simple means for combining representations is the direct sum:

**Definition 6.2.10.** A *direct sum* of representations $D_1$ and $D_2$ of a Lie algebra $\mathcal{G}$ on vector spaces $V_1$ and $V_2$ respectively is the representation $D = D_1 \oplus D_2$ of $\mathcal{G}$ on $V_1 \oplus V_2$ given by

$$D(X) = \big(D_1(X), D_2(X)\big).$$

Seen in reverse, if we start with one representation $D$ and find a way of expressing it as a direct sum of two representations, then we have a split $D$:

**Definition 6.2.11.** A representation $D$ of $\mathcal{G}$ on $V$ *splits* if it is isomorphic to the direct sum of two non-trivial representations.

When a representation $D$ of $\mathcal{G}$ on $V$ splits, it means that $V$ splits into a direct sum of two subspaces $V = V_1 + V_2$ in such a way that the action of $\mathcal{G}$ on $V$ does not mix $V_1$ with $V_2$:

$$v_r \in V_r \quad \Longrightarrow \quad D(X)v_r \in V_r \quad \text{for } r = 1, 2 \text{ and all } X \in \mathcal{G}. \qquad (6.2.12)$$

In other words, with respect to the decomposition $V = V_1 + V_2$, the linear transformations $D(X)$ are block diagonal for all $X \in \mathcal{G}$. Reversing our perspective again, if we are trying to find a splitting of $D$, it is often simplest to construct the corresponding splitting of $V$; and to split $V$, we begin by finding candidates for the $V_r$:

**Definition 6.2.13.** An *invariant* subspace of a representation $D$ of $\mathcal{G}$ on a vector space $V$ is a subspace $U$ of $V$ which has the property that

$$u \in U \quad \Longrightarrow \quad D(X)u \in U.$$

When the representation is obvious from the context, one just refers to 'invariant subspaces' and omits the qualifiers. The trivial invariant subspaces of a representation $D$ of $\mathcal{G}$ on $V$ are the two improper subspaces $\{0\}$ and $V$ of $V$.

Once we have an invariant subspace, we also have a natural representation of the Lie algebra on it:

**Definition 6.2.14.** A *subrepresentation* of a representation $D$ of a Lie algebra $\mathcal{G}$ on a vector space $V$ is the representation $D_U$ of $\mathcal{G}$ on an invariant subspace $U$ of $D$ obtained by restricting the linear transformations $D(X)$ to $U$:

$$u \in U \quad \Longrightarrow \quad D_U(u) = D(u).$$

We have described splitting of a representation in terms of direct sum of representations and direct sum of invariant subspaces. We can also describe splitting in terms of direct sum of subrepresentations. Clearly, all three descriptions are equivalent:

**Theorem 6.2.15.** *A representation $D$ of $\mathcal{G}$ on $V$ splits if and only if $D$ has a pair of non-trivial invariant subspaces $U$ and $U'$ which are complementary (i.e. $V = U + U'$ and $U \cap U' = \{0\}$).*

**Homework 6.2.16.** Define a representation $D$ of $su(2)$ on $V = M_4(\mathbf{C})$ by $D(X)M = XM - MX$. Use symmetry of $M$ to split $D$ into a sum of three subrepresentations.

It can happen that there is one invariant subspace $U$ for a representation $D$ of $\mathcal{G}$ on $V$, but no complementary invariant subspace. Consider, for example, the representation of $\mathcal{G} = \mathbf{R}$ on $V = \mathbf{R}^2$ given by

$$D(x) = \begin{pmatrix} 1 & x \\ 0 & 1 \end{pmatrix}. \tag{6.2.17}$$

Representations in which this happens can be very hard to analyze. To avoid this situation without seriously restricting the range of application in physics, we shall focus on unitary representations, defined as follows:

**Definition 6.2.18.** A representation $D$ of $\mathcal{G}$ on $V$ is a *unitary* representation if firstly $V$ is a complex vector space, secondly there is an inner product $\langle \, | \, \rangle$ on $V$, and thirdly the linear transformations $D(X)$ are anti-hermitian with respect to this inner product:

$$\langle D(X)u|v \rangle + \langle u|D(X)v \rangle = 0,$$

for all $u$, $v$ in $V$ and $X$ in $\mathcal{G}$.

**Homework 6.2.19.** Show that, if $D$ is a unitary representation of $\mathcal{G}$ on $V$, and if $v_1, \ldots, v_n$ is an orthonormal basis for $V$, then the matrix representing $D(X)$ with respect to the given basis is anti-hermitian for all elements $X$ in the Lie algebra.

**Homework 6.2.20.** Show that there is no non-trivial unitary representation of $sl(2, \mathbf{R})$, the Lie algebra of $2 \times 2$ real, traceless matrices.

It is useful to know that any representation of $so(n)$ or $su(n)$ on any vector space $V$ can be made unitary by constructing the right inner product on $V$. The relevance of unitarity to splitting is summarized in the following result:

**Theorem 6.2.21.** *If $D$ is a unitary representation of $\mathcal{G}$ on $V$, and if $U$ is an invariant subspace of $V$, then the orthogonal complement $U^\perp$ of $U$ in $V$ is a complementary invariant subspace.*

**Homework 6.2.22.** Use trace to construct an inner product on $su(n)$ which makes the adjoint representation unitary.

The concept of splitting a representation of $\mathcal{G}$ is central to classifying representations. If we start with an arbitrary representation and keep splitting it until the subrepresentations will split no more, we have expressed the original representation as a sum of elementary building blocks, the irreducible representations:

**Definition 6.2.23.** An *irreducible* representation is a representation with no non-trivial invariant subspaces.

**Homework 6.2.24.** Show that the adjoint representation of $so(n)$ is irreducible.

The simplest Lie algebras are the commutative ones:

**Definition 6.2.25.** An *abelian* or *commutative* Lie algebra is a Lie algebra $\mathcal{G}$ in which the bracket is trivial: $[X, Y] = 0$ for all $X$ and $Y$ in $\mathcal{G}$.

A representation $D$ of a commutative Lie algebra $\mathcal{G}$ on a vector space $V$, given bases $X_1, \ldots, X_k$ for the Lie algebra and $v_1, \ldots, v_n$ for $V$, is determined by an indexed set of commuting matrices, $M_r = \left[ D(X_r) \right]$. Matrices are not in general diagonalizable, but an anti-hermitian matrix can be diagonalized; indeed, any commuting family of anti-hermitian matrices can be simultaneously diagonalized. Hence, if $D$ is a unitary representation, and if the basis $v_1, \ldots, v_n$ is such that all the matrices $M_r$ are diagonal, then the one-dimensional subspaces $V_r$ determined by the basis vectors $v_r$ are invariant, and so we have a splitting of $D$ into $n$ one-dimensional irreducible subrepresentations:

**Theorem 6.2.26.** *Any unitary representation of a commutative Lie algebra is a direct sum of one-dimensional subrepresentations.*

**Homework 6.2.27.** Find the one-dimensional subrepresentations of the representation of $\mathbf{R}$ on $\mathbf{C}^2$ given by

$$D(t) = \begin{pmatrix} 0 & -t \\ t & 0 \end{pmatrix}.$$

Since the unitary representations of commutative Lie algebras are so simple, it makes sense to develop representation theory by connecting arbitrary Lie algebras to commutative ones. The essential concept is the Cartan subalgebra:

**Definition 6.2.28.** A *Cartan subalgebra* of a Lie algebra $\mathcal{G}$ is a maximal commuting Lie subalgebra of $\mathcal{G}$.

For any one of the Lie algebras of immediate interest to us, $u(n)$, $su(n)$, $so(n)$, and $so(1, 3)$, it is easy to show that its Cartan subalgebras all have the same dimension. This result shows that the dimension of a Cartan subalgebra often depends only on the Lie algebra and is therefore characteristic of the Lie algebra.

**Definition 6.2.29.** If all the Cartan subalgebras of a Lie algebra have the same dimension, then this dimension is called the *rank* of the Lie algebra.

**Homework 6.2.30.** Find Cartan subalgebras of $u(n)$, $su(n)$, $so(n)$, and $so(1,3)$. Show that these subalgebras have dimensions $n$, $n-1$, $[n/2]$ (the greatest integer less than $n/2$), and 2 respectively.

Suppose we have a representation $D$ of a Lie algebra $\mathcal{G}$ on a vector space $V$ and a Cartan subalgebra $\mathcal{H}$ of $\mathcal{G}$. The next two definitions show how the weight diagram of $D$ emerges from $D$ itself when we focus attention on the behavior of $D$ on $\mathcal{H}$.

First, the process of taking attention from the representation $D$ to the corresponding representation $D'$ of $\mathcal{H}$ can be made precise as follows:

**Definition 6.2.31.** The *restriction* of a representation $D$ of a Lie algebra $\mathcal{G}$ to a subalgebra $\mathcal{G}'$ of $\mathcal{G}$ is the representation $D'$ of $\mathcal{G}'$ defined by $D'(X) = D(X)$ for $X \in \mathcal{G}'$.

Second, we can summarize the consequences of applying the splitting theorem 6.2.26 to the restriction $D'$ of $D$ to $\mathcal{H}$ in a definition of a weight diagram:

**Definition 6.2.32.** Let $D$ be a unitary representation of $\mathcal{G}$ on $V$, let $\mathcal{H}$ be a Cartan subalgebra of $\mathcal{G}$ with basis $h_1, \ldots, h_k$, and let $v_1, \ldots, v_n$ be a basis of $V$ which diagonalizes all the $D(h)$'s simultaneously:

$$D(h_j)v_r = \lambda_{jr}v_r.$$

Then the *weights* of $D$ are the $k$-vectors $\lambda_r = (\lambda_{1r}, \ldots, \lambda_{kr})$. The set of these $n$ weights is called the *weight diagram* of the representation $D$.

In Section 3.11, we saw that a choice of Cartan subalgebra corresponded to a choice of commuting conserved quantities in field theory, and the diagonalizing vectors ($v_r$ above) corresponded to particle states or fields with well-defined quantum numbers. Now we see further, that a weight in representation theory corresponds to a set of quantum numbers of a quantum field, or of a particle state in quantum field theory.

Notice that the detailed structure of the weight diagram, its scale and orientation for example, depends on how we choose the bases for the Cartan subalgebra $\mathcal{H}$ and the vector space $V$; but the geometry of the whole weight diagram is really independent of these choices.

**Homework 6.2.33.** Find the weights of the representation of $so(4)$ on symmetric, traceless complex matrices $\mathbf{C}^9 \subset M_4(\mathbf{C})$ given by $D(X): M \mapsto XM - MX$.

The central concept in splitting a weight diagram is that of raising and lowering operators:

**Definition 6.2.34.** The *raising* and *lowering operators* of a Lie algebra $\mathcal{G}$ are the eigenvectors of the adjoint representation of the Lie algebra $\mathcal{G}$ restricted to the Cartan subalgebra $\mathcal{H}$.

The raising and lowering are generally complex linear combinations of elements of $\mathcal{G}$, so at this point we need $\mathcal{G}$ to be a complex vector space. We can compute the eigenvectors of the adjoint representation restricted to $\mathcal{H}$ either in $\mathcal{G}$ or in a representation $D$. Generally it is convenient to work in a representation and think of all operators as linear transformations of $V$ or even as matrices. To distinguish raising from lowering we say that one weight $w = (w_1, \ldots, w_k)$ is larger than another $w' = (w'_1, \ldots, w'_k)$ if the first component where $w$ and $w'$ are different satisfies $iw_r > iw'_r$. Raising operators raise weights, and lowering operators lower them, behaving like the raising and lowering operators of a quantum oscillator.

**Homework 6.2.35.** Find the raising and lowering operators for $su(3)$ with respect to the Cartan subalgebra of diagonal matrices in $su(3)$. Find the weights of the vector representation of $su(3)$ on $\mathbf{C}^3 - D(X) = X -$ and check that your raising and lowering operators work correctly!

The technique for splitting a representation on a vector space $V$ is to construct an invariant subspace from a vector $v \in V$ with maximal weight by applying all possible sequences of lowering operators to it and taking the span $U$ of the resulting vectors. Since our representation is unitary, the orthogonal complement $U^\perp$ is also invariant. If $U^\perp$ is not just the zero vector, we have a splitting and can apply the same steps to $U^\perp$. If $U^\perp$ is just the zero vector, then $U = V$ and $V$ is irreducible.

Because of the central role of the adjoint representation in representation theory, the weights of the adjoint representation have a distinguishing name:

**Definition 6.2.36.** The *roots* of a Lie algebra are the weights of the adjoint representation. The *root diagram* of a Lie algebra is the weight diagram of its adjoint representation.

These definitions give us the necessary language for a discussion of representations. As we analyze $su(2)$ and $so(1,3)$, we shall unfold the details of meaning as we need them.

The definitions above are strictly mathematical, intended to bring out the linear algebra underlying Lie algebra representation theory. Before we can comfortably apply these definitions to physics, we have to address two concerns.

The first concern comes from the use of anti-hermitian matrices: if we choose an orthogonal basis for the vector space $V$ in Definition 6.2.18 of a unitary representation, then a unitary representation $D$ would give rise to anti-hermitian matrices $\big[ D(X) \big]$. This is natural because the anti-hermitian matrices $u(n)$ are the generators of the unitary group $U(n)$. But anti-hermitian matrices have complex eigenvalues and are therefore a factor of $i$ away from quantum observables. To obtain real eigenvalues, it is necessary to work with hermitian matrices. To obtain hermitian matrices from anti-hermitian ones, it is necessary to multiply by $i$. The question is, where should we introduce this factor?

The second concern is that our presentation is coordinate-free, but physicists generally work in coordinates. Thus, a physicist views an abstract Lie algebra $\mathcal{G}$ through a basis $X_1, \ldots, X_n$, and specifies the bracket operation by a tensor:

**Definition 6.2.37.** The *structure constants* of a Lie algebra $\mathcal{G}$ with respect to a chosen basis $X_1, \ldots, X_l$ constitute the tensor $f_{rst}$ defined by the equation:

$$[X_r, X_s] = \sum_t f_{rst} X_t.$$

Furthermore, since a Lie algebra $\mathcal{G}$ is a vector space and its representations are linear transformations, a representation of a Lie algebra is determined by its values on a basis for the Lie algebra. Thus, to determine a representation $D$ of $\mathcal{G}$ on $V$ it is enough to specify the linear transformations $D(X_r)$ for a basis $X_1,\ldots,X_l$ of $\mathcal{G}$. If we choose a basis $v_1,\ldots,v_n$ for $V$, then $D(X_r)$ is represented by a matrix $M_r = [D(X_r)]$. Now the Lie algebra representation condition becomes a condition on the matrices $M_r$:

$$[X_r,X_s] = \sum_t f_{rst} X_t \quad \Longrightarrow \quad M_r M_s - M_s M_r = \sum_t f_{rst} M_t. \qquad (6.2.38)$$

With this coordinate viewpoint on a representation, it is easy to introduce the factor of $i$ which will resolve the first concern – multiply $M_r$ by $i$ and require a modified representation condition:

**Definition 6.2.39.** A *hermitian* representation of a Lie algebra defined by structure constants $f_{rst}$ is a set of hermitian matrices $M_r$ which satisfy the conditions

$$M_r M_s - M_s M_r = i \sum_t f_{rst} M_t.$$

Note that, since the structure constants are basis dependent, a hermitian representation cannot be defined without a choice of basis for the Lie algebra.

**Homework 6.2.40.** What are the structure constants of $su(2)$ with respect to the basis $-i\sigma_r$, where the $\sigma$'s are the Pauli matrices defined in the next section? Show that the Pauli matrices are a hermitian representation of $su(2)$.

The last definition is the starting point of most discussions in physics texts and the foundation for the rest of the chapter. When necessary, it is a simple matter to introduce bases and coordinates, and thereby translate all the definitions and all the little results of this section into the language of matrices. Yet it is advantageous to understand the coordinate-free basis for representation theory as set out above, because such a level of understanding will keep the mind clear and purposeful even in the midst of complicated calculations.

## 6.3 Detailed Structure of $su(2)$ and $so(1,3)$

The Lie algebra $su(2)$ is central in the analysis of all the Lie algebras derived from matrix groups. The key to understanding $su(2)$ is knowing the structure of its bracket action on itself. For our analysis of this action, and for connecting $so(1,3)$ with $su(2)$, it will be helpful to specify a real basis for the $2 \times 2$ complex matrices:

$$\begin{pmatrix} -i & 0 \\ 0 & -i \end{pmatrix} \quad \boxed{ v_1 = \begin{pmatrix} 0 & -i \\ -i & 0 \end{pmatrix} \quad v_2 = \begin{pmatrix} 0 & -1 \\ 1 & 0 \end{pmatrix} \quad v_3 = \begin{pmatrix} -i & 0 \\ 0 & i \end{pmatrix} } \quad su(2)$$

$$\boxed{ \sigma_0 = \begin{pmatrix} 1 & 0 \\ 0 & 1 \end{pmatrix} } \quad \boxed{ \sigma_1 = \begin{pmatrix} 0 & 1 \\ 1 & 0 \end{pmatrix} \quad \sigma_2 = \begin{pmatrix} 0 & -i \\ i & 0 \end{pmatrix} \quad \sigma_3 = \begin{pmatrix} 1 & 0 \\ 0 & -1 \end{pmatrix} } \quad \mathbf{R}^{1,3}$$

$$sl(2,\mathbf{C})$$

The top row of this table is a basis for the anti-hermitian matrices, while the bottom row is a basis for the hermitian ones. The bottom row is obtained from the top by multiplying by $i$. $\bar{\sigma}$ are the Pauli matrices. The value of representing Minkowski space by the vector space of hermitian matrices will be developed in Section 7.2. The three matrices $v_1$, $v_2$, $v_3$ satisfy:

$$v_r v_s + v_s v_r = -2\delta_{rs} 1_2 \qquad \text{and} \qquad v_r v_s - v_s v_r = 2\epsilon_{rs}{}^t v_t. \qquad (6.3.1)$$

Since $\sigma_r$ is $iv_r$, it follows that:

$$\sigma_r \sigma_s + \sigma_s \sigma_r = 2\delta_{rs} 1_2 \qquad \text{and} \qquad \sigma_r \sigma_s - \sigma_s \sigma_r = 2i\epsilon_{rs}{}^t \sigma_t. \qquad (6.3.2)$$

Absorb the factors of two by defining $\rho_r$ and $\tau$ by $\rho_r = \frac{1}{2}\sigma_r$ and $\tau_r = \frac{1}{2}v_r$.

To make the connection between $su(2)$ and $so(3)$, we shall first show that the adjoint action of $su(2)$ on itself is actually an isomorphism from $su(2)$ to $so(3)$. Indeed, since $su(2)$ has dimension 3, in the adjoint action of $su(2)$ on itself, the $2 \times 2$ matrices of $su(2)$ are transformed into $3 \times 3$ matrices. The Lie algebra structure, however, does not change. In detail, $\tau_r$ acts on generators $X$ in $su(2)$ by

$$ad(\tau_r)X = [\tau_r, X]. \qquad (6.3.3)$$

Let $X = x^s \tau_s$. Then:

$$\begin{aligned} ad(\tau_r)X = [\tau_r, X] &= [\tau_r, x^s \tau_s] \\ &= x^s [\tau_r, \tau_s] = x^s \epsilon_{rs}{}^t \tau_t. \end{aligned} \qquad (6.3.4)$$

Writing out the details for $r = 1$, we find:

$$\begin{aligned} ad(\tau_1)\tau_1 &= 0\tau_1 + 0\tau_2 + 0\tau_3 \\ ad(\tau_1)\tau_2 &= 0\tau_1 + 0\tau_2 + 1\tau_3 \\ ad(\tau_1)\tau_3 &= 0\tau_1 - 1\tau_2 + 0\tau_3 \end{aligned} \qquad (6.3.5)$$

from which we see that, with respect to the basis $\tau_r$, $ad(\tau_r)$ is represented by $3 \times 3$ matrices:

$$ad(\tau_1) = \begin{pmatrix} 0 & 0 & 0 \\ 0 & 0 & -1 \\ 0 & 1 & 0 \end{pmatrix}, \quad ad(\tau_2) = \begin{pmatrix} 0 & 0 & 1 \\ 0 & 0 & 0 \\ -1 & 0 & 0 \end{pmatrix}, \quad ad(\tau_3) = \begin{pmatrix} 0 & -1 & 0 \\ 1 & 0 & 0 \\ 0 & 0 & 0 \end{pmatrix}. \qquad (6.3.6)$$

Notice that $su(2)$ is a real vector space with basis $\tau_1$, $\tau_2$, and $\tau_3$. If we multiply these matrices by complex numbers and take sums, then we get the $sl(2, \mathbf{C})$ matrices. $sl(2, \mathbf{C})$ is the Lie algebra of $SL(2, \mathbf{C})$, the group of unit-determinant $2 \times 2$ complex matrices. $\{\tau_1, \tau_2, \tau_3\}$ is a basis for $sl(2, \mathbf{C})$ as a complex vector space. In general, given a real vector space $V$, multiplication by complex numbers has no meaning. However, since $\mathbf{C}$ is a two-dimensional real vector, we can always form the tensor product of real vector spaces $\mathbf{C} \otimes V$ (which has real dimension $2 \times \dim V$) and multiply vectors in this new vector space by complex numbers using the definition

$$\alpha(\beta \otimes v) \overset{\text{def}}{=} (\alpha\beta) \otimes v. \qquad (6.3.7)$$

Thus $\mathbf{C} \otimes V$ is a complex vector space with complex dimension dim $V$. It is called the *complexification* of $V$. Using this definition, we see that $sl(2, \mathbf{C})$ is the complexification of $su(2)$, but that the complexification of $sl(2, \mathbf{C})$ has real dimension 12 and is something new.

If we have a representation $D$ of a real Lie algebra $\mathcal{G}$ on a complex vector space $V$, then $D$ extends to a unique complex linear representation $\hat{D}$ of the complexification $\mathbf{C} \otimes \mathcal{G}$ of $\mathcal{G}$:

$$\hat{D}(X + iY) \stackrel{\text{def}}{=} D(X) + iD(Y). \tag{6.3.8}$$

If the vector space $V$ is real, it too will have to be complexified to $\mathbf{C} \otimes V$. In either case, $\hat{D}$ is called the *holomorphic extension* of $D$. The characteristic properties of $\hat{D}$ are, first, it agrees with $D$ on $\mathcal{G}$, and second, it is complex linear.

The process of complexification and holomorphic extension is usually so simple that one does not notice that anything has been done. One effectively identifies $V$ with $\mathbf{C} \otimes V$ and $D$ with $\hat{D}$. Thus $su(2)$ is commonly treated as its complexification $sl(2, \mathbf{C})$. In the case of $\mathbf{C} \otimes sl(2, \mathbf{C})$, however, it is essential to avoid mixing the complex numbers in $\mathbf{C}$ with those in $sl(2, \mathbf{C})$.

Our interest in complexifying $su(2)$ arises from the fact that the complex Lie algebra $\mathbf{C} \otimes so(1,3)$ is isomorphic to the direct sum $\mathbf{C} \otimes su(2) \oplus \mathbf{C} \otimes su(2)$. Define a basis for $so(1,3)$ by:

$$X_1 = \begin{pmatrix} 0 & 0 & 0 & 0 \\ 0 & 0 & 0 & 0 \\ 0 & 0 & 0 & -1 \\ 0 & 0 & 1 & 0 \end{pmatrix} \quad X_2 = \begin{pmatrix} 0 & 0 & 0 & 0 \\ 0 & 0 & 0 & 1 \\ 0 & 0 & 0 & 0 \\ 0 & -1 & 0 & 0 \end{pmatrix} \quad X_3 = \begin{pmatrix} 0 & 0 & 0 & 0 \\ 0 & 0 & -1 & 0 \\ 0 & 1 & 0 & 0 \\ 0 & 0 & 0 & 0 \end{pmatrix}$$

$$B_1 = \begin{pmatrix} 0 & 1 & 0 & 0 \\ 1 & 0 & 0 & 0 \\ 0 & 0 & 0 & 0 \\ 0 & 0 & 0 & 0 \end{pmatrix} \quad B_2 = \begin{pmatrix} 0 & 0 & 1 & 0 \\ 0 & 0 & 0 & 0 \\ 1 & 0 & 0 & 0 \\ 0 & 0 & 0 & 0 \end{pmatrix} \quad B_3 = \begin{pmatrix} 0 & 0 & 0 & 1 \\ 0 & 0 & 0 & 0 \\ 0 & 0 & 0 & 0 \\ 1 & 0 & 0 & 0 \end{pmatrix}$$

The splitting of $\mathbf{C} \otimes so(1,3)$ into $\mathbf{C} \otimes su(2) \oplus \mathbf{C} \otimes su(2)$ is obvious in the complex basis $T_r = \frac{1}{2}(X_r + iB_r)$, $\bar{T}_r = \frac{1}{2}(X_r - iB_r)$.

**Homework** 6.3.9. Verify that the $T$'s are anti-hermitian, that they satisfy the $su(2)$ commutation laws, and that $[T_r, \bar{T}_s] = 0$ for all $r$ and $s$.

Henceforth we shall be working in the world of complex Lie algebras and complex linear representations. To simplify notation and to conform to physics conventions, we shall therefore write $su(2)$ for the complex Lie algebra more correctly written $\mathbf{C} \otimes su(2)$; and we shall assume that all our complex representations are holomorphic extensions of real ones. When it is necessary to be careful, we can always return to the more precise notation.

This section has defined some notation, some specific bases for various Lie algebras, and the notion of complexification of Lie algebras and their representations. Complexification provides a link between the Lorentz Lie algebra and $su(2)$. Using this link, we can pursue our quest for representations of $so(1,3)$ by studying those of $su(2)$.

# 6.4  $su(2)$ in Particular

In this section, we will analyze the representations of the complex Lie algebra $su(2)$. Choose the matrix $\rho_3$ as the basis vector for a Cartan subalgebra. The analysis of a representation $D$ of $su(2)$ begins with finding the eigenvectors of $D(\rho_3)$. The procedure for analyzing the representation is to find the raising and lowering operators, to find a vector with maximal weight, and to apply the lowering operators to it to generate a basis for an invariant subspace of the representation. Repeating this procedure splits the representation into a direct sum of irreducible subrepresentations.

The raising and lowering operators for $su(2)$ satisfy

$$[\rho_3, X] = \lambda X. \tag{6.4.1}$$

The eigenvectors for this equation form a basis for the complex vector space $su(2)$:

$$R = \rho_1 + i\rho_2, \quad \rho_3, \quad L = \rho_1 - i\rho_2. \tag{6.4.2}$$

The eigenvalues associated with $R$, $\rho_3$, and $L$ are 1, 0, and $-1$. Notice how complexifying $su(2)$ has made it possible to have real eigenvalues and to find the corresponding eigenvectors.

Now we will turn our attention to the weight diagram of an arbitrary unitary representation $D$ of $su(2)$ on some vector space $V$. Let $\bar{\rho}_3$, $\bar{R}$, and $\bar{L}$ be the linear transformations $D(\rho_3)$, $D(R)$, and $D(L)$ respectively. Then, because $D$ is a representation, the Lie algebra relations between $\rho_3$, $R$, and $L$ will also hold for $\bar{\rho}_3$, $\bar{R}$, and $\bar{L}$.

Since $D$ is unitary, there exists a basis for $V$ in which $\bar{\rho}_3$ is hermitian. Hence $\bar{\rho}_3$ is diagonalizable. Let $v_1, \ldots, v_n$ be a basis of $\bar{\rho}_3$-eigenvectors for $V$. Let $w_r$ be the eigenvalue associated with $v_r$. Then $w_1, \ldots, w_n$ is the set of weights of $D$. Indeed, since our Cartan subalgebra is one-dimensional, $su(2)$ has rank one and the weight diagram is simply the set of numbers $\{\, w_r \mid 1 \leq r \leq n \,\}$ plotted on a line.

Since the Lie algebra element $R$ satisfies $[\rho_3, R] = R$, if $v$ is an eigenvector of $\bar{\rho}_3$ with weight $w$, we have

$$\begin{aligned}
[\rho_3, R] = R \quad &\Longrightarrow \quad [\bar{\rho}_3, \bar{R}] = \bar{R} \\
&\Longrightarrow \quad \bar{\rho}_3 \bar{R} = \bar{R}(\bar{\rho}_3 + 1) \\
&\Longrightarrow \quad \bar{\rho}_3(\bar{R}v) = (w+1)(\bar{R}v),
\end{aligned} \tag{6.4.3}$$

which means that $\bar{R}v$ has weight $w + 1$, and so $\bar{R}$ raises weights.

Similarly, $L = \rho_1 - i\rho_2$ satisfies $[\rho_3, L] = -L$. Then $\bar{\rho}_3 v = w_3 v$ implies that $\bar{L}v$ has weight $w_3 - 1$, and so $\bar{L}$ lowers weights.

Now we will proceed with the analysis of the representation $D$. First we prove a technical lemma to aid computation in the Lie algebra $D\big(su(2)\big)$.

**Lemma 6.4.4.** *If $R$ and $L$ are the raising and lowering operators of $su(2)$, if $D$ is a representation of $su(2)$ on a vector space $V$, if $\bar{R} = D(R)$, $\bar{L} = D(L)$, and $\bar{\rho}_3 = D(\rho_3)$, and if $v \in V$ is a vector of maximal weight $j$, then:*

$$\bar{R}\bar{L}^r v = r(2j + 1 - r)\bar{L}^{r-1}v \quad \text{for all } r \geq 0.$$

**Proof.** Push the $\bar{R}$ through the $\bar{L}$'s using the Lie bracket of $\bar{R}$ and $\bar{L}$:

$$\bar{R}\bar{L}^r = \bar{R}\bar{L}^r - \bar{L}\bar{R}\bar{L}^{r-1} + \bar{L}\bar{R}\bar{L}^{r-1} - \bar{L}^2\bar{R}\bar{L}^{r-2} + \bar{L}^2\bar{R}\bar{L}^{r-2} - \cdots - \bar{L}^r\bar{R} + \bar{L}^r\bar{R}$$
$$= [\bar{R},\bar{L}]\bar{L}^{r-1} + \bar{L}[\bar{R},\bar{L}]\bar{L}^{r-2} + \cdots + \bar{L}^{r-1}[\bar{R},\bar{L}] + \bar{L}^r\bar{R}$$
$$= 2(\bar{\rho}_3\bar{L}^{r-1} + \bar{L}\bar{\rho}_3\bar{L}^{r-2} + \cdots + \bar{L}^{r-1}\bar{\rho}_3) + \bar{L}^r\bar{R}.$$

Apply this equation to the vector $v$, and use $\bar{R}v = 0$ and $\bar{\rho}_3\bar{L}^q v = (j-q)\bar{L}^q v$ to find:

$$\bar{R}\bar{L}^r v = 2(\bar{\rho}_3\bar{L}^{r-1} + \bar{L}\bar{\rho}_3\bar{L}^{r-2} + \cdots + \bar{L}^{r-1}\bar{\rho}_3)v$$
$$= 2\Big((j-(r-1))\bar{L}^{r-1}v + \bar{L}(j-(r-2))\bar{L}^{r-2}v + \cdots + \bar{L}^{r-1}jv\Big)$$
$$= 2\Big((j-(r-1)) + (j-(r-2)) + \cdots + j\Big)\bar{L}^{r-1}v$$
$$= 2(rj - \tfrac{1}{2}r(r-1))\bar{L}^{r-1}v$$
$$= r(2j+1-r)\bar{L}^{r-1}v. \qquad \square$$

Next we show that the highest weight $j$ must be a positive multiple of $1/2$, and that this number is the highest weight for a unique irreducible unitary representation. Start by choosing a vector $v$ in the basis $v_1, \ldots, v_n$ whose weight $j$ is maximal in the set of weights of $D$. Since $\bar{L}^r v$ has weight $j - r$, the vector $\bar{L}^r v$ is either the zero vector or linearly independent of $\{\bar{L}^s v \mid 0 \le s < r\}$, $\bar{L}^{r+1}v = 0$ for large enough $r$. Determine $k$ from the conditions $\bar{L}^k v \ne 0$ and $\bar{L}^{k+1}v = 0$, then define a subspace $U$ of $V$ by

$$U = \text{Span}\{v, \bar{L}v, \ldots, \bar{L}^k v\}. \qquad (6.4.5)$$

Clearly $U$ is invariant under $\bar{L}$. Furthermore, $\bar{L}^r v$ is an eigenvector of $\bar{\rho}_3$ (with weight $j - r$), so $U$ is invariant under $\bar{\rho}_3$. Now apply the raising Lemma 6.4.4 to see that $\bar{R}\bar{L}^m v$ is in $U$, and so $U$ is also invariant under $\bar{R}$. Finally, since $\{\rho_3, R, L\}$ is a basis for $su(2)$, we conclude that $U$ is an invariant subspace for $D$.

Obviously, given any non-zero vector $u$ in $U$, we can raise it until it is a multiple of $v$ and then lower it to obtain a basis for $U$. Hence the representation $D$ has no non-trivial invariant subspaces in $U$, and is therefore irreducible.

Now we see that the weight diagram for $D$ restricted to act on $U$ is the set of numbers $j - r$ for $0 \le r \le n - 1$. A lemma on the sum of the weights will determine the relationship between $n$ and $j$:

**Lemma 6.4.6.** *The sum of the weights in any weight diagram is zero.*

**Proof.** The weights of a representation $D$ are simply the eigenvalues of $\bar{\rho}_3 = D(\rho_3)$ with multiplicity accounted for by repetition. The sum of these eigenvalues is therefore the trace of $\bar{\rho}_3$. But $i\bar{\rho}_3 = [\bar{\rho}_1, \bar{\rho}_2]$, and so

$$\text{tr}(i\bar{\rho}_3) = \text{tr}([\bar{\rho}_1, \bar{\rho}_2]) = \text{tr}(\bar{\rho}_1\bar{\rho}_2 - \bar{\rho}_2\bar{\rho}_1) = 0. \qquad \square$$

Since the sum of the weights is zero, $n = 2j + 1$. In particular, this shows that $j$ is a positive multiple of $1/2$. Finally, by the construction of $U$ from its vector with weight $j$, it is clear that the magnitude of the highest weight completely determines the representation of $su(2)$ on $U$, and that this subrepresentation on $U$ is irreducible.

Since $D$ is unitary, we see that the invariance of $U$ implies that $U^\perp$ is also invariant. Applying the same lowering process to a vector in $U^\perp$ with highest weight among all vectors in $U^\perp$ extracts another invariant subspace $U'$ corresponding to another irreducible subrepresentation. Repeated application leads to the complete decomposition of $D$ into a direct sum of irreducible subrepresentations.

Since the highest weight in an irreducible representation of $su(2)$ identifies the representation, the following definition is meaningful:

**Definition 6.4.7.** Write $D_j$ for the irreducible representation of $su(2)$ with highest weight $j$.

The representation $D_j$ is a Lie algebra homomorphism from $su(2)$ to $gl(\mathbf{C}^{2j+1})$. In particular, we have the following specific maps:

$$
\begin{aligned}
D_0 &: su(2) \to so(1) \subset gl(\mathbf{C}), & D_0(\rho_r) &= 0 \\
D_{1/2} &: su(2) \to su(2) \subset gl(\mathbf{C}^2), & D_{1/2}(\rho_r) &= \rho_r \\
D_1 &: su(2) \to so(3) \subset gl(\mathbf{C}^3), & D_1(\rho_r) &= iX_r
\end{aligned}
\tag{6.4.8}
$$

which illustrate the fact that only the integral weight representations can be written in terms of real, anti-symmetric matrices.

**Remark 6.4.9.** Write $L_j$ and $R_j$ for $D_j(L)$ and $D_j(R)$ respectively. We can use the data above to compute the matrix forms $[L_j]$ and $[R_j]$ of $L_j$ and $R_j$ with respect to the basis $\{v, \bar{L}v, \ldots, \bar{L}^{2j}v\}$ for $U$. Clearly $[L_j]$ follows directly from the structure of the basis, and $[R_j]$ may be written down from Lemma 6.4.4:

$$
[L_j] = \begin{pmatrix} 0 & & & & \\ 1 & 0 & & & \\ & & \ddots & & \\ & & & 0 & \\ & & & 1 & 0 \end{pmatrix}, \quad [R_j] = \begin{pmatrix} 0 & 1(2j) & & & \\ & 0 & 2(2j-1) & & \\ & & 0 & & \\ & & & \ddots & \\ & & & & 0 & 2j(1) \\ & & & & & 0 \end{pmatrix}.
\tag{6.4.10}
$$

If we scale the basis vectors $v_r = L_j^r v$ by setting $v_r = N_r v'_r$, and if we choose $N_r > 0$ such that

$$
r(2j + 1 - r)N_{r-1}^2 = N_r^2,
\tag{6.4.11}
$$

then

$$
\left.
\begin{aligned}
L_j v'_{r-1} &= \sqrt{r(2j + 1 - r)}\, v'_r \\
R_j v'_r &= \sqrt{r(2j + 1 - r)}\, v'_{r-1}
\end{aligned}
\right\}
\tag{6.4.12}
$$

which at once yields the new matrices for $R_j$ and $L_j$:

$$
[R_j] = \begin{pmatrix} 0 & \sqrt{1(2j)} & & & \\ & 0 & \sqrt{2(2j-1)} & & \\ & & 0 & & \\ & & & \ddots & \\ & & & & 0 & \sqrt{2j(1)} \\ & & & & & 0 \end{pmatrix} = [L_j]^\mathsf{T}.
\tag{6.4.13}
$$

Notice that $D_j(\rho_1) = \frac{1}{2}(R_j + L_j)$ and $D_j(\rho_2) = \frac{1}{2i}(R_j - L_j)$ are represented by hermitian matrices with respect to the basis $v'_0, \ldots, v'_{2j}$. This implies that, if we take $v'_0, \ldots, v'_{2j}$ to be orthonormal vectors, $D_j$ is a hermitian representation.

Generally in physics, representations are hermitian, and basis vectors are orthonormal and labeled by their weights. Thus, for the representation $D_j$ of $su(2)$ on $V = \mathbf{C}^{2j+1}$, the basis of eigenvectors of $D_j(\rho_3)$ will be labeled by the eigenvalues $j - r$ for $0 \leq r \leq 2j$. These eigenvalues $m$ may be integers or half integers; they start with $m = j$ and decrease by unity at each step. As the weight $m$ occurs in all the $D_j$ with $j > m$, it is sensible to include the index $j$ in the label.

Concretely, let $|j, m\rangle$ be the unit eigenvector of $D_j(\rho_3)$ with eigenvalue $m$:

$$|j, m\rangle \stackrel{\text{def}}{=} \frac{v_r}{|v_r|} \quad \text{for } m = j - r. \tag{6.4.14}$$

This choice of basis transforms the raising and lowering equations 6.4.12 into their common form:

$$\left.\begin{aligned} \bar{R}|j, m\rangle &= \sqrt{(j - m)(j + m + 1)}\,|j, m + 1\rangle \\ \bar{\rho}_3|m\rangle &= m|m\rangle \\ \bar{L}|j, m\rangle &= \sqrt{(j + m)(j - m + 1)}\,|j, m - 1\rangle \end{aligned}\right\} \tag{6.4.15}$$

In applications of $SU(2)$ symmetry to quantum theory, the quantity $j$ is also observable. From the perspective of the Lie algebra, it is determined by a non-linear combination of Lie algebra elements:

$$C_D \stackrel{\text{def}}{=} D(\rho_1)^2 + D(\rho_2)^2 + D(\rho_3)^2. \tag{6.4.16}$$

This linear operator $C_D$ is a *Casimir operator*, that is, it is a homogeneous polynomial in Lie algebra elements which commutes with every element in the Lie algebra:

$$[C_D, D(\rho_r)] = 0 \quad \text{for all } r. \tag{6.4.17}$$

The Casimir operator $C_j$ for $D = D_j$ determines $j$ through the equation

$$C_j|j, m\rangle = j(j + 1)|k\rangle. \tag{6.4.18}$$

**Homework** 6.4.19. Prove that the linear operator $C_D$ defined by 6.4.16 satisfies the equations 6.4.17. Using the facts that every linear operator on a complex vector space has an eigenvalue and that $D_j$ is irreducible, show that the Casimir operator $C_j$ is constant on $D_j$. By evaluating $C_j$ on $|j, j\rangle$, prove 6.4.18.

The classification of the irreducible representations of $su(2)$ by highest weight is the achievement of this section. The method of analysis used here – application of raising and lowering operators to weight diagrams – is basically the same for all unitary representations of $so(n)$ and $su(n)$. We will use it again when we investigate flavor symmetry of hadrons.

# 6.5 The Tensor Product

Our goal is to understand the range of options available for constructing Lorentz-invariant Lagrangians. At this point, we see that the representations $D_s$ of $su(2)$ are part of the possible action of the Lorentz Lie algebra on the column vector of fields in the theory. However, to form terms in the Lagrangian density, it is necessary to multiply these fields. The question is, how do we organize sums of products of fields in such a way that the resulting Lagrangian density is invariant? A product of fields is an element in a tensor product of the vector space of fields with itself. We need to know how to find invariant vectors in such tensor products.

In this section, we first derive the tensor product of Lie algebra representations from the tensor product of group representations. In the next, we show how to split tensor products of $su(2)$ representations into direct sums of irreducible representations. Finally, in Section 6.7, we extend this knowledge of splitting tensor products of $su(2)$ representations to $so(1,3)$ representations.

To understand a tensor product of Lie algebra representations, it is helpful to see its relationship to the simpler form of the tensor product of group representations. If $D$ and $E$ are representations of a group $G$ on vector spaces $U$ and $V$, then, following the definition 6.1.8, we define the Lie algebra representations $\bar{D} : \mathcal{G} \to gl(U)$ and $\bar{E} : \mathcal{G} \to gl(V)$ by

$$\bar{D}(X) = \frac{\partial}{\partial t} D\big(\exp(tX)\big) \Big|_{t=0},$$
$$\bar{E}(X) = \frac{\partial}{\partial t} E\big(\exp(tX)\big) \Big|_{t=0}. \tag{6.5.1}$$

The tensor product of $D$ and $E$ is the representation $F = D \otimes E$ of $G$ on $U \otimes V$ defined by

$$F(u \otimes v) = D(u) \otimes E(v). \tag{6.5.2}$$

Applying differentiation procedure above to this tensor product, we find:

$$\begin{aligned}
\bar{F}(X) &= \frac{\partial}{\partial t}\Big( D\big(\exp(tX)\big) \otimes E\big(\exp(tX)\big) \Big) \Big|_{t=0} \\
&= \Big( \frac{\partial}{\partial t} D\big(\exp(tX)\big) \Big) \otimes E\big(\exp(tX)\big) \Big|_{t=0} \\
&\quad + D\big(\exp(tX)\big) \otimes \Big( \frac{\partial}{\partial t} E\big(\exp(tX)\big) \Big) \Big|_{t=0} \\
&= \bar{D}(X) \otimes 1_V + 1_U \otimes \bar{E}(X).
\end{aligned} \tag{6.5.3}$$

This equation shows that, if $\bar{D}$ and $\bar{E}$ are representations of a Lie algebra $\mathcal{G}$ on vector spaces $U$ and $V$, the way to obtain a representation on the tensor product $U \otimes V$ is to form this combination of $\bar{D}$ and $\bar{E}$. As this combination is not the usual tensor product of matrices but is generally referred to as the tensor product of the representations, we shall write it with the symbol $\oslash$:

**Definition 6.5.4.** The *tensor product* of two representations $\bar{D}$ and $\bar{E}$ of a Lie algebra $\mathcal{G}$ on vector spaces $U$ and $V$ respectively is the representation $\bar{D} \oslash \bar{E}$ of $\mathcal{G}$ on $U \otimes V$ defined by

$$\bar{D} \oslash \bar{E} = \bar{D} \otimes 1_V + 1_U \otimes \bar{E}.$$

**Homework 6.5.5.** Verify the assumption hidden in this definition: if $\bar{D}$ and $\bar{E}$ are representations of a Lie algebra as linear transformations on vector spaces $U$ and $V$, then $\bar{D} \oslash \bar{E}$ is a representation of the Lie algebra on $U \otimes V$.

**Homework 6.5.6.** Let $\Phi$ be a column vector of $n$ complex scalar fields, and $\Psi$ be a square matrix of $n^2$ complex scalar fields. Let $X \in u(n)$ act on $\Phi$ and $\Psi$ by

$$X: \Phi \mapsto X\Phi \quad \text{and} \quad X: \Psi \mapsto X\Psi - \Psi X.$$

Show that $\Phi^\dagger \Phi$, $\Phi^\dagger \Psi \Phi$, and $\mathrm{tr}(\Psi^\dagger \Psi)$ are invariant under this action.

The conclusion, then, is that a representation of a Lie algebra on the vector of fields in a field theory breaks up into a sum of irreducible subrepresentations, and Definition 6.5.4 shows how to compute the action of the Lie algebra on a product of fields.

# 6.6 Weights and Tensor Products

Having clarified the notion of a tensor product of representations, we now discuss tensor products of representations in terms of weight diagrams. This leads to the decomposition of the tensor product of irreducible $su(2)$ representations into a direct sum of the same.

The first fact to understand is that the weights in a tensor product of representations are sums of the weights in the two representations. To see this, let $D$ and $F$ be two unitary representations of a Lie algebra $\mathcal{G}$ on vector spaces $U$ and $V$. Let $\mathcal{H}$ be a Cartan subalgebra of $\mathcal{G}$, and let $H_1, \ldots, H_k$ be a basis for $\mathcal{H}$. Restricting $D$ to $\mathcal{H}$, we find that there is a basis $u_1, \ldots, u_m$ for $U$ in which all the linear transformations $D(H_i)$ for $1 \le i \le k$ are represented by diagonal matrices. Then all the $u_r$ are eigenvectors of all the $D(H_i)$, and there exist eigenvalues $\alpha_{ir}$ such that:

$$D(H_i)u_r = \alpha_{ir}u_r. \tag{6.6.1}$$

Similarly, restricting $F$ to $\mathcal{H}$, we can choose a basis $v_1, \ldots, v_n$ with respect to which the matrices representing $H_i$ are all diagonal, so that there exist eigenvalues $\beta_{is}$ such that

$$D(H_i)v_s = \beta_{is}v_s. \tag{6.6.2}$$

The $m$ $k$-vectors $(\alpha_{1r}, \ldots, \alpha_{kr})$ are the weights of the representation $D$ with respect to the basis $H_i$ for a Cartan subalgebra of $\mathcal{G}$, and the $n$ $k$-vectors $(\beta_{1s}, \ldots, \beta_{ks})$ are the weights of the representation $F$. The weight diagrams of $D$ and $F$ consist respectively of $m$ and $n$ points in $\mathbf{R}^k$.

Now choose the vectors $u_r \otimes v_s$ as a basis for $U \otimes V$, and use Definition 6.5.4 for the action of the tensor product representation to see the additivity of the weights:

$$\begin{aligned} (D \oslash F)(H_i)\, u_r \otimes v_s &= \big(D(H_i)u_r\big) \otimes v_s + u_r \otimes \big(D(H_i)v_s\big) \\ &= \alpha_{ir}u_r \otimes v_s + \beta_{is}u_r \otimes v_s \\ &= (\alpha_{ir} + \beta_{is})\, u_r \otimes v_s. \end{aligned} \tag{6.6.3}$$

This shows that the weight of $u_r \otimes v_s$ is the sum of the weight of $u_r$ and the weight of $v_s$. In terms of field theory, this result effectively asserts that the charge of a product of fields is the sum of the charges of the fields, or that the charge of a particle pair is the sum of the charges of the particles. Thus:

**Theorem 6.6.4.** *If $D$ and $F$ are representations of a Lie algebra $\mathcal{G}$ on vector spaces $U$ and $V$, and if $\bar{\alpha}_1, \ldots, \bar{\alpha}_k$ and $\bar{\beta}_1, \ldots, \bar{\beta}_k$ are the weights of $D$ and $F$ with respect to a basis $H_1, \ldots, H_k$ for a Cartan subalgebra of $\mathcal{G}$, then the weights for the tensor representation $D \oslash F$ are the $mn$ $k$-vectors $\bar{\gamma}_{rs}$ defined by*

$$\bar{\gamma}_{rs} = \bar{\alpha}_r + \bar{\beta}_s.$$

To make this theorem more concrete and to discover its value, we analyze the tensor product $D_{1/2} \oslash D_1$ in detail. Figure 6.6 presents the structure of this example at a glance. It depicts the tensor product of the weight diagrams for $D_{1/2}$ and $D_1$, the direct sum of those for $D_{1/2}$ and $D_{3/2}$, and the relationship between the tensor product and the direct sum.

The $D_{1/2}$ representation acts on $V_{1/2} = \mathbf{C}^2$, and the eigenvectors of $\rho_3 = D_{1/2}(\rho_3)$ are

$$\rho_3 \begin{pmatrix} 1 \\ 0 \end{pmatrix} = \tfrac{1}{2} \begin{pmatrix} 1 \\ 0 \end{pmatrix} \quad \text{and} \quad \rho_3 \begin{pmatrix} 0 \\ 1 \end{pmatrix} = -\tfrac{1}{2} \begin{pmatrix} 0 \\ 1 \end{pmatrix}. \tag{6.6.5}$$

Call these eigenvectors $u_{1/2}$ and $u_{-1/2}$.

The representation $D_1$ acts on $V_1 = \mathbf{C}^3$, and the eigenvectors of $\rho_{3,1} = D_1(\rho_3)$ are

$$\rho_{3,1} \begin{pmatrix} 1 \\ i \\ 0 \end{pmatrix} = \begin{pmatrix} 1 \\ i \\ 0 \end{pmatrix}, \quad \rho_{3,1} \begin{pmatrix} 0 \\ 0 \\ 1 \end{pmatrix} = \begin{pmatrix} 0 \\ 0 \\ 0 \end{pmatrix}, \quad \rho_{3,1} \begin{pmatrix} 1 \\ -i \\ 0 \end{pmatrix} = -\begin{pmatrix} 1 \\ -i \\ 0 \end{pmatrix}. \tag{6.6.6}$$

Call these eigenvectors $v_1$, $v_0$, and $v_{-1}$. Of course, $v_1$, $v_0$, and $v_{-1}$ are simply the raising operator $R$, $\rho_3$ itself, and the lowering operator $L$ in coordinates with respect to the basis $\rho_r$.

In tensor products of representations, weights add. Hence the basis $u_\alpha \otimes v_\beta$ can be arranged according to the weights of these vectors:

$$
\begin{array}{cccc}
 & u_{-1/2} \otimes v_0 & u_{-1/2} \otimes v_1 & \\
u_{-1/2} \otimes v_{-1} & & & u_{1/2} \otimes v_1 \\
 & u_{1/2} \otimes v_{-1} & u_{1/2} \otimes v_0 & \\
\hline
-3/2 & -1/2 & 1/2 & 3/2
\end{array} \tag{6.6.7}
$$

By inspecting the weights of the tensor product, we see at once that the space of weight-3/2 vectors is one-dimensional, and the space of weight-1/2 vectors is two-dimensional. Hence $D_{1/2} \oslash D_1$ splits into a direct sum of $D_{3/2}$ and $D_{1/2}$. The invariant subspace $V_{3/2}$ for $D_{3/2}$ is obtained by lowering any weight-3/2 vector. To determine the invariant subspace $V_{1/2}$ for $D_{1/2}$, we can either take the orthogonal complement of $V_{3/2}$ or determine the weight-1/2 vector from its property of being annihilated by the raising operator. Raising and lowering operators are of significant practical value in physics, so we shall provide the details of this splitting.

Let $u = u_{1/2}$ and $v = v_1$ so that $u \otimes v$ is a vector with maximal weight. Let $L_1 = D_1(L)$ and recall that $L^2 = 0$ and $L_1^3 = 0$. Write $\bar{L}$ for $L \oslash L_1$, and use $\bar{L}$ to

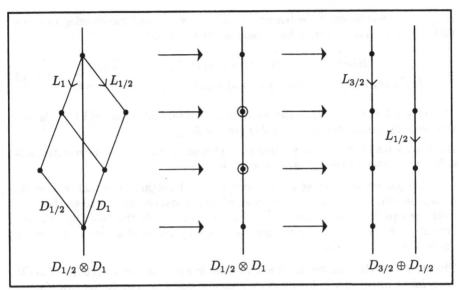

**Fig. 6.6.** Reading from left to right, the first half of the diagram shows how the two weights of $D_{1/2}$ combine with the three weights of $D_1$ to form the six of $D_{1/2} \otimes D_1$, and the second half shows how these six split into the four of $D_{3/2}$ and the two of $D_{1/2}$. Note that the initial and final sets of six weights are associated with different vectors in $\mathbf{C}^2 \otimes \mathbf{C}^4 = \mathbf{C}^4 \oplus \mathbf{C}^2$.

lower $u \otimes v$:

$$\bar{L}(u \otimes v) = (Lu) \otimes v + u \otimes (L_1 v), \tag{6.6.8}$$

$$\bar{L}^2(u \otimes v) = (L^2 u) \otimes v + 2(Lu) \otimes (L_1 v) + u \otimes (L_1^2 v)$$
$$= 2(Lu) \otimes (L_1 v) + u \otimes (L_1^2 v),$$

$$\bar{L}^3(u \otimes v) = (L^3 u) \otimes v + 3(L^2 u) \otimes (L_1 v) + 3(Lu) \otimes (L_1^2 v) + u \otimes (L_1^3 v)$$
$$= 3(Lu) \otimes (L_1^2 v),$$

$$\bar{L}^4(u \otimes v) = 0.$$

From the proof of Lemma 6.4.4, we see that the vectors above span a four-dimensional invariant subspace, and from the classification of irreducible representations of $su(2)$, we deduce that on this subspace the tensor representation $D_{1/2} \oslash D_1$ acts as $D_{3/2}$.

We can find the $D_{1/2}$ subrepresentation of the tensor product by finding its highest weight vector. Writing $R_1 = D_1(R)$ and $\bar{R} = R \oslash R_1$, this vector is a weight-1/2 vector in the kernel of $\bar{R}$:

$$\bar{R}\big(a(Lu) \otimes v + bu \otimes (L_1 v)\big) = a(RLu) \otimes v + bu \otimes (R_1 L_1 v) \tag{6.6.9}$$
$$= a\big((LR + 2\rho_3)u\big) \otimes v + bu \otimes \big((L_1 R_1 + 2\rho_{3,1})\big)v$$
$$= 2a(\rho_3 u) \otimes v_1 + 2bu \otimes (\rho_{3,1} v)$$
$$= (a + 2b)u \otimes v.$$

The kernel of the raising operator is therefore the set of vectors determined by $a + 2b = 0$. Choosing $a = 2$ and $b = -1$, we have a highest weight vector for the

$D_{1/2}$ subrepresentation of the tensor product. We can do a little lowering to check that this vector lies in a two-dimensional invariant subspace:

$$\bar{L}\big(2(Lu) \otimes v - u \otimes (L_1 v)\big) = (Lu) \otimes (L_1 v) - u \otimes (L_1^2 v),$$
$$\bar{L}\big((Lu) \otimes (L_1 v) - u \otimes (L_1^2 v)\big) = (Lu) \otimes (L_1^2 v) - (Lu) \otimes (L_1^2 v) = 0. \qquad (6.6.10)$$

The representation of $su(2)$ on the two-dimensional subspace spanned by this ladder is necessarily irreducible and must therefore be $D_{1/2}$.

**Homework 6.6.11.** Show that the invariant subspace for the $D_{1/2}$ subrepresentation is orthogonal to that for the $D_{3/2}$ subrepresentation.

This process of lowering vectors with maximal weights to extract irreducible subrepresentations from a tensor product of representations applies to all the symmetry groups in common use. When the group is $su(2)$, the following theorem presents a simple formula for the resulting decomposition of a tensor product of irreducible representations.

**Theorem 6.6.12.** *The tensor product of the irreducible representations of $su(2)$ splits into a direct sum of irreducible representations according to the formula*

$$D_s \oslash D_t = \bigoplus_{r=|s-t|}^{s+t} D_r.$$

**Proof.** The result follows at once from the methods illustrated above: given the classification of the irreducible representations of $su(2)$ in Section 6.4, analysis of the weight diagram is sufficient. $\qquad\square$

**Homework 6.6.13.** Let $u_r$ satisfy $D_{1/2} u_r = r u_r$ for $r = \pm\frac{1}{2}$, and $v_s$ satisfy $D_{3/2} v_s = s v_s$ for $s = 1, 0, -1$. Follow the argument above to split $D_{1/2} \oslash D_{3/2}$, but use the matrix form of the raising and lowering operators from 6.4.13 to express all vectors in terms of the basis $u_r \otimes v_s$.

Assuming that $u_r$, $v_s$, and $u_r \otimes v_s$ are orthonormal bases for their respective vector spaces, show that the two invariant subspaces $V_{3/2}$ and $V_{1/2}$ are orthogonal.

Note that, since $su(2)$ and $so(3)$ are isomorphic, all the points above about the structure and representations of $su(2)$ can be translated into knowledge of $so(3)$.

The formula of Theorem 6.6.12 for splitting tensor products of $su(2)$ representations is the main point of this section. The proof by example illustrates the structure of tensor representations and the use of raising and lowering operators in the analysis of representations. This method of analysis is of such general application in Lie algebra representation theory that anyone who will be going further in the study of symmetries in physics will do well to study its details now.

## 6.7 The Representations of $so(1,3)$

The goal now is to extend what we know of the representations for $su(2)$ to discover the structure of the representations of the Lorentz Lie algebra $so(1,3)$. We have seen that complex representations $D$ of the real Lie algebra $so(1,3)$ may be identified with their holomorphic extensions to the complexified Lie algebra $\mathbf{C} \otimes so(1,3)$, and from Homework 6.3.9 that the complexified Lie algebra splits into a direct sum of Lie subalgebras:

$$\left. \begin{array}{rcccc} S : \mathbf{C} \otimes so(1,3) & \to & \mathbf{C} \otimes su(2) & \oplus & \mathbf{C} \otimes su(2) \\ T_r & \mapsto & ( \quad \tau_r & , & 0 \quad ) \\ \bar{T}_r & \mapsto & ( \quad 0 & , & \tau_r \quad ) \end{array} \right\} \qquad (6.7.1)$$

Thus we need to investigate the relationship between representations of a direct sum of Lie algebras and representations of the summands.

In general, if $D$ and $D'$ are representations of the two Lie algebras $\mathcal{G}$ and $\mathcal{G}'$ on two vector spaces $V$ and $V'$, then there are natural constructions for representations of the direct sum $\mathcal{G} \oplus \mathcal{G}'$ on both the direct sum $V \oplus V'$ and on the tensor product $V \otimes V'$. However, we want the construction to build irreducible representations $F$ out of irreducible representations $D$ and $D'$. For this purpose, we define a representation $F$ of the direct sum $\mathcal{G} \oplus \mathcal{G}'$ on the tensor product of the vector spaces:

$$F(X, X')v \otimes v' \stackrel{\text{def}}{=} D(X)v \otimes v' + v \otimes D'(X')v'. \qquad (6.7.2)$$

This is not the tensor product of the representations $D$ and $D'$ since these representations act on different Lie algebras, so we will introduce a new symbol for this construction:

$$(D \odot D')(X, X') \stackrel{\text{def}}{=} D(X) \otimes 1_{V'} + 1_V \otimes D'(X'). \qquad (6.7.3)$$

This shows us how to obtain representations of direct sums of Lie algebras from representations of the summands.

**Homework 6.7.4.** Show that, if $D$ and $D'$ are irreducible representations of Lie algebras $\mathcal{G}$ and $\mathcal{G}'$ on vector spaces $V$ and $V'$, then the representation $D \odot D'$ of $\mathcal{G} \oplus \mathcal{G}'$ on $V \otimes V'$ is irreducible.

Applying this construction to the direct sum $\mathbf{C} \otimes su(2) \oplus \mathbf{C} \otimes su(2)$, we can define representations $D_{s,t}$ of $\mathbf{C} \otimes so(1,3)$ on $\mathbf{C}^{(2s+1)(2t+1)}$ in terms of the holomorphic extensions of the irreducible representations of $su(2)$ and the splitting map $S$:

$$D_{s,t}(X) \stackrel{\text{def}}{=} D_s \odot D_t(SX). \qquad (6.7.5)$$

From this definition, it follows that $D_{s,t}$ is complex linear, hence a holomorphic extension of a representation $D_{s,t}$ of $so(1,3)$. The equations for $D_{s,t}(X_r)$ and $D_{s,t}(B_r)$ are:

$$\begin{aligned} D_{s,t}(X_r) &= D_{s,t}(T_r + \bar{T}_r) \\ &= D_s(\tau_r) \otimes 1_{2t+1} + 1_{2s+1} \otimes D_t(\tau_r); \\ D_{s,t}(B_r) &= D_{s,t}(-iT_r + i\bar{T}_r) \\ &= D_s(-i\tau_r) \otimes 1_{2t+1} + 1_{2s+1} \otimes D_t(i\tau_r) \\ &= -iD_s(\tau_r) \otimes 1_{2t+1} + i1_{2s+1} \otimes D_t(\tau_r). \end{aligned} \qquad (6.7.6)$$

This completes our construction of irreducible representations of the Lorentz Lie algebra $so(1,3)$.

Now we need to prove that there are no other irreducible unitary representations of $so(1,3)$. Actually, we prove:

**Theorem 6.7.7.** *Any irreducible representation of $so(1,3)$ is isomorphic to $D_{s,t}$ for some $s$ and $t$.*

**Proof.** Let $F$ be any irreducible representation of $so(1,3)$ on any complex vector space $W$. Make $F$ complex linear, and define representations $D$ and $D'$ of $\mathbf{C} \otimes su(2)$ by

$$D(X) = F\big(S^{-1}(X,0)\big), \quad D'(X') = F\big(S^{-1}(0,X')\big), \tag{6.7.8}$$

so that

$$F(Y) = D \odot D'(SY).$$

Now for some half integers $s$ and $t$, $D$ and $D'$ contain subrepresentations $D_s$ and $D_t$. Hence $(D_s \odot D_t) \circ S$ is a subrepresentation of $F$. Since $F$ is irreducible, this implies that $F = D_{s,t}$. $\qquad\square$

Using the formulae 6.7.6, some of the representations $D_{s,t}$ may be written concretely:

$$
\begin{aligned}
&D_{0,0} \colon so(1,3) \to so(1), &&D_{0,0}(X) = 0; \\
&D_{0,1/2} \colon so(1,3) \to sl(2,\mathbf{C}), &&D_{0,1/2}(X_r) = \tau_r, \quad D_{0,1/2}(B_r) = \rho_r; \\
&D_{1/2,0} \colon so(1,3) \to sl(2,\mathbf{C},) &&D_{1/2,0}(X_r) = \tau_r, \quad D_{1/2,0}(B_r) = -\rho_r; \\
&D_{1/2,1/2} \colon so(1,3) \to so(1,3), &&D_{1/2,1/2}(X) = X.
\end{aligned}
\tag{6.7.9}
$$

$D_{0,1/2}$ and $D_{1/2,0}$ are both isomorphisms from $so(1,3)$ to $sl(2,\mathbf{C})$.

**Homework 6.7.10.** Show that

$$D_{1/2,0}(X) = -D_{0,1/2}(X)^\dagger$$

for all $X$ in $so(1,3)$.

**Homework 6.7.11.** Show that, if we try to define a representation of the Lorentz group $SO(1,3)$ as $2 \times 2$ complex matrices by defining

$$f\big(\exp(X)\big) = \exp\big(D_{1/2,0}(X)\big)$$

for $X$ in $so(1,3)$, then, by choosing two different logarithms $X$ for $\mathbf{1}_4$, we can make $f(\mathbf{1})$ both $\mathbf{1}_2$ and $-\mathbf{1}_2$. Use the tensor product splitting formula to show that $D_{s,t}$ only defines a representation of the Lorentz group through exponentiation when $s+t$ is a whole number.

The formula 6.7.5 for the irreducible representations of the Lorentz Lie algebra is the key result of this section. For future reference, it is also worth understanding that the $\odot$ product of irreducible representations is itself irreducible.

# 6.8 Tensor Products of $so(1,3)$ Representations

By using our knowledge of the tensor products $D_p \oslash D_q$, we can reduce tensor products $D_{s,t} \oslash D_{u,v}$ to sums of $D_{s,t}$:

$$D_{s,t} \oslash D_{u,v} = (D_s \odot D_t) \oslash (D_u \odot D_v) = (D_s \oslash D_u) \odot (D_t \oslash D_v). \qquad (6.8.1)$$

For example:

$$
\begin{aligned}
D_{1/2,1} \oslash D_{1,2} &= (D_{1/2} \oslash D_1) \odot (D_1 \oslash D_2) \\
&= (D_{3/2} \oplus D_{1/2}) \odot (D_3 \oplus D_2 \oplus D_1) \\
&= (D_{3/2} \odot D_3) \oplus (D_{3/2} \odot D_2) \oplus (D_{3/2} \odot D_1) \\
&\quad \oplus (D_{1/2} \odot D_3) \oplus (D_{1/2} \odot D_2) \oplus (D_{1/2} \odot D_1) \\
&= D_{3/2,3} \oplus D_{3/2,2} \oplus D_{3/2,1} \oplus D_{1/2,3} \oplus D_{1/2,2} \oplus D_{1/2,1}.
\end{aligned}
\qquad (6.8.2)
$$

This line of argument is sufficient to prove the following general statement:

**Theorem 6.8.3.** *The general formula for splitting a tensor product $D_{s,t} \oslash D_{u,v}$ into a direct sum of irreducible representations is:*

$$D_{s,t} \oslash D_{u,v} = \bigoplus_{p=|s-u|}^{s+u} \bigoplus_{q=|t-v|}^{t+v} D_{p,q}.$$

Note that the scalar representation $D_{0,0}$ occurs in this tensor product if and only if $s = u$ and $t = u$, and in this case the scalar occurs with multiplicity one.

# 6.9 Complex Conjugation, Parity, and Restriction to Rotations

If $D$ is a matrix representation of a group $G$, then the map $D^*$ defined by $D^*(g) = D(g)^*$ is also a representation. Clearly a similarity transformation of $D$ induces one of $D^*$, and so for arbitrary representations $D$, $D^*$ is well defined up to equivalence. Clearly also the dimension of $D$ is equal to that of $D^*$, and either both or neither are reducible. Hence, in particular, $D_s^*$ is equivalent to $D_s$.

**Theorem 6.9.1.** *The effect of complex conjugation on $D_{s,t}$ is given by:*

$$D_{s,t}^* \sim D_{t,s}.$$

**Proof.** We need only evaluate $D_{s,t}^*$ on the basis of rotation and boost generators, using first the formula 6.7.6, second the similarity of $D_s^* \sim D_s$, and third the similarity of linear transformations generated by the isomorphism between $\mathbf{C}^{2s+1} \otimes$

$\mathbf{C}^{2t+1}$ and $\mathbf{C}^{2t+1} \otimes \mathbf{C}^{2s+1}$:

$$\begin{aligned}
\left(D_{s,t}(X_r)\right)^* &= \left(D_s(\tau_r) \otimes \mathbf{1}_{2t+1} + \mathbf{1}_{2s+1} \otimes D_t(\tau_r)\right)^* \\
&\sim D_s(\tau_r) \otimes \mathbf{1}_{2t+1} + \mathbf{1}_{2s+1} \otimes D_t(\tau_r) \\
&\sim D_t(\tau_r) \otimes \mathbf{1}_{2s+1} + \mathbf{1}_{2t+1} \otimes D_s(\tau_r) \\
&\sim D_{t,s}(X_r);
\end{aligned}$$

$$\begin{aligned}
\left(D_{s,t}(B_r)\right)^* &= \left(-iD_s(\tau_r) \otimes \mathbf{1}_{2t+1} + i\mathbf{1}_{2s+1} \otimes D_t(\tau_r)\right)^* \\
&\sim iD_s(\tau_r) \otimes \mathbf{1}_{2t+1} - i\mathbf{1}_{2s+1} \otimes D_t(\tau_r) \\
&\sim -iD_t(\tau_r) \otimes \mathbf{1}_{2s+1} + i\mathbf{1}_{2t+1} \otimes D_s(\tau_r) \\
&\sim D_{t,s}(B_r).
\end{aligned}$$ $\qquad \square$

**Homework 6.9.2.** Show that $D_{1/2}^*$ is equivalent to $D_{1/2}$ by constructing the similarity matrix $S$ such that
$$D_{1/2}(X)^* = SD_{1/2}(X)S^{-1}$$
for all $X$ in $su(2)$. Use the matrix $S$ to demonstrate concretely the similarity between $D_{0,1/2}^*$ and $D_{1/2,0}$.

**Theorem 6.9.3.** *Let $P$ be the space reflection matrix. Then parity changes the representation $D_{s,t}$ into $D_{t,s}$:*

$$D_{s,t}(PXP^\top) \sim D_{t,s}(X),$$

*where the similarity is independent of the element $X$ in $so(1,3)$.*

**Proof.** The action of parity $\mathcal{P}$ on $so(1,3)$ is represented by the space reflection matrix $P$, so that $\mathcal{P} \cdot X = PXP^\top$. If $X$ is a rotation generator, then $PXP^\top = X$. If $B$ is a boost generator, then $PBP^\top = -B$. Therefore parity interchanges $T_r$ and $\bar{T}_r$. Now use the definition 6.7.5 of $D_{s,t}$ to compute the effect of parity on $D_{s,t}$:

$$\begin{aligned}
D_{s,t}\left(\mathcal{P} \cdot (\alpha^r T_r + \beta^r \bar{T}_r)\right) &= D_{s,t}(\alpha^r \bar{T}_r + \beta^r T_r) \\
&= \alpha^r \mathbf{1}_{2s+1} \otimes D_t(\tau_r) + \beta^r D_s(\tau_r) \otimes \mathbf{1}_{2t+1} \\
&\sim \alpha^r D_t(\tau_r) \otimes \mathbf{1}_{2s+1} + \beta^r \mathbf{1}_{2t+1} \otimes D_s(\tau_r) \\
&\sim D_{t,s}(\alpha^r T_r + \beta^r \bar{T}_r),
\end{aligned} \qquad (6.9.4)$$

where $\sim$ is the similarity that relates linear transformations of $\mathbf{C}^{2s+1} \otimes \mathbf{C}^{2t+1}$ to those of $\mathbf{C}^{2t+1} \otimes \mathbf{C}^{2s+1}$. $\qquad \square$

One instance of this result will be particularly useful in our discussion of fermions:
$$D_{0,1/2}(\mathcal{P} \cdot X) = D_{1/2,0}(X) = -D_{0,1/2}(X)^\dagger. \qquad (6.9.5)$$

**Theorem 6.9.6.** *The restriction of $D_{s,t}$ to the Lie subalgebra of rotation generators in $so(1,3)$ gives rise to the splitting formula:*

$$D_{s,t}\Big|_{\text{rotations}} = \bigoplus_{r=|s-t|}^{s+t} D_r.$$

**Proof.** When we restrict $D_{s,t}$ to act on rotation generators $X_r$, then, since $su(2)$ is isomorphic to $so(3)$, $D_{s,t}$ must split into a direct sum of $D_u$'s. From the definition of $D_{s,t}$,

$$D_{s,t}(X_r) = D_s(\tau_r) \otimes 1_{2t+1} + 1_{2s+1} \otimes D_t(\tau_r) = (D_s \oslash D_t)(X_r), \qquad (6.9.7)$$

and so we can use the splitting formula of Theorem 6.6.12 to complete the proof.□

## 6.10 Vectors, Tensors, and Symmetry

Since the highest weight $(s,t)$ uniquely identifies an irreducible representation of $so(1,3)$, we shall refer to the representation $D_{s,t}$ as $(s,t)$. Recall that the dimension of $(s,t)$ is $(2s+1)(2t+1)$. The vector representation has dimension four. The only four-dimensional irreducible representations of $so(1,3)$ have weights $(3/2,0)$, $(1/2,1/2)$, and $(0,3/2)$. Since the vector representation is real (equivalent to its complex conjugate), it must be the $(1/2,1/2)$ representation.

From Theorem 6.8.3, the tensor product of two vector representations of $so(1,3)$ decomposes into a sum

$$(1/2,1/2) \otimes (1/2,1/2) = (0,0) + (1,0) + (0,1) + (1,1), \qquad (6.10.1)$$

where we have followed the custom of writing '$\otimes$' for '$\oslash$' and '$+$' for '$\oplus$'. To identify these four classes of tensor, either regard the tensors as matrices $T^{\mu\nu}$, in which case the Lorentz group action is $\Lambda^{\top} T \Lambda$; or as $T^{\mu}{}_{\nu}$, in which case the group action is $\Lambda T \Lambda^{-1}$. The $(0,0)$ piece comes from the multiples of the metric matrix $g^{\mu\nu}$ or from the identity matrix $\delta^{\mu}_{\nu}$. The sum $(1,0)+(0,1)$ comes from anti-symmetric matrices $T^{\mu\nu}$ or from $T^{\mu}{}_{\nu}$ in $\mathbf{C} \otimes so(1,3)$. The $(1,1)$ piece is the set of symmetric matrices $T^{\mu\nu}$ which satisfy $g_{\mu\nu}T^{\mu\nu} = 0$. Check the dimensions: $(0,0)$ has dimension one, which corresponds to the dimension of the vector space with basis $g$; $(1,0)+(0,1)$ has dimension six, which corresponds to the dimension of the space of anti-symmetric matrices; $(1,1)$ has dimension nine, which corresponds to the dimension of the space of symmetric traceless matrices.

In general, in the splitting

$$(s,t) \otimes (s,t) = \overset{2s}{\underset{p=0}{\bigoplus}} \overset{2t}{\underset{q=0}{\bigoplus}} (2s-p, 2t-q), \qquad (6.10.2)$$

we can enquire which of the summands $(2s-p, 2t-q)$ is symmetric and which anti-symmetric as functions on the tensor product of $\mathbf{C}^{(2s+1)(2t+1)}$ with itself:

$$D_{(2s-p,2t-q)}(u \otimes v) \overset{?}{=} \pm D_{(2s-p,2t-q)}(v \otimes u).$$

In fact, the highest degree piece, $(2s,2t)$, is symmetric under interchange of variables, and as the sum of the indices goes down by one, we obtain anti-symmetric subrepresentations and so on down, alternating symmetric with anti-symmetric. Thus, in the equation above, the terms where $p+q$ are even are symmetric under interchange of factors, while the terms where $p+q$ are odd are anti-symmetric.

For example, in the product of $(0, 1/2)$ with itself,

$$(0, 1/2) \otimes (0, 1/2) = (0, 1) + (0, 0), \tag{6.10.3}$$

the $(0, 1)$ part is symmetric and the $(0, 0)$ part is anti-symmetric. Therefore the $(0, 0)$ subrepresentation will vanish on symmetric tensors $u \otimes u$. This suggests that $(0, 1/2)$ should be associated with fermions.

Similarly, in the product of $(1/2, 1/2)$ with itself,

$$(1/2, 1/2) \otimes (1/2, 1/2) = (1, 1) + (1, 0) + (0, 1) + (0, 0), \tag{6.10.4}$$

the $(0, 0)$ subrepresentation is symmetric in the fields and hence will vanish on anti-symmetric tensors; so we may associate $(1/2, 1/2)$ with bosons.

**Homework** 6.10.5. Write down the explicit form of the invariant terms above. How many linearly independent invariant cubic terms can be constructed from fields in $(s, t)$ representations with $s + t \leq 1$?

As we started from unitary and matrix representations of the Lorentz group in 6.1.3 and 6.1.7 respectively, and as we saw that any representation $D$ of the Lorentz group gives rise to a representation $\bar{D}$ of the Lorentz Lie algebra through differentiation, the final stroke of analysis in the chapter brings us back to the Lorentz group with the question, which of the representations $(s, t)$ of the Lorentz Lie algebra come from representations of the Lorentz group?

We begin by answering the simpler question, what is the relationship between the representations of $so(3)$ and of $SO(3)$? Since $D_{1/2}$ defines an isomorphism between $su(2)$ and $so(3)$, $D_j$ may be regarded as a representation of $so(3)$. Since the eigenvalues of $D_j(X_r)$ are $m$ for $-j \leq m \leq j$, and since $\exp(\tau X_r) = \mathbf{1}$ has solutions $\tau = 2\pi k$, the equation

$$E_j\big(\exp(\tau X_r)\big) \stackrel{\text{def}}{=} \exp\big(\tau D_j(X_r)\big) \tag{6.10.6}$$

defines a single-valued function $E_j$ if and only if $j$ is an integer. (Otherwise, there is a sign ambiguity.) To show that in this case $E_j$ is a representation, we could use the Campbell–Hausdorff formula as explained in Remark 3.8.21.

The action of $(s, t)$ on rotation generators is given by Theorem 6.9.6. Clearly, the integral-spin condition is satisfied if and only if $s + t$ is an integer. Hence:

**Theorem 6.10.7.** *The representations of the Lorentz group are determined by their derivatives $(s, t)$ at the origin. The Lorentz Lie algebra representations $(s, t)$ which arise as derivatives of Lorentz group representations are those for which $s + t$ is an integer.*

By convention, when $s + t$ is an integer, the notation $(s, t)$ is used for both the representation of the Lorentz Lie algebra and the associated representation of the Lorentz group.

This concludes our analysis of the representations of the Lorentz Lie algebra $so(1, 3)$. The main ideas and results have been presented as definitions and theorems so that they can easily be picked out from the background discussion.

Apropos of the original purpose to develop the possibility of non-scalar field theories, we see now an abundance of irreducible representations, each one of which can be taken as a single field, and we see in the tensor product splitting formula the means for identifying the invariant products of the fields which could be terms in the Lagrangian density. We shall see later that consistency of the quantum field theory actually restricts us to the $(0, 1/2)$, $(1/2, 0)$, and $(1/2, 1/2)$ representations. Supergravity theories, however, use more complicated representations, and the principles brought out here for forming Lorentz invariant terms are also used for forming terms invariant under internal symmetries. So, despite the elementary nature of the direct applications of the results of this chapter, we have not wasted our time by going this deeply into representation theory!

## 6.11 Summary

The analysis of representations of the Lorentz group presented here is far longer than strictly necessary. However, it is in the interests of future applications of symmetry in field theory to make a careful analysis of this group, since the terminology and methods of this analysis are so broadly applicable to the popular symmetry groups. In particular, the principles introduced in this chapter elucidate flavor symmetries and the structure of non-abelian gauge field theories.

   The conclusion of this chapter is that there are many representations of the Lorentz group that might appear in a quantum field theory and many ways of forming Lorentz-invariant terms from products of these representations. The next two chapters investigate the possibilities of the $(1/2, 0)$ and $(0, 1/2)$ representations for fermionic fields and the $(1/2, 1/2)$ representation for the electromagnetic field. In supergravity theories, the quantized gravitational field, the graviton, has spin 2 and its superpartner, the gravitino, has spin 3/2; higher spin fields are found among the massive excitations of the superstring. This chapter therefore provides the foundation for a profound expansion of quantum field theory.

# Chapter 7

# Two-Component Spinor Fields

The two two-dimensional representations of the Lorentz group give rise to left and right two-component spinor fields; on the basis of a thorough understanding of these representations, the procedure of perturbative canonical quantization developed for scalar fields can be applied to two-component spinor fields.

## Introduction

Having classified the representations of the Lorentz group in the last chapter, this chapter returns to the theme of Chapter 4, namely perturbative canonical quantization. Here and in the two subsequent chapters, we extend the principles of Chapter 4 to two-component Fermi fields, four-component Fermi fields, and vector fields.

This chapter has a purpose to present the quantum theory of Weyl fields and make the connection between Weyl fields and Dirac fields. By the end of this chapter, we shall be able to compute scattering amplitudes for Feynman diagrams with external and internal Weyl lines.

We begin in Sections 7.1 and 7.2 by finding a convenient description of $(0, 1/2)$ and $(1/2, 0)$, the smallest non-trivial representations of the Lorentz group. These spinor representations respectively determine the right and left two-component spinor or Weyl fields. Section 7.3 then tabulates all the Lorentz-invariant Lagrangian terms which can be formed from these Weyl fields. The anti-symmetric character of these terms implies that the Weyl fields must be fermionic. Section 7.4 introduces fermionic calculus so that Section 7.5 can derive canonical momenta and equations of motion from the Weyl Lagrangians. Section 7.6 uses the plane wave solutions of the equations of motion to build the free Weyl quantum field. At this point, canonical quantization of the free fields has been accomplished.

Sections 7.7, 7.8, and 7.9 extend the conservation results of Chapter 3 to Weyl fields. We find, in particular, that the conserved charge operator needs to be normally ordered and introduce the helicity quantum number.

In Section 7.10, we compute the contraction of Weyl fields, and in Sections 7.11 to 7.15, we use the Wick expansion of the $S$ matrix to derive Feynman diagrams and rules for a theory with interacting Weyl fields. This is perturbative canonical quantization of the interacting fields. Section 7.16 describes a common index notation for Weyl fields. Using this notation, Sections 7.17 and 7.18 show how the introduction of mass terms in a theory of Weyl fields makes it convenient and natural to introduce four-component or Dirac spinors. This leaves us ready for the next chapter on Dirac fields.

For a concrete interpretation, note that left Weyl fields can be used to represent neutrinos. Right Weyl fields do not represent any known low-energy particle, but both types of Weyl field can be used to describe high-energy particles.

# 7.1 The Lorentz Group Action on Spinors

In this section, we make concrete the implications of Chapter 6 for the two spinor representations $(0, 1/2)$ and $(1/2, 0)$ of the Lorentz Lie algebra. First, we exponentiate the Lie algebra map to obtain a map on the Lorentz group. Second, we provide specific formulae for the spinor actions of the Lorentz group.

Since $(0, 1/2)$ is a representation of the Lorentz Lie algebra that does not extend to a representation of the Lorentz group, it is best to think of change of Lorentz frame as a continuous motion $\Lambda(\theta) = \exp(\theta X)$ generated by a Lie algebra element $X$. In general, the result of this continuous transformation of frames on a spinor can be unambiguously defined by exponentiating the representation map:

$$D_{s,t}\big(\Lambda(\theta)\big) \overset{\text{def}}{=} \exp\big(\theta D_{s,t}(X)\big). \tag{7.1.1}$$

Here the symbol $D_{s,t}$ is used for the representations of the Lie algebra and of the group. It is conventional to forget about the $\theta$ parameter in this definition and write $D_{s,t}(\Lambda)$ even though this notation has a sign ambiguity whenever $s + t$ is not an integer. We also note in passing that the exponential map is not surjective, that is, there are elements in the Lorentz group which are not of the form $e^X$ for any $X$ in $so(1,3)$. However, every element in the Lorentz group is a product of elements of the form $e^X$, so knowledge of $D_{s,t}$ on the elements $e^X$ determines $D_{s,t}$ on the whole group.

Since, as explained in 6.7.9 and Section 6.3, the spinor representations map the Lorentz Lie algebra of $4 \times 4$ matrices to the Lie algebra $sl(2, \mathbf{C})$ of $2 \times 2$ matrices, a spinor is a two-component field. Write $\psi_R$ and $\psi_L$ respectively for the right and left spinor field. Following the pattern 6.1.3 of the action of a unitary representation on fields, the transformation rule for Weyl spinors can therefore be written as a $2 \times 2$ matrix multiplying two-component vectors, together with a change of evaluation point:

$$\begin{aligned}
\Lambda \cdot \psi_R(x) &\overset{\text{def}}{=} U(\Lambda)^\dagger \psi_R(x) U(\Lambda) = D_{0,1/2}(\Lambda)\psi_R(\Lambda^{-1}x); \\
\Lambda \cdot \psi_L(x) &\overset{\text{def}}{=} U(\Lambda)^\dagger \psi_L(x) U(\Lambda) = D_{1/2,0}(\Lambda)\psi_L(\Lambda^{-1}x),
\end{aligned} \tag{7.1.2}$$

where $\Lambda$ is any element of the Lorentz group.

If we let $\Lambda$ be a rotation or a boost in the definition 7.1.2 above, then we see at once that the effect of rotation on the two spinors is identical, but the effect of boosts is opposite:

**Theorem 7.1.3.** *Let $R(\bar{e}, \theta)$ be the right-hand rotation about an axis $\bar{e}$ through an angle $\theta$, and let $B(\bar{e}, \phi)$ be the boost in the $\bar{e}$ direction with velocity parameter $\phi$. Then:*

$$\begin{aligned}
D_{0,1/2}\big(R(\bar{e}, \theta)\big) &= \exp(\bar{e}\cdot\bar{\tau}\theta), & D_{0,1/2}\big(B(\bar{e}, \phi)\big) &= \exp(\bar{e}\cdot\bar{\rho}\phi); \\
D_{1/2,0}\big(R(\bar{e}, \theta)\big) &= \exp(\bar{e}\cdot\bar{\tau}\theta), & D_{1/2,0}\big(B(\bar{e}, \phi)\big) &= \exp(-\bar{e}\cdot\bar{\rho}\phi). \qquad \square
\end{aligned}$$

This concrete formula for the spinor actions of the Lorentz group provides the first step towards constructing Lagrangian densities with Weyl spinors.

## 7.2 The Lorentz Group and $SL(2, \mathbb{C})$

In this section, we show how the vector and spinor representations of the Lorentz group can be written in terms of a group of $2 \times 2$ matrices, $SL(2, \mathbb{C})$.

Remember that $D_{0,1/2}$ and $D_{1/2,0}$ map the Lorentz Lie algebra into $sl(2, \mathbb{C})$, and that according to Homework 6.7.10, the relationship between the two maps is given by

$$D_{0,1/2}(X) = -D_{1/2,0}(X)^\dagger. \tag{7.2.1}$$

Exponentiating these points, we find that the spinor representations of the Lorentz group take values in $SL(2, \mathbb{C})$, the group of $2 \times 2$ complex matrices with unit determinant, and, on continuous changes of Lorentz frames, the relationship between left and right spin representations of the Lorentz group is given by:

$$D_{0,1/2}(\Lambda) = D_{1/2,0}(\Lambda)^{\dagger\,-1}. \tag{7.2.2}$$

This relationship between $SO(1,3)$ and $SL(2, \mathbb{C})$ is a little abstract, depending as it does on exponentiation of generators. There is, however, a natural map from $SL(2, \mathbb{C})$ to the Lorentz group which makes the relationship more concrete. To provide the foundation for this map, we first define a map [ ] which represents Minkowski space by $2 \times 2$ hermitian matrices:

$$[\,]: \mathbb{R}^{1,3} \to M_2(\mathbb{C}),$$
$$p \mapsto [p] = p \cdot \sigma = p^\mu \sigma_\mu. \tag{7.2.3}$$

From this definition, we find

$$[p] = \begin{pmatrix} p^0 + p^3 & p^1 - ip^2 \\ p^1 + ip^2 & p^0 - p^3 \end{pmatrix} = \begin{pmatrix} p_0 - p_3 & -p_1 + ip_2 \\ -p_1 - ip_2 & p_0 + p_3 \end{pmatrix}, \tag{7.2.4}$$

and so

$$\det[p] = |p|^2. \tag{7.2.5}$$

Since every $2 \times 2$ hermitian matrix is in the image of [ ], we can identify this space of matrices with Minkowski space.

Now, $SL(2, \mathbb{C})$ acts on the space of $2 \times 2$ matrices by

$$g: M \mapsto gMg^\dagger. \tag{7.2.6}$$

Clearly, if $M$ is hermitian, then so is $gMg^\dagger$, so this action of $SL(2, \mathbb{C})$ preserves Minkowski space. Furthermore:

$$\det(gMg^\dagger) = \det(M), \tag{7.2.7}$$

and so this action of $SL(2, \mathbb{C})$ preserves the Minkowski length of vectors. Therefore the matrix $D$ defined by

$$g[p]g^\dagger = [Dp] \quad \text{for all } p \tag{7.2.8}$$

is a Lorentz transformation. Thus 7.2.8 defines a map $D(g)$ from $SL(2, \mathbb{C})$ to $SO(1,3)$.

**Homework 7.2.9.** Show that $D$ is a group representation, i.e. $D(gh) = D(g)D(h)$.

By setting $g = \exp(\tau X)$, differentiating with respect to $\tau$ and evaluating at $\tau = 0$, we find that

$$D(\tau_r) = X_r \quad \text{and} \quad D(\rho_r) = B_r, \tag{7.2.10}$$

where the boost $B_r$ is taken as acting on $p^\mu$. Thus $D$ on the Lie algebras is the inverse of $D_{0,1/2}$. Since the action of $D$ on the groups satisfies $D(1_2) = D(-1_2) = 1_4$, the inverse of $D$ on the groups has a sign ambiguity, as we have noted before. Thus:

**Theorem 7.2.11.** *Let $D$ be the map from $SL(2, \mathbf{C})$ to $4 \times 4$ matrices defined by the condition $[D(g)p] = g[p]g^\dagger$ for all $p$. Then $D$ takes values in the Lorentz group, is multiplicative, and satisfies*

$$D_{0,1/2}(\Lambda) = \pm g \quad \Longleftrightarrow \quad D(g) = \Lambda.$$

The map from Minkowski space to $2 \times 2$ hermitian matrices is not unique. Another natural possibility is to define $[\;]_P$ by

$$[k]_P \stackrel{\text{def}}{=} [Pk], \tag{7.2.12}$$

where $P$ is the parity operator on Minkowski space. (The parity-transformed vector $Pk$ is commonly written $\tilde{k}$, a notation we shall use later.) Thus

$$[k]_P = \begin{pmatrix} k^0 - k^3 & -k^1 + ik^2 \\ -k^1 - ik^2 & k^0 + k^3 \end{pmatrix} = \begin{pmatrix} k_0 + k_3 & k_1 - ik_2 \\ k_1 + ik_2 & k_0 - k_3 \end{pmatrix}. \tag{7.2.13}$$

**Homework 7.2.14.** Show that $\text{tr}[p][\tilde{q}] = 2p\cdot q$.

We have already seen in Theorem 6.9.3 that parity transforms $(0, 1/2)$ into $(1/2, 0)$, and so

$$D_{0,1/2}(\Lambda) = D_{1/2,0}(P\Lambda P). \tag{7.2.15}$$

Linking this with 7.2.2, we find that

$$\begin{aligned} D_{0,1/2}(P\Lambda P) &= D_{0,1/2}(\Lambda)^{\dagger^{-1}}; \\ D_{1/2,0}(P\Lambda P) &= D_{1/2,0}(\Lambda)^{\dagger^{-1}}. \end{aligned} \tag{7.2.16}$$

Apply $D$ to the first equation and substitute $\Lambda = D(g)$ to find

$$PD(g)P = D(g^{\dagger^{-1}}). \tag{7.2.17}$$

Now the action of $SL(2, \mathbf{C})$ on $[k]_P$ can easily be computed in terms of $D$:

$$g[k]_P g^\dagger = g[Pk]g^\dagger = [D(g)Pk] = [PD(g)Pk]_P = [D(g^{\dagger^{-1}})k]_P. \tag{7.2.18}$$

Substituting $(g^\dagger)^{-1}$ for $g$ we find another form of this equation:

$$g^{\dagger^{-1}}[k]_P g^{-1} = [D(g)k]_P. \tag{7.2.19}$$

We write a theorem to summarize the results so far:

**Theorem 7.2.20.** *Using the map $D$ of Theorem 7.2.11 and writing $\Lambda = D(g)$, the Lorentz transformation of vectors and spinors determined by $\Lambda$,*

$$\Lambda \cdot \psi_R = D_{0,1/2}(\Lambda)\psi_R$$
$$\Lambda \cdot \psi_L = D_{1/2,0}(\Lambda)\psi_L \qquad \Lambda \cdot k = \Lambda k,$$

*can be expressed as follows:*

$$\Lambda \cdot \psi_R = g\psi_R, \qquad [\Lambda k]_P = g^{\dagger^{-1}}[k]_P g^{-1},$$
$$\Lambda \cdot \psi_L = g^{\dagger^{-1}}\psi_L, \qquad [\Lambda k] = g[k]g^{\dagger}. \qquad \square$$

With this theorem, it is obvious how to arrange spinors and vectors in order to make all the $g$'s cancel out.

**Homework 7.2.21.** Show that from $g[k]g^{\dagger} = [\Lambda k]$ we can deduce the coordinate expression

$$g^{-1}\sigma^{\mu}g^{\dagger^{-1}} = \Lambda^{\mu}{}_{\nu}\sigma^{\nu}.$$

# 7.3 Invariant Terms from Spinors and Vectors

On the foundation provided by the two previous sections, we will now give one more route to a spinor invariant term and then list all the possible renormalizable invariant terms involving spinors and vectors.

Remember that

$$(0,1/2) \otimes (0,1/2) = (0,0) + (0,1). \tag{7.3.1}$$

Clearly, the scalar must have the form $\psi_R^{\mathsf{T}}\varepsilon\psi_R$ for some $2\times 2$ matrix $\varepsilon$. The condition for this term to be invariant is

$$g: \psi_R^{\mathsf{T}}\varepsilon\psi_R \mapsto \psi_R^{\mathsf{T}}g^{\mathsf{T}}\varepsilon g\psi_R = \psi_R^{\mathsf{T}}\varepsilon\psi_R. \tag{7.3.2}$$

This implies that $\varepsilon$ satisfies $g^{\mathsf{T}}\varepsilon g = \varepsilon$.

**Remark 7.3.3.** If we rewrite this condition as $g\varepsilon = \varepsilon g^{\mathsf{T}^{-1}}$, then we see that $\varepsilon$ is a similarity between two representations of $SL(2,\mathbf{C})$. This has the significant consequence that $\varepsilon\psi_L^*$ tranforms as a *right* Weyl spinor:

$$\Lambda \cdot \varepsilon\psi_L^* = \varepsilon g^{\mathsf{T}^{-1}}\psi_L^* = g\varepsilon\psi_L^*. \tag{7.3.4}$$

It is therefore possible to write any theory of Weyl spinors in terms of left Weyl spinors alone. Because the transformation of $\psi_L$ into $\varepsilon\psi_L^*$ reverses the signs of all internal quantum numbers associated with the field, it can be regarded as charge conjugation.

Following Homework 6.9.2, we see that we can take $\varepsilon = i\sigma_2$ and so obtain the scalar

$$\psi_R^{\mathsf{T}}\varepsilon\psi_R' = (\psi_R^1 \quad \psi_R^2)\begin{pmatrix} 0 & 1 \\ -1 & 0 \end{pmatrix}\begin{pmatrix} \psi_R'^1 \\ \psi_R'^2 \end{pmatrix}$$
$$= \psi_R^1\psi_R'^2 - \psi_R^2\psi_R'^1. \tag{7.3.5}$$

Naturally the same construction works just as well for left spinors, so that $\psi_L^\top \varepsilon \psi_L$ is also a Lorentz scalar.

For Lagrangian building, the only vector we currently have is the vector of partial derivative operators. Using this for $k$ in the equations above, we can immediately write out all the invariant terms that can be constructed out of a left spinor, a right spinor, and the partial derivative vector:

$$\psi_R^\dagger [\partial]_P \psi_R, \quad \psi_L^\dagger [\partial] \psi_L, \quad \psi_R^\dagger \psi_L, \quad \psi_L^\dagger \psi_R, \quad \psi_R^\top \varepsilon \psi_R, \quad \psi_L^\top \varepsilon \psi_L. \qquad (7.3.6)$$

The first two terms are free-field Lagrangian densities for left- and right-handed Weyl spinors. The next two terms mix left and right. They are called Dirac mass terms and could be part of the free electron Lagrangian density. The last two terms conserve charge only for neutral spinors. They are called Majorana mass terms and can be part of a free massive neutrino Lagrangian density. The Majorana mass terms have the form $\psi^1 \psi^2 - \psi^2 \psi^1$. Thus, if these terms are to contribute, the fields $\psi^r$ will have to anticommute. This is our first glimpse of fermionic properties.

**Homework 7.3.7.** Following the argument of Section 4.13 on mass dimensions of renormalizable interactions, show that the mass dimension of a Weyl spinor is 3/2.

**Homework 7.3.8.** Verify that the products in 7.3.6 are Lorentz invariant, and check that the list is a complete list of renormalizable terms constructed from Weyl spinors alone. Extend the list to include renormalizable couplings to scalar fields.

Note that the specific forms of the invariant terms given here are basis dependent. We are assuming throughout that the $(0, 1/2)$ and $(1/2, 0)$ representations are given by the specific matrices used to define them in Chapter 6.

This section concludes our treatment of spinor management techniques. The final list 7.3.6 of possible terms for a Lagrangian density is the first application. The construction of polarization spinors for the free spinor quantum fields will be another, and the computation of scattering amplitudes will be a third.

## 7.4 Fermionic Calculus

As there are indications already that spinor fields are fermionic, even in preparation for the classical treatment of spinor fields, it will be useful to understand calculus of fermions.

If $\psi_r$ are components of fermionic fields which satisfy $\psi_r \psi_s + \psi_s \psi_r = 0$ whenever $r \neq s$, then the differential of a product is given as usual by

$$\begin{aligned}
d(\psi_1 \cdots \psi_n) &= \sum_{r=1}^n \psi_1 \cdots d\psi_r \cdots \psi_n \\
&= \sum_{r=1}^n (-1)^{n-r} \psi_1 \cdots \hat{\psi}_r \cdots \psi_n d\psi_r,
\end{aligned} \qquad (7.4.1)$$

where $\hat{\psi}_r$ stands for the term omitted from the product.

To establish the meaning of differentiation with respect to a Fermi field, we postulate the property

$$\frac{\partial}{\partial \psi_r}(\psi_1 \cdots \psi_n) d\psi_r = d(\psi_1 \cdots \psi_n). \qquad (7.4.2)$$

It is clear from comparison with the formula for the differential that we should define the fermionic derivative operator by

$$\frac{\partial}{\partial \psi_r}(\psi_1 \cdots \psi_n) \overset{\text{def}}{=} (-1)^{n-r} \psi_1 \cdots \hat{\psi}_r \cdots \psi_n. \tag{7.4.3}$$

In general, if $f$ is any polynomial function of $k$ variables, then we shall now find

$$\frac{\partial}{\partial \psi_r} f(\psi_1, \ldots, \psi_n) d\psi_r = df(\psi_1, \ldots, \psi_n). \tag{7.4.4}$$

**Homework 7.4.5.** Prove that fermionic derivatives anti-commute:

$$\frac{\partial}{\partial \psi_r} \frac{\partial}{\partial \psi_s} + \frac{\partial}{\partial \psi_s} \frac{\partial}{\partial \psi_r} = 0.$$

For a complex scalar field $\phi$, we saw in Section 3.5 that we could vary $\phi$ and $\phi^\dagger$ separately to get the Euler–Lagrange equations. For Fermi fields, it is also useful to consider $\psi^\dagger$ and $\psi$ as distinct variables. In the definition above, it was a matter of choice whether we put the differential $d\psi_r$ on the right or the left of the expressions. However, the Lagrangian density and conserved currents are built around expressions of the form $\psi^\dagger M \psi$, where $M$ is a matrix. Consequently, it is convenient to put $d\psi$ on the right, and $d\psi^\dagger$ on the left.

$$df(\psi^\dagger_1, \ldots, \psi^\dagger_m, \psi_1, \ldots, \psi_n) \overset{\text{def}}{=} d\psi^\dagger_r \frac{\partial f}{\partial \psi^\dagger_r} + \frac{\partial f}{\partial \psi_s} d\psi_s. \tag{7.4.6}$$

From this definition we can deduce the commutation properties of the operators and the fields:

|             | $\partial_r$ | $\partial_s^\dagger$ | $\psi_u^\dagger$ | $\psi_v$ |
|-------------|--------------|----------------------|------------------|----------|
| $\partial_r$ | $A$ | $C$ | $C$ | $C$ |
| $\partial_s^\dagger$ | $C$ | $A$ | $A$ | $A$ |
| $\psi_u^\dagger$ | $C$ | $A$ | $A$ | $A$ |
| $\psi_v$ | $C$ | $A$ | $A$ | $A$ |

$$(7.4.7)$$

where $\partial_r$ and $\partial_s^\dagger$ stand for $\partial/\partial\psi_r$ and $\partial/\partial\psi_s^\dagger$ respectively, and the entries $C$ and $A$ indicate commutation and anti-commutation (up to a possible additive constant in the case of equal indices).

**Remark 7.4.8.** This convention is unusual. Commonly, one puts all the differentials on the right. Then all the fermionic derivatives anti-commute. The price for this slight simplification of the algebra is the appearance of odd minus signs in the definitions of canonical momenta and Noether currents and also the destruction of natural matrix structure.

It is generally possible to suppress indices on the fermionic fields and derivatives. Taking $\psi$ to be a column vector of fields implies that $\psi^\dagger$ is a row vector. Since differentiating the scalar $\psi^\dagger\psi$ with respect to $\psi$ leaves a row, it is natural to regard $\partial/\partial\psi$ as a row vector. Similarly, $\partial/\partial\psi^\dagger$ may be treated as a column vector. These conventions lead to expressions of the following type:

$$\frac{\partial}{\partial\psi^\dagger}\psi^\dagger = 1 \quad \text{and} \quad \frac{\partial}{\partial\psi} A \frac{\partial}{\partial\psi^\dagger} \psi^\dagger B \psi = \text{tr}(AB). \tag{7.4.9}$$

**Homework 7.4.10.** Evaluate $\dfrac{\partial}{\partial \psi} A \dfrac{\partial}{\partial \psi^\dagger} \big( (\psi^\dagger B \psi)(\psi^\dagger C \psi) \big)$.

Just as it is convenient to regard a Lagrangian density for a complex scalar field as a polynomial in the field, its conjugate, and their derivatives, just so a Lagrangian density $\mathcal{L}$ for a Fermi field $\psi$ is a function

$$\mathcal{L} = \mathcal{L}(\psi^\dagger, \psi, \partial_\mu \psi^\dagger, \partial_\mu \psi). \tag{7.4.11}$$

The action $S$ is defined as usual by integrating $\mathcal{L}$ over a block of space-time. Variations of the action are given in terms of variations of $\mathcal{L}$ according to the formula 7.4.4. Straight away we see that the Euler–Lagrange equations for $\psi$ and $\psi^\dagger$ are

$$\partial_\mu \Big( \frac{\partial \mathcal{L}}{\partial (\partial_\mu \psi^\dagger)} \Big) = \frac{\partial \mathcal{L}}{\partial \psi^\dagger} \quad \text{and} \quad \partial_\mu \Big( \frac{\partial \mathcal{L}}{\partial (\partial_\mu \psi)} \Big) = \frac{\partial \mathcal{L}}{\partial \psi}. \tag{7.4.12}$$

We therefore define the canonical momenta $\Pi$ and $\Pi^\dagger$ by

$$\Pi \overset{\text{def}}{=} \frac{\partial \mathcal{L}}{\partial (\partial_0 \psi)} \quad \text{and} \quad \Pi^\dagger \overset{\text{def}}{=} \frac{\partial \mathcal{L}}{\partial (\partial_0 \psi^\dagger)}, \tag{7.4.13}$$

with the understanding that $\Pi$ is a row vector and $\Pi^\dagger$ is a column vector. Then we can define the Hamiltonian $\mathcal{H}$ as the Legendre transform of $\mathcal{L}$:

$$\mathcal{H}(\Pi, \Pi^\dagger, \psi^\dagger, \psi, \partial_r \psi^\dagger, \partial_r \psi) \overset{\text{def}}{=} \Pi \partial_0 \psi + \partial_0 \psi^\dagger \Pi^\dagger - \mathcal{L}(\psi^\dagger, \psi, \partial_\mu \psi^\dagger, \partial_\mu \psi). \tag{7.4.14}$$

The natural structure of row and column vectors combined with our convention for the placement of fermionic differentials suggests this definition. One can easily show that this transformation does indeed eliminate the differentials of time derivatives in favor of the differentials of momenta.

**Homework 7.4.15.** Show that $d\mathcal{H}$ does not depend on $d(\partial_0 \psi)$.

## 7.5 Classical Massless Weyl Spinors

The purpose of this section is to set up a covariant family of bases $\{u_L(p), u_R(p)\}$ of two-component polarization spinors by finding all the plane wave solutions for the classical, massless spinor equations and to present the matrix properties of this basis. These classical spinors will be used as coefficients for the creation and annihilation operators in the free spinor fields.

A Lagrangian density must have a kinetic term. Hence, writing $\bar{\partial}$ for $P\partial$, the only possibilities for Lagrangian densities for right and left free, massless Weyl spinors are:

$$\begin{aligned} \mathcal{L}_R &= c\psi_R^\dagger [\bar{\partial}] \psi_R = c(\psi_R^\dagger \partial_0 \psi_R + \psi_R^\dagger \bar{\sigma} \cdot \bar{\partial} \psi_R); \\ \mathcal{L}_L &= c\psi_L^\dagger [\partial] \psi_L = c(\psi_L^\dagger \partial_0 \psi_L - \psi_L^\dagger \bar{\sigma} \cdot \bar{\partial} \psi_L). \end{aligned} \tag{7.5.1}$$

The constant $c$ will be fixed later to $c = i$ when we compute the Hamiltonian. Note that these Lagrangian densities are not hermitian, but with this value of $c$, the Lagrangian density is hermitian up to a total divergence. For example:

$$\begin{aligned} \mathcal{L}_R - \mathcal{L}_R^\dagger &= i\psi_R^\dagger \delta^{\mu\nu} \sigma_\mu (\partial_\nu \psi_R) + i(\partial_\nu \psi_R^\dagger) \delta^{\mu\nu} \sigma_\mu \psi_R \\ &= \partial_\mu (\psi_R^\dagger \delta^{\mu\nu} \sigma_\nu \psi_R). \end{aligned} \tag{7.5.2}$$

The Euler–Lagrange equation derived from $\mathcal{L}_R$ is

$$\frac{d}{dx^\mu}\frac{\partial \mathcal{L}_R}{\partial(\partial_\mu \psi_R^\dagger)} = \frac{\partial \mathcal{L}_R}{\partial \psi_R^\dagger} \quad \Longrightarrow \quad [\tilde{\partial}]\psi_R = 0. \tag{7.5.3}$$

Working similarly on $\mathcal{L}_L$, we find

$$\begin{aligned}[\tilde{\partial}]\psi_R &= (\partial_0 + \bar{\sigma}\cdot\tilde{\partial})\psi_R = 0; \\ [\partial]\psi_L &= (\partial_0 - \bar{\sigma}\cdot\tilde{\partial})\psi_L = 0. \end{aligned} \tag{7.5.4}$$

These equations are actually 'square roots' of the Klein–Gordon equation, as we now show. By direct computation, for any vector $k$,

$$[\tilde{k}][k] = [k][\tilde{k}] = k^2 1_2. \tag{7.5.5}$$

Applying this to the equations of motion gives the desired result:

$$\begin{aligned}\partial_\mu \partial^\mu \psi_R &= [\partial][\tilde{\partial}]\psi_R = 0, \\ \partial_\mu \partial^\mu \psi_L &= [\tilde{\partial}][\partial]\psi_L = 0. \end{aligned} \tag{7.5.6}$$

Since $k^2 = \det[k] = \det[\tilde{k}]$, classical plane wave solutions of the equations of motion exist for all $k$ which satisfy $k^2 = 0$:

$$k^2 = 0 \quad \Longrightarrow \quad \begin{cases} \psi_R^k(x) = u_R \exp(\pm ix\cdot k), & \text{for } [\tilde{k}]u_R = 0; \\ \psi_L^k(x) = u_L \exp(\pm ix\cdot k), & \text{for } [k]u_L = 0. \end{cases} \tag{7.5.7}$$

Consider, for example, the solutions associated with motion in the $z$-direction. Let $k_z = (1,0,0,1)$ (so that $k_z^3 = 1$) and the equations become:

$$\begin{aligned}[\tilde{k}_z] = \sigma_0 - \sigma_3 = \begin{pmatrix} 0 & 0 \\ 0 & 2 \end{pmatrix} &\Rightarrow u_R = \begin{pmatrix} \sqrt{2} \\ 0 \end{pmatrix}; \\ [k_z] = \sigma_0 + \sigma_3 = \begin{pmatrix} 2 & 0 \\ 0 & 0 \end{pmatrix} &\Rightarrow u_L = \begin{pmatrix} 0 \\ \sqrt{2} \end{pmatrix}. \end{aligned} \tag{7.5.8}$$

The two-component vectors $u_R$ and $u_L$ are called Weyl polarization spinors, two-component spinors, or simply spinors.

This shows that, in general, $u_R$ is an eigenvector of $\rho_3$ with eigenvalue $+1/2$ for the component of spin in the direction of motion of the particle, while $u_L$ is an eigenvector with eigenvalue $-1/2$. These eigenvalues are the spins of the two waves in the direction of motion $\bar{k}/k^0 = (0,0,1)$. This definite relationship of spin to wave depends on the fact that the wave, being massless, travels at the speed of light, and so the component of spin in the direction of motion is a Lorentz invariant of the particle. This Lorentz invariant is called helicity. The fixed relationship of spin to wave also indicates that the free left and right Lagrangian densities are not separately parity invariant. Note that, despite being a two-component field, the Weyl spinors $\psi_R$ and $\psi_L$ actually have only one complex degree (two real degrees) of freedom each.

In order to write down the free Weyl fields in terms of creation and annihilation operators we shall need polarization spinors $u_R(k)$ and $u_L(k)$ for all plane waves. We can define these spinors up to sign by boosting the special case of $u_R$ and $u_L$:

$$u_R(\Lambda k_z) \stackrel{\text{def}}{=} D_{0,1/2}(\Lambda)u_R;$$
$$u_L(\Lambda k_z) \stackrel{\text{def}}{=} D_{1/2,0}(\Lambda)u_L. \qquad (7.5.9)$$

From this definition, writing $g$ for $D_{0,1/2}(\Lambda)$, we can show that the conditions for the solutions 7.5.7 are preserved:

$$\begin{aligned}
[\Lambda k_z]_P u_R(\Lambda k_z) &= \left((g^{-1})^\dagger [\tilde{k}_z] g^{-1}\right) u_R \\
&= (g^{-1})^\dagger [\tilde{k}_z] g u_R = 0; \\
[\Lambda k_z]_L u_L(\Lambda k_z) &= \left(g[\tilde{k}_z]g^\dagger\right)(g^{-1})^\dagger u_L \\
&= g[\tilde{k}_z] u_L = 0.
\end{aligned} \qquad (7.5.10)$$

The fundamental properties of the polarization spinors are summarized in the equations of motion condition 7.5.7, and in the normalization and completeness relations that follow.

**Homework 7.5.11.** Let $\mathcal{G} \subset so(1,3)$ be the vector space of solutions to $X k_z = 0$. Using the notation of Section 6.3, show that $V$ has a basis $\{A, B, R\}$ where

$$A \stackrel{\text{def}}{=} B_1 - X_2, \quad B \stackrel{\text{def}}{=} B_2 + X_1, \quad R \stackrel{\text{def}}{=} X_3.$$

By computing the $3 \times 3$ matrices $\text{ad}(A)$, $\text{ad}(B)$, and $\text{ad}(R)$ or otherwise, show that the Lie algebra $\mathcal{G}$ is isomorphic to the Lie algebra $e_2$ of the group of Euclidean transformations of $\mathbf{R}^2$.

For $X \in \mathcal{G}$, find the values of $D_{0,1/2}(X)u_R(k_z)$ and $D_{1/2,0}(X)u_L(k_z)$. Conclude that the definition 7.5.9 makes the phases of the polarization spinors undefined at $k_z$.

**Remark 7.5.12.** This way of defining $u_R(k)$ and $u_L(k)$ is ambiguous because $\Lambda k_z = k$ has many solutions. Since any two solutions $\Lambda_r$ of this equation are related by $\Lambda_1^{-1}\Lambda_2 k_z = k_z$, the extent of the ambiguity is the centralizer $C_z$ of $k_z$.

By working on the Lie algebra level, we find that $C_z$ is a three-dimensional group isomorphic to $E_2$, the group of Euclidean motions of $\mathbf{R}^2$. From the spinor representations of Theorem 7.1.3, we discover that the motions in $C_z$ modify $u_R(k_z)$ and $u_L(k_z)$ by phases only.

We could eliminate this phase ambiguity by specifying a multiplicative complement $C_z'$ of $C_z$ in the Lorentz group. A choice of $C_z'$ is in effect a convention for choosing a unique solution $\Lambda_k$ of $\Lambda k_z = k$. However, $C_z' = \{ \Lambda_k \mid k^2 = 0, k_0 > 0 \}$ cannot be a group, and the resulting spinor functions $u_R(k)$ and $u_L(k)$ cannot be continuous. For example, we might choose $\Lambda_k$ to be a minimal rotation of $\bar{k}_z$ into the direction $\bar{k}$ followed by a boost; the discontinuity is then at $\bar{k} = -\bar{k}_z$.

It is natural to want $u_R(\Lambda k) = D_{0,1/2}(\Lambda)u_R(k)$. However, since spinor representations have a sign ambiguity, this equality can only hold up to a sign. In fact, since $C_z'$ is not a group, we only have equality up to a phase:

$$u_H(\Lambda k) = e^{-i\theta_H(\Lambda, k)} D_H(\Lambda) u_H(k), \qquad (7.5.13)$$

where $H$ is $R$ or $L$.

In practice, since all these phases cancel in cross sections and decay rates, it is possible not only to leave $C_z'$ unspecified but even to suppress the phases.

**Theorem 7.5.14.** *The polarization spinors satisfy the normalization relations:*

$$u_R(k)^\dagger \sigma^\mu u_R(k) = 2(Pk)^\mu;$$
$$u_L(k)^\dagger \sigma^\mu u_L(k) = 2k^\mu.$$

**Proof.** The way to see this is to start from the case where $k = k_z$,

$$u_R(k_z)^\dagger \sigma u_R(k_z) = 2(1, 0, 0, -1) = 2Pk_z \quad \text{where } \sigma = (\sigma^0, \sigma^1, \sigma^2, \sigma^3), \quad (7.5.15)$$

take the inner product with $Pq$,

$$u_R(k_z)^\dagger [q]_P u_R(k_z) = 2Pq \cdot Pk_z = 2q \cdot k_z, \quad (7.5.16)$$

and apply a Lorentz transformation to all the vectors in this equation to find:

$$
\begin{aligned}
u_R(k)^\dagger [p]_P u_R(k) &= u_R(\Lambda k_z)^\dagger [\Lambda q]_P u_R(\Lambda k_z) \\
&= u_R(k_z)^\dagger g^\dagger \left( (g^\dagger)^{-1} [q]_P g^{-1} \right) g u_R(k_z) \\
&= u_R(k_z)^\dagger [q]_P u_R(k_z) \\
&= 2q \cdot k_z \\
&= 2p \cdot k.
\end{aligned}
\quad (7.5.17)
$$

Substitute $Pq$ for $p$ to find

$$u_R(k)^\dagger [q] u_R(k) = 2(Pq) \cdot k = 2q \cdot (Pk), \quad (7.5.18)$$

from which we obtain the statement of the theorem by equating coefficients of $q_\mu$.□

**Theorem 7.5.19.** *The polarization spinors satisfy the completeness relations:*

$$u_R(k) u_R(k)^\dagger = [k];$$
$$u_L(k) u_L(k)^\dagger = [k]_P. \qquad\qquad\qquad □$$

This proposition may easily be proved by applying a Lorentz transformation to the $k = k_z$ case.

**Homework 7.5.20.** Prove the left spinor normalization and completeness formulae in Theorems 7.5.14 and 7.5.19. Also show that right and left spinors are orthogonal:

$$u_R(k)^\dagger u_L(k) = 0 \quad \text{for all } k.$$

**Homework 7.5.21.** Show that polarization spinors $u_R(k)$ and $u_L(k)$ are eigenvectors of the matrix $u_R(k) u_R(k)^\dagger$. Find the corresponding eigenvalues.

**Homework 7.5.22.** Show that $u_R(Pk) = i u_L(k)$ for all $k$ on the lightcone.

In this section, we have proposed Lagrangian densities for left and right Weyl spinors, deduced the equations of motion, and found their plane wave solutions.

These solutions contain two-component polarization spinors, that is, covariant functions of the energy-momentum four-vector of the wave which satisfy the normalization and completeness relations above. When we quantize the free, massless spinor theories in the next section, we shall use these polarization spinors to build the free quantum field.

## 7.6 Free Massless Weyl Quantum Fields

Now we shall quantize the massless Weyl spinor theory. First, we guess the form of the free fields, then compute the Hamiltonian. The result shows that the constant $c$ in the Lagrangian density must be $i$, and that the creation and annihilation operators must anti-commute. Second, from these free-field results, we deduce the equal-time commutation relations that an interacting Weyl spinor field should satisfy. This sequence of steps is canonical quantization in reverse, but, being a reversible sequence, it is effectively equivalent to canonical quantization. The advantage of proceeding in this sequence is that it is easier to guess the form of the free spinor field than to guess canonical commutation relations for it.

The free-field expansion for the Weyl spinors takes the usual form for charged fields, but with the addition of the polarization spinors:

$$
\begin{aligned}
\psi_R(x) &= \int \frac{d^3\bar{k}}{(2\pi)^3\, 2\omega(\bar{k})}\; u_R(k)\big(e^{ix\cdot k}\beta_L(k)^\dagger + e^{-ix\cdot k}\alpha_R(k)\big); \\
\psi_L(x) &= \int \frac{d^3\bar{k}}{(2\pi)^3\, 2\omega(\bar{k})}\; u_L(k)\big(e^{ix\cdot k}\beta_R(k)^\dagger + e^{-ix\cdot k}\alpha_L(k)\big).
\end{aligned}
\tag{7.6.1}
$$

The notation $\beta_L$ in the expansion of $\psi_R$ is used since, as we shall show in Section 7.8, this operator destroys a left-handed anti-particle.

**Remark 7.6.2.** If the free spinor field is to transform according to 7.1.2, then we must balance the phase in the transformation law 7.5.13 for polarization spinors

$$
u_H(\Lambda k) = e^{-i\theta_H(\Lambda, k)} D_H(\Lambda) u_H(k)
\tag{7.6.3}
$$

with a phase in the transformation law for creation and annihilation operators:

$$
\left.
\begin{aligned}
U^\dagger(\Lambda)\alpha_H(\Lambda k)U(\Lambda) &= e^{i\theta_H(\Lambda,k)}\alpha(k) \\
U^\dagger(\Lambda)\beta_H(\Lambda k)U(\Lambda) &= e^{-i\theta_H(\Lambda,k)}\beta(k)
\end{aligned}
\right\}
\tag{7.6.4}
$$

These phases never appear in predictions and can therefore be suppressed.

To determine the properties of the creation and annihilation operators more fully, we shall compute the Hamiltonian for the free right Weyl field. We shall find that these operators must satisfy anti-commutation relations if the Hamiltonian is to be bounded below.

Applying the conventions of Section 7.4 to the Lagrangian density 7.5.1, the canonical momentum of the free right Weyl spinor is

$$
\Pi = (\Pi_1, \Pi_2) = c\big((\psi_R^1)^\dagger, (\psi_R^2)^\dagger\big) = c\psi_R^\dagger,
\tag{7.6.5}
$$

while the momentum $\Pi^\dagger$ associated with $\psi_R^\dagger$ is zero. Hence, taking the evolution equations into account, the Hamiltonian density for the free right Weyl spinor is:

$$\mathcal{H} = c :\Pi \dot\psi_R - \mathcal{L}: = -c :\psi_R^\dagger \bar\sigma \cdot \bar\partial \psi_R: = c :\psi_R^\dagger \partial_0 \psi_R: \tag{7.6.6}$$

Substituting the free-field expansion into this result and integrating over space yields the Hamiltonian:

$$H = c \int \frac{d^3\bar p}{(2\pi)^3 2\omega(\bar p)} \frac{d^3\bar q}{(2\pi)^3 2\omega(\bar q)} d^3\bar x \tag{7.6.7}$$

$$:(e^{-ix\cdot p}\beta_L(p) + e^{ix\cdot p}\alpha_R(p)^\dagger)u_R(p)^\dagger u_R(q)\partial_0\big(e^{ix\cdot q}\beta_L(q)^\dagger + e^{-ix\cdot q}\alpha_R(q)\big):$$

$$= c \int \frac{d^3\bar p}{(2\pi)^3 2\omega(\bar p)} \frac{d^3\bar q}{(2\pi)^3 2\omega(\bar q)} d^3\bar x \; u_R(p)^\dagger u_R(q) i\omega(\bar q)$$

$$:e^{-ix\cdot(p-q)}\beta_L(p)\beta_L(q)^\dagger + e^{ix\cdot(p+q)}\alpha_R(p)^\dagger\beta_L(q)^\dagger$$

$$- e^{-ix\cdot(p+q)}\beta_L(p)\alpha_R(q) - e^{ix\cdot(p-q)}\alpha_R(p)^\dagger\alpha_R(q):$$

$$= \frac{ic}{2} \int \frac{d^3\bar p}{(2\pi)^3 2\omega(\bar p)} d^3\bar q\; u_R(p)^\dagger u_R(q)\; \big(\delta^{(3)}(\bar p - \bar q):\beta_L(p)\beta_L(q)^\dagger - \alpha_R(p)^\dagger\alpha_R(q):$$

$$+ \delta^{(3)}(\bar p + \bar q):e^{ix^0\omega(\bar p)+ix^0\omega(\bar q)}\alpha_R(p)^\dagger\beta_L(q)^\dagger - e^{-ix^0\omega(\bar p)-ix^0\omega(\bar q)}\beta_L(p)\alpha_R(q):\big)$$

$$= \frac{ic}{2} \int \frac{d^3\bar p}{(2\pi)^3 2\omega(\bar p)} \; u_R(p)^\dagger u_R(p):\beta_L(p)\beta_L(p)^\dagger - \alpha_R(p)^\dagger\alpha_R(p):$$

$$+ u_R(p)^\dagger u_R(\tilde p):e^{2ix^0\omega(\bar p)}\alpha_R(p)^\dagger\beta_L(\tilde p)^\dagger - e^{-2ix^0\omega(\bar p)}\beta_L(p)\alpha_R(\tilde p):$$

The Homeworks 7.5.20 and 7.5.22 show that $u_R(p)$ is orthogonal to $u_R(\tilde p)$. Using the spinor normalization rule of Theorem 7.5.14, we therefore obtain

$$H = \frac{ic}{2} \int \frac{d^3\bar p}{(2\pi)^3} :\beta_L(p)\beta_L(p)^\dagger - \alpha_R(p)^\dagger\alpha_R(p):$$

$$= -ic \int d^3\bar p\; \omega(\bar p):-b_L(p)b_L(p)^\dagger + a_R(p)^\dagger a_R(p): \tag{7.6.8}$$

From this equation for the Hamiltonian, we see that if the Hamiltonian is to be bounded below, then $b_L(\bar p)$ must anti-commute with $b_L(\bar p)^\dagger$, and $c = i$. The remainder of this section deals with the canonical anti-commutation relations for Fermi operators and fields.

Anti-commutation relations are written using

$$\{A, B\} \stackrel{\text{def}}{=} AB + BA. \tag{7.6.9}$$

Prompted by the need for $b_L(\bar p)$ to anti-commute with $b_L(\bar q)^\dagger$, we now assume that creation and annihilation operators associated with spinor fields satisfy anti-commutation relations of the form

$$\{a(\bar p), a(\bar q)^\dagger\} = \delta^{(3)}(\bar p - \bar q), \tag{7.6.10}$$

with all other anti-commutators vanishing. We also assume that all fermionic creation and annihilation operators commute with all bosonic creation and annihilation operators.

Since creation operators anti-commute, the particle states that they create satisfy Fermi statistics. For example,

$$|\bar{p}, \bar{q}\rangle \stackrel{\text{def}}{=} a(\bar{p})^{\dagger} a(\bar{q})^{\dagger} |0\rangle = -a(\bar{q})^{\dagger} a(\bar{p})^{\dagger} |0\rangle = -|\bar{q}, \bar{p}\rangle. \qquad (7.6.11)$$

Bras are defined by

$$\langle \bar{p}, \bar{q}| \stackrel{\text{def}}{=} = \langle 0| a(\bar{p}) a(\bar{q}). \qquad (7.6.12)$$

The normalization of the states follows from the anti-commutation relations (see the normalization equation 2.1.3 and Homework 2.3.5 for the Bose equivalent):

$$\langle \bar{p}', \bar{q}' | \bar{p}, \bar{q}\rangle = \delta^{(3)}(\bar{p}' - \bar{q})\delta^{(3)}(\bar{q}' - \bar{p}) - \delta^{(3)}(\bar{p}' - \bar{p})\delta^{(3)}(\bar{q}' - \bar{q}). \qquad (7.6.13)$$

Application of these creation operators to the vacuum builds up a fermionic Fock space.

**Homework 7.6.14.** Prove this normalization relation for two-particle fermionic states.

**Remark 7.6.15.** Obviously, the phases in the transformation law 7.6.4 for creation and annihilation operators will come into the transformation laws for the particle states. These phases will factor out of interfering amplitudes and not be present in cross sections or decay rates.

From the free right Weyl field formula and the anti-commutation of the creation and annihilation operators we deduce the appropriate form of the equal-time anti-commutation relations for an interacting Weyl spinor field $\psi$:

$$\{\psi^{r}(t, \bar{x}), \psi^{s}(t, \bar{y})\} = 0, \qquad \{\psi^{r}(t, \bar{x})^{\dagger}, \psi^{s}(t, \bar{y})^{\dagger}\} = 0,$$
$$\{\psi^{r}(t, \bar{x}), \psi^{s}(t, \bar{y})^{\dagger}\} = \delta^{rs}\delta^{(3)}(\bar{x} - \bar{y}). \qquad (7.6.16)$$

These anti-commutators can also be written in matrix form following the pattern

$$\left\{ (a \quad b), \begin{pmatrix} c \\ d \end{pmatrix} \right\} \stackrel{\text{def}}{=} \begin{pmatrix} \{a, c\} & \{b, c\} \\ \{a, d\} & \{b, d\} \end{pmatrix}. \qquad (7.6.17)$$

For example,

$$\{\psi(t, \bar{x}), \psi(t, \bar{y})^{\dagger}\} = \delta^{(3)}(\bar{x} - \bar{y})\mathbf{1}. \qquad (7.6.18)$$

**Homework 7.6.19.** Assuming the structure 7.6.1 of the free Weyl field $\psi_R$, using the normalization and completeness results 7.5.14 and 7.5.19 for the classical spinor $u_R(k)$, prove that the anti-commutation relations 7.6.10 on $a_R$ and $b_R$ are equivalent to the canonical anti-commutation relations 7.6.16 on the field and its momentum.

Note that there is no classical limit for this quantum field: nothing classical has anti-commutation relations. Furthermore, single Fermi fields are not self-adjoint, so they are not observables. To be self-adjoint, an observable must contain a product of an even number of Fermi fields, and therefore observables still satisfy commutation relations. It is the observables, then, that have a sensible classical limit, not the field theory.

On the basis of the Weyl polarization spinors, we can write down the formula for the free Weyl fields. The assumption of canonical anti-commutation relations for the creation and annihilation operators leads to a meaningful Hamiltonian, and so we now have a quantum theory of the free Weyl fields. As in the case of the scalar field, this is enough for perturbative canonical quantization of a theory with interacting spinors.

The next section applies Chapter 3 to bring a conserved current and quantity out of the $U(1)$ symmetry of the free spinor Lagrangians. Then follow two sections which clarify our understanding of angular momentum for fields with intrinsic spin. After that we return to our standard program for perturbative canonical quantization by unfolding in sequence contractions, Wick operators, and Feynman diagrams and rules.

## 7.7  Conserved Current and Quantity

The Lagrangian densities 7.5.1 have $U(1)$ symmetries

$$\psi_R \to e^{-i\tau q}\psi_R \quad \text{and} \quad \psi_L \to e^{-i\tau q}\psi_L, \tag{7.7.1}$$

where $q$ is arbitrary and represents the charge of the fields. One can derive the associated conserved currents and quantities from these Lagrangian densities using the classical formulae 3.4.10 and 3.4.11 and the fermionic calculus conventions of Section 7.4. For example,

$$j_R^\mu \overset{\text{def}}{=} \Pi \mathbf{D}\psi_R = (i\psi_R^\dagger \tilde\sigma^\mu)(-iq\psi_R) = q\psi_R^\dagger \tilde\sigma^\mu \psi_R. \tag{7.7.2}$$

**Remark 7.7.3.** Properly, one should make the Lagrangian density hermitian; this creates a term with a $\partial_0\psi_R^\dagger$ factor, and hence a canonical momentum for $\psi_R^\dagger$. The resulting conserved current is as above. An outline follows.

The hermitian form of the Lagrangian density is

$$\mathcal{L}_R \overset{\text{def}}{=} \frac{i}{2}\psi_R^\dagger \delta^{\mu\nu}\sigma_\mu\partial_\nu\psi_R - \frac{i}{2}(\partial_\nu\psi_R^\dagger)\delta^{\mu\nu}\sigma_\mu\psi_R. \tag{7.7.4}$$

Let $\psi_{cL} = \varepsilon\psi_R^*$ (the subscript $c$ stands for the charge-conjugate, and the subscript $L$ indicates that this is a left spinor field; one could also write $\psi_{Rc}$ for this field). Using the relation

$$\sigma_\mu^\top = -\varepsilon\tilde\sigma_\mu\varepsilon, \tag{7.7.5}$$

we can put $\mathcal{L}_R$ into the form

$$\mathcal{L}_R = \frac{i}{2}\psi_R^\dagger \tilde\sigma^\mu\partial_\mu\psi_R + \frac{i}{2}\psi_{cL}^\dagger \sigma^\mu\partial_\mu\psi_{cL}. \tag{7.7.6}$$

The canonical momenta are now

$$\Pi_R^\mu \overset{\text{def}}{=} \frac{i}{2}\psi_R^\dagger \tilde\sigma^\mu \quad \text{and} \quad \Pi_L^\mu \overset{\text{def}}{=} \frac{i}{2}\psi_{cL}^\dagger \sigma^\mu, \tag{7.7.7}$$

the infinitesimal symmetry acts as

$$\mathbf{D}\psi_R = -iq\psi_R \quad \text{and} \quad \mathbf{D}\psi_{cL} = iq\psi_{cL}, \tag{7.7.8}$$

and the conserved current is

$$j^\mu \overset{\text{def}}{=} \Pi_R^\mu \mathbf{D}\psi_R + \Pi_{cL}^\mu \mathbf{D}\psi_{cL}$$
$$= \frac{q}{2}\psi_R^\dagger \tilde{\sigma}^\mu \psi_R - \frac{q}{2}\psi_{cL}^\dagger \sigma^\mu \psi_{cL}. \tag{7.7.9}$$

Here the first and second terms are separately hermitian; it is clear that the two terms are equal. Hence we regain the earlier value for the conserved current.

And the moral of that is, though the original Lagrangian densities are not hermitian, they work!

In contrast to the scalar case, Noether's formula inevitably yields a conserved quantity which does not annihilate the vacuum; normal ordering is necessary:

$$Q_R \overset{\text{def}}{=} \int d^3\bar{x} \, {:}j_R^0(t,\bar{x}){:} = q \int d^3\bar{x} \, {:}\psi_R^\dagger \psi_R{:} \tag{7.7.10}$$

since $\tilde{\sigma}^0 = 1$. This conserved quantity will generate the symmetry through commutation.

**Homework 7.7.11.** Verify the assertion above that the conserved quantity defined by Noether's formula does not annihilate the vacuum.

**Homework 7.7.12.** Use the canonical anti-commutation relations 7.6.16 to show that

$$[\psi_R, Q_R] = q\psi_R.$$

**Remark 7.7.13.** In this case of free Weyl fields, since the conserved quantity must be normally ordered, and since normal ordering should in principle modify an expression in interacting fields by more than a constant, there is as yet no reason to suppose that Noether's conservation law survives quantization. We shall resolve this doubt in Chapter 13, and in the meantime be optimistic.

On the basis of our fermionic calculus conventions, the derivation of conserved currents and quantities for theories with Fermi fields follows the pattern set out for scalar fields.

# 7.8 Helicity

A massive particle can be considered in its rest frame, and the Lorentz representation $D_{s,t}$ which acts on it can be restricted to the rotations, thus splitting into a sum of $D_r$'s with highest weight $s + t$. This highest weight is called the spin of the particle. Spin is a Lorentz invariant which helps to identify a particle. For a massless particle, however, there is no rest frame. The corresponding Lorentz invariant is the helicity, defined as the component of spin in the direction of motion:

$$\text{Helicity} \overset{\text{def}}{=} \frac{\bar{k} \cdot \bar{J}}{|\bar{k}|}. \tag{7.8.1}$$

Now that we have quantized our spinor field we can investigate its helicity more thoroughly. It is convenient to specify $k = k_z = (1,0,0,1)$ in the definition of helicity above. Then we only need to compute the angular momentum $J_z$ of spinor particles moving along the $z$-axis. We proceed in three steps.

First, we specify the particle in terms of the spinor field by letting $|k\rangle$ be the state created by $\alpha_R(k)^\dagger$, and observing that

$$\langle 0|\psi_R(0)|k\rangle = \int \frac{d^3\bar{p}}{(2\pi)^3 \, 2\omega(\bar{p})} \, \langle 0|(\beta_L(p)^\dagger + \alpha_R(p))\alpha_R(k)^\dagger|0\rangle u_R(p) \tag{7.8.2}$$
$$= u_R(k).$$

Second, we connect the angular-momentum operator $J_z$ to the field $\psi_R$ through the characteristic property of the conserved quantity:

$$e^{i\theta J_z}\psi_R(x)e^{-i\theta J_z} = U(e^{-\theta R_z})\psi_R(x)U(e^{\theta R_z})$$
$$= D_{0,1/2}(e^{\theta R_z})\psi_R(e^{-\theta R_z}x) \tag{7.8.3}$$
$$= e^{-i\theta\rho_3}\psi_R(e^{-\theta R_z}x).$$

Then, differentiating with respect to $\theta$ and evaluating at $\theta = 0$ and $x = 0$, we obtain the commutation equation

$$[\psi_R(0), J_z] = \rho_3\psi_R(0). \tag{7.8.4}$$

Third, we assume that $|k_z\rangle$ is an eigenstate of $J_z$ and that $J_z$ annihilates the vacuum, and then evaluate the eigenvalue $\lambda$ as follows:

$$\frac{1}{2}u_R(k_z) = \rho_3 u_R(k_z) = \langle 0|\rho_3\psi_R(0)|k_z\rangle$$
$$= \langle 0|[\psi_R(0), J_z]|k_z\rangle$$
$$= \langle 0|\psi_R(0)J_z|k_z\rangle \tag{7.8.5}$$
$$= \langle 0|\psi_R(0)\lambda|k_z\rangle$$
$$= \lambda u_R(k_z),$$

which implies that $\lambda = 1/2$, and so the helicity of $|k_z\rangle$ is $1/2$.

Applying this logic to all our creation operators yields:

$$\begin{array}{llll}
\alpha_R(k)^\dagger|0\rangle & J_k = \tfrac{1}{2} & \text{particle} & (\nu_R) \\
\beta_L(k)^\dagger|0\rangle & J_k = -\tfrac{1}{2} & \text{anti-particle} & (\bar{\nu}_L) \\
\alpha_L(k)^\dagger|0\rangle & J_k = -\tfrac{1}{2} & \text{particle} & (\nu_L) \\
\beta_R(k)^\dagger|0\rangle & J_k = \tfrac{1}{2} & \text{anti-particle} & (\bar{\nu}_R)
\end{array} \tag{7.8.6}$$

The right-hand column gives a possible interpretation of the state in terms of neutrinos. As we would expect, helicity determines handedness, and our convention for the subscripts on the annihilation operators is natural.

**Homework** 7.8.7. The proof in the text covers the first line in this table. Verify that the other lines are correct.

Assuming reasonable behavior of the angular-momentum operator, this section has shown that left-handed spinors have helicity $-1/2$ and right-handed ones have helicity $1/2$. The next section shows that the angular-momentum operator has the desired reasonable behavior.

# 7.9 Angular Momentum of a Free Spinor

The purpose of this section is to investigate the structure of the angular-momentum operator $\bar{J}$ in detail, and to probe the assumptions made in the computation of helicity that this operator annihilates the vacuum and has $|k, \pm\frac{1}{2}\rangle$ for eigenstates.

To probe these assumptions, we need a formula for $\bar{J}$, and for this, we go back to the methods of Section 3.8. The real action determined by a fixed Lorentz generator $X$ on the spinor field $\psi_R$ is given by

$$\psi'_R(\theta, \psi_R)(x) \stackrel{\text{def}}{=} D_{0,1/2}(e^{\theta X})\psi_R(e^{-\theta X}x). \tag{7.9.1}$$

Differentiating with respect to $\theta$, setting $\theta = 0$, writing $\tau(X)$ for $D_{0,1/2}$ on the Lorentz Lie algebra, and using the anti-symmetry of $X^{\alpha\beta}$ gives us the velocity of the action:

$$\mathbf{D}\psi_R(x) = \tau(X)\psi_R(x) + x_\alpha X^\alpha{}_\beta \partial^\beta \psi_R(x);$$

$$TD\partial^\mu \psi_R(x) = \partial^\mu \mathbf{D}\psi_R(x) \tag{7.9.2}$$

$$= \tau(X)\partial^\mu \psi_R(x) + X^\mu{}_\beta \partial^\beta \psi_R(x) + x_\alpha X^\alpha{}_\beta \partial^\beta \partial^\mu \psi_R(x).$$

Since the Lagrangian density is identically zero on solutions of the equations of motion, on this set of fields we have

$$\mathbf{D}\mathcal{L}_R = 0. \tag{7.9.3}$$

The formula for the conserved current associated with a solution $\psi_R$ of the field equations therefore yields:

$$j^\mu = \Pi^\mu \mathbf{D}\psi_R = \psi_R^\dagger i\tilde{\sigma}^\mu \tau(X)\psi_R + x_\beta X^\beta{}_\alpha(\psi_R^\dagger i\tilde{\sigma}^\mu \partial^\alpha \psi_R). \tag{7.9.4}$$

Clearly, the first term is due to the spinor structure of the fields, while the second is orbital angular-momentum density. Remarkably, as the following homeworks show, these two currents are not separately conserved.

**Homework 7.9.5.** Show that for arbitrary $\psi_R$, $\mathbf{D}\mathcal{L}_R = \partial^\mu(x_\alpha X^\alpha{}_\mu \mathcal{L})$.

When we integrate to form the angular-momentum conserved quantities, the angular momentum is therefore a sum of orbital angular momentum and something new, the internal angular momentum associated with the spinor structure. Even in the case of the free field, these two types of angular momentum are not separately conserved.

**Homework 7.9.6.** Writing $\tau(X)$ for $D_{0,1/2}(X)$, use the Lorentz group action 7.2.20 on vectors as $2 \times 2$ matrices to show that

$$\tilde{\sigma}_\mu \tau(X) + \tau(X)^\dagger \tilde{\sigma}_\mu + \tilde{\sigma}_\nu X^\nu{}_\mu = 0.$$

**Homework 7.9.7.** Using the previous homework, show that internal and orbital angular momentum are not separately conserved, but that their sum is conserved.

The angular momentum $\bar{J}$ may now be defined as usual. For concreteness we shall focus on $J_z$. Recalling that $iD(X_3) = \rho_3$, we find:

$$J_z \stackrel{\text{def}}{=} \int d^3\bar{x}\, \psi^\dagger(x)\sigma^0 \rho_3 \psi(x) + \psi^\dagger(x)i(x^2\partial_1 - x^1\partial_2)\psi(x). \tag{7.9.8}$$

If we try to write this operator in terms of creation and annihilation operators, the integrals are too ill-defined to make sense. There are two reasons for this: first, the angular momentum of a plane wave state is ill-defined; and second, the Lorentz invariance of the fields forces them to be non-vanishing at infinity, so the integral over space cannot converge.

It is remarkable that, though the second reason applies to all conserved quantities in quantum field theory, we have not met convergence troubles until now. Former divergences were modest, and we buried them by glibly setting

$$\int dx \, \exp(-ixp) \stackrel{?}{=} 2\pi\delta(p). \qquad (7.9.9)$$

Actually, this 'equality' arises from changing the order of integration when such a change is not mathematically possible. In detail, seeing a connection between the Fourier identity and the delta function,

$$\begin{aligned} F(q) &= \int dx \left( \int \frac{dp}{2\pi} \, e^{-i(q-p)x} F(p) \right) \\ &= \int dp \, \delta(q-p) F(p), \end{aligned} \qquad (7.9.10)$$

one is tempted to change the order of integration in the Fourier identity and deduce the equality in question. In practice, no problems will arise so long as the delta function is only used in situations where the $\int dp$ and the original order of integration can be restored, and the conditions of the Fourier Theorem hold.

In our current situation, the appearance of an extra factor of $x$ in the angular-momentum operator 7.9.8 aggravates the situation. Indeed, we must now propose 'equalities'

$$\int dx \, xe^{-ixp} \stackrel{?}{=} i\frac{\partial}{\partial p} \int dx \, e^{-ixp} \stackrel{?}{=} 2\pi i\delta'(p). \qquad (7.9.11)$$

These 'equalities' will not lead one astray as long as there is a well-behaved integral over $p$ in the background. In such cases one can integrate by parts to remove the derivative from the delta function. Thus, in effect

$$\delta'(p)F(p) = -\delta(p)F'(p). \qquad (7.9.12)$$

The integrals defining $\psi$, however, are not sufficiently well behaved, and the factors corresponding to $F$ above contain creation or annihilation operators, which are not even continuous (let alone differentiable) as operator functions of momenta. It is therefore no surprise that $J_z$ cannot be expressed in terms of creation and annihilation operators.

**Remark** 7.9.13. Actually, there is a branch of mathematics, *distribution theory*, which puts the manipulations above on an axiomatic footing.

Instead of trying to represent $J_z$ concretely in terms of creation and annihilation operators, we must therefore use the operator in the form 7.9.8 above. For example, to check that it annihilates the vacuum, we first observe that the only danger is of particle pair production, and second evaluate the matrix element for this possibility:

$$\langle l, -\tfrac{1}{2}; k, \tfrac{1}{2} | J_z | 0 \rangle \qquad (7.9.14)$$

$$= \int d^3\bar{x} \, \frac{d^3\bar{p}}{(2\pi)^3 \, 2\omega(\bar{p})}$$
$$\times \langle l, -\tfrac{1}{2}; k, \tfrac{1}{2} | \psi^\dagger(x) \big(\rho_3 + (x^1 p_2 - x^2 p_1)\big) e^{ip\cdot x} u_R(p) | p, -\tfrac{1}{2} \rangle$$

$$= \int d^3\bar{x} \, \frac{d^3\bar{p}}{(2\pi)^3 \, 2\omega(\bar{p})} \frac{1}{(2\pi)^3 \, 2\omega(\bar{k})}$$
$$\times \langle l, -\tfrac{1}{2} | e^{ik\cdot x} u_R^\dagger(k) \big(\rho_3 + (x^1 p_2 - x^2 p_1)\big) e^{ip\cdot x} u_R(p) | p, -\tfrac{1}{2} \rangle$$

$$= \int d^3\bar{x} \, \frac{1}{(2\pi)^3 \, 2\omega(\bar{k})} e^{i(k+l)\cdot x} u_R^\dagger(k) \big(\rho_3 + (x^1 l_2 - x^2 l_1)\big) u_R(l)$$

$$= \frac{1}{2\omega(\bar{k})} e^{2i\omega(\bar{k})t} u_R^\dagger(k) \big( \delta^{(3)}(\bar{k}+\bar{l}) \rho_3$$
$$- i\delta'(k_1 + l_1) \delta(k_2 + l_2) \delta(k_3 + l_3) l_2$$
$$+ i\delta(k_1 + l_1) \delta'(k_2 + l_2) \delta(k_3 + l_3) l_1 \big) u_R(l)$$

$$= \frac{1}{2\omega(\bar{k})} e^{2i\omega(\bar{k})t} \delta^{(3)}(\bar{k}+\bar{l}) u_R^\dagger(k) \big(\rho_3 + i(l_2\partial^1 - l_1\partial^2)\big) u_R(l).$$

Next, to evaluate the derivatives acting on the polarization spinor $u$, we use the Lorentz equivariance of $u_R$. Let $R(\theta) = \exp(\theta X_3)$ be rotation through $\theta$ about the $z$-axis. Then, by differentiating

$$u_R\big(R(\theta)l\big) = D_{0,1/2}\big(R(\theta)\big) u_R(l) \qquad (7.9.15)$$

with respect to $\theta$ and evaluating at $\theta = 0$ we find

$$(l_1\partial^2 - l_2\partial^1) u_R(l) = D_{0,1/2}(X_3) u_R(l) = \tau_3 u_R(l). \qquad (7.9.16)$$

Substituting this in the final expression above yields zero, showing that $J_z$ does indeed annihilate the vacuum.

**Homework 7.9.17.** Define the internal angular-momentum operators $J_r^{\text{int}}$ by

$$J_r^{\text{int}} \overset{\text{def}}{=} \int d^3\bar{x} \, \psi^\dagger(x) i\bar{\sigma}^0 D(X_r) \psi(x)$$

$$= \int d^3\bar{x} \, \psi^\dagger(x) \rho_r \psi(x).$$

Use the canonical anti-commutation relations for the field to show that $J_r^{\text{int}}$ generates the associated internal Lorentz transformation:

$$[\psi(t,\bar{y}), J_r^{\text{int}}] = \rho_r \psi(t,\bar{y}).$$

**Homework 7.9.18.** Show that the states $|k_z, \tfrac{1}{2}\rangle$ created by $a^\dagger(k_z)$ are eigenstates of the angular-momentum operator $J_z$.

From the delicate points raised above on the convergence of integrals for conserved quantities, we see that conserved quantities operators often have a somewhat tenuous existence. Therefore, we shall not be surprised to discover quantum effects in interacting fields which violate the quantum version of classical conservation laws. However, even when the defining integrals fail to converge, the associated matrix elements are generally well defined. This completes our diversion into angular momentum, and we return in the next section to the business of perturbative canonical quantization.

## 7.10  Contractions for Massless Weyl Spinors

In this section, we shall take the first step towards Feynman rules for spinor fields by extending time ordering and normal ordering to Fermi fields. The bridge between Dyson's formula and the Feynman rules will again be the contraction of fields that results from expressing a time-ordered product as a sum of normally ordered products, so we conclude this section with the formulae for the contractions of the various combinations of Weyl fields that we shall use later.

The time-ordered product of Bose fields is independent of the order of the fields, but changing the order of components of Fermi fields introduces a factor of minus one for each interchange of Fermi fields:

$$T\big(A(x)B(y)\big) = -T\big(B(y)A(x)\big) \quad \text{for Fermi fields.} \tag{7.10.1}$$

In the interaction Hamiltonian, all Fermi fields occur in pairs, so time ordering of the terms in Dyson's formula exchanges pairs of Fermi fields, leaving the signs of the terms unchanged.

The normally ordered product of free Bose fields does not depend on the order of the fields in the product, but a normally ordered product of components of Fermi fields changes sign whenever any two neighboring fields are interchanged:

$$:A(x)B(y): = -:B(y)A(x): \quad \text{for Fermi fields.} \tag{7.10.2}$$

As for Bose fields, the contraction of components of Fermi fields is the difference between time and normal ordering:

$$\acute{A}(x)\grave{B}(y) \stackrel{\text{def}}{=} T\big(A(x)B(y)\big) - :A(x)B(y): = -\grave{B}(y)\acute{A}(x). \tag{7.10.3}$$

If we assume that $x^0 > y^0$, then we can express the right-hand side in terms of the anti-commutator of the Fermi fields:

$$T\big(A(x)B(y)\big) - :A(x)B(y): \tag{7.10.4}$$
$$= \big(A^+(x) + A^-(x)\big)\big(B^+(y) + B^-(y)\big)$$
$$\quad - \big(A^+(x)B^+(y) + A^+(x)B^-(y) - B^+(y)A^-(x) + A^-(x)B^-(y)\big)$$
$$= \big\{A^-(x), B^+(y)\big\}, \quad \text{when } x^0 > y^0.$$

Thus, for components $A$ and $B$ of two free Fermi fields $\psi$ and $\psi'$, we have:

$$\acute{A}(x)\grave{B}(y) = \begin{cases} \{A(x)^-, B(y)^+\}, & \text{if } x^0 > y^0; \\ -\{B(y)^-, A(x)^+\}, & \text{if } y^0 > x^0. \end{cases} \tag{7.10.5}$$

Since we now see that the contraction of free Fermi fields is a scalar, not an operator, we can evaluate the contraction by taking the vacuum expectation of the definition 7.10.3:

$$\acute{A}(x)\grave{B}(y) = \langle 0|T(A(x)B(y))|0\rangle. \qquad (7.10.6)$$

From the anti-commutation relations for the Fermi creation and annihilation operators, only when $\psi' = \psi^*$ is this contraction non-zero; a contraction of Fermi fields related to different particle types is zero.

As Weyl fields have two components, it will be convenient to extend the definition of contraction to Fermi fields $\psi$ as follows:

$$\acute{\psi}(x)\grave{\psi}^\dagger(y) \overset{\text{def}}{=} \begin{pmatrix} \acute{\psi}_1\grave{\psi}_1^\dagger & \acute{\psi}_1\grave{\psi}_2^\dagger \\ \acute{\psi}_2\grave{\psi}_1^\dagger & \acute{\psi}_2\grave{\psi}_2^\dagger \end{pmatrix} \overset{\text{def}}{=} -\,\grave{\psi}^\dagger(y)\acute{\psi}(x). \qquad (7.10.7)$$

The following theorem summarizes all we need to know about contractions of Weyl fields.

**Theorem 7.10.8.** *The non-zero contractions between left and right Weyl fields are given by*

$$\left.\begin{aligned}
\acute{\psi}_R(x)\grave{\psi}_R^\dagger(y) &= \int \frac{d^4k}{(2\pi)^4}\, \frac{i[k]}{k^2 + i\epsilon}\, e^{-i(x-y)\cdot k} \\
\acute{\psi}_R(x)^*\grave{\psi}_R(y)^\top &= \int \frac{d^4k}{(2\pi)^4}\, \frac{i\epsilon[\tilde{k}]\epsilon^{-1}}{k^2 + i\epsilon}\, e^{-i(x-y)\cdot k} \\
\acute{\psi}_L(x)\grave{\psi}_L^\dagger(y) &= \int \frac{d^4k}{(2\pi)^4}\, \frac{i[\tilde{k}]}{k^2 + i\epsilon}\, e^{-i(x-y)\cdot k} \\
\acute{\psi}_L(x)^*\grave{\psi}_L(y)^\top &= \int \frac{d^4k}{(2\pi)^4}\, \frac{i\epsilon[k]\epsilon^{-1}}{k^2 + i\epsilon}\, e^{-i(x-y)\cdot k}
\end{aligned}\right\}$$

*on account of the identity* $[k]^* = \epsilon[\tilde{k}]\epsilon^{-1}$. *In these formulae, the parameter $k$ represents momentum flowing from $y$ to $x$.*

**Proof.** Again assuming that $x^0 > y^0$, we can compute the contraction by using the free-field expansion of $\psi_L$:

$$\acute{\psi}_L(x)\grave{\psi}_L^\dagger(y) \qquad\qquad\qquad\qquad\qquad\qquad\qquad (7.10.9)$$

$$= \langle 0|\{\psi_L(x)^-, \psi_L^\dagger(y)^+\}|0\rangle$$

$$= \int \frac{d^3\bar{p}}{(2\pi)^3\, 2\omega(\bar{p})}\, \frac{d^3\bar{q}}{(2\pi)^3\, 2\omega(\bar{q})}\, u_L(p)u_L(q)^\dagger\, \langle 0|\{e^{-ix\cdot p}\alpha_L(p), e^{iy\cdot q}\alpha_L^\dagger(q)\}|0\rangle$$

$$= \int \frac{d^3\bar{p}}{(2\pi)^3\, 2\omega(\bar{p})}\, \frac{d^3\bar{q}}{(2\pi)^3\, 2\omega(\bar{q})}\, u_L(p)u_L(q)^\dagger e^{-ix\cdot p}e^{iy\cdot q}(2\pi)^3\, 2\omega(\bar{p})\delta^{(3)}(\bar{p} - \bar{q})$$

$$= \int \frac{d^3\bar{p}}{(2\pi)^3\, 2\omega(\bar{p})}\, u_L(p)u_L(p)^\dagger e^{-i(x-y)\cdot p}.$$

Now we use the method of Homework 2.4.30 and the normalization condition $u_L(p)u_L(p)^\dagger = [\bar{p}]$ to obtain a standard integral expression for the contraction of Weyl fields.

Similar reasoning applies to the other three contractions.    □

For example, with the interaction term

$$\mathcal{H}_I = \frac{g}{2}\psi_L^\mathsf{T}\varepsilon\psi_L\phi - \frac{g^*}{2}\psi_L^\dagger\varepsilon\psi_L^*\phi, \tag{7.10.10}$$

there are four non-zero contractions which connect one field in the first term to one in the second:

$$\overset{\frown}{\psi_L(x)^\mathsf{T}\psi_L(y)^\dagger}, \quad \overset{\frown}{\psi_L(x)^\mathsf{T}\psi_L(y)^*}, \quad \overset{\frown}{\psi_L(x)\psi_L(y)^\dagger}, \quad \overset{\frown}{\psi_L(x)\psi_L(y)^*}. \tag{7.10.11}$$

The matrix structure of the first and fourth options is not good. However, due to the invariance of each term under transpose, when constructing a Wick operator from a Wick diagram, we can always choose a contraction pattern that uses only the second and third options. Note that the factors of $\frac{1}{2}$ are necessary if our Wick and Feynman rules for combinatoric factors are to work.

As in the scalar case, the relationship between time ordering, normal ordering, and contraction supports the application of Wick's Theorem for converting time-ordered products into normal-ordered ones. We shall therefore find Feynman rules as before from Dyson's formula and Wick's Theorem. The only novelties will be, first, the overall sign of a Wick operator or invariant amplitude arising from the anti-commutation of the fields, and second, the concern for ordering and transposing the fields in order to preserve matrix multiplication in computations involving spinors.

## 7.11  Defining an Example

To illustrate Feynman rules for Fermi fields, we take the simplest possible Lagrangian density, one which couples a left Weyl spinor field $\psi_L$ to a scalar field $\phi$ using the Majorana-mass construction:

$$\mathcal{L} = \frac{1}{2}\partial_\mu\phi\,\partial^\mu\phi + \psi_L^\dagger i[\tilde{\partial}]\psi_L - \frac{\mu^2}{2}\phi^2 - \frac{g}{2}\psi_L^\mathsf{T}\varepsilon\psi_L\phi + \frac{g^*}{2}\psi_L^\dagger\varepsilon\psi_L^*\phi. \tag{7.11.1}$$

The two parts of the coupling are necessary to make the Lagrangian hermitian (up to a divergence). This Lagrangian density describes an interaction of a massless neutrino with a massive scalar field. The massive scalar could represent the $Z$ boson of the Standard Model of electroweak interactions, or more realistically, it could be a Higgs field which would, in a more complete theory, gain a vacuum expectation, and thus give the neutrino a Majorana mass $g\langle\phi\rangle$.

**Remark 7.11.2.** We shall use $g$ and $-g^*$ as coupling constants in order to distinguish between the two interaction monomials. In fact, it is sensible to choose the phase for $\psi_L$ which makes $g = i\lambda$ for real, positive $\lambda$. Since $\varepsilon = i\sigma_2$, the interaction Lagrangian density can now be written

$$\mathcal{L}' = -\frac{\lambda}{2}\psi_L^\mathsf{T}\sigma_2\psi_L\phi - \frac{\lambda}{2}\psi_L^\dagger\sigma_2\psi_L^*\phi.$$

The second-order term in Dyson's formula is:

$$S_2 = \frac{(-i)^2}{2!}\int d^4x\,d^4y \tag{7.11.3}$$

$$\times T\left(\frac{g}{2}\psi_L^\mathsf{T}\varepsilon\psi_L\phi + \frac{-g^*}{2}\psi_L^\dagger\varepsilon\psi_L^*\phi\right)_x\left(\frac{g}{2}\psi_L^\mathsf{T}\varepsilon\psi_L\phi + \frac{-g^*}{2}\psi_L^\dagger\varepsilon\psi_L^*\phi\right)_y.$$

When we normal order this term we get contractions of fields. When we split the uncontracted fields into creation and annihilation parts, then we see that $S_2$ finds expression in many different scattering and decay processes. To make the link between the Weyl fields and particles concrete, we convert the table of helicities 7.8.6 into the following table of fields and particles:

| Field | Destroys Outgoing | Destroys Incoming | Polarization Spinor | |
|-------|-------------------|-------------------|---------------------|---|
| $\psi_L$ | $\bar{\nu}_R, \frac{1}{2}$ | $\nu_L, -\frac{1}{2}$ | $u_L$ | (7.11.4) |
| $\psi_L^*$ | $\nu_L, -\frac{1}{2}$ | $\bar{\nu}_R, \frac{1}{2}$ | $u_L^*$ | |

The right-handed particle $\bar{\nu}_R$ is the anti-particle of $\nu_L$. The table shows how to match fields in a Wick operator to incoming and outgoing states. Of course, 'destroys outgoing' is equivalent to 'creates incoming,' and 'destroys incoming' is equivalent to 'creates outgoing,' but these other terms have no application in our current formulation of scattering theory.

This establishes a theory. The next three section bring out the Feynman rules for it by comparing specific amplitudes with the appropriate Feynman diagrams.

## 7.12 – First Application

In this section, we shall compute a simple scattering amplitude on the basis of the Lagrangian density 7.11.1, set up Wick vertices, Wick diagrams and Feynman diagrams, and, by comparison of the amplitude with the appropriate Feynman diagrams, deduce the Feynman rules for vertices and for external neutrinos.

Suppose we focus on the contribution of the the second-order term to $\nu_L \nu_L \rightarrow \nu_L \nu_L$ scattering:

$$\mathcal{M} = \langle q', -\tfrac{1}{2}; p', -\tfrac{1}{2} | S_2 | p, -\tfrac{1}{2}; q, -\tfrac{1}{2} \rangle. \tag{7.12.1}$$

We have labeled the bras and kets according to our convention for fermionic Fock space, namely, that the particle nearest the vertical line was created last:

$$\langle q', -\tfrac{1}{2}; p', -\tfrac{1}{2} | p, -\tfrac{1}{2}; q, -\tfrac{1}{2} \rangle \tag{7.12.2}$$

$$\stackrel{\text{def}}{=} \langle 0 | \alpha(q')\alpha(p')\alpha(p)^\dagger \alpha(q)^\dagger | 0 \rangle$$

$$= (2\pi)^3 \, 2\omega(\bar{p})(2\pi)^3 \, 2\omega(\bar{q})$$

$$\times \left( \delta^{(3)}(\bar{p}' - \bar{p})\delta^{(3)}(\bar{q}' - \bar{q}) - \delta^{(3)}(\bar{p}' - \bar{q})\delta^{(3)}(\bar{q}' - \bar{p}) \right).$$

To destroy the incoming neutrinos, we need two $\psi_L$'s; to destroy the outgoing neutrinos, we need two $\psi_L^*$'s. The part of $S_2$ that contributes must be the cross term in the product of the two brackets in 7.11.3 with the $\phi$ fields contracted:

$$\mathcal{M} = \frac{1}{2!} \int d^4x \, d^4y \, 2 \frac{-ig}{2} \frac{ig^*}{2} \overset{\centerdot}{\phi}(x)\overset{\centerdot}{\phi}(y)\langle q', -\tfrac{1}{2}; p', -\tfrac{1}{2} | \tag{7.12.3}$$

$$\times \psi_L^\dagger(x)_+ \varepsilon \psi_L^*(x)_+ \psi_L^\mathsf{T}(y)_- \varepsilon \psi_L(y)_- | p, -\tfrac{1}{2}; q, -\tfrac{1}{2} \rangle.$$

Noting that

$$\langle 0 | \alpha_L(k)\alpha_L(l) | p, -\tfrac{1}{2}; q, -\tfrac{1}{2} \rangle \tag{7.12.4}$$

$$= (2\pi)^3 \, 2\omega(\bar{p})(2\pi)^3 \, 2\omega(\bar{q})\left( \delta^{(3)}(\bar{l} - \bar{p})\delta^{(3)}(\bar{k} - \bar{q}) - \delta^{(3)}(\bar{l} - \bar{q})\delta^{(3)}(\bar{k} - \bar{p}) \right),$$

we can see that the integrand contains a sum of four products of delta functions, each product containing four delta functions. Changing integration parameters and using the anti-symmetry of $u_L(q)^{\mathsf{T}} \varepsilon u_L(p)$ and $u_L(q')^{\dagger} \varepsilon u_L(p')^*$, we can see that all four terms in this sum are equal. Hence:

$$
\begin{aligned}
\mathcal{M} = gg^* & \int \frac{d^4k}{(2\pi)^4} d^4x\, d^4y \, \frac{i}{k^2 - \mu^2 + i\epsilon} e^{-i(x-y)\cdot k} \\
& \times u_L(q)^{\mathsf{T}} \varepsilon u_L(p) e^{-iy\cdot(q+p)} u_L(q')^{\dagger} \varepsilon u_L(p')^* e^{ix\cdot(p'+q')} \\
= gg^* & \int \frac{d^4k}{(2\pi)^4} (2\pi)^4 \delta^{(4)}(k - p' - q')(2\pi)^4 \delta^{(4)}(p + q - k) \\
& \times u_L(q)^{\mathsf{T}} \varepsilon u_L(p) u_L(q')^{\dagger} \varepsilon u_L(p')^* \\
= (2\pi)^4 & \delta^{(4)}(p' + q' - p - q) \\
& \times \frac{i}{(p+q)^2 - \mu^2 + i\epsilon} (gg^*) u_L(q)^{\mathsf{T}} \varepsilon u_L(p) u_L(q')^{\dagger} \varepsilon u_L(p')^*,
\end{aligned}
\tag{7.12.5}
$$

from which we obtain the invariant amplitude

$$
i\mathcal{A} = \frac{i}{(p+q)^2 - \mu^2 + i\epsilon} (gg^*) u_L(q)^{\mathsf{T}} \varepsilon u_L(p) u_L(q')^{\dagger} \varepsilon u_L(p')^*.
\tag{7.12.6}
$$

The preceding points may be represented in terms of Wick and Feynman diagrams. First, the two Wick vertices representing the two terms in the interaction Hamiltonian density are

**Wick Vertices:** $\psi_L \underset{}{\searrow} \,{-ig\varepsilon}\, \underset{}{\swarrow} \psi_L^{\mathsf{T}}$   and   $\psi_L^* \underset{}{\searrow} \,{ig^*\varepsilon}\, \underset{}{\swarrow} \psi_L^{\dagger}$
$$\hspace{3.3cm} |\phi \hspace{3.6cm} |\phi \tag{7.12.7}$$

The arrows on the Fermi lines of these Wick vertices indicate left-handed helicity. As this quantity is not conserved, the two arrows at a vertex can both point inwards (or outwards). The joining constraint for lines with arrows, that an incoming arrow should be matched with an outgoing one, correctly disallows the construction of Wick diagrams which contain contractions which evaluate to zero. Furthermore, it is clear that the connected components of the set of Fermi lines in a Wick diagram are either Fermi lines which pass through the diagram or Fermi loops.

As there are two Fermi lines at each vertex and as Fermi lines are only joined to Fermi lines, clearly the Fermi lines in any Wick diagram join together to form either loops or lines which pass through the diagram. We shall refer to these loops or lines as the *maximal Fermi lines* in the diagram.

Second, the Wick operator which mediates $\nu_L \nu_L \to \nu_L \nu_L$ at second order is given by the diagram

**Wick Diagram:** $\quad$
$$
\begin{array}{ccc}
\psi_L^{\dagger} & & \psi_L^{\mathsf{T}} \\
& \diagdown \diagup & \\
ig^*\varepsilon & \diagup\diagdown & -ig\varepsilon \\
& & \\
\psi_L^* & & \psi_L
\end{array}
\tag{7.12.8}
$$

The contraction number for this Wick diagram is $c(D) = 1$, the symmetry number is $s(D) = 4$, and our standard rule $1/s(D)$ gives the correct combinatoric factor for $S(D)$ in the second-order operator 7.11.3.

Third, the corresponding Feynman diagram is

**Feynman Diagram:** $\qquad q', u_L^\dagger \diagdown \qquad \diagup u_L^\top, q$
$$ig^*\varepsilon \quad —igε \tag{7.12.9}$$
$\qquad\qquad\qquad\qquad\qquad p', u_L^* \diagup \qquad \diagdown u_L, p$

The symmetry number for this diagram is $s(D) = 1$, and so the absence of numerical factors in the final amplitude 7.12.6 can be predicted from the diagram alone.

To deduce Feynman rules which connect the Feynman diagram to the integral 7.12.5 for the amplitude, we observe that the matrix products in the integrand may be associated with the two Fermi lines in this diagram. The unusual feature of this particular theory is that the neutrino, having no charge, is its own charge conjugate. Hence the arrows do not give a consistent direction to Fermi lines in a Feynman diagram. Therefore the Feynman rule for Fermi lines is

1. Choose an orientation for each maximal Fermi line;
2. Multiply factors as they occur along each Fermi line according to the rule 'later on the left.'

We have already indicated a choice of orientation for the maximal Fermi lines in the diagrams above by putting fields or polarization spinors at the beginnings of the lines and transposes of fields and polarization spinors at the ends. In 7.12.9, for example, the 'later on the left' rule applied to the Fermi line on the right yields

$$\text{Orientation } p \to q \quad \Longrightarrow \quad \text{Product } u_L^\top(q)(-i\varepsilon)u_L(p). \tag{7.12.10}$$

By further inspection of the integrand, we find the Feynman rules for external neutrino lines:

$$\begin{aligned}\text{Incoming } \nu_L \text{ at momentum } p: &\quad u_L(p)\\ \text{Outgoing } \nu_L \text{ at momentum } p': &\quad u_L(p')^\dagger\end{aligned} \tag{7.12.11}$$

where it is understood that, when orientations have been chosen, we transpose these polarization spinors if necessary so that a column vector occurs at the beginning of a Fermi line and a row vector at the end.

Finally, the Feynman rules for the vertex factors clearly come directly from the interaction Lagrangian density:

$$\begin{aligned}\psi_L^\top \varepsilon \psi_L \phi \text{ vertex:} &\quad -ig\varepsilon(2\pi)^4\delta^{(4)}(P_{\text{in}})\\ \psi_L^\dagger \varepsilon \psi_L^* \phi \text{ vertex:} &\quad ig^*\varepsilon(2\pi)^4\delta^{(4)}(P_{\text{in}})\end{aligned} \tag{7.12.12}$$

The Feynman rules listed so far will not enable us to obtain the correct sign for the amplitude. The problem does not arise on the level of the Wick operator because $\psi_L^\top \varepsilon \psi_L$ is invariant under transpose. It arises on the level of the Feynman amplitude because firstly, $u_L^\top \varepsilon u_L$ changes sign under transpose, and secondly, the associated interchange of fermions in the initial or final state is overlooked.

Since we are only interested in the absolute values of amplitudes, the absolute sign of terms in the amplitude is of no interest. It is therefore more practical to have

Feynman rules that will give the correct relative sign between Feynman integrals that contribute to the same process. The next example makes this point in detail.

**Remark 7.12.13.** Indeed, it is obvious that the sign of an amplitude changes when we exchange fermions in either the initial or the final state, but that this exchange does not affect the Feynman diagram. To get the correct sign for an amplitude, we must therefore build the structure of the initial and final state into either the Feynman diagram or the Feynman rules. It is simplest to add an element of interpretation to the Feynman diagram which determines these states as follows.

In assigning polarization spinors and their transposes to external lines in a Feynman diagram, we have in effect chosen an orientation for each maximal Fermi line. Remembering the association between $u_L^\top \varepsilon u_L$ and $\psi_L^\top \varepsilon \psi_L$, we build an initial and final state from $|0\rangle$ and $\langle 0|$ by creating the pair of particles associated with each maximal Fermi line in turn, subject to the rule particles associated with row spinors precede those associated with column spinors.

If we associate initial and final states with a Feynman diagram in this way, then the rules above (and their extension to cover other types of fermion) will always give the correct sign. Interfering diagrams, however, could still have negative relative signs because of implied exchange of particles.

**Homework 7.12.14.** Instead of determining an initial and final state from the Feynman diagram as in the remark above, determine a rule for the sign factor as follows. Assume that the sequence of particles in the initial and final states corresponds to the sequence of subscripts in their momenta $p_r$. Using this data and the maximal Fermi lines from the Feynman diagram, give a rule for the sign factor that should multiply the amplitude derived from the rules given above.

This section has set up Wick vertices so that the natural joining constraint matches the non-zero contractions of the spinor fields, derived Feynman diagrams from here, and obtained the Feynman rules for vertices and for external neutrinos.

# 7.13 – Second Application

By computing a more complicated scattering on the basis of the Lagrangian density of 7.11.1, we will discover in this section the Feynman rules for external antineutrinos and relative signs.

Let us now consider the second-order contribution to $\nu_L \nu_L \to \bar{\nu}_R \bar{\nu}_R$. Since every annihilation and every creation in this scattering is mediated by the $\psi_L$ field, not the $\psi_L^*$ field, the matrix element is given by:

$$\mathcal{M} = \langle q', \tfrac{1}{2}; p', \tfrac{1}{2} | S_2 | p, -\tfrac{1}{2}; q, -\tfrac{1}{2} \rangle \tag{7.13.1}$$

$$= \frac{(-i)^2}{2!} \int d^4x\, d^4y\; \acute{\phi}(x) \acute{\phi}(y) \left( -\frac{g}{2} \right)^2$$
$$\times \langle q', \tfrac{1}{2}; p', \tfrac{1}{2} | {:} \psi_L(x)^\top \varepsilon \psi_L(x) \psi_L(y)^\top \varepsilon \psi_L(y) {:} | p, -\tfrac{1}{2}; q, -\tfrac{1}{2} \rangle.$$

The operator contains sixteen monomials in creation and annihilation operators. By inspection of the integrand, we see that only six of these monomials contribute to the amplitude. Since $\psi_L(x)^\top \varepsilon \psi_L(x)$ commutes with $\psi_L(y)^\top \varepsilon \psi_L(y)$, these six terms form three equivalent pairs; since also

$$ {:}(\psi_L^\top)^+ \varepsilon (\psi_L)^- {:} = {:}(\psi_L^\top)^- \varepsilon (\psi_L)^+ {:} \tag{7.13.2}$$

two of the three pairs are equal. The matrix element in 7.13.1 therefore expands to:

$$\langle q', \tfrac{1}{2}; p', \tfrac{1}{2} | {:} \psi_L(x)^\top \varepsilon \psi_L(x) \psi_L(y)^\top \varepsilon \psi_L(y) {:} | p, -\tfrac{1}{2}; q, -\tfrac{1}{2} \rangle \qquad (7.13.3)$$

$$= 2\langle q', \tfrac{1}{2}; p', \tfrac{1}{2} | \psi_L^\top(x)^+ \varepsilon \psi_L(x)^+ \psi_L^\top(y)^- \varepsilon \psi_L(y)^- | p, -\tfrac{1}{2}; q, -\tfrac{1}{2} \rangle$$

$$+ 4\langle q', \tfrac{1}{2}; p', \tfrac{1}{2} | {:} \psi_L^\top(x)^+ \varepsilon \psi_L(x)^- \psi_L^\top(y)^+ \varepsilon \psi_L(y)^- {:} | p, -\tfrac{1}{2}; q, -\tfrac{1}{2} \rangle.$$

Write $\mathcal{M}_M$ for the first term, the mass-pole term, and $\mathcal{M}_Y$ for the second, Yukawa-type term. Pursuing our detailed computation, we find the associated amplitudes $i\mathcal{A}_M$ and $i\mathcal{A}_Y$:

$$i\mathcal{A}_M = (-ig)^2 \frac{i}{(p+q)^2 - \mu^2 + i\epsilon} u_L(p')^\top \varepsilon u_L(q') u_L(q)^\top \varepsilon u_L(p), \qquad (7.13.4)$$

and

$$i\mathcal{A}_Y = (-ig)^2 \frac{i}{(p'-p)^2 - \mu^2 + i\epsilon} u_L(p')^\top \varepsilon u_L(p) u_L(q')^\top \varepsilon u_L(q) \qquad (7.13.5)$$

$$- (-ig)^2 \frac{i}{(p'-q)^2 - \mu^2 + i\epsilon} u_L(p')^\top \varepsilon u_L(q) u_L(q')^\top \varepsilon u_L(p).$$

The Wick operator in the integral 7.13.1 corresponds to the following Wick diagram:

**Wick Diagram:**     $\psi_L^\top \diagdown \qquad \diagup \psi_L^\top$     $-ig\varepsilon \diagdown\!\!\!\!\diagup -ig\varepsilon$     $\psi_L \diagup \qquad \diagdown \psi_L$     $(7.13.6)$

The contraction and symmetry numbers for this diagram are 1 and 8, and indeed the combinatoric coefficient of $1/8$ is evident in the Wick operator contained in the matrix element 7.13.1.

This Wick operator gives rise to three different processes, as represented by the following Feynman diagrams:

**Feynman Diagrams**    (7.13.7)

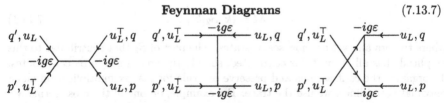

The internal symmetry numbers for these diagrams are all 1. The first diagram gives rise to the mass-pole amplitude $i\mathcal{A}_M$ of 7.13.4, and the other two give rise to the Yukawa-type amplitude $i\mathcal{A}_Y$ of 7.13.5. Again we find that symmetry of the Wick diagram converts into multiplicity of the Feynman diagrams: each combinatoric factor of $1/8$ in the operator is canceled in the amplitude. Hence the combinatoric devices of Section 4.4, though proved for scalar fields, are equally applicable here.

Comparing the amplitudes with the diagrams, we see how to extend our table of Feynman rules to cover anti-neutrinos:

$$\begin{array}{lll} \textbf{Incoming } \bar{\nu}_R \text{ at momentum } p{:} & u_L(p)^* & \\ \textbf{Outgoing } \bar{\nu}_R \text{ at momentum } q{:} & u_L(q)^\top & \end{array} \qquad (7.13.8)$$

Note that we favor orientations of maximal Fermi lines in which incoming particles are at the beginning and outgoing ones at the end.

When we apply these rules to compute an amplitude made up of interfering terms, we have to get the relative signs of the terms right, as we have done here. A rule for the relative signs of the amplitudes can be deduced from comparison of the Wick-operator matrix elements 7.13.3, the amplitudes 7.13.4 and 7.13.5, and the Feynman diagrams 7.13.7. First, we write down the amplitudes $i\mathcal{A}_r$ determined by application of the Feynman rules already proposed to our three Feynman diagrams. Second, we write down the sequences in which the external momenta appear in the resulting spinor products:

$$i\mathcal{A}_1: \quad (p',q',q,p); \qquad i\mathcal{A}_2: \quad (p',p,q',q); \qquad i\mathcal{A}_3: \quad (p',q,q',p). \qquad (7.13.9)$$

Third, we take any one of these sequences as a standard, for example $(p',q',p,q)$, and assign sign factors to the amplitudes $i\mathcal{A}_r$ according as the permutation required to put their sequence of momenta into the standard order is odd or even. This yields the signs

$$i\mathcal{A}_M = i\mathcal{A}_1, \qquad i\mathcal{A}_Y = i\mathcal{A}_2 - i\mathcal{A}_3, \qquad (7.13.10)$$

as desired.

The calculations of this section reveal the Feynman rule for relative signs between amplitudes which interfere and the Feynman rule for external anti-neutrinos.

## 7.14 – Third Application

We conclude the discovery of Feynman rules for the theory defined by the Lagrangian density 7.11.1 by computing a fermion loop amplitude. Comparison with the amplitude with the associated Feynman diagrams yields the Feynman rules for internal Weyl fermions and the sign factor associated with a Fermi loop.

Our final example using this model illustrates the Feynman rule that associates a minus sign with a fermion loop. For this purpose, we work out the neutrino loop correction to the scalar propagator:

$$\mathcal{M} = \langle k'|S_2|k \rangle, \qquad (7.14.1)$$

where the bra and ket are now scalar states. The part of $S_2$ that contributes to this amplitude has all Fermi fields contracted, so only the cross term in $S_2$ contributes. To organize the contractions and preserve our rule for matrix multiplication along maximal Fermi lines, we use the trace. For example, on one of the cross terms, we have:

$$\begin{aligned}
&\left(\psi_L(x)^\dagger \varepsilon \psi_L(x)^*\right)\left(\psi_L(y)^\top \varepsilon \psi_L(y)\right) \qquad (7.14.2) \\
&= \mathrm{tr}\left(\psi_L(x)^\dagger \varepsilon \psi_L(x)^* \psi_L(y)^\top \varepsilon \psi_L(y)\right) \\
&= -\mathrm{tr}\left(\psi_L(y)\psi_L(x)^\dagger \varepsilon \psi_L(x)^* \psi_L(y)^\top \varepsilon\right).
\end{aligned}$$

Notice how putting the operator product into a form suitable for contractions has introduced a factor of $-1$ as a result of Fermi exchange. Notice further that this same exchange has reversed the order of $x$ and $y$ also, so that application of the contraction formula will in effect restore the sign.

Since there are two cross terms and since the contractions can be done in two equivalent ways, the matrix element is given by

$$\mathcal{M} = \frac{gg^*}{8} \int d^4x \, d^4y \, (-4) \tag{7.14.3}$$
$$\times \, \mathrm{tr}\big(\dot\psi_L(y)\dot\psi_L(x)^\dagger \varepsilon \dot\psi_L(x)^* \dot\psi_L(y)^\mathsf{T} \varepsilon\big) \langle k'|{:}\phi(x)\phi(y){:}|k\rangle.$$

Two of the four creation-annihilation monomials contribute to this matrix element, and their contributions are equal:

$$\langle k'|{:}\phi(x)\phi(y){:}|k\rangle = 2\langle k'|\phi(x)_+\phi(y)_-|k\rangle = 2e^{ix\cdot k'}e^{-iy\cdot k}. \tag{7.14.4}$$

Substituting the formulae for the contractions from Theorem 7.10.8 yields:

$$i\mathcal{A} = -gg^* \int d^4p \, d^4q \; \mathrm{tr}\!\left(\frac{i\varepsilon[p]\varepsilon^{-1}}{p^2+i\epsilon} \, \varepsilon \, \frac{i[\tilde q]}{q^2+i\epsilon} \, \varepsilon\right) \delta^{(4)}(p-q+k). \tag{7.14.5}$$

The Wick diagram for the operator involved in this process is:

**Wick Diagram:** $\tag{7.14.6}$

Here, the symmetry number is 2, and the symmetry factor $1/2$ is as usual evident in the matrix element 7.14.3.

The Wick operator above generates two Feynman diagrams according to which of $\phi(x)$ and $\phi(y)$ annihilate the incoming particle. Furthermore, we must choose an orientation of the Fermi loop:

**Feynman Diagrams** $\tag{7.14.7}$

Notice that we have chosen $q$ to be oriented with the arrow and $p$ to be oriented in the same sense (clockwise or counter-clockwise) as $q$.

We deduce that the Feynman rule for internal Fermi lines depends on the relative orientation of the momentum and the arrow on the Fermi line:

$$
\begin{aligned}
&\textbf{Momentum } q \textbf{ parallel to arrow:} && \int \frac{d^4q}{(2\pi)^4} \frac{i[\tilde q]}{q^2+i\epsilon} \\[2mm]
&\textbf{Momentum } p \textbf{ opposed to arrow:} && \int \frac{d^4p}{(2\pi)^4} \frac{i\varepsilon[p]\varepsilon^{-1}}{p^2+i\epsilon}
\end{aligned} \tag{7.14.8}
$$

We note again the Fermi loop rule: compute the matrix element for a diagram using the Feynman rules listed above and then multiply by a factor of $(-1)$ for every Fermi loop in the diagram.

Since the symmetry number for both loop diagrams is 2, each loop integral has a combinatoric factor of $1/2$. However, the two Feynman integrals derived from the two loop diagrams are equal, and so no combinatoric factor is evident in the amplitude 7.14.5. We conclude that our universal combinatoric Feynman rule is valid here.

We can simplify the loop integral a bit by canceling $\varepsilon$'s and doing the $q$ integration:

$$
\begin{aligned}
i\mathcal{A} &= -gg^* \int d^4p \, \frac{\mathrm{tr}([p][\tilde{p}+\tilde{k}])}{(p^2+i\epsilon)((p+k)^2+i\epsilon)} \\
&= -2gg^* \int d^4p \, \frac{p\cdot(p+k)}{(p^2+i\epsilon)((p+k)^2+i\epsilon)}.
\end{aligned}
\tag{7.14.9}
$$

This integral is divergent. When we renormalize, it will be modified to a convergent combination of Feynman integrals by the $A\phi^2$ counterterm.

In this section, the calculation of an amplitude for a loop diagram brings out the Feynman rule for internal Weyl lines and the Feynman rule for fermion loops. This completes the derivation of the Feynman rules for the Lagrangian density 7.11.1.

## 7.15 – Conclusion of Example

In this section, we summarize the Feynman rules for Weyl fields derived in the three previous sections and introduce the use of the trace in computing squares of amplitudes.

The Feynman rules for the theory defined by the Lagrangian density 7.11.1 follow.

1. Feynman diagrams are based on the two vertices

$$\psi_L^T \varepsilon \psi_L \phi: \qquad \text{and} \qquad \psi_L^\dagger \varepsilon \psi_L^* \phi: \tag{7.15.1}$$

2. In constructing a Feynman diagram from a set of vertices, one may only join lines with incoming arrows to lines with outgoing arrows.
3. In a Feynman diagram, an orientation for each maximal Fermi line must be specified.
4. The Feynman rules assign factors to each vertex, external and internal Weyl line as follows:

| | |
|---|---|
| $\psi_L^T \varepsilon \psi_L \phi$ **vertex:** | $-ig\varepsilon(2\pi)^4 \delta^{(4)}(P_{\mathrm{in}})$ |
| $\psi_L^\dagger \varepsilon \psi_L^* \phi$ **vertex:** | $ig^*\varepsilon(2\pi)^4 \delta^{(4)}(P_{\mathrm{in}})$ |
| **Incoming** $\nu_L$ **at momentum** $p$: | $u_L(p)$ |
| **Outgoing** $\nu_L$ **at momentum** $p'$: | $u_L(p')^\dagger$ |
| **Incoming** $\bar{\nu}_R$ **at momentum** $q$: | $u_L(q)^*$ |
| **Outgoing** $\bar{\nu}_R$ **at momentum** $q'$: | $u_L(q')^T$ |

$$\tag{7.15.2}$$

| | |
|---|---|
| **Momentum** $p$ **parallel to arrow:** | $\displaystyle\int \frac{d^4p}{(2\pi)^4} \frac{i[\tilde{p}]}{p^2+i\epsilon}$ |
| **Momentum** $q$ **opposed to arrow:** | $\displaystyle\int \frac{d^4q}{(2\pi)^4} \frac{i\varepsilon[q]\varepsilon^{-1}}{q^2+i\epsilon}$ |

5. Matrices associated with fermions and vertices in a given diagram are grouped according to the maximal Fermi lines in that diagram and multiplied according to the 'later on the left' rule where later is defined by the chosen orientation of these lines.

6. Include an extra factor of $(-1)$ for every Fermi loop in a diagram.

7. Compute the relative sign between amplitudes which interfere by a assigning sign factor $(-1)^k$ to each amplitude $\mathcal{A}$, where $k$ is the number of transpositions needed to put the sequence of momenta in the polarization spinors of $\mathcal{A}$ into any predetermined order.

These seven items enable us to write down the Feynman diagrams and the Feynman integrals for the associated matrix elements in the theory with Lagrangian density 7.11.1.

The applications of these Feynman rules to decay and scattering processes follow the presentation of Chapter 5. The only new point is the computation of products of spinor terms. The following computation shows how to use the cyclic property of the trace in computing a square:

$$
\begin{aligned}
\left|u_L(q)^\top \varepsilon u_L(p)\right|^2 &= \left(u_L(q)^\top \varepsilon u_L(p)\right)\left(u_L(q)^\top \varepsilon u_L(p)\right)^\dagger \\
&= -\left(u_L(q)^\top \varepsilon u_L(p)\right)\left(u_L(p)^\dagger \varepsilon u_L(q)^*\right) \\
&= -\operatorname{tr}\left(\varepsilon u_L(q)^* u_L(q)^\top \varepsilon u_L(p) u_L(p)^\dagger\right) \\
&= -\operatorname{tr}\left(\varepsilon[\tilde{q}]^* \varepsilon[\tilde{p}]\right) \\
&= \operatorname{tr}\left([q][\tilde{p}]\right) \\
&= 2p{\cdot}q.
\end{aligned}
\tag{7.15.3}
$$

In general, interference of amplitudes such as the two in $\mathcal{A}_Y$ of 7.13.5 will lead to products of the following type:

$$
P \stackrel{\text{def}}{=} \left(u_L(p')^\top \varepsilon u_L(p) u_L(q')^\top \varepsilon u_L(q)\right)\left(u_L(p')^\top \varepsilon u_L(q) u_L(q')^\top \varepsilon u_L(p)\right)^\dagger. \tag{7.15.4}
$$

The four scalars of the form $u^\top \varepsilon u$ change sign when transposed and commute with each other. It is therefore possible to rearrange them (using the trace trick again) so that the polarization spinors form pairs of the form $u(k)u(k)^\dagger$:

$$
\begin{aligned}
P &= \operatorname{tr}\left(u_L(p')^\top \varepsilon u_L(p) u_L(p)^\dagger \varepsilon u_L(q')^* u_L(q')^\top \varepsilon u_L(q) u_L(q)^\dagger \varepsilon u_L(p')^*\right) \\
&= \operatorname{tr}\left(\varepsilon u_L(p')^* u_L(p')^\top \varepsilon u_L(p) u_L(p)^\dagger \varepsilon u_L(q')^* u_L(q')^\top \varepsilon u_L(q) u_L(q)^\dagger\right) \tag{7.15.5} \\
&= \operatorname{tr}\left([p'][\tilde{p}][q'][\tilde{q}]\right).
\end{aligned}
$$

To complete the reduction, we observe that the transformation properties of $\sigma_\mu$ and $\tilde{\sigma}_\mu$ make

$$
S_{\alpha\beta\gamma\delta} \stackrel{\text{def}}{=} \operatorname{tr}(\sigma_\alpha \tilde{\sigma}_\beta \sigma_\gamma \tilde{\sigma}_\delta) \tag{7.15.6}
$$

a Lorentz tensor, and investigate the case when a pair of indices are equal and the case when no two indices are equal to find that

$$
S_{\alpha\beta\gamma\delta} = 2g_{\alpha\beta}g_{\gamma\delta} - 2g_{\alpha\gamma}g_{\beta\delta} + 2g_{\alpha\delta}g_{\beta\gamma} - 2i\epsilon_{\alpha\beta\gamma\delta}. \tag{7.15.7}
$$

Thus we find that

$$P = 2\left(p'\cdot p\, q'\cdot q - p'\cdot q'\, p\cdot q + p'\cdot q\, p\cdot q' - i\epsilon(p'pq'q)\right). \qquad (7.15.8)$$

**Homework 7.15.9.** Show that, to lowest non-trivial order, the decay width for $\phi \to \nu_L \nu_L$ is $\Gamma = gg^*\mu/32\pi$.

**Homework 7.15.10.** Compute to second order the amplitude for $\nu_L \bar\nu_R \to \nu_L \bar\nu_R$. Show that the differential cross section for this process is

$$d\sigma = \frac{(gg^*)^2}{32\pi}\left(\frac{E_T \cos^2\frac{\theta}{2}}{\mu^2 + E_T^2 \cos^2\frac{\theta}{2}}\right)^2 d\cos\theta.$$

**Homework 7.15.11.** At second order there are two diagrams that contribute to $\nu_L \bar\nu_R \to \Phi\Phi$. Compute the amplitudes, their relative sign, and the differential cross section for this scattering.

We have carefully applied Dyson's formula to compute matrix elements and thereby deduced the Feynman rules for massless spinors. There are three new points. First, we compute fermion-line factors in an amplitude by following the fermion line through a diagram, starting with a spinor, multiplying the coupling matrices in sequence, and ending with a spinor. Second, for every fermion loop in the diagram, we add a factor of $(-1)$ to the amplitude. Third, we need a Feynman rule for relative signs of interfering amplitudes.

## 7.16 Two-Component Spinor Notation

In this section, we will introduce the index conventions for two-component spinors. The key point in understanding this notation is the relationship between the four ways in which an $SL(2,\mathbf{C})$ matrix $g$ can act on a two-component vector:

$$g\cdot\psi = \quad \text{one of} \quad g\psi,\ (g^{-1})^\top\psi,\ g^*\psi,\ (g^{-1})^\dagger\psi. \qquad (7.16.1)$$

Though two pairs of these actions are equivalent,

$$\varepsilon g\varepsilon^{-1} = (g^{-1})^\top \quad \text{and} \quad \varepsilon g^*\varepsilon^{-1} = (g^{-1})^\dagger, \qquad (7.16.2)$$

it is convenient for two-component computations to have notation that distinguishes all four. To reduce the clutter of indices, we shall write $\psi$ for $\psi_L$ and $\bar\psi$ for $\psi_L^*$. Now we associate upper and lower undotted and dotted indices with the four actions above as follows:

$$
\begin{array}{ccc}
g\cdot\psi = g\psi & \leftrightarrow \psi_\alpha \xrightarrow{\ *\ } \bar\psi_{\dot\alpha} \leftrightarrow & g\cdot\bar\psi = g^*\bar\psi \\
(1/2,0) & \Big\downarrow \varepsilon \qquad \Big\downarrow \dot\varepsilon & (0,1/2) \qquad (7.16.3) \\
g\cdot\psi = (g^{-1})^\top\psi \leftrightarrow \psi^\alpha & \xrightarrow{\ *\ } \bar\psi^{\dot\alpha} \leftrightarrow & g\cdot\bar\psi = (g^{-1})^\dagger\bar\psi
\end{array}
$$

where $g$ is now the $SL(2,\mathbf{C})$ element $D_{1/2,0}(\Lambda)$.

In this diagram, $\varepsilon$ is used to raise undotted indices while $\dot\varepsilon$ is used to raise dotted indices:

$$
\varepsilon = (\varepsilon^{\alpha\beta}) = \begin{pmatrix} 0 & 1 \\ -1 & 0 \end{pmatrix}, \quad \psi^\alpha = \varepsilon^{\alpha\beta}\psi_\beta;
$$

$$
\dot\varepsilon = (\varepsilon^{\dot\alpha\dot\beta}) = \begin{pmatrix} 0 & 1 \\ -1 & 0 \end{pmatrix}, \quad \bar\psi^{\dot\alpha} = \varepsilon^{\dot\alpha\dot\beta}\bar\psi_{\dot\beta}.
$$

$$(7.16.4)$$

The inverse matrices are used to lower indices:

$$(\varepsilon_{\alpha\beta}) = \begin{pmatrix} 0 & -1 \\ 1 & 0 \end{pmatrix}, \quad \psi_\alpha = \varepsilon_{\alpha\beta}\psi^\beta;$$

$$(\varepsilon_{\dot\alpha\dot\beta}) = \begin{pmatrix} 0 & -1 \\ 1 & 0 \end{pmatrix}, \quad \bar\psi_{\dot\alpha} = \varepsilon_{\dot\alpha\dot\beta}\bar\psi^{\dot\beta}. \tag{7.16.5}$$

Since the action of $SL(2,\mathbf{C})$ on hermitian matrices that corresponds to Lorentz transformations is

$$[k] \mapsto g[k]g^\dagger = [\Lambda(g)k], \tag{7.16.6}$$

the $\sigma$ matrices should have indices $\sigma_{\mu\alpha\dot\alpha}$. We can raise the spinor indices using $\varepsilon$ and $\dot\varepsilon$ to obtain:

$$\tilde\sigma_\mu^{\dot\alpha\alpha} \overset{\text{def}}{=} \varepsilon^{\dot\alpha\dot\beta}\varepsilon^{\alpha\beta}\sigma_{\mu\beta\dot\beta}. \tag{7.16.7}$$

We have reversed the spinor indices on $\tilde\sigma$ so that when we contract spinor indices between $\tilde\sigma$ and $\sigma$ the contraction will be between adjacent indices. As matrices, the $\tilde\sigma_\mu$ are related to the $\sigma_\mu$ by parity:

$$\tilde\sigma_0 = \sigma_0, \quad \tilde\sigma_r = -\sigma_r. \tag{7.16.8}$$

The four-vectors of components of $\sigma$ and $\tilde\sigma$ satisfy the orthonormality relations:

$$\sigma^\mu_{\alpha\dot\alpha}\tilde\sigma_\mu^{\dot\beta\beta} = -2\delta_\alpha^\beta\delta_{\dot\alpha}^{\dot\beta}. \tag{7.16.9}$$

These relations enables us to convert pairs of spinor indices into vector indices and vice versa:

$$v_{\alpha\dot\alpha} = \sigma^\mu_{\alpha\dot\alpha}v_\mu \iff v_\mu = -\frac{1}{2}\tilde\sigma_\mu^{\dot\alpha\alpha}v_{\alpha\dot\alpha}. \tag{7.16.10}$$

Using this notation we can write our Lorentz scalars just by contracting indices:

$$\psi_L^\top\varepsilon\psi_L = -\psi^\alpha\psi_\alpha,$$
$$\psi_R^\top\varepsilon\psi_R = \bar\psi_{\dot\alpha}\bar\psi^{\dot\alpha}, \tag{7.16.11}$$
$$\psi_L^\dagger[\partial]_P\psi_L = \tilde\sigma^{\mu\dot\alpha\alpha}\bar\psi_{\dot\alpha}\partial_\mu\psi_\alpha.$$

In index notation, an element $g$ in $SL(2,\mathbf{C})$ may be written $g = (g_\alpha{}^\beta)$. The relationship between $g$ and $\varepsilon$ is summarized in

$$g_\alpha{}^\beta g_\gamma{}^\delta\varepsilon^{\alpha\gamma} = \det(g)\varepsilon^{\beta\delta} = \varepsilon^{\beta\delta}. \tag{7.16.12}$$

From this equation we could develop notation for $(g^{-1})^\top$ and so on in order to verify Lorentz invariance in index notation, but we find it less confusing to use the index free notation for that kind of work.

Finally, there are conventions for suppressing indices:

$$\psi\chi \overset{\text{def}}{=} \psi^\alpha\chi_\alpha = -\psi_\alpha\chi^\alpha = \chi^\alpha\psi_\alpha = \chi\psi;$$
$$\bar\psi\bar\chi \overset{\text{def}}{=} \bar\psi_{\dot\alpha}\bar\chi^{\dot\alpha} = -\bar\psi^{\dot\alpha}\bar\chi_{\dot\alpha} = \bar\chi_{\dot\alpha}\bar\psi^{\dot\alpha} = \bar\chi\bar\psi. \tag{7.16.13}$$

Note the surprising consequence of these definitions, that this product of Fermi fields is commutative. Since $(\psi_\alpha)^\dagger = \bar\psi_{\dot\alpha}$, we can see that the adjoint converts the first of these products into the second:

$$(\psi\chi)^\dagger = (\psi^\alpha \chi_\alpha)^\dagger = \bar\chi_{\dot\alpha}\bar\psi^{\dot\alpha} = \bar\chi\bar\psi. \tag{7.16.14}$$

Weyl spinors are not commonly used, for most physicists are more accustomed to work in the Dirac notation presented in the next chapter. However, the notation presented here is fairly standard, and finds application in supersymmetric field theories. It has the advantage of compactness and makes Lorentz invariants obvious, but it tends to obscure the underlying Lorentz actions with a mess of dotted and undotted upper and lower indices.

## 7.17 Massive Weyl Spinors

Low energy physics is parity invariant. Since parity transforms the $(0, 1/2)$ representation into the $(1/2, 0)$ representation, a parity-invariant theory must have left and right Weyl spinors, $\psi_R$ and $\psi_L$. Since $\psi_L^\dagger \psi_R$ is an invariant term, a parity-invariant theory brings up the possibility of mixing between left and right Weyl spinors. If the fields mix in this way, it makes no sense to regard $\psi_R$ and $\psi_L$ as two separate fields. Instead one combines them into a single four-component field called a Dirac spinor. We shall arrive at the same conclusion in two other ways, first by analyzing mass terms in this section, second by analyzing the Lagrangian density and equations of motion in the following section.

In the following investigation into the structure of mass terms for Weyl spinors, we shall find that though all mass terms can be written as sums of Majorana mass terms, that is, terms of the form $\psi_L\psi_L$ or its adjoint $\psi_R\psi_R$, this is only natural for neutral fields. The mass terms for charged fields are best left in the form $\psi_L^\dagger \psi_R$ and its hermitian conjugate $\psi_R^\dagger \psi_L$. A mass term like this is called a *Dirac mass term*, and the two fields linked by the Dirac mass term are best written in a four-component notation which we present below.

Since the conjugate $\bar\psi_R$ to a right Weyl spinor $\psi_R$ is a left Weyl spinor, it is possible to write any theory involving left and right Weyl spinors in terms of left Weyl spinors and their conjugates. We shall therefore illustrate the general analysis of mass terms by studying a theory with two left Weyl spinors $\psi_1$ and $\psi_2$. The most general mass term in $\psi_1$, $\psi_2$, and their conjugates can be written in the two-component spinor notation as follows:

$$\text{Mass Term} = \frac{a}{2}\psi_1\psi_1 + b\psi_1\psi_2 + \frac{c}{2}\psi_2\psi_2 + \frac{a^*}{2}\bar\psi_1\bar\psi_1 + b^*\bar\psi_1\bar\psi_2 + \frac{c^*}{2}\bar\psi_2\bar\psi_2. \tag{7.17.1}$$

We can write this mass term using matrices as follows:

$$\text{Mass Term} = \frac{1}{2}\begin{pmatrix} \psi_1 & \psi_2 \end{pmatrix}\begin{pmatrix} a & b \\ b & c \end{pmatrix}\begin{pmatrix} \psi_1 \\ \psi_2 \end{pmatrix} + \text{h.c.}, \tag{7.17.2}$$

where 'h.c.' stands for 'hermitian conjugate'. This expression can be simplified by a change of basis. Note that, in order to preserve the form of the kinetic terms in the Lagrangian density, the change of basis should be unitary. To handle situations like this, it is useful to know the following two theorems:

**Theorem 7.17.3.** *If $M$ is an arbitrary $n \times n$ complex matrix, then there exist unitary matrices $U$ and $V$, and a real, diagonal, non-negative matrix $m$ such that $M = VmU^\dagger$.*

**Proof.** The proof has two steps. The first step is to reduce the general case to the case where $M$ is invertible. The second step is to prove this proposition in this special case.

Let $A$ and $B$ be complex vector spaces with inner products, both isomorphic to $\mathbf{C}^m$. Think of $M$ as a matrix which represents a linear transformation $T$ from $A$ to $B$ with respect to some choice of orthonormal bases for $A$ and $B$. Let $A_1 = \ker T$ and $B_2 = \operatorname{im} T$ be the kernel and image of $T$. Let $A_2 = A_1^\perp$ and $B_1 = B_2^\perp$ be the orthogonal complements of $A_1$ and $B_2$ in $A$ and $B$ respectively. Then $T$ is an isomorphism from $A_2$ to $B_2$. Choose orthonormal bases for $A_1$, $A_2$, $B_1$, and $B_2$. Then $T$ is represented by a matrix with block structure

$$[T] = \begin{pmatrix} N & 0 \\ 0 & 0 \end{pmatrix}, \tag{7.17.4}$$

where $N$ is invertible. This completes the first step.

Since $NN^\dagger$ is a positive-definite, hermitian matrix there is a unitary transformation $V$ such that $NN^\dagger = VDV^\dagger$ where $D$ is a diagonal matrix with positive diagonal elements. Thus $D$ has a unique, positive, diagonal square root $P$, and $NN^\dagger = VP^2V^\dagger$. Now let $U = N^\dagger VP^{-1}$. Then $U$ is unitary:

$$U^\dagger U = P^{-1}V^\dagger NN^\dagger VP^{-1} = 1, \tag{7.17.5}$$

and $N = VPU^\dagger$. This completes the second step.   $\square$

**Theorem 7.17.6.** *If $M$ is a complex, symmetric matrix, then there exists a unitary matrix $U$ and a real, diagonal, non-negative matrix $m$ such that $M = (U^\dagger)^\mathsf{T} mU^\dagger$.*

**Proof.** Applying the previous theorem gives us a factorization $M = VmU^\dagger$. We have to find $W$ such that $M = W^*mW^\dagger$. The idea of the proof is to extract a square root $S$ of the difference between $U^*$ and $V$ which will commute with $m$, so that we can adjust $U$ to $W$ and $V$ to $W^*$, while maintaining the equation with $M$.

Assume that we are in a basis in which all sets of equal eigenvalues in $m$ are together on the diagonal. Since $M = M^\mathsf{T}$,

$$VmU^\dagger = U^*mV^\mathsf{T} \quad \Rightarrow \quad UmV^\dagger = V^*mU^\mathsf{T}$$
$$\Rightarrow \quad VmU^\dagger UmV^\dagger = U^*mV^\mathsf{T}V^*mU^\mathsf{T} \tag{7.17.7}$$
$$\Rightarrow \quad U^\mathsf{T}Vm^2 = m^2U^\mathsf{T}V.$$

Thus $U^\mathsf{T}V$ is a unitary matrix which commutes with a diagonal matrix. This implies that $U^\mathsf{T}V$ is in block diagonal form, and so commutes with $m$. However,

$$VmU^\dagger = U^*mV^\mathsf{T} \quad \Longrightarrow \quad U^\mathsf{T}Vm = mV^\mathsf{T}U = m(U^\mathsf{T}V)^\mathsf{T}, \tag{7.17.8}$$

and so $U^\mathsf{T}V$ must be a symmetric unitary matrix. This gives us all the data we can deduce about the difference between $U^*$ and $V$.

The next step is to find a symmetric, unitary square root of $B = U^T V$. To preserve the block structure of $B$, we shall extract a symmetric, unitary square root for each block $B_r$ separately. Any unitary matrix is the exponential of an anti-hermitian matrix, so we can write $B_r$ in the form $\exp(X)$ where $X + X^\dagger = 0$. Any anti-hermitian matrix can be diagonalized by a unitary change of basis to a purely imaginary, diagonal matrix, and so we can write $X = AYA^\dagger$ where $Y$ is of this type. Now

$$B_r = \exp(AYA^\dagger) = A\exp(Y)A^\dagger \qquad (7.17.9)$$

is symmetric, so

$$A\exp(Y)A^\dagger = A^*\exp(Y)A^T, \qquad (7.17.10)$$

which is equivalent to

$$A^T A\exp(Y) = \exp(Y)A^T A. \qquad (7.17.11)$$

This shows that $A^T A$ is in block diagonal form with respect to the eigenspaces of $\exp(Y)$. Now, the eigenspaces of $\exp(Y)$ may not correspond to the eigenspaces of $Y$: consider

$$Y = \begin{pmatrix} \pi i & 0 \\ 0 & -\pi i \end{pmatrix}. \qquad (7.17.12)$$

Clearly there is a diagonal matrix $N$ with integer elements such that $Z = Y + 2\pi i N$ satisfies $\exp(Z) = \exp(Y)$ and the eigenspaces of $Z$ correspond to those of $\exp(Z)$. Now we can see that a symmetric, unitary square root of $B_r$ is given by

$$\sqrt{B_r} = A\exp(\tfrac{1}{2}Z)A^\dagger. \qquad (7.17.13)$$

Proceeding block by block in this way, we obtain a symmetric, unitary square root $S$ for $B$ itself. Since $S$ has the same block structure as $B$, $S$ also commutes with $m$.

Now we can use $S$ to balance $U^*$ with $V$. $S^2 = U^T V$ implies that $V = U^* S^2$, and so:

$$M = VmU^T = U^* S^2 mU^\dagger = U^* SmSU^\dagger = U^* SmS^T U^\dagger \qquad (7.17.14)$$
$$= (US^*)^* m(US^*)^\dagger. \qquad \square$$

Applying this theorem to 7.17.2, we find a new basis of fields, which we shall call $\eta_1$ and $\eta_2$, with respect to which the mass term is

$$\text{Mass Term} = \frac{1}{2}\begin{pmatrix} \eta_1 & \eta_2 \end{pmatrix}\begin{pmatrix} M_1 & 0 \\ 0 & M_2 \end{pmatrix}\begin{pmatrix} \eta_1 \\ \eta_2 \end{pmatrix} + \text{h.c.}, \qquad (7.17.15)$$

where $M_1$ and $M_2$ are both non-negative. In this diagonalization, we have actually reduced an arbitrary sum of fermion mass terms to a sum of Majorana mass terms:

**Corollary 7.17.16.** *If $\Psi = (\psi_1, \ldots, \psi_n)$ is a vector of Weyl spinors, and if $M$ is a symmetric complex matrix, then there exists a unitary change of basis for the space of Weyl spinors such that the mass term $\Psi^T M\Psi + \text{h.c.}$ becomes a sum of Majorana mass terms $\sum m_r \psi'_r \psi'_r$ with non-negative masses $m_r$.*

As an example, consider the mass matrix

$$M = \sigma_1 = \begin{pmatrix} 0 & 1 \\ 1 & 0 \end{pmatrix}. \qquad (7.17.17)$$

When we diagonalize using an orthogonal matrix, we obtain $m = \sigma_3$. Now we can change the phase of the second field to make the second mass positive. In terms of fields, this is equivalent to the change of basis

$$\eta_1 = \frac{1}{2}(\psi_1 + \psi_2), \quad \eta_2 = \frac{i}{2}(\psi_1 - \psi_2). \tag{7.17.18}$$

This conclusion alerts us to the fact that there is no physics in the sign of a Majorana mass. It is simply a self-interaction of the field.

So far, preserving the form of the kinetic terms in the Lagrangian density has provided the only constraint on the matrices used for diagonalizing the mass matrix. Conserved charges effectively provide more constraints, as we now show.

A Majorana mass term $\psi\psi + \bar{\psi}\bar{\psi}$ can only occur in a Lagrangian density if $\psi$ has no charge, but a Dirac mass term $\psi_1\psi_2 + \bar{\psi}_1\bar{\psi}_2$ can link fields with opposite charges. In the case of a Dirac mass involving charged fields $\psi_r$, though we can formally rewrite the mass matrix in diagonal form, the new basis of fields does not have definite charge. However, all observed particles have definite quantum numbers for all measurable conserved quantities, so it is commonly assumed that nature obeys superselection rules which compel the fundamental fields and particles to be eigenvectors for the measurable conserved quantity operators. We conclude then that our diagonalization matrix should not mix fields with different quantum numbers. For example, if we write $\eta_+$ for the left positron field and $\eta_-$ for the left electron field, then the Dirac mass term $\eta_-\eta_+ + \bar{\eta}_-\bar{\eta}_+$ can convert an incoming $e^+$ into an outgoing anti-$e^-$. Hence it makes no sense to regard these as two distinct particles: the positron is the anti-particle to the electron. Note, however, that this Dirac mass term does not occur in the high-energy, Standard Model Lagrangian density; there the couplings of the two fields are different, and these fields do represent distinct particles.

Since Dirac mass terms are unavoidable at low energies, we need a notation for combining the two Weyl spinors that mix in such a term into a single four-component Dirac field. Let $\eta$ and $\zeta$ be left Weyl spinors and rearrange their Dirac mass term as follows:

$$\text{Dirac Mass Term} = \frac{1}{2}(\eta \ \ \zeta)\begin{pmatrix} 0 & m \\ m & 0 \end{pmatrix}\begin{pmatrix} \eta \\ \zeta \end{pmatrix} + \text{h.c.}$$

$$= m(\eta\zeta + \bar{\eta}\bar{\zeta}) = (\eta \ \ \bar{\zeta})\begin{pmatrix} 0 & m \\ m & 0 \end{pmatrix}\begin{pmatrix} \bar{\eta} \\ \zeta \end{pmatrix}. \tag{7.17.19}$$

The final column vector is the Dirac four-component spinor $\psi$ which combines the two Weyl spinors into a single entity:

$$\psi \overset{\text{def}}{=} \begin{pmatrix} \bar{\eta} \\ \zeta \end{pmatrix} = \begin{pmatrix} \bar{\eta}^{\dot{1}} \\ \bar{\eta}^{\dot{2}} \\ \zeta_1 \\ \zeta_2 \end{pmatrix}. \tag{7.17.20}$$

For the sake of a simple mass term, we define the *Dirac adjoint* $\bar{\psi}$ of $\psi$ by:

$$\bar{\psi} \overset{\text{def}}{=} \psi^\dagger \begin{pmatrix} 0 & 1 \\ 1 & 0 \end{pmatrix} = (\bar{\zeta}_{\dot{1}} \ \ \bar{\zeta}_{\dot{2}} \ \ \eta^1 \ \ \eta^2). \tag{7.17.21}$$

Now the mass term is simply the matrix product $m\bar{\psi}\psi$.

Sometimes one needs to represent a single Weyl spinor in Dirac notation. This is accomplished by setting $\psi_L = \eta = \zeta$ in the definitions above. The resulting four-component spinor is called a *Majorana spinor*. The ensuing relationship between the top and bottom components of the four-component spinor is summarized in the constraint $\bar{\psi} = \psi$. Now the Dirac mass term $m\bar{\psi}\psi$ reduces to the Majorana mass term $m\psi_L\psi_L + m\bar{\psi}_L\bar{\psi}_L$.

**Homework** 7.17.22. How does the Lorentz group act on four-component spinors in the Weyl basis? Is $\bar{\psi}_W\psi_W$ really a Lorentz scalar?

At high energies, it is generally convenient to work with left Weyl spinors and fermion mass matrices because the properties of each Weyl field – its couplings and quantum numbers – effectively give it a unique individuality. At low energies, however, the breaking of symmetry and the loss of the associated high-energy conservation laws makes it convenient to simplify the fermion mass matrices by combining pairs of Weyl spinors into Dirac spinors.

Now we have seen how mass terms make it natural to introduce four-component notation, we shall approach the same point from the side of the Lagrangian density and the equations of motion. The approach will bring out the Dirac algebra of $4 \times 4$ matrices.

# 7.18 – Equations of Motion, and Dirac Algebra

In this section, we shall derive the Lagrangian density, Hamiltonian density, and equations of motion for the massive left and right Weyl fields. The form of the equations of motion not only provides a second motivation for four-component notation, but also serves to introduce the Dirac algebra of $4 \times 4$ matrices.

The Majorana mass term does not mix left Weyl fields with right. Thus we can write a Lagrangian density for a free, massive left Weyl field:

$$\mathcal{L}_L = \psi_L^\dagger i[\partial]\psi_L + g\psi_L^\mathsf{T}\varepsilon\psi_L - g^*\psi_L^\dagger\varepsilon\psi_L^*. \tag{7.18.1}$$

We have used $g$ here because we do not yet know how this coupling constant will relate to the mass as determined by the equations of motion. We shall find that $g = \frac{1}{2}m$ is correct.

Following the rules of fermionic calculus as set out in Section 7.4 and using the principle of independent variation of $\psi_L$ and $\psi_L^\dagger$, we find the equations of motion by varying $\psi_L^\dagger$:

$$\frac{d}{dx^\mu}\frac{\partial\mathcal{L}_L}{\partial(\partial_\mu\psi_L^\dagger)} = \frac{\partial\mathcal{L}_L}{\partial\psi_L^\dagger} \quad\Longrightarrow\quad i[\partial]\psi_L = 2g^*\varepsilon\psi_L^*. \tag{7.18.2}$$

If we conjugate this equation and rearrange the $\varepsilon$'s using $\varepsilon^{-1}[\partial]^*\varepsilon = [\tilde{\partial}]$, then we obtain an equation of motion for $\varepsilon\psi_L^*$:

$$i[\tilde{\partial}]\varepsilon\psi_L^* = 2g\psi_L. \tag{7.18.3}$$

Combining this with the equation of motion for $\psi_L$, we find the Klein–Gordon equation again:

$$\partial_\mu \partial^\mu \psi_L = [\tilde{\partial}][\partial]\psi_L = -2g^* i[\tilde{\partial}]\varepsilon \psi_L^* = -|2g|^2 \psi_L$$
$$\Rightarrow \quad (\partial_\mu \partial^\mu + 4|g|^2)\psi_L = 0. \tag{7.18.4}$$

From the Klein–Gordon equation, we can see that the classical mass of the left Weyl field is $m = 2|g|$. The phase in $g$ can be absorbed into $\psi_L$, so we can take $g = \frac{1}{2}m$ in the Lagrangian density.

**Homework 7.18.5.** Show that the Hamiltonian for a massive left Weyl field with the Lagrangian density above is bounded below.

The equations of motion mix the Weyl field $\psi_L$ with its conjugate $\psi_{Lc} = \varepsilon \psi_L^*$. It is convenient to rewrite the Lagrangian density and the equations of motion in terms of $\psi_L$ and $\psi_{Lc}$, as if these were two independent spinor fields. Thus:

$$\mathcal{L} = \frac{1}{2}\psi_L^\dagger i[\tilde{\partial}]\psi_L + \frac{1}{2}\psi_{Lc}^\dagger i[\partial]\psi_{Lc} - \frac{1}{2}m\psi_{Lc}^\dagger \psi_L - \frac{1}{2}m\psi_L^\dagger \psi_{Lc} \tag{7.18.6}$$

$$= \frac{1}{2}\begin{pmatrix} \psi_{Lc}^\dagger & \psi_L^\dagger \end{pmatrix}\begin{pmatrix} i[\tilde{\partial}] & 0 \\ 0 & i[\partial] \end{pmatrix}\begin{pmatrix} \psi_{Lc} \\ \psi_L \end{pmatrix} - \frac{1}{2}m\begin{pmatrix} \psi_{Lc}^\dagger & \psi_L^\dagger \end{pmatrix}\begin{pmatrix} 0 & 1 \\ 1 & 0 \end{pmatrix}\begin{pmatrix} \psi_{Lc} \\ \psi_L \end{pmatrix}.$$

The equations of motion are

$$\begin{pmatrix} i[\tilde{\partial}] & 0 \\ 0 & i[\partial] \end{pmatrix}\begin{pmatrix} \psi_{Lc} \\ \psi_L \end{pmatrix} = m\begin{pmatrix} 0 & 1 \\ 1 & 0 \end{pmatrix}\begin{pmatrix} \psi_{Lc} \\ \psi_L \end{pmatrix}. \tag{7.18.7}$$

**Homework 7.18.8.** Assuming the equal-time anti-commutation relations for the spinor fields, show that the Lagrangian density 7.18.6 is hermitian.

We can write these equations in the form

$$i(\partial_0 - \bar{\alpha}\cdot\bar{\partial})\psi = m\beta\psi, \tag{7.18.9}$$

where $\psi$ and the matrices $\alpha_r$ and $\beta$ are given by

$$\psi = \begin{pmatrix} \psi_{Lc} \\ \psi_L \end{pmatrix}, \quad \alpha_r \overset{\text{def}}{=} \begin{pmatrix} -\sigma_r & 0 \\ 0 & \sigma_r \end{pmatrix}, \quad \beta \overset{\text{def}}{=} \begin{pmatrix} 0 & 1 \\ 1 & 0 \end{pmatrix}. \tag{7.18.10}$$

Note that, according to the definitions of the previous sections, $\psi$ is a Majorana spinor. The matrices $\alpha_r$ and $\beta$ are hermitian and satisfy the algebraic equations

$$\{\alpha_r, \alpha_s\} = 2\delta_{rs}, \quad \{\alpha_r, \beta\} = 0, \quad \beta^2 = 1. \tag{7.18.11}$$

Since $\partial_r$ is associated with $\alpha_r$ and $\partial_0$ with 1, the Lorentz transformation properties of this formalism are not good. Remarkably, we can restore Lorentz covariance by the simple expedient of shifting $\beta$ as follows. From the definition 7.17.21 of the Dirac adjoint $\bar{\psi}$, we find that

$$\bar{\psi} = \psi^\dagger \beta. \tag{7.18.12}$$

Hence, if we define matrices $\gamma_\mu$ by

$$\gamma_0 = \beta = \begin{pmatrix} 0 & 1 \\ 1 & 0 \end{pmatrix}, \quad \gamma_r = \beta\alpha_r = \begin{pmatrix} 0 & \sigma_r \\ -\sigma_r & 0 \end{pmatrix}, \tag{7.18.13}$$

the Lagrangian density and equations of motion become the standard Dirac equation:

$$\mathcal{L} = \bar{\psi} i \gamma^\mu \partial_\mu \psi - m \bar{\psi} \psi \quad \Longrightarrow \quad (i\gamma^\mu \partial_\mu - m)\psi = 0. \qquad (7.18.14)$$

The $\gamma$ matrices satisfy the Lorentz covariant identity

$$\{\gamma_\mu, \gamma_\nu\} = g_{\mu\nu}, \qquad (7.18.15)$$

which is the defining relation for *Dirac algebra*.

In the previous section, we showed that Dirac four-component notation was convenient when fermion mass terms mixed left spinors with opposite charges. Here we have shown that the same notation is convenient for uniting the two versions of the equations of motion for a massless left Weyl spinor. The equations of motion also introduce the Dirac $\gamma$ matrices, the properties of which provide the practical basis for efficient computations with Dirac spinors.

We have at this point completed our discussion of two-component spinors and covered the transition to conventional four-component spinor notation. The following chapter provides an account of four-component spinor theory.

# 7.19 Summary

This chapter has pursued the lead provided by Chapter 6 and developed the quantum theory of fields which transform as $(1/2, 0)$ and $(0, 1/2)$ under the Lorentz group. The sequence of steps leading to the perturbative theory of quantum Weyl fields is close to that used in Chapter 2 for the scalar field: first, write down a classical Lagrangian density for the free-field theory; second, find the plane wave solutions of the classical equations of motion; third, introduce polarization spinors and particle creation and annihilation operators; fourth, define the free quantum field as an integral; fifth, compute the Hamiltonian density and verify that this operator is bounded below; sixth, compute the contraction of fields to determine the Feynman propagator; seventh, use the Wick expansion to find the vertex rules. Steps three and five introduced new elements, namely the polarization spinors and the anti-commutativity appropriate to fermions.

Weyl fields are regarded as fundamental at high energies where there are no quadratic interactions to create particle masses. At low energies, however, quadratic interactions arising from spontaneous symmetry breaking and renormalization group effects mix the Weyl fields and make them massive. This mixing makes it appropriate to use Dirac four-component notation at low energies.

Having brought out the Feynman rules for a Fermi field in the simplest case, in the following chapter, we turn to the development of Feynman rules for Dirac spinors. Indeed, since a Weyl field is a special case of a Dirac field, one can use the Dirac notation for both fields. The following chapter, though it draws on the lines of argument presented here for its proofs, is self-sufficient for its conclusions.

# Chapter 8

# Four-Component Spinor Fields

Applying perturbative canonical quantization to four-component spinor fields: developing the quantum theory of the free field on the basis of classical polarization spinors and Dirac algebra to obtain the Feynman diagrams and rules for the interacting field.

## Introduction

This chapter presents the standard quantum theory of Dirac fields. It implements canonical quantization perturbatively, following the principles laid out in Chapter 4 but in the style appropriate to fermions, as developed in Chapter 7. The advantages of the Dirac four-component notation even for Weyl-type fields like the neutrino originate in the many convenient features of the algebra of Dirac matrices. This chapter covers these features in detail, providing all the knowledge necessary for computing with fermions.

Sections 8.1 and 8.2 introduce Dirac algebra, its common representations and their basic properties, thereby providing a firm foundation for computation with four-component spinors. Section 8.3 uses the plane-wave solutions of the Dirac equation to set up classical polarization spinors for each on-shell momentum. In Section 8.4, on the basis of the Weyl spinor theory, we simply write down the equal-time anti-commutation relations for the Dirac field and give a free-field solution in terms of creation and annihilation operators. Again using our experience with Weyl fields, Section 8.5 quickly deduces Feynman rules from Dyson's formula and the Wick contraction of free Dirac fields. This completes the perturbative canonical quantization of the Dirac field.

Sections 8.6 and 8.7 go more deeply into properties of Dirac matrices, presenting just about every formula one will ever need for computing decay rates and cross sections. Section 8.8 integrates the previous sections by going through a scattering example in detail.

The last five sections, 8.9 to 8.13, cover the three discrete symmetries, charge conjugation, parity, and time reversal, both separately and in all combinations. We conclude with the $CPT$ invariance of Feynman amplitudes.

## 8.1 Representations of the Dirac Algebra

In this section, we introduce the Dirac equation, derive abstract Dirac algebra from it, and then present the three principle representations of Dirac algebra in $4 \times 4$ matrices.

The starting point for the theory of the Dirac field is the Dirac equation

$$(i\gamma^\mu \partial_\mu - m)\psi = 0. \tag{8.1.1}$$

Dirac regarded the symbols $\gamma^\mu$ as constants determined by the condition that the square of the operator $\gamma^\mu \partial_\mu$ should be the Klein–Gordon operator $\partial^\mu \partial_\mu$. As he did not assume that the $\gamma$'s commuted, this condition becomes

$$(\gamma^\mu \partial_\mu)^2 = \sum_{\mu=0}^{3}(\gamma^\mu)^2 \partial_\mu \partial_\mu + \sum_{\mu<\nu}(\gamma^\mu\gamma^\nu + \gamma^\nu\gamma^\mu)\partial_\mu \partial_\nu = \sum_{\mu=0}^{3} \partial_\mu \partial_\mu, \tag{8.1.2}$$

from which it is clear that

$$\{\gamma^\mu, \gamma^\nu\} = 2g^{\mu\nu}. \tag{8.1.3}$$

These constraints on the algebra generated by the $\gamma$'s define the *Dirac algebra*. We note here that the contraction of $\gamma^\mu$ with a vector such as $p_\mu$ is so common that there is a special notation 'slash $p$' for it:

$$p\!\!\!/ \overset{\text{def}}{=} \gamma^\mu p_\mu. \tag{8.1.4}$$

When working with the Dirac algebra, it is convenient to represent the $\gamma$'s by specific matrices. The smallest matrices for which this is possible are $4 \times 4$ complex matrices. Many of the properties of the $\gamma$ matrices depend only on the defining conditions of the Dirac algebra, but others are representation dependent. There are three principle representations of the Dirac algebra in $4 \times 4$ matrices. The Weyl representation for the Dirac algebra is the one in which the Lorentz action on the four-component spinors is most simple. The standard or Dirac representation is the one in which the non-relativistic limit is most simple. The *Majorana representation* is the one in which all the $\gamma$ matrices are purely imaginary and the Euler–Lagrange equations are real, which makes charge conjugation as simple as possible.

A theorem of Pauli states that the algebraic conditions 8.1.3 specify the $4 \times 4$ matrices $\gamma^\mu$ up to similarity. A consequence of this theorem is that two distinct choices of matrices that satisfy this condition are always related by a change of coordinates on the four-component spinors. In fact, the three principle representations are related by unitary changes of coordinates.

In the Weyl representation, the $\gamma$ matrices are defined as in Section 7.18:

$$\gamma^0 = \beta = \begin{pmatrix} 0 & 1 \\ 1 & 0 \end{pmatrix}, \quad \gamma^r = \beta\alpha^r = \begin{pmatrix} 0 & \sigma^r \\ -\sigma^r & 0 \end{pmatrix},$$

$$\text{or} \quad \gamma^\mu = \begin{pmatrix} 0 & \sigma^\mu \\ \tilde\sigma^\mu & 0 \end{pmatrix}. \tag{8.1.5}$$

The Weyl representation is the most convenient one for computing the effects of Lorentz transformations, since in this representation there is no mixing between the $(0, 1/2)$ representation and the $(1/2, 0)$ one.

For computing in the non-relativistic approximation, the standard or *Dirac representation* is more convenient. In the standard representation, the $\gamma$ matrices are given by:

$$\gamma_\mu = \frac{1}{2} \begin{pmatrix} \sigma_\mu + \tilde\sigma_\mu & -\sigma_\mu + \tilde\sigma_\mu \\ \sigma_\mu - \tilde\sigma_\mu & -\sigma_\mu - \tilde\sigma_\mu \end{pmatrix}$$

$$\Rightarrow \quad \gamma_0 = \begin{pmatrix} 1 & 0 \\ 0 & -1 \end{pmatrix}, \quad \gamma_r = \begin{pmatrix} 0 & -\sigma_r \\ \sigma_r & 0 \end{pmatrix}. \tag{8.1.6}$$

The final representation is designed to make charge conjugation simple. From the equation of motion $(i\partial\!\!\!/ - m)\psi = 0$, we see that $\psi^*$ will also satisfy the equation of motion if all the $\gamma$ matrices are purely imaginary. The following assignments show that this is possible:

$$\gamma_0 = \begin{pmatrix} 0 & \sigma_2 \\ \sigma_2 & 0 \end{pmatrix} \qquad \gamma_1 = \begin{pmatrix} -i\sigma_3 & 0 \\ 0 & -i\sigma_3 \end{pmatrix}$$

$$\gamma_2 = \begin{pmatrix} 0 & \sigma_2 \\ -\sigma_2 & 0 \end{pmatrix} \qquad \gamma_3 = \begin{pmatrix} i\sigma_1 & 0 \\ 0 & i\sigma_1 \end{pmatrix} \tag{8.1.7}$$

A representation of the Dirac algebra in which all the $\gamma$ matrices are purely imaginary is called a Majorana representation. In such a representation $\psi \to i\psi^*$ is the charge conjugation map: it changes the sign of all charges.

Following up on the consequence of Pauli's theorem, we will conclude by writing down the similarity transformations between the three principle representations.

$$S_{WD} \stackrel{\text{def}}{=} \frac{1}{\sqrt{2}} \begin{pmatrix} 1 & 1 \\ 1 & -1 \end{pmatrix} \stackrel{\text{def}}{=} S_{DW};$$

$$S_{DM} \stackrel{\text{def}}{=} \frac{1}{\sqrt{2}} \begin{pmatrix} 1 & \sigma_2 \\ \sigma_2 & -1 \end{pmatrix} \stackrel{\text{def}}{=} S_{MD}; \tag{8.1.8}$$

$$S_{MW} \stackrel{\text{def}}{=} \frac{1}{2} \begin{pmatrix} 1+\sigma_2 & 1-\sigma_2 \\ -1+\sigma_2 & 1+\sigma_2 \end{pmatrix}, \quad S_{WM} \stackrel{\text{def}}{=} \frac{1}{2} \begin{pmatrix} 1+\sigma_2 & -1+\sigma_2 \\ 1-\sigma_2 & 1+\sigma_2 \end{pmatrix};$$

where the subscripts indicate the application:

$$S_{AB}S_{BA} = 1, \quad \psi_A = S_{AB}\psi_B, \quad \gamma_A^\mu = S_{AB}\gamma_B^\mu S_{BA}. \tag{8.1.9}$$

Note that the matrices here are all unitary.

A Majorana spinor is simply a Weyl spinor in Dirac notation. In the Weyl representation of four-component spinors, this means that a *Majorana field* $\psi$ is made up from just one left Weyl field $\psi_L$ as follows:

$$\textbf{Majorana Field}: \quad \psi = \begin{pmatrix} \psi_{Lc} \\ \psi_L \end{pmatrix}. \tag{8.1.10}$$

Using the two-component notation of Section 7.17, since $\varepsilon = i\sigma_2$, an arbitrary Dirac field $\psi_W$ is a Majorana field if it satisfies the constraint

$$i\gamma_2^W \psi_W^* = \begin{pmatrix} 0 & i\sigma_2 \\ -i\sigma_2 & 0 \end{pmatrix} \begin{pmatrix} \bar{\psi}_1 \\ \psi_2 \end{pmatrix}^* = \begin{pmatrix} \bar{\psi}_2 \\ \psi_1 \end{pmatrix} \stackrel{?}{=} \begin{pmatrix} \bar{\psi}_1 \\ \psi_2 \end{pmatrix}. \tag{8.1.11}$$

The map $\psi_W \to i\gamma_2\psi_W^*$ is a charge conjugation map; it reverses internal quantum numbers and its square is the identity. Thus a Majorana field is an eigenfield of charge conjugation. In the Dirac and Majorana representations, this charge conjugation map is given by

$$\psi_D \to i\gamma_2^D \psi_D^* \quad \text{and} \quad \psi_M \to i\psi_M^*. \tag{8.1.12}$$

(Note that, despite appearances, the charge conjugation map is linear in the fields.)

We have defined the Dirac algebra, presented its three principle representations, and identified the benefit of these representations. Since the representations are connected by unitary changes of basis, they have many properties in common that are not intrinsic to Dirac algebra. The next section derives some of these in preparation for quantizing the Dirac field.

## 8.2 Basic Features of Dirac Algebra

Using the concept of abstract Dirac algebra and its three principle representations, this section derives the basic features of the abstract algebra and a few properties held in common by these representations.

In the Weyl representation, the transformation rules for Weyl spinors in Section 7.1 determine the $4 \times 4$ matrices which represent rotation and boost generators. These matrices can be defined in terms of the $\gamma$ matrices, indicating that there is a representation of $so(1,3)$ in $\gamma$ matricesrepresentation of the Lorentz Lie algebra in abstract Dirac algebra:

$$\text{Boosts:} \quad \frac{1}{2}\gamma_r\gamma_0 = \frac{1}{2}\begin{pmatrix} \sigma_r & 0 \\ 0 & -\sigma_r \end{pmatrix} = \begin{pmatrix} \rho_r & 0 \\ 0 & -\rho_r \end{pmatrix};$$

$$\text{Rotations:} \quad \frac{1}{2}\gamma_r\gamma_s = -\frac{i}{2}\epsilon_{rs}^t\begin{pmatrix} \sigma_t & 0 \\ 0 & \sigma_t \end{pmatrix} = \epsilon_{rs}^t\begin{pmatrix} \tau_t & 0 \\ 0 & \tau_t \end{pmatrix}.$$

(8.2.1)

**Homework** 8.2.2. Use the defining relation 8.1.3 of Dirac algebra to show that the products of $\gamma$'s in 8.2.1 form a representation of the Lorentz Lie algebra.

**Remark** 8.2.3. Using the notation of Section 7.2, the equations 8.2.1 show that the action of the Lorentz group on four-component spinors takes the following simple form in the Weyl representation:

$$D(\Lambda) \stackrel{\text{def}}{=} \begin{pmatrix} D_{0,1/2}(\Lambda) & 0 \\ 0 & D_{1/2,0}(\Lambda) \end{pmatrix} = \begin{pmatrix} g & 0 \\ 0 & (g^{-1})^\dagger \end{pmatrix}.$$

(8.2.4)

**Homework** 8.2.5. Using the anti-commutation property 8.1.3 of the $\gamma$'s, show that

$$\Lambda = e^{\theta\gamma_r\gamma_s} \implies D(\Lambda) = \mathbf{1}\cos\frac{\theta}{2} + \gamma_r\gamma_s\sin\frac{\theta}{2},$$

$$\Lambda = e^{\theta\gamma_r\gamma_0} \implies D(\Lambda) = \mathbf{1}\cosh\frac{\theta}{2} + \gamma_r\gamma_0\sinh\frac{\theta}{2}.$$

Following the discussion in Sections 7.17 and 7.18, it is convenient to introduce the Dirac adjoint $\bar{\psi}$ of a four-component spinor $\psi$:

$$\bar{\psi} \stackrel{\text{def}}{=} \psi^\dagger\gamma_0.$$

(8.2.6)

This notation permits us, for example, to write the Lagrangian for a free Dirac field $\psi$ in a compact form:

$$\mathcal{L}_{\text{Free}} \stackrel{\text{def}}{=} \bar{\psi}(i\not{\partial} - m)\psi.$$

(8.2.7)

Along with the Dirac adjoint of a four-component spinor, it is convenient to define a Dirac adjoint $\overline{M}$ of a $4 \times 4$ matrix $M$ so that

$$\overline{M\psi} = (M\psi)^\dagger\gamma_0 = \bar{\psi}\overline{M}.$$

(8.2.8)

The definition of $\overline{M}$ is deduced from this property:

$$\overline{M} = \gamma_0^{-1}M^\dagger\gamma_0 = \gamma_0 M^\dagger\gamma_0.$$

(8.2.9)

Applying this operation to the representation $D$, we find that:

$$\bar{D}(\Lambda) = \begin{pmatrix} g^{-1} & 0 \\ 0 & g^{\dagger} \end{pmatrix} = D(\Lambda)^{-1}. \tag{8.2.10}$$

**Homework 8.2.11.** Use the Dirac algebra condition 8.1.3 and Homework 8.2.5 to show that the matrices $\gamma^{\mu}$ transform as a Lorentz vector: $\bar{D}(\Lambda)\gamma^{\mu} D(\Lambda) = \Lambda^{\mu}{}_{\nu}\gamma^{\nu}$.

**Homework 8.2.12.** Following Homework 8.2.5 and using only Dirac algebra, show that $\bar{D}(\Lambda)D(\Lambda) = 1$.

Since both the algebra of the $\gamma$ matrices and the relationship between $\gamma$ matrices and Lorentz generators are independent of choice of basis for the four-component spinors, the matrix $\gamma^5$ defined by

$$\gamma^5 \overset{\text{def}}{=} i\gamma^0\gamma^1\gamma^2\gamma^3 \overset{\text{def}}{=} \gamma_5$$

$$= \begin{pmatrix} 1 & 0 \\ 0 & -1 \end{pmatrix} \quad \text{in the Weyl representation,} \tag{8.2.13}$$

can be used in any representation of the Dirac algebra to construct the projections $P_L$ and $P_R$ of a four-component spinor onto the left and right two-component spinors:

$$P_L \overset{\text{def}}{=} \frac{1}{2}(1 - \gamma^5), \quad P_R \overset{\text{def}}{=} \frac{1}{2}(1 + \gamma^5). \tag{8.2.14}$$

**Homework 8.2.15.** Show that in abstract Dirac algebra $\gamma^5\gamma^5 = 1$ and $\{\gamma^{\mu}, \gamma^5\} = 0$.

**Homework 8.2.16.** Show that in the Dirac and Majorana representations, $\gamma^5$ is given by

$$\gamma_D^5 = \begin{pmatrix} 0 & 1 \\ 1 & 0 \end{pmatrix} \quad \text{and} \quad \gamma_M^5 = \begin{pmatrix} -\sigma_2 & 0 \\ 0 & \sigma_2 \end{pmatrix}.$$

**Remark 8.2.17.** In the three principle representations, we have

$$\gamma_0 = \gamma_0^{\dagger}, \quad \gamma_r = -\gamma_r^{\dagger}, \quad \gamma_5^{\dagger} = \gamma_5. \tag{8.2.18}$$

This implies that $\bar{\gamma}_{\mu} = \gamma_{\mu}$, $\bar{\gamma}_5 = -\gamma_5$, and

$$(\bar{\psi}_1 M \psi_2)^{\dagger} = \bar{\psi}_2 \overline{M} \psi_1. \tag{8.2.19}$$

The basic properties of abstract Dirac algebra and its three principle representations presented in this section will be sufficient for the quantization of the Dirac field. More specific features of these representations will be developed as needed for the computation of invariant amplitudes. In the next section, we start the quantization process by finding classical polarization spinors from the plane wave solutions of the Dirac equation, and in the one after that we introduce the free Dirac field.

## 8.3  Polarization Spinors in the Standard Representation

From the equations of motion 8.1.1 for the free Dirac spinor field we find plane wave solutions of the form

$$\psi(x) = u(p)e^{-ix\cdot p}, \quad \text{where } \not{p}u(p) = mu(p). \tag{8.3.1}$$

The choice of sign in the exponent indicates association between these plane waves and annihilation operators. We will complete the analysis of this case first and then change the sign. Note that it is the mass that compels us to distinguish between the two cases. Following the terminology of the previous chapter, we shall refer to the classical spinor $u(p)$ as a Dirac polarization spinor, four-component spinor, or a polarization spinor if the context determines the number of components.

The equation $\not{p}u(p) = mu(p)$ has a non-zero solution $u(p)$ if and only if $m$ is an eigenvalue of $\not{p}$. From the algebraic conditions 8.1.3 we see that $\not{p}^2 = p^2\mathbf{1}$, and so

$$(\not{p} - m)(\not{p} + m) = (p^2 - m^2)\mathbf{1}, \tag{8.3.2}$$

which shows that $\not{p} - m$ is invertible unless $p^2 = m^2$. If $p^2 = m^2$, then we can Lorentz-transform $p$ into $p_m = (m, 0, 0, 0)$, in which case $\not{p}$ becomes $m\gamma^0$. Therefore $u(p_m) = (a, b, 0, 0)$ with $a$ and $b$ to be fixed by some convention. To obtain $u(p)$ over the whole mass hyperboloid we use

$$D(\Lambda)u(p_m) \overset{\text{def}}{=} u(\Lambda p_m). \tag{8.3.3}$$

As in the case of Weyl polarization spinors (Section 7.5), this definition is ambiguous; we shall investigate the ambiguity at the end of this section after presenting the generic properties of polarization spinors.

We would like to choose a basis $u_1(p)$, $u_2(p)$ for $\ker(\not{p} - m)$ which matches the $u_L(p)$, $u_R(p)$ basis in the limit as $m \to 0$. In the Weyl representation of Dirac algebra, we build a four-component spinor by putting $\psi_R$ on top and $\psi_L$ on bottom, so in the standard representation the corresponding Dirac polarization spinors are

$$u_R(k_z) = \begin{pmatrix} 1 \\ 0 \\ 1 \\ 0 \end{pmatrix}, \quad u_L(k_z) = \begin{pmatrix} 0 \\ 1 \\ 0 \\ -1 \end{pmatrix}. \tag{8.3.4}$$

To match the massive theory to the massless one, we choose a $z$-boost $\Lambda = \exp(2\theta B_3)$, write $p_z = \Lambda p_m$, and apply $\Lambda$ to $u(p_m)$:

$$p_z = (m\cosh 2\theta, 0, 0, m\sinh 2\theta)^\top,$$

$$\begin{aligned} \implies \quad u(p_z) &= \exp(\theta\gamma_3\gamma_0)u(p_m) \\ &= (1 + \theta\gamma_3\gamma_0 + \tfrac{1}{2!}\theta^2(\gamma_3\gamma_0)^2 + \cdots)u(p_m) \\ &= (1\cosh\theta + \gamma_3\gamma_0\sinh\theta)u(p_m) \\ &= (a\cosh\theta, b\cosh\theta, a\sinh\theta, -b\sinh\theta)^\top. \end{aligned} \tag{8.3.5}$$

Next, to match $p_z^3$ to the lightlike reference vector $k_z^3$, we choose the boost parameter $\theta$ so that $m \sinh 2\theta = 1$. Then we find:

$$
\begin{aligned}
u(p_z)^{\mathsf{T}} = a &\left( \sqrt{\frac{p_z^0 + m}{2m}}, 0, \sqrt{\frac{p_z^0 - m}{2m}}, 0 \right) \\
+ b &\left( 0, \sqrt{\frac{p_z^0 + m}{2m}}, 0, -\sqrt{\frac{p_z^0 - m}{2m}} \right).
\end{aligned}
\tag{8.3.6}
$$

Finally, we take the limit as $m \to 0$. In this limit, $\theta \to \infty$ and $p_z^0 \to 1$. Now we can see that, if we write $u_1(p_z)$ and $u_2(p_z)$ respectively for the two vectors on the right of 8.3.6, and if we choose their normalizations by $a = b = \sqrt{2m}$, then

$$
\lim_{m \to 0} u_1(p_z) = u_R(k_z), \quad \lim_{m \to 0} u_2(p_z) = u_L(k_z).
\tag{8.3.7}
$$

We therefore define the four-component polarization spinors by

$$
u_1(p_m) = \sqrt{2m} \begin{pmatrix} 1 \\ 0 \\ 0 \\ 0 \end{pmatrix}, \quad u_2(p_m) = \sqrt{2m} \begin{pmatrix} 0 \\ 1 \\ 0 \\ 0 \end{pmatrix}.
\tag{8.3.8}
$$

Since these classical spinors are derived from the $\exp(-ix \cdot k)$ solution, they are conventionally associated with particle annihilation operators.

Following the same line of analysis for the plane waves

$$
v(p) e^{ix \cdot p}, \quad \text{where } \not{p} v(p) = -m v(p),
\tag{8.3.9}
$$

leads to a basis $v_1(p)$, $v_2(p)$ for the space of solutions. The simplest derivation of this basis, however, uses the fact that $\gamma_5$ anti-commutes with all the other $\gamma$ matrices, and so:

$$
\gamma_5 (\not{p} + m) v(p) = (-\not{p} + m) \gamma_5 v(p),
\tag{8.3.10}
$$

which implies that we can define the $v$'s by $v_r(p) = \gamma_5 u_r(p)$. The result is:

$$
v_1(p_m) = \sqrt{2m} \begin{pmatrix} 0 \\ 0 \\ 1 \\ 0 \end{pmatrix} \quad \text{and} \quad v_2(p_m) = \sqrt{2m} \begin{pmatrix} 0 \\ 0 \\ 0 \\ 1 \end{pmatrix}.
\tag{8.3.11}
$$

The values of $v_r(p)$ are obtained by boosting:

$$
v_r(\Lambda p_m) \overset{\text{def}}{=} D(\Lambda) v_r(p_m).
\tag{8.3.12}
$$

**Homework 8.3.13.** Show that

$$
\lim_{m \to 0} v_1(p_z) = u_R(k_z) \quad \text{and} \quad \lim_{m \to 0} v_2(p_z) = u_L(k_z).
$$

Now we can prove polarization spinor normalization and completeness formulae using the same method that worked for the two-component case. In fact, we hardly ever need to know the specific values of the polarization spinors; the formulae below are sufficient to eliminate the polarization spinors in most computations.

**Theorem 8.3.14.** *The Dirac polarization spinors $u_r(p)$ and $v_r(p)$ satisfy*

| | | |
|---|---|---|
| *Equation of Motion:* | $\not{p}u_r(p) = mu_r(p),$ | $\not{p}v_r(p) = -mv_r(p);$ |
| *Normalization:* | $\bar{u}_r(p)u_s(p) = 2m\delta_{rs},$ | $\bar{v}_r(p)v_s(p) = -2m\delta_{rs};$ |
| *Completeness:* | $u^r(p)\bar{u}_r(p) = \not{p} + m,$ | $v^r(p)\bar{v}_r(p) = \not{p} - m.$ $\quad\square$ |

Notice how, because of the form of the $u$'s and $v$'s at $p_m$ in the standard representation of the Dirac algebra, in the non-relativistic limit, the four types of spinor are effectively just the standard basis for $\mathbf{C}^4$.

**Homework 8.3.15.** Prove the normalization and completeness equations in Theorem 8.3.14.

**Homework 8.3.16.** Show that $\bar{u}_r(p)\gamma^\mu u_s(p) = 2p^\mu \delta_{rs}$.

**Homework 8.3.17.** Show that $\bar{u}_r(p)v_s(p) = 0$.

Massive particles do not have Lorentz-invariant helicity: indeed, in the rest frame of the particle there is no preferred direction in which to measure spin. But if we think of the masses as 'turning on' as a consequence of some symmetry-breaking mechanism, then we can display the evolution and interpretation of the polarization spinors as follows:

**Helicity** $\qquad\qquad\qquad\qquad\qquad\qquad\qquad\qquad$ **Spin**

$$
\begin{array}{cccc}
1/2 & u_R(k_z) \to \begin{cases} u_1(p_z) & \text{particle} & 1/2 \\ v_1(p_z) & \text{anti-particle} & 1/2 \end{cases} \\
-1/2 & u_L(k_z) \to \begin{cases} u_2(p_z) & \text{particle} & -1/2 \\ v_2(p_z) & \text{anti-particle} & -1/2 \end{cases}
\end{array} \qquad (8.3.18)
$$

The properties of polarization spinors listed in Theorem 8.3.14 are invariant under the action of the Lorentz group, even when different elements act on $u$'s and $v$'s. If we write these properties in terms of matrices of polarization spinors $u = (u_1, u_2)$, $\bar{u} = (\bar{u}_1, \bar{u}_2)^\top$, and so on, then it is clear that they are invariant under arbitrary unitary change of basis of both $u$-space and $v$-space.

We noted above that the definition of $u_r(p)$ in terms of $u_r(p_m)$ is ambiguous. If $\Lambda_1$ and $\Lambda_2$ are two Lorentz transformations which satisfy $\Lambda p_m = p$, then $\Lambda_1^{-1}\Lambda_2$ is a rotation. Through the representation $D$, $SO(3)$ acts as $SU(2)$ on the space spanned by $u_1(p_m)$ and $u_2(p_m)$. Hence the ambiguity allows for arbitrary change of basis by $SU(2)$ at every point on the mass hyperboloid. The same remarks could be applied to $v_r$ if we relax the condition $v_r = \gamma_5 u_r$. We have, however, just noted that this degree of ambiguity does not affect the properties of Theorem 8.3.14, and so, since almost all computations with spinors eventually use only these generic properties, the ambiguity is of no significance.

As in the two-component case, to obtain well-defined polarization spinors, it is necessary to choose a specific basis at each point on the mass hyperboloid. In the two-component case, the effect of such a choice was to bring a phase into the Lorentz transformation law for polarization spinors. In the four-component case, however, the action of the Lorentz group will actually mix polarization spinors.

For the Dirac equation, as for the equation of motion for Weyl spinors, there are two types of classical solution, one associated with annihilation and the other with creation. The new element of mass in the Dirac equation destroys helicity as a Lorentz invariant and gives us spin instead; it lifts the degeneracy that associates $u_R$ or $u_L$ with both particle and anti-particle and gives us four polarization spinors to distinguish particle from anti-particle, spin up from spin down. Furthermore, with the definitions above for Dirac polarization vectors, our computations will depend continuously on spinor masses.

## 8.4 Canonical Quantization of Dirac Spinors

With our experience of canonical quantization for scalar and two-component spinor fields, the canonical quantization of the Dirac field will be entirely routine. In this section, we shall briefly outline the sequence of steps involved, making the conclusions clear at each step.

The free Dirac field Lagrangian density is

$$\mathcal{L} = \bar{\psi}(i\partial\!\!\!/ - m)\psi. \tag{8.4.1}$$

The fermionic calculus conventions of Section 7.4 imply that the canonical momentum $\Pi$ of the Dirac field $\psi$ is $i\psi^\dagger$. The canonical momentum for the adjoint field $\bar{\psi}$ is zero. (We could make the Lagrangian density hermitian and restore symmetry between $\psi$ and $\bar{\psi}$, but the results will be the same.) Since $\mathcal{L}$ is zero when $\psi$ is a solution of the Dirac equation, the Hamiltonian density on solutions is

$$\mathcal{H} = \bar{\psi}i\gamma^0\partial_0\psi - \mathcal{L} = \psi^\dagger i\partial_0\psi. \tag{8.4.2}$$

Following our analysis of the Weyl spinor case, we shall assume that the four components $\psi_i$ of a Dirac field $\psi$ satisfy equal-time anti-commutation relations:

$$\left\{\psi_i(t,\bar{x}), \psi_j(t,\bar{y})\right\} = 0, \quad \left\{\psi_i^\dagger(t,\bar{x}), \psi_j^\dagger(t,\bar{y})\right\} = 0,$$
$$\left\{\psi_i^\dagger(t,\bar{x}), \psi_j(t,\bar{y})\right\} = \delta_{ij}\delta^{(3)}(\bar{x}-\bar{y}); \tag{8.4.3}$$

or, suppressing the indices:

$$\left\{\psi(t,\bar{x}), \psi(t,\bar{y})\right\} = 0, \quad \left\{\bar{\psi}(t,\bar{x}), \bar{\psi}(t,\bar{y})\right\} = 0,$$
$$\left\{\bar{\psi}(t,\bar{x}), \psi(t,\bar{y})\right\} = \gamma_0\delta^{(3)}(\bar{x}-\bar{y}), \tag{8.4.4}$$

and that the integral form of the free field is:

$$\psi(x) = \int \frac{d^3\bar{p}}{(2\pi)^3\,2\omega(\bar{p})} \left(e^{ix\cdot p}v_r(p)\beta^r(p)^\dagger + e^{-ix\cdot p}u_r(p)\alpha^r(p)\right). \tag{8.4.5}$$

Assuming the Lorentz transformation law for the free field is

$$U^\dagger(\Lambda)\psi(x)U(\Lambda) = D(\Lambda)\psi(\Lambda^{-1}x), \tag{8.4.6}$$

knowledge of the specific Lorentz transformation law of the polarization spinors will imply knowledge of the transformation law for the creation and annihilation operators, and hence for the particle states. As noted in the previous section, Lorentz transformations generally mix polarization spinors, and hence will also mix particle states.

**Remark 8.4.7.** Since a rotation acts on both the momentum of a particle and the axis with respect to which spin up and down are observed, clearly we will avoid spin mixing by rotations if we choose this axis to coincide with the momentum. Such a choice excludes $\bar{p} = \bar{0}$ and determines the *helicity basis* for polarization spinors elsewhere.

The definition so far does not completely determine the polarization spinors. For uniqueness, we could choose $u$'s and $v$'s subject to the normalization and completeness conditions 8.3.14 at some point $p_z$ where $p^1 = p^2 = 0$ and $p^3 > 0$, and define the polarization spinors elsewhere by boosting to the correct energy and then rotating through a minimal angle. Then $u^r(p)$ is connected to $u^r(\bar{p})$ by the rotation which takes $p$ into $\bar{p}$ through the $p^3$ axis. This will leave the polarization spinors undefined at points with $p^1 = p^2 = 0$ and $p^3 \le 0$; there is a discontinuity along this half line.

If the particle is massive, there is a boost which will reverse the momentum without changing the spin. This boost will therefore interchange positive and negative helicity states. Boosts generally will generate more general mixtures. (Note that the Dirac mass term will cause dynamical mixing of positive and negative helicity states.)

**Homework 8.4.8.** Take $p \cdot p = 0$, let $X_p$ be the generator of right-handed rotations about $(p^1, p^2, p^3)$, assume that $\not{p}u = 0$ and for $\epsilon = \pm 1$ define $u_\epsilon = \frac{1}{2}(1 + \epsilon \gamma_5)u$. Show that $iX_p u_\epsilon = \frac{1}{2}\epsilon u$. (Establish the conclusion for $p^1 = p^2 = 0$ and then rotate.)

The equal-time anti-commutation relations on the free field are equivalent to the creation and annihilation operator anti-commutation relations:

$$
\begin{aligned}
\{\alpha^r(p), \alpha^s(q)^\dagger\} &= (2\pi)^3 2\omega(\bar{p})\delta^{rs}\delta^{(3)}(\bar{p} - \bar{q}), \\
\{\beta^r(p), \beta^s(q)^\dagger\} &= (2\pi)^3 2\omega(\bar{p})\delta^{rs}\delta^{(3)}(\bar{p} - \bar{q}).
\end{aligned}
\tag{8.4.9}
$$

As in Section 7.6 for the Weyl field, we use these creation operators and their anti-commutation relations to define Fock space for Dirac fermions.

**Homework 8.4.10.** Show that, given the form 8.4.5 of the Dirac field, the equal time anti-commutation relations 8.4.4 are equivalent to the creation and annihilation operator anti-commutation relations 8.4.9.

The anti-commutation relations lead to the correct form for the Hamiltonian:

$$
H = \int d^3\bar{x}\, \psi^\dagger i\partial_0 \psi = \int d^3\bar{p}\, \omega(\bar{p})\big(a_r(\bar{p})^\dagger a^r(\bar{p}) + b_r(\bar{p})^\dagger b^r(\bar{p})\big),
\tag{8.4.11}
$$

thus confirming the validity of the assumptions above.

**Homework 8.4.12.** Substitute the integral form of the free Dirac field into the Hamiltonian 8.4.2 and reduce the result to the expression above.

**Homework 8.4.13.** Show that $b_r(p)^\dagger$ is a particle creation operator by computing the energy of $b_r(p)^\dagger |i\rangle$ for an arbitrary initial state $|i\rangle$.

Since Lorentz invariance requires that every Lagrangian term contains an even number of spinors, in the absence of Majorana fermions, the Lagrangian density will always have a $U(1)$ symmetry

$$
\psi'(\tau, \psi) \stackrel{\text{def}}{=} e^{-i\tau}\psi.
\tag{8.4.14}
$$

Following Section 7.4 on fermionic calculus and Section 7.7 on the conservation law for Weyl spinors, we see at once that the Noether conserved current is $\bar{\psi}\gamma^\mu\psi$, but normal ordering will be necessary if the current is to be well defined even on the vacuum:

$$j^\mu \overset{\text{def}}{=} \Pi^\mu \mathbf{D}\psi = :\bar{\psi}\gamma^\mu\psi: \tag{8.4.15}$$

We shall generally be a bit loose and write such currents and their conserved quantities without the normal ordering.

**Homework 8.4.16.** Show that $\langle 0|\bar{\psi}\gamma^\mu\psi|0\rangle$ diverges. (Formally, it is infinite for $\mu = 0$ and zero otherwise, thereby challenging Lorentz covariance.)

As we have not scaled $\tau$ differently for different fermions, the conserved quantity $Q$ derived from this symmetry is simply fermion number. We exclude Majorana fermions because the Majorana condition 8.1.11 is not invariant under the transformation above. The couplings of Majorana fermions may give rise to processes in which fermion number is not conserved. Clearly, if we introduce charges for the fermions, then we can if necessary assign charge zero to the Majorana fermions and thereby obtain a $U(1)$ symmetry of the Lagrangian density.

**Homework 8.4.17.** Use the canonical anti-commutation relations 8.4.4 and 8.4.9 to show that

$$[\psi, Q] = \psi \quad \text{and} \quad Q|p\rangle = |p\rangle,$$

where $|p\rangle$ is the state created by $\alpha_r^\dagger(p)$.

We have defined the free Dirac field from its anti-commutation relations and presented the integral form of it. A theory with Dirac fields but no Majorana fields always has a symmetry whose conserved quantity measures fermion number. From this foundation we can obtain formulae for contractions between spinors and derive Feynman rules for any theory involving spinor fields.

## 8.5 Feynman Rules for Dirac Spinors

Following the derivation of Feynman rules for two-component spinors in Section 7.10, for Dirac spinors we also start from Dyson's formula and conclude with an analysis of the amplitude for a process. The line of argument is so similar for Weyl and Dirac spinors that we shall generally be content simply to state the conclusions.

The contractions of the spinors can be computed in the following way. The Dirac fields occur in the Lagrangian density in the forms $\psi$ and $\bar{\psi}$. The free-field expansions for these forms are:

$$
\begin{aligned}
\psi(x) &= \int \frac{d^3\bar{p}}{(2\pi)^3\, 2\omega(\bar{p})} \left( e^{ix\cdot p} v_r(p)\beta^r(p)^\dagger + e^{-ix\cdot p} u_r(p)\alpha^r(p) \right), \\
\bar{\psi}(y) &= \int \frac{d^3\bar{q}}{(2\pi)^3\, 2\omega(\bar{q})} \left( e^{iy\cdot q} \bar{u}_r(q)\alpha^r(q)^\dagger + e^{-iy\cdot q} \bar{v}_r(q)\beta^r(q) \right).
\end{aligned}
\tag{8.5.1}
$$

The relationship between fields and initial and final state particles that connects Wick operators with uncontracted $\psi$'s and $\bar{\psi}$'s is one of destruction; the fields in the operator must destroy both initial-state and final-state particles. The following table summarizes this relationship and includes the polarization spinor left behind by the annihilation:

| | **Destroys Outgoing** | | **Destroys Incoming** | | |
|---|---|---|---|---|---|
| $\psi$ | Anti-Particle | $v$ | Particle | $u$ | (8.5.2) |
| $\bar{\psi}$ | Particle | $\bar{u}$ | Anti-Particle | $\bar{v}$ | |

When $x^0 > y^0$, the contraction of $\psi$ with $\bar{\psi}$ is:

$$
\begin{aligned}
\overset{\smile}{\psi(x)\bar{\psi}(y)} &= \langle 0|T\psi(x)\bar{\psi}(y)|0\rangle \\
&= \int \frac{d^3\bar{p}}{(2\pi)^3 \, 2\omega(\bar{p})} \, e^{-ix\cdot p}e^{iy\cdot p}u^r(p)\bar{u}_r(p) \\
&= \int \frac{d^3\bar{p}}{(2\pi)^3 \, 2\omega(\bar{p})} \, e^{-ix\cdot p}e^{iy\cdot p}(\not{p}+m) \\
&= (i\not{\partial}_x + m)\int \frac{d^3\bar{p}}{(2\pi)^3 \, 2\omega(\bar{p})} \, e^{-ix\cdot p}e^{iy\cdot p}.
\end{aligned}
\tag{8.5.3}
$$

When $y^0 > x^0$, the same conclusion holds. The integral that is to be differentiated is just the contraction for a free scalar field with mass $m$. Hence:

$$
\begin{aligned}
\overset{\smile}{\psi(x)\bar{\psi}(y)} &= (i\not{\partial}_x + m)\overset{\smile}{\phi(x)\phi(y)} \\
&= (i\not{\partial}_x + m)\int \frac{d^4p}{(2\pi)^4} \, \frac{i}{p^2 - m^2 + i\epsilon}e^{-i(x-y)\cdot p} \\
&= \int \frac{d^4p}{(2\pi)^4} \, \frac{i(\not{p}+m)}{p^2 - m^2 + i\epsilon}e^{-i(x-y)\cdot p}.
\end{aligned}
\tag{8.5.4}
$$

Noting that $(\not{p}+m)(\not{p}-m) = p^2 - m^2$ and writing

$$
(\not{p} - m + i\epsilon)^{-1} = \frac{1}{\not{p} - m + i\epsilon},
\tag{8.5.5}
$$

we can rewrite the contraction as

$$
\overset{\smile}{\psi(x)\bar{\psi}(y)} = \int \frac{d^4p}{(2\pi)^4} \, \frac{i}{\not{p} - m + i\epsilon}e^{-i(x-y)\cdot p}.
\tag{8.5.6}
$$

The contraction between two $\psi$ or two $\bar{\psi}$ fields is zero.

With the only non-zero contraction in hand, we can write down the Feynman rules for internal lines. Note that, for Dirac spinors, we can generally assume that the field is charged, and therefore put arrows on the internal lines to indicate the flow of that charge. From the kinetic term in the Lagrangian density, we know that a spinor field has mass dimension 3/2. Hence a renormalizable Lorentz invariant interaction term contains either zero or two spinors. In particular, the renormalizable interactions between spinor and scalar fields have the forms

$$
\bar{\psi}_1\psi_2\phi \quad \text{or} \quad \bar{\psi}_1i\gamma^5\psi_2\phi.
\tag{8.5.7}
$$

Assuming that the Dirac field is charged, the vertex for these interactions is

$$\text{Feynman Vertex for } \bar{\psi}\Gamma\psi: \qquad \diagdown\!\!\!\!\!\!^{-i\Gamma}\!\!\!\!\diagup \qquad (8.5.8)$$

where the arrow indicates the flow of particles. The $n$th order Feynman diagrams are constructed from $n$ copies of this vertex with an arbitrary number of joinings subject only to the usual joining constraint for lines with arrows.

From the contraction formula, we can guess the Feynman rules for internal lines and vertices:

**Scalar, momentum $p$:** $\qquad\qquad \dfrac{i}{p^2 - \mu^2 + i\epsilon} \dfrac{d^4p}{(2\pi)^4}$

**Spinor, momentum $q$ along arrow:** $\qquad \dfrac{i(\slashed{q} + m)}{q^2 - m^2 + i\epsilon} \dfrac{d^4q}{(2\pi)^4}$ $\qquad (8.5.9)$

$\bar{\psi}_1\Gamma\psi_2\phi$**, incoming momentum $P_{\text{in}}$:** $\qquad (-i\lambda)\Gamma(2\pi)^4\delta^{(4)}(P_{\text{in}})$

Since the direction of the arrow on a Fermi line indicates the flow of particles, we can distinguish between particle and anti-particle external lines. This makes the corresponding Feynman rules simpler than those for the Weyl fermions in Section 7.15. Assuming that the external lines have momentum $p$ with $p^0 > 0$ and spin $r$, on the basis of the relationship between the fields and particles 8.5.2, we assign polarization spinors as follows:

| | |
|---|---|
| **Incoming spinor particle:** | $u^r(p)$ |
| **Incoming spinor anti-particle:** | $\bar{v}^r(p)$ |
| **Outgoing spinor particle:** | $\bar{u}^r(p)$ |
| **Outgoing spinor anti-particle:** | $v^r(p)$ |

$\qquad (8.5.10)$

As with Weyl spinors, we apply the Feynman rules by multiplying the matrices along maximal Fermi lines following the arrow and using the 'later on the left' rule. For the relative sign between amplitudes which interfere, we count the number of transpositions required to put the sequence of momenta in their spinor products into a standard order. If there is a Fermi loop rulefermion loop, then we take minus the trace of the product of all the matrices around the loop.

The following homework shows that these Feynman rules do not necessarily give the absolute sign of the amplitude.

**Homework 8.5.11.** Starting from Dyson's formula with interaction Lagrangian density

$$\mathcal{L}' = -\lambda\bar{\psi}_2\psi_1\phi - \lambda\bar{\psi}_1\psi_2\phi,$$

derive the tree-level amplitude for a $\psi_1$-fermion to decay into a $\psi_2$-fermion and a scalar. Show that the result agrees with that given by the Feynman rules above up to a $-1$ coming from anti-commutation of Fermi operators.

External spinor lines in Feynman diagrams are generally labeled by the appropriate polarization spinor according to the rules above. Thus, for example, taking $\psi$

and $\phi$ to represent a proton and a $\pi^0$ respectively, we have the following contribution to $P\pi^0$ elastic scattering:

$$D = \begin{array}{c} \bar{u}^{r'}(p') \\ q' \end{array} \!\!\!\!\!\!\!\! \begin{array}{c} q \\ u^r(p) \end{array} \tag{8.5.12}$$

$$\implies \quad i\mathcal{A}(D) = (-i\lambda)^2 \, \bar{u}^{r'}(p')\Gamma(\not{p}+\not{q}+m)\Gamma u^r(p) \, \frac{i}{(p+q)^2 - m^2 + i\epsilon}.$$

There are two second-order diagrams which contribute to $P\bar{P}$ elastic scattering:

$$D_1 = \begin{array}{c} \bar{u}^{r'}(p') \quad u^r(p) \\ v^{s'}(q') \quad \bar{v}^s(q) \end{array} \qquad D_2 = \begin{array}{c} \bar{u}^{r'}(p') \quad u^r(p) \\ v^{s'}(q') \quad \bar{v}^s(q) \end{array} \tag{8.5.13}$$

Since all the spins of the external fermions are the same in both diagrams, these diagrams represent amplitudes which interfere. The amplitudes for these diagrams are

$$i\mathcal{A}_1 = (-i\lambda)^2 \, \bar{u}^{r'}(p')\Gamma u^r(p) \, \bar{v}^s(q)\Gamma v^{s'}(q') \, \frac{i}{(p-p')^2 - \mu^2 + i\epsilon},$$
$$\tag{8.5.14}$$
$$i\mathcal{A}_2 = (-i\lambda)^2 \, \bar{u}^{r'}(p')\Gamma v^{s'}(q') \, \bar{v}^s(q)\Gamma u^r(p) \, \frac{i}{(p+q)^2 - \mu^2 + i\epsilon}.$$

Since the sequence of momenta in the spinor products differ by a single transposition, there is a relative sign of $(-1)$ in the total amplitude:

$$\mathcal{A} = \mathcal{A}_1 - \mathcal{A}_2. \tag{8.5.15}$$

The two Fermi lines in the vertex above can be contracted to create a diagram with a single external line. Such diagrams are called *tadpoles*. Following the principle of renormalization introduced in Section 4.13, tadpole graphs contribute to the vacuum expectation of a field and are eliminated by renormalization. This elimination of tadpoles also eliminates all graphs which have tadpole subgraphs:

$$\begin{array}{c} \text{(diagrams)} \end{array} + \begin{array}{c} \text{(diagram)} \end{array} = 0 \implies \begin{array}{c} \text{(diagram)} \end{array} + \begin{array}{c} \text{(diagram)} \end{array} = 0. \tag{8.5.16}$$

For the sake of an example, we give the loop integral for the tadpole:

$$i\mathcal{A}(\text{Tadpole}) = (-i\lambda)(-1) \int \frac{d^4q}{(2\pi)^4} \, \frac{i\,\mathrm{tr}\big(\Gamma(\not{q}+m)\big)}{q^2 - m^2 + i\epsilon}. \tag{8.5.17}$$

**Homework 8.5.18.** Following Section 4.10, write down the 18 connected second-order, tadpole-free Feynman diagrams that can be constructed from the vertex 8.5.8. Note which ones interfere. Using the approaches of Section 4.10 and Section 4.13 for tree and loop diagrams respectively, write down the amplitudes for these 19 diagrams. (No need to do the loop integrals.)

In practice, when we have a beam of Fermi particles, we do not generally know what their spins are. This means that, in our theory, we should average the differential cross sections over incoming spins. Also, when fermions are detected, they are generally detected without reference to their spins. This means that we should sum the differential cross sections over outgoing spins. Thus, starting from an invariant amplitude of the form

$$i\mathcal{A}_{sr} = F\bar{u}_s(p')\not{a}u_r(p) \tag{8.5.19}$$

which could represent a fermion with momentum $p$ and spin $r$ scattering off a scalar into a fermion with momentum $p'$ and spin $s$, we form the spin-sum, spin-average of the squares of the amplitudes using the trace trick of Section 7.15:

$$
\begin{aligned}
A^2 &= \frac{1}{2}\sum_{r,s}|\mathcal{A}_{sr}|^2 = \frac{1}{2}|F|^2\sum_{r,s}\bar{u}_s(p')\not{a}u_r(p)\bar{u}_r(p)\not{a}u_s(p') \\
&= \frac{1}{2}|F|^2\sum_{r,s}\mathrm{tr}\big(\bar{u}_s(p')\not{a}u_r(p)\bar{u}_r(p)\not{a}u_s(p')\big) \\
&= \frac{1}{2}|F|^2\sum_{r,s}\mathrm{tr}\big(\not{a}u_r(p)\bar{u}_r(p)\not{a}u_s(p')\bar{u}_s(p')\big) \\
&= \frac{1}{2}|F|^2\,\mathrm{tr}\bigg(\sum_{r,s}\big(\not{a}u_r(p)\bar{u}_r(p)\not{a}u_s(p')\bar{u}_s(p')\big)\bigg) \\
&= \frac{1}{2}|F|^2\,\mathrm{tr}\big(\not{a}(\not{p}+m)\not{a}(\not{p}'+m)\big),
\end{aligned}
\tag{8.5.20}
$$

which we can simplify when we have the Dirac algebra machinery which is the topic of the next section.

The formulae for the differential cross section $d\sigma$ developed in Chapter 5 are valid here. Where the spin-sum, spin-average rule applies, the formula is

$$d\sigma = \frac{1}{4E_T p_{\mathrm{in}}}A^2\mathcal{D}, \tag{8.5.21}$$

where $p_{\mathrm{in}} = |\bar{p}|$ is the magnitude of the momentum of a single incoming particle in the center of mass frame, and $\mathcal{D}$ is the invariant density of final states.

Having derived the formula for the contraction of Dirac spinors from the integral form of the free spinor field, the Feynman rules can be guessed without going through the application of Wick's Theorem to Dyson's formula. The application of the Feynman rules requires attention to the matrix products and signs, but is otherwise routine. Now in order to compute any invariant amplitude for a theory of scalars and spinors, all we need is the knowledge of Dirac algebra presented in the two following sections.

## 8.6 Dirac Algebra

When we apply the Feynman rules for spinors, we have to multiply $\gamma$ matrices along fermion lines and take traces of products of $\gamma$ matrices. We will now present a few rules for simplifying such products and traces. Because the formulae which follow are so generally sufficient to our purposes, it is not worthwhile to go into the details of proof. We shall be content to present them in a logical sequence.

An excellent general reduction formula for products of $\gamma$ matrices is:

**Lemma 8.6.1.** $\gamma^\alpha \gamma^\beta \gamma^\gamma = g^{\alpha\beta}\gamma^\gamma - g^{\alpha\gamma}\gamma^\beta + g^{\beta\gamma}\gamma^\alpha + i\epsilon^{\alpha\beta\mu\gamma}\gamma_\mu\gamma^5.$

**Proof.** To verify this formula, we consider the cases $\alpha = \beta = \gamma$, $\alpha = \beta \neq \gamma$, and $\alpha \neq \beta \neq \gamma \neq \alpha$ separately. The truth of the formula is obvious in each case. $\square$

The trace of a product of an odd number of $\gamma$ matrices is always zero:

**Lemma 8.6.2.** $\mathrm{tr}(\gamma^{\mu_1} \dots \gamma^{\mu_{2k+1}}) = 0.$

**Proof.** This result follows from the properties of $\gamma_5$ and the cyclic property of the trace. For example, with $k = 1$ we have

$$
\begin{aligned}
\mathrm{tr}(\gamma^\alpha \gamma^\beta \gamma^\gamma) &= \mathrm{tr}(\gamma_5 \gamma_5 \gamma^\alpha \gamma^\beta \gamma^\gamma) && \text{since } \gamma_5^2 = 1 \\
&= \mathrm{tr}(\gamma_5 \gamma^\alpha \gamma^\beta \gamma^\gamma \gamma_5) && \text{cyclic property of trace} \\
&= -\mathrm{tr}(\gamma_5 \gamma_5 \gamma^\alpha \gamma^\beta \gamma^\gamma) && \text{since } \gamma^\mu \gamma_5 = -\gamma_5 \gamma^\mu \\
&= -\mathrm{tr}(\gamma^\alpha \gamma^\beta \gamma^\gamma),
\end{aligned}
$$

and so $\mathrm{tr}(\gamma^\alpha \gamma^\beta \gamma^\gamma) = 0.$ $\square$

The trace of a product of an even number of $\gamma$ matrices can be computed from the reduction formula

**Lemma 8.6.3.** *Using a hat to indicate a matrix omitted from the product, we have*

$$
\mathrm{tr}(\gamma^{\mu_1} \dots \gamma^{\mu_{2k}}) = \sum_{r=1}^{2k-1} (-1)^{r+1} g^{\mu_r \mu_{2k}} \, \mathrm{tr}(\gamma^{\mu_1} \dots \widehat{\gamma^{\mu_r}} \dots \gamma^{\mu_{2k-1}}).
$$

**Proof.** This formula is a consequence of the defining property of Dirac algebra and the cyclic property of the trace. With $k = 2$ for example, we deduce the result from the two identities

$$
\mathrm{tr}(ABCD) = \mathrm{tr}(DABC), \tag{8.6.4}
$$

$$
ABCD + DABC = AB(CD + DC) - A(BD + DB)C + (AD + DA)BC. \quad \square
$$

In particular, this lemma implies:

**Corollary 8.6.5.** $\mathrm{tr}(\gamma^\mu \gamma^\nu) = 4g^{\mu\nu}$ and

$$
\mathrm{tr}(\gamma^\mu \gamma^\nu \gamma^\sigma \gamma^\tau) = 4(g^{\mu\nu}g^{\sigma\tau} - g^{\mu\sigma}g^{\nu\tau} + g^{\mu\tau}g^{\nu\sigma}).
$$

By considering the various cases of equal and unequal indices, from these formulae for traces of even products of $\gamma$'s we can deduce the results when we include a factor of $\gamma_5$:

**Corollary 8.6.6.** $\operatorname{tr}(\gamma_5\gamma^\mu\gamma^\nu) = 0$ and $\operatorname{tr}(\gamma_5\gamma^\mu\gamma^\nu\gamma^\sigma\gamma^\tau) = -4i\epsilon^{\mu\nu\sigma\tau}$.

Contracted products of $\gamma$ matrices also arise. The generic inductive step for simplifying such products is

$$\gamma_\alpha\gamma^{\mu_1} = 2g_\alpha^{\mu_1} - \gamma^{\mu_1}\gamma_\alpha \tag{8.6.7}$$
$$\implies \quad \gamma_\alpha\gamma^{\mu_1}\cdots\gamma^{\mu_n}\gamma^\alpha = 2\gamma^{\mu_2}\cdots\gamma^{\mu_n}\gamma^{\mu_1} - \gamma^{\mu_1}\gamma_\alpha\gamma^{\mu_2}\cdots\gamma^{\mu_n}\gamma^\alpha.$$

Repeated use of this step yields a general result and (after some rearrangements) the commonly used formulae:

**Lemma 8.6.8.** *Writing* $\hat{\gamma}$ *for a dropped factor,*

$$\gamma_\alpha\gamma^{\mu_1}\cdots\gamma^{\mu_n}\gamma^\alpha = 4(-1)^n\gamma^{\mu_1}\cdots\gamma^{\mu_n} - 2\sum_{r=1}^{n}(-1)^r\gamma^{\mu_1}\cdots\widehat{\gamma^{\mu_r}}\cdots\gamma^{\mu_n}\gamma^{\mu_r}.$$

**Corollary 8.6.9.** $\gamma_\alpha\gamma^\alpha = 4\mathbb{1}$, $\gamma_\alpha\gamma^\mu\gamma^\alpha = -2\gamma^\mu$, $\gamma_\alpha\gamma^\mu\gamma^\nu\gamma^\alpha = 4g^{\mu\nu}$, *and*

$$\gamma_\alpha\gamma^\mu\gamma^\nu\gamma^\rho\gamma^\alpha = -2\gamma^\rho\gamma^\nu\gamma^\mu,$$
$$\gamma_\alpha\gamma^\mu\gamma^\nu\gamma^\rho\gamma^\sigma\gamma^\alpha = 2(\gamma^\sigma\gamma^\mu\gamma^\nu\gamma^\rho + \gamma^\rho\gamma^\nu\gamma^\mu\gamma^\sigma). \qquad \square$$

From the formulae above for traces of four $\gamma$'s with and without $\gamma_5$ factors, we find the following commonly used contractions between traces:

**Corollary 8.6.10.** $\operatorname{tr}(\gamma_5\not{p}_1\gamma^\mu\not{p}_2\gamma^\nu)\operatorname{tr}(\not{p}_3\gamma_\mu\not{p}_4\gamma_\nu) = 0,$

$$\operatorname{tr}(\not{p}_1\gamma^\mu\not{p}_2\gamma^\nu)\operatorname{tr}(\not{p}_3\gamma_\mu\not{p}_4\gamma_\nu) = 32(p_1\cdot p_3\, p_2\cdot p_4 + p_1\cdot p_4\, p_2\cdot p_3),$$
$$\operatorname{tr}(\gamma_5\not{p}_1\gamma^\mu\not{p}_2\gamma^\nu)\operatorname{tr}(\gamma_5\not{p}_3\gamma_\mu\not{p}_4\gamma_\nu) = 32(p_1\cdot p_3\, p_2\cdot p_4 - p_1\cdot p_4\, p_2\cdot p_3). \qquad \square$$

So far in this section, the formulae have derived from the properties of abstract Dirac algebra. In the three principle representations, as noted above in Remark 8.2.17, we also have the following frequently used properties of the $\gamma$'s:

**Lemma 8.6.11.** *In any representation of Dirac algebra related to the Weyl representation by a unitary change of basis, we find that* $\gamma_\mu^\dagger = \tilde{\gamma}_\mu$, $\gamma_5^\dagger = \gamma_5$, $\bar{\gamma}_\mu = \gamma_\mu$, *and* $\bar{\gamma}_5 = -\gamma_5$.

The action of transposing is representation dependent:

$$
\begin{pmatrix} \gamma_0^{\mathsf{T}} \\ \gamma_1^{\mathsf{T}} \\ \gamma_2^{\mathsf{T}} \\ \gamma_3^{\mathsf{T}} \\ \gamma_4^{\mathsf{T}} \\ \gamma_5^{\mathsf{T}} \end{pmatrix} =
\begin{array}{ccc}
\mathbf{W} & \mathbf{D} & \mathbf{M} \\
\left\{\begin{matrix} + \\ - \\ + \\ - \\ + \\ + \end{matrix}\right\} &
\left\{\begin{matrix} + \\ - \\ + \\ - \\ + \\ + \end{matrix}\right\} &
\left\{\begin{matrix} - \\ + \\ + \\ + \\ - \\ - \end{matrix}\right\}
\end{array}
\begin{pmatrix} \gamma_0 \\ \gamma_1 \\ \gamma_2 \\ \gamma_3 \\ \gamma_4 \\ \gamma_5 \end{pmatrix}. \tag{8.6.12}
$$

Thus, for example, in the Dirac representation, $\gamma_3^{\mathsf{T}} = -\gamma_3$.

**Homework 8.6.13.** Show that $\operatorname{tr}(\gamma^{\mu_1}\cdots\gamma^{\mu_n}) = \operatorname{tr}(\gamma^{\mu_n}\cdots\gamma^{\mu_1})$. Deduce that equality is still true if we replace $\gamma^{\mu_r}$ by $\gamma_5$.

The conclusions of this section are generally sufficient for the computation of decays and cross sections with external fermions. In a few situations, one more trick is very useful, namely the Fierz transformation presented in the next section.

## 8.7 Fierz Transformations

When the product of matrices involves both $\gamma$ matrices and spinors, it is often useful to rearrange the sequence of the terms using the Fierz transformations. One purpose of the Fierz transformation is to get rid of quantities like $\not{p}u^r(q)$, which are not simple, by transforming them into quantities like $\not{p}u^r(p)$, which can be evaluated from the very definition of $u^r(p)$. Another purpose is to bring like spinors together by transformations of the following type:

$$\left(\bar{u}_e \gamma^\alpha (1-\gamma^5) u_{\nu_e}\right)\left(\bar{u}_{\nu_e} \gamma_\alpha (1-\gamma^5) u_e\right) \tag{8.7.1}$$
$$= -\left(\bar{u}_e \gamma_\alpha (1-\gamma^5) u_e\right)\left(\bar{u}_{\nu_e} \gamma^\alpha (1-\gamma^5) u_{\nu_e}\right).$$

The first step and foundation for the Fierz transformations is the use of bilinear form $\text{tr}(MN)$ on $4 \times 4$ matrices to set up dual bases for this 16-dimensional vector space. Thus, if $\mathcal{B} = (B^1, \ldots, B^{16})$ is any linear basis for the space of $4 \times 4$ complex matrices, and if the $16 \times 16$ matrix $\left[\text{tr}(B^i B^j)\right]$ is non-singular, then there exists a unique dual basis $\mathcal{B}' = (B_1, \ldots, B_{16})$ defined by the property

$$\text{tr}(B^i B_j) = \delta^i_j. \tag{8.7.2}$$

The second step is to use this duality result to write any $4 \times 4$ complex matrix $M$ as a linear combination of basis elements:

$$M \equiv \text{tr}(MB_i)B^i \equiv \text{tr}(MB^i)B_i. \tag{8.7.3}$$

Given two spinors $u$ and $v$, we can take $M = u\bar{v}$ and use the cyclic property of trace to find the fundamental expansions

$$u\bar{v} = \text{tr}(u\bar{v}B_i)B^i = (\bar{v}B_i u)B^i = (\bar{v}B^i u)B_i. \tag{8.7.4}$$

This equation is the essence of the Fierz transformations, as it effectively reverses the order of the spinors $u$ and $\bar{v}$.

To illustrate its application, let $u(n)$ stand for $u^{r_n}(p_n)$ or $v^{r_n}(p_n)$, define $(M, N)$ and its flipped version $(M, N)^f$ by

$$(M, N) \stackrel{\text{def}}{=} \bar{u}(4)Mu(2)\bar{u}(3)Nu(1),$$
$$(M, N)^f \stackrel{\text{def}}{=} \bar{u}(4)Mu(1)\bar{u}(3)Nu(2), \tag{8.7.5}$$

and transform the product of $(M, N)$'s using the fundamental expansion 8.7.4:

$$\begin{aligned}
(M, N) &= \bar{u}(4)Mu(2)\bar{u}(3)Nu(1) \\
&= \bar{u}(4)M\left(\bar{u}(3)B_k u(2)\right)B^k Nu(1) \\
&= \left(\bar{u}(4)MB^k Nu(1)\right)\left(\bar{u}(3)B_k u(2)\right) \\
&= (MB^k N, B_k)^f. \tag{8.7.6}
\end{aligned}$$

If we take $M = B^i$ and $N = B^j$, and if we express $B_k$ and $B^i B^k B^j$ in terms of $B^l$'s, then we see that $(B^i, B^j)$ is a linear combination of $(B^r, B^s)^f$'s. The resulting $256 \times 256$ matrix of coefficients may be called the Fierz matrix for the basis $\mathcal{B}$.

The formula above is too general for practical needs, and also too cumbersome. In applications, however, the matrices $B^i$ will be chosen to have reasonable Lorentz transformation properties, and the sums of products that arise will be Lorentz scalars. Since Dirac algebra is the algebra of $4 \times 4$ complex matrices, this naturally suggests building a basis from products of $\gamma$ matrices. The standard second-order products are the $\sigma^{\mu\nu}$'s defined by

$$\sigma^{\mu\nu} = \frac{i}{2}[\gamma^\mu, \gamma^\nu] = \begin{cases} i\gamma^\mu\gamma^\nu, & \text{if } \mu \neq \nu; \\ 0, & \text{if } \mu = \nu. \end{cases} \tag{8.7.7}$$

Note that $\sigma^{\mu\nu} = -\sigma^{\nu\mu}$, so that a basis should not contain more than the six $\sigma$'s with $\mu < \nu$. Now the standard basis $\mathcal{B}_S$ and its dual $\mathcal{B}'_S$ can be defined:

$$\mathcal{B}_S \stackrel{\text{def}}{=} \{1, \gamma^\mu, \sigma^{\mu\nu}, \gamma^5\gamma^\mu, \gamma^5\},$$
$$\mathcal{B}'_S \stackrel{\text{def}}{=} \{\tfrac{1}{4}1, \tfrac{1}{4}\gamma_\mu, \tfrac{1}{4}\sigma_{\mu\nu}, \tfrac{1}{4}\gamma_\mu\gamma_5, \tfrac{1}{4}\gamma_5\}. \tag{8.7.8}$$

**Homework 8.7.9.** Verify the duality condition 8.7.2 for $\mathcal{B}_S$ and $\mathcal{B}'_S$.

**Remark 8.7.10.** It is worth noting that each element of $\mathcal{B}_S$ is determined up to a scalar factor by its commutation/anti-commutation with the four $\gamma_\mu$. For example, $\gamma^5\gamma^0$ anti-commutes with $\gamma^0$ and commutes with the three $\gamma^r$.

The Lorentz group action on Dirac algebra is derived from the transformation of the term $\bar{\psi}M\psi$:

$$\Lambda \cdot M = \bar{D}(\Lambda)MD(\Lambda), \tag{8.7.11}$$

where $D(\Lambda)$ is specified in a representation independent way by 8.2.1. By direct computation using the algebra condition, we can verify that

$$\bar{D}(\Lambda)\gamma^\mu D(\Lambda) = \Lambda^\mu{}_\nu\gamma^\nu;$$
$$\bar{D}(\Lambda)\gamma^5 D(\Lambda) = \gamma^5. \tag{8.7.12}$$

(Strictly speaking, we verify this first for $\Lambda = \exp(X)$ for $X$ in the Lie algebra $so(1,3)$, and then extend to the connected Lorentz group by multiplying these $\Lambda$'s.) Thus our basis $\mathcal{B}$ is a basis of tensors, and this action of $SO(1,3)$ on the space of $4 \times 4$ complex matrices splits into a direct sum of five actions:

$$\begin{array}{cccccc} 1 & \gamma^\mu & \sigma^{\mu\nu}P_L & \sigma^{\mu\nu}P_R & \gamma^5\gamma^\mu & \gamma^5 \end{array}$$
$$\left((0,\tfrac{1}{2}) + (\tfrac{1}{2},0)\right)^2 = (0,0) + (\tfrac{1}{2},\tfrac{1}{2}) + (1,0) + (0,1) + (\tfrac{1}{2},\tfrac{1}{2}) + (0,0). \tag{8.7.13}$$

We recognize that $\sigma^{\mu\nu}P_R$ is associated with $(0,1)$ because $P_R$ projects onto the $(0,1/2)$ component of a Dirac field.

Up to a sign, parity $P$ is that element of the extended Lorentz group which satisfies $P^2 = 1$, commutes with rotation generators, and anti-commutes with boost generators. These properties must be shared by its representation $D(P)$ in relation to the representations of rotation and boost generators. We deduce that $D(P) = \gamma_0$

up to a sign. The action of parity on $\mathcal{B}_S$ distinguishes the two scalar and the two vector representations of the Lorentz group:

$$\gamma_0 1 \gamma_0 = 1 \quad \text{and} \quad \gamma_0 \gamma_5 \gamma_0 = -\gamma_5,$$

$$\gamma_0 \gamma^\mu \gamma_0 = \gamma_\mu \quad \text{and} \quad \gamma_0 \gamma_5 \gamma^\mu \gamma_0 = -\gamma_5 \gamma_\mu. \tag{8.7.14}$$

A Lorentz scalar which changes sign under parity is called a *pseudoscalar*; a Lorentz vector whose time component changes sign under parity is called an *axial vector*.

**Homework 8.7.15.** Show that the $\sigma$'s transform as $(1,0)+(0,1)$ under the Lorentz action of Homework 8.2.11. Deduce that there are two independent Lorentz scalars that are quadratic in the $\sigma$'s, and write down the general form of this scalar.

In applications of the Fierz transformations, the spinor products $(M, N)$ are generally Lorentz scalars. Hence, taking the result of Homework 8.7.15 into account, the $16 \times 16$ products permitted in the universal Fierz formula 8.7.6 reduce in the basis $\mathcal{B}_S$ to eight. The consequences of 8.7.6 could be expressed as an $8 \times 8$ matrix, but the result would indicate that even this reduced matrix is more complicated than necessary, and so instead of completing this calculation, we will change the basis $\mathcal{B}_S$ in order to find a simpler Fierz matrix.

A second, more convenient basis uses the spin projection matrices $P_- = P_L$ and $P_+ = P_R$. (This basis is not only convenient for simplifying the Fierz matrix, but also arises naturally in the context of the Standard Model of electroweak and strong interactions.) Noting that

$$\sigma^{01} P_\pm = \sigma^{01} \pm i\sigma^{23} = i\sigma^{23} P_\pm,$$

$$\sigma^{02} P_\pm = \sigma^{02} \pm i\sigma^{31} = i\sigma^{31} P_\pm, \tag{8.7.16}$$

$$\sigma^{03} P_\pm = \sigma^{03} \pm i\sigma^{12} = i\sigma^{12} P_\pm,$$

we can choose a better basis and determine its dual:

$$\mathcal{B}_P \overset{\text{def}}{=} \{P_+, P_-, \gamma^\mu P_+, \gamma^\mu P_-, \sigma^{0r} P_+, \sigma^{0r} P_-\},$$

$$\mathcal{B}_P' \overset{\text{def}}{=} \{\tfrac{1}{2}P_+, \tfrac{1}{2}P_-, \tfrac{1}{2}\gamma_\mu P_-, \tfrac{1}{2}\gamma_\mu P_+, \tfrac{1}{2}\sigma_{0r} P_+, \tfrac{1}{2}\sigma_{0r} P_-\}. \tag{8.7.17}$$

**Homework 8.7.18.** Show that the relationship of duality 8.7.2 holds between $\mathcal{B}_P$ and $\mathcal{B}_P'$.

The Fierz transformations for the Lorentz invariant products can now be summarized as follows:

$$(P_\pm, P_\pm) = \tfrac{1}{2}(P_\pm, P_\pm)^f + \tfrac{1}{8}(\sigma^{\mu\nu} P_\pm, \sigma_{\mu\nu} P_\pm)^f;$$

$$(P_\pm, P_\mp) = \tfrac{1}{2}(\gamma^\mu P_\mp, \gamma^\mu P_\pm)^f;$$

$$(\gamma^\mu P_\pm, \gamma_\mu P_\pm) = -(\gamma^\mu P_\pm, \gamma_\mu P_\pm)^f; \tag{8.7.19}$$

$$(\sigma^{\mu\nu} P_\pm, \sigma_{\mu\nu} P_\pm) = 6(P_\pm, P_\pm)^f - \tfrac{1}{2}(\sigma^{\mu\nu} P_\pm, \sigma_{\mu\nu} P_\pm)^f;$$

$$(\sigma^{\mu\nu} P_\pm, \sigma_{\mu\nu} P_\mp) = 0.$$

Since there are only eight Lorentz scalars, only eight of these ten expressions are linearly independent; indeed, the fourth pair is equivalent to the first.

**Homework 8.7.20.** Prove the Fierz formulae.

One should not worry about the details of the formulae above: they are not to be learned or proved. The purpose of presenting them here is just that one should know what is available and where to look it up!

## 8.8 Dirac Spinor Scattering Example

In this section, we shall integrate the Feynman rules of Section 8.5 and the formulae of Section 8.6 and illustrate their application in computations of scattering amplitudes involving Dirac spinors. For the sake of concreteness, we shall take the interaction Hamiltonian density

$$\mathcal{H}_I = \lambda \bar{\psi}\psi\phi, \tag{8.8.1}$$

where $\psi$ is a Dirac spinor with mass $m$ and $\phi$ is a scalar with mass $\mu$ and we shall compute the spin-sum, spin-average cross section from the amplitudes $\mathcal{M}^{rs}$ of fermion-scalar scatteringscattering aa fermion-scalar:

$$\mathcal{M}^{rs} = \langle p', r; q' | S - 1 | p, s; q \rangle. \tag{8.8.2}$$

To second order in $g$, there are two diagrams to consider, both containing a single fermion line connecting incoming fermion to outgoing fermion. Let $D_1$ be the diagram in which the fermion line absorbs the incoming scalar before emitting the outgoing scalar, and let $D_2$ be the other:

$$\mathcal{D}_1 = \begin{array}{c} p', r \\ \\ q' \end{array} \overbrace{\phantom{xxx}}^{\phantom{x}} \begin{array}{c} q \\ \\ p, s \end{array} \quad \text{and} \quad D_2 = \begin{array}{c} p', r \\ \\ q' \end{array} \overbrace{\phantom{xxx}}^{\phantom{x}} \begin{array}{c} q \\ \\ p, s \end{array} \tag{8.8.3}$$

(We have labeled the eternal Fermi lines with momenta and spins here, rather than with spinors.)

The symmetry factor for both diagrams is 1. Applying the Feynman rules, integrating over the internal line momentum to eliminate one vertex delta function, and factoring out a $(2\pi)^4 \delta^{(4)}(P_{\text{in}} - P_{\text{out}})$, we obtain the invariant amplitudes:

$$
\begin{aligned}
i\mathcal{A}_1^{rs} &= (-i\lambda)^2 \bar{u}^r(p') \frac{i((\not{p}+\not{q})+m)}{(p+q)^2 - m^2 + i\epsilon} u^s(p); \\[2mm]
i\mathcal{A}_2^{rs} &= (-i\lambda)^2 \bar{u}^r(p') \frac{i((\not{p}-\not{q}')+m)}{(p-q')^2 - m^2 + i\epsilon} u^s(p).
\end{aligned}
\tag{8.8.4}
$$

To compute the differential cross section we must add these two terms and take the spin-sum, spin-average of the square. Note that $\mathcal{A}_1^{rs}$ and $\mathcal{A}_2^{uv}$ interfere if and only if $r = u$ and $s = v$, so we must write out the squares before summing and averaging over spins. Hence we need to evaluate:

$$
\begin{aligned}
A^2 &\stackrel{\text{def}}{=} \frac{1}{2} \sum_{r,s} |\mathcal{A}_1^{rs} + \mathcal{A}_2^{rs}|^2 \\[2mm]
&= \frac{1}{2} \sum_{r,s} |\mathcal{A}_1^{rs}|^2 + \frac{1}{2} \sum_{r,s} (\mathcal{A}_1^{rs} \bar{\mathcal{A}}_2^{rs} + \bar{\mathcal{A}}_1^{rs} \mathcal{A}_2^{rs}) + \frac{1}{2} \sum_{r,s} |\mathcal{A}_2^{rs}|^2.
\end{aligned}
\tag{8.8.5}
$$

We shall refer to these three terms as $A_{11}^2$, $A_{12}^2$, and $A_{22}^2$ respectively.

Consider the first sum, the sum of the squares of the first-type amplitudes. Noting that $\not{p}u^s(p) = mu^s(p)$, we find that

$$A_{11}^2 \overset{\text{def}}{=} \frac{1}{2}\sum_{r,s}|A_1^{rs}|^2$$

$$= \frac{\lambda^4}{((p+q)^2 - m^2)^2}\frac{1}{2}\sum_{r,s}\left|\bar{u}^r(p')\not{q}u^s(p) + 2m\bar{u}^r(p')u^s(p)\right|^2. \tag{8.8.6}$$

Next, using the identity $(\bar{u}Mv)^\dagger = \bar{v}\overline{M}u$ from Remark 8.2.17, we simplify the spinor term using the trace procedure:

$$S_{11} \overset{\text{def}}{=} \frac{1}{2}\sum_{r,s}\left|\bar{u}^r(p')\not{q}u^s(p) + 2m\bar{u}^r(p')u^s(p)\right|^2 \tag{8.8.7}$$

$$= \frac{1}{2}\sum_{r,s}\text{tr}\big(u^r(p')\bar{u}^r(p')\not{q}u^s(p)\bar{u}^s(p)\not{q}$$
$$+ 4mu^r(p')\bar{u}^r(p')\not{q}u^s(p)\bar{u}^s(p) + 4m^2 u^r(p')\bar{u}^r(p')u^s(p)\bar{u}^s(p)\big)$$

$$= \frac{1}{2}\text{tr}\big((\not{p}' + m)\not{q}(\not{p} + m)\not{q} + 4m(\not{p}' + m)\not{q}(\not{p} + m) + 4m^2(\not{p}' + m)(\not{p} + m)\big)$$

$$= \frac{1}{2}\text{tr}\big(\not{p}'\not{q}\not{p}\not{q} + m^2(\not{q}\not{q} + 4\not{p}'\not{q} + 4\not{q}\not{p} + 4\not{p}'\not{p}) + 4m^4\big).$$

Considering that this is only one-third of the total computation, it looks bad. To simplify the equations, we shall now assume that this is a high-energy scattering process, so that the masses of the particles are negligible in comparison to their energies. The reduced expression is now simply

$$S_{11} = \frac{1}{2}\text{tr}(\not{p}'\not{q}\not{p}\not{q}). \tag{8.8.8}$$

We have presented the details above to indicate the kind of work that goes into computing low-energy cross sections accurately. If we are to take the high-energy approximation, it makes sense to begin by simplifying the two amplitudes $A_n^{rs}$ in 8.8.4. Setting $m = \mu = 0$ and using the spinor constraint $\not{p}u^s(p) = 0$, we find:

$$i A_1^{rs} \approx (-i\lambda)^2\frac{\bar{u}^r(p')i\not{q}u^s(p)}{(p+q)^2} = \frac{-i\lambda^2}{(p+q)^2}\bar{u}^r(p')\not{q}u^s(p);$$

$$i A_2^{rs} \approx (-i\lambda)^2\frac{\bar{u}^r(p')(-i\not{q}')u^s(p)}{(p-q')^2} = \frac{i\lambda^2}{(p-q')^2}\bar{u}^r(p')\not{q}'u^s(p). \tag{8.8.9}$$

Again using the trace trick, we can calculate the relevant squares and cross products of the spinor factors in the $A_{ij}^2$'s:

$$S_{11} \overset{\text{def}}{=} \frac{1}{2}\sum_{r,s}|\bar{u}^r(p')\not{q}u^s(p)|^2 = \frac{1}{2}\text{tr}\,\not{p}'\not{q}\not{p}\not{q};$$

$$S_{12} \overset{\text{def}}{=} \frac{1}{2}\sum_{r,s}\bar{u}^r(p')\not{q}u^s(p)\bar{u}^s(p)\not{q}'u^r(p') + \text{h.c.} = \frac{1}{2}\text{tr}(\not{p}'\not{q}\not{p}\not{q}') + \text{h.c.}; \tag{8.8.10}$$

$$S_{22} \overset{\text{def}}{=} \frac{1}{2}\sum_{r,s}|\bar{u}^r(p')\not{q}'u^s(p)|^2 = \frac{1}{2}\text{tr}\,\not{p}'\not{q}'\not{p}\not{q}'.$$

Now we use the results of the previous section to evaluate these traces. The formula we need is

$$\text{tr}(\gamma^\alpha \gamma^\beta \gamma^\gamma \gamma^\delta) = 4(g^{\alpha\beta} g^{\gamma\delta} - g^{\alpha\gamma} g^{\beta\delta} + g^{\alpha\delta} g^{\beta\gamma}). \tag{8.8.11}$$

We apply this by contracting it with the appropriate momentum vectors, and thereby obtain

$$
\begin{aligned}
S_{11} &= 2(p' \cdot q\, p \cdot q - p' \cdot p\, q \cdot q + p' \cdot q\, q \cdot p); \\
S_{12} &= 4(p' \cdot q\, p \cdot q' - p' \cdot p\, q \cdot q' + p' \cdot q'\, q \cdot p); \\
S_{22} &= 2(p' \cdot q'\, p \cdot q' - p' \cdot p\, q' \cdot q' + p' \cdot q'\, q' \cdot p).
\end{aligned}
\tag{8.8.12}
$$

To further simplify the computation, we now move to the center of momentum frame defined by the conditions $\bar{q} = -\bar{p}$ and $\bar{q}' = -\bar{p}'$. In this frame, using conservation of energy and the on-shell conditions for the external momenta, we find that

$$\sqrt{|\bar{p}|^2 + m^2} + \sqrt{|\bar{p}|^2 + \mu^2} = \sqrt{|\bar{p}'|^2 + m^2} + \sqrt{|\bar{p}'|^2 + \mu^2}, \tag{8.8.13}$$

which implies that $|\bar{p}| = |\bar{p}'|$. Set the $z$-axis along the direction of the incoming fermion, and choose the $y$-axis so that the scattering takes place in the $yz$-plane. Let $\theta$ be the angle through which the fermion is scattered. Then $p$, $q$, $p'$, and $q'$ can be concretely written out:

**Incoming:** $\quad p = (p_0, 0, 0, p_3), \qquad\qquad q = (q_0, 0, 0, -p_3); \tag{8.8.14}$

**Outgoing:** $\quad p' = (p_0, 0, p_3 \sin\theta, p_3 \cos\theta), \quad q' = (q_0, 0, -p_3 \sin\theta, -p_3 \cos\theta).$

From these specific forms of the momenta we can give concrete values for all inner products of these four vectors in terms of the parameters $p_3$ and $\theta$.

The high-energy assumption makes $p_0$, $q_0$, and $p_3$ all effectively equal to $\frac{1}{2} E_T$, where $E_T$ is the total energy of the collision. Hence:

**Incoming:** $\quad p \approx (p_0, 0, 0, p_0), \qquad\qquad q \approx (p_0, 0, 0, -p_0); \tag{8.8.15}$

**Outgoing:** $\quad p' \approx (p_0, 0, p_0 \sin\theta, p_0 \cos\theta), \quad q' \approx (p_0, 0, -p_0 \sin\theta, -p_0 \cos\theta).$

Applying these conclusions to the simplified expressions 8.8.12 for the spinor contributions to the $A_{ij}^2$'s, and noting that $q \cdot q = q' \cdot q' = 0$, we find:

$$
\begin{aligned}
S_{11} &= 4p_0^4(1 + \cos\theta)(2) = E_T^4 \cos^2 \tfrac{\theta}{2}; \\
S_{12} &= 4p_0^4\left((1 + \cos\theta)^2 - (1 - \cos\theta)^2 + (2)^2\right) \\
&= E_T^4 (1 + \cos\theta) = 2E_T^4 \cos^2 \tfrac{\theta}{2}; \\
S_{22} &= 4p_0^4(2)(1 + \cos\theta) = E_T^4 \cos^2 \tfrac{\theta}{2}.
\end{aligned}
\tag{8.8.16}
$$

The denominators of the amplitudes also simplify at high energies in the center of momentum frame:

$$
\begin{aligned}
(p + q)^2 &= E_T^2, \\
(p - q')^2 &= -p_0^2\left(\sin^2\theta + (1 + \cos\theta)^2\right) \\
&= -E_T^2 \cos^2 \tfrac{\theta}{2}.
\end{aligned}
\tag{8.8.17}
$$

This minus sign can be absorbed into the coupling factor $ig^2$ of the second family of amplitudes $\mathcal{A}_2^{rs}$ in 8.8.9, making the whole $-ig^2$ agree with the coupling factor of the first family, $\mathcal{A}_1^{rs}$. When we take the square of the sum, there will therefore be an overall $g^4$ and constructive interference.

Hence, multiplying the spinor factors 8.8.16 by the appropriate factors from 8.8.9, we obtain the sum of the squares of the amplitudes:

$$
\begin{aligned}
|\mathcal{A}|^2 &= \left(\frac{\lambda^2}{E_T^2}\right)^2 S_{11} + \left(\frac{\lambda^2}{E_T^2}\right)\left(\frac{\lambda^2}{E_T^2 \cos^2 \frac{\theta}{2}}\right) S_{12} + \left(\frac{\lambda^2}{E_T^2 \cos^2 \frac{\theta}{2}}\right)^2 S_{22} \\
&= \lambda^4 (\cos^2 \tfrac{\theta}{2} + 2 + \sec^2 \tfrac{\theta}{2}).
\end{aligned}
\tag{8.8.18}
$$

To complete the example, we return to the formula for the differential cross section 8.5.21, look up the form of the two-particle final state invariant density $\mathcal{D}$ in the center of momentum frame in Chapter 5, and so arrive at

$$
\begin{aligned}
d\sigma &= \frac{\lambda^4}{4 E_T p_{\text{in}}} (\cos \tfrac{\theta}{2} + \sec \tfrac{\theta}{2})^2 \frac{1}{16\pi^2} \frac{p_{\text{out}}}{E_T} \, d\Omega \\
&= \frac{g^4}{32\pi E_T^2} (\cos \tfrac{\theta}{2} + \sec \tfrac{\theta}{2})^2 \, d\theta,
\end{aligned}
\tag{8.8.19}
$$

since $p_{\text{out}} = p_3 = p_{\text{in}}$.

**Homework 8.8.20.** Take an interaction term $\lambda \bar{\psi} i \gamma_5 \psi \phi$ between a Fermi field $\psi$ with mass $m$ and a scalar field $\phi$ with mass $\mu$. Compute the high-energy approximation to the differential cross section for fermion-fermion scattering to fourth order in $\lambda$.

**Homework 8.8.21.** Let $\psi_r$ be two Dirac spinor fields with masses $m_r$, let $\phi$ be a scalar field with mass $\mu$, and let the interaction Hamiltonian density be $\mathcal{H}_I = \lambda(\bar{\psi}_1 \psi_2 \phi + \bar{\psi}_2 \psi_1 \phi)$. Compute the high-energy approximation to the differential cross section for $\Psi_1 \bar{\Psi}_2 \to \bar{\Psi}_1 \Psi_2$ scattering to fourth order in $\lambda$.

The computations above serve to integrate the concepts and notation of the chapter so far and give a clear idea of the steps one takes to work out a cross section in a theory with spinors. To complete our introduction to the theory of Dirac spinors, this chapter closes with a discussion of discrete symmetries in theories with such fields.

# 8.9  Charge Conjugation

Following the presentation of discrete symmetries for scalar field theories at the end of Chapter 3, we now begin the extension of our knowledge of discrete symmetries to Dirac fields. This section covers charge conjugation. The following sections cover parity, time reversal, action on bilinears, and an extension of $CPT$ invariance to theories with Dirac fermions.

In this section, although charge conjugation is not in the extended Lorentz group, we propose a general form for charge conjugation on fields which resembles the Lorentz action. From here, we deduce a specific definition of charge conjugation on fields, a formula for its action on creation and annihilation operators, and its action on particles. We find that the action on particles depends on the choice of polarization spinors.

Charge conjugation reverses charge but leaves energy and momentum unchanged. Therefore charge conjugation must transform particle creation operators into anti-particle creation operators and *vice versa*. This implies that charge conjugation will in effect mix polarization spinors.

Charge conjugation in field theory should map one identifiable field into another. Taking into account the action of $U_C$ on creation operators and the implied mixing of polarization spinors, we see that the general form of charge conjugation on a free field $\psi$ is

$$U_C^\dagger \psi(x) U_C \overset{\text{def}}{=} C\psi^*(x) \tag{8.9.1}$$

$$= \int \frac{d^3\bar{k}}{(2\pi)^3 \, 2\omega(\bar{k})} \left( e^{-ix\cdot k} C v^r(k)^* \beta_r(k) + e^{ix\cdot k} C u^r(k)^* \alpha_r^\dagger(k) \right),$$

where $C$ is some unitary matrix.

Substituting the free-field expansion for $\psi$ in this definition and separately equating coefficients of $\exp(ix\cdot k)$ and $\exp(-ix\cdot k)$, we find that

$$\left. \begin{array}{l} C u^r(k)^* \alpha_r^\dagger(k) = v^s(k) U_C^\dagger \beta_s^\dagger U_C \\[2mm] C v_r(k)^* \beta_r(k) = u^s(k) U_C^\dagger \alpha_s(k) U_C \end{array} \right\} \tag{8.9.2}$$

Now $u^1(k)$ and $u^2(k)$ form a basis for the kernel of $\not{k} - m$, and $\not{k} + m$ is a rank two matrix which satisfies $(\not{k} - m)(\not{k} + m) = 0$ when $k$ is on shell. Hence $u^1(k)$ and $u^2(k)$ form a basis for the image of $\not{k} + m$. Applying the same reasoning to the $v$'s, we find that $v^1(k)$ and $v^2(k)$ are in the kernel of $\not{k} + m$ and span the image of $\not{k} - m$. Hence the equations 8.9.2 for $C$ imply that

$$(\not{k} + m)C(\not{k} + m)^* = 0 \quad \text{and} \quad (\not{k} - m)C(\not{k} - m)^* = 0. \tag{8.9.3}$$

Subtracting one equation from the other, we find that

$$\not{k} C + C \not{k}^* = 0. \tag{8.9.4}$$

In the standard representation, $\gamma_0$, $\gamma_1$, and $\gamma_3$ are real while $\gamma_2$ is purely imaginary. In this case, evaluating 8.9.2 on the four basis vectors for energy-momentum space, we find that $C$ anti-commutes with $\gamma_0$, $\gamma_1$, and $\gamma_3$ and commutes with $\gamma_2$. Taking advantage of Remark 8.7.10 on commutation properties of the basis $\mathcal{B}_S$, we conclude that $C \propto \gamma_2$. To conform with our convention for the relationship between Dirac and Weyl spinors, as noted following the Majorana condition 8.1.11, it is convenient to take $C = i\gamma_2$. Then we have the specific form of charge conjugation

$$U_C^\dagger \psi(x) U_C \overset{\text{def}}{=} i\gamma_2 \psi^*(x). \tag{8.9.5}$$

Substituting $C = i\gamma_2$ in the equations 8.9.2 and using the completeness relations of Theorem 8.3.14, we conclude that

$$\left. \begin{array}{l} U_C^\dagger \alpha_r(k) U_C = \dfrac{1}{2m} \bar{u}_r(k) i\gamma_2 v^s(k)^* \beta_s(k) \\[4mm] U_C^\dagger \beta_r^\dagger U_C = -\dfrac{1}{2m} \bar{v}_r(k) i\gamma_2 u^s(k)^* \alpha_s^\dagger(k) \end{array} \right\} \tag{8.9.6}$$

This shows how charge conjugation on creation and annihilation operators, and hence on particles, will depend on the choice of polarization spinors.

From the commutation properties of $i\gamma_2$ with $\gamma_\mu$ noted above, we can deduce its commutation properties with boost and rotation generators and thereby discover that:

$$v^r(m,\bar{0}) \stackrel{\text{def}}{=} i\gamma_2 u^r(m,\bar{0})^* \implies v^r(k) \stackrel{\text{def}}{=} i\gamma_2 u^r(k)^*, \qquad (8.9.7)$$

whatever convention we have for defining the $u$'s. With this definition of the $v$'s, we find that

$$U_C^\dagger \alpha_s(k) U_C = \beta_s(k) \quad \text{and} \quad U_C^\dagger \beta_s^\dagger U_C = \alpha_s^\dagger(k). \qquad (8.9.8)$$

This relationship between $u$'s and $v$'s is particularly transparent in the Majorana representation. In general, the polarization spinors $u_r$ and $v_s$ are distinguished by their eigenvalues under $\not{p}$. Focus on $\bar{p} = \bar{0}$, where $\not{p} = m\gamma_0$. Since $\gamma_0$ is purely imaginary, $\gamma_0 u_r(0) = u_r(0)$ implies that $\gamma_0 u_r(0)^* = -u_r(0)^*$. We can therefore define $v_r(0)$ as $u_r(0)^*$. It is a curious feature of the Majorana representation that, since all the $\gamma$ matrices are purely imaginary, all the Lorentz generators are real, so that $D(\Lambda)$ is real for all $\Lambda$. We can therefore boost to find $v_r(p) = u_r(p)^*$ for all $p$.

**Homework 8.9.9.** From the property 8.9.4 of $C$, show that in the Majorana representation $C \propto 1$. Show that if $C = i\gamma_2$ in the standard representation, then $C = i$ in the Majorana representation.

**Homework 8.9.10.** Starting from the $u$'s of Section 8.3 at $\bar{p} = \bar{0}$, follow the construction above to find the $v$'s at $\bar{p} = \bar{0}$. Show that these $v$'s differ from those defined in Section 8.3 by a factor of $\sigma_2$.

In light of the foregoing homework, charge conjugation in the Majorana representation of the Dirac algebra takes the simple form

$$U_C^\dagger \psi U_C \stackrel{\text{def}}{=} i\psi^*. \qquad (8.9.11)$$

**Homework 8.9.12.** Show that charge conjugation reverses handedness, that is, the charge conjugate of a left-handed field is right handed and vice versa.

**Homework 8.9.13.** Define a Fermi wave packet by

$$|i\rangle = \int d^3\bar{p}\, d^3\bar{q}\, f_{rs}(\bar{p},\bar{q})\, a^r(\bar{p})^\dagger b^s(\bar{q})^\dagger |0\rangle.$$

Show that the structure of $U_C|i\rangle$ depends on that of $f_{rs}$ as follows:

$$f_{rs}(\bar{p},\bar{q}) = f_{sr}(\bar{q},\bar{p}) \implies U_C|i\rangle = -|i\rangle;$$
$$f_{rs}(\bar{p},\bar{q}) = -f_{sr}(\bar{q},\bar{p}) \implies U_C|i\rangle = |i\rangle.$$

The following example illustrates the physical consequences of charge conjugation invariance for the decay of positronium into two photons. The ground state of positronium $e^+e^-$ has $l = 0$. Since

$$D_{1/2} \otimes D_{1/2} = D_0^{(a)} \oplus D_1^{(s)}, \qquad (8.9.14)$$

where the superscripts stand for anti-symmetric and symmetric under exchange of factors, the spin states $s = 0$ and $s = 1$ respectively have odd and even wave functions. Now Homework 8.9.13 implies that

$$U_C|s = 0\rangle = |s = 0\rangle \quad \text{and} \quad U_C|s = 1\rangle = -|s = 1\rangle. \qquad (8.9.15)$$

Since a two-photon final state must satisfy $U_C|f\rangle = |f\rangle$, only the $s = 0$ state can decay into two photons.

Since $U_C^2 = 1$, if we write $\psi_c$ for $U_C^\dagger \psi U_C$, then obviously $U_C^\dagger \psi_c U_C = \psi$. However, because charge conjugation is linear in the fields, the value of $U_C^\dagger \psi^* U_C$ is representation dependent:

$$U_C^\dagger \psi^* U_C = \begin{cases} i\gamma_2\psi, & \text{standard representation;} \\ -i\psi, & \text{Majorana representation.} \end{cases} \tag{8.9.16}$$

**Homework 8.9.17.** Show that the Lagrangian for the free Dirac field is invariant under charge conjugation.

**Homework 8.9.18.** Show that the interaction

$$\mathcal{L}' = \lambda_1\bar{\psi}_2\psi_3 + \lambda_2\bar{\psi}_3\psi_1 + \lambda_3\bar{\psi}_1\psi_2 + \text{h.c.}$$

can be made charge-conjugation invariant by absorbing phase factors into the fields if and only if the product of the $\lambda$'s is real.

Charge conjugation is not an extension of the Lorentz group, but we have argued that the general form of charge conjugation on fields resembles the Lorentz action. From here, we have determined a specific definition of charge conjugation and determined its action on particles. This action on particles will in general mix spins, but this can be avoided by an intelligent choice of the polarization spinors.

# 8.10 Parity

Parity is in the extended Lorentz group, and so we can write down the general form of parity on fields straight away. From here, following the logic of the previous section, we deduce a specific definition of parity and a formula for the action of parity on particles. As for charge conjugation, the effect of parity on particles is found to depend on the choice of polarization spinors. We conclude that parity on fermion particle states reverses momentum, leaves energy and internal charges unchanged, need not change spins, but must change some phases so that Fermi particles and their anti-particles have opposite intrinsic parity.

When we extend the Lorentz group to include parity $P$, we must also choose some value $M$ for the representation $D(P)$ of parity on Dirac spinors. Since $P^2 = 1$, $M^2 = 1$ also. Hence $M$ is diagonalizable with eigenvalues $\pm 1$.

The action of parity on the free Dirac field now follows from the generic transformation formula for fields:

$$U_P^\dagger \psi(x) U_P = D(P)\psi(\tilde{x}) \tag{8.10.1}$$

$$= \int \frac{d^3\bar{k}}{(2\pi)^3\, 2\omega(\bar{k})} \left( e^{i\tilde{x}\cdot k} M v^r(k)\beta_r^\dagger(k) + e^{-i\tilde{x}\cdot k} M u^r(k)\alpha_r(k) \right)$$

$$= \int \frac{d^3\bar{p}}{(2\pi)^3\, 2\omega(\bar{p})} \left( e^{ix\cdot p} M v^r(\tilde{p})\beta_r^\dagger(\tilde{p}) + e^{-ix\cdot p} M u^r(\tilde{p})\alpha_r(\tilde{p}) \right),$$

where $p = \tilde{k}$.

Since $\psi(x)$ is a solution of the free Dirac equation, and since the Dirac operator commutes with $U_P$, $M\psi(\tilde{x})$ must also be a solution. Hence

$$(\slashed{p} - m)Mu^r(\tilde{p}) = 0 \quad \text{and} \quad (\slashed{p} + m)Mv^r(\tilde{p}) = 0. \tag{8.10.2}$$

Since $u(k)$'s and $v(k)$'s respectively span the images of $\slashed{k} + m$ and $\slashed{k} - m$, we conclude that

$$(\slashed{p} - m)M(\slashed{\tilde{p}} + m) = 0 \quad \text{and} \quad (\slashed{p} + m)M(\slashed{\tilde{p}} - m) = 0. \tag{8.10.3}$$

By subtracting these two equations we find that

$$\slashed{p}M - M\slashed{\tilde{p}} = 0. \tag{8.10.4}$$

Evaluating $p$ at the four basis vectors yields four equations which imply that $M$ commutes with $\gamma_0$ and anti-commutes with $\gamma^r$. As noted in Remark 8.7.10, we can deduce that $M \propto \gamma_0$. Since the eigenvalues of $M$ are $\pm 1$, $M = \pm\gamma_0$. It is conventional to take $M = \gamma_0$. Then the parity transformation of a Dirac field is given by

$$U_P^\dagger\psi(x)U_P \stackrel{\text{def}}{=} \gamma_0\psi(\tilde{x}). \tag{8.10.5}$$

**Homework 8.10.6.** Show that parity preserves the canonical anti-commutation relations for Dirac fields.

We can compute the action of parity on the free field another way:

$$U_P^\dagger\psi(x)U_P \tag{8.10.7}$$

$$= \int \frac{d^3\tilde{p}}{(2\pi)^3\, 2\omega(\tilde{p})} \left(e^{ix\cdot p}v^r(p)U_P^\dagger\beta_r^\dagger(p)U_P + e^{-ix\cdot p}u^r(p)U_P^\dagger\alpha_r^\dagger(p)U_P\right).$$

By comparison with the previous computation 8.10.1, and by using the normalization relations of Theorem 8.3.14, we find that

$$\left.\begin{array}{l} U_P^\dagger\alpha_r(p)U_P = \dfrac{1}{2m}\bar{u}_r(p)\gamma_0 u^s(\tilde{p})\alpha_s(\tilde{p}) \\[2ex] U_P^\dagger\beta_r^\dagger(p)U_P = -\dfrac{1}{2m}\bar{v}_r(p)\gamma_0 v^s(\tilde{p})\beta_s^\dagger(\tilde{p}) \end{array}\right\} \tag{8.10.8}$$

From these formulae, it is clear that the effect of parity on the creation and annihilation operators, and therefore on the particle states, depends on the choice of polarization spinors.

If we are to avoid mixing spins, then we must have $\gamma_0 u^s(\tilde{p}) = \eta u^s(p)$ and $\gamma_0 v^s(\tilde{p}) = \zeta v^s(p)$, where $\eta$ and $\zeta$ are phase factors. If we want the phases to be independent of $p$, then we can evaluate at $p = (m, \bar{0})$ to find $\eta = 1$ and $\zeta = -1$. If we now assume that the polarization spinors at points $p$ are defined in terms of those at $(m, \bar{0})$ by some Lorentz transformation $\Lambda_p$, then we can transform these spinor conditions into an equivalent condition on the $\Lambda$'s:

$$u^s(p) = \gamma_0 u^s(\tilde{p}) \quad \text{and} \quad v^s(p) = -\gamma_0 v^s(\tilde{p})$$

$$\Longleftrightarrow \quad \left.\begin{array}{l} D(\Lambda_p)u^s(m, \bar{0}) = \gamma_0 D(\Lambda_{\tilde{p}})u^s(m, \bar{0}) \\[1ex] D(\Lambda_p)v^s(m, \bar{0}) = -\gamma_0 D(\Lambda_{\tilde{p}})v^s(m, \bar{0}) \end{array}\right\} \tag{8.10.9}$$

$$\Longleftrightarrow \quad D(\Lambda_p) = \gamma_0 D(\Lambda_{\tilde{p}})\gamma_0$$

$$\Longleftrightarrow \quad \Lambda_{\tilde{p}} = P\Lambda_p P,$$

up to factor of $\pm 1$.

Hence, if we construct the polarization spinors by Lorentz transformations $\Lambda_p$ subject to the condition 8.10.9, then we find

$$U_P^\dagger \alpha_r^\dagger(p)U_P = \alpha_r^\dagger(\tilde{p}) \quad \text{and} \quad U_P^\dagger \beta_r^\dagger(p)U_P = -\beta_r^\dagger(\tilde{p}). \tag{8.10.10}$$

**Remark 8.10.11.** Since, for any boost $B$ and rotation $R$, $\widetilde{Bp} = B^{-1}\tilde{p}$ and $\widetilde{Rp} = R\tilde{p}$, and since $\gamma_0$ anti-commutes with boost generators and commutes with rotation generators, the condition 8.10.9 on the $\Lambda$'s is equivalent to defining the $u$'s and $v$'s by boosting the equations $\gamma_0 u_r(m, \bar{0}) = u_r(m, \bar{0})$ and $\gamma_0 v_r(m, \bar{0}) = -v_r(m, \bar{0})$.

**Remark 8.10.12.** In the helicity basis of Remark 8.4.7, the construction does not satisfy the condition 8.10.9, and indeed, by direct computation one can show that in this case parity generally mixes spins.

Thus parity need not change the spin of a fermion, but reverses its momentum and may change its phase. This phase change is called the *intrinsic parity* of the particle. Note that the intrinsic parities of Fermi particles and their anti-particles are opposite, while the intrinsic parities of Bose particles and their anti-particles are the same.

As we know from Section 6.9, parity should interchange left and right fields. If we use the projections

$$P_L = \frac{1}{2}(1 - \gamma^5), \quad P_R = \frac{1}{2}(1 + \gamma^5) \tag{8.10.13}$$

to separate out left and right two-component fields from a four-component field, then our parity transformation does indeed reverse the handedness:

$$Pu_L \stackrel{\text{def}}{=} \gamma_0 u_L = \gamma_0 \frac{1}{2}(1 - \gamma^5)u = \frac{1}{2}(1 + \gamma^5)\gamma_0 u = (\gamma_0 u)_R. \tag{8.10.14}$$

**Homework 8.10.15.** Show that the Lagrangian for a free Dirac field is parity invariant.

**Homework 8.10.16.** Let $\phi$ be a real scalar field with intrinsic parity $\varepsilon$. Show that an interaction $\lambda\bar{\psi}P_L\psi\phi + \text{h.c.}$ is parity invariant if and only if $\lambda^* = \varepsilon\lambda$.

Let the orbital angular momentum, internal angular momentum, and total angular momentum belong to representations $D_l$, $D_s$, and $D_j$ respectively. Then $D_j$ is a subrepresentation of $D_s \otimes D_l$, and so:

$$|l - s| \le j \le l + s. \tag{8.10.17}$$

The total parity of a state depends on both the intrinsic parity of the particles in it and on the decomposition of its wave function into spherical harmonics. The total parity of a state is defined by:

$$\text{Parity } P = \text{Intrinsic Parity} \times (-1)^l. \tag{8.10.18}$$

The physical consequences of parity conservation are illustrated by the following example of $p\bar{p} \to \pi\pi$.

An initial state consisting of a proton and an anti-proton at rest has $l = 0$, and $s = 0$ or $s = 1$. Experimental evidence indicates that the final state could be $\pi^+\pi^-$ but not $\pi^0\pi^0$. The explanation follows.

Since $l = 0$, the total parity of the initial state is the product of the intrinsic parities, which is $-1$ since fermions and their anti-particles have opposite intrinsic parity. Now the $\pi^+\pi^-$ state could have $l = 0$ or $l = 1$, but since the $\pi^0\pi^0$ state is bosonic, it must have $l = 0$. Also, since the intrinsic parity of a boson is the same as that of its anti-particle, the intrinsic parity factor for both these states is 1. The following chart relates this data to our two conserved quantities:

|  | $J_{\text{in}}$ | $P_{\text{in}}$ | $J^P$ |
|---|---|---|---|
| $p\bar{p}$ | $l = 0, s = 0$ | $(+1)(-1)$ | $0^-$ |
|  | $l = 0, s = 1$ | $(+1)(-1)$ | $1^-$ |

|  | $J_{\text{out}}$ | $P_{\text{out}}$ | $J^P$ |
|---|---|---|---|
| $\pi^+\pi^-$ | $l = 0, s = 0$ | $(\pm 1)(\pm 1)$ | $0^+$ |
|  | $l = 1, s = 0$ | $(\pm 1)(\pm 1)(-1)$ | $1^-$ |
| $\pi^0\pi^0$ | $l = 0, s = 0$ | $(\pm 1)(\pm 1)$ | $0^+$ |

$$(8.10.19)$$

Clearly the only possibility is that $p\bar{p}$ in the $l = 0$, $s = 1$ state decays into $\pi^+\pi^-$ in the $l = 1$ state.

**Homework 8.10.20.** Consider the decay $\pi^0 \to e^-e^+$. Using the fact that pions have intrinsic parity $-1$, show that the final fermions are in an $l = 0$, $s = 0$ state.

This section has shown how to extend the action of the Lorentz group on fields to include parity. The result, $\psi(x) \to \gamma_0\psi(\tilde{x})$, maps solutions of the Dirac equation into solutions, preserves the canonical anti-commutation relations, reverses handedness, and implies that Fermi particles have opposite intrinsic parity to their anti-particles. Finally, we saw that in applications of parity invariance, we must remember to include orbital angular momentum.

## 8.11 Time Reversal

As is our custom for discrete symmetries, we propose a general form for time reversal on fields, deduce a specific definition, and then investigate the relationship between polarization spinors and the effect of time reversal on particle states.

Time reversal is an element $T$ in the extended Lorentz group which, by definition, does not change the internal quantum numbers of a particle. It must therefore map a Dirac field $\psi(t, \tilde{x})$ into some multiple of $\psi(-t, \tilde{x})$. We may therefore extend the standard form for a Lorentz transformation of a field to time reversal:

$$\Omega_T \psi(t, \tilde{x})\Omega_T^{-1} \overset{\text{def}}{=} N\psi(-t, \tilde{x}) \tag{8.11.1}$$

$$= \int \frac{d^3\bar{k}}{(2\pi)^3 \, 2\omega(\bar{k})} \left( e^{-i\tilde{x}\cdot k} N v^r(k)\beta_r^\dagger(k) + e^{i\tilde{x}\cdot k} N u^r(k)\alpha_r(k) \right)$$

$$= \int \frac{d^3\bar{p}}{(2\pi)^3 \, 2\omega(\bar{p})} \left( e^{-ix\cdot p} N v^r(\tilde{p})\beta_r^\dagger(\tilde{p}) + e^{ix\cdot p} N u^r(\tilde{p})\alpha_r(\tilde{p}) \right).$$

Here $N$ stands for $D(T)$. Hence, since $T^2 = 1$, we also have $N^2 = 1$.

Substituting the free-field expansion for $\psi$ into the left-hand side also, we find, by comparing coefficients of the exponentials, that

$$\left.\begin{array}{l} u^s(p)^* \Omega_T \alpha_s(p) \Omega_T^{-1} = N u^r(\tilde{p}) \alpha_r(\tilde{p}) \\ v^s(p)^* \Omega_T \beta_s^\dagger(p) \Omega_T^{-1} = N v^r(\tilde{p}) \beta_r^\dagger(\tilde{p}) \end{array}\right\} \tag{8.11.2}$$

From here, as in the previous two sections, we use the relationships between $u$'s, $v$'s, and $\not{p} \pm m$ to obtain

$$(\not{p} - m)^* N(\not{\tilde{p}} + m) = 0 \quad \text{and} \quad (\not{p} + m)^* N(\not{\tilde{p}} - m) = 0. \tag{8.11.3}$$

Subtracting these two equations, we find that

$$\not{p}^* N - N \not{\tilde{p}} = 0. \tag{8.11.4}$$

In the standard representation, this implies that $N$ commutes with $\gamma_0$ and $\gamma_2$ but anti-commutes with $\gamma_1$ and $\gamma_3$ (see Section 8.6 for details). Consequently, $N \propto \gamma_1 \gamma_3$. It is conventional to take

$$N = i\gamma_1 \gamma_3, \tag{8.11.5}$$

in which case the definition of time reversal on fields becomes

$$\Omega_T \psi(x) \Omega_T^{-1} \stackrel{\text{def}}{=} i\gamma_1 \gamma_3 \psi(-t, \bar{x}). \tag{8.11.6}$$

This definition leads inevitably to an unexpected oddity:

$$\Omega_T^2 \psi(t, \bar{x}) \Omega_T^{-2} = \Omega_T N \psi(-t, \bar{x}) \Omega_T^{-1} = N^* N \psi(t, \bar{x}) = -\psi(t, \bar{x}). \tag{8.11.7}$$

This implies that $\Omega_T^2$ is the Fermi number operator $(-1)^F$, and $\Omega_T^4 = 1$.

**Homework** 8.11:8. From the properties of $N$ given in 8.11.4, deduce that $N \propto \gamma_0 \gamma_5$ in the Majorana representation. Show that $N = i\gamma_1 \gamma_3$ in the standard representation implies that $N = \gamma_0 \gamma_5$ in the Majorana representation.

Since $\Omega_T^\dagger$ does not exist, we cannot deduce the effect of time reversal on a conjugate field from its effect on a field. The natural criterion which determines time reversal of a conjugate field is invariance of the Dirac kinetic term. Suppose

$$\Omega_T \bar{\psi}(t, \bar{x}) \Omega_T^{-1} = \bar{\psi}(-t, \bar{x}) M. \tag{8.11.9}$$

Then the kinetic term transforms as follows:

$$\bar{\psi} i \not{\partial} \psi \xrightarrow{\;\;T\;\;} \bar{\psi} M (-i)(-\tilde{\not{\partial}}^*) N \psi, \tag{8.11.10}$$

and is invariant if and only if

$$M \tilde{\not{\partial}}^* N = \not{\partial}. \tag{8.11.11}$$

Using the commutation relation 8.11.4 for $N$, we see at once that $MN = \mathbf{1}$, and so $M = N$. Finally, we deduce the action of time reversal on the conjugate field:

$$\bar{\psi} \xrightarrow{\;\;T\;\;} N\bar{\psi} \quad \Longrightarrow \quad \psi^* \xrightarrow{\;\;T\;\;} \gamma_0^\top N^\top \gamma_0^\top \psi^* = -N\psi^*. \tag{8.11.12}$$

Now that we have a specific value for $N$, we can use the normalization relations of Theorem 8.3.14 to solve for the action of time reversal on creation and annihilation operators in the expression 8.11.2. Working similarly with the conjugate field, we can find its action on the two remaining operators:

$$\left.\begin{aligned}
\Omega_T \alpha_r(p)\Omega_T^{-1} &= \frac{1}{2m}\bar{u}_r(p)^* N u^s(\tilde{p})\alpha_s(\tilde{p}) \\
\Omega_T \beta_r^\dagger(p)\Omega_T^{-1} &= -\frac{1}{2m}\bar{v}_r(p)^* N v^s(\tilde{p})\beta_s^\dagger(\tilde{p}) \\
\Omega_T \alpha_r^\dagger(p)\Omega_T^{-1} &= -\frac{1}{2m}\bar{u}_r(p)N u^s(\tilde{p})^*\alpha_s^\dagger(\tilde{p}) \\
\Omega_T \beta_r(p)\Omega_T^{-1} &= \frac{1}{2m}\bar{v}_r(p)N v^s(\tilde{p})^*\beta_s(\tilde{p})
\end{aligned}\right\} \qquad (8.11.13)$$

**Homework** 8.11.14. Show directly from the action 8.11.13 of time reversal on creation operators that the square of time reversal is $-\mathbf{1}$.

Time reversal does not mix spins if and only if there exist phase factors $\eta$ and $\zeta$ such that

$$N u^s(\tilde{p})^* = \eta u^s(p) \quad \text{and} \quad N v^s(\tilde{p}) = \zeta v^s(p)^*, \qquad (8.11.15)$$

where $\eta$ and $\zeta$ could potentially be dependent on $s$ and $p$.

On the positive side, these equations are Lorentz covariant for constant phases, as we now show. The commutation properties of $\gamma_1\gamma_3$ with the boost and rotation generators of the Dirac-spinor representation imply that for a boost $B$, rotation $R$, and general Lorentz transformation $\Lambda = BR$, we have

$$\left.\begin{aligned}
D(B)\gamma_1\gamma_3 &= \gamma_1\gamma_3 D(B^{-1})^* \\
D(R)\gamma_1\gamma_3 &= \gamma_1\gamma_3 D(R)^*
\end{aligned}\right\} \implies D(\Lambda)\gamma_1\gamma_3 = \gamma_1\gamma_3 D(P\Lambda P)^*. \qquad (8.11.16)$$

Consequently:

$$\begin{aligned}
N u^s(\tilde{p})^* = \eta u^s(p) &\implies N u^s(\widetilde{\Lambda p})^* = \eta u^s(\Lambda p); \\
N v^s(\tilde{p}) = \zeta v^s(p)^* &\implies N v^s(\widetilde{\Lambda p}) = \zeta v^s(\Lambda p)^*.
\end{aligned} \qquad (8.11.17)$$

This covariance implies that a solution anywhere can be spread everywhere by Lorentz transformations.

On the negative side, the conditions 8.11.15 are inconsistent at $p = (m, \bar{0})$ for any choice of phases, as we may see by substituting each equation into itself and using the property $NN^* = -1$. If the phases are constants, then the same argument shows that these equations are inconsistent at all momenta.

We see therefore that if time reversal is to avoid mixing spins, then the phases cannot be constant and $p = (m, \bar{0})$ cannot be used as a reference momentum for the polarization spinors. This suggests setting up a global solution to the conditions 8.11.15 by choosing a particular solution using two reference momenta, $p$ and $\tilde{p}$.

Specifically, we could leave the polarization spinors undefined on the plane $p^3 = 0$ and define their values for $p^3 < 0$ in terms of their values for $p^3 > 0$ by

$$\left.\begin{aligned}
u^s(p) &\stackrel{\text{def}}{=} -N u^s(\tilde{p})^* \\
v^s(p) &\stackrel{\text{def}}{=} N v^s(\tilde{p})^*
\end{aligned}\right\} \quad \text{for } p^3 < 0. \qquad (8.11.18)$$

With this choice of polarization spinors, we have $\eta = \zeta = \text{sign}(p^3)$ and a basis of particle states whose spins are not mixed by time reversal. Substituting for the polarization spinors in the general time-reversal formulae 8.11.13 yields the specific formulae

$$\Omega_T \alpha_r(p)\Omega_T^{-1} = \eta(p_3)\alpha_r(\tilde{p}), \qquad \Omega_T \beta_r^\dagger(p)\Omega_T^{-1} = -\eta(p_3)\beta_r^\dagger(\tilde{p});$$
$$\Omega_T \alpha_r^\dagger(p)\Omega_T^{-1} = \eta(p_3)\alpha_r^\dagger(\tilde{p}), \qquad \Omega_T \beta_r(p)\Omega_T^{-1} = -\eta(p_3)\beta_r(\tilde{p}). \tag{8.11.19}$$

This solution to the conditions for non-mixing of spins makes it very clear that $\Omega_T^2 = -1$.

**Remark 8.11.20.** Another more common solution is to modify the helicity basis of Remark 8.4.7 to make it satisfy the constraints 8.11.15.

**Homework 8.11.21.** Find a basis of polarization spinors with respect to which time reversal interchanges spins. Verify that $\Omega_T^2 = -1$.

Finally, a quantum theory is invariant under time reversal if its Lagrangian and canonical commutation relations are both invariant. Firstly, as discussed in Section 3.14, invariance for the Lagrangian density means

$$\Omega_T \mathcal{L} \Omega_T^{-1} = \mathcal{L}^\dagger, \tag{8.11.22}$$

up to a divergence. Secondly, from the general formula

$$\{\psi^\dagger, \psi\} = C \implies \{\psi^\dagger A, B\psi\} = BCA, \tag{8.11.23}$$

we deduce that time reversal preserves the canonical anti-commutation relations for a Dirac field:

$$\Omega_T \{\psi^\dagger(t, \bar{x}), \psi(t, \bar{y})\}\Omega_T^{-1} = \{\psi^\dagger(-t, \bar{y})N, N\psi(-t, \bar{x})\} = 1\delta^{(3)}(\bar{x} - \bar{y}). \tag{8.11.24}$$

**Homework 8.11.25.** Show that the Lagrangian for the free Dirac field is time-reversal invariant.

Starting from a natural form of time reversal on fields, we have deduced a specific definition, found the general action of time reversal on operators and particles, and given a construction for a basis of polarization spinors with respect to which time reversal does not mix spins.

## 8.12 The Action of Discrete Symmetries on Bilinear Terms

In this section, we shall work out the action of $C$, $P$, and $T$ on bilinear functions $\bar{\eta}\Gamma\zeta$ of Fermi fields. Our results will allow us to check discrete symmetries of Lagrangians at a glance, and to demonstrate $CPT$ invariance of Lagrangian theories.

Writing $\eta$ and $\zeta$ for two Fermi fields, noting the factors of $\pm i$ in the action of $U_C$ as set out in 8.9.11 and 8.9.16, the action of charge conjugation on bilinear terms in Fermi fields may be worked out in the Majorana representation as follows:

$$\bar{\eta}\Gamma\zeta = \eta^\dagger \gamma_0 \Gamma\zeta \xrightarrow{\;C\;} \eta^\top \gamma_0 \Gamma\zeta^* = -\zeta^\dagger \Gamma^\top \gamma_0^\top \eta = -\bar{\zeta}\gamma_0 \Gamma^\top \gamma_0^\top \eta$$
$$= \bar{\zeta}\gamma_0 \Gamma^\top \gamma_0 \eta = \bar{\zeta}\gamma_0^* \Gamma^\top \gamma_0^* \eta \tag{8.12.1}$$
$$= \bar{\zeta}\bar{\Gamma}^* \eta.$$

Note that, although the steps in this computation are only valid for the Majorana representation, the conclusion is valid for all representations.

When parity acts on a bilinear term $\bar{\eta}\Gamma\zeta$, it induces a different transformation of the matrix $\Gamma$:

$$P\colon \bar{\eta}\Gamma\zeta \to \bar{\eta}\bar{\gamma}_0\Gamma\gamma_0\zeta \quad \implies \quad P\colon \Gamma \to \gamma_0\Gamma\gamma_0, \qquad (8.12.2)$$

since $\bar{\gamma}_0 = \gamma_0$ in any representation of Dirac algebra related to the Weyl one by a unitary change of basis.

Applying the definitions 8.11.6 and 8.11.12, the action of time reversal on fields induces the following action on a bilinear expression $\bar{\eta}\Gamma\zeta$:

$$\begin{aligned}
\Omega_T \bar{\eta}\Gamma\zeta\Omega_T^{-1} &= \Omega_T\bar{\eta}\Omega_T^{-1}\Omega_T\Gamma\Omega_T^{-1}\Omega_T\zeta\Omega_T^{-1} \\
&= \bar{\eta}N\Gamma^*N\zeta.
\end{aligned} \qquad (8.12.3)$$

The induced action on the matrix $\Gamma$ reduces to

$$\Gamma \xrightarrow{\ \ T\ \ } \gamma_3\gamma_1\Gamma^*\gamma_1\gamma_3. \qquad (8.12.4)$$

From these general formulae, it is easy to compute the effects of $C$, $P$, and $T$ on the five types of term $\bar{\eta}\Gamma\zeta$ identified by the parity properties of the basis $\mathcal{B}_S$ in 8.7.14:

|  | $C$ | $P$ | $T$ | $CPT$ |
|---|---|---|---|---|
| $\bar{\eta}\zeta$ | $\bar{\zeta}\eta$ | $\bar{\eta}\zeta$ | $\bar{\eta}\zeta$ | $\bar{\zeta}\eta$ |
| $\bar{\eta}\gamma^\mu\zeta$ | $-\bar{\zeta}\gamma^\mu\eta$ | $\bar{\eta}\gamma_\mu\zeta$ | $\bar{\eta}\gamma_\mu\zeta$ | $-\bar{\zeta}\gamma^\mu\eta$ |
| $\bar{\eta}\sigma^{\mu\nu}\zeta$ | $-\bar{\zeta}\sigma^{\mu\nu}\eta$ | $\bar{\eta}\sigma_{\mu\nu}\zeta$ | $-\bar{\eta}\sigma_{\mu\nu}\zeta$ | $\bar{\zeta}\sigma^{\mu\nu}\eta$ |
| $\bar{\eta}\gamma^5\gamma^\mu\zeta$ | $\bar{\zeta}\gamma^5\gamma^\mu\eta$ | $-\bar{\eta}\gamma^5\gamma_\mu\zeta$ | $\bar{\eta}\gamma^5\gamma_\mu\zeta$ | $-\bar{\zeta}\gamma^5\gamma^\mu\eta$ |
| $\bar{\eta}i\gamma^5\zeta$ | $\bar{\zeta}i\gamma^5\eta$ | $-\bar{\eta}\gamma^5 i\zeta$ | $-\bar{\eta}i\gamma^5\zeta$ | $\bar{\zeta}i\gamma^5\eta$ |

$$(8.12.5)$$

Thus, for example, $\bar{\psi}\psi$ is even under charge conjugation, and $\bar{\psi}\gamma^\mu\psi$ is odd; $\bar{\psi}\psi\phi$ is parity invariant if $\phi$ is a scalar field, while $\bar{\psi}i\gamma^5\psi\phi$ is parity invariant if $\phi$ is a pseudoscalar field.

**Homework 8.12.6.** Let $\phi$ be a neutral scalar field and $\psi$ a Dirac field. Show that the interactions $\bar{\psi}\psi\phi$ and $\bar{\psi}\gamma^\mu\psi\partial_\mu\phi$ (which is a dimension five non-renormalizable operator) are hermitian. Show further that they could separately be invariant under charge conjugation, but their sum could not be invariant.

**Homework 8.12.7.** Construct a hermitian Lagrangian density which cannot be made time-reversal invariant by any redefinition of the fields.

It is obvious from the table of the effects of the discrete symmetries on bilinear terms that, if we make all real fields scalar with respect to parity, then $CPT$ transforms a Lorentz-invariant Lagrangian density $\mathcal{L}(x)$ into $\mathcal{L}^\dagger(-x)$ – when all the Lorentz indices are contracted, the extra minus signs cancel out. When we include vector fields $A_\mu$, the same result holds as long as we arrange $CPT$ to transform $A_\mu(x)$ into $-A_\mu(-x)$.

# 8.13 Combinations of $C$, $P$, and $T$

In this section, we consider the products $CP$, $CT$, $PT$, and $CPT$, concluding our discussion of discrete symmetries with an application of $CPT$ to a Feynman integral.

Since the intrinsic parity of a fermion is opposite to the intrinsic parity of its anti-particle, charge conjugation and parity do not commute. In fact

$$U_C U_P = (-1)^F U_P U_C, \tag{8.13.1}$$

where $F$ is the fermion number operator.

The constructions for polarization spinors that work well for parity (boost from the rest frame) and for charge conjugation (obtain $v$'s by conjugating $u$'s) are compatible because boosts are real in the Majorana representation.

**Homework 8.13.2.** Show formally that charge conjugation and parity anti-commute on Dirac fields.

Time reversal and charge conjugation also anti-commute on Fermi fields. From the transformations 8.9.11, 8.9.16, 8.11.6, and 8.11.12, we find that

$$\psi(x) \xrightarrow{\;T\;} i\gamma_1\gamma_3\psi(-t,\bar{x}) \xrightarrow{\;C\;} (i\gamma_1\gamma_3)(i\gamma_2)\psi^*(-t,\bar{x}), \tag{8.13.3}$$

because $U_C$ commutes with $\gamma$ matrices, whereas

$$\psi(x) \xrightarrow{\;C\;} i\gamma_2\psi^*(x) \xrightarrow{\;T\;} (i\gamma_2)(-i\gamma_1\gamma_3)\psi^*(-t,\bar{x}), \tag{8.13.4}$$

because $i\gamma_2$ is real. Consequently,

$$\Omega_T U_C = (-1)^F U_C \Omega_T. \tag{8.13.5}$$

The final pair, parity and time reversal commute:

$$\Omega_T U_P = U_P \Omega_T. \tag{8.13.6}$$

**Homework 8.13.7.** Prove that parity and time reversal commute.

**Remark 8.13.8.** Writing $C$, $P$, and $T$ for $U_C$, $U_P$, and $\Omega_T$ respectively, the observations made above show that these operators generate a finite group $G$. In order to structure a distinction between unitary operators and anti-unitary operator, it is necessary to include a representation $I$ of $i$. The group $G$ is defined by the following relations:

$$C^2 = P^2 = T^4 = 1,$$
$$PC = CPT^2, \qquad TC = CT^3, \qquad TP = PT, \tag{8.13.9}$$
$$IC = CI, \qquad IP = PI, \qquad IT = TI^3.$$

Since $I^2$ commutes with all the generators, we can eliminate the representations in which $I = \pm 1$ by imposing the further condition

$$I^2 = -1. \tag{8.13.10}$$

The relations between the generators enable us to put any product of generators into a standard form $\pm C^a P^b T^c I^d$. Clearly, $G$ has 64 elements. Its center $Z$ is generated by $T^2$ and $I^2$. Since $G/Z$ is commutative and both $T^2$ and $I^2$ are commutators, $Z$ is also the commutator subgroup of $G$. Application of these operators to particle states provides real representations of $G$ in which $I$ acts as $i$.

It is natural to enquire whether there is a basis of polarization spinors in which $C$, $P$, and $T$ do not mix spins. Bringing the conditions forward from the three previous section, this would require

### Conditions for Non-Mixing of Spins

$$
\begin{array}{lll}
C: & u_r(p) = i\gamma_2 v_r^*(p) & \Longleftrightarrow & v_r(p) = i\gamma_2 u_r^*(p) \\
P: & u_r(p) = \gamma_0 u_r(\tilde{p}) & \Longleftrightarrow & v_r(p) = -\gamma_0 v_r(\tilde{p}) \\
T: & u_r(p) = \eta N u_r^*(\tilde{p}) & \Longleftrightarrow & v_r(p) = -\eta N v_r^*(\tilde{p})
\end{array}
\qquad (8.13.11)
$$

Here the logic is that the two aspects of the conditions for charge conjugation are equivalent, and if we use these conditions to define $v$'s in terms of $u$'s, then the two aspects of the conditions for parity and for time reversal are also equivalent. If we take $\eta = \text{sign}(p^3)$ as before, then these three conditions are compatible.

**Homework 8.13.12.** Show that the parity and time-reversal conditions above are incompatible if we take $\eta = 1$. In this case, construct a basis of polarization spinors in which $PT$ flips spins.

We have seen how $CPT$ transforms a Lagrangian into its hermitian conjugate. We shall now apply $CPT$ to Feynman diagrams and the associated Feynman integrals, taking for example a Feynman diagram with one fermion line passing through it. We shall show that the Feynman integral associated with this diagram is related to that for the $CPT$-transformed diagram by analytic continuation of external momenta and a sign coming from the interchange of the outgoing fermion with the incoming one. Figure 8.13 shows the structure of this example in detail, revealing in particular the origin of an anomalous minus sign.

Suppose that the incoming and outgoing fermion lines have momenta $p_1$ and $p_2$ and spinors $u_1(p_1)$ and $u_2(p_2)$. Write $p$ for the vector of external momenta $(p_1, \ldots, p_n)$. Then the form of the invariant amplitude is:

$$
i\mathcal{A}(p) = \bar{u}_2(p_2)\mathcal{I}(p)u_1(p_1). \qquad (8.13.13)
$$

When we use the Feynman rules to compute $i\mathcal{A}$, we can use the vertex delta functions to eliminate integrals over internal momenta until only integrals over loop momenta $l = l_1, \ldots, l_a$ remain. Then, since the numerator $N$ and denominator $D$ of the integrand come from products of propagators, the form of $i\mathcal{A}$ is given by:

$$
\begin{aligned}
i\mathcal{A}(p) &= \int \frac{d^{4a}l}{(2\pi)^{4a}} \frac{\bar{u}_2(p_2)N(p,l)u_1(p_1)}{D(p,l)} \\
&= \int \frac{d^{4a}l}{(2\pi)^{4a}} \frac{\bar{u}_2(\Lambda p_2)N(\Lambda p, \Lambda l)u_1(\Lambda p_1)}{D(\Lambda p, \Lambda l)} \\
&= \int \frac{d^{4a}l}{(2\pi)^{4a}} \frac{\bar{u}_2(p_2)\bar{D}(\Lambda)N(\Lambda p, \Lambda l)D(\Lambda)u_1(p_1)}{D(\Lambda p, \Lambda l)}.
\end{aligned}
\qquad (8.13.14)
$$

(Here $D$ is used for both the denominator and the Dirac representation of the Lorentz group.)

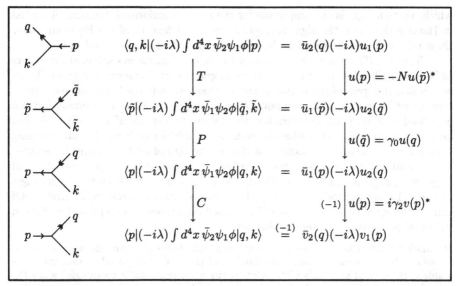

**Fig. 8.13.** This figure displays the effects of $T$, $P$, and $C$ acting in sequence on a Feynman diagram and the associated matrix element. For simplicity, time-reversal phases and fermion spins are not shown. The amplitudes on the right are derived from the diagrams by direct application of the Feynman rules. Note that, though the four matrix elements are equal, the bottom Feynman amplitude has the wrong sign.

Now we will use the analyticity of the Feynman rules in the momenta and the analyticity of the representation $D$ as a function of $\Lambda$ to substitute

$$\Lambda = \exp(\pi X_3 + i\pi B_3) = \exp(\pi X_3)\exp(i\pi B_3) = -1 \qquad (8.13.15)$$

and

$$D(\Lambda) = \exp(\tfrac{\pi}{2}\gamma_1\gamma_2 + \tfrac{i\pi}{2}\gamma_0\gamma_3) = \exp(\tfrac{\pi}{2}\gamma_1\gamma_2)\exp(\tfrac{i\pi}{2}\gamma_0\gamma_3)$$

$$= (\cos\tfrac{\pi}{2} + \gamma_1\gamma_2\sin\tfrac{\pi}{2})(\cosh\tfrac{i\pi}{2} + \gamma_0\gamma_3\sinh\tfrac{i\pi}{2}) \qquad (8.13.16)$$

$$= \gamma_5$$

into the formula above. The result is:

$$i\mathcal{A}(p) = \int \frac{d^{4a}l}{(2\pi)^{4a}} \frac{\bar{u}_2(p_2)\gamma_5 N(-p,-l)\gamma_5 u_1(p_1)}{D(-p,-l)}. \qquad (8.13.17)$$

If we define $u_r$ on negative-energy momenta by $u_r(-p_r) = \gamma_5 u_r(p_r)$, then the analyticity of the invariant amplitude implies that $\mathcal{A}(p) = \mathcal{A}(-p)$.

The conditions 8.13.11 on polarization spinors imply that

$$v^r(p) = i\gamma_2 u^r(p)^* = i\gamma_2\gamma_0 u^r(\tilde{p})^* = -i\gamma_2\gamma_0 N u^r(p) = i\gamma_5 u^r(p). \qquad (8.13.18)$$

The amplitude equality 8.13.17 can therefore be written

$$i\mathcal{A}(p) = i\mathcal{A}(-p) = -\bar{v}_2(p_2)\mathcal{I}(-p)v_1(p_1), \qquad (8.13.19)$$

which, up to a sign, is the amplitude for the $CPT$-transformed diagram. As noted in Homework 8.5.11, the sign discrepancy here is a feature of the Feynman rules: these rules can only be relied on for relative signs of interfering amplitudes.

Clearly, $CPT$ causes a crossing of a Feynman diagram as explained for scalars in Section 4.12. In extending crossing to diagrams with fermions, one generally assumes that the spins of the fermions are not changed, only the four-momenta. However, as we have seen in the computation 8.13.19 above, when an incoming fermion is crossed to an outgoing anti-fermion, the polarization spinor $u^r$ is finally evaluated on a negative-energy vector where its value is not initially defined. To make crossing work on an amplitude, one must first change the sign of all four-momenta, second, restore evaluation at positive energies using the conventions $u_r(-p) = \gamma_5 u_r(p)$ and $v_r(p) = i\gamma_5 u_r(p)$ as needed, and third, add factors of $-1$ for each fermi exchange. (These $-1$'s are generally irrelevant since crossed processes seldom interfere.) Of course, the equality up to sign of $CPT$-related amplitudes is exceptional and not a universal property of crossed amplitudes.

**Remark 8.13.20.** Clearly, if we apply $CPT$ to a particle decay amplitude, we can demonstrate equality between the decay amplitude and the related amplitude representing unstable particle creation. This $CPT$ relationship is experimentally meaningful when the particle decays into a pair state, but when the particle decays into three or more particles, it transcends experiment. This gap between theory and experiment arises because theory uses plane wave states, while experimental particle decay creates a highly structured spherical wave. It is this extra structure that makes the experimental situation non-reversible; two-particle collisions effectively run the universe, while three-particle collisions are so rare that they play no role in particle experiments.

We have shown how to define time reversal on Fermi fields and seen how the choice of polarization spinors affects the associated transformation of particle states. Putting the three common discrete symmetries together, we have shown that $CPT$ is a symmetry of any Lagrangian field theory and deduced that crossing can be extended to $CPT$-related processes.

The discrete symmetries are a little more intricate for Dirac spinors than they are for scalars. The best approach is (1) start from the form of the action on fields, (2) use the free-field expansion to find the specific form, (3) deduce the action on the creation and annihilation operators and particle states, and (4) find a convenient set of polarization spinors to simplify the action on particle states. The discrete symmetries are used to exclude physically inaccessible classes of final states from the range of evolution of an initial state.

# 8.14 Summary

This chapter has followed the outline of the chapter on Weyl spinors, taking care to present the logic of the transition from two-component notation to four-component notation. Technically, the main feature of the chapter is its presentation of the Dirac algebra of $4 \times 4$ matrices. This algebra enjoys so many useful mathematical properties that it is regarded as the essential tool for computation with spinors, so much so that even Weyl spinors are generally represented by four-component fields. Indeed, since all observed matter is either fermionic or bound states of fermions, the Dirac algebra techniques presented here are absolutely indispensable to a theorist.

The addition of mass terms to a theory of Weyl spinors destroys helicity as a Lorentz invariant of spinor particles, providing instead spin-up and spin-down

Lorentz eigenstates. As preparation of an initial state does not generally fix spins, and as the spins of particles in a final state are not generally observed, amplitudes must be averaged over initial spins and summed over final spins before squaring.

This concludes the investigation into spin-1/2 fields motivated by Chapter 6. There are theories, notably theories of hadrons, which use only scalar and fermion fields, but for modeling the electroweak and strong forces it is essential to have vector fields. Hence we shall not pause to discuss theories of scalars and fermions here but shall at once move on to study vector fields.

# Chapter 9

# Vector Fields and Gauge Invariance

Presenting the canonical quantization of a massive vector field, the existence of a massless limit when this field is coupled to a conserved current, and the gauge invariance of this massless limit; a preliminary understanding of QED.

## Introduction

Now that we have found fermionic fields, the need is for polarized fields that can represent the photon. Experimentally, photons have only the two transverse polarizations and no possibility of polarization along the direction of motion. Since these polarizations are vectors in Minkowski space, we shall begin our investigation using a vector field, a set of four hermitian fields $A_\mu$ that transform as a vector under the Lorentz-group action on the fields.

The most general Lagrangian density for a free vector field contains two free parameters, $a$ and $b$. After careful analysis of this Lagrangian density, we find in Section 9.1 that the classical theory is only free from off-shell modes and modes with longitudinal polarizations when $a = -1$ and $b \neq 0$, in which case the vector field is massive. We also observe that when $a = -1$ and $b = 0$, the vector field is massless and has a gauge symmetry which makes the longitudinal modes non-physical.

Rather than set up the functional integral machinery necessary for quantizing the gauge field theory, in this chapter we treat the gauge theory as the massless limit of a massive vector field theory. Using the plane wave solutions of the classical equations of motion, Section 9.2 presents canonical quantization of the free massive vector field, Section 9.3 derives Feynman rules for an interacting massive vector field, and Section 9.4 demonstrates that if the massive vector field is coupled to a conserved current, then the massless limit of the theory exists. Section 9.5 concludes the first part of the chapter by using a massive vector field theory to compute the Compton scattering of photons off electrons.

The next three sections investigate the massless limit. Section 9.6 brings out the geometrical background of gauge theory. Section 9.7 shows how any theory with a global $U(1)$ symmetry can be promoted to a gauge theory, and Section 9.8 gives a short proof that the addition of a mass term to a $U(1)$ gauge theory leaves the massive vector field coupled to a conserved current. This establishes the general validity of the technique used in Section 9.5 for computing Compton scattering. Section 9.9 closes the chapter by extending the discrete symmetries to a massive vector field.

Note that using massive vector fields to compute gauge-field amplitudes only works for vector fields which are not self-interacting. Thus, for non-abelian gauge fields, we shall need the fresh angle provided by the following chapters.

# 9.1 The Lagrangian Density and Plane Wave Solutions

The purpose of this section is to identify which classical Lagrangian densities can be subjected to canonical quantization and to bring out the classical significance of such densities. The possible terms in a Lagrangian density for a free vector field $A^\mu$ are obtained simply by contracting Lorentz indices:

$$A_\mu A^\mu, \quad \partial_\mu A_\nu \partial^\mu A^\nu, \quad \partial_\mu A_\nu \partial^\nu A^\mu, \quad \partial_\mu A^\mu \partial_\nu A^\nu. \tag{9.1.1}$$

The last two candidates are related by integration by parts, and so give the same contribution to the Lagrangian. Therefore we try setting

$$\mathcal{L} = \pm\frac{1}{2}(\partial_\mu A_\nu \partial^\mu A^\nu + a\partial_\mu A^\mu \partial_\nu A^\nu + bA_\mu A^\mu). \tag{9.1.2}$$

We shall use the Hamiltonian to find that the overall sign should be negative.

The equation of motion is:

$$-\partial_\nu \partial^\nu A_\mu - a\partial_\mu \partial_\nu A^\nu + bA_\mu = 0. \tag{9.1.3}$$

Substitute $A_\mu = e_\mu \exp(-ix\cdot k)$ to find the condition that the momentum $k$ and the polarization $e$ are parameters for a plane wave solution:

$$\begin{aligned} k^2 e_\mu + ae\cdot k\, k_\mu + be_\mu &= 0, \\ \Rightarrow \quad ((1+a)k^2 + b)e\cdot k &= 0. \end{aligned} \tag{9.1.4}$$

We need to understand how these equations constrain the momentum to be on a mass shell and how the polarization is related to the momentum.

If $a = 0$ in 9.1.4, then we obtain a shell condition $k^2 + b = 0$ but no condition on the polarization. This is no use for modeling photons, so we shall proceed on the assumption that $a \neq 0$. For the plane wave solution to be non-trivial we also require $e \neq 0$ and $k \neq 0$. The following table presents the consequences of 9.1.4 according to the various regions of the $ab$-plane:

| | | |
|---|---|---|
| $b = 0$ | $k^2 = 0 \implies e\cdot k = 0$ <br> $k^2 \neq 0 \implies e = \dfrac{e\cdot k}{k^2}k$ | $k^2 = 0 \implies e\cdot k = 0$ <br> $k^2 \neq 0 \implies e = 0$ |
| $b \neq 0$ | $k^2 + b = 0$ <br><br> and $e\cdot k = 0$ | $k^2 + \dfrac{b}{1+a} = 0$ <br><br> and $e = \dfrac{e\cdot k}{k^2}k$ |
| | $a = -1$ | $-1 \neq a \neq 0$ |

$$(9.1.5)$$

From this table, we see that only in the case $a = -1$, $b \neq 0$ are we free from longitudinal and off-shell modes. If we take $b = -\mu^2$ then we have a physically sensible theory of a vector field with mass $\mu$. A massive vector field has three transverse polarizations. Such a field could represent the $W$ and $Z$ bosons of electroweak theory after symmetry breaking has given them masses. We shall pursue this case first, as it is classically the simplest option.

For a massive vector field we take $a = -1$ and $b = -\mu^2$. Our massive vector field Lagrangian density is now

$$\mathcal{L} = \pm\frac{1}{2}(\partial_\mu A_\nu \partial^\mu A^\nu - \partial_\mu A^\mu \partial_\nu A^\nu - \mu^2 A_\mu A^\mu). \tag{9.1.6}$$

The generalized momenta and equations of motion are given by:

$$\frac{\partial \mathcal{L}}{\partial \partial^\mu A^\nu} = \mp(\partial_\mu A_\nu - \partial_\nu A_\mu);$$

$$\partial^\mu(\partial_\mu A_\nu - \partial_\nu A_\mu) + \mu^2 A_\nu = 0. \tag{9.1.7}$$

This is called the *Proca equation*. Because the first bracket in the Proca equation is anti-symmetric, taking the divergence of the Proca equation yields:

$$\mu^2 \partial^\nu A_\nu = 0. \tag{9.1.8}$$

Hence a massive vector field automatically has zero divergence. Using this to simplify the Proca equation, we find the Klein–Gordon equation again:

$$(\partial_\mu \partial^\mu + \mu^2)A_\nu = 0. \tag{9.1.9}$$

The emergence of the standard Klein–Gordon operator confirms the assignment of $-\mu^2$ to the parameter $b$. The two equations 9.1.8 and 9.1.9 are together equivalent to the Proca equation.

The form of the canonical momenta and the Proca equation in 9.1.7 suggest defining $F_{\mu\nu}$ by

$$F_{\mu\nu} \overset{\text{def}}{=} \partial_\mu A_\nu - \partial_\nu A_\mu. \tag{9.1.10}$$

We can rewrite the Lagrangian density in terms of $F_{\mu\nu}$ if we adjust it by a total divergence:

$$\mathcal{L} = \pm\left(\frac{1}{4}F_{\mu\nu}F^{\mu\nu} - \frac{1}{2}\mu^2 A_\nu A^\nu\right). \tag{9.1.11}$$

In this notation, the generalized momenta and the equations of motion are given by:

$$\frac{\partial \mathcal{L}}{\partial \partial^\mu A^\nu} = \mp F_{\mu\nu};$$

$$\partial^\mu F_{\mu\nu} + \mu^2 A_\nu = 0. \tag{9.1.12}$$

Now we consider obtaining the massless vector field as a limit of the massive one. In the limit, as the mass goes to zero, two things happen at once. First, one of the three transverse modes for on-shell momenta becomes longitudinal. Second, the theory suddenly develops a gauge symmetry and becomes a *gauge theory*,

$$A_\mu \to A_\mu + \partial_\mu \chi \quad \Longrightarrow \quad F_{\mu\nu} \to F_{\mu\nu}. \tag{9.1.13}$$

Note that the Lagrangian density is only gauge symmetric for the parameter values $a = -1$ and $b = 0$. The two events associated with the massless limit are tied together by the observation that the gauge symmetry allows us to choose a gauge in which there are no longitudinal photons. In detail, if $A_\mu$ is a classical wave packet of longitudinal modes,

$$A_\mu(x) = \int \frac{d^4k}{(2\pi)^4} \, e^{-ix\cdot k} \rho(k) k_\mu, \tag{9.1.14}$$

then

$$A_\mu(x) = \partial_\mu\left(i \int \frac{d^4k}{(2\pi)^4} \, e^{-ix\cdot k} \rho(k)\right), \tag{9.1.15}$$

and so $A_\mu(x)$ can be completely eliminated by a change of gauge.

This phenomenon of the massless vector field $A_\mu$ being gauge dependent is already familiar from classical electromagnetism, where $A^\mu = (\phi, \vec{A})$ is the four-vector potential for the electromagnetic field. The classical physical field is described by $F^{\mu\nu}$:

$$\vec{E} = -(F^{01}, F^{02}, F^{03}), \quad \vec{B} = -(F^{23}, F^{31}, F^{12}). \tag{9.1.16}$$

Maxwell's equations are:

$$\vec{\nabla} \cdot \vec{B} = 0, \qquad\qquad \vec{\nabla} \cdot \vec{E} = \rho,$$
$$\vec{\nabla} \times \vec{E} + \frac{\partial \vec{B}}{\partial t} = \vec{0}, \qquad \vec{\nabla} \times \vec{B} - \frac{\partial \vec{E}}{\partial t} = \vec{j}. \tag{9.1.17}$$

The homogeneous equations are equivalent to the existence of a four-vector potential $A = (\phi, \vec{A})$ for $F^{\mu\nu}$:

$$\vec{\nabla} \cdot \vec{B} = 0 \quad\Longleftrightarrow\quad \vec{B} = \nabla \times \vec{A};$$
$$\vec{\nabla} \times \left(\vec{E} + \frac{\partial \vec{A}}{\partial t}\right) = \vec{0} \quad\Longleftrightarrow\quad \vec{E} + \frac{\partial \vec{A}}{\partial t} = -\vec{\nabla}\phi. \tag{9.1.18}$$

The inhomogeneous equations in empty space correspond to the massless version $\partial_\mu F^{\mu\nu} = 0$ of the Proca equation 9.1.12. Thus the gauge freedom and elimination of longitudinal modes in the massless limit of our vector field theory correspond to the same concepts in classical electromagnetism.

The novel structure of the massless vector field Lagrangian density is a reflection of the fact that this field corresponds to a vector potential in classical field theory, not to a classical field. It turns out, however, that there are physical effects that require us to regard the potential itself as real. In the Bohm–Aharanov effect, for example, an electron beam is affected by the presence of a non-zero electromagnetic potential in an area where the electromagnetic field vanishes. Thus it makes sense to regard the potential $A_\mu$, modulo gauge transformations, as the physical field and the field strength $F^{\mu\nu}$ as a secondary or derived field.

In summary, we see that there are two potentially quantizable classical vector field Lagrangian densities. The first is the massive field for which $a = -1$ and $b = -\mu^2$, the second is the massless field for which $a = -1$ and $b = 0$. The massive theory is free from longitudinal modes and should be easy to quantize. The massless theory has longitudinal modes for all momenta, which are non-physical because of the gauge symmetry of the theory. Quantizing a theory with gauge symmetry will need some new techniques. Therefore we shall first quantize the massive theory, and specify conditions under which the massless limit is a theory of photons and then in the next chapter set up the machinery for quantizing gauge theories.

## 9.2 Quantization of the Massive Vector Field

We will now take the Lagrangian density

$$
\begin{aligned}
\mathcal{L} &= \pm\left(\frac{1}{4}F_{\mu\nu}F^{\mu\nu} - \frac{1}{2}\mu^2 A_\nu A^\nu\right) \\
&\Rightarrow \quad \partial_\mu F^{\mu\nu} + \mu^2 A^\nu = 0,
\end{aligned}
\tag{9.2.1}
$$

for a free, massive vector field and find the Hamiltonian, the canonical commutation relations, and the integral form of the free field. When we have found the Hamiltonian, we shall see that the overall sign must be a minus.

In order to find the Hamiltonian as the Legendre transform of the Lagrangian density, we need to write the Lagrangian density using a complete set of independent fields and their canonical momenta. The test for completeness of a set of fields is that the values of those fields at any initial time form a complete initial condition for the equations of motion. Thus the set of eight fields $\{A, \partial_0 A\}$ is complete. The condition for independence is that each field in the set can be specified at any initial time independently from the others. Since $\partial_\mu A^\mu = 0$, the set of fields $\{A, \partial_0 A\}$ is not independent. The six fields $A^i$ and $F^{0i}$, however, clearly form an independent set. We now show that this set is also complete. From the equations of motion we find

$$
A^0 = -\mu^{-2}\partial_\mu F^{\mu\nu} = \mu^{-2}\partial_i F^{0i}.
\tag{9.2.2}
$$

From the definition of $F^{0i}$ we find

$$
\partial^0 A^i = F^{0i} + \partial^i A^0,
\tag{9.2.3}
$$

and from the constraint $\partial_\mu A^\mu = 0$ we find

$$
\partial_0 A^0 = -\partial_i A^i.
\tag{9.2.4}
$$

Thus initial values of $A^i$ and $F^{0i}$ determine initial values of $\{A, \partial_0 A\}$, and so the set

$$
\{A^1, A^2, A^3, F^{01}, F^{02}, F^{03}\}
\tag{9.2.5}
$$

is indeed complete as well as independent.

Now we can write down the Hamiltonian density by writing the Lagrangian density in terms of this complete and independent set of fields and performing the Legendre transform as usual. The result is:

$$
\mathcal{H} = \pm\left(F^{0i}\partial_0 A_i - \frac{1}{2}F_{0i}F^{0i} - \frac{1}{4}F_{ij}F^{ij} + \frac{1}{2}\mu^2 A_\mu A^\mu\right).
\tag{9.2.6}
$$

To get a good form of the Hamiltonian, note that

$$
\begin{aligned}
F^{0i}\partial_0 A_i &= F^{0i}F_{0i} + F^{0i}\partial_i A_0 = F^{0i}F_{0i} - (\partial_i F^{0i})A_0 + \partial_i(F^{0i}A_0) \\
&= F^{0i}F_{0i} - \mu^2 A^0 A_0 + \partial_i(F^{0i}A_0).
\end{aligned}
\tag{9.2.7}
$$

The total divergence does not contribute to the Hamiltonian, so:

$$H = \pm \int d^3\bar{x} \, \frac{1}{2} F^{0i} F_{0i} - \frac{1}{4} F^{ij} F_{ij} - \frac{1}{2}\mu^2 A^0 A_0 + \frac{1}{2}\mu^2 A^i A_i. \qquad (9.2.8)$$

Since all of these terms are negative definite, the overall sign must be a minus.

Now that we have the overall sign, the canonical momenta are $\Pi^i = F^{i0}$, and we can write the equal-time commutation relations for the massive vector field:

$$\left[A_i(t,\bar{x}), A_j(t,\bar{y})\right] = 0, \quad \left[F_{i0}(t,\bar{x}), F_{j0}(t,\bar{y})\right] = 0,$$

$$\left[A_i(t,\bar{x}), F_{j0}(t,\bar{y})\right] = i\delta_{ij}\delta^{(3)}(\bar{x} - \bar{y}). \qquad (9.2.9)$$

To set up a free vector field as an integral of creation and annihilation operators, we need polarization vectors for all on-shell momenta. The plane wave solutions for the massive vector field are of the form $A = e^r \exp(-ix{\cdot}k)$ where $e^r{\cdot}k = 0$ for $r = 1, 2, 3$. As usual for constructing fields of polarization vectors on the mass hyperboloid, we choose a basis in the rest frame $k_{\mathrm{rf}} = (\mu, \bar{0})$ and boost. In this case, as opposed to the spinor situation, there is no preferred basis, so we simply choose $e^r(k_{\mathrm{rf}})$ to be a Lorentz orthonormal set of vectors transverse to $k_{\mathrm{rf}}$:

$$e^r(k_{\mathrm{rf}}) \cdot e^s(k_{\mathrm{rf}}) = -\delta^{rs}, \quad e^r(k_{\mathrm{rf}}) \cdot k_{\mathrm{rf}} = 0. \qquad (9.2.10)$$

Since the inner product is Lorentz invariant, we can use arbitrary solutions $\Lambda_k$ of $k = \Lambda k_{\mathrm{rf}}$ to define polarization vectors on the mass hyperboloid:

$$e^r(k) \overset{\mathrm{def}}{=} \Lambda_k e^r(k_{\mathrm{rf}}) \qquad (9.2.11)$$

implies that

$$e^r(k) \cdot e^s(k) = -\delta^{rs} \quad \text{and} \quad e^r(k) \cdot k = 0, \qquad (9.2.12)$$

for all $k$.

**Remark 9.2.13.** As noted Sections 8.3 and 8.4 in regard to polarization spinors, if we choose $\Lambda_k$ to be boosts, then rotations will mix the polarization vectors. If $\Lambda_k$ involve rotations, then boosts will do the mixing. As a consequence, Lorentz transformations will in general mix particle states of the vector field.

**Homework 9.2.14.** Choose $k_{\mathrm{rf}} = (\mu \cosh\theta, 0, 0, \mu \sinh\theta)$ for $\theta > 0$. Find the polarization vectors at $k_{\mathrm{rf}}$ which have spins $\pm 1$.

Photons are commonly either plain polarized or circularly polarized. The notation above is adequate for plane polarization, but as the previous homework shows, circular polarization introduces complex linear combinations of polarization vectors. To cover this case, we extend the normalization condition above to the case of complex polarization vectors as follows:

$$e^r(k)^* \cdot e^s(k) = -\delta^{rs} \quad \text{and} \quad e^r(k) \cdot k = 0. \qquad (9.2.15)$$

To obtain a completeness relation for $e^r(k)$, we begin from the rest frame. Let $X$ and $Y$ be the $3 \times 3$ matrices $X^r{}_i = e^r(k_{\mathrm{rf}})_i$ and $Y^j{}_s = e_s(k_{\mathrm{rf}})^j$. Then the normalization condition on the polarization vectors may be written $X^* Y = 1_3$.

This implies that $Y^*X = 1_3$ also. Writing this out with indices, together with the fact that $e^r(k_{\mathrm{rf}})_0 = 0$, gives us the following equality:

$$\sum_{r=1}^{3} e^r(k_{\mathrm{rf}})_\mu^* e^r(k_{\mathrm{rf}})_\nu = \begin{cases} 0, & \text{if } \mu\nu = 0; \\ \delta_{ij}, & \text{if } \mu = i \text{ and } \nu = j. \end{cases} \tag{9.2.16}$$

Before boosting we must rewrite the right-hand side as a tensor:

$$\sum_{r=1}^{3} e^r(k_{\mathrm{rf}})_\mu^* e^r(k_{\mathrm{rf}})_\nu = -g_{\mu\nu} + \frac{(k_{\mathrm{rf}})_\mu (k_{\mathrm{rf}})_\nu}{k_{\mathrm{rf}}^2}. \tag{9.2.17}$$

Now, since $\Lambda^* = \Lambda$, this formula readily spreads over the mass shell. We bring out this conclusion in a theorem:

**Theorem 9.2.18.** *Let $k_{\mathrm{rf}}$ be a reference vector and $\Lambda_k$ a solution of $\Lambda k_{\mathrm{rf}} = k$. If the three polarization vectors $e^r(k)$ satisfy $e^r(k)^* \cdot e^s(k) = -\delta^{rs}$ and $e^r(k) \cdot k = 0$ at $k = k_{\mathrm{rf}}$ and $e^r(k) = \Lambda_k e^r(k_{\mathrm{rf}})$ for all $k$, then the polarization vectors satisfy the normalization and completeness relations*

$$e^r(k)^* \cdot e^s(k) = -\delta^{rs} \quad \text{and} \quad e^r(k) \cdot k = 0;$$

$$\sum_{r=1}^{3} e^r(k)_\mu^* e^r(k)_\nu = -g_{\mu\nu} + \frac{k_\mu k_\nu}{\mu^2}. \qquad \square$$

Using these polarization vectors, we can write the free massive vector field as an integral of creation and annihilation operators:

$$A_\mu(x) = \int \frac{d^3\bar{k}}{(2\pi)^3 \, 2\omega(\bar{k})} \left( e^{ix\cdot k} e^r(k)_\mu^* a_r(k)^\dagger + e^{-ix\cdot k} e^r(k)_\mu a_r(k) \right). \tag{9.2.19}$$

Substituting the integral form of the free field into the Hamiltonian would confirm that the vector field must be bosonic.

**Homework 9.2.20.** Using the normalization and completeness relations for the polarization vectors, show that the free field satisfies the equal-time commutation relations if and only if the creation and annihilation operators satisfy:

$$\left[a_r(k), a_s(k')\right] = 0, \quad \left[a_r(k)^\dagger, a_s(k')^\dagger\right] = 0,$$
$$\left[a_r(k), a_s(k')^\dagger\right] = (2\pi)^3 2\omega(\bar{k}) \delta_{rs} \delta^{(3)}(\bar{k} - \bar{k}').$$

**Homework 9.2.21.** Substitute the free field 9.2.19 into the Hamiltonian 9.2.8 and verify that $H \geq 0$.

The crucial step in applying canonical quantization to the massive vector field is the determination of a complete set of independent fields for the theory. Once such a set has been identified, then canonical commutators can be imposed, and the fields can be expressed in terms of canonical creation and annihilation operators. The careful choice of Lagrangian parameters $a = -1$ and $b = -\mu^2 \neq 0$ eliminates all obstacles in defining polarization vectors. In the massless theory, any procedure for determining a complete set of independent fields must rest on a choice of gauge. The obvious problem now is to show that the resulting quantum theory is independent of this choice. The technique of functional integral quantization developed in the following chapters solves this problem gracefully.

## 9.3 Feynman Rules for the Massive Vector Field

In this section, we shall introduce two Lagrangian densities in which the massive vector field is respectively coupled to a charged spinor and to a charged scalar. The charged spinor theory could be a massive version of QED; there is no problem in deriving its Feynman rules and computing its scattering amplitudes. The charged scalar theory, however, introduces derivative couplings; these cause severe problems.

An example of a Lagrangian density which couples the massive vector field to a negatively charged spinor like the electron is:

$$\mathcal{L} = -\frac{1}{4}F_{\mu\nu}F^{\mu\nu} + \frac{1}{2}\mu^2 A_\mu A^\mu + \bar{\psi}(i\slashed{\partial} - e\slashed{A} - m)\psi. \tag{9.3.1}$$

**Homework 9.3.2.** Show that when $\mu = 0$, this Lagrangian density is invariant under the gauge transformation

$$\psi(x) \to e^{ie\theta(x)}\psi(x) \quad \text{and} \quad A_\mu(x) \to A_\mu(x) - \partial_\mu\theta(x).$$

When we apply Dyson's formula to a theory with a massive vector field, work out the Wick expansion of the $S$ matrix, and deduce the Feynman rules from there, then we find (as we found before for Fermi fields) that the polarization vectors in the free vector field should be associated with external lines: when an incoming vector particle with polarization $r$ and momentum $k$ is annihilated by $A_\mu$, a factor $e_\mu^r(k)$ is left behind. The Feynman rules for external photons follow from this:

$$
\begin{aligned}
&\textbf{Incoming photon:} \quad \text{Contract } \mathcal{A}_\mu \text{ with } e_r^\mu(k) \\
&\textbf{Outgoing photon:} \quad \text{Contract } \mathcal{A}_\mu \text{ with } e_r^\mu(k)^*
\end{aligned}
\tag{9.3.3}
$$

To compute an amplitude, we regard the $A_\mu$ as four separate fields and attach a Lorentz index to each external photon line. Thus, for a single incoming photon in a process, we would obtain four amplitudes $\mathcal{A}^\mu$. However, because there is no longitudinal mode, the polarization implied by the superscript must be regarded as fictitious. To obtain the amplitudes $\mathcal{A}^r$ for physical polarizations, we simply contract $\mathcal{A}^\mu$ with the appropriate polarization vector. Thus, for an incoming photon, we have

$$\mathcal{A}^r = \mathcal{A}^\mu e^r(k)_\mu. \tag{9.3.4}$$

Using the commutation rules for the creation and annihilation operators, we can readily show that the contraction of vector fields is given by:

$$\contraction{}{\hat{A}_\mu(x)}{}{\hat{A}_\nu(y)}\hat{A}_\mu(x)\hat{A}_\nu(y) = \int \frac{d^4k}{(2\pi)^4} \left(-g_{\mu\nu} + \frac{k_\mu k_\nu}{\mu^2}\right) \frac{ie^{-i(x-y)\cdot k}}{k^2 - \mu^2 + i\epsilon}. \tag{9.3.5}$$

This implies that the propagator for the massive vector field is

$$\textbf{Propagator:} \quad \left(-g_{\mu\nu} + \frac{k_\mu k_\nu}{\mu^2}\right)\frac{i}{k^2 - \mu^2 + i\epsilon}\frac{d^4k}{(2\pi)^4}. \tag{9.3.6}$$

The fundamental vertex for Wick diagrams is

**Wick Vertex:**                                (9.3.7)

From our experience with Dyson's formula, we can now guess the Feynman rule for the vector-spinor vertex:

$$\bar{\psi}e\gamma^\mu A_\mu \psi: \qquad (-ie)(2\pi)^4\delta^{(4)}(P_{\text{in}})\gamma^\mu. \qquad (9.3.8)$$

As we do not generally know the polarization of the vector fields, in computing cross sections we should average over incoming polarizations and sum over outgoing ones. For example, if we have a process with a single outgoing vector boson with momentum $k$, then we first compute the amplitudes $\mathcal{A}^\mu$ for a Lorentz index $\mu$ on the vector field line. Since $\partial_\mu A^\mu = 0$, we have a transversality condition $k_\mu \mathcal{A}^\mu = 0$. Now the sum over outgoing polarizations is:

$$\sum_{r=1}^{3} |e^r(k)^*_\mu \mathcal{A}^\mu|^2 = \sum_{r=1}^{3} (\mathcal{A}^\mu)^* e^r(k)_\mu e^r(k)^*_\nu \mathcal{A}^\nu$$

$$= (\mathcal{A}^\mu)^*\left(-g_{\mu\nu} + \frac{k_\mu k_\nu}{\mu^2}\right)\mathcal{A}^\nu \qquad (9.3.9)$$

$$= -\mathcal{A}^*_\mu \mathcal{A}^\mu.$$

To take an average over polarization, it is necessary to know the distribution of the polarizations in a beam. If the momentum is small in comparison to the mass of the vector particle, then we expect all three polarizations to be roughly equally represented. If the momentum is much larger than the mass, then (as we shall see in the next section) the approximately longitudinal mode effectively decouples, and we expect this mode to be a negligible proportion of a beam. Therefore, the polarization average depends on the momentum, the limiting cases being:

$$\text{Polarization Average} = \begin{cases} -\frac{1}{3}\mathcal{A}^*_\mu \mathcal{A}^\mu & \text{if } |\bar{k}| \ll \mu; \\ -\frac{1}{2}\mathcal{A}^*_\mu \mathcal{A}^\mu & \text{if } |\bar{k}| \gg \mu. \end{cases} \qquad (9.3.10)$$

This device works for each vector index on the amplitude.

**Homework 9.3.11.** Using the Lagrangian density 9.3.1 for massive QED, compute to lowest non-trivial order the invariant amplitudes for $\Psi\Psi \to \Psi\Psi$ scattering in the limit $\mu \to 0$. (Observe that the term $k^\mu k^\nu/\mu^2$ in the massive photon propagator does not contribute to these amplitudes, so that the limit $\mu^2 \to 0$ is physically sensible.) Find the differential cross section for $\Psi\Psi \to \Psi\Psi$ when $\mu = 0$.

**Homework 9.3.12.** In the style of the previous homework, find the amplitude and differential cross section for $\bar{\Psi}\Psi \to \bar{\Psi}\Psi$ (Bhabha) scattering.

Applying these Feynman rules to loop diagrams is simple enough. For example, from the diagrams

$$D_1 \overset{\text{def}}{=} k, \bar{u}^s \text{—} k, u^r \quad \text{and} \quad D_2 \overset{\text{def}}{=} k, \nu \text{—} k, \mu \qquad (9.3.13)$$

recalling that the elimination of vertex delta functions by integration in effect determines the internal momenta up to loop variables, we find amplitudes

$$i\mathcal{A}_1^{sr}(k) = (-ie)^2 \int \frac{d^4q}{(2\pi)^4} \left(-g_{\mu\nu} + \frac{q_\mu q_\nu}{\mu^2}\right) \frac{i}{q^2 - \mu^2 + i\epsilon} \qquad (9.3.14)$$

$$\times \bar{u}^s(k)\gamma^\nu \frac{i(\slashed{q} + \slashed{k} + m)}{(q+k)^2 - m^2 + i\epsilon} \gamma^\mu u^r(k)$$

and

$$i\mathcal{A}_2^{\nu\mu}(k) = -(-ie)^2 \int \frac{d^4q}{(2\pi)^4} \, \mathrm{tr}\left(\frac{i(\slashed{q}+m)}{q^2 - m^2 + i\epsilon} \gamma^\nu \frac{i(\slashed{q}+\slashed{k}+m)}{(q+k)^2 - m^2 + i\epsilon} \gamma^\mu\right). \qquad (9.3.15)$$

These integrals diverge. Chapters 17 and 18 introduce the regularization and renormalization techniques for bringing finite values out of such integrals. With a regularization procedure to express a divergent integral as a limit of convergent ones, we can manipulate the integrals in the ordinary way. In the contribution of the $q_\mu q_\nu/\mu^2$ term to $\mathcal{A}_1$, for example, we can use $(\slashed{k} - m)u^r(k) = 0$ to simplify the Dirac factors:

$$(\slashed{q} + \slashed{k} + m)\slashed{q}u^r(k) = (\slashed{q} + \slashed{k} + m)(\slashed{q} + \slashed{k} - m)u^r(k)$$
$$= ((q+k)^2 - m^2)u^r(k). \qquad (9.3.16)$$

Hence this contribution reduces to the integral of $-\bar{u}^s(k)\slashed{q}u^r(k)/(q^2 - \mu^2 + i\epsilon)$. Since the range of integration is symmetric and this integrand is odd in $q$, we conclude that the $q_\mu q_\nu/\mu^2$ term in fact evaluates to zero. This is significant because it implies that the massless limit of $\mathcal{A}_1$ is manageable.

**Homework 9.3.17.** Assuming that regularization will justify manipulating the integrand in $\mathcal{A}_2$, use Dirac algebra to evaluate the trace.

We have considered the interactions of the electromagnetic field (perhaps we should write 'potential' here, but convention is against it) with charged spinors. Now we will consider the possibility of interactions with charged scalars like pions or Higgses. We could take, for example, the most general dimension-four Lagrangian density which has a gauge-invariant massless limit:

$$\mathcal{L} = -\frac{1}{4}F_{\mu\nu}F^{\mu\nu} + \frac{1}{2}\mu^2 A_\mu A^\mu \qquad (9.3.18)$$

$$+ (\partial_\mu + ieA_\mu)\phi^\dagger(\partial^\mu - ieA^\mu)\phi - m^2\phi^\dagger\phi - \frac{\lambda}{4}(\phi^\dagger\phi)^2.$$

(Here we have assumed that $\phi$ has charge $+1$.)

**Homework 9.3.19.** Show that when $\mu = 0$, this Lagrangian density is invariant under the gauge transformation

$$\phi(x) \to e^{-ie\theta(x)}\phi(x) \quad \text{and} \quad A_\mu(x) \to A_\mu(x) - \partial_\mu\theta(x).$$

The interaction terms in this Lagrangian density contain derivatives of the scalar. Hence the canonical momenta for $\phi$ take a novel form:

$$\Pi_\phi^\mu = (\partial^\mu + ieA^\mu)\phi^\dagger. \qquad (9.3.20)$$

When we form the Hamiltonian following the previous derivation 9.2.8 but using the new equations of motion and momenta, we find the interaction Hamiltonian density

$$\mathcal{H}_I = (ieA^0\phi^\dagger\partial_0\phi - ieA^r\phi^\dagger\partial_r\phi + \text{h.c.}) + e^2(A_0A^0 - A_rA^r)\phi^\dagger\phi, \qquad (9.3.21)$$

which is not Lorentz covariant.

Furthermore, because the interactions involve derived fields, Wick's Theorem will bring the contraction of derived fields into the perturbation expansion, and we must therefore compute this contraction. From the definition of time ordering,

$$T\big(A(x)B(y)\big) = \theta(x^0 - y^0)A(x)B(y) + \theta(y^0 - x^0)B(y)A(x), \qquad (9.3.22)$$

we deduce that

$$\partial_0^x T\big(A(x)B(y)\big) = T\big(\partial_0^x A(x)B(y)\big) + \delta(x^0 - y^0)\big[A(x), B(y)\big]. \qquad (9.3.23)$$

Working now in the interaction picture where canonical commutation conditions determine the brackets that arise, we find:

$$\begin{aligned}
\partial_0^x\partial_0^y T\big(\phi(x)\phi^\dagger(y)\big) &= \partial_0^x T\big(\phi(x)\partial_0^y\phi^\dagger(y)\big) \\
&= T\big(\partial_0^x\phi(x)\partial_0^y\phi(y)^\dagger\big) + i\delta(x^0 - y^0)\big[\phi(x), \partial_0^y\phi^\dagger(y)\big] \qquad (9.3.24) \\
&= T\big(\partial_0^x\phi(x)\partial_0^y\phi^\dagger(y)\big) + i\delta^{(4)}(x - y),
\end{aligned}$$

which implies

$$T\big(\partial_\mu^x\phi(x)\partial_\nu^y\phi^\dagger(y)\big) = \partial_\mu^x\partial_\nu^y T\big(\phi(x)\phi^\dagger(y)\big) - i\delta_{0\mu}\delta_{0\nu}\delta^{(4)}(x - y). \qquad (9.3.25)$$

The contraction of derived fields may be evaluated by taking the vacuum expectation of this equation. Clearly the delta terms break covariance in the contraction.

In fact, the effect of covariance breakdown in the interaction Hamiltonian density cancels the effect of covariance breakdown in the contraction of derived fields, and (though this is hard to prove) the theory is perturbatively Lorentz covariant. However, the complexities are sufficiently formidable that no-one applies canonical quantization to a classical theory which contains derivative interactions, but turns instead to functional integral quantization.

Despite the non-covariant form of the canonical commutation relations for a massive vector field, the propagator is covariant, and the Feynman rules for massive QED are simple. In scalar QED, however, derivative couplings are necessary for the gauge invariance of the massless limit. Such couplings destroy covariance in the interaction Hamiltonian and the Feynman rules and thereby create unmanageable computational complexities. Clearly, we need a more powerful quantization technique.

## 9.4 The Massless Limit of the Massive Theory

To mimic Maxwell's inhomogeneous equations, we think of $(\rho, \vec{\jmath})$ as a four-vector current $J^\mu$ and couple this current to the massive vector field $A_\mu$ in such a way that in the massless limit the classical equations of motion correspond to Maxwell's equations:

$$\mathcal{L} = -\frac{1}{4}F_{\mu\nu}F^{\mu\nu} + \frac{1}{2}\mu^2 A_\nu A^\nu + A_\mu J^\mu$$
$$\implies \partial_\mu F^{\mu\nu} + \mu^2 A^\nu + J^\nu = 0. \tag{9.4.1}$$

Taking $J^\mu$ as a classical source for photons, we find the Feynman rule for the term $\mathcal{H}_I = -A_\mu J^\mu$ in the usual way from Dyson's formula:

$$A_\mu J^\mu, \text{ outgoing momentum } k, \text{ polarization } \mu: \quad i\hat{J}^\mu(-k), \tag{9.4.2}$$

where $\hat{J}^\mu$ is the Fourier transform of $J^\mu$. This Feynman rule may be expressed diagrammatically as follows:

$$\mathbf{Vertex} = \overset{i\hat{J}^\mu(-k) \quad k,\mu}{\underset{\longrightarrow}{\text{wwwwwwwwww}}} \tag{9.4.3}$$

This vertex may also be interpreted as the diagram for the emission of a massive photon, and the vertex factor as the invariant amplitude for this process.

Now hold the external current $J^\mu$ and the momentum $\bar{k} \neq 0$ fixed, and let the mass $\mu$ tend to zero. In an appropriate Lorentz frame,

$$(k^\mu) = (\omega, 0, 0, |\bar{k}|), \tag{9.4.4}$$

where

$$\omega = (\bar{k}^2 + \mu^2)^{1/2}. \tag{9.4.5}$$

The polarization vectors $e_r(k)^\mu$ may be chosen as follows:

$$e_1 = \frac{1}{\sqrt{2}}(0, 1, i, 0), \quad e_2 = \frac{1}{\sqrt{2}}(0, 1, -i, 0), \quad e_3 = \frac{1}{\mu}(|\bar{k}|, 0, 0, \omega). \tag{9.4.6}$$

In the massless limit, $e_3$ becomes longitudinal.

The amplitude for emitting a massive photon with $e_3$ polarization is

$$i\mathcal{A}_3 = -ie_3(k)^* \cdot \hat{J}(-k) = \frac{i}{\mu}\big(|\bar{k}|\hat{J}^0(-k) - \omega\hat{J}^3(-k)\big). \tag{9.4.7}$$

As $\mu \to 0$, this amplitude becomes the amplitude for emitting longitudinal photons and will in general diverge. The only chance for avoiding this divergence is to make $|\bar{k}|\hat{J}^0(-k) - \omega\hat{J}^3(-k)$ small compared to the mass. The simplest assumption that has this consequence is $k \cdot \hat{J}(-k) = 0$, or equivalently $\partial_\mu J^\mu = 0$. To show that this assumption is sufficient, estimate $i\mathcal{A}_3$ as follows:

$$i\mathcal{A}_3 = \frac{i}{\mu}\Big(\big(\omega\hat{J}^0(-k) - |\bar{k}|\hat{J}^3(-k)\big) + (|\bar{k}| - \omega)\big(\hat{J}^0(-k) + \hat{J}^3(-k)\big)\Big)$$
$$= \frac{i}{\mu}\big(\hat{J}^0(-k) + \hat{J}^3(-k)\big)O\Big(\frac{\mu^2}{|\bar{k}|}\Big) \tag{9.4.8}$$
$$= i\big(\hat{J}^0(-k) + \hat{J}^3(-k)\big)O\Big(\frac{\mu}{|\bar{k}|}\Big) \to 0 \quad \text{as } \mu \to 0.$$

We conclude that, if the massive vector theory is coupled to a conserved current, then, in the massless limit, the longitudinal modes that arise will be decoupled from the rest of the fields in the theory.

**Remark 9.4.9.** The gauge invariance of the massless limit implies that the $k_\mu k_\nu/\mu^2$ term in the propagator does not contribute to gauge-invariant sums of amplitudes.

We can also see that the conserved current hypothesis gives rise to repulsion between like charges. The vacuum to vacuum amplitude is given by:

$$
\begin{aligned}
i\mathcal{A} &= -\frac{1}{2} \int \frac{d^4k}{(2\pi)^4} \ \hat{J}(k)^\mu \hat{J}(-k)^\nu \left( -g_{\mu\nu} + \frac{k_\mu k_\nu}{\mu^2} \right) \frac{i}{k^2 - \mu^2 + i\epsilon} \\
&= \frac{1}{2} \int \frac{d^4k}{(2\pi)^4} \ \hat{J}(k)^\mu \hat{J}(-k)_\mu \frac{i}{k^2 - \mu^2 + i\epsilon},
\end{aligned}
\tag{9.4.10}
$$

where we have used current conservation $k \cdot \hat{J}(-k) = 0$. We will take a static charge distribution $\hat{J} = (\hat{\rho}, \bar{0})$.

We found in Section 4.7 that scalar exchange generates a purely attractive classical force governed by a Yukawa potential. Clearly the method of Section 4.7 applies to the vacuum to vacuum amplitude above to bring out the vacuum energy and the underlying pair potential. This potential is again a Yukawa potential; it becomes the Coulomb potential as $\mu \to 0$.

In the case of scalar exchange, the coupling to fermions has the form $\bar{\psi}\psi\phi$. Since the density $\rho = \bar{\psi}\psi$ is invariant under charge conjugation, scalar exchange generates a purely attractive force. In the case of vector exchange, however, firstly the vacuum to vacuum amplitude has the opposite sign because of the $-g^{\mu\nu}$ in the vector field propagator, and secondly the coupling has the form $\bar{\psi}\gamma^\mu\psi A_\mu$ causing charge conjugation to reverse the sign of the density $\rho = \bar{\psi}\gamma^0\psi$. Hence vector exchange leads to repulsion between like charges and attraction between opposite charges. This confirms that we are on the right track in using a massive vector field with a conserved current for modeling photons.

**Homework 9.4.11.** Using the Lagrangian density

$$
\mathcal{L}_F = -\frac{1}{4} F_{\mu\nu} F^{\mu\nu} + \frac{1}{2} \mu^2 A_\mu A^\mu + \bar{\psi}(i\slashed{\partial} - e\slashed{A} - m)\psi,
$$

find the invariant amplitude for $A\Psi \to A\Psi$ (Compton scattering) to lowest non-trivial order. Show that this amplitude vanishes if either massive photon polarization vector is longitudinal. (This is a consequence of $\partial_\mu A^\mu = 0$.)

**Remark 9.4.12.** From the repulsion/attraction structure of vector-particle exchange, we can deduce that the perturbation series for QED (whether the photon is massive or not) is an asymptotic series. For if the perturbation series converged for some value of the coupling $e_0$, then it would be absolutely convergent and analytic in $e$ for all values $|e| < |e_0|$. Since the vacuum is stable for real $e$, the perturbation series for vacuum decay must converge to zero for real $e$. Being analytic in $e$, the perturbation series for vacuum decay must therefore converge to zero for all values $|e| < |e_0|$. But rotating a real positive coupling $e$ to $ie$ reverses the character of the electromagnetic force, making like charges attract and opposite charges repel. As a consequence, the vacuum becomes unstable, cheerfully falling apart into opposite charges which pool together into separate clumps of positive and negative particles. Since the perturbation series fails to describe this phenomenon, we conclude that it is not convergent for any non-zero value of $e$.

The belief is that the perturbation series is an asymptotic series for real $e$ at $e = 0$. Thus for any particular value of $e$, there will be an optimum order in perturbation theory which will match the non-perturbative dynamics better than either lower or higher order computations. As $e \to 0$, the optimum order should become infinite. The particular optimum order for the $e$ of QED is not known.

The conclusion of this section is that if the massive vector field theory of the previous sections is coupled to a conserved current, then in the limit as its mass tends to zero, the conserved current will not create or absorb longitudinal vector particles. Hence, though longitudinal modes of the massless vector field are allowed on internal lines, as particles they can never interact with the observable universe.

## 9.5 Compton Scattering

In this section, we shall compute Compton scattering for polarized photons off unpolarized electrons in the lab frame in which the electrons are initially at rest. This lab frame corresponds to the common experimental set-up in which the electrons are initially in a stationary target. The section illustrates the techniques for computing with polarization vectors.

The two diagrams for Compton scattering at lowest order are

$$D_1 = \quad\text{(diagram)}\quad \text{and} \quad D_2 = \quad\text{(diagram)} \tag{9.5.1}$$

Omitting the $+i\epsilon$ to save space, the corresponding amplitudes are:

$$i\mathcal{A}_1^{r's'rs} \overset{\text{def}}{=} (-ie)^2 \bar{u}^{r'}(p') \not{\epsilon}'^{s'}(q')^* (\not{p} + \not{q} + m) \not{\epsilon}^s(q) u^r(p) \frac{i}{(p+q)^2 - m^2},$$

$$i\mathcal{A}_2^{r's'rs} \overset{\text{def}}{=} (-ie)^2 \bar{u}^{r'}(p') \not{\epsilon}^s(q)(\not{p} - \not{q}' + m) \not{\epsilon}'^{s'}(q')^* u^r(p) \frac{i}{(p-q')^2 - m^2}. \tag{9.5.2}$$

Suppressing polarization indices, assuming plane polarization so that the polarization vectors are real, and with a natural simplification of notation, we can write these expressions briefly as:

$$i\mathcal{A}_1 = -\frac{ie^2}{2p \cdot q} \bar{u}' \not{\epsilon}'(\not{p} + \not{q} + m)\not{\epsilon} u;$$

$$i\mathcal{A}_2 = \frac{ie^2}{2p \cdot q'} \bar{u}' \not{\epsilon}(\not{p} - \not{q}' + m)\not{\epsilon}' u. \tag{9.5.3}$$

We shall now assume that the photon is massless and, anticipating Chapter 12, that the resulting quantum theory is gauge invariant. These assumptions permit us to use Coulomb gauge and polarizations transverse to the *electron* momenta. Working in the lab frame $p = (m, \bar{0})$, this means:

$$\varepsilon \cdot p = 0 \quad \text{and} \quad \bar{\varepsilon} \cdot \bar{p} = 0;$$

$$\varepsilon' \cdot p = 0 \quad \text{and} \quad \bar{\varepsilon}' \cdot \bar{p}' = 0, \quad \text{and so} \quad \varepsilon' \cdot p' = 0. \tag{9.5.4}$$

With these assumptions, $\not{p}$ anti-commutes with $\not{\varepsilon}$ and $\not{\varepsilon}'$, and so we can use the Dirac equation $(\not{p} - m)u = 0$ to simplify the two amplitudes:

$$
\begin{aligned}
i\mathcal{A}_1 &= -\frac{ie^2}{2p\cdot q}\bar{u}'\not{\varepsilon}'\not{q}\not{\varepsilon}u; \\
i\mathcal{A}_2 &= -\frac{ie^2}{2p\cdot q'}\bar{u}'\not{\varepsilon}\not{q}\not{\varepsilon}'u.
\end{aligned}
\tag{9.5.5}
$$

(These expressions, of course, are no longer gauge invariant.)

Defining $S_a$ to be the spinor factor in $i\mathcal{A}_a$, the square of the total amplitude takes the form

$$
|i\mathcal{A}_1 + i\mathcal{A}_2|^2 = \frac{e^4}{4}\left(\frac{S_1\bar{S}_1}{p\cdot q\,p\cdot q} + \frac{S_1\bar{S}_2}{p\cdot q\,p\cdot q'} + \frac{S_2\bar{S}_1}{p\cdot q'\,p\cdot q} + \frac{S_2\bar{S}_2}{p\cdot q'\,p\cdot q'}\right). \tag{9.5.6}
$$

Taking the spin-sum, spin-average converts the $S_r\bar{S}_s$ into the $T_{r\bar{s}}$ defined by the following traces:

$$
\begin{aligned}
2T_{1\bar{1}} &\overset{\text{def}}{=} \operatorname{tr}\big((\not{p}' + m)\not{\varepsilon}'\not{q}\not{\varepsilon}(\not{p} + m)\not{\varepsilon}\not{q}\not{\varepsilon}'\big); \\
2T_{1\bar{2}} &\overset{\text{def}}{=} \operatorname{tr}\big((\not{p}' + m)\not{\varepsilon}'\not{q}\not{\varepsilon}(\not{p} + m)\not{\varepsilon}'\not{q}'\not{\varepsilon}\big); \\
2T_{2\bar{1}} &\overset{\text{def}}{=} \operatorname{tr}\big((\not{p}' + m)\not{\varepsilon}\not{q}'\not{\varepsilon}'(\not{p} + m)\not{\varepsilon}\not{q}'\not{\varepsilon}\big); \\
2T_{2\bar{2}} &\overset{\text{def}}{=} \operatorname{tr}\big((\not{p}' + m)\not{\varepsilon}\not{q}'\not{\varepsilon}'(\not{p} + m)\not{\varepsilon}'\not{q}'\not{\varepsilon}\big).
\end{aligned}
\tag{9.5.7}
$$

For all four traces, the sequence of steps in computation is similar. We illustrate them with $T_{1\bar{1}}$. Since the trace of an odd number of $\gamma$'s is zero, the term linear in $m$ vanishes. The term quadratic in $m$ involves a product $\not{\varepsilon}'\not{q}\not{\varepsilon}\not{\varepsilon}\not{q}\not{\varepsilon}'$. Since $\not{\varepsilon}\not{\varepsilon} = -1$, this product reduces to $\not{q}\not{q}$, which is zero. We therefore drop all terms in $m$ and $m^2$.

Next, since $\varepsilon\cdot p = 0$, $\not{\varepsilon}\not{p}\not{\varepsilon} = -\not{p}\not{\varepsilon}\not{\varepsilon} = \not{p}$, and so $T_{1\bar{1}}$ reduces to

$$
2T_{1\bar{1}} = \operatorname{tr}(\not{p}'\not{\varepsilon}'\not{q}\not{p}\not{q}\not{\varepsilon}'). \tag{9.5.8}
$$

Then, since $\not{q}\not{q} = 0$, we have $\not{q}\not{p}\not{q} = 2p\cdot q\,\not{q}$. Hence, using the normalization of $\varepsilon'$ and dropping a factor of 2,

$$
\begin{aligned}
T_{1\bar{1}} &= p\cdot q\,\operatorname{tr}(\not{p}'\not{\varepsilon}'\not{q}\not{\varepsilon}') \\
&= 4p\cdot q(2p'\cdot\varepsilon'\,q\cdot\varepsilon' + p'\cdot q).
\end{aligned}
\tag{9.5.9}
$$

The method applies to $T_{2\bar{2}}$; $T_{2\bar{2}}$ can also be evaluated using the substitution $q \leftrightarrow -q'$ and $\varepsilon \leftrightarrow \varepsilon'$ in $T_{1\bar{1}}$. To evaluate $T_{1\bar{2}}$ and $T_{2\bar{1}}$, it is best to eliminate $q'$ using $q' = q + p - p'$. Finally, if we use $p'\cdot\varepsilon' = q\cdot\varepsilon'$ and $p'\cdot q = p\cdot q'$ (which follow from $p' - q = p - q'$) to eliminate $p'\cdot\varepsilon'$ and $p'\cdot q$, the resulting values for the $T$'s are:

$$
\begin{aligned}
T_{1\bar{1}} &= 4p\cdot q\big(2(q\cdot\varepsilon')^2 + p\cdot q'\big); \\
T_{2\bar{2}} &= -4p\cdot q'\big(2(q'\cdot\varepsilon)^2 - p\cdot q\big); \\
T_{1\bar{2}} &= T_{2\bar{1}} = 4p\cdot q\,p\cdot q'\big(2(\varepsilon\cdot\varepsilon')^2 - 1\big) - 4(q\cdot\varepsilon')^2\,p\cdot q' + 4(q'\cdot\varepsilon)^2\,p\cdot q.
\end{aligned}
\tag{9.5.10}
$$

**Homework 9.5.11.** Verify the value given above for $T_{1\bar{2}}$. (The terms in $m^2$ really do cancel out.)

Replacing the $S$'s by these values for the $T$'s in the square of the amplitude 9.5.6 gives the spin-sum, spin-average $|\mathcal{A}|^2$ of the square of the amplitude:

$$|\mathcal{A}|^2 \overset{\text{def}}{=} \frac{1}{2} \sum_{\text{spins}} |i\mathcal{A}_1 + i\mathcal{A}_2|^2 = e^4 \left( \frac{p\cdot q}{p\cdot q'} + \frac{p\cdot q'}{p\cdot q} + 4(\varepsilon\cdot\varepsilon')^2 - 2 \right). \qquad (9.5.12)$$

**Homework 9.5.13.** How would this final expression look if we had used complex polarization vectors $\varepsilon$ and $\varepsilon'$?

If we write $q = (\omega, \bar{\omega})$ and $q' = (\omega', \bar{\omega}')$, and if we define the scattering angle $\theta$ by $\bar{\omega}\cdot\bar{\omega}' = \omega\omega' \cos\theta$, then we can use the lab-frame value of the invariant density of final states from Homework 5.2.9,

$$\mathcal{D} = \frac{\omega'^2}{16\pi^2 m\omega} d\Omega, \qquad (9.5.14)$$

and apply the formula 5.5.1 for the differential cross section with $|\bar{v}_1 - \bar{v}_2| = 1$ and $E_1 E_2 = m\omega$ to find the Klein–Nishina formula for the differential cross section of Compton scattering:

$$\begin{aligned}
\frac{\partial\sigma}{\partial\Omega} &= \frac{1}{4m\omega} \frac{\omega'^2}{16\pi^2 m\omega} e^4 \left( \frac{\omega}{\omega'} + \frac{\omega'}{\omega} + 4(\varepsilon\cdot\varepsilon')^2 - 2 \right) \\
&= \frac{e^4}{64\pi^2 m^2} \left( \frac{\omega'}{\omega} \right)^2 \left( \frac{\omega}{\omega'} + \frac{\omega'}{\omega} + 4(\varepsilon\cdot\varepsilon')^2 - 2 \right).
\end{aligned} \qquad (9.5.15)$$

**Homework 9.5.16.** Show that the polarization-sum, polarization-average of the term $(\varepsilon\cdot\varepsilon')^2$ works out simply enough:

$$\frac{1}{2} \sum_{\text{Polarizations}} (\varepsilon\cdot\varepsilon')^2 = \frac{1}{2}(1 + \cos^2\theta).$$

(Hint: note that the three vectors $\bar{\varepsilon}_1$, $\bar{\varepsilon}_2$, and $\bar{q}$ form an orthonormal basis for $\mathbf{R}^3$.)

**Homework 9.5.17.** Work out the differential cross section for unpolarized Compton scattering directly from the amplitudes 9.5.3. (Use the formulae of Section 8.6 to simplify contractions in products of $\gamma$ matrices.) The result,

$$\frac{\partial\sigma}{\partial\Omega} = \frac{e^4}{32\pi^2 m^2} \left( \frac{\omega'}{\omega} \right)^2 \left( \frac{\omega}{\omega'} + \frac{\omega'}{\omega} - \sin^2\theta \right), \qquad (9.5.18)$$

should agree with that derived by substituting the formula of the previous homework in 9.5.15.

# 9.6 The Gauge Principle

We have seen that, in order to remove the longitudinal modes of the massless vector field, we need the massless theory to be gauge symmetric. Thus the massless limit of a massive theory, if it is to be of any use, must be gauge invariant. Hence the only term in the interacting, massive vector field Lagrangian density that is not gauge invariant must be the mass term itself. In this section, we shall explain why gauge invariance is natural, and in the following sections we shall show how it constrains the form of the interaction terms and how it relates to current conservation in the massive theory.

The source of gauge invariance is the phase of a charged field. If $\psi$ is a classical charged scalar field, then its value $\psi(x)$ at each point $x$ in space-time is a complex number. We have so far assumed that there is a natural coordinate system for these phases which fixes the association between the physical state of the field and the complex number used to represent the state. In reality, there is a complex plane of values for $\psi$ to take at each point $x$, but no preferred coordinates for these complex planes. Geometrically, the total space we are using here is a product $M \times V$ of Minkowski space $M$ with a one-dimensional complex vector space $V = \mathbf{C}$. Such a structure, in which a vector space is hidden in each point of space-time, is called a *vector bundle*.

Write $V_x$ for the copy of the vector space $V$ at the point $x$. A choice of basis $v(x)$ for the spaces $V_x$ can be made in many ways. Indeed, for any particular basis $v(x)$ and any non-vanishing, complex-valued function $z(x)$, $v'(x) = v(x)z(x)$ is another basis. (Note that, for this section, we have adopted a convention of writing scalars on the right of the vectors that they multiply.) Since $V_x$ does have a modulus function $|v(x)|^2$ and a multiplication by $i$, the function $z$ may properly be restricted to the form $z = \exp(i\theta(x))$.

A choice of basis is called a *gauge*, and the set of all change-of-basis functions the *gauge group*. If we write $\theta(x) = \lambda\chi(x)$, then we can differentiate the group elements with respect to $\lambda$ and find that the arbitrary real-valued function $\chi$ is a Lie algebra element for the gauge group.

Now that we have seen the geometry underlying an internal $U(1)$ symmetry, it is natural to assert that it is the geometry of the situation that determines the physics, not the choice of basis. This assertion is a special case of the gauge principle, whose general statement proposes that, if a Lagrangian has an internal symmetry implemented through constant matrices, then it must also be invariant when the same matrices become position dependent.

To bring out the geometry, one removes indices by contracting the coordinates with the appropriate basis vectors. In this case, we simply multiply the basis by the coordinate, $v(x)\psi(x)$. For this vector in $V_x$ to be independent of choice of gauge, if $v' = vz$, then we must have $\psi' = z^*\psi$. Hence the gauge principle proposes that, if a Lagrangian is invariant under the global transformation $\psi'(x) = z^*\psi(x)$, then it must be invariant under the local transformation $\psi'(x) = z^*(x)\psi(x)$. This *local* or *gauge invariance* is just position-dependent charge symmetry, hence it is obvious that charge invariant terms like $\psi^\dagger\psi\phi$ are also gauge invariant. The derivative terms, however, break gauge invariance:

$$\partial_\mu\psi'^\dagger\partial^\mu\psi' = (\partial_\mu + i\partial_\mu\theta)\psi^\dagger(\partial^\mu - i\partial^\mu\theta)\psi, \tag{9.6.1}$$

which is not at all the same as the original term. The derivatives fail to transform well because our concept of differentiation is not geometric, but basis dependent.

Think of $\psi$ as a vector-valued function in coordinates which takes values in **C** and of $v\psi$ as the underlying geometric vector-valued function which takes values in the vector spaces $V_x$. Then it is $v\psi$ that we wish to differentiate, and it is clear that, if we walk along a path in space-time, then we will be able to use the modulus to detect changes in the length of $v\psi$, but we will not have any reference direction for detecting change in orientation of $v\psi$ in successive $V_x$'s.

In order to differentiate, we must actually choose a derivative operator from many equally good possibilities. Let us begin by specifying the abstract properties that such an operator should have. Writing $D_X$ for the act of differentiating a vector field in the $X$ direction in space-time, the three properties that make $D_X$ a derivative operator are:

$$\left. \begin{array}{c} D_{fX+gY} = fD_X + gD_Y \\ D_X(u+v) = D_X u + D_X v \\ D_X(vf) = (D_X v)f + v(X^\mu \partial_\mu f) \end{array} \right\} \qquad (9.6.2)$$

where $f$ and $g$ are scalar-valued functions and $u$ and $v$ are vector-valued functions. A direction-dependent operator $D$ on vector fields which satisfies these three conditions is called a *covariant derivative* or an *affine connection*.

The first axiom implies that differentiation in any direction can be computed from a knowledge of derivatives in the four coordinate directions:

$$D_X = X^\mu D_\mu, \qquad (9.6.3)$$

where $D_\mu$ is differentiation in the $\hat{x}_\mu$ direction.

To study a covariant derivative $D$, we choose basis vectors $v(x)$ for $V_x$ and define a scalar function $B$ of direction $X$ and position $x$ by

$$D_X v(x) = v(x)B_X(x) \quad \text{or} \quad D_\mu v(x) = v(x)B_\mu(x). \qquad (9.6.4)$$

The function $B_X$ tells us how fast our basis vector $v(x)$ is changing from the perspective of the connection $D$ as we move in the $X$ direction at a point $x$. In differential geometry, $B$ is called a *connection one-form*. The connection one-form is effectively the coordinate representation of the covariant derivative in the basis $v$.

To differentiate an arbitrary vector field $v'$, express $v'$ in terms of $v$ by an equation $v'(x) = v(x)z(x)$, and then differentiate this equation:

$$\begin{aligned} D_\mu v' &= (D_\mu v)z + v\partial_\mu z \\ &= vB_\mu z + v\partial_\mu z. \end{aligned} \qquad (9.6.5)$$

Writing the covariant derivative in coordinates implies suppressing the basis vectors $v(x)$. The result is the commonly used formula $D_\mu = \partial_\mu + B_\mu$.

However, if $v'$ is a basis, then the covariant derivative determines a connection one-form $B'$ with respect to $v'$:

$$D_\mu v' = v'B'_\mu. \qquad (9.6.6)$$

Combining this with the previous equation determines the transformation law for the connection one-form:

$$v' = vz \quad \Longrightarrow \quad B'_\mu = z^{-1} B_\mu z + z^{-1} \partial_\mu z \qquad (9.6.7)$$

In particular, if we substitute $z = \exp i\theta$, then we obtain the familiar formula:

$$B'_\mu = B_\mu + i \partial_\mu \theta. \qquad (9.6.8)$$

**Remark 9.6.9.** Another perspective on the covariant derivative comes through considering the vector fields that are annihilated by the differentiation operator. Suppose that $x(t)$ is a differentiable arc in space-time. Let $X(t)$ be the velocity of the point $x(t)$ at time $t$. Suppose $v(x)$ satisfies

$$D_{X(t)} v\big(x(t)\big) = 0 \qquad (9.6.10)$$

for all $t$. Then we say that the vectors $v(x)$ are *parallel* along the arc $x(t)$ (with respect to $D$, of course).

The idea of parallel vector fields may easily be understood from an example. Suppose that we are standing at the equator pointing East with a long stick. If we walk due North, we can keep the stick pointing perpendicular to the direction of motion. This is parallel transport, transport which generates parallel vectors. Suppose that, when we reach the North pole, we turn right and follow the direction of the stick down to the equator, keeping the stick pointing South. Then the stick continues to define parallel vectors. When we arrive at the equator, we turn right again and walk back to where we started, keeping the stick pointing South. When we arrive at our starting point, the stick has undergone parallel transport around a spherical triangle, but its final direction is perpendicular to its initial direction.

Besides illustrating the concept of parallel transport, this example shows that parallel transport must be associated with a path and cannot in general be extended to areas.

We have described a very general concept for differentiating vector fields. In the specific case of physical gauge symmetries, the vector spaces $V_x$ also have inner products. We can then constrain the differentiation rule by requiring that the inner product is a covariant constant:

$$\partial_\mu (v, v') = \big(D_\mu v, v'\big) + \big(v, D_\mu v'\big). \qquad (9.6.11)$$

Note how this equation achieves its object: the term $(v, v')$ is a product of three functions, the vector $v$, the vector $v'$, and the inner product itself, but the right-hand side contains derivatives of the first two only. A covariant derivative with the property 9.6.11 is called a *hermitian covariant derivative*.

When we have hermitian inner products on the vector spaces $V_x$ it simplifies computations to choose a field of unit basis vectors $v(x)$. Then the connection one-form $B$ satisfies:

$$
\begin{aligned}
B_\mu + B_\mu^* &= (B_\mu v, v) + (v, B_\mu v) \\
&= (D_\mu v, v) + (v, D_\mu v) \\
&= \partial_\mu (v, v) \\
&= 0,
\end{aligned}
\qquad (9.6.12)
$$

which shows that $B_\mu$ is purely imaginary.

Having investigated the structure of covariant derivatives, we can now see that the gauge principle forces all derivatives in a Lagrangian density with $U(1)$ symmetry to be covariant derivatives. The associated connection one-forms must therefore be interpreted as physical fields. Such fields are known as *gauge fields*.

The points above merely set out the geometry intrinsic to global symmetry. In short, if there is internal symmetry, then there must be degrees of freedom hidden in every point of space-time, derivatives must be covariant derivatives, and the emergent connection one-forms must be physical fields.

The detailed description of the geometry of gauge invariance presented in this section generalizes easily to arbitrary internal symmetry groups and forms the basis for understanding non-abelian gauge fields like gluons. The next section takes the first step in this direction by deriving quantum electrodynamics from a Lagrangian for charged fields.

## 9.7 Gauge Invariance and Minimal Coupling

To derive practical benefit from the insights of the previous section, in this section we first show how to bring the general discussion of gauge fields to a focus on the electromagnetic potential, and second describe the procedure for transforming a Lagrangian with a global $U(1)$ symmetry into one with a local $U(1)$ symmetry.

Let $\mathcal{L}(\phi, \psi)$ be a Lagrangian density with a $U(1)$ symmetry which distinguishes the neutral fields $\phi = (\phi^1, \ldots, \phi^m)$ from the charged fields $(\psi^1, \ldots, \psi^n)$. Writing $q_r$ for the charge of $\psi^r$, the $U(1)$ symmetry is implemented by multiplying $\psi^r$ by $\exp(-ieq_r\theta)$, where $\theta$ is a constant. Now each field $\psi^r$ is associated with a different vector space $V^r_x$ and some specific basis $v_r(x)$. The global symmetry is simply a coordinated, simultaneous change of bases from $v_r(x)$ to

$$v'_r(x) = v_r(x)e^{ieq_r\theta}. \tag{9.7.1}$$

From the gauge principle we understand that all derivatives in $\mathcal{L}$ must be covariant derivatives. Hence we must have a connection one-form $B^r$ for each vector bundle $V^r$. From the transformation law 9.6.8 for connection one-forms we see that the $B^r$ transform as follows:

$$B'^r_\mu = B^r_\mu + ieq_r\partial_\mu\theta. \tag{9.7.2}$$

Since there is only one photon, this family of transformation laws must be multiples of the transformation law for the photon, $A'_\mu = A_\mu - \partial_\mu\theta$. Hence $B^r = -ieq_r A$, and the covariant derivatives $D^r_\mu$ may be written in terms of $A_\mu$:

$$D^r_\mu = \partial_\mu - ieq^r A_\mu. \tag{9.7.3}$$

Thus the electromagnetic potential is, in effect, the connection one-form for all the vector bundles: however many matter fields there are, one symmetry generator only gives rise to one gauge field. Indeed, using the conserved quantity $Q$ of the original global symmetry, there is in fact only one covariant derivative:

$$D_\mu = \partial_\mu - ieA_\mu Q. \tag{9.7.4}$$

The conclusion of this geometric analysis is that we expect Lagrangian densities with global internal symmetries to have gauge symmetries and, for this to be the case, it is necessary and sufficient that all the derivatives in the Lagrangian density be covariant derivatives. It remains to comment on the construction of gauge invariant Lagrangian densities.

The simplest way of creating a gauge invariant Lagrangian density is to start from a Lagrangian density with a global symmetry and replace all the derivatives with covariant derivatives. This introduces the gauge field $A$ through the connection one-form, so we must also add the kinetic term for $A$. This method of creating gauge invariant Lagrangians is called *minimal coupling*, because it introduces only those interaction terms between the gauge field and the matter fields that are essential for gauge invariance. Thus, for example, from the Lagrangian density

$$\mathcal{L} = \bar{\psi}(i\partial\!\!\!/ - m)\psi, \tag{9.7.5}$$

we obtain the gauge invariant Lagrangian density of quantum electrodynamics:

$$\mathcal{L}_G = -\frac{1}{4}F_{\mu\nu}F^{\mu\nu} + \bar{\psi}(i\partial\!\!\!/ - e A\!\!\!/ - m)\psi, \tag{9.7.6}$$

where $\psi$ is the electron field.

**Homework 9.7.7.** The curvature $R_{\mu\nu}$ of a covariant derivative is defined by

$$R_{\mu\nu} \overset{\text{def}}{=} [D_\mu, D_\nu].$$

Using just the defining properties of the covariant derivative show that, for any function $z$ and any vector-valued field $v$,

$$R_{\mu\nu}(vz) = (R_{\mu\nu}v)z.$$

(This shows that the curvature is a tensor, not a differential operator.)

**Homework 9.7.8.** Show that the curvature tensor $R_{\mu\nu}$ is invariant under gauge transformations. Now write the curvature in terms of the connection one-form and conclude that

$$R_{\mu\nu} = -ieF_{\mu\nu},$$

where $F_{\mu\nu}$ is the electromagnetic field strength.

## 9.8 Mass Term and Current Conservation

We have seen that gauge invariance is a natural requirement on a theory with an internal symmetry, and that it is necessary in a theory with massless vector fields in order to eliminate the longitudinal modes. Now, in order to obtain a gauge invariant massless limit, we clearly must start from a massive vector field theory in which only the mass term breaks the gauge invariance. We shall now demonstrate that this assumption leads to current conservation, and thus, as we have seen, leads to the decoupling of the longitudinal mode in the massless limit.

Let $\mathcal{L}_0(A, \psi)$ be a gauge invariant Lagrangian density for a massless vector field coupled to some charged field $\psi$. Let $\mathcal{L}_\mu$ be the Lagrangian density

$$\mathcal{L}_\mu = \mathcal{L}_0 + \frac{1}{2}\mu^2 A_\mu A^\mu. \tag{9.8.1}$$

Now, by using the gauge group to vary the fields, and by applying the principle of least action to these variations, we shall show that the current is conserved in the massive theory.

For a field $\psi$ with charge $q$, the covariant derivative is

$$D_\mu = \partial_\mu - ieqA_\mu, \tag{9.8.2}$$

and the action of the gauge group is given by

$$\psi(x) \mapsto e^{-ieq\theta(x)}\psi(x), \quad \text{and} \quad A_\mu(x) \mapsto A_\mu(x) - \partial_\mu\theta(x). \tag{9.8.3}$$

To find the infinitesimal action we substitute $\theta(x) = \theta\chi(x)$, differentiate with respect to $\theta$, and evaluate at $\theta = 0$. This gives us the action of the Lie algebra element $\chi$ on the fields:

$$\psi(x) \mapsto -ieq\chi(x)\psi(x), \quad A_\mu(x) \mapsto A_\mu(x) - \partial_\mu\chi(x). \tag{9.8.4}$$

We are assuming that $\mathcal{L}_0$ is invariant under this action, so the effect on $\mathcal{L}_\mu$ is simply

$$\mathcal{L}_\mu \mapsto -\mu^2 \partial_\mu \chi A^\mu. \tag{9.8.5}$$

Integrating over space-time gives us the variation of the action:

$$\frac{\partial S}{\partial \theta}\Big|_{\theta=0} = -\int d^4x\, \mu^2 \partial_\mu \chi A^\mu = \int d^4x\, \mu^2 \chi \partial_\mu A^\mu. \tag{9.8.6}$$

According to the principle of least action, the evolution of the fields is such that $\delta S$ vanishes for all variations, hence $\partial_\mu A^\mu = 0$. The equations of motion are

$$\partial_\mu F^{\mu\nu} + \mu^2 A^\nu + J^\nu = 0, \tag{9.8.7}$$

and so we find that

$$\partial_\nu J^\nu = 0. \tag{9.8.8}$$

Hence, if we take a gauge invariant massless vector field theory, add a mass term, compute amplitudes, and then take the limit as the mass tends to zero, we will indeed find that the longitudinal mode decouples, and the limit is a sensible result for the massless vector field theory.

## 9.9 Discrete Symmetries

After the detailed treatment of discrete symmetries presented in the last sections of Chapters 3 and 8, we can afford to be brief here. The basis hypothesis is that QED is invariant under $C$, $P$, and $T$. The only term in the QED Lagrangian density to pose a threat is the interaction $\bar{\psi}\!\!\not{A}\psi$. We therefore define the discrete symmetries on $A$ in such a way that this term is invariant. From the tables of the actions of $C$, $P$, and $T$ on bilinears, we see that the following definitions are necessary and sufficient for the invariance of QED:

$$U_C^\dagger A(x) U_C \overset{\text{def}}{=} -A(x), \quad U_P^\dagger A(x) U_P \overset{\text{def}}{=} PA(\tilde{x}), \quad \Omega_T A(x) \Omega_T^{-1} \overset{\text{def}}{=} PA(-\tilde{x}). \tag{9.9.1}$$

To find the actions on the creation and annihilation operators, we follow the methods of Chapter 8: substitute the free-field expansion of $A$ into both sides of these definitions, equate coefficients of the exponentials, and then use the normalization of the polarization vectors. The result is

$$U_C^\dagger \alpha_r(k) U_C = -\alpha_r(\tilde{k}),$$
$$U_P^\dagger \alpha_r(k) U_P = -e_r(k)^* \cdot Pe^s(\tilde{k}) \alpha_s(\tilde{k}),$$
$$\Omega_T \alpha_r^\dagger(k) \Omega_T^{-1} = -e_r(k)^* \cdot Pe^s(\tilde{k})^* \alpha_s^\dagger(\tilde{k}).$$

(9.9.2)

Parity and time reversal will not mix polarizations if we use real polarization vectors which satisfy $Pe^s(\tilde{k}) = e^s(k)$. In the case of a massive vector field, if the polarization vectors at $k$ are defined by applying a solution $\Lambda_k$ of $\Lambda k_{\text{rf}} = k$ to their values at the vector $k_{\text{rf}}^\top = (\mu, \vec{0})$, then we need only choose $\Lambda_{\tilde{k}} = P\Lambda_k P$ and this relation will be satisfied. In any case, we could use the relation $Pe^s(\tilde{k}) = e^s(k)$ to define polarization vectors for $k^3 < 0$ in terms of those at $k^3 > 0$ and leave their values on $k^3 = 0$ undefined. With such choices of polarization vectors, we can ensure that none of the discrete symmetries mix polarization when applied to particle states of a vector field.

This section concludes our theoretical development of the common discrete symmetries, charge conjugation, parity, and time reversal. We find in particular that QED is invariant under all of them.

# 9.10  Summary

The free Lagrangian density for vector fields is more intricate than that for scalars or spinors; it gives rise to two different theories, one with a mass term and one without. The most important facts about these theories are that the massless theory has a gauge symmetry and the massive theory is continuous in the mass parameter at mass zero if and only if the vector field is only coupled to conserved currents. With the massive photon, we have managed to obtain a form of quantum electrodynamics, but we still need a means for quantizing a theory with derivative interactions and a means for either directly quantizing massless QED or of demonstrating gauge invariance for the massless limit of massive QED.

This chapter has also introduced the gauge principle which finds on geometric grounds that global symmetries should be associated with local symmetries and that local symmetries are upheld by massless gauge fields. The photon, the three $W$ bosons of the weak interactions, and the eight gluons of the strong interactions are all gauge bosons.

Chapter 9 brings Part 2 of the text to a conclusion with a glimpse of QED. We have now pushed perturbative canonical quantization about as far as it can comfortably go. With the gauge principle as a source of gauge-symmetric quantum field theories, it is really necessary to find a quantization technique that applies to theories with gauge symmetries. It is time to return to the roots of our quantization procedure and reformulate quantization on a more profound basis. This is the purpose of Part 3 – Chapters 10 to 13 – which develops functional calculus quantization from a new foundation in the interacting field.

# Chapter 10

# Reformulating Scattering Theory

Developing a new scattering theory from a foundation in the Green functions of the interacting field in order to provide a basis for renormalization, unstable particle theory, and in the next chapter, functional integral quantization.

## Introduction

Chapters 10 to 13 constitute the third part of this text. Where Chapter 4 offered perturbative canonical quantization and scattering theory based on the free field, these chapters offer functional calculus quantization and scattering theory based on the interacting field. This change in viewpoint enables us to quantize the Standard Model of the electroweak and strong interactions.

At this point, our scattering theory still depends on the concept of an adiabatic turning on and off of the interactions. This concept served to connect interacting fields to free fields and to identify particles in incoming and outgoing states. However, when we formulate scattering theory in this way, it is impossible to include unstable particles and bound states in the theory. (We have computed decay amplitudes at tree level and found them sensible; it is in the annihilation scattering diagram that the possibility of decay makes even the tree level computation meaningless – see Remark 5.5.9.) Furthermore, gauge symmetries are broken by turning off the interactions. Hence, as it is our purpose to develop gauge theories and theories with unstable particles and bound states, we must now eliminate the turning on and off function from scattering theory and reformulate scattering theory in terms of interacting fields.

The foundation for managing the new perturbation theory is a non-perturbative formalism for scattering. In this formalism, the basic unit of consideration is not an operator as in the Wick expansion, nor a process as in the Feynman expansion, but a function of external momenta, a Green function, which represents the total potential of natural law for transforming the incoming state into the outgoing one. Section 10.1 introduces Green functions and their generating functional. Section 10.2 describes the relationship between the generating functional and perturbation theory on a heuristic level, and Section 10.3 shows that the Green functions can indeed be computed by summing Feynman diagrams.

In the computations so far, we have assumed that the quantum field obeys the canonical commutation relations, but we have not used any of the particle states in Fock space. Therefore, Section 10.4 uses the quantum field to construct in and out states, and Section 10.5 applies this construction to obtain $S$-matrix elements from our Green functions. This completes the first part of the chapter. This material forms the basis for functional integral quantization in Chapter 11.

In order to construct in and out states, we had to rescale or renormalize the original or bare quantum fields to normalize their relation to one-particle states. We foresee that mass and coupling parameters will be subject to a similar renormalization. The remainder of this chapter investigates this renormalization.

Section 10.6 establishes a change in notation to make the renormalized field the fundamental symbol in the Lagrangian density. Section 10.7 comments on the difficulties of computing with the original bare parameters, and Section 10.8 explains how to compute with renormalized parameters by splitting the Lagrangian density into a physical or finite parameter part and a counterterm or divergent part. Pursuing the direction of Section 10.8, Section 10.9 presents the defining principle of counterterm renormalization – how to set renormalization conditions to fix the counterterm coefficients in terms of the physical parameters. Then Sections 10.10 and 10.11 provide a detailed example of counterterm renormalization. Up to this point, we have been working with scalar fields only; Sections 10.12 to 10.14 quickly extend the whole formalism of the previous sections to spinor and vector fields. This ends the second part of the chapter.

The purpose of the third and final part of the chapter – Sections 10.15 to 10.18 – is to extend the principles of earlier sections to unstable particles. We find that the principles do not extend to this case, but the conclusions can be used to provide an approximate model.

## 10.1  Generating Functionals and Green Functions

In this section, we shall introduce generating functionals and Green functions, a non-perturbative language for physical processes; in the next we shall show how these concepts relate to perturbation theory. The first step is to perturb an interacting field from its vacuum configuration with a classical source $j$. The generating functional $Z(j)$ is effectively the total response of all the fields in the theory. The degree of coupling to the source (which corresponds to the number of source vertices in a diagram) determines natural fragments of the total response. These natural fragments are the Green functions. Because the notation can get complicated even though the ideas are straightforward, we shall set up the language for a theory of a single scalar field. The notation extends naturally to other theories.

We shall take $\phi^4$-theory to illustrate the construction and properties of Green functions. The Lagrangian density for this theory is

$$\mathcal{L} = \frac{1}{2}\partial_\mu\phi\,\partial^\mu\phi - \frac{1}{2}\mu^2\phi^2 - \frac{\lambda}{4!}\phi^4, \tag{10.1.1}$$

giving a theory of a single self-interacting hermitian field.

To awaken the full potential of the $\phi^4$ field, we add a source term $j(x)\phi(x)$ to the Lagrangian density. Assuming that the *background field* or *classical source* $j(x)$ is zero at space-time infinity, the presence of $j$ will not affect the physical vacuum, $|0\rangle_\mathrm{P}$. The *generating functional* $Z(j)$ is the vacuum to vacuum amplitude of the total response of the field to the source:

$$Z(j) \stackrel{\text{def}}{=} {}_\mathrm{P}\langle 0|S_j|0\rangle_\mathrm{P}. \tag{10.1.2}$$

Since we are dealing with a vacuum expectation, there are no incoming or outgoing particles to identify, so $S_j = U_j(\infty, -\infty)$ may be computed without the turning on and off function.

Expanding the generating functional as a power series in $j$ brings out a set of symmetrical coefficient functions, the *Green functions* of $\phi^4$ theory:

$$Z(j) = \hat{G}_0 + \sum_{n=1}^{\infty} \frac{i^n}{n!} \int \frac{d^4 k_1}{(2\pi)^4} \cdots \frac{d^4 k_n}{(2\pi)^4} \, \hat{j}(-k_1) \cdots \hat{j}(-k_n) \hat{G}_n(k_1, \ldots, k_n)$$

$$= G_0 + \sum_{n=1}^{\infty} \frac{i^n}{n!} \int d^4 x_1 \cdots d^4 x_n \, j(x_1) \cdots j(x_n) G_n(x_1, \ldots, x_n),$$

(10.1.3)

where $G_n$ and $\hat{G}_n$ are related by

$$G_n(x_1, \ldots, x_n) = \int \frac{d^4 k_1}{(2\pi)^4} \cdots \frac{d^4 k_n}{(2\pi)^4} \, e^{ix_1 \cdot k_1} \cdots e^{ix_n \cdot k_n} \hat{G}_n(k_1, \ldots, k_n). \quad (10.1.4)$$

We will assume, as always, that the vacuum energy counterterm has been added to the Hamiltonian, so that in fact

$$Z(0) = \hat{G}_0 = G_0 = 1. \tag{10.1.5}$$

**Remark** 10.1.6. A Green function is generally the kernel of an integral operator which is itself the inverse of a *linear* differential operator in the context of a boundary value problem. Such a Green function gives the solution of the boundary value problem as a linear function of the forcing function $j$. In field theory, however, we have a non-linear system of partial differential equations, so the solutions are expressed in terms of a power series in the forcing function.

Thinking in Feynman diagrams for a moment, the generating functional is the sum of all vacuum bubbles in the $\phi^4$-theory with source. Notice how the presence of the source enables every physical diagram to have a unique representation as a vacuum bubble. Thus $Z(j)$ represents all possible physical processes.

Continuing with the Feynman diagram viewpoint, we see that all physical diagrams with $n$ external lines are represented by vacuum diagrams in $Z(j)$ with $n$ source vertices. Thus the generalized physical process with $n$ external lines is represented as a whole by that fragment of the generating functional which is coupled $n$ times to the source. This fragment, $G_n$ or $\hat{G}_n$ above, is called the *n-point Green function*.

Returning to analysis, the proper way to extract the Green functions from the generating functional is by functional differentiation as defined in Section 2.6:

$$\hat{G}_n(k_1, \ldots, k_n) = (2\pi)^{4n}(-i)^n \frac{\delta}{\delta \hat{j}(-k_1)} \cdots \frac{\delta}{\delta \hat{j}(-k_n)} Z(j) \Big|_{j=0},$$

$$G_n(x_1, \ldots, x_n) = (-i)^n \frac{\delta}{\delta j(x_1)} \cdots \frac{\delta}{\delta j(x_n)} Z(j) \Big|_{j=0}.$$

(10.1.7)

The classical background $j$ is merely a device which enables us to construct a generating functional for the Green functions. This generating functional simplifies functional integral quantization and creates a pleasing impression of the unity underlying the activity of the fields in a theory. Green functions, however, do not depend on the concept of a classical background, and functional integral quantization can be formulated directly in terms of Green functions. The classical background is technically and philosophically attractive and yet unnecessary.

With the concepts of a generating functional and Green functions, we have established a non-perturbative view of physical processes on the unmanifest or vacuum level. Of course, since many Feynman diagrams give rise to divergent integrals, the existence of these theoretical entities is in question. However, since the $n$-point Green function is closely related to the total amplitude for $k$ particles scattering into $n - k$ particles, the Green functions are more closely related to experiment than any individual Feynman diagram. We should therefore think of the generating functional and the Green functions as well defined by nature, whereas the Feynman diagrams, as they do not represent anything physical, are not subject to any such philosophical constraint.

The viewpoint on the generating functional presented in this section follows naturally from what we know of $\phi^4$ perturbation theory with a source. For the purposes of computation, however, it is necessary to see the Green functions in terms of Feynman diagrams. In the following section, we therefore explain the connection between the Green functions and sums of Feynman diagrams for $\phi^4$ without a source.

## 10.2 The Generating Functional in Perturbation Theory

In this section, we shall show that the relationship between $\hat{G}_n$ and $Z$ given by the functional derivative formula 10.1.7 is consistent with perturbative definitions of $\hat{G}_n$ and $Z$. Having thereby gained a general feeling for the relationship of Green functions and their generating functional to Feynman diagrams, we shall be ready for the more formal argument of the next section.

The Feynman diagram for the source term $j\phi$ in $\mathcal{L}$ is a dot with a line emerging from it. The associated Feynman rule is to put a factor of $i\hat{j}(-k)$ into the Feynman integral, where $k$ is the momentum coming from the source. Though every diagram that contributes non-trivially to $Z(j)$ has no external lines, we can convert those lines that terminate in source vertices into external lines with incoming momenta $k$ by functionally differentiating with respect to $i\hat{j}(-k)/(2\pi)^4$. Note that this functional differentiation removes the integration associated with the line attached to the source vertex, but does not remove the propagator factor.

This suggests that, when we compute $\hat{G}_n$ in perturbation theory, we modify the Feynman rules to include a propagator on each external line. For the resulting Green function to be meaningful, it will be necessary to assume that all momenta are off-shell.

If a vacuum bubble in $j$-theory has $n$ source vertices, then we must differentiate $n$ times to remove all the source factors. Doing this enables us to introduce $n$ different momenta, creating $n!$ Feynman diagrams in $\phi^4$-theory, each with $n$ external lines, and all related by crossing symmetry. Exactly the same differentiation procedure converts the generating functional into the $n$-point Green function. Therefore, if $Z(j)$ is a sum of $j$-theory Feynman integrals, then the Green functions are sums of $\phi^4$-theory Feynman integrals (with the caveat that propagator factors for external lines must be included). It remains to see if the combinatoric factors are correct. For this, it will be helpful to have an example.

Working under the convention that all momenta are incoming, consider the diagram

$$D(k_1, k_2, k_3, k_4) \overset{\text{def}}{=} \quad \begin{array}{c} k_1 \quad\quad k_2 \\ k_3 \quad\quad k_4 \end{array} \tag{10.2.1}$$

in $\phi^4$-theory without a source. The associated vacuum diagram in the theory with a source is

$$D_j \overset{\text{def}}{=} \tag{10.2.2}$$

We can write the amplitudes for these diagrams as Feynman integrals multiplied by symmetry factors:

$$i\mathcal{A}(D) \overset{\text{def}}{=} \frac{1}{s(D)} I(D) = \frac{1}{2} I(D) \quad \text{and} \quad i\mathcal{A}(D_j) \overset{\text{def}}{=} \frac{1}{s(D_j)} I(D_j) = \frac{1}{16} I(D_j). \tag{10.2.3}$$

From the relation 10.1.7 which expresses the Green functions as derivatives of the generating functional, the operator

$$\mathbf{D} \overset{\text{def}}{=} (2\pi)^{16} (-i)^4 \frac{\delta}{\delta \hat{j}(-k_1)} \frac{\delta}{\delta \hat{j}(-k_2)} \frac{\delta}{\delta \hat{j}(-k_3)} \frac{\delta}{\delta \hat{j}(-k_4)} \tag{10.2.4}$$

should transform $\mathcal{A}(D_j)$ into $\mathcal{A}(D)$. Actually, when we apply $\mathbf{D}$ to $I(D_j)$, we find that there are 4! terms corresponding to the 4! ways of assigning momenta to the lines in $D$. Now, since there are 2 symmetries of $D$, the 16 symmetries of $D_j$ group these 4! terms into groups of 8 which all come from the same Feynman diagram:

$$\mathbf{D}I(D_j) = 8I\big(D(k_1, k_2, k_3, k_4)\big) \tag{10.2.5}$$
$$+ 8I\big(D(k_1, k_3, k_2, k_4)\big) + 8I\big(D(k_1, k_4, k_3, k_2)\big).$$

Restoring symmetry factors, we find that

$$\mathbf{D}\mathcal{A}(D_j) = \mathcal{A}\big(D(k_1, k_2, k_3, k_4)\big) \tag{10.2.6}$$
$$+ \mathcal{A}\big(D(k_1, k_3, k_2, k_4)\big) + \mathcal{A}\big(D(k_1, k_4, k_3, k_2)\big).$$

Hence the contribution of the $D$-type diagrams to the Green functions matches the contribution of the associated vacuum diagram $D_j$ to the generating functional.

From this example, it is clear that the functional differentiation in the relationship 10.1.7 between $\hat{G}_n$ and $Z$ (1) effectively converts a vacuum diagram into a complete family of distinct diagrams related by crossing and (2) introduces a multiplicity factor which generates the correct symmetry factors in the final amplitudes.

**Remark 10.2.7.** The combinatorics here is identical to that in the relationship between Wick diagrams and Feynman diagrams. The symmetry factor of $D_j$ is the same as that in the Wick operator $S(D)$. The action of $\mathbf{D}$ on $\mathcal{A}(D_j)$ corresponds precisely to contraction of $S(D)$ with incoming and outgoing particle states. In Section 4.9, we were primarily interested in the multiplicity $s(D_j)/s(D)$ of a specific Feynman diagram derived from $S(D)$. Here, we have also considered the diversity of Feynman diagrams:

$$\#\{\textbf{Crossed Family of } D\} = n! \times \frac{s(D)}{s(D_j)}. \tag{10.2.8}$$

(There are $n!$ ways of labeling $n$ external lines with incoming momenta, and the number of ways of rearranging the labels without changing the diagram is $s(D_j)/s(D)$.) The transformation of symmetry factors is

$$\frac{1}{s(D_j)} \xrightarrow{\quad D \quad} \frac{1}{s(D_j)} \quad \text{for } n! \text{ diagrams}$$

$$\xrightarrow{\qquad} \frac{1}{s(D)} \quad \text{for } n! \times \frac{s(D)}{s(D_j)} \text{ distinct diagrams.}$$

(10.2.9)

We have demonstrated that if the ordinary Feynman rules, applied to the sum of vacuum diagrams in $\phi^4$-theory with a classical source $j$, give rise to $Z(j)$, and if the same Feynman rules modified to include propagator factors for external lines, applied to the sum of diagrams with $n$ external lines in $\phi^4$-theory, give rise to $\hat{G}_n$, then the functional derivative relation 10.1.7 holds order by order in perturbation theory.

The weakness in this argument is that the vacuum for the generating functional is the physical interacting-field vacuum, whereas the vacuum for Dyson's formula (the basis for the Feynman rules) is the fictitious free-field vacuum. Hence we do not yet know whether or not the generating functional is a sum of Feynman diagrams. The following section eliminates this weakness by presenting an argument based on Dyson's formula which includes a formally careful treatment of the vacua.

## 10.3  Green Functions and Feynman Diagrams

In this section, we establish the equality of the non-perturbative Green function defined in terms of the interacting field and the corresponding Green function of perturbation theory defined, as in the previous section, as a sum of Feynman integrals. The central principles of the demonstration are Dyson's formula, which provides expansions to compare, and the Riemann–Lebesgue Lemma, which in effect suppresses the difference between the physical and free vacua.

Let $\mathcal{H} = \mathcal{H}_0 + \mathcal{H}'$ be the physical Hamiltonian density and $H$ the corresponding Hamiltonian. (In the example of $\phi^4$-theory, $\mathcal{H}' = \lambda\phi^4/4!$.) Then the physical vacuum satisfies $H|0\rangle_P = 0$. Let $\mathcal{H}_j = \mathcal{H} - j\phi$ be the perturbed Hamiltonian density. Assume that $j(x) = 0$ for $x$ outside some space-time sphere so that the presence of $j$ does not change the physical vacuum. Apply Dyson's formula to the perturbed density by taking $-j\phi$ as the interaction. Then the interaction Hamiltonian density is simply $-j\phi_H$ where $\phi_H$ is the Heisenberg field in the unperturbed theory. Therefore

$$Z(j) = {}_P\langle 0|T \exp\left(i \int d^4x\, j(x)\phi_H(x)\right)|0\rangle_P \qquad (10.3.1)$$

$$= {}_P\langle 0|T \sum_{n=0}^{\infty} \frac{i^n}{n!} \int d^4x_1 \cdots d^4x_n\, j(x_1) \cdots j(x_n)\phi_H(x_1) \cdots \phi_H(x_n)|0\rangle_P,$$

from which we see a useful relation

**Theorem 10.3.2.** *The Green functions are related to the interacting field in the Heisenberg picture by*

$$G_n(x_1,\ldots,x_n) = {}_P\langle 0|T\phi_H(x_1) \cdots \phi_H(x_n)|0\rangle_P.$$

Now the sum of Feynman diagrams is given by Dyson's formula also, but with the interaction terms $\mathcal{H}' - j\phi$. Define $Z^F(j)$ to be the sum of Feynman diagrams. Then

$$Z^F(j) = \lim_{t_+ \to \infty} \lim_{t_- \to -\infty} \frac{1}{N} \langle 0 | T \exp\left(-i \int_{t_-}^{t_+} d^4x \, (\mathcal{H}_I - j\phi_I)\right) | 0 \rangle, \qquad (10.3.3)$$

where the vacuum energy normalization constant $N$ is given by

$$N = \lim_{t_+ \to \infty} \lim_{t_- \to -\infty} \langle 0 | T \exp\left(-i \int_{t_-}^{t_+} d^4x \, \mathcal{H}_I\right) | 0 \rangle. \qquad (10.3.4)$$

Note that the vacuum here is not the physical vacuum but the $H_0$ vacuum, and that $\phi_I$ are free fields.

**Remark** 10.3.5. Certainly $\phi_I$ is a free field, but there is a doubt whether it is correctly normalized. If we had one-particle states, we could check it, but we have not yet constructed these states. Since the free Hamiltonian is a homogeneous quadratic in $\phi_I$, the normalization of $\phi_I$ is determined by its canonical commutator with $\partial_0 \phi_I$. If we assume that the transformation from the Heisenberg to the interaction picture preserves the equal-time commutation relations, then we can deduce that $\phi_I$ is properly normalized and that its contraction $\dot{\phi_I}\phi_I$ gives rise to a correctly normalized propagator.

From this formula for $Z^F$, we use functional differentiation to obtain a formula for the Feynman Green functions $G_n^F$:

$$G_n^F(x_1, \ldots, x_n) \qquad (10.3.6)$$
$$= \lim_{t_+ \to \infty} \lim_{t_- \to -\infty} \frac{1}{N} \langle 0 | T \exp\left(-i \int_{t_-}^{t_+} d^4x \, \mathcal{H}_I\right) \phi_I(x_1) \cdots \phi_I(x_n) | 0 \rangle.$$

The right-hand side of 10.3.6 is very similar to Dyson's formula, only with $n$ extra free fields to contract. When we expand this vacuum expectation into Wick operators and then into Feynman diagrams, we therefore find that the Feynman Green function $\hat{G}_n^F$ is the sum of the Feynman diagrams with $n$ external lines multiplied by the propagators on those external lines.

Assuming that $t_1 \geq \cdots \geq t_n$, we can write out the time ordering in the Feynman Green function:

$$G_n^F(x_1, \ldots, x_n) \qquad (10.3.7)$$
$$= \lim_{t_+ \to \infty} \lim_{t_- \to -\infty} \frac{1}{N} \langle 0 | U_I(t_+, t_1)\phi_I(x_1)U_I(t_1, t_2)\phi_I(x_2) \cdots \phi_I(x_n)U_I(t_n, t_-) | 0 \rangle.$$

Now the relationship between the Schrödinger, Heisenberg, and interaction pictures implies

$$U_I(0, t)\phi_I(x)U_I(t, 0) = U(0, t)e^{-iH_0 t}\phi_I(x)e^{iH_0 t}U(t, 0)$$
$$= U(0, t)\phi_S(0, \bar{x})U(t, 0) \qquad (10.3.8)$$
$$= \phi_H(x),$$

and so the Feynman Green function is

$$G_n^F(x_1, \ldots, x_n) \tag{10.3.9}$$

$$= \lim_{t_+ \to \infty} \lim_{t_- \to -\infty} \frac{1}{N} \langle 0 | U_I(t_+, 0) \phi_H(x_1) \cdots \phi_H(x_n) U_I(0, t_-) | 0 \rangle$$

$$= \lim_{t_+ \to \infty} \lim_{t_- \to -\infty} \frac{\langle 0 | U_I(t_+, 0) \phi_H(x_1) \cdots \phi_H(x_n) U_I(0, t_-) | 0 \rangle}{\langle 0 | U_I(t_+, 0) U_I(0, t_-) | 0 \rangle}.$$

We can evaluate the limits by inserting a complete set of $H$ eigenstates between the unitary operators and the fields. Working on the ket $U_I(0, t_-)|0\rangle$, for example, we find:

$$\lim_{t \to -\infty} \langle \psi | U_I(0, t) | 0 \rangle = \lim_{t \to -\infty} \langle \psi | U(0, t) | 0 \rangle \quad \text{because } H_0 | 0 \rangle = 0$$

$$= \lim_{t \to -\infty} \langle \psi | e^{iHt} | 0 \rangle \tag{10.3.10}$$

$$= \lim_{t \to -\infty} \left( \langle \psi | 0 \rangle_P \, _P\langle 0 | 0 \rangle + \sum_n \int d\mu(n) \, \langle \psi | n \rangle_{PP} \langle n | 0 \rangle e^{iE_n t} \right)$$

$$= \langle \psi | 0 \rangle_{PP} \langle 0 | 0 \rangle,$$

since $E_n = 0$ for at most finitely many states and (by the Riemann–Lebesgue Lemma) the limit $t \to -\infty$ eliminates all other contributions.

Applying similar logic to the bra in 10.3.9, we find that the Feynman Green function is indeed the physical Green function:

$$G_n^F(x_1, \ldots, x_n) = \frac{\langle 0 | 0 \rangle_P \, _P\langle 0 | \phi_H(x_1) \cdots \phi_H(x_n) | 0 \rangle_P \, _P\langle 0 | 0 \rangle}{\langle 0 | 0 \rangle_P \, _P\langle 0 | 0 \rangle_P \, _P\langle 0 | 0 \rangle}$$

$$= \, _P\langle 0 | \phi_H(x_1) \cdots \phi_H(x_n) | 0 \rangle_P \tag{10.3.11}$$

$$= G_n(x_1, \ldots, x_n).$$

This argument shows that we can use the Feynman diagrams to compute the Green functions. Actually, Green functions are the only physically meaningful things we can compute using Feynman diagrams; we shall see that all $S$-matrix elements can be obtained from Green functions simply by removing the propagators from the external lines and rescaling. The only problem is that our Feynman diagrams arise from vacuum expectations; without the adiabatic turning on and off function, in and out states are undefined, and so the connection between our Feynman diagrams and particle processes is not at all clear. In and out states are the topic of the next section; the link between Green functions and $S$-matrix elements is the topic of the one after that.

# 10.4 In and Out States

Since we no longer have a turning on and off function, we cannot usefully define in and out states in terms of the free field, but must construct these states from the interacting field itself. In rising to this challenge, we shall find it necessary to rescale the fields and maybe to shift their vacuum expectations. This is field renormalization. The renormalized interacting field has the same one-particle to vacuum matrix elements as the free field. This makes it possible to define particle creation operators in terms of the renormalized interacting field. These particle creation operators are, however, time dependent.

In the interests of a meaningful perturbation theory, we adjust the field by a constant to make its vacuum-to-vacuum amplitude vanish. The vacuum expectation of a scalar field is position independent:

$$\langle 0|\phi(x)|0\rangle = \langle 0|e^{ix\cdot P}\phi(0)e^{-ix\cdot P}|0\rangle = \langle 0|\phi(0)|0\rangle, \tag{10.4.1}$$

hence we write the shift as

$$\phi(x) \to \phi(x) - \langle 0|\phi(0)|0\rangle. \tag{10.4.2}$$

If the Lagrangian has a symmetry like $\phi(x) \to -\phi(x)$, then $\langle 0|\phi(0)|0\rangle$ will be zero.

**Homework** 10.4.3. Show that, because the vacuum is a Lorentz scalar, the vacuum expectations of Fermi and vector fields must be zero.

Writing $\phi$ for $\phi_H$ and $|0\rangle$ for the physical vacuum $|0\rangle_P$, assume that we shall create one-particle states from $\phi$ with the following properties:

$$\langle k|k'\rangle = (2\pi)^3 \, 2\omega(\bar{k}) \, \delta^{(3)}(\bar{k} - \bar{k}'),$$
$$P^\mu|k\rangle = k^\mu|k\rangle, \quad \text{and} \quad U(\Lambda)|k\rangle = |\Lambda k\rangle. \tag{10.4.4}$$

The relationship between the interacting field $\phi$ and the one-particle states has simple dependence on position and momentum:

$$\begin{aligned}
\langle k|\phi(x)|0\rangle &= \langle k|e^{ix\cdot P}\phi(0)e^{-ix\cdot P}|0\rangle \\
&= e^{ix\cdot k}\langle k|\phi(0)|0\rangle; \\
\langle k|\phi(0)|0\rangle &= \langle k|U(\Lambda)^\dagger \phi(0)U(\Lambda)|0\rangle \\
&= \langle \Lambda k|\phi(0)|0\rangle.
\end{aligned} \tag{10.4.5}$$

We see in these expressions a field (or wave function) renormalization factor $z_\phi$:

$$z_\phi^{1/2} \stackrel{\text{def}}{=} \langle k|\phi(0)|0\rangle. \tag{10.4.6}$$

We shall assume that the phases of the one-particle states are normalized in such a way that $z_\phi$ is a positive real number.

The field renormalization factor represents the power of the interacting field to produce single-particle states out of the vacuum. For the free field, $z_\phi = 1$. Since an interacting field can also produce many-particle states, in this case $z_\phi$ will be less than unity. To obtain the normalized particle production characteristic of the free field, we simply divide the interacting field by its field renormalization factor.

Applying both vacuum-expectation subtraction and field renormalization to an interacting field $\phi$ gives us a renormalized field $\phi'$:

$$\phi' \stackrel{\text{def}}{=} z_\phi^{-1/2}\left(\phi - \langle 0|\phi(x)|0\rangle\right). \tag{10.4.7}$$

There are just a few essential properties of the renormalized field that we use repeatedly:

**Lemma 10.4.8.** *The characteristic properties of the renormalized field are*

$$\langle 0|\phi'(x)|0\rangle = 0,$$

$$\langle 0|\phi'(x)|k\rangle = e^{-ix\cdot k} \quad\text{and}\quad \langle k|\phi'(x)|0\rangle = e^{ix\cdot k}. \qquad \square$$

Next we need the physical analogue of in and out states. The presumption is that distant particles are effectively non-interacting, so if we can define single-particle states, then we can combine them for large $|t|$ to get many-particle states. To organize separation of particles, the construction of a single particle state must use wave packets rather than plane waves. We shall label our wave packets with solutions of the Klein–Gordon equation.

For every solution $g(x)$ of the Klein–Gordon equation, there exist unique functions $c_+(\bar{k})$ and $c_-(\bar{k})$ for which

$$g(x) = \int \frac{d^3\bar{k}}{(2\pi)^3\,2\omega(\bar{k})}\left(e^{ix\cdot k}c_+(\bar{k}) + e^{-ix\cdot k}c_-(\bar{k})\right), \tag{10.4.9}$$

where $k^0 = \omega(\bar{k})$ in the exponentials. The convention is to call the solutions associated with $c_-$ and $c_+$ *positive energy* and *negative energy* solutions respectively. We shall write $f$ for a positive energy solution:

$$f(x) \stackrel{\text{def}}{=} \int \frac{d^3\bar{k}}{(2\pi)^3\,2\omega(\bar{k})}\, c(\bar{k})e^{-ix\cdot k}, \quad\text{where } k^0 = \omega(\bar{k}). \tag{10.4.10}$$

We define a localized state $|f\rangle$ by

$$|f\rangle \stackrel{\text{def}}{=} \int \frac{d^3\bar{k}}{(2\pi)^3\,2\omega(\bar{k})}\, c(\bar{k})|k\rangle, \tag{10.4.11}$$

and note the implication $\langle q|f\rangle = c(\bar{q})$.

**Homework 10.4.12.** Show that

$$c_-(\bar{k}) = 2\omega(\bar{k})\int d^3\bar{x}\, e^{ix\cdot k}g(x),$$

where $k^0 = \omega(\bar{k})$. Deduce that

$$f(x) = e^{-ix\cdot l} \quad\Longrightarrow\quad |f\rangle = |l\rangle.$$

From Remark 2.4.24, we see how to obtain a wave-packet creation operator from the free field $\phi_F$:

$$i \int d^3\bar{x} \left(\phi_F(x)\partial_0 f(x) - f(x)\partial_0 \phi_F(x)\right) = \int \frac{d^3\bar{k}}{(2\pi)^3 \, 2\omega(\bar{k})} \, c(\bar{k})a^\dagger(k). \quad (10.4.13)$$

The same construction can be used to obtain a time-dependent creation operator $\phi'(f,t)$ for the one-particle state $|f\rangle$ from the interacting field:

$$\phi'(f,t) \overset{\text{def}}{=} i \int d^3\bar{x} \left(\phi'\partial_0 f - f\partial_0 \phi'\right). \quad (10.4.14)$$

The following lemma presents the fundamental properties of this operator.

**Lemma 10.4.15.** *If $f$ is as in 10.4.10 a positive energy solution to the Klein–Gordon equation, then the operator $\phi'(f,t)$ has the following properties:*

$$\langle 0|\phi'(f,t)|0\rangle = 0,$$

$$\langle k|\phi'(f,t)|0\rangle = \langle k|f\rangle \quad \text{and} \quad \langle 0|\phi'(f,t)|k\rangle = 0.$$

**Proof.** The first assertion follows directly from the definition of $\phi'(f,t)$ and the properties of $\phi'$. The second or particle-creation property is demonstrated by direct computation:

$$\langle k|\phi'(f,t)|0\rangle = i \int d^3\bar{x} \, \langle k|(\phi'\partial_0 f - f\partial_0\phi')|0\rangle \quad (10.4.16)$$

$$= i \int d^3\bar{x} \left(e^{ix\cdot k}\partial_0 f - f\partial_0 e^{ix\cdot k}\right)$$

$$= i \int d^3\bar{x} \int \frac{d^3\bar{q}}{(2\pi)^3 \, 2\omega(\bar{q})} \left(e^{ix\cdot k}\partial_0 e^{-ix\cdot q} - e^{-ix\cdot q}\partial_0 e^{ix\cdot k}\right)\langle q|f\rangle$$

$$= \int d^3\bar{x} \int \frac{d^3\bar{q}}{(2\pi)^3 \, 2\omega(\bar{q})} \left(\omega(\bar{q}) + \omega(\bar{k})\right)e^{ix\cdot(k-q)}\langle q|f\rangle$$

$$= \int \frac{d^3\bar{q}}{2\omega(\bar{q})} \left(\omega(\bar{q}) + \omega(\bar{k})\right)e^{it(\omega(\bar{k})-\omega(\bar{q}))}\delta^{(3)}(\bar{k} - \bar{q})\langle q|f\rangle$$

$$= \langle k|f\rangle.$$

The third property follows the same line of calculation, but with $\exp(-ix\cdot k)$ instead of $\exp(ix\cdot k)$. This sign change propagates, leading to $\omega(\bar{q}) - \omega(\bar{k})$ in the penultimate line, and hence finally to a zero. $\qquad\square$

To analyze the $n$-particle content of $\phi'(f,t)|0\rangle$, let $|n\rangle$ be a state orthogonal to all one-particle states. Inserting a resolution of the identity into the crucial bracket, we find that

$$\lim_{t\to\pm\infty} \langle\psi|\phi'(f,t)|0\rangle = \lim_{t\to\pm\infty}\left(\langle\psi|0\rangle\langle 0|\phi'(f,t)|0\rangle + \int \frac{d^3\bar{k}}{(2\pi)^3 \, 2\omega(\bar{k})} \, \langle\psi|k\rangle\langle k|\phi'(f,t)|0\rangle\right.$$

$$\left. + \sum_n \int d\mu(n) \, \langle\psi|n\rangle\langle n|\phi'(f,t)|0\rangle\right). \quad (10.4.17)$$

From the properties of $\phi'(f,t)$, we see that the first term is zero and the second term reduces to $\langle\psi|f\rangle$. In the third term, because in any $r$-particle state

$$E^2 > |\bar{P}|^2 + r\mu^2, \tag{10.4.18}$$

except on the set of measure zero where all momenta are equal, and because $|n\rangle$ is a combination of many-particle states, we see that $E_n - \omega(\bar{P}_n)$ is positive. Hence

$$\langle n|\phi'(f,t)|0\rangle \propto e^{i(E_n - \omega(\bar{P}_n))t} \tag{10.4.19}$$

oscillates fast as a function of $|n\rangle$ when $|t|$ is large. Using the stationary phase asymptotic estimate, we therefore find that the third term is zero.

A similar argument applies to $\langle 0|\phi'(f,t)$, and so in summary:

**Lemma 10.4.20.** *If $f$ is as in 10.4.10 a positive energy solution of the Klein–Gordon equation, then for any state $|\psi\rangle$, the operator $\phi'(f,t)$ satisfies*

$$\lim_{t\to\pm\infty} \langle\psi|\phi'(f,t)|0\rangle = \langle\psi|f\rangle \quad and \quad \lim_{t\to\pm\infty} \langle 0|\phi'(f,t)|\psi\rangle = 0.$$

**Homework** 10.4.21. Use the facts that

$$\phi'(x) = e^{ix\cdot P}\phi'(0)e^{-ix\cdot P} \quad and \quad \partial_0\phi(x) = e^{ix\cdot P}\big[iH, \phi'(0)\big]e^{-ix\cdot P}$$

to show that

$$\langle n|\phi'(f,t)|0\rangle = e^{i(E_n - \omega(\bar{P}_n))}\langle n|\phi'(0)|0\rangle\big\langle\big((\omega(\bar{P}_n), \bar{P}_n)\big|f\big\rangle.$$

By conjugating the definition 10.4.14 of $\phi'(f,t)$, we find an expression for the wave-packet annihilation operators:

$$\phi'(f,t)^\dagger = \phi'(-f^*,t). \tag{10.4.22}$$

If $f$ is made up of positive energy solutions to the Klein–Gordon equation, then $-f^*$ is made up of negative energy solutions. The properties of $\phi'(-f^*,t)$ follow at once from Lemma 10.4.20 and confirm that $\phi'(-f^*,t)$ is an annihilation operator.

It is possible to prove that these localized particle creation operators interrelate well to create many particle states, but we will simply assume good behavior. In the Heisenberg picture the limits of the states as $t\to\pm\infty$ are well defined, and so we can use $\phi'(f,t)$ and $\phi'(f,t)^\dagger$ respectively to create one-particle states. In the one-particle case, the two limits in fact create identical states. However, when we create two-particle states, then the sign in the limit is important. Let $f$ and $g$ be disjoint wave packets,

$$f(x)g(x) = 0 \quad \text{for all } x, \tag{10.4.23}$$

then we can define and interpret two types of two-particle states $|f,g\rangle^\pm$ as follows:

**Prepared State :** $\quad \langle\psi|f,g\rangle^- \overset{\text{def}}{=} \lim_{t\to-\infty} \langle\psi|\phi'(f,t)|g\rangle;$

**Detected State :** $\quad \langle\psi|f,g\rangle^+ \overset{\text{def}}{=} \lim_{t\to\infty} \langle\psi|\phi'(f,t)|g\rangle.$

$$\tag{10.4.24}$$

We shall assume, despite the obvious physical problems with interactions, that $f$ and $g$ can be plane waves, so that the states $|f\rangle$, $|g\rangle$ and $|f,g\rangle$ are momentum eigenstates. Then the Lorentz transformation and normalization properties of one-particle states assumed at the beginning of this section can easily be proved.

**Homework 10.4.25.** Use the $\phi^4$-theory equal-time commutation relations to show that

$$[P_r, \phi'(f,t)] = \phi'(i\partial_r f, t).$$

Deduce that, when $f$ is a plane wave, $\phi'(f,t)$ creates momentum eigenstates.

**Homework 10.4.26.** By computing the commutator of $\phi'(f_1, t)$ with $\phi'(-f_2^*, t)$, show that the plane-wave limit states $|k\rangle$ are correctly normalized.

Since we are working in the Heisenberg picture, the conceptual in and out states $|f,g\rangle^{\text{in}}$ and $|f,g\rangle^{\text{out}}$ are actually physical states. We will use the labels of the physical in states for the conceptual in and out states. Thus

$$|f,g\rangle \stackrel{\text{def}}{=} |f,g\rangle^{\text{in}} \stackrel{\text{def}}{=} |f,g\rangle^{\text{out}} \stackrel{\text{def}}{=} |f,g\rangle^-. \tag{10.4.27}$$

Define the $S$ matrix as an operator from conceptual in states to conceptual out states by

$$\langle f_1, f_2 | S | g_1, g_2 \rangle = {}^+\langle f_1, f_2 | g_1, g_2 \rangle^-. \tag{10.4.28}$$

Thus in effect $S|f_1, f_2\rangle^+ = |f_1, f_2\rangle^-$. We shall assume that this definition also makes sense for plane waves:

$$\langle p_1, p_2 | S | q_1, q_2 \rangle = {}^+\langle p_1, p_2 | q_1, q_2 \rangle^-. \tag{10.4.29}$$

The conserved quantity formulae of Chapter 3 and the results of that chapter for the action of a conserved quantity on other conserved quantities, on fields, and on particle states depend only on the canonical commutation relations. Assuming that quantum dynamics do not break the conservation law, the conserved quantity is a constant and will therefore commute with the elements of calculus present in the definition 10.4.14 of $\phi'(f,t)$. We conclude that in this case the results of Chapter 3 will apply to our current in and out states.

Now that we have seen how to construct in states, out states, and the $S$ matrix from the interacting field, it remains to link $S$-matrix elements to Green functions and the perturbative style of computation.

# 10.5 The LSZ Reduction Formula for Scattering Amplitudes

Since the particle states relate well to the renormalized fields, it is natural to expect that we will need Green functions defined in terms of these renormalized fields in order to obtain the $S$-matrix elements. In light of Theorem 10.3.2 on the bare Green functions, we define the renormalized Green functions $G'_n$ by

$$G'_n(x_1, \ldots, x_n) \stackrel{\text{def}}{=} \langle 0 | T\phi'(x_1) \cdots \phi'(x_n) | 0 \rangle. \tag{10.5.1}$$

Clearly, $G'_n$ is simply $G_n$ rescaled:

$$G'_n = z_\phi^{-n/2} G_n. \tag{10.5.2}$$

In this section, we will show how to relate general matrix elements to renormalized Green functions. The essence of the relationship is an operator that can convert fields $\phi'$ into particle creation or annihilation operators $\phi'(f, t)$. Successive application of such operators can convert a vacuum expectation of a time-ordered product of fields into a matrix element of a product of fields or into an $S$-matrix element. We will see that these operators, which appear complex in position space, reduce in momentum space to the process of removing propagators from the external lines of the renormalized Green function. This general conclusion is called the LSZ reduction formula after Lehmann, Symanzik, and Zimmermann.

For any solution $f$ of the Klein–Gordon equation, the operator $\phi'(g, t)$ satisfies:

$$
\begin{aligned}
\left(\lim_{t\to\infty} - \lim_{t\to-\infty}\right)\phi'(f, t) &= \int dt\, \partial_0 \phi'(f, t) \\
&= i\int d^4x\, \left(\phi'(x)\partial_0^2 f(x) - f(x)\partial_0^2 \phi'(x)\right) \\
&= i\int d^4x\, \left(\phi'(x)(\nabla^2 - \mu^2)f(x) - f(x)\partial_0^2 \phi'(x)\right) \\
&= \int d^4x\, f(x)(-i)(\partial_\mu \partial^\mu + \mu^2)\phi'(x).
\end{aligned}
\tag{10.5.3}
$$

Because of this equality, the operators on the right should be able to (1) convert the interacting fields in the renormalized Green function 10.5.1 into creation and annihilation operators and (2) take advantage of the time ordering to put these operators into the correct place to create incoming or outgoing particles. As we shall see, this relationship between the Klein–Gordon operator and the $\phi'(f, t)$ is the essence of the connection between renormalized Green functions and the $S$ matrix.

To reduce the length of the notation, let us give a name to the transformation which converts $\phi'$ into $\phi'(f, t)$:

$$
P(f, x_r)F(x_1, \ldots, x_n) \stackrel{\text{def}}{=} i\int d^4x_r\, f(x_r)\mathbf{D}_{x_r} F(x_1, \ldots, x_n),
\tag{10.5.4}
$$

where $\mathbf{D}$ is the Klein–Gordon operator and $F$ is any function of $n$ vectors. Then the previous equation becomes

$$
P(f, x)\phi'(x) = \left(\lim_{t\to-\infty} - \lim_{t\to\infty}\right)\phi'(f, t).
\tag{10.5.5}
$$

It is not entirely trivial to apply this formula to a Green function since neither $\partial_\mu \partial^\mu + \mu^2$ nor $\int dx^0$ commutes with $T$. When we apply both the differential and the integral operators, however, then the problems cancel out. To see this, write $\mathbf{D}$ for $\partial_\mu \partial^\mu + \mu^2$, let $B_r(t)$ be any operator-valued functions of time, let $t_n \geq \cdots \geq t_1$ be any sequence of times, and write $t_{n+1}$ and $t_0$ for $+\infty$ and $-\infty$ respectively. Now write the time-ordered product in terms of $\theta$ functions:

$$
T\phi'(x)B_n(t_n)\cdots B_1(t_1)
\tag{10.5.6}
$$
$$
= \sum_{r=0}^{n} B_n(t_n)\cdots B_{r+1}(t_{r+1})\theta(t_{r+1} - x^0)\theta(x^0 - t_r)\phi(x)B_r(t_r)\cdots B_1(t_1).
$$

The operator $\mathbf{D}$ acts on the terms in this sum as follows:

$$\mathbf{D}\theta(t_{r+1} - x^0)\theta(x^0 - t_r)\phi = \theta(t_{r+1} - x^0)\theta(x^0 - t_r)\mathbf{D}\phi'$$
$$- 2\big(\delta(t_{r+1} - x^0) - \delta(x^0 - t_r)\big)\partial_0\phi' \qquad (10.5.7)$$
$$+ \big(\delta'(t_{r+1} - x^0) + \delta'(x^0 - t_r)\big)\phi',$$

where a delta function or its derivative with an infinite argument is understood to be zero. This shows how much $\mathbf{D}$ fails to commute with time-ordering.

The integrals of these terms work out simply:

$$i\int d^4x\, f(x)\theta(t_{r+1} - x^0)\theta(x^0 - t_r)\mathbf{D}\phi(x) = \phi'(f, t_r) - \phi'(f, t_{r+1}),$$

$$i\int d^4x\, f(x)\big(2\delta(x^0 - t_r)\partial_0\phi' + \delta'(x^0 - t_r)\phi'\big)$$
$$= i\int d^4x\, \delta(x^0 - t_r)\big(2f\partial_0\phi' - \partial_0(f\phi')\big)$$
$$= i\int d^4x\, \delta(x^0 - t_r)(f\partial_0\phi' - \phi'\partial_0 f) \qquad (10.5.8)$$
$$= -\phi'(f, t_r),$$

$$i\int d^4x\, f(x)\big(-2\delta(t_{r+1} - x_0)\partial_0\phi' + \delta'(t_{r+1} - x_0)\phi'\big) = \phi'(f, t_{r+1}).$$

Now it is clear that these two integrals cancel each other term by term except at the boundaries of the sum over $r$, where there is no delta function contribution at times $\pm\infty$. Hence the operators $P(f, x)$ in effect commute with time-ordering and the equation 10.5.5 for the action of $P(f, x)$ on $\phi'$ generalizes to cover the action of $P(f, x)$ on a time-ordered product:

**Lemma 10.5.9.** *The interaction of integration over time, differentiation with respect to time, and time-ordering are such that*

$$P(f, x)\, T\big[\phi'(x)B_k(t_k)\cdots B_1(t_1)\big]$$
$$= T\big[B_k(t_k)\cdots B_1(t_1)\big]\phi'(f, -\infty) - \phi'(f, \infty)T\big[B_k(t_k)\cdots B_1(t_1)\big]. \qquad \square$$

Thus, for example, combining the previous lemma with the creation-operator properties of $\phi'(f, t)$ established in Lemma 10.4.20, we find

$$P(f, x_r)G'(x_1, \ldots, x_n) = \langle 0|T\phi'(x_1)\cdots\widehat{\phi'(x_r)}\cdots\phi'(x_n)|f\rangle, \qquad (10.5.10)$$

where the wide hat indicates the term to omit from the product. In light of the definition 10.4.22 of the annihilation operators, taking the adjoint of this equation yields an annihilation-operator example:

$$P(f^*, x_r)G'(x_1, \ldots, x_n) = \Big(\lim_{t\to-\infty} - \lim_{t\to\infty}\Big)\langle 0|T\phi'(x_1)\cdots\phi'(f^*, x_r)\cdots\phi'(x_n)|0\rangle$$
$$= \Big(\lim_{t\to\infty} - \lim_{t\to-\infty}\Big)\langle 0|T\phi'(x_1)\cdots\phi'(-f^*, x_r)\cdots\phi'(x_n)|0\rangle$$
$$= \langle f|T\phi'(x_1)\cdots\widehat{\phi'(x_r)}\cdots\phi'(x_n)|0\rangle. \qquad (10.5.11)$$

For a further example, we apply these $P$ operators to a four-point renormalized Green function to obtain an $S$-matrix element for scattering of wave packets. Assume that $\hat{f}_1 \hat{f}_2 = \hat{f}_3 \hat{f}_4 = 0$, so that our particles are ultimately non-interacting, write $\Delta\lim$ for the difference between the two time limits, and then analyze:

$$P(f_1^*, x_1)P(f_2^*, x_2)P(f_3, x_3)P(f_4, x_4)\langle 0|T\phi'(x_1)\phi'(x_2)\phi'(x_3)\phi'(x_4)|0\rangle \quad (10.5.12)$$

$$= \Delta\lim_{t_1} \Delta\lim_{t_2} \Delta\lim_{t_3} \Delta\lim_{t_4} \langle 0|T\phi'_{f_1}(t_1)^\dagger \phi'_{f_2}(t_2)^\dagger \phi'_{f_3}(t_3)\phi'_{f_4}(t_4)|0\rangle$$

$$= \Big(\lim_{t_1\to\infty} - \lim_{t_1\to-\infty}\Big)\Big(\lim_{t_2\to\infty} - \lim_{t_2\to-\infty}\Big)\langle 0|T\phi'_{f_1}(t_1)^\dagger \phi'_{f_2}(t_2)^\dagger|f_3, f_4\rangle^-$$

$$= \Big(\lim_{t_1\to\infty} - \lim_{t_1\to-\infty}\Big)\Big(\langle f_2|\phi'_{f_1}(t_1)|f_3, f_4\rangle^- - \langle 0|\phi'_{f_1}(t_1)|\psi\rangle\Big)$$

$$= {}^+\langle f_1, f_2|f_3, f_4\rangle^- - {}^-\langle f_1, f_2|f_3, f_4\rangle^- - \langle f_1|\psi\rangle + \langle f_1|\psi\rangle$$

$$= \langle f_1, f_2|S - 1|f_3, f_4\rangle.$$

The $|\psi\rangle$ enters because it is not clear what the effect of a wave-packet annihilation operator will be on an incoming pair of wave packets.

**Homework** 10.5.13. Experiment: reduce a three-point Green function to a $S$-matrix element for particle decay. Try different orders of application of the $\Delta\lim$'s. Your results should indicate that the creation operator $\phi_f(t)$ for an unstable field $\phi$ creates different one-particle states at different times.

Taking the plane-wave limit, $f_r(x) = \exp(-ix\cdot k_r)$, we find a simpler form of this result, as sketched in the steps below. Integrating by parts in the application of $P(f, x)$, we find:

$$P(e^{-ix\cdot k}, x)G'(x) = i\int d^4x\, e^{-ix\cdot k}\mathbf{D}G'(x)$$

$$= i(-k^2 + \mu^2)\int d^4x\, e^{-ix\cdot k}G'(x) \quad (10.5.14)$$

$$= (-i)(k^2 - \mu^2)\hat{G}'(k).$$

Conjugating this result and multiplying by $-1$, we find a similar expression for $f_r^*$:

$$P(e^{ix\cdot k}, x)G'(x) = (-i)(k^2 - \mu^2)\hat{G}'(-k). \quad (10.5.15)$$

Multiplying by $-1$ reverses the associated limit from Lemma 10.5.9, so that

$$\left.\begin{array}{l} P(e^{-ix\cdot k}, x) \to \big(\lim_{t\to-\infty} - \lim_{t\to\infty}\big)\phi'(e^{-ix\cdot k}; t) \\[2mm] P(e^{ix\cdot k}, x) \to \big(\lim_{t\to\infty} - \lim_{t\to-\infty}\big)\phi'(-e^{ix\cdot k}; t) \end{array}\right\} \quad (10.5.16)$$

Applying these two formulae to the four variables of our previous Green function we obtain:

$$\langle k_1, k_2|S - 1|k_3, k_4\rangle \quad (10.5.17)$$

$$= P(e^{ix_1\cdot k_1}, x_1)P(e^{ix_2\cdot k_2}, x_2)P(e^{-ix_3\cdot k_3}, x_3)P(e^{-ix_4\cdot k_4}, x_4)G'_4(x_1, x_2, x_3, x_4)$$

$$= \left(\prod_{r=1}^4 (-i)(k_r^2 - \mu^2)\right)\hat{G}'_4(-k_1, -k_2, k_3, k_4).$$

The first step, integration by parts in all four variables, is not justifiable. The final answer also does not make sense since we are multiplying an infinite quantity by zeros. The remedy is to allow the frequencies to be a little off-shell and build wave packets that are zero at space-time infinity out of mixed frequencies. Then the integration by parts is permissible, and the equality holds in the limit as the wave packets tend to on-shell plane waves. There is no benefit in probing the details. The outcome is that we can obtain any $S$-matrix element by removing the propagators from the external lines of the appropriate Green function. The final formula for the scattering amplitude is an example of the LSZ reduction formula. The obvious generalization to diverse in and out states and diverse particle types constitutes the reduction formula in full. Working with plane waves at different four-momenta avoids the possibility of interference between creation and annihilation operators which gave rise to the $|\psi\rangle$ above.

Notice that the only means of computation that we have is the Feynman diagram method, and the only physically significant things we can compute are the Green functions. Thus the importance of LSZ reduction is that it enables us to relate matrix elements of arbitrary products of fields to Green functions.

## 10.6 Change in Notation

At this stage in the development of a perturbation theory based on the interacting field, it is clear that the renormalized interacting field is the more practical entity to work with. We shall therefore change notation at this point as follows:

|  | **Bare**<br>**Interacting Field** | **Renormalized**<br>**Interacting Field** |
|---|:---:|:---:|
| **Old Notation:** | $\phi$ | $\phi'$ |
| **New Notation:** | $\phi_B$ | $\phi$ |

In short, bare fields and parameters will henceforth carry subscript B's while renormalized fields and parameters will go unadorned.

Now that we have established the non-perturbative perspective of the generating functional and Green functions, discovered how to create in and out states from the renormalized interacting field, and enjoyed the final simplicity of the LSZ reduction formula for converting Green functions to matrix elements, the question arises, how should we compute amplitudes?

There are basically two possibilities. The first is to use the bare fields and parameters and the Feynman rules derived from the bare Lagrangian density. The problem of computing the $z$ factors from these bare-theory Feynman rules is the topic of the next section. The second is to change variables in the Lagrangian density from bare fields and parameters to rescaled or renormalized fields and parameters and use the Feynman rules derived from the Lagrangian density regarded as a function of these renormalized entities. These renormalized-theory Feynman rules lead directly to the renormalized Green functions (now simply written $G$, with no prime), as we shall see. This second approach is more convenient, and so it is further developed in the remainder of this chapter and forms the basis of our treatment of renormalization.

## 10.7 Computation with Bare Fields and Parameters

In order to compute any physically significant quantity, we must use the LSZ reduction formula. The LSZ reduction formula applies to renormalized Green functions, which are derived from bare Green functions by adding a factor of $z_\phi^{-1/2}$ for each external $\phi$ line. The mode of computation that we have used so far determines the bare Green functions order by order in coupling constants. Hence, in order to make predictions, it is only necessary to have a formula for $z_\phi$.

From the definition of $z_\phi$, we can use the LSZ reduction formula to connect $z_\phi$ to the bare two-point Green function $G_B(x, y)$. We just need to apply the LSZ operator to $G_B(x, y)$ in order to transform one field into a particle creation operator:

$$G_B(x, y) \stackrel{\text{def}}{=} \langle 0|T\phi_B(x)\phi_B(y)|0\rangle$$
$$= z_\phi \langle 0|T\phi(x)\phi(y)|0\rangle \tag{10.7.1}$$

implies that

$$i \int d^4x \, e^{ik \cdot x} \mathbf{D}_x G_B(x, y) = z_\phi \big( \lim_{t \to \infty} - \lim_{t \to -\infty} \big) \langle 0|T\phi_k(t)\phi(y)|0\rangle$$
$$= z_\phi \langle k|\phi(y)|0\rangle \tag{10.7.2}$$
$$= e^{iy \cdot k} z_\phi.$$

To carry out the differentiation by $\mathbf{D}_x$, we represent $G_B(x, y)$ as the Fourier transform of $\hat{G}_B(p, q)$; we also simplify by setting $y = 0$:

$$z_\phi = i \int d^4x \, e^{ik \cdot x} \mathbf{D}_x G_B(x, 0)$$
$$= i \int d^4x \, e^{ik \cdot x} \mathbf{D}_x \int \frac{d^4p}{(2\pi)^4} \frac{d^4q}{(2\pi)^4} \, e^{ix \cdot p} e^{i0 \cdot q} \hat{G}_B(p, q)$$
$$= -i \int \frac{d^4p}{(2\pi)^4} \frac{d^4q}{(2\pi)^4} \, \delta^{(4)}(k + p)(p^2 - \mu^2) \hat{G}_B(p, q) \tag{10.7.3}$$
$$= -i \int \frac{d^4q}{(2\pi)^4} \, (k^2 - \mu^2) \hat{G}_B(-k, q).$$

As with our gesture at a proof of the LSZ reduction formula, so here too we have been loose; the last line is true in the limit $k^2 \to m^2$ and the derivation should generate this limit.

Now though $G_B(x, y)$ is a function of two variables, its Fourier transform $\hat{G}_B(p, q)$ (as we know from computing Feynman integrals) contains an energy-momentum conserving delta function. If we factor this out, what remains is the bare Feynman propagator plus all higher-order corrections as computed in the bare theory. We shall call this sum $D_B$:

$$(2\pi)^4 \delta^{(4)}(p + q) D_B(q^2) \stackrel{\text{def}}{=} \hat{G}_B(p, q). \tag{10.7.4}$$

Note that $D_B$ is a function of $q^2$ on account of Lorentz invariance.

Finally, we use this definition to eliminate the integral from the last expression for $z_\phi$ in 10.7.3 and make explicit the implied limit operator:

$$
\begin{aligned}
z_\phi &= \lim_{k^2 \to \mu^2} -i \int \frac{d^4q}{(2\pi)^4} \, (k^2 - \mu^2)(2\pi)^4 \delta^{(4)}(-k+q) D_B(q^2) \\
&= \lim_{k^2 \to \mu^2} -i(k^2 - \mu^2) D_B(k^2).
\end{aligned}
\tag{10.7.5}
$$

To compute $z_\phi$ then, we sum the invariant amplitudes for diagrams with two external lines, divide by the bare propagator, and take the limit as $k$ goes onto the mass shell.

There are some subtleties here. The mass $\mu$ in the formula for $z_\phi$ is the measured mass of a particle. It is not the bare mass $\mu_B$. To obtain predictions in the bare theory, we must specify some conditions which will determine the bare parameters as functions of measurable parameters. Such conditions are called *renormalization conditions*. An interacting theory must include a renormalization condition for each bare mass and each bare coupling parameter.

For example, the measured mass can be identified from the location of the pole in $D_B$. Hence, if we have computed $D_B$ to some order in perturbation theory, then we can set $\mu_B$ to that value which puts the pole at $\mu$. The result is that $\mu_B$ is different at different levels of perturbation theory. In general, every renormalization condition sets a function of all the bare parameters equal to a measurable parameter, and at the $n$th level of perturbation theory, these functions are generally linear in the $n$th-order part of the bare parameters. At zeroth order, $\mu_B$ is equal to $\mu$, and similar equalities will generally link the other bare parameters with measurable ones. At higher order, loop diagrams contribute, and $\mu_B - \mu$ commonly becomes infinite. To manage such infinities, we shall need some new procedures, namely regularization and renormalization.

Although we have shown how in principle to compute $z_\phi$, it is tiresome to work with bare parameters which are not only complex functions of measurable parameters but also dependent on the order of perturbation theory at which we are working. Furthermore, the diagrams involved are loop diagrams and will often generate divergent Feynman integrals. Thus we will need to cover renormalization before we can use our formula for $z_\phi$. Overall, we find that computing with bare parameters is very trying, and set up a better machinery in the next section.

One good point, however, comes from this section: from the formula 10.7.5 and the remarks which follow it, we can see that at tree level $z_\phi$ is simply 1. Hence our tradition of ignoring $z_\phi$ and renormalization when computing scattering and decay amplitudes at tree level is justified.

# 10.8 Computing with Renormalized Fields and Parameters

In this section, we shall show both from the generating functional and by analysis of Feynman diagrams that the Feynman rules of the renormalized theory directly support computation of the renormalized Green functions. To bring out the ideas without excess notation, we shall work with $\phi^4$-theory.

The bare Lagrangian density for $\phi^4$-theory is converted into the renormalized Lagrangian density as follows:

$$
\begin{aligned}
\mathcal{L} &\stackrel{\text{def}}{=} \frac{1}{2}\partial_\mu \phi_B \, \partial^\mu \phi_B - \frac{\mu_B^2}{2}\phi_B^2 - \frac{\lambda_B}{4!}\phi_B^4 \\
&= \frac{1}{2}\partial_\mu \phi \, \partial^\mu \phi - \frac{z_\phi \mu_B^2}{2}\phi^2 - \frac{z_\phi^2 \lambda_B}{4!}\phi^4 + \frac{1}{2}(z_\phi - 1)\partial_\mu \phi \, \partial^\mu \phi \\
&= \frac{1}{2}\partial_\mu \phi \, \partial^\mu \phi - \frac{\mu^2}{2}\phi^2 - \frac{\lambda}{4!}\phi^4 \\
&\quad + \frac{1}{2}(z_\phi - 1)\partial_\mu \phi \, \partial^\mu \phi - \frac{1}{2}\mu^2(z_\mu - 1)\phi^2 - \frac{1}{4!}\lambda(z_\lambda - 1)\phi^4,
\end{aligned}
\tag{10.8.1}
$$

where the physical parameters $\mu^2$ and $\lambda$ are related to the bare parameters by

$$
\mu^2 = \frac{z_\phi}{z_\mu}\mu_B^2 \quad \text{and} \quad \lambda = \frac{z_\phi^2}{z_\lambda}\lambda_B.
\tag{10.8.2}
$$

The change of variables above effectively splits the bare Lagrangian density $\mathcal{L}$ into a physical part $\mathcal{L}_{\text{ph}}$ and a counterterm part $\mathcal{L}_{\text{ct}}$ defined as follows:

$$
\begin{aligned}
\mathcal{L}_{\text{ph}} &\stackrel{\text{def}}{=} \frac{1}{2}\partial_\mu \phi \, \partial^\mu \phi - \frac{\mu^2}{2}\phi^2 - \frac{\lambda}{4!}\phi^4, \\
\mathcal{L}_{\text{ct}} &\stackrel{\text{def}}{=} \frac{1}{2}(z_\phi - 1)\partial_\mu \phi \, \partial^\mu \phi - \frac{1}{2}\mu^2(z_\mu - 1)\phi^2 - \frac{1}{4!}\lambda(z_\lambda - 1)\phi^4.
\end{aligned}
\tag{10.8.3}
$$

From the perspective of the adiabatic turning on and off function, we see that if we start with the physical density and turn the interactions on, the interactions will affect the relationship of the field to the particle states, change the mass of the field and its particle states, and introduce higher-order effects to modify the relationship between $\lambda$ and particle interactions. The counterterm density is added simply to preserve the intended physical significance of the original parameters $\phi$, $\mu$, and $\lambda$.

The bare generating functional is defined as the vacuum-to-vacuum expectation of the interacting theory augmented by the coupling $\mathcal{L}'_B = j_B \phi_B$:

$$
Z_B(j_B) \stackrel{\text{def}}{=} \langle 0|S_B(j_B)|0\rangle.
\tag{10.8.4}
$$

We define the renormalized generating functional as the vacuum-to-vacuum expectation of the renormalized interacting theory augmented with the coupling $\mathcal{L}' = j\phi$:

$$
Z(j) \stackrel{\text{def}}{=} \langle 0|S(j)|0\rangle.
\tag{10.8.5}
$$

If we set $\mathcal{L}' = \mathcal{L}'_B$, then

$$
j_B \phi_B = j\phi, \quad \text{and so} \quad j_B = z_\phi^{-1/2} j.
\tag{10.8.6}
$$

Since the bare and renormalized Lagrangian densities are now equal in value, and since rescaling the variables affects the field and its canonical momentum inversely, the bare and renormalized Hamiltonian densities are also equal, and so

$$
Z(j) = Z_B(z_\phi^{-1/2} j).
\tag{10.8.7}
$$

From this relationship between $Z$ and $Z_B$ and the relationship 10.5.2 of the bare and renormalized Green functions, it is clear that, just as functional differentiation of $Z_B(j_B)$ with respect to $j_B$ yields bare Green functions, just so functional differentiation of $Z(j)$ with respect to $j$ yields the renormalized Green functions.

Finally, we see that the argument of Section 10.3 which linked the bare Green functions to Feynman Green functions in the bare theory applies equally well to link renormalized Green functions to Feynman Green functions in the renormalized theory. We conclude that perturbation theory based on the renormalized Lagrangian density 10.8.1 will directly yield the renormalized Green functions.

It is instructive to see this same result by direct analysis of Feynman diagrams and rules. In order to complete such an analysis, it is necessary to know the Feynman rule for the kinetic counterterm in the renormalized Lagrangian density. We know from Section 9.3 that the contraction of derived fields is not covariant, but there is a possibility of associating the derivative with the vertex and thereby avoiding the need for a special propagator. For the free field, the presence of the derivative $\partial_\mu$ adds a factor of $-ip_\mu$ to the annihilation operator and a factor of $ip_\mu$ to the creation operator. This suggests adding a vertex factor of $-ip_\mu$ for the incoming momentum associated with a derived field. We shall confirm this tentative result in the next chapter when we have developed the more powerful analytical techniques of functional integral quantization. In the meantime, we propose the Feynman rule

**Vertex Rule for** $\frac{1}{2}(z_\phi - 1)\partial_\mu\phi\,\partial^\mu\phi$: $\quad i(z_\phi - 1)k^2.$ $\qquad$ (10.8.8)

Note that $1/2$ is the correct symmetry factor for this vertex and that we have left out the $2\pi$'s and delta function.

The Feynman rule for the mass counterterm is

**Vertex Rule for** $-\frac{1}{2}\mu^2(z_\mu - 1)\phi^2$: $\quad -i(z_\mu - 1)\mu^2.$ $\qquad$ (10.8.9)

Since wherever we insert a mass counterterm in a $\phi$ line, we can also insert an energy counterterm and *vice versa*, it is convenient to use the sum of the vertex rules to represent the sum of the counterterms:

$$
\begin{aligned}
V_{ct} &\overset{\text{def}}{=} -i(1 - z_\phi)k^2 + i(1 - z_\mu)\mu^2 \\
&= -i\big((k^2 - \mu^2) - z_\phi(k^2 - \mu_B^2)\big),
\end{aligned}
\qquad (10.8.10)
$$

since $z_\mu\mu^2 = z_\phi\mu_B^2$.

Write $P$ for the propagator $i/(k^2 - \mu^2)$. Then, after canceling integrals against delta functions, the sum of zero, one, two, etc. counterterm insertions in the propagator gives rise to the *effective propagator* $P'$:

$$
\begin{aligned}
P' &\overset{\text{def}}{=} P + PV_{ct}P + PV_{ct}PV_{ct}P + \cdots \\
&= P\big(1 + (V_{ct}P) + (V_{ct}P)^2 + \cdots\big) \\
&= P(1 - V_{ct}P)^{-1} \\
&= (P^{-1} - V_{ct})^{-1} \\
&= i\big(z_\phi(k^2 - \mu_B^2)\big)^{-1} \\
&= z_\phi^{-1} P_B,
\end{aligned}
\qquad (10.8.11)
$$

where $P_B$ is the propagator of the bare theory.

**Remark 10.8.12.** Such summation of divergent quantities is supported by a principle of perturbation theory; since $1 - z_\phi$ is zero at first order, its powers form a summable series. However, we are in trouble with another principle of perturbation theory; one should sum all contributions to a certain order before proceeding to higher-order contributions. In defining $P'$, however, we have formed a geometric series out of the lowest-order contribution, but neglected the loop contributions altogether. In so doing, we have effectively changed the order of summation in a divergent series. Such reorganization cannot be justified, but appears in this case to lead to a correct conclusion.

Finally, when we use this modified propagator to compute Green functions in the renormalized theory, the $z_\phi^{-1}$ in each propagator may be seen as a $z_\phi^{-1/2}$ associated with each end of every line in a diagram. The result is that the $\phi^4$ vertex factor $z_\lambda \lambda$ is modified by a factor of $z_\phi^{-2}$ and becomes $\lambda_B$, the effective propagator is $P_B$, and there is an extra factor of $z_\phi^{1/2}$ associated with each external line. Thus again we see that the computation of a Green function in the renormalized theory yields the renormalized Green function.

**Remark 10.8.13.** In Remark 10.3.5, we raised the question of the normalization of the free field $\phi_B^I$ obtained by viewing the bare interacting field in the interaction picture and proposed as a working hypothesis that the contraction $\overset{\frown}{\phi_B^I \phi_B^I}$ gives rise to the standard propagator. We now justify this hypothesis.

The basic relation between bare and renormalized Green functions is based on Theorem 10.3.2, which was derived in the Heisenberg picture:

$$\left.\begin{array}{l} G_B(x_1, \ldots, x_n) = \langle 0|T\phi_B(x_1) \cdots \phi_B(x_n)|0\rangle \\ G(x_1, \ldots, x_n) = \langle 0|T\phi(x_1) \cdots \phi(x_n)|0\rangle \end{array}\right\} \implies G = z_\phi^{-n/2} G_B. \tag{10.8.14}$$

The argument of Section 10.3 shows that these Green functions are equal to the corresponding Feynman Green functions

$$\left.\begin{array}{l} G_B^F = {}_0\langle 0|Te^{-i\int d^4x\, \mathcal{H}_B^I}\phi_B^I(x_1) \cdots \phi_B^I(x_n)|0\rangle_0 \\ G^F = {}_0\langle 0|Te^{-i\int d^4x\, \mathcal{H}^I}\phi^I(x_1) \cdots \phi^I(x_n)|0\rangle_0 \end{array}\right\} \tag{10.8.15}$$

The two interaction Hamiltonians differ by the rescaling of the field and the kinetic and mass counterterms which are only present in $\mathcal{H}^I$.

By direct analysis of Feynman diagrams and rules, we saw in this section that Feynman rules based on ordinary propagators yield an equality

$$G^F = z_\phi^{-n/2} G_B^F, \tag{10.8.16}$$

thereby confirming the consistency of our assumptions and indicating that the propagator for the bare field cannot be modified by a scale factor.

We are left with the paradox that rescaling the interacting field does not lead to rescaling the propagator. The resolution is that the two free fields $\phi_B^I$ and $\phi^I$ are not related by rescaling; they are defined by the choice of free Hamiltonian and have different mass parameters. The bare field is fictitious, a computational device. On account of its non-physical mass, its normalization cannot be computed from its relationship to the physical particles of the theory. Though we have no a priori reason for assuming that $\phi_B^I$ should have the usual normalization of a free field, we can nevertheless conclude that this assumption is necessary and consistent.

As we saw in the last section, so we have seen again in this section that our tree-level computations of the first nine chapters are valid, and can be regarded as computations in the renormalized theory where the $z$'s are all used at zeroth order and the counterterms therefore make no contribution. However, the renormalized Lagrangian density has two advantages. First, it is based on measurable parameters; the counterterm coefficients absorb all the divergent contributions of loop diagrams. Second, its Feynman diagrams and rules yield the renormalized Green functions directly.

The next step in using the renormalized Lagrangian density is to specify renormalization conditions which will determine the counterterm coefficients as functions of the physical parameters. A renormalization condition basically matches experimental values of mass and charge with predictions, which are functions of the counterterm coefficients. The simplest formulation of the renormalization conditions uses a new type of Green function, the one-particle irreducible Green functions, which we introduce next.

## 10.9 1PI Green Functions and Counterterm Renormalization

The renormalized $\phi^4$ Lagrangian density 10.8.1 contains two parameters, $m$ and $\lambda$, to be determined by experiment and three counterterm parameters, $z_\phi$, $z_m$, and $z_\lambda$, to be determined by renormalization conditions.

The first renormalization condition merely sets $G_0 = 1$. Since the perturbation expansion computes $S - 1$, this corresponds to adding a counterterm to the Hamiltonian to make the sum of vacuum bubbles zero at every order of perturbation theory. Since the Feynman integral corresponding to a disconnected diagram factorizes into the integrals for its connected components, this condition implies that any diagram with a vacuum bubble is part of a sum which evaluates to zero. Consequently, all diagrams which contain a vacuum bubble can be dropped from the perturbation expansion.

The second renormalization condition sets the sum of tadpole diagrams (diagrams with one external line) to zero. We follow the common convention of representing sums of all diagrams with a given set of external lines by a shaded circle. Thus the second renormalization condition is

$$\raisebox{-1em}{} = 0 \quad \Longleftrightarrow \quad \langle 0|\phi(x)|0\rangle = 0 \quad \Longleftrightarrow \quad G_1(x) = 0. \tag{10.9.1}$$

In general, from its very definition, the renormalized field satisfies this condition. In $\phi^4$-theory, the symmetry of the Lagrangian density under $\phi \to -\phi$ implies that even the bare field has zero vacuum expectation.

The vanishing of $G_1$ implies the vanishing of any diagram with a $G_1$ factor. For example, in $\phi^3$-theory, we have

$$\raisebox{-1em}{} = 0 \quad \Longrightarrow \quad \raisebox{-1em}{} = 0, \quad \raisebox{-1em}{} = 0, \quad \text{etc.} \tag{10.9.2}$$

In a renormalized theory, we therefore ignore all diagrams with tadpole factors.

**Homework 10.9.3.** In the $\phi^3$ example above, apply the Feynman rules to the diagrams with tadpoles to show that they give rise to integrals with $G_1$ factors.

From our experience with scattering theory based on an adiabatic turning on and off function, we can see that the pole in a propagator is at $m^2$ for bosons and $m$ for fermions, where $m$ is the physical mass of the one-particle state associated with the propagator. Similarly we can see that the normalized relationship between the free field and its one-particle states is reflected in the residue $i$ at this pole. The mass and field renormalization conditions in effect impose this pole and residue on the two-point Green function. We shall prepare for a more convenient formulation of these renormalization conditions by introducing the self-energy function.

We saw in developing the Wick expansion that the logarithm of the $S$ matrix is a sum corresponding to connected diagrams. Similarly, we can show that the logarithm of the generating functional,

$$iW(j) \stackrel{\text{def}}{=} \ln Z(j), \tag{10.9.4}$$

is the generating functional for *connected Green functions*:

$$\hat{G}_{\text{con}}(p_1,\ldots,p_n) \stackrel{\text{def}}{=} \left(-i\frac{\delta}{\delta\hat{j}(-p_1)}\right)\cdots\left(-i\frac{\delta}{\delta\hat{j}(-p_n)}\right)iW(j). \tag{10.9.5}$$

The $n$-point connected Green function is the sum of the Feynman integrals for all connected diagrams with $n$ external lines at specified momenta. The Feynman rules for connected Green functions include propagator factors for external lines.

As we have seen from numerous examples, scattering and decay diagrams are generally connected. The only notable exceptions are those hadron decays which result from the decay of a single constituent quark. We shall discuss these 'spectator processes' in Chapter 17.

The connected diagrams are themselves structured from simpler units, the one-particle irreducible diagrams defined as follows. A diagram is *one-particle irreducible* or *1PI* if it is a connected diagram which remains connected when any one internal line is cut. Topologically, this criterion indicates that the diagram is held together by loops.

To reduce a connected diagram to 1PI diagrams, simply mark each internal line which if cut would disconnect the original diagram, and then cut all marked lines. The remaining connected fragments are clearly 1PI diagrams. Furthermore, the original connected diagram may be seen as a tree diagram with these 1PI diagrams as vertices. Thus the 1PI Green functions (denoted by $\pm i\Gamma$'s) serve as vertex factors. Hence, in computing 1PI diagrams, we use the ordinary Feynman rules (no propagator factors for external lines) and drop the overall $(2\pi)^4\delta^{(4)}(P_{\text{in}})$. The functions $\pm i\Gamma$ are therefore sums of invariant amplitudes $i\mathcal{A}$.

The diagrams for Green functions, connected Green functions, and 1PI Green functions are generally of the forms

**Green Function:**

**Connected Green Function:**

**1PI Green Function:**

$$\tag{10.9.6}$$

As for generic amplitudes in Section 4.12, so for these Green functions it is generally convenient to label all lines with *incoming* momenta, and then to distinguish incoming from outgoing particles by the sign of the energy.

From the definition of the 1PI diagrams, it is clear that any diagram may be viewed as a tree diagram with 1PI diagrams for vertices. This implies that any Green function may be viewed as a sum of tree diagrams in 1PI Green functions. For example, assuming the vanishing of $G_1$, the two-point Green function may be expressed as a sum

$$\tag{10.9.7}$$

**Homework 10.9.8.** Prove this statement by showing that every diagram with two external lines and no tadpoles is a string of 1PI diagrams.

The two-point Green function is actually equal to the two-point connected Green function. This is because a disconnected diagram with two external lines is either a connected diagram plus a vacuum bubble or a sum of two tadpole (one external line) diagrams. However, on account of the first and second renormalization conditions, these disconnected diagrams sum to zero.

The two-point Green function is proportional to the *renormalized propagator* $D$ as follows:

$$(2\pi)^4 \delta^{(4)}(p+q) D(q^2) \stackrel{\text{def}}{=} G(p,q). \tag{10.9.9}$$

Since the value of $G_2$ at tree or zeroth level comes from a single line,

$$G(p,q) \stackrel{(0)}{=} \frac{i}{q^2 - \mu^2 + i\epsilon}(2\pi)^4 \delta^{(4)}(p+q), \tag{10.9.10}$$

the tree-level value of the renormalized propagator is simply the propagator:

$$D(q^2) \stackrel{(0)}{=} \frac{i}{q^2 - \mu^2 + i\epsilon}. \tag{10.9.11}$$

The renormalized propagator is the function which should have its pole and residue fixed by the field and mass renormalization conditions. However, as we show next, the introduction of a more fundamental function, the self-energy function, reduces these two conditions to vanishing of the self-energy and its first derivative at $q^2 = \mu^2$.

**Homework 10.9.12.** In a theory of a charge scalar $\psi$ interacting with a neutral scalar $\phi$ through $\mathcal{L}' = -\lambda\psi^\dagger\psi\phi$, using the fact that any Feynman diagram is a tree diagram in 1PI diagrams, write the reduced four-point Green function $G'_{\Psi\Phi\Psi\Phi}$ (no propagator factors on external lines) for $\Psi\Phi$ scattering in terms of renormalized propagators, $\Gamma_3$ and $\Gamma_4$.

**Homework 10.9.13.** Repeat the above homework for $\Psi\bar{\Psi}$ scattering.

**Homework 10.9.14.** Assume that $\lambda$ is related to $\Gamma_3$ by the renormalization condition

$$\Gamma_3(m^2, m^2, \mu^2) = \lambda.$$

Find the location and residue of the poles in the $G'_{\Psi\Phi}$ and $G'_{\Psi\bar{\Psi}}$.

**Remark 10.9.15.** The homework questions above show how the single theoretical coupling $\lambda$ occurs in two very different scattering processes. There is no simple reason for supposing that the two processes are related in this way. However, in the case of nucleon-pion scattering, experimental results confirm the predicted relationship of the two residues. This indicates that $\lambda$ is not merely a theoretical construct but actually represents an aspect of nature's functioning.

From the expansion 10.9.7 of the two-point Green function, when we write the renormalized propagator in terms of 1PI Green functions, only the two-point 1PI Green function is used. In general, a two-point 1PI Green function is called a *self-energy function*. In detail, writing $P$ for the propagator and $S$ for the self-energy function, applying the Feynman rules to the expansion 10.9.7 and dropping the overall $2\pi$'s and delta function, the renormalized propagator can be expanded as a sum

$$
\begin{aligned}
D &= P + PSP + PSPSP + \cdots \\
&= P(1 - SP)^{-1} \\
&= ((1 - SP)P^{-1})^{-1} \\
&= (P^{-1} - S)^{-1}.
\end{aligned}
\tag{10.9.16}
$$

(Note the similarity to the expansion of the effective propagator in 10.8.11.) Writing $-i\Pi(q^2)$ for the scalar self-energy function, we find that

$$
D(q^2) = \frac{i}{q^2 - \mu^2 - \Pi(q^2) + i\epsilon}.
\tag{10.9.17}
$$

From this relationship between $D$ and $\Pi$, it is clear that $D$ has a pole at $q^2 = \mu^2$ if and only if $\Pi$ has a zero there and that the residue of $D$ at this pole is $i$ if and only if the first derivative of $\Pi$ at this pole vanishes. These then are the final form of the mass and field renormalization conditions.

**Remark 10.9.18.** Though as noted above, 1PI Green functions are generally represented by $\Gamma$'s, the notation $\Pi$ is common for the two-point 1PI Green function. In Chapter 20, we will find that setting $-i\Gamma_2 = D^{-1}$ will prove convenient for defining a formal field theory with $\Gamma_2$ as the Fourier transform of the kinetic operator and the other $\Gamma$'s as vertex rules.

**Homework 10.9.19.** Show that the four-point connected Green function in $\phi^3$-theory can be expressed as a sum of four tree diagrams in 1PI Green functions and the renormalized propagator. What happens if we drop the term 'connected'?

There is no such canonical renormalization condition for the coupling factor $z_\lambda$. Hence we take $\lambda$ to be a finite parameter only indirectly determined by experiment and frame the coupling renormalization condition as an equality between $\lambda$ and $\Gamma$ evaluated at some convenient point. One common choice is the *symmetric point* defined by

$$
\textbf{Symmetric Point:} \quad p_r{\cdot}p_s = \frac{1}{3}(4\delta_{rs} - 1)\mu^2,
\tag{10.9.20}
$$

where $1 \leq r, s \leq 4$.

In summary, the renormalization conditions for $\phi^4$-theory may be written

### Renormalization Conditions

| | |
|---|---|
| **Mass:** | $\Pi(\mu^2) = 0;$ |
| **Field:** | $\left.\dfrac{\partial \Pi}{\partial q^2}\right|_{q^2 = \mu^2} = 0;$ |
| **Coupling:** | $\Gamma(\text{s.p.}) = \lambda.$ |

$$\tag{10.9.21}$$

The scattering formalism developed in these first eight sections of this chapter applies virtually unchanged to charged scalar fields. The background field $j$ is now charged, the coupling to this field is

$$\mathcal{L}' = j^\dagger \psi + \psi^\dagger j, \tag{10.9.22}$$

and the generating functional is a function of $j$ and $j^\dagger$; the LSZ reduction formula applies separately to $\psi$ and $\psi^\dagger$; the Green function for particle propagation and the Green function for anti-particle propagation differ only by an exchange of $x$ and $y$ and are therefore equal, the two renormalized propagators are equal, and the two self-energy functions are equal; and the renormalization conditions for this charged-scalar self-energy function are the same as those for the neutral scalar self-energy function. Really, the only new point is that the vacuum expectation for a charged scalar is always zero.

Renormalization conditions serve to connect theory to experiment. For convenience, renormalization conditions are framed in terms of the most fundamental components of Green functions, the 1PI Green functions. Following the division 10.8.3 of the Lagrangian density into physical and counterterm parts, the renormalization conditions, by simply imposing the naive interpretation of the parameters in the physical part on the theory as a whole, determine those unique values of the counterterm coefficients which alone can dynamically uphold this interpretation of these parameters.

Having completed this description of the renormalization of a scalar field theory, the immediate need is for an example. The next two sections serve this purpose. Then we shall extend the new perturbation theory to Fermi and vector fields.

# 10.10 Example: Field and Mass Renormalization Conditions

The precise meaning of the renormalization conditions of the last section is best made clear with an example. For simplicity, we will use $\phi^3$-theory. As for $\phi^4$-theory, we first express the bare Lagrangian density as a sum of physical and counterterm parts:

$$\mathcal{L} = \mathcal{L}_{\text{ph}} + \mathcal{L}_{\text{ct}}, \tag{10.10.1}$$

where

$$\mathcal{L}_{\text{ph}} \stackrel{\text{def}}{=} \frac{1}{2}\partial_\mu \phi\, \partial^\mu \phi - \frac{\mu^2}{2}\phi^2 - \frac{\lambda}{3!}\phi^3 \quad \text{and}$$

$$\mathcal{L}_{\text{ct}} \stackrel{\text{def}}{=} \frac{1}{2}(z_\phi - 1)\partial_\mu \phi\, \partial^\mu \phi - \frac{\mu^2}{2}(z_\mu - 1)\phi^2 - \frac{\lambda}{3!}(z_\lambda - 1)\phi^3. \tag{10.10.2}$$

Consider first the field and mass renormalization conditions. Since these are conditions on the self-energy $\Pi$, we must first compute this function. Assume that we have chosen a renormalization condition on the three-point 1PI Green function which makes $z_\lambda - 1$ vanish at zeroth order. (This condition is presented in the next section.) Then the lowest-order contribution comes from a loop diagram with $\lambda$ vertex factors and the sum of the kinetic and mass counterterms:

$$D = -\!\bigcirc\!- + -\!\!\times\!\!- \tag{10.10.3}$$

Recalling the vertex rule 10.8.8 for the kinetic counterterm, we see that the amplitude for the counterterm contribution at momentum $k$ is

$$i\mathcal{A}_{ct} = i(z_\phi - 1)k^2 - i(z_\mu - 1)\mu^2. \qquad (10.10.4)$$

Since we are intending to work to second order in $\lambda$, higher-order contributions are not included in these $z$'s. The amplitude for the loop diagram is a divergent integral:

$$i\mathcal{A}_{lp} = \frac{(-i\lambda)^2}{2} \int \frac{d^4p}{(2\pi)^4} \frac{i}{p^2 - \mu^2 + i\epsilon} \frac{i}{(p+k)^2 - \mu^2 + i\epsilon}. \qquad (10.10.5)$$

Note the symmetry factor.

Before tackling the loop integral, we bring out the pattern of renormalization in two paragraphs. As we shall see, making sense of a divergent Feynman integral introduces a regularization parameter $\Lambda$ with dimensions of mass. The self-energy function $\Pi$ should be as physically meaningful as the renormalized propagator and Green functions generally, so we assume that it is finite and independent of $\Lambda$. The $\Lambda$ dependence of the loop amplitude must therefore be absorbed by the counterterm coefficients. Consequently, we write the self-energy at second order as follows:

$$-i\Pi(k^2) \stackrel{(2)}{=} \lim_{\Lambda \to \infty} i\mathcal{A}_{lp}(k^2; \Lambda^2) + i\mathcal{A}_{ct}(k^2; \Lambda^2). \qquad (10.10.6)$$

Substituting the value found above for the counterterm contribution, we find:

$$\Pi(k^2) \stackrel{(2)}{=} \lim_{\Lambda \to \infty} -\mathcal{A}_{lp}(k^2; \Lambda^2) + \left(z_\phi(\Lambda^2) - 1\right)k^2 - \left(z_\mu(\Lambda^2) - 1\right)\mu^2. \qquad (10.10.7)$$

Application of the renormalization conditions now yields equations for the $z$'s in terms of the loop amplitude:

$$\left.\begin{array}{ll} \Pi(\mu^2) = 0 \implies & \left(z_\phi(\Lambda^2) - 1\right)\mu^2 - \left(z_\mu(\Lambda^2) - 1\right)\mu^2 = \mathcal{A}_{lp}(\mu^2; \Lambda^2) \\ \Pi'(\mu^2) = 0 \implies & \left(z_\phi(\Lambda^2) - 1\right) = \mathcal{A}'_{lp}(\mu^2; \Lambda^2) \end{array}\right\} \qquad (10.10.8)$$

where the prime means differentiation with respect to $p^2$. (Actually, the renormalization conditions only imply these equations up to terms which go to zero as $\Lambda \to \infty$.) Hence the values of the $z$'s are given by

$$z_\phi(\Lambda^2) = 1 + \mathcal{A}'_{lp}(\mu^2; \Lambda^2)$$

$$\text{and} \quad z_\mu(\Lambda^2) = 1 - \frac{1}{\mu^2}\mathcal{A}_{lp}(\mu^2; \Lambda^2) - \mathcal{A}'_{lp}(\mu^2; \Lambda^2). \qquad (10.10.9)$$

Eliminating the $z$'s from the equation for the self-energy, we find that

$$\Pi(k^2) = -\lim_{\Lambda \to \infty} \mathcal{A}_{lp}(k^2; \Lambda^2) - \mathcal{A}_{lp}(\mu^2; \Lambda^2) - (k^2 - \mu^2)\mathcal{A}'_{lp}(\mu^2; \Lambda^2). \qquad (10.10.10)$$

Thus we have to subtract the first two terms in the Taylor expansion of the amplitude from the amplitude in order to find the self-energy function. Such subtraction automatically reduces the high-energy divergence of an amplitude and is the basis for the renormalization procedure of Bogoliubov, Parasiuk, Hepp, and Zimmermann described in Chapter 20. Note that, since $\Pi$ is independent of $\Lambda$, we can drop all terms in $\mathcal{A}_{\mathrm{lp}}$ which go to zero as $\Lambda \to \infty$.

Now that we have seen the pattern of renormalization, we return to the problem of managing the divergent integral. First, before modifying the integral to make it convergent, we use the Feynman integration trick to put the integral into a standard form where the denominator is a power of $q^2 + a$ with $q$ the integration parameter and $a$ independent of $q$:

$$\frac{1}{uv} = \int_0^1 dx\, \frac{1}{\left((1-x)u + xv\right)^2} \tag{10.10.11}$$

implies that

$$\frac{1}{p^2 - \mu^2 + i\epsilon} \frac{1}{(p+k)^2 - \mu^2 + i\epsilon} \tag{10.10.12}$$

$$= \int_0^1 dx\, \left(p^2 + 2xp{\cdot}k + xk^2 - \mu^2 + i\epsilon\right)^{-2}$$

$$= \int_0^1 dx\, \left((p + xk)^2 + x(1-x)k^2 - \mu^2 + i\epsilon\right)^{-2}.$$

Next we change integration variable from $p$ to $q = p + xk$ so that the Feynman integral becomes

$$i\mathcal{A}_{\mathrm{lp}}(k^2) = \frac{\lambda^2}{2} \int_0^1 dx \int \frac{d^4q}{(2\pi)^4}\, \left(q^2 + x(1-x)k^2 - \mu^2 + i\epsilon\right)^{-2}. \tag{10.10.13}$$

Second, to make this integral converge, we shall modify it with a *Wick rotation* $q_0 \to iq_0$ and put an energy-momentum cutoff on the resulting Euclidean integral. The integral diverges in part because the integrand is constant on mass hyperboloids and in part because the integrand contains a small power of the integration variable. The Wick rotation eliminates the first cause of divergence, and the cutoff eliminates the second. At this point, the cutoff is a sensible way of giving character to the divergence, but the Wick rotation appears to be an extension of the Feynman rules. (We shall be able to justify Wick rotation in the functional integral framework.)

Writing $q_0 = iq_4$, Wick rotation transforms a Minkowski vector $q$ into a Euclidean vector $q_{\mathrm{E}} = (q_1, \ldots, q_4)$, $q^2$ into $-q_{\mathrm{E}}^2$, and $d^4q$ into $id^4q_{\mathrm{E}}$. Writing $r$ for $|q_{\mathrm{E}}|$ and $\Omega$ for the result of integrating over the angular variables in $d^4q_{\mathrm{E}}$, we find that the result of the Wick rotation and cutoff is a cutoff dependent amplitude

$$i\mathcal{A}_{\mathrm{lp}}(k^2; \Lambda) = \frac{i\Omega\lambda^2}{2(2\pi)^4} \int_0^1 dx \int_0^\Lambda dr\, r^3 \left(-r^2 + x(1-x)k^2 - \mu^2 + i\epsilon\right)^{-2}. \tag{10.10.14}$$

By considering the integral of $\exp(-r^2)$ over $\mathbf{R}^4$ in cartesian and polar coordinates, it is easy to show that $\Omega = 2\pi^2$. Changing variables from $r$ to $s = r^2$ then yields

$$\mathcal{A}_{\mathrm{lp}}(k^2; \Lambda) = \frac{\lambda^2}{32\pi^2} \int_0^1 dx \int_0^{\Lambda^2} ds\, s\left(s + \mu^2 - x(1-x)k^2 - i\epsilon\right)^{-2}. \tag{10.10.15}$$

The integral over $s$ is easily calculated. It has the form

$$\int_0^{\Lambda^2} ds \frac{s}{(s+a)^2} = \ln \frac{\Lambda^2 + a}{a} - \frac{\Lambda^2}{\Lambda^2 + a}. \tag{10.10.16}$$

Making $a$ a function of $k^2$ and writing $A(k^2; \Lambda^2)$ for the right-hand side, we find that

$$A(\mu^2; \Lambda^2) - A(k^2; \Lambda^2) = \ln\left(\frac{\Lambda^2 + a(\mu^2)}{a(\mu^2)} \frac{a(k^2)}{\Lambda^2 + a(k^2)}\right) - \frac{\Lambda^2}{\Lambda^2 + a(\mu^2)} + \frac{\Lambda^2}{\Lambda^2 + a(k^2)}$$

$$\rightarrow \ln \frac{a(k^2)}{a(\mu^2)} \quad \text{as } \Lambda \rightarrow \infty, \tag{10.10.17}$$

and

$$A'(\mu^2; \Lambda) = \frac{-x(1-x)}{\Lambda^2 + a(\mu^2)} + \frac{x(1-x)}{a(\mu^2)} - \frac{\Lambda^2}{(\Lambda^2 + a(\mu^2))^2}$$

$$\rightarrow \frac{x(1-x)}{a(\mu^2)} \quad \text{as } \Lambda \rightarrow \infty. \tag{10.10.18}$$

Restoring the integral over $x$ and the constant factors, we conclude that the self-energy to second order is

$$\Pi(k^2) \overset{(2)}{=} \frac{\lambda^2}{32\pi^2} \int_0^1 dx \ln \frac{a(k^2)}{a(\mu^2)} + (k^2 - \mu^2) \frac{x(1-x)}{a(\mu^2)} \tag{10.10.19}$$

$$\overset{(2)}{=} \frac{\lambda^2}{32\pi^2} \int_0^1 dx \ln \frac{\mu^2 - x(1-x)k^2 - i\epsilon}{\mu^2 - x(1-x)\mu^2 - i\epsilon} + (k^2 - \mu^2) \frac{x(1-x)}{\mu^2 - x(1-x)\mu^2 - i\epsilon}.$$

**Homework 10.10.20.** Show that the integrand in the formula for $\Pi$ above has a singularity if $k^2 \geq 4\mu^2$. Deduce that $\Pi$ is real if $k^2 < 4\mu^2$. If $k^2 \geq 4\mu^2$, then there is a possibility for the off-shell meson to decay into a pair of mesons; this leads to an imaginary part in $\Pi$. In this case, show that

$$\text{Im } \Pi(k^2) = -\frac{\lambda^2}{32\pi} \sqrt{1 - \frac{4\mu^2}{k^2}}.$$

(This illustrates the resolution of the meaningless pole problem discussed in Remark 5.5.9 along the lines of Remark 5.6.16.)

Suppose we forget the limit $\Lambda \rightarrow \infty$. Since the counterterms are specified at the renormalization point $p^2 = \mu^2$, the cancellation of $\Lambda$ dependence order by order in perturbation theory may only be complete at that point. In other words, as a function of $p^2$, $\Pi$ could have finite $\Lambda$ dependence. Hence the limit $\Lambda \rightarrow \infty$ is necessary. We can therefore generally afford to ignore all terms which go to zero in this limit. If, however, we are going to multiply anything by a divergent function like a $z$, then we may need these lower-order terms in order to get the finite parts of the product correct. For example,

$$(a\Lambda^2 + b + c\Lambda^{-2} + \cdots)(b' + c'\Lambda^{-2} + \cdots) = ab'\Lambda^2 + (ac' + bb') + O(\Lambda^{-2}). \tag{10.10.21}$$

If we had simply dropped the $c'$ term, the product would have had a finite error.

Having made these cautionary remarks, we shall now proceed to be cavalier with this limit, suppressing it as we have suppressed its companion $\epsilon \to 0$.

We see from this example how to use the field and mass renormalization conditions. In practice, since loop integrals often diverge, it is necessary to introduce a regularization parameter like $\Lambda$. Then the counterterm coefficients are functions of $\Lambda$ while the physically meaningful sums of diagrams like the self-energy function are independent of it. This implies that the total Lagrangian density is dependent both on the order of perturbation theory to which we are working and on the renormalization parameter. However, in the approach to renormalization developed here, at least the physical part of the Lagrangian density is constant and serves as a foundation from which the rest can be determined.

In the next section, we conclude this example by discussing the coupling renormalization condition.

# 10.11 Example: Coupling Renormalization Condition

The coupling renormalization condition for $\phi^3$-theory is a constraint on the three-point 1PI Green function $\Gamma_3$. This function depends on three external momenta subject to overall energy-momentum conservation. For a symmetrical vertex like $\phi^3$, it would be convenient to choose a symmetrical renormalization point. At first this appears impossible since the three momenta cannot be equal, but there is a trick. We know from our discussion of Lorentz-invariant functions in Section 4.11 that any invariant function of three Lorentz vectors $k_r$ can be expressed as a function of the six quadratic terms $k_r \cdot k_s$. If we assume energy-momentum conservation, then we can reduce the number of variables to the three squares $k_r^2$. Now in the case of $\Gamma_3$, the energy-momentum delta function has been factored out. So, having expressed $\Gamma_3$ as a function of the $k_r^2$, we can forget energy-momentum conservation and choose the renormalization point $k_r^2 = \mu^2$ for all $r$. The vertex renormalization condition can now be written

**Vertex Renormalization Condition:** $\qquad \Gamma_3(\mu^2, \mu^2, \mu^2) = \lambda.$ $\qquad$ (10.11.1)

The lowest-order contribution to $\Gamma_3$ from $\mathcal{L}_{\text{ph}}$ is the $\lambda$ vertex itself, while the next is the vertex loop correction:

$$D = \qquad\qquad\qquad\qquad (10.11.2)$$

The loop corrections to the external lines do not form 1PI diagrams and so are not included here. Rearranging the three vertex delta functions so as to extract the overall energy-momentum conserving delta function and then performing two of the three internal-line integrals eats up the remaining delta functions and leaves one loop integral, namely:

$$i\mathcal{A} = \frac{(-i\lambda)^3}{3!} \int \frac{d^4p}{(2\pi)^4} \frac{i}{p^2 - \mu^2} \frac{i}{(p + k_2)^2 - \mu^2} \frac{i}{(p - k_1)^2 - \mu^2}, \qquad (10.11.3)$$

where we have taken the $k_r$ to be incoming momenta and suppressed the $+i\epsilon$'s to save space.

Taking the vertex counterterm into account, we find that $\Gamma_3$ is given at third order by

$$-i\Gamma_3^{(3)} \stackrel{?}{=} -i\lambda + i\mathcal{A} - i(z_\lambda - 1)\lambda. \tag{10.11.4}$$

As in the previous section, so here the $z$ is taken to second order only. Since $\mathcal{A}$ will be finite after Wick rotation, neither $\mathcal{A}$ nor $z_\lambda$ will depend on the renormalization parameter $\Lambda$. The renormalization condition at first order in $\lambda$ sets $\Gamma_3$ equal to $\lambda$, so at *zeroth* order, $z_\lambda = 1$. At second order, we have

$$z_\lambda = 1 + \lambda^{-1}\mathcal{A}. \tag{10.11.5}$$

To proceed with the loop integral, we need the three-propagator Feynman integration trick:

$$\frac{1}{abc} = 2 \int_0^1 dx \int_0^{1-x} dy \left((1 - x - y)a + xb + yc\right)^{-3} \tag{10.11.6}$$

implies that

$$\frac{1}{p^2 - \mu^2} \frac{1}{(p + k_2)^2 - \mu^2} \frac{1}{(p - k_1)^2 - \mu^2} \tag{10.11.7}$$

$$= 2 \int_0^1 dx \int_0^{1-x} dy \left((1 - x - y)p^2 + x(p + k_2)^2 + y(p - k_1)^2 - \mu^2\right)^{-3}.$$

We can now use energy-momentum conservation to express the denominator as a function of the $k_r^2$'s:

$$(1 - x - y)p^2 + x(p + k_2)^2 + y(p - k_1)^2 - \mu^2 \tag{10.11.8}$$
$$= (p + xk_2 - yk_1)^2 - \mu^2 + yk_1^2 + xk_2^2 - (xk_2 - yk_1)^2$$
$$= (p + xk_2 - yk_1)^2 - \mu^2 + y(1 - x - y)k_1^2 + x(1 - x - y)k_2^2 + xyk_3^2.$$

Applying the Wick rotation, we find that

$$\mathcal{A}(k_1^2, k_2^2, k_3^2) = (-1)^3 \frac{\lambda^3}{3!} \frac{\pi^2}{(2\pi)^4} 2 \int_0^1 dx \int_0^{1-x} dy \int_0^\infty ds \frac{s}{(s + a)^3} \tag{10.11.9}$$

$$\text{where} \quad a = \mu^2 - y(1 - x - y)k_1^2 - x(1 - x - y)k_2^2 - xyk_3^2.$$

The integral over $s$ is easily performed,

$$\int_0^\infty ds \frac{s}{(s + a)^3} = \frac{1}{2a}, \tag{10.11.10}$$

yielding

$$\mathcal{A}(k_1^2, k_2^2, k_3^2) \tag{10.11.11}$$
$$= -\frac{\lambda^3}{96\pi^2} \int_0^1 dx \int_0^{1-x} dy \frac{1}{\mu^2 - y(1 - x - y)k_1^2 - x(1 - x - y)k_2^2 - xyk_3^2}.$$

We see at once that $\mathcal{A}$ is an analytic function of its three variables, and so we can evaluate $\mathcal{A}$ at the renormalization point $k_r^2 = \mu^2$, although this is inconsistent with energy-momentum conservation:

$$\mathcal{A}(\mu^2, \mu^2, \mu^2) = -\frac{\lambda^3}{96\pi^2\mu^2} \int_0^1 dx \int_0^{1-x} dy \, \frac{1}{1 - x - y + x^2 + xy + y^2}$$

$$\simeq -\frac{\lambda^3}{96\pi^2\mu^2} \times 0.67. \tag{10.11.12}$$

Hence $z_\lambda < 1$; the increase in the manifest coupling strength due to the self-interaction of the field at third order is compensated for by a decrease in the coupling renormalization factor.

**Homework** 10.11.13. Let $\phi$ be a neutral scalar field governed by an interaction Lagrangian density

$$\mathcal{L}' = -\frac{c}{3!}\phi^3 - \frac{d}{4!}\phi^4.$$

Find renormalization conditions for $z_c$ and $z_d$ which yield $z_c = z_d = 1$ at lowest order.

This section concludes our example on the meaning of the renormalization conditions. It shows how the Lorentz invariance of the three-point 1PI Green function can be used to select a convenient renormalization point, and extends the integral techniques of the last section to integrals with three propagator factors. The vertex renormalization factor $z_\lambda$ is finite because the associated operator has mass-dimension three, or equivalently, the coupling constant has mass-dimension one. In general, interactions have dimension four, coupling constants have dimension zero, and their renormalization factors are functions of a renormalization scale $\Lambda$.

With this understanding of the content and application of the renormalization conditions, we return to developing perturbation theory by extending our new formalism to spinor and vector fields.

# 10.12 Extending the Formalism to Dirac Fields

Before we start the extension of the LSZ reduction formula to Dirac fields, be warned! This section is primarily for theoretical completeness. The practical conclusion is that the usual Feynman formalism works as well for spinors as it does for scalars, and commonly no more need be known.

Green functions for theories with Dirac fields may conveniently be defined as vacuum expectations of time-ordered products of fields. This definition automatically provides the correct indices. The generating functional which sums the Green functions is now a functional of background Dirac fields, a $\bar{\jmath}$ for each $\psi$ and a $j$ for each $\bar{\psi}$. The connection between Green functions and perturbation theory is as before; use the Feynman formalism in the usual way but remember to include the propagators on the external lines.

Vacuum energy renormalization again as before simply involves adding a counterterm to the Hamiltonian which cancels all vacuum bubbles. This makes $G_0 = 1$. There is nothing special about the types of field involved in this.

The vacuum expectation of a Fermi field is zero since the vacuum is a scalar. This corresponds to the non-existence of tadpole graphs with an external Fermi line. Hence for each component $\psi_a$ of a Fermi field,

$$G_a(x) \stackrel{\text{def}}{=} \langle 0|\psi_a(x)|0\rangle = 0. \tag{10.12.1}$$

Field renormalization matches the one-particle absorption rate for an interacting field to that of the corresponding free field. We shall assume that we have one-particle states which are properly normalized and transform correctly under the Lorentz group, fit the field to them, and then show how to construct them from the field.

Since angular momentum is conserved, the amplitude for a bare interacting field to absorb a spin-1/2 particle at rest $p_m = (m, \bar{0})$ has the form

$$\langle 0|\psi_B(0)|p_m; \tfrac{1}{2}\rangle = au_1(p_m) + bv_1(p_m) = \begin{pmatrix} a \\ 0 \\ b \\ 0 \end{pmatrix} \tag{10.12.2}$$

in the standard representation.

**Homework** 10.12.3. Show that $\langle 0|\psi_B(0)|p_m; \tfrac{1}{2}\rangle$ transforms as a spin-1/2 spinor with respect to rotations about the $z$-axis.

If parity is conserved, then

$$\begin{aligned}
au_1(p_m) + bv_1(p_m) &= \langle 0|\psi_B(0)|p_m; \tfrac{1}{2}\rangle \\
&= \langle 0|U_P^\dagger \psi_B(0) U_P |p_m; \tfrac{1}{2}\rangle \\
&= \langle 0|\gamma_0 \psi_B(0)|p_m; \tfrac{1}{2}\rangle \\
&= \gamma_0 \left( au_1(p_m) + bv_1(p_m) \right) \\
&= au_1(p_m) - bv_1(p_m),
\end{aligned} \tag{10.12.4}$$

which implies that $b = 0$. In this case, we can renormalize the Fermi field by

$$\psi \stackrel{\text{def}}{=} z_\psi^{-1/2} \psi_B \quad \text{where} \quad z_\psi^{1/2} \stackrel{\text{def}}{=} \langle 0|\psi_B(0)|p_m; \tfrac{1}{2}\rangle. \tag{10.12.5}$$

**Homework** 10.12.6. Use the Lorentz group to show that $z_\psi$ works for both particle spins:

$$\langle 0|\psi_B(0)|p_m; -\tfrac{1}{2}\rangle = \langle 0|\psi_B(0)|p_m; \tfrac{1}{2}\rangle.$$

If parity is not conserved, then the right and left fields interact differently, like the left and right electron in the Standard Model. In this case, we can renormalize left and right fields with separate kinetic counterterms and renormalize any allowed mass terms as interactions.

Now that we have constructed renormalized Fermi fields, the new scattering formalism leading to the LSZ reduction formula is developed in the following steps.

1. Choose a classical solution $f$ of the free-field equations made up of positive energy plane waves.
2. Define a one-particle ket $|f\rangle$ in terms of this function.

3. Find the operator which converts the free field $\phi$ into the creation operator $\phi_f$ for $|f\rangle$.
4. Apply this operator to the renormalized interacting field to construct time-dependent wave-packet creation operators.
5. Integrate the time derivative of this operator and show that the resulting operator $P(f)$ effectively commutes with time ordering.
6. Apply $P(f)$ to the Green functions.
7. Take $f$ to the plane-wave limit, and deduce the LSZ reduction formula.

We shall actually skip step (2) and use the creation operator $\phi_f$ of step (3) to define the state $|f\rangle$.

The classical solutions of the Dirac equation are linear combinations of the general positive and negative frequency solutions

$$
f_+^r(x) = \int \frac{d^3\bar{p}}{(2\pi)^3\, 2\omega(\bar{p})}\, c_+(p) v^r(p) e^{ix\cdot p}
$$
$$
\text{and} \quad f_-^r(x) = \int \frac{d^3\bar{p}}{(2\pi)^3\, 2\omega(\bar{p})}\, c_-(p) v^r(p) e^{-ix\cdot p},
\tag{10.12.7}
$$

where $c_\pm(p)$ are arbitrary functions of momentum.

From the structure of the free Dirac field $\psi_F$,

$$
\psi_F(x) = \int \frac{d^3\bar{q}}{(2\pi)^3\, 2\omega(\bar{q})}\, v^r(q)\beta_r^\dagger(q)e^{ix\cdot q} + u^r(q)\alpha_r e^{-ix\cdot q},
\tag{10.12.8}
$$

we find spinor wave-packet creation and annihilation operators for both particles and anti-particles:

$$
\left.
\begin{aligned}
\bar{\psi}_F(f_-^r, t) &\overset{\text{def}}{=} \int d^3\bar{x}\, \bar{\psi}_F(x)\gamma^0 f_-^r(x) = \int \frac{d^3\bar{p}}{(2\pi)^3\, 2\omega(\bar{p})}\, c_-(p)\alpha_r^\dagger(p) \\
\psi_F(\bar{f}_+^r, t) &\overset{\text{def}}{=} \int d^3\bar{x}\, \bar{f}_+^r(x)\gamma^0 \psi_F(x) = \int \frac{d^3\bar{p}}{(2\pi)^3\, 2\omega(\bar{p})}\, c_+^\dagger(p)\beta_r^\dagger(p) \\
\psi_F(\bar{f}_-^r, t) &\overset{\text{def}}{=} \int d^3\bar{x}\, \bar{f}_-^r(x)\gamma^0 \psi_F(x) = \int \frac{d^3\bar{p}}{(2\pi)^3\, 2\omega(\bar{p})}\, c_-^\dagger(p)\alpha_r(p) \\
\bar{\psi}_F(f_+^r, t) &\overset{\text{def}}{=} \int d^3\bar{x}\, \bar{\psi}_F(x)\gamma^0 f_+^r(x) = \int \frac{d^3\bar{p}}{(2\pi)^3\, 2\omega(\bar{p})}\, c_+(p)\beta_r(p)
\end{aligned}
\right\}
\tag{10.12.9}
$$

From these formulae, we see at once what one-particle states to associate with the classical solutions $f_\pm^r$ and their adjoints $\bar{f}_\pm^r$:

$$
\begin{array}{ll}
\textbf{Outgoing Particle} \;\;\leftrightarrow \bar{f}_-^r & \textbf{Incoming Particle} \;\;\leftrightarrow f_-^r \\
\textbf{Outgoing Anti-Particle} \leftrightarrow f_+^r & \textbf{Incoming Anti-Particle} \leftrightarrow \bar{f}_+^r
\end{array}
\tag{10.12.10}
$$

We can apply these definitions to the renormalized interacting field, show that the resulting operators are indeed wave-packet creation and annihilation operators in the sense of Section 10.4, and justify the assumption made at the beginning of this section regarding the existence of one-particle states which are properly normalized and transform appropriately under the Lorentz action. We shall not go through the details.

**Homework** 10.12.11. Show that if $f$ and $g$ are solutions of the Dirac equation, then

$$\{\psi(\bar{f}, t), \bar{\psi}(g, t)\} = \int d^3\bar{x}\, \bar{f}(x)\gamma^0 g(x).$$

Derive a normalization relation for the wave-packet states.

The time derivative of these wave-packet creation and annihilation operators brings out the Dirac operator, for example:

$$
\begin{aligned}
\partial_0 \bar{\psi}(f_\pm^r, t) &= \int d^3\bar{x}\, (\partial_0 \bar{\psi})\gamma^0 f_\pm^r + \bar{\psi}(\gamma^0 \partial_0 f_\pm^r) \\
&= \int d^3\bar{x}\, (\partial_0 \bar{\psi})\gamma^0 f_\pm^r - \bar{\psi}(\gamma^j \partial_j - im) f_\pm^r \qquad (10.12.12) \\
&= i \int d^3\bar{x}\, \overline{(i\slashed{\partial} - m)\psi}\, f_\pm^r.
\end{aligned}
$$

Integrating over time brings out the operator $P(f_\pm^r, x)$ which we shall apply to the Green functions in order to create $S$-matrix elements:

$$
\begin{aligned}
P(f_\pm^r, x)\bar{\psi}(x) &\overset{\text{def}}{=} \pm i \int d^4x\, \overline{(i\slashed{\partial} - m)\psi(x)} f_\pm^r(x) \\
&= \pm(\lim_{t\to\infty} - \lim_{t\to-\infty})\bar{\psi}(f_\pm^r, t).
\end{aligned}
\qquad (10.12.13)
$$

The $\pm$ has been inserted so the annihilation option associated with $f_+^r$ has the $t \to \infty$ limit positive while the creation option associated with $f_-^r$ has the $t \to -\infty$ limit positive.

Similar reasoning applied to the $\bar{f}_\pm^r$ operators leads to the conclusion

$$
\begin{aligned}
P(\bar{f}_\pm^r, x)\psi(x) &\overset{\text{def}}{=} \pm i \int d^4x\, \bar{f}_\pm^r(x)(i\slashed{\partial} - m)\psi(x) \\
&= \pm(\lim_{t\to-\infty} - \lim_{t\to\infty})\bar{\psi}(\bar{f}_\pm^r, t).
\end{aligned}
\qquad (10.12.14)
$$

Now we must check that the $P$ operators commute with time ordering. We can indicate how the general proof should go by using the time-ordered product of two components $\psi_a$ and $\psi_b$ of Fermi fields. First, we write out the time-ordered product using Heaviside functions:

$$T\psi_a(t, \bar{x})\psi_b(u, \bar{y}) = \psi_a(t, \bar{x})\psi_b(u, \bar{y})\theta(t - u) - \psi_b(u, \bar{y})\psi_a(t, \bar{x})\theta(u - t). \quad (10.12.15)$$

Second, we compute the derivative of this with respect to time $t$:

$$
\begin{aligned}
\partial_t T\psi_a\psi_b &= (\partial_t\psi_a)\psi_b\theta(t - u) - \psi_b(\partial_t\psi_a)\theta(u - t) \\
&\quad + \psi_a\psi_b\delta(t - u) + \psi_b\psi_a\delta(u - t).
\end{aligned}
\qquad (10.12.16)
$$

Third, we multiply by any function $f$ of time, and integrate:

$$\int_{-\infty}^{\infty} dt\, f(t)\partial_t T\psi_a(x)\psi_b(y) \qquad (10.12.17)$$

$$= \left(f\psi_a\big|_u + \int_u^{\infty} dt\, f\partial_t\psi_a\right)\psi_b + \psi_b\left(f\psi_a\big|_u - \int_{-\infty}^u dt\, f\partial_t\psi_a\right).$$

Fourth, when we apply the full operator $P$ to the time-ordered product, since by 10.12.12 the integrand is a time derivative, the final integrals can be evaluated. The value at one limit of the integral cancels with the other term in its bracket, and we find for example that

$$P(\bar{f}^r_\pm, x)\, T\psi(x)\psi_b(y) = \mp\psi(\bar{f}^r_\pm, \infty)\psi_b(y) \mp \psi_b(y)\psi(\bar{f}^r_\pm, -\infty)$$
$$= \pm(\lim_{t\to-\infty} - \lim_{t\to\infty})\, T\psi(\bar{f}^r_\pm, t)\psi_b(y). \tag{10.12.18}$$

Similar reasoning applied to the $f^r_\pm$ operators yields

$$P(f^r_\pm, x)\, T\bar{\psi}(x)\psi_b(y) = \pm(\lim_{t\to\infty} - \lim_{t\to-\infty})\, T\bar{\psi}(f^r_\pm, t)\psi_b(y). \tag{10.12.19}$$

From here the LSZ reduction formula is evident: the application of the $P$ operators to a Green function will yield matrix elements between wave-packet states.

Finally, to obtain a form appropriate to momentum eigenstates, we take the limit as the functions $c_\pm(p)$ become delta functions. For example, from the definition 10.12.14 of $P(\bar{f}^r_-, x)$, integrating by parts in all four dimensions, we see that

$$P(\bar{u}^r(p)e^{ix\cdot p}, x)\, \psi(x) = -i \int d^4x\, \bar{u}^r(p)e^{ix\cdot p}(i\slashed{\partial} - m)\psi(x)$$
$$= \bar{u}^r(p)\,(-i)(\slashed{p} - m) \int d^4x\, e^{ix\cdot p}\psi(x). \tag{10.12.20}$$

The result looks silly because the $\slashed{p} - m$ will annihilate the $\bar{u}^r(p)$. Indeed, as in the scalar case, integrating by parts is not justified here. We should perform the calculation with wave packets before taking the plane-wave limit. However, we can interpret this result as follows: use the inverse propagator to cancel the pole in the integral as $p$ goes on shell and then multiply by the polarization spinor.

This yields the LSZ reduction formula for momentum eigenstates: assuming that no two particle momenta are equal, we can ignore the action of annihilation operators at $t = -\infty$ and the action of creation operators at $t = \infty$, and act on the fields in the Green function as follows to create matrix elements:

| **Desired Operator** | | **Transformation of Field** |
|---|---|---|
| $a^\dagger_r$ | $\leftrightarrow\ \ \bar{\psi}(f^r_-, -\infty)\ \ \leftrightarrow$ | $\int d^4x\, \bar{\psi}(x)e^{-ix\cdot p}(-i)(\slashed{p} - m)u^r(p)$ |
| $\beta^\dagger_r$ | $\leftrightarrow\ \ \psi(\bar{f}^r_+, -\infty)\ \ \leftrightarrow$ | $\bar{v}^r(p)(-i)(\slashed{p} - m)\int d^4x\, e^{-ix\cdot p}\psi(x)$ |
| $\alpha_r$ | $\leftrightarrow\ \ \psi(\bar{f}^r_-, \infty)\ \ \leftrightarrow$ | $\bar{u}^r(p)(-i)(\slashed{p} - m)\int d^4x\, e^{ix\cdot p}\psi(x)$ |
| $\beta_r$ | $\leftrightarrow\ \ \bar{\psi}(f^r_+, \infty)\ \ \leftrightarrow$ | $\int d^4x\, \bar{\psi}(x)e^{ix\cdot p}(-i)(\slashed{p} - m)u^r(p)$ |

$$\tag{10.12.21}$$

In the context of a Green function, the matrix multiplication would in general have to be represented by a mess of spinor indices. In applications, however, this is not necessary.

Consider the hadron decay $\Lambda \to N\pi$, where $N$ is a nucleon and $\pi$ a pion with appropriate charge. Let $p$, $q$, and $k$ be the momenta of the $\Lambda$, $N$, and $\pi$ respectively. Write the relevant three-point Green function as a $4 \times 4$ matrix,

$$G(x, y, z) \stackrel{\text{def}}{=} \langle 0|T\pi(x)N(y)\bar{\Lambda}(z)|0\rangle. \tag{10.12.22}$$

As suggested by Homework 10.5.13 on the creation operator associated with an unstable scalar field, so here we assume that the $t = \pm\infty$ states $|\Lambda\rangle^{\pm}$ are different; since the initial $\Lambda$ is distinguishable from its decay products, we must have

$$\left. \begin{aligned} {}^{+}\langle N, \pi|\Lambda\rangle^{+} &= {}^{-}\langle N, \pi|\Lambda\rangle^{-} = 0 \\ \langle N, \pi|S|\Lambda\rangle &\stackrel{\text{def}}{=} {}^{+}\langle N, \pi|\Lambda\rangle^{-} \neq 0 \end{aligned} \right\} \tag{10.12.23}$$

The matrix element for the decay is obtained from $G(x, y, z)$ as follows:

$$\langle N^{s}, q; \pi, k|S - \mathbf{1}|\Lambda^{r}, p\rangle \tag{10.12.24}$$

$$= P(u^{r}(p)e^{-iz\cdot p}, z)P(\bar{u}^{s}(q)e^{iy\cdot q}, y)P(e^{ix\cdot k}, x)G(x, y, z)$$

$$= (-i)(k^{2} - m_{\pi}^{2}) \times \bar{u}^{s}(q)(-i)(\slashed{q} - m_{N})\hat{G}(p, -q, -k)(-i)(\slashed{p} - m_{\Lambda})u^{r}(p).$$

Modulo delicate points in the management of unstable fields, this is encouraging: it shows that we can compute the $S$-matrix element by computing the $\hat{G}$, removing the propagators from the external lines, and then multiplying by the appropriate polarization spinors. Since our ordinary Feynman rules never put the propagators on the external lines in the first place, they gracefully embody this conclusion.

The theoretical conclusion of this section is simple: the LSZ reduction formula can be extended from scalar fields to Dirac fields and continues to transform Green functions into $S$-matrix elements. The practical conclusion is even simpler: the ordinary Feynman rules work for Dirac fields. Having established this link between renormalized Green functions, $S$-matrix elements, and Feynman rules, it remains to present the field and mass renormalization conditions for Dirac fields.

## 10.13  Dirac Renormalization Conditions

As for the scalar field, the field and mass renormalization conditions for a Dirac field will at first be constraints on the two-point Green function, but will be more conveniently expressed in terms of the field's self-energy function.

The two-point Green function for a Dirac field $\psi$ is defined by

$$G(x, y) \stackrel{\text{def}}{=} \langle 0|T\psi(x)\bar{\psi}(y)|0\rangle. \tag{10.13.1}$$

This definition makes $G(x, y)$ a $4 \times 4$ matrix. The LSZ reduction formula applies to the Fourier transform $\hat{G}(p, q)$. This function contains an overall energy-momentum conserving delta function. If we factor this out (along with its $(2\pi)^{4}$), then we obtain the renormalized propagator $S$:

$$(2\pi)^{4}\delta^{(4)}(p + q)S(\slashed{p}) \stackrel{\text{def}}{=} \hat{G}(p, q). \tag{10.13.2}$$

On the basis of the Feynman rules, we see that $S$ is a linear combination of powers of $\not{q}$ and the vertex matrices and use the notation $S(\not{q})$ to indicate this fact.

The Dirac self-energy function $-i\Sigma(\not{p})$ is again defined as the 1PI two-point Green function. Clearly, it is a function of $\not{p}$. The familiar device of inserting zero, one, two, and so on self-energy graphs into the propagator again expresses the renormalized propagator in terms of the self-energy function:

$$S(\not{p}) = \frac{i}{\not{p} - m - \Sigma(\not{p}) + i\epsilon}. \tag{10.13.3}$$

**Homework** 10.13.4. Show that the matrix structure does not invalidate the summation argument even if $\Sigma$ does not commute with $\not{p}$. (Formally, this amounts to showing that the scalar calculation 10.9.16 makes no assumptions regarding commutation.)

The vertex matrices may carry Lorentz indices, but in a diagram that contributes to $\Sigma$, all Lorentz indices will be contracted. Since $\sigma_{\mu\nu}p^\mu p^\nu = 0$, the general form of $\Sigma$ is given by

$$\Sigma(\not{p}) = a(p^2)\mathbf{1} + b(p^2)\not{p} + c(p^2)\gamma_5 + d(p^2)\gamma_5\not{p}. \tag{10.13.5}$$

Parity takes $\psi(x)$ into $\gamma_0\psi(\tilde{x})$, and so, from the definition of the Green function 10.13.1, the effect of parity on $\Sigma$ is

$$\Sigma(\not{p}) \xrightarrow{\text{Parity}} \gamma_0\Sigma(\tilde{\not{p}})\gamma_0. \tag{10.13.6}$$

Hence if the theory is parity invariant, then

$$\textbf{Parity} \quad \Longrightarrow \quad c(p^2) = d(p^2) = 0 \quad \Longrightarrow \quad \Sigma = a\mathbf{1} + b\not{p}. \tag{10.13.7}$$

In this case, the renormalized propagator takes the form

$$\begin{aligned} S(\not{p}) &= \frac{i}{(1-b)\not{p} - (m+a)} \\ &= \frac{i\big((1-b)\not{p} + (m+a)\big)}{(1-b)^2 p^2 - (m+a)^2} \\ &= \frac{i(\not{p} + \alpha)}{(1-b)(p^2 - \alpha^2)}, \end{aligned} \tag{10.13.8}$$

where the function $\alpha$ is defined by

$$\alpha(p^2) \overset{\text{def}}{=} \frac{m + a(p^2)}{1 - b(p^2)}. \tag{10.13.9}$$

Clearly the pole will be at $p^2 = m^2$ if we impose a mass renormalization condition

$$\alpha(m^2) = m, \quad \text{or} \quad a(m^2) + mb(m^2) = 0. \tag{10.13.10}$$

Using the condition $\alpha(m^2) = m$, we can rewrite $S$ in the environment of this pole to find the residue there:

$$S \overset{(2)}{=} \frac{i(\not{p} + \alpha)}{(1-b)(1 - 2m\alpha')(p^2 - m^2)}. \tag{10.13.11}$$

If the residue is to have the free-field value $i(\not{p} + m)$, then

$$\left(1 - b(m^2)\right)\left(1 - 2m\alpha'(m^2)\right) = 1. \tag{10.13.12}$$

Differentiating $\alpha$ from its definition and using the mass renormalization condition $\alpha(m^2) = m$, we find this simplifies to a field renormalization condition

$$2m\alpha'(m^2) + 2m^2 b'(m^2) + b(m^2) = 0. \tag{10.13.13}$$

The mass renormalization condition 10.13.10 and the field renormalization condition 10.13.13 can be succinctly expressed in terms of the self-energy $\Sigma(\not{p})$ if we think of the symbol $\not{p}$ as representing an analytic variable and substitute $\not{p}^2$ for all occurrences of $p^2$ in $\Sigma$:

$$
\begin{aligned}
\textbf{Mass Condition:} \quad & \Sigma(m) = 0; \\
\textbf{Field Condition:} \quad & \Sigma'(m) = 0.
\end{aligned}
\tag{10.13.14}
$$

**Homework** 10.13.15. Let $\mathcal{L}$ be a parity-invariant Lagrangian density in which $\psi$ is an interacting Fermi field. Show that rescaling $\psi$ leads to fermion kinetic and mass counterterms

$$\mathcal{L}_{\text{ct}}^F = (z_\psi - 1)\bar{\psi}i\not{\partial}\psi - (z_m - 1)m\bar{\psi}\psi.$$

Show that the self-energy function $\Sigma$ for $\psi$ satisfies

$$
\begin{aligned}
\Sigma(\not{k}) &\overset{(2)}{=} -\mathcal{A}(\not{k}) + (z_\psi - 1)\not{k} - (z_m - 1)m \\
&\overset{(2)}{=} -\mathcal{A}(\not{k}) + \mathcal{A}(m) + \mathcal{A}'(m)(\not{k} - m).
\end{aligned}
$$

where $\mathcal{A}$ is the amplitude for the 1PI diagram which contributes to $\Sigma$ at second order.

**Homework** 10.13.16. Building on the previous homework, let $\mathcal{L}$ be the theory defined by the interaction $\mathcal{L}' = -\lambda\bar{\psi}\psi\phi$, where $\phi$ is a neutral scalar. Compute the amplitude $\mathcal{A}(\not{k}; \Lambda^2)$ and thereby show that

$$\Sigma(\not{k}) = -\frac{\lambda^2}{16\pi^2}\int_0^1 dx\,(x\not{k} - m)\ln\frac{a(m^2, x)}{a(k^2, x)} - (\not{k} - m)\frac{m^2 x(1 - x^2)}{a(m^2, x)},$$

where

$$a(k^2, x) = x\mu^2 + (1 - x)m^2 - x(1 - x)k^2 - i\epsilon.$$

Show that $\Sigma$ is real for $k^2 \le (m+\mu)^2$ and develops an imaginary part when $k^2 > (m+\mu)^2$.

If the underlying theory is not parity invariant, then the left- and right-handed fields will have different couplings and commonly different charges. If the charges of the left and right fields do not match, then there will be no Dirac mass terms; if the charges are non-zero, there will be no Majorana mass terms either. Generally then, a parity-violating theory like the Standard Model will have no fermion mass terms, and the couplings in the theory will not generate any Feynman diagrams for such mass terms. We must therefore consider the left and right fields separately, defining two-point Green functions, renormalized propagators, self-energy functions, and renormalization conditions for each. Hence in this case our $\Sigma$ above combines the left-to-left and the right-to-right self-energy functions $\Sigma_L$ and $\Sigma_R$,

$$\Sigma_L = \bar{P}_L\Sigma P_L = P_R\Sigma P_L \quad \text{and} \quad \Sigma_R = \bar{P}_R\Sigma P_R = P_L\Sigma P_R, \tag{10.13.17}$$

will commonly have no left-right mixing terms,

$$P_L \Sigma P_L = P_R \Sigma P_R = 0, \tag{10.13.18}$$

and so has the form

**No Parity** $\implies$ $a(p^2) = c(p^2) = 0$ $\implies$ $\Sigma = b\not{p} + d\gamma_5\not{p}.$ $\tag{10.13.19}$

Finally, we note that a three-point 1PI Green function $\Gamma$ which has two external Fermi lines and one external Bose line is a $4 \times 4$ matrix. It may not be simply proportional to its lowest-order part. The specification of a renormalization condition for $\Gamma$ will depend on the theory, in particular, whether the Bose line represents a scalar or a vector particle. In general, the value of $\Gamma$ for Fermi particles will be different from its value for their anti-particles. Hence we must use Dirac projection operators to separate the two cases, thus giving a form like

$$(\not{p}' + m)\Gamma(p'^2, p^2, q^2)(\not{p} + m) = g(\not{p}' + m)M(\not{p} + m) \tag{10.13.20}$$

to the renormalization condition, where all particles are on-shell and $M$ is the coupling matrix in the interaction term.

The renormalization conditions for fermions have the same form as those for scalars. In a parity-invariant theory, we apply these conditions to Dirac fields, while in a parity-violating one, we must apply them to Weyl fields. This concludes the extension of the new scattering formalism to fermions. Now we turn our attention to vector fields.

## 10.14 Extending the Formalism to Vector Fields

The vacuum expectation of a vector field must itself be a vector. Lorentz covariance then implies that this VEV vanishes. From the perspective of Feynman diagrams, this corresponds to the non-existence of vector tadpoles.

Assume that we can create a set of one-particle states with appropriate Lorentz transformation properties. Writing $e^\varepsilon$ for the three $J_z$ eigen-polarizations of a massive vector particle with momentum $p_m = (m, \vec{0})$, we find that

$$\langle 0|A_B(0)|p_m; e^\varepsilon\rangle = ae^\varepsilon + bp_m. \tag{10.14.1}$$

Conservation of angular momentum for the case $\varepsilon = -1$ implies that $b = 0$. The Lorentz covariance of the equation implies that $a$ is independent of $\varepsilon$. Hence the renormalized vector field can be defined by rescaling:

$$A = z_A^{-1/2}A_B \quad \text{where} \quad z_A^{1/2} \stackrel{\text{def}}{=} \langle 0|A_B(0)|k; e^\varepsilon(k)\rangle. \tag{10.14.2}$$

(For the massless vector field, we simply repeat the above steps using helicity instead of angular momentum and restricting $\varepsilon$ to $\pm 1$.)

Following the steps for developing the LSZ reduction formula outlined in the section on fermions, we first identify the positive energy solutions to the field equations:

$$f_r(x) = \int \frac{d^3\bar{k}}{(2\pi)^3 \, 2\omega(\bar{k})} \, c(k)e_r(k)e^{-ix\cdot k} \quad \text{where} \quad k^0 = \omega(\bar{k}). \tag{10.14.3}$$

A general solution $g$ has the form $g = a^r f_r + b^r f_r^*$.

The creation and annihilation operators are extracted from the free vector field $A_F$ by the following transformation:

$$A_F(g, t) \stackrel{\text{def}}{=} i \int d^3\bar{x}\, A_F(x) \cdot \partial_0 g(x) - g(x) \cdot \partial_0 A_F(x)$$

$$\left. \begin{aligned} A_F(f_r, t) &= \int \frac{d^3\bar{k}}{(2\pi)^3\, 2\omega(\bar{k})}\; c(k)\alpha_r^\dagger(k) \\ A_F(-f_r^*, t) &= \int \frac{d^3\bar{k}}{(2\pi)^3\, 2\omega(\bar{k})}\; c^*(k)\alpha_r(k) \end{aligned} \right\} \qquad (10.14.4)$$

These results indicate what particle states to associate with the $f_r^\pm$'s.

Applying these transformations to the interacting field $A$ and differentiating with respect to time brings out the Klein–Gordon operator:

$$\partial_0 A(g, t) = -i \int d^3\bar{x}\, g \cdot (\Box + \mu^2)A. \qquad (10.14.5)$$

Integrating over time now yields the projection operators $P$ which convert fields in Green functions into creation or annihilation operators:

$$\begin{aligned} P(g, x)A(x) &\stackrel{\text{def}}{=} -i \int d^4x\, g \cdot (\Box + \mu^2)A \\ &= (\lim_{t \to \infty} - \lim_{t \to -\infty})A(g; t). \end{aligned} \qquad (10.14.6)$$

The application of these $P$'s to the Green functions yields the LSZ reduction formula for wave packets. Taking the plane-wave limit, we find for example that

$$P\big(e_r(p)e^{-ix \cdot k}, x\big)A(x) = \int d^4x\, e_r(k) \cdot i(k^2 - \mu^2)e^{-ix \cdot k} A(x). \qquad (10.14.7)$$

Eliminating $P$ between this and the previous equation yields the practical formulae

$$\begin{aligned} i(k^2 - \mu^2)e_r(k) \cdot \int d^4x\, e^{-ix \cdot k} A(x) &= (\lim_{t \to \infty} - \lim_{t \to -\infty})A\big(e_r(k)e^{-ix \cdot k}, t\big); \\ i(k^2 - \mu^2)e_r(k) \cdot \int d^4x\, e^{ix \cdot k} A(x) &= (\lim_{t \to -\infty} - \lim_{t \to \infty})A\big(-e_r^*(k)e^{ix \cdot k}, t\big). \end{aligned} \qquad (10.14.8)$$

The effective commutation of the $P$'s with time ordering enables us to extend these formulae to fields in time-ordered products. Application to a Green function yields the usual Feynman rules for vector external lines.

Note that the LSZ reduction formula yields more than $S$-matrix elements. It allows us to use Feynman diagrams to compute matrix elements for time-ordered products of fields. For example,

$$G_\mu(x, y, z) \stackrel{\text{def}}{=} \langle 0|T\phi(x)A_\mu(y)\phi(z)|0\rangle \qquad (10.14.9)$$

$$\implies \lim_{p^2 \to \mu^2} ie_r^\mu(p)(p^2 - \mu^2)\hat{G}_\mu(x, p, z) = \langle 0|T\phi(x)\phi(z)|p^{(r)}\rangle,$$

where $\hat{G}_\mu$ is the Fourier transform of $G_\mu$ in the second variable only, and $p^0 > 0$. The difference of a minus sign from the scalar formula comes from the $-g^{\mu\nu}$ in the Feynman propagator.

Note that the LSZ reduction formula for vector fields does not explicitly involve the inverse of the propagator. However, when $k$ is on-shell, since $k \cdot e_r(k) = 0$, the matrix factor acts as the identity on the polarization vectors:

$$\left(1 - \frac{kk^\top}{\mu^2}\right)e_r(k) = e_r(k). \tag{10.14.10}$$

The renormalized propagator $V$ for a vector field is a $4 \times 4$ matrix defined by

$$(2\pi)^4 \delta^{(4)}(p+q)V(q) \stackrel{\text{def}}{=} \hat{G}(p,q), \tag{10.14.11}$$

where

$$G(x,y) = \langle 0|TA(x)A^\top(y)|0\rangle. \tag{10.14.12}$$

Write $i\Pi$ for the sum of 1PI corrections to the vector field propagator. (The extra minus sign relative to the scalar case is again due to the $-g^{\mu\nu}$ in the vector field propagator.) This self-energy function $\Pi$ is a $4 \times 4$ matrix which has the form

$$\Pi(p) = a(p^2)\mathbf{1} + b(p^2)\frac{pp^\top}{\mu^2}. \tag{10.14.13}$$

Following the calculation 10.9.16 for summing the geometric series of corrections to the propagator, we find that

$$V(p) = (P^{-1} - i\Pi)^{-1}, \tag{10.14.14}$$

where $P$ is the propagator for the vector field. From the formula

$$\left(-\alpha\mathbf{1} + \beta\frac{pp^\top}{\mu^2}\right)^{-1} = \frac{1}{\alpha}\left(-\mathbf{1} + \frac{\beta\mu^2}{\beta p^2 - \alpha\mu^2}\frac{pp^\top}{\mu^2}\right), \tag{10.14.15}$$

we find that

$$V(p) = \frac{i}{p^2 - \mu^2 - a}\left(-\mathbf{1} + \frac{(\mu^2 + b)\mu^2}{\mu^4 + bp^2 + a\mu^2}\frac{pp^\top}{\mu^2}\right). \tag{10.14.16}$$

If $V(p)$ is to have a pole at $p^2 = \mu^2$, then we must have the mass renormalization condition $a(\mu^2) = 0$, and if the residue at this pole is to have the correct value, then we must have the field renormalization condition $a'(\mu^2) = 0$.

The curious feature of these conditions is that they place no constraint on $b$. Of course, in assessing the impact of the self-energy function on the renormalized propagator, we are breaking the fundamental rule of perturbation theory that demands complete analysis order by order. It appears natural to propose that the propagator is correct for on-shell vector particles and to deduce that the second-order contribution of the self-energy itself vanishes on-shell. This would imply that $b(\mu^2) = 0$. However, as we only have two quadratic counterterms, we cannot impose this third renormalization condition on the self-energy function. Hence, in light of the first two points, if an amplitude contributes a divergence to $b$, the massive vector theory will not be renormalizable.

The possibility of a non-vanishing $b$ even on-shell is associated with the following facts. (1) If the massive vector field is coupled to a conserved current, then the massless limit exists and the $pp^\top$ term in the propagator does not contribute to physically meaningful sums of Feynman integrals. The arbitrariness of the coefficient of the $pp^\top$ term reflects the freedom to choose a gauge in the massless limit. (2) If $b$ were to be divergent, the appropriate counterterm would be $(\partial_\mu A^\mu)^2$. Since current conservation implies $\partial_\mu A^\mu = 0$, this counterterm is only available when the current is not conserved. (3) If the current to which the massive vector field is coupled is not conserved, then the $pp^\top$ term does contribute, the high-energy behavior of the propagator fails to assist loop integrals to converge, there are infinitely many different types of divergent diagram, and renormalization fails. In summary, arbitrariness of $b$ is acceptable but divergence is not.

We cannot conveniently represent these renormalization conditions in terms of the self-energy function. It makes sense here to identify the function $a(p^2)$ as the *scalar self-energy function* and to use the notation $\Pi_S$ for it. Then the renormalization conditions take the usual form:

$$\text{Mass Condition:} \quad \Pi_S(\mu^2) = 0;$$
$$\text{Field Condition:} \quad \Pi_S'(\mu^2) = 0. \tag{10.14.17}$$

**Homework** 10.14.18. Show that the kinetic and mass counterterms for a massive vector field are

$$\mathcal{L}_{ct} = -\frac{1}{4}(z_A - 1)F_{\mu\nu}F^{\mu\nu} + \frac{1}{2}(z_\mu - 1)\mu^2 A_\mu A^\mu.$$

Using the derivative vertex rule to replace $\partial_\mu$ by $-ip_\mu$ where $p$ is the incoming momentum associated with the derived line, show that the vertex factor associated with $\mathcal{L}_{ct}$ is

$$\text{Vertex Factor:} \quad i(z_A - 1)(-p^2 g^{\mu\nu} + p^\mu p^\nu) + i(z_\mu - 1)\mu^2 g^{\mu\nu}.$$

**Homework** 10.14.19. Assume that we are working in spinor QED with a massive photon. Let $\mathcal{A} = \mathcal{A}_S \mathbf{1} + \mathcal{A}_M pp^\top$ be the amplitude for the fermion loop correction to the vector field propagator. From the previous homework we see that up to second order

$$i\Pi = i\mathcal{A} + i(z_A - 1)(-p^2\mathbf{1} + pp^\top) + i(z_\mu - 1)\mu^2\mathbf{1}.$$

Apply the renormalization conditions to conclude that

$$\left.\begin{array}{l}\Pi_S(p^2) = \mathcal{A}_S(p^2) - \mathcal{A}_S(\mu^2) - (p^2 - \mu^2)\mathcal{A}_S'(\mu^2) \\ \Pi_M(p^2) = \mathcal{A}_M(p^2) + \mathcal{A}_S'(\mu^2)\end{array}\right\}$$

**Homework** 10.14.20. Evaluate the amplitude $\mathcal{A}(p; \Lambda)$ and show that

$$\Pi_S(p^2) = \frac{\lambda^2}{2\pi^2}\left(c(p^2)p^2 + d(p^2 - \mu^2)\right) \quad \text{and} \quad \Pi_M(p^2) = -\frac{\lambda^2}{2\pi^2}\left(c(p^2) + d + \frac{1}{24}\right),$$

where the function $c(p^2)$ and constant $d$ are defined by

$$c(p^2) = \int_0^1 dx\, x(1-x)p^2 \ln\frac{m^2 - x(1-x)p^2 - i\epsilon}{m^2 - x(1-x)\mu^2 - i\epsilon}$$

$$\text{and} \quad d = \int_0^1 dx\, x^2(1-x)^2\frac{\mu^2}{m^2 - x(1-x)\mu^2 - i\epsilon}.$$

(Note that if we have shifted variables to put the denominator of a Feynman integral into the standard form $(q^2 + a)^{-n}$, then we can replace a $q^\mu q^\nu$ in the integrand with $q^2 g^{\mu\nu}/4$.)

The results of the preceding homework questions show that a pole in a Feynman diagram which contributes to $b$ in $\Pi$ can be canceled by our renormalization procedure despite the fact that $b$ is not directly the subject of a renormalization condition. We also find that the computation of the amplitude $\mathcal{A}$ for the Fermi loop correction to the vector field propagator makes no reference to the nature of the vector field, whether it is massive or massless. The same computation must arise in massless QED. This is disturbing because in this case we only have a kinetic counterterm and the divergence in $\mathcal{A}$ is not proportional to $-p^2 g^{\mu\nu} + p^\mu p^\nu$. We shall solve this problem in Chapter 19 by using dimensional regularization instead of cutoff regularization in the renormalization of QED.

The scattering and renormalization program of this chapter extends readily from scalar and spinor fields to include the massive vector field. The new feature of the vector field case is that only the scalar self-energy function is subject to renormalization conditions; if the matrix part contains a divergence, then the renormalization will fail and the theory will not be consistent.

This completes our treatment of the old field types. It remains to introduce unstable particles. In preparation for that final item, we have first a section on the analytic properties of the renormalized propagator and self-energy function.

## 10.15 Lehmann–Källén Spectral Representation

By analyzing the two-point Green function, we find an illuminating expression for the renormalized propagator. In particular, this formula sheds light on the analytic properties of $D(z)$ as a function of $z = p^2$ and enables us to prove that the imaginary part of the self-energy function $\Pi(z)$ is negative when the associated particles are unstable.

We begin by expressing the vacuum expectation of a product of renormalized scalar fields as an integral over energy. The vacuum expectation may be broken up by inserting a resolution of the identity as follows:

$$\langle 0|\phi(x)\phi(y)|0\rangle = \langle 0|\phi(x)|0\rangle\langle 0|\phi(y)|0\rangle \tag{10.15.1}$$
$$+ \int \frac{d^3\bar{k}}{(2\pi)^3 \, 2\omega(\bar{k})} \, \langle 0|\phi(x)|k\rangle\langle k|\phi(y)|0\rangle + \sum_n{}' \langle 0|\phi(x)|n\rangle\langle n|\phi(y)|0\rangle.$$

Because of the renormalization conditions, the first term in this sum is zero and the second term has the value

$$\textbf{2nd Term} = \int \frac{d^3\bar{k}}{(2\pi)^3 \, 2\omega(\bar{k})} \, e^{-ik\cdot(x-y)} \stackrel{\text{def}}{=} \Delta_+(x-y,\mu^2). \tag{10.15.2}$$

The third term can be progressively simplified by identifying and naming the contributions at different momenta. First we make the momentum dependence manifest:

$$\textbf{3rd Term} = \sum_n{}' \left|\langle n|\phi(0)|0\rangle\right|^2 e^{-iP_n\cdot(x-y)}$$
$$= \sum_n{}' \left|\langle n|\phi(0)|0\rangle\right|^2 \int d^4q\,\delta^{(4)}(P_n - q)e^{-iq\cdot(x-y)}. \tag{10.15.3}$$

Next, following Lehmann and Källén, we define the *spectral function* $\sigma(q^2)$ of $\phi$ by

$$\sigma(q^2)\theta(q^0) \overset{\text{def}}{=} {\sum_n}' \left|\langle n|\phi(0)|0\rangle\right|^2 (2\pi)^3 \delta^{(4)}(P_n - q). \qquad (10.15.4)$$

The spectral function gives the total possibility for $\phi(0)$ to produce a many-particle state at momentum $q$ from the vacuum.

Proceeding with the analysis of the third term, we find that

$$\begin{aligned}
\textbf{3rd Term} &= \int \frac{d^4q}{(2\pi)^3}\, \sigma(q^2)\theta(q^0)e^{-iq\cdot(x-y)} \\
&= \int \frac{d^4q}{(2\pi)^3} \int_0^\infty da^2\, \sigma(a^2)\delta(a^2 - q^2)\theta(q^0)e^{-iq\cdot(x-y)} \qquad (10.15.5) \\
&= \int_0^\infty da^2\, \sigma(a^2)\Delta_+(x - y, a^2).
\end{aligned}$$

Putting the parts together, we find the spectral representation of the vacuum expectation of a product of renormalized fields:

$$\langle 0|\phi(x)\phi(y)|0\rangle = \Delta_+(x - y, \mu^2) + \int_0^\infty da^2\, \sigma(a^2)\Delta_+(x - y, a^2). \qquad (10.15.6)$$

The basis for computing the spectral function perturbatively is a relation between $\langle n|\phi(0)|0\rangle$ and the Green functions. To link $\langle n|\phi(0)|0\rangle$ to Feynman rules, we start from $G_{n+1}$, apply the LSZ reduction operator to $n$ variables to create a wave-packet approximation to $\langle n|$, and then take the plane-wave limit:

$$\begin{aligned}
G(x, x_1, \ldots, x_n) &= \langle 0|T\phi(x)\phi(x_1)\cdots\phi(x_n)|0\rangle \\
&\Longrightarrow \prod_{r=1}^n P(f_r^*, x_r) \times G(x, x_1, \ldots, x_n) = \langle f_1, \ldots, f_n|\phi(x)|0\rangle \qquad (10.15.7) \\
&\Longrightarrow \prod_{r=1}^n (-i)(k_r^2 - \mu^2) \times \hat{G}(x, -k_1, \ldots, -k_n) = \langle n|\phi(x)|0\rangle.
\end{aligned}$$

Since the remaining field is no longer mixed with the others by time ordering, we now see how to write the final matrix element in terms of $\hat{G}_{n+1}$:

$$\langle n|\phi(x)|0\rangle = \prod_{r=1}^n (-i)(k_r^2 - \mu^2) \times \int \frac{d^4k}{(2\pi)^4}\, e^{ix\cdot k}\hat{G}(k, -k_1, \ldots, -k_n). \qquad (10.15.8)$$

Hence $\langle n|\phi(0)|0\rangle$ can be computed by summing Feynman diagrams with $n$ external lines at momenta $k, -k_1, \ldots, -k_n$, adding a propagator factor for the external line with momentum $k$, and then integrating over $k$.

From the definition of the spectral function, $\sigma(a^2)$ is non-negative. From the above procedure for computing $\sigma$, it is apparent that if there is a final state $\langle n|$ with energy-momentum $P_n = k$ which contributes to $\sigma$ through some diagram $D$, then more energy can be pumped through $D$ to make non-zero contributions for all energies greater than $k^0$. Hence, if $\sigma$ is non-zero at $a^2 = P_n^2$, then $\sigma$ will be non-zero for $a^2 > P_n^2$. Consequently, $\sigma$ is non-zero on a half line. (In the case of $\phi^4$-theory, the threshold is at $k^0 = 3\mu$.)

Recalling the function $\Delta(x - y, a^2)$ defined by

$$i\Delta(x - y, a^2) \stackrel{\text{def}}{=} \Delta_+(x - y, a^2) - \Delta_+(y - x, a^2), \qquad (10.15.9)$$

we see at once that the vacuum expectation of the commutator is given by

$$\langle 0 | [\phi(x), \phi(y)] | 0 \rangle = i\Delta(x - y, \mu^2) + \int_0^\infty da^2 \, i\sigma(a^2)\Delta(x - y, a^2). \qquad (10.15.10)$$

Applying $\partial_y^0$ and evaluating at $x^0 = y^0$ has the effect of turning all the functions into delta functions:

$$\partial_y^0 [\phi(x), \phi(y)] \Big|_{x^0 = y^0} = [\phi(x), \partial_y^0 \phi(y)] \Big|_{x^0 = y^0} = i z_\phi^{-1} \delta^{(3)}(\bar{x} - \bar{y}),$$
$$\partial_y^0 \Delta(x - y, a^2) \Big|_{x^0 = y^0} = \delta^{(3)}(\bar{x} - \bar{y}). \qquad (10.15.11)$$

Hence:

$$z_\phi^{-1} = 1 + \int_0^\infty da^2 \, \sigma(a^2). \qquad (10.15.12)$$

This formula shows that $0 \leq z_\phi \leq 1$, with $z_\phi = 1$ if and only if $\phi$ is a free field. Actually, for the interacting field, as we saw in Section 10.10, $z_\phi$ is a divergent function of a regularization parameter. This indicates that the integral of $\sigma$ is also divergent, and that the equality holds perturbatively subject to interpretation of $z_\phi^{-1}$ as the formal inverse of a formal power series. Non-perturbatively, we expect the integral over $a^2$ to diverge, so that $z_\phi$ will have to be zero.

The two-point Green function is related to the vacuum expectation of the product of fields as follows:

$$G(x, y) \stackrel{\text{def}}{=} \langle 0 | T\phi(x)\phi(y) | 0 \rangle \qquad (10.15.13)$$
$$= \theta(x^0 - y^0)\langle 0 | \phi(x)\phi(y) | 0 \rangle + \theta(y^0 - x^0)\langle 0 | \phi(y)\phi(x) | 0 \rangle.$$

We need to substitute the spectral representation for the vacuum expectations and take the Fourier transform of this expression. To simplify the work, note that if $\phi$ is the free field with mass $a$, then we will obtain the equality

$$\hat{G}_{\text{Free}}(p, q) = (2\pi)^4 \delta^{(4)}(p + q)\frac{i}{p^2 - a^2 + i\epsilon}. \qquad (10.15.14)$$

This enables us to deduce from 10.15.13 that

$$\int d^4x \, d^4y \, e^{-ix \cdot p} e^{-iy \cdot q} \left( \theta(x^0 - y^0)\Delta_+(x - y, a^2) + \theta(y^0 - x^0)\Delta_+(y - x, a^2) \right)$$
$$= (2\pi)^4 \delta^{(4)}(p + q)\frac{i}{p^2 - a^2 + i\epsilon}. \qquad (10.15.15)$$

If $\phi$ is now an interacting field, then we must substitute the spectral representation 10.15.6 in the formula 10.15.13 for the Green function and then take the Fourier transform. Since the integral over $a^2$ commutes with these operations, it is clear that the result is

$$D(p^2) = \frac{i}{p^2 - \mu^2 + i\epsilon} + \int_0^\infty da^2 \, \frac{i\sigma(a^2)}{p^2 - a^2 + i\epsilon}. \tag{10.15.16}$$

This expression shows that $D$ is an analytic function of $p^2$ considered as a single complex variable with domain the whole complex plane except $\mu^2$ and the real half line from the minimal energy of a possible many-particle decay up to infinity. We can absorb the $+i\epsilon$ into $p^2$, and define $D$ on the real axis by

$$D(p^2) \stackrel{\text{def}}{=} \lim_{\epsilon \to 0} D(p^2 + i\epsilon). \tag{10.15.17}$$

Notice that, if $\phi$ has unstable particles, then the minimal energy for decay is less than $\mu$, and the cut in the plane associated with the integral obliterates the isolated pole which determines $\mu$ as the mass of the particles. The next section investigates this case.

Since $\sigma$ is a real-valued function, $-iD$ has the Schwartz reflection property

$$(-iD(z))^* = -iD(z^*). \tag{10.15.18}$$

Sorting our real and imaginary parts, we find that

$$D(z^*) = -D(z)^* \quad \Longrightarrow \quad \left.\begin{array}{l} \text{Re}\, D(z^*) = -\,\text{Re}\, D(z) \\ \text{Im}\, D(z^*) = \text{Im}\, D(z) \end{array}\right\} \tag{10.15.19}$$

Taking $z = p^2 + i\epsilon$ we see that $\text{Im}\, D$ is continuous across the real axis, but that $\text{Re}\, D$ has a jump discontinuity whenever it is non-zero.

In fact, the spectral function is proportional to the discontinuity $\Delta D$ in $D$ across the real axis:

$$\Delta D(p^2) \stackrel{\text{def}}{=} \lim_{\epsilon \to 0^+} \left( D(p^2 + i\epsilon) - D(p^2 - i\epsilon) \right) \tag{10.15.20}$$

$$= \lim_{\epsilon \to 0^+} \left( \frac{i}{p^2 - \mu^2 + i\epsilon} - \frac{i}{p^2 - \mu^2 - i\epsilon} \right.$$

$$\left. + \int da^2 \, \sigma(a^2) \left( \frac{i}{p^2 - a^2 + i\epsilon} - \frac{i}{p^2 - a^2 - i\epsilon} \right) \right)$$

$$= \int da^2 \, \sigma(a^2) \times 2\pi\delta(p^2 - a^2)$$

$$= 2\pi\sigma(p^2).$$

Combining this result with the consequences 10.15.19 of the Schwartz reflection property, this implies that

$$\text{Re}\, D(p^2) = \pi\sigma(p^2). \tag{10.15.21}$$

Finally, we relate the imaginary part of the self-energy function $\Pi$ to the spectral function. As we saw above in 10.9.17, $\Pi$ satisfies

$$p^2 - \mu^2 - \Pi(p^2) = (-iD(p^2))^{-1}, \tag{10.15.22}$$

and so for real $p^2$ the imaginary part of $\Pi(p^2)$ is also proportional to $\sigma(p^2)$:

$$\operatorname{Im}\Pi(p^2) = \operatorname{Im}\big(iD(p^2)\big)^{-1} = \frac{\operatorname{Im}\big(-iD(p^2)\big)}{\big|D(p^2)\big|^2} = -\frac{\pi\sigma(p^2)}{\big|D(p^2)\big|^2}. \qquad (10.15.23)$$

From this expression, we deduce that the imaginary part of $\Pi(p^2)$ is negative on the half line where $\sigma(p^2)$ is non-zero.

The Lehmann–Källén spectral representation neatly highlights the difference between the free and interacting fields. It provides a foundation for discussion of the analytic properties of Green functions and thereby for the development of dispersion relations. For the purposes of this book, however, the main result of this section is that the imaginary part of the self-energy goes negative at the threshold for decay. We shall use this result in the next section to extend the new perturbation theory to unstable particles.

## 10.16 Extending the Formalism to Unstable Particles

The theory of perturbative canonical quantization with an adiabatic turning on and off function requires the assumption that all particles are stable. The tree-level homeworks of Chapter 5 have shown that decay rates can make sense anyway, but that the scattering amplitudes suffer from meaningless poles when an unstable-particle internal line has on-shell momentum. This section shows how to extend the new perturbation theory to unstable particles, and the next brings out some delicate points in this extension. The logic of this section will therefore be heuristic.

Since the error in measuring energy is inversely proportional to the duration of the measurement process, a short-lived particle does not have a well-defined mass. Thus the phrase 'unstable particle' is useful in proportion to the lifetime of the 'particle'. Ways of fixing masses of unstable particles are matters of convention. Different conventions can produce different masses for the same physical entity.

In our scalar nucleon-meson theory, we assumed that the meson was a stable particle. If we adjust the mass $m$ of the charged particle from $2m < \mu$ to $2m > \mu$, then the meson can decay into a nucleon anti-nucleon pair, the cut in the domain of analyticity of the renormalized propagator moves towards the origin and swamps the mass pole, and the self-energy function $\Pi(p^2)$ develops an imaginary part. These are the mathematical phenomena that correspond to the lack of a definite physical mass for the meson.

In more detail, the contribution of decay diagrams ensures that $\operatorname{Im}\Pi(k^2)$ is negative for $k^2 > 4m^2$. The counterterm that we would need to enforce our old renormalization conditions would have to have a complex part, and that would make the Hamiltonian non-hermitian. Hence we must abandon the old renormalization conditions and choose new ones; for example, one on the real part of the meson self-energy:

$$\operatorname{Re}\Pi(\mu^2) = 0 \quad \text{and} \quad \operatorname{Re}\frac{\partial\Pi}{\partial k^2}\bigg|_{k^2=\mu^2} = 0. \qquad (10.16.1)$$

This convention has the advantage of being continuous in the meson mass $\mu$.

As a consequence, the Feynman propagator for an unstable particle may not be a good approximation to its dynamics. Worse, if we try to compute the renormalized propagator perturbatively, the fact that $\Pi \neq 0$ on-shell can cause diagrams which are not 1PI to have meaningless Feynman integrals:

$$D = -\!\!\boxed{\text{1PI}}\!\!-\!\!\boxed{\text{1PI}}\!\!- \quad \Longrightarrow \quad i\mathcal{A}(D) = -i\Pi(k^2) \times \frac{i}{k^2 - \mu^2 + i\epsilon} \times -i\Pi(k^2),$$

$$\Pi(\mu^2) \neq 0 \quad \Longrightarrow \quad \lim_{k^2 \to \mu^2} i\mathcal{A}(D) \quad \text{diverges.} \tag{10.16.2}$$

Furthermore, since the sum that defines the renormalized propagator perturbatively is basically a geometric series in $\Pi/(k^2 - \mu^2)$, and since this fraction diverges as $k^2 \to \mu^2$, this geometric series is also divergent.

Since the ordinary Feynman propagator will not give a consistent perturbative theory, we must find a new propagator from the Green-function formalism. The renormalized propagator $D(k^2)$ is an analytic function of the mass $\mu$. This suggests defining the renormalized propagator in a theory with an unstable particle by analytic continuation in $\mu$ from a theory in which the particle is stable. The following items show how to arrive at a modified set of Feynman rules while avoiding the divergences noted above:

1. Compute $\Pi(k^2)$ with the usual Feynman rules.
2. Use this value of $\Pi$ to evaluate the renormalized propagator.
3. In computing any amplitude, discard all diagrams with subdiagrams which contribute to $\Pi$; use the renormalized propagator for the internal line in all other diagrams.

After $\Pi(k^2)$ has been computed, rules (2) and (3) can be implemented at lowest non-trivial order by modifying the physical and counterterm Lagrangian densities as follows:

$$\mathcal{L}_{\text{ph}} \longrightarrow \mathcal{L}_{\text{ph}} - \frac{1}{2}\Pi(\mu^2)\phi^2 \quad \text{and} \quad \mathcal{L}_{\text{ct}} \longrightarrow \mathcal{L}_{\text{ct}} + \frac{1}{2}\Pi(\mu^2)\phi^2. \tag{10.16.3}$$

This modification leaves the total Lagrangian density unaltered, but the propagator for the unstable field derived from the quadratic terms in the physical Lagrangian density has the correct pole, and the new counterterm cancels loop contributions at on-shell momenta.

Since we are using the renormalized propagator on the external lines in computing Green functions, the LSZ reduction formula should remove these propagators before putting the momenta on-shell. (If we multiply the Green function by $-i(k^2 - \mu^2)$ and move $k$ onto shell, the result will be zero because the Green function does not have a pole at $k^2 = \mu^2$.) However, since if $f$ is associated with creation of an incoming wave packet $|f\rangle$, then $-f^*$ is associated with creation of an outgoing wave packet $\langle f|$, we need to modify this rule:

$$\textbf{LSZ Factor} = \begin{cases} -i(k^2 - \mu^2 - \Pi), & \text{for } k_0 > 0; \\ -i(k^2 - \mu^2 - \Pi^*), & \text{for } k_0 < 0. \end{cases} \tag{10.16.4}$$

**Homework** 10.16.5. Find the LSZ factors for an unstable Dirac fermion.

From Homework 5.6.16 we know that the imaginary part of the forward scattering amplitude $\mathcal{A}$ for a single-particle state with energy $E$ is related to the decay width $\Gamma$ by $\text{Im}\,\mathcal{A} = E\Gamma$. The amplitude $\mathcal{A}$ is derived from the two-point Green function by application of the LSZ reduction formula. Thus, from

$$\hat{G}(k, k') = (2\pi)^4 \delta^{(4)}(k - k')D(k^2)$$
$$= (2\pi)^4 \delta^{(4)}(k - k') \times \frac{i}{k^2 - \mu^2 - \Pi(k^2)}, \qquad (10.16.6)$$

we find that

$$\langle k'|S - 1|k \rangle = (2\pi)^4 \delta^{(4)}(k - k')(-i)\big(k^2 - \mu^2 - \Pi^*(k^2)\big)\big|_{k^2=\mu^2}$$
$$= (2\pi)^4 \delta^{(4)}(k - k')i\Pi^*(\mu^2). \qquad (10.16.7)$$

Since $\Pi(\mu^2)$ is purely imaginary, we see that $\mathcal{A} = -\Pi(\mu^2)$ and $\text{Im}\,\Pi = -E\Gamma$. As we know from the previous section that $\text{Im}\,\Pi \le 0$, this conclusion verifies the correctness of our LSZ factors.

**Homework** 10.16.8. Using the interaction $\mathcal{L}' = -\lambda\psi^\dagger\psi\phi$ between a charged scalar $\psi$ and a neutral scalar $\phi$, show that to lowest non-trivial order

$$\text{Im}\,\Pi(\mu^2) = -\mu\Gamma = -\frac{\lambda^2}{16\pi}(\mu^2 - 4m^2)^{1/2}.$$

(Homework 10.10.20 effectively gives $\text{Im}\,\Pi$.)

The fact that $\Pi$ does not vanish for an unstable particle on-shell causes the perturbative definition of the renormalized propagator to become meaningless. We circumvent this obstruction to perturbation theory by modifying the Feynman rules on the basis of the analyticity of the underlying Green functions. Nevertheless, the very need to modify the Feynman rules indicates a weakness in the derivation of the Feynman rules.

In this section, we have presented a practical extension of the new scattering formalism to unstable particles. As the next section shows, however, the LSZ formalism does not extend to unstable particles, and the detailed structure of a model for them is still subject to active debate.

## 10.17 Modeling Unstable Particles

Before remarking on unstable particles in field theory, let us first review a few basic properties of unstable particle in particle mechanics.

One basic function characteristic of an unstable state is its *survival amplitude*:

$$S(t) \overset{\text{def}}{=} \langle \psi | e^{-iHt} | \psi \rangle. \qquad (10.17.1)$$

The survival probability function $P(t)$ is simply $\big|S(t)\big|^2$.

An unstable particle considered without reference to its decay products is modeled by an eigenstate $|\psi\rangle$ of $H$ for which $S(t)$ obeys an exponential decay law. This decay law implies that the eigenvalue associated with $|\psi\rangle$ must have an imaginary component:

$$H|\psi\rangle = \Big(\mu - \frac{i}{2}\tau\Big)|\psi\rangle. \qquad (10.17.2)$$

Since $S(-t)$ is the conjugate of $S(t)$, we conclude that the evolution of $|\psi\rangle$ is given by

$$|\psi(t)\rangle = \exp\left(-i\mu t - \frac{1}{2}\tau|t|\right)|\psi(0)\rangle. \tag{10.17.3}$$

The Fourier transform $\hat{\psi}$ of the exponential factor is

$$\hat{\psi}(\omega) = \frac{1}{2\pi}\frac{i}{(\omega - \mu) + \frac{i}{2}\tau} + \frac{1}{2\pi}\frac{-i}{(\omega - \mu) - \frac{i}{2}\tau} = \frac{1}{\pi}\frac{\frac{1}{2}\tau}{(\omega - \mu)^2 + \frac{1}{4}\tau^2}. \tag{10.17.4}$$

Thus $\hat{\psi}$ is the sum of two functions $\hat{\psi}_u$ and $\hat{\psi}_l$, one analytic in the upper half plane, the other analytic in the lower half plane. Because all three functions, $\hat{\psi}$, $\hat{\psi}_u$ and $\hat{\psi}_l$, depend on negative values of the energy $\omega$, they cannot be built up from physical states. Furthermore, for $\hat{\psi}_l$ and $\hat{\psi}$, the pole in the upper half plane leads to an advanced Green function, and hence violates causality.

These points indicate that a naive formulation of decay in quantum mechanics leads to a Hamiltonian which is not hermitian and violations of causality. However, if we model an unstable particle more carefully, we can easily prove deviations from the exponential decay law.

For simplicity, let us assume that we are describing a system consisting of a single unstable particle and its decay products. Let $\mathbf{H}$ and $H$ be respectively the Hilbert space and Hamiltonian for this system. Because $|\psi\rangle$ decays, firstly, it is not an element of the discrete spectrum of $H$, and secondly, it must be orthogonal to all the stable states. Consequently, $|\psi\rangle$ is associated with the continuous spectrum. This implies that there is a continuous function $\rho(E)$ with the property that $\rho(E)\,\delta E$ is the probability that $|\psi\rangle$ has energy in the range $[E, E + \delta E]$. Thus

$$\rho(E) = \langle\psi|E\rangle\langle E|\psi\rangle, \tag{10.17.5}$$

where $|E\rangle$ represents the eigenstates of the Hamiltonian with energy $E$. Clearly, $\rho(E)$ is zero outside the continuous spectrum of $H$.

The survival amplitude $S$ is simply the Fourier transform of the energy distribution function $\rho$:

$$S(t) = \langle\psi|e^{-iHt}|\psi\rangle = \int dE\, e^{-iEt}\rho(E). \tag{10.17.6}$$

We know that $H$ is bounded below, and so $\rho$ vanishes for all sufficiently large negative energies. According to the Paley–Wiener Theorem, it follows that the Fourier transform $S$ of $\rho$ vanishes as $t \to \infty$ more slowly than any exponential decay $e^{-at}$. The same conclusion holds for the survival probability function $P(t)$, of course.

Since the expectation of energy for $|\psi\rangle$ is finite, the derivative of the survival amplitude exists and is continuous:

$$|\dot{S}(t)| = \left|\int dE - iEe^{-iEt}\rho(E)\right| \leq \int dE\,|E|\rho(E). \tag{10.17.7}$$

Since $S(t)^* = S(-t)$, we find that the survival probability function $P(t)$ satisfies

$$\dot{P}(t) = \frac{d}{dt}\big(S(t)S(-t)\big) = \dot{S}(t)S(-t) - S(t)\dot{S}(-t), \tag{10.17.8}$$

and so, in particular,

$$\dot{P}(0) = 0. \tag{10.17.9}$$

Obviously, this implies that the survival probability is larger than any exponential decay function for sufficiently small positive times.

**Homework 10.17.10.** Suppose that the unstable state is observed $n$ times at intervals of $\Delta t = t/n$. Show that, in the limit as $n \to \infty$, the probability for the particle to decay becomes zero.

**Remark 10.17.11.** Deviations from exponential decay for small and large times are not experimentally accessible for the following reasons. In simple models for decaying particles, it can be shown that the effective limits of the exponential decay region are roughly at

$$t_1 \sim 10^{-14}\Gamma \quad \text{and} \quad t_2 \sim 190/\Gamma. \tag{10.17.12}$$

Taking $\pi \to \mu\bar{\nu}$ for example, $\tau \sim 3 \times 10^{-8}$ sec and $\Gamma \sim 3 \times 10^7 \text{sec}^{-1}$. This makes $t_1 \sim 10^{-21}$ sec and $t_2 \sim 6 \times 10^{-6}$ sec. Clearly, $t_1$ is inaccessibly small, while $t_2$ is so much larger than the half life that effectively no pions are left.

Because the main features of the above analysis of unstable particles are so fundamental, we expect them to be present in field theory also. In particular, we do not expect an unstable particle to have a well-defined mass. Consequently, an unstable particle cannot be accurately represented by a ket $|k\rangle$, where $k$ is subject to the usual shell condition $k^2 = \mu^2$. There are three general classes of approach to this problem: (1) integrate over momentum and energy to form wave packets; (2) let $\mu$ be an imaginary mass and supplement the theory with appropriate interpretations of imaginary stuff; (3) treat $\mu$ as real but unphysical (as we did in the previous section).

In the first solution, if a one-particle state $|f\rangle$ does not have definite energy and momentum, then the old particle-production renormalization condition

$$\langle k|\phi(x)|0\rangle = e^{ix \cdot k} \tag{10.17.13}$$

must be reformulated. Because of the complex Lorentz-transformation properties of $|f\rangle$, naturalness will not be much of a guide here. It is no longer clear even that the field renormalization factor $z_\phi$ is constant. Furthermore, the construction of the wave-packet creation operator $\phi_f$ would appear to need integration over space and time. Integration over time will muddy the whole LSZ formalism; it is not easy to see how to set up the renormalization condition and the definition of $\phi_f$ in such a way that the one-particle creation property can be proved.

In the second solution, the basic idea is that the momenta should satisfy a shell condition

$$k^2 = \mu^2 + \Pi(\mu^2). \tag{10.17.14}$$

The tricky point here is that, when any component of $k$ is not real, then Lorentz covariance implies that none of the energy-momentum operators are hermitian. There will be states $|k\rangle$ for which translation introduces exponential blow-up. Such disasters can be rendered harmless if we interpret $|k\rangle$ as an unphysical element in the spectrum of the Hamiltonian. In quantum mechanics, this has been accomplished by (1) choosing a dense subset of analytic functions in the Hilbert space of quantum states, and completing this with respect to a sesquilinear inner product to form a new state space, a Hardy space of analytic functions, and (2) analytically extending the Hamiltonian and $S$ matrix to the Hardy space. For field theory, such a formalism has yet to be developed.

In the third solution, as described in the previous section, we have a heuristic extension of a sound, practical formalism, marred by obvious modeling weaknesses such as the use of a field with definite mass for an unstable field and the representation of unstable particle states by kets with definite energy and momentum. This third approach is most common, but its weaknesses must lead to some small errors in predictions. In the following paragraphs, we indicate the need for an improved theoretical foundation for this approach.

The whole purpose of this chapter has been to develop a scattering theory based on the interacting field. The crucial step in the theory is the LSZ reduction formula which transforms Green functions into $S$-matrix elements. In the previous section, we proposed an LSZ-style reduction formula with inverse-propagator factors for an unstable field. To justify this formula, the inverse-propagator factors should originate in a classical field equation whose solutions $f$ are used to build the wave-packet creation operators $\phi(f, t)$. Consequently, the $f$'s for an unstable field should satisfy the equation

$$(\Box + \mu^2 + \Pi(\mu^2))f = 0. \tag{10.17.15}$$

Plane wave solutions of this equation have the form

$$e_k(x) \stackrel{\text{def}}{=} e^{-x \cdot k}, \quad \text{where} \quad k^2 = \mu^2 + \Pi(\mu^2). \tag{10.17.16}$$

Obviously, the four-vector $k$ cannot be real, and so the third approach really depends on the second.

Writing $k = p + iq$ for real $p$ and $q$, we can classify the Lorentz orbits according to the nature of $q$: positive energy and lightlike, forward lightcone, spacelike, and so on. If $q$ has positive energy and is lightlike, then there is a Lorentz transformation which puts $q$ into its rest frame. In this case, the energy $k_0$ always has a positive imaginary component. However, making $\bar{k}$ real breaks Lorentz symmetry down to the rotation group. Consequently, two vectors $k$ and $k'$ with real momenta are in the same orbit of the Lorentz group if and only if $k_0 = k_0'$ and $|\bar{k}| = |\bar{k}'|$. Similar remarks hold for the other cases.

If we interpret an imaginary component $k_\mu$ of $k$ as signifying an expectation $\text{Re} \, k_\mu$ and a standard deviation $\text{Im} \, k_\mu$ for the observation of $P_\mu$ in the associated one-particle state, then it appears most natural to assume that the orbit which corresponds to one-particle states for an unstable particle contains a vector $k_\omega$ defined by

$$k_\omega \stackrel{\text{def}}{=} (\omega, \bar{0}), \quad \text{where} \quad \omega^2 = \mu^2 + \Pi(\mu^2) \quad \text{and} \quad \text{Re} \, \omega > 0. \tag{10.17.17}$$

Of course, this interpretation of complex energy-momentum provides a close connection between the first two approaches to modeling an unstable particle.

The relationship between $\omega$ and the original parameters $\mu$ and $\Pi(\mu^2) = -i\mu\Gamma$ is given by

$$\begin{aligned}
\omega &= \sqrt{\mu^2 - i\mu\Gamma} \\
&= \mu\left(1 - \frac{1}{2}\left(\frac{i\Gamma}{\mu}\right) - \frac{1}{8}\left(\frac{i\Gamma}{\mu}\right)^2 - \frac{1}{16}\left(\frac{i\Gamma}{\mu}\right)^3 + \cdots\right) \\
&\simeq \mu\left(1 + \frac{\Gamma^2}{8\mu^2}\right) - \frac{i}{2}\Gamma\left(1 - \frac{\Gamma^2}{8\mu^2}\right).
\end{aligned} \tag{10.17.18}$$

Generally, $\Gamma/8\mu$ is so small that the differences $\operatorname{Re}\omega - \mu$ and $2\operatorname{Im}\omega + \Gamma$ are insignificant in comparison with the experimental errors in the values of $\mu$ and $\Gamma$. Only in the case of the $W$ and $Z$ masses is the uncertainty in the mass even of order $\Gamma/8\mu$.

**Homework 10.17.19.** From the translation axiom

$$e^{-ia\cdot P}\phi(x)e^{ia\cdot P} = \phi(x - a),$$

and the definition of $\phi(f, t)$, show that

$$\left[P_r, \phi(f, t)\right] = \phi(i\partial_r f, t) \quad \text{for } r = 1, 2, 3.$$

**Remark 10.17.20.** For the energy operator, however, the logic of the previous homework breaks down because we have no integral over time in $\phi(f, t)$. For a stable field, we find the corresponding result for energy by going to the plane-wave limit, imposing the mass-shell condition, and returning to the wave-packet case. The conclusion for the energy operator therefore depends on the mass renormalization condition, not merely on canonical commutation rules.

**Remark 10.17.21.** As a consequence of the previous homework, if $f$ is a plane wave $e_k(x)$, then the state created by $\phi(f, t)$ is a momentum eigenstate with momentum $\bar{k}$. Since $P_r$ must be hermitian, this implies that $\bar{k}$ must be real. This argument breaks down when $\bar{k}$ is not real because the plane-wave limit causes exponential blow-up in the defining integral for $\phi(f, t)$. On the orbit of $k_\omega$, we conclude that the plane-wave limit only exists when $\bar{k} = \bar{0}$. Thus the only momentum eigenstate we can define is the state $|k_\omega\rangle$ given by

$$|k_\omega\rangle \stackrel{\text{def}}{=} \phi(e^{-ix\cdot k_\omega}, t)|0\rangle. \tag{10.17.22}$$

Developing Homework 10.5.13 and the example 10.12.22 on LSZ reduction for a decay amplitude, it is worth noting that our definition of the $S$ matrix depends on the time-dependence of the creation operators:

$$\langle f|S|i\rangle \stackrel{\text{def}}{=} {}^+\langle f|i\rangle^-. \tag{10.17.23}$$

In the case of $\pi^0 \to 2\gamma$, this implies that $|\pi^0\rangle^-$ cannot be orthogonal to $|2\gamma\rangle^+$. Yet the definition of the state space requires

$$^-\langle 2\gamma|\pi^0\rangle^- = {}^+\langle 2\gamma|\pi^0\rangle^+ = 0. \tag{10.17.24}$$

Consequently, $|\pi^0\rangle^-$ must be different from $|\pi^0\rangle^+$. This is a property of unstable particle states which is not found in the states of stable particles. In a complete extension of the LSZ formalism to unstable particles, the proof of the one-particle creation property for $\phi(f, t)$ must somehow generate this property.

We find then that there is as yet no satisfactory account of unstable particles in field theory. The very concept is only poorly defined. The usual heuristic approach which makes a minimal modification in the Feynman rules and renormalization conditions has obvious weaknesses and lacks a theoretical foundation. However, this approach should be accurate to within the ambiguity $\sim (\Gamma/\mu)^2$ of its mass and decay width parameters, and so its failings have escaped experimental detection to date.

Working with the heuristic formalism, the following section shows how the non-zero value of $\Pi(\mu^2)$ gives rise to a Breit–Wigner peak in scattering amplitudes and the characteristic evolution of an unstable particle.

## 10.18 Propagation of Unstable Particles

In this section, we show how the shift in the pole of the renormalized propagator for an unstable particle is reflected in (1) particle production as a function of energy and (2) the exponential decay of an initial one-particle state.

The previous sections have shown that $\operatorname{Im}\Pi(k^2) = -\mu\Gamma$ for a particle at rest. Using the fact that $\Pi$ is at least second order in coupling constants, we can estimate $D$ to second order:

$$
\begin{aligned}
-iD(k^2)^{-1} &= k^2 - \mu^2 - \Pi(k^2) \\
&\stackrel{(2)}{=} k^2 - \mu^2 - \Pi(\mu^2) - (k^2 - \mu^2)\frac{\partial\Pi}{\partial k^2}\Big|_{k^2=\mu^2} \\
&\stackrel{(2)}{=} k^2 - \mu^2 + i\mu\Gamma \qquad \text{for } k^2 - \mu^2 \sim O(g^2), \\
&\stackrel{(2)}{=} k^2 - (\mu - \tfrac{1}{2}i\Gamma)^2.
\end{aligned}
\tag{10.18.1}
$$

This implies that $D$ has a pole near $k^2 = (\mu - \tfrac{1}{2}i\Gamma)^2$. This pole is in the analytic continuation of $D$ across the cut, and so has no direct physical significance. However, we do see now where the poles go as $\mu$ increases through the decay threshold.

If we perturb our meson field with a classical source $\mathcal{L}' = -\lambda\delta^{(4)}(x)\phi(x)$, then $\Gamma$ is the total width at half the maximum of the probability density for the momentum of a single meson final state. To see this, define $\mathcal{A}_{n0}$ to be the amplitude for producing the state $|n\rangle$ from the vacuum:

$$
\begin{aligned}
\mathcal{A}_{n0} &= -i\lambda \int d^4x\, \delta^{(4)}(x)\langle n|\phi(x)|0\rangle + O(\lambda^2) \\
&\stackrel{(1)}{=} -i\lambda\langle n|\phi(0)|0\rangle.
\end{aligned}
\tag{10.18.2}
$$

The probability density on momentum space is given by summing the squares of the amplitudes. Working to lowest non-trivial order:

$$
\begin{aligned}
\frac{d\text{Prob}}{d^4k} &\propto \sum_n \lambda^2 |\langle n|\phi(0)|0\rangle|^2 \delta^{(4)}(P_n - k) \\
&\propto \sigma(k^2) \\
&\propto \operatorname{Re} D(k^2) \\
&\propto \operatorname{Im}\left(\frac{1}{k^2 - \mu^2 + i\mu\Gamma}\right) \\
&\propto \frac{\mu\Gamma}{(k^2 - \mu^2)^2 + \mu^2\Gamma^2}.
\end{aligned}
\tag{10.18.3}
$$

To simplify this expression, let $k = (E, 0, 0, 0)$, and focus on the energy range $E \simeq \mu$. In this range

$$
k^2 - \mu^2 = (E + \mu)(E - \mu) \simeq 2\mu(E - \mu),
\tag{10.18.4}
$$

and so the probability density reduces to

$$
\frac{d\text{Prob}}{d^4k} \propto \frac{1}{(E - \mu)^2 + \tfrac{1}{4}\Gamma^2}.
\tag{10.18.5}
$$

This response of the meson field to a delta function external source is called a Breit–Wigner peak. The interpretation is that the field has approximate particle states reflected by the proximity of the pole in the renormalized propagator to the real axis.

The width of a Breit–Wigner peak is also the characteristic exponent in the decay of the unstable particle. Consider an initial state

$$|i\rangle \overset{\text{def}}{=} \int d^4x\, f(x)\phi(x)|0\rangle, \tag{10.18.6}$$

where $f$ is localized in position space and its Fourier transform $\hat{f}$ is localized in momentum space near $k^2 = \mu^2$. Let the particle detector be modeled by

$$\langle f| \overset{\text{def}}{=} \langle 0| \int d^4x'\, f(x'-y)\phi(x'). \tag{10.18.7}$$

Then, assuming that $y^2 \gg 0$ so that initial and detected states do not overlap, the amplitude for detection is

$$
\begin{aligned}
A(y) &= \langle 0| \int d^4x'\, f(x'-y)\phi(x') \int d^4x\, f(x)\phi(x)|0\rangle \\
&= \int d^4x'\, d^4x\, f(x'-y)f(x)\langle 0|T\phi(x')\phi(x)|0\rangle \\
&\propto \int d^4k\, |\hat{f}(k)|^2 e^{-iy\cdot k} D(k^2) \\
&\propto \int d^4k\, |\hat{f}(k)|^2 \frac{ie^{-iy\cdot k}}{k^2 - \mu^2 + i\mu\Gamma}.
\end{aligned}
\tag{10.18.8}
$$

Rewrite the fraction as the integral of an exponential

$$\frac{i}{k^2 - \mu^2 + i\mu\Gamma} = \int_0^\infty \frac{ds}{2\mu}\, e^{is(k^2-\mu^2+i\mu\Gamma)/2\mu} \tag{10.18.9}$$

so that

$$A(y) \propto \int_0^\infty \frac{ds}{2\mu} \int d^4k\, ie^{-iy\cdot k+is(k^2-\mu^2+i\mu\Gamma)/2\mu}|\hat{f}(k)|^2. \tag{10.18.10}$$

Now use the stationary phase formula

$$\int dt\, e^{ig(t)} f(t) \sim e^{ig(t_0)} f(t_0) \left|\frac{2\pi}{g''(t_0)}\right|^{\frac{1}{2}} e^{\pm \frac{i\pi}{4}}, \tag{10.18.11}$$

(where $t_0$ is the location of a stationary value of $g$ and the sign is the sign of $g''(t_0)$), to do the $k$ integrals. The stationary phase point is given by

$$\frac{\partial}{\partial k}\left(\frac{s}{2\mu}(k^2 - \mu^2 + i\mu\Gamma) - y\cdot k\right) = \frac{s}{\mu}k - y = 0 \quad \Rightarrow \quad k = \frac{\mu y}{s}. \tag{10.18.12}$$

At this point the Hessian of second derivatives is non-singular:

$$\frac{\partial}{\partial k^\mu}\frac{\partial}{\partial k^\nu}\left(\frac{s}{2\mu}(k^2 - \mu^2 + i\mu\Gamma) - y\cdot k\right) = \frac{s}{\mu}g^{\mu\nu}, \tag{10.18.13}$$

and so the four integrals will introduce factors

$$\left|\frac{2\pi}{f_0''}\right|^{\frac{1}{2}}\left|\frac{2\pi}{f_1''}\right|^{\frac{1}{2}}\left|\frac{2\pi}{f_2''}\right|^{\frac{1}{2}}\left|\frac{2\pi}{f_3''}\right|^{\frac{1}{2}} = (2\pi)^2\left(\frac{\mu}{s}\right)^2 \quad \text{and} \quad e^{-i\pi/4}e^{i\pi/4}e^{i\pi/4}e^{i\pi/4} = e^{i\pi/2} = i$$

(10.18.14)

into the integral. Thus we find

$$A(y) \propto \mu \int_0^\infty \frac{ds}{2s^2}\, e^{-\frac{i}{2}((\mu y^2/s)+\mu s)}e^{-s\Gamma/2}|\hat{f}(\mu y/s)|^2.$$

(10.18.15)

Apply the stationary phase formula again. The stationary point is given by

$$\frac{\partial}{\partial s}\left(\frac{\mu y^2}{s}+\mu s\right) = -\frac{\mu y^2}{s^2}+\mu = 0 \quad \Rightarrow \quad s = \sqrt{y^2}.$$

(10.18.16)

This makes $s$ the proper time between production and detection. The second derivative at this point is $2\mu/s$, which is positive. Hence

$$A(y) \propto e^{-i\mu s}e^{-s\Gamma/2}|\hat{f}(\mu y/s)|^2 \frac{\mu}{2s^2}\left(\frac{\pi s}{\mu}\right)^{\frac{1}{2}}e^{i\pi/4}$$

$$\propto |\hat{f}(\mu y/s)|^2 e^{-i\mu s}e^{-\frac{1}{2}s\Gamma}s^{-3/2}.$$

(10.18.17)

The first factor is the product of the production amplitude and the detection amplitude, the second is simply a proper-time dependent phase, the third is the exponential decay rate of the particles, and the fourth accounts for the spreading of the wave packet over time.

This section shows how the imaginary part of the self-energy function generates the characteristic estimates for production and propagation of unstable particles, thereby connecting the abstract structures developed in the earlier sections to experimental results. In light of the previous section, the emergence of an exponential decay law indicates the theoretical inadequacy of our model for unstable particles. However, as experiment is not yet demanding a better one, this model is sufficient for all practical purposes.

## 10.19 Summary

This chapter has transformed our understanding of perturbation theory by providing a foundation for it in the interacting field.

The first section introduced a non-perturbative concept, the Green functions, and the unified expression for the total dynamics of the interacting fields, the generating functional for the Green functions. The second and third sections showed how these Green functions could be computed using the usual Feynman diagrams and rules.

Having established a non-perturbative viewpoint and its connection to perturbation theory, the fourth and fifth sections provided the link to the $S$ matrix. First, it was necessary to identify in and out particle states in terms of the interacting field, and second, we had to connect the Green functions to those in and out states through the LSZ reduction formula.

In connecting the non-perturbative reality of the interacting field to the $S$ matrix, we discovered the field renormalization factor $z_\phi$. Section 10.7 presented a formula for this constant in the context of working with bare fields and parameters. However, it was found preferable to express the total Lagrangian density in terms of physical or renormalized fields and parameters and counterterms. Sections 8–11 presented the technique of counterterm renormalization which perturbatively sets the counterterms to those unique values which can uphold the physical interpretation of the renormalized fields and physical parameters. The first ten sections thus establish a new formalism for scattering theory based on an interacting scalar field which has stable particles. Sections 12–17 extend this formalism to spinor fields, vector fields, and fields with unstable particles.

The presence of unstable particles in a theory causes the Feynman rules derived from Dyson's formula to be somewhat meaningless. The underlying Green functions, however, are much less dramatically affected by the transition from stable to unstable particles. On this foundation of Green functions, it is natural to propose a modification of the Feynman rules to favor the renormalized propagator and thereby to restore meaning to the perturbation expansion. We note that this weakness of perturbative canonical quantization is also found in functional integral quantization.

This chapter provides a practical foundation for renormalization. As we noted at the end of Chapter 4, there are theories in which the divergences are so diverse that they cannot be absorbed by a finite number of counterterms. We have also noted that when the interactions have mass dimension at most four and the vector fields are massive and coupled to conserved currents, then we expect a finite number of counterterms to suffice. It remains to indicate the logic behind these rules of thumb.

At this point we have a choice, whether to turn to functional integral quantization and tree-level perturbation theory of gauge theories or to go more deeply into renormalization. However, since counterterm renormalization introduces derivative interactions, since we need functional integral quantization to derive covariant Feynman rules for these, and since the interesting theories are mostly gauge theories for which again we need functional integral quantization, we choose to develop functional integral quantization first and to postpone further investigation of renormalization until Chapter 20.

In the following chapter, we shall generate the path or functional integral from the completeness relation in quantum mechanics and then extend functional integration to fields in order to obtain a functional integral formula for the generating functional of a field theory. This formula constitutes functional integral quantization. Subsequent chapters will show how to quantize QED and non-abelian gauge theories in preparation for a presentation of the Standard Model of electroweak and strong interactions.

# Chapter 11
# Functional Integral Quantization

Providing a classical functional calculus for investigating interacting quantum fields from the level of the generating functional, proving the equivalence of this new quantization to perturbative canonical quantization, developing an efficient procedure for obtaining Feynman rules directly from a Lagrangian density, and preparing for functional quantization of gauge theories.

## Introduction

So far, we have used the perturbative canonical quantization of Chapter 4 as the method for transforming classical field theories into quantum field theories. By the end of Chapter 9, this program had encountered three problems. First, the nature of the adiabatic turning on and off function makes it theoretically unsound to apply that method to a theory with unstable particles or bound states or gauge invariance. Second, derivative interactions in massive scalar QED gave rise to non-covariant interactions and Feynman rules. Third, we cannot readily see that the gauge invariance of QED is preserved by this method of quantization. These troubles are greatly aggravated when we attempt to apply perturbative canonical quantization to non-abelian gauge theories. Clearly, a new and more powerful approach to quantization is needed.

Chapter 10 provided the first step, the theory of the generating functional. Taking the generating functional as the source of the perturbation expansion, we were able to eliminate the adiabatic turning on and off function, propose convenient renormalization conditions, make sense of divergent Feynman integrals, and extend perturbation theory to include fields with unstable particles.

Renormalization, however, generates kinetic counterterms, and therefore makes it imperative to develop a means for quantizing theories with derivative interactions. The purpose of this chapter is to solve the outstanding problems: quantization of theories with derivative interactions and theories with gauge symmetries.

The solution for these remaining problems is a new technique, functional integral quantization. The central theme of functional integral quantization is to express the dynamics of the quantum field from the viewpoint of classical fields. As we saw in our discussion of the vacuum state of a free scalar quantum field, the states of a quantum field can be regarded as a weighted average of classical fields. The functional integral is that notion of integration over a space of classical fields which enables us to write and manipulate these weighted averages. The emergence of quantum field theory from the functional integral follows naturally from the choice of weight function, $\exp\big(iS(\phi)\big)$, where $S(\phi)$ is the classical action of the classical field $\phi$.

Functional integral quantization is basically a non-perturbative viewpoint. It provides an integral formula for the generating functional and a variety of procedures for manipulating the integral. This chapter establishes the technique of functional integral quantization, and verifies its equivalence to perturbative canonical quantization for the theories treated in earlier chapters. Chapter 12 develops the application to gauge field theories, and Chapter 13 completes the theory of quantization of gauge theories with a discussion of anomalies – breaking of classical conservation laws by quantum effects – which could make a quantum gauge theory inconsistent. Further material on preservation of gauge invariance by renormalization is presented in Chapter 20.

The first five sections of this chapter investigate the functional integral derived from the completeness relation of quantum mechanics, Section 11.6 prepares for application to field theory, and Section 11.7 boldly applies the formalism to a scalar field, formally confirming its consistency with Dyson's formula. Section 11.8 derives the perturbation expansion from the functional integral form of the generating functional. This derivation yields a universal procedure for obtaining Feynman rules from an action. Section 11.9 provides a further consistency check. Section 11.10 extends functional integral quantization to the massive vector field. Finally, the last two sections develop and check the analogous procedure for fermions, namely functional derivative quantization.

## 11.1 Deriving the Functional Integral from Quantum Mechanics

To make the connection between functional integral quantization and canonical quantization we shall locate the source of the functional integral viewpoint in quantum principles, and we shall discover in the final structure of functional integral quantization the familiar notions of Dyson's formula, Green functions, and Feynman rules.

The foundation for the functional integral viewpoint is the quantum-mechanical completeness relation:

$$\langle q', T'|q, T\rangle = \int dq_{n-1} \cdots dq_1 \, \langle q', T'|q_{n-1}, t_{n-1}\rangle \cdots \langle q_1, t_1|q, T\rangle, \qquad (11.1.1)$$

where

$$T' = t_n > t_{n-1} > \cdots > t_1 > t_0 = T, \qquad (11.1.2)$$

and the brackets give the probability of evolving from the earlier state to the later. Since this equation expresses the probability of evolving from a state $|q\rangle$ at time $T$ to a state $|q'\rangle$ at time $T'$ in terms of all the paths $q_1, \ldots, q_n$ that the evolution could take, as we increase the number of intermediate states, the integrals approach a functional or path integral, an integral over all functions or paths $q(t)$. We shall make this point clear in detail in the following paragraphs.

Let $P$ and $Q$ be respectively the momentum and position operators for a quantum-mechanical particle, and let $|p\rangle$ and $|q\rangle$ be the corresponding eigenstates. Then the usual representation of quantum mechanics in one dimension is:

$$\begin{aligned} P &= -i\frac{\partial}{\partial x}, & Q &= x; \\ |p\rangle &= e^{ipx}, & |q\rangle &= \delta(x - q); \end{aligned} \qquad (11.1.3)$$

so that

$$\begin{aligned} P|p\rangle &= p|p\rangle, & Q|q\rangle &= q|q\rangle; \\ \langle q|p\rangle &= e^{ipq}, & \langle p|q\rangle &= e^{-ipq}. \end{aligned} \qquad (11.1.4)$$

Let $H(p, q)$ be a classical Hamiltonian of the form

$$H(p, q) = \frac{1}{2}p^2 + V(q), \tag{11.1.5}$$

and let $H(P, Q)$ be the corresponding quantum Hamiltonian. Then

$$\langle p|H(P, Q)|q \rangle = H(p, q)\langle p|q \rangle = H(p, q)e^{-ipq}, \tag{11.1.6}$$

and, to make the connection to the time-states notation in the completeness relation explicit,

$$\langle q', T'|q, T \rangle = \langle q'|e^{-iH(P,Q)(T'-T)}|q \rangle. \tag{11.1.7}$$

The completeness of the momentum eigenstates enables us to write the position eigenstates in terms of the momentum eigenstates:

$$\langle q'| = \int \frac{dp'}{2\pi} \langle q'|p' \rangle \langle p'| = \int \frac{dp'}{2\pi} e^{-ip'q'} \langle p'|. \tag{11.1.8}$$

Therefore the expectation of the Hamiltonian between position eigenstates is

$$\begin{aligned}
\langle q'|H(P, Q)|q \rangle &= \int \frac{dp'}{2\pi} e^{ip'q'} \langle p'|H(P, Q)|q \rangle \\
&= \int \frac{dp'}{2\pi} e^{ip'(q'-q)} H(p', q).
\end{aligned} \tag{11.1.9}$$

If we assume that the time intervals $\Delta t_r = t_r - t_{r-1}$ are all small, then we can use this formula in order to approximate each factor in the completeness relation above:

$$\begin{aligned}
\langle q_r, t_r|q_{r-1}, t_{r-1} \rangle &= \langle q_r|e^{-iH(P,Q)\Delta t_r}|q_{r-1} \rangle \\
&\simeq \langle q_r|(1 - iH(P, Q)\Delta t_r)|q_{r-1} \rangle \\
&\simeq \int \frac{dp_r}{2\pi} e^{ip_r q_r} \langle p_r|(1 - iH(P, Q)\Delta t_r)|q_{r-1} \rangle \\
&\simeq \int \frac{dp_r}{2\pi} e^{ip_r \Delta q_r} (1 - iH(p_r, q_{r-1})\Delta t_r) \\
&\simeq \int \frac{dp_r}{2\pi} e^{ip_r \Delta q_r - iH(p_r, q_{r-1})\Delta t_r},
\end{aligned} \tag{11.1.10}$$

where $\Delta q_r = q_r - q_{r-1}$, and the $\simeq$ indicates error of order $(\Delta t_r)^2$. Putting the factors together, we obtain another form of the completeness relation:

$$\langle q', T'|q, T \rangle \simeq \int \frac{dp_n}{2\pi} \frac{dp_{n-1} \, dq_{n-1}}{2\pi} \cdots \frac{dp_1 \, dq_1}{2\pi} \tag{11.1.11}$$

$$\times e^{ip_n \Delta q_n - iH(p_n, q_{n-1})\Delta t_n} \cdots e^{ip_1 \Delta q_1 - iH(p_1, q_0)\Delta t_1}.$$

Rewriting the product of exponentials as the exponential of a sum, we obtain:

$$\langle q', T'|q, T \rangle \simeq \int \frac{dp_n}{2\pi} \frac{dp_{n-1} \, dq_{n-1}}{2\pi} \cdots \frac{dp_1 \, dq_1}{2\pi} \tag{11.1.12}$$

$$\times \exp\left( i \sum_{r=1}^{n} p_r \Delta q_r - H(p_r, q_{r-1})\Delta t_r \right).$$

Now we can see the time-integral of the Lagrangian density emerging in the exponent as $\sup\{\Delta t_r\} \to 0$. In order to bring this integral out clearly, we need to change variables from the product of phase spaces to the space of discrete paths. We do this by interpreting the subscript $r$ as the discrete time variable for a momentum function $p$ and a position function $q$:

$$
\left.
\begin{aligned}
\mathbf{R}^n &\quad \to \quad \operatorname{Map}\big((t_1, \ldots, t_n), \mathbf{R}\big) \\
(p_1, \ldots, p_n) &\quad \mapsto \quad (p\colon t_r \mapsto p_r)
\end{aligned}
\right\}
$$

$$
\left.
\begin{aligned}
\mathbf{R}^{n-1} &\quad \to \operatorname{Map}\big((t_1, \ldots, t_{n-1}), \mathbf{R}\big) \\
(q_1, \ldots, q_{n-1}) &\quad \mapsto \quad (q\colon t_r \mapsto q_r)
\end{aligned}
\right\}
\tag{11.1.13}
$$

The natural basis for the space of discrete paths is the set of $n$ momentum paths $\hat{p}^r$ together with the set of $n-1$ position paths $\hat{q}^r$ defined as follows:

$$
\begin{aligned}
\hat{p}^r\colon t_s &\mapsto \delta_s^r, \quad \text{where } s = 1, \ldots, n; \\
\hat{q}^r\colon t_s &\mapsto \delta_s^r, \quad \text{where } s = 1, \ldots, n-1.
\end{aligned}
\tag{11.1.14}
$$

With this choice of basis, we may interpret the old variables $p_r$ and $q_r$ as coordinates on the space of discrete paths and the phase-space integral 11.1.12 as an integral over path-space, that is, as a discrete-time functional integral. This is the type of functional integral used in computer simulations of quantum field theory.

The continuous-time functional integral is obtained from the discrete-time version by taking the limit $n \to \infty$ in such a way that $\sup\{\Delta t_r\} \to 0$. Writing

$$
[dp] = \lim_{n \to \infty} \frac{dp^1}{2\pi} \cdots \frac{dp^n}{2\pi} \quad \text{and} \quad [dq] = \lim_{n \to \infty} dq^1 \cdots dq^{n-1},
\tag{11.1.15}
$$

we arrive at the continuous version of the completeness relation, a functional integral formula for the probability of evolution from $|q\rangle$ at time $T$ to $|q'\rangle$ at time $T'$:

**Theorem 11.1.16.** *If the evolution of a quantum particle is governed by a Hamiltonian $H = \frac{1}{2}p^2 + V(q)$, then the amplitude for finding a particle localized at $q$ at time $T$ at a position $q'$ at time $T'$ is given by the functional integral formula*

$$
\langle q', T' | q, T \rangle = \int [dq][dp] \, \exp\Big(i \int_T^{T'} dt \, p\dot{q} - H(p, q)\Big).
$$

The main conclusion of this section is that the completeness condition of quantum mechanics leads naturally to the notion that quantum-mechanical evolution simultaneously considers all classical paths. This notion finds formal expression in the functional integral formula for quantum amplitudes. The continuum functional integral is not very well defined from a mathematical perspective, yet in physical applications, its definition as a limit of the discrete functional integral is quite well behaved.

## 11.2 An Example: the Free Particle

Consider a free quantum particle. The Hamiltonian, Hamilton equations, and Lagrangian are

$$H(p,q) = \frac{1}{2}p^2, \quad \dot{q} = p \text{ and } \dot{p} = 0, \quad L(\dot{q},q) = \frac{1}{2}\dot{q}^2. \tag{11.2.1}$$

The functional integral equation for the quantum-mechanical evolution of the particle state is

$$\langle q', T'|q, T\rangle = \int [dq][dp] \, \exp\left(i\int_T^{T'} dt \, p\dot{q} - \frac{1}{2}p^2\right). \tag{11.2.2}$$

We will evaluate this functional integral using the discrete approximation of the original definition.

To prepare for the integral over $[dp]$, we first evaluate the standard complex Gaussian functional integral

$$I_p = \int [dp] \exp\left(i\int_T^{T'} dt \, -\frac{1}{2}p^2\right). \tag{11.2.3}$$

To evaluate this integral as a limit of discrete-time approximations, we break the interval $[T, T']$ into $n$ equal parts and use the basis $\hat{p}^r$ and coordinates $p_r$ for path space from 11.1.14. Since the path $\hat{p}^r$ has the constant value unity for $1/n^{\text{th}}$ the total time, we find:

$$\begin{aligned}
I_p &= \lim_{n\to\infty} \prod_{r=1}^{n} \int \frac{dp_r}{2\pi} \exp\left(-\frac{i(T'-T)}{2n}p_r^2\right) \\
&= \lim_{n\to\infty} \left(\frac{n}{2\pi(T'-T)}\right)^{n/2} e^{-n\pi i/4}.
\end{aligned} \tag{11.2.4}$$

Note that, though the integrands have unit modulus, they are the usual Fresnel integrals, and do actually converge. It is therefore unnecessary to rotate to Euclidean time, $t_{\text{E}} = it$. The limit, however, is definitely divergent; it is commonly referred to as an 'undefined constant'. We shall write $N$ for this limit.

Returning to the free particle, the function in the exponent is not $p^2/2$ but:

$$p\dot{q} - \frac{1}{2}p^2 = -\frac{1}{2}(p - \dot{q})^2 + \frac{1}{2}\dot{q}^2. \tag{11.2.5}$$

Hence, applying the complex Gaussian result above with translated variable $p - \dot{q}$, we find the result of the $[dp]$ integration in the free particle functional integral:

$$\langle q', T'|q, T\rangle = N\int [dq] \exp\left(i\int_T^{T'} dt \, \frac{1}{2}\dot{q}^2\right). \tag{11.2.6}$$

This integral, unlike the momentum integral, has fixed endpoints. Thus the discrete approximation $I_q^n$ is obtained by choosing $n-1$ equally spaced times and using the corresponding $q_r$ as integration variables:

$$I_q^n(q', T'; q, T) \overset{\text{def}}{=} \int dq_1 \cdots dq_{n-1} \, \exp\left(i\sum_{r=1}^{n} \frac{(q_r - q_{r-1})^2}{2\tau_n}\right), \tag{11.2.7}$$

where the interval $\tau_n$ is defined by

$$\tau_n \overset{\text{def}}{=} \frac{T' - T}{n}. \tag{11.2.8}$$

The integrals $I_q^n$ can be evaluated by induction. To simplify the computation, define a unit-interval form of $I_q^n$, $J_q^n$, as follows:

$$J_q^n(q', q) \overset{\text{def}}{=} \int dq_1 \cdots dq_{n-1} \, \exp\Big(\frac{i}{2} \sum_{r=1}^{n} (q_r - q_{r-1})^2\Big). \tag{11.2.9}$$

Now we can use the recursion relation

$$J_{n+1}(q', q) = (2\pi)^{1/2} \int dq_n \, e^{\frac{i}{2}(q' - q_n)^2} J_n(q_n, q) \tag{11.2.10}$$

to show inductively that

$$J_n(q', q) = \frac{(2\pi)^{(n-1)/2}}{\sqrt{n}} \exp\Big(-\frac{1}{2n}(q' - q)^2\Big) e^{(n-1)\pi i/4}. \tag{11.2.11}$$

Scaling this result leads to a formula for $I_q^n$:

$$
\begin{aligned}
I_q^n(q', T'; q, T) &= \tau_n^{(n-1)/2} J_q^n\Big(\frac{q'}{\sqrt{\tau_n}}, \frac{q}{\sqrt{\tau_n}}\Big) \\
&= \frac{(2\pi(T' - T))^{(n-1)/2}}{n^{n/2}} \exp\Big(\frac{i}{2(T' - T)}(q' - q)^2\Big) e^{(n-1)\pi i/4}.
\end{aligned}
\tag{11.2.12}
$$

The final answer for the functional integral $I_q$ follows directly:

$$I_q(q', T'; q, T) = N' \exp\Big(\frac{i}{2(T' - T)}(q' - q)^2\Big), \tag{11.2.13}$$

$$\text{where} \quad N' = \lim_{n \to \infty} \frac{(2\pi(T' - T))^{(n-1)/2}}{n^{n/2}} e^{(n-1)\pi i/4}.$$

Clearly $N'$ is another 'undefined constant', and so the functional integral over $[dq]$ diverges.

From this result for $I_q$ we obtain a value for the functional integral representing the evolution of a free particle in Euclidean time:

$$\langle q', T' | q, T \rangle = NN' \exp\Big(\frac{i}{2(T' - T)}(q' - q)^2\Big). \tag{11.2.14}$$

This is indeed the right formula; this is the complex Gaussian wave function usually derived directly from the Schrödinger equation:

$$i\frac{\partial}{\partial t}\psi = -\frac{1}{2}\frac{\partial^2}{\partial q^2}\psi \quad \text{and} \quad \psi(q, T) = \delta(q)$$

$$\implies \psi(q', T') = \frac{1}{\sqrt{2\pi(T' - T)}} \exp\Big(\frac{i}{2(T' - T)}(q' - q)^2\Big). \tag{11.2.15}$$

The product of the undefined constants $NN'$ is at first sight a meaningless infinity times zero. Going back to the defining limits, however, we see that the product should really be definite and finite:

$$NN' = \lim_{n\to\infty} \left(\frac{n}{2\pi(T'-T)}\right)^{n/2} e^{-n\pi i/4} \frac{\left(2\pi(T'-T)\right)^{(n-1)/2}}{n^{n/2}} e^{(n-1)\pi i/4}$$
$$= \left(2\pi(T'-T)\right)^{-1/2} e^{-\pi i/4}, \qquad (11.2.16)$$

which, up to a phase, is the proper normalization factor for the complex Gaussian. The moral is that $NN'$ exists as a limit even though $N$ and $N'$ do not. Thus, when the functional integral is used in physics, it is generally safe to use undefined constants like $N$ and $N'$ in an argument and then at the end replace the final undefined constant by a normalization factor.

This example shows that both the functional integral over $[dp]$ and that over $[dq]$ diverge, but that the double integral can be expressed as a finite limit of finite products. The existence of the double integral indicates that it is possible to redefine the measures $[dp]$ and $[dq]$ to make the integrals above exist. Indeed, the mathematical approach to functional integration uses Euclidean time and absorbs the standard Gaussian and the undefined constant into the measure, defining, for example,

$$d\mu(p) \stackrel{\text{def}}{=} \lim_{n\to\infty} \left(\frac{T'-T}{2\pi n}\right)^{n/2} dp_1 \cdots dp_n \, \exp\left(-\frac{n}{2(T'-T)} \sum_{r=1}^{n} p_r^2\right), \qquad (11.2.17)$$

where the $p$ in $d\mu(p)$ is a function, and the $p_r$ are the values of the function $p$ evaluated at the time $T + r(T'-T)/n$. With this measure one can evaluate integrals like

$$\int d\mu(p) \, F(p) \qquad (11.2.18)$$

for polynomial or exponential functions $F$ which are sufficiently small not to overcome the Gaussian structure of the measure.

Finally, to locate a familiar concept among the novelties of this example, we observe that the integral over $[dp]$ is a functional Fourier transform of a Hamiltonian Gaussian $G_H(p,q)$, and that the resulting function is the corresponding Lagrangian Gaussian $G_L(\dot{q},q)$:

$$G_H(p,q) \stackrel{\text{def}}{=} \exp\left(-\frac{i}{2}\int_T^{T'} dt\, p^2\right) \quad \text{and} \quad G_L(\dot{q},q) \stackrel{\text{def}}{=} \exp\left(\frac{i}{2}\int_T^{T'} dt\, \dot{q}^2\right)$$
$$\implies \int [dp] \, \exp\left(i\int_T^{T'} dt\, p\dot{q}\right) G_H(p,q) = G_L(\dot{q},q). \qquad (11.2.19)$$

If we were to work in Euclidean time, we would substitute $\dot{q} = i\dot{q}_{\rm E}$ in $G_L$ and thereby introduce an extra minus sign to restore symmetry between Hamiltonian and Lagrangian Gaussians.

**Homework** 11.2.20. Evaluate the free particle functional integral 11.2.2 by doing the integral over $[dq]$ first.

These detailed computations reveal the mathematical delicacy of the functional integral and its physical validity. The generalization to $n$ dimensions is presented in the next section.

## 11.3 Details of Functional Integration: Operator Ordering

The main result of the last section, Theorem 11.1.12, is limited to Hamiltonians which are free from operator ordering ambiguities. In this section, we show that the result can be extended to other Hamiltonians with the understanding that the operator ordering in the Hamiltonian must be reflected in the definition of the functional integral.

When quantizing a classical Hamiltonian, we can specify the ordering of operator products in a variety of ways. The simplest is symmetrical ordering:

$$(p^m q^n)_S \stackrel{\text{def}}{=} \frac{1}{2}(P^m Q^n + Q^n P^m), \qquad (11.3.1)$$

where $p$ and $q$ are the classical variables and $P$ and $Q$ are the associated quantum operators. Another common example is binomial ordering, of which there are two flavors:

$$(p^m q^n)_P \stackrel{\text{def}}{=} \frac{1}{2^m} \sum_{r=0}^{m} \binom{m}{r} P^{m-r} Q^n P^r;$$

$$(p^m q^n)_Q \stackrel{\text{def}}{=} \frac{1}{2^n} \sum_{s=0}^{n} \binom{n}{s} Q^s P^m Q^{n-s}. \qquad (11.3.2)$$

We shall write $H_S(P, Q)$, $H_P(P, Q)$, and $H_Q(P, Q)$ for the results of applying these transformations to the monomials in a function $H(p, q)$.

The most general ordering ordinarily considered uses an almost arbitrary function $f(u, v)$ and the Fourier transform:

$$H_f(P, Q) \stackrel{\text{def}}{=} \int \frac{dp\, du}{2\pi\, 2\pi} dq\, dv\, e^{ivP + iuQ} e^{-ivp - iuq} f(u, v) H(x, y). \qquad (11.3.3)$$

The case $f \equiv 1$ yields Weyl ordering $H_W(P, Q)$. Somewhat surprisingly,

$$H_W(P, Q) = H_P(P, Q) = H_Q(P, Q). \qquad (11.3.4)$$

We can obtain a generating functional $G_f$ for $f$-ordered products by substituting $\exp(iyp + ixq)$ for $H(p, q)$. Performing the integrals yields

$$G_f(P, Q; x, y) = e^{iyP + ixQ} f(x, y). \qquad (11.3.5)$$

**Homework** 11.3.6. By evaluating $[P^{m+1}, Q^{n+1}]$ in two different ways, show that

$$(P^m Q^n)_P = (P^m Q^n)_Q.$$

**Homework** 11.3.7. Let $R_r$ be equal to $P$ if $1 \le r \le m$ and equal to $Q$ if $m+1 \le r \le m+n$. Then

$$(p^m q^n)_W = \frac{1}{(m+n)!} \sum_{\pi} R_{\pi(1)} \cdots R_{\pi((m+n))}.$$

**Homework 11.3.8.** Prove that $H_W(P, Q) = H_P(P, Q)$. (Induction provides a workable strategy.)

**Homework 11.3.9.** Let $A$ and $B$ be operators, and write $C = [A, B]$. Prove that if $C$ commutes with $A$ and $B$, then
$$e^{A+B} = e^A e^B e^C.$$
This is a special case of the Campbell–Hausdorff formula.

The argument of the previous section converted the completeness relation into a functional integral on the basis of the evaluation 11.1.9 of the quantum Hamiltonian between position eigenstates. The three basic evaluation formulae are

$$\langle q'|P^m Q^n|q\rangle = \langle q'|P^m|q\rangle q^n = \int \frac{dp'}{2\pi} e^{ip'q' - ip'q} p'^m q^n;$$

$$\langle q'|Q^n P^m|q\rangle = \langle q'|P^m|q\rangle q'^n = \int \frac{dp'}{2\pi} e^{ip'q' - ip'q} p'^m q'^n; \qquad (11.3.10)$$

$$\langle q'|Q^r P^m Q^{n-r}|q\rangle = q'^r q^{n-r} \langle q'|P^m|q\rangle = \int \frac{dp'}{2\pi} e^{ip'q' - ip'q} q'^r p'^m q^{n-r}.$$

Hence, if we derived the quantum Hamiltonian by applying symmetrical or Weyl ordering to a classical Hamiltonian, then the classical Hamiltonian must be evaluated in the functional integral as follows:

$$H_S(P, Q) \leftrightarrow H_{\text{eff}} = \frac{1}{2}\big(H(p', q) + H(p', q')\big);$$
$$H_W(P, Q) \leftrightarrow H_{\text{eff}} = H\Big(p', \frac{q' + q}{2}\Big). \qquad (11.3.11)$$

If we use an arbitrary ordering function to quantize a classical theory, then we can write the resulting quantum Hamiltonian as a sum of symmetrical or Weyl ordered operators. However, this sum will in general contain $O(\hbar)$ terms in the quantum Hamiltonian, and these will generate $O(\hbar)$ terms in $H_{\text{eff}}$. In such cases, no reasonable definition of the functional integral will link the original classical Hamiltonian to its quantized counterpart.

The results of this section imply that we can drop the restriction on the form of the Hamiltonian in Theorem 11.1.16 to find:

**Theorem 11.3.12.** *Assume that the evolution of a quantum particle is governed by a Hamiltonian $H$ which can be expressed as a sum of symmetric or Weyl ordered operators without introducing $O(\hbar)$ terms. Then there is a definition of the functional integral for which the amplitude $\langle q', T'|q, T\rangle$ is given by the formula*

$$\langle q', T'|q, T\rangle = \int [dq][dp] \exp\Big(i \int_T^{T'} dt\, p\dot{q} - H(p, q)\Big).$$

If we introduce the usual low-energy, non-relativistic units for mass, length, and time, we will find a factor of $\hbar^{-1}$ in the exponent. The ordinary change of units that accompanies the movement of attention from high-energy particle phenomena to classical mechanics makes this factor increase, making it formally possible to evaluate the integral by the method of stationary phase. Thus, at low energy and small momenta, the dominant contribution to the integral comes from a region in path space about the point of stationary phase for the integrand. The point of stationary phase is the classical path determined by Hamilton's equations, as we show next.

To derive Hamilton's equations as the point of stationary phase from the Hamiltonian form of the action $S(p, q)$, we note that $\dot{p}$ and $\dot{q}$ are determined solely by $p$ and $q$ respectively, and use

$$S(p,q) = \int_T^{T'} dt \left(p\dot{q} - H(p,q)\right) = [pq]_T^{T'} + \int_T^{T'} dt \left(-q\dot{p} - H(p,q)\right) \qquad (11.3.13)$$

to find for fixed-endpoint variations that:

$$\left.\begin{aligned}
\frac{\delta}{\delta p}S(p,q) &= \int_T^{T'} dt \left(\dot{q} - \frac{\partial}{\partial p}H(p,q)\right)\delta p \\
\frac{\delta}{\delta q}S(p,q) &= \int_T^{T'} dt \left(-\dot{p} - \frac{\partial}{\partial q}H(p,q)\right)\delta q
\end{aligned}\right\} \qquad (11.3.14)$$

The arbitrariness of $\delta p$ and $\delta q$ as functions of time now imply Hamilton's equations,

$$\dot{q} = \frac{\partial}{\partial p}H(p,q) \quad \text{and} \quad \dot{p} = -\frac{\partial}{\partial q}H(p,q). \qquad (11.3.15)$$

This emergence of classical mechanics as the stationary phase approximation to quantum mechanics gives us a feeling for the difference between classical and quantum evolution. Due to the fact that $\hbar \neq 0$, the contribution of all the non-classical paths to the evolution of the quantum state can become significant at high energies.

With this strong connection between classical and quantum particle theories, it seems as though all the techniques of classical mechanics could be translated into techniques of quantum mechanics. We saw in Section 2.5, however, that there is no map from the Poisson algebra of classical observables to the commutator algebra of quantum observables. Despite this obstruction, the functional integral formula for quantum evolution provides a foundation for a comparison of concepts; in particular, it shows us that where the extrema of the classical action are sufficient to determine evolution in classical mechanics, quantum-mechanical evolution is sensitive to the entire action function.

Where the previous section showed how the concept of the path integral is implicit in the quantum-mechanical completeness relation, this section has indicated how the fine points in the definition of the path integral correspond to the choice of operator ordering in the quantum Hamiltonian. Understanding of the path integral in quantum theory would not be complete without this point. Further subtleties on the mathematics of the path integral tend to be increasingly less illuminating in their physical implications, so we shall stop the descent into details at this point and turn to the application in field theory. The first step in this direction is to find a functional integral formula for vacuum-to-vacuum expectations.

## 11.4 – Wick Rotation of Time

Theorem 11.3.12 expresses the evolution from one position eigenstate to another as a path integral. In our application to field theory, we shall want a path-integral expression for the generating functional, which is a vacuum-to-vacuum expectation. The following points indicate that we can accomplish this by rotating $t$ to $e^{-i\delta}t$.

In order to mimic the conditions of the generating functional in field theory, we assume that the quantum-mechanical Hamiltonian $H$ is modified to $H_J = H + qJ$, where $J$ is a time-dependent source and $q$ is the position operator. In the following argument, we shall assume that

$$T' > t' > t_0 > -t_0 > t > T \tag{11.4.1}$$

and

$$|t| > t_0 \quad \Longrightarrow \quad J(t) = 0. \tag{11.4.2}$$

First, we introduce energy eigenstates $|\phi_n\rangle$ of $H$ with energy $E_n$, where the function $\phi_n$ satisfies

$$
\begin{aligned}
\phi_n(q, t_2, t_1) &\overset{\text{def}}{=} \langle q, t_2 | \phi_n, t_1 \rangle \\
&= \langle q | e^{-iH(t_2 - t_1)} | \phi_n \rangle \quad \text{if } J = 0 \text{ for } t_1 > t > t_2 \\
&= e^{-iE_n(t_2 - t_1)} \langle q | \phi_n \rangle \\
&= e^{-iE_n(t_2 - t_1)} \phi_n(q).
\end{aligned}
\tag{11.4.3}
$$

(When $J$ is non-vanishing on the interval $[t_1, t_2]$, the Hamiltonian is time-dependent and only the first line, the definition, is valid.) We can now express the brackets of position eigenstates in terms of these energy eigenstates:

$$
\begin{aligned}
\langle Q', T' | q', t' \rangle &= \sum_m \langle Q', T' | \phi_m, T' \rangle \langle \phi_m, T' | q', t' \rangle \\
&= \sum_m \phi_m(Q') \phi_m^*(q') e^{-iE_m(T' - t')}, \\
\langle q, t | Q, T \rangle &= \sum_n \langle q, t | \phi_n, T \rangle \langle \phi_n, T | Q, T \rangle \\
&= \sum_n \phi_n(q) \phi_n^*(Q) e^{-iE_n(t - T)}.
\end{aligned}
\tag{11.4.4}
$$

Second, we represent the vacuum-to-vacuum amplitude in terms of position eigenstates:

$$
\begin{aligned}
\langle \Omega, T' | \Omega, T \rangle &= \int dq' \, dq \, \langle \Omega, T' | q', t' \rangle \langle q', t' | q, t \rangle \langle q, t | \Omega, T \rangle \\
&= \int dq' \, dq \, \phi_0^*(q') e^{-iE_0(T' - t')} \langle q', t' | q, t \rangle \phi_0(q) e^{-iE_0(t - T)}.
\end{aligned}
\tag{11.4.5}
$$

Third, we represent the evolution of a position eigenstate in a parallel manner:

$$\langle Q', T' | Q, T \rangle \tag{11.4.6}$$

$$
\begin{aligned}
&= \int dq' \, dq \, \langle Q', T' | q', t' \rangle \langle q', t' | q, t \rangle \langle q, t | Q, T \rangle \\
&= \sum_{m,n} \int dq' \, dq \, \phi_m(Q') \phi_m^*(q') e^{-iE_m(T' - t')} \langle q', t' | q, t \rangle \phi_n(q) e^{-iE_n(t - T)} \phi_n^*(Q) \\
&= \phi_0(Q') \phi_0^*(Q) \langle \Omega, T' | \Omega, T \rangle + \cdots
\end{aligned}
$$

Now we must suppress the terms with $m$ or $n$ greater than zero. Because the vacuum is the state with lowest energy, this can be accomplished by introducing imaginary components in $T$ and $T'$ in such a way that the exponentials become damping factors:

$$\left. \begin{array}{l} T \sim -e^{-i\delta}\infty \\ T' \sim e^{-i\delta}\infty \end{array} \right\} \quad \Longrightarrow \quad \langle Q', T'|Q, T\rangle \simeq \phi_0(Q')\phi_0^*(Q)\langle \Omega, T'|\Omega, T\rangle. \qquad (11.4.7)$$

This shows that, in order to obtain a vacuum-to-vacuum amplitude from a functional integral, we only need to suppress the contributions of higher energy states to the integral by making time complex. This rotation of time corresponds to a rotation $p_0 \to e^{i\delta}p_0$ of energy, which we found in Section 10.9 to be essential to computing Feynman loop integrals. Note that this small rotation is equivalent to the Wick rotation for a convergent Feynman integral. The next step towards application of the functional integral in field theory is to develop some basic formulae for doing Gaussian functional integrals.

## 11.5 – Evaluating Gaussians

The only functional integral which can be directly evaluated is the complex Gaussian. Thinking of this Gaussian as a probability measure, we can obtain expectation values for polynomials by differentiating the formula for the value of a complex Gaussian functional integral. The details for finite-dimensional functional integrals follow.

Let $\alpha$ be the constant $\exp(-\pi i/4)$. The ordinary Fresnel integral formula

$$\int dq \, \exp\left(-\frac{i}{2}cq^2\right) = \alpha\sqrt{\frac{2\pi}{c}} \qquad (11.5.1)$$

implies the product version

$$\int dq_1 \cdots dq_n \, \exp\left(-\frac{i}{2}\sum_{r=1}^{n} c_r q_r^2\right) = \alpha^n\sqrt{\frac{(2\pi)^n}{c_1 \cdots c_n}}, \qquad (11.5.2)$$

from which we obtain the matrix version, good for symmetric invertible matrices $C$:

$$\int d^n q \, \exp\left(-\frac{i}{2}q^\top C q\right) = \alpha^n\sqrt{\frac{(2\pi)^n}{\det C}}. \qquad (11.5.3)$$

Clearly, in the limit $n \to \infty$ the product of $2\pi\alpha^2$ factors diverges. We therefore define

$$(dq) \stackrel{\text{def}}{=} \frac{dq_1 \cdots dq_n}{(2\pi)^{n/2}\alpha^n} \qquad (11.5.4)$$

(note that $[dq]$ and $[dp]$ have different normalizations, but these measures will not be used again) and rewrite the matrix form of the Gaussian integral as

$$\int (dq) \, \exp\left(-\frac{i}{2}q^\top C q\right) = (\det C)^{-1/2}. \qquad (11.5.5)$$

The matrix $C$ in this expression is called the *covariance* of the complex Gaussian.

If we have a linear term in the exponent, then we can change variables to eliminate it, and use the formulae above to evaluate the resulting pure Gaussian integral. For the one-dimensional case, let $Q$ be a quadratic function of $q$:

$$
\begin{aligned}
Q(q) &= -\frac{1}{2}cq^2 + bq + a \\
&= -\frac{1}{2}c(q - \bar{q})^2 + \frac{1}{2}c\bar{q}^2 + a \\
&= -\frac{1}{2}c(q - \bar{q})^2 + Q(\bar{q}),
\end{aligned}
\tag{11.5.6}
$$

where $\bar{q} = b/c$ is the location of the stationary value of $Q$. Writing $q'$ for the shifted variable $q - \bar{q}$, the value of the corresponding integral is given by:

$$
\begin{aligned}
\int (dq)\, e^{iQ(q)} &= \int (dq')\, \exp\left(\frac{-icq'^2}{2} + iQ(\bar{q})\right) \\
&= c^{-1/2} e^{iQ(\bar{q})}.
\end{aligned}
\tag{11.5.7}
$$

Following the logic above, this formula generalizes easily to $n$ dimensions:

$$
\begin{aligned}
Q(q) &= -\frac{1}{2}q^\top C q + b^\top q + a \\
&= -\frac{1}{2}(q - \bar{q})^\top C(q - \bar{q}) + \frac{1}{2}\bar{q}^\top C\bar{q} + c \\
&= -\frac{1}{2}(q - \bar{q})^\top C(q - \bar{q}) + Q(\bar{q}),
\end{aligned}
\tag{11.5.8}
$$

where $C$ is invertible and symmetric, and $\bar{q} = C^{-1}b$ is again the location of the stationary value of the quadratic form. The formula for the $n$-dimensional complex Gaussian integral with linear terms is therefore:

$$
\int (dq)\, e^{iQ(q)} = (\det C)^{-1/2} e^{iQ(\bar{q})}.
\tag{11.5.9}
$$

Differentiating this last formula with respect to the parameters $b$ gives rise to a formula for the expectation of a polynomial in $q$ with respect to the generalized complex Gaussian distribution:

**Theorem 11.5.10.** *Let $C$ be an $n \times n$ invertible, symmetric matrix, let $Q(q) = -\frac{1}{2}q^\top C q + b^\top q + a$ be a quadratic form on $\mathbf{R}^n$. Then, for any polynomial $P$ in $n$ variables,*

$$
\int (dq)\, P(q) e^{iQ(q)} = \frac{1}{\sqrt{\det(C)}} P\left(-i\frac{\partial}{\partial b}\right) e^{iQ(\bar{q})},
$$

*where $\bar{q} = C^{-1}b$.*

We now bravely assume that the same formula holds for the continuum limit, the functional integral, where we have replaced the vector $q$ by the function $q(t)$, the vector inner product by the $L^2$ inner product

$$
(q, q') \overset{\text{def}}{=} \int_T^{T'} dt\, q(t)^\top q'(t),
\tag{11.5.11}
$$

the $\partial/\partial b$ by a functional derivative, and the matrix $C$ by a differential operator like $\partial^2/\partial t^2$. With these substitutions in the equation of Theorem 11.5.10, we find the central principle for evaluating functional integrals:

**Principle 11.5.12.** *Let $C$ be an invertible, symmetric differential operator, let $Q(q) = -\frac{1}{2}(q, Cq) + (b, q) + a$ be a quadratic form on $\mathbf{R}^n$-valued functions of time. Then, for any polynomial $P$ in $n$ variables,*

$$\int (dq)\, P(q) e^{iQ(q)} = \frac{1}{\sqrt{\det(C)}} P\left(-i\frac{\delta}{\delta b(t)}\right) e^{iQ(\bar{q})},$$

*where $\bar{q} = C^{-1}b$. Note that the stationary value of $Q$ is given by*

$$Q(\bar{q}) = \frac{1}{2}(b, C^{-1}b) + a.$$

This formula is a principle, not a theorem, because the conditions for its validity have not been made explicit. We will simply have to hope that in all reasonable applications, these implicit conditions are met. Certainly, the principle appears sensible when the operator $A$ has a well-defined determinant, but it takes a little trickery, a kind of regularization, even to interpret the formula for the operators we will be encountering. The upshot, in any case, is that this assumption provides the basis for evaluating functional integrals that are complex Gaussian in form. This is actually the only type of functional integral to which we can assign a numerical value without using a perturbation expansion.

**Remark 11.5.13.** It is impossible to define the determinant of an operator like $\Box + \mu^2$ because this operator has a continuous spectrum. To bring out a discrete spectrum, it is necessary to make space-time compact. The usual device is to add a point at infinity, thereby converting $\mathbf{R}^4$ into the four-dimensional sphere $S^4$. This process of compactification destroys the relativistic structure of Minkowski space. However, we saw in the previous section that there is a Wick rotation of time in the functional integral formula for a vacuum-to-vacuum expectation. Consequently, we can consider replacing $\Box + \mu^2$ by the Euclidean operator $D = -\Box_{\mathrm{E}} + \mu^2$. This Euclidean operator is elliptic, and so has a discrete spectrum with finite multiplicities of each eigenvalue. Despite these good properties, the spectrum diverges, and we need a regularization procedure if we are to define the determinant of $D$. One popular regularization uses the $\zeta$ function:

$$\zeta_D(s) \stackrel{\text{def}}{=} \sideset{}{'}\sum_n \lambda_n^{-s}. \tag{11.5.14}$$

The prime on the sum restricts to summation over non-zero eigenvalues. The determinant is now defined by

$$\det{}'(D) \stackrel{\text{def}}{=} \exp\left(-\frac{\partial}{\partial s}\zeta_D(s)\right)\Big|_{s=0}. \tag{11.5.15}$$

The prime on the det indicates omission of the zero eigenvalues. For the particular case of $D = -\Box_{\mathrm{E}} + \mu^2$, the lowest eigenvalue is associated with the constant functions, and so if $\mu^2 > 0$, the eigenvalues are all positive and the prime is not needed.

It is easy to check that this formula gives the usual determinant on matrices. Furthermore, if $D$ is any second-order elliptic operator on a compact $d$-dimensional manifold, classical results on the asymptotic values of the spectrum show that $\zeta_D$ is analytic on $\mathrm{Re}\, s > d/2$ and analytic extension yields a meromorphic function with only simple poles. If $d$ is even, then the poles are at $1, 2, \ldots, d/2$; if $d$ is odd, then they are at $d/2, d/2 - 1$, and so on down to $-\infty$. In either case, $\zeta_D$ is analytic at $s = 0$, and so the determinant is well defined and finite.

The derivation of the functional integral from the completeness relation of quantum mechanics leads to the Hamiltonian form of the functional integral. It is often convenient to perform the integration over momenta (as we did above for the free particle) in order to obtain the Lagrangian form of the functional integral, an integral over $(dq)$ alone. We shall show how this works out for one-dimensional quantum mechanics.

The form which arises in 11.2.2, the functional integral expression for the evolution of a quantum particle, is

$$Q(p) = -H(p, q) + p\dot{q}. \tag{11.5.16}$$

Assuming that the Hamiltonian is a non-degenerate quadratic in momenta, $Q$ is of the type considered above. Hence $Q$ has a stationary value when the momenta are related to the velocities by Hamilton's equations

$$\dot{q} = \frac{\partial}{\partial p} H(p, q), \tag{11.5.17}$$

and these equations have a unique solution $\bar{p} = \bar{p}(q, \dot{q})$. Using Principle 11.5.12 with $P \equiv 1$ to evaluate the $(dp)$ integral, we find:

$$\langle q', T' | q, T \rangle = \int (dq)(dp) \exp\left(i \int_T^{T'} dt\, p\dot{q} - H(p, q)\right)$$
$$= \int (dq)\, N(q) \exp\left(i \int_T^{T'} dt\, \bar{p}\dot{q} - H(\bar{p}, q)\right), \tag{11.5.18}$$

where the function $N(q)$ is the divergent factor corresponding to $(\det C)^{-1/2}$. If, as often happens, the quadratic momentum term in $H$ is independent of position, then $N$ will be a constant which can be factored out to give:

$$\langle q', T' | q, T \rangle = N \int (dq) \exp\left(i \int_T^{T'} dt\, L(\dot{q}, q)\right), \tag{11.5.19}$$

which is the Lagrangian form of the functional integral.

The existence and value of the Lagrangian form of the functional integral are therefore in question. If the Hamiltonian is either not quadratic in momenta or is a degenerate quadratic, then we cannot perform the integral over momenta. A change of variables might remove this obstruction, but this remedy will not always be available. If the Hamiltonian is a non-degenerate quadratic in momenta, and if the coefficient matrix for the quadratic terms is a function of position, then the normalization factor $N$ will be a non-trivial function of $q$, and it will not be easy to use the Lagrangian form of the functional integral. One approach is the use of unphysical or ghost fields to express $N(q)$ as a Gaussian integral; this technique effectively modifies the Lagrangian by a ghost term. Because of these obstacles in obtaining a Lagrangian form, we see again that the Hamiltonian is primary in the functional integral approach to quantum theory, the Lagrangian secondary.

Complex analysis provides the formulae for one-dimensional complex Gaussian integrals. The generalization to vector-valued functions is straightforward, but the extension to functional integrals, the continuum limit of the discrete or vector case, is extremely tricky. However, the principle for evaluating complex Gaussian expectations of functions, 11.5.12, is generally given the status of a theorem in physics because it has been found, when used with care, to give sensible results.

This completes the development of formulae for evaluating functional integrals which we shall use in the context of field theory. In this context, however, the evaluation principle 11.5.12 leads to the inverse of a differential operator. It will add greatly to our understanding if we convert our formula for the stationary values of action functionals from position space to momentum space. This is the purpose of the next section.

# 11.6 Fourier Transform of Inner Products and Operators

The following section introduces functional integral quantization. The proof of its key formula involves application of our Gaussian integral formulae to the action functional. This application uses the inverse of the covariance. The covariances that arise in field theory are linear differential operators with constant coefficients. Working with their inverses is made easier by expressing the action as a function of Fourier-transformed fields. This section provides the definitions and transformations used in the next section.

Let $\hat{f}$ be the Fourier transform of a function $f$ of position $x$. Then we can transform the position-space inner product $(f, g)$ into a momentum-space inner product as follows:

$$
\begin{aligned}
(f, g) &\overset{\text{def}}{=} \int d^4x\, f(x)g(x) \\
&= \int d^4x\, d^4y\, f(x)\delta^{(4)}(x - y)g(y) \\
&= \int \frac{d^4k}{(2\pi)^4}\, d^4x\, d^4y\, f(x)e^{-i(x-y)\cdot k}g(y) \qquad (11.6.1) \\
&= \int \frac{d^4k}{(2\pi)^4}\, \hat{f}(k)\hat{g}(-k) \\
&\overset{\text{def}}{=} (\hat{f}, \hat{g}).
\end{aligned}
$$

(The notation $(\hat{f}, \hat{g})$ is loose because $\hat{f}$ is in the same Hilbert space as $f$, but it is adequate in practice as the variables indicate when to include the $(2\pi)^{-4}$ factor.) Notice that both inner products are bilinear; the definitions do not include complex conjugation. In light of the Feynman rule for a classical source, we interpret $\hat{g}(-k)$ as creation of activity at momentum $k$ and $\hat{f}(k)$ as destruction of that activity.

Let $A(x)$ be a linear differential operator. Then the Fourier transform $\hat{A}$ of $A$ is an integral operator with kernel $\hat{A}(k, l)$ as set out in the following definitions:

$$
\hat{A}\hat{g}(k) \overset{\text{def}}{=} \int \frac{d^4l}{(2\pi)^4}\, \hat{A}(k, l)\hat{g}(l) \overset{\text{def}}{=} \widehat{Ag}(k). \qquad (11.6.2)
$$

From this definition we find that:

$$\hat{A}\hat{g}(k) = \int d^4x\, e^{-ix \cdot k} A(x) g(x)$$

$$= \int d^4x\, d^4y\, e^{-ix \cdot k} A(x) \delta^{(4)}(x - y) g(y)$$

$$= \int d^4x\, d^4y\, \frac{d^4l}{(2\pi)^4} e^{-ix \cdot k} A(x) e^{i(x-y) \cdot l} g(y) \qquad (11.6.3)$$

$$= \int \frac{d^4l}{(2\pi)^4} d^4x\, e^{-ix \cdot k} A(x) e^{ix \cdot l} \hat{g}(l).$$

Hence the kernel $\hat{A}(k, l)$ is given by

$$\hat{A}(k, l) = \int d^4x\, e^{-ix \cdot k} A(x) e^{ix \cdot l}. \qquad (11.6.4)$$

If the differential operator $A(x)$ has constant coefficients, then firstly we can write $A$ as a function of the partial differentiation operator, $A = A(\partial_x)$, and secondly we can perform the integration in the formula 11.6.4 to find:

$$\hat{A}(k, l) = (2\pi)^4 \delta^{(4)}(k - l) A(\partial_x) e^{ix \cdot l} \big|_{x=0} \qquad (11.6.5)$$

$$= (2\pi)^4 \delta^{(4)}(k - l) A(ik).$$

Now the integration in the integral operator $\hat{A}$ can also be performed, and so $\hat{A}$ reduces to a multiplication operator:

$$\hat{A}\hat{g}(k) = A(ik)\hat{g}(k) \quad \Longrightarrow \quad \hat{A} = A(ik). \qquad (11.6.6)$$

From the definition of $\hat{A}$, clearly

$$(f, Ag) = (\hat{f}, \widehat{Ag}) = (\hat{f}, \hat{A}\hat{g}). \qquad (11.6.7)$$

This equality obviously implies that if $A$ is symmetric, then so is $\hat{A}$:

$$(f, Ag) = (Af, g) \quad \Longrightarrow \quad (\hat{f}, \hat{A}\hat{g}) = (\hat{A}\hat{f}, \hat{g}). \qquad (11.6.8)$$

One important consequence of using bilinear inner products is that if $A$ is symmetric, then a perturbed operator like $A - i\epsilon$ will also be symmetric. This kind of perturbation destroys the property of being hermitian.

The free-field action functional takes the form

$$S(\phi, j) = \int d^4x\, \frac{1}{2} \partial_\mu \phi\, \partial^\mu \phi - \frac{\mu^2}{2} \phi^2 - j\phi$$

$$= \int d^4x\, -\frac{1}{2} \phi (\Box + \mu^2) \phi - j\phi. \qquad (11.6.9)$$

Notice that, since we are concerned with a vacuum-to-vacuum amplitude, the classical fields $\phi$ which will arise in the functional integral vanish at space-time infinity. Hence the integration by parts above does not create a surface term.

Clearly, we can regard this action as a quadratic form in $\phi$ with covariance $C = \square + \mu^2$. Now the Klein–Gordon operator is not invertible, so we generally need to perturb the covariance, for example, define $C_\epsilon$ and $S_\epsilon$ by

$$C_\epsilon \overset{\text{def}}{=} C - i\epsilon \quad \text{and} \quad S_\epsilon \overset{\text{def}}{=} -\frac{1}{2}(\phi, C_\epsilon\phi) + (j, \phi). \qquad (11.6.10)$$

We can foresee from Principle 11.5.12 for evaluating Gaussian functional integrals that we shall be interested in the stationary value of $S_\epsilon$. This stationary value occurs at $\phi_{\text{cl}} = C_\epsilon^{-1} j$. Writing $\phi' = \phi - \phi_{\text{cl}}$, we can eliminate the term linear in $\phi$:

$$S_\epsilon = -\frac{1}{2}(\phi - \phi_{\text{cl}}, C_\epsilon\phi - C_\epsilon\phi_{\text{cl}}) - \frac{1}{2}(\phi, C_\epsilon\phi_{\text{cl}}) - \frac{1}{2}(\phi_{\text{cl}}, C_\epsilon\phi) + \frac{1}{2}(\phi_{\text{cl}}, C_\epsilon\phi_{\text{cl}}) + (j, \phi)$$

$$= -\frac{1}{2}(\phi', C_\epsilon\phi') + \frac{1}{2}(j, C_\epsilon^{-1} j). \qquad (11.6.11)$$

Notice how the cancellation of the term linear in $\phi$ depends on the symmetry of $C_\epsilon$.

**Homework 11.6.12.** Check that, if $j(x) = 0$ for $t^2 + \bar{x}^2 \geq a^2$, then the integration by parts in 11.6.9 is valid for $\phi = \phi_{\text{cl}}$.

Converting to momentum space, the stationary value is given by

$$\text{stat}(S_\epsilon) = \frac{1}{2}(j, C_\epsilon^{-1} j)$$

$$= \frac{1}{2}(\hat{\jmath}, \hat{C}_\epsilon^{-1} \hat{\jmath}) \qquad (11.6.13)$$

$$= \frac{1}{2} \int \frac{d^4 k}{(2\pi)^4} \, \hat{\jmath}(k) C_\epsilon(-ik)^{-1} \hat{\jmath}(-k).$$

We note that we are not forced to have a $-k$ on the right here, but as noted following the definition 11.6.1 of the inner products, this sign convention leads to an interpretation that momentum originates on the right, propagates through $C_\epsilon(-ik)^{-1}$, and is absorbed on the left. Furthermore, this convention will be forced on us when we consider fermions.

While it is hard to compute directly with $C_\epsilon^{-1}$, it is easy to evaluate the appropriate transform:

$$C_\epsilon(-ik)^{-1} = \frac{1}{(-ik)^2 + \mu^2 - i\epsilon} = -\frac{1}{k^2 - \mu^2 + i\epsilon}. \qquad (11.6.14)$$

This result is just a factor of $-i$ away from the propagator. We can postulate that quite generally the propagator is related to the covariance by

$$\textbf{Propagator} = -iC_\epsilon(-ik)^{-1}. \qquad (11.6.15)$$

This section shows how to use the Fourier transform to find concretely the stationary value of a quadratic action functional. From the result for the free scalar field, we propose a formula for the propagator in terms of the covariance of the action. With these ideas and results in hand, we are ready for functional integral quantization.

# 11.7 Functional Integral Quantization: Lagrangian Form

Having seen how the functional integral arises from the completeness relation in quantum mechanics, and having worked out a formula for evaluating complex Gaussian functional integrals, we are ready to apply the functional integral formalism to quantum field theory. From the functional integral formula 11.3.12 for the evolution of a quantum-mechanical particle, we guess that the evolution of a quantum field $\phi$ should be given by a formula

$$\langle f', T' | f, T \rangle \overset{?}{=} \int (d\Pi)(d\phi) \, \exp\!\Big(i \int dt \, d^3\bar{x} \, \Pi \partial_0 \phi - \mathcal{H}(\Pi, \phi)\Big), \qquad (11.7.1)$$

where the states $|f, T\rangle$ are shape states of the quantum field. These states, however, are so inconvenient in practice that we prefer to compute vacuum-to-vacuum expectations over infinite time, to make these significant by introducing a classical source, and to connect these amplitudes to the functional integral through a Wick rotation of time. In short, we need a functional integral formula for generating functionals $Z(j)$ and therefore aim to justify the following functional integral quantization principle:

**Principle 11.7.2.** *The generating functional $Z(j)$ for a scalar field $\phi$ is determined by the Hamiltonian functional integral*

$$Z(j) = \int (d\Pi)(d\phi) \, \exp\!\Big(i \int d^4x \, \Pi \partial_0 \phi - \mathcal{H}(\Pi, \phi) + j\phi\Big).$$

For a scalar field theory defined by a Lagrangian density $\mathcal{L}$, Lorentz invariance ensures that the derivatives of the field occur only in the combination $\partial_\mu \phi \, \partial^\mu \phi$. If the theory is to be renormalizable, then in fact the only such term is the canonical kinetic term. In this case, the Hamiltonian as a function of the momentum is a non-degenerate quadratic form with constant coefficients. Hence the integral over momenta in the Hamiltonian form introduces the usual infinite constant $N$, leaving the Lagrangian form. Therefore, for renormalizable scalar field theories, it is sufficient to demonstrate the validity of the Lagrangian form:

**Principle 11.7.3.** *Let $\mathcal{L}$ be the Lagrangian density for a theory of a scalar field. If the derivatives of the field occur only in the canonical kinetic term of $\mathcal{L}$, then the generating function $Z(j)$ for a scalar field $\phi$ is determined by the Lagrangian functional integral*

$$Z(j) = N \int (d\phi) \, \exp\!\Big(i \int d^4x \, \mathcal{L}(\phi, \partial\phi) + j\phi\Big),$$

*where $N$ is chosen to make $Z(0) = 1$.*

**Remark 11.7.4.** In Section 11.3 we showed how the details of the definition of the functional integral in quantum mechanics corresponded to a specification of operator ordering in the Hamiltonian. Here, such details are simply not available. In quantum mechanics, moreover, the relationship between different orderings in $p^m q^n$ terms is always finite, whereas in field theory, changing the order of a product $\Pi(x)\phi(x)$ leads to a summand with a factor of $\delta^{(3)}(0)$.

From the operator perspective, normal ordering must eliminate such divergences – a normally ordered Hamiltonian must correspond to a unique operator expression. From the perspective of Feynman perturbation theory, these divergent terms are indistinguishable from contributions of loop diagrams, and so the combined force of regularization and renormalization eliminates the ambiguity. Finally, from the perspective of functional integral quantization, the ambiguity is first swamped by the lack of precise definition of the functional measure and then eliminated by collapse of non-perturbative pretensions into the usual perturbation series.

It is interesting to note that in field theory on a discrete space-time lattice, it is natural to give the matter fields values on the lattice vertices and the gauge fields values on the lattice edges.

To justify the Lagrangian functional integral principle 11.7.3, we begin with a few definitions. Let the classical action $S$ be the space integral of the total Lagrangian,

$$S(\phi, j) \overset{\text{def}}{=} \int d^4x \, \mathcal{L}(\phi, \partial\phi) + j\phi, \tag{11.7.5}$$

and define the interacting functional integral generating functional by:

$$Z_{\text{FI}}(j) \overset{\text{def}}{=} N \int (d\phi) \, \exp\big(iS(\phi, j)\big), \tag{11.7.6}$$

where $N$ is chosen to make $Z_{\text{FI}}(0) = 1$. Let $\mathcal{L}^0$ be the free-field Lagrangian density:

$$\mathcal{L}^0 = \frac{1}{2}\partial_\mu \, \partial^\mu \phi - \frac{1}{2}\mu^2\phi^2, \tag{11.7.7}$$

and let $S^0$ and $Z_{\text{FI}}^0(j)$ be respectively the action and the functional-integral generating function associated with $\mathcal{L}^0$.

In the proof that follows, first we evaluate the functional integral in $Z_{\text{FI}}^0(j)$. Second, we use Wick diagrams to evaluate $Z^0(j)$, and thereby establish Principle 11.7.3 for the free field. Third, applying a functional differentiation trick to the free-field result, we deduce the result for interacting fields.

The action for the free scalar field, perturbed to make its covariance invertible, is given by

$$S_\epsilon(\phi, j) \overset{\text{def}}{=} \int d^4x \, -\frac{1}{2}\phi(\Box + \mu^2 - i\epsilon)\phi + j\phi. \tag{11.7.8}$$

We saw in 11.6.9 that this action is a quadratic form in $\phi$, and we found a formula 11.6.13 for its stationary value. This formula provides the basis for a swift application of the Gaussian integration principle 11.5.12 to the functional integral $Z_{\text{FI}}^0$:

$$\begin{aligned}
Z_{\text{FI}}^0(j) &= N \int (d\phi) \, \exp(iS_\epsilon) \\
&= N \det(\Box + \mu^2 - i\epsilon)^{-1/2} \exp\big(i\operatorname{stat}(S_\epsilon)\big) \\
&= \exp\left(-\frac{i}{2} \int \frac{d^4k}{(2\pi)^4} \frac{\hat{j}(k)\hat{j}(-k)}{k^2 - \mu^2 + i\epsilon}\right),
\end{aligned} \tag{11.7.9}$$

by choice of normalization factor $N$.

As we saw in Section 4.6, the theory of a free scalar field with source term has only one connected Wick diagram which is a vacuum bubble, namely the diagram $C_2$ consisting of a single line with the source at each end. The contribution $S(C_2)$ of $C_2$ to the $S$ matrix is given by:

$$S(C_2) = \frac{1}{2} \int \frac{d^4 k}{(2\pi)^4} \, (i\hat{j}(k)) \frac{i}{k^2 - \mu^2 + i\epsilon} (i\hat{j}(-k)), \qquad (11.7.10)$$

and so, by the cluster expansion theorem, the vacuum-to-vacuum amplitude is the exponential of this diagram:

$$Z^0(j) = \langle 0|S(j)|0\rangle = \exp\big(S(C_2)\big), \qquad (11.7.11)$$

which matches the final expression above for $Z^0_{\mathrm{FI}}(j)$. We conclude that the Lagrangian functional integral principle is true for the free scalar field.

Now that we have established the basic connection between functional integration and the theory of the free scalar quantum field, it is an easy matter to extend the formalism to the interacting scalar quantum field. The trick is to differentiate the free field formula

$$Z^0(j) = Z^0_{\mathrm{FI}}(j) \qquad (11.7.12)$$

with respect to the classical source $j$ in such a way that we reconstitute the generating functional $Z(j)$ from the left-hand side and the functional integral $Z_{\mathrm{FI}}(j)$ from the right-hand side. The appropriate differential operator $\mathbf{D}_j$ is defined in terms of the interaction Lagrangian density $\mathcal{L}'$:

$$\mathbf{D}_j \stackrel{\text{def}}{=} \exp\left(i \int d^4 y \, \mathcal{L}'\left(-i\frac{\delta}{\delta j(y)}\right)\right). \qquad (11.7.13)$$

In the following argument, we demonstrate directly that

$$\mathbf{D}_j Z^0_{\mathrm{FI}}(j) \propto Z_{\mathrm{FI}}(j), \qquad (11.7.14)$$

and use Dyson's formula to show that

$$\mathbf{D}_j Z^0(j) \propto Z(j). \qquad (11.7.15)$$

In the application of $\mathbf{D}_j$ to the two forms of the generating function, the heart of the computation is the basic identity,

$$-i\frac{\delta}{\delta j(y)}\left(i \int d^4 x \, j(x)\phi(x)\right) = \phi(y), \qquad (11.7.16)$$

from which we deduce at once that

$$\mathbf{D}_j\left(i \int d^4 x \, j(x)\phi(x)\right) = \exp\left(i \int d^4 x \, \mathcal{L}'(\phi)\right). \qquad (11.7.17)$$

The effect of differential operator $\mathbf{D}_j$ on the functional integral $Z_{\mathrm{FI}}^0$ can be worked out in just a few lines:

$$
\begin{aligned}
\mathbf{D}_j Z_{\mathrm{FI}}^0(j) &= \mathbf{D}_j N \int (d\phi) \exp\left(i \int d^4x \, \mathcal{L}^0(\phi) + j\phi\right) \\
&= N \int (d\phi) \, \mathbf{D}_j \exp\left(i \int d^4x \, \mathcal{L}^0(\phi) + j\phi\right) \\
&= N \int (d\phi) \exp\left(i \int d^4x \, \mathcal{L}^0(\phi) + j\phi\right) \mathbf{D}_j\left(i \int d^4x \, j\phi\right) \\
&= N \int (d\phi) \exp\left(i \int d^4x \, \mathcal{L}^0(\phi) + \mathcal{L}'(\phi) + j\phi\right) \\
&= NN' Z_{\mathrm{FI}}(j).
\end{aligned}
\tag{11.7.18}
$$

Similarly, the differentiation of the time-ordered product is straightforward. The field $\phi$ here is the interaction picture field $\phi_I$, but $j$ is still classical, and hence unaffected by time ordering:

$$
\begin{aligned}
\mathbf{D}_j Z^0(j) &= \mathbf{D}_j \langle 0|T \exp\left(i \int d^4x \, j(x)\phi(x)\right)|0\rangle \\
&= \langle 0|T \mathbf{D}_j \exp\left(i \int d^4x \, j(x)\phi(x)\right)|0\rangle \\
&= \langle 0|T \exp\left(i \int d^4x \, j(x)\phi(x)\right) \mathbf{D}_j\left(i \int d^4x \, j(x)\phi(x)\right)|0\rangle \\
&= \langle 0|T \exp\left(i \int d^4x \, j(x)\phi(x)\right) \exp\left(i \int d^4x \, \mathcal{L}'(\phi)\right)|0\rangle \\
&= \langle 0|T \exp\left(i \int d^4x \, \mathcal{L}'(\phi(x)) + j(x)\phi(x)\right)|0\rangle \\
&= N'' Z(j).
\end{aligned}
\tag{11.7.19}
$$

Putting the parts together, we conclude that the application of the functional differentiation operator $\mathbf{D}_j$ to the free-field equation yields the interacting field equation:

$$
NN' Z_{\mathrm{FI}}(j) = \mathbf{D}_j Z_{\mathrm{FI}}^0(j) = \mathbf{D}_j Z^0(j) = N'' Z(j).
\tag{11.7.20}
$$

The normalization of the generating functionals implies that $NN' = N''$.

Having established Principle 11.7.3 formally without worrying about the as yet unavoidable absence of detailed definitions and technicalities of real analysis, we note that this new form of the generating functional yields a new expression for the Green functions:

$$
G(x_1, \ldots, x_n) = \int (d\phi) \, e^{iS(\phi)} \phi(x_1) \cdots \phi(x_n).
\tag{11.7.21}
$$

It is remarkable that the vacuum expectation of a time-ordered product of interacting fields can be computed using commuting classical fields alone.

Complex scalar fields $\psi$ are handled formally by the usual device of treating $\psi$ and $\psi^\dagger$ as independent variables, and formulae are verified by returning to a basis of hermitian fields. Thus for a finite-dimensional, complex vector $\psi$ we have:

$$
\begin{aligned}
\int (d\psi^\dagger)(d\psi) \exp(-i\psi^\dagger A \psi) &= \int (d\phi_1)(d\phi_2) \exp\left(-\frac{i}{2}\phi_1 C\phi_1 - \frac{i}{2}\phi_2 C\phi_2\right) \\
&= (\det C)^{-1},
\end{aligned}
\tag{11.7.22}
$$

and so the generating functional for the free, complex field $\psi$ may be expressed as the functional integral:

$$Z_0(j^\dagger, j) = N \int (d\psi)(d\psi^\dagger) \tag{11.7.23}$$

$$\times \exp\left(i \int d^4x \, \partial_\mu \psi^\dagger \partial^\mu \psi - \mu^2 \psi^\dagger \psi + j^\dagger \psi + j\psi^\dagger\right).$$

Note that, in order to make differentiation with respect to $j$ and $j^\dagger$ independent, we must list $j$ and $j^\dagger$ as separate variables in $Z_0$.

We have set up our analysis of action functionals using bilinear inner products. This is essential if we are to preserve the symmetry of the covariance and the ready application of our Gaussian integral evaluation principle. In the case of complex fields, we must therefore include the complex conjugates explicitly:

$$S_\epsilon = -(\psi^*, C_\epsilon \psi) + (j^*, \psi) + (\psi^*, j). \tag{11.7.24}$$

Now the conditions for a stationary value involve separate variation of $\psi$ and $\psi^*$. The result is

$$\psi_{\mathrm{cl}} = C_\epsilon^{-1} j \quad \text{and} \quad \psi_{\mathrm{cl}}^* = C_\epsilon^{-1} j^*. \tag{11.7.25}$$

Note that $\psi_{\mathrm{cl}}^*$ is not equal to $(\psi_{\mathrm{cl}})^\dagger$. This discrepancy is permissible because we are integrating over $\psi$ and $\psi^*$ as separate variables. Working with the underlying real fields confirms that this discrepancy is necessary. Because of the symmetry of $C_\epsilon$ with respect to the bilinear inner product, we find the expected formula for the stationary value:

$$S_\epsilon = -(\psi^* - \psi_{\mathrm{cl}}^*, C_\epsilon(\psi - \psi_{\mathrm{cl}})) + (\psi_{\mathrm{cl}}^*, C_\epsilon \psi_{\mathrm{cl}})$$
$$\implies \quad \mathrm{stat}(S_\epsilon) = (j^*, C_\epsilon^{-1} j). \tag{11.7.26}$$

Clearly, the proof of the validity of the Lagrangian form of the functional integral form of the generating functional for complex fields will follow the form of the proof for real fields. Indeed, we can see that reduction to the underlying real fields obviates the need for a separate proof.

We have proved the fundamental relationship 11.7.3 between functional integration and canonical quantization in the context of the scalar field. With a view to making functional integral quantization self-sufficient, the next section uses a scalar field theory example to make a direct link between functional integration and Feynman rules. After this example, we shall prove the validity of the Hamiltonian form of the generating function 11.7.2.

# 11.8 – Perturbation Expansion

We have derived the Feynman–Kac formula for the generating functional using Dyson's formula for the $S$ matrix. Previously, we derived the Feynman rules for perturbation theory from Dyson's formula. To complete our understanding of the functional integral formulation of scalar field theory, we now recover perturbation theory from the functional integral.

It is not obvious how to unfold the functional integral $Z_{\mathrm{FI}}$ into any kind of series. A trick, a transformation in viewpoint is necessary. We start from an intermediate form of the functional integral formula with a new $N$:

$$Z_{\mathrm{FI}}(j) = N\mathbf{D}_j Z^0(j) \tag{11.8.1}$$

$$= N \exp\left(i \int d^4 y\, \mathcal{L}'\left(-i\frac{\delta}{\delta j(y)}\right)\right) \exp\left(-\frac{1}{2} \int \frac{d^4 k}{(2\pi)^4} \frac{\hat{j}(k) i \hat{j}(-k)}{k^2 - \mu^2 + i\epsilon}\right).$$

The appearance of the Feynman propagator in the operand $Z^0$ suggests that we should associate each term

$$-\frac{1}{2} \int \frac{d^4 k}{(2\pi)^4} \frac{\hat{j}(k) i \hat{j}(-k)}{k^2 - \mu^2 + i\epsilon} \tag{11.8.2}$$

with an internal line in a Feynman diagram. To bring out the value of contraction between vertices associated with an internal line, we would like to interchange the operator and the operand. This can be accomplished by extending a curious Euclidean-space formula, valid for Euclidean-vector variables $j$ and $\phi$ and functions $D$ and $Z$ of Euclidean vectors,

$$D\left(-i\frac{\partial}{\partial j}\right) Z(j) \equiv Z\left(-i\frac{\partial}{\partial \phi}\right)\left(D(\phi) e^{ij\cdot\phi}\right)\Big|_{\phi=0}, \tag{11.8.3}$$

to the equivalent expression for functions and functionals:

**Principle 11.8.4.** *If $j$ and $\phi$ are functions on Minkowski space, and $D$ and $Z$ are functionals, then we have the following identity:*

$$D\left(-i\frac{\delta}{\delta j(y)}\right) Z(j) \equiv Z\left(-i\frac{\delta}{\delta \phi(x)}\right)\left(D(\phi) \exp\left(i \int d^4 z\, \phi(z) j(z)\right)\right)\Big|_{\phi=0}.$$

**Homework** 11.8.5. Verify the Euclidean space formula 11.8.3 in the case where $D$ and $Z$ are monomials in $n$ variables. Deduce that the formula is true when $D$ and $Z$ are analytic functions.

For the interchange principle above to be true, clearly we need some constraints on the functionals $D$ and $Z$. Rather than give precise general conditions, we shall merely observe that, in the restricted class of $D$'s and $Z$'s we shall encounter, the principle applies. Notice also that $\phi$ here is a dummy variable: setting $\phi = 0$ at the end of the differentiation eliminates $\phi$ from the final expression.

To apply this formula to $Z_{\mathrm{FI}}(j)$ as presented in 11.8.1, we first convert the $\hat{j}$'s in $Z^0(j)$ to $j$'s and then perform the interchange of operator and operand:

$$Z_{\mathrm{FI}}(j) = N \exp\left(i \int d^4z\, \mathcal{L}'\left(-i\frac{\delta}{\delta j(z)}\right)\right) \tag{11.8.6}$$

$$\times \exp\left(\frac{1}{2}\int \frac{d^4k}{(2\pi)^4}\, d^4x\, d^4y\, \frac{(ij(x)e^{-ix\cdot k})i(ij(y)e^{iy\cdot k})}{k^2 - \mu^2 + i\epsilon}\right)$$

$$= N \exp\left(\frac{1}{2}\int \frac{d^4k}{(2\pi)^4}\, d^4x\, d^4y\, \left(e^{-ix\cdot k}\frac{\delta}{\delta\phi(x)}\right)\frac{i}{k^2 - \mu^2 + i\epsilon}\left(e^{iy\cdot k}\frac{\delta}{\delta\phi(y)}\right)\right)$$

$$\times \left(\exp\left(i\int d^4z\, \mathcal{L}'(\phi(z)) + j(z)\phi(z)\right)\right)\bigg|_{\phi=0}.$$

Clearly the second exponential provides the vertices, while the first performs the contractions. Notice that performing the integral over $d^4k$ in the first term effectively replaces the propagator by the Green function for the covariance, so the link between the Lagrangian and the propagator is still clearly evident.

Now we want to identify the Feynman diagrams in the formula above for $Z_{\mathrm{FI}}(j)$. The central point is, as noted before, that the Feynman propagator for the scalar field is already conspicuous. To make the details more clear, let us assume that

$$\mathcal{L}'(\phi) = -\frac{g}{3!}\phi^3. \tag{11.8.7}$$

Now expand the exponentials in the formula for $Z_{\mathrm{FI}}(j)$ into a sum of powers of the contraction differential operator

$$\mathbf{C} \stackrel{\text{def}}{=} \int \frac{d^4k}{(2\pi)^4}\, d^4x\, d^4y\, \left(e^{-ix\cdot k}\frac{\delta}{\delta\phi(x)}\right)\frac{i}{k^2 - \mu^2 + i\epsilon}\left(e^{iy\cdot k}\frac{\delta}{\delta\phi(y)}\right) \tag{11.8.8}$$

acting on a sum of products of interaction and source terms

$$I(\phi^3) \stackrel{\text{def}}{=} -ig \int d^4z\, \phi(z)^3,$$
$$I(\phi j) \stackrel{\text{def}}{=} i \int d^4z\, \phi(z)j(z). \tag{11.8.9}$$

Using the shorthand for the terms in $Z_{\mathrm{FI}}$ defined above, we can expand the exponentials in $Z_{\mathrm{FI}}$ and multiply the expansions out into a sum of operators acting on field monomials:

$$Z_{\mathrm{FI}}(j) = \exp\left(\frac{1}{2}\mathbf{C}\right)\exp\left(\frac{1}{3!}I(\phi^3) + I(\phi j)\right)\bigg|_{\phi=0}$$

$$= \sum_{r=0}^{\infty}\frac{1}{2^r\, r!}\mathbf{C}^r \sum_{s=0}^{\infty}\frac{1}{s!}\left(\frac{1}{3!}I(\phi^3) + I(\phi j)\right)^s\bigg|_{\phi=0} \tag{11.8.10}$$

$$= \sum_{r=0}^{\infty}\sum_{s=0}^{\infty}\sum_{t=0}^{s}\frac{1}{2^r\, r!\, s!}\frac{s!}{t!\,(s-t)!}\frac{1}{3!^t}\mathbf{C}^r I(\phi^3)^t I(\phi j)^{s-t}\bigg|_{\phi=0}.$$

The final evaluation at $\phi = 0$ will eliminate any terms in which the order of the differential operator does not exactly match the degree of the monomial in $\phi$. Therefore we may assume that

$$2r = 3t + (s - t), \tag{11.8.11}$$

and eliminate $s$ from the summation using the general rule

$$\sum_{s=0}^{\infty} \sum_{t=0}^{s} \equiv \sum_{t=0}^{\infty} \sum_{s=t}^{\infty}, \tag{11.8.12}$$

together with the constraint that $r$ is non-negative, to obtain the upper bound for $t$ in the final formula for $Z_{\text{FI}}$:

$$Z_{\text{FI}}(j) = \sum_{r=0}^{\infty} \sum_{t=0}^{[2r/3]} \frac{1}{2^r\, 3!^t\, r!\, t!\, (2r - 3t)!} \mathbf{C}^r I(\phi^3)^t I(\phi j)^{2r-3t} \bigg|_{\phi=0}. \tag{11.8.13}$$

We construct diagrams for the terms which are non-zero after differentiation and evaluation at $\phi = 0$ as follows. Represent the interaction factor $I(\phi^3)$ by a vertex with three lines and the source factor $I(\phi j)$ by a vertex with one line. Represent the differential operator $\mathbf{C}$ by a line. When $\mathbf{C}$ acts on the interactions and source factors, contract the ends of its line with those of the vertices representing the terms on which the operator acts. Note that the non-vanishing terms are precisely those in which every $\phi$ in the interactions and source terms has been differentiated away, leaving, as we would expect, a diagram with no external lines.

Having represented the non-zero terms with diagrams, it remains to describe the Feynman rules. The Feynman rules must reconstruct from the diagram the value of the appropriate part of a perturbation monomial. A diagram constructed according to the rules above does not show which operator acts on which term in the interactions and source factors, nor how the two linear operators in $\mathbf{C}$ act. Thus the $r!$ ways of associating $\mathbf{C}$'s from $\mathbf{C}^r$ with $r$ lines, and the $2^r$ ways in which the individual $\mathbf{C}$'s can act correspond to $2^r\, r!$ separate, identical terms that arise in the perturbation expansion after computing the derivatives. Furthermore, differentiating the $I(\phi^3)^t$ term will produce a factor of $(3!)^t\, t!$, and differentiating the $I(\phi j)$ term will produce a factor of $(2r - 3t)!$. Therefore the combinatorics of differentiation eliminate the combinatorial factor

$$\frac{1}{2^r\, 3!^t\, r!\, t!\, (2r - 3t)!} \tag{11.8.14}$$

in the perturbation monomial from the Feynman value associated with the diagram. This simplification is a special feature of $\phi^3$-theory; more generally, in $\phi^4$-theory for example, the combinatoric factors reduce to the reciprocal of a non-trivial symmetry factor for the diagram.

Now we will break up $\mathbf{C}$ into parts associated with the vertices with which the line is contracted and a part associated with the line itself. The Feynman rule for the line itself is just the usual one:

**Internal line with momentum $k$:** $\qquad \displaystyle\int \frac{d^4k}{(2\pi)^4} \frac{i}{k^2 - \mu^2 + i\epsilon}. \tag{11.8.15}$

Making this choice for the propagator part of 11.8.8 leaves factors

$$\mathbf{C}_- \stackrel{\text{def}}{=} \int d^4x\, e^{-ix\cdot k}\frac{\delta}{\delta\phi(x)} \quad \text{and} \quad \mathbf{C}_+ \stackrel{\text{def}}{=} \int d^4y\, e^{iy\cdot k}\frac{\delta}{\delta\phi(y)} \tag{11.8.16}$$

to associate with the vertices. Comparison of the exponentials in $\mathbf{C}$ with those in the integral form of the free field suggests the following interpretation:

**Principle 11.8.17.** *Regard $\mathbf{C}_-$ as annihilation of a particle and $\mathbf{C}_+$ as creation so that differentiation of an interaction by $\mathbf{C}_-$ leads to incoming momentum and differentiation by $\mathbf{C}_+$ leads to outgoing momentum.*

Thus application of $\mathbf{C}_-$ to an interaction leads to a vertex with incoming momenta, while application of $\mathbf{C}_+$ leads to outgoing momenta. Hence, when we create an internal line by applying $\mathbf{C}$, momentum flows from the $\mathbf{C}_+$ end to the $\mathbf{C}_-$ end.

In diagrams, in light of the computations which follow, this principle becomes

$$\tag{11.8.18}$$

The vertex rule arises from application of $\mathbf{C}_-^3$ to $I(\phi^3)$. When we differentiate $I(\phi^3)$ three times, we obtain two delta functions:

$$\frac{\delta}{\delta\phi(x)}\frac{\delta}{\delta\phi(y)}\frac{\delta}{\delta\phi(z)}\int d^4z\,\phi(z)^3 = 3\frac{\delta}{\delta\phi(x)}\frac{\delta}{\delta\phi(y)}\phi(z)^2$$

$$= 6\delta^{(4)}(x-z)\delta^{(4)}(y-z). \tag{11.8.19}$$

Notice that this result is symmetrical in $x$, $y$, and $z$. Therefore, putting all the factors together, we obtain a Feynman factor for the interaction vertex with incoming momenta $k$, $p$, and $q$:

$$\textbf{Vertex Factor} = \mathbf{C}_-^3\frac{1}{3!}I(\phi^3) \tag{11.8.20}$$

$$= -ig\int d^4x\, d^4y\, d^4z\, e^{-ix\cdot k}e^{-iy\cdot p}e^{-iz\cdot q}\delta^{(4)}(x-z)\delta^{(4)}(y-z)$$

$$= -ig(2\pi)^4\delta^{(4)}(k+p+q),$$

which is the ordinary Feynman rule for the $\phi^3$ vertex. Note that the $-iH_I$ of Dyson's formula corresponds to the $iS$ of the functional integral, so in either case the coupling constants in the Lagrangian density pick up a factor of $i$.

Finally, when we differentiate $I(\phi j)$, taking the classical source as the source of momentum $k$ which flows along the line, we obtain the source factor

$$\textbf{Source Factor} = \mathbf{C}_+I(\phi j)$$

$$= \int d^4y\, e^{iy\cdot k}\frac{\delta}{\delta\phi(y)}I(\phi j)$$

$$= \int d^4y\, ij(y)e^{iy\cdot k} \tag{11.8.21}$$

$$= i\hat{j}(-k),$$

which is the correct Feynman rule for the current vertex.

This example shows that when we have a Lagrangian form of the generating functional, then the whole of perturbation theory is available in the functional integral form of the generating functional. The key to discerning the ordinary Feynman rules in the functional integral is the relationship between the covariance $C = \Box + \mu^2$ and the propagator. Having located the propagator in the generating functional, vertex rules are obtained directly from the Lagrangian by functional differentiation.

We have not yet shown that the Lagrangian form is valid in the presence of derivative interactions, but if we assume that it is valid, then for each derivative in an interaction we must compute

$$
\begin{aligned}
\mathbf{C}_- \partial^\mu \phi &= \int d^4x\, e^{-ix\cdot k} \frac{\delta}{\delta\phi(x)} \partial^\mu_y \phi(y) \\
&= \int d^4x\, e^{-ix\cdot k} \partial^\mu_y \delta^{(4)}(x-y) \\
&= \int d^4x\, - e^{-ix\cdot k} \partial^\mu_x \delta^{(4)}(x-y) \qquad (11.8.22)\\
&= \int d^4x\, \delta^{(4)}(x-y) \partial^\mu_x e^{-ix\cdot k} \\
&= -ik^\mu e^{-iy\cdot k},
\end{aligned}
$$

where $k$ is the incoming momentum associated with the derived field. This shows that the vertex rule prescribes an extra factor of $-ik^\mu$ for each derivative in the interaction monomial.

**Homework** 11.8.23. Use the Lagrangian form of the functional integral to derive the Feynman rules for the interactions $\psi^\dagger \psi \phi$, $\frac{1}{4}\phi_1^2\phi_2^2$, and $\psi^\dagger\psi\partial_\mu\phi\,\partial^\mu\phi$ where $\psi$ is a complex scalar and the $\phi$'s are real scalars.

In light of Section 10.16, the generalization of functional integral quantization to unstable fields is based on the transfer of the imaginary mass term from the counterterm Lagrangian density to the physical Lagrangian density:

$$
\mathcal{L}_{\text{ph}} \longrightarrow \mathcal{L}_{\text{ph}} - \frac{i}{2}\mu\Gamma\phi^2 \quad \text{and} \quad \mathcal{L}_{\text{ct}} \longrightarrow \mathcal{L}_{\text{ct}} + \frac{i}{2}\mu\Gamma\phi^2. \qquad (11.8.24)
$$

(The parameter $\Gamma$ is given by $\text{Im}\,\Pi(\mu^2)$, after $\Pi$ has been computed using $\Gamma = 0$.) With the Lagrangian density so organized, we can use the form 11.8.1 of the generating functional and the operator exchange principle 11.8.4 to extract the correct propagator for the unstable field.

From the detailed computations of this section, we see that functional integral quantization yields the same result as canonical quantization for scalar field theories; from either viewpoint, we can obtain the Feynman rules directly from the Lagrangian. As long as we take care to start from the Hamiltonian form of the generating functional, it is reasonable to generalize this relationship between the Lagrangian form and Feynman rules to other theories. Furthermore, if we have a theory with derivative interactions which has a valid Lagrangian form, then the procedure presented above for obtaining the vertex rule will yield covariant Feynman rules. This is the first indication that functional integration quantization will prove to be more powerful than canonical quantization.

## 11.9 – Hamiltonian Form

Now that we have shown how perturbation theory may be derived from the Lagrangian form of the generating function, we recall the more fundamental Hamiltonian form 11.7.2 and proceed with a direct justification for this form.

Express the Hamiltonian density $\mathcal{H}$ for a theory of a scalar field $\phi$ in terms of the canonical momentum $\Pi$, an interaction Hamiltonian density $\mathcal{H}'$, and source terms for $\phi$ and $\Pi$:

$$\mathcal{H} = \tfrac{1}{2}\Pi^2 - \tfrac{1}{2}\partial_r\phi\partial^r\phi + \tfrac{1}{2}\mu^2\phi^2 + \mathcal{H}'(\Pi,\phi) - J\phi - K\Pi. \tag{11.9.1}$$

The Hamiltonian form of the action $S_H$ is given by

$$S_H(\phi,\Pi;J,K) = \int d^4\bar{x}\,\Pi\partial_0\phi - \mathcal{H}(\Pi,\phi;J,K). \tag{11.9.2}$$

Define the functional integral generating functional $Z_{\text{FI}}$ by

$$Z_{\text{FI}}(J,K) \stackrel{\text{def}}{=} \int (d\Pi)(d\phi)\, e^{iS_H(\phi,\Pi;J,K)}. \tag{11.9.3}$$

The generating functional $Z(J,K)$ is defined in terms of the interaction-picture fields $\phi_I$ and $\Pi_I$ by Dyson's formula:

$$Z(J,K) \stackrel{\text{def}}{=} \langle 0|T\exp\left(-i\int d^4x\,\mathcal{H}'(\Pi_I,\phi_I) - J\phi_I - K\Pi_I\right)|0\rangle. \tag{11.9.4}$$

Our goal is to justify the equality of these two generating functions, that is, $Z_{\text{FI}} = Z$.

Following the steps used in the justification of the Lagrangian form, we first justify the equality for the free field by evaluation of the two generating functions and second apply a differential operator to bring out the interaction Hamiltonian on both sides of the equality. Because of the similarity with the previous justification, we shall not present all the details in this one.

The free-field generating function $Z^0(J,K)$ is a sum of vacuum diagrams, which is equal to the exponential of the sum of connected vacuum diagrams. For the free-field theory with sources, there are two Wick vertices:

**Wick Vertices** :    $J\bullet\!\!-\!\!-\phi$   and   $K\bullet\!\!-\!\!-\Pi$ \qquad (11.9.5)

There are three connected vacuum diagrams derived from the three non-zero contractions between $\phi$'s and $\Pi$'s. Their contributions to the $S$ matrix are given by:

$$S_{JJ} = -\frac{1}{2}\int \frac{d^4k}{(2\pi)^4}\,\frac{i}{k^2-\mu^2+i\epsilon}\,\hat{J}(k)\hat{J}(-k),$$

$$S_{JK} = -\int \frac{d^4k}{(2\pi)^4}\,\frac{k^0}{k^2-\mu^2+i\epsilon}\,\hat{J}(k)\hat{K}(-k), \tag{11.9.6}$$

$$S_{KK} = -\frac{1}{2}\int \frac{d^4k}{(2\pi)^4}\,\frac{i(\bar{k}^2+\mu^2)}{k^2-\mu^2+i\epsilon}\,\hat{K}(k)\hat{K}(-k).$$

Notice the symmetry factors. Hence $Z^0$ is given by

$$Z^0(J, K) = \exp\left(S_{JJ} + S_{JK} + S_{KK}\right). \tag{11.9.7}$$

The free-field action in Hamiltonian form, with a little integration by parts, may be written in terms of a covariance matrix:

$$S_H^0(\phi, \Pi; J, K) \tag{11.9.8}$$

$$= \int d^4x \, \frac{1}{2}\left(\Pi\partial_0\phi - \phi\partial_0\Pi - \Pi^2 - \phi\partial_r\partial^r\phi - \mu^2\phi\right) + J\phi + K\Pi$$

$$= \int d^4x \, -\frac{1}{2}(\phi \quad \Pi)\begin{pmatrix} \partial_r\partial^r + \mu^2 & \partial_0 \\ -\partial_0 & 1 \end{pmatrix}\begin{pmatrix} \phi \\ \Pi \end{pmatrix} + J\phi + K\Pi.$$

Write $C$ for the covariance, the matrix of differential operators. To evaluate the functional integral of $\exp(S_H^0)$, we need the matrix of operators to be invertible. Hence we regularize $C$ to $C_\epsilon$ defined by

$$C_\epsilon(\partial) \stackrel{\text{def}}{=} \begin{pmatrix} \partial_r\partial^r + \mu^2 - i\epsilon & \partial_0 \\ -\partial_0 & 1 \end{pmatrix}. \tag{11.9.9}$$

Using the obvious extension of the formula 11.6.13 for the stationary value of the action for a free scalar field to vectors of fields, we can find the stationary value of $S_H^0$. First, from

$$C_\epsilon(-ik) = \begin{pmatrix} \bar{k}^2 + \mu^2 - i\epsilon & -ik_0 \\ ik_0 & 1 \end{pmatrix}, \tag{11.9.10}$$

compute

$$\hat{C}_\epsilon^{-1} = -\frac{1}{k^2 - \mu^2 + i\epsilon}\begin{pmatrix} 1 & ik_0 \\ -ik_0 & \bar{k}^2 + \mu^2 - i\epsilon \end{pmatrix}. \tag{11.9.11}$$

Second, writing $q$ and $b$ respectively for the column vectors $(\phi, \Pi)^\top$ and $(J, K)^\top$, substitute in 11.6.13 to obtain the stationary value

$$\text{stat}(S_H^0) = \frac{1}{2}(\hat{b}, \hat{C}_\epsilon^{-1}\hat{b}) \tag{11.9.12}$$

$$= -\frac{1}{2}\int \frac{d^4k}{(2\pi)^4}\frac{1}{k^2 - \mu^2 + i\epsilon}(\hat{J} \quad \hat{K})\begin{pmatrix} 1 & ik_0 \\ -ik_0 & \bar{k}^2 + \mu^2 - i\epsilon \end{pmatrix}\begin{pmatrix} \hat{J}^* \\ \hat{K}^* \end{pmatrix}$$

$$= -i(S_{JJ} + S_{JK} + S_{KK}),$$

where we have written $\hat{J}^*$ and $\hat{K}^*$ to indicate $\hat{J}(-k)$ and $\hat{K}(-k)$ respectively. Third, we conclude that

$$Z_{\text{FI}}^0(J, K) \stackrel{\text{def}}{=} \int (d\phi)(d\Pi)\, e^{iS_H^0} = \exp\left(i\,\text{stat}(S_H^0)\right) = Z^0(J, K). \tag{11.9.13}$$

**Homework** 11.9.14. Show that making a Wick rotation $k_0 \to k_0(1 + i\epsilon)$ in $\hat{C}$ leads to an equivalent value for $\hat{C}_\epsilon^{-1}$.

For the interacting fields, we use the interaction Hamiltonian $\mathcal{H}'(\phi, \Pi)$ density to define a differential operator in just the way that we used the interaction Lagrangian density in Section 11.7. The appropriate differential operator $\mathbf{D}$ is

$$\mathbf{D} \overset{\text{def}}{=} \exp\left(-i \int d^4x\, \mathcal{H}'\left(-i\frac{\delta}{\delta J(x)}, -i\frac{\delta}{\delta K(x)}\right)\right). \tag{11.9.15}$$

It is now a routine matter to verify that $\mathbf{D}Z^0 \propto Z$ and $\mathbf{D}Z^0_{\text{FI}} \propto Z_{\text{FI}}$, and so to arrive at the desired conclusion:

$$Z^0 = Z^0_{\text{FI}} \quad \Longrightarrow \quad Z = Z_{\text{FI}}. \tag{11.9.16}$$

Following 11.8.6, we see how to derive the propagators from the free-field generating functional by the substitution

$$Z^0(J, K) \to Z^0\left(-i\frac{\delta}{\delta J}, -i\frac{\delta}{\delta K}\right). \tag{11.9.17}$$

From this computation we obtain the propagator matrix $-iC^{-1}(-ik)$ as expected (refer to 11.9.11), and, recalling the creation-annihilation rule 11.8.17, we deduce the four propagators:

$$
\left.
\begin{array}{ll}
\overset{\phi \quad \overset{\leftarrow}{k} \quad \phi}{} : & \dfrac{i}{k^2 - \mu^2 + i\epsilon} \\[2mm]
\overset{\phi \quad \overset{\leftarrow}{k} \quad \Pi}{} : & -\dfrac{k^0}{k^2 - \mu^2 + i\epsilon} \\[2mm]
\overset{\Pi \quad \overset{\leftarrow}{k} \quad \phi}{} : & \dfrac{k^0}{k^2 - \mu^2 + i\epsilon} \\[2mm]
\overset{\Pi \quad \overset{\leftarrow}{k} \quad \Pi}{} : & \dfrac{i(\bar{k}^2 + \mu^2)}{k^2 - \mu^2 + i\epsilon}
\end{array}
\right\} \tag{11.9.18}
$$

Having confirmed the correctness of these propagators through the following homework exercises, we see that the sign convention in the definition of the momentum-space inner product 11.6.1 leads to the correct sign of $k^0$ in $\Pi\phi$ and $\phi\Pi$ propagators.

**Homework** 11.9.19. Starting from the equation 11.9.13 for $Z^0(J, K)$, write out the details of the transformation 11.9.17 of $Z^0(J, K)$ into a partial derivative operator and derive the four propagators above.

**Homework** 11.9.20. We already derived contractions for derived fields in Section 9.3. Show that these contractions contain the four propagators 11.9.18.

The justification above for the validity of the Hamiltonian form of the generating functional applies to Hamiltonian densities which are independent of some canonical momenta or contain derivative interactions, Hamiltonian densities which do not directly give rise to Lagrangian densities. It also introduces the matrix of propagators and confirms the relationship between covariance and propagators proposed in 11.6.15:

Matrix of Propagators $= -i$(Momentum Space Covariance at $-ik$)$^{-1}$.   (11.9.21)

# 11.10 Extension to the Free Massive Vector Field

The functional integral quantization technique described above for scalar fields extends readily to the massive vector field. Functional integration for the massive vector field is completely analogous to scalar functional integration. We shall cover the idea briefly and leave verification of details as exercises.

For the free-field theory we have

$$
\begin{aligned}
\mathcal{L} &= -\tfrac{1}{4} F_{\mu\nu} F^{\mu\nu} + \tfrac{1}{2} \mu^2 A_\mu A^\mu \\
&= -\tfrac{1}{4}(\partial_\mu A_\nu - \partial_\nu A_\mu)(\partial^\mu A^\nu - \partial^\nu A^\mu) + \tfrac{1}{2}\mu^2 A_\mu A^\mu \\
&= -\tfrac{1}{2}(\partial_\mu A_\nu \partial^\mu A^\nu - \partial_\nu A_\mu \partial^\mu A^\nu) + \tfrac{1}{2}\mu^2 A_\mu A^\mu.
\end{aligned}
\tag{11.10.1}
$$

When we integrate over space to form the action $S^0$, we can integrate by parts to obtain:

$$
\begin{aligned}
S^0 &= \frac{1}{2} \int d^4x \, (A_\nu \partial_\mu \partial^\mu A^\nu - A_\mu \partial^\mu \partial_\nu A^\nu) + \mu^2 A_\mu A^\mu \\
&= -\frac{1}{2} \int d^4x \, A_\nu \big(\partial^\nu \partial_\mu - (\Box + \mu^2) g^\nu{}_\mu \big) A^\mu.
\end{aligned}
\tag{11.10.2}
$$

The covariance operator $C$ is a matrix. Writing $\mathbf{1}$ for the matrix $g^\nu{}_\mu$, we find

$$
C(-ik) = (k^2 - \mu^2)\mathbf{1} - kk^\top.
\tag{11.10.3}
$$

To compute the inverse of this matrix, we try multiplying it by the projection operator $P(k) = \mathbf{1} - \mu^{-2} kk^\top$ which annihilates $k$ when $k^2 = \mu^2$:

$$
\begin{aligned}
PC &= (k^2 - \mu^2)\mathbf{1} - kk^\top - \frac{k^2 - \mu^2}{\mu^2} kk^\top + \frac{k^2}{\mu^2} kk^\top \\
&= (k^2 - \mu^2)\mathbf{1}.
\end{aligned}
\tag{11.10.4}
$$

Since $C(-ik)$ is singular for on-shell momenta $k$, we must perturb our operator to $C_\epsilon$ by substituting $\mu^2 + i\epsilon$ for $\mu^2$. Then

$$
\begin{aligned}
P_\epsilon C_\epsilon &= (k^2 - \mu^2 + i\epsilon)\mathbf{1} \\
\implies\quad C_\epsilon^{-1} &= \frac{1}{k^2 - \mu^2 + i\epsilon} P_\epsilon \\
\implies\quad -iC_\epsilon^{-1} &= \left(-\mathbf{1} + \frac{kk^\top}{\mu^2 - i\epsilon}\right) \frac{i}{k^2 - \mu^2 + i\epsilon}.
\end{aligned}
\tag{11.10.5}
$$

Clearly, since $P_\epsilon$ has no singularities, we may set $P_\epsilon = P$ in the final expression and thereby find anew the propagator obtained in Chapter 9.

**Homework 11.10.6.** Show that $P_T = \mathbf{1} - (kk^\top/k^2)$ and $P_L = kk^\top/k^2$ are complementary orthogonal projection operators. Show that the $C(-ik)$ of 11.10.3 is diagonal with respect to the basis

$$
\hat{A}_T = P_T \hat{A}, \qquad \hat{A}_L = P_L \hat{A},
$$

and so verify the formula above for $-iC_\epsilon^{-1}$.

From this relationship between the covariance of functional integral quantization and the propagator of canonical quantization, it is obvious that the Lagrangian form of the generating functional is valid for the free vector field:

$$Z^0(J_\mu) = N \int (dA) \exp\left(i \int d^4x \ -\frac{1}{4}F_{\mu\nu}F^{\mu\nu} + \frac{1}{2}\mu^2 A_\mu A^\mu + J_\mu A^\mu\right), \quad (11.10.7)$$

where $(dA)$ means $(dA_0)(dA_1)(dA_2)(dA_3)$. Since we intend $J^\mu$ to be conserved and $A^\mu$ therefore to satisfy the constraint $\partial_\mu A^\mu = 0$, it is not a priori obvious that we can obtain the generating functional by integrating freely over all four components of the vector field.

**Homework 11.10.8.** Following the steps of Section 11.7, show that the generating functional $Z^0_{\rm FI}$ defined above by functional integration is equal to the generating $Z^0$ defined by Dyson's formula.

**Homework 11.10.9.** Show that the Feynman rule for the interaction $\mathcal{L}' = \frac{1}{2}E(\partial_\mu A^\mu)^2$ is

$$\nu \ \overset{\overleftarrow{q}}{\underset{}{\rule{1.2cm}{0.4pt}}}\!\!\times\!\!\overset{\overleftarrow{p}}{\underset{}{\rule{1.2cm}{0.4pt}}} \ \mu : \qquad iEp_\mu p_\nu (2\pi)^4 \delta^{(4)}(p-q).$$

As in Section 11.7, we can define a differential operator which will convert $Z^0_{\rm FI}$ and $Z^0$ into their interacting counterparts and thereby demonstrate the validity of the Lagrangian form of the functional integral for the massive vector field. (We shall prove this again in the next chapter in order to introduce a useful technique for comparing different forms of functional integral formulae.) We can now therefore quantize massive scalar QED and obtain covariant Feynman rules directly from the Lagrangian density:

$$\mathcal{L} \overset{\rm def}{=} -\frac{1}{4}F_{\mu\nu}F^{\mu\nu} + \frac{1}{2}\mu^2 A_\mu A^\mu + (\partial_\mu + ieA_\mu)\phi^\dagger(\partial^\mu - ieA^\mu)\phi + m^2\phi^\dagger\phi, \quad (11.10.10)$$

where $\phi$ is a complex scalar field with charge $+1$.

The propagators are simply those that we extract from the free-field generating functional, and these will be the usual Feynman propagators with no reference to derived fields. The vertex rules are obtained by applying the creation and annihilation factors (as in Principle 11.8.17) in $Z^0$ considered as a functional derivative operator. We find the usual Feynman rules for the familiar vertices.

For a derivative vertex, as noted in 11.8.22, we obtain an extra factor of $-ip^\mu$ for incoming momentum $k$ along a line corresponding to $\partial^\mu\phi$. Similarly, we obtain a factor of $iq_\mu$ for outgoing momentum $q$ along a line corresponding to $\partial_\mu\phi^\dagger$. Thus, from the interaction Lagrangian density

$$\mathcal{L}_I \overset{\rm def}{=} ieA_\mu(\phi^\dagger\partial_\mu\phi - \phi\partial_\mu\phi^\dagger) + e^2 A_\mu A^\mu \phi^\dagger\phi, \quad (11.10.11)$$

we obtain the vertex rules

$$i(ie)(-ip_\mu - iq_\mu) = ie(p_\mu + q_\mu),$$

$$2ie^2 g^{\mu\nu},$$

(11.10.12)

where the usual $(2\pi)^4\delta^{(4)}(P_{\text{in}})$'s are implicit.

**Homework 11.10.13.** Show that the operator-operand exchange identity 11.8.4 extends to complex scalar and massive vector fields. Work out the analogue of the real-field operators $\mathbf{C}_{\pm}$ and Principle 11.8.17 for complex scalar and massive vector fields. Verify the vertex rules above.

This section has shown that the Lagrangian form of the functional integral applies to free and interacting massive vector fields. The procedures of Section 11.8 then enable us to develop a perturbation theory directly from the Lagrangian density. This leads to covariant Feynman rules for massive scalar QED. In effect, the purposeful extraction of the ordinary covariant propagator guarantees covariant Feynman rules for the derived vertices. Now that we have seen one useful application of functional integral quantization, we take the next step towards quantization of useful theories by extending this quantization procedure to fermions.

# 11.11 Fermionic Functional Differentiation

Functional integral quantization uses classical fields to compute quantum evolution. To extend the functional integral to fermionic fields, we need some classical fermionic fields. We introduced fermionic calculus in Section 7.4 without worrying whether the fermionic variables were operators or classical objects. Now we need a little mathematical structure to describe honest classical fermions. This structure is derived from Grassmann algebra, the algebra generated by imposing an anti-commutative product on a vector space.

To define complex Grassmann algebra, we start with a complex vector space $V$ and define the tensor algebra of $V$ as the vector space

$$T(V) \stackrel{\text{def}}{=} \mathbf{C} \oplus V \oplus (V \otimes V) \oplus (V \otimes V \otimes V) \oplus \cdots \tag{11.11.1}$$

equipped with the tensor product as its multiplication law. Then the Grassmann algebra $\Lambda^*(V)$ of $V$ is the subspace of completely anti-symmetric tensors in $T(V)$ equipped with the anti-symmetric Grassmann product. The elements of the Grassmann algebra are called *alternating tensors*.

To define the Grassmann or anti-commutative product properly, let $\theta^1, \ldots, \theta^n$ be a basis for $V$ and define $\theta^I$ for $I = (i_1, \ldots, i_k)$ by

$$\theta^I \stackrel{\text{def}}{=} \frac{1}{k!} \sum_{\pi} \text{sign}(\pi)\theta^{i_{\pi(1)}} \otimes \cdots \otimes \theta^{i_{\pi(k)}}, \tag{11.11.2}$$

where the sum is taken over all permutations of the indices. By convention, $\theta^{\emptyset} = 1$. Clearly, if $I$ has a repeated index, then $\theta^I = 0$. The Grassmann product may now be defined by

$$\theta^I \theta^J \stackrel{\text{def}}{=} \theta^{IJ}. \tag{11.11.3}$$

We see at once that this product is bilinear and associative. Let the *degree* of $\theta^I$ be the number of indices in $I$, and write $\deg(\theta^I) = |I|$. Then we find that the Grassmann product satisfies the following analogue of the commutative law:

$$\theta^I \theta^J = (-1)^{|I||J|}\theta^J \theta^I. \tag{11.11.4}$$

**Remark** 11.11.5. Mathematicians call this the *wedge product*, and write $\theta^I \wedge \theta^J$ for $\theta^I \theta^J$.

The Grassmann algebra $\Lambda^*(V)$ is a direct sum of vector subspaces

$$\Lambda^k(V) \overset{\text{def}}{=} \text{span}\{\, \theta^I \mid |I| = k \,\}. \tag{11.11.6}$$

(The operator span constructs the vector space of all linear combinations of vectors in the set to which it is applied.) By convention,

$$\Lambda^0(V) = \mathbf{C}. \tag{11.11.7}$$

An element of $\Lambda^k(V)$ is known as an alternating tensor of degree $k$. The decomposition of $\Lambda^*(V)$ into a direct sum of $\Lambda^k(V)$ is called a *grading* of the algebra. The relationship of the grading to the algebra structure is simple: a sum of alternating tensors of degree $k$ has degree $k$, and a product of alternating tensors of degree $k$ and $l$ has degree $k+l$, and we have the graded commutativity law 11.11.4. These properties of the Grassmann algebra make it a *graded algebra*. One may also sum the even components and the odd components separately to form $\Lambda^+(V)$ and $\Lambda^-(V)$ respectively. Now the commutation law 11.11.4 states that alternating tensors of even degree commute with everything, but alternating tensors of odd degree anti-commute with each other.

Since $\theta^I = 0$ whenever two indices in $I$ are equal, the number of linearly independent $\theta^I$'s of degree $k$ is simply the number of ways of choosing $k$ distinct indices from a set of $n$ indices. Hence

$$\dim\big(\Lambda^k(V)\big) = \frac{n!}{k!\,(n-k)!} \quad \text{and} \quad \dim\big(\Lambda^*(V)\big) = 2^n. \tag{11.11.8}$$

In particular, $\Lambda^n(V)$ has dimension 1.

**Homework** 11.11.9. Let $V$ be an $n$-dimensional vector space with basis $\{\theta^1, \ldots, \theta^n\}$. Let $C$ be a change of basis matrix on $V$ so that the new basis $\{\phi^1, \ldots, \phi^n\}$ is given by $\phi = C\theta$. Show that the induced change of basis on $\Lambda^n(V)$ is simply multiplication by $\det(C)$:

$$\phi^1 \cdots \phi^n = \det(C)\theta^1 \cdots \theta^n.$$

Because of the fundamental anti-commutative property of the Grassmann algebra, namely $\theta\phi = -\phi\theta$ for tensors of degree 1, this algebra provides a classical, finite-dimensional analogue of Fermi fields. Fermi fields, however, come in pairs $\psi$ and $\bar\psi$. Hence, from a real basis $\phi^1, \ldots, \phi^{2n}$, we define a new basis $\theta^1, \bar\theta^1, \ldots, \theta^n, \bar\theta^n$ by

$$\left.\begin{aligned} \theta^r &\overset{\text{def}}{=} \phi^r + i\phi^{n+r} \\ \bar\theta^r &\overset{\text{def}}{=} \phi^r - i\phi^{n+r} \end{aligned}\right\} \tag{11.11.10}$$

The quadratic form on the Grassmann variables that mimics a fermion kinetic energy term is $Q = -\bar\theta^r C_{rs}\theta^s$. In the functional integral, the quadratic form is exponentiated. When we consider $\exp(iQ)$, we find that the term of highest degree has coefficient proportional to $\det C$:

$$e^{-i\bar\theta^r C_{rs}\theta^s} = 1 + (-iC_{rs}\bar\theta^r\theta^s) + \frac{1}{2!}(-iC_{rs}\bar\theta^r\theta^s)^2 + \cdots + \frac{1}{n!}(-iC_{rs}\bar\theta^r\theta^s)^n$$

$$= 1 + \cdots + \frac{(-i)^n}{n!} \sum_{\pi,\pi'} C_{\pi(1)\pi'(1)} \cdots C_{\pi(n)\pi'(n)} \bar{\theta}^{\pi(1)} \theta^{\pi'(1)} \ldots \bar{\theta}^{\pi(n)} \theta^{\pi'(n)}$$

$$= 1 + \cdots + (-i)^n \sum_{\pi} C_{\pi(1)1} \cdots C_{\pi(n)n} \bar{\theta}^{\pi(1)} \theta^1 \ldots \bar{\theta}^{\pi(n)} \theta^n \qquad (11.11.11)$$

$$= 1 + \cdots + (-i)^n \sum_{\pi} \operatorname{sign}(\pi) C_{\pi(1)1} \cdots C_{\pi(n)n} \bar{\theta}^1 \theta^1 \ldots \bar{\theta}^n \theta^n$$

$$= 1 + \cdots + (-i)^n \det(C) \bar{\theta}^1 \theta^1 \ldots \bar{\theta}^n \theta^n .$$

In analogy with the bosonic case, we call the matrix $C$ the *covariance* of the quadratic form $Q$.

**Homework 11.11.12.** Verify the argument 11.11.11 for $n = 3$.

The analogue of functional integration, which extracts $\det(C)^{-1}$ from the exponential of a quadratic form, is a linear operator which extracts $\det(C)$ from the sum above. As we merely want to remove the maximal product of $\bar{\theta}\theta$'s and eliminate lower order terms, the natural operator for this task is a fermionic differential operator:

$$\left( \frac{\partial}{\partial\theta} \frac{\partial}{\partial\bar{\theta}} \right) \overset{\text{def}}{=} i \frac{\partial}{\partial\theta^n} \frac{\partial}{\partial\bar{\theta}^n} \cdots i \frac{\partial}{\partial\theta^1} \frac{\partial}{\partial\bar{\theta}^1}$$

$$\implies \quad \left( \frac{\partial}{\partial\theta} \frac{\partial}{\partial\bar{\theta}} \right) e^{-i\bar{\theta}^r C_{rs} \theta^s} = \det C. \qquad (11.11.13)$$

A linear change of variables $\phi = A^{-1}\theta$ extends to the whole basis and the corresponding differential operators:

$$\left. \begin{array}{l} \theta = A\phi \\ \bar{\theta} = \bar{\phi}A^\dagger \end{array} \right\} \quad \implies \quad \left. \begin{array}{l} \partial_\theta = \partial_\phi A^{-1} \\ \partial_{\bar{\theta}} = (A^\dagger)^{-1}\partial_{\bar{\phi}} \end{array} \right\} \qquad (11.11.14)$$

Since the derivatives here are fermionic, the result of Homework 11.11.9 applies to show that

$$\left( \frac{\partial}{\partial\theta} \frac{\partial}{\partial\bar{\theta}} \right) = \det(AA^\dagger)^{-1} \left( \frac{\partial}{\partial\phi} \frac{\partial}{\partial\bar{\phi}} \right). \qquad (11.11.15)$$

As a consistency check, we can evaluate the fermionic Gaussian after changing variables:

$$\left( \frac{\partial}{\partial\theta} \frac{\partial}{\partial\bar{\theta}} \right) e^{-i\bar{\theta}C\theta} = \det(AA^\dagger)^{-1} \left( \frac{\partial}{\partial\phi} \frac{\partial}{\partial\bar{\phi}} \right) e^{-i\bar{\phi}A^\dagger CA\phi}$$

$$= \det(AA^\dagger)^{-1} \det(A^\dagger CA) \qquad (11.11.16)$$

$$= \det C.$$

This fermionic differential operator is generally called a fermionic integration operator:

$$`\int (d\psi)(d\bar{\psi})\,` \overset{\text{def}}{=} \left( \frac{\partial}{\partial\theta} \frac{\partial}{\partial\bar{\theta}} \right). \qquad (11.11.17)$$

However, since there is no underlying measure space and since the Jacobian for a change of variables is appropriate to a differential operator rather than an integral operator, we find this traditional nomenclature confusing.

Differentiation is applied to variables. Hence we refer to the elements of the basis as *Grassmann* or *anti-commuting* variables and the Grassmann algebra elements as *functions* of Grassmann variables.

**Remark 11.11.18.** Actually, it is more correct to call Grassmann variables Grassmann *coordinates*, as we shall briefly explain in the context of real Grassmann algebras.

A variable in mathematics is a symbol representing an arbitrary element of a set. Thus the usual real variables $x^r$ represent arbitrary elements of $\mathbf{R}$. However, in a vector $x = (x^1, \dots, x^n)$ there is an additional order structure. Thus the component $x^r$ is not simply a real variable, but rather it is the $r^{\text{th}}$ coordinate function on $\mathbf{R}^n$, that is, it is that function from the vector space $\mathbf{R}^n$ to its field of scalars $\mathbf{R}$ whose value is the $r^{\text{th}}$ component of the vector.

Let $V^*$ be the vector space of linear functions on $V$. The convention in mathematics is to regard each element $\theta$ in $\Lambda^1(V)$ as a coordinate function on $V^*$. When $V = \mathbf{R}^n$, then $\theta^r$ is identical as a linear function to $x^r$; only the multiplication is different. The product $x^I = x^{r_1} \cdots x^{r_k}$ is generally evaluated on a vector and could be evaluated on any *symmetric $k$-tensors*, while $\theta^I = \theta^{r_1} \cdots \theta^{r_k}$ is evaluated on *anti-symmetric $k$-tensors*.

To extend the finite-dimensional Grassmann integral to a functional integral, we must first consider what space of functions we want to integrate over. Since Fermi fields anti-commute, it is natural to use $\Lambda^1$-valued functions in the functional integral, but since Fermi fields anti-commute even when evaluated at distinct points in space-time, it appears that we need Grassmann basis elements $\theta_x$ and $\bar{\theta}_x$ for each space-time point $x$. The integrand in the functional integral, however, is an exponential of the Dirac action, and our desire is to bring the determinant of the covariance out of the exponential. It is not feasible to extract an uncountable product $\prod \bar{\theta}_x \theta_x$ from the exponential. Hence, instead of using a basis $\theta_x$, $\bar{\theta}_x$ and functions $\psi(x) = \varphi(x)\theta_x$, $\bar{\psi}(x) = \varphi^*(x)\bar{\theta}_x$, we start from a Hilbert space of complex-valued functions with countable basis $\varphi_n$ together with a countable set of Grassmann variables $\theta^n$, $\bar{\theta}^n$ and build classical Fermi functions

$$\psi(x) = \theta^n \varphi_n(x) \quad \text{and} \quad \bar{\psi}(x) = \bar{\theta}^n \varphi_n^*(x). \tag{11.11.19}$$

These functions do in fact anti-commute whether or not they are evaluated at the same point in space-time. We shall refer to functions of this type as *classical Fermi fields*.

To model fermions, we must model not only the anti-commuting property, but also the Lorentz transformation properties. For Weyl spinors, for example, we assume that the $\varphi_n$ form a basis for the Hilbert space of two-component classical spinor fields. For Dirac spinors, of course, each $\varphi_n$ must be a four-component classical spinor field. Thus $\bar{\theta}$ here is not the Dirac adjoint of $\theta$. Classical Fermi fields are used as source terms for Fermi fields in the definition of a generating functional for a theory with fermions. The source Lagrangian density for a single fermion $\psi$ has the form

$$\mathcal{L}_{\text{src}} \stackrel{\text{def}}{=} \bar{\jmath}\psi + \bar{\psi}j. \tag{11.11.20}$$

**Remark 11.11.21.** This use of Grassmann variables in classical Fermi fields differs from their use in supersymmetry where the $\theta$'s and $\bar{\theta}$'s carry spinor indices.

Functional differentiation with respect to a classical Fermi field is defined like that with respect to a Bose field:

$$
\left.
\begin{aligned}
\frac{\delta}{\delta j(x)} f(j) &= \lim_{\epsilon \to 0} \big( f(j + \epsilon \delta_x) - f(j) \big) \frac{1}{\epsilon} \\
\frac{\delta}{\delta \bar{\jmath}(x)} f(\bar{\jmath}) &= \lim_{\bar{\epsilon} \to 0} \frac{1}{\bar{\epsilon}} \big( f(\bar{\jmath} + \bar{\epsilon} \delta_x) - f(\bar{\jmath}) \big)
\end{aligned}
\right\}
\tag{11.11.22}
$$

but with the additional consideration that $\epsilon$ and $\bar{\epsilon}$ are Grassmann parameters. The positions of the factors $1/\epsilon$ and $1/\bar{\epsilon}$ uphold our convention of putting $d\psi$ on the right and $d\bar{\psi}$ on the left of a product. This definition implies, for example, that

$$
\frac{\delta}{\delta j(x)} \int d^4y\, \mathcal{L}_{\mathrm{src}} = \bar{\psi}(x) \quad \text{and} \quad \frac{\delta}{\delta \bar{\jmath}(x)} \int d^4y\, \mathcal{L}_{\mathrm{src}} = \psi(x).
\tag{11.11.23}
$$

Now we are ready for the leap of faith, extending the finite-dimensional Grassmann differentiation formula 11.11.16 to the Grassmann functional differentiation principle:

**Principle 11.11.24.** *If $\psi$ and $\bar{\psi}$ are classical Fermi fields, and if $C$ is a differential operator, then*

$$
\left( \frac{\partial}{\partial \psi} \frac{\partial}{\partial \bar{\psi}} \right) e^{-i(\bar{\psi}, C\psi)} = \det(C).
$$

**Remark 11.11.25.** Actually, there is a lot of solid mathematics supporting this principle. First, as for the Klein–Gordon operator in Remark 11.5.13, so here we must subject the central operator, $i\slashed{D} = i\slashed{\partial} + e\slashed{A} - m$, to Wick rotation and make it act on spinor fields over the 4-sphere, $S^4$. Second, the Grassmann differentiation formula can be extended to infinite dimensions.

The Wick rotation $k_0 \to ik_0$ is best understood in this context as making the $\gamma$'s Euclidean:

$$
(\gamma_{\mathrm{E}}^0, \gamma_{\mathrm{E}}^1, \gamma_{\mathrm{E}}^2, \gamma_{\mathrm{E}}^3) \stackrel{\text{def}}{=} (i\gamma^0, \gamma^1, \gamma^2, \gamma^3) \quad \Longrightarrow \quad \{\gamma_{\mathrm{E}}^\mu, \gamma_{\mathrm{E}}^\nu\} = -2\delta^{\mu\nu}.
\tag{11.11.26}
$$

The Euclidean operator $\slashed{D}_{\mathrm{E}} = \gamma_{\mathrm{E}}^\mu D_\mu$ is hermitian and a square root of the Laplacian,

$$
\slashed{D}_{\mathrm{E}}^2 = -\delta^{\mu\nu} \partial_\mu \partial_\nu,
\tag{11.11.27}
$$

and so has a discrete spectrum and finite-dimensional eigenspaces. Its eigenvalues $\lambda_n$, however, will be both positive and negative:

$$
\slashed{D}_{\mathrm{E}} \varphi_n = \lambda_n \varphi_n \quad \Longrightarrow \quad \slashed{D}_{\mathrm{E}} \gamma_5 \varphi_n = -(\lambda_n + 2m)\gamma_5 \varphi_n,
\tag{11.11.28}
$$

where $\varphi_n$ is a classical spinor field. We could define the determinant by using zeta-function regularization of the product of the absolute values of the eigenvalues, or, if we want to be more careful, we could use the Atiyah–Patodi invariant $\eta_{\slashed{D}}$ to take care of the signs:

$$
\eta_{\slashed{D}}(s) \stackrel{\text{def}}{=} \sideset{}{'}\sum \frac{\mathrm{sgn}(\lambda)}{|\lambda|^s},
$$

where the sum runs over the non-zero eigenvalues $\lambda$.

**Remark 11.11.29.** This definition of the determinant in the previous remark is consistent with our definition of classical Dirac fields as classical spinor fields multiplied by scalar Grassmann variables. If we had used classical scalar fields multiplied by Grassmann spinor variables, we would have obtained the fourth power of this determinant.

Once an orthonormal basis $\varphi_n$ has been chosen, the Grassmann variables $\theta^n$ become coordinates for the classical Dirac fields $\psi$. Then we can write

$$\left(\frac{\partial}{\partial\psi}\frac{\partial}{\partial\bar\psi}\right) = \left(\frac{\partial}{\partial\theta}\frac{\partial}{\partial\bar\theta}\right). \tag{11.11.30}$$

If the $\varphi_n$ are eigenfunctions of $\slashed{D}_{\mathrm E}$, then the action becomes

$$\begin{aligned}
S &\overset{\text{def}}{=} \int d^4x\, \bar\psi\, i\slashed{D}\psi \\
&= \sum_{m,n} \bar\theta^m\theta^n \int (-i)d^4x_{\mathrm E}\, \varphi_m^\dagger i\slashed{D}_{\mathrm E}\varphi_n \\
&= \sum_n \lambda_n\bar\theta^n\theta^n.
\end{aligned} \tag{11.11.31}$$

This form of $S$ indicates the possibility of justifying the Grassmann functional derivative principle 11.11.24 in the Euclidean case:

$$\det(C) = \det(-i\slashed{D}) = \det(-\slashed{D}_{\mathrm E}) = \prod\left(i\frac{\partial}{\partial\theta^n}\frac{\partial}{\partial\bar\theta^n}\right)e^{iS} = \prod(-\lambda_n).$$

**Homework 11.11.32.** Let $C$ and $A$ be $n \times n$ matrices. Show that the derivative of $\det(C)$ in the direction of $A$ is given by

$$A\cdot\partial \det(C) = \operatorname{tr}(AC^{-1})\det(C).$$

Let $\varphi = \theta^r\varphi_r$ be the expansion of the Euclidean-space classical Fermi field $\varphi$ with respect to an orthonormal basis for Hilbert space. Deduce that

$$\left(\frac{\partial}{\partial\theta}\frac{\partial}{\partial\bar\theta}\right) - i(\varphi^\dagger, A\varphi)e^{-i(\varphi^\dagger, C\varphi)} = \operatorname{Tr}(AC^{-1})\det(C),$$

where the functional trace is given by

$$\operatorname{Tr}(B) \overset{\text{def}}{=} \sum_r \int d^4x\, \varphi_r^\dagger(x)B\varphi_r(x).$$

**Remark 11.11.33.** We are often interested in chiral aa covariant derivatives

$$i\slashed{D}_\pm = (i\slashed{\partial} + e\slashed{A})P_\pm. \tag{11.11.34}$$

A chiral operator maps left-handed fields into right-handed ones and *vice versa*. Hence eigenvalues and eigenvectors can only be defined if we choose an identification between the spaces $H_\pm$ of left and right fields. The definition of the determinant as a number depends on this identification. (We note in passing that there is a definition of the determinant as a linear transformation in the sense of Homework 11.11.9 which is intrinsic to the operator.) In general, the determinant of a Dirac operator is positive by definition, and the details of the definition eliminate unimodular complex factors.

Principle 11.11.24 is closely related to the complex scalar functional integral 11.7.22. The relative inverse on the $\det A$ will support the relationship between the Grassmann functional derivative and quantum spinor field theory. Thus, if we think of $-\bar\psi C\psi$ as an interaction term, then we postulate

$$\det(A) \overset{?}{=} \sum_r S(D_r) = \exp\left(\sum_s S(C_s)\right), \tag{11.11.35}$$

where $D_r$ and $C_s$ are the Feynman diagrams and connected Feynman diagrams respectively. Now the connected diagram $C_s$ is a loop with $s$ vertices, so if $\psi$ is fermionic, then each $S(C_s)$ contains a loop-factor of $-1$ relative to the complex scalar case. Hence the relative inverse on the $\det C$ is essential to our purpose.

This section has briefly introduced the mathematics of classical anti-commuting fields and the determinant of the covariant Dirac operator. There is a wealth of mathematics in this area, but for most applications of field theory, the outline presented here will be quite sufficient. In particular, it provides the foundation for the extension of functional quantization to Fermi fields.

## 11.12 Functional Derivative Quantization for Fermions

It would be satisfying if we could motivate a Hamiltonian form for the Grassmann functional derivative by analyzing the quantum-mechanical completeness relation. In quantum mechanics, however, it is not possible to form position eigenstates with positive energy solutions of the Dirac equation. The logic that leads to the Hamiltonian form for the scalar field therefore does not extend to the Fermi field. The convention is therefore to use the Lagrangian form of the generating functional for fermions. There is no conflict with the use of the Hamiltonian form for scalars, and more generally, for bosons, since one can readily introduce an intermediate or Routhian form of the generating functional when both fermions and bosons are involved.

Thus, on the basis of the functional derivative principle 11.11.24 and the functional integral principle 11.7.3, we formulate the following principle regarding the generating functional for a free spinor field:

**Principle 11.12.1.** *The generating functional $Z^0$ for the free Dirac spinor can be written in terms of the functional differential as follows:*

$$Z^0(\bar{\jmath}, \jmath) = N\left(\frac{\partial}{\partial \psi} \frac{\partial}{\partial \bar{\psi}}\right) e^{iS^0(\psi, \bar{\psi}; \bar{\jmath}, \jmath)},$$

*where the action $S^0$ is given by*

$$S^0(\bar{\jmath}, \jmath) \stackrel{\text{def}}{=} \int d^4x\, \bar{\psi}(i\slashed{\partial} - m)\psi + \bar{\jmath}\psi + \bar{\psi}\jmath$$

*and $N$ is a normalization constant defined by $Z^0(0, 0) = 1$.*

To justify that the Lagrangian form of the generating functional really matches with Dyson's formula, we follow the logic of previous arguments. To simplify the presentation, we apply our real, symmetric bilinear form to classical Dirac fields:

$$(\bar{\psi}^\top, \psi) \stackrel{\text{def}}{=} \int d^4x\, \bar{\psi}\psi. \tag{11.12.2}$$

Write $C$ for the covariance $-i\slashed{\partial} + m$ and make $C$ invertible by perturbing it to $C_\epsilon = -i\slashed{\partial} + m - i\epsilon$. If we define $C_\epsilon'$ by $C_\epsilon' = i\slashed{\partial}^\top + m - i\epsilon$, then

$$(\bar{\psi}^\top, C_\epsilon\psi) = (C_\epsilon'\bar{\psi}^\top, \psi). \tag{11.12.3}$$

Next we eliminate the term linear in $\psi$ from the action in order to reduce the integrand to the standard form of Principle 11.11.24. Defining $\psi_{\mathrm{cl}}$ and $\bar{\psi}_{\mathrm{cl}}$ by

$$\psi_{\mathrm{cl}} \overset{\mathrm{def}}{=} C_\epsilon^{-1} j \quad \text{and} \quad \bar{\psi}_{\mathrm{cl}}^{\mathsf{T}} \overset{\mathrm{def}}{=} C_\epsilon'^{-1} \bar{j}^{\mathsf{T}}, \tag{11.12.4}$$

we find that

$$\begin{aligned}
S^0 &= -\left(\bar{\psi} - \bar{\psi}_{\mathrm{cl}}, C_\epsilon(\psi - \psi_{\mathrm{cl}})\right) - (\bar{\psi}, C_\epsilon \psi_{\mathrm{cl}}) - (\bar{\psi}_{\mathrm{cl}}, C_\epsilon \psi) \\
&\quad + (\bar{\psi}_{\mathrm{cl}}, C_\epsilon \psi_{\mathrm{cl}}) + (\bar{j}, \psi) + (\bar{\psi}, j) \\
&= -\left(\bar{\psi} - \bar{\psi}_{\mathrm{cl}}, C_\epsilon(\psi - \psi_{\mathrm{cl}})\right) + (\bar{j}, C_\epsilon^{-1} j).
\end{aligned} \tag{11.12.5}$$

Since we can translate Grassmann variables, we can apply the differentiation principle 11.11.24 to find

$$Z_{\mathrm{FD}}^0 = N \det(-i\slashed{\partial} + m - i\epsilon) e^{i(\bar{j}, C_\epsilon^{-1} j)} = e^{i(\bar{j}, C_\epsilon^{-1} j)}, \tag{11.12.6}$$

by definition of $N$. Following the method of 11.6.13 by which we expressed the stationary value of the scalar action as a momentum-space integral, we can transform the final integral in this expression from position space to momentum space:

$$\begin{aligned}
i(\bar{j}, C_\epsilon^{-1} j) &= -i \int d^4 z \, \bar{j}(z)(i\slashed{\partial} - m + i\epsilon)^{-1} j(z) \\
&= \int \frac{d^4 k}{(2\pi)^4} \left(i\hat{\bar{j}}(k)\right) \frac{i}{\slashed{k} - m + i\epsilon} \left(i\hat{j}(-k)\right).
\end{aligned} \tag{11.12.7}$$

Theorem 4.5.2 shows Dyson's formula for the generating functional reduces to the exponential of the connected vacuum bubbles. The Wick vertices for the free spinor theory with sources are

**Wick Vertices:**    $\bar{j} \bullet\!\!-\!\!\leftarrow\!\!-\!\bullet \psi$  and  $\bar{\psi} \bullet\!\!-\!\!\leftarrow\!\!-\!\bullet j$ $\tag{11.12.8}$

The only vacuum bubble diagram arises from the contraction of these two:

$$C \overset{\mathrm{def}}{=} \bar{j} \bullet\!\!-\!\!\leftarrow\!\!-\!\bullet j \tag{11.12.9}$$

The contribution of this operator to the $S$ matrix is given by the Feynman rules, and from 11.12.7, we see at once that

$$S(C) = i(\bar{j}, C_\epsilon^{-1} j). \tag{11.12.10}$$

Hence $Z^0(\bar{j}, j) = Z_{\mathrm{FD}}^0(\bar{j}, j)$.

Having established the equality of $Z^0$ and $Z_{\mathrm{FD}}^0$, we can follow the strategy of Section 11.7 and demonstrate the equality of $Z$ and $Z_{\mathrm{FD}}$ for interacting fermions. Since there are no non-trivial renormalizable interactions of fermions amongst themselves, any realistic theory will involve bosons. Suppose, for example, that we have a fermion $\psi$ interacting with a neutral scalar $\phi$ through $\mathcal{L}' = -\lambda\bar{\psi}\psi\phi$. Writing $\mathcal{L}^0$

for the sum of the free fermion and free scalar Lagrangian densities, the differential operator which constructs this interaction from $Z^0_{\text{FD}}$ is

$$\mathbf{D}\Big(-i\frac{\delta}{\delta j},\,-i\frac{\delta}{\delta \bar{j}},\,-i\frac{\delta}{\delta J}\Big) \overset{\text{def}}{=} \exp\Big(\int d^4x \, -\lambda\Big(-i\frac{\delta}{\delta j(x)}\Big)\Big(-i\frac{\delta}{\delta \bar{j}(x)}\Big)\Big(-i\frac{\delta}{\delta J(x)}\Big)\Big)$$

$$\implies \quad \mathbf{D}\Big(-i\frac{\delta}{\delta j},\,-i\frac{\delta}{\delta \bar{j}},\,-i\frac{\delta}{\delta J}\Big) \exp\Big(i\int d^4x \, \mathcal{L}^0 + \bar{j}\psi + \bar{\psi}j + J\phi\Big) \qquad (11.12.11)$$

$$= \exp\Big(i\int d^4x \, \mathcal{L}^0 - \lambda\bar{\psi}\psi\phi + \bar{j}\psi + \bar{\psi}j + J\phi\Big).$$

Clearly the arguments of Section 11.7 generalize, and we can conclude that the generating functional defined by the functional derivative is equal to the one defined by Dyson's formula.

It remains to obtain Feynman rules for spinors from the generating functional expressed in functional derivative form. We follow the logic of Section 11.8. First, suppressing the bosons, we write the generating functional in the form

$$Z(\bar{j}, j) = \mathbf{D}\Big(-i\frac{\delta}{\delta j},\,-i\frac{\delta}{\delta \bar{j}}\Big) e^{i(\bar{j}, C_\epsilon^{-1}j)}. \qquad (11.12.12)$$

Second, we generalize the operator-operand exchange principle 11.8.4 to classical fermions:

$$\mathbf{D}\Big(-i\frac{\delta}{\delta j},\,-i\frac{\delta}{\delta \bar{j}}\Big) Z^0(\bar{j}, j) \qquad (11.12.13)$$

$$= Z^0\Big(-i\frac{\delta}{\delta \psi},\,-i\frac{\delta}{\delta \bar{\psi}}\Big) \mathbf{D}(\bar{\psi}, \psi) \exp\Big(i\int d^4x \, \bar{j}\psi + \bar{\psi}j\Big) \Big|_{\psi=0,\bar{\psi}=0}.$$

Third, we identify the propagator from the new form of $Z^0$ and see how the action of $Z^0$ determines the vertex rules.

**Homework** 11.12.14. Verify the identity 11.12.13 for $\mathbf{D} = \bar{\psi}M\psi$ and $Z^0 = (\bar{j}Aj)(\bar{j}Bj)$.

Following 11.8.6, we combine the formula 11.12.7 for the stationary value of the action with the interchange formula 11.12.13 to bring out the propagator in $Z^0$ considered as an operator:

$$Z^0\Big(-i\frac{\delta}{\delta \psi},\,-i\frac{\delta}{\delta \bar{\psi}}\Big) \qquad (11.12.15)$$

$$= \exp\Big(\int \frac{d^4k}{(2\pi)^4} \, d^4x \, d^4y \, \big(e^{-ix\cdot k}\frac{\delta}{\delta \psi(x)}\big)\frac{i}{\not{k} - m + i\epsilon}\big(e^{iy\cdot k}\frac{\delta}{\delta \bar{\psi}(y)}\big)\Big).$$

The result is in agreement with our convention

$$\textbf{Propagator} = -iC_\epsilon^{-1}(-ik). \qquad (11.12.16)$$

Having extracted the Feynman propagator factor from $Z^0$, the vertex rules depend on the identification of a contraction and an annihilation operator in what remains. As in Principle 11.8.17, we shall therefore define $\mathbf{C}_-$ and $\mathbf{C}_+$ by

$$\mathbf{C}_- \overset{\text{def}}{=} \int d^4x \, \big(e^{-ix\cdot k}\frac{\delta}{\delta \psi(x)}\big) \quad \text{and} \quad \mathbf{C}_+ \overset{\text{def}}{=} \int d^4y \, \big(e^{iy\cdot k}\frac{\delta}{\delta \bar{\psi}(y)}\big). \qquad (11.12.17)$$

The vertex rules do not follow directly from this form of the $Z^0$ operator. We must decide how a power of the contraction operator should act on a product of vertices. The overall result will be independent of the order of this action since the operator is bosonic, but the vertex rules will be affected. The simplest rules follow from organizing dummy variables to make the contraction operators act in sequence along each Fermi line. For example, consider the action represented schematically by

$$ i\mathcal{A} = \frac{1}{2!} \left( \frac{\delta}{\delta\psi} P \frac{\delta}{\delta\bar{\psi}} \right)^{(1)} \left( \frac{\delta}{\delta\psi} P \frac{\delta}{\delta\bar{\psi}} \right)^{(2)} \left( (i\bar{\jmath}\psi)(-i\lambda\bar{\psi}\Gamma\psi\phi)(i\bar{\psi}j) \right). \tag{11.12.18} $$

If we make operator 1 contract the first $\psi$–$\bar{\psi}$ pair and operator 2 contract the second, then we can drop the $1/2!$ and find the vertex rules for the interactions in the Hamiltonian density:

$$ \bar{\psi}j: \qquad\qquad \mathbf{C}_+^{(2)}(i\bar{\psi}j) = i\hat{j}(k); $$

$$ \lambda\bar{\psi}\phi\psi: \qquad \mathbf{C}_+^{(1)}\mathbf{C}_-^{(2)}(-i\lambda\bar{\psi}\Gamma\psi\phi) = -i\lambda\Gamma\phi; \tag{11.12.19} $$

$$ \bar{\jmath}\psi: \qquad\qquad \mathbf{C}_-^{(1)}(i\bar{\jmath}\psi) = i\hat{\bar{\jmath}}(-k). $$

The vertex rule follows from an operator sign convention and a convention for applying powers of the contraction operator to a Fermi line. The Fermi loop rule is a logical consequence of these conventions.

**Homework** 11.12.20. Explain with an example how the minus sign emerges when we have a Fermi loop.

**Homework** 11.12.21. Check that functional calculus quantization applies equally well to Weyl spinors and verify the Feynman rules of Section 7.14.

On the foundation of the Fermi derivative formula for Gaussians derived in the previous section, in this section we have proposed a functional derivative formula for the fermionic generating functional and verified that this formula is consistent with Dyson's formula. Furthermore, the techniques of Section 11.8 generalize easily enough, and we find the usual Feynman rules for fermion propagators and interactions can be derived directly from this functional derivative formula. The main difference between the Fermi and Bose cases is that the Lagrangian form of the action is primary for fermions, not the Hamiltonian.

# 11.13 Summary

The discrete functional integral arises naturally from the completeness relation of quantum mechanics when we transform our viewpoint from integration over phase-space variables at intermediate times to integration over discrete paths. This form of the functional integral is a well-defined approximation to quantum processes which provides the basis for computer simulations. The functional integral itself, either as a limit of the discrete case or as an integral over path space, is less well defined, but provides physics with a principle of functional integral quantization.

The origin of the functional integral in quantum mechanics suggests that the Hamiltonian form of the action should be used for quantizing Bose fields. Extending the functional integral to fermions requires the use of Grassmann variables and classical Dirac fields. The process of extracting the determinant of the covariance from the exponential of the fermion action is simply an application of a linear differential operator.

On account of the different origin of functional differentiation over Fermi fields, there is no reason to prefer the Hamiltonian form of the action in this case. Hence, when a theory contains fermionic and bosonic fields, then the preferred action is based on the Routhian density.

On the basis of the bosonic functional integral and the fermionic functional derivative, it is possible to find a functional formula for any generating functional. The whole of perturbation theory can be unfolded from this functional formula. The covariance determines the propagators, and the creation and annihilation operators accompanying the covariance determine the vertex rules.

The functional calculus provides great flexibility in passing from one form of a theory to another. One may define the quantum field theory from its Routhian density and then change variables to obtain a Lagrangian form of the generating functional. The Feynman rules derived from a Lagrangian form are satisfactorily covariant.

It might appear that our new non-perturbative viewpoint could provide a framework in which renormalization is not necessary. Certainly, discrete space-time functional calculus does not need renormalization when considered in isolation, but renormalization is essential to the existence of a continuum limit. In the continuum formalism in common use, renormalization of the parameters in the Lagrangian compensates for the divergence intrinsic to function-space integration and differentiation.

We note also that the detailed definition of functional quantization determines the operator ordering in the quantum Hamiltonian. In field theory, the resulting Hamiltonian differs from its normal-ordered counterpart by an undefined constant. This constant is hidden in the normalization of the function-space measure.

Further discussion will be presented in the later chapters devoted to renormalization. Our current understanding is sufficient for our immediate purpose, namely, the extension of perturbation theory to gauge theories.

# Chapter 12

# Quantization of Gauge Theories

Extending the technique of functional integral quantization to gauge field theories through the Faddeev–Popov principle; quantization of QED and preparation for quantization of the Standard Model of the electroweak and strong interactions.

## Introduction

The previous chapter explained functional integral quantization of scalar, spinor, and massive vector fields and demonstrated that the Feynman rule for a derivative interactions proposed in Section 10.8 was correct. In this chapter, we extend functional integral quantization to gauge field theories. There are two elements in this extension. The first element is the Faddeev–Popov principle of integrating over orbits of the gauge group. This technique proposes a functional integral formula for the generating functional which is based on the Lagrangian form of the action. As a consequence, the Feynman rules are covariant and gauge invariance of the quantum theory is obvious. The second element is the use of Lagrangians which are first order in derivatives to prove equivalence between Faddeev–Popov quantization and canonical quantization in axial gauge. After this proof, we cheerfully accept the Faddeev–Popov principle for quantization of gauge theories.

Section 11.10 established the Lagrangian form of the generating functional for the free vector field. In Sections 12.1 and 12.2, we explain the use of first-order Lagrangians in the context of spinor and scalar QED with a massive photon. Section 12.3 introduces the Faddeev–Popov principle for quantizing gauge field theories, and Section 12.4 applies this principle to QED. Having mastered the basic concepts of quantization of gauge theories, we turn in Sections 12.5 and 12.6 to the structure of gauge theories built about the common symmetry groups, and then in Section 12.7 apply Faddeev–Popov quantization to non-abelian gauge theories to find the standard effective Lagrangian density which contains gauge-fixing and ghost-field terms. At this point, we can quantize the electromagnetic and strong interactions, but no the weak interactions, so Sections 12.8 to 12.10 introduce spontaneous symmetry breaking and extend Faddeev–Popov quantization to gauge theories with spontaneously broken symmetries. Finally, in Section 12.11 we indicate that the Faddeev–Popov technique is mathematically delicate and should be taken *cum grano salis*.

## 12.1 Massive Spinor QED

When we couple the vector field to a Dirac spinor to make massive QED, the canonical momenta of the vector and spinor fields are unaffected by this coupling, and so the Lagrangian form is still valid. It is, however, illuminating to see the connection between Lagrangian and Hamiltonian forms in this simplest case of dependent fields ($\partial_\mu A^\mu = 0$).

The Lagrangian density for massive spinor QED is

$$\mathcal{L} = -\frac{1}{4}F_{\mu\nu}F^{\mu\nu} + \frac{1}{2}\mu^2 A_\mu A^\mu + \bar{\psi}(i\slashed{\partial} - e\slashed{A} - m)\psi. \tag{12.1.1}$$

Adding a source term Lagrangian density

$$\mathcal{L}^{\text{src}} \overset{\text{def}}{=} J_\mu A^\mu + \bar{\jmath}\psi + \bar{\psi}j, \tag{12.1.2}$$

and integrating over space-time gives us the action for the Lagrangian form of the functional integral:

$$S_L(A, \psi, \bar{\psi}; J, \bar{\jmath}, j) = \int d^4 x\, \mathcal{L} + \mathcal{L}^{\text{src}}. \tag{12.1.3}$$

The Lagrangian form of the generating functional is therefore

$$Z_L(J, \bar{\jmath}, j) = N\left(\frac{\partial}{\partial \psi} \frac{\partial}{\partial \bar{\psi}}\right) \int (dA)\, e^{iS_L}. \tag{12.1.4}$$

The Routhian form of the functional integral starts from a Routhian density $\mathcal{R}$ which is Hamiltonian for the vector field and Lagrangian for the spinor field:

$$\begin{aligned}
\mathcal{R} &\overset{\text{def}}{=} F^{i0} \partial_0 A_i - \mathcal{L} \\
&= F_{i0} \partial^0 A^i + \frac{1}{2} F_{i0} F^{i0} + \frac{1}{4} F_{ij} F^{ij} - \frac{1}{2}\mu^2 A_\mu A^\mu - \bar{\psi}(i\slashed{\partial} - e\slashed{A} - m)\psi,
\end{aligned} \tag{12.1.5}$$

where $F_{ij} = \partial_i A_j - \partial_j A_i$ by definition and $A^0$ is determined by the Lagrangian equations of motion:

$$\partial_\mu F^{\mu\nu} + \mu^2 A^\nu = e\bar{\psi}i\gamma^\nu\psi \implies A^0 = \frac{e}{\mu^2}(\bar{\psi}i\gamma^0\psi - \partial_i F^{i0}). \tag{12.1.6}$$

(Note that, in the Routhian theory, the equations of motion are Hamiltonian in the conjugate variables $A_i$ and $F^{0i}$, and Lagrangian in the spinor field.)

The Routhian form of the action is

$$S_R(A^i, F^{i0}, \psi, \bar{\psi}; J_i, K_i, \bar{\jmath}, j) = \int d^4 x\, F^{i0} \partial_0 A_i - \mathcal{R} + \mathcal{R}^{\text{src}}, \tag{12.1.7}$$

where the density of source terms is defined by

$$\mathcal{R}^{\text{src}} \overset{\text{def}}{=} J_i A^i + K_i F^{i0} + \bar{\jmath}\psi + \bar{\psi}j. \tag{12.1.8}$$

The corresponding generating functional $Z_R$ is defined by

$$Z_R(J_i, K_i, \bar{\jmath}, j) \overset{\text{def}}{=} \left(\frac{\partial}{\partial \psi} \frac{\partial}{\partial \bar{\psi}}\right) \int \prod_i (dA^i)(dF^{i0})\, e^{iS_R}. \tag{12.1.9}$$

Notice that, the Feynman rules derived from the Routhian theory will be so far from covariant that computations will be very tedious. Furthermore, it is not obvious that the Routhian theory will make the same predictions as the Lagrangian theory. We are therefore faced with a conflict of interests: in theory, the Routhian form of the generating functional is preferable; for computations, the Lagrangian. Fortunately there exists an intermediate theory from which both forms can be derived.

Since $F_{\mu\nu} - (\partial_\mu A_\nu - \partial_\nu A_\mu) = 0$, we may add the square of this term to the Lagrangian density $\mathcal{L}$ without changing the classical theory. The result is a Lagrangian density which is first-order in derivatives:

$$\mathcal{L}_{\text{FO}} \overset{\text{def}}{=} \frac{1}{4}\left(F_{\mu\nu} - (\partial_\mu A_\nu - \partial_\nu A_\mu)\right)^2 + \mathcal{L} \tag{12.1.10}$$
$$= \frac{1}{4} F_{\mu\nu} F^{\mu\nu} - \frac{1}{2} F_{\mu\nu}(\partial^\mu A^\nu - \partial^\nu A^\mu) + \frac{1}{2}\mu^2 A_\mu A^\mu + \bar\psi(i\slashed\partial - e\slashed A - m)\psi.$$

If we now regard the $F$'s and $A$'s as independent variables, the equations of motion will establish the relationship between them.

Clearly, functional integration over $F$'s will effectively complete the square, restoring the classical definition of $F_{\mu\nu}$. Hence we may take the $F$'s as independent fields, and define the appropriate source Lagrangian density by:

$$\mathcal{L}_{\text{FO}}^{\text{src}} \overset{\text{def}}{=} J_\mu A^\mu + \frac{1}{2} K_{\mu\nu} F^{\mu\nu} + \bar\jmath\psi + \bar\psi j. \tag{12.1.11}$$

The action is given by

$$S_{\text{FO}} \overset{\text{def}}{=} \int d^4x\, \mathcal{L}_{\text{FO}} + \mathcal{L}_{\text{FO}}^{\text{src}} \tag{12.1.12}$$

and the generating functional by

$$Z_{\text{FO}}(J_\mu, K_{\mu\nu}, \bar\jmath, j) \overset{\text{def}}{=} N\left(\frac{\partial}{\partial\psi} \frac{\partial}{\partial\bar\psi}\right) \int (dA)(dF)\, e^{iS_{\text{FO}}}. \tag{12.1.13}$$

Notice that, if the action is quadratic in a field, functional integration over that field effectively substitutes the classical value for it into the action. This indirect use of the classical equations of motion is therefore valid even though we have not yet demonstrated that the evolution of the quantum operators is given by the classical equations.

Now we are in a position to demonstrate the equality of the three generating functionals $Z_{\text{L}}$, $Z_R$, and $Z_{\text{FO}}$. Firstly, integrating over $(dF)$ in $Z_{\text{FO}}$ and setting the extra sources to zero clearly gives $Z_{\text{L}}$. Indeed, the covariance is $\mathbf{1}$, so even the $\det(A)$ factor for this integration is trivial. Secondly, integrating over $(dF^{ij})$ and $(dA^0)$ in $Z_{\text{FO}}$ and setting their source terms to zero reduces $Z_{\text{FO}}$ to $Z_R$.

The conclusion of this argument is that, since the Lagrangian form of the generating functional for spinor QED with a massive photon is equivalent to the Routhian form, we can obtain the Feynman rules for this theory from the Lagrangian density. This corroborates the equivalence of functional integral and canonical quantization when both apply.

As noted in the introduction to this section, this result was expected because the canonical momenta associated with the vector and spinor fields are formally those of the free-field theories. However, the section demonstrates the use of the first-order formalism for connecting Lagrangian and Routhian forms of the generating functional.

# 12.2 Massive Scalar QED

One of the motivations for introducing a functional integral quantization was the problems posed by derivative couplings in the context of canonical quantization. We already know from Section 11.10 that the Feynman rules for scalar QED with a massive photon can be deduced from the Lagrangian. Now we confirm this conclusion using the first-order Lagrangian technique.

From Chapter 9, we recall that minimal coupling between a charged scalar and a massive vector field yields the Lagrangian density

$$\mathcal{L} \stackrel{\text{def}}{=} -\frac{1}{2}F_{\mu\nu}F^{\mu\nu} + \frac{1}{2}\mu^2 A_\mu A^\mu + (\partial_\mu + ieA_\mu)\phi^\dagger(\partial^\mu - ieA^\mu)\phi + m^2\phi^\dagger\phi. \quad (12.2.1)$$

To form the functional integral for the generating functional we must of course add source terms, but these follow the patterns presented above and therefore do not merit further comment.

The principle of quadratic extension must be applied here to both the vector and the scalar fields, so we write down the first-order, classically equivalent Lagrangian density

$$\mathcal{L}_{\text{FO}} \stackrel{\text{def}}{=} \mathcal{L} + \frac{1}{4}\big(F_{\mu\nu} - (\partial_\mu A_\nu - \partial_\nu A_\mu)\big)^2 \quad (12.2.2)$$
$$- \big(\Pi_\mu^\dagger - (\partial_\mu + ieA_\mu)\phi^\dagger\big)\big(\Pi^\mu - (\partial^\mu - ieA^\mu)\phi\big).$$

As we saw in 11.7.23, functional integral quantization of a complex scalar $\phi$ or $\Pi$ merely involves integrating over $\phi$ and $\phi^\dagger$ or $\Pi$ and $\Pi^\dagger$ as independent fields. Since we have here a theory of bosons, the correct classical form to quantize is the Hamiltonian one. However, there is no need to write out the Hamiltonian. Clearly, when we integrate the extended form over $(dF^{ij})$, $(dA^0)$, $(d\Pi_i)$, and $(d\Pi_i^\dagger)$ and drop their source terms, it will reduce to the Hamiltonian form. Similarly, when we integrate the extended form over $(dF^{\mu\nu})$, $(d\Pi_\mu)$, and $(d\Pi_\mu^\dagger)$ and drop their source terms, it will reduce to the Lagrangian form.

From this section, we see that a first-order Lagrangian density can be used to show the equivalence of the Hamiltonian and Lagrangian forms of the generating functional for scalar QED with a massive photon.

At this point, we know how first-order Lagrangians are used to connect Lagrangian and Routhian forms of the generating functional and appreciate the value of this technique for demonstrating the equivalence of Feynman rules derived by functional calculus based on the Lagrangian form to those derived by canonical quantization based on the Routhian form. The next step on the road to quantization of gauge theories is to introduce the Faddeev–Popov principle of integrating over the orbits of the gauge group in the function space of fields.

## 12.3 Outline of Faddeev–Popov Quantization of Gauge Theories

As we noted in Chapter 9, the quantizing a gauge theory depends on finding a complete set of independent fields which may be constrained by canonical, equal-time commutation relations and with which the Hamiltonian density may be defined. Even if we do find such a set of fields, it must depend on a choice of gauge, and then it is not at all obvious that the resulting quantum theory will be gauge invariant.

Functional integral quantization provides a very flexible means of gauge fixing and a method for demonstrating the gauge invariance of the resulting quantum theory. The demonstration that the generating functional is gauge invariant rests on a delicate computation of divergent determinants. In practice, one generally uses the functional integral to fix the gauge and define the Feynman rules and then verifies that the Feynman diagrams which could give rise to gauge-symmetry breaking do not in fact do so. It is worth noting, however, that this preservation of gauge symmetry is not automatic; as we shall see in the next chapter, it is a constraint on the fermion content (or matter representations) of the gauge theory.

Faddeev and Popov proposed that functional integral quantization could be applied to gauge theories as long as, instead of integrating over the space of classical fields and counting gauge-equivalent evolutions as distinct, one integrates over the space of equivalence classes in which gauge-equivalent evolutions are regarded as indistinguishable. Such a space of equivalence classes is called a *moduli space* because it is a parameter space for the equivalence classes. In the remainder of this section, we present Faddeev–Popov integration on the assumption of maximal simplicity. The result is good enough for applications presented in the three sections which follow. After that we pause to reflect on the delicate points.

In what follows, we first derive the Faddeev–Popov principle in two dimensions and second, extend the principle to field theory. The geometry of both steps is separately depicted in Figure 12.3.

For a finite-dimensional illustration of the Faddeev–Popov principle, consider a real action on the plane:

$$(x, y) \to t \cdot (x, y) = \sigma_t(x, y). \tag{12.3.1}$$

Let $f$ be a function on the plane which is constant on the orbits:

$$f\big(\sigma_t(x, t)\big) = f(x, y) \quad \text{for all } t. \tag{12.3.2}$$

Then $f$ is essentially a function on the space of orbits or moduli space of the real action. Now the question arises, how to integrate $f$? We need a measure on moduli space.

The first step towards a measure on moduli space is to parameterize it by choosing a function $c(x, y)$ which has the orbit-intersection and gauge-fixing properties:

(1) For all $(x, y)$ there exists $t$ such that $c\big(\sigma_t(x, y)\big) = 0$;

(2) $c(x, y) = c\big(\sigma_t(x, y)\big) \quad \Longleftrightarrow \quad t = 0$.

Now the graph $\Gamma_c$ of solutions to $c(x, y) = 0$ is obviously in 1–1 correspondence with the moduli space. Hence if we multiply $f$ by the gauge-fixing delta function $\delta\big(c(x, y)\big)$, the integral $dx\, dy$ will effectively be an integral over moduli space.

This integral, however, will depend on the detailed structure of $c$, not simply on $\Gamma_c$. For example, using $2c(x,y)$ would give a different value for the integral of $f$. We need a further factor in the integrand to cancel this dependence on $c$, a factor which for simplicity does not itself contribute to gauge-fixing. Following Faddeev and Popov, we propose the function $\Delta_c(x,y)$ defined by

$$\Delta_c(x,y) \stackrel{\text{def}}{=} \left| \frac{\partial}{\partial t} c\big(\sigma_t(x,y)\big) \right|_{t=t_0}, \tag{12.3.3}$$

where $t_0$ is defined by $c\big(\sigma_{t_0}(x,y)\big) = 0$, or equivalently by

$$\Delta_c(x,y)^{-1} = \int dt\, \delta\Big(c\big(\sigma_t(x,y)\big)\Big). \tag{12.3.4}$$

The function $\Delta_c$ is known as the *Faddeev–Popov determinant*.

**Homework 12.3.5.** Using the definition of $\Delta_c$ as an integral and the real-action property of $\sigma_t$, show that $\Delta_c\big(\sigma_t(x,y)\big) = \Delta_c(x,y)$.

Our candidate for an integral of $f$ over moduli space is thus

$$I_M(f) \stackrel{\text{def}}{=} \int dx\, dy\, \delta\big(c(x,y)\big) \Delta_c(x,y) f(x,y). \tag{12.3.6}$$

It remains to verify that this integral is independent of $c$. There are basically two steps, the first to show that the $\Delta_c$ factor makes the integral independent of the $c$ used to define $\Gamma_c$, and the second to show that the integral is independent of $\Gamma_c$. The first step is ordinary real analysis, and the second almost follows by making a position-dependent application of the real action to shift $\Gamma_c$ into some other zero set $\Gamma_{c'}$. The only hitch is that we need the Lebesgue measure $dx\, dy$ to be invariant under the real action. With these observations, we feel comfortable concluding our finite-dimensional illustration of the Faddeev–Popov principle with an example.

Define an area-preserving real action on the plane by

$$t \cdot (x,y) \stackrel{\text{def}}{=} (e^t x, e^{-t} y). \tag{12.3.7}$$

Clearly the moduli space of this action consists of the hyperbolae $xy = c$, the four half lines lying along the axes, and the origin. A continuous function on the plane which is constant on these orbits is a function $f(xy)$ of the single variable $xy$. (A continuous function cannot distinguish between the five orbits which taken together form the coordinate axes.)

Geometrically, choosing a gauge-fixing function $c$ means choosing a curve in the plane which intersects every orbit exactly once. Inspection of the graph makes it clear that such a curve cannot be connected. We choose the condition $y^2 = x^2$ which defines a pair of lines which intersects every orbit except the four half lines in a unique point. Dropping four points from the moduli space in this way will not affect the integral. (Figure 12.3 represents this particular real action and gauge condition.)

The Faddeev–Popov determinant $\Delta_c$ associated with this real action and gauge-fixing function is given by

$$\Delta_c(x,y)^{-1} = \left| \int dt\, \delta(e^{2t}x^2 - e^{-2t}y^2) \right| = \left| \frac{1}{4xy} \right|, \tag{12.3.8}$$

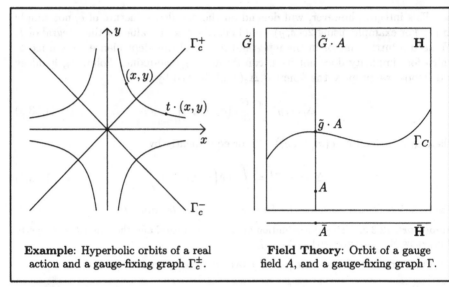

**Example:** Hyperbolic orbits of a real action and a gauge-fixing graph $\Gamma_c^\pm$.

**Field Theory:** Orbit of a gauge field $A$, and a gauge-fixing graph $\Gamma$.

**Fig. 12.3.** This figure presents the geometry of gauge fixing. On the left is a two-dimensional example, a plane divided into hyperbolic orbits and a graph $\Gamma_c^\pm$ of a gauge-fixing function $c$. On the right, the function space $\mathbf{H}$ is represented as a product of the gauge group $\tilde{G}$ and the moduli space $\bar{\mathbf{H}}$ of its orbits $\bar{A} = \tilde{G} \cdot A$, and $\Gamma_C$ is the graph of a gauge condition $C = 0$ which effectively determines $\tilde{g}$ as a function of $A$.

and so the integral over moduli space is

$$I_M(f) = \int dx\, dy\, 4|xy|\delta(x^2 - y^2)f(xy)$$

$$= \int_{-\infty}^{\infty} dx\, 2|x|\big(f(x^2) + f(-x^2)\big) \tag{12.3.9}$$

$$= 2\int_0^\infty ds\, \big(f(s) + f(-s)\big).$$

From the penultimate line, we see that the measure on moduli space is $2|x|\, dx$. We could have arrived at the same conclusion by making a change of variables

$$(X, t) \to (x, y) = (e^t X, e^{-t} X) \implies \frac{\partial(x, y)}{\partial(X, t)} = 2|X| \tag{12.3.10}$$

$$\implies dx\, dy = 2|X|\, dX\, dt.$$

**Homework 12.3.11.** Show that we arrive at the same value for $I_M(f)$ if we use the gauge-fixing function $x^2 - a^2$. Find the measure on moduli space associated with this gauge condition, and show that it is consistent with the measure $2|X|\, dX$ derived above.

From this example, we obtain a general form 12.3.6 for the integral over moduli space in terms of Lebesgue measure, a gauge-fixing delta function, and a Faddeev–Popov determinant. We also see that this approach to defining integration over moduli space depends on the invariance of Lebesgue measure under the action of the gauge group. Having understood the Faddeev–Popov construction in finite dimensions, we proceed to extend it formally to field theory.

Returning to field theory, the results above suggest the Faddeev–Popov principle for functional integral quantization of gauge theories:

**Principle 12.3.12.** *The generating functional of a quantized gauge field theory is given by the following functional integral:*

$$Z(J) \overset{\text{def}}{=} N \int (dA)\, \delta\big(C(A)\big) \Delta_C(A) e^{iS(A,J)},$$

*where $J$ represents classical sources for gauge invariant terms.*

To clarify the meaning of this formula, we take a moment to discuss delta functionals and gauge fixing.

Let $A$ be an element of a function space $\mathbf{H}$, let $\tilde{G}$ be a gauge group which acts on $\mathbf{H}$, and let $\bar{\mathbf{H}}$ be the moduli space of orbits of $\tilde{G}$ in $\mathbf{H}$. The purpose of a gauge-fixing condition $C(A) = 0$ is to determine a unique $A$ in every orbit of the gauge group. (All this structure is represented in Figure 12.3.)

The same idea can be expressed in another way: given an arbitrary $A$, there should be a unique gauge transformation $\tilde{g}$ such that $C(\tilde{g} \cdot A) = 0$. The gauge-fixing condition $C = 0$ should thereby determine $\tilde{g}$ as a function of $A$:

$$C(\tilde{g} \cdot A) = 0 \quad \Longleftrightarrow \quad \tilde{g}(x) = \tilde{g}(A, x). \tag{12.3.13}$$

Thus the gauge-fixing function $C$ has domain and range

$$C{:}\tilde{G} \times \mathbf{H} \to \mathbf{H}', \tag{12.3.14}$$

where $\mathbf{H}'$ is a function space determined by $C$.

With this perspective on $C$, we can consider the delta functional $\delta(C)$ as determining the gauge-group element $\tilde{g}(A)$. To interpret this assertion, let us introduce the functional measure $(d\tilde{g})$ for the gauge group $\tilde{G}$. Then the basic property of a delta functional on the gauge group is

$$\int (d\tilde{g})\, \delta(\tilde{g} - \tilde{g}_0) F(\tilde{g}) \equiv F(\tilde{g}_0), \tag{12.3.15}$$

where $\tilde{g}_0$ is a fixed element in the gauge group and $F$ is any continuous functional on the gauge group.

Just as the ordinary delta function $\delta(x - x_0)$ is related to $\delta\big(f(x)\big)$ by a Jacobian, so also the gauge-fixing delta functional is related to $\delta(\tilde{g} - \tilde{g}(A))$ by a functional Jacobian:

$$\Delta_C(A) \overset{\text{def}}{=} \left\| \frac{\delta C}{\delta \tilde{g}} \right\|_{\tilde{g}=\tilde{g}(A)} \quad \Longrightarrow \quad \Delta_C(A) \int (d\tilde{g})\, \delta\big(C(\tilde{g} \cdot A)\big) = 1. \tag{12.3.16}$$

(The notation $\|K\|$ stands for the absolute value of the determinant of $K$.)

For $C = 0$ to be a good gauge-fixing condition, clearly the condition 12.3.13 specifying the uniqueness of $\tilde{g}$ should be supplemented by the transversality condition $\Delta_C(A) > 0$.

**Remark** 12.3.17. Note that, because of the invariance of the gauge-group measure $(d\tilde{g})$ under the action of gauge-group elements $\tilde{g}'$, the Jacobian $\Delta_C(A)$ depends only on the gauge orbit of $A$:

$$\Delta_C(\tilde{g}' \cdot A)^{-1} = \int (d\tilde{g}) \, \delta\big(C(\tilde{g}\tilde{g}' \cdot A)\big)$$

$$= \int (d\tilde{g}\tilde{g}') \, \delta\big(C(\tilde{g}\tilde{g}' \cdot A)\big) \tag{12.3.18}$$

$$= \Delta_C(A)^{-1}.$$

In order to define the determinant here, it is necessary to choose some identification between the Lie algebra of $\tilde{G}$ and $\mathbf{H}'$. Different choices will affect the value of $\Delta_C(A)$ by a constant factor. In the integral definition of $\Delta_C$, this ambiguity originates in the undefined normalization of the functional measures on $\tilde{G}$ and $\mathbf{H}'$. Thus $\Delta_C(A)$ is not unique, but the ratio $\Delta_C(A)/\Delta_C(A_0)$ is.

In practice, as we noted in Remark 11.5.13, to define the determinant of an operator, we must make a Wick rotation and compactify space (or space-time). In this drastic transformation, distinctions between the original function spaces are destroyed.

From our example, we know that gauge-invariance of the Faddeev–Popov functional integral 12.3.12 depends on the measure on function space being invariant under gauge transformations. Since this measure is not precisely defined, any demonstration of invariance will be at best formal. As we shall see in Section 12.5, a gauge transformation acting on a vector made up of all the gauge fields combines a translation and a space-time dependent rotation. The rotation generates a unitary transformation of the Hilbert space of square-integrable functions. Our functional measure is based on Lebesgue measure and is therefore assumed to be invariant under translations and unitary transformations. Similarly, Section 12.5 shows that a gauge transformation acting on a vector of matter fields is a space-time dependent unitary transformation. Chapter 13 finds that the functional measure for the scalar fields in invariant under this transformation, but the functional measure for the Dirac fields may not be invariant. If the Dirac functional measure is not invariant, then (as we shall see) quantum dynamics causes conservation breakdown in currents coupled to gauge fields. Since this in turn causes breakdown of gauge invariance, we must conclude that Faddeev–Popov quantization breaks down.

Let $\Gamma_C$ be the set of solutions to $C(A) = 0$ in $\mathbf{H}$. Since a good gauge condition provides an identification of points $\bar{A}$ in $\bar{\mathbf{H}}$ with points $A_C$ in $\Gamma_C$, it determines an identification of $\tilde{G} \times \bar{\mathbf{H}}$ with $\mathbf{H}$:

$$\left. \begin{array}{ccccc} \tilde{G} \times \bar{\mathbf{H}} & \xrightarrow{\ \bar{F}_C\ } & \tilde{G} \times \Gamma_C & \xrightarrow{\ F_C\ } & \mathbf{H} \\ \tilde{g}\ ,\ \bar{A} & \longmapsto & \tilde{g}\ ,\ A_C & \longmapsto & \tilde{g} \cdot A_C \end{array} \right\} \tag{12.3.19}$$

The Faddeev–Popov construction assumes that $\bar{F}_C$ preserves measures, so that the measure on $\bar{\mathbf{H}}$ can be deduced from $F_C$:

$$Z = N \int (dA) \, \delta\big(C(A)\big) \Delta_C(A) e^{iS(A)} \tag{12.3.20}$$

$$= N \int (dA_C)(d\tilde{g}) \left\| \frac{\delta(\tilde{g} \cdot A_C)}{\delta\tilde{g} \, \delta A_C} \right\| \delta\big(C(\tilde{g} \cdot A_C)\big) \Delta_C(\tilde{g} \cdot A_C) e^{iS(\tilde{g} \cdot A_C)}$$

$$= N \int (dA_C) \left\| \frac{\delta(\tilde{g} \cdot A_C)}{\delta\tilde{g} \, \delta A_C} \right\|_{\tilde{g}=1} e^{iS(A_C)}$$

$$= N \int (d\bar{A}) \, e^{iS(\bar{A})},$$

from which we conclude that the measure $(d\bar{A})$ on the moduli space is determined by the parameterization $A_C$ of $\Gamma_C$ according to the relation

$$(d\bar{A}) \stackrel{\text{def}}{=} (dA_C) \left\| \frac{\delta(\tilde{g} \cdot A_C)}{\delta \tilde{g} \, \delta A_C} \right\|_{\tilde{g}=1}. \tag{12.3.21}$$

Reducing the integral over function space to an integral over moduli space depends on including in the Lagrangian only classical backgrounds which are coupled to gauge-invariant terms. In practice, we generally break gauge invariance by coupling background fields to all the fields of the theory, compute Green functions which are not gauge invariant, and finally restore gauge invariance by removing gauge-dependent propagators with the LSZ reduction formula.

In applying the Faddeev–Popov principle to gauge theories, it is necessary to obtain perturbative formulae for the delta function and the Faddeev–Popov determinant. In the following sections, we shall introduce tricks to move these factors into the action, creating a *Faddeev–Popov action* and a *Faddeev–Popov Lagrangian density* which differ from the originals by a gauge-fixing term and a ghost term.

In this section, we have laid out the details of Faddeev–Popov integration in finite dimensions and thereby seen the underlying assumptions clearly: how the gauge group should act and what the properties of a good gauge-fixing condition are. Extending this knowledge to field theory, we have shown that there is a meaningful Lagrangian form of the generating functional which is independent of the choice of gauge-fixing condition. Thus the Faddeev–Popov principle provides a foundation for quantizing gauge theories in arbitrary gauges. The following sections develop the applications of this principle, and the closing section of the chapter returns to the principle itself to clarify delicate points arising from these applications.

## 12.4 QED in Covariant Gauge

The application of the Faddeev–Popov principle to QED is completely routine. We choose a gauge condition, define the generating functional, lift the gauge-fixing delta functional into the action, and then obtain the Feynman rules. To close the section, we use the first-order Lagrangian technique to compare the Faddeev–Popov quantum theory with the canonical quantum theory in axial gauge. Remember that we are taking a rough and ready approach to application of the Faddeev–Popov principle, saving a discussion of fine points for the last section.

For QED, we choose $C(A) = \partial_\mu A^\mu - f(x)$, where $f$ is any function of space-time which vanishes at space-time infinity. Given an arbitrary potential $A_\mu$, there is a unique gauge $\chi$ with zero initial condition at $t = -\infty$ for which $A'_\mu = A_\mu - \partial_\mu \chi$ satisfies the gauge condition. If we use the Lie algebra elements $\chi$ as parameters for the gauge group, then we can evaluate the Faddeev–Popov determinant by functional differentiation:

$$\Delta_C(A) = \left\| \frac{\delta C(A)}{\delta \chi} \right\| = \| \partial_\mu \partial^\mu \|, \tag{12.4.1}$$

which is an undefined constant to be absorbed into the overall normalization factor $N$ of the Faddeev–Popov formula 12.3.12:

$$Z_f \overset{\text{def}}{=} N_f \left(\frac{\partial}{\partial \psi} \frac{\partial}{\partial \bar{\psi}}\right) \int (dA)\, e^{iS_{\text{QED}}} \delta(\partial_\mu A^\mu - f). \tag{12.4.2}$$

Now the $f$ in the definition of the generating functional can be removed by a gauge transformation, so $Z = Z_f$ and $N = N_f$ are independent of $f$. Because of this, we can insert an $f$-dependent factor in the integrand and integrate over $(df)$ in order to transform the gauge-fixing delta functional into a gauge-fixing term in the Lagrangian density:

$$Z = N' \int (df)\, \exp\left(-\frac{i}{2\alpha} \int d^4x\, f^2(x)\right) Z \tag{12.4.3}$$

$$= NN' \left(\frac{\partial}{\partial \psi} \frac{\partial}{\partial \bar{\psi}}\right) \int (df)(dA)\, \exp\left(-\frac{i}{2\alpha} \int d^4x\, f^2(x)\right) e^{iS(A,\bar{\psi},\psi)} \delta(\partial_\mu A^\mu - f)$$

$$= NN' \left(\frac{\partial}{\partial \psi} \frac{\partial}{\partial \bar{\psi}}\right) \int (dA)\psi\, \exp\left(iS(A,\bar{\psi},\psi) - \frac{i}{2\alpha} \int d^4x\, (\partial_\mu A^\mu)^2\right).$$

(For brevity, we have suppressed the classical sources.)

Thus the original gauge-invariant QED Lagrangian density $\mathcal{L}_{\text{QED}}$ is modified to a *Faddeev–Popov Lagrangian density* $\mathcal{L}_{\text{FP}}$ by the addition of a gauge-fixing term:

$$\mathcal{L}_{\text{FP}} \overset{\text{def}}{=} \mathcal{L}_{\text{QED}} - \frac{1}{2\alpha}(\partial_\mu A^\mu)^2. \tag{12.4.4}$$

It is neither obvious a priori that the $\mathcal{L}_{\text{FP}}$ theories are classically equivalent to the $\mathcal{L}_{\text{QED}}$ theory, nor that the extra term will effectively enforce the gauge condition, nor that the $\mathcal{L}_{\text{FP}}$ theories are independent of $\alpha$, nor that quantization will preserve all these non-obvious properties of $\mathcal{L}_{\text{FP}}$. It is the merit of the Faddeev–Popov principle that leads us to suppose that $\mathcal{L}_{\text{FP}}$ will provide a good QED.

**Homework 12.4.5.** Show that the gauge-fixing term does what it should on the classical level, i.e. that the equations of motion derived from $\mathcal{L}_{\text{FP}}$ imply that if $\partial_\mu A^\mu$ vanishes somewhere, then it is identically zero.

Now that we have a Faddeev–Popov gauge-fixed Lagrangian density, we could follow the logic of Section 11.10 to show from Dyson's formula that the Lagrangian form of the generating function is valid. The vertex Feynman rules are unchanged from QED with a massive photon; only the photon propagator is new. Using the projection operators

$$P_T \overset{\text{def}}{=} 1 - \frac{kk^\top}{k^2} \quad \text{and} \quad P_L \overset{\text{def}}{=} \frac{kk^\top}{k^2}, \tag{12.4.6}$$

we see that that part $S(A)$ of the action $S(A, \bar{\psi}, \psi)$ which is quadratic in $A$ may be written in terms of transversal and longitudinal components of $A$:

$$S(A) = \frac{1}{2} \int d^4x\, -(\partial_\mu A_\nu^T)(\partial^\mu A_T^\nu) - \frac{1}{\alpha}(\partial_\mu A_\nu^L)(\partial^\mu A_L^\nu), \tag{12.4.7}$$

so that the momentum-space covariance $\hat{C}$ is given by

$$\hat{C} = k^2 P_T + \frac{1}{\alpha} k^2 P_L. \tag{12.4.8}$$

Regularizing the factor of $k^2$ to $k^2 + i\epsilon$, we find the propagator:

$$\begin{aligned}
\text{Propagator} &= (-i)\frac{1}{k^2 + i\epsilon}\left(P_T + \frac{1}{\alpha}P_L\right)^{-1} \\
&= -\frac{i}{k^2 + i\epsilon}(P_T + \alpha P_L) \\
&= \frac{i}{k^2 + i\epsilon}\left(-\mathbf{1} + (1 - \alpha)\frac{kk^{\top}}{k^2}\right).
\end{aligned} \tag{12.4.9}$$

The components of the propagator are given by

$$\overline{\quad\mu \overset{\overleftarrow{\phantom{k}}}{\quad k \quad} \nu \quad} = \frac{i}{k^2 + i\epsilon}\left(-g_{\mu\nu} + (1 - \alpha)\frac{k_\mu k_\nu}{k^2}\right). \tag{12.4.10}$$

When $\alpha$ is set to zero, the propagator is totally transverse; this is *Landau gauge*. When $\alpha$ is set to unity, then the propagator reduces to $-ig_{\mu\nu}/k^2$; this is *Feynman gauge*.

Note that there is no value of $\alpha$ for which this propagator is the limit as $\mu^2 \to 0$ of the massive vector field propagator. The difference, however, is in the longitudinal terms $kk^{\top}$, which in both massive and massless theories do not contribute to amplitudes of measurable cross sections.

At this point, though it appears that the Faddeev–Popov principle has given us a sensible QED, considering the tenuous existence of the integrals involved it is wise to seek confirmation of the principle. One method for confirming the application to QED is to connect our results so far to some Routhian form of the generating functional. The logic is as follows: if there is a gauge in which the generating functional has a Routhian form, and if the Faddeev–Popov principle in that gauge yields the same generating functional, then we may have confidence that the principle applies to QED and that the quantum theory is also gauge invariant.

In fact, QED in axial gauge $A_3 = 0$ has a Routhian. The canonical variables are $A_1$, $A_2$, $F_{01}$, $F_{02}$, $\psi$, and $\bar{\psi}$, and the others are constrained by

$$\begin{aligned}
A_0: &\quad \partial_3\partial^3 A^0 = -\partial_3 F^{03} = \partial_1 F^{01} + \partial_2 F^{02} - e\bar{\psi}\gamma^0\psi; \\
F_{03}: &\quad F_{03} = \partial_0 A_3 - \partial_3 A_0 = -\partial_3 A_0; \\
F_{ij}: &\quad F_{ij} = \partial_i A_j - \partial_j A_i.
\end{aligned} \tag{12.4.11}$$

Letting $S_\mathcal{R}$ be the Routhian action plus source terms, we can write the generating functional

$$Z_\mathcal{R} \overset{\text{def}}{=} \left(\frac{\partial}{\partial\psi}\frac{\partial}{\partial\bar{\psi}}\right)\int (dA_1)(dA_2)(dF_{01})(dF_{02})\, e^{iS_\mathcal{R}}. \tag{12.4.12}$$

Since a gauge variation of $A_3$ by $\delta\chi$ has effect $\delta A_3 = \partial_3\delta\chi$, the operator $\delta A_3/\delta\chi = \partial_3$ has constant determinant. Hence applying the Faddeev–Popov principle to QED in axial gauge directly yields

$$Z_{\text{FP}} \overset{\text{def}}{=} N\left(\frac{\partial}{\partial\psi}\frac{\partial}{\partial\bar{\psi}}\right)\int (dA)\,\delta(A_3)e^{iS}, \tag{12.4.13}$$

where $S$ is the usual QED action plus source terms (12.1.3 with $\mu^2 = 0$).

To connect $Z_R$ to $Z_{FP}$ we need the extended form $\mathcal{L}_{FO}$ of the Lagrangian density (12.1.10 with $\mu^2 = 0$ plus source terms). Applying the Faddeev–Popov principle to $\mathcal{L}_{FO}$ determines a third generating functional:

$$Z_{FO} \overset{\text{def}}{=} N'\left(\frac{\partial}{\partial\psi}\,\frac{\partial}{\partial\bar{\psi}}\right) \int (dA)(dF)\,\delta(A_3)e^{iS_{FO}}. \tag{12.4.14}$$

Now we can obtain $Z_{FP}$ and $Z_R$ from $Z_{FO}$ simply by integrating over the appropriate constrained variables and setting source terms to zero. We conclude that first, in axial gauge, canonical quantization and the Faddeev–Popov principle define the same quantum theory and second, the Faddeev–Popov principle can therefore be used in QED to show that this quantum theory is gauge invariant.

The field theories of primary interest are gauge field theories. In general, to apply functional integral quantization to a field theory, it is necessary to have a Routhian density for it. For gauge theories, one must choose a gauge in order to obtain a Routhian density. Then, however, the resulting quantum theory may not be gauge invariant; even if it is, the Feynman rules will not be covariant, so computation will be very trying. These problems are solved by the Faddeev–Popov principle – integration over moduli space – and the use of first-order Lagrangians. With these two devices, it is easy to show that the Routhian form is equivalent to a family of Lagrangian forms, at least one of which has covariant Feynman rules. In the example of QED above, the covariant Lagrangian is simply the original, minimally-coupled one modified by a gauge-fixing term.

## 12.5  The Geometry of Gauge Field Theories

Now that we have understood the Faddeev–Popov method for quantizing gauge theories, in preparation for application, we need to explain the structure of classical gauge theories in detail. As with QED, we find it best to introduce the concepts from a geometrical standpoint.

A classical theory of matter fields comprising real scalars $\phi = (\phi_1, \ldots, \phi_m)^{\mathsf{T}}$ and Dirac fermions $\psi = (\psi_1, \ldots, \psi_n)^{\mathsf{T}}$ determined by a Lagrangian density $\mathcal{L}_M$ has a global symmetry group $G$ if there are representations $M$ and $N$ of $G$ on $\mathbf{R}^m$ and $\mathbf{C}^n$ respectively for which

$$\mathcal{L}_M\big(M(g)\phi, N(g)\psi\big) \equiv \mathcal{L}_M(\phi, \psi), \quad \text{for all } g \text{ in } G. \tag{12.5.1}$$

(A divergence could be added on the right, but for internal symmetries this is not necessary.)

As pointed out before in Section 9.5, there is a geometric or coordinate-free viewpoint on the classical theory described by $\mathcal{L}_M$. To make the transformation to the geometric perspective, it is only necessary to contract every coordinate vector with the appropriate basis vectors.

The fields $\phi$ and $\psi$ are coordinate-vector functions of space-time. Writing $U_x = \mathbf{R}^m$ and $V_x = \mathbf{C}^n$ for the target vector spaces at every point $x$ in space-time, we see that the standard bases $u = (u_1, \ldots, u_m)$ and $v = (v_1, \ldots, v_n)$ for $\mathbf{R}^m$ and $\mathbf{C}^n$ respectively underlie the coordinate representation of the fields. The geometric realities are the vector-valued functions $\Phi$ and $\Psi$ defined by

$$\Phi(x) \overset{\text{def}}{=} u_r(x)\phi^r(x) \quad \text{and} \quad \Psi(x) \overset{\text{def}}{=} v_s(x)\psi^s(x), \tag{12.5.2}$$

where we have promoted $u$ and $v$ to functions because these bases are selected independently at each point of space-time.

A change of coordinates is defined by a function $R(x) \times C(x)$ from space-time to $GL(m, \mathbf{R}) \times GL(n, \mathbf{C})$. Since the vector $\Phi$ is independent of coordinates, if we change coordinates by $M$, then we must also change bases by $M^{-1}$ as follows:

$$\left. \begin{array}{l} \phi' = M\phi \\ u' = uM^{-1} \end{array} \right\} \quad \Longleftrightarrow \quad u'\phi' = u\phi. \tag{12.5.3}$$

Similar points apply to $\Psi$, of course. Note that we are writing vectors to the right of scalars so that a vector like $\Phi$ may be written as a matrix product like $u\phi$.

The kinetic terms in the Lagrangian indicate the need for inner products on $U_x$ and $V_x$, and so we can restrict to orthonormal changes of coordinates determined by matrix-valued functions $O(x) \times U(x)$ from space-time to $SO(m) \times U(n)$. Finally, the coordinate changes which leave the Lagrangian invariant form a set of constant functions, namely $M(g) \times N(g)$, where $g$ is some fixed element in the symmetry group $G$.

The gauge principle proposes that the physics should depend only on the geometry of the fields and not at all on the coordinates used to describe them. Indeed, if the $G$ symmetry is unbroken, it must be impossible to determine preferred bases experimentally. Hence, this final collapse in permissible changes of coordinates is unacceptable, and the physical symmetry group should be the *gauge group* $\tilde{G}$ of *gauge transformations* $M(\tilde{g}(x)) \times N(\tilde{g}(x))$, where $\tilde{g}(x)$ is an arbitrary function from space-time to $G$.

Derivatives, however, do not commute with gauge transformations. Thus if $\phi' = M\phi$, then the form of the derivative term is changed by the gauge transformation:

$$\begin{aligned} \partial_\mu \phi' &= \partial_\mu (M\phi) \\ &= (\partial_\mu M)\phi + M\partial_\mu \phi. \end{aligned} \tag{12.5.4}$$

Hence, for example, the kinetic terms in $\mathcal{L}_M$ will not be gauge invariant.

To restore good gauge transformation properties to the derived fields, we must introduce a gauge multiplet $A_\mu$ of fields whose gauge transformation properties enable them to mop up the extra term $(\partial_\mu M)\phi$. Therefore, following Section 9.5, the effect of a gauge transformation on the four-vectors $A$ must be:

$$\begin{aligned} A'_\mu \phi' &\stackrel{\text{def}}{=} -(\partial_\mu M)\phi + MA_\mu \phi \\ \Longrightarrow \quad A'_\mu &= MA_\mu M^{-1} - (\partial_\mu M)M^{-1}. \end{aligned} \tag{12.5.5}$$

Notice that if $M$ is a function with values in any group $G$ of $m \times m$ matrices, then

$$(\partial_\mu M)M^{-1}\big|_x = \partial_\mu^y M(x+y)M(x)^{-1}\big|_{y=0}, \tag{12.5.6}$$

which shows that $(\partial_\mu M)M^{-1}$ is the derivative at the identity of a path through the identity in $G$, and hence a function with values in the Lie algebra $\mathcal{G}$ of $G$. Therefore, if $A_\mu$ takes values in $\mathcal{G}$, then so do $MA_\mu M^{-1}$ and $A'_\mu$. Hence it is natural and consistent to assume that $A_\mu$ takes values in $\mathcal{G}$.

From Section 9.5, we know that this transformation rule is consistent with a geometrical interpretation of $A$ as a connection one-form on the vector bundle made up of the $U_x$'s. Recall that the covariant derivative $D_\mu$ acts on vectors, and the connection one-form $A_\mu$ is defined by the action of $D_\mu$ on a basis:

$$D_\mu u_r \stackrel{\text{def}}{=} u_s A_\mu{}^s{}_r. \tag{12.5.7}$$

The properties of $D$ imply that the connection one-form obeys the transformation law 12.5.5:

$$\begin{aligned}
D_\mu \Phi &= u' A'_\mu \phi' + u' \partial_\mu \phi' \\
&= u A_\mu \phi + u \partial_\mu \phi \\
&= u' M A_\mu M^{-1} \phi' + u' M \partial_\mu M^{-1} \phi' \\
&= u' M A_\mu M^{-1} \phi' - u' (\partial_\mu M) M^{-1} \phi' + u' \partial_\mu \phi' \\
\implies A'_\mu &= M A_\mu M^{-1} - (\partial_\mu M) M^{-1},
\end{aligned} \tag{12.5.8}$$

where we have used the formula

$$\partial_\mu (M M^{-1}) = 0 \implies \partial_\mu M^{-1} = -M^{-1} (\partial_\mu M) M^{-1}. \tag{12.5.9}$$

The introduction of basis vectors enabled us to represent and use the geometric reality of $D_\mu \Phi$. Writing the basis vectors on the left makes it easy to suppress them, leaving the covariant derivative to act on the coordinate vector $\phi$:

$$D_\mu u_r \phi^r = u_r \partial_\mu \phi^r + u_r A^r{}_s \phi^s \quad \text{leads to} \quad D_\mu \phi = \partial_\mu \phi + A_\mu \phi. \tag{12.5.10}$$

In physics, the basis is usually suppressed in this way.

Since the derived fields $\partial_\mu \psi$ present the same transformation problem, there are also terms $(\partial_\mu N) N^{-1} \psi'$ to absorb. To cover both sets of terms with one set of fields, we must take our attention to the common level of $G$ which underlies both $M$ and $N$. First, define actions of the group $G$ on the vectors $\phi$ and $\psi$ through the representations $M$ and $N$:

$$\begin{aligned}
g \cdot \phi &\stackrel{\text{def}}{=} M(g)\phi; \\
g \cdot \psi &\stackrel{\text{def}}{=} N(g)\psi.
\end{aligned} \tag{12.5.11}$$

Second, extend these actions to the Lie algebra $\mathcal{G}$ of $G$ by substituting $g = \exp(\theta X)$, differentiating with respect to $\theta$, and evaluating at $\theta = 0$:

$$\begin{aligned}
X \cdot \phi &\stackrel{\text{def}}{=} \frac{\partial}{\partial \theta} M(e^{\theta X}) \phi \Big|_{\theta=0}; \\
X \cdot \psi &\stackrel{\text{def}}{=} \frac{\partial}{\partial \theta} N(e^{\theta X}) \psi \Big|_{\theta=0}.
\end{aligned} \tag{12.5.12}$$

If we now take $A_\mu$ to be an $\mathcal{G}$-valued function, then these two actions of the Lie algebra $\mathcal{G}$ enable $A_\mu$ to act naturally on the fields $\phi$ and $\psi$, and enable a single covariant derivative $D$ to act on both fields:

$$D_\mu \phi = \partial_\mu \phi + A_\mu \cdot \phi \quad \text{and} \quad D_\mu \psi = \partial_\mu \psi + A_\mu \cdot \psi. \tag{12.5.13}$$

Also, given a gauge transformation $\tilde{g}(x)$, the special-case transformation rule 12.5.5 becomes the universal rule

$$A'_\mu = \tilde{g} A_\mu \tilde{g}^{-1} - (\partial_\mu \tilde{g}) \tilde{g}^{-1}. \tag{12.5.14}$$

Now that we have the covariant derivative, we can solve the problem posed by the derived fields. Since the covariantly derived fields $D_\mu \phi$ and $D_\mu \psi$ transform as vectors under gauge transformations, if $\mathcal{L}_M(\phi, \partial\phi, \psi, \partial\psi)$ is invariant under the global symmetry group $G$, then $\mathcal{L}_M(\phi, D\phi, \psi, D\psi)$ is invariant under the gauge group $\tilde{G}$. The new Lagrangian density provides the *minimal coupling* between gauge and matter fields needed for gauge invariance.

To complete the minimal coupling procedure, we must add a kinetic term for the gauge fields. We recall from Section 9.5 that there is a curvature tensor $G_{\mu\nu}$ associated with a covariant derivative:

$$G_{\mu\nu}(A) \stackrel{\text{def}}{=} [D_\mu, D_\nu] = \partial_\mu A_\nu - \partial_\nu A_\mu + [A_\mu, A_\nu]. \tag{12.5.15}$$

The geometric Lagrangian density derived from the covariant derivative is

$$\mathcal{L}(A) = \text{tr}(G_{\mu\nu} G^{\mu\nu}). \tag{12.5.16}$$

**Homework 12.5.17.** Show that, under a gauge transformation $\tilde{g}(x)$, $G_{\mu\nu}$ transforms as a tensor representation: $G'_{\mu\nu} = \tilde{g} G_{\mu\nu} \tilde{g}^{-1}$.

**Remark 12.5.18.** In differential geometry, the operators $\partial_\mu$ are regarded as a basis for the vector space of linear differential operators $X(x) = f^\mu(x)\partial_\mu$, and the differentials $dx^\mu$ are regarded as a basis for the vector space of differentials $\theta(x) = a_\mu(x) \, dx^\mu$. These two vector spaces have a natural duality determined by $dx^\mu(\partial_\nu) = \delta^\mu_\nu$. The differentials generate the Grassmann algebra of *forms*. There is an *exterior derivative* $d$ defined on the forms by $d\theta = dx^\mu \wedge \partial_\mu \theta$. Thus the connection one-form, the covariant derivative, and the curvature may be written $A = A_\mu \, dx^\mu$, $D = d + A$, and $G = G_{\mu\nu} \, dx^\mu \wedge dx^\nu$. If we let $A \wedge A = [A_\mu, A_\nu] \, dx^\mu \wedge dx^\nu$, then we find the popular formula $G = dA + A \wedge A$.

The gauge fields $A^a_\mu$ are essentially the coefficient functions of the connection one-form $A$ with respect to some basis $X^1, \ldots, X^k$ for the Lie algebra $\mathcal{G}$. In quantum field theory, the standard normalization for vector-field kinetic terms is $-\frac{1}{4} F^a_{\mu\nu} F^{a\mu\nu}$, where $F^a_{\mu\nu}$ is $\partial_\mu A^a_\nu - \partial_\nu A^a_\mu$. To bring this term out of the geometric Lagrangian density 12.5.16 we need appropriate normalization of the basis $X^1, \ldots, X^k$.

Since $A_\mu$ transforms to $g A_\mu g^{-1}$ under $G$, it transforms to $[X, A_\mu]$ under $\mathcal{G}$, i.e. $A_\mu(x)$ transforms according to the adjoint representations of $G$ and $\mathcal{G}$ on $\mathcal{G}$. Hence, to normalize the basis, we need an inner product on $\mathcal{G}$ which is invariant under the adjoint.

Some groups, like the group $GL(2, \mathbf{R})$ of $2 \times 2$ real invertible matrices, possess no such invariant inner products. Ones that do are called *semi-simple* Lie groups. If the adjoint representation of a semi-simple Lie group is irreducible, then it is called *simple*. Semi-simple Lie groups are direct products of simple Lie groups. The same terms are used for the corresponding Lie algebras. Note furthermore that the invariant inner product on a simple Lie algebra is unique up to scale. Fortunately, the internal symmetry groups commonly used in physics are all semi-simple up to factors of $U(1)$.

**Homework** 12.5.19. Show that the group $SL(2, \mathbf{R})$ of $2 \times 2$ real matrices with unit determinant is not semi-simple.

Assume therefore that $G$ is simple. Let ad be the adjoint representation of $G$ so that $\text{ad}(X)Y = [X, Y]$. Then the Jacobi identity implies that the bilinear form

$$K(Y, Y') \overset{\text{def}}{=} - \text{tr}(\text{ad } Y \text{ ad } Y') \qquad (12.5.20)$$

is an invariant inner product on $\mathcal{G}$:

$$K(gYg^{-1}, gY'g^{-1}) = K(Y, Y'), \qquad (12.5.21)$$

or equivalently

$$K([X, Y], Y') + K(X, [X, Y']) = 0. \qquad (12.5.22)$$

Due to the size of the matrices $\text{ad}(X)$, this inner product is not convenient in practice. However, any simple Lie group $G$ may be regarded as a subgroup of a unitary group. Taking this stand, we can define a real inner product on $\mathcal{G}$ by

$$(X, X') \overset{\text{def}}{=} \text{tr}(X^\dagger X'). \qquad (12.5.23)$$

Since invariant inner products on simple Lie algebras are unique up to scale, this inner product may be used instead of $K$. Now we can impose an invariant normalization on the basis, and assume that for some constant $\kappa$ the $X$'s satisfy

$$\text{tr}\big((X^a)^\dagger X^b\big) = \frac{\kappa^2}{2}\delta^{ab}. \qquad (12.5.24)$$

Note that this normalization is not preserved by representations like $M$ and $N$, but depends upon the particular embedding of $G$ in a unitary group.

**Homework** 12.5.25. Show that if $X^1, \ldots, X^k$ is an orthonormal basis for a simple Lie algebra with respect to an invariant inner product, then the matrix representing $\text{ad}(X^a)$ is real and anti-symmetric for all $a$.

**Homework** 12.5.26. Compute the $K$ inner product of 12.5.20 for the generators $\tau^r = -\frac{i}{2}\sigma^r$ of $SU(2)$. Compute the $( \, , \, )$ inner product of 12.5.23 for spin 0, 1/2, and 1 representations of $su(2)$.

For the purposes of Feynman rules, we will need to expand the quadratic term in the field strength 12.5.15. Since $\mathcal{G}$ is a Lie algebra, the bracket of basis elements $X^a$ is a linear combination of basis elements:

$$[X^a, X^b] = f^{abc}X^c. \qquad (12.5.27)$$

The constants $f^{abc}$ defined by this equation are called the *structure constants* of $\mathcal{G}$.

The convention in physics is to define hermitian generators $T^a = iX^a$ in the complexified Lie algebra $\mathbf{C} \otimes \mathcal{G}$. The trace inner product 12.5.23 extends naturally to $\mathbf{C} \otimes \mathcal{G}$, so the normalization of the $T$'s is identical to that of the $X$'s:

$$\text{tr}(T^a T^b) = \frac{\kappa^2}{2}\delta^{ab}, \qquad (12.5.28)$$

and the structure constants determine the bracket of the $T$'s:

$$[T^a, T^b] = if^{abc}T^c. \tag{12.5.29}$$

In QED, the normalization of the $U(1)$ generator gives $\kappa^2 = 2$ in 12.5.28, while for simple Lie groups, $\kappa^2 = 1$ is a common choice.

Combining these two properties of the hermitian generators, we find a formula for the structure constants,

$$\frac{i\kappa^2}{2} f^{abc} = \text{tr}\big([T^a, T^b]T^c\big) = \text{tr}(T^aT^bT^c - T^aT^cT^b), \tag{12.5.30}$$

from which it is clear that $f^{abc}$ is a totally anti-symmetric tensor.

**Remark** 12.5.31. The Lie bracket [ , ] and the inner product $K$ are intrinsic to the Lie algebra $\mathcal{G}$. The equation 12.5.29 for the bracket of the hermitian generators is dependent on the choice of basis for $\mathcal{G}$, and the equation 12.5.28 depends on the choice of basis and the representation of these basis elements as matrices.

Finally, having defined a normalized hermitian basis for $\mathcal{G}$, we can express the connection one-form $A$ in terms of gauge fields $A^a$ and a coupling constant $g$:

$$A = -igA^aT^a. \tag{12.5.32}$$

The generic covariant derivative may now by written

$$D = \partial - igA^aT^a \tag{12.5.33}$$

and freely applied to each multiplet of matter fields using the notion of Lie algebra action as defined in 12.5.12.

If we define the field strengths $G^a_{\mu\nu}$ in terms of the curvature $G_{\mu\nu}$ by

$$G_{\mu\nu} \overset{\text{def}}{=} -igG^a_{\mu\nu}T^a, \tag{12.5.34}$$

then

$$G^a_{\mu\nu} = \partial_\mu A^a_\nu - \partial_\nu A^a_\mu + gf^{abc}A^b_\mu A^c_\nu, \tag{12.5.35}$$

and we have the appropriate normalization for the kinetic term. The Yang–Mills Lagrangian density for the gauge fields is therefore

$$\mathcal{L}_{\text{YM}}(A) \overset{\text{def}}{=} -\frac{1}{4}G^a_{\mu\nu}G^{a\mu\nu} = \frac{1}{2g^2\kappa^2}\text{tr}(G_{\mu\nu}G^{\mu\nu}). \tag{12.5.36}$$

**Homework** 12.5.37. Show that the generator $T^a$ determines a Noether current from $\mathcal{L}_{\text{YM}}$ which is given by

$$j^\mu \overset{\text{def}}{=} \frac{\partial \mathcal{L}}{\partial(\partial_\mu A^a_\nu)}\mathbf{D}A^a_\nu = gf^{abc}G^{b\mu\nu}A^c_\nu.$$

Verify that this current is conserved.

Through orthogonal transformation of $\phi$ and unitary transformation of $\psi$, we can split the representations $M$ and $N$ into direct sums of irreducible subrepresentations. Such a splitting also presents the elements in the image of $M$ and $N$ in block-diagonal form:

$$M = \bigoplus_{\alpha=1}^{\kappa} M_\alpha \quad\Longrightarrow\quad M(X) = \begin{pmatrix} M_1(X) & \cdots & 0 \\ \vdots & \ddots & \vdots \\ 0 & \cdots & M_\kappa(X) \end{pmatrix}, \tag{12.5.38}$$

and similarly for $N$ and for evaluation on a matrix $g$ in the symmetry group $G$. Each irreducible subrepresentation corresponds to a *multiplet* of matter fields; one labeled by $\alpha$ corresponds to a scalar multiplet $\phi^\alpha$ within the $\phi$ vector, and one labeled by $\beta$ corresponds to a Fermi multiplet $\psi^\beta$ within the $\psi$ vector.

The guiding light of this section is the gauge principle, which asserts that physics should depend on geometric structure rather than coordinate-dependent structure. On the basis of this principle, the generalization from $U(1)$ gauge symmetry to simple Lie group gauge symmetry is straightforward as long as we hold on to geometric principles. The essential concept from geometry is the covariant derivative, from which we derive the connection one-form and the curvature tensor. The need for canonical normalization of the gauge field kinetic term introduces an unexpected technicality, namely that the symmetry group should possess an inner product which is invariant under the adjoint representation. Such groups are products of simple groups and $U(1)$'s.

For a simple group, choice of a properly normalized basis for the Lie algebra and choice of a coupling constant determine in turn the gauge fields, the coupling to the matter fields, and the Yang–Mills Lagrangian density for the gauge fields. It is a special feature of the simple groups that the minimal coupling of the gauge fields to themselves and to the matter fields depends on only this one coupling constant. (Contrast the case of QED in which every one-dimensional representation has its own coupling constant.)

For semi-simple groups like the $SU(3) \times SU(2) \times U(1)$ of the Standard Model, one treats each factor independently, adding the Yang–Mills Lagrangian densities to the original matter Lagrangian density, and adding the connection one-forms to the partial derivatives in the original Lagrangian density.

## 12.6  Classical Minimally-Coupled Gauge Theories

Following the geometrical introduction to Yang–Mills theories of the last section, this section summarizes the structure and dynamics of a Yang–Mills theory. There are two notations, depending on whether one wants to introduce a basis for the Lie algebra of symmetry generators or not. We present the basis-dependent viewpoint first.

We start by choosing an orthonormal basis $T^a$ of symmetry generators and defining structure constants $f^{abc}$ to satisfy the two conditions

$$\text{tr}(T^a T^b) = \frac{\delta^{ab}}{2} \quad \text{and} \quad [T^a, T^b] = i f^{abc} T^c. \tag{12.6.1}$$

From the Jacobi identity, we find that the adjoint representation of $T^a$ is the matrix $F^a$ defined as follows:

$$(F^a)^{bc} \overset{\text{def}}{=} -i f^{abc} \quad \Longrightarrow \quad [F^a, F^b] = i f^{abc} F^c. \tag{12.6.2}$$

The symmetry generators $T^a$ act on the vector, spinor, and scalar fields through the representations $F^a$, $M^a$, and $N^a$. Writing $A_\mu$ for the column vector of the $A_\mu^a$'s to distinguish this vector from $A_\mu = -ig A_\mu^a T^a$, we have

$$T^a \cdot A_\mu = F^a A_\mu \quad \text{or} \quad (T^a \cdot A_\mu)^b = -i f^{abc} A_\mu^c,$$

$$T^a \cdot \psi = M^a \psi, \quad \text{and} \quad T^a \cdot \phi = N^a \phi. \tag{12.6.3}$$

The covariant derivative is

$$D_\mu = \partial_\mu - igA_\mu^a T^a. \tag{12.6.4}$$

The generator $T^a$ acts on the vector, spinor, and scalar fields as above to give:

$$\left. \begin{array}{l} D_\mu A_\nu^a = \partial_\mu A_\nu^a + gf^{abc}A_\mu^b A_\nu^c \\ D_\mu \psi = (\partial_\mu - igA_\mu^a M^a)\psi \\ D_\mu \phi = (\partial_\mu - igA_\mu^a N^a)\phi \end{array} \right\} \tag{12.6.5}$$

(Mathematically, the $a$ index in the first line is somewhat improper; one cannot differentiate the single component $A_\mu^a$, only the vector $A_{\cdot\mu}^\cdot$.) Note that the gauge fields $A_\mu^a$ are real fields.

The field strength is determined by

$$G_{\mu\nu}^a = \partial_\mu A_\nu^a - \partial_\nu A_\mu^a + gf^{abc}A_\mu^b A_\nu^c, \tag{12.6.6}$$

and the Yang–Mills Lagrangian density minimally coupled to matter fields has the form

$$\mathcal{L} = -\frac{1}{4}G_{\mu\nu}^a G^{a\mu\nu} + \bar\psi i \slashed{D}\psi + (D_\mu\phi)^\dagger(D^\mu\phi) + \mathcal{L}'(\bar\psi, \psi, \phi^\dagger, \phi), \tag{12.6.7}$$

where the possibilities for $\mathcal{L}'$ are determined by the requirement of global symmetry. From the density $\mathcal{L}$, we find the following canonical momenta:

$$\Pi^{a\mu\nu} = \frac{\partial\mathcal{L}}{\partial(\partial_\mu A_\nu^a)} = -G^{a\mu\nu};$$

$$\Pi_\psi^\mu = \bar\psi i\gamma^\mu, \qquad\qquad \Pi_{\bar\psi}^\mu = 0; \tag{12.6.8}$$

$$\Pi_\phi^\mu = (D^\mu\phi)^\dagger, \qquad\qquad \Pi_{\phi^\dagger}^\mu = D^\mu\phi.$$

From the canonical momenta and the action of the symmetry generators, we see that the matter current $J^{a\mu}$ associated with the generator $T^a$ is

$$J^{a\mu} \stackrel{\text{def}}{=} g\bar\psi M^a \gamma^\mu \psi \tag{12.6.9}$$
$$+ ig(\phi^\dagger N^a \partial^\mu \phi - \partial^\mu \phi^\dagger N^a \phi) + g^2 A^{b\mu}\phi^\dagger(N^b N^a + N^a N^b)\phi,$$

and the Noether conserved current is

$$j^{a\mu} = gf^{abc}G^{b\mu\nu}A_\nu^c + J^{a\mu}. \tag{12.6.10}$$

For a minimally-coupled Lagrangian density, the Noether current is the current coupled to the gauge field:

$$j^{a\mu} = \frac{\partial\mathcal{L}}{\partial A_\mu^a}. \tag{12.6.11}$$

This relationship would break down if $\mathcal{L}'$ involved the gauge fields.

**Homework** 12.6.12. Verify the formula for the Noether current. (The complex-scalar, Dirac-fermion, and gauge-field currents are given by 3.8.16, 8.4.15, and Homework 12.5.37 respectively.)

The equations of motion for the fields are:

$$\partial_\mu G^{a\mu\nu} + j^{a\nu} = 0, \quad i\not{D}\psi + \frac{\partial\mathcal{L}'}{\partial\bar\psi} = 0, \quad \partial_\mu D^\mu\phi - \frac{\partial\mathcal{L}'}{\partial\phi^\dagger} = 0. \tag{12.6.13}$$

From these equations, one can deduce conservation of the Noether current and covariant conservation for the matter current:

$$D_\mu J^\mu = 0 \quad \text{or} \quad \partial_\mu J^{a\mu} + gf^{abc}A^b_\mu J^{c\mu} = 0. \tag{12.6.14}$$

An infinitesimal gauge transformation is given by the action of a Lie-algebra element $\Theta$ on the fields:

$$\Theta(x) = -ig\theta^a(x)T^a \quad \Longrightarrow \quad \left.\begin{array}{c} (\Theta \cdot A'_\mu)^a = gf^{abc}\theta^b A^c_\mu - \partial_\mu\theta^a \\[4pt] \Theta \cdot \psi = -ig\theta^a M^a\psi \\[4pt] \Theta \cdot \phi = -ig\theta^a N^a\phi \end{array}\right\} \tag{12.6.15}$$

This action of the Lie algebra is commonly expressed in terms of small variations of the parameters, for example

$$\delta A^a_\mu \overset{\text{def}}{=} (\delta\Theta \cdot A'_\mu)^a = gf^{abc}\delta\theta^b A^c_\mu - \delta^{ab}\partial_\mu\delta\theta^b. \tag{12.6.16}$$

This concludes the presentation of the structure of a minimally-coupled gauge theory in coordinate notation.

To express the foregoing points in matrix notation, we contract all the group indices with the basis $T^a$, thereby avoiding dependence on the choice of basis. The following definitions are convenient:

$$A_\mu = -igA^a_\mu T^a, \quad G_{\mu\nu} = -igG^a_{\mu\nu}T^a, \quad \Pi_{\mu\nu} = -ig\Pi^a_{\mu\nu}T^a;$$
$$j^\mu = -igj^{a\mu}T^a, \quad \text{and} \quad J^\mu = -igJ^{a\mu}T^a. \tag{12.6.17}$$

These definitions support a variety of compact expressions:

$$D_\mu = \partial_\mu + A_\mu,$$
$$G_{\mu\nu} = [D_\mu, D_\nu] = \partial_\mu A_\nu - \partial_\nu A_\mu + [A_\mu, A_\nu],$$
$$j^\mu = [G^{\mu\nu}, A_\nu] + J^\mu,$$
$$\partial_\mu G^{\mu\nu} + j^\nu = 0, \tag{12.6.18}$$
$$D_\mu J^\mu = \partial_\mu J^\mu + [A_\mu, J^\mu] = 0,$$
$$D_\mu G^{\mu\nu} = \partial_\mu G^{\mu\nu} + [A_\mu, G^{\mu\nu}] = -J^\nu.$$

The structure of a gauge transformation by a gauge group element $\tilde{g}$ may be summarized as follows:

$$A' = \tilde{g}A\tilde{g}^{-1} - (\partial\tilde{g})\tilde{g}^{-1}, \quad \psi' = \tilde{g}\cdot\psi, \quad \phi' = \tilde{g}\cdot\phi,$$
$$\tilde{g} = e^\Theta \quad \Longrightarrow \quad \Theta \cdot A = [\Theta, A] - \partial\Theta. \tag{12.6.19}$$

**Homework 12.6.20.** Prove the Bianchi identity, $\epsilon^{\mu\nu\rho\sigma}D_\nu G_{\rho\sigma} = 0$.

This section is almost an appendix; it has simply listed the commonly used definitions and equations for a generic gauge theory. It is, however, worthwhile to go through the items above, demonstrating each to one's satisfaction. Sometimes it is best to start from the coordinate formulation, sometimes from the matrix formulation.

**Homework 12.6.21.** Prove every assertion in this section.

## 12.7  Quantization of Non-Abelian Gauge Theories

The significant novelty of the non-abelian gauge theories is the non-trivial self-coupling arising from the quadratic terms in 12.5.35. On account of these terms, there is no free-field theory. Hence perturbative canonical quantization simply does not apply. The Faddeev–Popov principle, however, gives us a covariant quantum theory easily enough. To verify the validity of the Lagrangian form, as before for QED, we use the first-order formalism to connect the Faddeev–Popov formula in axial-gauge to the corresponding Routhian form.

From the Lagrangian form of the action,

$$S_{\mathrm{L}}(A, J) = \int d^4x \; -\frac{1}{4}G^a_{\mu\nu}G^{a\mu\nu} + J^a_\mu A^{a\mu}, \tag{12.7.1}$$

and a choice of gauge fixing condition $C(A)$, the Faddeev–Popov principle yields the following Lagrangian form of the functional integral:

$$Z_{\mathrm{L}} = N \int (dA)\, e^{iS_{\mathrm{L}}(A,J)} \delta\big(C(A)\big) \Delta_C(A). \tag{12.7.2}$$

We can check this equation against a Routhian form in the case of axial gauge, where $C(A) = -igA^a_3 T^a$. With this value of $C$, the infinitesimal gauge transformation at a solution of $C(A) = 0$ is

$$(\Theta \cdot A)_3 = gf^{abc}\theta^b A^c_3 - \partial_3\theta^a = -\partial_3\theta^a, \tag{12.7.3}$$

and so the Faddeev–Popov determinant $\Delta_C$ is the usual undefined constant,

$$\Delta_C(A) = \left\|\frac{\delta C(\Theta \cdot A)}{\delta\Theta}\right\|_{\Theta=0} = \|\partial_3\|. \tag{12.7.4}$$

Absorbing this constant into the normalization $N$, the functional integral 12.7.2 becomes

$$Z^3_{\mathrm{L}} = N \int (dA)\, e^{iS_{\mathrm{L}}(A,J)} \delta(A_3). \tag{12.7.5}$$

Performing the integral over $(dA_3)$ effectively substitutes $A_3 = 0$ in the Lagrangian density, reducing it to the Lagrangian density for axial gauge.

The link between the Lagrangian form of the generating functional in axial gauge and the Routhian form in the same gauge rests on a first-order Lagrangian density defined by

$$\mathcal{L}_{\text{FO}}(G, A) \overset{\text{def}}{=} \frac{1}{4}\left(G^a_{\mu\nu} - (\partial_\mu A^a_\nu - \partial_\nu A^a_\mu + gf^{abc}A^b_\mu A^c_\nu)\right)^2 + \mathcal{L}. \tag{12.7.6}$$

Integrating $\mathcal{L}_{\text{FO}}$ over space defines the first-order action:

$$S_{\text{FO}}(G, A; J) = \int d^4x \, \mathcal{L}_{\text{FO}}(G, A) + J^a_\mu A^{a\mu}. \tag{12.7.7}$$

Then the first-order form of the generating functional $Z_{\text{FO}}$ is given by

$$Z_{\text{FO}}(J) = N \int (dG)(dA) \, e^{iS_{\text{FO}}(G,A;J)} \delta(A_3). \tag{12.7.8}$$

We see at once that integration over $(dG)$ yields the Faddeev–Popov formula in axial gauge, 12.7.5, and integration over $(dA_3)$ again reduces the Lagrangian density to the axial gauge Lagrangian density.

It remains to find a Routhian form. This is where axial gauge is such a convenient choice, for it permits the easy elimination of constrained fields. The Euler–Lagrange equations derived from the first-order Lagrangian in axial gauge,

$$\partial_\lambda \frac{\partial \mathcal{L}_{\text{FO}}}{\partial(\partial_\lambda G^a_{\mu\nu})} = \frac{\partial \mathcal{L}_{\text{FO}}}{\partial G^a_{\mu\nu}} \quad \text{and} \quad \partial_\lambda \frac{\partial \mathcal{L}_{\text{FO}}}{\partial(\partial_\lambda A^a_\mu)} = \frac{\partial \mathcal{L}_{\text{FO}}}{\partial A^a_\mu}, \tag{12.7.9}$$

are of two types; firstly, constraint equations that involve no time derivatives:

$$\left.\begin{array}{l} G^a_{rs} = \partial_r A^a_s - \partial_s A^a_r + gf^{abc}A^b_r A^c_s \\ G^a_{r3} = -\partial_3 A^a_r \\ \partial^3 G^a_{03} = -\partial^r G^a_{0r} + gf^{abc}G^b_{0r}A^{cr} \\ \partial_3 A^a_0 = -G^a_{03} \end{array}\right\} \quad \text{where } r, s = 1, 2. \tag{12.7.10}$$

and secondly evolution equations:

$$\left.\begin{array}{l} G^a_{0r} = \partial_0 A_r - \partial_r A_0 + gf^{abc}A^b_0 A^c_r \\ \partial^\mu G^a_{\mu r} = -gf^{abc}(G^b_{rs}A^{cs} + G^b_{0r}A^{c0}) \end{array}\right\} \quad \text{where } r, s = 1, 2. \tag{12.7.11}$$

Hence we may take the $A^a_r$ and $G^a_{0r}$ as dynamical variables and use the constraints to determine the $G^a_{rs}$, $G^a_{r3}$, $G^a_{03}$, and $A^a_0$. Thus the $A^a_r$ and $G^a_{0r}$ can be subjected to canonical quantization relations, and the $A^a_r$ can be used to make the Legendre transform to the axial-gauge Routhian density.

Now it is apparent that we can perform the functional integral over the constrained variables in the first-order form 12.7.8. Because the first-order Lagrangian density $\mathcal{L}_{\text{FO}}$ is quadratic in each field separately, integration will eliminate the constrained fields according to the constraint equations above, leaving the desired axial-gauge Routhian form for the generating functional.

Since in axial gauge, the Faddeev–Popov principle yields a quantum theory which is consistent with canonical quantization and with a Routhian form of the generating functional, we assume that the principle is valid in all gauges. For convenience in the Feynman rules, we choose to work in the covariant gauge, determined by $C(A) = \partial_\mu A^\mu$. The product of delta functionals $\delta(\partial_\mu A^{a\mu})$, as in the case of QED, can be converted into a gauge fixing term in the Lagrangian.

Now, however, the determinant $\Delta_C$ is not constant. Indeed, since 12.6.16 provides the expression

$$\delta A_\mu^a = g f^{abc} \delta\theta^b A_\mu^c - \partial_\mu \delta\theta^a, \tag{12.7.12}$$

the functional derivative evaluates to

$$\frac{\delta C}{\delta \tilde{g}} = \frac{\partial}{\partial(\delta\theta)} \partial^\mu (\delta A_\mu^a) = \partial^\mu \left( g f^{abc} A_\mu^c - \delta^{ab} \partial_\mu \right). \tag{12.7.13}$$

The determinant of the resulting operator is impossible to evaluate directly. It is necessary to express it as an exponent, so that the determinant factor effectively becomes part of the action, and the Feynman rules for the extended action include perturbative evaluation of this determinant.

The trick is to recall that functional integration over fermions yields the determinant of the covariance, not its inverse. Hence we may represent the determinant as follows:

$$\Delta_C = \left( \frac{\partial}{\partial \eta} \frac{\partial}{\partial \bar{\eta}} \right) \exp\left( i \int d^4x \, \bar{\eta}^a \partial^\mu (\delta^{ac} \partial_\mu + g f^{abc} A_\mu^b) \eta^c \right), \tag{12.7.14}$$

where $\eta$ (due to the absence of $\gamma$ matrices and the presence of two derivatives) is a *fermionic* complex scalar. The spin-statistics structure of the field $\eta$ prohibits direct physical interpretation for it, so $\eta$ is known as a *ghost field*. The absence of external ghost lines implies that these lines only occur in loops. Consequently the number of propagators is equal to the number of vertices and the overall sign of the ghost term (which has a factor of $+1$ to conform with the fermionic functional quantization principle 11.11.24) does not affect measurable matrix elements.

Ghost fields are present because functional integral quantization is defined by the Routhian form of the generating functional, and, as noted in the context of quantum mechanics in Section 11.5, the Lagrangian form will in general contain an operator determinant.

Adding the ghost term and the gauge-fixing term to the original Lagrangian density yields the Faddeev–Popov Lagrangian density:

$$\mathcal{L}_{\mathrm{FP}}(A, \eta, \bar{\eta}) \stackrel{\text{def}}{=} -\frac{1}{4} G_{\mu\nu}^a G^{a\mu\nu} \tag{12.7.15}$$

$$+ \bar{\eta}^a \partial^\mu (\delta^{ac} \partial_\mu + g f^{abc} A_\mu^b) \eta^c - \frac{1}{2\alpha} (\partial_\mu A^{a\mu})^2,$$

which in turn gives rise to the Faddeev–Popov action:

$$S_{\mathrm{FP}}(A, \eta, \bar{\eta}; J) \stackrel{\text{def}}{=} \int d^4x \, \mathcal{L}_{\mathrm{FP}} + J_\mu^a A^{a\mu} \tag{12.7.16}$$

and finally to the functional integral formula for the generating functional:

$$Z(J) = N\left(\frac{\partial}{\partial\eta}\frac{\partial}{\partial\bar\eta}\right)\int (dA)\, e^{iS_{\rm FP}(A,\eta,\bar\eta;J)}. \tag{12.7.17}$$

Since this final expression is in Lagrangian form, we can read off the Feynman rules by the usual processes of computing $-i\hat A^{-1}$ for covariance operators $A$ and differentiating vertices. Following the general principles of Sections 11.8 and 11.12, the first step is to pull contraction operators out of the functional-integral formula for the generating functional:

**Boson:** $\quad\dfrac{1}{2}\displaystyle\int\frac{d^4k}{(2\pi)^4}\,d^4x\,d^4y\left(e^{-ix\cdot k}\frac{\delta}{\delta\phi(x)}\right)\frac{i}{k^2-\mu^2+i\epsilon}\left(e^{iy\cdot k}\frac{\delta}{\delta\phi(y)}\right);$

$$\tag{12.7.18}$$

**Fermion:** $\quad\displaystyle\int\frac{d^4k}{(2\pi)^4}\,d^4x\,d^4y\left(e^{-ix\cdot k}\frac{\delta}{\delta\psi(x)}\right)\frac{i}{\not k-m+i\epsilon}\left(e^{iy\cdot k}\frac{\delta}{\delta\bar\psi(y)}\right).$

The second step is to choose a convention for the action of a product of these operators on a product of interactions. Following Sections 11.8 and 11.12, we define operators

$$\mathbf{C}_\phi^- \stackrel{\rm def}{=} \int d^4x\, e^{-ix\cdot k}\frac{\delta}{\delta\phi(x)} \quad\text{and}\quad \mathbf{C}_\phi^+ \stackrel{\rm def}{=} \int d^4y\, e^{iy\cdot k}\frac{\delta}{\delta\phi(y)}; \tag{12.7.19}$$

$$\mathbf{C}_\psi^- \stackrel{\rm def}{=} \int d^4x\, e^{-ix\cdot k}\frac{\delta}{\delta\psi(x)} \quad\text{and}\quad \mathbf{C}_{\bar\psi}^+ \stackrel{\rm def}{=} \int d^4y\, e^{iy\cdot k}\frac{\delta}{\delta\bar\psi(y)}. \tag{12.7.20}$$

and make the fermion operators act in the order $\mathbf{C}_\psi^+\mathbf{C}_\psi^-$ on an interaction.

For example, after integrating the ghost-ghost-vector interaction by parts, application of these principles yields:

$$\mathbf{C}_{\bar\eta^a}^+(k)\mathbf{C}_{A_\mu^b}^+(p)\mathbf{C}_{\eta^c}^-(l)\int d^4x\, gf^{a'b'c'}(-i\partial^\mu\bar\eta^{a'})A_\mu^{b'}\eta^{c'} \tag{12.7.21}$$

$$= \mathbf{C}_{\bar\eta^a}^+(k)\mathbf{C}_{A_\mu^b}^+(p)\int d^4x\, gf^{a'b'c}(-i\partial^\mu\bar\eta^{a'})A_\mu^{b'}e^{-ix\cdot l}$$

$$= \mathbf{C}_{\bar\eta^a}^+(k)\int d^4x\, gf^{a'bc}(-i\partial^\mu\bar\eta^{a'})e^{ix\cdot(p-l)}$$

$$= \mathbf{C}_{\bar\eta^a}^+(k)\int d^4x\, gf^{abc}\bar\eta^{a'}(l^\mu-p^\mu)e^{ix\cdot(p-l)}$$

$$= gf^{abc}k^\mu(2\pi)^4\delta^{(4)}(k+p-l).$$

The ghost here goes from $c$ to $a$, and $k^\mu$ is the outgoing momentum on the $a$ line.

The same procedure applies equally to the three-boson and four-boson vertices. As long as we use distinct symbols for the dummy suffices in the vertex and the propagator indices which go on both the lines at the vertex and the $\mathbf{C}$ operators, the calculation is routine.

The propagators and vertex rules in our pure Yang–Mills theory turn out to be:

**Ghost Propagator:**
$$-\frac{i\delta_{ab}}{k^2 + i\epsilon}$$

**Vector Boson Propagator :**
$$\left(-g_{\mu\nu} + (1-\alpha)\frac{k_\mu k_\nu}{k^2}\right)\frac{i\delta_{ab}}{k^2 + i\epsilon}$$

**Boson-Ghost Vertex:**
$$g f^{abc} k_\mu^{(a)} \tag{12.7.22}$$

**Four-Boson Vertex:**
$$ig^2(A_{\mu\nu\rho\sigma}^{abcd} + A_{\mu\rho\sigma\nu}^{acdb} + A_{\mu\sigma\nu\rho}^{adbc}),$$

$$\text{where} \quad A_{\mu\nu\rho\sigma}^{abcd} = f^{abe}f^{cde}(g_{\mu\rho}g_{\nu\sigma} - g_{\mu\sigma}g_{\nu\rho})$$

**Three-Boson Vertex:** $-igf^{abc}\left((k_1 - k_2)_\lambda g_{\mu\nu} + (k_2 - k_3)_\mu g_{\nu\lambda} + (k_3 - k_1)_\nu g_{\lambda\mu}\right)$

(We have omitted the integrals over propagators and the $(2\pi)^4\delta^{(4)}(P_{\text{in}})$ factors in the vertex rules.)

**Homework** 12.7.23. Verify the three-boson vertex rule.

Note the asymmetric character of the boson-ghost vertex rule; $k^{(a)}$ is the outgoing ghost momentum. Since in any physically meaningful Feynman diagram, the ghosts form simple loops, this asymmetry is not apparent in the Feynman integrals derived from such diagrams. This indicates that the overall sign of the ghost term in the Lagrangian density $\mathcal{L}_{\text{FP}}$ of 12.7.15 does not influence predictions. The sign in 12.7.15 is determined by careful normalization of the fermionic functional quantization operator in Section 11.11, but this sign would only be apparent in a computation involving ghost particles.

Note also that one may substitute bosons for ghosts in any ghost loop of any Feynman diagram to obtain a related Feynman diagram. Thus the ghost field acts as a regulator for loops of three-boson vertices, providing a further indication that it is the derivative couplings that make the naive Lagrangian form of the generating functional incorrect.

Finally, note that, if we take the optical theorem of Section 5.6 as the criterion for perturbative unitarity of a quantum theory, then we find that the ghost contributions are necessary for unitarity of the $S$ matrix.

If we begin from a theory of fermions and scalars with a global symmetry and then gauge this symmetry by introducing the appropriate gauge bosons and changing derivatives to covariant derivatives, then in the resulting minimally coupled theory there will be no new types of interaction. The quantization prescriptions above will cover all cases: the Feynman propagators will be the ones we have presented, and the vertex rules may be deduced directly from the interactions in the Lagrangian density.

**Remark** 12.7.24. Actually, in Coulomb and axial gauges it is not obvious how to remove the singularity of the covariance. The naive choices of propagator for the vector field and ghost-ghost-gluon vertex factor are constant as functions of one momentum coordinate, and therefore give rise to singular Feynman integrals. Hence, in order to interpret the Feynman integrals, it is necessary first to suppress these Feynman factors at high energy with some regularization procedure, second to sum the integrals which contribute to a Green function up to some order in the perturbation expansion, and third to remove the regularization.

There are regularization procedures for Coulomb and axial gauges which yield results consistent with covariant gauge, but these procedures are not determined by functional integral methods.

In 1962, Schwinger showed how to construct a quantum Hamiltonian for non-abelian gauge theories in Coulomb gauge. He found that operator ordering problems give rise to terms in the quantum Hamiltonian which are not present in the classical Hamiltonian. The Feynman rules derived from Schwinger's Hamiltonian agree up to fourth order with the regularized functional integral theory.

We conclude that functional integral quantization is not quite self-sufficient, but at times needs to be guided by some form of canonical quantization.

In conclusion, then, the Faddeev–Popov principle enables us to quantize gauge theories without reference to a Hamiltonian. This return of power to the Lagrangian brings with it the modest price of ghost fields, so that it is not the original classical Lagrangian that governs the quantum theory but the Faddeev–Popov Lagrangian. For perturbation theory, the pleasure of working with covariant Feynman rules far outweighs the horror of ghosts.

# 12.8 Global and Local Spontaneous Symmetry Breaking

As we shall see in Chapter 20, the quantum gauge theory of the previous section is renormalizable and (as we shall see in Chapter 15) provides models for the electromagnetic and strong interactions. The weak interactions, however, have a short range as indicated by the smallness of the four-fermi coupling constant $G_F$. They could be modeled by massive vector bosons. Since the weak interactions involve transfer of isotopic spin, they must carry this quantum number and therefore must be self-interacting. On account of the high-energy structure of the propagators for massive vector fields, a theory of self-interacting massive vector bosons is generally not renormalizable. Spontaneous symmetry breaking (SSB) provides a mechanism for transferring the renormalizability of gauge theories to theories with interacting massive vector bosons.

In this section, we present examples of the simplest mechanism for spontaneously breaking either global or gauge symmetry. In breaking a global symmetry, we shall find a massless scalar emerges, the *Goldstone boson*. In breaking a gauge symmetry, the *Higgs mechanism* takes effect: the degree of freedom associated with this massless particle becomes the third degree of freedom for a gauge field which has become massive.

The principle of SSB is that the fields in any theory will necessarily display quantum fluctuations about a vacuum state represented by a minimal-energy configuration of the classical limit. Hence one must choose such a minimal-energy configuration and shift the fields to the new origin determined by this configuration before quantizing. This process of shifting the fields will generally be found to break symmetries.

In classical theories, shifting the fields generally does not lead to symmetry breakdown because the spaces of fields commonly used are translation invariant. However, if we are describing small vibrations, then this translation invariance breaks down, the sets of small vibrations associated with distinct minimal-energy configurations are different function spaces, and there may be symmetries of the

Lagrangian which, by transforming the minimal-energy configurations into one another, likewise transform the associated function spaces.

In a theory with Fermi fields, the Hamiltonian does not evaluate to a real number and minimization makes no sense. Since we want to preserve Lorentz invariance, we assume that all quantum Fermi fields have zero vacuum expectation values (VEV's) and therefore set their classical counterparts to zero before minimizing.

Having determined the minimal-energy configuration of classical fields, one returns to quantum theory by first identifying the classical values as VEV's, and then expressing the original fields as perturbations from these VEV's.

It is helpful to bear a simple example in mind, namely the case of a potential $V(r) = (r^2 - v^2)^2$, where $r$ is the radial variable in $\mathbf{R}^2$. This potential has $SO(2)$ symmetry, a maximum at the origin, and minima on the circle $r = v$. The total set of minima is $SO(2)$-symmetric, but when one specific minimum is selected, then the symmetry is broken. One may, however, use the action of $SO(2)$ to change coordinates, so that the minimum has cartesian coordinates $(v, 0)$. It is clear also that small radial oscillations are harmonic, while angular perturbations are not controlled by a restoring force.

If we now consider a theory of an $\mathbf{R}^2$-valued field $\phi = (\phi_1, \phi_2)$ with the same potential $V$, we can intuit at once that a minimal-energy configuration of the field is constant and after a change of coordinates may be written $\bar{\phi} = (v, 0)$. Furthermore, the radial oscillations represented by the shifted field $\phi_1' = \phi_1 - v$ will be massive, and the angular oscillations represented by $\phi_2' = \phi_2$ will be massless. This massless field $\phi_2'$, as it is associated with the direction in which the broken symmetry generator would move $\bar{\phi}$, is a Goldstone boson.

Finally, if the theory is a gauge theory, then we see that we can use the angular displacements $\phi_2'$ to define a gauge, that is, the equation $\phi_2' = 0$ can be interpreted as a gauge condition. This absorption of the Goldstone boson by a gauge symmetry is part of the Higgs mechanism (which we present in detail below).

For an example of the spontaneous breaking of a global symmetry, we start from a Lagrangian density $\mathcal{L}_M$ for a charged scalar $\phi$,

$$\mathcal{L}_M \overset{\text{def}}{=} \partial_\mu \phi^\dagger \partial^\mu \phi - \frac{\kappa^2}{4}(\phi^\dagger \phi - v^2)^2, \tag{12.8.1}$$

make a change of variables to real fields $r$ and $\theta$ through

$$\phi(x) = e^{-i\theta(x)}\big(v + r(x)\big), \tag{12.8.2}$$

and write the Lagrangian density $\mathcal{L}_M$ as a function $\mathcal{L}_M'$ of these real fields:

$$\partial_\mu \phi = e^{-i\theta}\partial_\mu r - i e^{-i\theta}(v + r)\partial_\mu \theta$$

$$\implies \mathcal{L}_M' = \partial_\mu r\, \partial^\mu r + (v + r)^2 \partial_\mu \theta\, \partial^\mu \theta - \frac{\kappa^2}{4}(2vr + r^2)^2. \tag{12.8.3}$$

Hence $\theta$ is massless (the Goldstone boson) and $r$ has mass $\kappa v$. (To bring out canonical kinetic terms, we must rescale $\theta$ and $r$, but this rescaling will not change the masses of the fields.)

The set of minimal-energy configurations for this example of global symmetry breaking is given by $\phi^\dagger \phi = v^2$ and $\partial_\mu \phi = 0$. Thus a solution $\bar\phi$ is a constant of the form $ve^{-i\alpha}$. In light of the change of variables 12.8.2, after quantization $\alpha$ will be the VEV of the $\theta$ field.

The original symmetry group $U(1)$ is a group of symmetries of the set of $\bar\phi$'s. Symmetry is broken by the fact that the universe has favored a particular $\bar\phi$ in this set. After symmetry breaking, we can use $U(1)$ to change bases for the fields in $\mathcal{L}'_M$. Because of the original symmetry, the net effect will be an apparent rotation of $\phi$:

$$
\begin{aligned}
\mathcal{L}'_M(\bar\phi; \phi', \partial\phi') &\stackrel{\text{def}}{=} \mathcal{L}_M(\bar\phi + \phi', \partial(\bar\phi + \phi')) \\
&= \mathcal{L}_M(e^{i\alpha}\bar\phi + e^{i\alpha}\phi', e^{i\alpha}\partial(\bar\phi + \phi')) \\
&= \mathcal{L}'_M(e^{i\alpha}\bar\phi; e^{i\alpha}\phi', e^{i\alpha}\partial\phi') \\
&= \mathcal{L}'_M(e^{i\alpha}\bar\phi; \phi'', \partial\phi'').
\end{aligned}
\tag{12.8.4}
$$

Hence there is a $U(1)$ change of basis which makes $\langle\theta\rangle = 0$, and the resulting value $v$ of $\bar\phi$ is unique.

If we promote the global symmetry of $\mathcal{L}_M$ to a local symmetry by minimally coupling $\phi$ to a gauge field $A$, then we obtain a Lagrangian density with local or gauge symmetry:

$$
\mathcal{L} \stackrel{\text{def}}{=} -\frac{1}{4}F_{\mu\nu}F^{\mu\nu} + (D_\mu\phi)^\dagger(D^\mu\phi) - \frac{\kappa^2}{4}(\phi^\dagger\phi - v^2)^2,
\tag{12.8.5}
$$

where the covariant derivative is given by

$$
D_\mu = \partial_\mu - ieA_\mu.
\tag{12.8.6}
$$

With the shift of fields to the minimum of the classical potential and the change of variables from $\phi$ to $\theta$ and $r$, the original Lagrangian density $\mathcal{L}$ becomes a function $\mathcal{L}'$ of these new variables, and we find in particular that

$$
\begin{aligned}
|D_\mu\phi|^2 &= |\partial_\mu\phi - ieA_\mu\phi|^2 \\
&= \left|\partial_\mu r - i(v + r)(eA_\mu + \partial_\mu\theta)\right|^2 \\
&= \partial_\mu r\,\partial^\mu r + (v + r)^2(eA_\mu + \partial_\mu\theta)(eA^\mu + \partial^\mu\theta).
\end{aligned}
\tag{12.8.7}
$$

In this expression, we see a mass term for $A$. Furthermore, this term contains the only occurrence of the $\theta$ field in $\mathcal{L}'$, and $\theta$ appears here as a gauge transformation of $A$. Because of the mass term, $\mathcal{L}'$ no longer has gauge symmetry, yet we can apply a gauge transformation to write $\mathcal{L}'$ in terms of gauge-transformed fields. Such gauge transformations affect $A$ and $\theta$ but not $v$ or $r$. The particular gauge transformation

$$
\phi \to e^{i\theta}\phi \quad \text{and} \quad A_\mu \to A_\mu + \frac{1}{e}\partial_\mu\theta
\tag{12.8.8}
$$

transfers the $\theta$ degree of freedom from $\phi$ to $A$ so that the Lagrangian density becomes:

$$
\mathcal{L}' = -\frac{1}{4}F_{\mu\nu}F^{\mu\nu} + \partial_\mu r\,\partial^\mu r + (v + r)^2 e^2 A_\mu A^\mu - \frac{\kappa^2}{4}(2vr + r^2)^2.
\tag{12.8.9}
$$

This transfer of the degree of freedom from the former Goldstone mode to the newly massive vector field is the Higgs mechanism.

The set of minimal-energy configurations for this example of local symmetry breaking is given by $\phi^\dagger\phi = v^2$, $F_{\mu\nu} = 0$ and $D_\mu\phi = 0$. The vanishing of $F_{\mu\nu}$ implies that $A$ is pure gauge. In the calculation above, the minimal-energy configuration is described by $\bar{\phi} = ve^{-i\theta}$ and $e\bar{A}_\mu = \partial_\mu\theta$. Interpreted as a vacuum state, this configuration appears to break Lorentz invariance.

Again, the set of minimal-energy configurations is invariant under the original group of gauge symmetries, and symmetry is broken because the universe has favored some particular configuration $\bar{\phi}$ and $\bar{A}_\mu$. After symmetry breaking, the original gauge group can be used to change bases with net effect being an apparent transformation of $(\bar{\phi}, \bar{A})$:

$$
\begin{aligned}
\mathcal{L}'(\bar{\phi}, \bar{A}; \phi', D\phi', A) &\stackrel{\text{def}}{=} \mathcal{L}\big(\bar{\phi} + \phi', D(\bar{\phi} + \phi'), \bar{A} + A\big) \\
&= \mathcal{L}\big(\tilde{g}\cdot\bar{\phi} + \tilde{g}\cdot\phi', \tilde{g}\cdot D(\bar{\phi} + \phi'), \tilde{g}\cdot\bar{A} + \tilde{g}\cdot A\big) \\
&= \mathcal{L}'\big(\tilde{g}\cdot\bar{\phi}, \tilde{g}\cdot\bar{A}; \tilde{g}\cdot\phi', \tilde{g}\cdot D\phi', \tilde{g}\cdot A\big) \\
&= \mathcal{L}'\big(\tilde{g}\cdot\bar{\phi}, \tilde{g}\cdot\bar{A}; \phi'', D''\phi'', A''\big).
\end{aligned}
\tag{12.8.10}
$$

Hence there is always a basis for the fields in which $\bar{\phi}$ is a positive real constant $v$ (which is unique) and $\bar{A}_\mu = 0$, making it clear that Lorentz invariance was never in danger.

The analysis above is purely classical. The connection to quantum field theory comes from interpreting a minimal-energy configuration as representing the quantum vacuum of the universe. In the context of canonical quantization, since we cannot eliminate a quantum field like $\theta$ by a gauge transformation, we must choose the basis of fields in the classical Lagrangian density before quantizing. In functional integral quantization, we can apply these classical transformations as long as we keep track of the function spaces and the measures on them. With these examples in hand, we can comfortably proceed to a more general presentation of Goldstone's Theorem and the Higgs mechanism.

## 12.9 Goldstone's Theorem and the Higgs Mechanism

The examples in the previous section show how scalar potentials can lead to spontaneous breaking of a $U(1)$ symmetry, generating in the global case a Goldstone boson and in the local case a massive vector field. In general, whatever the group and whatever the symmetry-breaking mechanism, in the case of a global symmetry, Goldstone's Theorem shows that for each broken $U(1)$ there is a massless boson, and in the case of a local symmetry, the Higgs mechanism transfers the associated degree of freedom to a gauge field which has become massive. We now present a brief proof of these assertions for the case of symmetry breaking by a scalar potential.

Let $\mathcal{L}$ be a Lagrangian density for a theory with $n$ real scalar fields $\phi = (\phi_1, \ldots, \phi_n)$ and possibly fermi and vector fields also. Let $G$ be the symmetry group of $\mathcal{L}$. As we are interested in the action of $G$ on the scalar fields, we may as well assume that $G$ is a subgroup of $SO(n)$. Let $V(\phi)$ be the *scalar potential*, that part of $\mathcal{L}$ which involves $\phi$ but not $\partial_\mu\phi$, fermions, or vector fields.

In a configuration with minimal energy, we know from the previous section that the fields will be constants with the only non-zero values given by the $n$ equations $\partial_k V(\phi) = 0$ for the scalars. Let $\bar{\phi}$ be a fixed solution to these equations. When we shift fields to $\phi' = \phi - \bar{\phi}$, then the symmetry group $\bar{G}$ of the $\phi'$ theory comprises all those elements of $G$ which leave $\bar{\phi}$ invariant:

$$\bar{G} \overset{\text{def}}{=} \{\, g \in G \mid g\bar{\phi} = \bar{\phi}\,\}. \tag{12.9.1}$$

Since we regard different values of $\bar{\phi}$ as representing different vacuum states, in reducing $G$ to $\bar{G}$, we are throwing away those transformations which connect quantum theories based on different vacua.

The goal is to understand the scalar masses after shifting. The Taylor expansion of $V$ in the shifted fields $\phi'$ contains no linear terms. The mass matrix $M^2$ for $\phi'$ comes from the second-order term:

$$(M^2)_{rs} = \frac{\partial^2 V}{\partial\phi^r\,\partial\phi^s}\Big|_{\phi=0}. \tag{12.9.2}$$

Now we shall use the original symmetry to find some null vectors for $M^2$. Choose hermitian generators $T^1,\ldots,T^k$ for $\bar{G}$, then extend to a family of generators $T^1,\ldots,T^l$ for $G$. Since $V$ is symmetric under $G$,

$$
\begin{aligned}
\frac{\partial}{\partial\theta}V(e^{-i\theta T^a}\phi)\Big|_{\theta=0} &= \frac{\partial V}{\partial\phi^s}\frac{\partial}{\partial\theta}\big(e^{-i\theta T^a}\phi\big)^s\Big|_{\theta=0}\\
&= -i\frac{\partial V}{\partial\phi^s}(T^a\phi)^s\\
&= 0,
\end{aligned}
\tag{12.9.3}
$$

and so

$$\frac{\partial}{\partial\phi^r}\Big(\frac{\partial V}{\partial\phi^s}(T^a\phi)^s\Big) = \frac{\partial^2 V}{\partial\phi^r\,\partial\phi^s}(T^a\phi)^s + \frac{\partial V}{\partial\phi^s}(T^a)^s{}_r = 0. \tag{12.9.4}$$

Evaluating at $\phi = \bar{\phi}$, the second term becomes zero. Hence, if $T^a\bar{\phi} \neq 0$, then $T^a\bar{\phi}$ is a null vector for the scalar mass matrix.

From the definition of $\bar{G}$, $T^a\bar{\phi} = 0$ for $1 \leq a \leq k$, and the set $\{\, T^b\bar{\phi} \mid k < b \leq l\,\}$ is linearly independent. Hence, if we make a unitary change of basis to diagonalize $M^2$, we will find at least $l - k$ massless bosons. Since the vectors $T^b\bar{\phi}$ are not necessarily orthogonal, it may not be possible to associate these vectors with elements of the new basis. Nevertheless, for $k < b \leq l$, we can associate a massless mode with the broken symmetry generator $T^b$:

$$\textbf{Goldstone Boson:}\quad \phi_G'^b \overset{\text{def}}{=} \bar{\phi}^\top X^b \phi', \tag{12.9.5}$$

where $X^b$ is the real generator $-iT^b$. This is the Goldstone theorem for SSB. Again, as shown in 12.8.4, there will be an element of $G$ which, acting as a change of basis transformation on the fields $\phi'$ in $\mathcal{L}'$, will make the VEV's of the Goldstone bosons zero. The resulting $\bar{\phi}$ will be unique.

**Remark** 12.9.6. In the example 12.8.1, we found a definite parameter $v$ to represent the VEV of the scalar fields. Generally, the action of $G$ on $\mathbf{R}^n$ defines orbits with dimension $l - k$, and so there is a moduli space of these orbits which has dimension $m = n - l + k$. In our $U(1)$ example then, the moduli space has dimension one and a particular orbit can be identified by a single parameter. More generally, it would be convenient to use $m$ algebraically independent $G$-invariant polynomials on $\mathbf{R}^n$. For example, if $G = SU(2)$ and $G$ acts on $\mathbf{R}^5$ through the representation $D_2$, then the moduli space has dimension two and can be parametrized by the two generators of the algebra of $SU(2)$-invariant polynomials:

$$\left. \begin{aligned} p_2 &= \phi_2^2 + \phi_1^2 + \phi_0^2 + \phi_{-1}^2 + \phi_{-2}^2 \\ p_3 &= 6\sqrt{2}\,\phi_2\phi_0\phi_{-2} - 3\sqrt{3}\,\phi_2\phi_{-1}^2 - 3\sqrt{3}\,\phi_1^2\phi_{-2} + 3\sqrt{2}\,\phi_1\phi_0\phi_{-1} - \sqrt{2}\,\phi_0^3 \end{aligned} \right\} \quad (12.9.7)$$

where we have labeled the fields by their $\sigma_3$ quantum numbers.

Goldstone's Theorem applies to global symmetries which are not gauge symmetries. Higgs observed that, when a symmetry generator is also a generator of a gauge symmetry, then the gauge transformations can be used after symmetry breaking to remove the Goldstone degree of freedom. The general form of the change of basis is given in 12.8.10. The upshot is that a change of basis by a gauge transformation $\tilde{g}$ reduces to a transformation of the minimal-energy configuration in $\mathcal{L}'$:

$$(\bar{\phi}, \bar{A}) \rightarrow (\tilde{g} \cdot \bar{\phi}, \tilde{g} \cdot \bar{A}). \quad (12.9.8)$$

Thus, where the set of minimal-energy configurations for global symmetry breaking has dimension $l - k$ and is invariant under $G$, the set for local symmetry breaking is infinite dimensional and invariant under the gauge group $\tilde{G}$ derived from $G$.

Now we will use the gauge transformations to eliminate the former Goldstone bosons. First, using $b$ as an index for the broken symmetry generators, we parameterize the scalar fields with reference to a constant minimal-energy configuration $\bar{\phi}$ by

$$\phi(x) = e^{-i\theta^b(x)T^b}\left(\bar{\phi} + \phi'(x)\right) \quad \text{where } \phi' \text{ satisfies } \quad \bar{\phi}^\top T^b \phi' = 0. \quad (12.9.9)$$

Second, we define a gauge transformation $\tilde{g}$ to remove the angular degrees of freedom from $\phi$,

$$\tilde{g}(x) \overset{\text{def}}{=} e^{i\theta^b(x)T^b}. \quad (12.9.10)$$

This gauge which eliminates all the Goldstone modes is called *unitarity gauge*. Third, we compute the effect of $\tilde{g}$ on the gauge fields:

$$A'^c T^c = \tilde{g}A^a T^a \tilde{g}^{-1} - \frac{i}{g}(\partial \tilde{g})\tilde{g}^{-1} = A^a T^a + \frac{1}{g}\partial\theta^b T^b + O(\theta^2), \quad (12.9.11)$$

and deduce that the gauge transformation transfers the degree of freedom $\theta^b$ to the gauge field $A^b$.

As in the Lagrangian density 12.8.5, so here the shift in the fields generates a mass term for the gauge fields. From

$$D_\mu\phi = \partial_\mu\phi' - igA_\mu^a T^a(\bar{\phi} + \phi'), \quad (12.9.12)$$

using the real, anti-symmetric generators $X^a = -iT^a$, we find that

$$|D_\mu\phi|^2 \quad \text{contains} \quad g^2|A_\mu^a X^a \bar{\phi}|^2, \quad (12.9.13)$$

and conclude that the Lagrangian density $\mathcal{L}'$ contains the vector-boson mass term

$$\textbf{Mass Term:} \qquad \frac{g^2}{2}(\bar{\phi}^\top X^a X^b \bar{\phi}) A^a_\mu A^{b\mu}. \qquad (12.9.14)$$

Clearly, this term gives masses to the $l - k$ fields $A^b$ where $k < b \leq l$, that is, the vector fields associated with the broken symmetry generators.

Since $\bar{\phi}^\top X Y \bar{\phi}$ is a symmetric bilinear form on $\mathcal{G}$ which vanishes if $X$ is an unbroken generator, there will be an orthonormal basis $Y^a$ for $\mathcal{G}$ which diagonalizes this bilinear form and satisfies $Y^a = X^a$ for the unbroken generators. (Such a change of basis could change the structure constants.) Clearly then, we may as well assume that our $X$'s already have the property

$$a \neq b \quad \Longrightarrow \quad \bar{\phi}^\top X^a X^b \bar{\phi} = 0. \qquad (12.9.15)$$

Then the mass matrix does not mix the gauge fields, and the mass $m_b$ of the field $A^b$ is given by

$$m_b^2 \overset{\text{def}}{=} - g^2 \bar{\phi}^\top X^b X^b \bar{\phi}. \qquad (12.9.16)$$

Since experiment has virtually eliminated the possibility of massless scalars or vector bosons (other than the photon and gluons), the Higgs mechanism is central to the formulation of realistic gauge field theories. The ideas are simple to state, even when the details become intricate. The geometry of the potential is the key concept.

## 12.10  Quantization after Spontaneous Symmetry Breaking

Quantization of a classical theory derived by spontaneous symmetry breaking from a Lagrangian density with gauge symmetry is a direct application of either canonical or functional integral quantization; the problem posed by a gauge symmetry is removed by the breaking of this symmetry, and so the Faddeev–Popov principle is neither needed nor applicable. The quantization follows in outline the quantization of QED with a massive photon. In particular, the propagator for the massive gauge fields has no convergence at infinity in momentum space. If the quantum theory is to be renormalizable, the massive gauge field must be coupled to a conserved current even after symmetry breaking.

On the classical level, a shift of the origin in function space will not affect the conservation of a current associated with a broken symmetry generator. However, when we break symmetry by choosing a function space of small vibrations around a minimal-energy configuration of the fields, then the conserved quantity derived from this current becomes an operator between distinct function spaces.

On the quantum level also, spontaneous symmetry breaking generally does not lead to a breakdown of conservation in the currents associated with the broken generators, but the integral defining the conserved quantity becomes divergent, indicating that it is in effect an operator which changes the state space. However, when gauge bosons are coupled to fermions there is a possibility of quantum dynamics leading to the breaking of current conservation for any current. We shall discuss this in detail in the next chapter.

The particular form of the propagator for the massive vector field after spontaneous symmetry breaking depends on the choice of basis for the fields. In the remarks above, we assumed unitary gauge, the gauge in which the massless scalars generated by the symmetry breaking are removed from the Lagrangian. Instead of using a gauge transformation to make a change of basis after symmetry breaking, one can add a gauge-fixing term to the gauge-invariant Lagrangian density and then break the symmetry by shifting scalar fields.

For example, we can define $R$ gauge by adding the gauge-fixing term

$$\mathcal{L}_{gf} \stackrel{\text{def}}{=} -\frac{1}{2\alpha}(\partial_\mu A^{a\mu})(\partial_\nu A^{a\nu}) \tag{12.10.1}$$

to the Lagrangian density $\mathcal{L}$. Then, writing $m_a$ for the mass of the gauge field $A^a$ after symmetry breaking as in 12.9.16, the covariance for this gauge field becomes

$$C_a \stackrel{\text{def}}{=} -g^{\mu\nu}(\square + m_a^2) + (1 - \alpha^{-1})\partial^\mu\partial^\nu, \tag{12.10.2}$$

and so the $R$-gauge propagator is

$$\textbf{Propagator:} \quad \left(-g^{\mu\nu} + (1-\alpha)\frac{k^\mu k^\nu}{k^2 - \alpha m_a^2 + i\epsilon}\right)\frac{i}{k^2 - m_a^2 + i\epsilon}. \tag{12.10.3}$$

Since $S$-matrix elements must be independent of $\alpha$, clearly, when an $S$-matrix element is computed, the pole in the propagator at $k^2 = \alpha m_a^2$ must be canceled by other terms.

In the limit $\alpha \to \infty$, as the momentum goes on-shell, this propagator has the projection operator onto transverse momenta as a factor. This indicates that the theory does not generate longitudinal outgoing vector particles or equivalently, that it describes a unitary theory of transversal vector particles. The asymptotic structure of this propagator is bad, however, suggesting that the theory is not renormalizable. In the limit $\alpha \to 1$, however, the $R$-gauge propagator behaves asymptotically like the scalar-field propagator, which is good for convergence of loop integrals, but has no projection operator to eliminate longitudinal vector bosons. Since the two limits describe the same theory, the existence of $R$ gauge suggests that a theory with massive vector fields generated by spontaneous symmetry breaking will be renormalizable and unitary. A slight modification of this approach to gauge fixing introduced by 't Hooft in fact makes a proof possible.

If instead of the parameterization 12.9.9 of $\phi$ in terms of angles, we simply translate the origin and split the radial fields $\phi'$ from the tangential fields $\phi''$ defined by the real broken generators $X^b$ so that

$$\phi(x) = \bar{\phi} + \phi'(x) + \phi''(x), \tag{12.10.4}$$

where

$$\bar{\phi}^\mathsf{T}T^b\phi' = 0 \quad \text{and} \quad \phi''(x) = f^b(x)X^b\bar{\phi}, \tag{12.10.5}$$

then the term $|D_\mu\phi|^2$ will contain a term which mixes $A$'s with $\partial_\mu f$'s:

$$\begin{aligned}|D_\mu\phi|^2 &= (\partial_\mu\phi + gA_\mu^a X^a\phi)^\mathsf{T}(\partial^\mu\phi + gA^{a\mu}X^a\phi)\\&= \cdots - 2g(\bar{\phi}^\mathsf{T}X^a X^b\bar{\phi})A_\mu^a\partial^\mu f^b + \cdots\end{aligned} \tag{12.10.6}$$

As in the computation 12.9.16 of the masses of the gauge fields after symmetry breaking, we can assume that the $X$'s are such that this term simply mixes $A^b$ with $\partial_\mu f^b$. This implies that $f^b$ is not an independent field.

To remove this mixing term from the Lagrangian density, 't Hooft proposed the gauge-fixing term

$$
\begin{aligned}
\mathcal{L}_{\mathrm{gf}} &\overset{\text{def}}{=} -\frac{1}{2\alpha g^2} \sum_a (g\partial_\mu A^{a\mu} + \alpha m_a^2 f^a)^2 \\
&= -\sum_a \left( \frac{1}{2\alpha}(\partial_\mu A^{a\mu})^2 + \frac{m_a^2}{g}(\partial_\mu A^{a\mu})f^a + \frac{\alpha m_a^4}{2g^2} f_a^2 \right).
\end{aligned}
\tag{12.10.7}
$$

The second term combines with the mixing term 12.10.6 to form a total derivative, which can be dropped without affecting the stationary points of the action.

After the addition of $\mathcal{L}_{\mathrm{gf}}$ to $\mathcal{L}'$, the covariance and propagator for the field $A^a$ are as before for the $R$ gauge. Since $\phi''$ has zero mass in $\mathcal{L}'$, the quadratic terms in $f^b$ are a kinetic term coming from $|D_\mu \phi|^2$ and a mass term coming from $\mathcal{L}_{\mathrm{gf}}$:

$$
\mathcal{L}' + \mathcal{L}_{\mathrm{gf}} = -\sum_b \left( \frac{m_b^2}{2g^2} \partial_\mu f^b \, \partial^\mu f^b + \frac{\alpha m_b^4}{2g^2} f_b^2 \right) + \cdots
\tag{12.10.8}
$$

Scaling $f^b$ to normalize the kinetic term,

$$
\phi_b'' \overset{\text{def}}{=} \frac{m_b}{g} f^b,
\tag{12.10.9}
$$

we find that $\phi_b''$ has a propagator

$$
\textbf{Propagator for } \phi_b'': \qquad \frac{i}{k^2 - \alpha m_b^2 + i\epsilon}.
\tag{12.10.10}
$$

Thus the pole in the $\phi_b''$ propagator matches the undesirable pole in the propagator of $A^b$, making it feasible to demonstrate unitarity and renormalizability of the quantum field theory defined by spontaneous symmetry breaking.

Interacting massive vector fields are needed to model the weak interactions. To obtain a renormalizable theory with interacting massive vector fields, we spontaneously break the symmetry in a gauge theory. The $R$ gauge indicate that symmetry breaking preserves both the renormalizability and the unitarity of the gauge theory, and the 't Hooft gauge provide the basis for a proof of this. Of course, underlying all these manipulations is the assumption that the currents associated with the former global symmetry are conserved. This assumption is discussed in the next chapter.

# 12.11 Faddeev–Popov Quantization Revisited

The first concern in setting up Faddeev–Popov quantization carefully is for a definition of the function spaces involved. First, the Lie algebra $\tilde{\mathcal{G}}$ of the gauge group $\tilde{G}$ cannot be the ordinary Hilbert space of square-integrable functions because such functions can become increasingly rough when multiplied together. The Lie algebra should consist of functions which are at least continuous, and so if a Hilbert space is desired, one should use a Hilbert space $\mathbf{H_3}$ of functions whose third derivatives are square integrable. Second, this choice implies that the pure-gauge configurations $A = -(\partial \tilde{g})\tilde{g}^{-1}$ has one less degree of differentiability and lie in a Hilbert space $\mathbf{H_2}$ of functions whose second derivatives are square integrable. Clearly, the degree of roughness of the pure-gauge configurations puts an upper limit on the differentiability of the $A$'s generally, and so we may as well assume that the $A$'s lie in a Hilbert space of the type $\mathbf{H_2}$.

The second concern is that the space of functions may not factorize into a product of gauge group and moduli space. Consider the example of a gauge group $\tilde{G}$ of phase functions $\exp\big(i\theta(x)\big)$ acting on a space $\mathbf{H}$ of complex-valued functions of a real variable. Then, if $\phi$ is an element of $\mathbf{H}$ which is zero on a set $Z$ of real numbers, the action of a gauge-group element $\tilde{g}$ on $\phi$ will not depend on the values of $\tilde{g}$ on $Z$. Consequently the space $\mathbf{H}$ cannot be factored into a product of the form $\tilde{G} \times \mathbf{H}'$ for any subset $\mathbf{H}'$ of $\mathbf{H}$.

Since a small perturbation of a $\phi$ with zeros will remove the zeros, the set of $\phi$'s which have zeros should as a whole have measure zero with respect to any reasonable measure on function space. It is therefore possible for factorization to fail only on a set with measure zero, in which case the obstruction to factorization may be solved by throwing away that set.

The third subtlety arises from gauge fixing. It appears that we could define a gauge Im $\phi = 0$, Re $\phi \geq 0$, and identify the moduli space $\overline{\mathbf{H}}$ with the subspace $\mathbf{H_0^+}$ of non-negative, real-valued functions in $\mathbf{H}$. However, a function $\phi$ can in effect wind the real line around the origin in the complex plane infinitely often. Consequently, the generator $\theta(x)$ required to make the value of such a $\phi$ real and positive will be unbounded. In this case, the lie algebra of the gauge groups has no metric and there is no measure on the gauge group.

From the points above, it is clearly a difficult technical task to choose function spaces $\mathbf{H}$ and $\tilde{\mathcal{G}}$ for the fields and the Lie algebra of the gauge group and a gauge-fixing condition $C(A) = 0$ for which the following desirable conditions hold:

1. $\mathbf{H}$ and $\tilde{\mathcal{G}}$ are Hilbert (or other standard) spaces;
2. The functions in $\tilde{\mathcal{G}}$ can be exponentiated to form a gauge group $\tilde{G}$.
3. $\tilde{G}$ acts on $\mathbf{H}$.
4. For each $A$, there exists a finite (or at worst countable) number $n_A$ of gauge-group elements $\tilde{g}$ such that $C(\tilde{g} \cdot A) = 0$.
5. The numbers $n_A$ are independent of $A$ except for a set with zero measure in function space.

A lot of work remains to be done to make the Faddeev–Popov principle acceptable mathematically.

Within physics, there are several gauges which should be linked by the Faddeev–Popov principle. For each gauge $C(A) = 0$ and each field configuration $A$, we must solve $C(\tilde{g} \cdot A) = 0$ for a gauge group element $\tilde{g}$, and we must show that $\tilde{g}$ is unique. The following table shows for four common gauges for the electromagnetic potential $A$, the name of the gauge, the gauge condition, the equation to solve, and the boundary conditions which make the solution unique:

| Gauge | Condition | Gauge Generator | Boundary Condition | |
|---|---|---|---|---|
| **Covariant:** | $\partial_\mu A^\mu = 0$ | $\Box \chi + \partial_\mu A^\mu = 0$ | $\chi(t_0, \bar{x}) = \partial_0 \chi(t_0, \bar{x}) = 0$ | (12.11.1) |
| **Coulomb:** | $\partial_r A^r = 0$ | $\Delta \chi + \partial_r A^r = 0$ | $\lim_{|\bar{x}| \to \infty} \chi(t, \bar{x}) = 0$ | |
| **Temporal:** | $A_0 = 0$ | $\partial_0 \chi + A_0 = 0$ | $\chi(t_0, \bar{x}) = 0$ | |
| **Axial:** | $A_3 = 0$ | $\partial_3 \chi + A_3 = 0$ | $\chi(x^0, x^1, x^2, x_0^3) = 0$ | |

In this table, $t_0$ and $x_0^3$ stand for fixed numbers, and any variables in $\chi$ not constrained by zero subscripts or limits in the right-hand column are subject to a 'for all' quantifier. We observe at once that solutions to the equation for the gauge generator will not generally be square-integrable over space or space-time.

Clearly, these boundary conditions cannot be combined. Hence, if we have a gauge group capable of transforming any $A$ into a solution of any one of these gauge conditions, then there will be an infinite-dimensional space of solutions $\chi$ to each gauge condition. In this case, the operators in the Faddeev–Popov determinants will have infinitely many zero modes and the determinants will evaluate to zero. Furthermore, if we start with a field $A$ which vanishes at infinity, then put $A$ into axial gauge, the new field $A'$ does not vanish at infinity. If we assume that space of $A$'s is Lorentz invariant, then we can rotate $A'$. The final field $A''$ has enough structure at infinity that we cannot put it into Coulomb gauge without violating the boundary condition given in the table above. Thus it becomes apparent that if we are to have all these gauges available at once, the space of $A$'s is so big that a functional integral over $A$ cannot possibly be approximated by a finite lattice calculation.

Since we cannot form one huge gauge group to simultaneously manage transformations into all gauges, it appears necessary to work with a variety of gauge groups, one for each gauge. In this case, even if we can find a good space of fields, the moduli space will be gauge dependent and the validity of the Faddeev–Popov principle will be in doubt.

In physics these mathematical concerns are generally suppressed by the drastic transformations that go into computing operator determinants, namely, making Minkowski space Euclidean by a Wick rotation and making Euclidean space compact by adding a point at infinity. These transformations make the original function spaces irrelevant, effectively replacing them by spaces of functions on the three-dimensional sphere. There is probably a real weakness in the non-perturbative formalism here, but as far as perturbation theory goes, the conclusions drawn from Faddeev–Popov quantization are consistent and match experiment well.

In the case of non-abelian gauge theories, the gauge conditions above give non-linear equations for the gauge-group element $\tilde{g}$. Taking Coulomb gauge on for example, the equations for $\tilde{g}$ to preserve the gauge-fixing condition at $A = 0$ are

$$\partial_r\big((\partial^r \tilde{g})\tilde{g}^{-1}\big) = 0. \qquad (12.11.2)$$

This set of non-linear differential equations has non-constant solutions. Such indeterminacy in $\tilde{g}$ is called a *Gribov ambiguity*. Notice that the equation 12.11.2 is invariant under the substitution $\tilde{g} \rightarrow g_0 \tilde{g} g_1$ where $g_0$ and $g_1$ are any global symmetry transformations.

The family of solutions $G = $ const has finite volume and determines the expected gauge-fixed field theories. The theories determined by similar families of non-constant solutions are called *Gribov copies*. The Faddeev–Popov functional integral includes contributions from the Gribov copies. There is as yet no technique for deriving amplitudes from such a functional integral. However, since the gauge transformation connecting the expected theory to a Gribov copy modifies the gauge fields by an amount proportional to the inverse of the coupling constant, the influence of the Gribov copies cannot be reflected in perturbation theory. Other arguments exist to confirm that perturbatively the Gribov copies decouple. (The occurrence of the reciprocal of the coupling here is reminiscent of its appearance in soliton scattering.)

**Remark 12.11.3.** An explicit solution of the gauge-fixing equation 12.11.2 can be found for gauged $SU(2)$. The group $SU(2)$ is easy to parameterize since the exponential map on its Lie algebra is 1–1 on the interior of the smallest circle where all points map to $-\mathbf{1}$. Thus every element $U$ except $-\mathbf{1}$ has a unique representation

$$U = e^{i\hat{n}\cdot\vec{\sigma}} = \mathbf{1}\cos|\bar{n}| + i\hat{n}\cdot\vec{\sigma}\sin|\bar{n}|, \qquad (12.11.4)$$

where $\hat{n} = \bar{n}/|\bar{n}|$ and $\bar{n}\cdot\bar{n} < \pi$. To obtain a gauge-group element $U(\bar{x})$ we only have to make $\bar{n}$ a function of $\bar{x}$. We take

$$\bar{n} = \frac{\omega(r)}{r}(y, x, z) \qquad (12.11.5)$$

(the exchange of $x$ and $y$ is intended) and change coordinates to spherical polars, thereby generating

$$U(r, \theta, \phi) = \begin{pmatrix} c_\omega + ic_\theta s_\omega & e^{i\phi}s_\theta s_\omega \\ -e^{-i\phi}s_\theta s_\omega & c_\omega - ic_\theta s_\omega \end{pmatrix}, \qquad (12.11.6)$$

where the notations $s_\alpha$ and $c_\alpha$ stand for $\sin \alpha$ and $\cos \alpha$ respectively.

Writing the divergence and gradient operators in spherical polars and changing coordinates from $r$ to $t = \ln r$, the equation 12.11.2 on $U$ yields

$$\left(U_{tt} + U_t + U_{\theta\theta} + \frac{c_\theta}{s_\theta}U_\theta + \frac{1}{s_\theta^2}U_{\phi\phi}\right)U^\dagger + U_t U_t^\dagger + U_\theta U_\theta^\dagger + \frac{1}{s_\theta^2}U_\phi U_\phi^\dagger = 0. \qquad (12.11.7)$$

When we work out the details of this, miraculous cancellations occur, leaving a product of a matrix function of $\theta$ and $\phi$ with the following differential equation in $\omega$:

$$\ddot{\omega} + \dot{\omega} - 2\sin\omega\,\cos\omega = 0. \qquad (12.11.8)$$

Changing variables from $\omega$ to $\varpi = 2\omega$, we find an equation which describes a damped pendulum in a constant gravitational field with $\varpi$ being the angle from the point of unstable equilibrium.

Since $U = \mathbf{1}$ at the origin, $\varpi(r = 0)$ must be $2k\pi$ for some integer $k$. Since $\varpi$ is only determined up to a multiple of $2\pi$, there is no loss of generality in taking $k = 0$. Hence we have a boundary condition

$$\lim_{t \to -\infty} \varpi(t) = 0. \qquad (12.11.9)$$

Clearly, there are three solutions with this initial condition: the pendulum never falls, or it falls clockwise, or it falls anti-clockwise. Checking the eigenvalues of the linearized differential equation in a neighborhood of the unstable point shows that each falling solution exists and is unique up to translation in time. Translation in time corresponds to rescaling the coordinate $r$, so we conclude that there are two one-parameter families of non-trivial solutions to the gauge-fixing condition 12.11.2 of the form 12.11.6.

Finally, note that, since $SU(2)$ is a subgroup of every simple group, the Gribov ambiguity is a feature of every non-abelian gauge theory.

**Remark 12.11.10.** The Faddeev–Popov functional integral is commonly derived by using the gauge-invariance of $(dA)$, $S(A)$, and $\Delta\big(C(A)\big)$ to factor the integral over the gauge group out of the integral over the configuration space:

$$
\begin{aligned}
Z &= \int (dA)\, e^{iS(A)} \\
&= \int (dA)(d\tilde{g})\, \Delta_C(A)\delta\big(C(\tilde{g}^{-1}\cdot A)\big)e^{iS(A)} \\
&= \int (dA)(d\tilde{g})\, \Delta_C(\tilde{g}\cdot A)\delta\big(C(A)\big)e^{iS(\tilde{g}\cdot A)} \qquad\text{(12.11.11)} \\
&= \int (dA)(d\tilde{g})\, \Delta_C(A)\delta\big(C(A)\big)e^{iS(A)} \\
&= N \int (dA)\, \Delta_C(A)\delta\big(C(A)\big)e^{iS(A)}.
\end{aligned}
$$

In this argument, it is apparent that the gauge group may be chosen to suit the gauge-fixing condition. Since no direct reference to the moduli space is made, the moduli space can be dependent on this choice. If it is so dependent, however, there is no reason why $Z$ should be gauge invariant. In a gauge with a Gribov ambiguity, the final delta functional gives rise to an unmanageable sum over the set of its zeros.

This section reveals some of the missing mathematics in the Faddeev–Popov procedure. Generally, the weakness of the foundation in functional analysis does not affect the physics because of the Wick rotation and compactification of space, and the existence of Gribov copies does not affect perturbation theory. However, the Gribov ambiguity suggests that the non-perturbative consequences of this quantization procedure (if such an interpretation can be found) will indicate significant departures from the conclusions of perturbation theory.

## 12.12 Summary

This chapter has developed the functional integral formalism introduced in the previous chapter by showing how to connect Routhian form to Lagrangian form through first-order form and by introducing the Faddeev–Popov principle for quantization of gauge theories.

The primary conclusion of the chapter is that the Faddeev–Popov principle gives correct Feynman rules for gauge theories. However, we note that, since the non-linearity of the Yang–Mills equations for non-abelian gauge fields provides a major obstruction to canonical quantization, and since, if the gauge fields were massive, then the gauge symmetry would break and functional integral quantization would also not apply, neither quantization technique applies to non-abelian gauge field theories with gauge-field mass terms.

Of equal importance with these conclusions are the geometric structure of gauge theories and the transformation of a classical, gauge-invariant Lagrangian into a Faddeev–Popov Lagrangian under Faddeev–Popov quantization.

With the Feynman rules for gauge theories in hand, we are in a position to understand the Standard Model of the electroweak and strong interactions at tree level. The Standard Model will be the topic for the next two chapters.

# Chapter 13

# Anomalies and Vacua in Gauge Theories

Completing the presentation of quantization of gauge theories with an investigation into the possibility that quantum effects might cause a classical conservation law to break down and thereby render the Faddeev–Popov procedure inapplicable.

## Introduction

The logic of the gauge principle, the simplicity of the Faddeev–Popov principle, and the absence of any competing means for constructing quantum field theories with non-abelian symmetries have together made the constructions of the previous chapter central to high-energy model building. Therefore, in this chapter, we pursue the analysis of gauge theories a little further.

Gauge invariance of a classical field theory does not imply gauge invariance of the corresponding quantum field theory. Invariance under gauge transformations is tested in two stages: first, infinitesimal gauge transformations, and second, finite gauge transformations. Of course, invariance under infinitesimal gauge transformations implies invariance under all exponentials of infinitesimal gauge transformations, but there may be gauge transformations that are not of this form. Hence the need for two steps.

A classical symmetry which does not survive quantization is said to be *anomalous*. When infinitesimal gauge symmetries are anomalous, then the quantum theory is inconsistent. But the finite gauge symmetries can be anomalous without making the quantum theory inconsistent.

The opening sections, Sections 13.1 to 13.5 discuss the first or local level of gauge invariance and clarify the relation between classical equations of motion and conservation laws and their quantum counterparts. This material culminates in the Section 13.6 which summarizes the basic theorems on local anomalies. The closing sections, Sections 13.7 to 13.10, describe the structure of the common gauge groups and investigates the connection between the second or global level of gauge invariance and the quantum vacuum. This theory of anomalies will be applied to the Standard Model of the electroweak and strong interactions in Chapter 15, and to the decay $\pi^0 \to 2\gamma$ in Chapter 17.

## 13.1 Preservation of Conservation Laws in Bosonic Theories

As we saw in Section 9.8, gauge invariance implies that gauge fields couple to conserved currents. Hence, if a classical conservation law breaks down for a current which is coupled to gauge fields, gauge invariance cannot be maintained. In this situation, there is no known renormalization procedure and the theory is declared inconsistent. Furthermore, even when the current does not couple to gauge fields, its divergence may make significant contributions to measurable phenomena. It is therefore important to know when a classical conservation law is broken by quantum effects and, if it is, to have a formula for the divergence of the current.

It is apparent from the comments in Chapters 11 and 12 that functional integration is a delicate tool. The comparative ease with which it quantizes gauge theories is balanced by a lack of mathematical support for the functional measure, and to a lesser extent for the operator determinants and the fermionic functional derivative. However, many errors of naive application can be remedied, and many insights may be gained by formal manipulations. In this section and the next, we use functional methods to discover when quantum effects break a classical conservation law. The fundamental idea is to use changes of variables in the functional integral formula for the generating functional to find quantum expectations of the classical equations of motion and currents. This section establishes current conservation in bosonic theories, and the next examines conservation breakdown in theories with fermions.

Assume that we have a theory of real bosonic fields $\phi$ defined by a Lagrangian density $\mathcal{L}(\phi, \partial\phi)$. We saw in Section 2.6 a direct relationship between the action and the equations of motion:

**Lemma 13.1.1.** *Functional differentiation of the action yields the equations of motion:*

$$\frac{\delta S}{\delta \phi_m(a)} = \frac{\partial \mathcal{L}}{\partial \phi_m}\bigg|_{x=a} - \partial_\mu \frac{\partial \mathcal{L}}{\partial(\partial_\mu \phi_m)}\bigg|_{x=a}.$$

In the formula for the generating functional, we can induce functional differentiation of the action by using the translation of a field $\phi_m$ defined by the change of basis

$$\phi_r'(x) \overset{\text{def}}{=} \phi_r(x) + \epsilon \delta_{rm} \delta^{(4)}(x-a). \tag{13.1.2}$$

A change of variables leaves the generating functional unchanged, of course. In the formalism of physics, the functional measure itself is translation invariant. Hence, only the action is affected:

$$\int (d\phi)\, e^{iS(\phi)} i \frac{\delta}{\delta \phi_m(a)} S(\phi) = 0. \tag{13.1.3}$$

Hence translation of variables in the generating functional does indeed bring out a factor of the classical equations of motion in the integrand.

If we define the expected value $\langle X \rangle$ of any function $X$ of the fields $\phi$ by

$$\langle X \rangle \overset{\text{def}}{=} N \int (d\phi)\, e^{iS(\phi)} X(\phi), \tag{13.1.4}$$

then we can conclude that:

**Corollary 13.1.5.** *The classical equation of motion is true in the average:*

$$\left\langle \frac{\partial \mathcal{L}}{\partial \phi_m} - \partial_\mu \frac{\partial \mathcal{L}}{\partial(\partial_\mu \phi_m)} \right\rangle = 0.$$

To test for a non-zero value of the classical equations in quantum theory, we take the expectation of the product of the classical equations with any other function of the fields. To evaluate such an expectation, we use the fact that, for any polynomial functional $F(\phi)$, the functional integral of a functional derivative vanishes,

$$\int (d\phi)\, \frac{\delta}{\delta \phi_m(x)} \big( F(\phi) e^{iS} \big) = 0, \tag{13.1.6}$$

to justify integrating by parts:

**Lemma 13.1.7.** *Expectations involving the equations of motion simplify:*

$$\left\langle F \frac{\delta S}{\delta \phi_m(x)} \right\rangle = i \left\langle \frac{\delta F}{\delta \phi_m(x)} \right\rangle.$$

In light of Lemma 13.1.1, the left-hand side represents the quantum expectation of the equations of motion multiplied by some polynomial $F$ in the fields.

In the operator perspective, combining the time-ordered product formula 10.3.2 for a Green function with the functional integral equation 11.7.21 for the same Green function, functional expectations of products of classical fields are vacuum expectations of time-ordered products of the corresponding quantum fields. Consequently, we can start from equations of the form

$$\langle 0|T\hat\phi(x)\hat\phi(x_1)\cdots\hat\phi(x_n)|0\rangle = \int (d\phi)\,\phi(x)\phi(x_1)\cdots\phi(x_n), \qquad (13.1.8)$$

differentiate both sides and add to build up the equations of motion on $\phi(x)$. Then we see at once that the terms on the right of Lemma 13.1.7 come from differentiating a time-ordered product with respect to time, and the remaining terms combine to give:

$$\left\langle 0 \left| T \frac{\delta S(\hat\phi)}{\delta \hat\phi_m(x)} \hat\phi(x_1)\cdots\hat\phi(x_n) \right| 0 \right\rangle = 0. \qquad (13.1.9)$$

From here, application of the LSZ reduction formula shows that all matrix elements of the equations of motion vanish. We conclude that:

**Lemma 13.1.10.** *A scalar quantum field satisfies the classical equations of motion.*

This justifies the naïve form of canonical quantization used in Chapter 3 and its subsequent applications. However, in preparation for the analysis of local anomalies, it will be worthwhile to develop the functional integral approach further with a view to deriving the same conclusion within that framework.

**Homework 13.1.11.** Show that Lemma 13.1.7 applied to the action for a free scalar field gives the result of Homework 2.4.29:

$$(\Box_x + m^2)\langle 0|T\phi(x)\phi(y)|0\rangle = -i\delta^{(4)}(x - y).$$

**Homework 13.1.12.** Show that Lemma 13.1.10 generalizes to massive vector fields (see Euler–Lagrange equations 9.1.9 and equal-time commutation relations 9.2.9). Show that it also extends to abelian and to non-abelian gauge fields.

In the classical proof of conservation, one needs to use the equations of motion multiplied by the action $\mathbf{D}\phi$ of the symmetry generator $T$ on the vector of fields:

$$(\partial_\mu \Pi^\mu)\mathbf{D}\phi = \frac{\partial \mathcal{L}}{\partial \phi}\mathbf{D}\phi. \qquad (13.1.13)$$

Since $T$ is proportional to an element in $so(n)$, $T_{mm} = 0$ for all $m$, and the VEV on the right of the expectation formula 13.1.7 evaluates to zero:

$$\left\langle \frac{\delta F}{\delta \phi_m(x)} \right\rangle = \left\langle \frac{\delta}{\delta \phi_m(x)} (T\phi(x))_m \right\rangle = 0. \qquad (13.1.14)$$

Consequently, the classical use of the equations of motion is valid for theories with only real scalar fields. Since a theory with complex scalar fields can be expressed in terms of real ones, and since this argument applies equally to vector fields, we can drop the restrictions 'real' and 'scalar' and deduce the same result for theories of scalar fields generally:

**Theorem 13.1.15.** *In theories of scalar fields, quantization never causes a break-down of current conservation.*

In preparation for the more delicate analysis of theories with fermions, we next approach the same conclusion from a second direction in order to establish a formula for the expectation of the divergence of the current.

Since a classical current $j^\mu$ is associated with a symmetry generator $T$, we make the change of variables defined by

$$\phi' = e^{-i\tau T}\phi. \tag{13.1.16}$$

Let $\mathbf{D}_\tau$ be the operation of differentiating with respect to $\tau$ and then evaluating at $\tau = 0$. Then:

$$\mathbf{D}_\tau \mathcal{L} = 0 \quad \Longrightarrow \quad \langle \mathbf{D}_\tau \mathcal{L} \rangle = \left\langle \frac{\partial \mathcal{L}}{\partial(\partial_\mu \phi)} \partial_\mu(-iT\phi) + \frac{\partial \mathcal{L}}{\partial \phi}(-iT\phi) \right\rangle \tag{13.1.17}$$

$$= \left\langle \left( \left( \frac{\partial \mathcal{L}}{\partial \phi} - \partial_\mu \frac{\partial \mathcal{L}}{\partial(\partial_\mu \phi)} \right)(-iT\phi) + \partial_\mu \left( \frac{\partial \mathcal{L}}{\partial(\partial_\mu \phi)}(-iT\phi) \right) \right) \right\rangle$$

$$= \left\langle \frac{\delta S}{\delta \phi_m(x)}(-iT\phi(x))_m \right\rangle + \langle \partial_\mu j^\mu(x) \rangle$$

$$= 0,$$

and so, using the evaluation formula of Lemma 13.1.7,

$$\langle \partial_\mu j^\mu(x) \rangle = -\left\langle \frac{\delta}{\delta \phi_m(x)}(T\phi(x))_m \right\rangle. \tag{13.1.18}$$

As before, the final expression evaluates to zero because the diagonal elements in $T$ are all zero.

These elegant functional arguments shed light on the character of quantum evolution, enabling us to see that the classical equations of motion and conservation laws hold for scalar fields. In the following section, we extend the analysis to fermions and confirm the possibility of current non-conservation.

# 13.2 Breakdown of Conservation Laws in Fermionic Theories

The techniques of the previous section apply with natural modifications to theories with fermions. First, as in Lemma 13.1.1, there is a direct relationship between the action and the equations of motion for a classical Fermi field:

**Lemma 13.2.1.** *Functional differentiation of the action yields the equations of motion:*

$$\frac{\delta S}{\delta \psi(x)} = \frac{\partial \mathcal{L}}{\partial \psi} - \partial_\mu \frac{\partial \mathcal{L}}{\partial(\partial_\mu \psi)} \quad \text{and} \quad \frac{\delta S}{\delta \bar{\psi}(x)} = \frac{\partial \mathcal{L}}{\partial \bar{\psi}} - \partial_\mu \frac{\partial \mathcal{L}}{\partial(\partial_\mu \bar{\psi})}.$$

Second, the functional derivative is invariant under translation of fields

$$\psi \to \psi + \epsilon \delta_x, \tag{13.2.2}$$

where $\epsilon$ is a Grassmann parameter, so we find that the expectation of the classical equations of motion is again zero.

**Homework** 13.2.3. Verify that the expected value of the fermion equations of motion vanishes by writing out the details.

Third, following 13.1.4, we define the expectation of a function $F$ of fermions by

$$\langle F \rangle \stackrel{\text{def}}{=} \left( \frac{\partial}{\partial \psi} \frac{\partial}{\partial \bar{\psi}} \right) F(\psi, \bar{\psi}) e^{iS(\psi, \bar{\psi})}. \tag{13.2.4}$$

Since the square of a fermionic derivative is zero, we find that expectation of a derivative is again zero:

$$\left. \begin{aligned} \left( \frac{\partial}{\partial \psi} \frac{\partial}{\partial \bar{\psi}} \right) \frac{\delta}{\delta \psi(x)} \left( F(\psi, \bar{\psi}) e^{iS(\psi, \bar{\psi})} \right) = 0 \\ \left( \frac{\partial}{\partial \psi} \frac{\partial}{\partial \bar{\psi}} \right) \frac{\delta}{\delta \bar{\psi}(x)} \left( \bar{F}(\psi, \bar{\psi}) e^{iS(\psi, \bar{\psi})} \right) = 0 \end{aligned} \right\} \tag{13.2.5}$$

Therefore the evaluation Lemma 13.1.7 also extends to the fermionic case.

Because of the commutation rules of fermion calculus set out in Section 7.4, it is inconvenient to write down a general formula for the fermionic result. Instead, we focus on the case at hand, in which

$$F = \mathbf{D}_\tau \psi = -iT\psi \quad \text{and} \quad \bar{F} = \mathbf{D}_\tau \bar{\psi} = \bar{\psi} \bar{T} i, \tag{13.2.6}$$

where $\bar{T} = \gamma^0 T^\dagger \gamma^0$. For such cases, a schematic computation shows how the signs work out:

$$\frac{\delta}{\delta \psi} (e^{iS} T\psi) = -i e^{iS} \frac{\delta S}{\delta \psi} T\psi + e^{iS} \frac{\delta}{\delta \psi} (T\psi). \tag{13.2.7}$$

The following lemma states the result we need.

**Lemma 13.2.8.** *The fermionic analogue of integration by parts yields equalities between expectations as follows:*

$$\left\langle \frac{\delta S}{\delta \psi(x)} (-iT\psi) \right\rangle = -i \left\langle \frac{\delta}{\delta \psi(x)} (-iT\psi) \right\rangle, \quad \left\langle (\bar{\psi} \bar{T} i) \frac{\delta S}{\delta \bar{\psi}(x)} \right\rangle = -i \left\langle \frac{\delta}{\delta \bar{\psi}(x)} (\bar{\psi} \bar{T} i) \right\rangle.$$

Fourth, as we would expect from the computation 13.1.17 of $\mathbf{D}_\tau \mathcal{L}$ for scalar fields, the computation of $\mathbf{D}_\tau \mathcal{L}$ for fermions links their Noether currents with their equations of motion:

$$\begin{aligned} \mathbf{D}_\tau \mathcal{L} &= \frac{\partial \mathcal{L}}{\partial \psi} \mathbf{D}_\tau \psi' + \frac{\partial \mathcal{L}}{\partial (\partial_\mu \psi)} \mathbf{D}_\tau \partial_\mu \psi' + \mathbf{D}_\tau \bar{\psi}' \frac{\partial \mathcal{L}}{\partial \bar{\psi}} + \mathbf{D}_\tau \partial_\mu \bar{\psi}' \frac{\partial \mathcal{L}}{\partial (\partial_\mu \bar{\psi})} \\ &= \left( \frac{\partial \mathcal{L}}{\partial \psi} - \partial_\mu \frac{\partial \mathcal{L}}{\partial (\partial_\mu \psi)} \right) \mathbf{D}_\tau \psi' + \mathbf{D}_\tau \bar{\psi}' \left( \frac{\partial \mathcal{L}}{\partial \bar{\psi}} - \partial_\mu \frac{\partial \mathcal{L}}{\partial (\partial_\mu \bar{\psi})} \right) \\ &\quad + \partial_\mu \left( \Pi^\mu \mathbf{D}_\tau \psi' + \mathbf{D}_\tau \bar{\psi}' \bar{\Pi}^\mu \right) \\ &= \frac{\delta S}{\delta \psi} \mathbf{D}_\tau \psi' + \mathbf{D}_\tau \bar{\psi}' \frac{\delta S}{\delta \bar{\psi}} + \partial_\mu \left( \Pi^\mu \mathbf{D}_\tau \psi' + \mathbf{D}_\tau \bar{\psi}' \bar{\Pi}^\mu \right). \end{aligned} \tag{13.2.9}$$

Assuming that $T$ is a symmetry generator for $\mathcal{L}$, $\mathbf{D}_\tau \mathcal{L} = 0$. In this case,

$$\partial_\mu j_T^\mu = -\frac{\delta S}{\delta \psi}(-iT\psi) - (\bar{\psi}\bar{T}i)\frac{\delta S}{\delta \bar{\psi}}. \tag{13.2.10}$$

Applying the Lemma 13.2.8 to this result, we find a formula for the expectation of $\partial_\mu j_T^\mu$ analogous to the scalar formula 13.1.18:

**Lemma 13.2.11.** *If a Lagrangian density $\mathcal{L}$ is invariant under $\psi \to e^{-i\tau T}\psi$, then the divergence of the current $j_T^\mu = \bar{\psi}\gamma^\mu T\psi$ is given by:*

$$\left\langle \partial_\mu j_T^\mu \right\rangle = \left\langle \frac{\delta}{\delta \psi}(T\psi) - \frac{\delta}{\delta \bar{\psi}}(\bar{\psi}\bar{T}) \right\rangle,$$

*where $\bar{T} = \gamma^0 T^\dagger \gamma^0$.*

The matching of Dirac indices between the functional derivative operators and their operands leads to a trace Tr over Dirac and group indices. In this fermionic case, the diagonal elements of $T$ are in general non-zero, so naive evaluation of the functional derivatives leads to the formula

$$\left\langle \partial_\mu j_T^\mu \right\rangle = \mathrm{Tr} \begin{pmatrix} T & 0 \\ 0 & \pm T \end{pmatrix} \delta(0), \tag{13.2.12}$$

where the sign is positive if there is a $\gamma_5$ in $T$ and negative otherwise. Now, even if the trace is zero, the delta factor is infinite, and we can no longer be sure that the expectation of the divergence is zero.

The factor of $\delta(0)$ indicates a formal weakness in our derivation. It is necessary to return to the definitions for fermion field calculus in Section 11.11 and regularize the functional derivative from that firmer foundation. The following section gives the details of the regularization, and the next computes the trace properly.

## 13.3 Wick Rotation and Fermionic Functional Calculus

The first step in framing an improved definition of fermionic functional derivative operators is to reduce their function space to a Hilbert space. For this, we need a genuine inner product; the Dirac form $\bar{\psi}\psi'$ yields an indefinite metric. For this we must make a Wick rotation.

The general spirit of the transformation is to change only the time/energy components of tensors and to leave the spatial structure unchanged. In the following display, the equalities are the substitutions which implement the Wick rotation. The essential definitions are on the left, and their consequences are on the right:

$$k_0 \overset{\mathrm{W}}{=} ik_\mathrm{E}^4 \quad \text{and} \quad x_0 \overset{\mathrm{W}}{=} -ix_\mathrm{E}^4 \quad \Longrightarrow \quad x{\cdot}k = x_\mathrm{E}{\cdot}k_\mathrm{E} \quad \text{and} \quad k^2 = -k_\mathrm{E}^2;$$

$$\gamma^0 \overset{\mathrm{W}}{=} -i\gamma_\mathrm{E}^4 \quad \Longrightarrow \quad \{\gamma_\mathrm{E}^\alpha, \gamma_\mathrm{E}^\beta\} = -2\delta^{\alpha\beta}, \quad \not{k} = \not{k}_\mathrm{E}, \quad \text{and} \quad \not{\partial} = \not{\partial}_\mathrm{E}; \tag{13.3.1}$$

$$\gamma^5 \overset{\mathrm{W}}{=} \gamma_\mathrm{E}^5 = -\gamma_\mathrm{E}^1 \gamma_\mathrm{E}^2 \gamma_\mathrm{E}^3 \gamma_\mathrm{E}^4, \quad \epsilon^{\mu\nu\rho\sigma} \overset{\mathrm{W}}{=} i\epsilon_\mathrm{E}^{\alpha\beta\gamma\delta} \quad \text{where } \epsilon^{1234} = 1.$$

These definitions also imply that $\gamma_E^{\mu\dagger} = -\gamma_E^\mu$, $\not{\partial}_E^2 = -\Delta$, and a Euclidean trace formula:

$$\text{tr}(\gamma_E^5 \gamma_E^\alpha \gamma_E^\beta \gamma_E^\gamma \gamma_E^\delta) = -4\epsilon_E^{\alpha\beta\gamma\delta}. \tag{13.3.2}$$

After Wick rotation, the operator $\not{D}$ is hermitian with respect to the usual $\varphi^\dagger \varphi'$ structure, so we can use its eigenfunctions $\varphi_k$ as a basis for the classical fermions:

$$\varphi_k \stackrel{\text{def}}{=} \begin{pmatrix} \varphi_{k,1} \\ \vdots \\ \varphi_{k,n} \end{pmatrix} \quad \text{such that} \quad \not{D}\varphi_k = \lambda_k \varphi_k, \tag{13.3.3}$$

where $n$ is the number of fermion fields in $\psi$ and each component field $\varphi_{k,r}$ itself has four components. Furthermore, we assume that these functions are normalized as follows:

$$\int d^4x\, \varphi_k^\dagger(x)\varphi_l(x) = \delta_{kl}. \tag{13.3.4}$$

From the orthonormality of the $\varphi$'s, we find that

$$K(x,y) \stackrel{\text{def}}{=} \sum_k \varphi_k(x)\varphi_k^\dagger(y) \quad \Longrightarrow \quad \int d^4y\, K(x,y)\varphi_l(y) = \varphi_l(x). \tag{13.3.5}$$

Assuming completeness of the $\varphi$'s, we infer that

$$\sum_k \varphi_k(x)\varphi_k^\dagger(y) = \delta^{(4)}(x-y)\mathbf{1}_4 \otimes \mathbf{1}_n. \tag{13.3.6}$$

Taking components of this equation leads to a relation on component fields:

$$\sum_k \varphi_{k,r}(x)\varphi_{k,s}^\dagger(y) = \delta^{(4)}(x-y)\mathbf{1}_4\delta_{rs}, \tag{13.3.7}$$

and taking the trace of this over group and Dirac indices provides a second normalization relation:

$$\sum_k \varphi_k^\dagger(x)\varphi_k(y) = 4n\delta^{(4)}(x-y)\delta_{rs}. \tag{13.3.8}$$

Following the principles of Section 11.11, on the basis of the completeness assumption, we can expand our vector $\psi$ of classical Fermi fields in terms of the basis $\varphi_k$ using Grassmann parameters as coefficients:

$$\psi(x) = \sum_k \theta_k \varphi_k. \tag{13.3.9}$$

Now any functional of $\psi$'s can be regarded as a function of the coefficients $\theta_k$, and so we can define functional derivative operators by:

$$\frac{\delta}{\delta\psi(x)} \stackrel{\text{def}}{=} \sum_k \varphi_k^\dagger(x)\frac{\partial}{\partial\theta_k}. \tag{13.3.10}$$

On account of the normalization relation 13.3.8, this definition has the defining property of the fermionic functional differentiation operator:

$$\frac{\delta}{\delta\psi(x)}\psi(y) = \sum_k \varphi_k^\dagger(x)\varphi_k(y) = 4n\delta^{(4)}(x-y). \qquad (13.3.11)$$

The factor of $4n$ reflects the total number of Dirac component fields in $\psi$.

**Homework 13.3.12.** We can verify the fermion functional calculus by working in the two-dimensional space with basis $\varphi_1$ and $\varphi_2$. Expand $\psi$, $S$, and $e^{iS}$ in terms of this basis and evaluate the following:

$$\left(\frac{\partial}{\partial\theta}\frac{\partial}{\partial\bar\theta}\right)e^{iS}, \quad \left(\frac{\partial}{\partial\theta}\frac{\partial}{\partial\bar{\bar\theta}}\right)e^{iS}\left(\bar\psi\bar T\frac{\delta S}{\delta\bar\psi}\right), \quad \text{and} \quad \left(\frac{\partial}{\partial\theta}\frac{\partial}{\partial\bar{\bar\theta}}\right)e^{iS}\left(\frac{\delta S}{\delta\psi}T\psi\right).$$

Having expressed the functional derivative as a sum of partial derivatives, one obvious possibility for regularization is to damp the high-eigenvalue contributions before summing:

$$\left(\frac{\delta}{\delta\psi(x)}\right)_{\mathrm M} \overset{\text{def}}{=} \sum_k e^{-\lambda_k^2/M^2}\varphi_k^\dagger(x)\frac{\partial}{\partial\theta_k}. \qquad (13.3.13)$$

The damping and the limit will cancel in the defining property because in that case the original series is summable.

Building on Section 11.11, this section has gone more deeply into the basic operations of fermionic functional calculus. With these details of Wick rotation and fermionic functional derivative regularization, we are ready to compute the traces in Lemma 13.2.11 and find the expected divergence of the axial currents.

## 13.4 The Abelian Anomaly

abelian anomalyIn this section, we illustrate the technique for computing the traces in Lemma 13.2.11 in the case of $T = \gamma_5$. In this case, the symmetry group associated with the anomalous current is abelian, and the expected divergence of the current is known as the *abelian anomaly*.

Consider a Lagrangian density for a Yang–Mills multiplet coupled to a vector $\psi$ of massless charged spinors,

$$\mathcal{L} = -\frac{1}{4}G_{\mu\nu}^a G^{a\mu\nu} + \bar\psi i\slashed{D}\psi, \qquad (13.4.1)$$

with its axial $U(1)$ symmetry and current

$$\psi' = e^{-i\tau\gamma_5}\psi \quad \text{and} \quad j_5^\mu = \bar\psi\gamma^\mu\gamma_5\psi. \qquad (13.4.2)$$

For this example, the divergence Lemma 13.2.11 implies that the Noether current $j_5^\mu$ satisfies

$$\langle\partial_\mu j_5^\mu(x)\rangle = \lim_{M\to\infty}\left\langle\left(\frac{\delta}{\delta\psi(x)}\right)_M(\gamma_5\psi(x)) - \left(\frac{\delta}{\delta\bar\psi(x)}\right)_M(\bar\psi(x)\bar\gamma_5)\right\rangle$$
$$= 2\lim_{M\to\infty}\sum_k e^{-\lambda_k^2/M^2}\varphi_k^\dagger(x)\gamma_5\varphi_k(x). \qquad (13.4.3)$$

(From Homework 11.11.32, we see at once that $\langle \partial_\mu j_5^\mu \rangle = 2\,\text{Tr}(\gamma_5 \delta_x)$, so we could consider this last expression as a regularization of the trace in function space.)

The presence of the $\varphi$'s in this formula indicates that we must now do all our computations in Euclidean space. We shall not put subscript E's on all the symbols because that would clutter up the formulae. Simply assume that they are there until a further note in the text brings us back to Minkowski space.

Because we have no specific information about the eigenvalues $\lambda_k$, the first step is to eliminate them in favor of the Euclidean operator $\not{D}$. The second step is to use the trace Tr over Dirac and group indices to manipulate the result into a convenient form:

$$\langle \partial_\mu j_5^\mu(x) \rangle = 2 \lim_{M \to \infty} \sum_k \bar\varphi_k^\dagger(x) \gamma_5 e^{-\not{D}^2/M^2} \varphi_k(x)$$

$$= 2 \lim_{M \to \infty} \sum_k \lim_{y \to x} \text{Tr}\left(\gamma_5 e^{-\not{D}^2/M^2} \varphi_k(x) \varphi_k^\dagger(y)\right)$$

$$= 2 \lim_{M \to \infty} \lim_{y \to x} \text{Tr}\left(\gamma_5 e^{-\not{D}^2/M^2} \delta^{(4)}(x-y)\right) \qquad (13.4.4)$$

$$= 2 \lim_{M \to \infty} \lim_{y \to x} \int \frac{d^4k}{(2\pi)^4} \, \text{Tr}\left(\gamma_5 e^{-\not{D}^2/M^2} e^{-ik\cdot(x-y)}\right)$$

$$= 2 \lim_{M \to \infty} \int \frac{d^4k}{(2\pi)^4} \, \text{Tr}\left(\gamma_5 e^{ik\cdot x} e^{-\not{D}^2/M^2} e^{-ik\cdot x}\right).$$

We simplify this expression in two steps. First we use the natural relationship between conjugation and exponentiation

$$g^{-1} \exp(C) g = \exp(g^{-1} C g) \qquad (13.4.5)$$

to transform the operator as follows:

$$e^{ik\cdot x} \exp(-\not{D}^2/M^2) e^{-ik\cdot x} = \exp(-e^{ik\cdot x} \not{D}^2 e^{-ik\cdot x})$$

$$= \exp\left(-(e^{ik\cdot x} \not{D} e^{-ik\cdot x})^2\right) \qquad (13.4.6)$$

$$= \exp\left(-(\not{D} - i\not{k})^2/M^2\right)$$

$$= \exp(-k^2/M^2) \exp\left(-(2ik_\alpha D^\alpha + \not{D}\not{D})/M^2\right),$$

where the final expression contains implicit evaluation on the constant function unity. Notice that $\{\not{k}, \not{l}\} = -2k\cdot l$ in Euclidean space. Second, we scale the integration parameter $k$ to bring its $M$'s out of the exponent:

$$\langle \partial_\mu j_5^\mu(x) \rangle = 2 \lim_{M \to \infty} M^4 \int \frac{d^4k}{(2\pi)^4} e^{-k^2} \, \text{Tr}\left(\gamma_5 \exp\left(-\frac{2ik_\mu D^\mu + \not{D}\not{D}}{M^2}\right)\right). \qquad (13.4.7)$$

In light of the trace over Dirac indices, when we expand the exponential in $D$ the terms with less than four gamma matrices evaluate to zero. In light of the limit over $M$, terms with more than four $D$'s will also drop out. The only term remaining is the one with $\not{D}^4$, and this gives rise to the following trace:

$$\text{Tr}(\gamma_5 \not{D}^4) = -4\epsilon^{\alpha\beta\gamma\delta} \, \text{tr}(D_\alpha D_\beta D_\gamma D_\delta)$$

$$= -\epsilon^{\alpha\beta\gamma\delta} \, \text{tr}([D_\alpha, D_\beta][D_\gamma, D_\delta]) \qquad (13.4.8)$$

$$= -\epsilon^{\alpha\beta\gamma\delta} \, \text{tr}(G_{\alpha\beta} G_{\gamma\delta}).$$

Finally, recalling that there is a factor of $\frac{1}{2}$ associated with the trace because it comes from the quadratic term in an exponential, and with the help of the formula

$$\int \frac{d^4k}{(2\pi)^4} e^{-k^2} = \frac{1}{16\pi^2}, \tag{13.4.9}$$

we can substitute into equation 13.4.7 for $\langle \partial_\mu j_5^\mu \rangle$ and obtain

$$\langle \partial_\mu j_5^\mu(x) \rangle = -\frac{1}{16\pi^2} \epsilon^{\alpha\beta\gamma\delta} \operatorname{tr}(G_{\alpha\beta} G_{\gamma\delta}). \tag{13.4.10}$$

Now we have completed our computation in Euclidean space, it is only necessary to transform the conclusion back into Minkowski space. According to the Wick substitutions 13.3.1 of the previous section, the Euclidean $\epsilon$ becomes $-i$ times the Minkowski $\epsilon$, and $G_{r4}$ becomes $iG_{0r}$ on account of the anti-symmetry of $G_{\mu\nu}$ and the substitutions $-i\partial^0$ for $\partial^4$ and $-iA^0$ for $A^4$. Surprisingly, then, there is no change in the form of the final formula:

**Theorem 13.4.11.** *In the gauge theory defined by the matter Lagrangian density* $\mathcal{L} = \bar\psi i \not D \psi$, *the divergence of the axial current* $j_5^\mu = \bar\psi \gamma^\mu \gamma_5 \psi$ *is given by*

$$\langle \partial_\mu j_5^\mu \rangle = -\frac{1}{8\pi^2} \operatorname{tr}(G_{\mu\nu} \tilde G^{\mu\nu}),$$

*where* $G_{\mu\nu} = [D_\mu, D_\nu]$ *and* $\tilde G^{\mu\nu} = \frac{1}{2} \epsilon^{\mu\nu\rho\sigma} G_{\rho\sigma}$.

Note that, if we apply this regularization procedure to the vector current $j^\mu$, we find that the overall factor of 2 in the formula 13.4.3 for the divergence becomes a factor of zero, so this current is conserved.

**Homework 13.4.12.** Show that, if we use $\exp(-\lambda_k^4/M^4)$ instead of $\exp(-\lambda_k^2/M^2)$ to regularize the functional derivative, we get the same formula for the anomaly.

The emergence of this anomaly indicates circumstances in which quantum currents do not obey classical conservation laws. Even adding a mass term for the fermion field and explicitly breaking the axial symmetry does not eliminate the anomaly. Indeed, we find in this case that

$$\langle \partial_\mu j_5^\mu \rangle = 2im \langle \bar\psi \gamma_5 \psi \rangle - \frac{1}{8\pi^2} \operatorname{tr}(G_{\mu\nu} \tilde G^{\mu\nu}). \tag{13.4.13}$$

**Homework 13.4.14.** Verify that adding a mass term to explicitly break the axial symmetry leads to equation 13.4.13.

At this point, it is natural to wonder how the anomaly emerges in the operator formalism from the sum of commutators $[iP_\mu, j_5^\mu]$. Do the Heisenberg equations depart from the classical ones? The argument above suggests that the canonical commutators should be regularized. However, there is no obvious Lorentz invariant regularization procedure for them. Sections 18.10 and 18.12 provide instead two different ways of regularizing the axial current itself. In both approaches, the Fermi field equations are valid and the regularization of the axial current causes $[iP_\mu, j_5^\mu]$ to be non-zero. Miraculously, all three approaches agree on the value of the anomaly.

**Homework** 13.4.15. Starting from the equation

$$\langle 0|T\bar{\psi}(y)\gamma_5\gamma^\mu\psi(x)|0\rangle = \left(\frac{\partial}{\partial\psi}\frac{\partial}{\partial\bar{\psi}}\right)\bar{\psi}(y)\gamma_5\gamma^\mu\psi(x)e^{iS}, \qquad (13.4.16)$$

show that

$$\langle 0|T\bar{\psi}(y)\gamma_5(i\slashed{D}-m)\psi(x)|0\rangle = 0.$$

Despite the apparent arbitrariness of regularizing the functional derivative operator through a discrete eigenexpansion of the covariant derivative, the form of the axial $U(1)$ anomaly given here agrees with results derived from perturbation theory. We conclude that the insights above into the properties of the classical equations of motion and the conservation laws which depend on them are valid and judge that the functional integral, when handled with due care, provides a remarkably robust formalism in physics.

This section, while greatly clarifying the relationship between classical and quantum evolution and between classical and quantum conservation laws, reveals the possibility that global symmetries may break down in gauge theories. Because this effect is an essential feature of the quantum theory (and may even make the theory inconsistent), it should be thoroughly understood. To help with this, the next section presents the same phenomenon from a different perspective.

# 13.5  Functional Jacobians

There is an alternative viewpoint on anomalies – the original method of Fujikawa – which locates their origin in functional Jacobians. This viewpoint is illuminating, so in this section, we shall use Fujikawa's technique to derive the abelian anomaly a second time.

Consider a gauge change of basis $\psi' = e^{-i\tau\chi T}\psi$, where $\chi$ is a function of spacetime. The generating functional must be invariant because this is a change of basis, but the functional measure may not be. We therefore introduce a determinant $D = D(\tau)$ for the functional Jacobian generated by this change of basis:

$$\left(\frac{\partial}{\partial\psi}\frac{\partial}{\partial\bar{\psi}}\right)D \overset{\text{def}}{=} \left(\frac{\partial}{\partial\psi'}\frac{\partial}{\partial\bar{\psi}'}\right). \qquad (13.5.1)$$

The dependence of $D$ on $\tau$ is multiplicative,

$$D(\sigma + \tau) = D(\sigma)D(\tau), \qquad (13.5.2)$$

so, differentiating with respect to $\sigma$ and evaluating at $\sigma = 0$, we find a differential equation for $D$:

$$\frac{\partial}{\partial\tau}D(\tau) = \dot{D}(0)D(\tau). \qquad (13.5.3)$$

Writing $\ln J$ for the constant $\dot{D}(0)$, we conclude that

$$D(\tau) = e^{\tau\ln J} = J^\tau, \qquad (13.5.4)$$

and so

$$\left(\frac{\partial}{\partial\psi'}\frac{\partial}{\partial\bar{\psi}'}\right) = \left(\frac{\partial}{\partial\psi}\frac{\partial}{\partial\bar{\psi}}\right)e^{\tau\ln J}. \qquad (13.5.5)$$

The function $J(\chi)$ is known as *Fujikawa's anomalous Jacobian*.

To make the functional Jacobian more concrete, we now compute it for the Lagrangian density 13.4.1 of the previous section. The gauged change of basis is axial,

$$\psi' = e^{-i\tau\chi\gamma_5}\psi \quad \text{and} \quad \bar{\psi}' = \bar{\psi}e^{-i\tau\chi\gamma_5}, \tag{13.5.6}$$

and so the contributions from $\psi$ and $\bar{\psi}$ add:

$$
\begin{aligned}
\left(\frac{\partial}{\partial\psi'}\frac{\partial}{\partial\bar{\psi}'}\right) &= \det\left(e^{2i\tau\chi\gamma_5}\right)\left(\frac{\partial}{\partial\psi}\frac{\partial}{\partial\bar{\psi}}\right) \\
&= \exp\left(2i\tau\,\mathrm{Tr}(\chi\gamma_5)\right)\left(\frac{\partial}{\partial\psi}\frac{\partial}{\partial\bar{\psi}}\right).
\end{aligned}
\tag{13.5.7}
$$

Comparing with the characteristic property 13.5.5 of $J$, obviously:

**Lemma 13.5.8.** *The anomalous Jacobian is given by*

$$\ln J(\chi) = 2i\,\mathrm{Tr}(\chi\gamma_5).$$

To compute the trace, we need a basis for the Hilbert space of classical Euclidean Fermi fields. Choosing the eigenfunctions $\varphi_k$ of $\slashed{D}_{\mathrm{E}}$ again, we find that

$$\mathrm{Tr}(\chi\gamma_5) = \sum_k \int d^4y\,\varphi_k^\dagger(y)\chi(y)\gamma_5\varphi_k(y). \tag{13.5.9}$$

This sum is divergent, and needs to be regularized. If, as before for the functional derivative, we damp the high frequency contributions by $\exp(-\lambda_k^2/M^2)$, we obtain:

$$\mathrm{Tr}(\chi\gamma_5) = \lim_{M\to\infty}\sum_k e^{-\lambda_k^2/M^2}\int d^4y\,\varphi_k^\dagger(y)\chi(y)\gamma_5\varphi_k(y). \tag{13.5.10}$$

This expression is linear in $\chi$, so we can set $\chi(y) = \delta^{(4)}(x-y)$ without loss of generality. The resulting right-hand side is exactly the expression evaluated in the last section in the proof of the abelian anomaly theorem. We conclude that

$$\mathrm{Tr}(\delta_x^{(4)}\gamma_5) = -\frac{1}{16\pi^2}\,\mathrm{tr}(G_{\mu\nu}\tilde{G}^{\mu\nu}), \tag{13.5.11}$$

and

**Theorem 13.5.12.** *The change of variables* $\psi'(y) = e^{-i\tau\delta^{(4)}(x-y)\gamma_5}\psi(y)$ *gives rise to a Jacobian* $(J_x)^\tau$ *in the change of functional measure, where the anomalous Jacobian* $J_x$ *is given by*

$$\ln J_x = -\frac{i}{8\pi^2}\,\mathrm{tr}(G_{\mu\nu}\tilde{G}^{\mu\nu}).$$

Differentiation of the generating functional also provides a means for linking $J_x$ and the Noether current. Again, let $\mathbf{D}_\tau$ be the operation of differentiating with respect to $\tau$ and then evaluating at $\tau = 0$. Then:

$$
\begin{aligned}
\mathbf{D}_\tau Z &= N\mathbf{D}_\tau\left(\frac{\partial}{\partial\psi'}\frac{\partial}{\partial\bar{\psi}'}\right)e^{iS(\psi')} \\
&= N\mathbf{D}_\tau\left(\frac{\partial}{\partial\psi}\frac{\partial}{\partial\bar{\psi}}\right)e^{iS(\psi')+\tau\ln J_x} \\
&= N\left(\frac{\partial}{\partial\psi}\frac{\partial}{\partial\bar{\psi}}\right)e^{iS(\phi)}\left(i\mathbf{D}_\tau S(\psi') + \ln J_x\right).
\end{aligned}
\tag{13.5.13}
$$

Since, of course, $Z$ is invariant, $\mathbf{D}_\tau Z = 0$, and we conclude that the functional Jacobian is determined by the $\tau$-dependence of the action:

$$\ln J_x = -i\langle \mathbf{D}_\tau S \rangle. \tag{13.5.14}$$

**Remark 13.5.15.** The relation between the anomalous Jacobian and the abelian anomaly is not a coincidence; it follows directly from differentiation of the generating functional with respect to the gauged axial change of basis 13.5.6.

Following the earlier result 13.2.9 on $\mathbf{D}_\tau \mathcal{L}$, we find that

$$i\mathbf{D}_\tau S = \int d^4 y \frac{\delta S}{\delta \psi} T \delta_x^{(4)} \psi - \bar{\psi} \delta_x^{(4)} \bar{T} \frac{\delta S}{\delta \bar{\psi}} + \partial_\mu (\chi j^\mu). \tag{13.5.16}$$

The integral of the divergence vanishes, and we find

$$\ln J_x = -\frac{\delta S}{\delta \psi} T \chi \psi + \bar{\psi} \chi \bar{T} \frac{\delta S}{\delta \bar{\psi}}. \tag{13.5.17}$$

Taking the expectation, using Lemma 13.2.8 to integrate the right-hand side by parts, and comparing the result with the anomaly formula in Lemma 13.2.11, we find that

$$\ln J_x = i\langle \partial_\mu j_5^\mu \rangle. \tag{13.5.18}$$

Thus Fujikawa's computation of the anomalous Jacobian is equivalent to and consistent with the calculation leading to Theorem 13.4.11 on the abelian anomaly.

Finally, we apply the anomalous Jacobian to show how an axial rotation $\psi' = e^{-i\tau\gamma_5}\psi$ effectively generates an anomalous term in the Lagrangian density. In this case, $\chi(y) = 1$, and the anomalous Jacobian $J_1$ is given by

$$\ln J_1 = \int d^4 x \ln J_x = -\frac{i}{8\pi^2} \int d^4 x \, \text{tr}(G_{\mu\nu}\tilde{G}^{\mu\nu}). \tag{13.5.19}$$

Since the action $S(\bar{\psi}, \psi)$ is invariant under this change of variables, there is no longer a term $\mathbf{D}_\tau S$ to uphold the invariance of the generating functional. Consequently,

$$\begin{aligned} Z_\tau &= N\Big(\frac{\partial}{\partial \psi'} \frac{\partial}{\partial \bar{\psi}'}\Big) e^{iS(\bar{\psi}', \psi')} \\ &= N\Big(\frac{\partial}{\partial \psi} \frac{\partial}{\partial \bar{\psi}}\Big) e^{iS(\bar{\psi}, \psi) + \tau \ln J_1}. \end{aligned} \tag{13.5.20}$$

On account of the form of $\ln J_1$, we can absorb its contribution to the action into the Lagrangian density:

**Theorem 13.5.21.** *If $\mathcal{L}(\bar{\psi}, \psi)$ is invariant under the axial rotation $\psi' = e^{-i\tau\gamma_5}\psi$, then the change of basis $\psi \to \psi'$ in the functional derivative formula for the generating functional is equivalent to replacing the Lagrangian density $\mathcal{L}$ by*

$$\mathcal{L}_\tau(\bar{\psi}, \psi) \stackrel{\text{def}}{=} \mathcal{L}(\bar{\psi}, \psi) - \frac{\tau}{8\pi^2} \text{tr}(G_{\mu\nu}\tilde{G}^{\mu\nu}).$$

The definition of the anomalous Jacobian shows how a simple rotation in a gauge theory can modify an action through the subtraction of a term $\tau \ln J$. Even when we can rotate the term away, the anomalous Jacobian, being proportional to the divergence of the quantum current, can still contribute to physical processes. For example, as we shall see in Section 16.12, the anomalous divergence of the axial current provides the primary contribution to the process $\pi^0 \to 2\gamma$.

In the previous section, anomalies emerged from the breakdown of the classical evolution equations. In this section, we have found the same anomalies emerging from the functional measure for the fermions. In both cases, the anomaly ultimately arises from regularization of the covariant derivative acting on the fermions. In the next section, we present a practical test for a theory to be free from anomalies.

## 13.6 Local Anomaly Theorems

From the previous sections, we see that it is the fermionic structure of a gauge theory which causes breakdown of classical symmetries, and that the essential problem is the regularization of the covariant derivative. Thus, in the fermionic functional differentiation principle 11.11.24,

$$\left(\frac{\partial}{\partial \psi} \frac{\partial}{\partial \bar\psi}\right) e^{-i\bar\psi C \psi} = \det(C), \tag{13.6.1}$$

the meaning of $\det(C)$ is actually unclear when $C = -i\not{D}$; we cannot treat this determinant as an undefined constant, because it is a non-trivial function of the gauge fields. However, any regularization of $-i\not{D}$ which makes the determinant meaningful also endangers the classical conservation laws. This explains why anomalies only arise in theories where fermions couple to gauge fields.

To conclude our presentation of local anomalies, it remains to generalize the abelian anomaly theorem to the non-abelian case, and, because of the threat anomalies pose to consistency, to provide a test for freedom from anomalies.

In this section, we present three anomaly theorems. The results depend on whether the left and right fermions have identical couplings to the gauge fields or not. In the two simplest cases, the gauge fields couple to the vector current $\bar\psi\gamma^\mu\psi$ and the chiral current $\bar\psi_L\gamma^\mu\psi_L$. The first theorem covers vector coupling, and the second covers chiral coupling. The third theorem, bringing out a point in common to the previous two, provides a general test for the presence of anomalies.

As explained in Section 12.6, for a minimally-coupled gauge theory, conservation of the Noether current is classically equivalent to covariant conservation of the matter current. To be in better accord with other texts, we modify the normalization 12.6.17 of the matrices of currents by a factor of $ig^{-1}$:

$$j^\mu \overset{\text{def}}{=} j_a^\mu T^a, \quad \text{and} \quad J^\mu \overset{\text{def}}{=} J_a^\mu T^a. \tag{13.6.2}$$

Then the classical evolution equations imply that

$$\partial_\mu j^\mu = D_\mu J^\mu \overset{\text{def}}{=} \partial_\mu J^\mu + [A_\mu, J^\mu]. \tag{13.6.3}$$

The first theorem covers theories like QED in which the gauge fields couples to a fermionic vector current. It is derived from the trace formula 13.2.11 using the techniques that lead to Theorem 13.4.11 on the abelian anomaly. In this theorem, the global symmetries are chiral while the potential gauge symmetries are the vector subgroup of the global symmetries.

**Theorem 13.6.4.** *Let* $\mathcal{L} = \bar{\psi}i\slashed{D}\psi$, *where* $\slashed{D}$ *is free from* $\gamma_5$'s. *The global symmetries of* $\mathcal{L}$ *are* $U(n)_L \times U(n)_R$. *Assume that the symmetry generators* $T^a$ *of* $U(n)$ *are normalized by* $\mathrm{tr}(T^aT^b) = \frac{1}{2}\delta^{ab}$. *Then, after quantization, the matrix* $J^\mu = (\bar{\psi}\gamma^\mu T^a\psi)T^a$ *of matter vector currents satisfies:*

$$D_\mu J^\mu = 0.$$

*The classical chiral symmetry defined by* $\psi \to e^{-i\tau T\gamma_5}\psi$ *determines a matrix* $J_5^\mu = (\bar{\psi}\gamma^\mu T^a\gamma_5\psi)T^a$ *of matter chiral currents which, after quantization, satisfies:*

$$(D_\mu J_5^\mu)^a = -\frac{1}{8\pi^2}\mathrm{tr}(T^a G_{\mu\nu}\tilde{G}^{\mu\nu}).$$

Thus the left and right contributions to the matter currents coupled to the gauge fields cancel, the anomalies in the axial currents are covariant, and the gauge symmetry is not broken.

Note that neither this theorem nor the following one specifies which generators are gauged. If none are gauged, then $G_{\mu\nu}^a$ is identically zero and even the axial currents are conserved.

**Homework 13.6.5.** Check that the abelian anomaly is correctly included in the conclusion of Theorem 13.6.4. Note the normalization of the generators.

The second anomaly theorem covers theories in which the gauge fields couple only to fermions of one chirality. Here all the global symmetries are potential gauge symmetries.

**Theorem 13.6.6.** *Let* $\mathcal{L}_L = \bar{\psi}_L i\slashed{D}\psi_L$. *Assume that the symmetry generators* $T^a$ *are normalized by* $\mathrm{tr}(T^aT^b) = \frac{1}{2}\delta^{ab}$. *Then, after quantization, the matrix* $J_L^\mu = (\bar{\psi}_L\gamma^\mu T^a\psi_L)T^a$ *of matter currents satisfies:*

$$(D_\mu J_L^\mu)^a = -\frac{1}{24\pi^2}\epsilon^{\mu\nu\sigma\tau}\partial_\mu \mathrm{tr}\left(T^a\left(A_\nu\partial_\sigma A_\tau + \tfrac{1}{2}A_\nu A_\sigma A_\tau\right)\right).$$

*(For the Lagrangian density for right-handed fermions, drop the factor of* $-1$ *on the right.)*

Note that the value of $D_\mu J_L^\mu$, the *chiral anomaly*, is not covariant, and if it does not vanish for group-theoretic (or other) reasons, then it will destroy gauge invariance and render the quantum theory inconsistent.

**Homework 13.6.7.** Convert the left-chiral anomaly above into the right-chiral anomaly. In detail, working in the Majorana representation of the Dirac algebra and defining $\psi_R$ by $\psi_R = i\psi_L^*$ enables us to write the Lagrangian density $\mathcal{L}_L$ in terms of right-handed fields. Deduce that:

1. $T_R^a = -T_L^{a*}$.

2. $J_L^{a\mu} \overset{\mathrm{def}}{=} \bar{\psi}_L\gamma^\mu T_L^a\psi_L = \bar{\psi}_R\gamma^\mu T_R^a\psi_R \overset{\mathrm{def}}{=} J_R^{a\mu}$.

3. $(D_\mu J_L^\mu)^a = (D_\mu J_R^\mu)^{a*}$.

4. $\mathrm{tr}\left(T_L^a(A_L^\nu\partial^\sigma A_L^\tau + \tfrac{1}{2}A_L^\nu A_L^\sigma A_L^\tau)\right) = -\mathrm{tr}\left(T_R^a(A_R^\nu\partial^\sigma A_R^\tau + \tfrac{1}{2}A_R^\nu A_R^\sigma A_R^\tau)\right)^*$.

This permits us to work with a Lagrangian density $\mathcal{L} = \mathcal{L}_L + \mathcal{L}_R$ and to apply Theorem 13.6.6 to both the left and right currents.

There appears to be some inconsistency between these two anomaly theorems. Though classically the axial current in the first theorem is the difference of the axial currents in the second, the axial anomaly in the first is not the difference of the axial anomalies in the second. The reason for this is that the regularization depends on the theory. The fundamental distinction between the various currents involved is whether or not they can be obtained by varying the action with respect to a field. If they can, then they must satisfy an integrability condition, first formulated by Wess and Zumino. In the first theorem, the gauge currents are obtained by varying the action with respect to a field, but the axial currents cannot be so obtained. Here, we use a regularization which preserves this property of the gauge fields and obtain a *covariant anomaly* in the axial currents. In the second theorem, the chiral currents can be obtained by varying the action, so we use a regularization which upholds this property and derive a *consistent anomaly* in the axial currents.

By inspection of the anomalies in the two theorems above, we can see that the symmetries of the operator terms enable us to use the Lie bracket to factor out a coefficient $\text{tr}(T^a\{T^b, T^c\})$. In the second theorem, taking advantage of Homework 13.6.7, we can combine left and right anomalies by writing a general Lagrangian density in terms of left-handed fields $\psi_L$ and $\psi_{cL}$. Consequently, we can summarize the conditions under which a gauge theory is anomaly free as follows:

**Theorem 13.6.8.** *Let $\mathcal{L}$ be a Lagrangian density for a classical gauge theory with gauge potential $A$, fermions $\psi$, and scalars $\phi$. Let $\mathcal{G}$ be the Lie algebra of generators of global, internal symmetries of $\mathcal{L}$, and let $T_L^a$ and $T_{cL}^a$ be the representations of $\mathcal{G}$ on the left Weyl spinors $\psi_L$ and $\psi_{cL}$ respectively. Then:*

1. *A vector quantum theory is free from chiral anomalies if and only if*

$$\text{tr}(T_L^a\{T_L^b, T_L^c\}) = 0;$$

2. *A chiral quantum theory is free from anomalies if and only if*

$$\text{tr}(T_L^a\{T_L^b, T_L^c\}) + \text{tr}(T_{cL}^a\{T_{cL}^b, T_{cL}^c\}) = 0;$$

*A vector or chiral theory is free from anomalies in the gauge currents if and only if the same condition holds for the gauged generators.*

**Homework** 13.6.9. Demonstrate that $\text{tr}(T^a\{T^b, T^c\})$ is a factor in both anomalies. In light of the normalization $\text{tr}(T^aT^b) = \frac{1}{2}\delta^{ab}$, the following *d-f* notation is convenient:

$$\left.\begin{array}{l} \text{tr}(T^a[T^b, T^c]) = 2if^{abc} \\ \text{tr}(T^a\{T^b, T^c\}) = 2d^{abc} \end{array}\right\} \quad\Longleftrightarrow\quad \text{tr}(T^aT^bT^c) = d^{abc} + if^{abc}.$$

There are representations for which the trace condition of Theorem 13.6.8 is automatically satisfied. In particular, the real representations have this property:

**Definition 13.6.10.** *A hermitian representation of a Lie algebra $\mathcal{G}$ is a real representation if there exists a similarity transform $S$ for which $STS^{-1} = -T^*$ for all representation matrices $T$.*

Clearly, for a real representation, $\text{tr}(T^a\{T^b, T^c\}) = 0$. All hermitian representations of the groups $SU(2)$, $SO(n)$ for $n \neq 4$, and the symplectic groups, are real.

For other groups, the trace condition is a constraint on the matter representations in the theory.

**Homework** 13.6.11. Show that when the gauge fields couple to vector currents, then the representation of the gauged symmetries on the left Weyl fields is real.

**Remark** 13.6.12. Perturbation theory confirms the structure of the anomalies presented here, at least at low order. For simplicity of exposition, suppose that we have gauged all the symmetries so that there are gauge fields $A_\mu^a$ and $A_{5\mu}^a$ associated with the generators $T^a$ and $\gamma_5 T^a$ respectively. Then the anomaly arises from the interaction of three gauge fields through loop diagrams. The simplest such diagrams are the triangles:

$$\text{(13.6.13)}$$

where the matrices $M^a$ could be either $-i\lambda T^a$ or $-i\lambda\gamma_5 T^a$.

These triangle diagrams give rise to divergent Feynman integrals. Renormalization conditions which support conservation of both vector and axial vector currents turn out to be inconsistent. In practice, generally only the generators $T^a$ are gauged, and so for the consistency of the gauge theory, it is necessary to maintain conservation of the vector currents. The outcome is an anomalous divergence in the axial currents.

If we write out the Feynman integrals, it is easy to show that the sum of the two integrals is proportional to $\text{Tr}\big(M^a\{M^b, M^c\}\big)$. Consequently, the analysis of triangle diagrams supports the test for anomalies proposed in Theorem 13.6.8.

The most significant implication of this section is that the first kind of gauge invariance, invariance under infinitesimal gauge transformations, may not survive quantization. If it does not survive, then the Faddeev–Popov principle does not apply and the theory is inconsistent. Thus it is essential to verify that all combinations of generators which could contribute to symmetry-breaking chiral anomalies satisfy the trace constraint presented in the final theorem above.

This concludes our introduction to anomalies generated by infinitesimal gauge transformations. The following sections consider gauge transformations that cannot be obtained by exponentiating infinitesimal ones, and show how they can be associated with what might be called 'anomalous vacuum states.'

# 13.7 Basic Topology for the Discussion of Global Anomalies

Modulo $U(1)$ factors, the common symmetry groups in field theory are products of simple groups like $SO(n)$ and $SU(n)$. In this and the following sections, we shall show that the gauge group $\tilde{G}$ and the space **A** of gauge potentials associated with a simple group $G$ both comprise countably many components, that the different components of **A** represent evolution between countably many distinct vacuum states, that the physical vacuum of a gauge invariant theory must be a superposition of these vacua, and that the composite structure of the physical vacuum brings an extra term, the global anomaly, into a Faddeev–Popov Lagrangian density.

Subsequent analysis depends upon a few notions of topology and global analysis. Topology is the study of continuity both on the level of topological spaces and on the level of functions between topological spaces. For the following discussion, think of a topological space as a subset of either Euclidean space or of a function space. The primary purpose of topology is to find numerical invariants of topological spaces on the basis of their continuous structure. Global analysis, however, adds differentiable structure to the topological spaces and has a primary purpose in finding integral formulae for topogical invariants. In this section, we shall first introduce a few concepts from topology and then develop some of the associated global analysis.

Connectedness is a fundamental notion in topology. For example, any point in the group $U(n)$ can be connected by a continuous arc to the identity element, any point in $O(n)$ can be so connected to $\pm 1$, but no continuous arc in $O(n)$ connects **1** to $-\mathbf{1}$. The associated definitions and propositions follow.

**Definition 13.7.1.** A topological space is *arcwise connected* if any two points in the space can be connected by a continuous arc.

**Definition 13.7.2.** An *arcwise-connected component* of a topological space $X$ is any set of points $C[x]$ in $X$ which are all connected by continuous arcs to any fixed point $x \in X$.

**Lemma 13.7.3.** $C[C[x]] = C[x]$.

**Corollary 13.7.4.** $C[x]$ and $C[y]$ intersect if and only if they are equal.

**Theorem 13.7.5.** *Any topological space is a disjoint union of its arcwise-connected components.*

Thus, for example, $U(n)$ has one arcwise-connected component, $O(n)$ has two, and $O(1,3)$ has four. The number of components of a topological space is a topological invariant, meaning that this number is not changed by any continuous, invertible transformation of the space.

Winding number is another fundamental topological concept. For example, if we throw a loop of string onto a plane and pick a point in the plane not on the string, then the string will wind around that point a definite number of times. This winding number will not change under continuous motions of the string as long as such motions do not take the string across the chosen point.

The winding number of a string about a point can also be defined analytically using the complex plane. Represent the string by a differentiable function $\gamma$ from the unit interval to $\mathbf{C}$ which satisfies $\gamma(0) = \gamma(1)$. Let $a \in \mathbf{C}$ be a point not in the image of $\gamma$. Then the winding number of $\gamma$ around $a$ is given by:

$$w(\gamma, a) \stackrel{\text{def}}{=} \frac{1}{2\pi i} \int_\gamma \frac{dz}{z - a}. \tag{13.7.6}$$

The winding number is a topological invariant which depends only on continuity; the analytic formula computes this topological index using differentiable functions. This example illustrates the general relationship between topology and global analysis.

The concept of winding number generalizes to higher dimensions. For example, any function from the $n$-dimensional sphere $S^n$ to itself can always be deformed continuously into one which clearly wraps the sphere around itself a definite number of times.

The natural concept of a deformation of a function $f$ into a function $g$ is represented by a choice of arc from $f$ to $g$ in the appropriate function space. Thus a function can be deformed into any other function in the same arcwise-connected component of the underlying function space. This understanding of deformations leads to the Hopf degree theorem which links winding numbers and components of function spaces as follows:

**Theorem 13.7.7.** *Let* $\text{Map}(S^n, S^n)$ *be the space of continuous functions from* $S^n$ *to itself; let* $[S^n, S^n]$ *be the set of arcwise-connected components of* $\text{Map}(S^n, S^n)$. *Then the winding number is constant on each arcwise-connected component of* $\text{Map}(S^n, S^n)$ *and provides a bijective map between* $[S^n, S^n]$ *and the integers* $\mathbf{Z}$.

The Hopf theorem 13.7.7 for functions from spheres to themselves extends to functions from the three-sphere to arcwise-connected simple groups as follows:

**Theorem 13.7.8.** *Let* $G$ *be an arcwise-connected simple group, let* $\text{Map}(S^3, G)$ *be the space of continuous functions from* $S^3$ *to* $G$, *and let* $[S^3, G]$ *be the set of arcwise-connected components in* $\text{Map}(S^3, G)$. *Then the winding number is constant on each arcwise-connected component of* $\text{Map}(S^3, G)$ *and provides a bijective map between* $[S^3, G]$ *and the integers* $\mathbf{Z}$.

It is in fact possible to deform any map from $S^3$ into an arcwise-connected simple group $G$ into a map from $S^3$ into an $SU(2)$ or $SO(3)$ subgroup of $G$. Thus the winding number of the theorem above is effectively the winding number of a map from $S^3$ to itself.

In particular, since $SU(2)$ is naturally the sphere $S^3$,

$$SU(2) = \left\{ \begin{pmatrix} w + ix & y + iz \\ -y + iz & w - ix \end{pmatrix} \;\middle|\; \begin{array}{l} (w, x, y, z) \in \mathbf{R}^4 \text{ and} \\ w^2 + x^2 + y^2 + z^2 = 1 \end{array} \right\}, \tag{13.7.9}$$

functions from $S^3$ to itself or to any simple group with any winding number can be constructed from the following simple fact:

**Theorem 13.7.10.** *For each integer* $k$, *the map* $g \mapsto g^k$ *from* $SU(2)$ *to itself has winding number* $k$.

This concludes the introduction to topology. The concepts of arcwise connectedness and winding number are necessary and sufficient foundation for the coming discussion of the gauge group and the space of gauge potentials. Now we turn to global analysis for some integral formulae for winding numbers and related topological invariants.

For the purposes of global analysis, it is convenient to regard the sphere $S^n$ as the union of Euclidean space $\mathbf{R}^n$ and a single point at infinity, usually identified with the north pole of the sphere. Indeed, stereographic projection from the north pole $N$ is a coordinate map from $S^n \setminus \{N\}$ to $\mathbf{R}^n$. In these coordinates, a continuous function on the sphere becomes a continuous function on $\mathbf{R}^n$ which has a direction-independent limit as the radius goes to infinity.

Using $\mathbf{R}^3$ as coordinates for $S^3$, global analysis provides a remarkable formula for the winding number of Theorem 13.7.8:

**Theorem 13.7.11.** *Let* $\tilde{g}$ *be a differentiable function from* $S^3$ *to an arcwise-connected simple group* $G$. *Then the winding number of* $\tilde{g}$ *is given by*

$$w(\tilde{g}) = \frac{1}{24\pi^2} \int d^3\bar{x} \, \epsilon^{rst} \, \text{tr}(A_r A_s A_t),$$

where $A_r = -(\partial_r \tilde{g})\tilde{g}^{-1}$.

To make the notion of winding number precise while avoiding an overdose of topological terminology, we can take the equation of this theorem as the definition of the winding number.

Some topological indices are harder to visualize than the winding number and are best defined through global analysis. Let $A$ be a gauge potential on $S^4 = \mathbf{R}^{1,3} \cup \{\infty\}$ with values in the Lie algebra of $G$, and let $G_{\mu\nu}(A)$ be the field strength of $A$. Then the following integral defines the Pontrjagin index $p(A)$ associated with $A$:

$$p(A) \overset{\text{def}}{=} -\frac{1}{16\pi^2} \int d^4x \, \text{tr}(G_{\mu\nu}\tilde{G}^{\mu\nu}). \tag{13.7.12}$$

The integrand may be called the *Pontrjagin class*. As $A$ is defined on $S^4$, the coordinate form of $A$ vanishes sufficiently fast at space-time infinity to make the integral converge. The following proposition brings out the significance of $p(A)$:

**Theorem 13.7.13.** *If $A(t)$ is a continuous function from $\mathbf{R}$ to gauge potentials on $S^4$, then the Pontrjagin index $p(A(t))$ is constant.*

**Homework 13.7.14.** Prove the preceding theorem by differentiating $p(A(t))$ with respect to time and using the Bianchi identities $\epsilon^{\mu\nu\rho\sigma}D_\nu G_{\rho\sigma} = 0$.

From the definition of $G_{\mu\nu}$ and the Bianchi identities (see previous homework), it is simple to verify that the Pontrjagin class is a divergence:

$$\text{tr}(G_{\mu\nu}\tilde{G}^{\mu\nu}) = \partial_\mu \epsilon^{\mu\nu\rho\sigma} \, \text{tr}\left(G_{\rho\sigma}A_\tau - \frac{2}{3}A_\rho A_\sigma A_\tau\right). \tag{13.7.15}$$

The operand of $\partial_\mu$ is called a *Chern–Simons term*. Non-zero values of $p(A)$ arise from surface integrals of the Chern–Simons term which do not necessarily vanish at space-time infinity.

These three points, the formulae for the winding number, the Pontrjagin index, and the relationship between the Chern–Simons term and the Pontrjagin class, are all we need from global analysis.

**Homework 13.7.16.** Prove the equality 13.7.15.

**Homework 13.7.17.** Show that the Chern–Simons term may also be written

$$2\epsilon^{\mu\rho\sigma\tau} \, \text{tr}\left(A_\rho \partial_\sigma A_\tau + \frac{2}{3}A_\rho A_\sigma A_\tau\right).$$

**Homework 13.7.18.** Let $\tilde{g}$ be an element of the gauge group. Let $\tilde{g}_t$ be the function on $\mathbf{R}^3$ determined by evaluating $\tilde{g}$ at a fixed time $t$. By inspection of the Chern–Simons term and the Pontrjagin class, show that $w(\tilde{g}_t)$ is independent of $t$.

This concludes the introduction to topology and global analysis. We are only interested in winding numbers of functions from $S^3$ to simple groups. These numbers we have seen from the perspectives of both topology and global analysis. The Chern–Simons term has been introduced because it provides a link between the winding number and the Pontrjagin index: compare Theorem 13.7.11 with 13.7.15, and note the possibility for expressing the Pontrjagin index as a three-dimensional integral of the Chern–Simons term. The application to quantum gauge theory in the following sections shows how the winding number characterizes initial and final vacuum configurations, how the Pontrjagin index represents the difference between the two, and how the Pontrjagin class becomes part of the Faddeev–Popov Lagrangian density.

## 13.8  The Topology of the Gauge Group

The first stage of application to gauge theories of the foregoing points from topology and global analysis is in the investigation into the topology of the gauge group.

The gauge group $\tilde{G}$ derived from a group $G$ consists of those functions from space-time to $G$ which are constant at infinity:

$$\tilde{G} \stackrel{\text{def}}{=} \left\{ g \colon \mathbf{R}^{1,3} \to G \;\middle|\; \lim_{|\bar{x}| \to \infty} g(t, \bar{x}) \text{ exists for all } t \right\}. \qquad (13.8.1)$$

Since the sphere $S^3$ may be regarded as $\mathbf{R}^3$ extended by a point at infinity, the condition on gauge group elements at infinity enables us to regard these elements as time-dependent functions on $S^3$, that is, as arcs in $\text{Map}(S^3, G)$. Since spacelike surfaces in space-time may well be spheres, the condition that gauge transformations be constant at infinity may be physically natural. Technically, however, this condition is essential if functions derived from the gauge group and its representations are to have convergent integrals.

The boundary condition in the definition of the gauge group is also responsible for some topological structure. Since a gauge group element $\tilde{g}$ is an arc in $\text{Map}(S^3, G)$, the image of $\tilde{g}$ lies entirely in some arcwise-connected component of this space. Hence the winding number of $\tilde{g}$ at time $t$ is independent of $t$. This observation suggests extending the notion of winding number to gauge group elements and introducing a decomposition of the gauge group as follows:

**Definition 13.8.2.** The *winding number* $w(\tilde{g})$ of a gauge group element $\tilde{g}$ is the number defined by evaluating the integral of Theorem 13.7.11 on $\tilde{g}$ at any fixed time.

**Definition 13.8.3.** Write $\tilde{G}_k$ for the subset of $\tilde{G}$ of gauge group elements with winding number $k$.

We now show that the $\tilde{G}_k$ are the arcwise-connected components of $\tilde{G}$:

**Lemma 13.8.4.** *Any gauge group element $\tilde{g}$ is arcwise-connected to the time-independent gauge group element $\tilde{g}_0$ defined by $\tilde{g}_0(x) = \tilde{g}(0, \bar{x})$.*

**Proof.** To establish the proposition, we simply need to define an arc leading from $\tilde{g}$ to $\tilde{g}_0$. Since $\tilde{g}_0^{-1} \tilde{g}$ is a gauge group element which is the identity at time zero, we can find a field of generators $\tilde{\chi}(t, \bar{x})$ such that

$$\tilde{g}_0^{-1}(x) \tilde{g}(x) = \exp\big(\tilde{\chi}(x)\big). \qquad (13.8.5)$$

The boundary conditions on $\tilde{g}$ imply that $\tilde{\chi}$ is a bounded function on Minkowski space. Hence we can define a continuous arc $\tilde{g}_\lambda$ connecting $\tilde{g}_0$ to $\tilde{g}$ as follows:

$$\tilde{g}_\lambda(x) \stackrel{\text{def}}{=} \tilde{g}_0(x) \exp\big(\lambda \tilde{\chi}(x)\big). \qquad \square$$

**Theorem 13.8.6.** *The subsets $\tilde{G}_k$ of $\tilde{G}$ comprising all gauge transformations $\tilde{g}$ which have winding number $k$ are the arcwise-connected components of $\tilde{G}$.*

**Proof.** Clearly, if $k \neq l$, no element of $\tilde{G}_k$ can be connected by an arc in $\tilde{G}$ to any element in $\tilde{G}_l$. Hence we need only show that if $\tilde{g}$ and $\tilde{g}'$ are elements of $\tilde{G}_k$, then there is an arc in $\tilde{G}$ connecting $\tilde{g}$ to $\tilde{g}'$. From the lemma above, we have arcs connecting $\tilde{g}$ and $\tilde{g}'$ to $\tilde{g}_0$ and $\tilde{g}'_0$ respectively. It therefore remains to connect $\tilde{g}_0$ to $\tilde{g}'_0$. However, $\tilde{g}_0$ and $\tilde{g}'_0$ are elements of $\mathrm{Map}(S^3, G)$, and, as they have the same winding number, Theorem 13.7.7 shows that they are in the same arcwise-connected component of $\mathrm{Map}(S^3, G)$. But $\mathrm{Map}(S^3, G)$ may be considered as the set of time-independent gauge group elements, so an arc in a component of $\mathrm{Map}(S^3, G)$ is also an arc in $\tilde{G}$. □

The winding-number decomposition

$$\tilde{G} = \bigcup_{k=-\infty}^{\infty} \tilde{G}_k \tag{13.8.7}$$

is therefore the topological decomposition of the gauge group $\tilde{G}$ into its arcwise-connected components. That is all we need to know.

# 13.9 The Topology of the Space of Gauge Potentials

The second stage of application of topology and global analysis to gauge theories is in the investigation into the topology of the space of gauge potentials.

From a gauge group element $\tilde{g}$ we can define a gauge potential or connection one-form $A(\tilde{g})$ by

$$A_\mu(\tilde{g}) \stackrel{\text{def}}{=} -(\partial_\mu \tilde{g})\tilde{g}^{-1}. \tag{13.9.1}$$

Since $A(\tilde{g})$ can be reduced to $A = 0$ by a gauge transformation, $A(\tilde{g})$ is said to be a *pure-gauge connection one-form*. The curvature (field strength) $G_{\mu\nu}$ of $A(\tilde{g})$ is zero. However, since $A(\tilde{g})$ determines the winding number $w(\tilde{g})$ of $\tilde{g}$ through the formula of Theorem 13.7.11, these pure-gauge connection one-forms do have some individual character.

To make the total energy associated with a gauge potential finite, we restrict the space of gauge potentials $A$ by the following boundary conditions:

**Definition 13.9.2.** The space **A** of connection one-forms $A$ is the subset of Lie algebra-valued functions on Minkowski space which vanish at spatial infinity and tend to pure gauge configurations at temporal infinity. Thus, for every $A \in \mathbf{A}$, we have two gauge-group elements $\tilde{g}_\pm$ such that

$$\left.\begin{array}{l} \displaystyle\lim_{t \to -\infty} A_\mu(t, \bar{x}) = -\big(\partial_\mu \tilde{g}_-(\bar{x})\big)\tilde{g}_-(\bar{x})^{-1} \\[2mm] \displaystyle\lim_{t \to \infty} A_\mu(t, \bar{x}) = -\big(\partial_\mu \tilde{g}_+(\bar{x})\big)\tilde{g}_+(\bar{x})^{-1} \end{array}\right\}$$

From the boundary conditions, it is clear that a gauge potential $A$ in the space **A** determines initial and final winding numbers:

**Definition 13.9.3.** Define $w_-(A)$ and $w_+(A)$ by $w_\pm(A) \overset{\text{def}}{=} w(\tilde{g}_\pm)$.

Since, as indicated by Homework 13.7.18, continuous deformation of $A$ cannot change these winding numbers, we find that the space of gauge potentials is a union of subspaces which cannot be connected by arcs:

**Definition 13.9.4.** Write $\mathbf{A}_{m,n}$ for the subspace of $\mathbf{A}$ whose elements have initial and final winding numbers $m$ and $n$ respectively.

Gauge potentials generally form arcwise-connected spaces. Here, however, the boundary conditions in Definition 13.9.2 force elements of $\mathbf{A}$ to be in one of the subspaces $\mathbf{A}_{m,n}$; linear combinations across these subspaces are not contained in $\mathbf{A}$. In fact, the sets $\mathbf{A}_{m,n}$ are the arcwise-connected components of $\mathbf{A}$:

**Theorem 13.9.5.** *The decomposition*

$$\mathbf{A} = \bigcup_{m=-\infty}^{\infty} \bigcup_{n=-\infty}^{\infty} \mathbf{A}_{m,n}$$

*is the topological decomposition of $\mathbf{A}$ into arcwise-connected components.*

**Proof.** Assume that $A^0$ and $A^1$ are both in $\mathbf{A}_{m,n}$. Let $\tilde{g}_\pm^0$ and $\tilde{g}_\pm^1$ be time-independent gauge transformations provided by the boundary conditions of Definition 13.9.2 on $A^0$ and $A^1$ respectively. Then $\tilde{g}_+^0$ and $\tilde{g}_+^1$ lie in the same component $\tilde{G}_n$ of the gauge group. Consequently, they can be connected by an arc $\tilde{g}_+^\tau$. Clearly, this arc can be an arc of time-independent gauge transformations. Similarly, we can construct an arc $\tilde{g}_-^\tau$ to connect $\tilde{g}_-^0$ and $\tilde{g}_-^1$.

Now $\tilde{g}_+^\tau(\tilde{g}_+^0)^{-1}$ and $\tilde{g}_-^\tau(\tilde{g}_-^0)^{-1}$ are arcs in the identity component $\tilde{G}_0$ of the gauge group which start at the identity element when $\tau = 0$. Obviously it is possible to find an arc of time-dependent gauge transformations $\tilde{g}_t^\tau$ which fills in the triangle between these two arcs:

$$\lim_{t \to -\infty} \tilde{g}_t^\tau = \tilde{g}_-^\tau(\tilde{g}_-^0)^1 \quad \text{and} \quad \lim_{t \to \infty} \tilde{g}_t^\tau = \tilde{g}_+^\tau(\tilde{g}_+^0)^1. \tag{13.9.6}$$

Let us choose to make $\tilde{g}_t^0 = 1$. Then, $\tilde{g}_t^\tau \cdot A^0$ takes the value $A^0$ at $\tau = 0$, and at $\tau = 1$ becomes a gauge potential $A^{1\prime}$ which has the same initial and final $\tilde{g}$'s as $A^1$.

Finally, the usual affine combination $A_\lambda = (1 - \lambda)A^{1\prime} + \lambda A^1$ can be used to connect $A^{1\prime}$ with $A^1$. $\qquad \square$

The gauge group acts on the gauge potentials in the usual way:

$$\tilde{g} \cdot A_\mu \overset{\text{def}}{=} \tilde{g} A_\mu \tilde{g}^{-1} - (\partial_\mu \tilde{g})\tilde{g}^{-1}. \tag{13.9.7}$$

When we consider the decomposition by winding numbers and recall Theorem 13.7.10 that the winding number of $g \mapsto g^k$ on $SU(2)$ is $k$, then we find that a gauge group element in $\tilde{G}_k$ transforms a gauge potential in $\mathbf{A}_{m,n}$ into one in $\mathbf{A}_{m+k,n+k}$:

$$\left. \begin{array}{ccc} \tilde{G}_k \times \mathbf{A}_{m,n} & \to & \mathbf{A}_{m+k,n+k} \\ \tilde{g} \ , \quad A & \mapsto & \tilde{g} \cdot A \end{array} \right\} \tag{13.9.8}$$

Hence a quantum field theory could be invariant under $\tilde{G}_0$ if we only integrate over $\mathbf{A}_{m,n}$, but if it is to be invariant under $\tilde{G}$, then we must integrate over $\bigcup_k \mathbf{A}_{m+k,n+k}$.

The structure of the space of gauge potentials is therefore simple: it is just a union of arcwise-connected components indexed by initial and final winding numbers. Now we see what it means to assert that the action of the gauge group can be considered on two levels. The local level, considered in the opening sections of this chapter, consists of infinitesimal gauge transformations and, through exponentiation, of $\tilde{G}_0$. The global level, considered in the next section, consists of what remains when these first level transformations are forgotten, namely the set of arcwise-connected components of $\tilde{G}$.

## 13.10 The Global Anomaly for Quantum Gauge Theories

Having presented the basic concepts of topology and global analysis, and having analyzed the topology of the gauge group and the space of gauge potentials, it remains to bring out the fruit of the line of thought, the consequences of this topological structure for quantum gauge theories. The point of focus which brings out the consequences is the physical vacuum.

Since the gauge potentials in $\mathbf{A}_{m,n}$ evolve from initial winding number $m$ to final winding number $n$, and since quantum fluctuations are small perturbations from a classical configuration, quantum fluctuations are presumably modeled by integration over $\mathbf{A}_{m,n}$ itself. However, if $m \neq n$, the initial and final state of the universe will be different even on the classical level. We therefore interpret a pure gauge configuration with winding number $n$ as an aspect of a vacuum state $|n\rangle$ for a quantum gauge theory.

From the action 13.9.8 of the components of the gauge group on the components of the space of gauge potentials, we see that

$$\tilde{G}_k|n\rangle = |n + k\rangle. \tag{13.10.1}$$

Note that the whole of $\tilde{G}_0$ acts trivially, so that the infinitesimal transformations are forgotten and we are properly on the global level of the action of the gauge group.

Therefore, no vacuum state is invariant under the whole gauge group. The best one can do is to construct $\theta$ vacua which are eigenstates of every gauge transformation:

$$|\theta\rangle \stackrel{\text{def}}{=} \sum_{n=-\infty}^{\infty} e^{in\theta}|n\rangle. \tag{13.10.2}$$

Then the elements of the gauge group act as follows:

$$\begin{aligned}
\tilde{G}_k|\theta\rangle &= \sum e^{in\theta}\tilde{G}_k|n\rangle \\
&= \sum e^{in\theta}|n + k\rangle \\
&= e^{-ik\theta}|\theta\rangle.
\end{aligned} \tag{13.10.3}$$

Since the $\theta$ vacua are the best candidates for physical vacua, we use one of these to define a fully gauge-invariant generating functional:

$$
\begin{aligned}
Z_\theta & \overset{\text{def}}{=} \lim_{t \to \infty} \langle \theta | e^{-iHt} | \theta \rangle \\
& = \sum_{m,n} e^{-i(n-m)\theta} \lim_{t \to \infty} \langle n | e^{-iHt} | m \rangle \\
& = N \sum_{m,n} e^{-i(n-m)\theta} \int (dA_{m,n}) \, \exp(iS_{\text{eff}}) \\
& = N \sum_{m,n} \int (dA_{m,n}) \, \exp\Big( iS_{\text{eff}} - i\theta \big( w_+(A_{m,n}) - w_-(A_{m,n}) \big) \Big).
\end{aligned}
\tag{13.10.4}
$$

Now finally, because of the zero boundary condition on $A_{m,n}$ at spatial infinity, we can use the winding-number formula 13.7.11, the Chern–Simons term 13.7.15, and the Pontrjagin index 13.7.12 to express $w_- - w_+$ as an integral over space-time:

$$
\begin{aligned}
w_+(A_{m,n}) - w_-(A_{m,n}) & = \frac{1}{24\pi^2} \int dt \, \partial_0 \int d^3\bar{x} \, \epsilon^{rst} \, \text{tr}(A_r A_s A_t) \\
& = -\frac{1}{16\pi^2} \int dt \int d^3\bar{x} \, \epsilon^{\mu\rho\sigma\tau} \partial_\mu \, \text{tr}\Big( G_{\rho\sigma} A_\tau - \frac{2}{3} A_\rho A_\sigma A_\tau \Big) \\
& = -\frac{1}{16\pi^2} \int d^4x \, \text{tr}(G_{\mu\nu} \tilde{G}^{\mu\nu}).
\end{aligned}
\tag{13.10.5}
$$

Thus the net effect of this analysis of the gauge group, the space of gauge potentials, and the vacuum state is that a Faddeev–Popov Lagrangian density must be modified to include a $\theta$-dependent *global anomaly*,

$$
\mathcal{L}_\theta \overset{\text{def}}{=} \mathcal{L}_{\text{FP}} + \frac{\theta}{16\pi^2} \, \text{tr}(G_{\mu\nu} \tilde{G}^{\mu\nu}).
\tag{13.10.6}
$$

We note that the global anomaly has the same structure as the axial anomaly in Theorem 13.6.4. Indeed, if there is a massless fermion which has vector couplings to the gauge bosons, then the axial rotation of Theorem 13.5.21 can be used to absorb the global anomaly.

Thus all the points in the section on topology and global analysis have been applied to support the conclusion that there is a one-parameter family of physical vacua and the Faddeev–Popov Lagrangian density depends on which vacuum our environment is in.

# 13.11 Summary

This chapter has probed quantum gauge theories using the formalism of functional integral quantization. The investigation followed the analysis of the gauge group: the first level of investigation was based on infinitesimal gauge transformations; the second, on arcwise-connected components of the gauge group.

The first conclusion is that a departure of operator evolution from the classical equations endanger classical conservation laws for theories in which gauge bosons are coupled to fermions. Remarkably, the Noether currents associated with gauge symmetries will nevertheless be conserved as long as the fermion representations in the theory satisfy a trace constraint. This allows for the existence of consistent, realistic gauge theories.

When a Noether current is not conserved, its divergence can be computed by regularizing either the functional derivative operator or equivalently the functional Jacobian. Such divergences can contribute to observable processes.

The second conclusion is that the global structure of any non-abelian gauge group compels us to model the physical vacuum by a superposition of vacua with pure winding numbers. The consequence of this is that the physical vacuum is not unique, and that the Faddeev–Popov Lagrangian density depends on the vacuum.

With this chapter, Part 3 of the text comes to rest. We have established the interacting field as the basis for quantization and succeeded in quantizing gauge theories. In Part 4, which comprises Chapters 14 to 17, we develop the Standard Model from its roots in the Weinberg–Salam Lagrangian density to its fruits in decay rates and scattering amplitudes.

# Chapter 14

# $SU(3)$ Representation Theory

Preparation for color symmetry of the strong interactions and flavor
symmetry of hadrons.

## Introduction

Having presented quantization of gauge field theories in Part 3, the primary purpose at this point is to introduce the Standard Model of electroweak and strong interactions. The Chapters 14 to 17 which constitute Part 4 introduce $SU(3)$ representation theory, the Standard Model of the strong and electroweak interactions, hadron physics, and applications of the Standard Model to decay rates and cross sections.

Since the strong interactions are described by an $SU(3)$ gauge theory, it is convenient to cover a small amount of $SU(3)$ representation theory now. The basic matter fields of the strong interactions are the six quarks, each of which is an $SU(3)$ vector of three Fermi fields. These quarks are never observed in isolation but only found in bound states called hadrons. Being bound states, the hadrons are far beyond the reach of perturbation theory. To a first approximation, their structure and properties are therefore deduced using representation theory as follows.

The quarks can be set in a sequence according to their effective masses in the hadrons; using the standard notation for the six quarks, the result is $u$, $d$, $s$, $c$, $b$, and $t$, representing the up, down, strange, charm, bottom, and top quarks respectively. The $u$ and $d$ quarks are nearly degenerate in mass, and so we can postulate an approximate $SU(2)$ symmetry of the strong interactions which acts on the doublet $(u, d)$; this symmetry is in fact evident in the properties of the hadrons. The $s$ quark is considerably more massive but is still light by hadron standards, and so we can also postulate a more approximate $SU(3)$ symmetry which acts on the vector $(u, d, s)$; this symmetry is also evident. The remaining quarks are too heavy, however, for the extended symmetries to be of much value for predicting hadron properties. Hadron structure, on the other hand, can be understood using any of these symmetries. If we select any set of $n$ quark types, then we find that the hadrons with zero orbital angular momentum made up of just these quarks form irreducible representations of $SU(n)$. We shall introduce the $uds$ hadrons in this way using $SU(3)$ symmetry.

Note that we have introduced two completely independent $SU(3)$'s here. They have nothing to do with each other! The gauge group is known as color $SU(3)$, written $SU(3)_c$, and mixes the three components (colors) of each quark field separately. The approximate $SU(n)$ symmetries are called flavor symmetries, written $SU(n)_f$, and mix the quark fields without mixing colors.

In preparation then for the strong interactions and for the subsequent investigation into the structure and properties of the hadrons, it will be convenient to establish a solid understanding of $SU(3)$ representation theory. For a concrete beginning, Section 14.1 introduces the topic using matrices, and Sections 14.2 and 14.3 explain the interpretation of $SU(3)$ as flavor and color. The remaining sections, Sections 14.4 to 14.7, provide the abstract details of three equivalent viewpoints:

weight diagrams, tensors, and Young diagrams. Weight diagrams provide the geometry of the representations and a basis for proof techniques which generalize to simple groups. Tensors provide a link between theory and applications and a basis for proof techniques which generalize to $SU(n)$. Young diagrams provides the most convenient algorithm for reducing tensor products of $SU(n)$ representations into direct sums of irreducible ones.

# 14.1 $SU(3)$ Representations of Small Degree

This section introduces $SU(3)$ representation theory on the basis of the vector and adjoint representations of $SU(3)$. Using these simplest representations as examples, it presents the Gell-Mann basis for $su(3)$ and the standard basis for a Cartan subalgebra, introduces weight diagrams, roots, conjugate representations, tensor products, and splitting of representations thereby showing how the concepts of Chapter 6 bear on $SU(3)$.

As with $SU(2)$ and the Lorentz group, analysis of representations is easier on the Lie algebra $su(3)$; fortunately representations $R$ of $SU(3)$ and $\rho$ of its Lie algebra are in one-to-one correspondence under the exponential map:

$$R\big(\exp(X)\big) = \exp\big(\rho(X)\big). \tag{14.1.1}$$

Following the method used to analyze $su(2)$ representations, we first construct a maximal commuting subalgebra (a Cartan subalgebra) of the Lie algebra $su(3)$ by choosing two hermitian generators of $SU(3)$ that commute:

$$\lambda_3 = \begin{pmatrix} 1 & 0 & 0 \\ 0 & -1 & 0 \\ 0 & 0 & 0 \end{pmatrix} \quad \text{and} \quad \lambda_8 = \frac{1}{\sqrt{3}} \begin{pmatrix} 1 & 0 & 0 \\ 0 & 1 & 0 \\ 0 & 0 & -2 \end{pmatrix}. \tag{14.1.2}$$

These generators correspond to the charge operators for the two independent $SU(3)$ quantum numbers. Second, we analyze any representation $\rho$ by diagonalizing the matrices $\rho(\lambda_3)$ and $\rho(\lambda_8)$. This simultaneous diagonalization corresponds to finding states or fields with pure $SU(3)$ quantum numbers in an $SU(3)$ multiplet.

For example, the identity (or vector) representation $\mathbf{1}(X) = X$ acts on the three-dimensional space $\mathbf{C}^3$ by matrix multiplication, and the eigenvectors of $\lambda_3$ and $\lambda_8$ are the standard basis $e_1$, $e_2$, $e_3$ for $\mathbf{C}^3$. The three pairs of eigenvalues, the weights of $\mathbf{1}$, form an equilateral triangle:

$$\text{Weights of } \mathbf{1}: \quad \left(1, \tfrac{1}{\sqrt{3}}\right), \ \left(-1, \tfrac{1}{\sqrt{3}}\right), \ \left(0, -\tfrac{2}{\sqrt{3}}\right). \tag{14.1.3}$$

(Of course, to obtain convenient quantum numbers we change the scale of $\lambda_8$.) This triangle uniquely identifies the vector representation of $su(3)$.

In general, if the structure constants $f_{abc}$ of a real Lie algebra $\mathcal{G}$ are real,

$$[X_a, X_b] = f_{abc} X_c \quad \Longrightarrow \quad [X_a^*, X_b^*] = f_{abc} X_c^*. \tag{14.1.4}$$

Hence, if $\rho$ is any representation of $\mathcal{G}$ on a multiplet of fields, then the conjugate map $\rho^*$ defined by $\rho^*(X) = \rho(X)^*$ is a representation of $\mathcal{G}$ on the multiplet of anti-fields.

We use the complex linearity of $\rho$ and $\rho^*$ to extend this observation to the complex Lie algebra $\mathbf{C} \otimes \mathcal{G}$. Thus, if $H = iX$ is a hermitian generator of $\mathcal{G}$, then the correct formula for $\rho^*(H)$ is

$$\rho^*(H) = \rho^*(iX) = i\rho^*(X) = i\rho(X)^*$$
$$= i\rho(-iH)^* = i\left(-i\rho(H)\right)^* = -\rho(H)^*. \tag{14.1.5}$$

Conjugation therefore changes the sign of all the quantum numbers (weights) in a representation, and the weight diagram of $\rho^*$ is obtained from the weight diagram of $\rho$ by reflecting all the weights in the origin.

Thus, for a second example, the weights of the $su(3)$ representation $\mathbf{1}^*$ are given by:

$$\text{Weights of } \mathbf{1}^*: \quad \left(-1, -\tfrac{1}{\sqrt{3}}\right), \ \left(1, -\tfrac{1}{\sqrt{3}}\right), \ \left(0, \tfrac{2}{\sqrt{3}}\right). \tag{14.1.6}$$

If $\rho$ and $\sigma$ are two representations of a Lie algebra $\mathcal{G}$ on two vector spaces $U$ and $V$ respectively, then, as in Section 6.5, we can define the tensor product representation $\rho \otimes \sigma$ of $\mathcal{G}$ on $U \otimes V$ by

$$\rho \otimes \sigma(X) \overset{\text{def}}{=} \rho(X) \otimes \mathbf{1}_V + \mathbf{1}_U \otimes \sigma(X). \tag{14.1.7}$$

The tensor product is the representation of $\mathcal{G}$ on pair states or quadratic terms in a Lagrangian density.

As we saw in Section 6.6, the weights of the tensor product representation $\rho \otimes \sigma$ are all possible sums of the weights of the two representations $\rho$ and $\sigma$. Thus, for example, the representation $\mathbf{1} \otimes \mathbf{1}^*$ has weights

| $\mathbf{1} \otimes \mathbf{1}^*$ | $\left(-1, -\tfrac{1}{\sqrt{3}}\right)$ | $\left(1, -\tfrac{1}{\sqrt{3}}\right)$ | $\left(0, \tfrac{2}{\sqrt{3}}\right)$ |
|---|---|---|---|
| $\left(1, \tfrac{1}{\sqrt{3}}\right)$ | $(0,0)$ | $(2,0)$ | $(1, \sqrt{3})$ |
| $\left(-1, \tfrac{1}{\sqrt{3}}\right)$ | $(-2,0)$ | $(0,0)$ | $(-1, \sqrt{3})$ |
| $\left(0, -\tfrac{2}{\sqrt{3}}\right)$ | $(-1, -\sqrt{3})$ | $(1, -\sqrt{3})$ | $(0,0)$ |

$$\tag{14.1.8}$$

These weights form a hexagon with three points at its center. Note that a weight diagram must indicate the multiplicity of the weights in it. Figure 14.1 shows the structure of this tensor product with a view to bringing out the relationship between the weight diagrams of the factors and the weight diagram of the tensor product. The figure also shows the interpretation in terms of quarks, the topic of Section 14.3.

The Gell-Mann basis for $su(3)$ is a standard extension of the Pauli basis for $su(2)$:

$$\lambda_1 = \begin{pmatrix} 0 & 1 & 0 \\ 1 & 0 & 0 \\ 0 & 0 & 0 \end{pmatrix} \quad \lambda_2 = \begin{pmatrix} 0 & -i & 0 \\ i & 0 & 0 \\ 0 & 0 & 0 \end{pmatrix} \quad \lambda_3 = \begin{pmatrix} 1 & 0 & 0 \\ 0 & -1 & 0 \\ 0 & 0 & 0 \end{pmatrix} \quad su(2)$$

$$\lambda_4 = \begin{pmatrix} 0 & 0 & 1 \\ 0 & 0 & 0 \\ 1 & 0 & 0 \end{pmatrix} \quad \lambda_5 = \begin{pmatrix} 0 & 0 & -i \\ 0 & 0 & 0 \\ i & 0 & 0 \end{pmatrix}$$

$$\lambda_6 = \begin{pmatrix} 0 & 0 & 0 \\ 0 & 0 & 1 \\ 0 & 1 & 0 \end{pmatrix} \quad \lambda_7 = \begin{pmatrix} 0 & 0 & 0 \\ 0 & 0 & -i \\ 0 & i & 0 \end{pmatrix} \quad \lambda_8 = \frac{1}{\sqrt{3}}\begin{pmatrix} 1 & 0 & 0 \\ 0 & 1 & 0 \\ 0 & 0 & -2 \end{pmatrix} \quad \Bigg\} \, su(3)$$

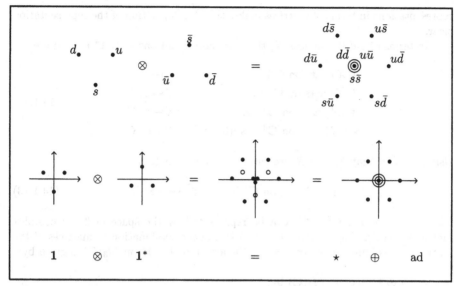

**Fig. 14.1.** The bottom line of this figure shows the tensor product under consideration. The middle line shows the related weight diagrams, and the upper line shows the interpretation in terms of quark pair states or mesons. The central diagram in the middle line depicts the structure of the weight diagram for the tensor product as the sum of the second weight diagram with each weight of the first.

Just as the $W$ bosons are associated with the Pauli generators of $su(2)$, so we shall associate the gluons $G$ with the Gell-Mann matrices. These matrices satisfy the normalization condition

$$\text{tr}(\lambda_a \lambda_b) = 2\delta_{ab}. \tag{14.1.9}$$

If we regard $su(3)$ as a real vector space isomorphic to $\mathbf{R}^8$, then the adjoint action of $su(3)$ on itself,

$$\text{ad}(X)Y = [X, Y], \tag{14.1.10}$$

is an eight-dimensional representation of $su(3)$. The following table gives a basis in which the eight-by-eight matrices $\text{ad}(\lambda_3)$ and $\text{ad}(\lambda_8)$ are diagonal, and the pair of eigenvalues (the weight) associated with each basis element:

### Weights of ad

$$
\begin{array}{llll}
\lambda_3: & (0,0) & \lambda_1 + i\lambda_2: (2,0) & \lambda_4 + i\lambda_5: (1,\sqrt{3}) \\
\lambda_1 - i\lambda_2: (-2,0) & & & \lambda_6 + i\lambda_7: (-1,\sqrt{3}) \\
\lambda_4 - i\lambda_5: (-1,-\sqrt{3}) & \lambda_6 - i\lambda_7: (1,-\sqrt{3}) & \lambda_8: & (0,0)
\end{array} \tag{14.1.11}
$$

According to the definition in Section 6.2, the weights of the adjoint representation are called roots. The associated eigenvectors are the raising and lowering operators of $su(3)$. From 14.1.11 above, we see that the eigenvectors for each nonzero root form a one-dimensional vector space. Assuming a choice of normalization condition such as $\text{tr}(E_\alpha^\dagger E_\alpha) = 1$, we can therefore label an eigenvector $E_\alpha$ with its root $\alpha$.

Comparison with the weights of $\mathbf{1} \otimes \mathbf{1}^*$ in 14.1.8 shows that $\mathbf{1} \otimes \mathbf{1}^*$ splits into a direct sum of the ad representation with the trivial representation $\star$. Rewriting these representations in terms of matrices makes this decomposition of the representation clear.

In terms of $su(3)$ matrices $X$, the four representations $\star$, **1**, **1**$^*$ and ad are:

$$\begin{aligned}
&\star(X) \text{ acts on } \mathbf{C} \text{ by:} && z \mapsto 0; \\
&\mathbf{1}(X) \text{ acts on } \mathbf{C}^3 \text{ by:} && \vec{z} \mapsto X\vec{z}; \\
&\mathbf{1}^*(X) \text{ acts on } \mathbf{C}^3 \text{ by:} && \vec{z} \mapsto X^*\vec{z}; \\
&\text{ad}(X) \text{ acts on } \mathbf{C}^8 \cong su(3) \text{ by:} && Y \mapsto [X, Y].
\end{aligned} \tag{14.1.12}$$

Using the fact that $X^\dagger = -X$, we can transpose the action of **1**$^*$:

$$\mathbf{1}^*(X) \text{ acts on } \mathbf{C}^3 \text{ by:} \qquad \vec{z}^\mathsf{T} \mapsto -\vec{z}^\mathsf{T} X. \tag{14.1.13}$$

The tensor product $\mathbf{C}^3 \otimes \mathbf{C}^3$ can be represented by the space of $3 \times 3$ complex matrices $M_3(\mathbf{C})$. Any matrix in $M_3(\mathbf{C})$ is a linear combination of matrices of the form $\vec{w}\vec{z}^\mathsf{T}$ – column times row. Hence the action of $\mathbf{1} \otimes \mathbf{1}^*$ on $M_3(\mathbf{C})$ is given by:

$\mathbf{1} \otimes \mathbf{1}^*$ acts on $M_3(\mathbf{C})$ by:

$$\begin{aligned}
&\vec{w}\vec{z}^\mathsf{T} \mapsto X\vec{w}\vec{z}^\mathsf{T} - \vec{w}\vec{z}^\mathsf{T} X \quad \text{on rank one matrices,} \tag{14.1.14} \\
&M \mapsto XM - MX = [X, M] \quad \text{in general.}
\end{aligned}$$

This action looks exactly like the adjoint action except for the dimension of the space acted on. However, from the analysis of weights above, we expect $M_3(\mathbf{C})$ to contain a one-dimensional invariant subspace on which $su(3)$ acts trivially. Indeed, $[X, M] = 0$ for all $X$ in $su(3)$ if and only if $M$ is a scalar multiple of the identity matrix, so the one-dimensional invariant subspace of $M_3(\mathbf{C})$ is $\{ z\mathbf{1}_3 \mid z \in \mathbf{C} \}$. The complementary invariant subspace is the space of traceless matrices $\{ zX \mid z \in \mathbf{C}, X \in su(3) \}$. Hence we can write

$$\begin{aligned}
M_3(\mathbf{C}) &\cong \{ z\mathbf{1}_3 \mid z \in \mathbf{C} \} \oplus \{ zX \mid z \in \mathbf{C}, X \in su(3) \} \\
\mathbf{1} \otimes \mathbf{1}^* &\cong \qquad \star \qquad \oplus \qquad\qquad \text{ad}
\end{aligned} \tag{14.1.15}$$

The singlet is extracted by the following decomposition of a matrix $M$ into scalar and traceless parts:

$$M \equiv \frac{1}{3} \text{Tr}(M)\mathbf{1}_3 + \left(M - \frac{1}{3}\text{Tr}(M)\right). \tag{14.1.16}$$

When $M = \vec{w}\vec{z}^\mathsf{T}$, then the singlet (no surprise) is $z_1 w_1 + z_2 w_2 + z_3 w_3$.

This section has revived the terminology of representation theory set out in Chapter 6 in the context of the simplest representations of $su(3)$, namely $\star$, **1**, **1**$^*$, and ad. To add a material value to the theory, the next section presents the interpretation of this material in the two applications of primary interest, namely to quarks through color and flavor symmetries. After that we shall be prepared for descriptions of $su(3)$ weight diagrams, tensor representations, and Young diagrams as set out in the subsequent three sections.

# 14.2  Interpretation of $SU(3)$ as Flavor

When $SU(3)$ acts as a flavor symmetry, the vector representation **1** is interpreted as acting on a column vector of the three light quark fields:

$$\psi \overset{\text{def}}{=} \begin{pmatrix} u \\ d \\ s \end{pmatrix}. \tag{14.2.1}$$

The conjugate representation $\mathbf{1}^*$ acts on the anti-quark fields $\bar{u}$, $\bar{d}$, and $\bar{s}$.

The conventional isospin and hypercharge quantum numbers associated with $SU(3)_f$ are defined by the Lie algebra elements

$$I_3 \overset{\text{def}}{=} \frac{1}{2}\lambda_3 = \begin{pmatrix} \frac{1}{2} & 0 & 0 \\ 0 & -\frac{1}{2} & 0 \\ 0 & 0 & 0 \end{pmatrix},$$

$$Y \overset{\text{def}}{=} \frac{1}{2\sqrt{3}}\lambda_8 = \begin{pmatrix} \frac{1}{6} & 0 & 0 \\ 0 & \frac{1}{6} & 0 \\ 0 & 0 & -\frac{1}{3} \end{pmatrix}. \tag{14.2.2}$$

Electric charge corresponds to the Lie algebra element

$$Q \overset{\text{def}}{=} I_3 + Y = \begin{pmatrix} \frac{2}{3} & 0 & 0 \\ 0 & -\frac{1}{3} & 0 \\ 0 & 0 & -\frac{1}{3} \end{pmatrix}. \tag{14.2.3}$$

Note that these Lie algebra elements give the quantum numbers of particle states; as we saw in Section 3.5, a field has quantum numbers opposite to those of the particle that it destroys. In detail, if $X = -iT$ is the real symmetry generator which acts on the column vector $\psi$ of quark fields, the associated conserved current is

$$j^\mu(X) = \Pi^\mu \mathbf{D}\psi = (\bar{\psi}\gamma^\mu i)(X\psi)$$
$$= \bar{\psi}\gamma^\mu T\psi, \tag{14.2.4}$$

and the associated conserved quantity is

$$\hat{Q}(X) = \int d^3\bar{x}\, \bar{\psi}\gamma^0 T\psi. \tag{14.2.5}$$

(The vector $\psi$ contains three four-component spinors, and so the notation $\gamma^\mu T$ stands for the tensor product $\gamma^\mu \otimes T$.) From the canonical anti-commutation relations, we deduce that the action of $\hat{Q}$ on the fields is

$$[\hat{Q}, \psi] = -T\psi. \tag{14.2.6}$$

The tensor product $\mathbf{1} \otimes \mathbf{1}^*$ is represented by the action of $SU(3)_f$ on $\psi\bar{\psi}$. The nine particle pairs in this product form a basis for the linear space of either the pseudoscalar or the axial vector mesons. Conservation of electric charge prohibits mixing between states with different charges. The identification for pseudoscalars follows:

$$\psi\bar{\psi} = \begin{pmatrix} u\bar{u} & u\bar{d} & u\bar{s} \\ d\bar{u} & d\bar{d} & d\bar{s} \\ s\bar{u} & s\bar{d} & s\bar{s} \end{pmatrix} \tag{14.2.7}$$

implies that

$$\psi\bar{\psi} = \frac{1}{\sqrt{3}}\eta_1 \mathbf{1} + \begin{pmatrix} \frac{1}{\sqrt{2}}\pi^0 + \frac{1}{\sqrt{6}}\eta_8 & \pi^+ & K^+ \\ \pi^- & -\frac{1}{\sqrt{2}}\pi^0 + \frac{1}{\sqrt{6}}\eta_8 & K^0 \\ K^- & \bar{K}^0 & -\frac{2}{\sqrt{6}}\eta_8 \end{pmatrix}. \qquad (14.2.8)$$

Notice that $\pi^0$ and $\eta_8$ are associated with the generators $\lambda_3$ and $\lambda_8$ respectively. The pure-charge states $\eta_1$ (singlet) and $\eta_8$ (member of octet) may be mixed in the mass eigenstates $\eta'$ and $\eta$, but $\eta'$ is mainly $\eta_1$, and $\eta$ is mainly $\eta_8$.

The action of $SU(3)$ as flavor symmetry on the light quarks organizes the mesons into singlets and octets. This identification of pair states with mesons is a first approximation: it gives a first approximation to the structure of the mesons but omits the spin of the quarks and the gluon and virtual particle content of the physical particles. In Section 16.1, we shall include spin and color in the analysis and extend the analysis to baryons.

## 14.3 Interpretation of $SU(3)$ as Color

When $SU(3)$ acts as color symmetry $SU(3)_c$, we must distinguish the three components of the quark fields by a color index, for example:

$$u = \begin{pmatrix} u^r \\ u^b \\ u^g \end{pmatrix}. \qquad (14.3.1)$$

The indices here stand for red, blue, and green. We write $u^a$ to indicate any one of the three colors of the up quark. The vector representation $\mathbf{1}$ of $SU(3)_c$ simply mixes the colors of each quark. The conjugate representation $\mathbf{1}^*$ mixes anti-color components of anti-quarks, for example:

$$\bar{u} = \begin{pmatrix} \bar{u}^{\bar{r}} \\ \bar{u}^{\bar{b}} \\ \bar{u}^{\bar{g}} \end{pmatrix}. \qquad (14.3.2)$$

The Cartan subalgebra elements $\lambda_3$ and $\lambda_8$ determine the two color quantum numbers which characterize each color and anti-color.

As $SU(3)_c$ is the gauge group of the strong interactions, the generators of $SU(3)_c$ correspond to the vector bosons of the strong interactions, that is, to the eight gluons. Just as the electromagnetic potential responds to electrical charge, so the gluons respond to color charge. In fact this response is so strong that colored particles are never observed further apart than a proton radius; they are confined in color-singlet combinations called hadrons. Thus, for example, the $\pi^+$ as a $u\bar{d}$ pair must have the color structure

$$\pi^+ \leftrightarrow u^r\bar{d}^{\bar{r}} + u^b\bar{d}^{\bar{b}} + u^g\bar{d}^{\bar{g}}. \qquad (14.3.3)$$

The other combinations of color and anti-color can only occur in virtual particles.

Since $SU(3)_c$ is an unbroken symmetry, there is actually no experiment capable of distinguishing one color from another. Hence the specification of the color of a quark field depends on an arbitrary choice of basis for the three-dimensional space of colors at each point in space-time. Physics should be independent of the choice of basis; change of basis is organized by the gauge group associated with $SU(3)_c$; hence the need for a gauge theory.

Consider that at some moment we see a $\pi^+$ made up of a $u^r \bar{d}^{\bar{r}}$ pair. Since the $u^r$ and the $\bar{d}^{\bar{r}}$ may be separated in space, someone else may choose a basis in which our $u^r$ is still a $u^r$ but our $\bar{d}^{\bar{r}}$ appears to be a $\bar{d}^{\bar{g}}$. To restore color balance, we must include the relative twisting between the two bases: the second observer will also see a $g^{\bar{r}}$ gluon. Since the existence of a gluon field should not be observer dependent, we deduce that this gauge force is a necessary physical reality.

Finally, the eight gluons carry color and anti-color according to their appearance in the octet in $u\bar{u}$:

$$\textbf{Gluon Octet} = \begin{pmatrix} g^{r\bar{r}} & g^{r\bar{b}} & g^{r\bar{g}} \\ g^{b\bar{r}} & g^{b\bar{b}} & g^{b\bar{g}} \\ g^{g\bar{r}} & g^{g\bar{b}} & g^{g\bar{g}} \end{pmatrix}. \tag{14.3.4}$$

Of course, as this is an octet, there is no gluon corresponding to $r\bar{r} + g\bar{g} + b\bar{b}$.

## 14.4 $SU(3)$ **Weight Diagrams**

Having introduced $SU(3)$ representations in this concrete way, we can now progress to some more abstract generalities. First, an irreducible representation is determined by its highest weight, where we identify the highest weight as that one which is furthest right in the weight diagram. (In general, we would order the weights as we order a dictionary, first by the first coordinate, then when first coordinates are equal by second coordinates, and so on.) Second, every highest weight can be found in the weight diagram of some tensor product of **1** and **1**\*. Third, the irreducible representation $(m, n)$ with highest weight

$$w = m\left(1, \tfrac{1}{\sqrt{3}}\right) + n\left(1, -\tfrac{1}{\sqrt{3}}\right) \tag{14.4.1}$$

occurs exactly once as a subrepresentation

$$(m, n) \subset \underbrace{\mathbf{1} \otimes \ldots \otimes \mathbf{1}}_{m \text{ times}} \otimes \underbrace{\mathbf{1}^* \otimes \ldots \otimes \mathbf{1}^*}_{n \text{ times}}. \tag{14.4.2}$$

Fourth, the multiplicities of the weights in $(m, n)$ are computed by applying the following algorithm to the weight diagram drawn with all multiplicities set to zero:

1. Add one to the multiplicity of the weights;
2. If the weights form a triangle, stop; otherwise forget the outermost weights and return to step (1).

Fifth, the degree of $(m, n)$ is given by

$$\deg(m, n) = \frac{1}{2}(m + n + 2)(m + 1)(n + 1). \tag{14.4.3}$$

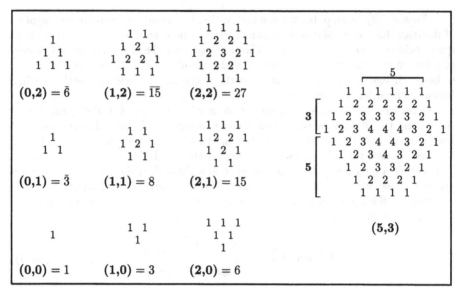

**Fig. 14.4.** Here we see the structure of the weight diagrams of irreducible representations at a glance. The location of a numeral indicates a weight and its value, the multiplicity of that weight. On the left are nine small representations. On the right, a diagram of $(5, 3)$ shows by example how to construct the weight diagram for $(m, n)$.

The small representations of $su(3)$ are generally specified by their dimension and, if $m < n$, a macron. Thus 3 stands for $(1, 0)$, $\bar{3}$ for $(0, 1)$, 8 for $(1, 1)$, and so on. Note that 15 would be ambiguous, being the dimension of both $(4, 0)$ and $(2, 1)$.

The weight diagrams of $SU(3)$ are based on the equilateral triangles corresponding to **1** and **1\***. If we write a weight diagram with numerals for multiplicities at the location of the corresponding weights, then the rows always have the form

$$\text{Row of Multiplicities} = 1\ 2\ \ldots\ k\ k\ \ldots\ k\ k\ \ldots\ 2\ 1: \qquad (14.4.4)$$

a row starts with a 1, counts up to a maximum $k$, and then counts down to 1. The best way to draw the weight diagram for $(m, n)$ follows:

1. Make a row of $m + 1$ 1's to represent multiplicity 1;
2. Add $n$ more rows of dots, increasing the number of numerals and the maximum multiplicity by one for each new row;
3. Add $m - n$ more rows, decreasing the number of numerals by one for each new row and maintaining the same maximum as the previous row;
4. Add $n$ more rows, decreasing the maximum by one for each new row.

This procedure is illustrated in Figure 14.4.

Since the weight diagram of a tensor product of irreducible representations is simply the sum of all the pairs of weights of the two representations (with multiplicities multiplied), with the algorithm above for drawing weight diagrams, it is possible to compute tensor products and split them into sums of irreducibles:

1. Draw the weight diagram for the tensor product;
2. Locate the highest weight and subtract the weight diagram for the corresponding irreducible representation;

3. Repeat step (2) until there are no weights left.

This graphical procedure is only practical for small representations of $SU(3)$, however, so we shall introduce tensor representations and the Young diagram algorithm in order to provide a more powerful and general method of splitting tensor representations.

**Homework 14.4.5.** Split the tensor products $3 \otimes 3$, $6 \otimes \bar{3}$, and $6 \otimes \bar{6}$ using weight diagrams.

This section gives mastery of $SU(3)$ weight diagrams. This geometric foundation provides a sense of familiarity with the smaller irreducible representations and insight into the structure and splitting of tensor products. On this basis, we introduce the algebraic approach to tensors in the next section.

## 14.5 Tensor Representations

In general, a representation matrix for $U$ acting in $(m, n)$ is useless not only because of its high degree but also because its elements are polynomials of degree $m$ and $n$ in the elements of $U$ and $U^*$ respectively. For $m+n > 2$, it is far more practical to work with tensors. In this section, we introduce the tensor form of $(m, n)$ and describe the weight diagram from this perspective. In the following section, we present an algorithm for splitting tensor products of irreducible $su(3)$ representations into direct sums of same, and analyze symmetry of tensors. This knowledge of tensors will also be useful when we apply flavor $SU(3)$ to baryons.

A tensor $T_J^I$ is a set of complex numbers labeled by indices $I = i_1 \cdots i_m$ and $J = j_1 \cdots j_n$, where each index ranges from 1 to 3. The simplest tensors $T^i$ and $T_j$ are simply the three components of a vector in $\mathbf{C}^3$. The upper and lower indices indicate different actions of $SU(3)$: $T^i$ and $T_j$ correspond to the representations **1** and **1**$^*$ respectively. Thus:

$$(U \cdot T)^i = U^i{}_r T^r \quad \text{and} \quad (U \cdot T)_j = T_s \bar{U}^s{}_j, \tag{14.5.1}$$

where $\bar{U} = U^*$. In general, a multi-index tensor $T_J^I$ is transformed by a product of $|I|$ $U$'s and $|J|$ $\bar{U}$'s. For example,

$$(U \cdot T)^{i_1 i_2}_j = U^{i_1}{}_{r_1} U^{i_2}{}_{r_2} T_s^{r_1 r_2} \bar{U}^s{}_j. \tag{14.5.2}$$

The splitting of tensor representations into irreducible components is achieved through contractions with the isotropic tensors $\delta_j^i$, $\epsilon^{ijk}$ and $\epsilon_{ijk}$. For an example of the use of $\delta_j^i$, the nine dimensions of $T_j^i$ may be split into the eight dimensions of traceless tensors and a scalar:

$$T_j^i \mapsto \left( T_j^i - \frac{1}{3} T_k^k \delta_j^i \right) \oplus \frac{1}{3} T_k^k. \tag{14.5.3}$$

The $\oplus$ is necessary here because the two summands are in different tensor spaces. Clearly, this map represents the splitting of $3 \otimes \bar{3}$ into $8 \oplus 1$.

For an example of the use of $\epsilon_{ijk}$, the nine dimensions of $T^{ij}$ naturally splits into six dimensions of symmetric tensors and three dimensions of anti-symmetric, and the $\epsilon_{ijk}$ may be used to lower the anti-symmetric indices:

$$T^{ij} \mapsto \frac{1}{2}(T^{ij} + T^{ji}) \oplus \frac{1}{2} \epsilon_{ijk}(T^{ij} - T^{ji}). \tag{14.5.4}$$

Here, we recognize the splitting of $3 \otimes 3$ into $6 \oplus \bar{3}$.

**Homework 14.5.5.** Split $T^{ijk}$ and $T^{ij}_k$ into irreducible components using $\epsilon$'s and $\delta$'s.

From these examples, it is clear that any tensor which does not vanish when two of its indices are contracted with $\delta^i_j$, $\epsilon^{ijk}$, or $\epsilon_{ijk}$ is necessarily reducible. It is also true (though less obvious) that this vanishing condition is sufficient for irreducibility.

A tensor which vanishes upon contraction with $\delta^i_j$ is called *traceless*. One which vanishes upon contraction of two indices with an $\epsilon$ is symmetric in those two indices. Consequently, the irreducible representations correspond to tensors $T^I_J$ which are traceless and totally symmetric in the indices of $I$ and $J$ separately. For example,

$$
\begin{aligned}
3 &\longleftrightarrow T^i, & \bar{3} &\longleftrightarrow T_i; \\
6 &\longleftrightarrow T^{ij} = T^{ji}, & \bar{6} &\longleftrightarrow T_{ij} = T_{ji}; \\
8 &\longleftrightarrow T^i_j, & \text{where} \quad T^i_i &= 0.
\end{aligned}
\tag{14.5.6}
$$

**Homework 14.5.7.** Verify the formula 14.4.3 for the degree of $(m, n)$ by working out the dimension of the space of traceless, totally symmetric tensors with $m$ upper indices and $n$ lower indices.

The tensors with pure weights are simultaneously eigentensors of $\lambda_3$ and $\lambda_8$. The determination of these tensors is trivial when we realize that the up quark is represented by $T^i = \delta^{1i}$, the down quark by $T^i = \delta^{2i}$, and the strange quark by $T^i = \delta^{3i}$. The anti-quarks are represented by $T_j$'s, of course. Consequently, the tensors with pure weights are those which vanish for all values of the indices except one. For example, remembering to change the quantum numbers for anti-particles,

$$
\lambda_3 \cdot T^{12}_3 = (1 - 1 + 0)T^{12}_3 = 0
\tag{14.5.8}
$$

and

$$
\lambda_8 \cdot T^{12}_3 = \frac{1}{\sqrt{3}}(1 + 1 + 2)T^{12}_3 = \frac{4}{\sqrt{3}}T^{12}_3.
\tag{14.5.9}
$$

(It is generally improper to make a Lie algebra element act on a single component of a tensor, but in this case it makes sense.)

Thus the tensor with highest weight has $T^{1 \cdots 1}_{2 \cdots 2} = 1$ and all other components zero; it stands for a product of $u$'s and $\bar{d}$'s. As for $SU(2)$ representations, the rest of the irreducible representation can be obtained by lowering this highest-weight tensor. However, because $SU(3)$ has rank two, the weight diagram is two dimensional, and we need to lower in two different directions. The basic lowering operators for $su(3)$ are those which lower the $\lambda_3$ quantum number minimally:

$$
L_{31} = \begin{pmatrix} 0 & 0 & 0 \\ 0 & 0 & 0 \\ 1 & 0 & 0 \end{pmatrix} \quad \text{and} \quad L_{23} = \begin{pmatrix} 0 & 0 & 0 \\ 0 & 0 & 1 \\ 0 & 0 & 0 \end{pmatrix}
\tag{14.5.10}
$$

which lower $1(u)$ to $3(s)$ and $3(s)$ to $2(d)$. The other lowering operator is the bracket of these two:

$$
L_{21} = [L_{23}, L_{31}] = \begin{pmatrix} 0 & 0 & 0 \\ 1 & 0 & 0 \\ 0 & 0 & 0 \end{pmatrix}.
\tag{14.5.11}
$$

Suppose, for example, that $T^i_j$ is the highest-weight tensor in the octet; that is, $T^1_2 = 1$ and other components vanish. Then $L_{31}$ applies as follows:

$$S^i_j \overset{\text{def}}{=} (L_{31} \cdot T)^i_j = (L_{31})^i_r T^r_j. \tag{14.5.12}$$

Clearly, $S^i_j$ is zero unless $i = 3$ and $j = 2$, in which case it is 1. Thus $L_{31}$ has effectively lowered $u\bar{d}$ to $s\bar{d}$.

**Homework** 14.5.13. Use the lowering operators to generate a pure weight basis for the tensors $T^{ij}$.

Since these lowering operators are elements of the complexified Lie algebra, they can act on any representation. However, when they act on $(0,1)$, we have to be careful to conjugate only the original real representation. For example,

$$L_{31} = \frac{1}{2} \begin{pmatrix} 0 & 0 & 1 \\ 0 & 0 & 0 \\ -1 & 0 & 0 \end{pmatrix} + \frac{i}{2} \begin{pmatrix} 0 & 0 & i \\ 0 & 0 & 0 \\ i & 0 & 0 \end{pmatrix} \tag{14.5.14}$$

implies that $L_{31}$ acts on $(0,1)$ as the matrix

$$\bar{L}_{31} = \frac{1}{2} \begin{pmatrix} 0 & 0 & 1 \\ 0 & 0 & 0 \\ -1 & 0 & 0 \end{pmatrix} + \frac{i}{2} \begin{pmatrix} 0 & 0 & -i \\ 0 & 0 & 0 \\ -i & 0 & 0 \end{pmatrix} = -L_{\bar{1}\bar{3}}. \tag{14.5.15}$$

Thus $L_{31}$ lowers $u$ to $s$ and $\bar{s}$ to $-\bar{u}$.

**Homework** 14.5.16. Use the lowering operators to generate a pure weight basis for the tensors $T^{ij}_k$.

The details in this section make it clear how to use tensors to represent products involving quarks and anti-quarks and how lowering operators act on such products. These details will be useful when we use flavor $SU(3)$ to predict properties of the baryon decuplet in Chapter 16. However, even to understand the structure of the hadrons from the perspective of quarks, we need a little more knowledge of tensor products. This is provided by the next section.

# 14.6 Splitting Tensor Products

The purpose of this section is to present an algorithm for decomposing tensor products of irreducible representations of $SU(3)$ into sums of irreducible representations, and to identify symmetric and anti-symmetric components in such sums.

The tensor product of tensors $T^I_J$ and $T^{I'}_{J'}$ is given by

$$T^{II'}_{JJ'} = T^I_J T^{I'}_{J'}. \tag{14.6.1}$$

The tensor product of the spaces of tensors $T^I_J$ and $T^{I'}_{J'}$ is the space of tensors $T^{II'}_{JJ'}$. We saw in the previous section how to use the isotropic tensors $\delta^i_j$, $\epsilon_{ijk}$, and $\epsilon^{ijk}$ to project a tensor onto its irreducible components. The first step uses $\delta$'s to remove the traces, and the second uses $\epsilon$'s to replace anti-symmetric pairs of indices by single indices. The following two-step algorithm summarizes the results.

Let $(m, m'; n, n')$ be the representation of $SU(3)$ on tensors $T^{II'}_{JJ'}$ where $|I| = m$, etc. and $T$ is traceless with respect to $IJ$ and $I'J'$ separately and symmetric in the indices in $I$, $I'$, $J$, and $J'$ separately.

An $SU(3)$ tensor product may now be split into irreducible representations through the following two steps:

$$(m, n) \otimes (m', n') = \bigoplus_{r=0}^{\min(m, n')} \bigoplus_{s=0}^{\min(m', n)} (m - r, m' - s; n - s, n' - r), \quad (14.6.2)$$

and

$$(m, m'; n, n') = (m + m', n + n') \quad (14.6.3)$$
$$\oplus \bigoplus_{t=1}^{\min(m, m')} (m + m' - 2t, n + n' + t)$$
$$\oplus \bigoplus_{t=1}^{\min(n, n')} (m + m' + t, n + n' - 2t).$$

Thus, for example:
$$3 \otimes 3 = (1, 0) \otimes (1, 0)$$
$$= (1, 1; 0, 0) \quad (14.6.4)$$
$$= (2, 0) \oplus (0, 1) = 6 \oplus \bar{3}.$$

When analyzing a tensor product of a representation with itself, we can first split it into even and odd subrepresentations with respect to interchange of factors, and then split these subrepresentations into irreducible representations. In the case of an $su(3)$ representation $(m, n)$, interchange of factors leads to interchange of $r$ and $s$ in 14.6.2. Hence, if $r = s$, the subrepresentation $(m - r, m - s; n - s, n - r)$ is invariant; but if $r \neq s$, then since as representations

$$(m - r, m - s; n - s, n - r) = (m - s, m - r; n - r, n - s), \quad (14.6.5)$$

diagonalization of the interchange operator leads to one even and one odd subrepresentation $(m - r, m - s; n - s, n - r)$. Finally, counting the first term in 14.6.3 as the $t = 0$ term, when we apply 14.6.3 to $(m - r, m - s; n - s, n - r)$, symmetry alternates with anti-symmetry as the index $t$ increases from zero.

Thus, for example,

$$(1, 1) \otimes (1, 1)$$
$$= (1, 1; 1, 1)_+ \oplus (1, 0; 0, 1)_+ \oplus (0, 1; 1, 0)_- \oplus (0, 0; 0, 0)_+ \quad (14.6.6)$$
$$= ((2, 2)_S \oplus (3, 0)_A \oplus (0, 3)_A) \oplus (1, 1)_S \oplus (1, 1)_A \oplus (0, 0)_S,$$

which becomes in the dimension notation:

$$8 \otimes 8 = 27_S \oplus 10_A \oplus \overline{10}_A \oplus 8_S \oplus 8_A \oplus 0_S. \quad (14.6.7)$$

Note that the subscript $+$ does not indicate that the subrepresentation is symmetric; it refers to an overall sign which indicates whether to start with $S$ or $A$ at $t = 0$.

**Homework 14.6.8.** Use the tensor decomposition algorithm to split the product $27 \otimes 27$ into a sum of irreducibles. Indicate which terms are even and which odd under exchange of factors.

Finally, from this decomposition formula, we can deduce a special case of Schur's Lemma:

**Theorem 14.6.9.** *There is a scalar subrepresentation in* $(m, n) \otimes (m', n')$ *if and only if* $(m, n) = (n', m')$, *in which case the scalar is unique.*

This section has introduced the algebraic description of $SU(3)$ representations as tensors and of the splitting of spaces of tensors into irreducible pieces through the use of isotropic tensors. The concepts presented here generalize readily to $SU(n)$ (and with a little more work to $SO(n)$). With this knowledge of tensor representations, we are ready to understand Young Diagrams.

# 14.7 Young Diagrams

The method of Young diagrams provides the most efficient device for splitting tensor products of $SU(n)$ representations. Each irreducible representation of $SU(n)$ may be concretely presented as a representation on a space of tensors $T^I$ with certain symmetry constraints on the index $I$. A Young diagram is an arrangement of $|I|$ boxes which depicts the defining symmetry of such a space of tensors. Writing loosely, boxes in a column represent anti-symmetrization of the corresponding indices, and boxes in a row represent symmetrization. For example, for $SU(n)$, a column of $n$ boxes stands for the space of totally anti-symmetric tensors with $n$ indices, that is, the singlet representation. A column with more than $n$ boxes will represent the zero tensor, and so such columns are not found in Young diagrams. The graphical convention is to align the tops of the columns and put taller columns to the left of shorter.

The degree of an irreducible representation is reckoned from a fraction. To find the numerator, fill the boxes with numbers starting with $n$'s along the diagonal, increasing by one with each step right, and decreasing by one with each step down, and multiply these numbers. Thus for $SU(3)$, we have an example:

$$\rightarrow \boxed{\begin{array}{ccc}3&4&5\\2\end{array}} \rightarrow 2 \times 3 \times 4 \times 5 \tag{14.7.1}$$

The denominator is computed by writing in each box one plus the total number of boxes directly below and directly to the right, and multiplying the results. Continuing the $SU(3)$ example, we find:

$$\rightarrow \boxed{\begin{array}{ccc}4&2&1\\1\end{array}} \rightarrow 1 \times 4 \times 2 \times 1 \tag{14.7.2}$$

For $SU(3)$, then, this Young diagram corresponds to an irreducible representation with degree 15.

The procedure for splitting a tensor product is straightforward. Take two irreducible factors at a time, write their Young diagrams, label the columns in the second diagram alphabetically from the top so that all the boxes in any one row contain the same distinguishing letter. Now write out all possible ways of moving the boxes of the second diagram to the first diagram which satisfy the following constraints:

1. Attach the boxes in alphabetical sequence so that at each step we have a Young diagram;
2. Never put two boxes with the same label in the same column;
3. When finished, read down the columns from the top right to form running totals for occurrences of each letter and check that a later letter never has a greater count than an earlier letter.

Rule (3) does not look so good in words, but is easy enough to work with. The following diagram illustrates the sequence in which to take the labels:

$$\textbf{Sequence for Rule (3)}: \quad \begin{array}{|c|c|c|} \hline 5 & 3 & 1 \\ \hline 6 & 4 & 2 \\ \hline 7 & & \\ \hline \end{array} \qquad (14.7.3)$$

and the next diagram gives an example in which Rule (3) is violated in the 4th box:

| Box No. | 1 | 2 | 3 | 4 | 5 | 6 | 7 |
|---------|---|---|---|---|---|---|---|
| a's     | 1 | 1 | 1 | 1 | 1 | 1 | 2 |
| b's     | 0 | 1 | 1 | 2 | 2 | 2 | 2 |

$$(14.7.4)$$

The following computation illustrates the outcome of formal use of the rules for splitting tensor products of Young diagrams:

$$(14.7.5)$$

This computation can be interpreted for $SU(n)$ for all $n$ with the understanding that if $n = 2$, then a diagram with a column of three boxes should be dropped.

This example illustrates a systematic approach to following Rule (3) insofar as the *a* box moves from left to right in the diagrams on the right-hand side. The general procedure follows. Suppose that we have constructed all consistent diagrams by moving boxes labeled $a_1$ to $a_{k-1}$. Then first attach the boxes labeled $a_k$ as far left as possible on each of these diagrams and then create new diagrams by moving these boxes one at a time to the right. In this process of moving a box to the right, as soon as a violation of Rule (3) occurs, that box has already been in all allowed locations, so we turn to the next box of type $a_k$ and begin moving it to the right.

The following silly alternative for splitting the tensor product of the previous example again illustrates the left to right movement of boxes:

$$(14.7.6)$$

In this computation, the following patterns were eliminated by the constraints:

$$(14.7.7)$$

The Young diagrams for the $SU(3)$ representations $(m, n)$ consist of $n$ columns of two boxes followed by $m$ columns of single boxes. For example:

$$(2,3) = \begin{array}{c}\boxed{\phantom{x}}\end{array} \longleftrightarrow T^{ij}_{rst} \tag{14.7.8}$$

On the basis of the geometric and algebraic approaches to $SU(3)$ representation theory presented in the preceding sections, it is easy to sense the logic behind the Young algorithm. With a little practice, this algorithm enables us to compute arbitrary $SU(n)$ tensor products efficiently. At this point, we have covered 95% of the $SU(3)$ representation theory that we need. Later, we will cover the remaining 5% in two more points. First, for the discussion of flavor $SU(3)$, we will explain the splitting of $(m, n)$ when it is restricted to an $su(2) \times u(1)$ subalgebra of $su(3)$. Second, for application to flavor and color, we will analyze the symmetry of a triple tensor product under interchange of factors.

# 14.8 Summary

This chapter has extended the abstract notions of Chapter 6 in the direction of $SU(3)$ representation theory. Section 14.1 offers a very concrete treatment of the four primary representations, the singlet or scalar representation $\star$, the vector representation $\mathbf{1}$, its conjugate $\mathbf{1}^*$, and the adjoint or octet representation ad. Sections 14.2 and 14.3 describe the interpretation of these representations in the context of flavor and color respectively. This level of understanding is already sufficient for QCD. Section 14.4 summarizes the geometric viewpoint on $SU(3)$ representation theory. Knowing this, one will be prepared for the classical representation theory of the semi-simple Lie groups. Section 14.5 brings out the same knowledge in the tensor notation; the tensor spaces of small dimension could represent quark or hadron multiplets and will be used in Chapter 16. Finally, Section 14.6 introduces the Young algorithm for splitting tensor products of $SU(n)$ representations. This is all the $SU(3)$ representation one generally needs. Two further topics, the restriction of $SU(3)$ to $SU(2) \times U(1)$ and the symmetry of triple tensor products, are developed in Chapter 16.

# Chapter 15

# The Structure of the Standard Model

Presenting the most successful theory in the history of science; its structure, some characteristic effects, anomalies, and overall consistency.

## Introduction

The Standard Model of electroweak and strong interactions is the basis for all high-energy particle theory. It has achieved such wide acceptance that the primary test for the validity of a model in high-energy physics is that it yields this theory at sufficiently low energies. The Standard Model is an extension of the Weinberg–Salam model of the electroweak interactions to include quantum chromodynamics (QCD), the quark-gluon theory of the strong interactions.

The Weinberg–Salam model is a gauge theory, built around the group $SU(2) \times U(1)$. As a gauge theory, it has massless vector bosons for each symmetry generator. However, massless bosons correspond to long-range forces and would therefore have been detected long ago. Hence the theory includes a means for giving the vector bosons masses; it proposes the existence of some scalar fields (Higgs fields) and a gauge-invariant potential for them whose classical minima are only invariant under a $U(1)$ subgroup of $SU(2) \times U(1)$. When the symmetry breaks, the vector field associated with the unbroken symmetry remains massless and is identified with the photon, but those associated with broken symmetries become massive, and we are reconciled with experimental evidence regarding massless vector bosons.

QCD, the quark-gluon theory of strong interactions, is a gauge theory based on $SU(3)$ color symmetry. The quarks were originally proposed to account for observed $SU(3)$ flavor multiplet structure in the hadrons. Color $SU(3)$ was introduced to restore Fermi statistics to the baryons regarded as bound states of quarks. Color symmetry is unbroken, but the emergence of bound states of quarks at the nuclear distance scale indicates that at this scale QCD is not a perturbative theory. At smaller distance scales, however, perturbative QCD works well.

The chapter has three parts: Sections 1 to 8 introduce the Weinberg–Salam theory of the electroweak interactions, Sections 9 and 10 extend the Weinberg–Salam to include quarks and strong interactions, and Sections 11 to 15 discuss current conservation in the Standard Model. Section 15.16 integrates the results of the previous sections in a brief presentation of the structure of the Standard Model.

In writing this chapter, we decided to go straight for established conclusions and not to worry about the sequence of steps by which the results were first discovered. Thus, we have sacrificed something of the flavor of model-building in order to present the Standard Model in simple stages.

## 15.1 The Electroweak Lagrangian

The Weinberg–Salam model proposes the existence of lepton families, each family comprising a pair of particles like the electron and neutrino. The theory puts no constraint on the number of families, but gives each family a standard structure. The electron-type particle is massive, hence a Dirac spinor; but since no right-handed neutrino has ever been observed, the theory includes only the left-handed neutrino-type particles. Thus a single lepton family is represented by three chiral spinors. In particular, for the electron-neutrino family, we write:

$$\text{Family:} \quad e_L, \, e_R, \, \nu_L, \tag{15.1.1}$$

where the $L$-$R$ subscripts indicate projections $P_{\pm}$ of the Dirac fields,

$$
\begin{aligned}
e_L &\overset{\text{def}}{=} P_- e = \tfrac{1}{2}(1 - \gamma_5)e; \\
e_R &\overset{\text{def}}{=} P_+ e = \tfrac{1}{2}(1 + \gamma_5)e.
\end{aligned}
\tag{15.1.2}
$$

To bring in the gauge group, the model proposes that the doublet field

$$\psi_L \overset{\text{def}}{=} \begin{pmatrix} \nu_L \\ e_L \end{pmatrix} \tag{15.1.3}$$

transforms as a vector under $SU(2)$, while the singlet $e_R$ transforms as a scalar. Experiment suggests that there are precisely three of these lepton families, comprising the electron, the muon, the tauon, and their neutrinos.

Since the $SU(2) \times U(1)$ symmetry which we are building into our theory must break to the $U(1)$ of electromagnetism, we include a symmetry-breaking doublet of complex scalar fields $\phi = (\phi^+, \phi^0)$, which we have labeled according to their electromagnetic charge. Since the $U(1)$ of electromagnetism is unbroken, this notation indicates that the $\phi^+$ is defined as that component of the doublet which has zero vacuum expectation value after the breaking of the $SU(2) \times U(1)$ symmetry. The fields $\phi$ are called *Higgs fields* or simply *higgses*. The symmetry-breaking mechanism will become clear as our presentation proceeds.

The Lagrangian density for a single lepton family is a sum of kinetic terms and all possible interactions between $\psi_L$, $e_R$, and $\phi$. Renormalizability indicates that the Lagrangian density for these matter fields must have the following form:

$$
\begin{aligned}
\mathcal{L}_{\text{M}} \overset{\text{def}}{=} \; &\bar{\psi}_L i \partial\!\!\!/ \psi_L + \bar{e}_R i \partial\!\!\!/ e_R + (\partial_\mu \phi)^\dagger (\partial^\mu \phi) \\
&- f \bar{e}_R \phi^\dagger \psi_L - f \bar{\psi}_L \phi e_R - \frac{\lambda}{2} \left( \phi^\dagger \phi - \frac{v^2}{2} \right)^2.
\end{aligned}
\tag{15.1.4}
$$

According to the minimal-coupling principle, to promote the global $SU(2) \times U(1)$ symmetry to a gauge symmetry, one must add a gauge field kinetic term for each symmetry generator, and replace all partial derivatives by covariant derivatives. Let $T^a = \tfrac{1}{2}\sigma^a$ and $Y$ be the hermitian generators of $SU(2)$ and $U(1)$ respectively, so that the $T$'s are normalized to satisfy:

$$[T^a, T^b] = i\epsilon^{abc} T^c \quad \text{and} \quad \text{Tr}(T^a T^b) = \frac{1}{2}\delta^{ab}. \tag{15.1.5}$$

To be proper, the generators of $SU(2) \times U(1)$ are

$$\hat{T}^a \stackrel{\text{def}}{=} T^a \otimes 1 \quad \text{and} \quad \hat{Y} \stackrel{\text{def}}{=} 1 \otimes Y, \tag{15.1.6}$$

but we shall identify $\hat{T}^a$ and $\hat{Y}$ with $T^a$ and $Y$ respectively.

Let $W_\mu^a$ and $B_\mu$ be the gauge fields corresponding to the generators $T^a$ and $Y$ respectively. Then action of $T^a$ on the various multiplets is

$$T^a \cdot \psi_L = T^a \psi_L, \quad T^a \cdot e_R = 0, \quad \text{and} \quad T^a \cdot W_\mu' = -i\epsilon^{abc} W_\mu^c, \tag{15.1.7}$$

and (following Section 12.6) the covariant derivative and the field strengths are:

$$D_\mu = \partial_\mu - igW_\mu^a T^a - ig' B_\mu Y;$$
$$W_{\mu\nu}^a = \partial_\mu W_\nu^a - \partial_\nu W_\mu^a + g\epsilon^{abc} W_\mu^b W_\nu^c; \tag{15.1.8}$$
$$B_{\mu\nu} = \partial_\mu B_\nu - \partial_\nu B_\mu.$$

Therefore, the complete Weinberg–Salam Lagrangian density for one lepton family is:

$$\mathcal{L}_{\text{EW}} = -\frac{1}{4} W_{\mu\nu}^a W^{a\mu\nu} - \frac{1}{4} B_{\mu\nu} B^{\mu\nu} + \bar{\psi}_L i \not{D} \psi_L + \bar{e}_R i \not{D} e_R$$
$$+ (D_\mu \phi)^\dagger (D^\mu \phi) - f\bar{e}_R \phi^\dagger \psi_L - \bar{f}\bar{\psi}_L \phi e_R - \frac{\lambda}{2}\left(\phi^\dagger \phi - \frac{v^2}{2}\right)^2. \tag{15.1.9}$$

The rank of $SU(2) \times U(1)$ is 2; the pair $T_3$, $Y$ span a maximal commuting subalgebra of $su(2) \times u(1)$. Each particle in the theory has therefore two quantum numbers. The $T_3$ quantum numbers, known as *third component of weak isospin*, have been assigned already by the choice of doublets and singlets. The $Y$ quantum number, called *hypercharge*, may be chosen to make the Lagrangian density $Y$-invariant, and to make electromagnetic charge $Q$ satisfy

$$Q = T_3 + Y. \tag{15.1.10}$$

Since the $Q$-charge of the electron, the neutrino, and the Higgs fields are known, this constraint 15.1.10 determines the hypercharge for these fields, and hence, from the couplings in 15.1.9, we can determine the hypercharges of the gauge fields. Actually, the hypercharge assignments to the Higgs fields are over-determined. The consistency of the theory is a consequence of the intelligent choice of doublets and singlets.

The following table summarizes the $T_3$ and $Y$ quantum numbers of all the fields:

|  | $W^+$ | $W^-$ | $W_3$ | $B$ | $\nu_e$ | $e_L^-$ | $e_R^-$ | $\phi^+$ | $\phi^0$ |
|---|---|---|---|---|---|---|---|---|---|
| $T_3$ | 1 | $-1$ | 0 | 0 | $\frac{1}{2}$ | $-\frac{1}{2}$ | 0 | $\frac{1}{2}$ | $-\frac{1}{2}$ |
| $Y$ | 0 | 0 | 0 | 0 | $-\frac{1}{2}$ | $-\frac{1}{2}$ | $-1$ | $\frac{1}{2}$ | $\frac{1}{2}$ |
| $Q = T_3 + Y$ | 1 | $-1$ | 0 | 0 | 0 | $-1$ | $-1$ | 1 | 0 |

where the $Q$-eigenfields $W^\pm$ are defined by

$$W^+ = \frac{W_1 - iW_2}{\sqrt{2}} \quad \text{and} \quad W^- = \frac{W_1 + iW_2}{\sqrt{2}}. \tag{15.1.11}$$

## 15.2 – Symmetry Breaking

We now apply the general description of spontaneous symmetry breakdown presented in Section 12.9 to the Standard Model.

The Higgs doublet $\phi$ has to gain a vacuum expectation value $\bar{\phi} = \langle \phi \rangle$ in order to minimize the potential

$$V(\phi) = \frac{\lambda}{2}\left(\phi^\dagger \phi - \frac{v^2}{2}\right)^2. \tag{15.2.1}$$

The minimum is achieved by any vector $\phi$ with length $v/\sqrt{2}$. We can use $SU(2)$ at each point in space-time to rotate $\langle \phi(x) \rangle$ into the constant field

$$\langle \phi(x) \rangle = \begin{pmatrix} 0 \\ v/\sqrt{2} \end{pmatrix}. \tag{15.2.2}$$

In unitarity gauge, the doublet $\phi$ may be written in terms of radial fluctuations around this minimum described by a real field $\phi'$:

$$\phi = \begin{pmatrix} \phi^+ \\ \phi^0 \end{pmatrix} = \begin{pmatrix} 0 \\ v/\sqrt{2} \end{pmatrix} + \begin{pmatrix} 0 \\ \phi' \end{pmatrix}. \tag{15.2.3}$$

Substituting this expression for $\phi$ in the Lagrangian 15.1.9 represents the shift from the unphysical vacuum $\langle \phi \rangle = 0$ to the physical vacuum 15.2.2.

This VEV breaks the $SU(2) \times U(1)$ symmetry down to $U(1)_Q$:

$$\begin{aligned}
(\epsilon_a T^a + \epsilon Y)\langle \phi \rangle &= \frac{1}{2}\begin{pmatrix} \epsilon_3 + \epsilon & \epsilon_1 - i\epsilon_2 \\ \epsilon_1 + i\epsilon_2 & -\epsilon_3 + \epsilon \end{pmatrix}\begin{pmatrix} 0 \\ v/\sqrt{2} \end{pmatrix} \\
&= \frac{v}{2\sqrt{2}}\begin{pmatrix} \epsilon_1 - i\epsilon_2 \\ -\epsilon_3 + \epsilon \end{pmatrix}
\end{aligned} \tag{15.2.4}$$

implies that

$$(\epsilon_a T^a + \epsilon Y)\langle \phi \rangle = 0 \quad \Longleftrightarrow \quad \epsilon_1 = \epsilon_2 = 0, \; \epsilon_3 = \epsilon, \tag{15.2.5}$$

in which case

$$(\epsilon_a T^a + \epsilon Y) = \epsilon Q. \tag{15.2.6}$$

Consequently the electromagnetic charge $Q$ is indeed the only unbroken symmetry generator.

## 15.3 – Identifying the Photon

The $Q = 0$ bosons are $W_3$ and $B$. The photon field $A$ must be a normalized linear combination of $W_3$ and $B$; the $Z$ field is the orthogonal linear combination:

$$\left. \begin{array}{l} A = \quad\cos\theta_W\, B + \sin\theta_W\, W_3 \\ Z = -\sin\theta_W\, B + \cos\theta_W\, W_3 \end{array} \right\} \quad \text{or} \quad \left. \begin{array}{l} B = \cos\theta_W\, A - \sin\theta_W\, Z \\ W_3 = \sin\theta_W\, A + \cos\theta_W\, Z \end{array} \right\} \quad (15.3.1)$$

The parameter $\theta_W$ is known as the *Weinberg angle*.

This definition of $A$ and $Z$ implies that the neutral vector boson contribution to the covariant derivative may be written

$$\begin{aligned} gW_3 T_3 + g'BY &= A(g\sin\theta_W\, T_3 + g'\cos\theta_W\, Y) \\ &\quad + Z(g\cos\theta_W\, T_3 - g'\sin\theta_W\, Y). \end{aligned} \quad (15.3.2)$$

The coupling of the photon field $A$ to matter must be $eAQ$, and so we obtain a constraint:

$$g\sin\theta_W\, T_3 + g'\cos\theta_W\, Y = eQ = eT_3 + eY; \quad (15.3.3)$$

which implies that the unknown gauge coupling constants $g$ and $g'$ can be expressed in terms of the unknown Weinberg angle and the known value of $e$:

$$g = \frac{e}{\sin\theta_W} \quad \text{and} \quad g' = \frac{e}{\cos\theta_W}. \quad (15.3.4)$$

Eliminating $g$ and $g'$ from the neutral part of the covariant derivative yields the useful expression

$$gW_3 T_3 + g'BY = AeQ + Z\frac{e}{\sin\theta_W \cos\theta_W}(T_3 - \sin^2\theta_W\, Q). \quad (15.3.5)$$

## 15.4 – Covariant Derivative

Now that we have identified the photon, the next step in developing a computational understanding of the Lagrangian 15.1.9 is to compute specific matrix forms for the actions of the covariant derivative $D_\mu$ on the various $SU(2)$ multiplets in the Lagrangian. These matrices should be expressed in terms of the preferred low-energy basis of vector fields, $W^\pm$, $Z$ and $A$.

On $SU(2)$ singlets $T_3 = 0$ and $Y = Q$, so the covariant derivative takes the form:

$$D_\mu = \partial_\mu - ig'B_\mu Q. \quad (15.4.1)$$

On $SU(2)$ doublets the covariant derivative $D_\mu$ takes the form (suppressing $\mu$'s):

$$D = \partial - i\begin{pmatrix} eAQ + \left(\dfrac{e}{\sin 2\theta_W} - \dfrac{e\sin\theta_W}{\cos\theta_W}Q\right)Z & \dfrac{e}{\sqrt{2}\sin\theta_W}W^+ \\[2ex] \dfrac{e}{\sqrt{2}\sin\theta_W}W^- & eAQ - \left(\dfrac{e}{\sin 2\theta_W} + \dfrac{e\sin\theta_W}{\cos\theta_W}Q\right)Z \end{pmatrix}. \quad (15.4.2)$$

## 15.5 – Particle Masses

After symmetry breaking, the Higgs kinetic term $(D_\mu \phi)^\dagger (D^\mu \phi)$ contains a product of vector fields and the Higgs VEV. This product is the mass term for the vector bosons.

In detail, after symmetry breaking,

$$D_\mu \phi = D_\mu \left( \begin{pmatrix} 0 \\ v/\sqrt{2} \end{pmatrix} + \begin{pmatrix} 0 \\ \phi' \end{pmatrix} \right) \tag{15.5.1}$$

which, according to 15.4.2 with the appropriate values of $Q$, contains the term

$$-i \begin{pmatrix} eA_\mu + \left( \dfrac{e}{\sin 2\theta_W} - \dfrac{e \sin \theta_W}{\cos \theta_W} \right) Z_\mu & \dfrac{e}{\sqrt{2} \sin \theta_W} W_\mu^+ \\ \dfrac{e}{\sqrt{2} \sin \theta_W} W_\mu^- & -\dfrac{e}{\sin 2\theta_W} Z_\mu \end{pmatrix} \begin{pmatrix} 0 \\ \dfrac{v}{\sqrt{2}} \end{pmatrix}$$

$$= -\dfrac{iv}{\sqrt{2}} \begin{pmatrix} \dfrac{e}{\sqrt{2} \sin \theta_W} W_\mu^+ \\ -\dfrac{e}{\sin 2\theta_W} Z_\mu \end{pmatrix}. \tag{15.5.2}$$

Hence the Higgs kinetic energy term $(D_\mu \phi)^\dagger (D^\mu \phi)$ contains the term

$$\dfrac{v^2}{2} \left\| \begin{matrix} \dfrac{e}{\sqrt{2} \sin \theta_W} W_\mu^+ \\ -\dfrac{e}{\sin 2\theta_W} Z_\mu \end{matrix} \right\|^2 = \dfrac{e^2 v^2}{4 \sin^2 \theta_W} W^{+\mu} W_\mu^- + \dfrac{e^2 v^2}{8 \sin^2 \theta_W \cos^2 \theta_W} Z^\mu Z_\mu. \tag{15.5.3}$$

These are the vector boson mass terms. The photon field $A$ is massless because it is associated with the unbroken symmetry. The masses of the other fields are

$$M_W = \dfrac{ev}{2 \sin \theta_W} \quad \text{and} \quad M_Z = \dfrac{ev}{2 \sin \theta_W \cos \theta_W}. \tag{15.5.4}$$

Notice that measurement of $M_W$ and $M_Z$ determines the Weinberg angle and the VEV $v$.

Absorbing a phase into $e_R$ if necessary, we can assume that the coupling constant $f$ in 15.1.9 is real and positive. Then, after symmetry breaking, the term

$$f \bar{e}_R \phi^\dagger \psi_L + f \bar{\psi}_L \phi e_R \tag{15.5.5}$$

contains the electron mass term

$$\dfrac{fv}{\sqrt{2}} (\bar{e}_R e_L + \bar{e}_L e_R). \tag{15.5.6}$$

Since $f$ is another unknown, there is no predictive power in this expression.

## 15.6 – The Couplings of the Vector Bosons to Leptons

Since lepton scattering and weak decay experiments are of immediate relevance to the Lagrangian density 15.1.9, we will be most interested in the coupling of the leptons to the vector bosons. For this purpose we will use the form 15.4.2 of the covariant derivative with appropriate values of $Q$ to analyze the terms $\bar{\psi}_L i\slashed{D}\psi_L$ and $\bar{e}_R i\slashed{D}e_R$ in the Lagrangian density:

$$\slashed{D}\psi_L = \slashed{\partial}\psi_L - i \begin{pmatrix} \dfrac{e}{\sin 2\theta_W}\slashed{Z} & \dfrac{e}{\sqrt{2}\sin\theta_W}\slashed{W}^+ \\[2ex] \dfrac{e}{\sqrt{2}\sin\theta_W}\slashed{W}^- & -e\slashed{A} - \dfrac{e\cos 2\theta_W}{\sin 2\theta_W}\slashed{Z} \end{pmatrix} \begin{pmatrix} \nu_e \\ e_L \end{pmatrix}. \qquad (15.6.1)$$

Hence the term $\bar{\psi}_L i\slashed{D}\psi_L$ contains the interaction

$$\begin{pmatrix} \bar{\nu}_e & \bar{e}_L \end{pmatrix} \begin{pmatrix} \dfrac{e}{\sin 2\theta_W}\slashed{Z} & \dfrac{e}{\sqrt{2}\sin\theta_W}\slashed{W}^+ \\[2ex] \dfrac{e}{\sqrt{2}\sin\theta_W}\slashed{W}^- & -e\slashed{A} - \dfrac{e\cos 2\theta_W}{\sin 2\theta_W}\slashed{Z} \end{pmatrix} \begin{pmatrix} \nu_e \\ e_L \end{pmatrix}, \qquad (15.6.2)$$

which multiplies out to

**Couplings to Left Leptons** $= \dfrac{e}{2\sin\theta_W \cos\theta_W}\bar{\nu}_e\slashed{Z}\nu_e$ $\qquad (15.6.3)$

$$+ \frac{e}{\sqrt{2}\sin\theta_W}(\bar{\nu}_e\slashed{W}^+ e_L + \bar{e}_L\slashed{W}^-\nu_e)$$

$$- \bar{e}_L\left(e\slashed{A} + \frac{e\cos 2\theta_W}{2\sin\theta_W \cos\theta_W}\slashed{Z}\right)e_L.$$

This expression contains all the couplings between the left leptons and the vector bosons.

The couplings to the right lepton are simple. Since $e_R$ is an $SU(2)$ singlet it is annihilated by the $T^a$ and does not couple to the $W^a$ at all. The vector boson coupling for $e_R$ comes only from the term $\bar{e}_R i\slashed{D}e_R$, and may be written down from the expression 15.4.1 for the covariant derivative on singlets:

**Couplings to Right Lepton** $= -\bar{e}_R(-g'\slashed{B}Y)e_R$

$$= -\bar{e}_R\big(-g'\slashed{B}(-1)\big)e_R \qquad (15.6.4)$$

$$= -\bar{e}_R(e\slashed{A} - e\tan\theta_W \slashed{Z})e_R.$$

Summing the expressions 15.6.3 and 15.6.4 for left and right couplings gives the expression for the total coupling of leptons to bosons:

**Lepton–Boson Couplings** $\qquad\qquad\qquad\qquad\qquad (15.6.5)$

$$= \frac{e}{\sqrt{2}\sin\theta_W}(\bar{\nu}_e\slashed{W}^+ e_L + \bar{e}_L\slashed{W}^-\nu_e)$$

$$+ \frac{e}{\sin 2\theta_W}(\bar{\nu}_e\slashed{Z}\nu_e - \cos 2\theta_W\,\bar{e}_L\slashed{Z}e_L + 2\sin^2\theta_W\,\bar{e}_R\slashed{Z}e_R)$$

$$- e(\bar{e}_L\slashed{A}e_L + \bar{e}_R\slashed{A}e_R).$$

All the tree-level scattering of leptons is summarized in this expression. Loop diagrams, of course, could also contain the cubic and quartic self-interactions of the vector fields.

## 15.7 – Effective Four-Fermi Interactions

Since the masses of the $W$ and $Z$ bosons are in fact very large compared to the energies of low-energy scattering experiments, it is sensible to build an effective Lagrangian density for low-energy processes which does not refer to these vector bosons. The processes in question are lepton scatterings and decays, both of which are described by diagrams with four external Fermi lines. Thus the effective Lagrangian density must have interactions of the form $\bar{\psi}_1 A \psi_2 \bar{\psi}_3 B \psi_4$, where $A$ and $B$ are arbitrary $4 \times 4$ matrices. These *four-Fermi* interactions are not renormalizable, and can only be used at tree level. The essential element in building the low-energy effective Lagrangian density is the computation of the coupling constants for its four-Fermi interactions.

Scattering amplitudes are functions of $E_T$ and the scattering angle. Thus the coupling 'constants' for the four-Fermi interactions will be energy dependent. However, if we assume that $E_T$ is well below the $W$ and $Z$ masses, then it is reasonable to compute to first order in these masses. In this approximation, the couplings are constant to order $(E_T/M_W)$.

The essence of the low-energy approximation lies in discarding loop diagrams and simplifying the propagators for the massive vector bosons according to the following schema:

$$
\left(-g_{\mu\nu} + \frac{k_\mu k_\nu}{\mu^2}\right) \frac{i}{k^2 - \mu^2 + i\epsilon} = \frac{ig_{\mu\nu}}{\mu^2} + \frac{k^2 g_{\mu\nu} - k_\mu k_\nu}{\mu^4} + \cdots
$$
$$
\simeq \frac{ig_{\mu\nu}}{\mu^2}.
$$

(15.7.1)

It is the constancy of the resulting approximation that makes the couplings in the effective Lagrangian density constant.

The lepton aspect of the scattering and decay diagrams is computed from 15.6.5, in which the heavy vector bosons couple to fermion currents as follows:

$$
\left.
\begin{aligned}
W_\mu^1: &\quad \frac{e}{2\sin\theta_W}(\bar{\nu}_e \gamma^\mu e_L + \bar{e}_L \gamma^\mu \nu_e) \\
W_\mu^2: &\quad -\frac{ie}{2\sin\theta_W}(\bar{\nu}_e \gamma^\mu e_L - \bar{e}_L \gamma^\mu \nu_e)
\end{aligned}
\right\}
$$

(15.7.2)

or

$$
\left.
\begin{aligned}
W_\mu^+: &\quad \frac{e}{\sqrt{2}\sin\theta_W}\bar{\nu}_e \gamma^\mu e_L \\
W_\mu^-: &\quad \frac{e}{\sqrt{2}\sin\theta_W}\bar{e}_L \gamma^\mu \nu_e
\end{aligned}
\right\}
$$

(15.7.3)

and

$$
Z_\mu: \quad \frac{e}{\sin 2\theta_W}(\bar{\nu}_e \gamma^\mu \nu_e - \cos 2\theta_W \, \bar{e}_L \gamma^\mu e_L + 2\sin^2\theta_W \bar{e}_R \gamma^\mu e_R).
$$

(15.7.4)

Combining these couplings with the simplified form of the propagators for the massive vector fields leads to low energy ($k^2 \ll M_W^2$) scattering amplitudes of the form:

$$iA = (-i)^2 (\text{coupling constant})^2 (\text{four spinor term}) \frac{ig_{\mu\nu}}{\text{boson mass}^2}, \qquad (15.7.5)$$

where (coupling constant)$^2$ represents the product of two coupling constants from 15.6.5.

To illustrate how these amplitudes can be represented by four-Fermi interactions, we work out the details for the $W^-$ contribution to $e_L \bar\nu$ scattering. The amplitude for this contribution is:

$$
\begin{aligned}
iA &= -\left(\frac{e}{\sqrt{2}\sin\theta_W}\right)^2 (\bar\nu_e \gamma^\mu e_L)(\bar e_L \gamma^\nu \nu_e)\left(\frac{ig_{\mu\nu}}{M_W^2}\right) \\
&= (-i)\left(\frac{e}{\sqrt{2}M_W \sin\theta_W}\right)^2 (\bar\nu_e \gamma^\mu e_L)(\bar e_L \gamma_\mu \nu_e),
\end{aligned}
\qquad (15.7.6)
$$

where the field symbols now represent the associated spinors.

Thus the $W^-$ contribution to $e_L\bar\nu$ scattering is duplicated by the four-fermi operator

$$\textbf{Four-Fermi Operator} = -\left(\frac{e}{\sqrt{2}M_W \sin\theta_W}\right)^2 (\bar\nu_e \gamma^\mu e_L)(\bar e_L \gamma_\mu \nu_e). \qquad (15.7.7)$$

Hence, as far as this process is concerned, we may drop the $W^-$ from the Lagrangian density as long as we add the four-Fermi operator above.

There are, however, many four-fermi operators arising from the couplings 15.6.5. It is convenient to sum them, and express the sum in terms of the self-interactions of the fermionic parts of the Noether currents. From the generic formula

$$J^\mu \overset{\text{def}}{=} \bar\psi \gamma^\mu T \psi \qquad (15.7.8)$$

we write down:

$$
\left.
\begin{aligned}
J_1^\mu &= \frac{1}{2}\bar\nu_e \gamma^\mu e_L + \frac{1}{2}\bar e_L \gamma^\mu \nu_e \\
J_2^\mu &= -\frac{i}{2}\bar\nu_e \gamma^\mu e_L + \frac{i}{2}\bar e_L \gamma^\mu \nu_e \\
J_3^\mu &= \frac{1}{2}\bar\nu_e \gamma^\mu \nu_e - \frac{1}{2}\bar e_L \gamma^\mu e_L
\end{aligned}
\right\}
\qquad (15.7.9)
$$

or

$$
\left.
\begin{aligned}
J_+^\mu &= J_1^\mu - iJ_2^\mu = \bar e_L \gamma^\mu \nu_e \\
J_-^\mu &= J_1^\mu + iJ_2^\mu = \bar\nu_e \gamma^\mu e_L
\end{aligned}
\right\}
\qquad (15.7.10)
$$

and

$$J_Y^\mu = -\frac{1}{2}\bar\psi_L \gamma^\mu \psi_L - \bar e_R \gamma^\mu e_R. \qquad (15.7.11)$$

From the Lie algebra equalities

$$Q = T_3 + Y \quad \text{and} \quad Q_Z = T_3 - \sin^2\theta_W Q, \qquad (15.7.12)$$

we deduce the currents associated with these generators:

$$J_{\text{EM}}^{\mu} = -\bar{e}_L \gamma^{\mu} e_L - \bar{e}_R \gamma^{\mu} e_R, \tag{15.7.13}$$

$$\text{and} \quad J_Z^{\mu} = \frac{1}{2}\bar{\nu}_e \gamma^{\mu} \nu_e - \frac{1}{2} \cos 2\theta_W \, \bar{e}_L \gamma^{\mu} e_L + \sin^2 \theta_W \, \bar{e}_R \gamma^{\mu} e_R.$$

These currents agree with those displayed in 15.7.2 and 15.7.4 up to coupling factors.

Clearly, since the current coupled to a massive vector boson is proportional to the corresponding conserved current, the sum of the four-Fermi interactions associated with a massive vector boson is proportional to the square of the current. Using the traditional parameter $G_F$ (known as *G-Fermi*), substituting the currents 15.7.10 in the operator 15.7.6, the four-Fermi interactions with intermediate $W^-$'s are given by

$$\textbf{Four-Fermi Operator} = -4\frac{G_F}{\sqrt{2}}\bar{e}_L \gamma_{\mu}\nu_e \, \bar{\nu}_e \gamma^{\mu} e_L = -4\frac{G_F}{\sqrt{2}}J_{+\mu}J_-^{\mu}. \tag{15.7.14}$$

The constant $G_F$ can be determined by comparing the leading monomial with the four-fermi interaction term 15.7.7:

$$\frac{G_F}{\sqrt{2}} = \frac{1}{4}\left(\frac{e}{\sqrt{2}M_W \sin \theta_W}\right)^2 = \frac{e^2}{8M_W^2 \sin^2 \theta_W}. \tag{15.7.15}$$

**Homework 15.7.16.** Show that $J_1^{\mu} J_{1\mu} + J_2^{\mu} J_{2\mu} = 4J_+^{\mu} J_{-\mu}$, so that the part of $J_1^{\mu} J_{1\mu}$ which does not conserve charge is canceled by the contribution from $J_2^{\mu} J_{2\mu}$.

For the $Z$ boson, notice that

$$\left(\frac{2e}{\sin 2\theta_W}\right)^2 \frac{1}{M_Z^2} = \left(\frac{e}{\sin \theta_W}\right)^2 \frac{1}{M_W^2}, \tag{15.7.17}$$

so that the product of the damping influence of a boson mass with the square of its coupling strength is independent of the vector boson. Hence the coefficient that connects the $Z$ current-current interaction $J_Z^{\mu} J_{Z\mu}$ with the effective four-fermi interactions mediated by the $Z$ boson is the same as the coefficient $4G_F/\sqrt{2}$ that connects the $W_-$ current-current interaction with the effective four-fermi interactions mediated by the $W_-$ boson.

In summary, we have determined in 15.7.6 and 15.7.7 the structure and coefficient of the four-Fermi operator that reproduces a tree-level lepton scattering diagram, seen that the currents 15.7.2 and 15.7.4 that couple to the vector bosons in the Lagrangian are simply the fermionic part 15.7.10 of the conserved currents, and have concluded that the sum of current-current interactions is proportional to the sum of the four-fermi operators of type 15.7.7. Comparison of coefficients of a single four-fermi operator fixes the constant of proportionality 15.7.15. Thus all possible tree-level lepton processes mediated by the heavy bosons $W^{\pm}$ or $Z$ may be computed from the four-fermi interaction Lagrangian density

$$\mathcal{L}' = -\frac{G_F}{\sqrt{2}}\left(4J_{+\mu}J_-^{\mu} + 4(J_3^{\mu} - \sin^2 \theta_W \, J_{\text{EM}}^{\mu})(J_{3\mu} - \sin^2 \theta_W \, J_{\text{EM}\mu})\right). \tag{15.7.18}$$

## 15.8 – Three Families of Leptons

We have presented the Standard Model for one family of leptons. The theory is not substantially changed by the addition of more families. As we shall show below, the families properly defined do not mix.

Experimental evidence for the number of families is based on lepton loop contributions to scatterings. It is now accepted that there are three families of leptons. We shall therefore analyze the three-family Standard Model.

The novel possibility is of family mixing. The vector bosons and Higgs fields are in common to all families, and so the three-family Lagrangian density comprises the vector boson and Higgs kinetic terms, kinetic terms for $\psi_L = (\psi_L^1, \psi_L^2, \psi_L^3)^\top$ and $e_R = (e_R^1, e_R^2, e_R^3)^\top$, the Higgs potential, and the interaction term

$$\mathcal{L}_I \stackrel{\text{def}}{=} \bar{\psi}_L \phi M e_R + \text{h.c.} = M_{ij} \bar{\psi}_L^i \phi e_R^j + \text{h.c.}, \tag{15.8.1}$$

where $M$ is an arbitrary $3 \times 3$ matrix.

Fortunately, the Lagrangian density without the interaction term has a symmetry $\psi_L \to U\psi_L$ and $e_R \to Ve_R$, where $U$ and $V$ are arbitrary elements $U(3)$. Symmetries of this kind are known as *flavor*, *family*, or *generation* symmetries. This particular example is a bit of a fake because it is not a symmetry of the Lagrangian as a whole. However, we can use $U$ and $V$ to diagonalize $M$, thereby finding a basis in which no family is directly coupled to any other family.

The connection between this theoretically preferable basis for the families and the basis of physical particles is good because the theoretical basis is the most stable over time. The decoupling of the families is reflected in a $U(1)^3$ symmetry,

$$\left.\begin{array}{l} \psi_L^j \to e^{-i\theta_j} \psi_L^j \\ e_R^j \to e^{-i\theta_j} e_R^j \end{array}\right\} \tag{15.8.2}$$

and in the consequent conservation of family numbers in electroweak processes. Thus, for example, we find decays like

$$\mu \to e + \bar{\nu}_e + \nu_\mu \tag{15.8.3}$$

are permitted, but $\mu \to e + \nu_e + \bar{\nu}_\mu$ cannot occur in the Standard Model.

**Homework 15.8.4.** The left Fermi fields $\psi^r$ for $1 \le r \le 9$ in the three-family theory are $e_R^{j*}$, $e_L^j$, and $\nu_L^j$ for $1 \le j \le 3$. The pure kinetic term for the Fermi fields is $\bar{\psi}^r i \partial\!\!\!/ \psi^r$. This term has $U(9)$ symmetry. Show how the gauge couplings reduce this $U(9)$ to the $U(1)^3$ symmetry of family conservation.

**Homework 15.8.5.** Using the four-Fermi interaction Lagrangian density 15.7.18 generalized to three generations of leptons, taking the electron mass to be zero, find the decay width for the muon decay process 15.8.3. Estimate the lifetime of a muon at rest.

As an example of a process which reveals the left-right asymmetry of the electroweak interactions, consider $e^- e^+ \to \mu^- \mu^+$. At tree level, only the annihilation diagram contributes, but the internal line in this diagram may be either a photon or a $Z$ boson. From the theory, the $Z$ has both vector ($V$) and axial vector ($A$) couplings while the photon has only vector couplings. Therefore, the amplitudes for the $Z$ and $\gamma$ contributions are of the forms

$$\mathcal{A}_Z = V_Z + A_Z \quad \text{and} \quad \mathcal{A}_\gamma = V_\gamma, \tag{15.8.6}$$

so the differential cross section

$$d\sigma \propto |A_Z + A_\gamma|^2 \qquad (15.8.7)$$

contains an interference term of the form

$$d\sigma_I \propto V_Z V_Z^* V_\gamma V_\gamma^* + V_Z A_Z^* V_\gamma V_\gamma^* + A_Z V_Z^* V_\gamma V_\gamma^* + A_Z A_Z^* V_\gamma V_\gamma^*. \qquad (15.8.8)$$

The $VVVV$ term is like the dominant $|A_\gamma|^2$ contribution, symmetric under interchange of the final state $\mu^+$ and $\mu^-$. It therefore contributes equally to $\mu^+$ scattering at angles $\theta$ and $\theta + \pi$. The $VAVV$ terms are helicity dependent and so violate parity, but these effects have not yet been observed. Finally, the $AAVV$ term is anti-symmetric under interchange of the final state muons, and gives rise therefore to an asymmetry in the $\mu^+$ scattering angles. This asymmetry has been observed and agrees well with the predicted effect.

**Homework 15.8.9.** In $e^-e^+ \to \mu^-\mu^+$ scattering, let $\theta$ be the angle between the incoming $e^-$ and the outgoing $\mu^-$ in the center of mass frame. Assuming that the energy of the electrons is large compared to the muon mass (so that lepton masses may be ignored) but small compared to the $Z$ mass (so that the effective interactions may be used and the purely weak contribution may be neglected), compute the differential cross section $d\sigma/dc$, where $c = \cos\theta$. Find the front-back asymmetry:

$$\frac{1}{\sigma}\left( \int_0^1 dc \frac{d\sigma}{dc} - \int_{-1}^0 dc \frac{d\sigma}{dc} \right) \simeq -\frac{3}{4\sqrt{2}} \frac{G_F}{e^2} E_T^2.$$

This concludes our presentation of the structure of the electroweak Lagrangian density. The theory is elegant, simple, and successful. Even at tree level, the theory generates a wide variety of characteristic effects. Beyond tree level lies the theory of loop effects, which rests on renormalization. From renormalization comes the renormalization group equations which determine the coupling constants in the effective theory at all different energy levels. With this level of understanding, the masses of the $W$ and $Z$ bosons can be correctly predicted, and indications for grand unification can be discovered.

# 15.9  The QCD Lagrangian

As electroweak theory does not explain hadrons, it is necessary to propose an extension of the Weinberg–Salam model to include hadrons. Hadrons, however, are understood to be bound states of quarks, and so instead of extending the Weinberg–Salam theory to include hadrons, one extends it to include quarks.

It is generally believed that quarks form hadrons under the influence of the strong force, and that this force is mediated by an $SU(3)$ octet of gauge fields, the gluons. From experiment, it appears that this $SU(3)$ symmetry is unbroken, and therefore not directly detectable. There are, however, a number of motivations and verifications for this color $SU(3)$.

The spin-statistics theorem provides the theoretical motivation for including in our theory a symmetry which in principle can never be observed. For example, it is thought that the $\Delta^{++}$ fermion is made of three $u$ quarks (and an unknown state of the gluon fields), and that the $3/2$ spin of the $\Delta^{++}$ comes from the alignment of the

1/2 spins of its constituent quarks. Thus the field for the $\Delta^{++}$ is symmetric under interchange of constituents. But the spin-statistics theorem asserts that half-integer spin particles must obey Fermi statistics. The remedy is to make each quark come in three colors and to require the $\Delta^{++}$ to be a color singlet state so that the $\Delta^{++}$ is automatically anti-symmetric in color indices. Therefore, in the interests of building a consistent theory of quarks as constituents of known particles, we assume that quarks come in three colors and hadrons are color singlets.

Verification comes from many decay and scattering experiments which indicate that if the quarks could be distinguished by observable quantum numbers, the quark contributions to the relevant amplitudes would be about three times too small. (When we wrote earlier that experiment indicates six quarks, we meant six $SU(3)$ vectors of quarks.)

Having indicated the case for an $SU(3)$ gauge theory of quarks, we can proceed to extend the Weinberg–Salam model to the quarks. To fit the quarks into the Weinberg–Salam model framework we assume *a priori* that the left-handed fields form $SU(2)$ doublets while the right-handed fields are singlets.

In QCD, there are six quark types. The kinetic term for the quarks does not mix either right fields with left or the three colors. Hence the kinetic term may be written in terms of 36 Weyl fermions. Since the kinetic term for a Dirac fermion $\psi$ can be written in terms of the left field $\psi_L$ and the charge conjugate of the right field $\psi_{cL} = U_C^\dagger \psi_R U_C$,

$$
\begin{aligned}
\textbf{Kinetic Term} &= \bar{\psi} i \slashed{\partial} \psi \\
&= \bar{\psi}_L i \slashed{\partial} \psi_L + \bar{\psi}_R i \slashed{\partial} \psi_R \\
&= \bar{\psi}_L i \slashed{\partial} \psi_L + \bar{\psi}_{cL} i \slashed{\partial} \psi_{cL} - \partial_\mu (\bar{\psi}_{cL} i \gamma^\mu \psi_{cL}),
\end{aligned}
\tag{15.9.1}
$$

we see that the quark kinetic term has the form $\bar{\psi}_L^r i \slashed{\partial} \psi_L^r$ where $1 \le r \le 36$. This term clearly has $U(36)$ symmetry.

As we are building an $SU(3) \times SU(2) \times U(1)$ gauge theory, we will want covariant derivatives in the kinetic energy term. Since the $U(36)$ does not act on the gauge fields, these couplings break $U(36)$ to the maximal subgroup that commutes with $SU(3) \times SU(2) \times U(1)$.

The 36 left quark fields form nine irreducible representations under the action of $SU(3) \times SU(2)$. Each irreducible representation of a product of simple groups is a tensor product of irreducible representations of the simple groups taken separately. Hence we can identify an irreducible representation of $SU(3) \times SU(2)$ by the degrees $m$ and $n$ of the factors in this tensor product. Thus, in this context, we write $(m, n)$ for the irreducible representation of $SU(3) \times SU(2)$ of degree $mn$ formed by tensoring the irreducible representations of $SU(3)$ and $SU(2)$ which have degrees $m$ and $n$ respectively. In this notation, the left up-type and down-type quarks form three $(3, 2)$'s, the left anti-up-type quarks and the left anti-down-type quarks form six $(3, 1)$'s:

$$
\begin{array}{ccc}
36 & \longrightarrow & 3 \times (3, 2) + 6 \times (3, 1) \\
\text{of } U(36) & & \text{of } SU(3) \times SU(2)
\end{array}
\tag{15.9.2}
$$

The charges of the quarks must also be considered. Up-type and down-type quarks have charges $2/3$ and $-1/3$ respectively, while the corresponding anti-quarks have the opposite charges. Hence the part of $U(36)$ which commutes with $SU(3) \times SU(2) \times U(1)$ must preserve the classification into $(3, 2)$'s, anti-up-type $(3, 1)$'s, and anti-down-type $(3,1)$'s. Since every field in a given multiplet is uniquely identified by $SU(3) \times SU(2) \times U(1)$ quantum numbers, the coupling to the gauge fields breaks the $U(36)$ symmetry of the kinetic terms to $U(3)^3$.

Since the remaining symmetry does not mix quarks with anti-quarks, we shall drop the charge-conjugate fields in favor of right-handed fields. Since we have three families of quarks, each set of quantum numbers identifies a three-dimensional space of quark fields. Writing superscripts and subscripts to identify color and family respectively, we may freely choose the nine orthonormal fields $U_{iL}^1$, $U_{iR}^1$, and $D_{iR}^1$. Then the $su(2)$ lowering operator determines $D_{iL}^1$ and the $su(3)$ lowering operators determine the other elements of the basis. Clearly, the freedom in the choice of basis is the $U(3)^3$ symmetry found above.

We have now specified the $SU(3) \times SU(2)$ transformation properties of the quarks and also their electromagnetic charge. The hypercharge assignments are therefore determined by the relation $Q = T^3 + Y$. The following table displays the quark isospin, hypercharge, and electromagnetic charges:

|  | $u_L$ | $d_L$ | $u_R$ | $d_R$ |
|---|---|---|---|---|
| $T_3$ | $\frac{1}{2}$ | $-\frac{1}{2}$ | $0$ | $0$ |
| $Y$ | $\frac{1}{6}$ | $\frac{1}{6}$ | $\frac{2}{3}$ | $-\frac{1}{3}$ |
| $Q$ | $\frac{2}{3}$ | $-\frac{1}{3}$ | $\frac{2}{3}$ | $-\frac{1}{3}$ |

(15.9.3)

We shall use the Gell-Mann matrices $\lambda^a$ for the generators of $SU(3)$, and write $f^{abc}$ for the associated structure constants. The generators of the product group $SU(3) \times SU(2) \times U(1)$ should be written as tensor products, for example

$$\hat{\lambda}^a \overset{\text{def}}{=} \lambda^a \otimes \mathbf{1}_2 \otimes 1, \tag{15.9.4}$$

but as before in Section 15.1, we shall suppress the identity elements.

Assuming only the leptons, vector bosons and Higgs scalars of the electroweak Lagrangian, we find that the fractional charges of the quarks (a consequence of the hypothesis that baryons comprise three quarks) make it impossible to couple quarks to leptons, but do not obstruct coupling to the Higgs fields. Hence the QCD Lagrangian density involves only the $SU(2)$ gauge bosons and the Higgs fields of the electroweak theory:

$$\begin{aligned}
\mathcal{L}_{\text{QCD}} = &-\frac{1}{4}G_{\mu\nu}^a G^{a\mu\nu} + \bar{\chi}_{jL} i \slashed{D} \chi_{jL} + \bar{D}_{jR} i \slashed{D} D_{jR} + \bar{U}_{jR} i \slashed{D} U_{jR} \\
&- g_{ij}\bar{\chi}_L^i \phi D_R^j - h_{ij}\bar{\chi}_L^i \tilde{\phi} U_R^j + \text{h.c.},
\end{aligned} \tag{15.9.5}$$

where

$$\begin{aligned}
D_\mu &= \partial_\mu - ig_s G_\mu^a \frac{\lambda^a}{2} - igW_\mu^a T^a - ig' B_\mu Y \\
G_{\mu\nu}^a &= \partial_\mu G_\nu^a - \partial_\nu G_\mu^a + g_s f^{abc} G_\mu^b G_\nu^c \\
\chi_{iL} &= (U_{iL}, D_{iL})^{\mathsf{T}},
\end{aligned} \tag{15.9.6}$$

and $\tilde{\phi}$ is another form of the Higgs doublet $\phi$:

$$\tilde{\phi} = \begin{pmatrix} (\phi^0)^* \\ -(\phi^+)^* \end{pmatrix} = \begin{pmatrix} 0 & 1 \\ -1 & 0 \end{pmatrix} \begin{pmatrix} \phi^+ \\ \phi^0 \end{pmatrix}^* = \epsilon \phi^*. \tag{15.9.7}$$

The matrix $\epsilon$ above is the similarity transformation that converts the identity representation of $SU(2)$ into its conjugate:

$$\epsilon(U\phi)^* = \epsilon U^*\phi^* = U\epsilon\phi^* = U\tilde{\phi},$$

showing that $\tilde{\phi}$ is an $SU(2)$ doublet.

The Higgs potential in the electroweak part of the total Lagrangian is still effective in giving the Higgs field a VEV. This VEV does not break the color symmetry of QCD, but provides a quark mass matrix. Recall that in unitary gauge the Higgs field $\phi$ becomes

$$\phi = \begin{pmatrix} \phi^+ \\ \phi^0 \end{pmatrix} = \begin{pmatrix} 0 \\ v/\sqrt{2} \end{pmatrix} + \begin{pmatrix} 0 \\ \phi' \end{pmatrix}, \tag{15.9.8}$$

causing $\tilde{\phi}$ to become

$$\tilde{\phi} = \epsilon\phi^*$$
$$= \begin{pmatrix} v/\sqrt{2} \\ 0 \end{pmatrix} + \begin{pmatrix} \phi' \\ 0 \end{pmatrix}. \tag{15.9.9}$$

Substituting these expressions for $\phi$ and $\tilde{\phi}$ into the QCD Lagrangian gives rise to quark mass terms:

$$\textbf{Quark Mass Term} = -\frac{g_{ij}v}{\sqrt{2}}\bar{D}_{iR}D_{jL} - \frac{h_{ij}v}{\sqrt{2}}\bar{U}_{iR}U_{jL} + \text{h.c.} \tag{15.9.10}$$

Note that, since the coefficients $g_{ij}$ and $h_{ij}$ are arbitrary, the symmetry breaking parameter $v \propto \|\langle\phi\rangle\|$ does not directly affect QCD and there is no need for more Higgs fields to generate greater diversity of masses.

**Homework 15.9.11.** Investigate the nature of the strong interactions at high energy by computing tree-level amplitudes for quark-quark and quark–anti-quark scattering by means of gluons.

## 15.10 – Diagonalizing the Quark Mass Matrices

We can now use the $U(3)^3$ symmetry of the Lagrangian density 15.9.5 to simplify the quark mass term 15.9.10. The governing principle of the ensuing simplification is expressed in the diagonalization theorem of Section 7.17, namely, if $M$ is an arbitrary complex matrix, then there exist unitary matrices $U$ and $V$ such that $UMV^\dagger$ is diagonal, real and non-negative.

Write $M_{ij}$ for $h_{ij}v/\sqrt{2}$, and select unitary change of basis matrices $U$ and $V$ for the $U_R$'s and $U_L$'s respectively to diagonalize $M$:

$$M_U \overset{\text{def}}{=} UMV^\dagger = \begin{pmatrix} m_u & 0 & 0 \\ 0 & m_c & 0 \\ 0 & 0 & m_t \end{pmatrix}. \tag{15.10.1}$$

This change of basis determines the up, charm, and top quark fields, and their masses.

Next write $M'_{ij}$ for $g_{ij}v/\sqrt{2}$, and select unitary change of basis matrices $U'$ and $V'$ for the $D_R$'s and $D_L$'s respectively to diagonalize $M'$:

$$M_D \overset{\text{def}}{=} U'M'V'^\dagger = \begin{pmatrix} m_d & 0 & 0 \\ 0 & m_s & 0 \\ 0 & 0 & m_b \end{pmatrix}. \tag{15.10.2}$$

This change of basis determines the down, strange, and bottom quark fields, and their masses.

There is a problem, however: we are only allowed to choose three unitary matrices, not four. It might seem unnatural to worry about preserving $SU(2)$ doublets after symmetry breaking, but even after symmetry breaking, the coupling of the $W^\pm$ bosons still reveals this pairing of the left-handed fields. To preserve the $SU(2)$ doublet structure we must have $V' = V$. Hence the mass terms can only be reduced to

$$-\bar{U}_R M_U U_L - \bar{D}_R M_D K^\dagger D_L + \text{h.c.}, \tag{15.10.3}$$

where

$$K^\dagger = V'V^\dagger. \tag{15.10.4}$$

The quark fields in this form of the mass term have pure $SU(3) \times SU(2) \times U(1)$ quantum numbers.

The matrix $K$ is a measure of the discrepancy between the mass and charge eigenquarks. It is a convention to make the up-type and right-handed down-type quark fields both mass and charge eigenfields, and to focus the discrepancy in the left-handed down-type quark fields.

We can redefine the phases of the left and right down-type quarks to simplify $K$. First we return to the older four-quark theory, and show how to simplify the $2 \times 2$ unitary matrix that arose in that theory.

Any $2 \times 2$ unitary matrix $C$ can be written in the form

$$C = \begin{pmatrix} a\cos\theta & -ac\sin\theta \\ b\sin\theta & bc\cos\theta \end{pmatrix}, \tag{15.10.5}$$

where $|a| = |b| = |c| = 1$. Hence we can use phase redefinitions as follows to make $C$ real:

$$\begin{aligned} C' &= \begin{pmatrix} a^* & 0 \\ 0 & b^* \end{pmatrix} \begin{pmatrix} a\cos\theta & -ac\sin\theta \\ b\sin\theta & bc\cos\theta \end{pmatrix} \begin{pmatrix} 1 & 0 \\ 0 & c^* \end{pmatrix} \\ &= \begin{pmatrix} \cos\theta & -\sin\theta \\ \sin\theta & \cos\theta \end{pmatrix}. \end{aligned} \tag{15.10.6}$$

The remaining parameter, $\theta = \theta_C$, is known as the *Cabibbo angle*.

This argument can be extended to $K$ by factoring $K$ as follows:

$$K = U_1 U_2 U_3, \tag{15.10.7}$$

where the $U$'s are unitary with the forms:

$$U_1 = \begin{pmatrix} x & 0 & 0 \\ 0 & x & x \\ 0 & x & x \end{pmatrix}, \quad U_2 = \begin{pmatrix} x & x & 0 \\ x & x & 0 \\ 0 & 0 & x \end{pmatrix}, \quad U_3 = \begin{pmatrix} x & 0 & 0 \\ 0 & x & x \\ 0 & x & x \end{pmatrix}. \tag{15.10.8}$$

Using the Cabibbo result for $2 \times 2$ unitary matrices, choose change of phase matrices $A_1$, $B_1$, $A_3$ and $B_3$ such that $U_1' = A_1^* U_1 B_1$ and $U_3' = A_3^* U_3 B_3$ are real, and apply these matrices to $K$ as follows:

$$
\begin{aligned}
K' &= A_1^* K B_3 \\
&= A_1^* U_1 B_1 B_1^* U_2 A_3 A_3^* U_3 B_3 \\
&= U_1' U_2' U_3',
\end{aligned}
\tag{15.10.9}
$$

where $U_2' = B_1^* U_2 A_3$. Since $U_2'$ has the same form as $U_2$, we can use phase matrices $A_2$ and $B_2$ of the form $\mathrm{diag}(x, x, y)$ which will commute with $U_1'$ and $U_3'$ to make the $2 \times 2$ block in $U_2'$ real:

$$
\begin{aligned}
K'' &= A_2^* K' B_2 \\
&= U_1' A_2^* U_2' B_2 U_3' \\
&= U_1' U_2'' U_3',
\end{aligned}
\tag{15.10.10}
$$

where $U_2''$ has the form

$$
U_2'' = \begin{pmatrix} \cos\theta & -\sin\theta & 0 \\ \sin\theta & \cos\theta & 0 \\ 0 & 0 & e^{i\delta} \end{pmatrix}.
\tag{15.10.11}
$$

Having selected phases for the down-type mass eigenquarks, the resulting unitary matrix $V = K''$ is known as the Kobayashi–Maskawa or KM matrix. There are various parameterizations of this matrix, but all have three angles and one complex phase. The phase in the Lagrangian density breaks $CP$ invariance.

It is conventional to use the symbols $d$, $s$ and $b$ for the mass eigenquarks, and to introduce new symbols, $d'$, $s'$ and $t'$, for the charge eigenquarks . With this convention, $V$ converts left-handed down-type mass eigenquarks into left-handed down-type charge eigenquarks:

$$
\begin{pmatrix} d' \\ s' \\ b' \end{pmatrix}_L = \begin{pmatrix} V_{ud} & V_{us} & V_{ub} \\ V_{cd} & V_{cs} & V_{cb} \\ V_{td} & V_{ts} & V_{tb} \end{pmatrix} \begin{pmatrix} d \\ s \\ b \end{pmatrix}_L.
\tag{15.10.12}
$$

If we fix a mass eigenstate basis for left and right quark fields in which the mass matrices are in their reduced forms $M_U$ and $M_D$, then the KM matrix is transferred from the mass terms to the currents that couple to the weak vector bosons. The $U(3)^3$ symmetry of the quark kinetic term with covariant derivatives is broken by the Higgs VEV's to a residual $U(1)$ arising from an overall quark phase freedom. This symmetry corresponds to conservation of baryon number.

**Homework 15.10.13.** Write out the neutral quark currents which couple to the photon and the $Z$ boson. Show that the KM matrix can be eliminated from these currents. This shows that at tree level, neutral currents do not generate any flavor-changing effects.

A family of leptons and quarks is made up of multiplets of particles which form irreducible representations of the gauge group $SU(3) \times SU(2) \times U(1)$. At this point, it does not appear necessary to associate the lepton multiplets with quark multiplets, but in fact there is a balancing of leptons and quarks, a *generation*, which avoids anomalies. A generation of lepto-quarks consists of the following particles:

**One Generation:**

$$
\begin{array}{|ccc|c|ccc|}
\hline
\nu_e & u_L^r & u_L^b & u_L^g & & u_R^r & u_R^b & u_R^g \\
e_L & d_L^r & d_L^b & d_L^g & e_R & d_R^r & d_R^b & d_R^g \\
\hline
\bar{\nu}_e & \bar{u}_L^{\bar{r}} & \bar{u}_L^{\bar{b}} & \bar{u}_L^{\bar{g}} & & \bar{u}_R^{\bar{r}} & \bar{u}_R^{\bar{b}} & \bar{u}_R^{\bar{g}} \\
\bar{e}_L & \bar{d}_L^{\bar{r}} & \bar{d}_L^{\bar{b}} & \bar{d}_L^{\bar{g}} & \bar{e}_R & \bar{d}_R^{\bar{r}} & \bar{d}_R^{\bar{b}} & \bar{d}_R^{\bar{g}} \\
\hline
\end{array}
$$

The three generations may be briefly written:

**The Three Generations:**

$$
\begin{array}{|cc|cc|cc|}
\hline
\nu_e & u & \nu_\mu & c & \nu_\tau & t \\
e & d & \mu & s & \tau & b \\
\hline
\end{array}
$$

The KM matrix is the only part of the total Lagrangian which mixes generations. In fact the dominant branching ratios preserve generation number, but processes which mix generations are not rare.

Flavor symmetries originate in small quark masses, and are useful in analysis of processes dominated by the strong interactions at energies where the quark masses are indeed negligible. Since flavor ignores electroweak effects, it is not influenced by the KM matrix. Flavor $SU(2)$ acts on $u$ and $d$, and flavor $SU(3)$ acts on $u$, $d$, and $s$. We shall see in the next chapter how these quarks combine into hadrons, and how flavor symmetry therefore becomes an approximate symmetry of hadronic physics.

This analysis makes clear the necessary structure of the six-quark QCD Lagrangian density. The notable novelty is KM matrix which provides flavor mixing of mass eigenstates and a $CP$ violating phase. At high energies, the coupling constant $g_s$ of QCD is small enough to permit computation by perturbation theory. The next section, passing over the non-perturbative QCD of the confinement scale, introduces the hadrons, and the final section presents basic techniques of perturbative QCD in the context of scattering processes involving hadrons.

## 15.11 The Standard Model: Consistency

From Chapter 13, we know that there are three types of anomaly:

1. Local anomalies in local symmetries – these break gauge symmetry,
2. Anomalies in global symmetries – these are gauge invariant but break the global symmetry,
3. Global anomalies in local symmetries – these are also gauge invariant.

The symmetries of the Standard Model before symmetry breaking are the gauged $SU(3) \times SU(2) \times U(1)$ and the global $U(1)^4$ of lepton flavor and baryon number symmetries. In this and the following sections, we will show that the gauge symmetries are free from local anomalies, $U(1)^4$ has local anomalies, and $SU(3)$ and $SU(2)$ have global anomalies.

To verify the consistency of the Standard Model, we first show that the currents associated with gauge-group generators are free from local anomalies. To illustrate the meaning of the local anomaly theorems of Section 13.4, we first consider each factor of the gauge group separately.

In the absence of $G$ and $B$ bosons, as the $W$ bosons couple only to left-handed fermions, Theorem 13.6.6 applies, indicating a danger of non-covariant anomalies. However, all representations of $SU(2)$ are real, and so the trace condition of Theorem 13.6.8 is satisfied, and we can conclude that the coefficient of the anomaly in Theorem 13.6.6 is zero. Hence there is no anomaly involving $W$'s alone.

Similarly, in the absence of $W$ and $B$ bosons, the gluons couple equally to left and right quarks, and so Theorem 13.6.4 applies, showing that there is no anomaly in the gauge currents involving only gluons.

The $B$ boson, however, couples to right and left fermions asymmetrically, so even in the absence of $G$ and $W$ bosons, neither theorem applies. However, the more general trace test of Theorem 13.6.8 does apply, and so, even though we do not know the hypercharge anomaly as a function of $B$, we can easily verify that its coefficient is zero:

$$\operatorname{tr}\big(Y\{Y,Y\}\big) = 2\operatorname{tr}Y^3$$

$$= 2(-\tfrac{1}{2})^3 + 6(\tfrac{1}{6})^3 + (1)^3 + 3(-\tfrac{2}{3})^3 + 3(\tfrac{1}{3})^3 \qquad (15.11.1)$$
$$\quad\;\; \nu_e, e_L \qquad u_L, d_L \qquad e_{cL} \qquad u_{cL} \qquad d_{cL}$$

$$= -\tfrac{9}{36} + \tfrac{1}{36} + 1 - \tfrac{32}{36} + \tfrac{4}{36} = 0.$$

These remarks illustrate the application of the computations of Chapter 13 to the present situation, but it is of course also necessary to show that cross terms between the gauge bosons have zero coefficients. In fact, we must take recourse to Theorem 13.6.8 and verify that all traces $\operatorname{tr}\big(A\{B,C\}\big)$ vanish for all generators $A$, $B$, and $C$ of $SU(3) \times SU(2) \times U(1)$. From the structure of the representations in the Standard Model, all these trace constraints reduce to two conditions, $\operatorname{tr}Y^3 = \operatorname{tr}Y = 0$, both of which are satisfied for each generation of leptons and quarks.

**Homework** 15.11.2. Verify the assertion made in the last paragraph. More precisely, show that, since $SU(3)$ couples only to vector currents, for all generators $A$, $B$, $C$ of $SU(3) \times SU(2) \times U(1)$ there exist scalars $\alpha$ and $\beta$ such that

$$\operatorname{tr}\big(A\{B,C\}\big) = \alpha\operatorname{tr}Y + \beta\operatorname{tr}Y^3,$$

where $Y$ is the generator of the $U(1)$. (Remember, the fermions are represented as left Weyl spinors, and the trace is taken over all fermions.)

Check that $\operatorname{tr}Y$ is zero on each generation of leptons and quarks.

The conclusion of this section is that the gauge currents in the Standard Model are free from anomalies. It is remarkable that this statement is not true for the Weinberg–Salam model of leptons; the quarks must be included for consistency.

To complete the discussion of anomalies in the Standard Model, we consider the local anomalies in the $U(1)$ global symmetries in the next section, and then discuss the $SU(3)$ and $SU(2)$ global anomalies.

# 15.12 – Lepton and Baryon Number Violation

In this section, we find that topologically non-trivial configurations of the $SU(2)$ gauge field lead to violation of lepton and baryon number. Estimating the magnitude of these effects from the generating functional shows that they are not significant at ordinary energies and temperatures.

Our analysis of the symmetries of the Standard Model showed that there is a global $U(1)^4$ symmetry remaining after diagonalizing the mass matrices for the leptons and up-type quarks, and removing three phases from the KM matrix. These are the lepton-flavor and baryon number symmetries. Because the couplings in the Standard Model uphold these symmetries, they cannot be broken in perturbation theory. However, following the test for anomalies of Theorem 13.6.8, the lepton numbers of $\nu_L$, $e_L$, and $e_{cL}$ are 1, 1, and $-1$ respectively, so $\text{tr}(L_e^3) \neq 0$. Similarly, the baryon numbers of $u_L$, $d_L$, $u_{cL}$, and $d_{cL}$ are proportional to 1, 1, $-1$, and $-1$ respectively, so, since $T_a$ annihilates right-handed fields, $\text{tr}(BT_a^2) \neq 0$. We conclude that lepton-flavor and baryon numbers are not conserved.

Following the abelian-anomaly Theorem 13.4.11, the divergence in the associated currents is proportional to $\text{tr}(G_{\mu\nu}\tilde{G}^{\mu\nu})$, where $G_{\mu\nu}$ is the $SU(2)$ field strength. The violation of particle number conservation laws therefore has the form

$$\Delta N \propto \Delta Q \propto \int d^4x \, \text{tr}(G_{\mu\nu}\tilde{G}^{\mu\nu}). \tag{15.12.1}$$

Complete analysis shows that $SU(2)$ gauge fields break the four global particle flavor symmetries down to the $U(1)$ of $L - B$. In fact, the changes $\Delta L$ and $\Delta B$ in lepton and baryon number satisfy

$$\Delta L = \Delta B = -3p, \tag{15.12.2}$$

where $p$ is the Pontrjagin index of the $SU(2)$ gauge fields:

$$p = -\frac{g^2}{32\pi^2} \int d^4x \, G_{\mu\nu}^a \tilde{G}^{a\mu\nu}. \tag{15.12.3}$$

Thus $B - L$ is conserved, but $B + L$ can change by any multiple of 6.

**Homework** 15.12.4. Work out the constants of proportionality in 15.12.1 and verify the conservation of $L - B$.

To find a bound for the contribution of such gauge field configurations to the functional integral, we first use the Wick rotation and introduce the Euclidean Lagrangian density $\mathcal{L}_E$ and action $S_E$:

$$iS = i \int d^4x \, \mathcal{L} \xrightarrow{x_0 \to -ix_4} -\int d^4x_E \, \mathcal{L}_E = -S_E. \tag{15.12.5}$$

Notice the signs: to get the Euclidean Lagrangian density, we substitute $i\partial_4$ for $\partial_0$ in $-\mathcal{L}$.

In Euclidean space, the operation $G_{\mu\nu} \to \tilde{G}_{\mu\nu}$ satisfies

$$\tilde{\tilde{G}}_{\mu\nu} = G_{\mu\nu}. \tag{15.12.6}$$

It will be convenient to define $G_{\mu\nu}^{\pm}$ by

$$G_{\mu\nu}^{\pm} \stackrel{\text{def}}{=} \frac{1}{2}(G_{\mu\nu} \pm \tilde{G}_{\mu\nu}). \tag{15.12.7}$$

In the case of a gauge field $G_{\mu\nu}^a$, the Euclidean action and the Pontrjagin index can be written in terms of $G_{\mu\nu}^{\pm}$:

$$
\begin{aligned}
S_{\text{E}} &= \frac{1}{4} \int d^4 x_{\text{E}} \, G_{\mu\nu}^a G^{a\mu\nu} \\
&= \frac{1}{2} \int d^4 x_{\text{E}} \sum_{\mu < \nu} \left( |G_{\mu\nu}^{+a}|^2 + |G_{\mu\nu}^{-a}|^2 \right), \\
p_{\text{E}} &= \frac{g^2}{32\pi^2} \int d^4 x_{\text{E}} \, G_{\mu\nu}^a \tilde{G}^{a\mu\nu} \\
&= \frac{g^2}{16\pi^2} \int d^4 x_{\text{E}} \sum_{\mu < \nu} \left( |G_{\mu\nu}^{+a}|^2 - |G_{\mu\nu}^{-a}|^2 \right).
\end{aligned}
\tag{15.12.8}
$$

It is clear that

$$S_{\text{E}} \geq \frac{8\pi^2}{g^2} p_{\text{E}}. \tag{15.12.9}$$

Consequently, the contribution of the topologically non-trivial $SU(2)$ gauge configurations to the generating functional is of order

$$e^{iS_{\min}} = \exp\left(-\frac{8\pi^2 p_{\text{E}}}{g^2}\right), \tag{15.12.10}$$

where $g$ is the weak coupling constant. Note that the $1/g^2$ dependence cannot arise perturbatively. When $p_{\text{E}} = 1$, this evaluates to about $e^{-78}$. This factor would have to be extracted before expanding the $p_{\text{E}} = 1$ part of the functional integral into Feynman diagrams, and so it effectively suppresses all amplitudes for lepton-flavor and baryon non-conservation.

**Remark** 15.12.11. The inequality 15.12.9 is an equality if and only if the gauge multiplet satisfies the self-duality condition

$$G_{\mu\nu}^{-a} = 0 \quad \text{or} \quad \tilde{G}^{a\mu\nu} = G^{a\mu\nu}. \tag{15.12.12}$$

Since a self-dual configuration $G_{\mu\nu}$ is a minimal value for $S$, such configurations are *instantons*, that is, solutions of the Euclidean Euler-Lagrange equations.

**Remark** 15.12.13. When we consider Yang–Mills theory coupled to Higgs fields, then there are degenerate absolute minima connected by saddle points. The solutions of the equations of motion corresponding to the saddle points are called *sphalerons*. The energy of the $SU(2)$ sphaleron is estimated to be 7–14 TeV. There is a possibility of tunneling between different vacua (absolute minima) at high temperatures. There is also a remote chance that many-particle processes could be sensitive to the sphaleron.

There is, therefore, a tiny contribution of topologically non-trivial $SU(2)$ gauge configurations to mixing of leptons and baryons. This contribution is well within the bounds set by experimental data on lepton flavor and baryon number non-conservation. There is a remote possibility that at high temperatures or in many-particles processes these contributions could become significant.

## 15.13 – Global Anomalies

This section covers the last of the three categories of anomaly, the global anomaly. According to the analysis in Section 13.8, every simple factor in the gauge group will give rise to a set of static solutions to the Euler-Lagrange equations which has countably many pathwise connected components. Each component corresponds to a quantum vacuum. Tunneling between these vacua permits the formation of $\theta$ vacua for each simple factor. Putting the simple factors together, one obtains a family of vacua labeled by a vector of $\theta$'s. In the case of the Standard Model, this implies that we have a set of vacua $|\theta_2, \theta_3\rangle$.

When the temperature of the early universe is high enough that $SU(2)$ is a symmetry, then the $SU(2)$ global anomaly could help create the observed preponderance of baryons in the universe. When $SU(2)$ is broken by the Higgs potential at low temperature, the tunneling between vacua associated with different pathwise-connected components of the $SU(2)$ gauge group is suppressed, and so the $SU(2)$ global anomaly effectively disappears from the Lagrangian density. (The perturbative influence of the global anomaly after symmetry breaking will be suppressed by a factor of $M_Z^{-4}$.)

The $SU(3)$ global anomaly introduces $CP$ violation into the strong interactions. This is serious, because the strong interactions are observed to be $CP$ invariant to a high degree of precision. Instead of discussing this anomaly in the context of the Standard Model as a whole, we therefore investigate it in the context of the strong interactions alone. This is the purpose of the next two sections.

## 15.14 Strong Interactions: Approximate Symmetries

In the context of the strong interactions alone, the global $SU(3)$ anomaly is indistinguishable from an axial $U(1)$ anomaly. In this section, we explain how to consider the strong interactions in isolation from the electroweak interactions, introduce the axial $U(1)$ anomaly of the strong interactions, and discuss the significance of the resulting $CP$ violation. We find that the coefficient of the anomaly must be ridiculously small to avoid conflict between predictions and experiments.

The Lagrangian density $\mathcal{L}_{\text{QCD}}$ has only the Standard Model symmetries. If, however, we conceptually isolate the strong interactions from the electroweak, and if we consider the strong interactions at high energies, then the masses of the $u$, $d$, and $s$ quarks are negligible and the strong interactions of these three quarks are determined by the minimal coupling between their free quark Lagrangian and the gluons. (Because the other three quarks are comparatively heavy, they can hardly be brought into this approximation scheme.)

The resulting three-quark Lagrangian density has a chiral symmetry, $U(3)_L \times U(3)_R$. If $T$ is a $U(3)$ generator, then we write $T_V$ and $T_A$ respectively for the generators $(T, T)$ and $(T, -T)$ of $U(3)_L \times U(3)_R$. If we consider the $3 \times 3$ matrices as acting on the column vector of quark fields, then we can write

$$\left.\begin{aligned} T_V &= T\left(\frac{1+\gamma_5}{2}\right) + T\left(\frac{1-\gamma_5}{2}\right) = T \\ T_A &= T\left(\frac{1+\gamma_5}{2}\right) - T\left(\frac{1-\gamma_5}{2}\right) = T\gamma_5 \end{aligned}\right\} \qquad (15.14.1)$$

Clearly, in the original Standard Model Lagrangian density $\mathcal{L}_{\text{EW}} + \mathcal{L}_{\text{QCD}}$, only $\mathbf{1}_V$ is a symmetry generator. Hence, from the perspective of the strong interactions of the $u$, $d$, and $s$ quarks at high energies, the primary effect of interactions with the electroweak theory is the generation of quark masses. This effect breaks $U(3)_L \times U(3)_R$ to the vector $U(1)$ symmetry and an approximate $SU(3)$ symmetry.

The Goldstone Theorem on the existence of a massless boson for each broken global symmetry generator can be extended to a principle: if a generator of a global approximate symmetry is broken, then there will be light scalars in the theory.

Taking this principle into account, we can tabulate the fate of the generators of $U(3)_L \times U(3)_R$ as follows:

| Symmetry | Interpretation | Observation | Conclusion |
|---|---|---|---|
| $u(1)_V$ | Baryon Number | Conserved | Unbroken |
| $su(3)_V$ | Flavor Symmetry | Hadrons form approximate flavor families | Unbroken approximate symmetry |
| $su(3)_A$ | Chiral Flavor Symmetry | Octet of light pseudoscalars | Spontaneously broken |
| $u(1)_A$ | Axial $U(1)$ | No $\gamma_5$ partners to baryon multiplets | Broken |
| | | No ninth light pseudoscalar | Not broken spontaneously |

$$(15.14.2)$$

The odd item is $u(1)_A$, which must be broken without being spontaneously broken. The resolution of this paradox is that, as we have seen, $U(1)_A$ is anomalous. In detail, following Theorem 13.4.11, the axial current $j_5^\mu$ is not conserved, but its divergence is given by the abelian anomaly:

$$\partial_\mu j_5^\mu = -\frac{n_F}{8\pi^2}\, \text{tr}(G_{\mu\nu}\tilde{G}_{\mu\nu}), \qquad (15.14.3)$$

where $n_F$ is the number of quark flavors and the trace is taken in the adjoint representation of $SU(3)$.

**Remark 15.14.4.** One can use the Chern–Simons term of 13.7.15 to regulate the current:

$$j_{5R}^\mu \stackrel{\text{def}}{=} j_5^\mu - \frac{n_F}{8\pi^2}\epsilon^{\mu\rho\sigma\tau}\, \text{tr}(G_{\rho\sigma}A_\tau - \tfrac{2}{3}A_\rho A_\sigma A_\tau), \qquad (15.14.5)$$

but then, though the conserved quantity $Q_{5R}$ is well defined, it is does not commute with the color-$SU(3)$ gauge transformations $\mathbf{G}_k$ which have winding number $k \neq 0$:

$$\mathbf{G}_k Q_{5R}\mathbf{G}_k^{-1} = Q_{5R} - 2n_F k. \qquad (15.14.6)$$

Hence $\exp(-i\tau Q_{5R})$ is not a symmetry of the vacuum $|\Theta\rangle$:

$$\exp(-i\tau Q_{5R})|\Theta\rangle = |\Theta - 2n_F\tau\rangle. \qquad (15.14.7)$$

From the definition of the anomalous Jacobian, we deduced in Theorem 13.5.21 that a $U(1)_A$ transformation modifies the Lagrangian density:

$$\psi \to e^{-i\tau\gamma_5}\psi \quad \Longrightarrow \quad \mathcal{L} \to \mathcal{L}_{\text{eff}} = \mathcal{L} - \frac{\tau n_F}{16\pi^2}\, \text{tr}(G_{\mu\nu}\tilde{G}_{\mu\nu}). \qquad (15.14.8)$$

The abelian anomaly is charge-conjugation invariant but, because of the $\epsilon$ tensor, odd under parity. The strong interactions, however, are observed to be $CP$ invariant. This raises the strong $CP$ problem: why is the coefficient of the abelian so small as to be unobserved? In more detail, since the axial $U(1)$ was part of the $U(3)^3$ used in the previous section to reduce the mass matrix for the down-type quarks to the KM matrix, the particular form of the KM matrix effectively sets an axial gauge. In this gauge, the parameter $\tau n_F$ takes on some definite value $\bar{\theta}$. A bound for $\bar{\theta}$ can be computed from experimental bounds on the dipole electric moment of the neutron. The conclusion is

$$\bar{\theta} \leq 10^{-9}.$$

The chiral $U(3)$ flavor symmetry of the strong interactions at high energy is broken to baryon-number $U(1)$ and $SU(3)$ approximate flavor symmetry by the weak interactions. The Goldstone principle requires light scalars for the broken generators. The pseudoscalar meson octet is available for axial $SU(3)$, but there no candidate for the Goldstone boson for axial $U(1)$ has been observed. However, invoking the anomaly theory of Chapter 13, we see that axial $U(1)$ is not a symmetry of the quantum theory and therefore is not subject to the Goldstone principle.

The anomaly itself violates $CP$, while the strong interactions appear to conserve $CP$. We conclude that the coefficient of the anomaly (after fixing the axial gauge in our choice of KM matrix) must be surprisingly small. The next section takes a first step in modifying the theory to excise this element of surprise.

## 15.15 – The Strong $CP$ Problem

One possible explanation for the unnatural smallness of $\bar{\theta}$ actually arranges that $\bar{\theta} = 0$. Following Peccei and Quinn, this can be achieved in the six-quark Lagrangian density by using two Higgs doublets to give masses to the quarks in order to make the axial $U(1)$ a valid symmetry:

$$\mathcal{L}' = g_{jk}(\bar{u}_{Lj}, \bar{d}_{Lj}) \begin{pmatrix} \phi_2^0 \\ \phi_2^- \end{pmatrix} u_{Rk} + h_{jk}(\bar{u}_{Lj}, \bar{d}_{Lj}) \begin{pmatrix} \phi_1^+ \\ \phi_1^0 \end{pmatrix} d_{Rk} + \text{h.c.} \qquad (15.15.1)$$

A general $U(1)$ transformation of these terms,

$$\begin{pmatrix} u_{Lj} \\ d_{Lj} \end{pmatrix} \to e^{i\alpha_L} \begin{pmatrix} u_{Lj} \\ d_{Lj} \end{pmatrix}, \quad \begin{pmatrix} \phi_1^0 \\ \phi_1^- \end{pmatrix} \to e^{i\alpha_1} \begin{pmatrix} \phi_1^0 \\ \phi_1^- \end{pmatrix}, \quad \begin{pmatrix} \phi_2^0 \\ \phi_2^- \end{pmatrix} \to e^{i\alpha_2} \begin{pmatrix} \phi_2^0 \\ \phi_2^- \end{pmatrix},$$

$$u_{Rk} \to e^{i\alpha_{Ru}} u_{Rk}, \quad \text{and} \quad d_{Rk} \to e^{i\alpha_{Rd}} d_{Rk}, \qquad (15.15.2)$$

is a symmetry if and only if

$$\left. \begin{array}{c} -\alpha_L + \alpha_2 + \alpha_{Ru} = 0 \\ -\alpha_L + \alpha_1 + \alpha_{Rd} = 0 \end{array} \right\} \qquad (15.15.3)$$

These two constraints have three independent solutions:

| Group | $\alpha_L$ | $\alpha_{Ru}$ | $\alpha_{Rd}$ | $\alpha_1$ | $\alpha_2$ |
|---|---|---|---|---|---|
| $U(1)_V$ | 1 | 1 | 1 | 0 | 0 |
| $U(1)_{PQ}$ | 0 | 1 | −1 | 1 | −1 |
| $U(1)_A$ | 1 | −1 | −1 | 2 | 2 |

$$(15.15.4)$$

Note that, when we consider all the quark fields, these generators are mutually orthogonal. We can now use $U(1)_A$ to set $\bar{\theta} = 0$, and then use $U(1)_V$ and $U(1)_{PQ}$ to reduce the down-type quark mass matrix to the normal form involving the KM matrix.

When the Higgs fields take on non-zero VEV's, then the Peccei–Quinn $U(1)$ is broken. Consequently there must be a corresponding Goldstone boson, the axion. Since the Peccei–Quinn symmetry is not gauged, there is no Higgs mechanism to absorb the axion. However, the confining effect of the strong interactions also breaks this symmetry by generating quark condensates:

$$\langle \bar{q}_L q_R + \bar{q}_R q_L \rangle \neq 0. \tag{15.15.5}$$

This dynamical symmetry breaking actually gives the axion a mass bounded above by about 1 Mev. This axion has not been observed.

The strong $CP$ problem and its companion the invisible axion problem cannot be resolved on the level of the Standard Model. This is one motive for considering grand unified theories (GUT's), theories with larger symmetry groups which contain the $SU(3) \times SU(2) \times U(1)$ of the Standard Model. One solution on the GUT level provides a mechanism which causes the axion to be so weakly coupled to ordinary matter that it is no longer a surprise that it has not been observed.

## 15.16 Summary

The purpose of the chapter is to establish clearly the structure of the Standard Model of the electroweak and strong interactions. The central concern addressed by this model is to find a quantizable vector-boson theory of these forces. Since the hadron families suggest an approximate $SU(3)$ flavor symmetry, the spin-statistics theorem indicates an underlying $SU(3)$ color symmetry. Since interactions dominated by strong and weak forces suggest the weak-isospin conserved quantity, it is natural to propose the rank-one group $SU(2)$ as an internal symmetry in the model of these forces. Finally, adding the independent generator required for the electric charge $U(1)$, we conclude that the theory should be built around a symmetry group $SU(3) \times SU(2) \times U(1)$. The gauge principle indicates that this group should be taken as a gauge group. Functional integral quantization now applies through the Faddeev-Popov principle.

Since the photon is the only observed massless vector boson, the theory must include some mechanism for making the other vector bosons either massive or unobservable. Spontaneous symmetry breaking of the $SU(2) \times U(1)$ is invoked to make the weak vector bosons massive, and the non-perturbative phenomenon of color confinement is called upon to make the gluons unobservable.

The starting point in building the Standard Model is the grouping of matter fields into left-handed $SU(2)$ doublets and right-handed $SU(2)$ singlets. Since left and right spinors have different transformation properties, the theory contains no fermion mass terms. The trick is to organize the singlets and doublets so that the symmetry-breaking mechanism can give masses to everything except the neutrinos. A single doublet of scalars is sufficient for this.

The resulting theory contains two unexpected features. First, for the quarks, the charge eigenfields are not identical to the mass eigenfields. The discrepancy is by convention focused in the left-handed $d$, $s$, and $b$ quarks, and normalized into the Kobayashi–Maskawa matrix which converts mass eigenfields into charge eigenfields.

Second, there is also a global anomaly due to the $SU(3)$ gauge symmetry. This adds a $CP$-violating term, a multiple of the Pontrjagin class, to the Lagrangian density. The physical effects of this operator are not observed, so its coefficient must be unreasonably small. This element of unreason can only be removed from the Standard Model by developing grand unified theories.

On the basis of the structure of the strong interactions presented in this chapter, the following chapter uses color confinement and flavor symmetry to develop the rudiments of hadron physics. When the structure and basic properties of the hadrons have been covered, then we can develop tree-level Standard Model scattering theory.

# Chapter 16

# Hadrons, Flavor Symmetry, and Nucleon-Pion Interactions

Connecting the strong interactions to observed particles in preparation for application of the Standard Model to low-energy scattering.

## Introduction

In this chapter, we begin the study of hadrons, the strongly interacting particles like neutrons, protons, and pions. The first level of study introduces the hadrons as bound states of quarks; the second level, the nucleon-pion theory of nuclear structure.

The concept of a quark arose from insight into experimental data on strong interactions. First, the data indicated approximate conservation of two quantum numbers, and second, when the known hadrons were plotted with respect to these two quantum numbers, the results resembled $SU(3)$ weight diagrams. Since all $SU(3)$ representations can be obtained as subrepresentations of tensor products of the vector representation and its conjugate, it was therefore natural (though bold) to propose that the vector representation of $SU(3)$ must correspond to the fundamental particles in the theory of hadrons. The basis elements for this representation are the $u$, $d$, and $s$ quarks, and the $SU(3)$ in question is the flavor symmetry.

Experiment indicates that there are at least six quarks. These are $u$, $d$, $c$, $s$, $b$, and $t$. Cosmology and the scale of quark confinement require that there are no more than eight quarks; particle data indicates that there are at most six quarks. As we pointed out in the last chapter, for the theory of quarks to be consistent with the spin-statistics theorem, it is necessary to introduce a second $SU(3)$ symmetry, color, and thus we have QCD.

In QCD, the coupling of the quarks to the color gauge fields, the gluons, is strongly energy dependent: at very high energies, it is effectively zero, while at nuclear energies it is effectively infinite. The distance scale at which the coupling becomes infinite is called the *confinement scale*. The existence of the confinement scale indicates that quarks cannot be isolated, and the fact that all detectable particles have zero color quantum numbers suggests that quarks bind together to produce color singlets. Thus mesons (spin-0 hadrons) are thought to comprise two quarks, and baryons (spin-1/2 hadrons other than quarks) are thought to comprise three.

Due to the non-perturbative nature of confinement-scale strong interactions, it is extremely hard to make predictions about hadrons on the basis of QCD. For example, the observed spin of the proton cannot be completely accounted for on the basis of the dynamics of its constituents, and the relative masses of the nucleons and the pions have not been explained. Hence, we need flavor symmetry, a level of study between the moderate-energy approximation to hadron physics provided by the nucleon-pion theory and the high-energy approximation afforded by QCD.

Section 16.1 provides the foundation of the chapter by using group theory to find the hadrons as combinations of quarks. Section 16.2 describes the machinery of flavor symmetry in detail, and Sections 16.3 to 16.8 investigate the consequences

| $Y$ \ $I_3$ | $-\frac{1}{2}$ | $0$ | $\frac{1}{2}$ |  | $-\frac{1}{2}$ | $0$ | $\frac{1}{2}$ |
|---|---|---|---|---|---|---|---|
|  | $-1$ | $0$ | $1$ |  | $-1$ | $0$ | $1$ |
| $\frac{1}{2}$ | $K^0$ |  | $K^+$ |  | $K^{*0}$ |  | $K^{*+}$ |
|  | 498 |  | 494 |  | 896 |  | 892 |
| $0$ | $\pi^-$ | $\pi^0\ \eta$ | $\pi^+$ |  | $\rho^-$ | $\rho^0\ \omega$ | $\rho^+$ |
|  | 140 | 135  549 | 140 |  | 767 | 768  782 | 767 |
| $-\frac{1}{2}$ | $K^-$ |  | $\bar{K}^0$ |  | $K^{*-}$ |  | $\bar{K}^{*0}$ |
|  | 494 |  | 498 |  | 892 |  | 896 |

|  |  |
|---|---|
| **Pseudoscalar** | **Axial-Vector** |
| **Meson Octet** | **Meson Octet** |

**Fig. 16.1a.** The mesons with orbital angular momentum $l = 0$ form the two $SU(3)$ octets shown here and two $SU(3)$ singlets, $\eta'$ and $\phi$, which are respectively a pseudoscalar and an axial vector. The particles in each octet are identified by two quantum numbers, the hypercharge $Y$ and the third component of isospin $I_3$. (Masses are in MeV.)

of flavor symmetry and flavor symmetry breaking for the hadron families. Finally, Sections 16.9 to 16.12 develop the $\sigma$ Model of nucleon-pion interactions, concluding with the computation of the $\pi^0 \to 2\gamma$ decay rate from the $\sigma$ Model abelian anomaly.

## 16.1 Hadrons

The $l = 0$ states of the hadrons fall naturally into groups according to their isospin, hypercharge, and mass quantum numbers. If we embed $SU(2)_I \times U(1)_Y$ in $SU(3)_f$ so that (as in 14.2.2) $I_3$ and $Y$ become the elements $I_3^f$ and $Y^f$ of $su(3)_f$ defined by

$$I_3^f \overset{\text{def}}{=} \begin{pmatrix} \frac{1}{2} & 0 & 0 \\ 0 & -\frac{1}{2} & 0 \\ 0 & 0 & 0 \end{pmatrix} \quad \text{and} \quad Y^f \overset{\text{def}}{=} \begin{pmatrix} \frac{1}{6} & 0 & 0 \\ 0 & \frac{1}{6} & 0 \\ 0 & 0 & -\frac{1}{3} \end{pmatrix}$$

$$\implies Q = I_3^f + Y^f = \begin{pmatrix} \frac{2}{3} & 0 & 0 \\ 0 & -\frac{1}{3} & 0 \\ 0 & 0 & -\frac{1}{3} \end{pmatrix},$$

(16.1.1)

then the hadrons with zero orbital angular momentum appear to form $SU(3)$ multiplets. The nine pseudoscalar mesons and the nine axial-vector mesons form two octets and two singlets as shown in Figure 16.1a. The baryons form a spin-1/2 octet and a spin-3/2 decuplet as shown in Figure 16.1b.

From the masses of the particles, it appears that $SU(3)$ is a decent approximate symmetry, and isospin is a good one. Having indicated that these patterns

| $\begin{matrix}\phantom{}& I_3 \\ Y & \end{matrix}$ | $-\frac{1}{2}$ | $\frac{1}{2}$ | | $-\frac{3}{2}$ | $-\frac{1}{2}$ | $\frac{1}{2}$ | $\frac{3}{2}$ |
|---|---|---|---|---|---|---|---|
| | $-1$ | $0$ | $1$ | $-1$ | $0$ | $1$ | |

| $\frac{1}{2}$ | | $\boldsymbol{n}$ | | $\boldsymbol{p}$ | | $\boldsymbol{\Delta^-}$ | $\boldsymbol{\Delta^0}$ | $\boldsymbol{\Delta^+}$ | $\boldsymbol{\Delta^{++}}$ |
| | | 940 | | 938 | | 1236 | 1233 | 1232 | 1231 |

| $0$ | $\boldsymbol{\Sigma^-}$ | $\boldsymbol{\Sigma^0}$ | $\boldsymbol{\Lambda}$ | $\boldsymbol{\Sigma^+}$ | | $\boldsymbol{\Sigma^-}$ | $\boldsymbol{\Sigma^0}$ | $\boldsymbol{\Sigma^+}$ |
| | 1198 | 1193 | 1116 | 1189 | | 1387 | 1384 | 1383 |

| $-\frac{1}{2}$ | | $\boldsymbol{\Xi^-}$ | | $\boldsymbol{\Xi^0}$ | | | $\boldsymbol{\Xi^-}$ | $\boldsymbol{\Xi^0}$ |
| | | 1321 | | 1315 | | | 1535 | 1532 |

| $-1$ | | | | | | | $\boldsymbol{\Omega^-}$ | |
| | | | | | | | 1672 | |

**Baryon Octet**          **Baryon Decuplet**

**Fig. 16.1b.** The baryons with orbital angular momentum $l = 0$ form a spin-1/2 octet and a spin-3/2 decuplet under flavor $SU(3)$. The masses of the $\Delta^-$ and $\Delta^+$ given here are tentative. In particular, the mass of the $\Delta^+$ is probably nearer 1235 MeV though some experiments indicate the value given here. All masses are in MeV's.

motivated the quark theory, we now proceed to show how QCD can be the source of the hadrons.

**Remark 16.1.2.** The isospin multiplets in the baryon octet are deduced from the form of $I_3^f$ in 16.1.1. The irreducible representations $(1,0)$ and $(0,1)$ split into irreducible representations of isospin and hypercharge as follows:

$$(1,0) \to \left(\tfrac{1}{2}\right)^{\frac{1}{6}} \oplus (0)^{-\frac{1}{3}},$$
$$(0,1) \to (0)^{\frac{1}{3}} \oplus \left(\tfrac{1}{2}\right)^{-\frac{1}{6}}. \tag{16.1.3}$$

Therefore, recalling our general points on the structure of weight diagrams for irreducible representations of $su(3)$, we see that:

1. Each row may be separately split into a direct sum of irreducible representations of isospin;
2. The hypercharges are constant on each row, and decrease by $\frac{1}{2}$ as we go from one row down to the next.

By inspection and reflection, we deduce the following formula:

$$(m,n) \to \bigoplus_{r=0}^{m} \bigoplus_{s=0}^{n} \left(\frac{r+s}{2}\right)^{\frac{n-m}{3} + \frac{r-s}{2}}. \tag{16.1.4}$$

Since experiment finds abundant hadrons and no isolated quarks, and since theory proposes that gluons couple to color charge, it is reasonable to assume that the energy dependence of the strong coupling constant forces quarks into color-singlet states at the nucleon scale. We begin this discussion of hadrons by showing that they have the same quantum numbers as color-singlet combinations of quarks.

The masses of the hadrons cannot be predicted from any reasonable hypothesis about quark masses. Nevertheless, the scattering of hadrons is reasonably well accounted for by the quark model, and so it assumed that the hadrons primarily comprise these quark color-singlet states.

As we intend to mix the six quarks types in the singlets, we can group the quarks into multiplets under color, flavor, and spin symmetry. For the purposes of this introduction, it will be sufficient to take the three lightest quarks and use $SU(3)_c \times SU(3)_f \times SU(2)_s$. The light quarks form a $(3,3,2)$ multiplet and their anti-quarks a $(\bar{3}, \bar{3}, 2)$ multiplet under this group; so we need to find color singlets among the tensor products of these representations. The flavor and spin quantum numbers of the color singlets will enable us to identify the singlets with particles. From what we know of $SU(3)$ representations, the smallest interesting tensor products will correspond to $q\bar{q}$ and $qqq$ bound states.

First we consider the $q\bar{q}$ combinations. The splitting of the tensor products is routine:

$$
\begin{aligned}
(3,3,2) \otimes (\bar{3}, \bar{3}, 2) &= (3 \otimes \bar{3}, 3 \otimes \bar{3}, 2 \otimes 2) \\
&= (1 \oplus 8, 1 \oplus 8, 1 \oplus 3) \\
&= \underbrace{(1,1,1) \oplus (1,8,1)}_{\text{spin } 0} \oplus \underbrace{(1,1,3) \oplus (1,8,3)}_{\text{spin } 1} \oplus \cdots ,
\end{aligned}
\tag{16.1.5}
$$

where the dots stand for non-trivial color multiplets. Since the color singlets both have spin $s = 0$, the parity of these singlets is $P = P_I(-1)^l$ where $P_I = -1$ is the intrinsic parity and $l$ is the orbital angular momentum. Taking $l = 0$, the first two representations describe pseudoscalars, the second two describe axial vectors.

To identify the corresponding particles, we use the isospin and flavor hyper-charge quantum numbers defined in 16.1.1 for these singlets. The quantum numbers $(I_3^f, Y^f)^Q$ for the flavor singlet are $(0,0)$, and those in a flavor 8 are:

$$
\text{Octet Quantum Numbers:} \quad
\left\{
\begin{array}{lll}
(0,0)^0 & (1,0)^+ & (\tfrac{1}{2}, \tfrac{1}{2})^+ \\[4pt]
(-1,0)^- & & (-\tfrac{1}{2}, \tfrac{1}{2})^0 \\[4pt]
(-\tfrac{1}{2}, -\tfrac{1}{2})^- & (\tfrac{1}{2}, -\tfrac{1}{2})^0 & (0,0)^0
\end{array}
\right.
\tag{16.1.6}
$$

From these quantum numbers, we identify the first group of nine spin-0 color singlets with the nine pseudoscalar mesons (the octet of Figure 16.1a together with the $\eta'$ particle), and the second group of nine spin-1 color singlets with the axial-vector mesons (the octet of Figure 16.1a together with the $\phi$ particle).

**Remark 16.1.7.** Since the quark vector $(u, d, s)$ is a vector with respect to $SU(3)_f$, we can represent the $q\bar{q}$ combinations by the following schema:

$$
\begin{array}{c}
d \quad u \\
s
\end{array}
\otimes
\begin{array}{c}
\bar{s} \\
\bar{u} \quad \bar{d}
\end{array}
=
\begin{array}{c}
d\bar{s} \quad u\bar{s} \\
d\bar{u} \quad xxx \quad u\bar{d} \\
s\bar{u} \quad s\bar{d}
\end{array}
\tag{16.1.8}
$$

The three $x$'s represent three linear combinations of $u\bar{u}$, $d\bar{d}$, and $s\bar{s}$ to be determined.

If we take the spin of the quark to be opposite to the spin of the anti-quark, then this indicates the predominant quark content of six of the nine pseudoscalar mesons. (If we assume that the spins are aligned, then we see the structure of six of the nine axial-vector mesons.)

In particular, $\pi^+$ is associated with $u\bar{d}$. Since the pions form an isospin triplet, the pair representing $\pi^0$ is obtained by lowering $u\bar{d}$. From the definition of the conjugate representation, if $Lu = d$, then $L\bar{d} = -\bar{u}$. Hence

$$\pi^0 \longleftrightarrow \frac{1}{\sqrt{2}}(u\bar{u} - d\bar{d}). \tag{16.1.9}$$

A good basis for the linear combinations of $u\bar{u}$, $d\bar{d}$, and $s\bar{s}$ orthogonal to $u\bar{u} - d\bar{d}$ is

$$\eta_1 \longleftrightarrow \frac{1}{\sqrt{3}}(u\bar{u} + d\bar{d} + s\bar{s}) \quad \text{and} \quad \eta_8 \longleftrightarrow \frac{1}{\sqrt{6}}(u\bar{u} + d\bar{d} - 2s\bar{s}). \tag{16.1.10}$$

These two states are both isospin singlets.

Since flavor symmetry is only an approximate symmetry, mesons with identical electric charge and spin may mix. Inspection of the $q\bar{q}$ content of the nine states identified above shows that states associated with the pions cannot mix with the others, $K^0 \leftrightarrow d\bar{s}$ can mix with $\bar{K}^0 \leftrightarrow s\bar{d}$, and $\eta_1$ can mix with $\eta_8$. In fact, the physical particles $K_L$ and $K_S$ (the kaons with long and short lifetimes respectively) are mixtures of $K^0$ and $\bar{K}^0$, and the physical particles $\eta$ and $\eta'$ could be mixtures of $\eta_8$ and $\eta_1$ with $\eta$ and $\eta'$ being predominantly $\eta_8$ and $\eta_1$ respectively.

Similar remarks apply to the axial-vector mesons.

Similarly, the $qqq$ tensor product may be analyzed. Baryons being fermions, however, it is necessary to find states which are anti-symmetric in all group indices. For this purpose we need to understand the representations of the group $S_3$ of permutations of $\{1, 2, 3\}$.

The group $S_3$ has three irreducible representations, $s$, $a$, and $m$, known as the symmetric, anti-symmetric, and mixed representations. Concretely, they may be defined by

$$s(\pi) \stackrel{\text{def}}{=} 1 \quad \text{and} \quad a(\pi) = \epsilon_{\pi(1)\pi(2)\pi(3)}, \tag{16.1.11}$$

and

$$m\big((123)\big) \stackrel{\text{def}}{=} \begin{pmatrix} \cos(2\pi/3) & -\sin(2\pi/3) \\ \sin(2\pi/3) & \cos(2\pi/3) \end{pmatrix} \quad \text{and} \quad m\big((12)\big) \stackrel{\text{def}}{=} \begin{pmatrix} 0 & 1 \\ 1 & 0 \end{pmatrix}. \tag{16.1.12}$$

The tensor products of these representations may be computed from the following table:

| $\otimes$ | $s$ | $a$ | $m$ |
|-----------|-----|-----|-----|
| $s$ | $s$ | $a$ | $m$ |
| $a$ | $a$ | $s$ | $m$ |
| $m$ | $m$ | $m$ | $s + a + m$ |

$$\tag{16.1.13}$$

Counting dimensions of the spaces of symmetric and anti-symmetric tensors shows which irreducible subrepresentations are of these types:

$$2 \otimes 2 \otimes 2 = 2 \oplus 2 \oplus 4_s,$$
$$3 \otimes 3 \otimes 3 = 1_a \oplus 8 \oplus 8 \oplus 10_s, \tag{16.1.14}$$

so the pair of 2's and the pair of 8's must be mixed by permutation of the factors.

**Homework** 16.1.15. Use Young diagrams to confirm the conclusions of the foregoing paragraph.

We can now use the tensor product table for $S_3$ to extract the anti-symmetric part of the $qqq$ representation. Writing $\rho^n$ for the tensor products of $\rho$ with itself $n$ times, we find:

$$(3, 3, 2)^3 \stackrel{\text{def}}{=} (3^3, 3^3, 2^3)$$
$$= (1_a \oplus 8 \oplus 8 \oplus 10_s, 1_a \oplus 8 \oplus 8 \oplus 10_s, 2 \oplus 2 \oplus 4_s) \qquad (16.1.16)$$
$$= (1_a, (8 \oplus 8)_m, (2 \oplus 2)_m) \oplus (1_a, 10_s, 4_s) \oplus \cdots$$

The two $m$ representations are in effect tensored together, and so

$$(1_a, (8 \oplus 8)_m, (2 \oplus 2)_m) = (1_a, \underbrace{8, 2}_{s}) \oplus (1_a, \underbrace{8, 2}_{a}) \oplus (1_a, \underbrace{(8, 2) \oplus (8, 2)}_{m}). \qquad (16.1.17)$$

Therefore the anti-symmetric color-singlets in $(3, 3, 2)^3$ are given by

$$(3, 3, 2)^3 = (1_a, 8, 2) \oplus (1_a, 10_s, 4_s) \oplus \cdots \qquad (16.1.18)$$

We use isospin and flavor hypercharge again to define and identify states with pure quantum numbers in these two fermionic representations. From these quantum numbers and from the spin, we conclude that $(1, 8, 2)$ can be identified with the baryon octet, and $(1, 10, 4)$ with the hyperon decuplet.

Hadrons cannot be understood purely in terms of quarks. Scattering data indicate that the structure of the gluon field cannot be ignored, and that gluon and quark contributions to hadron structure are not sufficient to explain the data. Computation of the strong coupling constant as a function of energy indicates the existence of a confinement scale for color. From this we feel comfortable taking color singlet combinations of quarks as a first approximation to hadrons. The resulting construction of hadrons from quarks is pure group theory.

Having made the transition from quarks to hadrons on the basis of flavor symmetry, in the next section we present a few consequences of flavor symmetry for the hadrons.

## 16.2 Flavor Symmetry: Action on Octets

Flavor symmetry originated as an insight which discerned flavor weight diagrams in hadron quantum numbers. The symmetry is approximate since the masses of the hadrons in the multiplets are not actually all equal, and flavor quantum numbers are not conserved in all observed processes. Indeed, in the theory of electroweak and strong interactions, flavor symmetry is broken by the quark mass matrices and the electroweak couplings. However, the $u$, $d$, and to a lesser extent the $s$ quarks have small masses by baryon standards, so $SU(3)$ flavor is a reasonable approximate symmetry of the strong interactions. We can therefore imagine that the mass splitting in each of the baryon multiplets of the previous section is due to those same electroweak interactions which violate the flavor conservation laws.

The purpose of this section is to develop a matrix version of the formalism of Chapter 3 appropriate to the application of flavor symmetry to the three hadron octets. In the following sections, we assume that the strong interactions dominate hadron structure, so that flavor symmetry is a good approximation, and then compute how symmetry breaking affects the magnetic moments, masses, and decays in the hadron families.

Let $\psi$ be the vector of quark fields $(u, d, s)^\top$. Write $\mathbf{M}$ for the matrix of paired fields defined by

$$\mathbf{M} \stackrel{\text{def}}{=} \psi\bar{\psi} - \frac{1}{3}\bar{\psi}\psi\mathbf{1}. \tag{16.2.1}$$

As a principle of the naive quark model which we are developing here, we assume that to a first approximation we may regard $\mathbf{M}$ as a matrix of meson fields, pseudoscalars or axial vectors depending on the orientation of the spins:

$$\mathbf{M}_0 \simeq \begin{pmatrix} \frac{1}{\sqrt{2}}\pi^0 + \frac{1}{\sqrt{6}}\eta & \pi^+ & K^+ \\ \pi^- & -\frac{1}{\sqrt{2}}\pi^0 + \frac{1}{\sqrt{6}}\eta & K^0 \\ K^- & \bar{K}^0 & -\frac{2}{\sqrt{6}}\eta \end{pmatrix}, \tag{16.2.2}$$

$$\mathbf{M}_1 \simeq \begin{pmatrix} \frac{1}{\sqrt{2}}\rho^0 + \frac{1}{\sqrt{6}}\omega & \rho^+ & K^{*+} \\ \rho^- & -\frac{1}{\sqrt{2}}\rho^0 + \frac{1}{\sqrt{6}}\omega & K^{*0} \\ K^{*-} & \bar{K}^{*0} & -\frac{2}{\sqrt{6}}\omega \end{pmatrix}. \tag{16.2.3}$$

To extract a specific paired field $\phi_M$ from this matrix, we use a numerical matrix $M$ as coordinates:

$$\phi_M \stackrel{\text{def}}{=} \text{tr}(M^\dagger\mathbf{M}). \tag{16.2.4}$$

Thus, for example,

$$M = \begin{pmatrix} 0 & 1 & 0 \\ 0 & 0 & 0 \\ 0 & 0 & 0 \end{pmatrix} \implies \phi_M = \pi^+. \tag{16.2.5}$$

Note that $\phi_M$ is anti-linear in $M$.

The baryon octet has a similar structure. Let $\mathbf{B}$ be the octet in $\psi \otimes \psi \otimes \psi$. Then we associate the entries in $\mathbf{B}$ with particles as follows:

$$\mathbf{B} \simeq \begin{pmatrix} \frac{1}{\sqrt{2}}\Sigma^0 + \frac{1}{\sqrt{6}}\Lambda^0 & \Sigma^+ & p \\ \Sigma^- & -\frac{1}{\sqrt{2}}\Sigma^0 + \frac{1}{\sqrt{6}}\Lambda^0 & n \\ \Xi^- & \Xi^0 & -\frac{2}{\sqrt{6}}\Lambda^0 \end{pmatrix}. \tag{16.2.6}$$

We extract specific elements from $\mathbf{B}$ using a matrix $B$ as coordinates:

$$\psi_B \stackrel{\text{def}}{=} \text{tr}(B^\dagger\mathbf{B}). \tag{16.2.7}$$

Suppressing spin and momentum, we define one-particle states $|M\rangle$ and $|B\rangle$ with $\phi_M$ and $\psi_B$ respectively by

$$\langle 0|\phi_M|M\rangle = 1 \quad \text{and} \quad \langle 0|\psi_B|B\rangle = 1. \tag{16.2.8}$$

These theoretical one-particle states may represent linear combinations of physical particles. Note that, since the fields are anti-linear in the coordinate matrices $M$ and $B$, the kets must be linear.

The currents $j_C^\mu$ associated with an element $C$ in $su(3)$ (complexified if necessary) are given by

$$j_C^\mu \overset{\text{def}}{=} \bar\psi \gamma^\mu C \psi. \tag{16.2.9}$$

We shall write $Q_C$ for the associated conserved quantity.

To put the currents into a matrix, we choose a basis $C_r$ for $su(3)$ and define a dual basis $C^s$ by the condition

$$\text{tr}(C_r C^s) = \delta_r^s. \tag{16.2.10}$$

Then we can define a matrix $\mathbf{J}^\mu$ of currents by

$$\mathbf{J}^\mu \overset{\text{def}}{=} \sum_r j_{C_r}^\mu C^r. \tag{16.2.11}$$

A change basis from $C_r$ to $D_r$ induces a change of dual basis:

$$D_r = S^u{}_r C_u \quad\Longrightarrow\quad D^s = (S^{-1})^s{}_v C^v. \tag{16.2.12}$$

It is therefore clear that $\mathbf{J}^\mu$ is independent of the choice of basis. We find that

$$\mathbf{J}^\mu = \begin{pmatrix} \bar u \gamma^\mu u & \bar d \gamma^\mu u & \bar s \gamma^\mu u \\ \bar u \gamma^\mu d & \bar d \gamma^\mu d & \bar s \gamma^\mu d \\ \bar u \gamma^\mu s & \bar d \gamma^\mu s & \bar s \gamma^\mu s \end{pmatrix} - \frac{1}{3}(\bar u \gamma^\mu u + \bar d \gamma^\mu d + \bar s \gamma^\mu s)\mathbf{1}. \tag{16.2.13}$$

The current $j_C^\mu$ is extracted from the octet of currents $\mathbf{J}^\mu$ by the trace:

$$j_C^\mu = \text{tr}(C\mathbf{J}^\mu). \tag{16.2.14}$$

Note that $j_C^\mu$ is linear in $C$, and there is a transpose relationship between the location of a current in $\mathbf{J}^\mu$ and its matrix coordinate, for example:

$$C = \begin{pmatrix} 0 & 0 & 1 \\ 0 & 0 & 0 \\ 0 & 0 & 0 \end{pmatrix} \quad\Longrightarrow\quad j_C^\mu = \bar u \gamma^\mu s. \tag{16.2.15}$$

To introduce the group action, we use the exponential map on real generators $X$:

$$g = e^X \quad\Longrightarrow\quad U(g) \overset{\text{def}}{=} e^{-iQ_X}. \tag{16.2.16}$$

From the action of $Q_X$ on $\psi$, we can deduce the action of $U(g)$ on everything. More directly, since the vector $\psi$ transforms as a 3 of $SU(3)$, the unitary representation of $SU(3)$ must satisfy

$$U(g)^\dagger \psi U(g) = g\psi. \tag{16.2.17}$$

Consequently:

$$U(g)^\dagger \mathbf{J}^\mu U(g) = g\mathbf{J}^\mu g^\dagger, \qquad U(g)^\dagger j_C^\mu U(g) = j_{g^\dagger C g}^\mu;$$

$$U(g)^\dagger M U(g) = gMg^\dagger, \qquad U(g)^\dagger \phi_M U(g) = \phi_{g^\dagger M g}; \tag{16.2.18}$$

$$U(g)^\dagger B U(g) = gBg^\dagger, \qquad U(g)^\dagger \psi_B U(g) = \psi_{g^\dagger B g}.$$

and
$$U(g)|M\rangle = |gMg^\dagger\rangle, \quad U(g)|B\rangle = |gBg^\dagger\rangle. \tag{16.2.19}$$

(The actions on the kets may be verified from the annihilation property 16.2.8 of the kets.)

Finally, we can define matrix elements of currents by

$$\mathcal{M}(M', C, M) \overset{\text{def}}{=} \langle M'|j_C^\mu|M\rangle, \tag{16.2.20}$$

and deduce that these matrix elements are invariant under the group action:

$$\mathcal{M}(gM'g^\dagger, gCg^\dagger, gMg^\dagger) = \mathcal{M}(M', C, M). \tag{16.2.21}$$

The concepts, notation, and transformation rules presented in this section provide the basis for the application of flavor symmetry which follow.

## 16.3 – Baryon Octet Magnetic Moments

In this section, we use flavor symmetry to predict differences between the magnetic moments of the baryons in the octet.

To the extent that flavor symmetry is good, the baryons in the octet **B** have equal magnetic moments and equal masses. A better approximation may be obtained from the interactions of the octet with the electromagnetic current $j_{\text{em}}^\mu$. To use flavor symmetry we must regard $j_{\text{em}}^\mu$ as an element in the flavor octet of currents $\mathbf{J}^\mu$:

$$j_{\text{em}}^\mu = \text{tr}(Q\mathbf{J}^\mu), \tag{16.3.1}$$

where the matrix $Q$ is defined as in 16.1.1 by its action on the vector $\psi$ of quark fields.

Writing $B$ and $B'$ for matrix coordinates of baryons in the baryon octet, the magnetic moment $\mu_{B'B}$ for the baryon transition from $B$ to $B'$ is the coefficient of $\bar{u}'i\sigma_{\mu\nu}q^\nu u$ in the amplitude associated with $\langle B'|j_{\text{em}}^\mu|B\rangle$. As the spinor terms are immune to flavor symmetry, we can abuse the notation a little and set

$$\mu_{B'B} \overset{\text{def}}{=} c\langle B'|j_{\text{em}}^\mu|B\rangle, \tag{16.3.2}$$

where $c$ is a constant.

The central hypothesis is that the matrix elements of the baryon octet with the current octet is itself an octet; that is,

$$\mathcal{M}(B', B) \overset{\text{def}}{=} \langle B'|\mathbf{J}^\mu|B\rangle \tag{16.3.3}$$

implies

$$\begin{aligned}
\mathcal{M}(gB'g^\dagger, gBg^\dagger) &= \langle B'|U(g)^\dagger\mathbf{J}^\mu U(g)|B\rangle \\
&= \langle B'|g\mathbf{J}^\mu g^\dagger|B\rangle \\
&= g\mathcal{M}(B', B)g^\dagger.
\end{aligned} \tag{16.3.4}$$

As we need to extract $j_{\text{em}}^\mu$, and as singlets are easier to work with than octets, we use the equivalent hypothesis that

$$\mathcal{M}(B', C, B) \overset{\text{def}}{=} \langle B'|\text{tr}(C\mathbf{J}^\mu)|B\rangle \tag{16.3.5}$$

is a singlet.

The matrix elements $\mathcal{M}(B', C, B)$ lie in a singlet subrepresentation of $8 \otimes 8 \otimes 8$. Since

$$8 \otimes 8 = 1 \oplus 8 \oplus 8 \oplus 10 \oplus \overline{10} \oplus 27, \tag{16.3.6}$$

there are only two singlets in $8 \otimes 8 \otimes 8$. Furthermore, $\mathcal{M}$ is linear in $B$ and $C$ and anti-linear in $B'$. We can easily write down two linearly independent, sesquilinear singlets, for example, $\mathrm{tr}(CBB'^\dagger)$ and $\mathrm{tr}(CB'^\dagger B)$. Hence, since $\mathcal{M}(B', C, B)$ is itself a singlet, it must be a linear combination of these two:

$$\mathcal{M}(B', C, B) = \alpha\,\mathrm{tr}(CBB'^\dagger) + \beta\,\mathrm{tr}(CB'^\dagger B). \tag{16.3.7}$$

This formula for $\mathcal{M}(B', C, B)$ implies that the magnetic moments $\mu_{B'B}$ satisfy

$$\mu_{B'B} = \alpha c\,\mathrm{tr}(QBB'^\dagger) + \beta c\,\mathrm{tr}(QB'^\dagger B). \tag{16.3.8}$$

Setting $B' = B$, and substituting appropriate values for $B$, we find eight magnetic moments in terms of two parameters, $\alpha c$ and $\beta c$. Using the experimental values of $\mu$ for the nucleons to determine the parameters, we are left with six predictions. The following table uses $a = \frac{1}{6}\alpha c$ and $b = \frac{1}{6}\beta c$ to summarize the results:

|  | | Theory | Experiment | |
|---|---|---|---|---|
| $\mu_p$ | $=4a - 2b$ | | $=\ \ 2.79$ | |
| $\mu_n$ | $=-2a - 2b$ | | $=-1.91$ | |
| $\mu_{\Xi^0}$ | $=-2a - 2b$ | $=-1.91$ | $-1.25 \pm .01$ | |
| $\mu_{\Xi^-}$ | $=-2a + 4b$ | $=-0.88$ | $-0.68 \pm .03$ | $(16.3.9)$ |
| $\mu_{\Sigma^+}$ | $=4a - 2b$ | $=\ \ 2.79$ | $2.42 \pm .05$ | |
| $\mu_{\Sigma^0}$ | $=a + b$ | $=\ \ 0.95$ | ?? | |
| $\mu_{\Sigma^-}$ | $=-2a + 4b$ | $=-0.88$ | $-1.16 \pm .03$ | |
| $\mu_\Lambda$ | $=-a - b$ | $=-0.95$ | $-0.61$ | |

The error percentages between theory and experiment indicate the extent to which the symmetry hypothesis is good. We find it appears correct to within 50%; not too bad considering how little computation we have done.

**Homework 16.3.10.** Use tensors to extend the methods of the previous section to flavor decuplets. Make a prediction for the magnetic moments in the baryon decuplet. (The lifetimes of these particles are so short that there is only experimental data for the $\Delta^{++}$.)

# 16.4 Constituent Quarks and Hadron Magnetic Moments

The $SU(3)$ transformations of the hadrons is only a part of the information about hadrons contained in the naive quark model. We can go through the details of representation theory and identify the structure of each hadron as a wave function of quarks with definite spin.

Writing $q_\pm$ for a quark $q$ with spin $\pm 1/2$ in the $z$ direction, it is obvious that the quark wave function for the $\Delta^{++}$ is given by:

$$\Delta^{++} \longleftrightarrow u_+ u_+ u_+. \tag{16.4.1}$$

Wave functions for other members of the baryon decuplet can be obtained from this one by lowering quarks, and wave functions for members of the baryon octet can be obtained from these by lowering spin. The results for the baryon octet follow.

The structures of the six baryons at the corners of the octet are basically identical. For brevity, we define generic spin-up wave functions $F$ and $F_{\text{sym}}$ by

$$F(x,y,z) \stackrel{\text{def}}{=} 2x_+ y_+ z_- - x_+ y_- z_+ - x_- y_+ z_+ \tag{16.4.2}$$
$$\text{and} \quad F_{\text{sym}}(x,x,y) = F(x,x,y) + F(x,y,x) + F(y,x,x).$$

The wave functions for corner elements of the baryon octet are obtained by substituting the appropriate quarks and normalizing, for example:

$$p \longleftrightarrow \frac{1}{3\sqrt{2}} F_{\text{sym}}(u,u,d) \quad \text{and} \quad n \longleftrightarrow \frac{1}{3\sqrt{2}} F_{\text{sym}}(d,d,u). \tag{16.4.3}$$

For $\Lambda$ and $\Sigma^0$, we take $F^\Lambda$ and $F^{\Sigma^0}$ to be

$$F^\Lambda \stackrel{\text{def}}{=} u_+ d_- s_+ - d_+ u_- s_+ - u_- d_+ s_+ + d_- u_+ s_+,$$
$$F^{\Sigma^0} \stackrel{\text{def}}{=} 2u_+ d_+ s_- - u_+ d_- s_+ - u_- d_+ s_+ + 2d_+ u_+ s_- - d_+ u_- s_+ - d_- u_+ s_+, \tag{16.4.4}$$

so that

$$\Lambda \longleftrightarrow \frac{1}{2\sqrt{3}} F^\Lambda_{\text{sym}} \quad \text{and} \quad \Sigma^0 \longleftrightarrow \frac{1}{6} F^{\Sigma^0}_{\text{sym}}. \tag{16.4.5}$$

For such wave functions, the magnetic moment operator (in nuclear magnetons) is given by

$$\mu_{\text{op}} \stackrel{\text{def}}{=} \frac{m_p}{m} Q\sigma_3, \tag{16.4.6}$$

where $m_p$ is the proton mass, and $m$ and $Q$ are the mass and electric charge operators respectively. One computes an expectation of this operator as follows:

$$\langle u_+ d_+ s_- | \mu_{\text{op}} | u_+ d_+ s_- \rangle = \frac{2}{3}\frac{m_p}{m_u} - \frac{1}{3}\frac{m_p}{m_d} + \frac{1}{3}\frac{m_p}{m_s}. \tag{16.4.7}$$

We can identify these terms with the magnetic moments of the quarks, with a plus or minus according as the quarks has spin up or down:

$$\mu_q \stackrel{\text{def}}{=} \frac{m_p}{m_q} Q_q. \tag{16.4.8}$$

Applying this style of computation to the particles in the baryon octet, we find that

$$\mu_p = \frac{1}{3}(4\mu_u - \mu_d), \quad \mu_\Lambda = \mu_s, \quad \mu_{\Sigma^0} = \frac{1}{3}(2\mu_u + 2\mu_d - \mu_s). \tag{16.4.9}$$

Finally, to get prediction from here, we must make some assignments for the quark masses. In the naive quark model, the masses of the baryons come from the masses of their constituent quarks. Since the masses of the proton and neutron are almost equal, we take

$$m_u = m_d = \frac{1}{3}m_p \simeq 310\,\text{MeV}. \tag{16.4.10}$$

The mass gaps between the isomultiplets in the baryon octet and in the baryon decuplet are quite diverse, ranging from 123 MeV to 258 MeV. Clearly, the constituent quark is a loose concept. However, we can make decent predictions with an estimate

$$m_s \simeq m_u + 180\,\text{MeV} \simeq 490\,\text{MeV}. \tag{16.4.11}$$

With these masses for the constituent quarks, we find magnetic moments

$$\mu_u \simeq 2, \quad \mu_d \simeq -1, \quad \mu_s \simeq -0.64. \tag{16.4.12}$$

The predictions for the magnetic moments in the baryon octet are:

|  |  | **Theory** | **Experiment** |  |
|---|---|---|---|---|
| $\mu_p$ | = | 3 | 2.79 | |
| $\mu_n$ | = | $-2$ | $-1.91$ | |
| $\mu_{\Xi^0}$ | = | $-1.52$ | $-1.25 \pm .01$ | |
| $\mu_{\Xi^-}$ | = | $-0.52$ | $-0.68 \pm .03$ | (16.4.13) |
| $\mu_{\Sigma^+}$ | = | 2.88 | $2.42 \pm .05$ | |
| $\mu_{\Sigma^0}$ | = | 0.88 | ?? | |
| $\mu_{\Sigma^-}$ | = | $-1.12$ | $-1.16 \pm .03$ | |
| $\mu_\Lambda$ | = | $-0.64$ | $-0.61$ | |

Note that these predictions are considerably better than those of the previous section, which were based on flavor symmetry alone. This is because the use of constituent quarks takes into account the primary effect of breaking of $SU(3)_f$, namely, the difference between quark masses.

**Remark 16.4.14.** There is a big difference between the concept of a constituent quark and a current quark. The former is supposed to be a primary component of a hadronic bound state, whereas the latter is supposed to be at high energy and enjoying asymptotic freedom. Current quarks are the appropriate concept whenever perturbation theory is applicable. Thus the masses in the quark propagators of the Standard Model are not the masses of constituent quarks, but far smaller. Weinberg's 1977 estimates are

$$m_u^0 \simeq 4.3\,\text{MeV}, \quad m_d^0 \simeq 7.5\,\text{MeV}, \quad m_s^0 \simeq 150\,\text{MeV}. \tag{16.4.15}$$

**Homework 16.4.16.** Find the wave functions for the two meson nonets.

**Homework 16.4.17.** Find the wave functions for the baryon decuplet and predict their magnetic moments.

# 16.5 Flavor Symmetry – Gell-Mann–Okubo Mass Formula

Since each hadron family contains particles with significantly different masses, flavor $SU(3)$ is obviously broken. Since the mass splitting within isomultiplets is significantly less than that between isomultiplets, it appears that there is an aspect of the strong-interaction Hamiltonian which breaks $SU(3)$ but preserves isospin. Specifically, following Gell-Mann and Okubo, we suppose that the Hamiltonian for the strong processes is a sum,

$$H_{\text{strong}} = H_{\text{very strong}} + H_{\text{medium strong}}, \tag{16.5.1}$$

where the first summand is $SU(3)$ invariant and the second summand transforms as an $I = 0$, $Y = 0$ element of a flavor octet, like the $\Lambda$ baryon.

The mass corrections in any flavor multiplet $(m, n)$ have the form:

$$(\text{Mass Correction})_T \propto \langle T|H_{\text{ms}}|T\rangle, \tag{16.5.2}$$

where $T$ is a tensor of type $(m, n)$. The hypothesis on $H_{\text{ms}}$ implies that the more general matrix element

$$\mathcal{M}(T', T) \stackrel{\text{def}}{=} \langle T'|H_{\text{ms}}|T\rangle. \tag{16.5.3}$$

transforms as an element of an octet. Since $(m, n)^* \otimes (m, n)$ contains two octets if $mn \neq 0$ and one otherwise, $\mathcal{M}$ is a linear combination of at most two quantities which transform like $H_{\text{ms}}$. Thus we can write the mass correction to the particle with tensor coordinate $T$ as follows:

$$(\text{Mass Correction})_T \tag{16.5.4}$$
$$= 2a(Y^f)^u_v T^{vr_2\cdots r_m}_{s_1\cdots s_n}(T^\dagger)^{s_1\cdots s_n}_{ur_2\cdots r_m} + 2b(Y^f)^u_v T^{r_1\cdots r_m}_{us_2\cdots s_n}(T^\dagger)^{vs_2\cdots s_n}_{r_1\cdots r_m}.$$

(We could have derived this by introducing a fictitious matrix of Hamiltonians and using $Y^f$ to extract $H_{\text{ms}}$ from this matrix.)

**Remark 16.5.5.** Gell-Mann and Okubo wrote $\mathcal{M}$ in terms of two conserved quantities which transform correctly. From the matrix of conserved quantities, we find $Y$ itself for one, and from the adjoint of this matrix, we find a second in $Y^2 - \bar{I}^2$. Thus

$$(\text{Mass Correction})_{I,Y} = aY + b\big(I(I + 1) - Y^2\big). \tag{16.5.6}$$

Actually, to interpret $Y^2 - \bar{I}^2$ we must leave the linear space $su(3)$, and use the space of tensors over this vector space, modulo a condition which connects the tensor product to the Lie bracket:

$$X \otimes Y - Y \otimes X = [X, Y].$$

This new algebraic structure $\mathcal{E}$ is called the *enveloping algebra* of $su(3)$. As the elements of $su(3)$ generate $\mathcal{E}$, $SU(3)$ acts on $\mathcal{E}$. The quadratic tensors in $su(3)$ elements transform as $8 \otimes 8$, but in $\mathcal{E}$ the anti-symmetric tensors are reduced by the Lie bracket. In any case, the symmetric 8 in $8 \otimes 8$ remains unreduced in $\mathcal{E}$ and therefore provides an octet of quadratic quantum numbers. The expression $Y^2 - \bar{I}^2$ is the $I = 0$, $Y = 0$ element of this octet.

When we apply a representation to $su(3)$, that representation extends uniquely to the whole of the enveloping algebra $\mathcal{E}$. Hence $Y^2 - \bar{I}^2$ is well defined on any representation of $su(3)$.

The mass correction is interpreted as a correction to the coefficient of the quadratic term in the Hamiltonian, that is, as a correction to $m$ for a fermion and a correction to $m^2$ for a boson. There is no compelling argument for this interpretation, but it works better than any other.

Defining a matrix $P$ by

$$2Y^f = \tfrac{1}{3}1 - P \quad \Longrightarrow \quad P = \begin{pmatrix} 0 & 0 & 0 \\ 0 & 0 & 0 \\ 0 & 0 & 1 \end{pmatrix}, \tag{16.5.7}$$

eliminating $Y^f$ in favor of $P$ in the mass correction formula 16.5.4, and writing a superscript (2) to indicate squaring for bosons, 16.5.4 yields a formula

$$M_B^{(2)} = M^{(2)} + a\,\mathrm{tr}(PBB^\dagger) + b\,\mathrm{tr}(PB^\dagger B). \tag{16.5.8}$$

The values of $M_B^{(2)}$ are constant on isomultiplets within each octet, so we can replace the label $B$ by the pair $(I, Y)$ and conclude that

$$M^{(2)}(1,0) = M^{(2)}, \qquad M^{(2)}(0,0) = M^{(2)} + \tfrac{2}{3}a + \tfrac{2}{3}b,$$
$$M^{(2)}(\tfrac{1}{2}, \tfrac{1}{2}) = M^{(2)} + b, \qquad M^{(2)}(\tfrac{1}{2}, -\tfrac{1}{2}) = M^{(2)} + a. \tag{16.5.9}$$

Finally, since there are four isospin multiplets in an octet and three parameters to eliminate, there is one predictive relation – the Gell-Mann–Okubo formula:

$$\frac{1}{2}M^{(2)}(1,0) - M^{(2)}(\tfrac{1}{2}, \tfrac{1}{2}) - M^{(2)}(\tfrac{1}{2}, -\tfrac{1}{2}) + \frac{3}{2}M^{(2)}(0,0) = 0. \tag{16.5.10}$$

Applying this equation to the baryon, pseudoscalar meson, and axial vector meson octets, we find:

| | $\frac{1}{2}$ | $-1$ | $-1$ | $\frac{3}{2}$ | **Experiment** | |
|---|---|---|---|---|---|---|
| **Baryons** | $\Sigma$ | $N$ | $\Xi$ | $\Lambda$ | | |
| | 1192 | 939 | 1318 | 1113 | 8.5 MeV | |
| **Pseudoscalars** | $\pi^2$ | $(K^+)^2$ | $(\bar{K}^0)^2$ | $\eta^2$ | | (16.5.11) |
| | 0.019 | 0.244 | 0.248 | 0.301 | $-0.03$ (BeV)$^2$ | |
| **Axial Vectors** | $\rho^2$ | $(K^{*+})^2$ | $(\bar{K}^{*0})^2$ | $\omega^2$ | | |
| | 0.590 | 0.796 | 0.802 | 0.612 | $-0.385$ (BeV)$^2$ | |

where the numbers at the top of each column are the coefficients in the Gell-Mann–Okubo relation 16.5.10. The experiment column lists the values of this relation on the three octets; the prediction is that these values should be zero. The first two predictions are good enough; the third is not.

**Homework** 16.5.12. Work out the consequences of the Gell-Mann–Okubo hypothesis 16.5.1 for the baryon decuplet. You should find two predictions, and they should be good to within 1%.

# 16.6 – Isosinglet Mixing

The unsatisfactory outcome of applying the Gell-Mann–Okubo mass formula 16.5.10 to the axial-vector mesons arises from a relationship between mass eigenstates and flavor eigenstates. The Gell-Mann–Okubo mass relation does not apply to mass eigenstates, but to charge eigenstates.

As we have noted, there is a flavor-singlet, axial-vector meson $\omega_1$. Writing $\omega_8$ for the $I = 0$, $Y = 0$ eigenstate, we find that since both $\omega_1$ and $\omega_8$ have $I = 0$ and $Y = 0$, these fields can mix without breaking isospin of hypercharge symmetries.

Suppose that the $\omega_8$ and the $\omega_1$ have a mass matrix (analogous to the KM matrix of QCD) with off-diagonal entries:

$$M^2 \stackrel{\text{def}}{=} \begin{pmatrix} m_1^2 & m^2 \\ m^2 & m_8^2 \end{pmatrix}. \tag{16.6.1}$$

Then the mass eigenstates of $M^2$, $|\omega\rangle$ and $|\phi\rangle$, can correspond to the observed particles, $\omega$ and $\phi$. Since $M^2$ is positive definite, and since $\text{tr}(M^2)$ is invariant under change of basis, the mass conditions

$$m_1^2 + m_8^2 = M_\omega^2 + M_\phi^2$$
$$\text{and} \quad |m_1^2 - m_8^2| \leq |M_\omega^2 - M_\phi^2| \tag{16.6.2}$$

are necessary and sufficient for the consistency of this mixing hypothesis.

In fact, the Gell-Mann–Okubo relation sets the mass of the charge eigenstate $\omega_8$ and experiment sets the mass of $\phi$, so that the mass of $\omega_1$ can be determined from the trace condition. In summary, the squared masses in units of $(\text{GeV})^2$ are

| | | |
|---|---|---|
| **Experiment:** | $M_\phi^2 = 1.004;$ | |
| **GM–O Relation:** | $m_8^2 \simeq 0.864;$ | |
| **Mixing Hypothesis:** | $m_1^2 \simeq 0.752;$ | |
| **Experiment:** | $M_\omega^2 = 0.612.$ | |

$$\tag{16.6.3}$$

Using a mixing angle $\theta$, we can write the mass eigenstates in terms of the charge eigenstates:

$$|\omega\rangle = \cos\theta|\omega_1\rangle - \sin\theta|\omega_8\rangle,$$
$$|\phi\rangle = \sin\theta|\omega_1\rangle + \cos\theta|\omega_8\rangle. \tag{16.6.4}$$

Then from the data above we can compute $\sin^2\theta$:

$$M^2 = M_\phi^2|\phi\rangle\langle\phi| + M_\omega^2|\omega\rangle\langle\omega| \tag{16.6.5}$$

$$\implies m_8^2 = \langle\omega_8|M^2|\omega_8\rangle = M_\phi^2\cos^2\theta + M_\omega^2\sin^2\theta$$

$$\implies \sin^2\theta = \frac{M_\phi^2 - m_8^2}{M_\phi^2 - M_\omega^2} \simeq 0.36.$$

**Remark 16.6.6.** We have explained isosinglet mixing with the benefit of hindsight. When Gell-Mann was first promoting flavor symmetry, rather than drop the idea in the face of the poor mass structure of the axial-vector octet, he used the lightness of the $\omega$ as the basis for predicting the existence of the $\phi$ particle.

**Homework 16.6.7.** We also have a pseudoscalar meson $\eta'$ with mass 957 MeV. From the Gell-Mann–Okubo relation for the pseudoscalar meson masses, what is the maximum mixing angle between the $SU(2)$ singlet in the octet and the $SU(3)$ singlet?

# 16.7 – Electromagnetic Mass Splitting

The Gell-Mann–Okubo analysis of mass splitting in the baryon octet intentionally neglects mass splitting within isomultiplets. Such mass splittings can result from electroweak effects. We expect that the mass of the $W^\pm$ and $Z$ bosons will make the weak contribution to the masses negligible and conclude that electromagnetic effects are primarily responsible for the mass splitting within hadron isomultiplets.

On the level of Feynman diagrams, the electromagnetic correction to masses is generated by emission and absorption of electromagnetic current. We therefore propose the mass correction formula

$$\delta M_B \propto \langle B | j_{\text{em}}^\mu j_{\text{em}\,\mu} | B \rangle. \tag{16.7.1}$$

Analysis begins from the flavor-symmetry hypothesis, which in this case implies that the matrix element

$$\mathcal{M}(B', C', C, B) \stackrel{\text{def}}{=} \langle B' | \operatorname{tr}(C' \mathbf{J}^\mu) \operatorname{tr}(C \mathbf{J}_\mu) | B \rangle \tag{16.7.2}$$

is a singlet.

This $\mathcal{M}$ is anti-linear in $B'$, and linear in $C'$, $C$, and $B$. Furthermore, since the currents are bosonic, $\mathcal{M}$ is symmetric in $C'$ and $C$. Hence $\mathcal{M}$ is a singlet formed from an $8 \otimes 8$ and the symmetric part of an $8 \otimes 8$. Since

$$8 \otimes 8 = 1_s \oplus 8_s \oplus 8_a \oplus 10_a \oplus \overline{10}_a \oplus 27_s, \tag{16.7.3}$$

there are four scalars coming from the terms $1 \otimes 1_s$, $(8 \oplus 8) \otimes 8_s$, and $27 \otimes 27_s$. Again, it is easy to write down four linearly independent singlets. At $C' = C$, we could use:

$$\operatorname{tr}(B'^\dagger C^2 B), \quad \operatorname{tr}(B'^\dagger C B C), \quad \operatorname{tr}(B'^\dagger B C^2), \quad \operatorname{tr}(B'^\dagger B) \operatorname{tr}(C^2). \tag{16.7.4}$$

Since $\mathcal{M}$ is a singlet, it must be a linear combination of these four.

Substitute $C = Q = P - \frac{1}{3}\mathbf{1}$ with $P = \operatorname{diag}(1,0,0)$ to obtain a formula for the mass splitting:

$$\delta M_B = a \operatorname{tr}(B^\dagger P B) + b \operatorname{tr}(B^\dagger B P) + c \operatorname{tr}(B^\dagger P B P) + d \operatorname{tr}(B^\dagger B). \tag{16.7.5}$$

Substituting the appropriate $B$'s and adding the fictitious octet mass $m$ yields expressions for the baryons masses:

$$
\begin{aligned}
M_p &= m + a + d & M_{\Sigma^+} &= m + a + d \\
M_n &= m + d & M_{\Sigma^0} &= m + \tfrac{1}{2}(a+b+c) + d \\
M_{\Xi^-} &= m + b + d & M_{\Sigma^-} &= m + b + d \\
M_{\Xi^0} &= m + d & M_\Lambda &= m + \tfrac{1}{6}(a+b+c) + d
\end{aligned}
\tag{16.7.6}
$$

Now we must take into account the medium-strong mass splitting which separates the isomultiplets: this introduces three more parameters, making a total of seven. The eight mass formulae above therefore reduce to one prediction. To eliminate the medium-strong effects, we first compute the mass splittings within the three non-trivial isomultiplets,

$$M_p - M_n = a, \qquad\qquad M_{\Xi^-} - M_{\Xi^0} = b,$$
$$M_{\Sigma^+} - M_{\Sigma^0} = \frac{1}{2}(a - b - c), \qquad M_{\Sigma^0} - M_{\Sigma^-} = \frac{1}{2}(a - b + c). \tag{16.7.7}$$

Finally, eliminating $a$, $b$, and $c$ we find a mass relation which is well supported by experiment:

$$
\begin{array}{cc}
\textbf{Theory} & \textbf{Experiment} \\
(M_p - M_n) + (M_{\Xi^0} - M_{\Xi^-}) - (M_{\Sigma^+} - M_{\Sigma^-}) = 0 & 0.4 \pm .7 \text{ (MeV)}
\end{array} \tag{16.7.8}
$$

**Homework 16.7.9.** Apply the method of this section to make four predictions relating the baryon decuplet mass splittings within isomultiplets. Check your result against experimental data. Note the magnitude of the experimental uncertainties.

# 16.8 – Leptonic Decays

A further application of flavor symmetry provides relationships between the decay rates of the axial-vector mesons. Using matrices $B$ in $su(3)$ as coordinates for the axial-vector meson octet, the leptonic decay amplitudes for the octet are given by

$$\mathcal{M}_B \propto \langle 0 | j_{\text{em}}^\mu | B \rangle, \tag{16.8.1}$$

and the symmetry hypothesis is that the more general amplitude

$$\mathcal{M}(C, B) \stackrel{\text{def}}{=} \langle 0 | \operatorname{tr}(C J^\mu) | B \rangle \tag{16.8.2}$$

is a flavor singlet. However, in $8 \otimes 8$ there is only one singlet, for example $\operatorname{tr}(CB)$. Hence:

$$\mathcal{M}(C, B) \propto \operatorname{tr}(CB). \tag{16.8.3}$$

As the octet includes $\omega_8$ but not the physical particles $\phi$ or $\omega$, we must use the $(\omega_8, \omega_1)$ basis for the isosinglets. However, since $1 \otimes 8$ contains no singlets, the leptonic decay amplitudes for $\omega_1$ are all zero. Hence we can relate $\phi$ and $\omega$ leptonic decays to $\omega_8$ ones:

$$\Gamma(\phi \to \text{leptons}) = \Gamma(\omega_8 \to \text{leptons}) \cos^2 \theta,$$
$$\Gamma(\omega \to \text{leptons}) = \Gamma(\omega_8 \to \text{leptons}) \sin^2 \theta, \tag{16.8.4}$$

where 'leptons' stands for either a specific set of leptons like $e^+ e^-$ or all possibilities together.

The trace formula enables us to compute ratios of amplitudes for leptonic decays for any pair of particles in the octet which have the same charge, for example:

$$\frac{\Gamma(\omega_8 \to \text{leptons})}{\Gamma(\rho^0 \to \text{leptons})} = \left( \frac{\operatorname{tr}(Q B_{\omega_8})}{\operatorname{tr}(Q B_{\rho^0})} \right)^2 = \frac{1}{3}. \tag{16.8.5}$$

In particular, then, we can make specific predictions like:

$$\frac{\Gamma(\phi \to e^+ e^-)}{\Gamma(\rho^0 \to e^+ e^-)} = \frac{1}{3} \cos^2 \theta = 0.21 \qquad \text{(Experiment: } 0.21 \pm 0.02\text{)}$$

$$\frac{\Gamma(\omega \to e^+ e^-)}{\Gamma(\rho^0 \to e^+ e^-)} = \frac{1}{3} \sin^2 \theta = 0.12 \qquad \text{(Experiment: } 0.09 \pm 0.01\text{)}$$

(16.8.6)

which compare well with the experimental values.

This section has avoided the difficult issue of explaining hadrons and their properties from the level of QCD by proposing a mechanism for color confinement and developing the consequences of this hypothesis through group theory. The results on hadron multiplets, magnetic moments, masses, and leptonic decays are in sufficiently good agreement with experiment that we feel to be on the right track.

Some applications of flavor symmetry to scattering and decay amplitudes will be found in Chapter 17.

## 16.9 Nucleon-Pion Theory

The previous sections cover the naive quark model and its immediate consequences for hadrons. This and the following sections introduce the Lagrangian theory of nucleon-pion scattering based on isospin symmetry. A more complete approach – chiral Lagrangian theory – proposes a non-renormalizable theory of hadron processes based on flavor symmetry.

Nuclear forces appear to be attractive only, and short range. The results of Chapter 4 on scalar exchange show that this mechanism could be responsible for the nuclear forces. The nuclei $B^{12}$ and $N^{12}$ have roughly the same energy levels, suggesting that in the absence of electromagnetic effects, the force between protons is equal to that between neutrons. The occurrence of some of these levels in the more elaborate spectrum of $C^{12}$ suggests that in the absence of the electromagnetic forces, the force between a proton and a neutron (in their anti-symmetric state) is equal to the force between protons or between neutrons. Since the mass difference between the proton and neutron could be due to electromagnetic effects, it is natural to consider that, were it not for electromagnetism, nuclear processes would have $SU(2)$ symmetry. Historically, this symmetry of the short range nuclear force was the first glimpse of isospin.

Write $p$ and $n$ for the Dirac fields representing the proton and neutron. From a consideration of the electromagnetic charges involved, a minimal set of scalars for the three forces would be the three pions, $\pi_0$, $\pi_-$, and $\pi_+ = (\pi_-)^\dagger$. In terms of hermitian fields $\pi_r$ we would have:

$$\pi_0 = \pi_3, \quad \sqrt{2}\,\pi_+ = \pi_1 - i\pi_2, \quad \sqrt{2}\,\pi_- = \pi_1 + i\pi_2. \qquad (16.9.1)$$

It is natural then to form $SU(2)$ multiplets from these fields as follows:

$$N \stackrel{\text{def}}{=} \begin{pmatrix} p \\ n \end{pmatrix} \quad \text{and} \quad \Pi \stackrel{\text{def}}{=} \sum_{r=1}^{3} \pi_r \sigma_r = \begin{pmatrix} \pi_0 & \pi_1 - i\pi_2 \\ \pi_1 + i\pi_2 & -\pi_0 \end{pmatrix}. \qquad (16.9.2)$$

The action of an element $g$ of $SU(2)$ on these multiplets is given by

$$N \to gN \quad \text{and} \quad \Pi \to g\Pi g^\dagger. \qquad (16.9.3)$$

With these multiplets, there is a natural Lagrangian density:

$$\mathcal{L} \stackrel{\text{def}}{=} \bar{N}(i\partial\!\!\!/ - m)N + \frac{1}{4}\operatorname{tr}(\partial_\mu \Pi \, \partial^\mu \Pi) - \frac{\mu^2}{4}\operatorname{tr}(\Pi^2) - g\bar{N}\Gamma\Pi N, \qquad (16.9.4)$$

where $\Gamma$ is some combination of $1$ and $\gamma_5$. (Note that $\gamma$ matrices must be understood to act on both spinor components of $N$, and that $\gamma$ stuff commutes with $SU(2)$ stuff.)

**Remark** 16.9.5. The constraint that the Lagrangian density be renormalizable is not to be taken too seriously, for the strength of the strong interaction is such that for many experiments $g$ is too large for perturbation theory. This Lagrangian density is really to be used at tree level, and to be regarded as a first approximation to a non-renormalizable effective Lagrangian density for hadrons.

The parity of $\pi^0$ was determined from its decay

$$\pi^0 \to 2\gamma \to e^+e^- + e^+e^-. \qquad (16.9.6)$$

Assuming that the decay is primarily due to electromagnetic and strong forces, this decay is parity preserving. Now we can connect the parity of the $\pi^0$ to the structure of the final state in two steps, as follows.

First, a parity invariant effective Lagrangian density for $\pi^0$ decay into two photons must contain one of the interactions:

$$\pi_0 F_{\mu\nu}F^{\mu\nu} \quad \text{or} \quad \pi_0 \epsilon^{\mu\nu\rho\sigma} F_{\mu\nu} F_{\rho\sigma}, \qquad (16.9.7)$$

or, in terms of the electric and magnetic fields,

$$\pi_0(\bar{E}^2 - \bar{B}^2) \quad \text{or} \quad \pi_0(2\bar{E}{\cdot}\bar{B}). \qquad (16.9.8)$$

Clearly, the first option corresponds to a scalar decaying into photons with parallel polarizations, the second to a pseudoscalar decaying into photons with perpendicular polarizations.

Second, the plane of an $e^+e^-$ pair contains the electric field vector of the parent photon. An analysis of the relative orientations of the two planes established that the $\pi^0$ is a pseudoscalar.

Since the neutral pion is a pseudoscalar, if the interaction $\pi_0\bar{p}\Gamma p$ is to be parity invariant, then the results of Section 8.10 indicate that $\Gamma$ must be proportional to $\gamma_5$. To make the term hermitian, we take $\Gamma = i\gamma_5$.

**Homework** 16.9.9. By an analysis of the two-photon final state in Coulomb gauge $\bar{\partial}\bar{A} = 0$, prove that, if a spin-0 particle which decays into two photons through parity preserving processes, then the two photons have parallel or perpendicular polarizations according as the initial particle is a scalar or a pseudoscalar.

To identify the coupling constant $g$ out of context, it is conventional to write

$$g_{\pi NN} = g.$$

Experimentally,

$$g_{\pi NN} \simeq 13.5, \qquad (16.9.10)$$

so the perturbation expansion based on this Lagrangian density is doomed. Only the tree-level diagrams should be considered meaningful.

**Homework 16.9.11.** Starting from the interaction Lagrangian density

$$\mathcal{L}_I = -g_p \pi_0 \bar{p} i \gamma_5 p - g_n \pi_0 \bar{n} i \gamma_5 n - g_c \pi_+ \bar{p} i \gamma_5 n - g_c^* \pi_- \bar{n} i \gamma_5 p,$$

show first that we may redefine the phases of the fields to make all the coupling constants real and $g_p$ and $g_c$ non-negative, and second that if the tree-level amplitudes for $pp$, $nn$, and $pn$ (in the anti-symmetric state) scattering are all equal, then

$$g_p^2 = g_n^2 = g_p g_n + g_c^2.$$

From scattering experiments we know that $g_c \neq 0$. Conclude that $\mathcal{L}_I$ can be written in the form $-g \bar{N} i \gamma_5 \pi N$ given above.

# 16.10 Nucleon-Pion Scattering

The existence of the $SU(2)$ symmetry places a constraint on nucleon-nucleon, and nucleon-pion amplitudes. (Pion-pion scattering, of course, is not governed by the Lagrangian density 16.9.4.) Considering, for example, nucleon-pion scattering, we note that there are six initial and six final states, hence 36 amplitudes. The symmetry, however, reduces these 36 unknowns to two unknown parameters.

The trick is to remember the decomposition of the tensor products of $SU(2)$, and to use the $SU(2)$ invariance of the matrix elements. First, the six states form two irreducible representations:

$$D_{1/2} \otimes D_1 = D_{1/2} + D_{3/2}, \tag{16.10.1}$$

and second, their tensor products contain just two $SU(2)$ singlets:

$$(D_{1/2} + D_{3/2}) \otimes (D_{1/2} + D_{3/2}) \tag{16.10.2}$$
$$= D_{1/2} \otimes D_{1/2} + D_{3/2} \otimes D_{3/2} + \text{cross terms}$$
$$= D_0 + D_0 + \text{non-singlets}.$$

Following the principles of Section 6.4, we can use the lowering operator to write explicit bases for $D_{1/2}$ and $D_{3/2}$:

| $D_{3/2}$ | $D_{1/2}$ |
|---|---|
| $|p\pi^+\rangle$ | |
| $|n\pi^+\rangle + \sqrt{2}|p\pi^0\rangle$ | $\sqrt{2}|n\pi^+\rangle - |p\pi^0\rangle$ |
| $2\sqrt{2}|n\pi^0\rangle + 2|p\pi^-\rangle$ | $|n\pi^0\rangle - \sqrt{2}|p\pi^-\rangle$ |
| $6|n\pi^-\rangle$ | |

$$(16.10.3)$$

Writing $|j, m\rangle$ for the $I_3 = m$ unit vector in the spin-$j$ representation, we have

$$\left|\tfrac{3}{2}, \tfrac{3}{2}\right\rangle = |p\pi^+\rangle \tag{16.10.4}$$

$$\left|\tfrac{3}{2}, \tfrac{1}{2}\right\rangle = \sqrt{\tfrac{1}{3}}\,|n\pi^+\rangle + \sqrt{\tfrac{2}{3}}\,|p\pi^0\rangle \qquad \left|\tfrac{1}{2}, \tfrac{1}{2}\right\rangle = \sqrt{\tfrac{2}{3}}\,|n\pi^+\rangle - \sqrt{\tfrac{1}{3}}\,|p\pi^0\rangle$$

$$\left|\tfrac{3}{2}, -\tfrac{1}{2}\right\rangle = \sqrt{\tfrac{2}{3}}\,|n\pi^0\rangle + \sqrt{\tfrac{1}{3}}\,|p\pi^-\rangle \qquad \left|\tfrac{1}{2}, -\tfrac{1}{2}\right\rangle = \sqrt{\tfrac{1}{3}}\,|n\pi^0\rangle - \sqrt{\tfrac{2}{3}}\,|p\pi^-\rangle$$

$$\left|\tfrac{3}{2}, -\tfrac{3}{2}\right\rangle = |n\pi^-\rangle$$

To make the connection to amplitudes explicit, let $J_r$ be the conserved quantity operator associated with the isospin generator $I_r = \sigma_r$, and note that

$$J^2 \stackrel{\text{def}}{=} J_1^2 + J_2^2 + J_3^2 \quad \Longrightarrow \quad [J^2, J_r] = 0. \tag{16.10.5}$$

Since the Hamiltonian is $SU(2)$ invariant, the $S$ matrix satisfies

$$[J_r, S] = 0, \tag{16.10.6}$$

and so both $J_3$ and $J^2$ are conserved. Hence the only non-zero amplitudes are of the form $\langle j, m|S|j, m \rangle$. Furthermore,

$$\begin{aligned}
\langle j, m|S|j, m \rangle &= \left((j+m)(j-m+1)\right)^{-1/2} \langle j, m|SR|j, m-1 \rangle \\
&= \left((j+m)(j-m+1)\right)^{-1/2} \langle j, m|RS|j, m-1 \rangle \\
&= \langle j, m-1|S|j, m-1 \rangle,
\end{aligned} \tag{16.10.7}$$

showing that all the non-zero amplitudes are independent of $m$.

**Remark 16.10.8.** The method used above to analyze nucleon-pion scattering amplitudes can be applied to the representations of any simple Lie group or algebra. Indeed, let $\mathcal{G}$ be a simple Lie algebra of rank $k$, let $H_1, \ldots, H_k$ be a maximal commuting set of hermitian generators in $\mathcal{G}$, and let $\rho$ be any representation of $\mathcal{G}$ on a complex vector space $U$. The weight diagram for $\rho$, that is, the set of quantum numbers $q = (q_1, \ldots, q_k)$ for all states in $U$ which have pure quantum numbers with respect to the generators $H_j$, is a finite set of points in $\mathbf{R}^k$.

Recall from Section 6.2 that the weights $\alpha$ of the adjoint representation are called roots. The states $E_\alpha$ in $\mathcal{G}$ which have pure $H$ quantum numbers under the adjoint representation of $\mathcal{G}$ are raising and lowering operators:

$$[H_j, E_\alpha] = \alpha_j E_\alpha. \tag{16.10.9}$$

Now we state the facts. First, there is an invariant inner product, on $U$. With respect to this inner product $\rho(H_j)$ is hermitian for all $1 \le j \le k$. Third, there is a decomposition of $U$ into invariant orthogonal subspaces $U_r$ such that the restriction $\rho_r$ of $\rho$ to $U_r$ is irreducible. Fourth, if the tensor product $\sigma_1 \otimes \sigma_2$ of two irreducible representations contains the scalar representation then $\sigma_1$ is equivalent to $\sigma_2^*$; if $\sigma_1$ is equivalent to $\sigma_2^*$, then the scalar representation occurs with multiplicity one in the tensor product. Fifth, each $\rho(E_\alpha)$ (properly normalized to $N_q E_\alpha$) determines a standard $su(2)$ ladder through every pure state $|q\rangle$ in $U$. Sixth, a representation $\sigma$ is irreducible if and only if every pair of pure states is connected by these $su(2)$ ladders.

The details of the promised generalization will depend on the irreducible representations $\sigma_s$ of the particles under consideration, which representations make up the in state, which the out state, and which the operators in a family of matrix elements. A formal statement of the general result is available in the Wigner–Eckart Theorem. If the operator is an $S$ matrix in a $\mathcal{G}$ symmetric theory, then we see from the fourth point that the incoming and the outgoing states must lie in the same irreducible representation of $\mathcal{G}$, from $q$ conservation that $|q\rangle$ can only scatter into itself, and from the sixth point that $\bar{q}S|q\rangle$ is independent of $q$ in the weight diagram of an irreducible representation.

Clearly then, the two singlets indicated by the analysis above of the tensor product are

$$\mathcal{A}_{3/2} \stackrel{\text{def}}{=} \langle \tfrac{3}{2}, \tfrac{3}{2}|S|\tfrac{3}{2}, \tfrac{3}{2} \rangle \quad \text{and} \quad \mathcal{A}_{1/2} \stackrel{\text{def}}{=} \langle \tfrac{1}{2}, \tfrac{1}{2}|S|\tfrac{1}{2}, \tfrac{1}{2} \rangle. \tag{16.10.10}$$

All other scattering amplitudes can be expressed as a linear combination of these. For example, the amplitude for $p\pi^-$ elastic scattering may be computed simply by change of basis:

$$|p\pi^-\rangle = \sqrt{\tfrac{1}{3}} \, |\tfrac{3}{2}, -\tfrac{1}{2}\rangle - \sqrt{\tfrac{2}{3}} \, |\tfrac{1}{2}, -\tfrac{1}{2}\rangle$$

$$\implies \quad \langle p\pi^-|S|p\pi^-\rangle = \frac{1}{3}\mathcal{A}_{3/2} + \frac{2}{3}\mathcal{A}_{1/2}.$$

(16.10.11)

Actually, the $p\pi^+$ elastic scattering amplitude $\mathcal{A}_{3/2}$ is dominated by a pole at 1236 MeV, indicating the existence of the $\Delta^{++}$ particle, whereas the processes that contribute to $\mathcal{A}_{1/2}$ have no pole. Hence around 1236 MeV, the amplitude for $p\pi^-$ should be about 1/3 the amplitude for $p\pi^+$. This has been verified experimentally.

**Homework** 16.10.12. Analyze the consequences of $SU(2)$ symmetry for nucleon-nucleon scattering.

This section has introduced a field theory of hadrons. The theory can only be used at tree level, but it does indicate how the pion exchange by nucleons can generate the attractive forces that bind the nucleons into a nucleus, and it provides a simple model in which to analyze the consequences of symmetry for matrix elements. The following section expands on this simple beginning.

## 16.11 The σ Model

The primitive nucleon-pion theory of the previous sections can be developed to a theory with four pions and a chiral $SU(2)$ symmetry. In this extended theory, the pion multiplet, acting as Higgs fields, breaks the chiral symmetry to the original vector $SU(2)$ and gives masses to the nucleons. The minimal coupling of this theory to the photon has an anomaly which provides the operator for the decay $\pi^0 \to 2\gamma$.

Actually, if we make the group $SU(2)_L \times SU(2)_R$ act on the nucleon-pion Lagrangian density 16.9.4 by allowing $SU(2)$ to act independently on the left and right spinors, then the interaction term mixes with the nucleon mass term. This suggests introducing a fourth 'pion' field, $\sigma$, and regarding the nucleon mass term as a part of the interaction which separated off on account of spontaneous symmetry breaking. When the chiral symmetry breaks to isospin, the pions are identified as the Goldstone bosons, and the $\sigma$ as the massive scalar. The next few paragraphs describe this $\sigma$ Model in detail.

Define the extended pion multiplet $\Sigma$ by

$$\Sigma \overset{\text{def}}{=} \sigma\sigma_0 + i\Pi = \begin{pmatrix} \sigma + i\pi_0 & i\pi_1 + \pi_2 \\ i\pi_1 - \pi_2 & \sigma - i\pi_0 \end{pmatrix}.$$

(16.11.1)

Since classically $\Sigma$ takes values in the real vector space $\mathbf{R} \times su(2)$, and since the group $SU(2)$ can be identified as the unit sphere in this space,

$$SU(2) = \{ a^0\mathbf{1} - ia^r\sigma_r \mid \delta_{\mu\nu}a^\mu a^\nu = 1 \},$$

(16.11.2)

it is obvious that the form of $\Sigma$ is preserved under multiplication by elements of $SU(2)$ on either the right or the left. Hence we may define $SU(2)_L \times SU(2)_R$ transformations of the nucleon and $\Sigma$ fields by:

$$N' = (g_L, g_R)N \overset{\text{def}}{=} g_L P_L N + g_R P_R N \quad \text{and} \quad \Sigma' = g_L \Sigma g_R^\dagger.$$

(16.11.3)

Noting that any invariant function of $\Sigma$ is a function of $\Sigma^\dagger \Sigma = (\sigma^2 + \bar{\pi}^2)\mathbf{1}$, an invariant Lagrangian density for the $\sigma$ Model may now be written down:

$$
\mathcal{L}(N, \Sigma) \overset{\text{def}}{=} \bar{N} i \not{\partial} N + \frac{1}{4} \text{tr}(\partial_\mu \Sigma^\dagger \partial^\mu \Sigma) \tag{16.11.4}
$$
$$
- g \bar{N} \Sigma P_R N - g \bar{N} \Sigma^\dagger P_L N - V(\text{tr}\, \Sigma^\dagger \Sigma),
$$

where $V$ is chosen to bring about symmetry breaking:

$$
V = \frac{\lambda}{4}(\sigma^2 + \bar{\pi}^2 - F_\pi^2)^2, \tag{16.11.5}
$$

for some constant $F_\pi$. This density defines the $\sigma$ Model.

Note that this Lagrangian density also has a hypercharge or lepton-number symmetry. The generator $Y$ of this symmetry combines with $I_3$ to give electric charge $Q = Y + I_3$. When this pion-nucleon theory is coupled to electromagnetism, the axial symmetries become anomalous.

Using the $SU(2)$ symmetry to rotate the classical minimum of $V$, we can assert that the effect of $V$ is to set $\langle \sigma \rangle = F_\pi$ and $\langle \bar{\pi} \rangle = 0$. Shifting fields to $\sigma' = \sigma - F_\pi$, we find the Lagrangian density becomes:

$$
\mathcal{L}(N, \Pi, \sigma') = \bar{N}(i\not{\partial} - gF_\pi)N + \frac{1}{2}\partial_\mu \sigma' \, \partial^\mu \sigma' + \frac{1}{4}\text{tr}(\partial_\mu \Pi \, \partial^\mu \Pi) \tag{16.11.6}
$$
$$
- g\sigma' \bar{N} N + g\bar{N} \Pi i \gamma_5 N - \frac{\lambda}{4}(\sigma'^2 + \bar{\pi}^2 + 2F_\pi \sigma')^2.
$$

This is basically the nucleon-pion Lagrangian density 16.9.4 with nucleon mass $m = gF_\pi$, but also with zero pion mass (after all, here the pions are Goldstone bosons), an extra massive scalar $\sigma'$, and a quartic interaction term for the pions.

Since pions are so much lighter than nucleons, neglecting the pion mass can give a good approximation to nucleon-pion dynamics. A more significant consequence of the $\sigma$ Model arises from the breaking of chiral symmetry. When we set $\langle \sigma \rangle = F_\pi$ and $\langle \bar{\pi} \rangle = 0$, the full $SU(2)_L \times SU(2)_R$ symmetry breaks to $SU(2)$, and so the currents $j_{5r}$ associated with the axial generators $\gamma_5 I_r$ are no longer conserved. This breakdown of conservation manifests first in a linear contribution of the pions to the currents:

$$
j_{5r}^\mu = -\sigma \partial^\mu \pi_r + \pi_r \partial^\mu \sigma - \bar{N} \gamma^\mu \gamma_5 I_r N \tag{16.11.7}
$$
$$
= -F_\pi \partial^\mu \pi_r + \text{bilinear terms},
$$

and second, in a non-vanishing matrix element:

$$
\langle 0 | j_{5r}^\mu(0) | \pi_s(p) \rangle = -F_\pi \partial^\mu \langle 0 | \pi_r(x) | \pi_s(p) \rangle \, |_{x=0}
$$
$$
= -F_\pi \partial^\mu e^{-ix \cdot p} \delta_{rs} \, |_{x=0} \tag{16.11.8}
$$
$$
= i F_\pi p^\mu \delta_{rs}.
$$

This equation indicates the possibility of measuring $F_\pi$ from pion decays such as $\pi^+ \rightarrow l^+ + \nu_l$ (see Section 17.5 for details). The coupling constant $g = g_{\pi NN}$ is determined experimentally from nucleon-nucleon scattering. The nucleon masses

$m_N$ are well known. Consequently, the nucleon mass formula $m = gF_\pi$ is a testable prediction:

$$\begin{array}{llll}
\textbf{Prediction:} & m_N & = g_{\pi NN} \times & F_\pi \\
\textbf{Experiment:} & 939 \text{ MeV} & 13.5 \times & 93 \text{ MeV} = 1256 \text{ MeV}
\end{array} \qquad (16.11.9)$$

Thus this weak version of the Goldberger–Treiman relation works fairly well. (The complete Goldberger–Treiman relation is presented in Section 17.6.)

In this section we have presented a more elaborate theory which aims to bring out the deeper structure of nucleon-pion interactions in the absence of other forces and particle types. The purpose of the theory is to exhibit the pions as Goldstone bosons, to generate nucleon masses, and to explain pion decay modes.

One important decay mode, $\pi^0 \to 2\gamma$, cannot be explained from the theory as set out above. Also, we have not yet covered the axial $U(1)$ and its anomaly. In the next section, we conclude our introduction to nucleon-pion theory with a discussion of the anomaly and its connection to this decay mode.

## 16.12 – Abelian Anomaly

Returning to the axial symmetries of the classical Lagrangian density 16.11.4, we see that, if we include minimal coupling of the $\sigma$ Model to the photon, then according to Theorem 13.6.4 there will be an anomaly in the Noether currents $j_{5,r}^\mu$ associated with the generators $\gamma_5 I_r$. Since the gauge group is abelian, the Noether current and the matter current are identical, and so the anomaly for $\gamma_5 I_3$ is:

$$\partial_\mu j_5^\mu = \frac{e^2}{8\pi^2} \operatorname{tr}(I_3 Q^2) F_{\mu\nu} \tilde{F}^{\mu\nu}. \qquad (16.12.1)$$

The trace here is taken over $p$ and $n$. Hence the trace evaluates to $1/2$ and the anomaly is given by:

$$\partial_\mu j_5^\mu = \frac{e^2}{16\pi^2} F_{\mu\nu} \tilde{F}^{\mu\nu}. \qquad (16.12.2)$$

**Remark 16.12.3.** Actually, the abelian anomaly in the $\sigma$ Model has its roots in QCD. If we consider the minimal coupling between the strong interactions and the photon at high energy, then we have an $SU(3) \times U(1)$ gauge theory with chiral symmetry. Theorem 13.6.4 now implies that the gluon anomaly for the axial $U(1)$ generated by $\gamma_5 I_3$ must be supplemented by an electromagnetic anomaly like 16.12.1 but with the trace taken over $u$ and $d$:

$$\operatorname{tr}_{u,d}(I_3 Q^2) = 3\left(\tfrac{1}{2}(\tfrac{2}{3})^2 - \tfrac{1}{2}(-\tfrac{1}{3})^2\right) = \tfrac{1}{2}, \qquad (16.12.4)$$

where the factor of 3 accounts for the three colors of each quark.

From the form of the axial current 16.11.7 and the formula for the divergence of the neutral current 16.12.2, we can calculate the amplitude for the decay $\pi^0 \to 2\gamma$. The strategy is to use both formulae to compute the value of $p_\mu \mathcal{M}^\mu$ near $p^2 = 0$, where

$$\mathcal{M}^\mu(k, l; p) \overset{\text{def}}{=} \langle \gamma(k), \gamma(l) | \hat{j}_5^\mu(p) | 0 \rangle. \qquad (16.12.5)$$

For the first phase in the argument, we substitute the specific form of the axial current into $\mathcal{M}^\mu$ from 16.11.7 and use the LSZ reduction formalism to link $\mathcal{M}^\mu$ to a pion-decay matrix element. From the current formula 16.11.7 we see at once that

$$\mathcal{M}^\mu = \langle \gamma(k), \gamma(l) | (-ip^\mu) F_\pi \hat{\pi}^0(p) | 0 \rangle + \cdots \qquad (16.12.6)$$

Now observe that, if we apply the appropriate LSZ reduction operators to the Green function

$$G_{\mu\nu}(x, y, z) \overset{\text{def}}{=} \langle 0|TA_\mu(x)A_\nu(y)\pi^0(z)|0\rangle, \qquad (16.12.7)$$

and if we reduce the $A$'s to outgoing photons first, the final step of reduction takes the form:

$$-ip^2\langle\gamma(k), \gamma(l)|\hat{\pi}^0(p)|0\rangle = \langle\gamma(k), \gamma(l)|\pi^0(p)\rangle, \qquad (16.12.8)$$

where the $-ip^2$ is the inverse propagator factor for the pion taken as a massless particle in the $\sigma$ Model. Multiplying by $i/p^2$ and using the result in our previous expression 16.12.6, we obtain

$$\mathcal{M}^\mu = \frac{F_\pi p^\mu}{p^2}\langle\gamma(k), \gamma(l)|\pi^0(p)\rangle + \cdots \qquad (16.12.9)$$

This shows that the $\sigma$ Model pion contributes a pole to $\mathcal{M}^\mu$ at $p^2 = 0$. The current $j_5^\mu$ cannot create any other one-particle state, and all many-particle states occur at $p^2 > 0$. Consequently, this pole is the distinctive contribution of the pion to $\mathcal{M}^\mu$ and the behavior of $\mathcal{M}^\mu$ near $p^2 = 0$ is dominated by this pole. Consequently,

$$p_\mu\mathcal{M}^\mu \simeq F_\pi\langle\gamma(k), \gamma(l)|\pi^0(p)\rangle. \qquad (16.12.10)$$

We assume now that giving the pion its small mass will not invalidate this equation.

For the second phase of the argument, since $p_\mu\mathcal{M}^\mu$ is determined by $\partial_\mu j_5^\mu(x)$, we can use the anomaly formula 16.12.2 to compute $p_\mu\mathcal{M}^\mu$. It will be helpful to work out the matrix elements of $F_{\mu\nu}(x)$ ahead of time.

Since the electromagnetic potential $A^\mu$ is an interacting field, a matrix element of $F_{\mu\nu}(x)$ must be evaluated perturbatively by Feynman diagrams. For our purposes, it is sufficient to take the lowest-order term. The lowest-order term, however, is actually nothing other than the free-field approximation. Hence

$$\begin{aligned}
\langle\gamma(k)|F_{\mu\nu}(x)|0\rangle &= \langle\gamma(k)|\partial_\mu A_\nu(x) - \partial_\nu A_\mu(x)|0\rangle \\
&= \partial_\mu e^{ix\cdot k}e_\nu - \partial_\nu e^{ix\cdot k}e_\mu \qquad (16.12.11) \\
&= e^{ix\cdot k}(ik_\mu e_\nu - ik_\nu e_\mu),
\end{aligned}$$

and

$$\begin{aligned}
\langle\gamma(k), \gamma(l)|F_{\mu\nu}(x)\tilde{F}^{\mu\nu}(x)|0\rangle &\simeq \frac{1}{2}\epsilon^{\mu\nu\rho\sigma}\langle\gamma(k), \gamma(l)|F_{\mu\nu}(x)F_{\rho\sigma}(x)|0\rangle \qquad (16.12.12) \\
&\simeq \epsilon^{\mu\nu\rho\sigma}e^{ix\cdot(k+l)}(ik_\mu e_\nu - ik_\nu e_\mu)(il_\rho e_\sigma - il_\sigma e_\rho) \\
&\simeq -4e^{ix\cdot(k+l)}\epsilon^{\mu\nu\rho\sigma}k_\mu e_\nu l_\rho e_\sigma.
\end{aligned}$$

Hence the matrix element $p_\mu\mathcal{M}^\mu$ can be determined from the anomaly formula 16.12.2 as follows:

$$\begin{aligned}
p_\mu\mathcal{M}^\mu &= \langle\gamma(k), \gamma(l)|p_\mu\hat{j}_5^\mu(p)|0\rangle \\
&= \langle\gamma(k), \gamma(l)|\int d^4x\, e^{-ix\cdot p}(-i)\partial_\mu j_5^\mu(x)|0\rangle \\
&= \langle\gamma(k), \gamma(l)|(-i)\frac{e^2}{16\pi^2}\int d^4x\, e^{-ix\cdot p}F_{\mu\nu}(x)\tilde{F}^{\mu\nu}|0\rangle \qquad (16.12.13) \\
&\simeq i\frac{e^2}{4\pi^2}\int d^4x\, e^{ix\cdot(k+l-p)}\epsilon^{\mu\nu\rho\sigma}k_\mu e_\nu l_\rho e_\sigma \\
&\simeq i\frac{e^2}{4\pi^2}\epsilon^{\mu\nu\rho\sigma}k_\mu e_\nu l_\rho e_\sigma(2\pi)^4\delta^{(4)}(k + l - p).
\end{aligned}$$

From the conclusions of the two steps, to derive the amplitude the $\pi^0 \rightarrow 2\gamma$ we have only to equate the two values of $p_\mu \mathcal{M}^\mu$:

$$\langle \gamma(k), \gamma(l) | \pi^0(p) \rangle = i \frac{e^2}{4\pi^2 F_\pi} \epsilon^{\mu\nu\rho\sigma} k_\mu e_\nu l_\rho e_\sigma (2\pi)^4 \delta^{(4)}(k + l - p). \qquad (16.12.14)$$

From this matrix element, we deduce the invariant amplitude $i\mathcal{A}$ for $\pi^0 \rightarrow 2\gamma$:

$$i\mathcal{A} \overset{\text{def}}{=} \langle \gamma(k), \gamma(l) | \pi^0(p) \rangle = i \frac{e^2}{4\pi^2 F_\pi} \epsilon^{\mu\nu\rho\sigma} k_\mu e_\nu l_\rho e_\sigma. \qquad (16.12.15)$$

From the amplitude, following Section 5.2 and Section 5.4, we derive the differential decay width $d\Gamma$. Summing over final state polarizations and inserting a factor of $\frac{1}{2}$ to avoid double-counting the final states, we find the decay width

$$\Gamma = \frac{\alpha^2 m_\pi^3}{64\pi^3 F_\pi^2} \simeq 7.63 \text{ eV}, \qquad (16.12.16)$$

where $\alpha = e^2/4\pi$. The experimental value is

$$\Gamma_{\text{exp}} = 7.3 \pm 0.2 \text{ eV}. \qquad (16.12.17)$$

The good agreement between the two indicates the validity of the anomaly analysis.

This concludes our discussion of nucleon-pion theory. The $\sigma$ Model indicates that the pions can be viewed as Goldstone bosons of a broken, approximate flavor symmetry, and that pion decay modes can, to a first approximation, be understood from the structure of the matter currents associated with broken generators. As foretold in Chapter 12, we also see how the $\pi^0 \rightarrow 2\gamma$ decay is fundamentally a consequence of an abelian anomaly.

At this point we shall turn our attention to understanding the connection between the hadrons and the quarks. The Lagrangian theory of hadrons can be developed further, but, for our purposes, the rudiments of this theory given above will be sufficient.

## 16.13 Summary

This chapter has shown how basic properties of hadrons can be understood in terms of their constituent quarks. The central assumption was that the strong coupling constant diverges at the nuclear scale due to loop effects, thereby giving rise to quark confinement in color singlet states. Taking flavor, spin, and the spin-statistics theorem into account, we used group theory to determine the possible hadrons. For orbital angular momentum $l = 0$, this approach lead directly to the pseudoscalar meson octet and singlet, to the axial vector meson octet and singlet, to the baryon octet, and to the baryon decuplet. It did not, however, give any insight into the state of the gluon field in the hadrons.

Having derived the $l = 0$ hadrons from QCD by symmetry arguments, we proceeded to work out the consequences of the underlying symmetry for magnetic moments, mass splitting, and leptonic decays. Comparison with experiment was convincing.

Finally, isospin symmetry was used to construct a model of nucleon-pion processes. The lightness of the pions and the proximity of a chiral symmetry in the nucleon-pion model suggested a model in which the pions are Goldstone bosons for a broken chiral symmetry, namely, the $\sigma$ Model. This model indicated the structure of the axial currents and presence of an abelian anomaly, all of which helped to explain pion decay modes.

Now that we understand the hadronic environment of the Standard Model, we are ready to begin Standard-Model perturbation theory. The next chapter uses the Standard Model at tree level to investigate decays and scattering processes involving hadrons.

# Chapter 17

# Tree-Level Applications of the Standard Model

Connecting the four-Fermi operators of the Standard Model to particle decays and scattering amplitudes through the quark model of hadrons.

## Introduction

The Standard Model can be applied to those particle phenomena in which nonperturbative aspects of QCD do not significantly affect the scattering or decay process. For example, because of the asymptotic freedom of QCD, a high energy electron scattering off a proton interacts with the constituent quarks and gluons of the proton as if they were free particles, and the Standard Model is competent to represent these interactions. On the other hand, scattering of hadrons is beyond the reach of perturbation theory.

The types of decay and scattering which we might be able to compute are those mediated by the four-Fermi interactions of Chapter 15. The following table summarizes the possibilities, writing $l$, $q$, and $h$ for a lepton, quark, and hadron respectively:

| Type | Decay | Example | Scattering | Example |
|------|-------|---------|------------|---------|
| Leptonic | $l \to l's$ | $\mu \to e^- \nu_e \bar{\nu}_\mu$ | $l_1 l_2 \to l_3 l_4$ | $e^+ e^- \to \mu^+ \mu^-$ |
| Lepto-quark | $l \to l + h's$ | $\tau \to \nu_\tau \pi^-$ | $l_1 l_2 \to h's$ | $e^+ e^- \to h's$ |
| | $q \to l_1 l_2 + h's$ | $n \to p e^- \bar{\nu}_e$ | $lq \to l + h's$ | Parton Model |
| | | | $q_1 q_2 \to l_1 l_2$ | $\pi^+ \to \mu^+ \nu_\mu$ |
| Quark | $q \to h's$ | $s, c, b, t$ decays | $q_1 q_2 \to h's$ | $\eta, \eta', \omega, \phi$ decays |

Note that it is generally impossible to predict what hadrons will be produced, and so even in the examples we have not listed any specific hadrons in the outgoing state.

The four-quark operators of the effective Standard Model permit decay of heavy quarks into hadrons and annihilation of quarks into quarks. Examples are $s$ decay in kaons and hyperons, and quark-annihilation decay of neutral mesons. Though symmetry arguments can constrain the simpler matrix elements, because of the propensity of pions to pop up in the final states, even the simplest significant analysis of these decays is not all that simple, but requires more specific techniques than we want to present in this text.

The purely leptonic operators are covered in Section 17.1; this section is short because there are so few leptons. Section 17.2 gives an overview of the applications of the lepto-quark operators, and Sections 17.3 to 17.6 unfold the details. The last four sections, Sections 17.7 to 17.10, aim at describing baryon $\beta$ decays and deriving the true Goldberger–Treiman relation, thereby correcting the $\sigma$-Model approximation given in Section 16.11. The quark parts of the lepto-quark operators are somewhat buried in the baryon structure, so we first derive form factors for baryon transitions using symmetry arguments and then apply these to $\beta$ decays.

# 17.1 Leptonic Interactions

The simplest applications of the Standard Model are the ones involving the four-Fermi interactions of leptons. These operators can mediate lepton decays into three-lepton final states and lepton scattering into lepton-pair final states. We have already seen the character of these applications in Homeworks 15.8.5 (muon decay) and 15.8.9 ($e^+ e^- \rightarrow \mu^+ \mu^-$).

Three common decays mediated by four-lepton operators are:

$$\begin{aligned}
\mu^- &\rightarrow e^- \bar{\nu}_e \nu_\mu; \\
\tau^- &\rightarrow e^- \bar{\nu}_e \nu_\tau; \\
\tau^- &\rightarrow \mu^- \bar{\nu}_\mu \nu_\tau.
\end{aligned} \tag{17.1.1}$$

Tree-level computations give reasonable predictions for these decays.

**Homework** 17.1.2. The decay width of the tauon is a sum of leptonic and hadronic parts. Since the muon is light relative to the tauon, the two leptonic decay modes for the tauon given above should have roughly equal rates. Building on Homework 15.8.5 on the leptonic decay of the muon, estimate the leptonic decay width of the tauon. From the experimental lifetime of the tauon, $\tau \simeq 0.3 \times 10^{-12}$ sec, estimate the ratio of leptonic and hadronic decay widths of the tauon.

Neutrino-electron scattering is observed: energetic $\nu_\mu$'s are produced by $\pi^\pm$ decay and $\bar{\nu}_e$'s are produced by reactors. The cross sections for these processes are small, however, since

$$\sigma \propto G_F^2 s \quad \text{where} \quad s \sim 2(\text{Beam Energy})m_e. \tag{17.1.3}$$

In $e^+ e^- \rightarrow \mu^+ \mu^-$, the largest amplitude comes from annihilation into a photon. When we include the contribution from annihilation into a $Z$ boson, then the interference between the amplitudes magnifies the $Z$ process contribution. Therefore the data for $e^+ e^- \rightarrow \mu^+ \mu^-$ provides a better test of Standard Model predictions.

# 17.2 Lepto-quark Operators

At tree level, all the operators in the Standard Model which mediate interactions between leptons and quarks have the form $\bar{l}_2 W_\mu P_L l_1 \, \bar{q}_2 W^\mu P_L q_1$. The derived four-Fermi operators have the form

**Four-Fermi Operator:** $\qquad \dfrac{G_F}{\sqrt{2}} \bar{l}_2 \gamma^\mu (1 - \gamma^5) l_1 \, \bar{q}_2 \gamma_\mu (1 - \gamma^5) q_1.$ $\qquad$ (17.2.1)

Each operator can mediate three types of scattering and two types of decay.

This gives rise to five classes of application of the four-Fermi operators:

|  | **Operator** | **Example** |
|---|---|---|
| **Parton Model:** | | $\nu_\mu p \to \mu + \text{hadrons}$ |
| **Lepton Annihilation:** | | $e^+ e^- \to \text{hadrons}$ |
| **Quark Annihilation:** | | $\pi^- \to e\bar{\nu}_e$ |
| **Quark Decay:** | | $\pi^- \to \pi^0 e\bar{\nu}_e$ |
| **Lepton Decay:** | | $\tau \to \pi^- \nu_\tau$ |

$$(17.2.2)$$

The following sections treat the parton model of hadrons, the mode-counting approach to hadron production by lepton annihilation and by lepton decay, hadron decay by quark annihilation, and hadron decay by quark decay. Lepton decay with creation of a single hadron is related by crossing to hadron decay by quark annihilation. Material on the lepton-decay process is contained in the sections on mode counting and quark annihilation. Each section develops a technique for computing the contribution of one four-Fermi operator to an observed process.

# 17.3 – The Parton Model

Feynman proposed that the baryons could be understood as collections of point-like particles contained in a bag. These constituents of baryons he called *partons*. (When Feynman introduced the parton model, he was not aware of quarks or gluons.) The concept was to be applied at high energies where the interactions between a probe and the baryon would happen so fast that the baryon as a whole could not respond, and the interaction as a whole would be dominated by the contributions of interactions between the probe and single partons. Of course, after this interaction, the parton would be so excited that the original baryon would disintegrate and the strong forces would act in uncomputable ways to combine the resulting strongly-interacting debris into baryons and mesons. However, if the probe were a lepton, then it should be possible to detect the outgoing lepton and to determine the total energy and momentum of the outgoing hadrons from energy-momentum conservation. In this case, some predictions could be made. In this section, we shall derive a typical cross section using the parton model.

Since leptons do not scatter off gluons, lepton-parton scattering means lepton-quark scattering. If we take the lepton to be an electron and assume that an electron is detected in the final state, then photon and $Z$ exchange contribute, but the dominant contribution to the amplitude comes from photon exchange. In this case, the diagram for lepton-parton scattering is the usual

$$D = \tag{17.3.1}$$

The invariant amplitude for this diagram is

$$i\mathcal{A}^{rs}_{r's'} \stackrel{\text{def}}{=} (-ie)(-ieQ)\frac{-i}{q^2}\bar{u}(k')\gamma^\mu u(k)\bar{u}(p')\gamma_\mu u(p), \tag{17.3.2}$$

where $Q$ is the charge of the quark. The spin-sum, spin-average of the squared amplitudes is given by

$$|\mathcal{A}|^2 \stackrel{\text{def}}{=} \frac{1}{4}\sum_{r,s}\sum_{r',s'}|\mathcal{A}^{rs}_{r's'}|^2$$

$$= \frac{e^4Q^2}{4q^4}\,\text{tr}(\not{k}'\gamma^\mu\not{k}\gamma^\nu)\,\text{tr}(\not{p}'\gamma_\mu\not{p}\gamma_\nu) \tag{17.3.3}$$

$$= \frac{8e^4Q^2}{q^4}(k'{\cdot}p'\,k{\cdot}p + k'{\cdot}p\,k{\cdot}p').$$

To proceed, we need to know something about the momenta $p$ and $p'$. If the virtual photon has large energy, then $p$ is effectively proportional to the initial hadron momentum $P$. We shall introduce some statistics for the constant of proportionality later when we combine the parton scattering cross sections to make up the hadron scattering cross section. A more subtle point is that, if the hadron is to disintegrate, then the excited parton must travel a considerable distance before the strong forces take over. Hence, if disintegration is observed, we may assume that the excited parton is a physical particle with on-shell momentum. This assumption enables us to compute the parton scattering cross section.

Working in the lab frame, that is, the rest frame of the hadron, and using the high energy of the scattering to drop small masses, we find that:

$$k^2 = k'^2 = p'^2 = 0; \quad p = (m, \bar{0});$$

$$p + k = p' + k' \implies 2p{\cdot}k + m^2 = 2p'{\cdot}k'; \tag{17.3.4}$$

$$p - k' = p' - k \implies 2p{\cdot}k' - m^2 = 2p'{\cdot}k.$$

It is convenient to parameterize the cross section by the change in electron energy. We therefore define parameters $E$, $E'$ and $y$ as follows:

$$E = k^0, \ E' = k'^0 \quad \text{and} \quad y = \frac{E - E'}{E}. \tag{17.3.5}$$

Then

$$k{\cdot}p = mE, \qquad k'{\cdot}p' = mE + \tfrac{1}{2}m^2,$$

$$k'{\cdot}p = mE', \qquad k{\cdot}p' = mE' - \tfrac{1}{2}m^2. \tag{17.3.6}$$

Substituting these values into the squared amplitude of 17.3.3 yields

$$
\begin{aligned}
|\mathcal{A}|^2 &= \frac{8e^4 Q^2}{q^4}\left((mE + \tfrac{1}{2}m^2)(mE) + (mE')(mE' - \tfrac{1}{2}m^2)\right) \\
&= \frac{8e^4 Q^2}{q^4}\left(m^2 E^2\left(1 + (1-y)^2\right) + \tfrac{1}{2}m^3 Ey\right).
\end{aligned}
\tag{17.3.7}
$$

As $m \ll E$, we shall drop the $m^3 E$ term.

With the squared amplitude in good form, it is now time to derive the form of the cross section in the lab frame as a function of $E'$. Beginning from the definitions of Section 5.5, we find that:

$$
\begin{aligned}
d\sigma &\overset{\text{def}}{=} \frac{1}{4|k^0 \bar{p} - p^0 \bar{k}|}|\mathcal{A}|^2 \mathcal{D} \\
&= \frac{|\mathcal{A}|^2}{4mE}(2\pi)^4 \delta^{(4)}(p + k - p' - k')\frac{d^3 \bar{k}'}{(2\pi)^3 2k'^0}\frac{d^3 \bar{p}'}{(2\pi)^3 2p'^0} \\
&= \frac{|\mathcal{A}|^2}{64\pi^2 mE}\delta(m + E - E' - p'^0)\frac{d^3 \bar{k}'}{E' p'^0}.
\end{aligned}
\tag{17.3.8}
$$

To eliminate the last delta function, we express $d^3 \bar{k}'$ in terms of $E'$ and spherical angular coordinates,

$$
d^3 \bar{k}' = E'^2 \, dE' \, dc \, d\phi.
\tag{17.3.9}
$$

In eliminating $\bar{p}'$ from the differentials, we have imposed the momentum conservation constraint. Consequently, since $p'^2 = 0$, the energy $p'^0$ is determined by the other parameters:

$$
p'^0 = |\bar{p}'| = |\bar{k} - \bar{k}'| = (E^2 + E'^2 - 2EE'c)^{1/2}.
\tag{17.3.10}
$$

Hence the derivative of the function in the energy delta function is given by

$$
\frac{\partial}{\partial c}(m + E - E' - p'^0) = \frac{EE'}{p'^0},
\tag{17.3.11}
$$

and so, when we integrate over $c$, we find:

$$
\begin{aligned}
d\sigma &= \frac{|\mathcal{A}|^2}{64\pi^2 mE}\frac{p'^0}{EE'}\frac{E'^2 \, dE' \, d\phi}{E' p'^0} \\
&= \frac{|\mathcal{A}|^2}{64\pi^2 mE^2} dE' \, d\phi \\
&= \frac{|\mathcal{A}|^2}{32\pi mE^2} dE' \\
&= \frac{|\mathcal{A}|^2}{32\pi mE} dy.
\end{aligned}
\tag{17.3.12}
$$

Finally, substituting in the value of the squared amplitude, we obtain the desired cross section:

$$
\frac{d\sigma}{dy} = \frac{e^4 Q^2}{4\pi q^4}mE\left(1 + (1-y)^2\right),
\tag{17.3.13}
$$

with the convention that $q^4 = (q^2)^2$.

To build up the hadronic cross section from its parton cross sections, we must connect the momenta $p_j$ for the $j$th parton to the momentum $P$ of the hadron. If the exchanged energy is large compared to the hadron mass, then we can assume that the lateral momenta of the partons in negligible. There are, however, no constraints either on how many partons there will be in a hadron or on the proportion of the hadron energy-momentum a particular parton will carry. We therefore introduce functions $f_j(\xi_j)$ which give the probability densities that the $j$th parton will have energy-momentum $p_j = \xi_j P$.

If the target is a proton, since the proton is the lightest baryon and since there will be a baryon in the final state, the final hadronic state momentum $P'$ will satisfy the constraint

$$P'^2 \geq P^2 = M^2. \tag{17.3.14}$$

Up to degree of accuracy that we are using here, the nucleons have equal masses and so, if the target is a neutron, then the same constraint will hold. Hence, to simplify computations, we will assume that the target is made of nucleons.

Since $P' = P + q$, we deduce that

$$2P\cdot q + q^2 \geq 0, \tag{17.3.15}$$

and define impact parameters $x_\mathrm{H}$ and $y_\mathrm{H}$ by

$$x_\mathrm{H} \overset{\text{def}}{=} -\frac{q^2}{2P\cdot q} \in [0,1];$$
$$y_\mathrm{H} \overset{\text{def}}{=} \frac{P\cdot q}{P\cdot k} = \frac{E - E'}{E} \in [0,1]. \tag{17.3.16}$$

Since $p_j = \xi_j P$, $q^2 \gg M^2$, and $p'^2 = 0$, we find that

$$p_j^2 + 2p_j\cdot q + q^2 = (p_j + q)^2 = p'^2 = 0$$
$$\text{and} \quad -\frac{q^2}{2p_j\cdot q} = 1 + \frac{p_j^2}{2p_j\cdot q} = 1 + O\!\left(\frac{M}{\sqrt{q^2}}\right), \tag{17.3.17}$$

and conclude

$$x_\mathrm{H} = -\frac{q^2}{2P\cdot q} = -\frac{\xi_j q^2}{2p_j\cdot q} \simeq \xi_j. \tag{17.3.18}$$

The nucleon parameters $y_\mathrm{H}$ and $M$ are simply related to the parton parameters $y$ and $m$:

$$y_\mathrm{H} = y \quad \text{and} \quad x_\mathrm{H} M = m. \tag{17.3.19}$$

Finally, the exchange momentum satisfies

$$-q^2 = 2P\cdot k\, x_\mathrm{H} y_\mathrm{H} = 2ME x_\mathrm{H} y_\mathrm{H}. \tag{17.3.20}$$

Taking $x_\mathrm{H}$ and $y_\mathrm{H}$ as the parameters in the nucleon differential cross section, we combine parton cross sections to form the nucleon differential cross section as follows:

$$\frac{d\sigma_\mathrm{H}^e}{dx_\mathrm{H}\, dy_\mathrm{H}} = \sum_j \frac{d\sigma_j}{dy_j} f_j(x_\mathrm{H})$$
$$= \sum_j \frac{e^4 Q_j^2}{4\pi q^4} ME\big(1 + (1 - y_\mathrm{H})^2\big) x_\mathrm{H} f_j(x_\mathrm{H}). \tag{17.3.21}$$

Experimental data on such scatterings as functions of the impact parameters provides data on the functions $f_j$. In fact, such data indicates a weak dependence on $q^2$, a result in agreement with predictions based on higher-order QCD effects.

Since the formula 17.3.21 for the electron-nucleon differential cross section is only valid for high values of $-q^2$, $x_H y_H$ is constrained by a lower bound. In practice, $ME$ is sufficiently large that we can permit $x_H$ and $y_H$ to range freely over the unit square as long as we suppress the meaningless pole due to the $q^{-4}$ factor. To a first approximation, this is accomplished by taking $q^4$ to be a constant.

In this approximation, we can define the expected value $X_j$ of $x_H$ for each parton

$$X_j \stackrel{\text{def}}{=} \int_0^1 dx_H \, x_H f_j(x_H) \tag{17.3.22}$$

and thereby reduce the differential cross section to

$$\frac{d\sigma_H^e}{dy_H} = \frac{e^4}{4\pi q^4} ME\big(1 + (1 - y_H)^2\big) \sum_j Q_j^2 X_j. \tag{17.3.23}$$

**Homework 17.3.24.** Apply the parton model to derive the differential cross section for the scattering of a muon neutrino with a nucleon (the final-state lepton will be a muon). Use the simplifying assumptions that (1) only the light quarks $u$, $\bar{u}$, $d$, and $\bar{d}$ contribute significantly, (2) the KM matrix is the identity, and (3) the exchange momentum satisfies $m_N^2 \ll -q^2 \ll M_Z^2$.

Write down the differential cross section for the scattering of the muon anti-neutrino with a nucleon.

Writing $X_q = n_q \int dx \, x f_q(x)$ for the expected proportion of hadron momentum carried by quarks of type $q$, the answer should be

$$\frac{d\sigma_H^\nu}{dy_H} = \frac{2mEG_F^2}{\pi} \big((X_d + (1 - y_H)^2 X_{\bar{u}}\big);$$

$$\frac{d\sigma_H^{\bar{\nu}}}{dy_H} = \frac{2mEG_F^2}{\pi} \big((X_{\bar{d}} + (1 - y_H)^2 X_u\big).$$

**Homework 17.3.25.** In the previous homework, what are the results if we use the correct KM matrix but continue to ignore heavy quarks?

If we take a target with equal numbers of neutrons and protons, then isospin symmetry will make

$$X_u = X_d \stackrel{\text{def}}{=} X \quad \text{and} \quad X_{\bar{u}} = X_{\bar{d}} \stackrel{\text{def}}{=} \bar{X}. \tag{17.3.26}$$

Integrating the differential cross section in Homework 17.3.24 over $y_H$, we find in this case that

$$\sigma^\nu = \frac{2mEG_F^2}{\pi} (X + \tfrac{1}{3}\bar{X});$$

$$\sigma^{\bar{\nu}} = \frac{2mEG_F^2}{\pi} (\tfrac{1}{3}X + \bar{X}). \tag{17.3.27}$$

Comparing these predictions with experimental data, we can obtain the values of $X$ and $\bar{X}$:

$$X \simeq 0.5 \quad \text{and} \quad \bar{X} \simeq 0.1. \tag{17.3.28}$$

This indicates that the quarks carry only about half the nucleon momentum. Anti-quarks (and heavy quarks and other particles) contribute a little, but the bulk of the other half is in fact carried by gluons.

In the previous chapter, we treated hadrons as tensor products of quark states. It is now clear that this model is inadequate; hadron structure is far more intricate. It fails to take into account either the gluon field or the sea of quark–anti-quark pairs in hadron. As yet, the Standard Model does not even provide an explanation of the mass and spin of a proton.

The cross section for electron scattering from the same target obtained by integrating 17.3.23 over $y_{\mathrm{H}}$ is

$$\sigma^e = \frac{e^4 mE}{4\pi q^4} \frac{4}{3} \frac{5}{9} (X + \bar{X}). \qquad (17.3.29)$$

Eliminating $X$ and $\bar{X}$ from these three cross-section formulae yields the relation

$$\sigma^e = \frac{5}{72} \frac{e^4}{G_{\mathrm{F}}^2 q^4} (\sigma^\nu + \sigma^{\bar\nu}). \qquad (17.3.30)$$

Comparison with experiment confirms the validity of the line of thought presented in this section.

The fundamental assumptions of the parton model are that:

1. If the energy exchanged between the probe particle and the target hadron is large compared to the mass of the hadron, then the interaction is at first dominated by the weak interaction of the probe with a parton;
2. The disintegration of the target signals that the excited parton was on-shell;
3. Hadron formation follows and does not significantly affect the total differential cross section.

In the computations above, to illustrate the parton model with a simple example, we took the probe to be an electron so that a virtual photon could mediate the interaction and we could select final states which contained an electron; and we took a proton for the target so that we would at least know of the final state that it contained a proton. The conclusion is a remarkably simple formula for the cross section. Next, we use the formulae of the Homework 17.3.24 first to determine the unknown expectations $X_q$ from experiment and second to generate predictions.

## 17.4 – The Parton Model and Mode Counting

The parton model can also be used to estimate branching ratios through mode counting. The essential new principles are:

1. All the various decay modes of a high-energy electroweak vector boson into lepton and light quark pairs are equally probable;
2. A light quark pair produces hadrons in the final state without affecting this probability.

When an amplitude is dominated by a single diagram, these principles apply directly to make the cross sections for the various modes roughly proportional to the products of the appropriate coupling constants. On the basis of this proportionality, we can use the computable cross section for a lepton option to make predictions about non-computable hadron options.

For example, when an $e^+e^-$ pair annihilates into a photon, the photon may decay into $e^+e^-$, $\mu^+\mu^-$, or $q\bar{q}$ where $q$ is any quark that is light enough to be produced. The difference between the electroweak amplitudes for these processes is a factor of the charge $Q$ of a final state particle. Hence we deduce that the branching ratio

$$R \overset{\text{def}}{=} \frac{\sigma(e^+e^- \rightarrow \text{hadrons})}{\sigma(e^+e^- \rightarrow e^+e^-)} \tag{17.4.1}$$

is given by

$$R \simeq \sum_{m_q^2 < E} n_c Q_q^2, \tag{17.4.2}$$

where $n_c$ is the number of quark colors, $E$ is the energy of the incoming electron in the c.o.m. frame, and the sum is taken over quarks $q$ with masses $m_q$ and charges $Q_q$.

**Remark 17.4.3.** Actually, the strong interactions enhance the cross section for hadron final states. The dominant contribution comes from diagrams in which one of the outgoing quarks emits a gluon. Taking these diagrams into account yields

$$R \simeq n_c \left(1 + \frac{g_s^2}{4\pi^2} + O(g_s^4)\right) \sum_{m_q^2 < E} Q_q^2, \tag{17.4.4}$$

where $g_s$ is the strong coupling constant. As QCD is asymptotically free, $g_s$ tends to zero as the energy becomes larger. Hence these QCD effects will not significantly affect the mode-counting prediction for $R$ at high energies. However, at thresholds for $q\bar{q}$ resonances like $\rho$, $\omega$, $\phi$, $J/\phi$, and $\Upsilon$, $R$ is observed to have sharp narrow peaks, indicating the inadequacy 17.4.4 at these energies.

The decays of the tauon provide a similar opportunity for mode counting on the basis of the parton model. The tauon decays into a $\nu_\tau$ and a $W^-$. Treating the KM matrix as the identity for a first approximation, we find that the $W^-$ can decay into one of $e^-\bar{\nu}_e$, $\mu^-\bar{\nu}_\mu$, or $d\bar{u}$ in three colors. Hence we predict that

$$\Gamma(\tau \rightarrow e^- + \bar{\nu}_e + \nu_\tau) = 0.2 \times \Gamma_{s=0},$$
$$\Gamma(\tau \rightarrow \mu^- + \bar{\nu}_\mu + \nu_\tau) = 0.2 \times \Gamma_{s=0}, \tag{17.4.5}$$
$$\Gamma(\tau \rightarrow (s = 0 \text{ hadrons}) + \nu_\tau) = 0.6 \times \Gamma_{s=0},$$

where $\Gamma_{s=0}$ is the decay width for all tauon decays whose final states contain no strange hadrons. These estimates are in good accord with experiment, though the leptonic modes appear experimentally to be relatively suppressed by a few percent. Note that we have squared the amplitudes before summing because the three color options for $d\bar{u}$ are orthogonal and do not interfere with each other.

Because of mass constraints, if the $W^-$ decays into $s\bar{c}$ or $b\bar{t}$, there must be further decay before a particle state is formed. This leads to a $G_F$ suppression. Hence the insistence on 'light' quark pairs in the principle of mode counting.

If we use the correct KM matrix, then $W^-$ decays into $d'\bar{u}$, and so we obtain some strange hadrons from the $d'$:

$$d' = V_{ud}d + V_{us}s + V_{ub}b$$
$$\simeq 0.97\,d + 0.22\,s + 0.004\,b. \tag{17.4.6}$$

The $b$ is too heavy to survive in the final state, and anyway its contribution is heavily suppressed by the 0.004 factor. Hence the ratio of non-strange hadron final states to comparable strange hadron final states should be about

$$\left(\frac{0.97}{0.22}\right)^2 \simeq 20. \tag{17.4.7}$$

In fact, observation of tauon decays yields percentages of the total decay width as follows:

$$\Gamma(\tau \to \rho^- \nu_\tau) \simeq 22.7\%,$$
$$\Gamma(\tau \to K^{*-} \nu_\tau) \simeq 1.4\%. \tag{17.4.8}$$

There are other decay modes with strange-hadron final states, but $\rho^-$ is a $\bar{u}d$ bound state and so should be compared with the $\bar{u}s$ bound state $K^{*-}$. Comparison between prediction and data indicates the merit of mode counting.

Mode counting is a very easy way of obtaining estimates for cross sections and decays. The method applies when there are a variety of processes which contain identical weak interaction components and irrelevant or comparable QCD components. It is of limited applicability, however, since these hypotheses are hard to satisfy.

# 17.5 – Quark Annihilation

So far, we have studied the application of the weak-interaction lepto-quark operators in situations where there is at least one incoming fermion. Now we turn to situations where the incoming particles are quarks.

There are two possibilities here: either two quarks interact (annihilate into a vector boson) or one quark decays. In a quark decay, following the parton principle, all other quarks are assumed to be non-participant, merely spectators, until the strong interactions take over in the final stage of hadronization. In this section, we shall show how to approach annihilation processes and in the next, spectator processes.

The simplest annihilation process is the decay $\pi^- \to e^- \bar{\nu}_e$. To find the amplitude for this process, we assume that it takes place in the following three steps:

1. The pion decays into a quark current through strong interactions;
2. The quarks in this current then annihilate into a $W^-$;
3. The $W^-$ then decays into the final state leptons.

This sequence of events may be represented by a diagram:

$$\tag{17.5.1}$$

With these assumptions, we can eliminate the $W^-$ in favor of the four-Fermi interaction between a quark current $j^\mu$ and a lepton current $l^\mu$:

$$\mathcal{L}' \overset{\text{def}}{=} -\frac{G_F}{\sqrt{2}} l^\mu j_\mu, \tag{17.5.2}$$

where

$$l^\mu = \bar{e}\gamma^\mu(1-\gamma^5)\nu_e \quad \text{and} \quad j^\mu = \bar{u}\gamma^\mu(1-\gamma^5)d'. \tag{17.5.3}$$

From Dyson's formula, the matrix element for the decay $\pi^- \to e^- \bar{\nu}_e$ is to lowest order

$$\mathcal{M} \overset{\text{def}}{=} \langle e^-, \bar{\nu}_e | (S-1) | \pi^- \rangle$$
$$= -\frac{iG_F}{\sqrt{2}} \int d^4x \, \langle e^-, \bar{\nu}_e | l^\mu(x) j_\mu(x) | \pi^- \rangle. \tag{17.5.4}$$

If we insert a resolution of the identity between the currents, the current matrix elements arising from the insertion of an $n$-particle state can be derived from Green functions which have $n+4$ variables. The vacuum insertion alone has a zeroth order component and should therefore dominate the expansion:

$$\mathcal{M} = -\frac{iG_F}{\sqrt{2}} \int d^4x \, \langle e^-, \bar{\nu}_e | l^\mu(x) | 0 \rangle \langle 0 | j_\mu(x) | \pi^- \rangle. \tag{17.5.5}$$

The modified process may be represented diagrammatically:

$$\tag{17.5.6}$$

The strong interactions are parity invariant. The value of the matrix element $\langle 0 | j_\mu(x) | \pi^- \rangle$ must be a function of the pion momentum alone, so the matrix element must be a vector. Since the pion is a pseudoscalar, the vector part of the quark current does not contribute.

The $\sigma$ Model of Section 16.8 introduced the link between the axial current and the pion in the context of pion-nucleon theory. To make the transition from that toy model to the Standard Model, we need only interpret the axial current as the light-quark axial current $j_5^\mu$:

$$\langle 0 | j_5^\mu(x) | \pi^- \rangle = i\sqrt{2} \, F_\pi e^{-ix \cdot p} p^\mu, \tag{17.5.7}$$

where $p$ is the momentum of the pion and the extra factor of $\sqrt{2}$ comes from the normalization of the current. There is at present no way of computing this matrix element, so $F_\pi$ is a free parameter to be determined from experiment.

Using 17.5.7 to eliminate the quark-current matrix element from $\mathcal{M}$ yields

$$\mathcal{M} = -G_F F_\pi p^\mu \int d^4x \, e^{-ix \cdot p} \langle e^-, \bar{\nu}_e | l^\mu(x) | 0 \rangle$$
$$= -G_F F_\pi p^\mu \langle e^-, \bar{\nu}_e | \hat{l}^\mu(p) | 0 \rangle. \tag{17.5.8}$$

In the spirit of the four-Fermi operator, we regard the fields in $\hat{l}^\mu$ as free. According to the following homework, the matrix element of the lepton current is

$$\langle e^-, \bar{\nu}_e | \hat{l}^\mu(p) | 0 \rangle = -(2\pi)^4 \delta^{(4)}(p - k - l) \bar{e}(k) \gamma^\mu (1 - \gamma^5) \nu_e(l), \tag{17.5.9}$$

where $e(k)$ and $\nu_e(l)$ are polarization spinors.

**Homework 17.5.10.** Show that $\langle f | \hat{A}(p) | i \rangle = (2\pi)^4 \delta^{(4)}(p_f - p - p_i) \langle f | A(0) | i \rangle$, where $A$ is any function of quantum fields.

**Remark 17.5.11.** The LSZ reduction formula implies that the lowest-order contribution (possibly zero) to a matrix element comes from the substitution of free fields for interacting fields. It is worthwhile to understand this in detail. We shall therefore derive the lepton-current matrix element 17.5.9 without assuming that the fields are free.

Clearly, the lepton-current matrix element 17.5.9 originates in a four-point Green function $G^{\alpha\beta\gamma\delta}$ with four spinor indices. Since we use momentum-space perturbation theory, we must connect the Fourier transform of the matrix element to $\hat{G}$. In the current, two fields are evaluated at the same point, so we introduce a lemma:

$$f(x) = g(x, x) \quad \Longrightarrow \quad \hat{f}(p) = \int \frac{d^4 q}{(2\pi)^4} \hat{g}(p - q, q), \qquad (17.5.12)$$

and apply the lemma through the definition

$$g^{\mu}(x, y) = \langle e^{-}, \bar{\nu}_e | T \bar{e}(x) \gamma^{\mu}(1 - \gamma^5) \nu_e(y) | 0 \rangle, \qquad (17.5.13)$$

which makes $\hat{f}^{\mu}(p)$ equal to the leptonic-current matrix element of 17.5.9.

Clearly, $\hat{g}^{\mu}$ can be derived from $\hat{G}$ by LSZ reduction of two fields and contraction of the remaining two with the Dirac indices of $\gamma^{\mu}(1 - \gamma^5)$. On the level of $\hat{g}^{\mu}$, the two fields in $l^{\mu}$ should be represented in diagrams by distinct external lines because they can carry independent momenta. On the level of $\hat{f}^{\mu}$, there is only one independent momentum, and it is better to represent this term by a source in a single line:

$$\textbf{Source Diagram:} \qquad \hat{l}^{\mu}(p) \longleftrightarrow p - q \; {\dashleftarrow} \; q$$
$$\qquad\qquad\qquad\qquad\qquad\qquad\qquad\qquad\qquad\qquad (17.5.14)$$
$$\textbf{Vertex Rule:} \qquad \hat{l}^{\mu}(p) \longleftrightarrow \gamma^{\mu}(1 - \gamma^5)$$

As usual, the arrow indicates flow of leptons, a neutrino coming in and an electron coming out.

On the basis of this link to an underlying Green function and with this diagram convention, we can write down Feynman diagrams which contribute to the leptonic-current matrix element $\hat{f}^{\mu}$:

$$\qquad\qquad\qquad\qquad\qquad\qquad\qquad\qquad\qquad\qquad\qquad (17.5.15)$$

The first diagram gives the free-field approximation, the unique lowest-order term.

The KM matrix expresses the charge eigenquark $d'$ as a sum of mass eigen-quarks. Since the $\pi^-$ is so light, the heavy quarks must be off-shell and must decay, generating higher-order processes which we shall neglect. Consequently, we can replace $d'$ in $j_5^{\mu}$ with $V_{ud}d$, and $\mathcal{M}$ evaluates to

$$\mathcal{M} = (2\pi)^4 \delta^{(4)}(p - k - l) G_{\mathrm{F}} F_{\pi} V_{ud}\, p_{\mu} \bar{e}(k) \gamma^{\mu}(1 - \gamma^5) \nu_e(l). \qquad (17.5.16)$$

The associated amplitude is

$$i\mathcal{A} = G_{\mathrm{F}} F_{\pi} V_{ud}\, \bar{e}(k) \slashed{p}(1 - \gamma^5) \nu_e(l). \qquad (17.5.17)$$

Since $\bar{e}\slashed{k} = m_e \bar{e}$, $\slashed{l}\nu_e = 0$, and $p = k + l$ this simplifies to

$$i\mathcal{A} = G_{\mathrm{F}} F_{\pi} V_{ud} m_e \, \bar{e}(k)(1 - \gamma^5) \nu_e(l). \qquad (17.5.18)$$

The mass suppression by $m_e$ in the amplitude arises because the current coupled to the $W^-$ contains a $P_L$, and so when the $W^-$ decays, it creates a right-handed electron and a left-handed neutrino. To conserve angular momentum, it is therefore necessary for the electron mass term to convert the right-handed electron into a left-handed one. Hence the factor of $m_e$.

The same analysis applies to the decay $\pi^- \to \mu \bar\nu_\mu$ to yield an amplitude

$$i\mathcal{A}' = G_F F_\pi V_{ud} m_\mu \, \bar u(k)(1 - \gamma^5)\nu_\mu(l). \tag{17.5.19}$$

We therefore expect the ratio of the decay rates to be roughly proportional to the square of the ratio of the masses. Experiment gives

$$\left.\begin{array}{l} \Gamma(\pi^- \to \mu \bar\nu_\mu) \simeq 1 \\ \Gamma(\pi^- \to e \bar\nu_e) \simeq 1.2 \times 10^{-4} \end{array}\right\} \quad \text{and} \quad \left(\frac{m_\mu}{m_e}\right)^2 \simeq 4 \times 10^4. \tag{17.5.20}$$

Our prediction is off by a factor of 4 because we have omitted the suppression factor associated with larger mass of final state particles. Generally, decay into lighter particles is favored. In this case, angular-momentum suppression of the lower-mass final state dominates, mass-suppression of the higher-mass final state is secondary, and, as the following homework shows, the two together give a prediction in good agreement with experiment.

**Homework 17.5.21.** Compute the decay widths for $\pi^- \to e \bar\nu_e$ and $\pi^- \to \mu \bar\nu_\mu$ from the amplitudes above. Show that their ratio is

$$\frac{\Gamma(\pi^- \to \mu \bar\nu_\mu)}{\Gamma(\pi^- \to e \bar\nu_e)} = \left(\frac{m_\mu}{m_e}\right)^2 \left(\frac{m_\pi^2 - m_\mu^2}{m_\pi^2 - m_e^2}\right)^2.$$

Compare the predicted decay widths and their predicted ratio with experiment.

Having done this reduction carefully once, one sees that the amplitude can be written down directly from the quark-current matrix element and the four-Fermi interaction:

$$i\mathcal{A} = \frac{G_F}{\sqrt{2}}(\text{Quark-Current Matrix Element})_\mu \tag{17.5.22}$$
$$\times (\text{Polarization Spinor Current})^\mu.$$

The matrix element for a $K^-$ to decay into the quark current $\bar u \gamma^\mu d$ introduces a further uncomputable strong interaction parameter, $F_K$:

$$\langle 0|\bar u(0)\gamma^\mu d(0)|K^-(p)\rangle = i\sqrt{2}\, F_K p^\mu. \tag{17.5.23}$$

In the limit of exact $SU(3)$ flavor symmetry, $F_K = F_\pi$, but in practice the value is larger on account of the greater instability of the strange quark:

$$F_K \simeq 1.2 F_\pi. \tag{17.5.24}$$

**Homework 17.5.25.** Compute the decay widths for $K^- \to e \bar\nu_e$ and $K^- \to \mu \bar\nu_\mu$. Compare with experiment.

**Homework 17.5.26.** The technique used above for computing pion and kaon decays works equally well for the tauon decays $\tau \to \pi^- \nu_\tau$ and $\tau \to K^- \nu_\tau$. Compute the decay widths for these two processes and compare with experiment.

This section has introduced the application of the Standard-Model lepto-quark operators to incoming quark states. The logic for this application depends (as usual) on separating strong interaction effects from the core electroweak process.

The analysis above shows how the underlying parton model supports a factorization of the decay matrix element into hadronic and leptonic factors. The hadronic factor cannot be computed; it must be parametrized. If the result is to be useful, clearly the same parameter must arise in modeling distinct experiments as happens with the various pion decays.

# 17.6 – Quark Decay

Spectator processes form the last category of applications of the effective Standard-Model lepto-quark operators. In these processes, a heavy quark decays into a quark and a lepton pair, and then the strong interactions assert themselves to produce color singlets.

We shall begin with the simplest example of a spectator process, the $d$ decay in the $\pi^-$ which causes the $\pi^-$ decay

$$\left.\begin{array}{c} d' \to u e^- \bar{\nu}_e \\ \implies \quad \pi^- \to \pi^0 e^- \bar{\nu}_e \end{array}\right\} \tag{17.6.1}$$

The up anti-quark in the $\pi^-$ is the spectator here.

The four-Fermi operator which mediates this decay is again

$$\textbf{Operator} = -\frac{G_F}{\sqrt{2}} \bar{e}\gamma^\mu(1-\gamma_5)\nu_e \, \bar{u}\gamma_\mu(1-\gamma_5)d'. \tag{17.6.2}$$

Following the factorization summary of the last section, we can write the amplitude for this decay as follows:

$$i\mathcal{A} = \frac{G_F}{\sqrt{2}} \bar{e}\gamma^\mu(1-\gamma_5)\nu_e \, \langle \pi^0 | j^\mu(0) - j_5^\mu(0) | \pi^- \rangle. \tag{17.6.3}$$

Since the matrix element of the quark current must be a function of the two pion momenta, the matrix element must be a vector. Since the pions are both pseudoscalars, the parity invariance of the strong interactions prohibits a contribution from the axial part of the quark current.

As in the analysis of pion decay through the annihilation process, we do not expect the heavy quarks to contribute significantly, and so we approximate $j^\mu$ using $V_{ud}$ from the KM matrix:

$$i\mathcal{A} = \frac{G_F V_{ud}}{\sqrt{2}} \bar{e}\gamma_\mu(1-\gamma_5)\nu_e \, \langle \pi^0 | \bar{u}(0)\gamma_\mu d(0) | \pi^- \rangle. \tag{17.6.4}$$

Since the pions are on-shell, the Lorentz covariance of the quark-current matrix elements implies that they have the form

$$\begin{aligned} \mathcal{M}^\mu &\overset{\text{def}}{=} \langle \pi^0 | \bar{u}(0)\gamma^\mu d(0) | \pi^- \rangle \\ &= A(q^2)(p^\mu + p'^\mu) + B(q^2)q^\mu, \end{aligned} \tag{17.6.5}$$

where $p$ and $p'$ are the momenta of $\pi^-$ and $\pi^0$ respectively, and $q = p - p'$ is the exchange momentum.

Now, if isospin symmetry is broken by the electroweak interactions, then we can investigate this strong-interaction matrix element in the limit of isospin symmetry. We first restore isospin symmetry by equating the pion masses. Since symmetry has been restored, the quark current $j^\mu$ is conserved and $q_\mu \mathcal{M}^\mu = 0$. Since the pion masses are equal, $q_\mu(p + p')^\mu = 0$. Hence we conclude that $B$ must be zero:

$$q^2 B(q^2) = 0 \quad \Longrightarrow \quad B(q^2) = 0. \tag{17.6.6}$$

Furthermore, $\bar{u}\gamma^\mu d$ is the current associated with the isospin raising operator, so the integral of $\bar{u}\gamma^0 d$ gives the hermitian operator $Q$ which represents raising on the state space:

$$Q = \int d^3\bar{x}\, \bar{u}(x)\gamma^0 d(x) \quad \Longrightarrow \quad Q|\pi^-\rangle = \sqrt{2}\,|\pi^0\rangle. \tag{17.6.7}$$

**Remark 17.6.8.** This action of $Q$ on the particles states is best remembered by the following simple logic based on the destruction and creation properties of the fields:

$$\left.\begin{array}{ll} Q \leftrightarrow \bar{u}d & \text{(fields)} \\ \pi^0 \leftrightarrow \dfrac{u\bar{u} - d\bar{d}}{\sqrt{2}} & \text{(particles)} \\ \pi^- \leftrightarrow d\bar{u} & \text{(particles)} \end{array}\right\} \quad \Longrightarrow \quad Q\pi^- = u\bar{u} - d\bar{d} = \sqrt{2}\,\pi^0. \tag{17.6.9}$$

Even from this viewpoint, the factor of $\sqrt{2}$ originates in the normalization of $Q$.

We find the value of $A$ by evaluating $\bar{\pi}^0 Q|\pi^-\rangle$ in two ways:

$$\langle \pi^0|Q|\pi^-\rangle = \sqrt{2}\,\langle \pi^0|\pi^0\rangle = \sqrt{2}\,(2\pi)^3\, 2p^0\delta^{(3)}(\bar{q}), \tag{17.6.10}$$

and

$$\begin{aligned} \langle \pi^0|Q|\pi^-\rangle &= \int d^3\bar{x}\, \langle \pi^0|\bar{u}(x)\gamma^0 d(x)|\pi^-\rangle \\ &= \int d^3\bar{x}\, e^{ix\cdot q}\langle \pi^0|\bar{u}(0)\gamma^0 d(0)|\pi^-\rangle \\ &= (2\pi)^3\delta^{(3)}(\bar{q})(p^0 + p'^0)A(q^2). \end{aligned} \tag{17.6.11}$$

Equating the two results, we find that

$$A(0) = \sqrt{2}. \tag{17.6.12}$$

If we maintain isospin symmetry, then the degeneracy of the pion masses makes this $\pi^-$ decay impossible. However, if we now include the electroweak symmetry-breaking effects, it is reasonable to assume that the current matrix element does not change much, and so the value of the amplitude is approximately

$$i\mathcal{A} \simeq G_F V_{ud}\bar{e}(\not{p} + \not{p}')\nu_e. \tag{17.6.13}$$

**Homework 17.6.14.** Taking $m_e = 0$, find the differential decay width $d\Gamma$ for $\pi^- \to \pi^0 e\bar{\nu}_e$ as a function of the energies $E_1$ and $E_2$ of the final-state leptons. Find the range of the variables $E_r$, and find $\Gamma$ by numerical integration. Check your result against experimental data.

**Remark** 17.6.15. If we take care of the Clebsch–Gordon coefficients, then we can use the same argument to find amplitudes for nuclei like $^{14}O$ or $^{34}Cl$ which suffer $d$ decays. Comparison with data for such decays supports the bounds on the KM matrix element

$$0.9747 < V_{ud} < 0.9759. \tag{17.6.16}$$

Hadron decays caused by $s$ decay can be analyzed in the same way. Parity eliminates the axial part of the quark current, and the KM matrix introduces a factor of $V_{us}$ into the effective light-quark current:

$$\bar{u}\gamma^\mu(1 - \gamma^5)d' \to V_{us}\bar{u}\gamma^\mu s. \tag{17.6.17}$$

For example, using the general form 17.5.22 for decay amplitudes of this type, the amplitude for the decay

$$K^- \to \pi^0 e\bar{\nu}_e \tag{17.6.18}$$

is

$$i\mathcal{A} = \frac{G_F V_{us}}{\sqrt{2}}\bar{\nu}_e\gamma_\mu e \, \langle\pi^0|j^\mu|K^-\rangle. \tag{17.6.19}$$

The current matrix element, on the assumption of restored $SU(3)$ flavor symmetry in the limit of equal quark masses, has the value

$$\begin{aligned}
\langle\pi^0|j^\mu|K^-\rangle &= C(q^2)(p^\mu + p'^\mu) + D(q^2)(p^\mu - p'^\mu) \\
&= C(0)(p^\mu + p'^\mu).
\end{aligned} \tag{17.6.20}$$

The value of $C(0)$ is $1/\sqrt{2}$.

**Homework** 17.6.21. We found $A(0)$ for pion decay by constructing and applying an $SU(2)$ raising operator. Construct and apply an $SU(3)$ raising operator to prove that $C(0)$ is equal to $1/\sqrt{2}$.

**Homework** 17.6.22. Taking $m_e = 0$, find the differential decay width $d\Gamma$ for $K^- \to \pi^0 e\bar{\nu}_e$ as a function of the energies $E_1$ and $E_2$ of the final-state leptons. For a first approximation, find $\Gamma$ under the assumption that the pion mass is negligible. For a better approximation, find $\Gamma$ by numerical integration of $d\Gamma$. Compare your answers with experiment. Is numerical integration worthwhile in this case?

**Remark** 17.6.23. Since the $u$, $d$, and $s$ masses are so far from degeneracy, it is actually rather surprising that the value of $V_{us}$ computed from comparing this amplitude with experiment is within the correct range:

$$0.218 < V_{us} < 0.224. \tag{17.6.24}$$

With this section, we conclude our presentation of measurable effects arising from the lepto-quark operators in the four-Fermi limit of the Standard Model. The next section investigates matrix elements of quark currents between baryon states.

# 17.7 Baryons: Form Factors for Current Matrix Elements

The analysis of pion and kaon decays presented in the previous sections depends on an analysis of quark-current matrix elements. The matrix elements we encountered there had simple forms because mesons particle states are characterized by momentum. When we attempt the same analysis for baryon decays, since a baryon cannot decay into a quark current, we must consider baryon-baryon transitions which generate a quark current. Then the spins of baryon particle states complexify the result. In this section, we investigate the structure of such matrix elements and then apply the results.

Let $j^\mu = \bar{\psi} T \gamma^\mu \psi$ and $j_5^\mu = \bar{\psi} T \gamma^5 \gamma^\mu \psi$ be vector currents and axial currents respectively. Define matrix elements $\mathcal{M}^\mu$ and $\mathcal{M}_5^\mu$ for these currents between baryon states as follows:

$$\mathcal{M}^\mu \stackrel{\text{def}}{=} \langle B', p', s'|j^\mu(0)|B, p, s\rangle \quad \text{and} \quad \mathcal{M}_5^\mu \stackrel{\text{def}}{=} \langle B', p', s'|j_5^\mu(0)|B, p, s\rangle. \quad (17.7.1)$$

In analyzing these matrix elements, we shall assume that the parities of the two baryons are equal. (In light of Homework 17.5.10, these matrix elements do not contain an energy-momentum conserving delta function).

The analysis begins with an application of Lorentz covariance to identify form factors:

$$\mathcal{M}^\mu = \bar{u}(p', s')(F_1\gamma^\mu + F_2 p^\mu + F_3 p'^\mu + F_4\sigma^{\mu\nu}p_\nu + F_5\sigma^{\mu\nu}p'_\nu)u(p, s),$$

$$\mathcal{M}_5^\mu = \bar{u}(p', s')(G_1\gamma^\mu + G_2 p^\mu + G_3 p'^\mu + G_4\sigma^{\mu\nu}p_\nu + G_5\sigma^{\mu\nu}p'_\nu)\gamma_5 u(p, s), \quad (17.7.2)$$

where the $F_r$ and $G_r$ are functions of $(p - p')^2$.

The Gordon relations

$$\left.\begin{array}{c} p^\mu = \gamma^\mu \not{p} + i\sigma^{\mu\nu}p_\nu \\ p'^\mu = \not{p'}\gamma^\mu - i\sigma^{\mu\nu}p'_\nu \end{array}\right\} \quad (17.7.3)$$

enable us to write both expressions in terms of the exchange momentum $q = p - p'$. Evaluated between polarization spinors, the $\not{p}$ and $\not{p'}$ terms reduce to masses, and so we find that the matrix elements have forms:

$$\mathcal{M}^\mu = \bar{u}(p', s')(f_1\gamma^\mu - if_2\sigma^{\mu\nu}q_\nu + f_3 q^\mu)u(p, s),$$

$$\mathcal{M}_5^\mu = \bar{u}(p', s')(g_1\gamma^\mu - ig_2\sigma^{\mu\nu}q_\nu + g_3 q^\mu)\gamma_5 u(p, s). \quad (17.7.4)$$

**Homework** 17.7.5. Fill in the details – derive 17.7.4 carefully.

Next we use $PT$ symmetry of the strong interactions to show that the coefficient functions $f_r$ and $g_r$ are real. In the Majorana representation, if we choose the polarization spinors so that

$$u(p, -s) = \gamma_5 u(p, s)^*, \quad (17.7.6)$$

then the spinor fields transform as

$$\psi(x) \to \psi_{PT}(x) \stackrel{\text{def}}{=} \gamma_5 \psi(-x)^*. \quad (17.7.7)$$

Since $PT$ reverses the order of fields, the currents transform as follows:

$$
\begin{aligned}
j_\mu^{PT}(0) &= (\psi'^\dagger \gamma_0 \gamma_\mu \psi)_{PT} \\
&= \psi_{PT}^T (\gamma_0 \gamma_\mu)^T (\psi'_{PT})^* \\
&= \psi^\dagger \gamma_5^T \gamma_\mu^T \gamma_0^T \gamma_5 \psi' \\
&= \bar\psi \gamma_0 \gamma_5^T \gamma_\mu^T \gamma_0^T \gamma_5 \psi' \\
&= \bar\psi \gamma_\mu \psi' \\
&= j_\mu(0)^*,
\end{aligned}
\tag{17.7.8}
$$

and similarly

$$
j_{5\mu}^{PT}(0) = -j_{5\mu}(0)^*.
\tag{17.7.9}
$$

Taking into account the assumption that the initial and final baryon states have equal parity, application of $PT$ to the matrix elements proceeds as follows:

$$
\begin{aligned}
\mathcal{M}_{PT}^\mu &= \langle B, p, -s | j^{\mu *}(0) | B', p', -s' \rangle \\
&= \Big( \langle B', p', -s' | j^\mu(0) | B, p, -s \rangle \Big)^*, \\
\mathcal{M}_{5\,PT}^\mu &= -\langle B, p, -s | j_5^{\mu *}(0) | B', p', -s' \rangle \\
&= -\Big( \langle B', p', -s' | j_5^\mu(0) | B, p, -s \rangle \Big)^*,
\end{aligned}
\tag{17.7.10}
$$

and so, upon substitution of 17.7.4, we find that the coefficient functions $f_r$ and $g_r$ are indeed real.

If we now assume that the currents involve only the $u$, $d$, and $s$ quarks and that flavor $SU(3)$ is an exact symmetry, then the functions $f_3$ and $g_2$ must vanish. We relegate the details to a homework and state the conclusion:

$$
\begin{aligned}
\mathcal{M}^\mu &\simeq \bar u(p', s')(f_1 \gamma^\mu - i f_2 \sigma^{\mu\nu} q_\nu) u(p, s), \\
\mathcal{M}_5^\mu &\simeq \bar u(p', s')(g_1 \gamma^\mu + g_3 q^\mu) \gamma_5 u(p, s).
\end{aligned}
\tag{17.7.11}
$$

**Homework 17.7.12.** Hermitian generators of $SU(3)$ give rise to a basis of hermitian currents $j^\mu$. Show that if $j^\mu$ is hermitian, then the definition 17.7.1 and the form 17.7.4 of the matrix elements $\mathcal{M}^\mu$ and $\mathcal{M}_5^\mu$ implies that $f_3 = g_2 = 0$.

**Homework 17.7.13.** If the initial and final baryon states had opposite parities, what final form would replace 17.7.11?

The final form of the quark-current matrix elements depends on the assumptions that the initial and final baryons have the same intrinsic parity, that the matrix elements are dominated by the strong interactions (which are $PT$ invariant), and that flavor-symmetry violating effects are of secondary importance. The first assumption can be met. The second and third are tree-level truths: they do not apply, for example, to processes involving flavor-changing neutral currents since these depend on loop diagrams and $W$ exchange.

The following two sections present an application of this reduced form for the quark-current matrix elements first, to $\beta$ decays in the baryon octet and second, to neutron decay in particular.

# 17.8 – β Decays

We focus now on those baryon $\beta$ decays $B \to B'e\bar{\nu}_e$ in which the two baryons have the same parity. The operators mediating these transitions have the usual current-current form and, as in the case of pion decay, give rise to an amplitude of the form 17.5.22. Since the weak interactions are involved, the quark-current matrix element in this amplitude is the sum of the vector and axial matrix elements given in 17.7.11.

When this matrix element is coupled to a lepton current, the $q$ in the $g_3$ term gives us the divergence of the lepton current, and this in turn introduces a factor of a lepton mass. For example:

$$q_\mu \, \bar{e}(k')\gamma^\mu(1 - \gamma_5)\nu_e(k) = \bar{e}(k')(\not{k}' + \not{k})(1 - \gamma_5)\nu_e(k)$$
$$= m_e\bar{e}(k')(1 - \gamma_5)\nu_e(k). \tag{17.8.1}$$

The contribution of the $g_3$ term in 17.7.11 is therefore negligible if there is a final-state electron.

The contribution of the $f_2$ term in the quark-current matrix element is dominated by those of the $f_1$ and $g_1$ terms. The reason for this is that there are no tree-level contributions to $f_2$: it is the anomalous magnetic moment. Tree-level diagrams contribute to the $f_1$ term and generate the Dirac magnetic moment through the following application of the Gordon relations:

$$\bar{u}'(p')\gamma_\mu u(p) = \bar{u}'(p)(p'_\mu + p_\mu + i\sigma_{\mu\nu}q^\nu)u(p), \tag{17.8.2}$$

but it takes a loop diagram to change the relative coefficients of the terms on the right. Furthermore, the vertex renormalization condition constrains only the loop contributions to $f_1$. Hence, assuming renormalizability, the contributions to $f_2$ must be of order

$$f_2 \sim \frac{e^2\Omega}{(2\pi)^4} = \frac{e^2}{8\pi^2}, \tag{17.8.3}$$

where $\Omega$ is the volume of the unit sphere in four dimensions. For our current concerns, we conclude that $f_2$ is negligible in comparison to $f_1$ and $g_1$.

On the basis of the foregoing discussion, the amplitude for a baryon $\beta$ decay may be written

$$i\mathcal{A} \stackrel{\text{def}}{=} i\frac{G_F V_{B'B}}{\sqrt{2}} \, \bar{e}^{r'}(k')\gamma^\mu(1 - \gamma^5)\nu_e^r(k) \, \bar{u}^{s'}(p')\gamma_\mu(g_V + g_A\gamma^5)u^s(p), \tag{17.8.4}$$

where $V_{B'B}$ is the appropriate element of the KM matrix, and $g_V = f_1$ and $g_A = g_1$ depend on $B$, $B'$, and the exchange momentum $q = p - p'$.

The spin-sum, spin-average of the squared amplitude works out to

$$|\mathcal{A}|^2 = 16G_F^2|V_{B'B}|^2\big((g_V^2 + g_A^2)(k{\cdot}p'\, k'{\cdot}p + k{\cdot}p\, k'{\cdot}p') \tag{17.8.5}$$
$$+ 2g_V g_A(k{\cdot}p\, k'{\cdot}p' - k{\cdot}p'\, k'{\cdot}p) - MM'(g_V^2 - g_A^2)k{\cdot}k'\big),$$

where $M$ and $M'$ are the masses of $B$ and $B'$ respectively. In terms of the energies $E$ and $E'$ of the electron and final state baryon, this formula becomes

$$|\mathcal{A}|^2 = 8G_F^2|V_{B'B}|^2\Big((g_V + g_A)^2(M^2 - M'^2 - 2M(M - E - E'))M(M - E - E')$$
$$+ (g_V - g_A)^2(M^2 - M'^2 - 2ME)ME \tag{17.8.6}$$
$$- (g_V^2 - g_A^2)MM'(M^2 + M'^2 - 2ME')\Big).$$

The boundary of the Dalitz plot is given by the constraints

$$0 \le E \le \frac{M^2 - M'^2}{2M} \tag{17.8.7}$$

and

$$\frac{M'^2 + (M - 2E)^2}{2(M - 2E)^2} \le E' \le \frac{M^2 + M'^2}{2M}. \tag{17.8.8}$$

The differential decay width is

$$d\Gamma = \frac{1}{64\pi^3} \frac{1}{M} |\mathcal{A}|^2 \, dE \, dE'. \tag{17.8.9}$$

For computation and estimation, it is convenient to eliminate $M$ and $M'$ in favor of the parameters

$$\Delta M \stackrel{\text{def}}{=} M - M' \quad \text{and} \quad \delta \stackrel{\text{def}}{=} \frac{M - M'}{M + M'}, \tag{17.8.10}$$

and to eliminate $E$ and $E'$ in favor of variables $x$ and $y$ given by

$$x \stackrel{\text{def}}{=} (1 + \delta) \frac{E}{\Delta M} \quad \text{and} \quad 1 - y \stackrel{\text{def}}{=} \frac{1 + \delta}{\delta} \frac{E' - M'}{\Delta M}. \tag{17.8.11}$$

The differential decay width becomes

$$d\Gamma = \frac{G_F^2 |V_{B'B}|^2}{8\pi^3} \frac{(\Delta M)^5}{(1 + \delta)^3} F(x, y) \, dx \, dy, \tag{17.8.12}$$

where

$$F(x, y) \stackrel{\text{def}}{=} 2(g_V^2 + g_A^2)(x - x^2) - \frac{1}{2}(g_V^2 - g_A^2)y \tag{17.8.13}$$

$$- \delta(g_V + g_A)^2(1 - 2x)y + \delta^2 \left( \frac{1}{2}(g_V^2 + g_A^2)y - (g_V + g_A)^2 y^2 \right),$$

and the boundary of the Dalitz plot becomes

$$0 \le x \le 1 \quad \text{and} \quad 0 \le y \le \frac{4(x - x^2)}{(1 + \delta)^2 - 4\delta x}. \tag{17.8.14}$$

To first order in $\delta$, integration yields a decay width

$$\Gamma = \frac{G_F^2 |V_{B'B}|^2}{60\pi^3} (\Delta M)^5 (g_V^2 + 3g_A^2)(1 - 3\delta). \tag{17.8.15}$$

The parameters $g_V$ and $g_A$ cannot as yet be predicted, but it is reasonable to suppose that they are of order unity. For example, in the decay

$$\Omega^- \to \Xi^0 e \bar{\nu}_e, \tag{17.8.16}$$

the predicted and experimental decay widths are

$$\textbf{Theory:} \qquad 1.3 \times 10^{-14} \times (g_V^2 + 3g_A^2)$$
$$\textbf{Experiment:} \qquad 4.5 \times 10^{-14} \tag{17.8.17}$$

Note that if we had not performed a spin-sum, spin-average but had instead kept track of the various spins, then we would have obtained a more specific prediction from which comparison with experiment could determine both $g_V$ and $g_A$ up to an overall phase.

**Homework 17.8.18.** Assuming the form 17.8.4 for the amplitude, provide the details of the computation which yields the decay width above.

**Homework 17.8.19.** From the formula 17.8.15 for the $\beta$ decay width, use experimental data to show that $g_V^2 + 3g_A^2$ is of order unity for the observed $\beta$ transitions in the baryon octet: $n \to pe\bar{\nu}_e$, $\Lambda \to pe\bar{\nu}_e$, $\Sigma^+ \to \Lambda e\bar{\nu}_e$, $\Sigma^- \to ne\bar{\nu}_e$, $\Sigma^- \to \Lambda e\bar{\nu}_e$, $\Xi^- \to \Lambda e\bar{\nu}_e$, and $\Xi^- \to \Sigma^0 e\bar{\nu}_e$.

# 17.9 – Flavor Symmetry in Octet $\beta$ Decays

We now focus on the $\beta$ decays which mediate transitions between elements of the baryon octet. In this case, applying the techniques of Chapter 16, we can use flavor $SU(3)$ to find the structure of $g_V$ and $g_A$ as functions of $B$ and $B'$ at zero momentum transfer.

Writing $\psi$ for the vector of light quarks, the elements of $su(3)$ are associated with vector and axial currents:

$$j_T^\mu \stackrel{\text{def}}{=} \bar{\psi}\gamma^\mu T\psi \quad \text{and} \quad j_{5T}^\mu \stackrel{\text{def}}{=} \bar{\psi}\gamma^\mu\gamma_5 T\psi. \tag{17.9.1}$$

These currents transform as two octets under flavor $SU(3)$ and so from Section 16.2 we find that:

$$\mathcal{M}_{(5)}^\mu \stackrel{\text{def}}{=} \langle B', p', s'|j_{(5)T}^\mu|B, p, s\rangle \tag{17.9.2}$$

$$\implies \left. \begin{array}{l} g_V = (D' + F')\,\text{tr}(B'^\dagger TB) + (D' - F')\,\text{tr}(BTB'^\dagger) \\[2mm] g_A = (D + F)\,\text{tr}(B'^\dagger TB) + (D - F)\,\text{tr}(BTB'^\dagger) \end{array} \right\}$$

Since flavor $SU(3)$ is an approximate symmetry, these formulae are only approximately correct. One can consider switching off the electroweak effects and making flavor symmetry exact, and consider these formulae to be valid in this limit. Exact flavor symmetry implies equal masses in particle multiplets and so zero momentum transfer in transitions. One can propose that these expressions for $g_V$ and $g_A$ are correct at zero momentum transfer, and by studying their dependence on momentum transfer, one could in principle estimate the error in the formulae 17.9.2.

The functions $D'$ and $F'$ can be evaluated using the $I_3$ current and conserved quantity. First, we find a formula for diagonal current matrix elements by evaluating the expectation of the conserved quantity $Q_3$ in two ways:

$$\langle p', s'|Q_3|p, s\rangle = \lambda_3\langle p', s'|p, s\rangle$$
$$= \lambda_3 2\omega(\bar{p})\delta_{s's}(2\pi)^3\delta^{(3)}(\bar{p} - \bar{p}'), \tag{17.9.3}$$

where $\lambda_3$ is the $I_3$ quantum number of the ket, and

$$
\begin{aligned}
\langle p', s'|Q_3|p, s\rangle &= \int d^3\bar{x}\, \langle p', s'|j_3^0(0, \bar{x})|p, s\rangle \\
&= \int d^3\bar{x}\, e^{i x \cdot (p' - p)} \langle p', s'|j_3^0(0)|p, s\rangle \\
&= \langle p', s'|j_3^0(0)|p, s\rangle (2\pi)^3 \delta^{(3)}(\bar{p} - \bar{p}').
\end{aligned}
\tag{17.9.4}
$$

Comparison of the last two results 17.9.3 and 17.9.4 shows that

$$
\begin{aligned}
\langle p', s'|j_3^0(0)|p, s\rangle &= \lambda_3\, 2\omega(\bar{p})\delta_{s's} \\
&= \lambda_3\, \bar{u}(p, s')\gamma^0 u(p, s),
\end{aligned}
\tag{17.9.5}
$$

where the last step follows from a normalization relation of the spinors.

Since $j_3$ is a vector current, we can use this formula to evaluate the matrix elements $\mathcal{M}$ and thereby determine $D'$ and $F'$. Writing $P$ and $\Xi^-$ for the matrix coordinates which identify the $P$ and $\Xi^-$ particles in the baryon octet,

$$
P = \Xi^{-\dagger} = \begin{pmatrix} 0 & 0 & 1 \\ 0 & 0 & 0 \\ 0 & 0 & 0 \end{pmatrix} \quad \text{and} \quad I_3 = \begin{pmatrix} \frac{1}{2} & 0 & 0 \\ 0 & -\frac{1}{2} & 0 \\ 0 & 0 & 0 \end{pmatrix},
\tag{17.9.6}
$$

we find from formula 17.9.5 that

$$
\begin{aligned}
\langle P|j_3(0)|P\rangle &= g_V^P \bar{u}\gamma^0 u = \frac{1}{2}\bar{u}\gamma^0 u, \\
\langle \Xi^-|j_3(0)|\Xi^-\rangle &= g_V^{\Xi^-} \bar{u}'\gamma^0 u' = -\frac{1}{2}\bar{u}'\gamma^0 u'.
\end{aligned}
\tag{17.9.7}
$$

Consequently,

$$
g_V^P = \frac{1}{2} \quad \text{and} \quad g_V^{\Xi^-} = -\frac{1}{2}.
\tag{17.9.8}
$$

To evaluate these $g_V$'s using the formula 17.9.2, we compute the traces

$$
\text{tr}(P^\dagger I_3 P) = \text{tr}(\Xi^- I_3 \Xi^{-\dagger}) = \frac{1}{2} \quad \text{and} \quad \text{tr}(P I_3 P^\dagger) = \text{tr}(\Xi^{-\dagger} I_3 \Xi^-) = 0, \tag{17.9.9}
$$

and so find that

$$
g_V^P = \frac{1}{2}(D' + F') \quad \text{and} \quad g_V^{\Xi^-} = \frac{1}{2}(D' - F').
\tag{17.9.10}
$$

Combining these two expressions 17.9.8 and 17.9.10 for the $g_V$'s shows that

$$
D' + F' = 1 \quad \text{and} \quad D' - F' = -1,
\tag{17.9.11}
$$

and so

$$
D' = 0 \quad \text{and} \quad F' = 1.
\tag{17.9.12}
$$

The angular distribution of final state leptons in octet $\beta$ decays provides some specific values for $g_V$ and $g_A$. Using the formulae for $g_V$ and $g_A$ presented in the table below, we can make a best-fit estimate of the parameters $D$ and $F$:

$$\left. \begin{array}{lcc}
 & \textbf{Exp.} & \textbf{Theory} \\
n \to p e \bar\nu_e & -1.257(3) & D+F \\
\Lambda \to p e \bar\nu_e & -0.718(15) & \frac{1}{3}(D+3F) \\
\Sigma^- \to n e \bar\nu_e & 0.340(17) & -D+F \\
\Xi^- \to \Lambda e \bar\nu_e & -0.25(5) & -\frac{1}{3}(D-3F)
\end{array} \right\} \implies \left. \begin{array}{l} D \simeq -0.82 \\ F \simeq -0.43 \end{array} \right\} \quad (17.9.13)$$

Substituting these values in the equations 17.9.2 for $g_V$ and $g_A$ and using the results in the formula 17.8.15 for the decay width, we finally obtain purely numerical predictions for the $\beta$ transitions:

| Process | $V_{B'B}$ | $g_V$ | $g_A$ | $\Gamma_{\text{Th}}$ | $\Gamma_{\text{Exp}}$ |
|---|---|---|---|---|---|
| $n \to p e \bar\nu_e$ | $V_{ud}$ | $1$ | $D+F$ | $1.4 \times 10^{-24}$ | $7.4 \times 10^{-25}$ |
| $\Lambda \to p e \bar\nu_e$ | $V_{us}$ | $-\frac{3}{\sqrt{6}}$ | $-\frac{1}{\sqrt{6}}(D+3F)$ | $1.1 \times 10^{-15}$ | $2.1 \times 10^{-15}$ |
| $\Sigma^+ \to \Lambda \bar e \nu_e$ | $V_{ud}$ | $0$ | $\frac{2}{\sqrt{6}}D$ | $1.6 \times 10^{-16}$ | $1.6 \times 10^{-16}$ |
| $\Sigma^- \to n e \bar\nu_e$ | $V_{us}$ | $-1$ | $D-F$ | $4.5 \times 10^{-15}$ | $4.4 \times 10^{-15}$ |
| $\Sigma^- \to \Lambda e \bar\nu_e$ | $V_{ud}$ | $0$ | $-\frac{2}{\sqrt{6}}D$ | $4.3 \times 10^{-16}$ | $2.5 \times 10^{-16}$ |
| $\Xi^- \to \Lambda e \bar\nu_e$ | $V_{us}$ | $\frac{3}{\sqrt{6}}$ | $-\frac{1}{\sqrt{6}}(D-3F)$ | $2.5 \times 10^{-15}$ | $2.3 \times 10^{-15}$ |
| $\Xi^- \to \Sigma^0 e \bar\nu_e$ | $V_{us}$ | $\frac{1}{\sqrt{2}}$ | $\frac{1}{\sqrt{2}}(D+F)$ | $2.9 \times 10^{-16}$ | $3.6 \times 10^{-16}$ |

Overall, the predictions are within a factor of two of the experimental values.

**Homework 17.9.14.** The experimental upper bound on the $\Xi^0 \to \Sigma^+$ $\beta$ transition is

$$\textbf{Experiment:} \quad \Gamma(\Xi^0 \to \Sigma^+ e \bar\nu_e) < 2.5 \times 10^{-15} \text{ MeV}.$$

Show that our theory is consistent with this upper bound.

**Homework 17.9.15.** The following octet transitions have been observed to yield pions:

$$\Lambda \to p\pi^-, \qquad \Lambda \to n\pi^0;$$
$$\Sigma^+ \to p\pi^0, \qquad \Sigma^+ \to n\pi^+, \qquad \Sigma^- \to n\pi^-;$$
$$\Xi^0 \to \Lambda\pi^0, \qquad \Xi^- \to \Lambda\pi^-.$$

Writing the amplitude for the process $A \to BC$ in the form

$$i\mathcal{A} \overset{\text{def}}{=} G_F m_C^2 \, \bar u_B (a - b\gamma^5) u_A,$$

show that the decay width is given by

$$\Gamma = \frac{a^2 + b^2}{4\pi^2} G_F^2 m_C^4 \, \varepsilon,$$

where

$$\varepsilon = \frac{M_A^2 + M_B^2 - m_C^2}{2M_A^2} \frac{\sqrt{(M_A + M_B)^2 - m_C^2}}{2M_A} \frac{\sqrt{(M_A - M_B)^2 - m_C^2}}{m_C}.$$

Show that for these transitions, $\frac{1}{2} < \varepsilon < 2$. By comparing predicted decay widths with experimental data, show that $a$ and $b$ are of order unity.

Flavor symmetry is valid in a theoretical limit in which the electroweak effects are turned off so that, for example, masses in particle multiplets are equal. It may seem odd to have used flavor symmetry to make predictions for phenomena which can only occur when this symmetry is broken. However, it appears that the dependence of our formulae on physical masses enables us to take into account the principle effects of symmetry breaking. In particular, the evaluation of $g_V$ and $g_A$ in the $SU(3)$-symmetry limit yields values at zero momentum transfer. The assumption that symmetry breaking does not significantly change their values has worked out to one significant figure.

The application of flavor symmetry to octet $\beta$ decays provides only a minor improvement in the estimated rates derived from the postulated form of the amplitude. However, some of the predictions for $g_V$ and $g_A$ can be verified experimentally from the angular distribution of the final state leptons.

# 17.10 – The Goldberger–Treiman Relation

In this section, starting from the final form 17.7.11 of the axial quark-current matrix element associated with neutron decay, we deduce the Goldberger–Treiman relation which links four parameters of hadron processes.

The proof depends on moving from physical reality to a theoretical reality and back. In the physical reality, chiral flavor $SU(3)$ is an approximate symmetry, the pions are massive, and computations are hard. In the theoretical reality, the pions are massless, chiral flavor symmetry is restored, and computations are effortless. We use the pion mass parameter to connect the two realities, derive the Goldberger–Treiman in the theoretical reality, and then transport it back to the physical reality.

If we assume that chiral flavor $SU(3)$ is a valid symmetry, then the axial currents are conserved. Contracting $\mathcal{M}_5^\mu$ with $q_\mu$, we therefore conclude that

$$(m' + m)g_1(q^2) = q^2 g_3(q^2) \tag{17.10.1}$$

in the limit $q^2 \to 0$. This relation implies that as $q \to 0$ either $g_1 \to 0$ or $g_3 \to \infty$. However, as we saw in the last section, measurements of $g_1$ at $q^2 \simeq m_e^2$ may be obtained from observations of neutron decays, and so we can deduce that

$$j_5^\mu = V_{ud}\, \bar{u}\gamma_5\gamma^\mu d \quad \Longrightarrow \quad g_A \overset{\text{def}}{=} g_1^{N \to P}(m_e^2) = -1.26. \tag{17.10.2}$$

This suggests that $g_1(0) \simeq -1.26$, and so $g_3$ has a pole at $q^2 = 0$.

Considering neutron decays in the context of axial $SU(3)$ symmetry, the pole may be understood as arising from the neutron emitting a $\pi^-$ which then decays into the axial current:

$$\tag{17.10.3}$$

where the couplings come from Section 16.8 on nucleon-pion theory and 17.5.7. In this context, the pions are massless, and so the contribution of pion emission to the axial-current matrix element between an initial neutron and a final proton is

$$\mathcal{M}^\mu_\pi \overset{\text{def}}{=} \frac{2F_\pi g_{\pi NN} q^\mu}{q^2 + i\epsilon} \bar{u}(p', s')\gamma_5 u(p, s), \qquad (17.10.4)$$

and so the contribution to $g_3$ is

$$g_3(q^2) = \frac{2F_\pi g_{\pi NN}}{q^2 + i\epsilon}. \qquad (17.10.5)$$

Finally, putting 17.10.1, 17.10.2, and 17.10.5 together, we find the

**Goldberger–Treiman Relation:** $\quad m_N |g_A| = F_\pi g_{\pi NN}. \qquad (17.10.6)$

With $m_N = 939\,\text{MeV}$, $g_A = -1.26$, $F_\pi = 93\,\text{MeV}$, and $g_{\pi NN} = 13.5$ the error in this relation is about 6%.

Just one caution: this relation was derived under the assumption of chiral flavor symmetry, and then checked in the experimental reality in which that symmetry is broken. For this check to be meaningful, we must bring forward the hidden assumptions that the breaking of this axial symmetry of the strong interactions by electroweak effects does not significantly change the parameters in the Goldberger–Treiman relation. As the magnitude of the pion masses indicates the extent of the symmetry breaking, we are in effect assuming that $g_A$ and $g_{\pi NN}$ do not vary rapidly as functions of exchange energy $q^2$ on the interval $0 \le q^2 \le m_\pi^2$.

Note that, in sliding from the theoretical reality back to the physical reality, the pion pole gets shifted from $q^2 = 0$ to $q^2 = m_\pi^2$. Since the neutron-proton mass difference is too small for neutron decay to create a pion, there is no pion-pole contribution to neutron decay in physical reality.

The Goldberger–Treiman relation fits the data well. It thereby indicates the value of chiral flavor symmetry, supporting the notion that the pions are Goldstone bosons for broken axial symmetries and the hypothesis that the axial currents are approximately conserved.

## 17.11 Summary

This chapter has briefly indicated how the Standard Model may be used at tree level to make interesting predictions. The applications have been classified on the basis of the structure of the effective four-Fermi operators involved. The four-quark operators are not much use since the non-perturbative effects of QCD are hard to factor out in the processes to which these operators contribute. The four-lepton operators are the easiest to use because with them the strong interactions are irrelevant. The most interesting applications come from the lepto-quark operators, where the parton model perspective enables us to distinguish the strong from the weak interaction effects.

This chapter brings us to the end of Part 4 of this text. We have now satisfactorily indicated what can be accomplished with tree-level perturbation theory. Further progress depends on loop effects. Loop effects allow us to refine predictions, understand asymptotic freedom of QCD, and also to glimpse the high-energy regime of grand unified theories. Building on the principles of renormalization set out in Chapter 10, Part 5, comprising Chapters 18 to 21, dives into the theory of renormalization.

# Chapter 18

# Regularization and Renormalization

Presenting the principles and procedures for identifying the character of a divergent integral with a view to systematically canceling the divergences that arise in perturbation theory.

## Introduction

This chapter begins Part 5, the last part of the book. As we have now pushed tree-level applications as far as we usefully can, Part 5 is concerned with higher-order corrections in theory and in practice. Chapter 18 sets the stage for the investigation by providing a standard regularization for Feynman integrals. Chapter 19 presents a detailed account of renormalization as it applies to the simplest divergences in QED. Chapter 20 goes into the theory of renormalization; it presents the forest formula for systematically matching counterterms with divergences and the proof that renormalization preserves gauge invariance in theories whose fermion multiplets pass the consistency test of Theorem 13.6.8. Finally, Chapter 21 brings the book to completion with a careful account of the renormalization group.

Many fundamental loop diagrams give rise to divergent integrals. The first step in defining higher-order corrections is to express divergent Feynman integrals as limits of convergent integrals, and the second step is to arrange for systematic cancellation of divergences. As we saw in Chapter 10, such cancellation can generally only be organized by putting similar divergences in the coefficients of the bare Lagrangian density, or equivalently in the coefficients of the counterterms. This is a drastic step, suggesting that the Lagrangian formulation of the laws of nature has been pushed beyond its intrinsic limitations. Part 5 shows that, despite the delicacy of the approach, renormalized Lagrangian theory has solid theoretical features and practical successes.

We noted in Section 10.14 that the simple regularization used in Chapter 10 would break gauge invariance. For gauge theories, dimensional regularization is widely preferred. This chapter provides the foundation for Part 5 by presenting a detailed account of computations with dimensional regularization.

Section 18.1 shows how to put a Feynman integral into standard form, and Section 18.2 computes the values of the resulting momentum integrals in the convergent case. Sections 18.3 and 18.4 explain the character of regularization and renormalization. The essential features of both procedures were presented in the context of examples in Chapter 10; here, we bring out the underlying structure of these operations. Section 18.5 describes three commonly used regularization procedures. Section 18.6 defines our preference, dimensional regularization, and Section 18.7 finds formulae for the regularized values of all standard momentum integrals, convergent and divergent. The next two sections, Sections 18.8 and 18.9, provide technical details which are frequently used in applications of dimensional regularization. Sections 18.10 to 18.12 present interesting examples of dimensional regularization at work.

# 18.1 Feynman Integral Technique for Loop Diagrams

To establish a foundation for our discussion of regularization, we begin by laying out the stages in putting a Feynman integral into a standard form characterized by a simple structure of the denominator:

$$\text{Denominator} = \left(\sum_c l_c^2 + C\right)^N, \tag{18.1.1}$$

where $l_c$ are loop momenta, and $C$ is a function of the external momenta. This is accomplished in three steps.

For a general Feynman integral $I$, we use a notation $k_a$, $p_b$ and $l_c$ for the momenta as follows:

$$\left.\begin{array}{llll}
k_a & - & \text{internal momenta,} & 1 \leq a \leq N \\
p_b & - & \text{external momenta,} & 1 \leq b \leq E \\
l_c & - & \text{loop momenta,} & 1 \leq c \leq L
\end{array}\right\} \tag{18.1.2}$$

Then the $k$'s are linear functions of the $p$'s and $l$'s and the integral as a whole is a function of the $p$'s:

$$I(p) = \int \frac{d^4 l_1}{(2\pi)^4} \cdots \frac{d^4 l_L}{(2\pi)^4} \frac{N(p,l)}{D(p,l)}, \tag{18.1.3}$$

where the numerator $N$ is determined by the choice of denominator,

$$D(p,l) = \prod_{a=1}^{N} (k_a^2 - m_a^2 + i\epsilon). \tag{18.1.4}$$

The first step towards evaluation is to transform the product of many propagators into a power of something. We saw how to accomplish this for two factors in 10.10.11 and for three factors in 10.11.6. The Feynman integration trick used there generalizes to many propagators as follows.

Let $P_a = k_a^2 - m_a^2 + i\epsilon$ be the denominator of the $a$th propagator factor. Then

$$\prod_{a=1}^{N} \frac{1}{P_a} = (-i)^N \int_0^\infty dz_1 \cdots dz_N \, \exp\left(i \sum_a z_a P_a\right) \tag{18.1.5}$$

$$= (-i)^N \int_0^\infty dz_1 \cdots dz_N \int_0^\infty \frac{d\lambda}{\lambda} \, \exp\left(i \sum_a z_a P_a\right) \delta\left(1 - \frac{1}{\lambda}\sum_a z_a\right).$$

Now changing variable from $z$'s to $x$'s defined by $z_a = \lambda x_a$, we find that

$$\prod_{a=1}^{N} \frac{1}{P_a} = (-i)^N \int_0^1 dx_1 \cdots dx_N \int_0^\infty d\lambda \, \lambda^{N-1} \exp\left(i\lambda \sum_a x_a P_a\right) \delta\left(1 - \sum_a x_a\right)$$

$$= (N-1)! \int_0^1 dx_1 \cdots dx_N \, \frac{\delta\left(1 - \sum x_a\right)}{\left(\sum x_a P_a\right)^N}. \tag{18.1.6}$$

The parameters $x_a$ are called *Feynman parameters*. Note that the delta function has allowed us to cut the range of integration for the Feynman parameters to the interval $[0,1]$.

**Remark** 18.1.7. The sequence of steps can be changed. Because $a$ is quadratic in momenta, after using the Schwinger representation of the propagator

$$\frac{1}{a+i\epsilon} = -i \int_0^\infty dx \, e^{ix(a+i\epsilon)}, \tag{18.1.8}$$

the integrand is a complex Gaussian in momenta, and so we can perform the momentum integrations. The $\lambda$ trick can then be used to simplify the parameter integrations.

Alternatively, we can make a Wick rotation of external and internal variables, then use the representation

$$\frac{1}{a} = \int_0^\infty dx \, e^{-sa} \tag{18.1.9}$$

to transform the Feynman integral into a Gaussian.

This insertion of Feynman parameters does not affect the numerator at all. The general form 18.1.3 of a Feynman integral becomes

$$I(p) = (N-1)! \int d^N x \int \frac{d^4 l_1}{(2\pi)^4} \cdots \frac{d^4 l_L}{(2\pi)^4} \frac{N(p,l)\delta(1 - \sum x_a)}{\left(\sum x_a P_a\right)^N}. \tag{18.1.10}$$

The second step is to change variables in order to remove mixing of loop momenta in the quadratic terms of the denominator $P = \sum x_a P_a$ of the effective propagator. Now $P$ must have a minimum at some point $l_c = \bar{l}_c$. Defining new variables $l_c^{(1)}$ by $l_c = \bar{l}_c + l_c^{(1)}$, we can eliminate terms in $P$ which are linear in loop variables:

$$P(p,x,l) = P^{(1)}(p,x,l^{(1)}) = \sum_{r,s} A^{rs}(x) l_r^{(1)} l_s^{(1)} + C(p,x). \tag{18.1.11}$$

This step brings apparent $x$-dependence to the numerator:

$$N(p,l) = N^{(1)}(p,x,l^{(1)}). \tag{18.1.12}$$

The third step is to put the quadratic terms into normal form. We make an orthogonal change of coordinates to diagonalize $A$,

$$P(p,x,l) = P^{(2)}(p,x,l^{(2)}) = \sum_c A_{cc}^{(1)}(x) l_c^{(2)} l_c^{(2)} + C(p,x), \tag{18.1.13}$$

and then normalize the $l^{(2)}$ by

$$l_c^{(3)} = \sqrt{A_{cc}^{(1)}(x)} \, l_c^{(2)} \tag{18.1.14}$$

in order to arrive at

$$P(p,x,l) = P^{(3)}(p,x,l^{(3)}) = \sum_c l_c^{(3)} \cdot l_c^{(3)} + C(x,p). \tag{18.1.15}$$

Finally, using the fact that the product of the eigenvalues of $A$ is equal to its determinant, we conclude that the integral 18.1.10 becomes

$$I(p) = (N-1)! \int d^N x \tag{18.1.16}$$

$$\int \frac{d^4 l_1^{(3)}}{(2\pi)^4} \cdots \frac{d^4 l_L^{(3)}}{(2\pi)^4} \left(\frac{1}{\det(A)}\right)^2 \frac{N^{(3)}(p,x,l^{(3)})\delta\left(1 - \sum x_a\right)}{\left(\sum l_c^{(3)} \cdot l_c^{(3)} + C(x,p)\right)^N},$$

where $N^{(3)}$ is simply the original numerator $N$ as a function of the last set of loop variables.

Since $C$ does not depend on the loop variables, this integral is simple enough. The integration over the Feynman parameters is more tricky. There are, however, many programs for doing these integrals, and doing complicated cases by hand is not worth the time.

**Homework 18.1.17.** Try this procedure out on the following loop diagrams in $\phi^3$-theory and QED respectively:

 and

Now that we have seen the outline of a Feynman integration, we are left with the problem of interpreting divergent cases. Indeed, in these cases, the translation of loop variables is not a *priori* justifiable. By the end of this chapter, however, we will see that the procedure presented here is effectively part of the dimensional regularization technique which we use in subsequent chapters.

## 18.2 Feynman Integrals in Euclidean Space

In this section, we prepare the ground for management of divergent Feynman integrals by discussing convergent Feynman integrals. We begin with the one-loop case, develop some formulae for the momentum integral, and then extend these formulae to the momentum integrals of the multi-loop cases.

For one-loop diagrams, the matrix $A$ in the standard form 18.1.16 for a Feynman integral is simply the identity matrix. Consequently, Feynman integrals for one-loop diagrams reduce to Minkowski-space integrals like

$$I_n = (N-1)! \int dx_1 \cdots dx_N \int \frac{d^4 l}{(2\pi)^4} \left(\frac{1}{l^2 - a(p,x)}\right)^N, \tag{18.2.1}$$

where $a$ contains a $-i\epsilon$, and the limit as $\epsilon \to 0$ is as usual implicit in the evaluation of the expression.

These integrals are all divergent because the integrands are constant on mass hyperbolae. In fact, all Feynman integrals in Minkowski space diverge, and the first step in all procedures for evaluating Feynman integrals is to take advantage of Section 11.4, where we showed that functional integral quantization provides a justification for perturbing the Feynman integrals in the direction of Wick rotation. Once we have rotated energy off the real axis, the rest of the Wick rotation does not change the convergence properties of Feynman integrals. Consequently, we may as well assume that evaluation in Euclidean space is part of the Feynman rules. With this understanding, the integral 18.2.1 converges for $n \geq 2$.

**Homework 18.2.2.** Define the $D$-dimensional integral $I_n^D$ by

$$I_n^D \overset{\text{def}}{=} \int \frac{d^D l}{(2\pi)^D} \left( \frac{1}{l^2 - a} \right)^n.$$

Prove that if $D \geq 2$, then the integral $I_n^D$ diverges for all $n$. (Hint: integrate over the angular variables in $\vec{l}$; change variables from $r = |\vec{l}|, l_0$ to $u = (l_0^2 - r^2)/2, v = l_0 r$.)

Suppressing Feynman parameters, rotation of $l_0$ to $il_4$ converts the Minkowski-space integral 18.2.1 above into the Euclidean-space integral

$$I_n = i \int \frac{d^4 l_{\mathrm{E}}}{(2\pi)^4} \left( \frac{1}{-l_{\mathrm{E}}^2 - a} \right)^n, \tag{18.2.3}$$

where

$$l_{\mathrm{E}} = (l_4, l_1, l_2, l_3) = (-il_0, l_1, l_2, l_3) \quad \text{and} \quad l_{\mathrm{E}}^2 = l_1^2 + l_2^2 + l_3^2 + l_4^2. \tag{18.2.4}$$

Notice that when we treat $l_0$ as a complex variable, then the poles in the integrand lie on $\mathrm{Im}(l_0^2) = -i\epsilon$, a hyperbola in the second and fourth quadrants of the $l_0$-plane. Hence, rotating the contour of integration counter-clockwise from the real axis to the imaginary axis avoids all possible poles.

Curiously enough, the favored role played by energy in the Euclidean-space integral does not cause the final integral to break Lorentz symmetry. This effect cannot come from any similarity between the Lorentz group and $SO(4)$ because no similarity could convert a group with infinite volume into a group with finite volume. The effect depends on the integration eliminating off-diagonal terms.

In more detail, a Feynman integral will depend on external momenta or polarization vectors $k = (k_1, \ldots, k_m)$ as follows:

$$I(p) = \int \frac{d^4 l}{(2\pi)^4} \frac{N(p, l)}{(l^2 - a(p))^n}. \tag{18.2.5}$$

When the integrand does not contain dot products between internal and external momenta or spinors corresponding to external fermions, then making internal momenta Euclidean will not interfere with the Lorentz action on the external momenta. If the integrand does contain dot products between internal and external momenta (but still contains no spinors), then $p$ will be contracted with integrals like

$$I_n^{\mu\nu}(p) = \int \frac{d^4 l}{(2\pi)^4} \frac{l^\mu l^\nu}{(l^2 - a(p))^n}. \tag{18.2.6}$$

We see at once that if the integrand had an odd number of $l$'s on top, the Euclidean integral would evaluate to zero because of the rotational symmetry of measure and denominator. The same argument shows us that if $\mu \neq \nu$ then $I_n^{\mu\nu} = 0$. In fact, writing $\eta$ for the change of coordinates matrix $\mathrm{Diag}(i, 1, 1, 1)$, we find that

$$
\begin{aligned}
I_n^{\mu\nu}(p) &= i\eta^\mu{}_\sigma \eta^\nu{}_\tau \int \frac{d^4 l_{\mathrm{E}}}{(2\pi)^4} \frac{l_{\mathrm{E}}^\sigma l_{\mathrm{E}}^\tau}{(-l_{\mathrm{E}}^2 - a)^n} \\
&= i\eta^\mu{}_\sigma \eta^\nu{}_\tau \int \frac{d^4 l_{\mathrm{E}}}{(2\pi)^4} \frac{\delta^{\sigma\tau}}{4(-l_{\mathrm{E}}^2 - a)^n} \\
&= i \int \frac{d^4 l_{\mathrm{E}}}{(2\pi)^4} \frac{g^{\mu\nu}}{4(-l_{\mathrm{E}}^2 - a)^n},
\end{aligned}
\tag{18.2.7}
$$

a Lorentz covariant result!

If the integrand also contains spinors then the spinor terms will couple to the internal momenta through tensor currents of the form

$$S^{\alpha_1 \cdots \alpha_m} = \bar{u}(p_1) \gamma^{\alpha_1} \cdots \gamma^{\alpha_m} u(p_2). \tag{18.2.8}$$

Such tensor currents will couple to integrals of the form just analyzed, and therefore will not break the Lorentz invariance of the Euclidean-space integral.

The basic form of the Euclidean-space integrals is

$$I = \int d^4x \, F(|x|). \tag{18.2.9}$$

The infinitesimal volume $d^4x$ can be written in spherical polars, and the angular integration can be performed at once to give

$$I = \Omega \int_0^\infty dr \, r^3 F(r). \tag{18.2.10}$$

The factor of $\Omega = 2\pi^2$ is found by taking $F = \exp(-x^2)$. Applying this formula to the basic integrand yields:

**Theorem 18.2.11.** *Wick rotation assigns a value to the simplest Feynman integral as follows:*

$$\int \frac{d^4l}{(2\pi)^4} \frac{1}{(l^2 - a)^n} = \frac{i}{16(n-1)(n-2)\pi^2} \left(-\frac{1}{a}\right)^{n-2}$$

for $n \geq 3$.

This is a one-loop result. Clearly, in the multi-loop case, if we first put our Feynman integral into standard form following Section 18.1, the result of performing the first integral sets up an integrand in standard form for the second, and so on. Consequently, when the integral as a whole is convergent, this one theorem is effectively a multi-loop result.

**Homework 18.2.12.** Perform the momentum integrals for the two-loop diagram in Homework 18.1.17.

If there are powers of $l^2$ in the numerator, then we can use partial fractions to evaluate the Euclidean-space integral. If there are products of $l^\mu$'s in the numerator, since the denominator is spherically symmetric after Wick rotation, then we use the argument of 18.2.7 above to reduce these products to powers of $l^2$ and $g^{\mu\nu}$'s:

**Theorem 18.2.13.** *In the Euclidean-space integral*

$$\int \frac{d^4lp}{(2\pi)^4} \frac{N(l)}{D(l^2)},$$

*the value of the renormalized integral will be unchanged if we make the following substitutions in the numerator $N(l)$ of the integrand:*

Terms of odd order in $l$: $\qquad l^\mu \to 0;$

Terms quadratic in $l$: $\qquad l^\mu l^\nu \to \frac{1}{4} l^2 g^{\mu\nu};$

Terms quartic in $l$: $\qquad l^\mu l^\nu l^\rho l^\sigma \to \frac{1}{24} l^2 l^2 (g^{\mu\nu} g^{\rho\sigma} + g^{\mu\rho} g^{\nu\sigma} + g^{\mu\sigma} g^{\nu\rho}). \qquad \square$

**Homework** 18.2.14. Evaluate the momentum integral for the box diagram

where the fermion and scalar have masses $m$ and $\mu$ respectively, and the coupling is $\mathcal{L}' = -\lambda \bar{\psi}\psi\phi$.

This section has explained the evaluation of convergent Feynman integrals in Euclidean space by Wick rotation. As indicated by the Euclidean-space integral formula 18.2.11 above, power-counting of the integration variable is enough to indicate when the integral is finite after Wick rotation. The next step to renormalization is regularization of the divergent cases.

## 18.3  The General Character of Regularization

As we learned in Section 10.10, after Wick rotation, there are two steps in any renormalization procedure. The first step expresses a divergent Euclidean-space integral as a limit of convergent integrals; this is regularization. The second step identifies and eliminates the divergent part of the limit; this is renormalization. In this section, we introduce the principles of regularization, and in the next, application to renormalization.

Wick rotation alone makes many Feynman integrals converge, but not the simplest and most significant loop integrals like the self-energy integrals of Chapter 10. As we saw in that chapter, such integrals must be subjected to a regularization procedure. A one-dimensional example will bring out the essence of this concept. Let $I$ be the integral

$$I = \int_{-\infty}^{\infty} dx \, \frac{2x}{x^2 + 1}. \tag{18.3.1}$$

The definition of an integral over the whole real line expresses $I$ as a double limit:

$$\begin{aligned} I &= \lim_{a \to -\infty} \lim_{b \to \infty} \int_a^b dx \, \frac{2x}{x^2 + 1} \\ &= \lim_{a \to -\infty} \lim_{b \to \infty} \left[ \ln(x^2 + 1) \right]_a^b \\ &= \lim_{a \to -\infty} \lim_{b \to \infty} \ln\left( \frac{b^2 + 1}{a^2 + 1} \right). \end{aligned} \tag{18.3.2}$$

At first it would appear that $I = 0$ because the integrand is an odd function while the range of integration is symmetric. Actually, since neither limit exists, $I$ has no definite value. To bring out the zero value, we can define a *regularized integral* $\text{Reg}_\Lambda(I)$ of $I$ as follows:

$$\text{Reg}_\Lambda(I) \stackrel{\text{def}}{=} \int_{-\Lambda}^{\Lambda} dx \, \frac{2x}{x^2 + 1}, \tag{18.3.3}$$

and assign to $I$ the *regularized value* given by the limit of the regularized integral:

$$\text{Reg}(I) \stackrel{\text{def}}{=} \lim_{\Lambda \to \infty} \text{Reg}_\Lambda(I). \tag{18.3.4}$$

Clearly $\text{Reg}(I) = 0$, and regularization has assigned a definite value to a divergent integral. This has happened primarily because the integrand was an odd function of the integration variable. Generally, regularization does not make even logarithmic divergences converge. However, as the following homework suggests, regularized logarithmic divergences are generally translation invariant.

**Homework 18.3.5.** Without translating the integration variable in a divergent integral, show that the following Euclidean integral vanishes:

$$\int d^2x \, \frac{1}{x^2 + a^2} - \frac{1}{(x+c)^2 + a^2} = 0.$$

In cases where the regularization limit diverges, regularization in effect quantifies the divergence of divergent integrals. For example, the integral

$$I' \overset{\text{def}}{=} \int dx \, \frac{e^x}{1 + e^x} \tag{18.3.6}$$

has a regularization

$$\text{Reg}_\Lambda(I') = \ln\left(\frac{1 + e^\Lambda}{1 + e^{-\Lambda}}\right). \tag{18.3.7}$$

The divergence survives the regularization, but at least it has been given a precise quantitative character. As explained in the next section, this enables renormalization to bring out some finite value.

Regularization has the strange side-effect of extending the domain of the integral operator. The symmetric regularization above, for example, makes all odd functions integrable, whatever their behavior at infinity:

$$f(-x) = -f(x) \quad \Longrightarrow \quad \text{Reg} \int dx \, f(x) = 0. \tag{18.3.8}$$

This extended domain is not translation invariant.

In perturbation theory with four dimensions of energy-momentum, divergences are either logarithmic, linear, or at worst quadratic. Regularization typically assigns vanishing regularized values to integrals of odd functions. Generally, regularized logarithmic divergences are translation invariant, but regularized linear divergences and quadratic divergences are not translation invariant. This makes it necessary to formulate a new Feynman rule which specifies the origin of loop momenta in Feynman integrals which diverge linearly or quadratically. Commonly, this rule is treated as part of the regularization procedure. (For example, in Homeworks 10.13.16 and 10.14.19 on self-energy functions, following the example in Section 10.10 and the principles of Section 18.1, one translates the integration variable to put the denominators of the integrands into standard form.)

This section has brought out the principles of the Euclidean cut-off regularization used in Chapter 10. Theoretically, there is great flexibility in defining a regularization procedure. Only a few, however, are practical.

# 18.4 The General Character of Renormalization

The advantage of quantifying the divergence is that we can now make renormalization, that is, the removal the divergences, a systematic, well-defined procedure. In the simplest case, we can renormalize sums of divergent integrals simply by exchanging the sum and the limit operators. For example, defining

$$I'(c) \stackrel{\text{def}}{=} \int dx \, \frac{e^{x+c}}{1 + e^{x+c}}, \tag{18.4.1}$$

we find that

$$\text{Reg}_\Lambda\big(I'(c)\big) = \ln\Big(\frac{1 + e^{\Lambda+c}}{1 + e^{-\Lambda+c}}\Big), \tag{18.4.2}$$

and can thereby define a reasonable renormalized value for the difference $I'(c) - I'(0)$ as follows:

$$\begin{aligned}
\text{Ren}\big(I'(c) - I'(0)\big) &\stackrel{\text{def}}{=} \lim_{\Lambda \to \infty} \Big(\text{Reg}_\Lambda\big(I'(c)\big) - \text{Reg}_\Lambda\big(I'(c)\big)\Big) \\
&= \lim_{\Lambda \to \infty} \Big(\ln\Big(\frac{1 + e^{\Lambda+c}}{1 + e^{-\Lambda+c}}\Big) - \ln\Big(\frac{1 + e^\Lambda}{1 + e^{-\Lambda}}\Big)\Big) \\
&= \lim_{\Lambda \to \infty} \ln\Big(\frac{1 + e^{\Lambda+c}}{1 + e^\Lambda} \frac{1 + e^{-\Lambda}}{1 + e^{-\Lambda+c}}\Big) \\
&= c.
\end{aligned} \tag{18.4.3}$$

This illustrates the breakdown of translation invariance in regularized linearly divergent integrals.

Here, the individual integrals diverge, but there is a linear combination of their integrands which is integrable. This exchange of linear combination and integral operators is effectively the technique of counterterm renormalization. For example, in 10.10.10, elimination of counterterms from a scalar self-energy function $\Pi$ using the field and mass renormalization conditions yielded

$$\Pi(p^2) = -\lim_{\Lambda \to \infty} \Big(\mathcal{A}_{\text{lp}}(p^2; \Lambda^2) - \mathcal{A}_{\text{lp}}(\mu^2; \Lambda^2) - (p^2 - \mu^2)\mathcal{A}'_{\text{lp}}(\mu^2; \Lambda^2)\Big). \tag{18.4.4}$$

The function $\Pi$ is finite because the linear combination of integrands on the right is in fact integrable.

**Remark** 18.4.5. This example suggests eliminating regularization by defining counterterm coefficients not as numbers, but as regularized divergent integrals, and modifying the Feynman rules: instead of integrating every diagram separately, add the integrands generated by all contributing processes before integrating. This is in effect the Zimmermann implementation of the Bogoliubov–Parasiuk–Hepp renormalization scheme, which we present in Chapter 20.

A more drastic type of renormalization uses regularization to separate out a divergence from an integral and assigns a finite value to the integral by simply throwing the divergence away. In field theory, the minimal subtraction or MS renormalization scheme works in this way.

For a one-dimensional example, if

$$I'' \overset{\text{def}}{=} \int_{-\infty}^{\infty} dx \, \frac{x^2}{x^2 + 1}, \tag{18.4.6}$$

then

$$
\begin{aligned}
\text{Reg}_\Lambda(I) &= \int_{-\Lambda}^{\Lambda} dx \, \frac{x^2}{x^2 + 1} \\
&= \left[ x - \tan^{-1}(x) \right]_{-\Lambda}^{\Lambda} \\
&= 2\left( \Lambda - \tan^{-1}(\Lambda) \right).
\end{aligned} \tag{18.4.7}
$$

Now we can see that, in comparison to powers of $\Lambda$, the integral diverges linearly, and that if we throw away the linear term in $\Lambda$ then the residue converges. We could therefore define the renormalized value $\text{Ren}(I)$ of $I$ by:

$$\text{Ren}(I) \overset{\text{def}}{=} \lim_{\Lambda \to \infty} -2\tan^{-1}(\Lambda) = -\pi. \tag{18.4.8}$$

Notice how surprising this answer is: the integrand is positive everywhere, yet the renormalized integral is negative! Upon reflection, however, it is clear that any process of removing infinities will yield this kind of paradox.

Regularization, as we have seen, is a procedure for expressing divergent integrals as limits of convergent ones. Renormalization is a procedure for removing divergences. Understanding the nature of regularization and renormalization and their relationship, we can now approach the details of these procedures.

## 18.5 Regularization

There are three principle methods for regularizing the divergent Feynman integrals: truncate the domain of integration in Euclidean space; introduce regulator fields; integrate in $D$ dimensions. We will briefly introduce all three techniques in this section.

The first method of regularization cuts the range of integration down to $-\Lambda^2 \leq l^2 \leq \Lambda^2$. It is natural in a theory which is only valid up to a certain energy-momentum threshold $\Lambda$, beyond which some more complete theory will take over. In this case, after Wick rotation, the energy constraint becomes $l_{\text{E}}^2 \leq \Lambda^2$. The bizarre way in which the Lorentz invariant, Minkowski-space constraint has become a rotation invariant, Euclidean-space constraint reveals the non-trivial nature of the perturbation $l_0 \to e^{i\delta} l_0$.

The Euclidean energy-momentum cut-off provides a natural regularization parameter $\Lambda$:

$$
\begin{aligned}
\text{Reg}_{\text{E}}^\Lambda \left( I(p) \right) &\overset{\text{def}}{=} i \int_{l_{\text{E}}^2 \leq \Lambda^2} \frac{d^4 l_{\text{E}}}{(2\pi)^4} \, \frac{N(p, l_{\text{E}})}{\left( -l_{\text{E}}^2 - a(p) \right)^n} \\
\implies \quad &\text{Reg} I(p) = \lim_{\Lambda \to \infty} \text{Reg}_{\text{E}}^\Lambda \left( I(p) \right),
\end{aligned} \tag{18.5.1}
$$

whenever the limit exists.

For Feynman integrals $I(p)$ of the type 18.2.5, the cut-off integral can be evaluated and written in the form

$$\text{Reg}_{\text{E}}^{\Lambda}\big(I(p)\big) = c(p) + d(p)\ln(\Lambda) + d_1(p)\Lambda + d_2(p)\Lambda^2. \qquad (18.5.2)$$

For the same reasons that Wick rotation preserves Lorentz invariant, so also does cutting off the domain of integration in Euclidean space. From the form of this final equation we see that the Euclidean energy-momentum cut-off enables us to isolate and identify divergences in Feynman integrals.

Cut-off regularization has the advantage of being the simplest regularization to implement, but the disadvantage of breaking gauge invariance. In QED, for example, as we noted following Homework 10.14.20, if we apply cut-off regularization to the electron-loop contribution to the photon propagator, we find a divergence that calls for a counterterm which breaks the gauge symmetry.

The second method is Pauli–Villars regularization which adds fictitious particles to the theory in order to modify every propagator in the following way:

$$\frac{i}{p^2 - m^2 + i\epsilon} \longrightarrow \frac{i}{p^2 - m^2 + i\epsilon} - \frac{i}{p^2 - M^2 + i\epsilon}$$
$$= \frac{-i(M^2 - m^2)}{(p^2 - m^2 + i\epsilon)(p^2 - M^2 + i\epsilon)}. \qquad (18.5.3)$$

The effect of this modification on the Euclidean-space integral is to double the rate at which the integrand falls off at infinity. The regularization parameter is the mass $M$ of the fictitious particle: as $M \to \infty$, we recover the old propagator.

In this regularization scheme, it is necessary to include a fictitious regulator field for every physical field that causes a divergence in some loop integral, and to organize that the sum of physical and regulated contributions is convergent. For this cancellation to work, firstly, the regulator field must have identical couplings to the physical field which it regulates, and hence identical Lorentz transformation properties. Secondly, to cancel a loop divergence, it is necessary to make the contribution of the corresponding regulator-field loop opposite in sign. Thus the scalar regulator fields have Fermi statistics, and spinor regulator fields have Bose statistics. For consistency, the original Lagrangian density should be modified by the inclusion of regulator fields. This has not been accomplished for a non-abelian gauge theory.

**Remark 18.5.4.**   For example, the Pauli-Villars bare Lagrangian density for QED has the form:

$$\mathcal{L}_{\text{PV}} = -\frac{1}{4}F_{\mu\nu}F^{\mu\nu}(1 + M^{-2}\Box)F^{\mu\nu} - \frac{1}{2\alpha}(\partial_\mu A^\mu)(1 + M^{-2}\Box)(\partial_\mu A^\mu)$$
$$+ \bar{\psi}(i\slashed{D} - m)\psi + \bar{\psi}'(i\slashed{D} - M)\psi',$$

where $\psi'$ is the regulator field for $\psi$ and the Klein–Gordon operator provides regularization of the photon propagator.

A scalar theory would be regularized differently, for example:

$$\mathcal{L} = \frac{1}{2}\partial_\mu\phi_1\,\partial^\mu\phi_1 + \frac{1}{2}\partial_\mu\phi_2\,\partial^\mu\phi_2 - \frac{1}{2}m_1^2\phi_1^2 - \frac{1}{2}m_2^2\phi_1^2 - \frac{\lambda}{2}\phi_1\phi_2^2 \qquad (18.5.5)$$

will be modified to

$$\mathcal{L}_{\text{PV}} = \frac{1}{2}\partial_\mu\phi_1\,\partial^\mu\phi_1 + \frac{1}{2}\partial_\mu\phi_2\,\partial^\mu\phi_2 + \frac{1}{2}\partial_\mu\phi\,\partial^\mu\phi$$
$$- \frac{1}{2}m_1^2\phi_1^2 - \frac{1}{2}m_2^2\phi_1^2 - \frac{1}{2}M^2\phi^2 \qquad (18.5.6)$$
$$- \frac{\lambda}{2}\phi_1(\phi_2 + \phi)^2,$$

where $\phi$ is a regulator field for $\phi_2$. We do not need a regulator field for $\phi_1$ because this field cannot form a loop on its own, and regulating the occurrences of $\phi_2$ alone will be enough to force convergence on loop integrals.

Our third method of regularization uses the dimension $D$ of space-time as the regularization parameter. The basic observation is that the $D$-dimensional Euclidean-space integral

$$I_n^D = i \int \frac{d^D l_{\text{E}}}{(2\pi)^D} \left(\frac{1}{-l_{\text{E}}^2 - a}\right)^n \qquad (18.5.7)$$

is convergent for all $n \geq 1$ if the dimension $D \leq 2$. Therefore, if we can make $D$ a continuous parameter, then we can isolate the divergence as a function of $D$ by letting $D \to 4$. The art enters into the method in seeing how to make the integral a continuous function of $D$.

Cut-off, propagator, and dimensional regularization are the three regularization procedures most commonly used. Cut-off regularization is convenient when there is no gauge symmetry. Propagator regularization is often used in QED, but becomes a little tricky in non-abelian gauge theories. Dimensional regularization is generally favored because it manages all cases efficiently. There are other regularization procedures, and new schemes which promise easier computation or better theoretical properties are discovered from time to time. We, however, shall use dimensional regularization, and therefore devote the next section to developing that concept.

## 18.6  Dimensional Regularization Defined

In order to define dimensional regularization, we must first obtain a formula for integrals in $D$ dimensions which is analytic in $D$. Such a formula cannot be defined for arbitrary integrands, but there is no problem in the special case of Feynman integrals.

Define an integral $I(f)$ by:

$$I(f) \overset{\text{def}}{=} \int \frac{d^4 l}{(2\pi)^4} \, f(l^2). \qquad (18.6.1)$$

Then Euclidean renormalization in $D$ dimensions provides a modified integral

$$I_D(f) \overset{\text{def}}{=} i \int \frac{d^D l_E}{(2\pi)^D} \, f(-l_E^2). \qquad (18.6.2)$$

We can simplify this integral using spherical polars and integrating over angular variables. Writing $A_D$ for the area of the unit sphere in $D$ dimensions, we find:

$$I_D(f) = \frac{i A_D}{(2\pi)^D} \int_0^\infty dr \, r^{D-1} f(-r^2). \qquad (18.6.3)$$

The neatest way of computing the area $A_D$ of the unit sphere in $D$-dimensional space is to evaluate the integral of $\exp(-r^2)$ in cartesian and spherical polar coordinates:

$$\int d^D x \, \exp\left(-|x|^2\right) = A_D \int dr \, r^{D-1} \exp(-r^2)$$

$$\Longrightarrow \quad \left(\int dt \, \exp(-t^2)\right)^D = \frac{A_D}{2} \int ds \, \exp(-s) s^{(D-2)/2} \qquad (18.6.4)$$

$$\Longrightarrow \quad \pi^{D/2} = \frac{A_D}{2} \Gamma(D/2),$$

where the function $\Gamma$ is as usual defined by

$$\Gamma(z) = \int_0^\infty dt \, t^{z-1} e^{-t} \quad \text{for Re}(z) > 0. \qquad (18.6.5)$$

Hence we conclude that

**Lemma 18.6.6.** *The area $A_D$ of the unit sphere in $D$ dimensions is given by*

$$A_D = \frac{2\pi^{D/2}}{\Gamma(D/2)}.$$

**Remark** 18.6.7. There are many delightful properties of the $\Gamma$ function. The most basic one is

$$\Gamma(z) = (z-1)\Gamma(z-1). \qquad (18.6.8)$$

This implies that $\Gamma(n) = (n-1)!$ whenever $n$ is a positive integer.

Substituting this expression for the area $A_D$ into the formula 18.6.3 for $I^D(f)$ provides a definition of the dimensional-regularized value $\text{Reg}_D(I(f))$ of $I(f)$:

**Definition 18.6.9.** Dimensional regularization assigns to the integral

$$I(f) = \int \frac{d^4 l}{(2\pi)^4} f(l^2)$$

the value

$$\text{Reg}_D(I(f)) \overset{\text{def}}{=} \frac{i}{2^{D-1}\pi^{D/2}\Gamma(D/2)} \int_0^\infty dr \, r^{D-1} f(-r^2).$$

Since $\Gamma(D/2)$ is analytic, and since the remaining integral is simply a Mellin transform, $\text{Reg}_D(I(f))$ is an analytic function of $D$ for a wide range of functions $f$.

With this definition of dimensional regularization for functions $f(l^2)$ in hand, the next step is to extend the definition to the slightly broader class of integrands encountered in perturbation theory, and find some formulae for regularized values of Feynman integrals in standard form.

# 18.7 Dimensional Regularization Applied

Note that we cannot simply integrate in dimension $D = 1$ or $D = 2$ and then deduce the dimensional regularization formulae. It is necessary to use the spherical symmetry of the integrand after Wick rotation in order to introduce $D$ as a complex variable. It will therefore be necessary to manipulate the integrands of Feynman integrals to bring out such symmetry before applying dimensional regularization. Specification of some such systematic manipulation must be included in the definition of dimensional regularization.

In practice, the steps of manipulation are standard: first, use Feynman parameters to combining propagators; second, shift variables to put the denominator in standard form; third, simplify the numerator using substitutions like those in Theorem 18.2.13 above.

The generalization of the substitution formulae in Theorem 18.2.13 to degree $2n$ in $D$ dimensions reads

**Lemma 18.7.1.** *For an integrand $l^{\mu_1} \cdots l^{\mu_{2n}} f(l^2)$, dimensional regularization generates the following substitutions:*

$$l^{\mu_1} \cdots l^{\mu_{2n}} \longrightarrow \frac{\Gamma(D/2)}{2^n \Gamma(n + D/2)} \underbrace{(g^{\mu_1\mu_2} \cdots g^{\mu_{2n-1}\mu_{2n}} + \cdots)}_{(2n-1)!! \text{ terms}} (l^2)^n,$$

*where $(2n - 1)!! = (2n - 1) \cdot \ldots \cdot 3 \cdot 1$ is the number of ways of splitting $2n$ indices into an unordered set of $n$ unordered pairs.*

**Proof.** The proof is three steps. First, any polynomial of degree $2n$ in $D$ variables which is invariant under $SO(D)$ is proportional to $r^{2n}$. Second, observe that there must therefore exist a constant $c$ such that

$$\int d^D x \, x^{\mu_1} \cdots x^{\mu_{2n}} f(x^2) = c(\delta^{\mu\nu'}\text{s}) \int d^D x \, (x^2)^n f(x^2), \qquad (18.7.2)$$

where $x$ is a Euclidean variable. Third, find $c$ by evaluating the two sides for $f(x^2) = \exp(-x^2)$. $\qquad \square$

One more fine but important point before deriving formulae: a dimensionless coupling constant in four dimensions will not be dimensionless in $D \neq 4$ dimensions. Indeed, in $D$ dimensions the Lagrangian will have the mass dimension of $M^D$, and so we can deduce the mass dimensions of the fields from the kinetic term:

$$\begin{aligned}
\textbf{Bose Field:} \quad & M^{\frac{D-2}{2}}; \\
\textbf{Fermi Field:} \quad & M^{\frac{D-1}{2}}.
\end{aligned} \qquad (18.7.3)$$

If we write $\lambda_F$ and $\lambda_B$ for generic couplings of one-Bose–two-Fermi and four-Bose operators in a Lagrangian density, then we can deduce the mass dimension of these formerly dimensionless coupling constants in dimension $D = 4 - 2\delta$:

$$\begin{aligned}
\lambda_F(\delta): \quad & M^{\frac{4-D}{2}} = M^\delta; \\
\lambda_B(\delta): \quad & M^{4-D} = M^{2\delta}.
\end{aligned} \qquad (18.7.4)$$

In order to maintain the dimensionless status of the coupling constants, we write

$$\lambda_F(\delta) = \lambda_F \mu^\delta \sim \lambda_F(1 + \delta \ln \mu),$$
$$\lambda_B(\delta) = \lambda_B \mu^{2\delta} \sim \lambda_B(1 + 2\delta \ln \mu),$$
(18.7.5)

where $\mu$ is some generic mass. As $D \to 4$, these $\mu$-terms can make finite contributions to divergent integrals. As we shall see in Chapter 21, this generic mass parameter $\mu$ is the fundamental parameter in the renormalization group equations.

**Homework 18.7.6.** What power of $\mu$ preserves the dimension of the coupling $\lambda(\delta)$ in $\lambda \phi^3$?

From this definition, we can provide formulae for the dimensional regularization of the standard Feynman integrals:

**Theorem 18.7.7.** *The values of generic Feynman integrals after dimensional regularization follow:*

$$\text{Reg}_D \int \frac{d^4 p}{(2\pi)^4} \frac{1}{(p^2 - a)^n} = \frac{i}{16\pi^2} \frac{\Gamma(n-2+\delta)}{\Gamma(n)} \left(-\frac{1}{a}\right)^{n-2} \left(\frac{4\pi}{a}\right)^\delta,$$

$$\text{Reg}_D \int \frac{d^4 p}{(2\pi)^4} \frac{p^\mu}{(p^2 - a)^n} = 0,$$

$$\text{Reg}_D \int \frac{d^4 p}{(2\pi)^4} \frac{p^2}{(p^2 - a)^n} = \frac{i(2-\delta)}{16\pi^2} \frac{\Gamma(n-3+\delta)}{\Gamma(n)} \left(-\frac{1}{a}\right)^{n-3} \left(\frac{4\pi}{a}\right)^\delta,$$

$$\text{Reg}_D \int \frac{d^4 p}{(2\pi)^4} \frac{p^\mu p^\nu}{(p^2 - a)^n} = \frac{i g^{\mu\nu}}{32\pi^2} \frac{\Gamma(n-3+\delta)}{\Gamma(n)} \left(-\frac{1}{a}\right)^{n-3} \left(\frac{4\pi}{a}\right)^\delta,$$

*where* $D = 4 - 2\delta$.

Note that, because the mass dimension of $a$ is $+2$, the Laurent expansion of $(4\pi/a)^\delta$ in powers of $\delta$ introduces a term $\ln(4\pi/a)$, which does not have well-defined mass dimension. In all applications of these formulae, there will be coupling factors like $\lambda_F^2$ which will modify this term to $(4\pi\mu^2/a)^\delta$.

The theorem above covers integration over a single loop variable. However, assuming that we have followed the procedure of Section 18.1 and put our Feynman integral into standard form, if there is a second momentum variable, then the formulae on the right have the same form in this second variable as the integrands on the left did in the first. Only the exponent $n - 2 + \delta$ is not an integer. Since $\Gamma$ is an analytic function, this poses no problem; the same formulae work equally well when $n$ is not an integer. Consequently, as before for the Euclidean integral formulae in Theorem 18.2.11, so here the one-loop formulae trivially generate multi-loop formulae. The following remark provides a proof for the case $n$ an integer, but it generalizes simply enough.

**Remark 18.7.8.** The first formula can be proved by contour integration and the others follow easily. For the first, we have shown that

$$I_n \overset{\text{def}}{=} \int \frac{d^D p}{(2\pi)^D} \frac{1}{(p^2 + a)^n} = \frac{i}{2^D \pi^{D/2}} \frac{(-1)^n}{\Gamma(D/2)} \int_0^\infty ds \frac{s^{D/2-1}}{(a+s)^n},$$
(18.7.9)

so it remains to evaluate the Mellin transform

$$J \overset{\text{def}}{=} \int_0^\infty ds \frac{s^\alpha}{(a+s)^n} \quad \text{where } \alpha = \frac{D}{2} - 1.$$
(18.7.10)

To define the integral in the complex plane, we need to choose a branch of $s^\alpha$. Since the value on the positive real axis is determined, we have only to choose a cut. We choose the positive real axis for the cut. Then we can evaluate $J$ by integrating along the two sides of the cut as follows.

Choose a contour $C = C_1 + C_2 + C_3$ consisting of the integral from zero to $\infty + i\epsilon$, from $\infty + i\epsilon$ round a circle to $\infty - i\epsilon$, and from $\infty - i\epsilon$ back to zero. The integral round the circle vanishes. The integral over $C_1$ is $J$, and that over $C_2$ is proportional to $J$:

$$\int_{C_2} ds \frac{s^\alpha}{(a+s)^n} = \int_\infty^0 ds \frac{e^{2\alpha\pi i} s^\alpha}{(a+s)^n} = -e^{2\pi\alpha i} J. \tag{18.7.11}$$

Hence

$$\int_C ds \frac{s^\alpha}{(a+s)^n} = (1 - e^{2\alpha\pi i})J = -2ie^{\alpha\pi i} \sin \pi\alpha\, J. \tag{18.7.12}$$

We can evaluate the integral round $C$ by the residue theorem. Since $a$ contains a $-i\epsilon$, there is no ambiguity in the branch of the integrand in a neighborhood of the pole. Hence we can expand the integrand as a power series in $z = a + s$:

$$\frac{s^\alpha}{(a+s)^n} = \frac{(z-a)^\alpha}{z^n}$$

$$= (-a)^\alpha z^{-n}\left(1 - \frac{z}{a}\right)^\alpha \tag{18.7.13}$$

$$= (-a)^\alpha z^{-n} \sum_{r=0}^\infty \frac{(\alpha)(\ldots)(\alpha+1-r)}{r!}\left(-\frac{z}{a}\right)^r,$$

and apply the residue theorem to find the value of the contour integral:

$$\int_C ds \frac{s^\alpha}{(a+s)^n} = 2\pi i\, \mathrm{Res}(-a)$$

$$= 2\pi i(-a)^\alpha \frac{(\alpha)(\ldots)(\alpha+2-n)}{(n-1)!}\left(-\frac{1}{a}\right)^{n-1} \tag{18.7.14}$$

$$= 2\pi i \frac{\Gamma(\alpha+1)}{\Gamma(\alpha+2-n)\Gamma(n)}\left(-\frac{1}{a}\right)^{n-1-\alpha}.$$

Setting the two values of the contour integral equal, we obtain a formula for $J$:

$$J = \frac{\pi}{\sin \pi\alpha} \frac{\Gamma(\alpha+1)}{\Gamma(\alpha+2-n)\Gamma(n)}\left(-\frac{1}{a}\right)^{n-2}\left(\frac{1}{a}\right)^{1-\alpha}. \tag{18.7.15}$$

Recalling the identity

$$\frac{\pi}{\sin \pi z} = \Gamma(z)\Gamma(1-z) = (-1)^k \Gamma(z-k)\Gamma(1+k-z), \tag{18.7.16}$$

we see that

$$\frac{\pi}{\sin \pi\alpha} = (-1)^{n-2}\Gamma(\alpha+2-n)\Gamma(n-1-\alpha). \tag{18.7.17}$$

Hence

$$J = (-1)^{n-2} \frac{\Gamma(\alpha+1)\Gamma(n-1-\alpha)}{\Gamma(n)}\left(-\frac{1}{a}\right)^{n-2}\left(\frac{1}{a}\right)^{1-\alpha}. \tag{18.7.18}$$

Substituting this value of $J$ into the formula for $I_n$ and using $\alpha = 1 - \delta$ yields the first formula, as desired. $\qquad\square$

Invariance of a $D$-dimensional Feynman integrand under $SO(1, D-1)$ transformations of the integration parameter leads to $SO(D)$ invariance of the Euclidean-space integrand, and this invariance allows us to integrate over angular variables.

The resulting formulae 18.7.7 are analytic in the dimension parameter $D$ and may therefore be used to define dimensional regularization. The extended definition of dimensional regularization and the formulae presented above are sufficient for most purposes.

In applications to divergent Feynman integrals, these formulae will contain $\Gamma$ functions which may have poles or zeroes at $D = 4$. The next section therefore summarizes the relevant properties of the $\Gamma$ function.

## 18.8 The $\Gamma$ Function

In computations, it will be necessary to understand the behavior of the $\Gamma$ function in the vicinity of its poles. The following Laurent expansions should be sufficient:

$$\Gamma(\delta) = \frac{1}{\delta} - \gamma + O(\delta)$$

$$\implies \quad \Gamma(\delta - n) = \frac{1}{(\delta - 1)(\ldots)(\delta - n)}\Gamma(\delta) \tag{18.8.1}$$

$$= \frac{(-1)^n}{n!}\left(\frac{1}{\delta} + 1 + \frac{1}{2} + \cdots + \frac{1}{n} - \gamma\right) + O(\delta),$$

where $\gamma$, Euler's constant, and $\gamma_n$ are given by

$$\gamma \simeq .5772, \quad \text{and}$$

$$\gamma_n \stackrel{\text{def}}{=} 1 + \frac{1}{2} + \cdots + \frac{1}{n} - \gamma. \tag{18.8.2}$$

A convenient convention defines $\gamma_0 = -\gamma$.

**Remark 18.8.3.** Using the relation $\Gamma(z)\Gamma(1-z) = \pi/\sin(\pi z)$ with $z = \delta$, we can compute the coefficients of positive powers of $\delta$ in the Laurent expansion of $\Gamma(\delta)$. For example,

$$\Gamma(\delta) = \frac{1}{\delta} - \gamma + \left(\frac{\pi^2}{12} + \frac{\gamma^2}{2}\right)\delta + O(\delta^2). \tag{18.8.4}$$

Since the $\Gamma$'s have simple poles, it is necessary to expand every other $\delta$-dependent factor up to order $\delta$ if we are to get the correct finite part as $\delta \to 0$. The following Laurent expansion will be sufficient:

$$z^\delta = 1 + \delta \ln(z) + O(\delta^2). \tag{18.8.5}$$

Commonly, we shall use these expansions in the combination

$$\Gamma(\delta - n)z^\delta = \frac{(-1)^n}{n!}\left(\frac{1}{\delta} + \gamma_n + O(\delta)\right)\left(1 + \delta \ln z + O(\delta)\right)$$

$$= \frac{(-1)^n}{n!}\left(\frac{1}{\delta} + \gamma_n + \ln z\right) + O(\delta). \tag{18.8.6}$$

The application of the Laurent expansions of this section to the dimensional regularization is best made clear by an example. Taking $1 \le n \le 2$,

$$\text{Reg}_D \int \frac{d^4p}{(2\pi)^4} \frac{\lambda^2}{(p^2 - a)^n} \tag{18.8.7}$$

$$= \frac{i\lambda^2}{16\pi^2\Gamma(n)}\left(-\frac{1}{a}\right)^{n-2}\Gamma(n-2+\delta)\left(\frac{4\pi\mu^2}{a}\right)^\delta$$

$$= \frac{i\lambda^2(-1)^{2-n}}{16\pi^2}\left(-\frac{1}{a}\right)^{n-2}\left(\frac{1}{\delta}+\gamma_{2-n}\right)\left(1+\delta\ln\left(\frac{4\pi\mu^2}{a}\right)\right)+O(\delta)$$

$$= \frac{i\lambda^2 a^{2-n}}{16\pi^2}\left(\frac{1}{\delta}+\gamma_{2-n}+\ln\left(\frac{4\pi\mu^2}{a}\right)\right)+O(\delta).$$

Renormalization will define a counterterm to cancel the pole so that the limit of the remainder will be finite as $\delta \to 0$.

**Homework** 18.8.8. Follow the argument of Section 10.10 using dimensional regularization instead of a cut-off. More precisely, compute the one-loop contributions to the self-energy and vertex functions for the interaction Lagrangian density $\mathcal{L}' = -\lambda\phi^3/3!$, and determine the three $z$'s and the coefficients of the kinetic, mass, and vertex counterterms.

The material of this section is trivial enough, but very practical: it is used in every application of dimensional regularization. The next section completes the theory of dimensional regularization by explaining how to manage Dirac algebra in $D$ dimensions.

## 18.9 Dirac Algebra in $D$ Dimensions

Since our integrals are being performed in $D$ dimensions, it is necessary to use Dirac algebra in $D$ dimensions. The axioms that define $D$-dimensional Dirac algebra are

$$\{\gamma^\alpha,\gamma^\beta\} = 2g^{\alpha\beta}\mathbf{1},$$

$$(\gamma^\alpha)^\dagger = \gamma_\alpha = \begin{cases} \gamma^\alpha, & \text{if } \alpha = 0; \\ -\gamma^\alpha, & \text{if } \alpha \geq 1. \end{cases} \tag{18.9.1}$$

A set of matrices $\gamma^1,\ldots,\gamma^D$ which satisfy these axioms are called a representation of the Dirac algebra. It is not hard to show that the complex Dirac algebras are simply the following matrix algebras:

$$\begin{array}{ll} \textbf{DirAlg}_0 = \mathbf{C} & \textbf{DirAlg}_1 = \mathbf{C} \oplus \mathbf{C} \\ \textbf{DirAlg}_2 = M_2(\mathbf{C}) & \textbf{DirAlg}_3 = M_2(\mathbf{C}) \oplus M_2(\mathbf{C}) \\ \textbf{DirAlg}_4 = M_4(\mathbf{C}) & \textbf{DirAlg}_5 = M_4(\mathbf{C}) \oplus M_4(\mathbf{C}) \end{array} \tag{18.9.2}$$
$$\cdots \qquad\qquad \cdots$$

Thus the fundamental representation of the Dirac algebra in dimension $D = 2k$ or $D = 2k+1$ has degree $d = 2^k$.

The common Dirac-algebra formulae depend on the representation of the Dirac algebra only through the degree of the representation. This comes in as the trace of the identity matrix. For the purposes of dimensional regularization, it is permissible to set this trace to its four-dimensional value 4. Using the proper value makes a finite difference in the counterterms and no difference in the predictions.

On the basis of the structure of the $D$-dimensional Dirac algebra alone, the usual contraction relations readily generalize. For example, writing

$$\gamma^{\alpha_1\ldots\alpha_s} \stackrel{\text{def}}{=} \gamma^{\alpha_1}\ldots\gamma^{\alpha_s}, \tag{18.9.3}$$

we find from the commutation relations that

$$\gamma^\beta \gamma^{\alpha_1 \cdots \alpha_s} \gamma_\beta = 2 \sum_{r=1}^{s} (-1)^{r-1} \gamma^{\alpha_1 \cdots \widehat{\alpha_r} \cdots \alpha_s \alpha_r} + (-1)^s D \gamma^{\alpha_1 \cdots \alpha_s}, \tag{18.9.4}$$

where the hat indicates an index to omit.

The usual trace operator is available, and with the standard convention that $\mathrm{Tr}(1) = 4$, traces of products of $\gamma$ matrices are independent of the choice of matrices that represent the Dirac algebra. Thus:

$$\mathrm{Tr}(\gamma^\alpha \gamma^\beta) = 4g^{\alpha\beta};$$
$$\mathrm{Tr}(\gamma^\alpha \gamma^\beta \gamma^\eta \gamma^\zeta) = 4(g^{\alpha\beta} g^{\eta\zeta} - g^{\alpha\eta} g^{\beta\zeta} + g^{\alpha\zeta} g^{\beta\eta}). \tag{18.9.5}$$

The trace of an odd number of $\gamma$ matrices is zero.

Unfortunately, the natural generalization of $\gamma^5$ to $D$ dimensions fails to reproduce the desired four-dimensional formulae in the limit $D \to 4$. If we define $\gamma_5$ in dimension $D = 2k$ by

$$\gamma_5 \overset{\text{def}}{=} i\gamma^0 \cdots \gamma^{2k-1}, \tag{18.9.6}$$

then $\gamma_5$ has the desirable properties

$$\gamma_5^2 = 1, \quad \gamma_5^\dagger = \gamma^5, \quad \{\gamma_5, \gamma^\alpha\} = 0, \quad \mathrm{Tr}(\gamma_5) = 0. \tag{18.9.7}$$

(For $k > 2$, the notation $\gamma_5$ is ambiguous, but as we never substitute 5 for $\mu$ in $\gamma_\mu$, in practice no confusion arises.)

**Remark** 18.9.8. In dimension $D = 2k + 1$, $\gamma_5$ is generally defined by

$$\gamma_5 \overset{\text{def}}{=} \gamma^0 \cdots \gamma^{2k}. \tag{18.9.9}$$

Then $\gamma_5$ commutes with the other $\gamma$ matrices, so the two matrices $P_\pm = \frac{1}{2}(1 \pm \gamma_5)$ are idempotents in the Dirac algebra, $P_\pm^2 = 1$, and thus split the algebra into two commuting subalgebras as shown in the list 18.9.2 of Dirac algebras above.

The problem lies in the formulae for traces involving $\gamma_5$ with other $\gamma$ matrices. We find that as a function of $D$, such traces are identically zero. First, for two other $\gamma$'s, the contraction relations show that

$$\begin{aligned}
D\,\mathrm{Tr}(\gamma_5 \gamma^{\alpha\beta}) &= \mathrm{Tr}(\gamma_5 \gamma_\epsilon \gamma^{\epsilon\alpha\beta}) \\
&= -\mathrm{Tr}(\gamma_\epsilon \gamma_5 \gamma^{\epsilon\alpha\beta}) \\
&= -\mathrm{Tr}(\gamma_5 \gamma^{\epsilon\alpha\beta} \gamma_\epsilon) \\
&= -\mathrm{Tr}(\gamma_5 (2\gamma^{\beta\alpha} - 2\gamma^{\alpha\beta} + D\gamma^{\alpha\beta})).
\end{aligned} \tag{18.9.10}$$

We deduce that

$$(D - 2)\,\mathrm{Tr}(\gamma_5 \gamma^{\alpha\beta}) = -\mathrm{Tr}(\gamma_5(\gamma^{\beta\alpha} + \gamma^{\alpha\beta})) = -2g^{\alpha\beta}\,\mathrm{Tr}(\gamma_5) = 0. \tag{18.9.11}$$

Consequently, if we promote $D$ to a complex variable and assume that $\mathrm{Tr}(\gamma_5 \gamma^{\alpha\beta})$ is meromorphic in $D$, then this trace must vanish identically.

Using the contraction relations on the four-$\gamma$ case, we find that

$$D \operatorname{Tr}(\gamma_5 \gamma^{\alpha\beta\gamma\delta}) = -\operatorname{Tr}\big(\gamma_5(2\gamma^{\beta\gamma\delta\alpha} - 2\gamma^{\alpha\gamma\delta\beta} + 2\gamma^{\alpha\beta\delta\gamma} - 2\gamma^{\alpha\beta\gamma\delta} + D\gamma^{\alpha\beta\gamma\delta})\big). \quad (18.9.12)$$

If we assume the vanishing of the trace in the two-$\gamma$ case, then this expression reduces to

$$(D-4) \operatorname{Tr}(\gamma_5 \gamma^{\alpha\beta\gamma\delta}) = 0. \qquad (18.9.13)$$

Again, analyticity of the trace implies that it vanishes identically. Consequently, the usual four-dimensional formula has no natural analytic extension.

One approach to interpreting $\gamma_5$ and its companion $\epsilon_{\mu\nu\rho\sigma}$ in dimensional regularization uses the four-dimensional definitions:

$$\gamma^5 = i\gamma^0\gamma^1\gamma^2\gamma^3;$$
$$\epsilon_{0123} = 1. \qquad (18.9.14)$$

These definitions break $D$-dimensional Lorentz covariance, but preserve Lorentz covariance in four dimensions and work well in the limit $D \to 4$.

**Remark** 18.9.15. Another approach defines a new trace operator tr$_5$ for traces with $\gamma_5$ factors. The operator tr$_5$ in this approach is not cyclic, and so avoids the trace catastrophe above.

A useful and fairly standard notation separates the range of indices into 0–3 and 4–$D$ using bars and hats respectively. In particular,

$$\gamma_\mu = \bar{\gamma}_\mu + \hat{\gamma}_\mu \quad \text{and} \quad g_{\mu\nu} = \bar{g}_{\mu\nu} + \hat{g}_{\mu\nu}, \qquad (18.9.16)$$

where $\bar{\gamma}_m u$ is zero for $\mu > 3$, $\hat{\gamma}_\mu$ is zero for $\mu < 4$, and similarly for $g_{\mu\nu}$. The splitting of $g_{\mu\nu}$ is valid only because the metric tensor vanishes for mixed indices. Note that

$$\{\bar{\gamma}_\mu, \gamma_5\} = 0 \quad \text{but} \quad [\hat{\gamma}_\mu, \gamma_5] = 0. \qquad (18.9.17)$$

(The use of the bar here conflicts with its earlier use as the Dirac adjoint operator, but in practice the context makes the intended meaning obvious.)

**Homework** 18.9.18. Use dimensional regularization to compute to one loop the self-energy functions for a fermion $\psi$ with mass $m$ and a scalar $\phi$ with mass $\mu$ which interact through $\mathcal{L}' = -\lambda\bar{\psi}\psi\phi$. Determine the two kinetic and mass renormalization factors and the coefficients of the corresponding counterterms.

**Homework** 18.9.19. Continuing with the example of the previous homework, use dimensional regularization to compute to one loop the vertex function, its renormalization factor and the coefficient of the vertex counterterm.

At this point, we have set out all the machinery necessary for applying dimensional regularization with confidence. In the concluding section of this chapter, we shall extend the concepts of regularization and renormalization to operators and in the process find opportunities to use dimensional regularization.

# 18.10 Regularization of Composite Operators

We noted in Section 2.4 that a quantum field is not an operator on Hilbert space. For the free field, we can readily see that its matrix element between physical states is well defined, but its value on a physical state is a distribution, not a physical state or even an element of Hilbert space. Consequently, the meaning of a product of fields at the same space-time point is like the square of a distribution not easy to define precisely. Such a product of fields is known as a *composite operator*. From the perspective of canonical quantization, as we noted in Section 4.13, composite operators in a Hamiltonian make normal ordering necessary. In functional integral quantization, the divergence problem is buried in the definition of the functional measure. In this section, we introduce the regularization and renormalization of composite operators.

We begin the study of composite operator regularization using the axial current as an example. Consider the following paradox. Directly from the canonical quantization relations, we find that

$$\left[ j^0(\bar{x}, t), j^r(\bar{y}, t) \right] = 0. \tag{18.10.1}$$

However, if we assume current conservation, then the space-divergence of the left-hand side yields a time-derivative of the current, and we obtain

$$\frac{\partial}{\partial y^r} \left[ j^0(\bar{x}, t), j^r(\bar{y}, t) \right] = -\left[ j^0(\bar{x}, t), \partial_t j^0(\bar{y}, t) \right]$$
$$= i \left[ j^0(\bar{x}, t), \left[ j^0(\bar{y}, t), H \right] \right]. \tag{18.10.2}$$

Now we take the vacuum expectation of the final expression and estimate its value by inserting a complete set of intermediate states:

$$\langle 0 | \left[ j^0(\bar{x}, t), \left[ j^0(\bar{y}, t), H \right] \right] | 0 \rangle \tag{18.10.3}$$
$$= \sum \!\!\!\!\!\!\int d\mu(n) \, \langle 0 | j^0(\bar{x}, t) | n \rangle \langle n | \left[ j^0(\bar{y}, t), H \right] | 0 \rangle - \langle 0 | \left[ j^0(\bar{y}, t), H \right] | n \rangle \langle n | j^0(\bar{x}, t) | 0 \rangle$$
$$= -2 \sum \!\!\!\!\!\!\int d\mu(n) \, e^{i(\bar{x} - \bar{y}) \cdot \bar{p}_n} E_n \left| \langle 0 | j^0(\bar{0}, t) | n \rangle \right|^2.$$

Clearly, when $\bar{x} = \bar{y}$, this final expression is definitely negative. Obviously, naive use of the canonical commutation relations with currents is simply wrong.

One obvious possibility for regularization of currents is simply to separate the points at which the operators are evaluated. For a current in a gauge theory, however, this would break gauge invariance. Thus, for the axial current, Schwinger proposed a regularization which included a compensating exponential in the gauge field:

$$j_5^\mu(x; \epsilon) \stackrel{\text{def}}{=} \bar{\psi}(x + \tfrac{\epsilon}{2}) \gamma^\mu \gamma_5 \psi(x - \tfrac{\epsilon}{2}) \exp\left( -ie \int_{-\epsilon/2}^{\epsilon/2} dy \cdot A(y) \right). \tag{18.10.4}$$

This regularized axial current is invariant under the transformation

$$\psi(x) \to e^{-ie\theta(x)} \psi(x) \quad \text{and} \quad A_\mu(x) \to A_\mu(x) + \partial_\mu \theta(x). \tag{18.10.5}$$

Similarly, the vector current $j^\mu(x)$, the mass term and its axial counterpart $j_5 = \bar\psi\gamma_5\psi$ can be regularized.

The basic properties of the regularized axial current are

$$\langle 0|j_5^\mu(x;\epsilon)|0\rangle = O\left(\frac{1}{\epsilon}\right), \tag{18.10.6}$$

and, assuming the equations of motion of QED, direct computation yields

$$\langle 0|\partial_\mu j_5^\mu(x;\epsilon)|0\rangle = ie\langle 0|\epsilon^\mu j_5^\nu(x;\epsilon)|0\rangle F_{\mu\nu}(x) + 2im\langle 0|j_5(x;\epsilon)|0\rangle + O(\epsilon)$$
$$= \frac{\alpha}{4\pi}\epsilon^{\mu\nu\rho\sigma}F_{\mu\nu}(x)F_{\rho\sigma}(x) + 2im\langle 0|j_5(x;\epsilon)|0\rangle + O(\epsilon). \tag{18.10.7}$$

Thus Schwinger's regularization leads directly to the anomaly in the divergence of the axial current.

With reference to the canonical-commutator paradox presented above, if we define an average of the regularized current over spatial $\epsilon$'s,

$$j_r(\bar y, t; \delta) = \frac{3}{4\pi\delta^3}\int_{|\bar\epsilon|\le\delta} d^3\bar\epsilon\, j_r(\bar y, t; \bar\epsilon), \tag{18.10.8}$$

we find that actually

$$[j^0(\bar x, t), j^r(\bar y, t; \delta)] = \partial^r\delta^{(3)}(\bar x - \bar y)\frac{c}{3\delta^2}, \tag{18.10.9}$$

for some constant $c$. The term on the right is known as a *Schwinger term*.

In general, regularization of currents leads to violation of the expected commutation relations:

$$\left[j_0^a(\bar x, t), j_r^b(\bar y, t)\right] = if^{abc}j_r^c(\bar x, t)\delta^{(3)}(\bar x - \bar y) + \mathcal{O}_{rs}^{ab}\frac{\partial}{\partial y_s}\delta^{(3)}(\bar x - \bar y), \tag{18.10.10}$$

where $\mathcal{O}_{rs}^{ab}$ is an operator. The divergence of the delta function shows that these Schwinger terms do not invalidate the results of Chapter 3, in which only the spatial integrals of these current-current commutators were computed.

More generally, whenever we have a product of operators evaluated at the same point, it is necessary to regularize. The most widely used technique is to separate the points of evaluation and use Wilson's operator product expansion (OPE), two forms of which follow:

1. **Short Distance Expansion:** As $x \to y$,

$$A(x)B(y) \simeq \sum_n C_n(x - y)\mathcal{O}\left(\frac{x + y}{2}\right); \tag{18.10.11}$$

2. **Lightcone Expansion:** As $x^2 \to 0$,

$$A(x)B(-x) \simeq \sum_n C_n(x)\mathcal{O}_n(x). \tag{18.10.12}$$

It is not at all obvious that the Wilson coefficients $C_n$ and local operators $\mathcal{O}_n$ exist, but there are proofs. The form of the OPE allows one to take higher-order products.

The non-existence of operator products like $\phi^2(x)$ and the resulting paradoxes in current algebra make operator regularization a necessity. The results of regularization of currents are the Schwinger terms in current commutators and the axial anomaly.

The modern approach to operator regularization is Wilson's operator product expansion. Two advantages of the OPE over earlier approaches are its manifest covariance and universality of form. One disadvantage is the abstract nature of the operators in the expansion; they cannot be expressed as polynomials in the fundamental fields since these are the very operator products we are busy regularizing. The character of the operators and coefficients is best revealed by computing the effect of the left-hand side in a Green function. Such an approach is similar to that used in operator renormalization, the topic of the next section.

## 18.11  Renormalization of Composite Operators

In this section, we regularize and renormalize Green functions involving composite operators and deduce some basic principles for renormalization of composite operators.

Currents are simply quadratic composite operators. We saw the value of current matrix elements in predicting hadronic decays and baryon transitions in the second half of Chapter 17. More generally, if $\Omega(x)$ is a product of two fields,

$$\Omega(x) = B_1(x)B_2(x), \qquad (18.11.1)$$

then we can construct the Green functions $G_{\Omega,n}$ by inserting the operator $\Omega$ in $G_n$:

$$G_{\Omega,n}(x; y_1, \ldots, y_n) \overset{\text{def}}{=} \langle 0|T\big(\Omega(x)\phi_1(y_1)\cdots\phi_n(y_n)\big)|0\rangle. \qquad (18.11.2)$$

In order to compute such Green functions, we need to express $\hat{G}_{\Omega,n}$ as a sum of Feynman amplitudes. We can derive the appropriate Feynman rules by separating the points of evaluation of $B_1$ and $B_2$ and relating $\hat{G}_{\Omega,n}$ to an ordinary $(n+2)$-point Green function $G_{2,n}$ defined as follows:

$$G_{2,n}(x_1, x_2; y) \overset{\text{def}}{=} \langle 0|T\big(B_1(x_1)B_2(x_2)\phi_1(y_1)\cdots\phi_n(y_n)\big)|0\rangle. \qquad (18.11.3)$$

Then

$$G_{\Omega,n}(x; y_1, \ldots, y_n) = \int \frac{d^4k_1}{(2\pi)^4}\frac{d^4k_2}{(2\pi)^4}\frac{d^{4n}l}{(2\pi)^{4n}} \, e^{ix\cdot k_1}e^{ix\cdot k_2}e^{iy\cdot l}G_{2,n}(k_1, k_2; l). \qquad (18.11.4)$$

Since we prefer momentum-space Feynman rules, we need to transform this last equation:

$$G_{\Omega,n}(p; q) = \int d^4x\, d^{4n}y \, e^{-ix\cdot p}e^{-iy\cdot q}G_{\Omega,n}(x; y_1, \ldots, y_n). \qquad (18.11.5)$$

Substituting for $G_{\Omega,n}(x; y_1, \ldots, y_n)$, performing the position space integrals to get delta functions, and then performing two momentum-space integrals, when the dust settles we find that:

$$G_{\Omega,n}(p; q) = \int \frac{d^4k}{(2\pi)^4} \, G_{2,n}(k, p-k; q). \qquad (18.11.6)$$

If we regard $G_{2,n}(k, p-k; q)$ as a sum of diagrams, then this integral formula implies that $G_{\Omega,n}$ is a sum of the same diagrams with the $k$ and $p-k$ lines joined to form a loop. Because the momentum does not match at the point of joining, we must treat that point as a vertex for a source which contributes momentum $p$ to the diagram. The outcome is that the ordinary Feynman rules will work if we introduce a new vertex for $\Omega$:

**Diagram for $\Omega$ Insertion:**   $k \overset{\longleftarrow}{\underset{\underset{p}{\uparrow\uparrow}}{\phantom{xx}}} \overset{\longrightarrow}{p - k}$   (18.11.7)

On the level of Green functions, taking $n = 2$ for example, we could represent the relation between $G_{2,n}(k, p-k; q)$ and $G_{\Omega,n}(p; q)$ by the following transformation of diagrams:

$$\qquad\qquad (18.11.8)$$

Notice that there is no $i\lambda$ coupling factor or external propagator factor associated with the operator insertion. Also, the rule for inserting the $\Omega$ vertex does not permit the vertex to form a tadpole.

Clearly, there will be divergent contributions to $G_{\Omega,n}$. Furthermore, since any diagram is a tree diagram in 1PI diagrams, it is also clear that divergent contributions originate in 1PI diagrams, and so for the purposes of operator renormalization, we may work with 1PI Green functions $\Gamma_{\Omega,n}$. To renormalize these functions, we can impose a renormalization condition and then define counterterm insertion operators to mop up divergences.

Consider the case of $\Gamma_{\Omega,2}$ for $\Omega = \frac{1}{2}\phi^2$ in $\phi^4$-theory. Here, the only possible counterterm is a multiple of $\phi^2$, so the renormalized 1PI Green function may be written

$$\Gamma'_{\Omega,2} = z_\Omega \Gamma_{\Omega,2}. \qquad (18.11.9)$$

The extra $z$ factor may be thought of as acting on the operator $\Omega$ itself, effectively renormalizing it:

$$\Omega' \overset{\text{def}}{=} z_\Omega \Omega. \qquad (18.11.10)$$

We can see at once that there is a parallel with Wilson's OPE: $\Omega'$ corresponds to an expansion operator $\mathcal{O}_1(x)$, and the pole in the perturbative expansion of $z_\Omega^{-1}$ corresponds to a delta function in the Wilson coefficient $C_1(x)$ of this operator.

Pursuing the example further, the two lowest-order diagrams which contribute to $\Gamma_{\Omega,2}$ are:

$$\begin{array}{ccc} {}^{q_1}\!\diagdown \\ \phantom{xx}\rangle\!\!=\!\!=\!p \\ {}_{q_2}\!\diagup \end{array} \quad \text{and} \quad \begin{array}{c} {}^{q_1}\!\diagdown \\ \phantom{xx}\rangle\!\!\supset\!\!=\!p \\ {}_{q_2}\!\diagup \end{array} \qquad (18.11.11)$$

If we write $\frac{1}{2}(z_\Omega - 1)\phi^2$ for the counterterm, the value of the tree diagram at zeroeth order in the coupling $\lambda$ is given by

$$i\mathcal{A}_{\text{tree}} = z_\Omega. \qquad (18.11.12)$$

To set the value of $z_\Omega$, we need a renormalization condition. Let us choose

$$\Gamma'_{\Omega,n}(0;q,-q) = 1. \tag{18.11.13}$$

Then, to zeroth order, $z_\Omega = 1$ and the counterterm vanishes. Note that because there is no $i\lambda$ factor associated with the insertion, we are dropping the conventional factor of $i$ in these $\Gamma$s.

To find $z_\Omega$ at first order, we compute the loop contribution:

$$
i\mathcal{A}_{\text{loop}} = (-i\lambda) \int \frac{d^4k}{(2\pi)^4} \frac{i}{k^2 - m^2} \frac{i}{(k-p)^2 - m^2}
$$
$$
= i\lambda \int_0^1 dx \int \frac{d^4k}{(2\pi)^4} \left( (1-x)(k^2 - m^2) + x\big((k-p)^2 - m^2\big) \right)^{-2}.
$$
$$\tag{18.11.14}$$

Simplifying the denominator and using the substitution $l = k - xp$, we are in a position to apply the dimensional regularization integral formulae of Theorem 18.7.7 to find $i\mathcal{A}_{\text{loop}}$ to $O(\delta)$:

$$
i\mathcal{A}_{\text{loop}} = i\lambda \int_0^1 dx \, \frac{i}{16\pi^2} \Gamma(\delta) \left( \frac{4\pi\mu^2}{m^2 - x(1-x)p^2} \right)^\delta \tag{18.11.15}
$$
$$
= -\frac{\lambda}{16\pi^2} \int_0^1 dx \left( \frac{1}{\delta} - \gamma + \ln \frac{4\pi\mu^2}{m^2} - \ln\left(1 - x(1-x)\frac{p^2}{m^2}\right) \right)
$$
$$
= -\frac{\lambda}{16\pi^2} \left( \frac{1}{\delta} - \gamma + \ln \frac{4\pi\mu^2}{m^2} \right) + \frac{\lambda}{16\pi^2} \int_0^1 dx \, \ln\left(1 - x(1-x)\frac{p^2}{m^2}\right).
$$

The renormalization condition implies that the counterterm cancels $i\mathcal{A}_{\text{loop}}$ at $p = 0$. Thus

$$
z_\Omega - 1 = \frac{\lambda}{16\pi^2} \left( \frac{1}{\delta} - \gamma + \ln \frac{4\pi\mu^2}{m^2} \right), \tag{18.11.16}
$$

and so we find concrete first-order expressions for the renormalized operator $\Omega' = z_\Omega\Omega$ and for $\Gamma'_{\Omega,n}$:

$$
\Gamma'_{\Omega,n}(p;q,-q) = \frac{\lambda}{16\pi^2} \int_0^1 dx \, \ln\left(1 - x(1-x)\frac{p^2}{m^2}\right). \tag{18.11.17}
$$

It is quite possible for composite operators to mix under renormalization. In the case of $\Omega = \phi^4$, the counterterms will have the form of $\phi^4$, $\partial_\mu\phi\,\partial^\mu\phi$, and $\phi^2$ insertions, and so the renormalization of $\phi^4$ involves not only rescaling but also subtractions of $\partial_\mu\phi\,\partial^\mu\phi$ and $\phi^2$. In general, if $\Omega_r$ are $n$ operators for which the counterterms $\Delta\Omega_r$ are given by

$$
\Delta\Omega_r = C_r^s \Omega_s, \tag{18.11.18}
$$

then we can form a source Lagrangian density $\mathcal{L}_{\text{src}}$ for these operators,

$$
\mathcal{L}_{\text{src}} = \sum \chi^r (\Omega_r + \Delta\Omega_r), \tag{18.11.19}
$$

and make unitary changes of basis in $\chi$'s and in $\Omega$'s to diagonalize the matrix $1 + C$. In the new basis of $\chi'_r$ and $\Omega'_r$, $1 \le r \le n$, we find that

$$\mathcal{L}_{\text{src}} = \sum z_r \chi'_r \Omega'_r, \tag{18.11.20}$$

so renormalization is again accomplished by simple rescaling in an appropriate basis.

**Homework** 18.11.21. Cheng & Li offer an example of operator mixing. Specify a theory of a neutral scalar $\phi$ interacting with a Fermi field $\psi$ by the interaction Lagrangian density

$$\mathcal{L}' = -\frac{\lambda}{3!}\phi^3 - \frac{\kappa}{4!}\phi^4 - g\bar{\psi}\psi\phi,$$

and choose $\Omega_2 = \bar{\psi}\psi$ and $\Omega_3 = \phi^3$. Find divergent diagrams with insertions of these operators which generate operator mixing.

At this point, we can surmise that the counterterms $\Delta\Omega$ can be determined from a finite set of 1PI Green functions. For example, the counterterms for $\Omega = \phi^4$ do not include any operators $\phi^n$ with $n > 4$, so the only relevant 1PI Green functions are $\Gamma_{\Omega,n}$ for $n \le 4$. More generally, we need only consider 1PI Green functions for which the operator representing the external lines has mass dimension at most the mass dimension of $\Omega$. Taking $\Omega = \frac{1}{2}\phi^2$ and $G_{\Omega,4}$ for example, consider the diagrams

$$\tag{18.11.22}$$

The diagram on the left is convergent, that on the right is divergent, but it only diverges because of a divergent subdiagram (the loop diagram of 18.11.11) which contributes to $G_{\Omega,2}$. Thus the renormalization of $\Omega$ based on $G_{\Omega,2}$ makes $G_{\Omega,4}$ finite.

Note that an operator like $\phi^6$ with mass dimension greater than four is renormalizable as an operator even though it is not a possible interaction in a renormalizable Lagrangian density. The reason for this is that non-renormalizability in a Lagrangian arises from multiple insertions of such operators; a single insertion is no problem. Note also that we have been content here to present examples and generalizations. The theory of renormalization introduced in Chapter 10 will be further developed in Chapter 20.

The concept of operator renormalization is fundamental in quantum field theory. It reveals a weakness in canonical quantization of a Lagrangian field theory and indicates the necessity of renormalization even on the level of the Lagrangian itself. (In functional integral quantization, the same necessity arises from the weakness of the Lebesgue functional measure.) In the next section, we put this new understanding of operators to work in a detailed study of the axial current.

## 18.12 The Abelian Anomaly Revisited

To integrate the material of this chapter, in this section we compute the axial-current anomaly of Theorem 13.4.11 using current algebra, operator insertions, and dimensional regularization.

We begin with the traditional use of current algebra to make a prediction about the divergence of the axial current. Let us define the following $j$'s and Green function:

$$j_5^\mu = \bar\psi\gamma^\mu\gamma_5\psi, \quad j^\mu = \bar\psi\gamma^\mu\psi, \quad j_5 = \bar\psi\gamma_5\psi;$$

$$G_5^{\lambda\mu\nu}(x;y,z) = \langle 0|T\big(j_5^\lambda(x)j^\mu(y)j^\nu(z)\big)|0\rangle; \qquad (18.12.1)$$

$$G_5^{\mu\nu}(x;y,z) = \langle 0|T\big(j_5(x)j^\mu(y)j^\nu(z)\big)|0\rangle.$$

Then, expressing the time-ordered product as a sum of six terms each with two $\theta$ functions, we can take the divergence of $G_5^{\lambda\mu\nu}$ and find

$$\partial_\lambda G_5^{\lambda\mu\nu}(x;y,z) = \langle 0|T\big(\partial_\lambda j_5^\lambda(x)j^\mu(y)j^\nu(z)\big)|0\rangle + \langle\text{four commutators}\rangle. \quad (18.12.2)$$

Now each commutator has the form $\big[j_5^0(x), j^0(y)\big]$. Generally, from the ETCR's we have

$$\big[\psi^\dagger(x)A\psi(x), \psi^\dagger(y)B\psi(y)\big]_{x^0=y^0} = (AB - BA)\delta^{(3)}(\bar x - \bar y), \qquad (18.12.3)$$

and so the current commutators vanish at equal times. However, in the divergence 18.12.2 of $G_5^{\lambda\mu\nu}$, each commutator is multiplied by a delta function of the relevant time difference. Consequently, the four commutator terms vanish. (As each commutator involves only the time components of the currents, there will be no Schwinger terms to disrupt this argument.)

Allowing for a fermion mass term, the divergence of the axial current is given by

$$\begin{aligned}
\partial_\mu j_5^\mu &\overset{?}{=} \bar\psi\gamma^\mu\gamma_5\partial_\mu\psi + \text{h.c.} \\
&= -\bar\psi\gamma_5\slashed{D}\psi + \text{h.c.} \\
&= i\bar\psi\gamma_5(i\slashed{D} - m)\psi + im\bar\psi\gamma_5\psi + \text{h.c.} \\
&= 2imj_5.
\end{aligned} \qquad (18.12.4)$$

Thus we expect

$$\frac{\partial}{\partial x^\lambda}G_5^{\lambda\mu\nu}(x;y,z) \overset{?}{=} 2imG_5^{\mu\nu}(x;y,z). \qquad (18.12.5)$$

An identity which, like this one, relates different Green functions is known as a *Ward identity*. We know from Sections 13.4 and 18.10 that there is an anomaly in this current, so this calculation is flawed at the outset. We will use dimensional regularization to correct it.

We begin the refinement of the calculation by promoting the whole QED Lagrangian density to $D$ dimensions. Now $\gamma_5$ is the same in $D$ dimensions as it is in four. Consequently, the axial transformation is no longer a symmetry of the kinetic term. Further, in order to make the axial current hermitian, it is necessary to modify its definition to

$$\begin{aligned}
j_5^\mu(x;\delta) &\overset{\text{def}}{=} \frac{1}{2}\bar\psi[\gamma^\mu, \gamma_5]\psi \\
&= \bar\psi\bar\gamma^\mu\gamma_5\psi,
\end{aligned} \qquad (18.12.6)$$

where we have used the terminology of Section 18.9, writing

$$\gamma^\mu = \bar{\gamma}^\mu + \hat{\gamma}^\mu \qquad (18.12.7)$$

to distinguish between the first four and the last $D - 4$ components. Now the divergence of the axial current may be written

$$\partial_\mu j_5^\mu(x; \delta) = -(\bar{\psi}\gamma_5 \overleftarrow{\partial}\psi + \text{h.c.}) \qquad (18.12.8)$$

$$= \underbrace{(i\bar{\psi}\gamma_5(i\overleftarrow{\partial} - m)\psi + \text{h.c.})}_{\Omega_{\text{eom}}} + \underbrace{(\bar{\psi}\gamma_5\hat{\partial}\psi + \text{h.c.})}_{\Omega_{\text{anom}}} + \underbrace{2im\bar{\psi}\gamma_5\psi}_{\Omega_{\text{mass}}}.$$

We shall show that the 'equation of motion' or EoM term $\Omega_{\text{eom}}$ does not contribute at the one-loop level and that the 'anomalous' term $\Omega_{\text{anom}}$ gives rise to the standard abelian anomaly. ($\Omega_{\text{anom}}$ could also be called *evanescent*, meaning that it disappears when $D \to 4$.)

In light of the previous section, we see that the Green functions involved in the proposed identity can be computed using the principle of operator insertions. Indeed, they are nothing but operator insertions, and the two one-loop diagrams for insertions of any one of the three terms in 18.12.8 have the same forms:

$$(18.12.9)$$

Each momentum here flows in the direction of the arrow on its Fermi line.

The vertex factor $V_{5L}$ and $V_{5R}$ depends on the operator inserted. For the mass term, it is simply $2im\gamma_5$, but for the others we have:

| Operator | $V_{5L}$ | $V_{5R}$ | |
|---|---|---|---|
| $\Omega_{\text{eom}}$ | $i\gamma_5(\slashed{k} - \slashed{q} - m)$ | $i\gamma_5(-\slashed{k} - \slashed{p} - m)$ | |
| $\Omega_{\text{eom}}^\dagger$ | $i(\slashed{k} + \slashed{p} - m)\gamma_5$ | $i(-\slashed{k} + \slashed{q} - m)\gamma_5$ | $(18.12.10)$ |
| $\Omega_{\text{anom}}$ | $-i\gamma_5(\hat{\slashed{k}} - \hat{\slashed{q}})$ | $-i\gamma_5(-\hat{\slashed{k}} - \hat{\slashed{p}})$ | |
| $\Omega_{\text{anom}}^\dagger$ | $-i(\hat{\slashed{k}} + \hat{\slashed{p}})\gamma_5$ | $-i(-\hat{\slashed{k}} + \hat{\slashed{q}})\gamma_5$ | |

The Feynman integrands for the left and right triangle diagrams are:

$$\mathcal{I}_L = -\text{tr}\left(V_{5L}\frac{i}{\slashed{k} - \slashed{q} - m}\gamma^\nu\frac{i}{\slashed{k} - m}\gamma^\mu\frac{i}{\slashed{k} + \slashed{p} - m}\right);$$

$$\mathcal{I}_R = -\text{tr}\left(V_{5R}\frac{i}{-\slashed{k} - \slashed{p} - m}\gamma^\mu\frac{i}{-\slashed{k} - m}\gamma^\nu\frac{i}{-\slashed{k} + \slashed{q} - m}\right). \qquad (18.12.11)$$

Substituting the value of $V_{5L}$ or $V_{5R}$ for the EoM terms leads to a cancellation between the vertex and one of the Fermi propagators, for example:

$$\mathcal{I}_L^{\text{eom}} = \text{tr}\left(\gamma_5\gamma^\nu\frac{i}{\slashed{k} - m}\gamma^\mu\frac{i}{\slashed{k} + \slashed{p} - m}\right). \qquad (18.12.12)$$

The trace with a $\gamma_5$ yields an $\epsilon$ tensor, and after integration this tensor must have two indices contracted with $p$. Consequently, the amplitude derived from $\mathcal{I}_L^{\text{eom}}$ is zero. Clearly, the same argument applies to all four integrals arising from the EoM terms.

The contribution of the mass term is accounted for in the proposed Ward identity, so there is no need to evaluate it. The departure from the proposed identity is therefore solely due to the anomalous term. We evaluate its contribution next.

Multiplying the Fermi-loop minus into the vertex factor for the anomalous operators and dropping a factor of $i^4$, the two diagrams and the two operators $\Omega_{\text{anom}}$ and $\Omega_{\text{anom}}^\dagger$ generate the following four Feynman integrands:

$$\mathcal{I}_{\Omega L}^{\lambda\mu\nu} = D^{-1}\,\text{tr}\big(\gamma_5(\hat{k} - \hat{q})(k\!\!\!/ - q\!\!\!/ + m)\gamma^\nu(k\!\!\!/ + m)\gamma^\mu(k\!\!\!/ + p\!\!\!/ + m)\big), \qquad (18.12.13)$$

$$\mathcal{I}_{\Omega^\dagger R}^{\lambda\mu\nu} = D^{-1}\,\text{tr}\big((-\hat{k} + \hat{q})\gamma_5(-k\!\!\!/ - p\!\!\!/ + m)\gamma^\mu(-k\!\!\!/ + m)\gamma^\nu(-k\!\!\!/ + q\!\!\!/ + m)\big),$$

$$\mathcal{I}_{\Omega R}^{\lambda\mu\nu} = D^{-1}\,\text{tr}\big(\gamma_5(-\hat{k} - \hat{p})(-k\!\!\!/ - p\!\!\!/ + m)\gamma^\mu(-k\!\!\!/ + m)\gamma^\nu(-k\!\!\!/ + q\!\!\!/ + m)\big),$$

$$\mathcal{I}_{\Omega^\dagger L}^{\lambda\mu\nu} = D^{-1}\,\text{tr}\big((\hat{k} - \hat{p})\gamma_5(k\!\!\!/ - q\!\!\!/ + m)\gamma^\nu(k\!\!\!/ + m)\gamma^\mu(k\!\!\!/ + p\!\!\!/ + m)\big),$$

where $D$ here is the denominator

$$D = (k^2 - m^2)\big((k - q)^2 - m^2\big)\big((k + p)^2 - m^2\big). \qquad (18.12.14)$$

The first pair and the second pair of integrands are actually equal. The trick is to apply the Dirac adjoint inside the trace and then take the complex conjugate. The trace is invariant under this operation, but it reverses the order of the $\gamma$ matrices and also, if we work in the Majorana representation, changes the sign of every $\gamma$ matrix.

The integrands simplify dramatically because of the hatted terms. First, the trace satisfies

$$\text{tr}(\gamma_5\hat{\gamma}^\rho\gamma^\sigma\gamma^\mu\gamma^\nu) = 0, \qquad (18.12.15)$$

and so the $m$'s in the numerator cannot contribute. Second, since the hatted terms vanish as the $\delta \to 0$, the integrals will vanish unless integration produces a $1/\delta$ factor. Clearly, the four-$k$ term in the numerator vanishes because of the $\text{tr}(\gamma_5 \ldots)$, so only the two-$k$ terms contribute. Therefore, if we differentiate with respect to $m^2$, we obtain a convergent integral which vanishes as $\delta \to 0$. Consequently, the integral is independent of $m$ and we can compute at $m = 0$. Third, the momenta $p$ and $q$ are external, so we can assume that

$$\hat{p} = \hat{q} = 0. \qquad (18.12.16)$$

Finally, applying the first principle in the form

$$\text{tr}(\gamma_5\hat{\gamma}^{\mu_1\cdots\mu_4}\gamma^\rho\gamma^\sigma) = -\,\text{tr}(\gamma_5\hat{\gamma}^{\mu_1\cdots\mu_4}\gamma^\sigma\gamma^\rho), \qquad (18.12.17)$$

we can make the numerator truly trivial:

$$
\begin{aligned}
\mathcal{I}_{\Omega L}^{\lambda\mu\nu} + \mathcal{I}_{\Omega\dagger L}^{\lambda\mu\nu} &= D^{-1}\,\mathrm{tr}\big(\gamma_5(2\hat{k})(\slashed{k}-\slashed{q})\gamma^\nu\slashed{k}\gamma^\mu(\slashed{k}+\slashed{p})\big) \\
&= -2D^{-1}\,\mathrm{tr}\big(\gamma_5\hat{k}(\slashed{k}-\slashed{q})\gamma^\nu\slashed{k}(\slashed{k}+\slashed{p})\gamma^\mu\big) \\
&= -2D^{-1}\,\mathrm{tr}\big(\gamma_5\hat{k}(\slashed{k}-\slashed{q})\gamma^\nu\slashed{k}\slashed{p}\gamma^\mu\big) \\
&= 2D^{-1}\,\mathrm{tr}\big(\gamma_5\hat{k}(\slashed{k}-\slashed{q})\slashed{k}\gamma^\nu\slashed{p}\gamma^\mu\big) \\
&= -2D^{-1}\,\mathrm{tr}\big(\gamma_5\hat{k}\slashed{q}\slashed{k}\gamma^\nu\slashed{p}\gamma^\mu\big) \\
&= 2D^{-1}\,\mathrm{tr}\big(\gamma_5\hat{k}\slashed{k}\slashed{q}\gamma^\nu\slashed{p}\gamma^\mu\big) \\
&= -8iD^{-1}(\hat{k})^2\epsilon^{\sigma\nu\rho\mu}q_\sigma p_\rho.
\end{aligned}
\tag{18.12.18}
$$

It remains to perform the integral. We substitute

$$
(\hat{k})^2 = k^\kappa k^\lambda \hat{g}_{\kappa\lambda},
\tag{18.12.19}
$$

and use Feynman parameters to rewrite the denominator

$$
\begin{aligned}
D^{-1} &= 2\int_\Delta dx\,dy\,\big(x(k+p)^2 + y(k-q)^2 + (1-x-y)k^2\big) \\
&= 2\int_\Delta dx\,dy\,\big((k+xp-yq)^2 + x(1-x)p^2 + y(1-y)q^2 + 2xyp\cdot q\big)^{-3} \\
&= 2\int_\Delta dx\,dy\,(l^2 - a)^{-3},
\end{aligned}
\tag{18.12.20}
$$

with the obvious meanings for $l$ and $a$. Remembering a factor of 2 to account for all four integrands in 18.12.13, application of the integration formulae of Theorem 18.7.7 yields an anomaly

$$
\begin{aligned}
I_{\mathrm{anom}} &= 4\int_\Delta dx\,dy\,\frac{ig^{\kappa\lambda}}{32\pi^2}\frac{\Gamma(\delta)}{\Gamma(3)}\left(\frac{4\pi}{a}\right)^\delta \times -8i\hat{g}_{\kappa\lambda}\epsilon^{\sigma\nu\rho\mu}q_\sigma p_\rho \\
&= \int_\Delta dx\,dy\,\frac{(-2\delta)}{\pi^2}\frac{1}{2}\left(\frac{1}{\delta}-\gamma+\ln\frac{4\pi}{a}\right) \times \epsilon^{\sigma\nu\rho\mu}q_\sigma p_\rho \\
&= -\frac{1}{2\pi^2}\epsilon^{\mu\nu\rho\sigma}p_\rho q_\sigma.
\end{aligned}
\tag{18.12.21}
$$

(The ln term should have a $\mu^2$ for dimensional validity. This would come from adding factors of $e$ to the electromagnetic currents.)

Thus, of the three component operators in the divergence of the regularized axial current, only $\Omega_{\mathrm{anom}}$ contributes at one-loop level to the Ward identity. Multiplying its contribution $I_{\mathrm{anom}}$ by an energy-momentum conserving delta function, we find that the proposed Ward identity is modified by the anomaly:

$$
k_\lambda \hat{G}_5^{\lambda\mu\nu}(k;p,q) = 2m\hat{G}_5^{\mu\nu}(k;p,q) - \frac{1}{2\pi^2}\epsilon^{\mu\nu\rho\sigma}p_\rho q_\sigma(2\pi)^4\delta^{(4)}(p+q-k).
\tag{18.12.22}
$$

Actually, Adler and Bardeen showed that there are no higher-order corrections to the anomaly. We can confirm this from Theorem 13.4.11 on the abelian anomaly.

To simplify the presentation, we set $m = 0$. The VEV of $\partial_\lambda j_5^\lambda$ was defined by the fermionic part of the QED action,

$$\langle \partial_\lambda j_5^\lambda(x) \rangle = \left( \frac{\partial}{\partial \psi} \frac{\partial}{\partial \bar{\psi}} \right) \partial_\lambda j_5^\lambda(x) e^{i(\bar{\psi}, i \slashed{D} \psi)}. \tag{18.12.23}$$

Differentiating both sides with respect to the photon field yields

$$i \frac{\delta}{\delta A_\mu(y)} \, i \frac{\delta}{\delta A_\nu(z)} \langle \partial_\lambda j_5^\lambda(x) \rangle = e^2 \left( \frac{\partial}{\partial \psi} \frac{\partial}{\partial \bar{\psi}} \right) \partial_\lambda j_5^\lambda(x) j^\mu(y) j^\nu(z) e^{i(\bar{\psi}, i \slashed{D} \psi)}. \tag{18.12.24}$$

If we now restore the action $S(A)$ arising from the photon kinetic term and integrate over the photon field, then we find that

$$e^2 \partial_\lambda G_5^{\lambda\mu\nu}(x; y, z) = N \int (dA) \, e^{iS(A)} i \frac{\delta}{\delta A_\mu(y)} \, i \frac{\delta}{\delta A_\nu(z)} \langle \partial_\lambda j_5^\lambda(x) \rangle. \tag{18.12.25}$$

To compute the functional derivatives on the right, we use the anomaly formula of Theorem 13.4.11, adjusted by a factor of 2 to account for the changed normalization of the current:

$$\langle \partial_\mu j_5^\mu(x) \rangle = \frac{e^2}{16\pi^2} \epsilon^{\mu\nu\rho\sigma} F_{\mu\nu}(x) F_{\sigma\rho}(x). \tag{18.12.26}$$

Writing $\mathcal{F}$ for the Fourier transform operator, a little computation yields

$$i \frac{\delta}{\delta \hat{A}_\mu(p)} \, i \frac{\delta}{\delta \hat{A}_\nu(q)} \mathcal{F} \left( \langle \partial_\mu j_5^\mu(x) \rangle \right)_k = -\frac{1}{2\pi^2} \epsilon^{\mu\nu\rho\sigma} p_\rho q_\sigma (2\pi)^4 \delta^{(4)}(p + q - k). \tag{18.12.27}$$

Therefore the integrand in 18.12.25 is constant, and the integral cancels against the normalization factor $N$. We conclude as before that

$$k_\lambda \hat{G}_5^{\lambda\mu\nu}(k; p, q) = -\frac{1}{2\pi^2} \epsilon^{\mu\nu\rho\sigma} p_\rho q_\sigma (2\pi)^4 \delta^{(4)}(p + q - k), \tag{18.12.28}$$

only now the result is non-perturbative.

**Homework** 18.12.29. Complete the calculation in 18.12.27.

**Homework** 18.12.30. Adjust the derivation of the abelian anomaly in Section 13.4 to allow for a fermion mass. Use your improved abelian anomaly theorem to show that the improved Ward identity 18.12.22 is true to all orders.

It is remarkable how the same value of the abelian anomaly arises from so many different directions. We discovered it first using functional calculus; in that approach the anomaly arose first from a breakdown of the classical equations of motion, and second from the transformation properties of the functional Jacobian. Underlying both arguments was the need for regularization in fermionic functional calculus. In Section 18.10, it cropped up as a result of Schwinger's regularization of the axial current. Now it has arisen from an evanescent contribution to the divergence of the axial current as computed in $D$ dimensions. The conclusion we draw is that regularization is inescapable, fairly arbitrary, and yet gives rise to physical effects which are independent of the particular regularization scheme chosen.

# 18.13 Summary

This chapter has explained the nature of renormalization and treated its foundation, the concept of regularization, in detail. In brief, renormalization is a system for transforming and combining the divergent Feynman integrals which arise from loop diagrams in order to bring out finite values. Such systems exist for gauge theories in which the coupling constants have non-negative mass dimension, but may not exist for other theories of interacting vector fields.

Assuming the infinitesimal Wick rotation which makes most Minkowski-space integrals converge, the foundation of any renormalization scheme is a regularization procedure which quantifies divergent Feynman integrals as limits of convergent integrals. The three most widely used regularization procedures are Euclidean, propagator, and dimensional regularization. Euclidean regularization is conceptually the simplest one, but it breaks gauge invariance. Propagator regularization is often used in QED, but has not been extended to non-abelian gauge theories. Dimensional regularization has found favor as the most powerful and computationally simple of the three.

The primary practical value of this chapter is the development of dimensional regularization to the point where it is ready for application to any divergent Feynman integral we may encounter. On this foundation, the next chapter begins the process of renormalizing QED.

# Chapter 19

# Renormalization of QED: Three Primitive Divergences

Applying dimensional regularization to divergent diagrams in QED, finding renormalization factors and counterterms, and making predictions at the one-loop level.

## Introduction

With dimensional regularization in hand, it is possible to use a bare Lagrangian density in $4 - 2\delta$ dimensions and avoid divergent integrals altogether by never taking the limit as $\delta \to 0$. The outcome is that the parameters in the bare Lagrangian become finite functions of $\delta$ which diverge in this limit. It is, however, awkward to be continually computing to extract finite predictions from a theory whose parameters are nearly infinite. Following Section 10.8, a more efficient approach to computation is to maintain the separation of finite and divergent coefficients by splitting the Lagrangian density into two parts:

$$\mathcal{L} = \mathcal{L}_{\mathrm{ph}} + \mathcal{L}_{\mathrm{ct}}, \tag{19.0.1}$$

where $\mathcal{L}_{\mathrm{ph}}$ has finite parameters based directly on experimentally determined quantities, and $\mathcal{L}_{\mathrm{ct}}$ is the counterterm Lagrangian density which absorbs all the divergences.

In QED, up to crossings, there are five one-loop diagrams which appear to have divergent integrals:

(19.0.2)

However, the diagram for a three-photon interaction cancels with its crossed version, and the diagram for photon-photon scattering is actually finite. This is just as well, because counterterms for these processes could not be gauge invariant. The three remaining diagrams are called *primitive divergences*. As we shall see in Chapter 20, all other divergent diagrams in QED are related to these so that once these are renormalized, the renormalization spreads automatically to all diagrams.

To lay the foundation for computation, Section 19.1 introduces the counterterm Lagrangian density and Section 19.2 proposes renormalization conditions to fix the counterterm coefficients. Section 19.3 points out a relation between the Feynman integrals for the electron self-energy and the vertex correction and demonstrates the equality $z_e = z_\psi$ between the electric charge and electron field renormalization factors. The next five sections, Sections 19.4 to 19.8, compute the Feynman integrals, counterterms, and renormalized values of the three primitive divergences identified above. The results, of course, confirm the equality of Section 19.3.

The photon internal lines in the primitively divergent diagrams cause infrared divergences. In Sections 19.3 to 19.8, we suppress these divergences by inserting a small mass into the photon propagator. Section 19.9 shows how the infrared divergences should be handled – the low energy contribution of a diagram with electrons in the final state is canceled by contributions from related diagrams with soft photon emission.

Having established sensible procedures for computing higher-order corrections to our old tree-level amplitudes, Section 19.10 indicates the greater accuracy of predictions which include loop contributions. To prepare for the renormalization group of Chapter 21, Section 19.11 uses QED as an example to introduce the minimal subtraction renormalization scheme.

## 19.1 The Counterterm Lagrangian Density

In this section, we shall consider two ways of arriving at a QED Lagrangian density $\mathcal{L}$ which is a sum of physical and counterterm parts. Comparing the two results will show how renormalization can break gauge invariance.

The first approach presents $\mathcal{L}$ as the most general gauge-invariant Lagrangian density together with a gauge-fixing term:

$$\mathcal{L} = -\frac{1}{4}F_{\text{B}\,\mu\nu}F_{\text{B}}^{\mu\nu} + \bar{\psi}_{\text{B}}(i\partial\!\!\!/ - e_{\text{B}}A\!\!\!/_{\text{B}})\psi_{\text{B}} - m_{\text{B}}\bar{\psi}_{\text{B}}\psi_{\text{B}} - \frac{1}{2\alpha_{\text{B}}}(\partial_\mu A_{\text{B}}^\mu)^2. \tag{19.1.1}$$

This total Lagrangian density can be split into physical-parameter and counterterm parts using the equations

$$A_{\text{B}} = z_A^{1/2}A, \quad \psi_{\text{B}} = z_\psi^{1/2}\psi$$

$$m_{\text{B}} = \frac{z_m}{z_\psi}m, \quad e_{\text{B}} = \frac{z_e}{z_\psi z_A^{1/2}}e, \quad \alpha_{\text{B}} = \frac{z_A}{z_\alpha} \tag{19.1.2}$$

to change variables from bare fields and parameters to renormalized fields and physical parameters.

The change of variables is accomplished in two steps, first the fields, second the coupling constants:

$$\begin{aligned}
\mathcal{L} = {}& -\frac{1}{4}z_A F_{\mu\nu}F^{\mu\nu} + z_\psi\bar{\psi}i\partial\!\!\!/\psi - e_{\text{B}}z_\psi z_A^{1/2}\bar{\psi}A\!\!\!/\psi - m_{\text{B}}z_\psi\bar{\psi}\psi - \frac{1}{2\alpha_{\text{B}}}z_A(\partial_\mu A^\mu)^2 \\
= {}& -\frac{1}{4}F_{\mu\nu}F^{\mu\nu} + \bar{\psi}i\partial\!\!\!/\psi - e_{\text{B}}z_\psi z_A^{1/2}\bar{\psi}A\!\!\!/\psi - m_{\text{B}}z_\psi\bar{\psi}\psi - \frac{1}{2\alpha_{\text{B}}}z_A(\partial_\mu A^\mu)^2 \\
& -\frac{1}{4}(z_A - 1)F_{\mu\nu}F^{\mu\nu} + (z_\psi - 1)\bar{\psi}i\partial\!\!\!/\psi \\
= {}& -\frac{1}{4}F_{\mu\nu}F^{\mu\nu} + \bar{\psi}i\partial\!\!\!/\psi - e\bar{\psi}A\!\!\!/\psi - m\bar{\psi}\psi - \frac{1}{2\alpha}(\partial_\mu A^\mu)^2 \\
& -\frac{1}{4}(z_A - 1)F_{\mu\nu}F^{\mu\nu} + (z_\psi - 1)\bar{\psi}i\partial\!\!\!/\psi - (z_e - 1)e\bar{\psi}A\!\!\!/\psi - (z_m - 1)m\bar{\psi}\psi \\
& -\frac{1}{2\alpha}(z_\alpha - 1)(\partial_\mu A^\mu)^2.
\end{aligned} \tag{19.1.3}$$

We can readily identify $\mathcal{L}_{\text{ph}}$ and $\mathcal{L}_{\text{ct}}$ as the first and following lines of this final expression for $\mathcal{L}$.

The second approach to defining the total Lagrangian density $\mathcal{L}$ starts from the physical-parameter Lagrangian density

$$\mathcal{L}_{\text{ph}} = -\frac{1}{4}F_{\mu\nu}F^{\mu\nu} + \bar{\psi}(i\slashed{\partial} - e\slashed{A})\psi - m\bar{\psi}\psi - \frac{1}{2\alpha}(\partial_\mu A^\mu)^2, \tag{19.1.4}$$

and adds the counterterm Lagrangian density in order to provide interactions which can cancel any divergences that may arise from diagrams computed with $\mathcal{L}_{\text{ph}}$. From this viewpoint, we need to consider all possible counterterms with mass-dimension at most four subject only to Lorentz and charge conjugation invariance. This leads to

$$\mathcal{L}_{\text{ct}} = -\frac{A}{4}F_{\mu\nu}F^{\mu\nu} + B\bar{\psi}i\slashed{\partial}\psi - Ce\bar{\psi}\slashed{A}\psi - Dm\bar{\psi}\psi$$
$$- \frac{E}{2\alpha}(\partial_\mu A^\mu)^2 + \frac{F}{2}A_\mu A^\mu + \frac{G}{4}(A_\mu A^\mu)^2. \tag{19.1.5}$$

Comparison of the conclusions of the two approaches to splitting the total Lagrangian density indicates that

$$\left.\begin{aligned} A &= z_A - 1 \\ B &= z_\psi - 1 \\ C &= z_e - 1 \\ D &= z_m - 1 \end{aligned}\right\} \qquad \left.\begin{aligned} E &= z_\alpha - 1 \\ F &= 0 \\ G &= 0 \end{aligned}\right\} \tag{19.1.6}$$

The five equations involving $z$'s may be regarded as defining the $z$'s in terms of the counterterms, but the other two equations have to be verified by actual computation (or abstract argument). The first four equations above are associated with the gauge-invariant part of the Lagrangian density.

The Feynman rules for the counterterms with derivatives follow from the universal rule for interactions with derived fields: for each $\partial_\mu$, multiply the usual vertex factor $-i\lambda/s(D)$ by $-iq_\mu$, where $q$ is the momentum coming in to the vertex along the line associated with the derived field. (This rule emerged from we derived the perturbation expansion from the functional integral form of the generating functional in Section 11.8.)

**Homework 19.1.7.** Verify the following Feynman rule for the $A$ counterterm:

$$-\frac{A}{4}F_{\mu\nu}F^{\mu\nu}: \qquad -iA(q^2 g^{\mu\nu} - q^\mu q^\nu).$$

The discrepancy between the two forms of $\mathcal{L}$ (the $F$ and $G$ terms) indicates the delicacy of renormalization. It is the need to cancel divergences that suggests introducing these terms, but for gauge invariance, we need them to be zero. If either $F$ or $G$ were non-zero, then gauge invariance would be broken by the quantum dynamics of QED, and Faddeev–Popov quantization would not apply to QED. In short, if $F$ or $G$ were non-zero, our current methods would fail to produce a meaningful theory. In fact, QED is strictly renormalizable, and $F$ and $G$ are zero to all orders of perturbation theory. In this chapter, we shall demonstrate that $F$ and $G$ vanish to second order in the coupling constant $e$; in the next, we shall prove that for a gauge theory there always exists a renormalization procedure which preserves gauge invariance.

The transformation from the bare Lagrangian density to the renormalized Lagrangian density is nothing more than a change of variables. This transformation is worthwhile because it makes finite, physical quantities directly available for tree-level computation and makes the divergences in the bare parameters available to cancel divergent loop contributions.

Having organized our Lagrangian density intelligently, we must next specify the renormalization conditions which will determine the counterterms as functions of physical parameters.

## 19.2 Renormalization Conditions

As counterterm renormalization has introduced seven new parameters into the Lagrangian, we must choose seven renormalization conditions to fix their values. (Of the parameters in $\mathcal{L}_{ph}$, $e$ and $m$ are determined by experiment and $\alpha$ is arbitrary.) In this section, we bring forward and modify the material in Chapter 10 in order to frame renormalization conditions appropriate to QED with a massless photon.

In Section 10.14, we saw that, for the massive vector field, transversality of the renormalized propagator $V^{\mu\nu}(q)$ to the momentum $q$ was not automatic; it was a consequence of field and mass renormalization conditions imposed on a scalar self-energy function. Furthermore, the self-energy function $\Pi^{\mu\nu}(q)$ was not transversal even when $q^2 = \mu^2$.

In the massless case, we shall work in Landau gauge, setting $\alpha = 0$ in the photon propagator. In this gauge, the structure of the propagator implies that whatever the value of the self-energy function $\Pi^{\mu\nu}(q)$, the renormalized propagator $V^{\mu\nu}(q)$ will be transversal. In detail, letting

$$\Pi^{\mu\nu}(q) = \Pi_{\mathrm{S}}(q^2)g^{\mu\nu} + \Pi_{\mathrm{M}}(q^2)q^\mu q^\nu,$$

$$\eta = \frac{i}{q^2 + i\epsilon}, \quad P_T = 1 - \frac{qq^\mathsf{T}}{q^2}, \quad \text{and} \quad P = -\eta P_T, \tag{19.2.1}$$

where the subscript M in $\Pi_{\mathrm{M}}$ indicates a coefficient of a non-trivial matrix, we find that

$$\begin{aligned}
V(q) &= P + P(i\Pi)P + \cdots \\
&= P\big(1 + (-i\eta\Pi_{\mathrm{S}}) + (-i\eta\Pi_{\mathrm{S}})^2 + \cdots\big) \\
&= -\eta(1 + i\eta\Pi_{\mathrm{S}})^{-1}P_T \\
&= -i(q^2 - \Pi_{\mathrm{S}} + i\epsilon)^{-1}P_T.
\end{aligned} \tag{19.2.2}$$

Thus, the field and mass renormalization conditions are as usual natural conditions on the scalar self-energy function $\Pi_{\mathrm{S}}$.

For the massless vector field, however, we must also uphold gauge invariance. This means that only gauge-invariant counterterms should be needed. For this purpose, we impose transversality on the self-energy function itself. We deduce at once

$$\textbf{Gauge Condition:} \quad q_\mu \Pi^{\mu\nu}(q) = 0$$

$$\begin{aligned}
\implies \quad & \Pi_{\mathrm{S}} + q^2\Pi_{\mathrm{M}} = 0 \\
\implies \quad & \Pi^{\mu\nu}(q) = \Pi_{\mathrm{S}}(q^2)P_T^{\mu\nu}.
\end{aligned} \tag{19.2.3}$$

Note that we are simply setting a foundation for computation here. We will demonstrate in Chapter 20 that QED is gauge invariant and that this gauge condition is appropriate.

Now we are in a position to specify the renormalization conditions which fix the values of the seven counterterm coefficients. Following Sections 10.10 to 10.14, we find two conditions on both the fermion self-energy function $\Sigma$ and the gauge-field scalar self-energy function $\Pi_S$. A fifth renormalization condition constrains the vertex function $\Gamma$ to uphold the experimental value of the electric charge. The sixth and seventh ones come from the gauge condition on the photon self-energy function $\Pi$. The following table shows which counterterms are determined by which conditions.

|  | **Condition** | **Constrains** |
|---|---|---|
| **Gauge:** | $q_\mu \Pi^{\mu\nu}(q) = 0$ | $A, E, F, G$ |
| **Photon:** | $\Pi_S(0) = 0, \quad \dfrac{\partial \Pi_S}{\partial p^2}\Big|_{p^2=0} = 0$ | $A, E, F, G$ |
| **Electron:** | $\Sigma(m) = 0, \quad \dfrac{\partial \Sigma}{\partial \not{p}}\Big|_{\not{p}=m} = 0$ | $B, D$ |
| **Coupling:** | $\bar{u}(p')\Gamma_\mu(p',p)u(p) = e\bar{u}(p')\gamma_\mu u(p)$ $(\text{at } p=p', \ p^2=m^2)$ | $C$ |

$$(19.2.4)$$

Note that, to differentiate $\Sigma$ with respect to $\not{p}$, we treat $\not{p}$ as a single complex variable and express $p^2$ as $\not{p}^2$. Note further the polarization spinors in the coupling constant condition; as we shall see, it is too much to expect that $\Gamma_\mu$ will be proportional to $\gamma_\mu$.

To use the renormalization conditions, we first expand each counterterm into a formal sum over powers of the coupling constant $e$, second impose the renormalization conditions separately at each order, and third solve the equations iteratively. Although every function in the renormalization conditions ultimately depends on all the counterterms, this approach is in principle simple because at any given order the renormalization conditions are linear in the counterterms at that order.

## 19.3  Charge Renormalization

Before doing detailed computations of the three primitive divergences and the associated counterterms, we will bring out a fundamental relationship between electron self-energy diagrams and vertex correction diagrams. We will introduce this relationship using the amplitudes derived from the relevant primitive divergences, and then establish it to all orders. We shall conclude that $z_e = z_\psi$.

The second-order vertex correction function $\Lambda_F^\mu$ is defined in terms of the primitively-divergent vertex diagram as follows:

$$-ie\Lambda_F^\mu(p',p) \qquad (19.3.1)$$

$$\overset{\text{def}}{=} (-ie)^3 \int \frac{d^4k}{(2\pi)^4} \, \gamma^\nu \, \frac{i}{\not{p}' - \not{k} - m} \gamma^\mu \frac{i}{\not{p} - \not{k} - m} \gamma_\nu \frac{-i}{k^2 - \lambda^2}.$$

Notice that we have introduced a photon mass $\lambda$ in order to remove the low energy divergence from the integral. (We have suppressed the $+i\epsilon$'s to save space. The subscript 'F' indicates a function derived from Feynman diagrams as opposed to one defined in terms of Green functions. In this chapter, subscript 'F' functions will generally be second-order.)

Clearly, this integral bears a striking resemblance to that of the electron self-energy function:

$$-i\Sigma_F(\not p) \overset{\text{def}}{=} (-ie)^2 \int \frac{d^4k}{(2\pi)^4} \gamma^\nu \frac{i}{\not p - \not k - m} \gamma_\nu \frac{-i}{k^2 - \lambda^2}, \qquad (19.3.2)$$

Directly from the integrals for $\Sigma_F$ and $\Lambda_F^\mu$, it appears that differentiating $\Sigma_F$ with respect to the external momentum will transform the integrand for $\Sigma_F$ into the integrand for $\Lambda_F^\mu$. To make this work, we observe that differentiating a Fermi propagator is effectively differentiating an inverse matrix, and the rule for that is

$$\partial(M^{-1}M) = \partial(M^{-1})M + M^{-1}\partial M = 0$$
$$\implies \quad \partial(M^{-1}) = -M^{-1}(\partial M)M^{-1}. \qquad (19.3.3)$$

Application to the Dirac propagator yields

$$\frac{\partial}{\partial p^\mu} \frac{i}{\not p - \not k - m} = -\frac{i}{\not p - \not k - m}(-i\gamma^\mu)\frac{i}{\not p - \not k - m}. \qquad (19.3.4)$$

From the dimensional regularization formulae of Theorem 18.7.7, we see that differentiation with respect to the parameter $a$ commutes with dimensional regularization. We deduce that differentiation with respect to external momenta also commutes with dimensional regularization. By inspection of the integrands in $\Lambda_F^\mu$ and $\Sigma_F$ (19.3.1 and 19.3.2 respectively), we see that the differentiation formula 19.3.4 implies:

**Lemma 19.3.5.** *After dimensional regularization, differentiation of the second-order contribution $\Sigma_F$ to the electron self-energy function with respect to the external momentum $p$ yields the second-order vertex correction function $\Lambda_F^\mu$ at zero photon momentum:*

$$\frac{\partial}{\partial p^\mu}\Sigma_F(\not p; \delta) = -\Lambda_F^\mu(p, p; \delta).$$

The second-order contributions to the electron self-energy function $\Sigma(\not p)$ are:

$$-i\Sigma(\not p) \overset{(2)}{=} -i\Sigma_F(\not p) + iB_2\not p - iD_2 m. \qquad (19.3.6)$$

When we have to compute the right-hand terms, we shall find that they are functions of the dimensional-regularization parameters $\delta$ and $\mu$. For this section, we suppress this dependence. The renormalization conditions on the electron self-energy now imply that the right-hand side and its derivative vanish at $\not p = m$:

$$\Sigma(\not p)\Big|_{\not p=m} = B_2 m - D_2 m \quad \text{and} \quad \frac{\partial}{\partial \not p}\Sigma(\not p)\Big|_{\not p=m} = B_2. \qquad (19.3.7)$$

Similarly, second-order contributions to the vertex correction function $\Gamma^\mu(p', p)$ (derived from third-order contributions to the vertex) yield

$$-ie\Gamma^\mu(p', p) \overset{(2)}{=} -ie\gamma^\mu - ie\Lambda_F(p', p) - ieC_2\gamma^\mu. \qquad (19.3.8)$$

The vertex function renormalization condition yields

$$C_2 \bar{u}(p)\gamma^\mu u(p) = -\bar{u}(p)\Lambda_F^\mu(p,p)u(p), \quad \text{where } p^2 = m^2. \tag{19.3.9}$$

Combining the formulae 19.3.7 and 19.3.9 for the $B_2$ and $C_2$ counterterms with the Feynman integral relationship of Lemma 19.3.5, we find that $C_2 = B_2$:

$$
\begin{aligned}
C_2 \bar{u}(p)\gamma^\mu u(p) &= -\bar{u}(p)\Lambda_F^\mu(p,p)u(p) \\
&= \bar{u}(p)\frac{\partial}{\partial p^\mu}\Sigma_F(\not{p})u(p) \\
&= \bar{u}(p)\gamma^\mu \frac{\partial}{\partial \not{p}}\Sigma_F(\not{p})u(p) \\
&= \frac{\partial}{\partial \not{p}}\Sigma_F(\not{p})\bigg|_{\not{p}=m} \bar{u}(p)\gamma^\mu u(p) \\
&= B_2 \bar{u}(p)\gamma^\mu u(p).
\end{aligned}
\tag{19.3.10}
$$

Consequently, the vertex and electron-field renormalization factors $z_e$ and $z_\psi$ are indeed equal at second order.

This relationship is so simple and inevitable, it seems possible to extract the principles of the second-order argument and thereby structure a proof to all orders. The rest of this section accomplishes this goal.

The fundamental principles in the above argument are:

1. Differentiating the integrand arising from the electron self-energy diagram at second order yields an integrand which arises from the vertex-correction diagram evaluated at zero photon momentum;
2. Dimensional regularization permits us to interchange the momentum integral with the differentiation operator;
3. The simple relationships between $B_2$ and $\Sigma_F$ and between $C_2$ and $\Lambda_F$ hold at all orders.

Clearly, if we can generalize the first principle to all orders, then principles (2) and (3) will bring out the desired conclusion.

The first problem is that, when we differentiate with respect to the external momentum, it is not at first clear which lines in a self-energy diagram are dependent on this momentum. The first lemma resolves this muddle by constraining the choice of loop momenta.

**Lemma 19.3.11.** *In an electron self-energy diagram, loop momenta can be assigned so that the only internal propagators whose momenta depend on the external momentum are those and only those on the Fermi line which passes through the diagram.*

**Proof.** Clearly, if $D$ is an electron self-energy diagram, then the incoming electron line must pass through $D$ and become the outgoing line. Write $\gamma$ for this Fermi line and its vertices. Let $D'$ be the diagram obtained by removing $\gamma$ from $D$, and let $D'_1, \ldots, D'_k$ be the connected components of $D'$. The external lines of $D'_r$ are all photon lines. Since the internal vertices in $D'_r$ provide only overall energy-momentum conservation as a constraint on its external momenta, we can assign arbitrary momenta to all but one external line of $D'_r$ and still solve its vertex delta functions. As internal lines in $D$, these external lines of $D'_r$ can therefore be assigned loop momenta subject to the constraint of overall energy-momentum conservation in $D'_r$; the additional vertex delta functions on $\gamma$ simply serve to determine the momenta on the segments of $\gamma$ itself. Thus the propagators in $\gamma$ and only these propagators depend on the external momentum. $\qquad\square$

As we shall be interested in operations on Feynman integrands, not merely on Feynman integrals, we introduce a notation for these integrands:

**Definition 19.3.12.** Write $\mathcal{I}(D)$ for the Feynman integrand derived from a diagram or sum of diagrams $D$.

From the formula 19.3.4 for the derivative of a Dirac propagator, we see that we can represent such differentiation in a Feynman integrand on the level of the underlying Feynman diagram:

**Lemma 19.3.13.** *Differentiating a fermion propagator in a Feynman integrand* $\mathcal{I}(D)$ *with respect to its momentum corresponds to inserting a vertex with a zero-momentum photon into its line in the diagram* $D$.

When we differentiate an electron self-energy integrand $\mathcal{I}(D)$, according to the previous lemma we shall not obtain the sum of the integrands for all possible zero-energy photon insertions in $D$ but only those associated with the Fermi line which passes through $D$. Thus differentiation of self-energy diagrams will leave many vertex-correction diagrams unaccounted for. The next lemma (which is due to Furry) provides the basis for eliminating the extra diagrams.

**Lemma 19.3.14.** *Let* $D$ *be a QED diagram containing a Fermi loop* $\gamma$ *with an odd number of vertices. Let* $\bar{D}$ *be the same diagram with the arrows on* $\gamma$ *reversed. Then the Feynman integrands* $\mathcal{I}(D)$ *and* $\mathcal{I}(\bar{D})$ *cancel.*

**Proof.** This result is a direct consequence of charge-conjugation invariance in QED. Since charge conjugation effectively changes $e$ to $-e$, charge conjugation changes the sign of the integrand arising from $\gamma$. Since the integrands associated with $\gamma$ and its charge conjugate $\bar{\gamma}$ are opposite, the same is true for the integrands arising from $D$ and $\bar{D}$. $\qquad\square$

**Homework 19.3.15.** Let $\gamma$ be a Fermi loop in QED with an odd number of vertices. Let $\bar{\gamma}$ be the loop with arrows reversed. Verify directly that the Feynman integrands for $\gamma$ and $\bar{\gamma}$ sum to zero.

On the basis of these lemmas, we can now prove a general relationship between electron self-energy and vertex-correction diagrams:

**Lemma 19.3.16.** *Let* $D$ *be an electron self-energy diagram, and let* $\partial D$ *be the set of vertex correction diagrams obtained from* $D$ *by inserting a single vertex with a zero-momentum external photon. Then*

$$e\frac{\partial}{\partial p^\mu}\mathcal{I}(D) = -\mathcal{I}(\partial D),$$

*where* $p$ *is the momentum of the incoming electron.*

**Proof.** Let $p$ be the momentum of the incoming electron in $D$. Let $\gamma$ be the Fermi line which passes through $D$. Solve the vertex delta functions in $D$ in order to place the $p$ dependence in the propagators of $\gamma$ alone. Then differentiating $\mathcal{I}(D)$ gives rise to a sum of terms $\mathcal{I}(D'_r)$ which correspond to the diagrams $D'_r$ obtained from $D$ by inserting a vertex in the $r$th segment of $\gamma$ with external photon at zero momentum. Thus

$$e\frac{\partial}{\partial p^\mu}\mathcal{I}(D) = -\sum_r \mathcal{I}(D'_r). \tag{19.3.17}$$

To prove the lemma, we must show that the contributions of $\partial D \setminus D'$ sum to zero. These contributions are of two types: those diagrams $D'_e$ which have only even Fermi loops, and those diagrams $D'_o$ which have at least one odd Fermi loop. From the previous lemma, we know that the sum of the integrands of the $D'_o$ is zero. For the $D'_e$, consider a Fermi loop $\gamma$ with $2n+1$ vertices. Differentiate $\gamma$ with respect to the loop momentum. The result will be a sum of $2n+1$ diagrams $\gamma'_s$ with an even number of vertices and one external photon at zero momentum. Apply the same operation to the loop $\bar\gamma$ obtained by reversing the arrows in $\gamma$ and sum. Since the integrands for $\gamma$ and $\bar\gamma$ sum to zero, so do the integrands for the $\gamma'$'s and $\bar\gamma'$'s. Clearly, the diagrams in $D'_e$ can be grouped in such a way that this cancellation takes place separately in each group. $\qquad\square$

This last lemma shows that differentiation sets up a correspondence between electron self-energy and vertex correction diagrams which will support the desired identity:

**Theorem 19.3.18.** *Using dimensional regularization on divergent integrals, the relationship between $\Lambda_F^\mu$ and $\Sigma_F$ established in Lemma 19.3.5 holds to all orders:*

$$\frac{\partial}{\partial p^\mu}\Sigma_F(p;\delta) = -\Lambda_F^\mu(p,p;\delta).$$

Consequently, the $B_2(\delta)$ and $C_2(\delta)$ counterterms are also equal to all orders, and we can conclude that:

**Corollary 19.3.19.** $z_e(\delta) = z_\psi(\delta)$ *to all orders.*

The significance of this result is that the renormalization of the electric charge depends only on the photon renormalization factor:

$$e = \frac{z_\psi(\delta)z_A^{1/2}(\delta)}{z_e(\delta)}e_B(\delta) = z_A^{1/2}(\delta)e_B(\delta). \tag{19.3.20}$$

This implies that the simple ratios of observed charges does not require fine-tuning of bare parameters:

**Corollary 19.3.21.** *The ratios of observed charges of fermions are equal to the ratios of their bare charges.*

**Homework** 19.3.22. Does the same result hold for scalar particles?

**Homework** 19.3.23. Derive the formula of Theorem 19.3.18 from the Ward identity

$$q_\mu\Lambda_F^\mu(p',p) = \Sigma_F(p\!\!\!/') - \Sigma_F(p\!\!\!/).$$

**Homework** 19.3.24. Following the proof of Theorem 19.3.18 given in this section, prove the Ward identity of the previous homework. (No integration is needed. First establish the Ward identity at second order, and then bring out the full value of the technique you used.)

In this section, spotting a simple relationship between the second-order contributions to electron self-energy and vertex correction functions $\Sigma_F$ and $\Lambda_F^\mu$, we have constructed a proof that links $\Sigma_F$ and $\Lambda_F^\mu$ at all orders, brought out the implied equality of $B$ and $C$ counterterms, deduced the equality of $z_e$ and $z_\psi$, and finally concluded that charge renormalization is independent of the Fermi fields.

The next five sections take up the detailed work of computing the primitive divergences, counterterms, and renormalization factors.

## 19.4 The Photon Self-Energy Function

The first loop diagram we compute is the second-order correction $\Pi_F^{\mu\nu}(q)$ to the photon propagator given by inserting an electron loop:

$$i\Pi_F^{\mu\nu}(q) = -(-ie)^2 \int \frac{d^4k}{(2\pi)^4} \,\text{Tr}\!\left(\frac{i(\slashed{k}+\slashed{q}+m)}{(k+q)^2-m^2+i\epsilon}\gamma^\mu \frac{i(\slashed{k}+m)}{k^2-m^2+i\epsilon}\gamma^\nu\right). \quad (19.4.1)$$

Following the procedure of Section 18.1, we introduce a Feynman parameter to put the denominator of $\Pi_F^{\mu\nu}$ into standard form:

$$\frac{1}{(k+q)^2-m^2+i\epsilon}\frac{1}{k^2-m^2+i\epsilon} \qquad\qquad (19.4.2)$$

$$= \int_0^1 dx \left(x((k+q)^2-m^2+i\epsilon)+(1-x)(k^2-m^2+i\epsilon)\right)^{-2}$$

$$= \int_0^1 dx \left(k^2+2xk\cdot q+xq^2-m^2+i\epsilon\right)^{-2}$$

$$= \int_0^1 dx \left((k+xq)^2+x(1-x)q^2-m^2+i\epsilon\right)^{-2}.$$

It is convenient to separate the Feynman-parameter integral from the momentum-space integral as follows:

$$\Pi_F^{\mu\nu}(q) = \int_0^1 dx\, I^{\mu\nu}(q,x) \qquad\qquad (19.4.3)$$

$$\text{for}\quad I^{\mu\nu}(q,x) = ie^2 \int \frac{d^4k}{(2\pi)^4} \frac{\text{tr}\big((\slashed{k}+\slashed{q}+m)\gamma^\mu(\slashed{k}+m)\gamma^\nu\big)}{\big((k+xq)^2+x(1-x)q^2-m^2+i\epsilon\big)^2}.$$

The next step is to make the substitution $l = k + xq$.
Writing $N(q,k)$ for the numerator, we work out $N(q, l-xq)$:

$$N^{\mu\nu}(q,k) = 4\big((k+q)^\mu k^\nu + (k+q)^\nu k^\mu - (k+q)\cdot k\, g^{\mu\nu} + m^2 g^{\mu\nu}\big)$$

$$= 4\big(2l^\mu l^\nu + (1-x)(q^\mu l^\nu + q^\nu l^\mu) - 2x(1-x)q^\mu q^\nu \qquad (19.4.4)$$

$$- (l^2 + (1-2x)q\cdot l - x(1-x)q^2)g^{\mu\nu} + m^2 g^{\mu\nu}\big).$$

Since the denominator is a function of $l^2$, we can drop terms linear in $l$ to obtain a reduced numerator, $N'(q, l, x)$:

$$N'^{\mu\nu}(q, l, x) = 8l^\mu l^\nu - 4(l^2 - a)g^{\mu\nu} + 8b^{\mu\nu}, \tag{19.4.5}$$

where

$$a = m^2 - x(1-x)q^2 \quad \text{and} \quad b^{\mu\nu} = x(1-x)(q^2 g^{\mu\nu} - q^\mu q^\nu). \tag{19.4.6}$$

Note that $b$ is proportional to the projection matrix $P_T$.

Now that the Feynman integrand is in standard form, we can use dimensional regularization to define the regularized integral $I(q, x; \delta)$:

$$I^{\mu\nu}(q, x; \delta) = \text{Reg}_D\, ie^2 \int \frac{d^4 l}{(2\pi)^4} \frac{8l^\mu l^\nu - 4(l^2 - a)g^{\mu\nu} + 8b^{\mu\nu}}{(l^2 - a + i\epsilon)^2}. \tag{19.4.7}$$

When we apply the formulae in Section 18.7, we find that the contributions of the first two terms cancel exactly. Inserting a factor of $\mu^\delta$ for each factor of the electric charge $e$ as explained in Section 18.7, the third term yields

$$I^{\mu\nu}(q, x; \delta) = -\frac{e^2 b^{\mu\nu}}{2\pi^2} \Gamma(\delta) \Big(\frac{4\pi\mu^2}{a}\Big)^\delta. \tag{19.4.8}$$

**Remark 19.4.9.** If we had replaced $l^\mu l^\nu$ by $l^2 g^{\mu\nu}/4$ in $N'$ above, the first two terms would not have canceled and the final value would not have satisfied the gauge condition $p_\mu I^{\mu\nu} = 0$. It is therefore necessary to keep the dimension dependence in such substitutions: $l^2 g^{\mu\nu}/D$ would give the same result as we have just obtained.

The integral of $I^{\mu\nu}$ in formula 19.4.8 cannot be performed as it stands. We must use the Laurent expansions of $\Gamma(\delta)$ and $z^\delta$,

$$\begin{aligned}
\Gamma(\delta) z^\delta &= \Big(\frac{1}{\delta} - \gamma\Big)(1 + \delta \ln z) + O(\delta) \\
&= \frac{1}{\delta} - \gamma + \ln z + O(\delta),
\end{aligned} \tag{19.4.10}$$

to find a simpler approximate value of the integrand:

$$I^{\mu\nu}(q, x; \delta) = -\frac{e^2}{2\pi^2} x(1-x) \Big(\frac{1}{\delta} - \gamma + \ln \frac{4\pi\mu^2}{a}\Big) q^2 P_T + O(\delta). \tag{19.4.11}$$

The factor of $q^2 P_T$ in $I^{\mu\nu}$ enables us to define the Feynman-amplitude analogues of $\Pi_S$ and $\Pi_M$. Summarizing our results so far, we state:

**Theorem 19.4.12.** Let $\alpha = e^2/4\pi$. The regularized value $\Pi_F(q; \delta)$ of the Feynman self-energy function for the photon at second order contains a factor of $q^2 P_T$, permitting the definition of functions $\Pi_{FS}$ and $\Pi_{FM}$ by

$$\Pi_F(q^2; \delta) = \Pi_{FS}(q^2; \delta) P_T = -\Pi_{FM}(q^2; \delta) q^2 P_T,$$

where

$$\Pi_{FM}(q^2; \delta) = \frac{\alpha}{3\pi} \Big(\frac{1}{\delta} - \gamma\Big) + \frac{2\alpha}{\pi} \int_0^1 dx\, x(1-x) \ln \frac{4\pi\mu^2}{m^2 - x(1-x)q^2}.$$

This expression shows that $\Pi_F$ is proportional to the transverse projection operator $P_T$. Renormalization can therefore be accomplished using the photon kinetic counterterm; no longitudinal or mass counterterm is needed. This hints at preservation of gauge invariance in renormalized QED.

At low or high energies, we can expand the logarithm as a Taylor series. With the low energy assumption $|q^2| \ll m^2$, we find:

$$\ln \frac{4\pi\mu^2}{m^2 - x(1-x)q^2} = \ln \frac{4\pi\mu^2}{m^2} - \ln\left(1 - x(1-x)\frac{q^2}{m^2}\right) \tag{19.4.13}$$

$$= \ln \frac{4\pi\mu^2}{m^2} + x(1-x)\left(\frac{q^2}{m^2}\right) + \frac{1}{2}x^2(1-x)^2\left(\frac{q^2}{m^2}\right)^2 + \cdots$$

Using the formula

$$\int_0^1 dx\, x^r (1-x)^s = \frac{r!\, s!}{(r+s+1)!}, \tag{19.4.14}$$

we can complete the integration over the Feynman parameter $x$ to obtain the following corollary:

**Corollary 19.4.15.** *Let $\alpha = e^2/4\pi$ and assume $|q^2| \ll m^2$. Then the function $\Pi_{FM}$ of Theorem 19.4.12 has an expansion in powers of $q^2/m^2$:*

$$\Pi_{FM}(q^2;\delta) \stackrel{(\delta^0)}{=} \frac{\alpha}{3\pi}\left(\frac{1}{\delta} - \gamma + \ln \frac{4\pi\mu^2}{m^2}\right) + \frac{\alpha}{15\pi}\left(\frac{q^2}{m^2}\right) + \frac{\alpha}{140\pi}\left(\frac{q^2}{m^2}\right)^2 + \cdots$$

The self-energy is generally most significant as a modification of the propagator in an internal line when the momentum on that line is far off-shell. For this case, we assume that $|q^2| \gg m^2$, and make the approximation

$$\ln \frac{4\pi\mu^2}{m^2 - x(1-x)q^2} \simeq \ln \frac{4\pi\mu^2}{-q^2 - i\epsilon} - \ln\left(x(1-x) + i\epsilon\right) \tag{19.4.16}$$

$$\simeq \ln \frac{4\pi\mu^2}{-q^2 - i\epsilon} - \ln x(1-x).$$

The logarithm is commonly defined with a branch cut along the negative real axis. The $i\epsilon$ in this formula determines which branch of the logarithm to take and enables us to change the sign of $q^2$ if $q$ is time-like:

$$\ln(-q^2 - i\epsilon) = \ln(q^2 + i\epsilon) + \ln(-1 - i\epsilon) = \ln(q^2 + i\epsilon) - \pi i. \tag{19.4.17}$$

Having made this remark, we will drop the $i\epsilon$ in future formulae.

Substituting the high energy approximation 19.4.16 in the formula 19.4.12 for $\Pi_{FM}$ and performing the necessary integration,

$$\int_0^1 dx\, x(1-x) \ln x(1-x) = -\frac{5}{18}, \tag{19.4.18}$$

we obtain a high energy formula for $\Pi_{FM}$:

**Corollary 19.4.19.** *Let $\alpha = e^2/4\pi$ and assume $|q^2| \gg m^2$. Then the function $\Pi_{\mathrm{FM}}$ of Theorem 19.4.12 has an approximation*

$$\Pi_{\mathrm{FM}}(q^2; \delta) = \frac{\alpha}{3\pi}\left(\frac{1}{\delta} - \gamma + \ln\frac{4\pi\mu^2}{-q^2} + \frac{5}{3}\right) + O(\delta).$$

The main points to note at this stage are the existence and definition of the Feynman scalar self-energy function 19.2.3, the value of it as presented in 19.4.12, and the clear identification of the divergence in the photon self-energy function in 19.4.12 and the subsequent corollaries. With all this data on the Feynman self-energy integral for the photon, we can proceed to apply the renormalization conditions, determine the counterterm coefficients, and find the renormalized value of the self-energy.

# 19.5  Gauge and Photon Propagator Renormalization Conditions

Now that we have a formula for the electron loop contribution to the photon propagator, we can apply the gauge and photon renormalization conditions to determine the counterterms at second order.

Using the counterterm Lagrangian density 19.1.5 at second order, we find that the photon self-energy function $\Pi$ is a sum of $\Pi_{\mathrm{F}}$ and the contributions from the kinetic energy counterterm, the gauge-fixing counterterm, and the photon mass counterterm:

$$i\Pi^{\mu\nu}(q) \overset{(2)}{=} \lim_{\delta \to 0}\left(i\Pi_{\mathrm{F}}^{\mu\nu} + iA_2(q^\mu q^\nu - q^2 g^{\mu\nu}) - \frac{i}{\alpha}E_2 q^\mu q^\nu + iF_2 g^{\mu\nu}\right), \qquad (19.5.1)$$

where, on account of dimensional regularization, every term on the right is a function of $\delta$.

Imposing the gauge condition $q_\mu \Pi^{\mu\nu}(q) = 0$ and recalling that there is a factor of $P_{\mathrm{T}}$ in $\Pi_{\mathrm{F}}^{\mu\nu}$, we deduce that

$$\textbf{Gauge Condition:} \quad \frac{i}{\alpha}q^2 E_2 q^\nu - iF_2 q^\nu = 0. \qquad (19.5.2)$$

Since the gauge condition is to be imposed for all momenta, not just on-shell momenta, we conclude that

$$E_2 = F_2 = 0. \qquad (19.5.3)$$

Furthermore, since the gauge condition implies $\Pi_{\mathrm{S}} = -q^2\Pi_{\mathrm{M}}$, the mass and field renormalization conditions $\Pi(0) = \Pi'(0) = 0$ reduce to $\Pi_{\mathrm{M}}(0) = 0$.

Substituting $\Pi = -q^2\Pi_{\mathrm{M}}P_T$ and $\Pi_{\mathrm{F}} = -q^2\Pi_{\mathrm{FM}}P_T$ in the equation 19.5.1, we find the equation

$$\Pi_{\mathrm{M}}(q^2) = \lim_{\delta \to 0}\Pi_{\mathrm{FM}}(q^2; \delta) + A_2(\delta). \qquad (19.5.4)$$

Consequently, suppressing the limit, the two renormalization conditions on the photon self-energy function reduce to the equality

$$A_2(\delta) = -\Pi_{\mathrm{FM}}(0; \delta). \qquad (19.5.5)$$

Eliminating $A_2$ from equation 19.5.4 yields

$$\Pi_M(q^2) = \lim_{\delta \to 0}\big(\Pi_{FM}(q^2; \delta) - \Pi_{FM}(0; \delta)\big). \tag{19.5.6}$$

From the formulae 19.4.12, 19.4.15, and 19.4.19 for $\Pi_F$, we obtain respectively general, low energy, and high energy expressions for the renormalized photon self-energy function:

**Theorem 19.5.7.** *Up to second order in perturbation theory, the renormalized scalar self-energy $\Pi_S$ is given by $\Pi_S(q^2) = -q^2\Pi_M$, where*

$$\Pi_M(q^2) \stackrel{(2)}{=} -\frac{e^2}{2\pi^2}\int_0^1 dx\, x(1-x)\ln\Big(1 - x(1-x)\frac{q^2}{m^2}\Big)$$

$$\stackrel{(2)}{=} \frac{\alpha}{15\pi}\Big(\frac{q^2}{m^2}\Big) + \frac{\alpha}{140\pi}\Big(\frac{q^2}{m^2}\Big)^2 + O\Big(\frac{q^2}{m^2}\Big)^3, \qquad \text{if } |q^2| \ll m^2;$$

$$\stackrel{(2)}{=} -\frac{\alpha}{3\pi}\Big(\ln\frac{-q^2}{m^2} - \frac{5}{3}\Big) + O\Big(\frac{m^2}{|q^2|}\ln\frac{|q^2|}{m^2}\Big), \qquad \text{if } |q^2| \gg m^2. \quad \square$$

Note that in the general case, the argument in the logarithm has a zero at $q^2 = 4m^2$. For larger values of $q^2$ the argument goes negative in the range of integration and the logarithm has an imaginary component. This is a general phenomenon: when energy on a propagator goes high enough to allow pair production, then its self-energy function develops a non-zero imaginary part.

The relationship 19.1.6 shows that the photon field renormalization factor is determined up to second order in perturbation theory by the value of $\Pi_{FM}$ given in 19.4.12:

$$\Pi_{FM}(q^2; \delta) \stackrel{(2)}{=} \frac{e^2}{2\pi^2}\Gamma(\delta)\int_0^1 dx\, x(1-x)\Big(\frac{4\pi\mu^2}{m^2 - x(1-x)q^2}\Big)^\delta$$

$$\implies A(\delta) = -\Pi_{FM}(0; \delta) \stackrel{(2)}{=} -\frac{\alpha}{3\pi}\Gamma(\delta)\Big(\frac{4\pi\mu^2}{m^2}\Big)^\delta \tag{19.5.8}$$

$$\implies z_A(\delta) = 1 + A(\delta) \stackrel{(2)}{=} 1 - \frac{\alpha}{3\pi}\Gamma(\delta)\Big(\frac{4\pi\mu^2}{m^2}\Big)^\delta.$$

From this section, we see how the renormalization conditions determine the counterterms, how gauge invariance is maintained at second order, and how a divergent Feynman integral is made to yield a finite contribution to a Green function. We also see that, though the counterterm $A$ and photon renormalization factor $z_A$ depend on the dimensional regularization parameters $\mu$ and $\delta$, the photon self-energy function is independent of these parameters.

## 19.6  Electron Self-Energy Function

The next primitively divergent diagram we analyze is the second-order correction to the electron self-energy function $\Sigma(\not{p})$. Curiously enough, the one-loop Feynman integral is infrared convergent, but the renormalization conditions make use of its derivative, which is infrared divergent. We shall see this clearly as the computation concludes. Because of this infrared divergence, we introduce a photon mass $\lambda$ as an infrared regulator.

The one-loop Feynman integral which contributes to $\Sigma$ is given by:

$$-i\Sigma_{\mathrm{F}}(\not{p}) \overset{\mathrm{def}}{=} (-ie)^2 \int \frac{d^4k}{(2\pi)^4} \gamma^\nu \frac{i(\not{p} - \not{k} + m)}{(p-k)^2 - m^2} \gamma^\mu \frac{-ig_{\mu\nu}}{k^2 - \lambda^2}, \tag{19.6.1}$$

where the $+i\epsilon$ has been suppressed to save space. Note that $\Sigma(\not{p})$ and $\Sigma_{\mathrm{F}}(\not{p})$, like the electron propagator, are $4 \times 4$ matrices of functions.

Introducing a Feynman parameter, we combine the propagators:

$$\frac{1}{D(p,k,x)} \overset{\mathrm{def}}{=} \frac{1}{(p-k)^2 - m^2} \frac{1}{k^2 - \lambda^2}$$

$$= \int_0^1 dx \left( (k - xp)^2 + x(1-x)p^2 - xm^2 - (1-x)\lambda^2 \right)^{-2}. \tag{19.6.2}$$

Writing

$$a = xm^2 + (1-x)\lambda^2 - x(1-x)p^2 \quad \text{and} \quad l = k - xp, \tag{19.6.3}$$

the denominator reduces to

$$D'(k,l,x) \overset{\mathrm{def}}{=} D(p, l + xp, x) = (l^2 - a)^2. \tag{19.6.4}$$

Using Dirac algebra in $D = 4 - 2\delta$ dimensions, the numerator simplifies to

$$N(p,k) \overset{\mathrm{def}}{=} -2(1-\delta)\not{p} + 2(1-\delta)\not{k} + (4 - 2\delta)m. \tag{19.6.5}$$

Shifting variables from $k$ to $l = k - xp$ and dropping linear terms in $l$ yields a new numerator

$$N'(p,x) \overset{\mathrm{def}}{=} N(p, l + xp) = -2(1-\delta)(1-x)\not{p} + (4 - 2\delta)m. \tag{19.6.6}$$

Applying dimensional regularization to $\Sigma_{\mathrm{F}}(\not{p})$, we obtain the regularized electron self-energy function $\Sigma_{\mathrm{F}}(\not{p}; \delta)$, where $D = 4 - 2\delta$:

$$\Sigma_{\mathrm{F}}(\not{p}; \delta) \overset{\mathrm{def}}{=} \mathrm{Reg}_D - ie^2 \int_0^1 dx \int \frac{d^4l}{(2\pi)^4} \frac{N'(p,x)}{(l^2 - a)^2}. \tag{19.6.7}$$

Substituting these values for the numerator and denominator into the definition of $\Sigma_{\mathrm{F}}(\not{p}; \delta)$ and interchanging the order of integration puts the loop integral into normal form:

$$\Sigma_F(\not p; \delta) = \text{Reg}_D - ie^2 \int_0^1 dx \int \frac{d^4 l}{(2\pi)^4} \frac{N'(p; x)}{(l^2 - a)^2} \tag{19.6.8}$$

$$= \frac{e^2}{16\pi^2} \int_0^1 dx\, N'(p; x) \Gamma(\delta) \left(\frac{4\pi\mu^2}{a}\right)^\delta$$

$$= \frac{\alpha}{4\pi} \int_0^1 dx \left(-2(1-\delta)(1-x)\not p + (4-2\delta)m\right) \left(\frac{1}{\delta} - \gamma + \ln\frac{4\pi\mu^2}{a}\right)$$

$$= \frac{\alpha}{4\pi} \int_0^1 dx \left( (-2(1-x)\not p + 4m) \left(\frac{1}{\delta} - \gamma + \ln\frac{4\pi\mu^2}{a}\right) + (2(1-x)\not p - 2m) \right),$$

where $O(\delta)$ terms have been dropped.

Integrating the simple stuff and extracting the $\mu$ dependence in this formula yields:

**Theorem 19.6.9.** *The regularized value of $\Sigma_F(\not p; \delta)$, the electron self-energy function, is given to order $\delta$ by*

$$\Sigma_F(\not p; \delta) = \frac{\alpha}{4\pi} \left(\frac{1}{\delta} - \gamma + \ln\frac{4\pi\mu^2}{m^2}\right)(-\not p + 4m) + \frac{\alpha}{4\pi}(\not p - 2m)$$

$$+ \frac{\alpha}{4\pi} \int_0^1 dx \left(2(1-x)\not p - 4m\right) \ln\left(x + (1-x)\frac{\lambda^2}{m^2} - x(1-x)\frac{p^2}{m^2}\right). \quad \square$$

Note that the integral converges for $p^2 = m^2$ and $\lambda = 0$; it is the derivative of the integral that suffers infrared divergence.

**Homework** 19.6.10. Show that the integral in Theorem 19.6.9 develops an imaginary part when $p^2 > (m + \lambda)^2$.

The outcome of our computation of the second-order correction to the electron propagator is a clear identification of the ultra-violet divergence. This will permit us to renormalize $\Sigma_F$ and interpret the remaining finite part.

# 19.7 Electron Propagator Renormalization Conditions

The formula for the second-order photon correction to the electron propagator enables us to deduce the consequences of the electron propagator renormalization conditions. First, we bring forward and regularize the electron self-energy formula 19.3.6 and the mass and field renormalization conditions 19.3.7:

$$\Sigma(\not p) \overset{(2)}{=} \lim_{\delta \to 0} \left(\Sigma_F(\not p; \delta) - B_2(\delta)\not p + D_2(\delta)m\right)$$

$$\Sigma(m) = 0 \quad \Longrightarrow \quad \Sigma_F(m; \delta) = B_2(\delta)m - D_2(\delta)m, \tag{19.7.1}$$

$$\text{and} \quad \Sigma'(m) = 0 \quad \Longrightarrow \quad \Sigma_F'(m; \delta) = B_2(\delta).$$

Eliminating the counterterms from the formula for $\Sigma$ yields

$$\Sigma(\not p) \overset{(2)}{=} \lim_{\delta \to 0} \left(\Sigma_F(\not p; \delta) - \Sigma_F(m; \delta) - \Sigma_F'(m; \delta)(\not p - m)\right). \tag{19.7.2}$$

Thus, the value of $\Sigma(\not{p})$ is simply $\Sigma_F(\not{p})$ minus the first two terms in its Taylor series about $\not{p} = m$. Clearly, in the formula 19.6.9 for $\Sigma_F(\not{p})$, the terms of degree 0 or 1 in $\not{p}$ do not contribute to $\Sigma(\not{p})$.

From the formula 19.6.9 for $\Sigma_F$,

$$\Sigma_F(\not{p};\delta) = \frac{\alpha}{4\pi}\left(\frac{1}{\delta} - \gamma + \ln\frac{4\pi\mu^2}{m^2}\right)(-\not{p} + 4m) + \frac{\alpha}{4\pi}(\not{p} - 2m) \tag{19.7.3}$$
$$+ \frac{\alpha}{2\pi}\int_0^1 dx\,((1-x)\not{p} - 2m)\ln\left(x + (1-x)\frac{\lambda^2}{m^2} - x(1-x)\frac{p^2}{m^2}\right),$$

we find that

$$\Sigma_F(m;\delta) = \frac{3m\alpha}{4\pi}\left(\frac{1}{\delta} - \gamma + \ln\frac{4\pi\mu^2}{m^2}\right) - \frac{m\alpha}{4\pi} \tag{19.7.4}$$
$$- \frac{m\alpha}{2\pi}\int_0^1 dx\,(1+x)\ln\left(x^2 + (1-x)\frac{\lambda^2}{m^2}\right),$$

and

$$B_2(\delta) = \Sigma_F'(m;\delta) \tag{19.7.5}$$
$$= -\frac{\alpha}{4\pi}\left(\frac{1}{\delta} - \gamma + \ln\frac{4\pi\mu^2}{m^2}\right) + \frac{\alpha}{4\pi}$$
$$+ \frac{\alpha}{2\pi}\int_0^1 dx\,(1-x)\ln\left(x^2 + (1-x)\frac{\lambda^2}{m^2}\right) + \frac{2m^2(x - x^3)}{x^2m^2 + (1-x)\lambda^2}.$$

The integrals of the logarithms are infrared finite. We can set $\lambda = 0$ in these and obtain values

$$\int_0^1 dx\,\ln x^2 = -2 \quad\text{and}\quad \int_0^1 dx\,x\ln x^2 = -\frac{1}{2}. \tag{19.7.6}$$

The integrals of the fractions are partly infrared convergent. Thus, up to infrared convergent terms, we have:

$$\int_0^1 dx\,\frac{(x - x^3)m^2}{x^2m^2 + (1-x)\lambda^2} = \int_0^1 dx\,\frac{xm^2}{x^2m^2 + (1-x)\lambda^2} - \frac{x^3m^2}{x^2m^2 + (1-x)\lambda^2}$$
$$= \frac{1}{2}\int_0^1 dx\,\frac{2xm^2 - \lambda^2}{x^2m^2 + (1-x)\lambda^2} - 2x$$
$$= \frac{1}{2}\left[\ln(x^2m^2 + (1-x)\lambda^2) - x^2\right]_0^1$$
$$= \frac{1}{2}\ln\frac{m^2}{\lambda^2} - \frac{1}{2}. \tag{19.7.7}$$

Putting the parts together, we find a formula for the renormalized self-energy function $\Sigma(\not{p})$:

**Theorem 19.7.8.** *The renormalized electron self-energy function* $\Sigma(\not{p})$ *is given by*

$$\Sigma(\not{p}) = \frac{\alpha}{2\pi}(\not{p} - m)\left(1 + \ln\frac{\lambda^2}{m^2}\right) + \frac{\alpha}{8\pi}(3\not{p} - 8m)$$

$$+ \frac{\alpha}{2\pi}\int_0^1 dx\,((1-x)\not{p} - 2m)\ln\left(1 - (1-x)\frac{p^2}{m^2}\right)$$

$$= \frac{\alpha}{2\pi}(\not{p} - m)\left(1 + \ln\frac{\lambda^2}{m^2}\right) + \frac{\alpha}{2\pi}(\not{p} - 2m)\frac{m^2 - p^2}{p^2}\ln\frac{m^2 - p^2}{p^2}, \quad \text{if } p^2 \simeq m^2;$$

$$= \frac{\alpha}{2\pi}(\not{p} - m)\left(1 + \ln\frac{\lambda^2}{m^2}\right) + \frac{\alpha}{4\pi}\not{p} + \frac{\alpha}{4\pi}(\not{p} - 4m)\ln\frac{-p^2}{m^2}, \quad \text{if } |p^2| \gg m^2. \quad \square$$

Note that in this approximation $\Sigma'(m)$ is non-zero because we dropped $\lambda$'s from the logarithm.

The divergent influence of the photon mass $\lambda$ in the renormalized electron self-energy function indicates some fundamental oversight in our approach to infrared regularization. We present the corrections in Section 19.9. In general, the first step is to transform our formulae from dependence on a photon mass to dependence on an infrared momentum cut-off. The second step is to add in the contribution of bremsstrahlung diagrams in which the energy of final-state photons is so low that the apparatus cannot resolve them. The outcome is dependence of renormalized amplitudes on an apparatus resolution parameter rather than an artificial infrared regulator.

From the relationship between renormalization factors and counterterms summarized in 19.1.6, from the expression 19.7.5 for $B_2$ and the integral formulae following 19.7.5, we find the following formulae for $B_2(\delta)$ and $z_\psi(\delta)$:

$$z_\psi(\delta) = 1 + B(\delta) \overset{(2)}{=} 1 - \frac{\alpha}{4\pi}\left(\frac{1}{\delta} - \gamma + 4 + \ln\frac{4\pi\mu^2}{m^2} + 4\ln\frac{\lambda}{m}\right). \tag{19.7.9}$$

**Homework 19.7.10.** Show that the electron mass counterterm $D_2$ is given by

$$D_2(\delta) = -\frac{\alpha}{\pi}\left(\frac{1}{\delta} - \gamma + 2 + \ln\frac{4\pi\mu^2}{m^2} + \ln\frac{\lambda}{m}\right).$$

**Homework 19.7.11.** The expression 19.6.1 for $\Sigma_F$ was derived in Feynman gauge. Show that $\Sigma_F$ is not gauge invariant. More precisely, evaluate the contribution of the $k_\mu k_\nu$ term in the general photon propagator when $\Sigma_F$ is evaluated between polarization spinors $\bar{u}(p)$ and $u(p)$. Conclude that change of gauge modifies $z_\psi$ by a constant.

Since both $z_m$ and $z_\psi$ are divergent functions of $\delta$ and dependent on $\mu$, the bare field and bare mass exist only in $4 - 2\delta$ dimensions and are themselves functions of $\delta$ and $\mu$.

## 19.8 The Vertex Correction

The second-order correction to the vertex is determined from the function $\Lambda_F^\mu$ defined in 19.3.1. Manipulating the definition into the usual computational form yields:

$$\Lambda_F^\mu(p',p) = -ie^2 \int \frac{d^4k}{(2\pi)^4} \frac{\gamma^\rho(p\!\!\!/' - k\!\!\!/ + m)\gamma^\mu(p\!\!\!/ - k\!\!\!/ + m)\gamma_\rho}{\left((p'-k)^2 - m^2\right)\left((p-k)^2 - m^2\right)(k^2 - \lambda^2)}. \qquad (19.8.1)$$

A photon mass is necessary here because the integral is infrared divergent. There is significant physics in this diagram, and so it is worthwhile to spend some space in calculating this integral. There is an analytic formula for this integral as it stands, but for our purposes it will be sufficient to perform the calculation under the simplifying assumptions that (1) $p$ and $p'$ are on-shell and (2) $\Lambda_F^\mu$ is evaluated between polarization spinors.

The propagators in the denominator $D(p', p, k)$ may be combined as follows:

$$\frac{1}{D} = 2 \int_0^1 \int_0^{1-x} dx\, dy\, \frac{1}{P^3}, \qquad (19.8.2)$$

where the effective propagator $P$ is defined and simplified as follows:

$$\begin{aligned}
P &\stackrel{\text{def}}{=} x(p'-k)^2 + y(p-k)^2 + (1-x-y)k^2 - (x+y)m^2 - (1-x-y)\lambda^2 \\
&= k^2 - 2xp'\!\cdot\!k - 2yp\!\cdot\!k + xp'^2 + yp^2 - (x+y)m^2 - (1-x-y)\lambda^2 \\
&= (k - xp' - yp)^2 + x(1-x)p'^2 + y(1-y)p^2 - 2xyp'\!\cdot\!p \\
&\quad - (x+y)m^2 - (1-x-y)\lambda^2.
\end{aligned} \qquad (19.8.3)$$

The first term determines the shifted loop variable:

$$l \stackrel{\text{def}}{=} k - xp' - yp. \qquad (19.8.4)$$

Since we are assuming shell conditions on $p$ and $p'$, $p\!\cdot\!p'$ can be conveniently expressed in terms of the photon momentum $q$:

$$-2p\!\cdot\!p' = (p'-p)^2 - 2m^2 = q^2 - 2m^2, \qquad (19.8.5)$$

and the expression for $P$ simplifies to

$$P = l^2 - (x+y)^2 m^2 - (1-x-y)\lambda^2 + xyq^2. \qquad (19.8.6)$$

Defining $a$ by

$$a \stackrel{\text{def}}{=} (x+y)^2 m^2 + (1-x-y)\lambda^2 - xyq^2, \qquad (19.8.7)$$

we finally arrive at the standard form of the denominator:

$$\frac{1}{D} = 2 \int_\Delta dx\, dy\, (l^2 - a)^{-3}. \qquad (19.8.8)$$

Now we must make the same translation of loop variable in the numerator $N^\mu$:

$$N^\mu(p',p,k) \stackrel{\text{def}}{=} \gamma^\rho(\not{p}' - \not{k} + m)\gamma^\mu(\not{p} - \not{k} + m)\gamma_\rho \qquad (19.8.9)$$
$$= \gamma^\rho(-\not{l} + (1-x)\not{p}' - y\not{p} + m)\gamma^\mu(-\not{l} - x\not{p}' + (1-y)\not{p} + m)\gamma_\rho.$$

Since the denominator is a function of $l^2$, we can drop terms linear in $l$ from the numerator to obtain a new numerator given by

$$N'^\mu = \gamma^\rho \not{l}\gamma^\mu \not{l}\gamma_\rho + \gamma^\rho((1-x)\not{p}' - y\not{p} + m)\gamma^\mu(-x\not{p}' + (1-y)\not{p} + m)\gamma_\rho. \quad (19.8.10)$$

The term in $l$ leads to a divergent integral, the other terms lead to a convergent one. It will be convenient to split $N'$ accordingly into a sum of $N'^\mu_{\text{dvgt}}$ and $N'^\mu_{\text{cvgt}}$. Applying the contraction formula 18.9.4 to $N'^\mu_{\text{dvgt}}$ yields

$$N'^\mu_{\text{dvgt}} = \gamma^\rho \not{l}\gamma^\mu \not{l}\gamma_\rho = -2(1-\delta)\not{l}\gamma^\mu \not{l}. \qquad (19.8.11)$$

At this point, the expressions for the denominator and the numerator have been adequately analyzed, and we can proceed to do the momentum integral.

The split of the numerator into $N'^\mu_{\text{dvgt}}$ and $N'^\mu_{\text{cvgt}}$ induces a split of $\Lambda^\mu_F$ into $\Lambda^\mu_{\text{Fdvgt}}$ and $\Lambda^\mu_{\text{Fcvgt}}$. From the $D$-dimensional regularization formulae of Theorem 18.7.7 and the $D$-dimensional trace formula 18.9.4, we find at once that:

$$\Lambda^\mu_{\text{Fdvgt}}(p',p;\delta) = -4(1-\delta)\gamma_\alpha\gamma^\mu\gamma_\beta \int_\Delta dx\,dy\,\text{Reg}_D(-ie^2)\int \frac{d^4l}{(2\pi)^4}\frac{l^\alpha l^\beta}{(l^2-a)^3}$$
$$= -4(1-\delta)\gamma_\alpha\gamma^\mu\gamma_\beta \int_\Delta dx\,dy\,(-ie^2)\frac{ig^{\alpha\beta}}{32\pi^2}\frac{\Gamma(\delta)}{\Gamma(3)}\left(\frac{4\pi\mu^2}{a}\right)^\delta$$
$$= \gamma^\mu\frac{e^2}{16\pi^2}\int_\Delta dx\,dy\,2(1-\delta)^2\Gamma(\delta)\left(\frac{4\pi\mu^2}{a}\right)^\delta \qquad (19.8.12)$$
$$= \gamma^\mu\frac{e^2}{8\pi^2}\int_\Delta dx\,dy\left(\frac{1}{\delta} - \gamma - 2 + \ln\frac{4\pi\mu^2}{a}\right).$$

For the convergent part $\Lambda^\mu_{\text{Fcvgt}}$, we simply use the Euclidean integral formula of Theorem 18.2.11 to find that

$$\Lambda^\mu_{\text{Fcvgt}}(p',p) = 2\int_\Delta dx\,dy(-ie^2)\int \frac{d^4l}{(2\pi)^4}\frac{N'^\mu_{\text{cvgt}}}{(l^2-a)^3}$$
$$= -2\int_\Delta dx\,dy\,(-ie^2)\frac{i}{32\pi^2}\frac{N'^\mu_{\text{cvgt}}}{a} \qquad (19.8.13)$$
$$= -\frac{e^2}{16\pi^2}\int_\Delta dx\,dy\,\frac{N'^\mu_{\text{cvgt}}}{a}.$$

To summarize our work so far, we state a lemma:

**Lemma 19.8.14.** *Taking $p$ and $p'$ on-shell in $\Lambda^\mu_F$ and evaluating the momentum integral in the definition 19.8.1 of $\Lambda^\mu_F$ subject to sandwiching between polarization spinors $\bar{u}(p')$ and $u(p)$ leads to the formula*

$$\Lambda^\mu_F(p',p;\delta) = \gamma^\mu\frac{\alpha}{2\pi}\int_\Delta dx\,dy\left(\frac{1}{\delta} - \gamma - 2 + \ln\left(\frac{4\pi\mu^2}{a}\right)\right) - \frac{\alpha}{4\pi}\int_\Delta dx\,dy\,\frac{N'^\mu_{\text{cvgt}}}{a},$$

where

$$N_{\text{cvgt}}^{\prime\mu} = \gamma^\rho\big((1-x)\slashed{p}' - y\slashed{p} + m\big)\gamma^\mu\big(-x\slashed{p}' + (1-y)\slashed{p} + m\big)\gamma_\rho,$$

and

$$a = (x+y)^2 m^2 + (1-x-y)\lambda^2 - xyq^2.$$

Next, we tackle the integration over Feynman parameters. The first step is to eliminate the excess $\gamma$'s from the numerator $N_{\text{cvgt}}^{\prime\mu}$. The second trivially eliminates one parameter using a scaling trick. Integration over the remaining parameter is then straightforward. We illustrate the steps for the divergent part.

The divergent part $\Lambda_{\text{Fdvgt}}^\mu$ of $\Lambda_{\text{F}}^\mu$ is not actually infrared divergent, so we can set $\lambda = 0$ in this function. Then the integral to complete is essentially

$$I(q^2) \overset{\text{def}}{=} \int_\Delta dx\, dy\; \ln\!\left(\frac{a}{m^2}\right)$$
$$= \int_\Delta dx\, dy\; \ln\big((x+y)^2 + 4xy s_2^2\big), \tag{19.8.15}$$

where we made the substitution

$$-q^2 = 4m^2 \sinh^2 \frac{\theta}{2} \tag{19.8.16}$$

and written $s_2$ for $\sinh(\theta/2)$. Note that this definition of $\theta$ implies that $p \cdot p' = m^2 \cosh\theta$.

We notice that the integrand is a function of a homogeneous quadratic in Feynman parameters. Consequently, we can transform integration over one Feynman parameter into a trivial integration over a scale variable $\alpha$:

$$\int dx\, dy\; f(x,y) = \int dx\, dy\, d\alpha\; \delta(\alpha - x - y) f(x,y)$$
$$= \int dX\, dY\, d\alpha\; \alpha^2 \delta(\alpha - \alpha X - \alpha Y) f(\alpha X, \alpha Y)$$
$$= \int dX\, dY\, d\alpha\; \delta(1 - X - Y)\alpha f(\alpha X, \alpha Y) \tag{19.8.17}$$
$$= \int dX\, d\alpha\; \alpha f(\alpha X, \alpha(1-X)).$$

This trick is particularly useful when $f$ is homogeneous, but may also help in other cases. Note that the substitution $x = \alpha X$, $y = \alpha Y$ sets limits of integration over $X$ and $Y$ in terms of the value of $\alpha$ and the limits of integration over $x$ and $y$. For Feynman parameters with original domain of integration the triangle $\Delta$, we naturally choose to use $\alpha$ in the interval $[0,1]$ so that, in light of the delta function, we can take $\Delta$ for the domain of $X$ and $Y$.

Applying this process to the integral $I$ of 19.8.15 and writing $x$ instead of $X$ for aesthetic reasons, we find

$$I = \int_0^1 dx \int_0^1 d\alpha\; \alpha \ln(\alpha^2) + \alpha \ln\big(1 + 4x(1-x)s_2^2\big)$$
$$= \int_0^1 dx\; -\frac{1}{2} + \frac{1}{2}\ln\big(1 + 4x(1-x)s_2^2\big). \tag{19.8.18}$$

The quadratic in the logarithm has roots

$$x_\pm = \frac{1}{2} \pm \frac{1}{2}\frac{c_2}{s_2}, \tag{19.8.19}$$

where $c_2$ stands for $\cosh(\theta/2)$. Both roots lie outside the range of integration. Factorizing the quadratic leads to a simple expression:

$$
\begin{aligned}
I &= -\frac{1}{2} + \frac{1}{2}\int_0^1 dx \ \ln(x - x_-) + \ln(x_+ - x) + \ln(4s_2^2) \\
&= -\frac{3}{2} + \frac{\theta/2}{\tanh(\theta/2)}.
\end{aligned} \tag{19.8.20}
$$

Multiplying $I$ by $-\gamma^\mu \alpha / 2\pi$ and substituting in the first integral of Lemma 19.8.14, we find:

**Lemma 19.8.21.** *The divergent part $\Lambda^\mu_{\text{Fdvgt}}$ of the vertex correction function $\Lambda^\mu_{\text{F}}$ defined in Lemma 19.8.14 is given by:*

$$\Lambda^\mu_{\text{Fdvgt}}(p', p; \delta) = \gamma^\mu \frac{\alpha}{4\pi}\left(\frac{1}{\delta} - \gamma + 1 + \ln\frac{4\pi\mu^2}{m^2} - \frac{\theta}{\tanh(\theta/2)}\right),$$

*where $\theta$ is determined by $-q^2 = 4m^2 \sinh^2(\theta/2)$.*

The same steps apply to the convergent part $\Lambda^\mu_{\text{Fcvgt}}$ of $\Lambda^\mu_{\text{F}}$. Only, in preparation for these steps, it is necessary to eliminate excess $\gamma$'s from the numerator $N'^\mu_{\text{cvgt}}$. In this case, there is no need to use dimensional regularization or $D$-dimensional Dirac algebra. The contraction formulae alone are good enough and yield

$$N'^\mu_{\text{cvgt}} = (Aq^2 + Bm^2)\gamma^\mu + Cmp^\mu + Dmp'^\mu, \tag{19.8.22}$$

where the coefficients are given by

$$
\begin{aligned}
A &= -2(1-x)(1-y), \\
B &= 4 - 8(x+y) + 2(x+y)^2, \\
C &= 4x - 4xy - 4y^2, \\
D &= 4y - 4xy - 4x^2.
\end{aligned} \tag{19.8.23}
$$

Note that, since the denominator and parameter integral are symmetric in $x$ and $y$, $C$ and $D$ are in effect equal.

Taking advantage of the Gordon relations,

$$
\left.
\begin{aligned}
p^\mu &= \gamma^\mu \not{p} + i\sigma^{\mu\nu}p_\nu \\
p'^\mu &= \not{p}'\gamma^\mu - i\sigma^{\mu\nu}p'_\mu
\end{aligned}
\right\} \tag{19.8.24}
$$

and setting $q = p' - p$ again, we can rewrite $N'^\mu_{\text{cvgt}}$ in a manner which brings out the contribution to the magnetic moment:

$$N'^\mu_{\text{cvgt}} = \left(Aq^2 + (B+C+D)m^2\right)\gamma^\mu - \frac{i}{2}(C+D)m\sigma^{\mu\nu}q_\nu. \tag{19.8.25}$$

Using the expressions for $A$, $B$, $C$, and $D$ above, the scaling trick makes $N'^\mu_{\text{cvgt}}/a$ simple to integrate, and we find:

**Lemma 19.8.26.** *The convergent part $\Lambda^\mu_{\text{Fcvgt}}$ of the vertex correction function $\Lambda^\mu_F$ defined in Lemma 19.8.14 is given by:*

$$\Lambda^\mu_{\text{Fcvgt}}(p',p) = \gamma^\mu \frac{\alpha}{\pi}\left(\theta\coth\theta\left(\ln\frac{\lambda}{m}+1\right) + \frac{1}{2}\left(\frac{\theta}{\sinh\theta}-1\right)+\frac{1}{4}\right)$$
$$- \gamma^\mu\frac{2\alpha}{\pi}\coth\theta\int_0^{\theta/2}d\phi\,\phi\tanh\phi + \frac{i\alpha}{4\pi m}\sigma^{\mu\nu}q_\nu\frac{\theta}{\sinh\theta},$$

*where $q = p' - p$ and $\theta$ is determined by $-q^2 = 4m^2\sinh^2(\theta/2)$.*

The vertex renormalization condition simply sets $C_2$ to cancel any non-zero value of $\Lambda^\mu_F$ evaluated between polarization spinors at zero photon momentum. By inspection of the last two lemmas, we find the following contributions at $q = 0$:

$$\Lambda^\mu_{\text{Fdvgt}}(q=0) = \gamma^\mu\frac{\alpha}{4\pi}\left(\frac{1}{\delta}-\gamma-1+\ln\frac{4\pi\mu^2}{m^2}\right),$$
$$\Lambda^\mu_{\text{Fcvgt}}(q=0) = \gamma^\mu\frac{\alpha}{\pi}\left(\ln\frac{\lambda}{m}+1+\frac{1}{4}\right). \tag{19.8.27}$$

This implies that the vertex renormalization factor $z_e$ is given by

$$z_e(\delta) = 1 + C(\delta)\overset{(2)}{=} 1 - \frac{\alpha}{4\pi}\left(\frac{1}{\delta}-\gamma+4+\ln\frac{4\pi\mu^2}{m^2}+4\ln\frac{\lambda}{m}\right). \tag{19.8.28}$$

Notice that no scalar counterterm can cancel the $\sigma^{\mu\nu}$ contribution to $\Lambda^\mu_{\text{Fcvgt}}$. It is therefore essential to evaluate our renormalization condition at $q = 0$, not merely at $q^2 = 0$. Notice further that this value of $z_e$ agrees with the value of $z_\psi$ obtained in 19.7.9.

Adding the counterterm $C_2\gamma^\mu$ to $\Lambda^\mu_F$ yields the renormalized value $\Lambda^\mu$ of the vertex correction function. Expressing $\Lambda^\mu$ as the sum of contributions $\Lambda^\mu_{\text{dvgt}}$ and $\Lambda^\mu_{\text{cvgt}}$ from the divergent and convergent parts respectively, we have:

$$\Lambda^\mu_{\text{dvgt}}(p',p) = \gamma^\mu\frac{\alpha}{2\pi}\left(1-\frac{\theta/2}{\tanh(\theta/2)}\right), \tag{19.8.29}$$

and

$$\Lambda^\mu_{\text{cvgt}} = \gamma^\mu\frac{\alpha}{2\pi}\left((\theta\coth\theta-1)\left(\ln\frac{\lambda}{m}+1\right)+\frac{1}{2}\left(\frac{\theta}{\sinh\theta}-1\right)\right)$$
$$- \gamma^\mu\frac{2\alpha}{\pi}\coth\theta\int_0^{\theta/2}d\phi\,\phi\tanh\phi+\frac{i\alpha}{4\pi m}\sigma^{\mu\nu}q_\nu\frac{\theta}{\sinh\theta}.$$

The final step in our computation is to extract the low and high energy approximations to $\Lambda^\mu$. For the low energy case, this is simply a matter of taking each product to order $\theta^2 = -q^2/m^2$. At high energy,

$$\frac{1}{4}(e^\theta - 2 + e^{-\theta}) = \sinh^2\frac{\theta}{2} = -\frac{q^2}{4m^2} \quad\Longrightarrow\quad \theta\simeq\ln\frac{-q^2}{m^2}. \tag{19.8.30}$$

With these ideas, it is a simple matter to obtain the following conclusions:

**Theorem 19.8.31.** *Writing $q = p' - p$, the renormalized value $\Lambda^\mu$ of the vertex correction $\Lambda^\mu_F$ is given by*

$$
\Lambda^\mu(p',p) = 
\begin{cases}
\gamma^\mu \dfrac{\alpha}{3\pi} \dfrac{q^2}{m^2}\left(\ln\dfrac{m}{\lambda} - \dfrac{3}{8}\right) + \dfrac{i\alpha}{4\pi m}\sigma^{\mu\nu}q_\nu, & \text{if } -q^2 \ll m^2; \\[3mm]
-\gamma^\mu \dfrac{\alpha}{\pi}\left(\left(\ln\dfrac{-q^2}{m^2} - 1\right)\ln\dfrac{m}{\lambda} + \dfrac{1}{4}\left(\ln\dfrac{-q^2}{m^2}\right)^2\right), & \text{if } -q^2 \gg m^2. \quad \square
\end{cases}
$$

This section has shown in detail how to compute the Feynman integral for $\Lambda^\mu_F$, how to renormalize $\Lambda^\mu_F$ and obtain the vertex correction $\Lambda^\mu$ at second order, and finally, how to find the low and high energy approximations to $\Lambda^\mu$. Because of the persistent infrared divergence, it has been necessary to keep the photon mass $\lambda$ away from zero. This unsatisfactory aspect of our work will be rectified in the following section. After that, we shall comment on the applications of this computation.

## 19.9 The Infrared Divergence

In the previous section, we eliminated the infrared divergence from the vertex correction function $\Lambda^\mu_F$ by inserting a small photon mass $\lambda$. If we are to be consistent, we have two options. First, we could use this mass everywhere, even in the infrared convergent electron self-energy function $\Sigma_F$. This is the approach of massive-photon QED described in Chapter 9. In this case, gauge invariance is restored in the massless limit. Second, we could improve our analysis of infrared divergences.

For the purposes of high energy approximations, the photon mass is negligible, and so the first approach, being simpler, is generally to be preferred. At low energies, however, it is necessary to take the second route. In this section, we outline the relevant theory and steps of computation in the simplest case.

Consider the phenomenon of Compton scattering off a classical source $J^0$ coupled to the electromagnetic potential $A$ by $\mathcal{L}' = eJ_\mu A^\mu$. The first-order diagram for this is:

$$
D^{(2)}_C = \bar{u}(p') \overset{-ie\gamma^0}{\longleftarrow} u(p) \tag{19.9.1}
$$

$$
ie\hat{J}^0(-q)
$$

The associated amplitude and differential cross section are given by

$$
i\mathcal{A}^{(2)}_C = -ie^2\hat{J}^0(-q)\bar{u}(p')\gamma^0 u(p)\frac{1}{q^2 + i\epsilon},
$$
$$
\left(\frac{d\sigma}{d\Omega}\right)^{(4)}_C = \frac{\alpha^2}{q^4}\left|\bar{u}(p')\gamma^0 u(p)\right|^2. \tag{19.9.2}
$$

**Remark 19.9.3.** The spin sum, spin average of this differential cross section yields the Mott formula for scattering of unpolarized electrons:

$$
\left(\frac{d\sigma}{d\Omega}\right)^{(4)}_M = \frac{\alpha^2}{q^4}(4E^2 + q^2). \tag{19.9.4}
$$

If we take the value of the vertex correction from the last section, then we find that in squaring the contributions of the second and fourth order diagrams, we obtain a sixth-order correction to the fourth-order Coulomb differential cross section above:

$$\left(\frac{d\sigma}{d\Omega}\right)^{(6)}_{\mathrm{C}} = \left(\frac{d\sigma}{d\Omega}\right)^{(4)}_{\mathrm{C}} \left(1 - \frac{2\alpha}{\pi} f(q^2) \ln\frac{m}{\lambda}\right), \qquad (19.9.5)$$

where $f(q^2)$ simplifies in the low and high energy regimes as follows:

$$f(q^2) = \begin{cases} \dfrac{-q^2}{3m^2}, & \text{if } -q^2 \ll m^2; \\[2ex] \ln\!\left(\dfrac{-q^2}{m^2}\right) - 1, & \text{if } -q^2 \gg m^2. \end{cases} \qquad (19.9.6)$$

Of course, $q$ is always space-like since the incoming and outgoing momenta $p$ and $p'$ are on the same mass shell. This corrected formula for the differential cross section cannot be correct because it diverges as $\lambda \to 0$.

Regulating the photon propagator with a mass preserves Lorentz invariance at the expense of gauge invariance. It is better for low energy approximations to preserve gauge invariance at the expense of Lorentz invariance and regulate the photon propagator with an infrared momentum cut-off $|\bar{l}| > l_{\min}$.

If we start from the integral 19.8.1 for the second-order vertex correction function $\Lambda_{\mathrm{F}}^{\mu}$, simplify as if it were sandwiched between $\bar{u}(p')$ and $u(p)$ factors, and assume that

$$\lambda \ll l_{\min} \ll m, \qquad (19.9.7)$$

then we find that the correction $\delta\Lambda_{\mathrm{F}}^{\mu}$ to $\Lambda_{\mathrm{F}}^{\mu}$ is given by

$$\delta\Lambda_{\mathrm{F}}^{\mu}(p',p) \stackrel{\text{def}}{=} \Lambda_{\mathrm{F}}^{\mu}(p',p)_{l_{\min}} - \Lambda_{\mathrm{F}}^{\mu}(p',p)_{\lambda}$$

$$\simeq e^2 \gamma^{\mu} \int_{|\bar{l}| < l_{\min}} \frac{d^3\bar{l}}{(2\pi)^3 \, 2\omega(\bar{l})} \, \frac{p' \cdot p}{p \cdot l \, p' \cdot l}. \qquad (19.9.8)$$

Because the momentum cut-off is not covariant, there is some delicacy in the renormalization of the new vertex function. Also, without a photon mass, the infrared singularity in the vertex correction at zero photon momentum makes it impossible to use our vertex renormalization condition. One escape route is to regard the massless theory as a perturbation of the massive theory in which the kinetic term for the electron is not properly normalized. As a result, the Green functions of the massless theory are to be obtained from those of the massive theory by multiplying the external electron lines by factors

$$\sqrt{z_{\psi}} \simeq 1 + \frac{1}{2}(z_{\psi} - 1). \qquad (19.9.9)$$

On the perturbative level, this means using $\delta z_{\psi}/2$ as a counterterm on external electron lines.

To find $\delta z_{\psi}$, we recall from Theorem 19.3.18 that

$$\frac{\partial}{\partial p^{\mu}} \Sigma_{\mathrm{F}}(p;\delta) = -\Lambda_{\mathrm{F}}(p,p;\delta). \qquad (19.9.10)$$

The proof of this equality was purely diagrammatic, and so the conclusion holds with or without a photon mass or infrared cut-off. Furthermore, from the renormalization conditions 19.3.7 on $\Sigma(\not p)$ and the link 19.1.6 between counterterms and $z$'s, we find that

$$z_e(\delta) - 1 = \frac{\partial}{\partial \not p}\Sigma_{\rm F}(\not p; \delta)\bigg|_{\not p = m}. \tag{19.9.11}$$

Consequently, we can deduce the value of $\delta z_\psi$ directly from formula 19.9.8 for $\delta\Lambda_{\rm F}^\mu$:

$$\delta z_\psi = \frac{\partial}{\partial \not p}\delta\Sigma_{\rm F} = -e^2 \int_{|\bar l| < l_{\min}} \frac{d^3\bar l}{(2\pi)^3\, 2\omega(\bar l)} \frac{m^2}{(l\cdot p)^2}. \tag{19.9.12}$$

**Remark 19.9.13.** Note that $\delta z_\psi$ and therefore $z_\psi$ now depend on $p$. Lorentz invariance is broken by the apparatus, and this is reflected in every aspect of the cut-off theory. In practice, this is quite acceptable. In theory, it is rather disturbing that Faddeev–Popov quantization of a gauge theory must be supplemented by an infrared regularization which either breaks gauge invariance with a photon mass or breaks Lorentz invariance with a momentum cut-off.

Adding to $\delta\Lambda_{\rm F}^\mu$ the extra terms coming from $1 + \frac{1}{2}\delta z_\psi$ factors on the external lines of the original Coulomb scattering diagram $D_{\rm C}^{(2)}$ leads to a renormalized correction formula:

$$\delta\Lambda^\mu(p', p) = e^2\gamma^\mu \int_{|\bar l| < l_{\min}} \frac{d^3\bar l}{(2\pi)^3\, 2\omega(\bar l)} \left(\frac{p'\cdot p}{p\cdot l\, p'\cdot l} - \frac{m^2}{2(l\cdot p)^2} - \frac{m^2}{2(l\cdot p')^2}\right), \tag{19.9.14}$$

which, being integrated at low and high energies, yields

$$\delta\Lambda^\mu(p', p) = \begin{cases} \gamma^\mu \dfrac{\alpha}{\pi}\dfrac{-q^2}{3m^2}\left(\ln\dfrac{2l_{\min}}{\lambda} - \dfrac{5}{6}\right), & \text{if } -q^2 \ll m^2; \\[2ex] \gamma^\mu \dfrac{\alpha}{\pi}\left(\ln\dfrac{-q^2}{m^2} - 1\right)\left(\ln\dfrac{m}{\lambda} - \ln\dfrac{E}{l_{\min}}\right), & \text{if } -q^2 \gg m^2; \end{cases} \tag{19.9.15}$$

where $E$ is the energy of the incoming electron.

This change in the vertex correction function leads to the following modification in the infrared divergence in the differential cross section 19.9.5:

$$\left(\frac{d\sigma}{d\Omega}\right)_{\rm C}^{(6)} = \left(\frac{d\sigma}{d\Omega}\right)_{\rm C}^{(4)}\left(1 - \frac{2\alpha}{\pi}f(q^2)\ln\frac{E}{l_{\min}}\right). \tag{19.9.16}$$

The problem with this last result is that when the infrared cut-off is removed, the differential cross section diverges. This happens because we have overlooked the finite resolution of the apparatus and the consequent contribution of bremsstrallung diagrams:

$$D_{\rm B}^{(2)} = \quad\text{+}\quad \tag{19.9.17}$$

If we assume that the energy $l^0$ of the final state photon lies in the range

$$l_{\min} < l^0 < l_{\max}, \tag{19.9.18}$$

then the differential cross section arising from these two diagrams is

$$\left(\frac{d\sigma}{d\Omega}\right)_{\mathrm{B}}^{(6)} = \left(\frac{d\sigma}{d\Omega}\right)_{\mathrm{C}}^{(4)} \left(\frac{2\alpha}{\pi} f(q^2) \ln \frac{l_{\max}}{l_{\min}}\right).$$    (19.9.19)

Adding this correction to our previous result 19.9.16 yields a differential cross section which depends naturally enough on the sensitivity of the apparatus:

$$\left(\frac{d\sigma}{d\Omega}\right)_{\mathrm{C}}^{(6)} = \left(\frac{d\sigma}{d\Omega}\right)_{\mathrm{C}}^{(4)} \left(1 - \frac{2\alpha}{\pi} f(q^2) \ln \frac{E}{l_{\max}}\right).$$    (19.9.20)

**Homework 19.9.21.** Check all the formulae in this section. Good luck!

This section shows how to make a low energy prediction in a theory with ultraviolet and infrared divergences. Miraculously, the infrared blowup is effectively renormalized by the bremsstrallung contribution which slips into any experiment due to the finite resolution of the apparatus. In fact, infrared divergences are never a threat to renormalizability, but they do throw a wrench into the renormalization machinery based on on-shell renormalization conditions. In a theory with massless particles, some renormalization prescription like minimal subtraction is more appropriate.

## 19.10 Lamb Shift and Anomalous Magnetic Moments

In this section, we add the renormalized vertex correction term from the previous section to our final formula for the vertex function from the one before, and we comment on the physical implications of the result.

Two sections back in Theorem 19.8.31, we established the formulae

$$\Lambda^\mu(p', p; \lambda) = \begin{cases} \gamma^\mu \dfrac{\alpha}{3\pi} \dfrac{q^2}{m^2} \left(\ln \dfrac{m}{\lambda} - \dfrac{3}{8}\right) + \dfrac{i\alpha}{4\pi m}\sigma^{\mu\nu} q_\nu, & \text{if } -q^2 \ll m^2; \\[2ex] -\gamma^\mu \dfrac{\alpha}{\pi}\left(\left(\ln \dfrac{-q^2}{m^2} - 1\right)\ln \dfrac{m}{\lambda} + \dfrac{1}{4}\left(\ln \dfrac{-q^2}{m^2}\right)^2\right), & \text{if } -q^2 \gg m^2. \end{cases}$$    (19.10.1)

In the last section, we found that proper treatment of infrared divergences generated a modification $\delta\Lambda_{\mathrm{F}}^\mu$ defined by

$$\delta\Lambda^\mu(p', p; l_{\min}, \lambda) = \begin{cases} -\gamma^\mu \dfrac{\alpha}{3\pi} \dfrac{q^2}{m^2} \left(\ln \dfrac{2l_{\min}}{\lambda} - \dfrac{5}{6}\right), & \text{if } -q^2 \ll m^2; \\[2ex] \gamma^\mu \dfrac{\alpha}{\pi}\left(\ln \dfrac{-q^2}{m^2} - 1\right)\left(\ln \dfrac{m}{\lambda} - \ln \dfrac{E}{l_{\min}}\right), & \text{if } -q^2 \gg m^2, \end{cases}$$    (19.10.2)

where $E$ is the energy of the incoming electron.

Adding the correction to the original yields the correctly regularized and renormalized vertex function

$$\Lambda^\mu(p', p; l_{\min}) = \begin{cases} \gamma^\mu \dfrac{\alpha}{3\pi} \dfrac{q^2}{m^2} \left(\ln \dfrac{m}{2l_{\min}} + \dfrac{5}{6} - \dfrac{3}{8}\right) + \dfrac{i\alpha}{4\pi m}\sigma^{\mu\nu} q_\nu, & \text{if } -q^2 \ll m^2; \\[2ex] -\gamma^\mu \dfrac{\alpha}{\pi}\left(\left(\ln \dfrac{-q^2}{m^2} - 1\right)\ln \dfrac{E}{l_{\min}} + \dfrac{1}{4}\left(\ln \dfrac{-q^2}{m^2}\right)^2\right), & \text{if } -q^2 \gg m^2. \end{cases}$$    (19.10.3)

We should also include the effects of correcting the photon propagator to second order. From Theorem 19.5.7 on the value of the renormalized photon scalar self-energy function $\Pi_S$, we find that

$$-\Pi_M(p^2) = \begin{cases} -\dfrac{\alpha}{15\pi}\dfrac{q^2}{m^2}, & \text{if } -q^2 \ll m^2; \\[2mm] -\dfrac{\alpha}{\pi}\left(\dfrac{1}{3}\ln\dfrac{-q^2}{m^2} - \dfrac{5}{3}\right), & \text{if } -q^2 \gg m^2. \end{cases} \qquad (19.10.4)$$

The renormalized propagator $V(q)$ has the form

$$\begin{aligned} V(q) &= \frac{i}{q^2 - \Pi_S + i\epsilon} P_T \\ &= \frac{i}{q^2(1 + \Pi_M) + i\epsilon} P_T \\ &\simeq \frac{i}{q^2 + i\epsilon} P_T(1 - \Pi_M), \end{aligned} \qquad (19.10.5)$$

so $-\Pi_M$ is the perturbative correction to the propagator.

Thus the effective vertex in Coulomb scattering is given by

$$\textbf{Effective Coulomb Vertex} = e\hat{J}_\mu(-q)\bar{u}(p')\Gamma^\mu_{\text{eff}}(q)u(p), \qquad (19.10.6)$$

where $\Gamma^\mu_{\text{eff}}$ is defined by summing the first-order term $\gamma^\mu$, the final value of the third-order correction $\Lambda^\mu$, and $\gamma^\mu\Pi_S$ at second order. For $-q^2 \ll m^2$, this gives

$$\Gamma^\mu_{\text{eff}} \overset{\text{def}}{=} \gamma^\mu + \gamma^\mu \frac{\alpha}{3\pi}\frac{q^2}{m^2}\left(\ln\frac{m}{2l_{\min}} + \frac{5}{6} - \frac{3}{8} - \frac{1}{5}\right) + \frac{i\alpha}{4\pi m}\sigma^{\mu\nu}q_\nu. \qquad (19.10.7)$$

The effective vertex is commonly written in terms of form factors:

$$\begin{aligned} \Gamma^\mu_{\text{eff}}(q) &= \gamma^\mu F_1(q^2) + \frac{i}{2m}\sigma^{\mu\nu}q_\nu F_2(q^2) \\ &= \frac{1}{2m}(p^\mu + p'^\mu)F_1(q^2) + \frac{i}{2m}\sigma^{\mu\nu}q_\nu\left(F_1(q^2) + F_2(q^2)\right), \end{aligned} \qquad (19.10.8)$$

where the last line follows by the Gordon relations when the effective vertex is evaluated between spinors. The magnetic moment is the coefficient $F_1 + F_2$ on the last line. From the form 19.10.7 of the effective Coulomb vertex,

$$F_1 \propto \frac{\alpha}{3\pi}\frac{q^2}{m^2} \quad \text{for } -q^2 \ll m^2, \qquad (19.10.9)$$

and so does not contribute significantly to the magnetic moment when $q^2 \sim 0$. However, even at $q^2 = 0$, $F_2$ does not vanish; its contribution is known as the *anomalous magnetic moment*. At second order, therefore, we have

$$\textbf{Anomalous Magnetic Moment} = \frac{\alpha}{2\pi}\mu_0, \qquad (19.10.10)$$

where $\mu_0 = 1/2m$ is the Dirac magnetic moment.

**Remark** 19.10.11. Actually, the anomalous magnetic moment of the electron has been calculated to sixth order. The coefficient of $\mu_0$ is predicted to be:

$$a_e^{\text{qed}} = 0.5\left(\frac{\alpha}{\pi}\right) - 0.328\,478\,445\left(\frac{\alpha}{\pi}\right)^2 + 1.183(11)\left(\frac{\alpha}{\pi}\right)^3. \tag{19.10.12}$$

Using $\alpha^{-1} = 137.035987(29)$, we find excellent agreement with experiment:

$$\left.\begin{array}{l} a_e^{\text{th}} = 0.001\,159\,652\,359(282) \\ a_e^{\text{exp}} = 0.001\,159\,652\,410(200) \end{array}\right\} \tag{19.10.13}$$

The theoretical value has been improved by going beyond QED to include the muon loop correction to the photon propagator ($\sim 10^{-12}$), hadronic ($\sim 10^{-12}$), and weak contributions ($\sim 10^{-14}$).

**Remark** 19.10.14. The same computation covers the bulk of the muon anomalous magnetic moment. The QED contribution here is

$$a_\mu^{\text{qed}} = 0.5\left(\frac{\alpha}{\pi}\right) + 0.765\,782\left(\frac{\alpha}{\pi}\right)^2 + 24.45(6)\left(\frac{\alpha}{\pi}\right)^3 + 128.3(71.4)\left(\frac{\alpha}{\pi}\right)^4 \tag{19.10.15}$$

$$= 0.001\,165\,851\,8(24).$$

The hadronic contribution is computed to be

$$a_\mu^{\text{had}} = 66.7(9.4) \times 10^{-9}. \tag{19.10.16}$$

The resulting theoretical value and the experimental value are as follows:

$$\left.\begin{array}{l} a_\mu^{\text{th}} = 0.001\,165\,919(10) \\ a_\mu^{\text{exp}} = 0.001\,165\,922(9) \end{array}\right\} \tag{19.10.17}$$

The weak interactions are expected to contribute on the order of $10^{-9}$.

The Lamb shift is the splitting of the levels $2S_{1/2}$ and $2P_{1/2}$ in the hydrogen atom. Since the electron in the atom is bound, there are some delicate points in applying our results to this case. Since atomic theory is not our focus, we shall be content to summarize the conclusions. Based on the effective Coulomb vertex above, the splitting $\Delta E$ is predicted to be

$$\Delta E^{\text{qed}} = 1052.1 \text{ MHz.} \tag{19.10.18}$$

Taking higher-order corrections and the recoil of the nucleus into account, we have the following predictions:

**Erickson (1971):**     $\Delta E^{\text{th}} = 1057.916(10)$
**Mohr (1975):**     $\Delta E^{\text{th}} = 1057.864(14)$
$\tag{19.10.19}$

Experimental values are in agreement up to the errors specified:

**Lundeen & Pipkin (1975):**     $\Delta E^{\text{exp}} = 1057.893(20)$
**Andrews & Newton (1976):**     $\Delta E^{\text{exp}} = 1057.862(20)$
$\tag{19.10.20}$

In the Introduction, we remarked that the loop diagram for photon-photon scattering actually gives rise to a convergent Feynman integral. With all the crossings possible, there are in fact six such diagrams. Here, we conclude the chapter by giving a low energy approximation for the differential cross section:

$$\frac{d\sigma}{d\Omega} = \frac{\alpha^4}{4\pi^2}\frac{139}{90^2}\left(\frac{\omega}{m}\right)^6\frac{1}{m^2}(3 + \cos^2\theta)^2 \quad \text{for } \omega \ll m. \tag{19.10.21}$$

The absence of tree-level contributions is reflected in the high power of $\alpha$. Photon-photon scattering, the scattering of light by itself, is a purely quantum-mechanical phenomenon.

The computation of the anomalous magnetic moments of the electron and muon, and the computation of the Lamb shift even in QED alone are remarkable successes of quantum field theory. Seeing how well this delicate structure matches nature's performances, we are motivated to study the underlying principles more thoroughly.

# 19.11  Renormalization by Minimal Subtraction

In this section, taking advantage of our familiarity with the $z$'s of QED, we return to the theme of the opening section and comment on the infinite variety of possible renormalization schemes.

The basis for counterterm renormalization are the renormalization conditions; these conditions determine those values of the counterterm coefficients which uphold the physical interpretation of the field and coupling parameters in the Lagrangian density. However, as we saw in Section 10.7, it is quite possible to establish a bare Lagrangian density at a fixed level of perturbation theory with $\delta \neq 0$ and fixed $\mu$, compute amplitudes with it, and add external-line renormalization factors to obtain contributions to Green functions or $S$-matrix elements. Consequently, the split of the bare Lagrangian density into finite and counterterm parts is a matter of convenience.

The approach which we have used so far is known as *mass-shell counterterm renormalization*. It has as its characteristic merit that the kinetic terms are normalized, so the renormalization factors for external lines reduce to unity. It is a good system for obtaining high-precision predictions.

Another possibility is to split the $z$'s into divergent or pole parts and finite parts with respect to $\delta$, and to use the pole parts for counterterm coefficients and add the finite parts back in to the physical parameters. This renormalization scheme is called *minimal subtraction* or simply MS. Thus, in MS renormalization of QED, we define the MS $z$'s as the poles in the on-shell $z$'s,

$$z'_A = 1 - \frac{\alpha}{3\pi}\frac{1}{\delta} \quad \text{and} \quad z'_e = z'_\psi = 1 - \frac{\alpha}{4\pi}\frac{1}{\delta}, \tag{19.11.1}$$

let $z''$'s stand for the finite parts of the on-shell $z$'s,

$$z_A = z'_A + z''_A, \quad z_\psi = z'_\psi + z''_\psi, \quad \text{and} \quad z_e = z'_e + z''_e, \tag{19.11.2}$$

and, assuming that $z_\alpha = 1$ and using $z_e = z_\psi$, use the $z''$'s to split the bare Lagrangian density:

$$\mathcal{L}_B = \mathcal{L}_{\text{fin}} + \mathcal{L}_{\text{div}}, \tag{19.11.3}$$

where

$$\mathcal{L}_{\text{fin}} = -\frac{1}{4}(1 + z''_A)F_{\mu\nu}F^{\mu\nu} \tag{19.11.4}$$

$$+ (1 + z''_\psi)\bar{\psi}(i\slashed{\partial} - e\slashed{A})\psi - (1 + z''_m)m\bar{\psi}\psi + \frac{1}{2\alpha}(\partial_\mu A^\mu)^2,$$

and

$$\mathcal{L}_{\text{div}} = -\frac{1}{4}(z'_A - 1)F_{\mu\nu}F^{\mu\nu} + (z'_\psi - 1)\bar{\psi}(i\partial\!\!\!/ - e A\!\!\!/)\psi - (z'_m - 1)m\bar{\psi}\psi. \quad (19.11.5)$$

The MS renormalization scheme has the advantage that the counterterms are extremely simple; the $z'$'s are independent of masses and $\mu$. This scheme is convenient for theoretical purposes, such as the proof that renormalization preserves gauge invariance of gauge theories (see Chapter 20), and the effects of changing $\mu$ – the renormalization group – (see Chapter 21). It is a good system for getting approximate predictions at high energies. For accuracy, one must however determine

$$\bar{m} = (1 + z''_m)m \quad (19.11.6)$$

by computing $z''_m$, and one must also include the renormalization factors

$$(1 + z''_A)^{-1/2} \quad \text{and} \quad (1 + z''_\psi)^{-1/2} \quad (19.11.7)$$

on the external photon and electron lines.

The MS scheme tends to produce large coefficients in the perturbation expansion. A refinement of the MS scheme, the $\overline{\text{MS}}$ scheme, modifies the $z'$'s to absorb the obvious finite part of the $z$'s and reduce these coefficients. This is accomplished by the simple expedient of substituting $\bar{\mu}$ for $\mu$, where $\bar{\mu}$ is given by

$$\ln \bar{\mu}^2 \stackrel{\text{def}}{=} \ln(4\pi\mu^2) - \gamma. \quad (19.11.8)$$

Clearly, any division of the bare Lagrangian density can be used. It is just a matter of picking the one which makes the task at hand as simple as possible.

## 19.12  Summary

In this chapter, on the basis of the dimensional regularization formulae presented in Chapter 18, we have unfolded the implications of Chapter 10 for renormalization in a useful theory, QED. On one hand, the procedure is automatic:

1. Write out the physical-parameter Lagrangian density;
2. Identify the possible counterterms and form a counterterm Lagrangian density;
3. Use Feynman rules as usual to find Feynman integrals which contribute to 1PI Green functions;
4. Regularize the divergent integrals;
5. Apply the renormalization conditions to fix the counterterm coefficients;
6. Compute renormalized 1PI Green functions;
7. Look for physical consequences of the renormalized 1PI Green functions.

On the other hand, the infrared divergences cause some problems. First, because the only renormalizable counterterm is structured around $\gamma^\mu$ not $\sigma^{\mu\nu}$, in order to find counterterms for the Feynman vertex function, it is necessary to eliminate the magnetic moment by setting the photon momentum $q$ to zero in the renormalization condition. Second, because of the infrared divergence of the Feynman vertex function, it is necessary to use either a photon mass or a momentum cut-off to regulate the low energy contributions to the integral. The photon mass upsets gauge invariance and distorts the relation $z_e = z_\psi$. The momentum cut-off makes the renormalization point inaccessible.

Fundamentally, infrared divergences are harmless. They are always eliminated when we estimate the finite resolution of the apparatus and compute the inclusive differential cross section. In QED, this means adding to the main process any indistinguishable bremsstrahlung processes.

It takes a little intelligence to work around these obstacles and obtain appropriately renormalized vertex corrections. Recognizing that the mass regularization had in effect distorted the field renormalization and therefore also the normalization of the kinetic term in the Lagrangian density, we found it necessary to return to the principles of Chapter 10 and introduce corrections on external electron lines in order to bring the vertex from the mass regularization to the momentum regularization. The triumphant outcome of matching prediction to experiment for the electron magnetic moment is a strong indication that this devious procedure for computing higher-order corrections is valid.

Having thus established renormalization on a practical footing, we turn in the next chapter to the theory of renormalization. In particular, we explain how to reduce the divergences of any diagram, no matter how complex, to the three primitive divergences described here, thereby extending our practical technique for computing higher-order contributions to the whole range of Feynman diagrams.

# Chapter 20

# Renormalization and Preservation of Symmetries

Developing the theory of renormalization in order to extend renormalization from primitive divergences to all diagrams and demonstrate preservation of global and local symmetries.

## Introduction

Chapter 4 opened the topic of renormalization, indicating the need for counterterms to balance divergent contributions to the scalar self-energy. There we also noted that interactions of mass dimension greater than four would generate an infinite tower of counterterms. Chapter 10 provided a foundation for renormalization in the theory of Green functions, and gave examples of the use of renormalization conditions in the computation of first-order corrections. Chapter 19 discussed the renormalization of the primitive divergences of QED in detail, introducing the use of a photon mass as an infrared regulator. In this chapter, we extend the theoretical foundation of renormalization and in particular justify the approach taken in Chapter 19.

Sections 20.1 and 20.2 present the renormalization procedure of Bogoliubov, Parasiuk, Hepp, and Zimmermann (BPHZ). This procedure is effectively the counterterm renormalization technique expressed in terms of Taylor series. It extends the renormalization techniques of the last chapter to all diagrams. Section 20.3 concludes the first unit of the chapter with some renormalization theorems which show that BPHZ renormalization preserves global symmetries.

At this point in the chapter, we still have no guarantee that renormalization will preserve a local symmetry. In the next two units, we demonstrate perturbative preservation of gauge invariance for first abelian and second non-abelian gauge theories.

Section 20.4 uses the effective interaction $\Gamma$ to show how the gauge invariance of QED is equivalent to a relation on the generator of 1PI Green functions, the Ward–Takahashi identity $\delta\bar{\Gamma} = 0$. Section 20.5 provides an interpretation of $\Gamma$ as the effective action for the classical field theory derived from a quantum field. Section 20.6 unfolds the identity $\delta\bar{\Gamma} = 0$ into perturbative Ward–Takahashi identities. In particular, this section establishes the consistency of the gauge condition $q_\mu \Pi^{\mu\nu}$ used in Chapter 19. Section 20.7 shows that counterterms generated by dimensional regularization and minimal subtraction preserve the gauge invariance of QED. Here we also discover that the photon-mass and gauge-fixing parameters $\lambda$ and $\alpha$ are not modified by radiative corrections, thus demonstrating the validity of the approach to computation used in Chapter 19.

Section 20.8 uses the Becchi–Rouet–Stora (BRS) symmetry of a Faddeev–Popov action to prove the Slavnov–Taylor identity for non-abelian gauge theories. Section 20.9 uses this identity to show that there is a way of organizing the counterterms which makes the renormalized action equal to the original action up to a rescaling of fields and coupling constants. This establishes perturbative gauge invariance of the renormalized theory. Since we know from Chapter 13 that gauge currents can be anomalous, the argument rests on a hidden assumption that the BRS current is conserved.

# 20.1 Superficial Degree of Divergence

In this section, we prepare for the analysis of renormalization and global symmetry by developing a simple formula for the mass dimension of a Feynman integral. The following section describes how this formula can be used to systematically renormalize any divergent Feynman integral. With this procedure of renormalization in hand, we can then summarize the conclusions on renormalization and preservation of global symmetries.

As we have seen from our loop examples, it is easy to determine the mass dimension of a Feynman integral directly from the corresponding Feynman diagram. The mass dimension of the integrand in a Feynman integral comes directly from the propagators, and the dimension of the integral operator comes from the internal lines.

The propagator dimension is the asymptotic degree of the propagator as the momentum goes to infinity, and these dimensions are as follows:

**Scalar:**  $p_s = -2;$                                                (20.1.1)

**Fermi:**  $p_f = -1;$

**Vector:**  $p_v = \begin{cases} 0 & \text{massive vector field in general,} \\ -2 & \text{massive vector field coupled to conserved current,} \\ -2 & \text{gauge field.} \end{cases}$

The two cases of the massive vector field arise because, when a vector field is coupled to a conserved current, the $k^\mu k^\nu$ term in the propagator can be omitted. In light of the foregoing remarks on divergences, the mass dimension of a Feynman integral is reasonably enough called the *superficial degree of divergence*. We aim to find a convenient formula for this number.

Suppose that the theory has a variety of interactions labeled by the index $i$, and that the $i$th interaction has $s_i$ scalar, $f_i$ fermion, $v_i$ vector fields, and $d_i$ derivatives. Write $c_i$ for the coupling constant in the Lagrangian density term for the $i$th interaction. Then the dimension of the coupling constant $c_i$ is given by:

$$\delta(c_i) = 4 - s_i - \tfrac{3}{2}f_i - v_i - d_i. \qquad (20.1.2)$$

For a specific Feynman diagram $D$, let $V_i$ be the number of vertices of type $i$ in the diagram, let $I_s$, $I_f$, and $I_v$ be the number of internal scalar, Fermi, and vector lines respectively, and let $E_s$, $E_f$, and $E_v$ be the number of external scalar, Fermi, and vector lines in the diagram. Then, since each internal line has two ends at vertices while each external line has only one, we can equate the number of line ends with the number of lines available at the vertices for each of the three line types:

$$\left. \begin{aligned} E_s + 2I_s &= \sum V_i s_i \\ E_f + 2I_f &= \sum V_i f_i \\ E_v + 2I_v &= \sum V_i v_i \end{aligned} \right\} \qquad (20.1.3)$$

Recalling that each derivative in a vertex adds a factor of momentum to the numerator, it is now a simple matter to write down and manipulate a formula for the superficial degree of divergence $\delta(D)$ for the Feynman integral $I(D)$:

$$\delta(D) = 4L + I_s p_s + I_f p_f + I_v p_v + \sum V_i d_i \qquad (20.1.4)$$

$$= 4(1 + I_s + I_f + I_v - \sum V_i) + I_s p_s + I_f p_f + I_v p_v + \sum V_i d_i$$

$$= 4 - \sum V_i(4 - d_i) + \tfrac{1}{2}(4 + p_s)(\sum V_i s_i - E_s)$$

$$+ \tfrac{1}{2}(4 + p_f)(\sum V_i f_i - E_f) + \tfrac{1}{2}(4 + p_v)(\sum V_i v_i - E_v)$$

$$= 4 - \tfrac{1}{2}\big((4 + p_s)E_s + (4 + p_f)E_f + (4 + p_v)E_v\big)$$

$$- \sum V_i\Big(4 - \tfrac{1}{2}\big((4 + p_s)s_i + (4 + p_f)f_i + (4 + p_v)v_i\big) - d_i\Big).$$

Superficial divergence and convergence correspond respectively to $\delta(D)$ positive and negative. The vanishing of $\delta(D)$ corresponds to superficial logarithmic divergence.

The first part in this expression for the superficial degree of divergence depends only on the external lines, and will be the same for all diagrams which contribute to the same scattering or decay process. The second part indicates how the superficial degree of divergence is affected when we increase the number of vertices while holding the numbers $E_s$, $E_f$, and $E_v$ of external lines fixed. The remarkable thing about the second part is that the contributions from the different vertices are entirely independent.

Thus a single vertex $i$ for which $\delta(c_i) < 0$ inevitably gives rise to infinitely many diagrams with $\delta(D) > 0$. In this case, there are in fact no miraculous cancellations, and infinitely many different types of counterterm are needed to absorb the tower of divergences. For example, in Section 4.14 we saw how a $\phi^5$ term lead to a breakdown of renormalizability in a scalar field theory.

There are now various cases depending on the nature of the vector fields. If we have a massive vector field which is not coupled to a conserved current, then we find:

$$\delta(D) = (4 - E_s - \tfrac{3}{2}E_f - 2E_v) - \sum V_i(4 - s_i - \tfrac{3}{2}f_i - 2v_i - d_i). \qquad (20.1.5)$$

Lorentz invariance implies that a vector field must couple to either a derivative of a scalar field, or to a vector field, or to a pair of Fermi fields, or to a product of more fields than this. Only in the first two cases will the resulting interaction make a non-positive contribution to $\delta(D)$. Thus a renormalizable theory with a massive vector field which is not coupled to a conserved current has only quadratic interactions and is effectively a theory of free fields.

If we have only gauge fields or massive vector fields which are coupled to conserved currents, then we find instead:

$$\delta(D) = (4 - E_s - \tfrac{3}{2}E_f - E_v) - \sum V_i(4 - s_i - \tfrac{3}{2}f_i - v_i - d_i). \qquad (20.1.6)$$

Restated, this becomes the main result of the section:

**Theorem 20.1.7.** *Assume that we are working in either a gauge theory or a theory of a massive vector field coupled to a conserved current. Let $D$ be a Feynman* diagram in which the coupling constant $c_i$ occurs at $V_i$ vertices, and let $c_D$ be a coupling constant appropriate to a Faddeev–Popov Lagrangian term whose vertex has the external lines of $D$. Then the superficial degree of divergence $\delta(D)$ of $D$ is related to the mass dimensions $\delta(c)$ of the couplings $c$ by

$$\delta(D) = \delta(c_D) - \sum V_i \delta(c_i).$$

Though we have not yet linked superficial divergence to actual divergence, we can foresee an implication of this theorem: if a theory of the specified type is to be renormalizable, then its coupling constants must have non-negative mass dimensions. Clearly, in this case, there are only a finite number of Green functions which have superficially divergent contributions.

In QED, QCD, and the Standard Model, the coupling constants satisfy $\delta(c) = 0$. In such theories, the superficial degree of divergence of a diagram $D$ depends only on the number of its external lines.

**Homework 20.1.8.** List the Green functions in QED which have contributions from diagrams which are superficially divergent.

From this analysis of superficial degree of divergence, we conclude that there are very few interactions which do not make a theory non-renormalizable. Here, up to integration by parts and total derivatives, is a table of the possibilities:

| | | |
|---|---|---|
| **Scalars:** | $\phi, \phi^2, \phi^3, \phi^4,$ | $\partial_\mu \phi \, \partial^\mu \phi;$ |
| **Spinors:** | $\bar{\psi}\psi, \bar{\psi}\gamma^5\psi,$ | $\bar{\psi}\slashed{\partial}\psi, \bar{\psi}\gamma^5\slashed{\partial}\psi;$ |
| **Vectors** $(p_v = 0)$: | $A_\mu A^\mu,$ | $\partial_\mu A^\mu, \partial_\mu A_\nu \partial^\mu A^\nu, \partial_\mu A^\mu \partial_\nu A^\nu;$ |
| **Vectors** $(p_v = -2)$: | $A_\mu A^\mu,$ | $\partial_\mu A^\mu, \partial_\mu A_\nu \partial^\mu A^\nu, \partial_\mu A^\mu \partial_\nu A^\nu,$ |
| | $(A_\mu A^\mu)^2,$ | $A_\mu A^\mu \partial_\nu A^\nu;$ |
| **Mixed:** | $\bar{\psi}\psi\phi, \bar{\psi}\gamma^5\psi\phi,$ | $\phi\partial_\mu A^\mu, \phi^2 \partial_\mu A^\mu,$ |
| | $A_\mu A^\mu \phi^2, \bar{\psi}\slashed{A}\psi,$ | $\phi A_\mu A^\mu.$ |

Note that $\bar{\psi}\gamma^5\slashed{\partial}\psi$, $\partial_\mu A^\mu$, and $\phi A_\mu A^\mu$ cannot occur in a renormalizable quantum field theory. The term $\phi\partial_\mu A^\mu$, if it occurs, can always be removed by a redefinition of fields. The term $\phi$ is a valid tadpole counterterm.

**Homework 20.1.9.** Show how to remove a term $\phi\partial_\mu A^\mu$ from a massive vector field theory.

**Homework 20.1.10.** It is clear from Lorentz invariance that tadpole diagrams with an external vector field vanish. Confirm this from the perspective of perturbation theory.

This section has defined and interpreted the superficial degree of divergence of a Feynman integral. The main conclusion at this stage is that interactions of mass dimension greater than four will give rise to infinitely many distinct divergences, each needing its own characteristic counterterm. Such theories are called *non-renormalizable* because they have infinitely many parameters and therefore no predictive power. The technique for using the superficial degree of divergence in renormalization will be described in the next section.

# 20.2  The Forest Formula

It is not clear whether or not an integral of positive mass dimension diverges because there may be cancellations in the numerator due to Dirac algebra and contractions with external polarization spinors and vectors. Neither is it clear that a negative mass dimension will force convergence, since the negative dimension may be focussed on one loop variable thereby leaving positive dimension for another. However, Bogoliubov and Parasiuk proposed a counterterm renormalization procedure based upon this parameter. Hepp proved that this approach worked, and Zimmermann provided the ultimate algorithm, the forest formula, for counterterm renormalization. We shall describe the BPHZ procedure now.

We have already noted that any diagram is a tree diagram in 1PI diagrams. Clearly, a divergence of any diagram must therefore originate in a 1PI subdiagram. The new insight behind the forest formula is that a systematic treatment of divergences in any Feynman integral can be based on an analysis of superficially divergent 1PI subdiagrams of the original Feynman diagram $D$. Here, the appropriate concept of a subdiagram is not just the graph-theoretic one, but the following modification:

**Definition 20.2.1.** A *subdiagram* of a Feynman diagram $D$ is any subset $V$ of the vertices of $D$ together with all the internal lines of $D$ that both begin and end at vertices of $V$. We shall write $1PI(D)$ for the set of 1PI subdiagrams of $D$.

For example, in the following diagram, the edges in bold face and the vertices they meet constitute a subdiagram:

$$\textbf{Subdiagrams:} \qquad \text{} \qquad\qquad \text{(20.2.2)}$$

A subdiagram $d$ of a diagram $D$ is defined as a diagram with no external lines so that we can collapse it to a point and so obtain a new diagram $D/d$. The Feynman rules for subdiagrams are the same as those for diagrams, only spinor factors for external Fermi lines are omitted.

As for notation, we shall use $D$ and $d$ respectively for an arbitrary diagram and subdiagram, $\Gamma$ and $\gamma$ respectively for a 1PI diagram and 1PI subdiagram, and $D'$ and $\Gamma'$ respectively for the subdiagrams of $D$ and $\Gamma$ which contain all the vertices and internal lines of the original. The subdiagrams of a diagram $D$ other than $D'$ are called *proper subdiagrams*.

Weinberg's convergence theorem establishes a foundation for the forest formula by establishing the utility of superficial degree of convergence as a test for genuine convergence:

**Theorem 20.2.3.** *In a theory with no massless fields, if $\delta(\gamma) < 0$ for all 1PI subdiagrams $\gamma$ of a Feynman diagram $D$, then the Euclidean integral $I_E(D)$ is absolutely convergent.*

Actually, Weinberg with characteristic thoroughness, investigated convergence of a class of Euclidean integrals wider than those derived from Feynman rules. The result stated here is a corollary of Weinberg's more general conclusion. Inspection of the derivation of the result stated above shows that it holds for any set of generalized Feynman diagrams and rules which satisfy the following constraints:

1. The momenta associated with the lines of the diagram satisfy momentum conservation at each vertex;
2. The generalized propagator for a line is a Lorentz-covariant function of that line's momentum;
3. The generalized vertex factor is a constant Lorentz tensor like $-ie$, $-ie\gamma^\mu$, or $-ieg^{\mu\nu}$.

The case of derivative couplings is included as follows. If, due to such a coupling, we need to put a $-ik^\mu$ at some vertex $V$, we can include $k_\nu$ in the appropriate generalized propagator and add a factor $-ig^{\mu\nu}$ at $V$.

**Remark 20.2.4.** If we have massless fields, then there may be infrared divergences. These are absent in diagrams involving only dimensionless coupling constants as long as no partial sum of external momenta vanishes.

From Weinberg's convergence criterion, it is clear that we need to work with Feynman integrands and superficially divergent 1PI subdiagrams, so we introduce some notation:

**Definition 20.2.5.** Write $(2\pi)^4\delta^{(4)}(P_n)\mathcal{I}(D)$ for the Feynman integrand associated with the Feynman diagram $D$ after the elimination of all internal delta functions against integrals. Thus the loop differentials are included in $\mathcal{I}(D)$ but not the loop integrals. Write $\mathcal{I}_E(D)$ for the Euclidean integrand obtained by substituting $p_0 \to ip_4$ so that the corresponding invariant amplitude is given by $i\mathcal{A}(D) = \int \mathcal{I}_E(D)$.

**Definition 20.2.6.** A *renormalization part* is a 1PI diagram which is superficially divergent. Write $\mathrm{RP}(D)$ for the set of renormalization parts in $1\mathrm{PI}(D)$.

Thus, for example, the diagram on the left of 20.2.2 has two renormalization parts, namely, the whole diagram and the electron loop, while the diagram on the right has only one, effectively the whole diagram.

The next step towards the forest formula is provided by superficially divergent 1PI diagrams whose 1PI subdiagrams meet the hypotheses of Weinberg's Theorem:

**Definition 20.2.7.** A *primitive divergence* or *primitively divergent diagram* is a Feynman diagram $D$ with $\delta(D) \geq 0$ and $\delta(\gamma) < 0$ for all $\gamma \in 1\mathrm{PI}(D)$.

The three divergent diagrams we investigated in Chapter 19 are of this type. There we saw that counterterm renormalization effectively modified each primitively divergent integrand $\mathcal{I}(p)$ by subtracting the first few terms in its Taylor expansion. The number of terms to subtract is at most the superficial degree of divergence of the primitively divergent diagram. Gauge invariance, as we saw in Chapter 19, often reduces the actual degree of divergence, but over-subtraction does not create any theoretical problems.

To write formulae, we introduce the Taylor expansion operator $T$:

**Definition 20.2.8.** Let $T$ be the operator on Feynman integrands $\mathcal{I}(D)$ which yields the Taylor series of $\mathcal{I}(D)$ in external momenta about the zero of momentum up to and including the terms of order $\delta(D)$.

Taking the expansion about zero momentum causes no problems if there are no massless particles. It merely shifts the renormalization point and thereby modifies the finite parameters in the Lagrangian density. If there are massless particles, our whole approach will need substantial refinement.

On the basis of our experience with loop integrals, we can guess that the integral $\int(1-T)\mathcal{I}(D)$ will be convergent. In fact, the Taylor expansion of the integrand in this expression starts with derivatives of order $\delta(D)+1$. As each derivative decreases the superficial degree of divergence by 1, clearly every term in the Taylor expansion is superficially convergent. To apply Weinberg's Theorem, we need only to extend our Feynman rules so that the terms in the Taylor expansion can be represented by diagrams.

In the Taylor series, derivatives are evaluated at the origin and the external variable is coupled to a differential operator. For example:

$$\frac{\partial}{\partial p^{\mu}}\left(\frac{i}{(p+k)^2 - m^2}\right)\bigg|_{p=0} p^{\mu} = \frac{i}{k^2-m^2}(-2ip{\cdot}k)\frac{i}{k^2-m^2}. \qquad (20.2.9)$$

Diagrammatically, this corresponds to inserting a vertex in the scalar propagator. For example:

$$(20.2.10)$$

This generalized Feynman rule cannot be derived from a Lagrangian, but it is good enough to establish the fundamental connection between diagrams and integrands which makes Weinberg's convergence theorem relevant.

Thus, once we have chosen loop momenta for the original diagram $D$, the $n$th term in the Taylor expansion can be represented by a sum of diagrams $\partial^n D$ constructed from the original diagram $D$ by putting the $n$ insertions in all possible combinations along the lines carrying external momenta. For $n > \delta(D)$, all the diagrams in $\partial^n D$ satisfy the criteria of Weinberg's Theorem, so the Euclidean Feynman integral for each diagram is absolutely convergent.

**Homework 20.2.11.** Draw the diagrams $\partial^n D$, $n = \delta(D) + 1$, for the following $D$'s:

Since the Taylor series itself is absolutely convergent, there is theoretically no problem in summing the contributions of all the diagrams $\partial^n D$ for $n > \delta(D)$ to obtain $(1-T)\delta(D)$ itself. In summary, we therefore have:

**Theorem 20.2.12.** *In a theory with no massless fields, if $D$ is a primitive divergence, then the Euclidean integral $\int(1-T)\mathcal{I}_E(D)$ is absolutely convergent.*

Thus, Taylor-series subtraction provides a renormalization of primitive divergences, providing the simplest case of the forest formula:

**Definition 20.2.13.** *The forest formula renormalized value* $\mathrm{Ren}_{FF}I(D)$ *of a primitive divergence $D$ is given by*

$$\mathrm{Ren}_{FF}I(D) = \int(1-T)\mathcal{I}_E(D).$$

The Taylor series $T\mathcal{I}(D)$ is a Lorentz-invariant polynomial $P(p_1,\ldots,p_n)$ in incoming external momenta. Thus, if we write $\mathcal{O} = \varphi_1\cdots\varphi_n$ for the operator which is represented by the external lines of $D$, $T\mathcal{I}(D)$ can be represented in the Lagrangian density by a counterterm $iP(i\partial_1, \ldots, i\partial_n)\mathcal{O}$, where the derivative $\partial_r$ acts only on the field $\varphi_r$ associated with $p_r$. Consequently, the forest formula renormalized value of a primitive divergence is simply a counterterm renormalization prescription. What the forest formula will add to our understanding of counterterm renormalization is a system for including lower-order counterterms in the computation of higher-order divergences.

The obvious way of representing the counterterms $iP(i\partial_1, \ldots, i\partial_n)\mathcal{O}$ diagrammatically is to introduce a vertex with the external lines of $\Gamma$. This vertex is obtained from $\Gamma$ by collapsing its maximal renormalization part $\Gamma'$. Clearly, if we have two 1PI subdiagrams with no edges or vertices in common, we can collapse both:

$$D/\{\gamma_1, \gamma_2\} \overset{\text{def}}{=} (D/\gamma_1)/\gamma_2. \tag{20.2.14}$$

In general, if $D$ is any diagram, it is meaningful to collapse any set $\{\gamma_1, \ldots, \gamma_n\}$ of pairwise-disjoint 1PI subdiagrams to dots and write $D/\{\gamma_1, \ldots, \gamma_n\}$ for the result.

At this point, it is clear that the general strategy for renormalization of a 1PI diagram is first to remove divergences from its 1PI subdiagrams, and then remove the divergence of the diagram itself. This is a recursive approach which starts with a complicated divergence and terminates with primitive divergences. In preparation for the general formula, the following paragraphs cover the three ways in which two subdivergences can be related.

Two renormalization parts $\gamma_1$ and $\gamma_2$ are said to be *nested* if they satisfy $\gamma_1 \supset \gamma_2$. In this case, we naturally renormalize $\gamma_2$ first. For example, we can take a 1PI diagram $\Gamma$ with a proper 1PI subdiagram $\gamma$ and identify $\gamma_1$ with the maximal subdiagram $\Gamma'$ and $\gamma_2$ with $\gamma$:

$$\Gamma \qquad\qquad \Gamma/\gamma \tag{20.2.15}$$

In this case, both diagrams have an overall divergence to eliminate, so the subtractions can be represented by a formula

$$\text{Ren}_{\text{FF}}\mathcal{I}(\Gamma) = (1 - T_\Gamma)(1 - T_\gamma)\mathcal{I}(\Gamma). \tag{20.2.16}$$

Note that the diagrams themselves establish the precise meaning of the operators, making clear for example how $T_\gamma$ acts on the subdivergence of $\mathcal{I}(\Gamma)$ associated with $\gamma$. Note further that there are no natural diagrams for the two overall divergences separately; both are represented by $\Gamma/\Gamma'$.

Two renormalization parts with no edges or vertices in common are said to be *disjoint*. In this case their counterterms can coexist. For example, the following sum of diagrams represents the removal of subdivergences from the first diagram:

$$\Gamma \qquad\quad \Gamma/\gamma_1 \qquad\quad \Gamma/\gamma_2 \qquad\quad \Gamma/\{\gamma_1, \gamma_2\} \tag{20.2.17}$$

As each of these four diagrams is in itself superficially divergent, to complete the renormalization we will also need a Taylor subtraction for the whole of each diagram. The result can be summarized in the formula

$$\mathrm{Ren_{FF}}\big(\mathcal{I}(\Gamma)\big) = (1 - T_\Gamma)(1 - T_{\gamma_1} - T_{\gamma_2} + T_{\gamma_1}T_{\gamma_2})\mathcal{I}(\Gamma)$$
$$= (1 - T_\Gamma)(1 - T_{\gamma_1})(1 - T_{\gamma_2})\mathcal{I}(\Gamma). \tag{20.2.18}$$

Here, as the diagrams unambiguously indicate, $T_{\gamma_1}$ and $T_{\gamma_2}$ act on the subintegrands associated with the diagrams $\gamma_1$ and $\gamma_2$ respectively.

Two renormalization parts which are neither nested nor disjoint are described as *overlapping*. In this case, we cannot collapse both at once and their counterterms cannot coexist. For example,

$$\Gamma \qquad\qquad \Gamma/\gamma_1 \qquad\qquad \Gamma/\gamma_2 \tag{20.2.19}$$

In this case, the formula for the renormalized value will be

$$\mathrm{Ren_{FF}}\mathcal{I}(\Gamma) = (1 - T_\Gamma)(1 - T_{\gamma_1} - T_{\gamma_2})\mathcal{I}(\Gamma). \tag{20.2.20}$$

It is fairly evident that this procedure of recursive renormalization of renormalization parts will result in integrands which meet the convergence criteria of Weinberg's Theorem. We now give a formula generalizing the examples above:

**Theorem 20.2.21.** *If $D$ is a Feynman diagram, and if $\mathrm{RP}(D)$ is the set of renormalization parts of $D$, then the following integrand has an absolutely convergent Euclidean integral:*

$$\mathrm{Ren_{FF}}\mathcal{I}(D) = \prod_{\gamma \in \mathrm{RP}(D)} (1 - T_\gamma)\mathcal{I}(D),$$

*where it is understood that*

  i) *If $\gamma$ and $\gamma'$ are overlapping, then $T_\gamma T_{\gamma'} = 0$;*
  ii) *If $\gamma \supset \gamma'$, then $T_\gamma$ is to be applied after $T_{\gamma'}$.*

The formula in this theorem is known as the *forest formula*. Now we must explain this term. The basic idea is to express the product over renormalization parts as a sum over subsets of $\mathrm{RP}(D)$. Given such a subset $S$, the corresponding product of Taylor operators will be non-zero if and only if there are no overlapping divergences in $S$. It is now natural to introduce a term *forest* for a set $S$ of non-overlapping diagrams in $\mathrm{RP}(D)$. Writing $\mathcal{F}(D)$ for the set of forests in $D$, we see that

$$\mathrm{Ren_{FF}}\mathcal{I}(D) = \sum_{F \in \mathcal{F}(D)} \prod_{\gamma \in F} (-T_\gamma)\mathcal{I}(D). \tag{20.2.22}$$

**Remark** 20.2.23. Naturally, one expects a forest to be a set of trees, and indeed it is, but a set of trees in a new graph whose vertices represent elements of $\mathrm{RP}(D)$ and whose directed edges represent the inclusion relation. Now a forest $F$ can be seen to be a set of trees in this graph, comprising one tree for every maximal renormalization part in $F$.

In practice, the use of Taylor series as a means for generating counterterms leads to rather horrendous integrands. The technique of dimensional regularization is more snappy. Although dimensional regularization cannot be applied to an integrand, we can still use the pattern of counterterm renormalization expressed by the forest formula to guide the use of counterterms in dimensional regularization. The two approaches, Taylor subtractions and dimensional regularization, lead to finite differences in the counterterms.

**Homework 20.2.24.** Use this technique of renormalization – dimensional regularization plus forest formula – to show that the operator counterterms $\Delta\Omega$ used to renormalize a composite operator $\Omega$ satisfy the mass dimension inequality $\dim(\Delta\Omega) < \dim(\Omega)$. (Refer to Section 18.11.)

The forest formula is an algorithm which caps the development of counterterm renormalization. With a cheerful disregard for history, the essential logic leading to it begins with Weinberg's convergence theorem because this theorem indicates the value of the superficial degree of divergence as a diagnostic device for divergences. The next main idea is the Taylor expansion of Feynman integrands to reduce primitively divergent integrands to convergent ones. Finally, investigation of the three relationships between renormalization parts – nested, disjoint, overlapped – leads to the forest formula.

In the forest formula we have a general procedure for applying lower-order counterterms to eliminate subdivergences of 1PI diagrams. This procedure satisfactorily extends and fulfills the renormalization technique we introduced in Chapter 10 and applied in Chapter 19.

# 20.3 Renormalization and Global Symmetry

The forest formula does indeed produce absolutely convergent integrands for theories with no massless fields. With more work, it can be extended to theories with massless fields also. There is, however, a concern whether or not renormalization preserves the symmetries of the original classical theory. Since diagrams are not symmetric, it may be that the counterterms break classical symmetries. In fact, we may distinguish several cases:

1. **Non-Renormalizable Theories**: theories with at least one interaction of dimension greater than four – such theories have infinitely many independent parameters and are therefore of no value for prediction.
2. **Renormalizable Theories**: theories in which all interactions have dimension at most four and at least one interaction has dimension four – such theories have infinitely many divergent diagrams.
3. **Super-Renormalizable Theories**: theories whose interactions have dimension at most three – these theories have only a finite number of divergent diagrams.
4. **Strictly Renormalizable Theories**: theories in which the counterterms have the same interactions as terms already in the Lagrangian.

We can use the forest formula to define a minimal subtraction renormalization scheme based on dimensional regularization. This scheme has the special property of preserving the split between the dimension-four and dimension-three interactions in the following sense. Writing $g$, $\lambda$, and $m$ for coupling constants of dimension zero, one, and two respectively, we can separate the action $S$ into two parts $S'$ and $S''$ determined by the coupling constants:

$$S(g, \lambda, m) = S'(g) + S''(\lambda, m). \tag{20.3.1}$$

Performing the analogous separation for the bare action $S_B$, we have:

$$S_B(g_B, \lambda_B, m_B) = S'_B(g_B) + S''_B(\lambda_B, m_B). \tag{20.3.2}$$

Then it can be proved that:

1. $g_B$ is a function $g_B(g; \delta, \mu)$;
2. $\lambda_B$ and $m_B$ are both polynomials in $\lambda$ and $m$ with coefficients which are functions of $g$, $\delta$, and $\mu$;
3. The global symmetries of $S'(g)$ are identical to those of $S'_B(g_B)$.

The last point enables us to construct a strictly renormalizable Lagrangian density $\mathcal{L} = \mathcal{L}' + \mathcal{L}''$ by choosing a set of fields and a group $G$ of global symmetries, building $\mathcal{L}'$ from all dimension-four interactions invariant under $G$, and choosing an arbitrary $\mathcal{L}''$ from interactions with dimension three or less. Furthermore, from a consideration of the bare Lagrangian density, it is clear that property (3) above will hold in any renormalization scheme based on dimensional regularization.

With the forest formula in hand, it is possible to make a detailed analysis of the counterterms and demonstrate the preservation of global symmetries. For gauge theories, however, we need some more powerful argument to conclude that renormalization can preserve gauge invariance. This is the topic of the following sections.

## 20.4 The Ward–Takahashi Identity

In gauge theories, the global aspect of the symmetry results in conserved currents as usual, but the local aspect yields identities among the Green functions. For QED, these identities are known as Ward–Takahashi identities; for non-abelian gauge theories, Slavnov–Taylor identities. In both cases, we shall first prove a compact non-perturbative form of these *identities* which we shall refer to as the related *identity*. To provide further insight into the functional integral formalism, we shall prove the two identities separately by different methods.

This section presents a proof of the Ward–Takahashi identity in two phases. First, we find an average-field condition for gauge invariance of the generating functional. Second, we transform this expression into a constraint on the generator of 1PI Green functions. The following section unpacks specific identities from this constraint.

We are familiar with the importance of gauge invariance after quantization of a gauge theory. Having computed a few loop effects in QED, we can now appreciate in detail how the gauge-fixing parameter $\alpha$ inevitably complexify every computation. It is far from obvious that the Green functions are independent of this parameter. Certainly, for canonical quantization, to find a complete set of independent canonical variables and a Hamiltonian, or for functional integral quantization, to integrate over moduli space, it is necessary to add a gauge-fixing term to the QED Lagrangian; but why should the effects of this extra term not permeate the Green functions? The

Ward–Takahashi identity which we prove in this section proclaims gauge invariance on the level of 1PI Green functions and remarkably limits the influence of the gauge-fixing term to the photon self-energy.

Now we begin the proof, aiming first at a condition on average values of classical fields which expresses gauge invariance. If $\chi$ is in the Lie algebra of the gauge group, then we can define an infinitesimal gauge transformation of a field $X$ in the direction $\chi$ by

$$\delta X \overset{\text{def}}{=} \frac{\partial}{\partial \theta} e^{-ie\theta\chi} \cdot X \Big|_{\theta=0}. \tag{20.4.1}$$

In QED, we have a photon $A_\mu$, fermions $\psi^a$ with charges $q^a$, and scalars $\phi^b$ with charges $q^b$. The covariant derivative is $D_\mu = \partial_\mu - ieqA_\mu$. Writing $\varphi$ for the vector of all these fields, then an infinitesimal gauge transformation determines a vector $M(\varphi)$ according to the following equalities:

$$\delta\varphi = \begin{pmatrix} \delta A_\mu \\ \delta\psi^a \\ \delta\phi^b \end{pmatrix} = \begin{pmatrix} -\partial_\mu\chi \\ -ieq^a\chi\psi^a \\ -ieq^b\chi\phi^b \end{pmatrix} = \begin{pmatrix} -\partial_\mu \\ -ieq^a\psi^a \\ -ieq^b\phi^b \end{pmatrix} \chi = M(\varphi)\chi. \tag{20.4.2}$$

Notice that $M(\varphi)$ is at most first-order in the fields.

The variation of the gauge-fixing term similarly defines an operator $N(\varphi)$ which is linear in $\varphi$:

$$\delta\left(-\frac{1}{2\alpha}\int d^4x\,(\partial_\mu A^\mu)^2\right) = \int d^4x\,\frac{1}{\alpha}(\partial_\mu A^\mu)\partial_\nu\partial^\nu\chi = \int d^4x\,N(\varphi)\chi. \tag{20.4.3}$$

Since gauge transformations translate the vector boson $A$ and rotate the matter fields, the functional measure loosely written $(d\varphi)$ is invariant under gauge transformations. Also, since the original action $S$ is gauge invariant, $\delta S = 0$. Hence the effect of such a transformation on the generating functional is simply:

$$\begin{aligned} \delta Z(J) &= \delta N \int (d\varphi)\, e^{iS_{\text{eff}}(\varphi,J)} \\ &= N \int (d\varphi)\, e^{iS_{\text{eff}}(\varphi,J)}\, i \int d^4x\, \big(J{\cdot}M(\varphi) + N(\varphi)\big)\chi, \end{aligned} \tag{20.4.4}$$

where the dot represents summation over fields. Of course, a gauge transformation of the integration variable is simply a change of coordinates in the functional integral, so

$$\delta Z(J) = 0. \tag{20.4.5}$$

Since $N$ and $M$ are at most first-order in $\varphi$, the final functional integral can be evaluated in terms of the average field $\bar\varphi$ defined by

$$\bar\varphi(x) \overset{\text{def}}{=} \frac{\int(d\varphi)\,\varphi(x)\exp\big(iS_{\text{eff}}(\varphi,J)\big)}{\int(d\varphi)\,\exp\big(iS_{\text{eff}}(\varphi,J)\big)}, \tag{20.4.6}$$

and so the gauge invariance of $Z$ implies that

**Condition on Average Fields:** $\quad \int d^4x\, \big(N(\bar\varphi) + J\cdot M(\bar\varphi)\big)\chi = 0. \tag{20.4.7}$

This equation marks the end of the first phase in our analysis of gauge invariance in QED. In the second phase, we transform this condition on average fields into a condition on the generator of 1PI Green functions.

The definition of the average field $\bar\varphi$ suggests introducing a logarithm $iW$ of the generating functional:

$$\bar\varphi(x) = -i\frac{\delta}{\delta J(x)}\ln Z(J) \stackrel{\text{def}}{=} \frac{\delta}{\delta J(x)}W(J). \tag{20.4.8}$$

Following Section 10.9, we identify $iW$ as the generating functional for connected Green functions. Recalling that the connected diagrams themselves are tree diagrams in 1PI diagrams, we next express $iW$ in terms of the generating functional of 1PI Green functions.

To build diagrams out of 1PI subdiagrams, we see the 1PI subdiagrams with three or more external lines as contributing to vertices and the 1PI subdiagrams with two external lines as contributing to the renormalized propagators. For $n > 2$, we therefore define $n$-point 1PI Green functions $i\hat\Gamma_n$ as sums of the Feynman amplitudes $i\mathcal{A}$ for 1PI diagrams, and for $n = 2$, we define $i\hat\Gamma_2 = -D^{-1}$.

Using multi-indices to refine our notation for the $\hat\Gamma$'s, assuming that our QED has $k$ distinct fields, we shall write $I$ for $(i_1, \ldots, i_k)$, $n = |I|$ for $\sum i_r$, $I!$ for $\prod i_r!$, $\hat\Gamma_I$ for the $n$-point function with $i_r$ external lines of field-type $r$, and $\varphi^I$ for $\prod \varphi_r^{i_r}$, with the understanding that this product of $n$ fields is to be evaluated on $n$ separate points $x^1, \ldots, x^n$. The significance of using $\varphi$ for the variable will emerge as we progress. Now we can define the *effective interaction*, the generating functional for the $\hat\Gamma$'s:

$$\Gamma(\varphi) \stackrel{\text{def}}{=} \sum_I \frac{1}{I!}\int d^4x^1 \cdots d^4x^n\, \varphi^I(x)\Gamma_I(x^1, \ldots, x^n), \tag{20.4.9}$$

where

$$\Gamma_I(x^1, \ldots, x^n) \stackrel{\text{def}}{=} \int \frac{d^4k_1}{(2\pi)^4}\cdots\frac{d^4k_n}{(2\pi)^4} \tag{20.4.10}$$
$$\times\, e^{ik_1\cdot x_1}\cdots e^{ik_n\cdot x_n}(2\pi)^4\delta^{(4)}(k_1 + \cdots + k_n)\hat\Gamma_I(k_1, \ldots, k_n).$$

The quantum theory defined by this effective interaction is known as *formal $\Gamma$ theory*. The generating functional for formal $\Gamma$ theory is:

$$Z_\Gamma(J) \stackrel{\text{def}}{=} N\int(d\varphi)\,\exp\Big(\frac{i}{\hbar}\big(\Gamma(\varphi) + (J, \varphi)\big)\Big). \tag{20.4.11}$$

From the principles of Section 11.8, the propagators are determined by the covariance $-\Gamma_2$ and are given by $i\hat\Gamma_2^{-1}$. Consequently, as desired, the renormalized propagators $D$ of QED are the tree-level propagators of formal $\Gamma$ theory. Now it is easy to see that connected Green functions in QED are finite sums of tree-level amplitudes in formal $\Gamma$ theory. For example, the Green function for electron-positron scattering may be split up as follows:

$$\tag{20.4.12}$$

Note that the lines in these diagrams stand for renormalized propagators.

To develop the connection between formal $\Gamma$ theory and QED, we show next that loop diagrams and tree diagrams are separated by powers of Planck's constant. Quite generally, a diagram with $I$ internal lines and $V$ vertices will have a factor of $\hbar^{I-V}$ in its Feynman integral. However, for a connected diagram, the number of loops $L$ is given by

$$L = I - V + 1, \tag{20.4.13}$$

so the associated Feynman integral contains a factor of $\hbar^{L-1}$. Thus amplitudes are dominated by tree-level diagrams when such exist, and amplitudes for purely quantum processes like photon-photon scattering are damped by relative factors of $\hbar$. Hence we can evaluate tree-level $\Gamma$ theory by taking $\hbar$ as a formal parameter and letting $\hbar \to 0$. Consequently, we find that:

$$e^{iW(J)/\hbar} = Z_\Gamma(J)e^{\mathcal{O}(1)} \quad \text{as } \hbar \to 0, \tag{20.4.14}$$

where $\mathcal{O}(1)$ stands for the error due to loop diagrams in formal $\Gamma$ theory.

We can now use the method of stationary phase to evaluate the integral for $Z_\Gamma(J)$ as $\hbar \to 0$:

$$W(J) = \Gamma(\bar{\varphi}) + \int d^4x\, J \cdot \bar{\varphi}, \tag{20.4.15}$$

where $\bar{\varphi} = \varphi(J)$ is defined by

$$\frac{\delta\Gamma}{\delta\bar{\varphi}} + J = 0. \tag{20.4.16}$$

Hence $W(J)$ is the Legendre transform of $\Gamma(\bar{\varphi})$, and when we evaluate $\delta W/\delta J$ we find $\bar{\varphi}$, in accord with the definition 20.4.8 of $\bar{\varphi}$ as the average field.

We can now eliminate the operator $N$ from the condition 20.4.7 for gauge invariance. Indeed, since

$$M(\bar{\varphi})\chi = \delta\bar{\varphi} \quad \text{and} \quad J = -\left.\frac{\delta\Gamma}{\delta\varphi}\right|_{\varphi=\bar{\varphi}}$$
$$\implies \int d^4x\, J \cdot M(\bar{\varphi})\chi = -\int d^4x\, \frac{\delta\Gamma}{\delta\varphi}\delta\bar{\varphi} = -\delta\Gamma(\bar{\varphi}), \tag{20.4.17}$$

we see at once that

$$\delta\Gamma(\bar{\varphi}) = \int d^4x\, N(\bar{\varphi})\chi = \delta\left(-\frac{1}{2\alpha}\int d^4x\,(\partial_\mu \bar{A}^\mu)^2\right). \tag{20.4.18}$$

Consequently, the failure of gauge invariance in the effective interaction is entirely due to the contribution of the gauge-fixing term to the 2-point 1PI Green function. Defining $\bar{\Gamma}$ by

$$\bar{\Gamma}(\bar{\varphi}) \overset{\text{def}}{=} \Gamma(\bar{\varphi}) + \frac{1}{2\alpha}\int d^4x\,(\partial_\mu \bar{A}^\mu)^2, \tag{20.4.19}$$

we find that $\bar{\Gamma}$ is gauge invariant:

**Ward–Takahashi Identity:**  $\delta\bar{\Gamma}(\bar{\varphi}) = 0. \tag{20.4.20}$

Note that the subtraction of the gauge-fixing action from $\Gamma$ only affects the photon two-point 1PI function, and that only at the zero-loop level. We shall use the bar in $\Gamma \to \bar{\Gamma}$ throughout the following sections to indicate removal of the gauge-fixing term from Lagrangian density, action, or effective interaction as needed.

The determination of the consequences of gauge invariance in QED involves an investigation into Green functions. The appearance of the average field $\bar{\varphi}$ in the gauge variation of the generating functional causes us to introduce the generating functional $W = -i \ln Z$ of connected Green functions. The investigation into the structure of connected Green functions leads to the identification of the 1PI Green functions as their essential vertices. Then the $\hbar$ dependence of Feynman integrals enables us to suppress loops in the formal $\Gamma$ theory of 1PI Green functions and hence to derive QED from the effective interaction. The outcome of this derivation is a second, implicit expression for the average field $\bar{\varphi}$ which can be used to eliminate $\bar{\varphi}$ from the gauge invariance condition. The resulting expression is the Ward–Takahashi identity, the expression of gauge-invariance in the 1PI Green functions.

The next section presents an illuminating insight into the role of $\Gamma$ as an effective action for the classical field $\bar{\varphi}$. After that, we return to the study of the Ward–Takahashi identity, describing the method for bringing specific identities out of $\delta\bar{\Gamma} = 0$, and giving applications of the first two identities.

## 20.5  Effective Classical Action

In this section, we show that the effective interaction $\Gamma$ of the previous section is an effective classical action for the VEV of $\phi$. This is naturally accomplished by taking the main points of the previous section pertaining to $\Gamma$ in reverse order. The material presented here constitutes a brief but delightful interlude in the discussion of the Ward–Takahashi identity.

The field $\bar{\varphi}$ defined in 20.4.6 is the 'classical' field associated with the quantum field $\phi$ in the presence of the background field $J$:

$$\bar{\varphi}(x) = \frac{\delta W(J)}{\delta J(x)} = \frac{\langle 0 | \phi(x) | 0 \rangle_J}{\langle 0 | 0 \rangle_J}. \tag{20.5.1}$$

Following the spirit of Lagrangian field theory, we next introduce an effective action $\Gamma(\bar{\varphi})$ to generate the classical equations of motion:

$$\frac{\delta\Gamma}{\delta\bar{\varphi}(x)} + J(x) = 0. \tag{20.5.2}$$

Comparison with the defining equation 20.5.1 for $\bar{\varphi}$ shows that this equation has a solution for $\Gamma$, namely the Legendre transform of $W$:

$$\Gamma(\bar{\varphi}) = W(J) - \int d^4x \, J(x)\bar{\varphi}(x). \tag{20.5.3}$$

To make this inquiry self-sufficient, we have only to show that $\Gamma$ is the generating functional of 1PI Green functions.

The technique is to differentiate the definition 20.5.1 of $\bar{\varphi}$ with respect to this field using the functional chain rule,

$$\frac{\delta}{\delta\bar{\varphi}(x)} = \int d^4y \, \frac{\delta J(y)}{\delta\bar{\varphi}(x)} \frac{\delta}{\delta J(y)}, \tag{20.5.4}$$

and use the definition 20.5.2 of $\Gamma$ to express derivatives of $J$ in terms of derivatives of $\Gamma$.

It helps to introduce a more efficient notation. We choose to write subscripts for partial derivatives and let repeated indices imply an integral as the following examples suggest:

$$\Gamma_{xy} \stackrel{\text{def}}{=} \frac{\delta^2 \Gamma}{\delta \bar{\varphi}(x)\, \delta \bar{\varphi}(y)} \tag{20.5.5}$$

and

$$W_{xy} \Gamma_{yz} \stackrel{\text{def}}{=} \int d^4 y\, \frac{\delta^2 W}{\delta \bar{\varphi}(x)\, \delta \bar{\varphi}(y)} \frac{\delta^2 \Gamma}{\delta J(y)\, \delta J(z)}. \tag{20.5.6}$$

With these definitions, the functional chain rule reduces to

$$A(\bar{\varphi}) = B(J) \quad \Longrightarrow \quad A_x + B_y \Gamma_{yx} = 0. \tag{20.5.7}$$

Finally, we define $\Gamma_n$ to be the coefficients in the functional expansion of $\Gamma$ in powers of $\bar{\varphi}$, and define $W_n$ in terms of $W$ as $G_n$ is defined in terms of $Z$:

$$\Gamma_n(x_1, \ldots, x_n) \stackrel{\text{def}}{=} \Gamma_{x_1 \cdots x_n}(\bar{\varphi}) \big|_{\bar{\varphi}=0}. \tag{20.5.8}$$

Now differentiation of 20.5.1 with respect to $\bar{\varphi}$ yields

$$\delta^{(4)}(y_1 - x_2) + W_{y_1 y_2} \Gamma_{y_2 x_2} = 0. \tag{20.5.9}$$

Evaluation at $\phi = J = 0$ shows that $i\Gamma_2^{-1} = iW_2$ is the renormalized propagator $G_2$. A second differentiation yields

$$W_{y_1 y_2} \Gamma_{y_2 x_2 x_3} - W_{y_1 y_2 y_3} \Gamma_{y_2 x_2} \Gamma_{y_3 x_3} = 0. \tag{20.5.10}$$

In light of the previous result, we can multiply by $\Gamma_{y_1 x_1}$ to find

$$\Gamma_{x_1 x_2 x_3} + W_{y_1 y_2 y_3} \Gamma_{y_1 x_1} \Gamma_{y_2 x_2} \Gamma_{y_3 x_3} = 0, \tag{20.5.11}$$

from which we conclude that $\Gamma_3$ is the three-point 1PI Green function – the diagrams which contribute to $W_3$ are automatically 1PI.

Differentiating this relation produces something less trivial,

$$\Gamma_{x_1 x_2 x_3 x_4} \tag{20.5.12}$$
$$= W_{y_1 y_2 y_3 y_4} \Gamma_{y_1 x_1} \Gamma_{y_2 x_2} \Gamma_{y_3 x_3} \Gamma_{y_4 x_4} - W_{y_1 y_2 y_3} \Gamma_{y_1 x_1 x_4} \Gamma_{y_2 x_2} \Gamma_{y_3 x_3}$$
$$- W_{y_1 y_2 y_3} \Gamma_{y_1 x_1} \Gamma_{y_2 x_2 x_4} \Gamma_{y_3 x_3} - W_{y_1 y_2 y_3} \Gamma_{y_1 x_1} \Gamma_{y_2 x_2} \Gamma_{y_3 x_3 x_4}.$$

We can use the previous relation to eliminate three-point $W$'s and thereby obtain

$$\Gamma_{x_1 x_2 x_3 x_4} = W_{y_1 y_2 y_3 y_4} \Gamma_{y_1 x_1} \Gamma_{y_2 x_2} \Gamma_{y_3 x_3} \Gamma_{y_4 x_4} - W_{y_1 y_2} \Gamma_{y_1 x_1 x_4} \Gamma_{y_2 x_2 x_3}$$
$$- W_{y_1 y_2} \Gamma_{y_1 x_1 x_3} \Gamma_{y_2 x_2 x_4} - W_{y_1 y_2} \Gamma_{y_1 x_1 x_2} \Gamma_{y_2 x_3 x_4}. \tag{20.5.13}$$

If we represent this by diagrams, then we can see that $\Gamma_4$ must be 1PI. It is rather remarkable that the 1PI property, a property of graphs, should emerge in a natural manner from Legendre relationship of $\Gamma$ and $W$.

**Remark 20.5.14.** In order to derive formulae linking 1PI Green functions $i\Gamma_n$ to connected Green functions $iW_n$, we need a $-i$ for every $\Gamma_{xx'}$ to get inverse propagators, a $-i$ for every subscript on a $W$ to get $W_n$'s, an extra $i$ for every $W$, and an extra $i$ for every $\Gamma$ with more than two suffices.

**Homework 20.5.15.** For $n > 2$, the expansion of a connected Green function $\hat{W}_n$ in terms of the 1PI green functions $\hat{\Gamma}_m$ can be expressed as a sum of tree diagrams with Feynman rules that associate $i\hat{\Gamma}_m$ with an $m$-point vertex and a renormalized propagator $\hat{D}$ with an internal line.

Represent the expansions 20.5.11 and 20.5.13 of $W_3 D^{-3}$ and $W_4 D^{-4}$ respectively in diagrams. Devise a diagrammatic representation of the differentiation operator acting recursively on these expansions. (This representation of the calculus above effectively proves that the $\Gamma_n$ are 1PI.)

At this point, we feel comfortable with the assertion that $\Gamma$ is the generating functional of 1PI Green functions. It is a bit tricky to complete the demonstration formally, especially when we consider different field types and interactions. For this reason, in the previous section we based our analysis on formal $\Gamma$ theory. This ends the interlude on the interpretation of $\Gamma$. We return in the next section to the theory of Ward–Takahashi identities.

# 20.6  Perturbative Ward–Takahashi Identities

The final step in understanding the Ward–Takahashi identity is to unfold the variation condition 20.4.20 and find specific identities between the various 1PI Green functions. For simplicity, we consider QED with a photon $A$ and single spinor field $\psi$ of charge $-1$ and write $(A_\mu, \bar{\psi}, \psi)$ for $\bar{\varphi}$. Then the infinitesimal gauge transformations 20.4.2 become

$$\delta A_\mu = -\partial_\mu \chi \quad \text{and} \quad \delta \psi = ie\chi\psi. \tag{20.6.1}$$

Labeling the 1PI Green functions by Lorentz subscripts indicating external $A$ lines, and two integer superscripts indicating numbers of external $\bar{\psi}$ and $\psi$ lines, we find that

$$\bar{\Gamma}(A_\mu, \bar{\psi}, \psi) = \int d^4x\, d^4y \left( \frac{1}{2} \bar{\Gamma}_{\mu\nu}(x, y) A^\mu(x) A^\nu(y) + \bar{\psi}(x) \Gamma^{1,1}(x, y) \psi(y) \right)$$
$$+ \int d^4x\, d^4y\, d^4z\, \bar{\psi}(x) \Gamma^{1,1}_\mu(x, y, z) \psi(y) A^\mu(z) + \cdots \tag{20.6.2}$$

Using the infinitesimal gauge transformations 20.6.1 and the symmetry of the 1PI Green functions under interchange of external lines of the same type, we find that

$$\delta\bar{\Gamma} = - \int d^4x\, d^4y\, \bar{\Gamma}_{\mu\nu}(x, y) A^\mu(x) \partial^\nu_y \chi(y) \tag{20.6.3}$$

$$- \int d^4x\, d^4y\, \bar{\psi}(x) \Gamma^{1,1}(x, y) \psi(y) \big( ie\chi(x) - ie\chi(y) \big)$$

$$- \int d^4x\, d^4y\, d^4z\, \bar{\psi}(x) \Gamma^{1,1}_\mu(x, y, z) \psi(z) \Big( \big( ie\chi(x) - ie\chi(y) \big) A^\mu(z) + \partial^\mu_z \chi(z) \Big)$$

$$- \cdots$$

Substituting $\chi(w) = \delta^{(4)}(w - a)$ yields:

$$
\begin{aligned}
\delta\bar{\Gamma}_a = {} & \int d^4x \; \partial_y^\nu \bar{\Gamma}_{\mu\nu}(x, a) A^\mu(x) \\
& - \int d^4y \; ie\bar{\psi}(a)\Gamma^{1,1}(a, y)\psi(y) + \int d^4x \; ie\bar{\psi}(x)\Gamma^{1,1}(x, a)\psi(a) \\
& - \int d^4x \, d^4z \; ie\bar{\psi}(a)\Gamma_\mu^{1,1}(a, y, z)\psi(y)A^\mu(z) \qquad (20.6.4) \\
& + \int d^4x \, d^4y \; ie\bar{\psi}(x)\Gamma_\mu^{1,1}(x, a, z)\psi(a)A^\mu(z) \\
& + \int d^4y \, d^4z \; \bar{\psi}(x)\partial_z^\mu\Gamma_\mu^{1,1}(x, y, a)\psi(y) + \cdots
\end{aligned}
$$

To find the Ward–Takahashi identities in position space, we use functional differentiation followed by evaluation at $\bar{\varphi} = 0$:

$$
\frac{\delta}{\delta A^\mu(b)} \delta\bar{\Gamma}_a \Big|_{\bar{\varphi}=0} = \partial_y^\nu \Gamma_{\mu\nu}(b, a) = 0,
$$

$$
\frac{\delta^2}{\delta\bar{\psi}(b)\,\delta\psi(c)} \delta\bar{\Gamma}_a \Big|_{\bar{\varphi}=0} = -ie\delta(a - b)\Gamma^{1,1}(a, c) \qquad (20.6.5)
$$

$$
+ ie\Gamma^{1,1}(b, a)\delta(a - c) + \partial_z^\mu\Gamma_\mu^{1,1}(b, c, a) = 0,
$$

and so on. These are the Ward–Takahashi identities for 1PI Green functions in position space.

The definition 20.4.10 of the $\Gamma$'s in terms of the momentum-space 1PI Green functions enables us to convert the identities above into momentum-space identities. The inversion formula is:

$$
\begin{aligned}
(2\pi)^4 \delta^{(4)}(k_1 + \cdots + k_n)\hat{\Gamma}_I(k_1, \ldots, k_n) \qquad & (20.6.6) \\
= \int d^4x_1 \cdots d^4x_n \; e^{-ik_1 \cdot x_1} \cdots e^{-ik_n \cdot x_n}\Gamma_I(x^1, \ldots, x^n). &
\end{aligned}
$$

We shall omit the hat in $\hat{\Gamma}$ and let the variables indicate whether we are considering $\Gamma$ in position or momentum space.

Applying this transformation to the first identity in 20.6.5 above yields

$$
k^\mu \bar{\Gamma}_{\mu\nu}(k) = 0 \quad \Longrightarrow \quad \bar{\Gamma}_{\mu\nu}(k) = F(k^2)P_T, \qquad (20.6.7)
$$

where $P_T$ is the transverse projection operator on momentum space. This transversality of $\bar{\Gamma}_{\mu\nu}(k)$ provides information about the photon self energy.

First, we find a formula for the renormalized propagator $V(k)$ is terms of the self-energy function $\Pi(k)$ by summing the series of self-energy insertions:

$$
\begin{aligned}
V &= P + P(i\Pi)P + P(i\Pi)P(i\Pi)P + \cdots \\
&= P(1 - i\Pi P)^{-1} \qquad (20.6.8) \\
&= (P^{-1} - i\Pi)^{-1}.
\end{aligned}
$$

Writing $\Pi$ and the Feynman propagator in terms of the transverse and longitudinal projections,

$$\Pi(k) = \Pi_T(k^2)P_T + \Pi_L(k^2)P_L$$

$$\text{and} \quad P = \frac{-i}{k^2 + i\epsilon}P_T + \frac{-i\alpha}{k^2 + i\epsilon}P_L, \tag{20.6.9}$$

and substituting in the formula for $V$ yields

$$V = \frac{-i}{k^2 - \Pi_T + i\epsilon}P_T + \frac{-i\alpha}{k^2 - \alpha\Pi_L + i\epsilon}P_L. \tag{20.6.10}$$

Second, from the definition of $\Gamma_{\mu\nu}$ and the Ward–Takahashi identity for $\Gamma$ we have

$$V^{-1}(k) = i\Gamma(k) = i\bar{\Gamma}(k) + \frac{i}{\alpha}k^2 P_L$$

$$= F(k^2)P_T + \frac{i}{\alpha}k^2 P_L. \tag{20.6.11}$$

Comparison of these two formulae for $V$ implies that $\Pi_L = 0$. Recall that in Chapter 19 we worked in Landau gauge ($\alpha = 0$) and proposed a gauge condition $k_\mu \Pi^{\mu\nu} = 0$. Now we can see that this gauge condition is consistent in any gauge.

To take the Fourier transform of the second identity 20.6.5 above, we multiply the identity by exponentials in $k \cdot a$, $p \cdot b$, and $q \cdot c$, integrate over $a$, $b$, and $c$, and drop the delta factor:

$$\delta(a - b)\Gamma^{1,1}(a,c) \ \rightarrow \ \Gamma^{1,1}(p + k, q);$$

$$\Gamma^{1,1}(b,a)\delta(a - c) \ \rightarrow \ \Gamma^{1,1}(p, q + k); \tag{20.6.12}$$

$$\partial_z^\mu \Gamma_\mu^{1,1}(b,c,a) \ \rightarrow \ ik^\mu \Gamma_\mu^{1,1}(p, q, k).$$

Finally, recalling that $i\Gamma^{1,1}(-q, q) = S(q)^{-1}$ and using the dropped delta function to eliminate $k$ everywhere, we obtain the original momentum-space Ward–Takahashi identity:

$$(p + q)^\mu \Gamma_\mu^{1,1}(p, q) = ieS(-p)^{-1} - ieS(q)^{-1}$$

$$= -e(\not{p} + \not{q}) - e\Sigma(-p) + e\Sigma(q). \tag{20.6.13}$$

As we saw in Section 19.3, this identity lies at the basis of the result $z_e = z_\psi$. (Note a change of sign convention: here in Chapter 20, the 1PI amplitudes $i\mathcal{A}$ are represented by $i\Gamma$, but in Chapter 19, the negative charge of the electron field makes it convenient to represent the three-point function in QED by $-i\Gamma$.)

**Homework 20.6.14.** Verify this identity 20.6.13 at lowest order in perturbation theory.

**Homework 20.6.15.** What is the general form of a Ward–Takahashi identity for QED?

**Homework 20.6.16.** Work out the first few Ward–Takahashi identities for scalar QED.

On the level of perturbative computations, the Ward–Takahashi identities are often invoked to explain the improved convergence of Feynman integrals in QED with reference to the superficial degree of divergence. Take the electron-loop contribution to the photon propagator, for example. From Lorentz covariance, the amplitude $\mathcal{A}^{\mu\nu}$ has the form

$$i\mathcal{A}^{\mu\nu}(p^2) = A(p^2)g^{\mu\nu} + B(p^2)p^\mu p^\nu. \tag{20.6.17}$$

Inspection of the Feynman integral for $\mathcal{A}^{\mu\nu}$ indicates that the integrands $\mathcal{I}_A$ and $\mathcal{I}_B$ for $A$ and $B$ respectively have superficial degrees of divergence given by

$$\delta(\mathcal{I}_A) = 2 \quad \text{and} \quad \delta(\mathcal{I}_B) = 0. \tag{20.6.18}$$

The Ward–Takahashi identity for $\Gamma'_{\mu\nu}$, as we have seen, implies the transversality of the photon self energy, so $p_\mu \mathcal{A}^{\mu\nu} = 0$ and we find

$$A(p^2)p^\nu + B(p^2)p^2 p^\nu = 0. \tag{20.6.19}$$

Consequently, there must be a cancellation in the numerator of $A$ which reduces the actual degree of divergence from 2 to 0. Indeed, this is what we found following equation 19.4.7.

The same analysis applied to the sum of the box diagrams for photon-photon scattering shows that this sum is convergent with effective $\delta = -2$, not logarithmically divergent as the superficial degrees of divergence would suggest. (The individual diagrams do diverge; the argument only applies to a Green function order by order, not to individual diagrams.)

This section has provided detailed understanding of the Ward–Takahashi identity. In practice, only the first two terms are commonly used, and the primary value of these terms is the applications given above, namely, in the transversality of the photon self-energy and the relation $z_e = z_\psi$. These points hint at the preservation of gauge invariance in renormalized QED, setting a direction for the next section where we show how the Ward–Takahashi identity can be used to demonstrate this point.

## 20.7 – Gauge Invariance of QED

We saw in Section 20.2 that the combination of dimensional regularization and counterterm renormalization preserves global symmetries of the physical-parameter action. With the Ward–Takahashi identity, we will now show that this same computational technique preserves gauge invariance in QED.

In practice, we start computing the effective interaction $\Gamma$ with the physical-parameter action $S_0$ derived from the physical-parameter Lagrangian density

$$\mathcal{L}_{\text{ph}} = -\frac{1}{4} F_{\mu\nu} F^{\mu\nu} + \bar\psi(i\slashed{D} - m)\psi - \frac{1}{2\alpha}(\partial_\mu A^\mu)^2, \tag{20.7.1}$$

and define counterterms order by order to cancel the poles in regularized Feynman integrals. Here, order by order means in powers of $\hbar$, or equivalently in number of loops. Thus, $S_0$ is the zero-loop value of the action, $S'_n$ is the $n$-loop counterterm contribution, and $S_n$ is the action at $n$-loop level, so that:

$$S_n = S_0 + \sum_{r=1}^{n} S'_r \quad \text{and} \quad S_B = S_\infty = S_0 + \sum_{r=1}^{\infty} S'_r. \tag{20.7.2}$$

For this to make sense, of course, we must hold on to dimension $D = 4 - 2\delta$ and not take $\delta = 0$.

In more detail, let $\Gamma'_n(S)$ be the 1PI generating functional computed at exactly $n$th order in $\hbar$ using the action $S$. Define $\Gamma_n(S)$ similarly to be the generating functional computed to order at most $n$. Then, in particular,

$$\Gamma_n(S'_n) = \Gamma'_n(S'_n) = S'_n. \tag{20.7.3}$$

We now define inductively a sequence of actions $S_n$ for which $\Gamma_n(S_n)$ is finite.

Certainly, there is no problem at the zero-loop level since $\Gamma_0(S_0) = S_0$ is obviously finite. We proceed to assume that $\Gamma_m(S_m)$ is finite for $0 < m < n$, compute $\Gamma'_n(S_{n-1})$, and find its pole terms. Remarkably enough, from the iterative use of our dimensional regularization formulae, one can see that the coefficients of the pole terms in any Feynman integral are polynomial in external momenta. This indicates that we can use local counterterms $S'_n$ to absorb the pole terms.

The finiteness condition for $\Gamma_n(S_n)$ can be expanded as follows:

$$\begin{aligned}\Gamma_n(S_{n-1} + S'_n) &= \Gamma_{n-1}(S_{n-1}) + \Gamma'_n(S_{n-1} + S'_n) \\ &= \Gamma_{n-1}(S_{n-1}) + \Gamma'_n(S_{n-1}) + \Gamma'_n(S'_n).\end{aligned} \tag{20.7.4}$$

Using the Laurent expansion in $\delta$, we can separate the finite and divergent parts of this equation. For simplicity, let us renormalize by minimal subtraction, that is, by choosing counterterms merely to cancel the divergent part:

$$S'_n \stackrel{\text{def}}{=} -\Gamma'^{\text{div}}_n(S_{n-1}). \tag{20.7.5}$$

Clearly, this is sufficient to make $\Gamma_n(S_n)$ finite as $\delta \to 0$. This procedure determines the counterterms $S'_n$ at every order.

Now we want to see whether the resulting $S_n$ are gauge invariant for all $n$. The previous section showed that the gauge-fixing term in the action generates an identical gauge-fixing term in the effective interaction. Thus, rather than gauge invariance (GI), we seek gauge invariance up to the original gauge-fixing term ($\text{GI}_\alpha$).

The argument is by induction on $n$. The base for the induction is that $S_0$ is $\text{GI}_\alpha$. The second step is to assume that $S_m$ is $\text{GI}_\alpha$ for $0 < m < n$. The third step is to show that $S_n$ is also $\text{GI}_\alpha$.

Since $S_{n-1}$ is $\text{GI}_\alpha$, $\Gamma(S_{n-1})$ satisfies the Ward–Takahashi identities. Clearly, this implies that the identities are satisfied at each level – tree, one-loop, two-loop, and so on. Furthermore, the divergent and finite parts of the Laurent expansions at each level are not mixed by a gauge transformation and must also be $\text{GI}_\alpha$. Consequently, the $n$-loop counterterm action $S'_n$ define by 20.7.5 is $\text{GI}_\alpha$.

**Remark** 20.7.6. Since there is only one fundamental vertex in QED, it is easy to show that

$$L - 1 = \frac{1}{2}(V - E),$$

where $L$, $V$, and $E$ are respectively the number of loops, vertices, and external lines in a QED diagram. This indicates that the argument above can be reformulated in terms of powers of $e$ in place of powers of $\hbar$.

We also see directly from the form 20.4.18 of $\delta\Gamma$ that the gauge parameter $\alpha$ is not modified by radiative corrections, divergent or otherwise. This is a relief, because it would be vexing if the Feynman propagator for the photon were dependent on the order in perturbation theory.

Another delicate point in Chapter 19 was the use of a photon mass $\lambda$ as an infrared regulator. It would be inconvenient if the radiative corrections to $\lambda$ were non-zero and disastrous if they were divergent. However, the proof of the Ward–Takahashi identity above rests solely on the linearity of the gauge variation $\delta\bar\varphi$ in $\varphi$. This holds true even if we add a mass term to the QED Lagrangian density. In this case, we would find the identity $\delta\bar\Gamma = 0$ for the modified effective interaction $\bar\Gamma$ defined by

$$\bar\Gamma = \Gamma + \int d^4x \left( \frac{1}{2\alpha}(\partial_\mu A^\mu)^2 - \frac{\lambda^2}{2} A_\mu A^\mu \right). \tag{20.7.7}$$

We conclude that $\lambda$ is not modified by radiative corrections, and the calculations of Chapter 19 are valid.

**Homework** 20.7.8. We have shown that one can add a gauge-fixing term to massive-photon QED. Find the resultant photon propagator. Deduce from the gauge-invariance of $\bar\Gamma$ that the $kk^\top$ term in the propagator does not contribute to 1PI Green functions.

This simple argument shows that, whatever terrible multi-loop integrals crop up in QED with or without a photon mass, the counterterms for them will always be gauge invariant. Thus Faddeev–Popov quantization is consistent with renormalization for QED. Furthermore, the gauge parameter $\alpha$ and the photon mass $\lambda$ are not subject to radiative corrections, so our use of these parameters in the photon propagator has been correct.

# 20.8 Slavnov–Taylor Identity

The Slavnov–Taylor identity is the equivalent in non-abelian gauge field theories of the Ward–Takahashi identity for abelian gauge field theories. The derivation of this identity can follow that used above for the Ward–Takahashi identity. However, as we already have the benefit of that derivation, we shall introduce another fundamental technique of field theory, the Becchi–Rouet–Stora (BRS) transformation, and derive the Slavnov–Taylor identity for a Yang–Mills theory using this.

There are three phases in this section. First, we define a BRS symmetry. Second, we find an expression for the BRS-invariance of the generating functional in terms of connected Green functions. Third, we transform this expression into the Slavnov–Taylor identity on the effective interaction.

A BRS transformation in a gauge theory captures the structural connection between the original Lagrangian density, the gauge-fixing terms, and the ghost terms with such precision that the Faddeev–Popov action is invariant under it. We shall now define a BRS transformation for a pure Yang–Mills theory.

For a pure Yang–Mills Lagrangian density $\mathcal{L}_{\text{YM}} = -\frac{1}{4}G^a_{\mu\nu}G^{a\mu\nu}$, the Faddeev–Popov action is

$$\mathcal{L}_{\text{FP}}(A, \bar\eta, \eta) \overset{\text{def}}{=} -\tfrac{1}{4}G^a_{\mu\nu}G^{a\mu\nu} + \bar\eta^a\partial^\mu D^{ac}_\mu\eta^c - \frac{1}{2\alpha}(\partial_\mu A^{a\mu})^2$$
$$\text{where} \quad D^{ac}_\mu \overset{\text{def}}{=} \delta^{ac}\partial_\mu + gf^{abc}A^b_\mu. \tag{20.8.1}$$

Note that $\bar\eta$ and $\eta$ are adjoints.

The generators $\chi = \chi^a T^a$ of infinitesimal gauge transformations act on the gauge fields $A_\mu = -igA^a_\mu T^a$ by:

$$\chi \cdot A_\mu = -\partial_\mu\chi + g[\chi, A_\mu] = -D_\mu\chi. \tag{20.8.2}$$

Though $\mathcal{L}_{\text{YM}}$ is invariant under this transformation, there seems at first no way to extend it to involve the ghost fields. The road to a BRS transformation is opened by the insight that we can substitute $\theta\eta$ for $\chi$ where $\theta$ is a constant fermionic factor. Thus the BRS transformation $\bar{\delta}$ is defined on the gauge fields by:

$$\bar{\delta}A_\mu \stackrel{\text{def}}{=} -\theta D_\mu \eta. \tag{20.8.3}$$

With this modification, $\mathcal{L}_{\text{YM}}$ is still invariant, but the fields are now mixing. The variation of the gauge-fixing term is non-zero:

$$\bar{\delta}\left(-\frac{1}{2\alpha}(\partial_\mu A^\mu)^2\right) = \frac{\theta}{\alpha}(\partial_\mu A^\mu)\partial_\nu D^\nu \eta, \tag{20.8.4}$$

which result reminds us so of the ghost term that we see how to remove it by defining the BRS transformation of $\bar{\eta}$:

$$\bar{\delta}\bar{\eta} \stackrel{\text{def}}{=} -\frac{\theta}{\alpha}(\partial_\mu A^\mu). \tag{20.8.5}$$

Now the variation of the $\bar{\eta}$ in the ghost term cancels the variation of the gauge-fixing term, and we can choose the BRS transformation of $\eta$ to make the remaining variation of the ghost term zero:

$$\begin{aligned}
\bar{\delta}(D_\mu \eta) &= D_\mu(\bar{\delta}\eta) + (\bar{\delta}D_\mu)\eta \\
&= D_\mu(\bar{\delta}\eta) + g[\bar{\delta}A_\mu, \eta] \\
&= D_\mu(\bar{\delta}\eta) - g\theta[D_\mu\eta, \eta] \\
&= D_\mu(\bar{\delta}\eta) - D_\mu\left(\tfrac{1}{2}g\theta[\eta, \eta]\right),
\end{aligned} \tag{20.8.6}$$

showing that $\bar{\delta}(D_\mu\eta) = 0$ if we define $\bar{\delta}\eta$ by:

$$\bar{\delta}\eta \stackrel{\text{def}}{=} \frac{\theta}{2}g[\eta, \eta]. \tag{20.8.7}$$

(Since $\theta$ is a ubiquitous constant factor, it is often simply dropped from the definition of $\bar{\delta}$.)

**Homework 20.8.8.** The definition of all the Lie brackets above is

$$[\alpha, \beta] \stackrel{\text{def}}{=} \alpha^a \beta^b [T^a, T^b].$$

Show that the bracket so defined has the following properties:

$$[\alpha, \beta] = \begin{cases} [\beta, \alpha], & \text{if both terms are fermionic;} \\ -[\beta, \alpha], & \text{if at least one term is bosonic;} \end{cases}$$

$$D_\mu[\alpha, \beta] = [D_\mu\alpha, \beta] + [\alpha, D_\mu\beta];$$

$$\bar{\delta}[\alpha, \beta] = [\bar{\delta}\alpha, \beta] + [\alpha, \bar{\delta}\beta].$$

Check the details of the derivation 20.8.6 of $\bar{\delta}\eta$.

**Homework 20.8.9.** Show that $(\partial_\theta\bar{\delta})^2 = 0$.

Putting the parts together, the transformation $\bar{\delta}$ is an infinitesimal symmetry of the Faddeev–Popov Lagrangian density. Since $\bar{\delta}^2 = 0$, the algebra generated by $\bar{\delta}$ is closed, and $\bar{\delta}$ does not give rise to any further linearly independent infinitesimal symmetries.

**Homework 20.8.10.** Define a ghost-number symmetry by

$$\eta \to e^{-i\tau}\eta \quad \text{and} \quad \bar{\eta} \to e^{i\tau}\bar{\eta}.$$

Find the associated Noether current and conserved quantity $Q_{\text{gh}}$. Propose canonical anti-commutation rules for the ghost fields which will enable $Q_{\text{gh}}$ to act properly on them:

$$[\eta, Q_{\text{gh}}] = \eta \quad \text{and} \quad [\bar{\eta}, Q_{\text{gh}}] = -\bar{\eta}.$$

**Homework 20.8.11.** Find the Noether current associated with the BRS symmetry. Using the ghost canonical quantization relations of the previous homework, verify that the associated conserved quantity generates the BRS transformation.

The Slavnov–Taylor identity now arises from application of $\bar{\delta}$ to the functional integral form of the generating functional for the gauge theory. As the variation creates non-linear terms $\bar{\delta}A_\mu$ and $\bar{\delta}\eta$, it is necessary to include terms for these in the source term Lagrangian density:

$$\mathcal{L}_{\text{src}} \overset{\text{def}}{=} J_\mu \cdot A^\mu + \bar{u} \cdot \eta + \bar{\eta} \cdot u - \bar{X}_\mu \cdot D^\mu \eta + \frac{1}{2} g Y \cdot [\eta, \eta], \tag{20.8.12}$$

where the dot represents summation over a group index. Note that $J$ and $Y$ are bosonic, while $\bar{u}$, $u$, and $\bar{X}$ are fermionic.

Consider now using the BRS transformation to define a change of variables $A' = A + \bar{\delta}A$, etc. in the functional integral form of the generating functional determined by the Lagrangian density $\mathcal{L} = \mathcal{L}_{\text{eff}} + \mathcal{L}_{\text{src}}$. The change of variables cannot affect the value of the generating functional, but we can gain information by equating the non-change of the whole to the sum of the changes in the parts. Since $\mathcal{L}_{\text{eff}}$ and the $X$ and $Y$ source terms are invariant (recall $\bar{\delta}^2 = 0$), the BRS variation only affects the functional measure and the $J$, $\bar{u}$, and $u$ source terms.

The functional measure, as we saw in Chapter 11, is a product of an integration measure and a differentiation measure. A change of variables will change the integration measure by a Jacobian and the differentiation measure by an inverse Jacobian. When the change of variables mixes the integration and differentiation variables, we will need to extend the old notion of Jacobian. The appropriate concept is Berezin's *superdeterminant* defined as follows:

$$M = \begin{pmatrix} A & B \\ C & D \end{pmatrix} \implies \text{sdet}(M) \overset{\text{def}}{=} \det(A - BD^{-1}C)\det(D)^{-1}. \tag{20.8.13}$$

The following homework shows how the superdeterminant naturally arises in functional calculus.

**Homework 20.8.14.** Let $z = (z_1, \ldots, z_m)$ be complex variables; let $\eta = (\eta^1, \ldots, \eta^n)$ be fermionic. Define superspace in coordinates by adjoining the variables into one vector, $(z, \eta)$. Let $M$ be a general linear transformation of superspace:

$$M = \begin{pmatrix} A & B \\ C & D \end{pmatrix}.$$

Define the superdeterminant or Berezinian sdet of $M$ by the value for the generalized Gaussian in $z$ and $\eta$:

$$\text{sdet}\begin{pmatrix} A & B \\ C & D \end{pmatrix}^{-1} \stackrel{\text{def}}{=} \left(\frac{\partial}{\partial\eta}\frac{\partial}{\partial\bar\eta}\right)\int (dz^*)(dz)\, e^{i(z^*Az + z^*B\eta + z C\bar\eta + \bar\eta D\eta)}.$$

By shifting variables or otherwise, show that

$$\text{sdet}(M) = \frac{\det(A)}{\det(D - CA^{-1}B)} = \frac{\det(A - BD^{-1}C)}{\det(D)}.$$

In our case, the Jacobian matrix $\mathbf{J}$ of the BRS change of variables $A' = A + \bar\delta A$, etc., is given by

$$\mathbf{J} = \frac{\partial(A', \bar\eta', \eta')}{\partial(A, \bar\eta, \eta)}, \tag{20.8.15}$$

where

$$\left.\begin{array}{l} A' = A'(A, \eta) \\ \bar\eta' = \bar\eta'(A, \bar\eta) \\ \eta' = \eta'(\eta) \end{array}\right\} \implies \mathbf{J} = \begin{pmatrix} dA'/dA & 0 & dA'/d\eta \\ d\bar\eta'/dA & d\bar\eta'/d\bar\eta & 0 \\ 0 & 0 & d\eta'/d\eta \end{pmatrix}. \tag{20.8.16}$$

Directly from the form of the Jacobian matrix, we see that the term $BD^{-1}C$ in its superdeterminant is zero. Consequently

$$\text{sdet}(\mathbf{J}) = \det\left(\frac{\partial A'}{\partial A}\right) \det\begin{pmatrix} d\bar\eta'/d\bar\eta & 0 \\ 0 & d\eta'/d\eta \end{pmatrix}^{-1}. \tag{20.8.17}$$

If the gauge group is $k$-dimensional, then each diagonal term can be expressed in block-diagonal form with each block being a $k \times k$ matrix. The structure of these blocks is given by:

$$\frac{\partial A_\mu'^a}{\partial A_\mu^b} = \delta^{ab} - g\theta f^{abc}\eta^c;$$

$$\frac{\partial \bar\eta'^a}{\partial \bar\eta^b} = \delta^{ab}; \tag{20.8.18}$$

$$\frac{\partial \eta'^a}{\partial \eta^b} = \delta^{ab} - g\theta f^{abc}\eta^c.$$

Since the $\theta$ terms are all off-diagonal, and since $\theta^2 = 0$, the determinant of each $k \times k$ block is unity. Thus the Jacobian for the change of variables is an infinite product of determinants of finite-dimensional Jacobian matrices, each of which is unity. Hence the BRS variation annihilates the functional measure:

$$\bar\delta\left(\left(\frac{\partial}{\partial\eta}\frac{\partial}{\partial\bar\eta}\right)(dA)\right) = 0. \tag{20.8.19}$$

Since the BRS variation of the generating functional therefore reduces to the variation of the $J$, $\bar u$, and $u$ source terms:

$$\bar\delta Z = \left(\frac{\partial}{\partial\eta}\frac{\partial}{\partial\bar\eta}\right)\int (dA)\, e^{iS_{\text{eff}} + iS_{\text{src}}}\, i\bar\delta S_{\text{src}}, \tag{20.8.20}$$

where the variation of the source action is given by

$$\bar{\delta}S_{\text{src}} = \int d^4x \, (J_\mu \cdot \bar{\delta}A^\mu + \bar{u} \cdot \bar{\delta}\eta + \bar{\delta}\bar{\eta} \cdot u)$$
$$= -\theta \int d^4x \left( J_\mu \cdot D^\mu\eta - \frac{1}{2}g\bar{u} \cdot [\eta, \eta] + \frac{1}{\alpha}(\partial_\mu A^\mu) \cdot u \right). \tag{20.8.21}$$

Since we have included extra sources for the non-linear terms $\bar{\delta}A$ and $\bar{\delta}\eta$, we see that the variation of the generating functional is linear in fields associated with source terms, and so a sum of average fields. Therefore, writing $W = -i \ln Z$ as before, we find that $\bar{\delta}Z = 0$ implies the following relation for the generating functional for connected Green functions:

**BRS Invariance:** $\int d^4x \left( J_\mu \cdot \frac{\delta W}{\delta \bar{X}_\mu} + \bar{u} \cdot \frac{\delta W}{\delta Y} + \frac{1}{\alpha}\left( \partial_\mu \frac{\delta W}{\delta J_\mu} \right) \cdot u \right) = 0.$ (20.8.22)

Define the 1PI Green functions and the effective interaction $\Gamma$ as in QED, and express the generating functional in terms of the $\hbar \to 0$ limit of the resulting formal $\Gamma$ theory. Then, noting that we do not integrate over $\bar{X}$ or $Y$ in formal $\Gamma$ theory, we find again a connection between $\Gamma$ and $W$:

$$W(J, \bar{u}, u, \bar{X}, Y) = \int d^4x \, (J_\mu \cdot B^\mu + \bar{u} \cdot \zeta + \bar{\zeta} \cdot u) + \Gamma(B, \zeta, \bar{\zeta}, \bar{X}, Y), \tag{20.8.23}$$

where $B$, $\zeta$, and $\bar{\zeta}$ are determined implicitly by the stationary phase condition:

$$\frac{\delta \Gamma}{\delta B^\mu} + J_\mu = 0, \quad \frac{\delta \Gamma}{\delta \zeta} + \bar{u} = 0, \quad \frac{\delta \Gamma}{\delta \bar{\zeta}} + u = 0. \tag{20.8.24}$$

The derivatives of $W$ with respect to the corresponding sources show that $B$, $\zeta$, and $\bar{\zeta}$ are the averages of $A$, $\eta$, and $\bar{\eta}$ respectively:

$$\frac{\delta W}{\delta J_\mu} = B^\mu, \quad \frac{\delta W}{\delta \bar{u}} = \zeta, \quad \frac{\delta W}{\delta u} = \bar{\zeta}. \tag{20.8.25}$$

Of course, the functional derivatives of $W$ and $\Gamma$ with respect to the sources $\bar{X}$ and $Y$ are equal:

$$\frac{\delta W}{\delta \bar{X}} = \frac{\delta \Gamma}{\delta \bar{X}}, \quad \frac{\delta W}{\delta Y} = \frac{\delta \Gamma}{\delta Y}. \tag{20.8.26}$$

Since the total action is linear in $\bar{\eta}$, we can extract a constraint by translating the $\bar{\eta}$ variable. Changing variable to $\bar{\eta}' = \bar{\eta} + \epsilon\bar{\xi}$, where $\bar{\xi}$ is a fermionic complex scalar and $\epsilon$ is a real number, does not change the fermionic measure. Writing $D_\epsilon$ for the operation of differentiating with respect to $\epsilon$ and then setting $\epsilon = 0$, we therefore find that

$$D_\epsilon Z = \left( \frac{\partial}{\partial\eta} \frac{\partial}{\partial\bar{\eta}} \right) \int (dA) \, e^{iS} \, iD_\epsilon S$$
$$= \left( \frac{\partial}{\partial\eta} \frac{\partial}{\partial\bar{\eta}} \right) \int (dA) \, e^{iS} \, i\bar{\xi}(u + \partial^\mu D_\mu \eta). \tag{20.8.27}$$

As this change of variables does not change the generating functional either, we see that $D_\epsilon Z = 0$. Consequently

$$Zu = -\partial_\mu \frac{\delta Z}{\delta \bar{X}_\mu} = -Z \partial_\mu \frac{\delta W}{\delta \bar{X}_\mu}, \tag{20.8.28}$$

and

**Constraint on $\bar{\zeta}$ Dependence :**    $\dfrac{\delta \Gamma}{\delta \bar{\zeta}} + \partial_\mu \dfrac{\delta \Gamma}{\delta \bar{X}_\mu} = 0.$    (20.8.29)

Now we can write the BRS invariance condition 20.8.22 in terms of $\Gamma$ and eliminate the variation in $\bar{\zeta}$ to obtain:

$$\int d^4x \left( \frac{\delta \Gamma}{\delta B^\mu} \cdot \frac{\delta \Gamma}{\delta \bar{X}_\mu} + \frac{\delta \Gamma}{\delta \zeta} \cdot \frac{\delta \Gamma}{\delta Y} + \frac{1}{\alpha} (\partial_\mu B^\mu) \cdot \partial_\nu \frac{\delta \Gamma}{\delta \bar{X}_\nu} \right) = 0. \tag{20.8.30}$$

**Remark 20.8.31.** At zeroth order, $\Gamma$ is simply the Faddeev–Popov action $S$ derived from $\mathcal{L}_{\mathrm{FP}}$, and this identity reduces to the condition for invariance of the Faddeev–Popov action under the BRS transformation. Indeed, we can use the following substitutions to transform BRS variations into variations with respect to sources:

$$\bar{\delta} B_\mu = -\theta D_\mu \zeta = \theta \frac{\delta S}{\delta \bar{X}_\mu}, \qquad \bar{\delta} \zeta = g\theta[\zeta, \zeta] = \theta \frac{\delta S}{\delta Y},$$

$$\bar{\delta} \bar{\zeta} = -\frac{\theta}{\alpha} \partial_\mu B^\mu, \qquad \frac{\delta S}{\delta \bar{\zeta}} = -\partial_\mu \frac{\delta S}{\delta \bar{X}_\mu}. \tag{20.8.32}$$

The outcome is

$$\frac{\delta S}{\delta B^\mu} \cdot \bar{\delta} B_\mu + \frac{\delta S}{\delta \zeta} \cdot \bar{\delta} \zeta + \bar{\delta} \bar{\zeta} \cdot \frac{\delta S}{\delta \bar{\zeta}} = 0$$

$$\implies \quad \frac{\delta S}{\delta B^\mu} \cdot \frac{\delta S}{\delta \bar{X}_\mu} + \frac{\delta S}{\delta \zeta} \cdot \frac{\delta S}{\delta Y} + \frac{1}{\alpha} (\partial_\mu B^\mu) \cdot \partial_\nu \frac{\delta S}{\delta \bar{X}_\nu} = 0. \tag{20.8.33}$$

As before for the Ward–Takahashi identity, gauge invariance of $\Gamma$ is only broken by the gauge-fixing term in the action. Defining the functional $\bar{\Gamma}$ by removing this term,

$$\bar{\Gamma} \overset{\text{def}}{=} \Gamma + \frac{1}{2\alpha} \int d^4x \, (\partial_\mu B^\mu)^2, \tag{20.8.34}$$

we find that $\bar{\Gamma}$ satisfies the homogeneous Slavnov–Taylor identity,

**Slavnov–Taylor Identity:**    $\displaystyle \int d^4x \left( \frac{\delta \bar{\Gamma}}{\delta B^\mu} \frac{\delta \bar{\Gamma}}{\delta \bar{X}_\mu} + \frac{\delta \bar{\Gamma}}{\delta \zeta} \frac{\delta \bar{\Gamma}}{\delta Y} \right) = 0.$    (20.8.35)

This is the dense version of the Slavnov–Taylor identities. We can expand it by choosing specific variations in the sources and using functional differentiation. Note, however, that the technical remedy to the non-linearity of the $\bar{\delta}$ variations has left us with five sources, and it is not easy to compute the dependence of $\Gamma'$ on $\bar{X}$ and $Y$.

**Homework 20.8.36.** Since QED is abelian, it does not need a ghost field. We can, however, add a decoupled ghost term to the Lagrangian density:

$$\mathcal{L}_{\mathrm{eff}}^{\mathrm{qed}} \overset{\text{def}}{=} -\frac{1}{4} F_{\mu\nu} F^{\mu\nu} + \bar{\psi}(i\slashed{\partial} - m)\psi - \frac{1}{2\alpha}(\partial_\mu A^\mu)^2 + \bar{c}\partial_\mu \partial^\mu c.$$

Find a BRS symmetry for this Lagrangian density. Use the argument of this section to obtain the Ward–Takahashi identity from this symmetry.

**Homework 20.8.37.** Consider QED with a gauge-fixing condition $\partial_\mu A^\mu + \frac{1}{2} A_\mu A^\mu = 0$. Find (1) the Faddeev–Popov Lagrangian density $\mathcal{L}$, (2) a BRS symmetry for $\mathcal{L}$, and (3) the condition for gauge invariance of the effective interaction.

At this point, the derivation of the Slavnov–Taylor identity may be more illuminating than the conclusion. Certainly, the existence of a BRS infinitesimal symmetry is remarkable. Such symmetries are invaluable for quantizing constrained field theories and string theories. The primary value of the Slavnov–Taylor identity is in the proof that renormalization of a gauge theory can preserve gauge invariance. This is the topic of the next section.

# 20.9 – Renormalization and Gauge Invariance

In the Slavnov–Taylor identity, we have a powerful constraint on the effective interaction $\Gamma$ computed from an action $S$ which is gauge invariant up to its gauge-fixing term. Using this constraint, we show in this section that the counterterms needed for perturbative renormalization may be chosen gauge invariantly. In preparation for this application, we first specify more precisely how $\Gamma$ is computed.

As in the case of QED, we start computing the effective interaction $\Gamma$ with the physical-parameter action $S_0$ derived from the Faddeev-Popov Lagrangian density

$$\mathcal{L}_{\text{FP}} = -\frac{1}{4} G_{\mu\nu} \cdot G^{\mu\nu} + \bar{\eta} \cdot \partial_\mu D^\mu \eta - \frac{1}{2\alpha} (\partial_\mu A^\mu)^2 - \bar{X}_\mu \cdot D^\mu \eta + \frac{1}{2} g Y \cdot [\eta, \eta], \quad (20.9.1)$$

where the dot again represents summation over a group index, and we define counterterms $S'_n$ order by order to cancel the poles in regularized Feynman integrals.

More precisely, we use dimensional regularization and minimal-subtraction counterterm renormalization. The peculiar advantage of dimensional regularization is that any amplitude $\mathcal{A}$ will always have a Laurent expansion in the two parameters $\hbar$ and $\delta$:

$$\mathcal{A} = \frac{1}{\hbar} \mathcal{A}_{-1} + \sum_{r=0}^{\infty} \sum_{s=1}^{r+1} \mathcal{A}_{r,s} \frac{\hbar^r}{\delta^s}. \quad (20.9.2)$$

Note that the power $s$ of $\delta^{-1}$ will never be greater than the number of loops $r+1$ and that dependence on positive powers of $\delta$ has been suppressed. Clearly, if we have products of amplitudes to consider, we should only suppress such dependence after taking the product. The existence of these expansions enables us to expand the Slavnov–Taylor identity into a family of identities bigraded by the powers of $\hbar$ and of $\delta$.

To use the grading over powers of $\hbar$, we let $\Gamma'_n$ be $\Gamma$ at $n$th order and $\Gamma_n$ be $\Gamma$ up to $n$th order. Thus $\Gamma'_0 = \Gamma_0 = S_0$ and:

$$\Gamma_n = \sum_{r=0}^{n} \Gamma'_n \quad \text{and} \quad \Gamma = \Gamma_\infty = \sum_{r=0}^{\infty} \Gamma'_r. \quad (20.9.3)$$

(The prime represents the discrete derivative.) Then, if we define the Slavnov–Taylor functional $\Theta \star \Phi$ by

$$\Theta \star \Phi \overset{\text{def}}{=} \int d^4 x \, \frac{\delta \Theta}{\delta A_\mu} \cdot \frac{\delta \Phi}{\delta \bar{X}\mu} + \frac{\delta \Theta}{\delta \eta} \cdot \frac{\delta \Phi}{\delta Y}, \quad (20.9.4)$$

the Slavnov–Taylor identity splits into a sequence of relations indexed by implicit power of $\hbar$:

$$\sum_{r+s=n} \bar{\Gamma}'_r \star \bar{\Gamma}'_s = 0. \tag{20.9.5}$$

Perturbative gauge invariance, however, requires $\bar{\Gamma}_n$ to satisfy the full Slavnov–Taylor identity:

$$\bar{\Gamma}_n \star \bar{\Gamma}_n = 0. \tag{20.9.6}$$

Clearly the quadratic nature of the identity makes this a stronger condition than the $n$th-order version 20.9.5. By construction, our $\Gamma'$'s counterterms satisfy the $n$th-order relation, but they may not be gauge invariant and so may fail the identity 20.9.6.

**Homework** 20.9.7. Verify by direct computation that $\bar{\Gamma}_0 \star \bar{\Gamma}_0 = 0$.

At first order, for example, using the Laurent expansion in $\delta$, we can separate $\Gamma'_1(S_0)$ into finite and divergent parts:

$$\Gamma'_1(S_0) = \Gamma'^{\text{fin}}_1(S_0) + \Gamma'^{\text{div}}_1(S_0). \tag{20.9.8}$$

To renormalize at the one-loop level, we must choose counterterms $S'_1$ to eliminate $\Gamma'^{\text{div}}_1(S_0)$:

$$S'_1 \overset{\text{def}}{=} -\Gamma'^{\text{div}}_1(S_0). \tag{20.9.9}$$

Then, to test for gauge invariance, we compute the Slavnov–Taylor functional for $\bar{\Gamma}_1(S_1)$ and find:

$$\bar{\Gamma}_1(S_1) \star \bar{\Gamma}_1(S_1) = \big(S + \Gamma'^{\text{fin}}_1(S_0)\big) \star \big(S + \Gamma'^{\text{fin}}_1(S_0)\big). \tag{20.9.10}$$

We now proceed to eliminate the gauge-invariant action $S$.

From the graded identities 20.9.5 we see that

$$\Gamma'_1(S_0) \star S + S \star \Gamma'_1(S_0) = 0. \tag{20.9.11}$$

Because $S$ has no $\delta$ dependence, the grading of the Slavnov–Taylor identity by powers of $\delta$ enables us to separate the finite and divergent parts of this identity:

$$\left. \begin{array}{r} \Gamma'^{\text{fin}}_1(S_0) \star S + S \star \Gamma'^{\text{fin}}_1(S_0) = 0 \\ \Gamma'^{\text{div}}_1(S_0) \star S + S \star \Gamma'^{\text{div}}_1(S_0) = 0 \end{array} \right\} \tag{20.9.12}$$

Consequently the Slavnov–Taylor functional for $\bar{\Gamma}_1(S_1)$ reduces to:

$$\bar{\Gamma}_1(S_1) \star \bar{\Gamma}_1(S_1) = \Gamma'^{\text{fin}}_1(S_0) \star \Gamma'^{\text{fin}}_1(S_0), \tag{20.9.13}$$

which will not be zero at second order in $\hbar$.

Generally, the $\bar{\Gamma}_n$ defined above are only solutions of the Slavnov–Taylor identity up to order $n$. To obtain a true solution, it is necessary to modify the $\Gamma$'s at higher order. To allow for this modification, we relax the constraint that $n$ is an upper bound on the order of $S_n$, $S'_n$, $\Gamma_n$, and $\Gamma'_n$ in $\hbar$.

Again, we intend to define $\Gamma_n$ inductively. As before, the base for the induction is that $\Gamma_0 = S_0$ satisfies $\bar{\Gamma}_0 \star \bar{\Gamma}_0 = 0$. For the second step, we assume that we have a renormalized action $S_{n-1}$ such that $\Gamma_{n-1}(S_{n-1})$ is finite up to $O(\hbar^n)$ and satisfies the Slavnov–Taylor identity. The third step is to construct $S'_n$ so that $\Gamma_n(S_n)$ is finite up to $O(\hbar^{n+1})$ and satisfies the identity.

We saw in the case of QED that the counterterms can be regarded as renormalization factors in the physical-parameter Lagrangian density. Here, we expect the same phenomenon. In detail, the action $\bar{S}_n$ derived from $S_n$ by removing the gauge-fixing term should basically be $\bar{S}_0$ with renormalized variables:

$$\bar{S}_n(A, \bar{\eta}, \eta, \bar{X}, Y; g) \overset{?}{=} \bar{S}_0(A_n, \bar{\eta}_n, \eta_n, \bar{X}_n, Y_n; g_n) + O(\hbar^{n+1}), \tag{20.9.14}$$

where

$$A_n = z_{A,n}^{1/2} A, \quad \bar{\eta}_n = z_{\bar{\eta},n}^{1/2} \bar{\eta}, \quad \eta_n = z_{\eta,n}^{1/2} \eta, \quad \text{etc.} \tag{20.9.15}$$

Writing $\bar{S}_{0,n}$ for the rescaled $\bar{S}_0$, the value of the Slavnov–Taylor functional on $\bar{S}_{0,n}$ is

$$\bar{S}_{0,n} \star \bar{S}_{0,n} = \int d^4x \left( z_{A,n}^{1/2} z_{\bar{X},n}^{1/2} \frac{\delta \bar{S}_{0,n}}{\delta A} \cdot \frac{\delta \bar{S}_{0,n}}{\delta \bar{X}} + z_{\eta,n}^{1/2} z_{Y,n}^{1/2} \frac{\delta \bar{S}_{0,n}}{\delta \eta} \cdot \frac{\delta \bar{S}_{0,n}}{\delta Y} \right). \tag{20.9.16}$$

By comparison with the $\bar{S}$ case, clearly this vanishes if

$$z_{A,n}^{1/2} z_{\bar{X},n}^{1/2} \overset{?}{=} z_{\eta,n}^{1/2} z_{Y,n}^{1/2}. \tag{20.9.17}$$

These postulates give concrete value to the detailed argument which follows.

In QED, our condition 20.7.4 that $S'_n$ be an effective set of counterterms completely determined $S'_n$. To allow for higher-order corrections in $S'_n$, we therefore relax this condition to

$$\Gamma_n'^{\text{div}}(S_{n-1}) + \Gamma'_n(S'_n) = O(\hbar^{n+1}), \tag{20.9.18}$$

together with the supplementary assumption that the higher-order contributions to $\Gamma'_n$ are simply the higher-order counterterms:

$$\Gamma'_n(S'_n) \overset{\text{def}}{=} S'_n. \tag{20.9.19}$$

Now the grading of the Slavnov–Taylor identity is thrown off. To direct our search for higher-order corrections while upholding the necessary cancellation property at $n$th order, we constrain the higher-order terms by imposing the graded identity 20.9.5 on $\Gamma'_n(S_{n-1})$:

$$\Gamma'_n \star \bar{S}_0 + \bar{S}_0 \star \Gamma'_n = -\Gamma'_{n-1} \star \Gamma'_1 - \cdots - \Gamma'_1 \star \Gamma'_{n-1}. \tag{20.9.20}$$

Dimensional regularization makes the pole terms on the left-hand side of this equation well defined, and the right-hand side is finite by hypothesis. So we can split $\Gamma'_n$ into finite and divergent parts again, and find a condition on the divergent part:

$$\Gamma_n'^{\text{div}} \star \bar{S}_0 + \bar{S}_0 \star \Gamma_n'^{\text{div}} = 0. \tag{20.9.21}$$

Now we seek a gauge-invariant solution to this equation.

**Homework** 20.9.22. Show that any functional $G(A)$ is gauge invariant if and only if

$$D_\mu \frac{\delta G}{\delta A_\mu} = 0.$$

Deduce that $G(A)$ is a solution of the condition 20.9.21.

The mass dimensions and ghost numbers of our fields and sources are

| | $A_\mu$ | $\bar\eta$ | $\eta$ | $\bar X_\mu$ | $Y$ | |
|---|---|---|---|---|---|---|
| **Ghost Number** | 0 | −1 | 1 | −1 | −2 | (20.9.23) |
| **Mass Dimension** | 1 | 1 | 1 | 2 | 2 | |

Taking mass dimensions, ghost-number conservation, and the constraint 20.8.29 on $\bar\eta$ dependence into account, one can verify that the general form of a solution to 20.9.21 which could come from counterterms in our gauge theory is

$$S'_n = \int d^4 \left( F(A) - (\bar X^a_\mu - \bar\eta^a \partial_\mu) C^{ab\mu}(A)\eta^b + \frac{1}{2} D^{abc} Y^a \eta^b \eta^c \right), \qquad (20.9.24)$$

where $F(A)$ has dimension four, $C(A)$ has dimension one, and $D^{abc}$ is a scalar anti-symmetric in $bc$.

Since our gauge-fixing term leaves the global symmetry unbroken, we can choose to make $S'_n$ invariant too. Thus,

$$C^{ab}_\mu(A) = c_1 \delta^{ab} \partial_\mu + c_2 g f^{abc} A^b_\mu \quad \text{and} \quad D^{abc} = c_3 g f^{abc}. \qquad (20.9.25)$$

Writing $G(A)$ for the integral of $F(A)$ and substituting the resulting general form into the invariance condition 20.9.21 leads to the further constraints $c_3 = c_2$ and

$$D^{ab}_\mu \frac{\delta G}{\delta A^b_\mu} + g(c_2 - c_1) f^{abc} A^b_\mu \frac{\delta S_{\text{YM}}}{\delta A^c_\mu} = 0, \qquad (20.9.26)$$

where $S_{\text{YM}}$ is Yang–Mills part of the action $S_0$.

As Homework 20.9.22 indicates, any gauge-invariant function will satisfy the homogeneous equation. Since we are restricted to mass dimension at most four, the only possibility is $S_{\text{YM}}(A)$. Adding a particular solution yields the general counterterm solution for $G(A)$:

$$G(A) = c_4 S_{\text{YM}}(A) + (c_2 - c_1) A^a_\mu \frac{\delta S_{\text{YM}}}{\delta A^a_\mu}. \qquad (20.9.27)$$

Putting the parts together, we find a general counterterm solution to the Slavnov–Taylor constraint 20.9.21:

$$S'_n = c_4 S_{\text{YM}}(A) + (c_2 - c_1) A^a_\mu \frac{\delta S_{\text{YM}}}{\delta A^a_\mu} \qquad (20.9.28)$$

$$- \int d^4 x \left( (\bar X^a_\mu - \bar\eta^a \partial_\mu)(c_1 \delta^{ac} \partial^\mu + c_2 g f^{abc} A^b_\mu)\eta^c - \frac{1}{2} c_2 g f^{abc} Y^a \eta^b \eta^c \right).$$

**Remark** 20.9.29. The condition 20.9.21 above can be written as a differential equation,

$$D\Gamma'^{\text{div}}_n = 0, \qquad (20.9.30)$$

where the differential operator $\mathbf{D}$ is defined by

$$\mathbf{D} \overset{\text{def}}{=} \int d^4x \left( \frac{\delta \bar{S}_0}{\delta A_\mu} \cdot \frac{\delta}{\delta \bar{X}_\mu} + \frac{\delta \bar{S}_0}{\delta \eta} \cdot \frac{\delta}{\delta Y} + \frac{\delta \bar{S}_0}{\delta \bar{X}_\mu} \cdot \frac{\delta}{\delta A_\mu} + \frac{\delta \bar{S}_0}{\delta Y} \cdot \frac{\delta}{\delta \eta} \right). \tag{20.9.31}$$

Since $\mathbf{D}$ is fermionic, $\mathbf{D}^2$ simplifies considerably independent of any special properties of $S_0$. In particular, the second-order part of the operator vanishes. The remaining terms can be collected into derivatives of $\bar{S}_0 \star \bar{S}_0$. Therefore $\mathbf{D}^2$ vanishes. Since $\mathbf{D}^2 = 0$, $\mathbf{D}F$ is a solution of $\mathbf{D}\Gamma = 0$ for any functional $F$.

From Homework 20.9.22, we know that any gauge-invariant functional $G(A)$ will also satisfy $\mathbf{D}G(A) = 0$.

In fact, though we have not proved it, the general solution is just the sum of these two options:

$$\mathbf{D}\Gamma_n^{\prime\text{div}} = 0 \quad \Longrightarrow \quad \Gamma_n^{\prime\text{div}} = G(A) + \mathbf{D}F(A, \bar{\eta}, \eta, \bar{X}, Y). \tag{20.9.32}$$

The particular solution found above comes from taking $G = S_{\text{YM}}$ and

$$F = cX_\mu \cdot A^\mu + c'Y \cdot \eta, \tag{20.9.33}$$

for some value of the constants $c$ and $c'$.

Having guessed the solution will be a rescaled version of $\bar{S}_0$, we can manipulate this last expression into an operator on $\bar{S}_0$ which will perform the rescaling. First, we use the homogeneity of $\mathcal{L}_{\text{YM}}(A, g)$ to eliminate $S_{\text{YM}}$ in favor of its derivatives:

$$\mathcal{L}_{\text{YM}}(A, g) = \frac{1}{2} A \frac{\delta S_{\text{YM}}}{\delta A} - \frac{1}{2} g \frac{\delta S_{\text{YM}}}{\delta g}. \tag{20.9.34}$$

Second, we compute the homogeneous derivatives of $S_0$:

$$\begin{cases} A_\mu \cdot \dfrac{\delta \bar{S}_0}{\delta A_\mu} = A_\mu \cdot \dfrac{\delta S_{\text{YM}}}{\delta A_\mu} - g(\bar{X} - \bar{\eta}\partial)_\mu \cdot [A^\mu, \eta] \\[2mm] Y \cdot \dfrac{\delta \bar{S}_0}{\delta Y} = \dfrac{1}{2} gY \cdot [\eta, \eta] \\[2mm] \bar{X}_\mu \cdot \dfrac{\delta \bar{S}_0}{\delta \bar{X}_\mu} = -\bar{X}_\mu \cdot D^\mu \eta \\[2mm] \bar{\eta} \cdot \dfrac{\delta \bar{S}_0}{\delta \bar{\eta}} = \bar{\eta} \cdot \partial_\mu D^\mu \eta \\[2mm] \eta \cdot \dfrac{\delta \bar{S}_0}{\delta \eta} = -(\bar{X} - \bar{\eta}\partial)_\mu \cdot D^\mu \eta + gY \cdot [\eta, \eta] \end{cases} \tag{20.9.35}$$

$$\begin{cases} g \dfrac{\delta \bar{S}_0}{\delta g} = g \dfrac{\delta S_{\text{YM}}}{\delta g} - g(\bar{X} - \bar{\eta}\partial)_\mu \cdot [A^\mu, \eta] + gY \cdot [\eta, \eta] \end{cases}$$

Third, we write $S_n'$ in terms of these expressions:

$$S_n' = \mathbf{C}(a_n, b_n, c_n) \bar{S}_0(A, \bar{\eta}, \eta, \bar{X}, Y; g), \tag{20.9.36}$$

where the counterterm operator $\mathbf{C}$ is given by

$$\mathbf{C}(a_n, b_n, c_n) \tag{20.9.37}$$

$$\overset{\text{def}}{=} -\int d^4x \left( a_n \left( A_\mu \cdot \frac{\delta}{\delta A_\mu} + Y \cdot \frac{\delta}{\delta Y} \right) + b_n \left( \bar{X}_\mu \cdot \frac{\delta}{\delta \bar{X}_\mu} + \bar{\eta} \cdot \frac{\delta}{\delta \bar{\eta}} + \eta \cdot \frac{\delta}{\delta \eta} \right) + c_n g \frac{\delta}{\delta g} \right),$$

and the parameters $a_n$, $b_n$, and $c_n$ are related to the $c$'s by

$$a_n = c_2 - c_1 + \frac{c_4}{2}, \quad b_n = \frac{c_1}{2}, \quad \text{and} \quad c_n = -\frac{c_4}{2}. \tag{20.9.38}$$

From this formula for $S'_n$, it is clear that the renormalized action generated by adding the $S'_n$ to $\bar{S}_0$ is given by

$$\bar{S}_n(A, \bar{\eta}, \eta, \bar{X}, Y; g) = \bar{S}_0(A_n, \bar{\eta}_n, \eta_n, \bar{X}_n, Y_n; g_n), \tag{20.9.39}$$

where

$$A_n = (1 + \cdots + a_n)A, \quad Y_n = (1 + \cdots + a_n)Y, \tag{20.9.40}$$

$$\bar{X}_n = (1 + \cdots + b_n)\bar{X}, \quad \bar{\eta}_n = (1 + \cdots + b_n)\bar{\eta}, \quad \eta_n = (1 + \cdots + b_n)\eta,$$

$$g_n = (1 + \cdots + c_n)g.$$

Note that, with this result, we have more than satisfied the conditions 20.9.14 and 20.9.17 of our 'inspired guess'.

**Homework** 20.9.41. Using the Feynman rules for a Yang–Mills theory (Section 12.7), show that the pole terms for the gauge-boson and ghost loops

and

are given respectively by

$$\Pi^{ab}_{\mu\nu}(p) = \frac{g^2}{8\pi^2\delta} f^{acd} f^{bcd} \left( \frac{19}{6} g_{\mu\nu} p^2 - \frac{11}{3} p_\mu p_\nu \right)$$

$$\text{and} \quad \tilde{\Pi}^{ab}_{\mu\nu}(p) = \frac{g^2}{8\pi^2\delta} f^{acd} f^{bcd} \left( \frac{1}{6} g_{\mu\nu} p^2 + \frac{1}{3} p_\mu p_\nu \right).$$

(This shows that the counterterms for these two diagrams separately break gauge invariance, but for the sum of the two, they preserve gauge invariance.)

The proof of the Slavnov–Taylor identity and the proof of gauge invariance of a renormalized Yang–Mills theory extends to gauge theories with matter fields. The basis for the extension is the following generalization of the BRS transformation.

The general principle of BRS transformations is to start from a gauge transformation $\tilde{g} = \exp(\Theta)$ for which

$$\delta\psi = \Theta\psi, \quad \delta\bar{\psi} = -\bar{\psi}\Theta, \quad \text{and} \quad \delta A_\mu = -D_\mu\Theta, \tag{20.9.42}$$

and substitute $\Theta = -ig\theta\eta$, where $\theta$ is a Grassmann parameter and $\eta$ is an adjoint of ghosts. This yields

$$\bar{\delta}\psi = -ig\theta\eta\psi, \quad \bar{\delta}\bar{\psi} = ig\bar{\psi}\theta\eta, \quad \text{and} \quad \bar{\delta}A_\mu = -\theta D_\mu\eta. \tag{20.9.43}$$

Note that we have $\eta$ not $\bar{\eta}$ in the transformation of $\bar{\psi}$. This transformation automatically annihilates the gauge-invariant terms. It remains to link the gauge-fixing and ghost terms with the choice

$$\bar{\delta}\bar{\eta} = -\frac{\theta}{\alpha}(\partial_\mu A^\mu) \quad \text{and} \quad \bar{\delta}\eta = \frac{\theta}{2}[\eta, \eta]. \tag{20.9.44}$$

**Homework 20.9.45.** Verify that the sum of gauge-fixing and ghost terms is annihilated by this transformation:

$$\bar{\delta}\left(\bar{\eta}\partial^{\mu} D_{\mu}\eta - \frac{1}{2\alpha}(\partial_{\mu}A^{\mu})^2\right) = 0.$$

With this symmetry in hand, we expect that all gauge theories will be renormalizable. We know, however, that theories in which the gauge fields couple to chiral currents rather than vector currents may be anomalous and therefore inconsistent. This indicates that the BRS transformation may in such theories itself be anomalous, its current may not be conserved, its conserved quantity operator may not annihilate the vacuum.

Actually, it is clear from the fermionic Jacobian matrix of the BRS transformation,

$$\frac{\partial\psi'}{\partial\psi} = 1 - ig\theta\eta = e^{-ig\theta\eta}, \tag{20.9.46}$$

that the Jacobian $\mathbf{J}_\psi$ is

$$\mathbf{J}_\psi \overset{\text{def}}{=} \det\left(\frac{\partial\psi'}{\partial\psi}\right) = \exp\big(\text{tr}(-ig\theta\eta)\big). \tag{20.9.47}$$

Hence if there are $\gamma_5$'s in the gauge couplings, there will be a $\gamma_5$ in the trace here, and we will be in the situation discussed in Section 13.5: the Jacobian will be anomalous unless the fermion representations in the theory pass the anomaly test of Theorem 13.6.8. If the fermion representations pass this test, then the BRS symmetry will be preserved by quantization, the Slavnov–Taylor identity will hold, and the renormalized theory will indeed be gauge invariant.

**Homework 20.9.48.** Consider QED with a photon mass $\lambda$. Add the usual gauge-fixing term and the ghost terms

$$\mathcal{L}_{\text{gh}} = \bar{\eta}\Box\eta - \alpha\lambda^2\bar{\eta}\eta.$$

Show that the theory is BRS-invariant. Show that the ghost mass renormalization factor satisfies $z_\lambda = 1$. Conclude that there is no higher-order correction to the photon mass.

This argument shows that the counterterms required by perturbative computations of $\Gamma$ build up the rescaled action order by order in powers of $\hbar$. At any particular order, the renormalized action is not gauge invariant, but if we allow the higher-order corrections coming from products of $z$'s to enter the renormalized action ahead of time, then we can restore gauge invariance at every order. We conclude that gauge invariance is not broken by renormalization of a Yang–Mills theory. The result extends to theories in which the gauge currents are not anomalous according to the trace criterion of Theorem 13.6.8.

# 20.10 Summary

This chapter has introduced the theory of renormalization. The first step was to extend the ideas of Chapter 10 and Chapter 19 to a complete system for counterterm renormalization. We did this by introducing the superficial degree of divergence of a Feynman diagram, and then showing that this index could be used to direct a recursive procedure for eliminating divergences from any Feynman integral. The final result, the forest formula, is of fundamental practical value, whether it is used in its original form with Taylor expansions, or in its other manifestations such as minimal subtraction.

Having established a recursive definition of counterterms, it is possible to raise the question whether they preserve global symmetries or not. The conclusion is that counterterms generated by the forest formula preserve global symmetries.

QED, QCD, and the Standard Model, however, depend on the preservation of gauge symmetries. To approach this question, we introduced the effective interaction, the generating functional for the 1PI Green functions, and transformed a gauge variation of the functional integral formula for the generating functional into a condition on this effective interaction. In the case of an abelian gauge theory, we thereby demonstrated the Ward–Takahashi identity and proved that QED with or without a photon mass is a consistent quantum theory. In the case of a non-abelian gauge theory, we derived the Slavnov–Taylor identity and showed that the counterterms in these theories can be organized in such a way that the gauge symmetry is preserved order by order.

With these achievements, we feel that the theory of renormalization has been adequately introduced. For the final topic of this text, we turn to the dependence of the Green functions on the renormalization mass parameter $\mu$, derive the renormalization group equations, and present some of their fundamental applications.

# Chapter 21

# The Renormalization Group Equations

The scaling of finite parameters with energy – providing an efficient approach to computing predictions at high energy and a link between high-energy theories and low energy phenomenology.

## Introduction

In this final chapter, we present the renormalization group equations. These equations arise from the freedom to reparameterize a Lagrangian density while preserving the $S$ matrix. We have seen in counterterm renormalization that coupling parameters like the electric charge are defined by experimental values with specific initial and final states. Clearly, we can change the experimental definition of the coupling parameter without changing the theory. To carry out such a program in detail is complicated. It is more efficient to define scaling in terms of the generic energy parameter introduced by regularization. Commonly, the renormalization group equations are differential equations which express the change of all finite parameters in the Lagrangian as a function of such a generic energy parameter.

The general framework of the renormalization group is developed in the first three sections. Section 21.1 gives a more detailed introduction to the concept of the renormalization group. Section 20.2 links dependence on the generic energy parameter to energy-dependence in Green functions, thereby setting the stage for interpretation of the renormalization group. Section 20.3 establishes the practical foundation of renormalization group theory: the $S$ matrix determines the bare Lagrangian density, so invariance of the $S$ matrix under reparameterization is equivalent to invariance of the bare Lagrangian density.

The renormalization group equations are generally very complicated. Minimal subtraction gives rise to significant simplifications. The remaining sections take this angle. Section 20.4 establishes the form of the renormalization group equations for $\phi^4$-theory with minimal subtraction. Section 20.5 computes the RGE coefficients at one-loop level and Section 20.6 solves these RGEs. Section 20.7 closes the example of $\phi^4$-theory with gives a brief argument to justify the use of the RGEs in perturbation theory: it shows how low-order computations of RGE coefficients can be used to obtain approximations to high-energy behavior which take into account the dominant contributions of all loops.

Having seen the nature of the RGEs in the simplest example, Sections 20.8 and 20.9 extend the theory to QED, pure Yang–Mills theory, and Yang–Mills theory coupled to matter fields. Here we glimpse the fundamental properties of QCD – the existence of a confinement scale and the validity of perturbative calculations at high energies. Section 20.10 describes some qualitative features of the RGEs and their solutions in $\phi^4$-theory and gauge theories. Section 20.11 describes the scaling of Green functions and the fundamental application of RGEs, namely, efficient computation of Green functions at high energy.

Finally, we bring the text to a close by extending our notion of a Lagrangian theory to include effective non-polynomial Lagrangians, a class of Lagrangians which arise naturally when we attempt to link the Standard Model to a unified field theory. Section 20.12 provides background information on the unification program, and Section 20.13 indicates how Wilson's non-perturbative or exact renormalization group suggests the consistency of perturbative theories defined by non-polynomial Lagrangians.

# 21.1  The Renormalization Group

Assume that we have (1) a finite-parameter Lagrangian density $\mathcal{L}_{\text{fp}}$, (2) renormalization conditions on the self-energy and vertex functions associated with every term in $\mathcal{L}_{\text{fp}}$, and (3) a regularization scheme. Then we can compute the counterterm Lagrangian density $\mathcal{L}_{\text{ct}}$ as a function of the regularization parameter to any desired order in perturbation theory.

We have seen that the renormalization conditions are somewhat arbitrary, particularly for vertex functions and for self-energy functions of unstable particles. It is clearly possible to change the renormalization conditions and compensate for the change by adjusting the parameters in $\mathcal{L}_{\text{fp}}$. These changes will of course generate changes in $\mathcal{L}_{\text{ct}}$. The family of such changes is called the *renormalization group*. It is simply the set of reparameterizations of a single theory.

Small deformations of the renormalization conditions naturally generate small changes in the finite parameters of $\mathcal{L}_{\text{fp}}$ and in the Green functions derived from the total Lagrangian density $\mathcal{L}$. From here we obtain the *renormalization group equations* or *RGEs*, the differential equations which govern the dependence of the finite parameters and Green functions on the renormalization conditions.

The renormalization conditions are imposed on 1PI Green functions. The parameters involved in defining a single renormalization condition fall into two categories, the renormalization point parameters ($a$'s) which specify a Lorentz orbit of the external momenta, and the renormalized value parameter ($b$'s) which sets the value of the 1PI Green function at the renormalization point. For example, for a three-point function for scalar fields we might have

$$\Gamma(p_1^2 = a_1^2, p_2^2 = a_2^2, p_3^3 = a_3^2) = b. \tag{21.1.1}$$

Generally, then, the full Lagrangian density is a function $\mathcal{L}(a, b)$. Given some specific renormalization point parameters and values $a_0$ and $b_0$, then we can use $\mathcal{L}(a_0, b_0)$ to compute the $b$'s as functions of the $a$'s. For values of $a$ far from $a_0$, such computation is lengthy because higher-order diagrams make significant contributions. The RGEs, however, enable the functions $b(a)$ to be computed more efficiently.

Geometrically, the complete solution $b$ with the specified initial condition at low energy is represented by a curved trajectory $B$. The generic one-loop estimates for the dependence of $b$ on $a$ determine curves which are tangential to $B$ at the one-loop level. Algebraically, when we integrate the one-loop RGEs we are using this one-loop information about the tangents to $B$ to generate an approximation to $B$ which (as we shall see in Section 20.7) actually includes the dominant contribution of all higher-order loops.

The RGEs, then, provide an efficient approach to computing the dependence of $\mathcal{L}_{\text{fp}}$ and Green functions on the renormalization point. By construction, the solution $b(a)$ of the RGEs for the renormalized value parameters with initial condition $b(a_0) = b_0$ has the property that $\mathcal{L}(a, b(a))$ defines the same theory as $\mathcal{L}(a_0, b_0)$ for all $a$. In Section 21.3, we shall see further that

$$\mathcal{L}\big(a, b(a)\big) = \mathcal{L}(a_0, b_0) \tag{21.1.2}$$

for all $a$; the RGEs simply determine a reparameterization of the bare Lagrangian density.

Generally, solutions to the RGEs will generate the dominant higher-order contribution up to terms of order the largest of the logs $\ln(a_{0r}/a_{0s})$ and $\ln(a_r/a_s)$. Thus, if the $a_0$'s and $a$'s are well clustered in comparison with the gap between the $a_0$'s and $a$'s, the approximation will be useful. It is therefore common practice to ignore differences between the components of $a$ and thereby reduce the number of independent variables to a single parameter $\mu$, the energy scale. The following section explains the relationship between RGEs and classical scale transformations.

## 21.2 Scale Invariance, Scale Covariance, and RGEs

Historically, the renormalization group had its origin in scale invariance. There is a viewpoint that fundamental Lagrangian densities of quantum field theories should not have dimensionful parameters. Masses and dimension-three interactions should be generated by spontaneous symmetry breaking or the phenomenon of dimensional transmutation. (Dimensional transmutation is the emergence of a mass scale in a massless theory due to quantum effects. The emergence of the confinement scale in QCD is a good example.) Classically, such Lagrangian densities are scale invariant. Quantum effects, however, make the associated current anomalous and break the symmetry. The scale transformation can still be applied to the Green functions of quantum field theories. Their effects are summarized in the Callan–Symanzik equation. In the limit of zero mass, the Callan–Symanzik equation effectively becomes the renormalization group equation.

Starting on the classical level, let $\mathcal{L}$ be a Lagrangian density and define a scale transformation by $x \mapsto e^s x$. Let $\mathbf{D}$ be the operator $d/ds$ followed by evaluation at $s = 0$. Then finite and infinitesimal transformations of the fields can be defined by

$$\begin{aligned}
s \cdot \phi(x) &\overset{\text{def}}{=} e^s \phi(e^s x) &&\implies& \mathbf{D}\phi(x) &= (1 + x \!\cdot\! \partial)\phi(x), \\
s \cdot \psi(x) &\overset{\text{def}}{=} e^{3s/2} \psi(e^s x) &&\implies& \mathbf{D}\psi(x) &= \Big(\tfrac{3}{2} + x \!\cdot\! \partial\Big)\psi(x).
\end{aligned} \tag{21.2.1}$$

Throwing in a partial derivative, we find, for example, that

$$s \cdot \partial_\mu \phi(x) \overset{\text{def}}{=} \partial_\mu s \cdot \phi(x) \implies \mathbf{D}\,\partial_\mu \phi(x) = (2 + x \!\cdot\! \partial)\partial_\mu \phi(x). \tag{21.2.2}$$

Consequently, the infinitesimal scale transformation of the partial derivative operator is simply

$$\mathbf{D}\partial_\mu = \partial_\mu. \tag{21.2.3}$$

**Remark 21.2.4.** Since $\mathbf{D}x^\mu = x^\mu$, we would expect $\mathbf{D}\partial_\mu = -\partial_\mu$. This would preserve the duality relation $\partial_\mu x^\nu = \delta_\mu^\nu$. Our definition above violates the duality relation in the interests of preserving the functional form of $\mathcal{L}$:

$$\mathcal{L}(\phi, \partial\phi) \to \mathcal{L}\big(s \cdot \phi, \partial(s \cdot \phi)\big). \tag{21.2.5}$$

Applying these transformations to an operator $A$ of mass-dimension $[A]$ and writing $N_\phi$, $N_\psi$, and $N_\partial$ for the number of $\phi$'s, $\psi$'s, and $\partial$'s in $A$, we find that

$$\mathbf{D}A = \left(N_\phi + \frac{3}{2}N_\psi + N_\partial + x\cdot\partial\right)A$$
$$= ([A] - 4)A + \partial\cdot(xA), \tag{21.2.6}$$

and so $\mathbf{D}A$ is a divergence if and only if $[A] = 4$. Consequently, mass terms and dimension-three operators in a Lagrangian break scale invariance even on the classical level.

Let $\mathcal{L}$ be the usual Lagrangian density of $\phi^4$-theory with mass parameter $m$. After quantization, we would expect that taking the limit as the mass $m$ goes to zero should restore scale invariance to the Green functions. Actually, as we shall see, quantum effects get in the way.

First, since the mass dimension of $G_n(x)$ is simply $n$, the mass dimension of $n$ fields $\phi$, we find the mass dimension of $G_n(p)$. Then, dropping $n$ propagators and a delta function gives the mass dimension of the $n$-point 1PI Green function:

$$[G_n(p)] = -3n \implies [\Gamma_n(p)] = 4 - n. \tag{21.2.7}$$

Second, we represent momenta by a dimensionless matrix $\Omega$ defined by

$$\Omega_{rs} = -\frac{p_r\cdot p_s}{E^2}, \tag{21.2.8}$$

where $E$ is an energy parameter. (If the theory were not parity invariant, dependence on the anti-symmetric combinations of $p$'s would have to be included separately.) Third, we can naively write $\Gamma_n$ as a function

$$\Gamma_n \overset{?}{=} \Gamma_n(E^2\Omega, m, g)$$
$$\overset{?}{=} E^{4-n}\Gamma_n\left(\Omega, \frac{m}{E}, g\right). \tag{21.2.9}$$

In the limit $m \to 0$, we expect the last function to reduce to $\Gamma_n(\Omega, 0, g)$. Actual computation, however, reveals that regularization introduces some other energy parameter $\mu$ which mediates the relationship between the momenta and the mass:

$$\Gamma_n = \Gamma_n(E^2\Omega, m, g; \mu)$$
$$= E^{4-n}\Gamma_n\left(\Omega, \frac{m}{E}, g; \frac{\mu}{E}\right) \tag{21.2.10}$$
$$= E^{4-n}\Gamma_n\left(\Omega, 0, g; \frac{\mu}{E}\right) \quad \text{when } m \to 0.$$

The energy dependence of $\Gamma_n$ is governed by the Callan–Symanzik equation, a topic with which we shall not deal, while the $\mu$ dependence is governed by the RGE. The result above indicates that (1) the $E$ dependence of $\Gamma_n$ is more complicated than the scale invariance of $\mathcal{L}$ would lead us to expect, (2) in the massless or high energy limit, the $\mu$ dependence of $\Gamma_n$ determines its energy dependence. Thus scale invariance in the classical theory becomes scale covariance in the quantum theory, and in scale-free theories this scale covariance is governed by the renormalization group equations.

Now that we have established the purpose and character of scale transformations in quantum theories, our next concern is to find an invariant with respect to which we can calculate the effects of a scale transformation. The bare Lagrangian density serves this purpose.

## 21.3 The Constancy of the Bare Lagrangian

In this section, we establish that a theory is determined by its bare Lagrangian density. At first thought, this appears obvious. However, our procedure for defining a theory starts from a finite-parameter Lagrangian density, a regularization procedure, and a set of renormalization conditions, and then defines counterterms order by order in powers of coupling constants. From this procedure, it is not at all clear that the resulting bare Lagrangian makes sense as a non-perturbative entity, so it is not even clear what the assertion means.

To make the discussion more concrete, assume that we have chosen a regularization scheme with parameter $\Lambda$, and the finite-parameter Lagrangian density and renormalization conditions of $\phi^3$-theory:

$$\mathcal{L}_{\text{fp}} = \frac{1}{2}\partial_\mu\phi\,\partial^\mu\phi - \frac{m^2}{2}\phi^2 - \frac{\lambda}{6}\phi^3,$$

$$\Pi(m^2) = \Pi'(m^2) = 0 \quad \text{and} \quad \Gamma(p_1^2 = p_2^2 = p_3^2 = m^2) = \lambda. \tag{21.3.1}$$

Then the counterterm Lagrangian density has the form

$$\mathcal{L}_{\text{ct}} = \frac{1}{2}(z^\phi - 1)\partial_\mu\phi\,\partial^\mu\phi - \frac{m^2}{2}(z^m - 1)\phi^2 - \frac{\lambda}{6}(z^\lambda - 1)\phi^3, \tag{21.3.2}$$

where the three renormalization factors are understood to be formal power series in the coupling constant with coefficients that are functions of $\Lambda$:

$$z(\Lambda) = 1 + \lambda^2 z_2(\Lambda) + \lambda^4 z_4(\Lambda) + \cdots \tag{21.3.3}$$

The total or bare Lagrangian density $\mathcal{L}$ is simply the sum of $\mathcal{L}_{\text{fp}}$ and $\mathcal{L}_{\text{ct}}$:

$$\mathcal{L} = \frac{1}{2}z^\phi\partial_\mu\phi\,\partial^\mu\phi - \frac{m^2}{2}z^m\phi^2 - \frac{\lambda}{6}z^\lambda\phi^3. \tag{21.3.4}$$

Note that, for the purposes of this chapter, we are regarding the bare Lagrangian as a function of the renormalized field.

This definition presents the bare Lagrangian density as a formal power series in finite parameters. (We write 'formal' because we are ignoring questions of convergence.) With this understanding, we can use functional integral quantization to derive Feynman rules from the bare Lagrangian. The amplitude coming from a Feynman diagram will be a formal power series in the finite parameters. To obtain the usual expressions of perturbation theory, we sum over the contributing diagrams and truncate the resulting power series at the desired order.

For a fixed set of finite parameters, because we are not rescaling the field, the amplitudes do not depend on how we split the bare Lagrangian into free and interaction parts. Change of finite parameters from $\lambda$ to $\bar{\lambda}$ is defined by writing the old parameters as unitary formal power series in the new:

$$\lambda = \bar{\lambda} + c_2 \bar{\lambda}^2 + c_3 \bar{\lambda}^3 + \cdots \tag{21.3.5}$$

To make the bare Lagrangian density invariant, we must also define new $z$'s by

$$\bar{\lambda} z_{\bar{\lambda}} = \lambda z_\lambda. \tag{21.3.6}$$

Then, for a fixed process, the amplitude $\bar{\mathcal{A}}(\bar{\lambda})$ computed from the bare Lagrangian with parameters $\bar{\lambda}$ is related to the amplitude $\mathcal{A}(\lambda)$ computed with parameters $\lambda$ by formal power series substitution:

$$\bar{\mathcal{A}}(\bar{\lambda}) = \mathcal{A}\big(\lambda(\bar{\lambda})\big). \tag{21.3.7}$$

Truncation is obviously a parameter-dependent operation. However, $\bar{\mathcal{A}}(\bar{\lambda})$ truncated at order $n$ in $\bar{\lambda}$ must equal $\mathcal{A}(\lambda)$ truncated at order $n$ in $\lambda$ up to $n$th order in either set of parameters. (Proof: a discrepancy at order $k \leq n$ cannot be corrected by adding terms of order $l > n$ and is therefore inconsistent with the power-series equality 21.3.7 above.)

If we also rescale the field, then our Green functions will pick up a scaling factor. Expressing this factor as a formal power series, we find that the same conclusion holds: the old and new power series coming from functional integral quantization are formally identical, and truncation at order $n$ will preserve the equality up to that order.

Splitting the bare Lagrangian up creates no combinatoric problems. The multinomial coefficients that arise in expanding the exponential of a split interaction Lagrangian in functional integral quantization are identical to the multiplicities that arise in the corresponding Feynman diagrams. For example, in a diagram with six identical vertices, there are $6!/1!3!2!$ ways of leaving one vertex unchanged, and putting in three first-order counterterms and two second-order ones. This is exactly the coefficient of $l_2^3 l_3^2$ in $(1 + l_2 + l_3)^6$.

This equality of combinatoric factors implies that the symmetry factor of a Feynman diagram depends only on the line structure, not on whether vertices are marked with crosses or orders.

For concreteness, we shall describe the computation of the self-energy function $\Pi$. First we compute the contribution $\Pi_{\text{fp}}$ to $\Pi$ from the finite parameter Lagrangian density alone:

$$\Pi_{\text{fp}}(p^2) = \lambda^2 F_2(p^2) + \lambda^4 F_4(p^2) + \cdots \tag{21.3.8}$$

Counterterm renormalization modifies $\Pi_{\text{fp}}$, taking vertex corrections and $\phi^2$ insertions into account order by order. Let us define $F_{n;r}$ to be the sum of $n$-vertex diagrams with one $r$th-order quadratic insertion; the insertion modifies a propagator in the Feynman integrand as follows:

$$\frac{i}{k^2 - m^2} \longrightarrow \frac{i}{k^2 - m^2}(iz_r^\phi k^2 - iz_r^m m^2)\frac{i}{k^2 - m^2}. \tag{21.3.9}$$

Similarly, $F_{n;r,s}$ is the sum of $n$-vertex diagrams with quadratic insertions of order $r$ and $s$, and so on.

With this notation, we find:

$$\Pi = \lambda^2 F_2 - \lambda^2(z_2^\phi p^2 - z_2^m m^2)$$
$$+ \lambda^4(F_{2;2} + F_4) + 2\lambda^4 z_2^\lambda F_2 - \lambda^4(z_4^\phi p^2 - z_4^m m^2)$$
$$+ \lambda^6(F_{2;4} + F_{2;2,2} + F_{4;2} + F_6) + 2\lambda^6 z_2^\lambda F_{2;2} \qquad (21.3.10)$$
$$+ \lambda^6(2z_4^\lambda + z_2^\lambda z_2^\lambda)F_2 + 4\lambda^6 z_2^\lambda(F_{2;2} + F_4) - \lambda^6(z_6^\phi p^2 - z_6^m m^2)$$
$$+ \cdots$$

Now we are ready to examine the computation of $\Pi$ from $\mathcal{L}$ in terms of bare parameters and verify that it yields this expansion.

Summing the Feynman diagrams for $\Pi$ using the Feynman rules for the bare Lagrangian yields an expression

$$\Pi_{\text{bp}} = (\lambda z^\lambda)^2 F_2^{\text{bp}} + (\lambda z^\lambda)^4 F_4^{\text{bp}} + \cdots \qquad (21.3.11)$$

When computing with bare parameters, the propagator expands into the renormalized propagator plus all the quadratic insertions:

$$\frac{i}{z^\phi k^2 - z^m m^2} = i\big((k^2 - m^2) + (z^\phi - 1)k^2 - (z^m - 1)m^2\big)^{-1} \qquad (21.3.12)$$
$$= \frac{i}{k^2 - m^2} + \frac{i}{k^2 - m^2}\big(i(z^\phi - 1)k^2 - i(z^m - 1)m^2\big)\frac{i}{k^2 - m^2} + \cdots$$

Clearly, substituting this expansion into the functions $F_n^{\text{bp}}$ establishes a formula for $F_n^{\text{bp}}$ in terms of the finite parameter functions $F_n$:

$$F_n^{\text{bp}} = F_n + F_{n;\times} + F_{n;\times,\times} + \cdots$$
$$= F_n + \lambda^2 F_{n;2} + \lambda^4 F_{n;4} + \lambda^6 F_{n;6} + \cdots$$
$$+ \lambda^4 F_{n;2,2} + \lambda^6 F_{n;2,4} + \cdots \qquad (21.3.13)$$
$$+ \lambda^6 F_{n;2,2,2} + \cdots$$
$$+ \cdots,$$

where the subscript $\times$ represents the quadratic insertion with coefficients $z^\phi - 1$ and $z^m - 1$.

From these expansions, it is clear that $\Pi_{\text{bp}}$ of 21.3.11 corresponds to $\Pi$ of 21.3.10 up to the terms $\lambda^n(z_n^\phi p^2 - z_n^m m^2)$. These terms come from the counterterm insertion in the one-propagator Feynman diagram; there is no associated diagram in the bare-parameter computation. Consequently, we conclude that

$$\Pi_{\text{bp}} = \Pi + (z^\phi - 1)p^2 - (z^m - 1)m^2. \qquad (21.3.14)$$

Rearranging the last equation, we effectively find equality of the renormalized propagators:

$$z^\phi p^2 - z^m m^2 - \Pi_{\text{bp}} = p^2 - m^2 - \Pi. \qquad (21.3.15)$$

This is natural since the quadratic terms are used to define the propagator. For $n > 2$, we simply have $\Gamma_n = \Gamma_n^{\text{bp}}$. (Note that these formulae usually have $z_\phi$ factors because the bare computation generally employs the bare field.)

Thus again we see that it is the bare Lagrangian which determines the Green functions $G_n$ and, for $n > 2$, the 1PI Green functions $\Gamma_n$, and again we conclude that using unitary formal power series like 21.3.5 and 21.3.6 to change finite parameters in the bare Lagrangian density leaves these Green functions formally invariant, with discrepancies arising solely due to truncation.

Finally, let us consider changing the scale of the field. If we substitute

$$\phi = \frac{1}{a}\bar{\phi}, \tag{21.3.16}$$

then the VEV of $\bar{\phi}$ is $a$, and so the Green functions $G_n$ and $\bar{G}_n$ computed with respect to $\phi$ and $\bar{\phi}$ respectively satisfy

$$G_n = \langle 0|\phi \ldots \phi|0\rangle = a^{-n}\langle 0|\bar{\phi} \ldots \bar{\phi}|0\rangle = a^{-n}\bar{G}_n. \tag{21.3.17}$$

The propagator is modified as follows:

$$\frac{i}{p^2 - m^2} \longrightarrow \frac{ia^2}{p^2 - m^2}, \tag{21.3.18}$$

and so on the one hand, dropping propagators from external lines to obtain 1PI Green functions will yield

$$\Gamma_n = a^n\bar{\Gamma}_n \quad \text{for } n \geq 2, \tag{21.3.19}$$

and on the other hand, the LSZ reduction formula will be changed by

$$\lim_{p^2 \to m^2} i(p^2 - m^2)G_n \longrightarrow \lim_{p^2 \to m^2} ia(p^2 - m^2)\bar{G}_n. \tag{21.3.20}$$

(There is only one power of $a$ because the factor comes from a single field.) Consequently, the scattering matrices $S$ and $\bar{S}$ derived from the $\phi$ and $\bar{\phi}$ Feynman rules will be identical:

$$\bar{S} = S. \tag{21.3.21}$$

Putting the parts together, we find that when the finite and divergent parameters undergo the general multiplicative transformation

$$\left.\begin{array}{ccc} \phi = \dfrac{1}{a}\bar{\phi} & m^2 = b\bar{m}^2 & \lambda = c\bar{\lambda} \\[2mm] z^{\phi} = a^2\bar{z}^{\phi} & z^m = \dfrac{a}{b}\bar{z}^m & z^{\lambda} = \dfrac{a^3}{c}\bar{z}^{\lambda} \end{array}\right\} \tag{21.3.22}$$

then the bare Lagrangian density is invariant, and the induced transformation of the Green functions is given by:

$$G_n = \frac{1}{a^n}\bar{G}_n \quad \text{for } n \geq 2, \quad \Gamma_n = a^n\bar{\Gamma}_n \quad \text{for } n > 2,$$

$$\text{and} \quad p^2 - m^2 - \Pi = \frac{1}{a^2}(p^2 - \bar{m}^2 - \bar{\Pi}). \tag{21.3.23}$$

Our conclusion is that parameter changes of this type leave the $S$ matrix formally invariant. For perturbative invariance, that is, invariance after truncation at $n$th order up to terms of order $n+1$, we need only make $a$, $b$, and $c$ formal power series in $\lambda$.

**Homework 21.3.24.** Show that, if we start from the finite parameter Lagrangian density

$$\bar{\mathcal{L}}_{\text{fp}} = \frac{1}{2}\partial_\mu\bar{\phi}\,\partial^\mu\bar{\phi} - \frac{1}{2}\bar{m}^2\bar{\phi}^2 - \frac{1}{6}\bar{\lambda}\bar{\phi}^3, \tag{21.3.25}$$

use the modified renormalization conditions

$$\bar{\Pi}(m^2) = m^2 - \bar{m}^2, \quad \bar{\Pi}'(m^2) = 1 - a^2, \quad \text{and} \quad \bar{\Gamma}(m^2, m^2, m^2) = a^3\lambda, \tag{21.3.26}$$

then compute the counterterm Lagrangian density $\bar{\mathcal{L}}_{\text{ct}}$, we shall find that

$$\bar{\mathcal{L}}_{\text{fp}} + \bar{\mathcal{L}}_{\text{ct}} = \mathcal{L}. \tag{21.3.27}$$

The argument of this section then implies $\bar{S} = S$.

The transformation 21.3.22 does not preserve the functional relation between the $z$'s and the finite parameters. In effect, there is an implicit change of renormalization scheme here. This is not unreasonable, for the $S$ matrix should not depend on the renormalization scheme.

In more detail, approaching the same idea from the other side, suppose we have computed a bare Lagrangian density $\mathcal{L}$ from a definite $\mathcal{L}_{\text{fp}}$ in a renormalization scheme $R$. Let $R'$ be another scheme. We must choose a new finite-parameter Lagrangian density $\mathcal{L}'_{\text{fp}}$ which will lead to an equality of bare Lagrangian densities, $\mathcal{L}' = \mathcal{L}$. The equations determining our choice are simply

$$z'_{\phi'}\phi'^2 = z_\phi\phi^2, \quad z'_{m'}m'^2 = z_m m^2, \quad \text{and} \quad z'_{g'}g' = z_g g. \tag{21.3.28}$$

Here the functions $z'$ depend on $R'$; they are not the same as the $z$'s. However, at lowest order we must have $z' = 1$ for all the $z'$'s. This enables us to find $\phi'$, $m'$, and $g'$ as formal power series in $\phi$ (trivial), $m$, and $g$. (The coefficients will be functions of regularization parameters.) Now, on the basis of $\mathcal{L}' = \mathcal{L}$, computing from $\mathcal{L}'$ in scheme $R'$ will yield the formal scattering matrix derived from $\mathcal{L}$ with scheme $R$: $S' = S$. If we compute to some given order, then we will obviously have equality to that order along with discrepancies at higher orders.

This section has established that the bare Lagrangian as a formal power series in finite parameters determines the $S$ matrix as a formal power series. On this basis, it is possible to change finite parameters and renormalization scheme in a theory using arbitrary formal power series.

In the next section, we take the angle that the renormalization scheme should be invariant, that is, the coefficients of the counterterms should be fixed functions of the finite parameters. This constraint leaves a one-parameter family of changes of variable, the renormalization group.

## 21.4 Minimal Subtraction and the Renormalization Group

Having established the constancy of the bare Lagrangian, we can now consider systematic variation of the finite parameters. Chapter 19 provides an example of the $z$'s obtained in on-shell renormalization. Section 19.11 shows how these $z$'s are modified in the minimal subtraction or MS renormalization scheme. The characteristic of the $z$'s generated by MS renormalization is their independence from masses and the regularization parameter $\mu$. It is therefore simplest to introduce variation of finite parameters in this scheme.

From the previous section, it would appear that we could change coupling, mass, and field parameters arbitrarily and independently. However, this freedom exists at best on the non-perturbative level. In perturbation theory, the $z$'s have extra perturbative structure which enables them to cancel divergences in Feynman diagrams order by order. If we change the value of the finite parameters indiscriminately, then this order by order cancellation will break down. Therefore it is necessary to adjust all the finite parameters in proper balance.

Let $\mathcal{L}_{\text{fp}}$ be a finite-parameter Lagrangian density, for example

$$\mathcal{L}_{\text{fp}} = \frac{1}{2}\partial_\nu \phi_0 \, \partial^\nu \phi_0 - \frac{1}{2}m_0^2\phi_0^2 - \frac{1}{4!}g_0\phi_0^4. \tag{21.4.1}$$

Since we intend to use dimensional regularization with parameters $\mu_0$ and $\delta$, $\mathcal{L}_{\text{fp}}$ needs to be adjusted to keep $g_0$ dimensionless:

$$\mathcal{L}_{\text{fp}}^{(\mu_0,\delta)} = \frac{1}{2}\partial_\nu \phi_0 \, \partial^\nu \phi_0 - \frac{1}{2}m_0^2\phi_0^2 - \frac{1}{4!}g_0\mu_0^{2\delta}\phi_0^4. \tag{21.4.2}$$

Let $\mathcal{L}^{(\mu_0,\delta)}$ be the bare Lagrangian density constructed from $\mathcal{L}_{\text{fp}}^{(\mu_0,\delta)}$ using dimensional regularization with parameters $\mu_0$ and $\delta$ and MS renormalization:

$$\mathcal{L}^{(\mu_0,\delta)} = \frac{1}{2}z_0^\phi \partial_\nu \phi_0 \, \partial^\nu \phi_0 - \frac{1}{2}z_0^m m_0^2\phi_0^2 - \frac{1}{4!}z_0^g g_0\mu_0^{2\delta}\phi_0^4. \tag{21.4.3}$$

The $z$'s here are dimensionless and by construction have the functional form

$$z_0 = z(g_0,\delta). \tag{21.4.4}$$

The explicit appearance of a $\mu_0$ in the Lagrangian density gives us a finite parameter to vary. However, we do not want to vary the regularization scheme. We therefore regard the $\mu_0$ in the Lagrangian density as a variable $\mu$ which satisfies a boundary condition $\mu = \mu_0$.

To obtain the perturbative RGEs, first we vary the $\mu$ in the Lagrangian density away from $\mu_0$ while holding $\mathcal{L}^{(\mu_0,\delta)}$ constant. This forces $g_0$ to become a function $g = g(\mu)$. Second, we maintain the functional dependence of the $z$'s on the coupling parameter. Thus, when $g_0$ is promoted to a variable $g$, the $z$'s become functions $z(g,\delta)$.

The functional form of the $z$'s expresses the generic relationships between $\mathcal{L}_{\text{fp}}$ and $\mathcal{L}_{\text{ct}}$ determined by the Feynman expansion. Preserving the functional form of the $z$'s as we vary $\mu$ guarantees perturbative finiteness of amplitudes calculated at any $\mu$.

Bringing in some notation for these ideas, we write

$$\mathcal{L}^{(\mu_0,\delta)} = \mathcal{L}^{(\mu_0,\delta)}\big(\phi(\mu), \partial\phi(\mu), m(\mu), g(\mu); \mu\big). \tag{21.4.5}$$

The explicit parameters determine the Lagrangian density on the basis of the generic $z$'s:

$$\mathcal{L}^{(\mu_0,\delta)} = \frac{1}{2}z_\phi \partial_\nu \phi\, \partial^\nu \phi - \frac{1}{2}z_m m^2 \phi^2 - \frac{1}{4!}z_g g \mu^{2\delta}\phi^4. \tag{21.4.6}$$

Equating terms in the two forms 21.4.3 and 21.4.6 of the Lagrangian density and defining $Z$'s to simplify future expressions, we find the three governing equations for the variation in parameters:

$$z_\phi(g,\delta)\phi^2(\mu,\delta) = z_\phi(g_0,\delta)\phi_0^2,$$

$$Z_m m \stackrel{\text{def}}{=} \frac{z_m(g,\delta)}{z_\phi(g,\delta)}m(\mu,\delta) = \frac{z_m(g_0,\delta)}{z_\phi(g_0,\delta)}m_0, \tag{21.4.7}$$

$$Z_g g \mu^{2\delta} \stackrel{\text{def}}{=} \frac{z_g(g,\delta)}{z_\phi(g,\delta)^2}g\Big(\frac{\mu}{\mu_0},\delta\Big)\mu^{2\delta} = \frac{z_g(g_0,\delta)}{z_\phi(g_0,\delta)^2}g_0\mu_0^{2\delta}.$$

(Since $g$ is dimensionless, we have written it as a function of dimensionless parameters.)

Now we need to see how to solve these equations under the assumption that the $z$'s are known functions. A direct solution is not available, but we can obtain differential equations for $\phi$, $m$, and $g$. The dependence of the $z$'s on $g$ makes it necessary to derive the equation for $g$ first.

It is convenient to write $t = \ln\mu$ and regard $g$ as a function of $t$. Then we differentiate the equation 21.4.7 with respect to $t$ and drop a $\mu^{2\delta}$ factor to obtain

$$2\delta g Z_g + \frac{\partial g}{\partial t}Z_g + g\frac{\partial Z_g}{\partial g}\frac{\partial g}{\partial t} = 0. \tag{21.4.8}$$

Writing

$$w = g\frac{\partial}{\partial g}\ln Z_g, \tag{21.4.9}$$

this implies

$$2\delta g + (1+w)\frac{\partial g}{\partial t} = 0. \tag{21.4.10}$$

To go any further we need to have more information about $w$ as a function of $\delta$. Since minimal subtraction only puts pole terms in $z$'s, clearly at any given order, $Z_g$ is simply a rational function in $1/\delta$. Furthermore, since $Z_g$ is 1 to zeroth order, $Z_g$ tends to 1 as $\delta$ tends to infinity. Consequently $\ln Z_g$ is analytic in $1/\delta$ for sufficiently large $\delta$. This implies that $\ln Z_g$ and its derivatives with respect to $g$ will have Taylor expansions in $1/\delta$. In particular, taking the order to infinity, we will obtain a formal sum

$$w(g,\delta) = \sum_{n=1}^{\infty}\frac{w_n(g)}{\delta^n}. \tag{21.4.11}$$

**Homework 21.4.12.** Show that in $\phi^4$-theory, $\phi^3$-theory, and QED the $k$th order contribution to $G_n$ has respectively $\frac{1}{2}(2k-n+2)$, $\frac{1}{2}(k-n+2)$, and $\frac{1}{2}(k-n+2)$ loops.

**Homework 21.4.13.** From the previous homework, prove that all $z$'s defined in $\phi^4$-theory by dimensional regularization and MS renormalization are of the form $z = 1 + t$, where

$$t(g, \delta) = \sum_{k=1}^{\infty} \sum_{n=1}^{m} t_{kn} g^k \delta^{-n}.$$

Show that $z^{-1}$ has the same form as $z$, and conclude that $Z_m$ and $Z_g$ have this form also.

Clearly with this form of $w$, taking $\delta$ to infinity in the differential equation 21.4.10 for $g$, we find that $\partial g / \partial t$ has the following form:

$$\frac{\partial g}{\partial t} = \beta(g) - 2\delta g. \tag{21.4.14}$$

This $\beta$ function is central to the renormalization group equations. (Since $g$ is finite, there cannot be any $1/\delta$ dependence in $\partial g / \partial t$.)

Directly from the equation 21.4.10, we now find that

$$w = \frac{\beta}{2\delta g - \beta} = \frac{\beta}{2\delta g} \left( 1 + \left( \frac{\beta}{2\delta g} \right) + \left( \frac{\beta}{2\delta g} \right)^2 + \cdots \right). \tag{21.4.15}$$

Comparison with the expansion 21.4.11 of $w$ shows that

$$\beta = 2g w_1. \tag{21.4.16}$$

This shows that $w_1$ determines all the other $w_n$'s. Furthermore, if we expand $Z_g$ in terms of $1/\delta$,

$$Z_g = 1 + \sum_{n=1}^{\infty} \frac{c_n(g)}{\delta^n}, \tag{21.4.17}$$

we find that

$$
\begin{aligned}
w &= g \frac{\partial}{\partial g} \ln Z_g \\
&= \sum_{n=1}^{\infty} \frac{1}{\delta^n} g \frac{\partial c_n}{\partial g} \left( 1 + \sum_{n=1}^{\infty} \frac{c_n}{\delta^n} \right)^{-1} \\
&= \frac{1}{\delta} g \frac{\partial c_1}{\partial g} + O\left( \frac{1}{\delta^2} \right).
\end{aligned}
\tag{21.4.18}
$$

Consequently

$$\beta = 2g w_1 = 2g^2 \frac{\partial c_1}{\partial g} = 2g^2 \frac{\partial}{\partial g} \operatorname{Res} Z_g, \tag{21.4.19}$$

where $\operatorname{Res} Z_g$ is the residue $c_1$ of $Z_g$ at $\delta = 0$.

Now that we have obtained a manifestly finite form for $\beta$, we can return to the differential equation 21.4.14 and take $\delta$ to zero. The result is the *renormalization group equation* for the coupling parameter $g$:

$$\textbf{RGE for } g: \quad \mu \frac{\partial g}{\partial \mu} = \beta(g), \quad \text{where} \quad \beta = 2g^2 \frac{\partial}{\partial g} \operatorname{Res} Z_g. \tag{21.4.20}$$

Now that we have a differential equation for the evolution of the coupling parameter, we can derive the corresponding equations for the field and mass parameters. The governing equation for the field is

$$z_\phi(g,\delta)\phi^2(\mu,\delta) = \text{const.} \qquad (21.4.21)$$

Differentiating with respect to $t = \ln\mu$ and dividing by $2z_\phi\phi$, we find that

$$\frac{\partial\phi}{\partial t} + \frac{\phi}{2z_\phi}\frac{\partial z_\phi}{\partial g}\frac{\partial g}{\partial t} = 0. \qquad (21.4.22)$$

Since $\phi$ is finite, there cannot be any $1/\delta$ dependence in $\phi$. Since the derivative of $\ln z_\phi$ will be a power series in $1/\delta$ without a constant term, the coefficient of $\phi$ in this equation is independent of $\delta$:

$$\gamma(g) \overset{\text{def}}{=} \frac{1}{z_\phi}\frac{\partial z_\phi}{\partial g}\frac{\partial g}{\partial t}. \qquad (21.4.23)$$

For this to be possible, we must use the $\delta$-dependent $\beta$ function of 21.4.14. Then we find that $\gamma$ is determined by the simple pole in $z_\phi$, the field $\phi$ is a function of $\mu$ but independent of $\delta$, and the renormalization group equation for $\phi(\mu)$ is

**RGE for $\phi$:** $\quad \dfrac{\partial\phi}{\partial t} + \dfrac{1}{2}\gamma(g)\phi = 0, \quad$ where $\quad \gamma = -2g\dfrac{\partial}{\partial g}\operatorname{Res}\ln z_\phi. \qquad (21.4.24)$

Similarly for the mass, we find a function $\gamma_m$ which is independent of $\delta$,

$$Z_m(g,\delta)m^2(\mu,\delta) = \text{const.}$$

$$\implies \frac{\partial m}{\partial t} + \frac{m}{2Z_m}\frac{\partial Z_m}{\partial g}\frac{\partial g}{\partial t} = 0 \qquad (21.4.25)$$

$$\implies \gamma_m(g) \overset{\text{def}}{=} \frac{1}{Z_m}\frac{\partial Z_m}{\partial g}\frac{\partial g}{\partial t},$$

deduce that $\gamma_m$ is given by the simple pole in $Z_m$, and find the renormalization group equation for the mass:

**RGE for $m$:** $\quad \dfrac{\partial m}{\partial t} + \dfrac{1}{2}\gamma_m(g)m = 0, \quad$ where $\quad \gamma_m = -2g\dfrac{\partial}{\partial g}\operatorname{Res}Z_m. \qquad (21.4.26)$

In this section, we have carefully derived the renormalization group equations in $\phi^4$-theory subject to dimensional regularization and renormalization by minimal subtraction. The resulting RGEs are particularly simple because the $z$'s do not depend on the mass $m$ or the regularization parameter $\mu$.

Since the $z$'s in the MS scheme are defined without reference to renormalization conditions, we cannot directly interpret the running parameters derived from these RGEs as parameters determined by a running renormalization point. However, the general argument 21.2.10 for linking $\mu$ to the energy dependence of the Green functions still applies, and so the link to renormalization point is in principle correct.

## 21.5 RGE Example – Scalar Theory

Now that we have established the procedure for finding renormalization group equations, we turn to presenting examples. In this section, we work out the details for $\phi^4$-theory.

The first step is to compute the values of the $z$'s to first order. We shall find that $z_\phi$ is trivial but $z_m$ and $z_g$ are not. The second step uses these first-order values to compute the $z$'s at second order. The third step extracts the RGEs from these values of the $z$'s. In the computations we shall make use of the dimensional regularization formulae of Theorem 18.7.7 and the expansions in $\delta$ given in Section 18.8.

There are three second-order diagrams which contribute to $G_4$:

$$D_1^g = \text{\Large$\bowtie$} \qquad D_2^g = \text{\Large$\ominus$} \qquad D_3^g = \text{\Large$\bowtie\!\!\ominus$} \tag{21.5.1}$$

Clearly, the last two can be obtained from the first by crossing. Since all three diagrams generate logarithmic divergences, the Taylor expansion in external momenta has a divergence in the constant term only. Consequently the residue at the pole does not involve momenta and the divergences in all three are equal. Note that the symmetry factors are 2 for all three diagrams.

The amplitude $\mathcal{A}_1^g$ associated with $D_1^g$ is given by

$$i\mathcal{A}_1^g = \frac{(-ig)^2}{2} \int \frac{d^4k}{(2\pi)^4} \frac{i}{k^2 - m^2} \frac{i}{(k + p_1 + p_2)^2 - m^2}. \tag{21.5.2}$$

Since we only want the pole, we can evaluate the integral at zero external momenta:

$$\begin{aligned}
i\mathcal{A}_1^g &= \frac{g^2}{2} \frac{i}{16\pi^2} \frac{\Gamma(\delta)}{\Gamma(2)} \left(\frac{4\pi\mu^2}{m^2}\right)^\delta + \text{finite} \\
&= i\frac{g^2}{32\pi^2} \frac{1}{\delta} + \text{finite}.
\end{aligned} \tag{21.5.3}$$

Consequently, the three diagrams $D_r^g$ contribute three times this pole to $G_4$.

The vertex counterterm and its Feynman rule are

**Vertex Counterterm** $-\frac{1}{4!}(z_g - 1)g\phi^4$:    $-i(z_g - 1)g$. $\tag{21.5.4}$

To absorb the second-order pole in $G_4$ we must take

$$z_g - 1 = \frac{3g}{32\pi^2} \frac{1}{\delta}. \tag{21.5.5}$$

In $\phi^4$-theory, there is one divergent first-order diagram and it contributes to the self-energy function:

$$D^{(1)} = \text{\Large$\frown$}\!\!-\!\!-\!\!- \tag{21.5.6}$$

The symmetry factor $s(D^{(1)})$ is 2. Dimensional regularization gives its amplitude $\mathcal{A}^{(1)}$ the value

$$\begin{aligned}
i\mathcal{A}^{(1)} &= \frac{-ig}{2} \int \frac{d^4k}{(2\pi)^4} \frac{i}{k^2 - m^2} \\
&= \frac{g}{2} \frac{i}{16\pi^2} \frac{\Gamma(\delta - 1)}{\Gamma(\delta)} \left(\frac{1}{m^2}\right)^{-1} \left(\frac{4\pi\mu^2}{m^2}\right)^\delta \\
&= im^2 \left(\frac{g}{32\pi^2}\right) \left(\frac{4\pi\mu^2}{m^2}\right)^\delta \Gamma(\delta - 1).
\end{aligned} \tag{21.5.7}$$

Expanding the functions of $\delta$ yields the pole term

$$i\mathcal{A}^{(1)} = -im^2 \left(\frac{g}{32\pi^2}\right)\frac{1}{\delta} + \text{finite.} \tag{21.5.8}$$

As there is no $p^2$ dependence in this pole, renormalization by minimal subtraction leaves $z_\phi$ at 1. The counterterm involving $z_m$ and its Feynman rule are

**Mass Counterterm** $-\frac{1}{2}(z_m - 1)m^2\phi^2$:    $-i(z_m - 1)m^2.$ $\qquad$ (21.5.9)

To absorb the pole in $\mathcal{A}^{(1)}$, we must set

$$z_m - 1 = -\left(\frac{g}{32\pi^2}\right)\frac{1}{\delta}. \tag{21.5.10}$$

The diagrams in finite parameters which contribute to the self-energy at second order are

$$D_1^{(2)} = \underset{\text{\Large 8}}{\rule{2cm}{0.4pt}} \quad \text{and} \quad D_2^{(2)} = \underset{\text{\Large O}}{\rule{2cm}{0.4pt}} \tag{21.5.11}$$

Also, since we have counterterms at first order, we have to include the counterterm insertions in the first-order divergence $D^{(1)}$:

$$D_3^{(2)} = \underset{\text{\Large O}}{\rule{2cm}{0.4pt}} \quad \text{and} \quad D_4^{(2)} = \underset{\text{\Large O}}{\rule{2cm}{0.4pt}} \tag{21.5.12}$$

The symmetry factor for $D_1^{(2)}$ is 4, so its amplitude $\mathcal{A}_1^{(2)}$ is given by:

$$i\mathcal{A}_1^{(2)} = \frac{(-ig)^2}{4} \int \frac{d^4k}{(2\pi)^4} \frac{d^4l}{(2\pi)^4} \left(\frac{i}{k^2 - m^2}\right)^2 \frac{i}{l^2 - m^2} \tag{21.5.13}$$

$$= \frac{i}{4}g^2 \left(\frac{i}{16\pi^2} \frac{\Gamma(\delta)}{\Gamma(2)} \left(\frac{4\pi\mu^2}{m^2}\right)^\delta\right) \left(\frac{i}{16\pi^2} \frac{\Gamma(\delta - 1)}{\Gamma(1)} \left(\frac{1}{m^2}\right)^{-1} \left(\frac{4\pi\mu^2}{m^2}\right)^\delta\right)$$

$$= im^2 \left(\frac{g}{32\pi^2}\right)^2 \left(\frac{4\pi\mu^2}{m^2}\right)^{2\delta} \Gamma(\delta - 1)^2 (1 - \delta).$$

Expanding the functions of $\delta$ yields the pole terms

$$i\mathcal{A}_1^{(2)} = im^2 \left(\frac{g}{32\pi^2}\right)^2 \frac{1}{\delta}\left(\frac{1}{\delta} + 1 - 2\gamma + 2\ln\frac{4\pi\mu^2}{m^2}\right) + \text{finite.} \tag{21.5.14}$$

Minimal subtraction produces $z$'s with polynomial dependence on $m^2$. The $\ln m^2$ in the residue must cancel when we add up all the second-order contributions.

Since $s(D_2^{(2)}) = 6$, the amplitude $\mathcal{A}_2^{(2)}$ for the $D_2^{(2)}$ is given by

$$i\mathcal{A}_2^{(2)}(p^2) = \frac{(-ig)^2}{6} \int \frac{d^4k}{(2\pi)^4} \frac{d^4l}{(2\pi)^4} \frac{i}{k^2 - m^2} \frac{i}{l^2 - m^2} \frac{i}{(p - k - l)^2 - m^2}. \tag{21.5.15}$$

It is not so easy to evaluate this. Fortunately, we only need the pole terms and these can be obtained without too much trouble.

To get started, we use Wick rotation (see Section 13.3 for details) and assume that all $x$'s and $p$'s are in Euclidean space:

$$\mathcal{A}_2^{(2)}(-p^2) = \frac{g^2}{6} \int \frac{d^4k}{(2\pi)^4} \frac{d^4l}{(2\pi)^4} \frac{1}{k^2+m^2} \frac{1}{l^2+m^2} \frac{1}{(p-k-l)^2+m^2}. \qquad (21.5.16)$$

To simplify the integration, we insert a Fourier transform and its inverse,

$$\mathcal{A}_2^{(2)}(-p^2) = \int d^4x\, e^{-ip\cdot x} \int \frac{d^4q}{(2\pi)^4} e^{ix\cdot q} \mathcal{A}_2^{(2)}(-q^2), \qquad (21.5.17)$$

distribute the $\exp(ix\cdot q)$ among the three propagators in $\mathcal{A}_2^{(2)}(q)$, and then translate the $q$ variable to find

$$\mathcal{A}_2^{(2)}(-p^2) = \frac{g^2}{6} \int d^4x\, e^{-ip\cdot x} \left( \int \frac{d^4q}{(2\pi)^4} \frac{e^{ix\cdot q}}{q^2+m^2} \right)^3. \qquad (21.5.18)$$

Having eliminated dot products of momenta, we can apply dimensional regularization to the $x$ and $q$ integrals.

To perform the $x$ integration, we represent the propagator by

$$\frac{1}{q^2+m^2} = \int_0^\infty ds\, e^{-s(q^2+m^2)}, \qquad (21.5.19)$$

perform the Fourier transform of the propagator in this representation,

$$\int \frac{d^Dq}{(2\pi)^D} \frac{e^{ix\cdot q}}{q^2+m^2} = \int_0^\infty ds \int \frac{d^Dq}{(2\pi)^D} e^{ix\cdot q} e^{-s(q^2+m^2)} \qquad (21.5.20)$$

$$= \int_0^\infty ds \int \frac{d^Dq}{(2\pi)^D} \exp\left( -s\left(q - \frac{ix}{2s}\right)^2 - \frac{x^2}{4s} - sm^2 \right)$$

$$= \frac{1}{(2\pi)^{D/2}} \int_0^\infty ds\, \frac{1}{(2s)^{D/2}} \exp\left( -\frac{x^2}{4s} - sm^2 \right),$$

introduce three separate parameters for the three momentum integrals in $\mathcal{A}_2^{(2)}(-p^2)$, so that the $x$ integral reduces to

$$\int d^Dx\, e^{-ip\cdot x} e^{-\sigma x^2} = \left( \frac{2\pi}{2\sigma} \right)^{D/2} \exp\left( -\frac{p^2}{4\sigma} \right), \qquad (21.5.21)$$

where

$$\sigma = \frac{1}{4}\left( \frac{1}{s_1} + \frac{1}{s_2} + \frac{1}{s_3} \right). \qquad (21.5.22)$$

Putting the parts together, we finally obtain a convenient expression for the amplitude:

$$\mathcal{A}_2^{(2)}(-p^2) = -\frac{1}{6}\left( \frac{g}{16\pi^2} \right)^2$$

$$\times (4\pi\mu^2)^{2\delta} \int ds_1\, ds_2\, ds_3\, (s_2 s_3 + s_1 s_3 + s_1 s_2)^{-2+\delta} \qquad (21.5.23)$$

$$\times \exp\left( -(s_1+s_2+s_3)m^2 - \frac{s_1 s_2 s_3}{s_2 s_3 + s_1 s_3 + s_1 s_2}p^2 \right).$$

We can now change the sign of $p^2$ and interpret $p$ as a Minkowski-space variable.

This integral is divergent at $D = 4$, but as usual differentiating with respect to $p^2$ reduces the degree of divergence; indeed, in the Taylor series about $p^2 = 0$, only the constant and the coefficient of $p^2$ diverge. It takes some intelligent manipulations to evaluate even these terms. Here we shall simply reproduce the result:

$$
\begin{aligned}
i\mathcal{A}_2^{(2)}(p^2) &= im^2\left(\frac{g}{32\pi^2}\right)^2\left(\frac{4\pi\mu^2}{m^2}\right)^{2\delta}\Gamma(\delta-1)^2(1+\delta) \\
&+ \frac{i}{6}\left(\frac{g}{32\pi^2}\right)^2\frac{p^2}{\delta} + \text{finite.}
\end{aligned}
\tag{21.5.24}
$$

Expanding in $\delta$ yields

$$
\begin{aligned}
i\mathcal{A}_2^{(2)}(p^2) &= im^2\left(\frac{g}{32\pi^2}\right)^2\frac{1}{\delta}\left(\frac{1}{\delta}+3-2\gamma+2\ln\frac{4\pi\mu^2}{m^2}\right) \\
&+ \frac{i}{6}\left(\frac{g}{32\pi^2}\right)^2\frac{p^2}{\delta} + \text{finite.}
\end{aligned}
\tag{21.5.25}
$$

The amplitude $\mathcal{A}_3^{(2)}$ for diagram $D_3^{(2)}$ is given by

$$
\begin{aligned}
i\mathcal{A}_3^{(2)} &= \frac{-ig}{2}\int\frac{d^4k}{(2\pi)^4}\frac{i}{k^2-m^2}\left(i\frac{gm^2}{32\pi^2}\frac{1}{\delta}\right)\frac{i}{k^2-m^2} \\
&= -m^2\frac{g^2}{64\pi^2}\frac{1}{\delta}\int\frac{d^4k}{(2\pi)^4}\frac{1}{(k^2-m^2)^2} \\
&= -m^2\frac{g^2}{64\pi^2}\frac{1}{\delta}\left(\frac{i}{16\pi^2}\frac{\Gamma(\delta)}{\Gamma(2)}\left(\frac{4\pi\mu^2}{m^2}\right)^{\delta}\right) \\
&= im^2\left(\frac{g}{32\pi^2}\right)^2\left(\frac{1}{\delta}-1\right)\left(\frac{4\pi\mu^2}{m^2}\right)^{\delta}\Gamma(\delta-1).
\end{aligned}
\tag{21.5.26}
$$

Note that dimensional regularization adds a factor of $\mu^{2\delta}$ for the original $g$ but not for the $g$ in $z_m - 1$ since the $z$'s are dimensionless in any dimension.

Expanding in $\delta$ we find the pole terms

$$
i\mathcal{A}_3^{(2)} = -im^2\left(\frac{g}{32\pi^2}\right)^2\frac{1}{\delta}\left(\frac{1}{\delta}-\gamma+\ln\frac{4\pi\mu^2}{m^2}\right) + \text{finite.}
\tag{21.5.27}
$$

The diagram $D_4^{(2)}$ is obtained from $D^{(1)}$ by substituting the counterterm vertex for the finite one. As a result, the amplitude $\mathcal{A}_4^{(2)}$ is just $\mathcal{A}^{(1)}$ with an extra factor of $z_g - 1$:

$$
\begin{aligned}
i\mathcal{A}_4^{(2)} &= \frac{-i(z_g-1)g}{2}\int\frac{d^4k}{(2\pi)^4}\frac{i}{k^2-m^2} \\
&= \frac{3g^2}{64\pi^2}\frac{1}{\delta}\left(\frac{i}{16\pi^2}\frac{\Gamma(\delta-1)}{\Gamma(1)}\left(\frac{1}{m^2}\right)^{-1}\left(\frac{4\pi\mu^2}{m^2}\right)^{\delta}\right) \\
&= im^2\left(\frac{g}{32\pi^2}\right)^2\frac{3}{\delta}\left(\frac{4\pi\mu^2}{m^2}\right)^{\delta}\Gamma(\delta-1).
\end{aligned}
\tag{21.5.28}
$$

Expanding in $\delta$ yields

$$i\mathcal{A}_4^{(2)} = -im^2 \left(\frac{g}{32\pi^2}\right)^2 \frac{3}{\delta}\left(\frac{1}{\delta} + 1 - \gamma + \ln\frac{4\pi\mu^2}{m^2}\right). \tag{21.5.29}$$

Summing the four $\mathcal{A}^{(2)}$'s we find that the $\ln m^2$ contributions to the poles cancel. We can see this directly from the expanded forms of the amplitudes, or we can use the unexpanded forms 21.5.13, 21.5.25, 21.5.26, and 21.5.28. In detail, define $\mathcal{B}$'s by

$$\mathcal{A}_r^{(2)}(p^2 = 0) = m^2\left(\frac{g}{32\pi^2}\right)^2 \mathcal{B}_r, \tag{21.5.30}$$

and set

$$R = \frac{4\pi\mu^2}{m^2}, \tag{21.5.31}$$

then the $\mathcal{B}$'s are given by

$$\mathcal{B}_1 = R^{2\delta}\Gamma(\delta-1)^2(1-\delta), \qquad \mathcal{B}_3 = \left(\frac{1}{\delta}-1\right)R^{\delta}\Gamma(\delta-1),$$

$$\mathcal{B}_2 = R^{2\delta}\Gamma(\delta-1)^2(1+\delta), \qquad \mathcal{B}_4 = \frac{3}{\delta}R^{\delta}\Gamma(\delta-1), \tag{21.5.32}$$

$$\mathcal{B}_1 + \mathcal{B}_2 = 2R^{2\delta}\Gamma(\delta-1)^2, \qquad \mathcal{B}_3 + \mathcal{B}_4 = \left(\frac{4}{\delta}-1\right)R^{\delta}\Gamma(\delta-1).$$

Adding the four $\mathcal{B}$'s together and completing the square, we find

$$\sum \mathcal{B}_r = 2\left(R\Gamma(\delta-1) + \left(\frac{1}{\delta}-\frac{1}{4}\right)\right)^2 - 2\left(\frac{1}{\delta}-\frac{1}{4}\right)^2 + \text{finite}$$

$$= -2\left(\frac{1}{\delta}-\frac{1}{4}\right)^2 + \text{finite} \tag{21.5.33}$$

$$= -\frac{2}{\delta^2} + \frac{1}{\delta} + \text{finite}.$$

This gives us the second-order contribution to $z_m$. Putting the parts together, we see that the counterterms are given by

$$z_g \overset{(1)}{=} 1 + \frac{3g}{32\pi^2}\frac{1}{\delta},$$

$$z_\phi \overset{(2)}{=} 1 - \frac{1}{6}\left(\frac{g}{32\pi^2}\right)^2 \frac{1}{\delta}, \tag{21.5.34}$$

$$z_m \overset{(2)}{=} 1 - \frac{g}{32\pi^2}\frac{1}{\delta} - \left(\frac{g}{32\pi^2}\right)^2\left(\frac{2}{\delta^2} - \frac{1}{\delta}\right).$$

Recalling that

$$Z_m = \frac{z_m}{z_\phi} \quad \text{and} \quad Z_g = \frac{z_g}{z_\phi^2}, \tag{21.5.35}$$

we note that $Z_g$ is equal to $z_g$ at first order, while

$$Z_m = 1 - \frac{g}{32\pi^2}\frac{1}{\delta} - \left(\frac{g}{32\pi^2}\right)^2\left(\frac{2}{\delta^2} - \frac{7}{6}\frac{1}{\delta}\right). \tag{21.5.36}$$

Now the formulae of the previous section yield the characteristic functions of the RGEs:

$$\beta(g) = 2g^2 \frac{\partial}{\partial g} \operatorname{Res} Z_g \stackrel{(2)}{=} \frac{3g^2}{16\pi^2},$$

$$\gamma_m(g) = -2g \frac{\partial}{\partial g} \operatorname{Res} Z_m \stackrel{(2)}{=} \frac{g}{16\pi^2} - \frac{5}{6}\left(\frac{g}{16\pi^2}\right)^2, \qquad (21.5.37)$$

$$\gamma(g) = -2g \frac{\partial}{\partial g} \operatorname{Res} z_\phi \stackrel{(2)}{=} \frac{1}{6}\left(\frac{g}{16\pi^2}\right)^2.$$

**Homework** 21.5.38. Using the value of $\beta$ in 21.4.14 and the values of the $z$'s in 21.5.34, show that the functions $\gamma(g)$ and $\gamma_m(g)$ defined by

$$\gamma(g) = \frac{\partial}{\partial t} \ln z_\phi \quad \text{and} \quad \gamma_m(g) = \frac{\partial}{\partial t} \ln Z_m$$

are independent of $\delta$ up to second order in $g$.

This section illustrates the computation of the RGE scaling functions $\beta$, $\gamma_m$, and $\gamma$. On the basis of the previous section, the procedure is routine but the details of extracting pole terms from Feynman integrals beyond one loop remain tricky. Fortunately, as we shall discover in the following sections, a good deal of useful information can be obtained from RGEs at first order.

## 21.6 – Solving the RGEs

To solve a system of RGEs, we begin with the equation for the running coupling parameter. In the case of $\phi^4$-theory, this means solving

$$\frac{\partial g}{\partial t} = \frac{3g^2}{16\pi^2}. \qquad (21.6.1)$$

This is easy because we can separate the variables:

$$\int \frac{dg}{g^2} = \int dt \, \frac{3}{16\pi^2} \quad \Longrightarrow \quad \frac{1}{g_0} - \frac{1}{g} = \frac{3}{16\pi^2} \ln \frac{\mu}{\mu_0}$$

$$\Longrightarrow \quad g(\mu) = \frac{16\pi^2 g_0}{16\pi^2 - 3g_0(\ln\mu - \ln\mu_0)}. \qquad (21.6.2)$$

The principle applies to higher-order RGEs as well, but the integration gets tougher as the degree of $\beta$ rises.

From the RGE for the scale of the field $\phi$,

$$\frac{\partial \phi}{\partial t} = \frac{1}{6}\left(\frac{g}{16\pi^2}\right)^2 \phi = \frac{1}{18} \frac{1}{16\pi^2} \frac{\partial g}{\partial t} \phi, \qquad (21.6.3)$$

we see that it is more natural to find $\phi$ as a function of $g$ than as a function of $\mu$. Indeed, we deduce at once that

$$\ln \frac{\phi}{\phi_0} = \frac{1}{18} \frac{g - g_0}{16\pi^2}, \qquad (21.6.4)$$

or more conveniently

$$\phi = \phi_0 \exp\left(\frac{1}{18}\frac{g-g_0}{16\pi^2}\right).\tag{21.6.5}$$

Finally, the RGE for the running mass is

$$\frac{\partial m^2}{\partial t} = \left(\frac{g}{16\pi^2} - \frac{5}{6}\left(\frac{g}{16\pi^2}\right)^2\right)m^2.\tag{21.6.6}$$

Again it is natural to take advantage of the absence of explicit $\mu$ dependence and solve for $m^2$ as a function of $g$:

$$\frac{1}{m^2}\frac{dm^2}{dg} = \frac{16\pi^2}{3g^2}\left(\frac{g}{16\pi^2} - \frac{5}{6}\left(\frac{g}{16\pi^2}\right)^2\right) = \frac{1}{3g} - \frac{5}{18}\frac{1}{16\pi^2}.\tag{21.6.7}$$

Integrating this we find that

$$\ln\frac{m^2}{m_0^2} = \frac{1}{3}\ln\frac{g}{g_0} - \frac{5}{18}\frac{g-g_0}{16\pi^2}.\tag{21.6.8}$$

Exponentiating yields the final result

$$m^2 = m_0^2\left(\frac{g}{g_0}\right)^{1/3}\exp\left(-\frac{5}{18}\frac{g-g_0}{16\pi^2}\right).\tag{21.6.9}$$

**Homework** 21.6.10. Show that minimal subtraction assigns to the diagram

$$D = \times\!\!\times$$

the renormalized amplitude $\mathcal{A}$ given by

$$\mathcal{A}((p+q)^2) = \frac{1}{16\pi^2}\left(\ln\frac{e^\gamma m^2}{4\pi\mu^2} + z\ln\frac{z+1}{z-1} - 2\right),$$

where

$$z^2 = 1 - \frac{4m^2}{(p+q)^2}.$$

(Note that external line corrections are not included here. They would be included if we were computing a contribution to a Green function.)

**Homework** 21.6.11. Define a renormalization point for scattering in $\phi^4$-theory by the values of the Mandelstam parameters

$$s = 4m^2, \quad u = 0, \quad t = 0.$$

Using the previous homework, show that coupling strength $g_{ph}$ at this renormalization point is given to second order by

$$g_{ph} = g + \frac{g^2}{16\pi^2}\left(3\ln\frac{e^\gamma m^2}{4\pi\mu^2} - 2\right).$$

Show that the running values of $g$ and $m$ found above make $g_{ph}$ independent of $\mu$ at second order. (Here corrections to the external lines are not included because $g_{ph}$ is an $S$-matrix element.)

From these expressions for the running parameters we conclude that increasing $\mu$ causes all three parameters $g$, $\phi$, and $m$ to increase. It would appear that there is a maximum value for $m$ and a singularity in $g$ when $\mu$ gets very large, but our formulae are meaningless for such values of $\mu$ because they are way beyond the Planck scale, and this kind of theory has no meaning there, and way beyond the range in which a first-order RGE can be relied upon. Because of these limitations on the range of $\mu$, the damping of $\ln \mu$ by $16\pi^2$ in the formula for $g$ makes the range of $g$ very small, and the damping of $g$ dependence in the exponential factors in $\phi$ and $m$ makes these exponentials completely negligible. In effect, $g$ increases a little with $\mu$, $\phi$ is constant and $m^2$ scales like $g^{1/3}$.

We introduced the RGEs with the promise that they would enable us to compute high-energy amplitudes with low-order effort. The following section indicates that we are on the right track by showing how low-order RGEs lead to excellent approximations to a high-energy effective coupling.

## 21.7 – The Leading-Logarithm Approximation

In this section, we will show how, in $\phi^4$-theory with dimensional regularization and MS renormalization, high-order contributions enter low-order RGEs.

First, we expand $\beta$ in the RGE as a power series in the running coupling, which we shall write as $\bar{g}$ in this section:

$$\frac{d\bar{g}}{dt} = \beta(\bar{g}) = b_2 \bar{g}^2 + b_3 \bar{g}^3 + \cdots, \tag{21.7.1}$$

where $t = \ln(\mu/\mu_0)$. (Of course, the odd $b$'s are zero, but the sums will look better if we leave such coefficients in.) The solution $\bar{g}$ is a function $\bar{g}(g,t)$ of $t$ and the initial value, $\bar{g} = g$ at $t = 0$. Note that, because we are using the MS scheme, there is no $t$ dependence in the coefficients $b_k$.

Second, we guess at the form of this solution, writing

$$\bar{g}(g,t) = \sum_{n=1}^{\infty} \sum_{r=1}^{n} c_{n,r} g^n t^{n-r}. \tag{21.7.2}$$

The terms in $\bar{g}$ are naturally graded by the index $r$ which indicates the deficit $\Delta$ in the power of $t$ relative to the power of $g$:

$$\Delta(g^n t^{n-r}) = n - (n-r) = r. \tag{21.7.3}$$

Third, we must see if we can solve the RGE by iteratively determining the coefficients $c_{n,r}$. If we let $\bar{g}_r$ be the sum of terms with constant deficit $r$,

$$\bar{g}_r \stackrel{\text{def}}{=} \sum_{n=r}^{\infty} c_{n,r} g^n t^{n-r}, \tag{21.7.4}$$

then, since the deficit function is clearly additive on products,

$$\Delta(\bar{g}_r \bar{g}_s) = r + s. \tag{21.7.5}$$

Consequently, we can use this grading to split the renormalization group equation for $\bar{g}$ into a 'triangular' set of linked equations:

$$\frac{d\bar{g}}{dt} = b_2\bar{g}^2 + b_3\bar{g}^3 + \cdots \tag{21.7.6}$$

$$\implies \quad \frac{d\bar{g}_1}{dt} + \frac{d\bar{g}_2}{dt} + \cdots = b_2(\bar{g}_1 + \bar{g}_2 + \cdots)^2 + b_3(\bar{g}_1 + \bar{g}_2 + \cdots)^3 + \cdots$$

$$\implies \quad \Delta = 2: \quad \frac{d\bar{g}_1}{dt} = b_2\bar{g}_1^2$$

$$\Delta = 3: \quad \frac{d\bar{g}_2}{dt} = 2b_2\bar{g}_1\bar{g}_2 + b_3\bar{g}_1^3$$

$$\Delta = 4: \quad \frac{d\bar{g}_3}{dt} = \cdots$$

Clearly, the functions $\bar{g}_r$ can be iteratively determined and then summed to solve the original RGE. In fact, if we take the power series expansions of the $\bar{g}_r$ and substitute them in each equation, each coefficient $c_{n,r}$ can be determined from coefficients with lower indices, and the values $c_{1,1} = 1$ and $c_{n,n} = 0$ for $n > 1$ determined by the initial condition completes the solution.

**Homework 21.7.7.** Why is $b_1 = 0$ in the expansion of $\beta$?

The consequence of this grading of the renormalization group equation is that the equation can be solved up to any deficit $\Delta$. In particular, the coefficient $c_2$ determines the leading logarithm approximation $\bar{g}_1$. This is remarkable because typically $c_2$ is computed from one-loop diagrams, yet $\bar{g}_1$ contains the leading logarithm contribution of all multi-loop diagrams.

Having seen the effectiveness of the RGEs in making the best use of low-order computations, we turn next to RGEs in QED and Yang–Mills theories.

## 21.8  Further Examples – QED

In QED, as we saw in Section 19.11, if we use an off-shell mass $m$, we can set up a Lagrangian density under the minimal subtraction renormalization scheme:

$$\mathcal{L} = -\frac{1}{4}z_A F_{\mu\nu}F^{\mu\nu} + z_\psi \bar{\psi}i\partial\!\!\!/\psi - z_m m\bar{\psi}\psi - z_e \mu^\delta e\bar{\psi}A\!\!\!/\psi. \tag{21.8.1}$$

From Section 19.11, writing $\alpha = e^2/4\pi$, we have the values of the $z$'s at second order in $e$:

$$z_A = 1 - \frac{\alpha}{3\pi}\frac{1}{\delta}, \quad z_m = 1 - \frac{\alpha}{\pi}\frac{1}{\delta}, \quad \text{and} \quad z_e = z_\psi = 1 - \frac{\alpha}{4\pi}\frac{1}{\delta}. \tag{21.8.2}$$

The scaling of $e$ comes from that component $Z_e$ of $z_e$ which is not due to the scaling of the fields:

$$Z_e = \frac{z_e}{z_\psi\sqrt{z_A}} = \frac{1}{\sqrt{z_A}} \overset{(2)}{=} 1 + \frac{\alpha}{6\pi}\frac{1}{\delta}. \tag{21.8.3}$$

The cancellation here is valid at all orders because of the Ward identity. (In general, derivation of the equality $z_e = z_\psi$ from the Ward identity 20.6.13 depends on the renormalization conditions. For minimal subtraction, however, we can deduce the equality directly from the analysis of diagrams presented in Section 19.3.)

Because the coupling parameter $e$ is multiplied by $\mu^\delta$ instead of the $\mu^{2\delta}$ appropriate in $\phi^4$-theory, we find that the methods of Section 21.4 apply with

$$\frac{\partial e}{\partial t} = \beta(e) - \delta e. \tag{21.8.4}$$

Straight away we find the following RGE for $e$:

$$\mu\frac{\partial}{\partial\mu}e = e^2\frac{\partial}{\partial e}\operatorname{Res}Z_e = \frac{e^3}{12\pi^2}, \tag{21.8.5}$$

or in terms of $\alpha$,

$$\mu\frac{\partial}{\partial\mu}\alpha = 4\alpha^2\frac{\partial}{\partial\alpha}\operatorname{Res}Z_e = \frac{2\alpha}{3\pi}. \tag{21.8.6}$$

The solution of the RGE is

$$\alpha(\mu) = \frac{\alpha_0}{1 - \dfrac{2\alpha_0}{3\pi}\ln\dfrac{\mu}{\mu_0}}. \tag{21.8.7}$$

Similarly, the scaling of the electron mass is governed by

$$Z_m = \frac{z_m}{z_\psi}\overset{(2)}{=}1 - \frac{3\alpha}{4\pi}\frac{1}{\delta}, \tag{21.8.8}$$

and the RGE for the mass is given by

$$\frac{\mu}{m}\frac{\partial}{\partial\mu}m = e\frac{\partial}{\partial e}\operatorname{Res}Z_m = -\frac{3\alpha}{2\pi}. \tag{21.8.9}$$

Using the RGE above for $\alpha(\mu)$, the solution of this equation is

$$m = m_0\exp\left(-\frac{4}{9}(\alpha - \alpha_0)\right). \tag{21.8.10}$$

The scaling of the fields is similar to the scaling of the mass. Since $m_0$ is very small, the change in mass as $\mu$ runs up to the Planck scale is insignificant.

The scaling laws above are fine for pure QED. When considering QED in the context of the known particles, clearly we must include the loop contributions of all charged particles in these calculations. In particular, the photon self-energy diagram can contain a particle–anti-particle pair for any charged field. However, the practical value of the RGEs will be lost if we simply include the poles from all fields right from the start. Instead, we include a field only when the energy of the incoming state is at the mass of that field. This will keep the log terms small.

Adding fields as the energy increases demands a new perspective on the role of Lagrangian densities in the definition of a theory. In particular, we must have a different Lagrangian density $\mathcal{L}_r^{(\mu)}$ for each energy range $[m_r, m_{r+1}]$, and on each interval $[m_r, m_{r+1}]$, the dependence of $\mathcal{L}_r^{(\mu)}$ on $\mu$ must be given by RGEs. The system is integrated by a boundary condition at the end of each interval: $\mathcal{L}_{r-1}^{(m_r)}$ is equal to $\mathcal{L}_r^{(m_r)}$ up to terms involving fields with mass $m_r$. Thus, instead of having one Lagrangian density for all energies, we have an energy-dependent family of *effective Lagrangian densities*. This scheme makes for more efficiency in computation.

In such a scheme, one can introduce non-renormalizable operators to represent the effect of heavy-particle exchange in light fields. The four-Fermi operators of the low-energy Standard Model provide an example. As long as one can obtain an RGE for the coefficients of these operators, they can be used at tree level over a range of energies. The drawback of using energy-dependent effective Lagrangian densities is the difficulty in estimating the accuracy of a calculation.

Returning to QED, writing $m_r$ and $Q_r$ for the mass and charge of the matter fields, we define $z_A$ by the energy-dependent expression

$$z_A(\mu) = 1 - \frac{\alpha}{3\pi} \frac{1}{\delta} \sum_{m_r < \mu} Q_r^2. \tag{21.8.11}$$

With this modification, the RGE becomes

$$\mu \frac{\partial}{\partial \mu} \alpha = \frac{2\alpha}{3\pi} \sum_{m_r < \mu} Q_r^2. \tag{21.8.12}$$

To solve this, we organize the fields so that $m_r \leq m_{r+1}$ for all $r$ and then integrate over each domain $[m_r, m_{r+1}]$ separately, using the end result of each integration as the initial condition for the next. The net result is

$$\frac{1}{\alpha_r} = \frac{1}{\alpha_0} - \frac{2}{3\pi} \Big( Q_1^2 \ln \frac{\mu_r}{\mu_0} + \cdots + Q_r^2 \ln \frac{\mu_r}{\mu_{r-1}} \Big). \tag{21.8.13}$$

(When $\mu$ reaches the mass of the $W$ vector bosons, their contribution should also be included.)

Note that the one-loop RGE is linear in contributions of the different particles. Hence we could obtain the same result by separately integrating the RGE for each contribution from the mass of its particle up to $\mu$ and then adding.

Applying this formula to the range from $\mu \sim m_e$ to $\mu \sim M_W$ with the quark masses given in Appendix A.1, we find that

$$9.7 < \frac{1}{\alpha(m_e)} - \frac{1}{\alpha(M_W)} < 10.3. \tag{21.8.14}$$

This fits well with the values

$$\alpha(m_e) = \frac{1}{137} \quad \text{and} \quad \alpha(M_W) = \frac{1}{127}. \tag{21.8.15}$$

Note that each quark comes in three colors; without this factor of three, the difference 21.8.14 would be about 6.5, which is simply too small.

In this section, we have seen that our theory of RGEs applies readily enough to QED at low energies. However, if we are to cover the range of energies from the mass of the electron to that of the $W$ boson, then we cannot keep the logarithms small with this approach. We must introduce an energy-dependent family of effective Lagrangian densities linked by the RGEs and boundary conditions. Such a structure takes maximal advantage of perturbative computations but makes it hard to estimate errors.

## 21.9 – RGEs in Yang–Mills Theories

In QED, the $\beta$ function is derived from the matter fields. By contrast, in Yang–Mills theories the $\beta$ function comes from gauge fields. In this section, we find the evolution of the coupling parameter in Yang–Mills theory. To minimize the notation, we shall assume that the global symmetry group $G$ of the theory is a simple group as defined in Section 12.6.

The basis for the RGE is the introduction of the $\mu$ parameter in the Lagrangian density to keep the coupling parameter $g$ dimensionless. In four dimensions, the Lagrangian density is given by

$$\mathcal{L} = -\frac{1}{4}(\partial_\mu A_\nu^a - \partial_\nu A_\mu^a + g f^{abc} A_\mu^b A_\nu^c)(\partial^\mu A^{a\nu} - \partial^\nu A^{a\mu} + g f^{ade} A^{d\mu} A^{e\nu}), \quad (21.9.1)$$

which implies that we must replace $g$ by $g\mu^\delta$ to obtain $\mathcal{L}^{(\mu,\delta)}$.

As we found in Section 12.8, quantization introduces a gauge-fixing term and ghosts. We write $\zeta$ for the gauge parameter. The ghost fields and $\zeta$ scale, but we shall not be interested in their scaling laws. To find $z_A$ at second-order, we must evaluate three diagrams:

$$\hspace{10cm} (21.9.2)$$

The value of $z_A$ is set by the poles in the first two; the third evaluates to zero.

The value of $z_g$ could be computed from either the three-bose or four-bose vertices. Taking the three-bose vertex, for the second-order contribution we must extract the poles from the bose and ghost loops diagrams:

$$\hspace{10cm} (21.9.3)$$

The computations resemble the QED computations of Chapter 19. They result in the following values for the $z$'s:

$$z_A = 1 + \frac{g^2}{32\pi^2}\left(\frac{13}{3} - \zeta\right)\frac{C_2(G)}{\delta},$$

$$z_g = 1 + \frac{g^2}{32\pi^2}\left(\frac{17}{6} - \frac{3}{2}\zeta\right)\frac{C_2(G)}{\delta}, \quad (21.9.4)$$

where $C_2(G)$ is a number defined by

$$\sum_{c,d} f^{acd} f^{bcd} = \delta^{ab} C_2(G). \quad (21.9.5)$$

The number $C_2(G)$ exists because $G$ is simple; $\delta^{ab} C_2(G)$ is the quadratic Casimir operator of $G$ in the adjoint representation. For the groups $SU(n)$, we have

$$C_2\big(SU(n)\big) = n. \quad (21.9.6)$$

**Homework 21.9.7.** Verify these values for $z_A$ and $z_g$.

The scaling of $g$ depends on the function $Z_g$ given by

$$Z_g \overset{\text{def}}{=} \frac{z_g}{z_A^{3/2}} = 1 - \frac{g^2}{32\pi^2} \frac{11}{3} \frac{C_2(G)}{\delta}. \tag{21.9.8}$$

From here we deduce the $\beta$ function as usual:

$$\beta \overset{\text{def}}{=} g^2 \frac{\partial}{\partial g} \operatorname{Res} Z_g = -\frac{g^3}{16\pi^2} \frac{11}{3} C_2(G). \tag{21.9.9}$$

Thus the RGE for $g$ is like that for $e$ in QED but with the opposite sign:

$$\mu \frac{\partial}{\partial \mu} g = -\frac{g^3}{16\pi^2} b \quad \text{where} \quad b = \frac{11}{3} C_2(G). \tag{21.9.10}$$

Introducing $\alpha_g$ for $g^2/4\pi$, we find that

$$\frac{1}{\alpha_g(\mu)} = \frac{1}{\alpha_g(\mu_0)} + \frac{b}{2\pi} C_2(G) \ln \frac{\mu}{\mu_0}. \tag{21.9.11}$$

The extreme values of $\alpha_g$ in this solution are as follows:

$$\lim_{\mu \to \infty} \alpha_g(\mu) = 0 \quad \text{and} \quad \lim_{\mu \to \Lambda} \alpha_g = \infty, \tag{21.9.12}$$

where

$$\Lambda = \mu_0 \exp\left(-\frac{2\pi}{b\alpha_g(\mu_0)}\right). \tag{21.9.13}$$

Clearly, taking $\mu$ beyond the Planck scale is meaningless. Furthermore, the low-energy limit is obviously beyond the range of perturbative techniques. Nevertheless, these limits indicate the possibility that a theory can be *asymptotically free* at high energies and *confining* at low energies.

Suppose now that we couple our Yang–Mills theory to a vector current of fermions through a hermitian representation $R^a$ of the Lie algebra of $G$. Because $G$ is a simple group, there is a number $T(R)$ which satisfies

$$\operatorname{tr}(R^a R^b) = \delta^{ab} T(R). \tag{21.9.14}$$

It will also be convenient to write $C_2(R)$ for the scalar part of the second-order Casimir operator computed in the fermion representation $R$:

$$R^a R^a = C_2(R)\mathbf{1}. \tag{21.9.15}$$

To second order, $z_A$ must be modified to include all the fermion loops,

$$\tag{21.9.16}$$

$z_g$ can be computed from the $\bar{\psi}A\psi$ vertex,

$$(21.9.17)$$

and the renormalization factor $z_\psi$ for the fermion kinetic term comes from the diagram

$$(21.9.18)$$

Again, except for the diagram with a three-bose vertex, the computations here are like those for QED up to a group-theoretic factor. We find the following values for the $z$'s:

$$z_A = 1 + \frac{g^2}{16\pi^2}\left(\left(\frac{13}{6} - \frac{\zeta}{2}\right)C_2(G) - \frac{4}{3}T(R)\right)\frac{1}{\delta},$$

$$z_\psi = 1 - \frac{g^2}{16\pi^2}\frac{C_2(R)}{\delta}, \qquad (21.9.19)$$

$$z_g = 1 - \frac{g^2}{16\pi^2}\left(\left(\frac{3}{4} + \frac{\zeta}{4}\right)C_2(G) + C_2(R)\right)\frac{1}{\delta}.$$

The scaling of the coupling $g$ is now governed by

$$Z_g = \frac{z_g}{z_\psi\sqrt{z_A}} \overset{(2)}{=} 1 - \frac{g^2}{32\pi^2}\left(\frac{11}{3}C_2(G) - \frac{4}{3}T(R)\right). \qquad (21.9.20)$$

**Homework** 21.9.21. Verify the values for these three $z$'s.

From here, we see that the RGE for $g$ is

$$\mu\frac{\partial}{\partial\mu}g = -\frac{g^3}{16\pi^2}b \quad \text{where} \quad b = \frac{11}{3}C_2(G) - \frac{4}{3}T(R), \qquad (21.9.22)$$

and the evolution of $g$ is given in terms of $\alpha_g$ by

$$\frac{1}{\alpha_g(\mu)} = \frac{1}{\alpha_g(\mu_0)} + \frac{b}{2\pi}\ln\frac{\mu}{\mu_0}. \qquad (21.9.23)$$

(The matter fields do not change the algebraic form of the RGE and its solution from 21.9.11, but merely change the value of $b$.) We conclude that there is asymptotic freedom as long as the matter representation satisfies $4T(R) < 11C_2(G)$, a condition which is amply satisfied in the case of QCD with six triplets of Dirac quarks. As before, there is a critical value $\Lambda$ of $\mu$ for which $\alpha_g$ blows up:

$$\Lambda = \mu_0\exp\left(-\frac{2\pi}{b\alpha_g(\mu_0)}\right) \implies \alpha_g(\Lambda) = \infty. \qquad (21.9.24)$$

Eliminating $\mu_0$ from 21.9.23 in terms of $\Lambda$ yields

$$\frac{1}{\alpha(\mu)} = \frac{b}{2\pi}\ln\frac{\mu}{\Lambda}. \qquad (21.9.25)$$

We can use the formulae above to predict the evolution of the strong coupling parameter $\alpha_s$ in terms of an initial parameter $\alpha_s(M_Z)$. A best fit of the resulting curves to experimental data can be used to estimate the actual value of $\alpha_s(M_Z)$ and fix the predicted values of $\alpha_s$ for all energies. From here, the confinement scale $\Lambda$ can be computed. The following homework outlines a simpler approach to estimating the confinement scale. The conclusion $\Lambda \sim 0.16$ GeV is lower than the experimental value of 0.2 GeV, but the result is pretty good for a one-loop estimate to a non-perturbative parameter.

**Homework 21.9.26.** In QCD with $N_F$ flavors of quarks,

$$b = 11 - \frac{2}{3}N_F. \tag{21.9.27}$$

Starting from $\mu_0 = 91$ GeV, the $Z$ mass, with the experimental value $\alpha_s(\mu_0) = 0.118$, and using the values 4.3 GeV and 1.3 GeV for the masses of the bottom and charm quarks respectively, predict values for $\alpha_s$ at $\mu_1 = 11$ GeV and $\mu_2 = 1.8$ GeV. Compare your results to the experimental values 0.16 and 0.31. From the theoretical value of $\alpha_s(\mu_2)$, find $\Lambda \sim 0.16$ GeV. (Note that the value of $\alpha_g(\mu_2)$ given here is the average of results 0.26 and 0.37 derived from two very different means of determining this parameter.)

**Homework 21.9.28.** Show that for QCD, in the basis for $su(3)$ which satisfies $T(3) = 1/2$ for the vector representation 3, we find $C_2(G) = 3$ and $C_2(3) = 4/3$. Conclude that for $N_F$ flavors of quarks, each of which comes in three colors, QCD is asymptotically free at one-loop level as long as $N_F < 17$.

**Homework 21.9.29.** Show that if we choose a basis for $su(n)$ in which $T(n) = 1/2$, ($n$ here is the vector representation of $SU(n)$), then $C_2\big(SU(n)\big) = n$ and $C_2(n) = (n^2-1)/2n$.

Continuing to work in Yang–Mills theory coupled to a vector current of fermions through a representation $R$, computing to two loops one finds that

$$\mu\frac{\partial}{\partial\mu}\alpha_g = -b_2\alpha_g^2 - b_3\alpha_g^3, \tag{21.9.30}$$

where

$$b_2 = \frac{11}{6}C_2(G) - \frac{2}{3}T(R)$$
$$\text{and} \quad b_3 = \frac{17}{12}C_2(G)^2 - \frac{1}{2}C_2(R)T(R) - \frac{5}{6}C_2(G)T(R). \tag{21.9.31}$$

Here it is possible for $b_2$ to be negative and $b_3$ to be positive.

**Homework 21.9.32.** Show that $T(R)$ and $C_2(R)$ are related by

$$\dim(G)\,T(R) = \deg(R)\,C_2(R).$$

Show that $T$ acts on the Lie-algebra direct sum and tensor product as follows:

$$T(R \oplus S) = T(R) + T(S) \quad \text{and} \quad T(R \otimes S) = \deg(S)\,T(R) + \deg(R)\,T(S).$$

(The meaning of "$\otimes$" here is given in Definition 6.5.4.)
    Compute $T(R)$ and $C_2(R)$ for the first few representations of $su(3)$: (0,0), (1,0), (2,0), (1,1), (3,0), (2,1).

**Homework 21.9.33.** Using the results of Homework 21.9.32, show that if $b_2 < 0$, then $b_3 < 0$.

   This section concludes our examples of RGEs for finite parameters. It indicates the variety of asymptotic behaviors of gauge theories. The following section presents the key points graphically.

## 21.10 – Qualitative Features of RGEs

The heart of the RGEs in the minimal subtraction scheme is the running of the coupling parameter. This continues until $\beta(g)$ arrives at a zero. A zero of $\beta$ arrived at by increasing $\mu$ is called an *ultra-violet fixed point* and a limit arrived at by decreasing $\mu$ is called an *infra-red fixed point*. The following diagram illustrates a hypothetical $\beta$ as a function of $g$ and the corresponding flow of $g$:

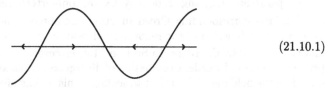

$$(21.10.1)$$

Of course, the domain $\mu < 0$ is unphysical, and an infra-red fixed point in a theory with massive fields will not generally provide any insight into the low-energy behavior of the theory.

**Homework** 21.10.2. Using the results of Homework 21.9.32, show that there is an infra-red fixed point $\alpha_g^{(\mathrm{IR})} = 1/57$ if $G = SU(3)$ with $R = (3, 0)$.

**Homework** 21.10.3. Show that in QCD with $n$ flavors of quark, there is an infra-red fixed point for $8 < n < 17$. For what range of $n$ is $\alpha_g^{(\mathrm{IR})}$ perturbative?

**Homework** 21.10.4. Show that, if $G = SU(2)$ and $R = D_{3/2}$, there is an infra-red fixed point $\alpha_g^{(\mathrm{IR})} \simeq 0.03$.

   From the previous sections, we find that the following graphs can arise for $\beta$ functions:

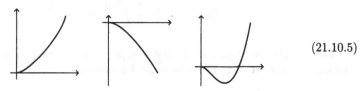

$$(21.10.5)$$

The first graph represents the $\beta$ functions for $g$ in $\phi^4$ and for $\alpha$ in QED. The second graph represents the $\beta$ function for $\alpha_g$ in pure Yang–Mills theory. As indicated by the two-loop RGE 21.9.31, the third graph arises in Yang–Mills theory with an appropriate matter representation. In this case, there is an infra-red fixed point $\alpha_g^{(\mathrm{IR})}$, and if $\alpha_g^{(\mathrm{IR})}$ is sufficiently small, $\alpha_g$ can be perturbative for the whole range of energy. In light of Homework 21.9.33, the reflection of the third graph in the $x$-axis is not possible.

   We have now covered the basic knowledge of running parameters. One of the most far-reaching implications of this running is the possibility that coupling parameters which appear distinct at low energy may become equal as the energy rises. Indeed, the three couplings of the Standard Model appear to converge on a single point. This motivated the Georgi–Glashow $SU(5)$ grand unified theory. More careful analysis of the trajectories indicates that the $SU(2)$ and $SU(3)$ couplings converge first and are then joined by the $U(1)$ coupling. This running leads to the flipped $SU(5)$ theory of Hagelin *et al.*

Next, to make a fundamental link to applications of the RGEs, the following section shows how these parameters arise naturally in solving the RGEs for Green functions.

## 21.11 – RGEs for Green Functions

In this section, we will find and solve the RGEs for Green functions. Continuing with the conventions of Section 21.3, let $G_n$ be a Green function computed from a finite-parameter Lagrangian density $\mathcal{L}_{\mathrm{fp}}$ by counterterm renormalization, and let $G_n^{(\mu_0)}$ be the corresponding Green function computed from the bare Lagrangian density $\mathcal{L}$ as a function of the renormalized field. Section 21.3 established that $G_n$ and $G_n^{(\mu_0)}$ agree to all orders. In Section 21.4, however, we noted that, if we are to preserve the order by order cancellation of divergences, then all the finite parameters including the field must be subject to scaling. This scaling gives rise to a Lagrangian density $\mathcal{L}^{(\mu)}$ which is just $\mathcal{L}$ considered as a function of finite parameters at scale $\mu$. Let $G_n^{(\mu)}$ be the $n$-point Green function computed from $\mathcal{L}^{(\mu)}$. Applying the scaling law 21.3.17, we find that

$$G_n^{(\mu)} = \langle \phi \rangle^n G_n^{(\mu_0)}. \tag{21.11.1}$$

The function $\langle \phi \rangle$ is commonly expressed as a ratio of $z$'s, but such a ratio needs interpretation. We find it more meaningful to use the analysis of Section 21.4 and define $\langle \phi \rangle$ as the solution to the RGE 21.4.24 for $\phi$:

$$\frac{\partial}{\partial t}\langle \phi \rangle + \frac{1}{2}\gamma(g)\langle \phi \rangle = 0, \quad \langle \phi(\mu_0) \rangle = 1. \tag{21.11.2}$$

The solution to this RGE is

$$\langle \phi(\mu) \rangle = \exp\left(-\frac{1}{2}\int_{\mu_0}^{\mu}\frac{d\nu}{\nu}\,\gamma\big(g(\nu)\big)\right). \tag{21.11.3}$$

Making the parameter dependence explicit in the context of $\phi^4$-theory, the relation between the Green functions can be written

$$G_n^{(\mu_0)}(m_0^2, g_0) = \langle \phi(\mu) \rangle^{-n} G_n^{(\mu)}(m^2, g). \tag{21.11.4}$$

Differentiating with respect to $t = \ln \mu$ yields the RGE for the Green function:

**RGE for $G_n$:**  $\left(\mu\dfrac{\partial}{\partial \mu} - \gamma_m(g)\dfrac{\partial}{\partial m^2} + \beta(g)\dfrac{\partial}{\partial g} + \dfrac{n}{2}\gamma(g)\right)G_n^{(\mu)}(m^2, g) = 0.$ (21.11.5)

**Remark** 21.11.6. If we apply the method of Cauchy characteristics to the RGE for $G_n^{(\mu)}$, we find that the evolution of the coordinate functions $m^2$ and $g$ along a Cauchy characteristic is given by their RGEs, and the evolution of the independent function $G_n^{(\mu)}$ is basically the RGE for $\langle \phi \rangle^n$,

$$\mu\frac{d}{d\mu}G^{(\mu)} + \frac{n}{2}\gamma(g)G^{(\mu)}. \tag{21.11.7}$$

For 1PI Green functions $\Gamma_n^{(\mu)}$, the same reasoning leads to the equality and RGE

$$\Gamma_n^{(\mu_0)}(m_0^2, g_0) = \exp\left(-\frac{n}{2}\int_{\mu_0}^{\mu}\frac{d\nu}{\nu}\,\gamma(g(\nu))\right)\Gamma_n^{(\mu)}(m^2, g; \mu),$$

$$\left(\mu\frac{\partial}{\partial\mu} - \gamma_m(g)\frac{\partial}{\partial m^2} + \beta(g)\frac{\partial}{\partial g} - \frac{n}{2}\gamma(g)\right)\Gamma_n^{(\mu)}(m^2, g) = 0. \tag{21.11.8}$$

We gave a general argument to link the dimensional regularization parameter $\mu$ to the high-energy behavior of Green functions. Here we make a practical extension of the argument using the running parameters.

Feynman integrals generally give rise to factors $\ln(p^2/\mu^2)$. (The photon self-energy function of Theorem 19.4.12 provides an example.) At high $p^2$, these factors enhance higher-order contributions relative to lower-order ones. Consequently, for accurate predictions, it often appears necessary to compute to several loops. The use of running parameters provides a technique for obtaining high-energy accuracy with low-order computations.

To bring out the details, let us write $G^{(\mu)}(p, m, g)$ for $G^{(\mu)}(p)$ in order to separate the finite-parameter dependence from the scale dependence in the Green functions, let $\kappa$ be a dimensionless parameter, and let $\mu = \kappa\mu_0$. Then, first applying the scaling law 21.11.4 and second using dimensional analysis (as in 21.2.10), we see that

$$G_n^{(\mu_0)}(p, m_0^2, g_0) = \langle\phi(\mu)\rangle^{-n} G_n^{(\mu)}(p, m^2, g)$$

$$= \kappa^{-3n}\langle\phi(\mu)\rangle^{-n} G_n^{(\mu_0)}\left(\frac{p}{\kappa}, \frac{m^2}{\kappa^2}, g\right). \tag{21.11.9}$$

The last expression means, (1) compute $n$-point amplitudes using a hybrid Lagrangian density whose finite parameters are $\phi(\mu)$, $m^2/\kappa^2$, and $g$, but whose explicit $\mu$ is set at $\mu_0$, (2) evaluate the result at $p/\kappa$, and (3) scale with $\langle\phi(\mu)\rangle$ and $\kappa$ factors.

If $g$ remains small as $\kappa \to \infty$, then we can choose $\kappa$ to make $p^2/\kappa^2\mu_0^2$ close to $\pm 1$. With this choice, a low-order calculation of $G_n^{(\mu_0)}$ using the hybrid Lagrangian yields a good approximation to high-energy behavior of $G_n^{(\mu_0)}$. Commonly, as $\kappa \to \infty$, we also have $m/\kappa \to 0$, so we can neglect masses in this calculation.

This section actually gives the practical culmination of the RGE technique. It has shown how to use the RGEs of the finite parameters to structure a 'hybrid' Lagrangian density, a density from which high-energy Green functions can be computed with high accuracy using only low-order diagrams. Indeed, this particular Lagrangian density is optimal for its tree-level approximations to high-energy processes.

**Homework 21.11.10.** By considering amplitudes of Feynman diagrams, verify that the product $\kappa^{-3n} G_n^{(\mu_0)}\left(\frac{p}{\kappa}, \frac{m^2}{\kappa^2}, g\right)$ is independent of $\kappa$.

## 21.12  The Path to Unified Field Theories

Recent thinking on the significance of gauge field theory and renormalizability is motivated by quantum theories of gravity and by unified field theories. Before closing the book with a final section on the current understanding of renormalizability, let us, in this section, sketch the path to unified field theories. Since the idea of unification is dull unless a rich diversity is first established, we shall start the story with classical physics.

The first taste of unification is found in Newton's laws of mechanics, which established that the law of gravity governs both terrestrial and celestial motions of bulk matter. From here it appeared possible, in principle, to reduce knowledge of the universe to knowledge of initial conditions and natural law. Then, over the following century, the field of natural law itself became diversified. In particular, laws of optics, thermodynamics, electricity, and magnetism were discovered. The first step of unification in the field of natural law was achieved by Maxwell, whose four equations of electromagnetism may be viewed as the source of the diverse laws of electricity, magnetism, and optics.

Meanwhile, between 1803 and 1808, Dalton developed a comprehensive atomic theory of matter. Einstein's 1905 paper on Brownian motion used statistical mechanics to link the Dalton's hypothetical molecules to an observable behavior of bulk matter. Interpreted in light of this atomic theory, Rutherford's experiments indicated that the positive charge and the bulk of the atomic mass concentrated in a very small part of the apparent volume of an atom. Rutherford proposed his planetary-system model of atomic structure in 1911. However, Maxwell's laws would predict for this structure rapid radiation of electromagnetic energy from the orbiting electrons and almost instantaneous collapse of the electron orbits into the nucleus.

The last great achievement in classical physics was Einstein's 1915 general theory of relativity, a field theory of gravity. Here, then, at the close of the classical era, we have two comprehensive field theories, gravity and electromagnetism, and a great muddle of laws classifying and governing the particles in the universe.

While classical thinking was reaching the limit of its effective domain, Planck had, in 1900, introduced particles or quanta of light as a possible explanation for the observed radiation spectra of hot objects, and Einstein had shown how this quantum principle could be used to explain the photoelectric effect. In 1913, Bohr then bravely extended the quantum principle to the atom, proposing that the electrons could only occupy certain specific orbits. And in the early 1920s, de Broglie integrated the two poles of the quantum principle – the application to light and the application to matter – into a universal principle of wave-particle duality.

Even at this early date, the quantum principle as presented by de Broglie promised to provide an integrating perspective on matter and electromagnetism. However, there was (and still is) no experimental indication of gravity waves or gravity particles, so there was neither experimental nor theoretical motivation to apply the quantum principle to gravity.

Developments on the path to unified field theories can (with a pinch of salt) be seen as technical implementations of de Broglie's insight. First came the quantum theory of particles – Schrödinger's wave mechanics, Heisenberg's matrix mechanics, and Dirac's canonical quantization. Then came Dirac's relativistic wave equations for the electron and the seeds of quantum field theory – particles, anti-particles, creation and annihilation operators.

For a long time, it seemed impossible to make field theory work. Calculations simply diverged. But finally, Schwinger's operator theory, Feynman's path-integral theory, and Dyson's perturbative expansion of the scattering matrix led to an integrated and practical understanding of relativistic fields. With this step, the myriads of particles of any given type could be viewed as activity of a single field. Diversity was reduced from the set of all particles to the set of particle types.

Now, in principle, with every type of matter being represented by a quantum field of its own, all fundamental transformations could be encapsulated in monomial interaction terms, and a universal Lagrangian field theory could be built.

This level of technical understanding proved inadequate for the task of modeling the weak interactions. It took many years to find conditions under which the renormalization program could be extended. First the four-Fermi interactions were replaced by interactions with a non-abelian gauge field, and second the gauge field was made massive by spontaneous symmetry breaking.

The discovery that massive vector fields could be renormalizable if they were derived in this way suggested a principle: the fundamental laws of nature are highly symmetrical, but the symmetry is 'hidden' at low energies. After the formulation of the Standard Model with its three coupling constants, it was therefore natural to consider running the energies up and seeing if the coupling constants converged. With the range of experimental error in the values of the couplings, it appeared at first that all three could converge at about $10^{15}$ GeV. Georgi proclaimed $SU(5)$ unification, a viewpoint in which $SU(5)$ was the 'real' symmetry of the three forces represented in the Standard Model and the low energy reality emerged through spontaneous symmetry breaking. Later, with better experimental bounds on the coupling constants, it appeared that the strong and electroweak couplings converged first and then the electromagnetic coupling joined them. This finding suggest the $SU(5) \times U(1)$ gauge theory of Hagelin and others.

While these adventures in coupling unification were going forward, another aspect of the Standard Model was being refined. For symmetry breaking to achieve a $Z$ mass of 91 GeV and an electron mass of 0.5 MeV requires a fine tuning of high-energy parameters. This fine tuning becomes especially acute in the context of renormalizing divergent contributions to the electron mass. This fine-tuning or 'mass hierarchy problem' is felt to be a weakness of the theory.

To soften the need for fine tuning, supersymmetry was introduced. A supersymmetry is an infinitesimal symmetry of a Lagrangian which exchanges fermions and bosons. A global supersymmetry has a fermionic generator $Q_\alpha$, where $\alpha$ is a Weyl spinor index. The benefit of supersymmetry in this context is that the crucial divergent loop diagrams are now completely canceled by diagrams in supersymmetric partners.

Of course, there is the drawback that supersymmetry does not pair up any two known particles, so supersymmetry must be broken somehow and the non-detection of the unobserved superpartners must be explained. As no other solution to the hierarchy problem presented itself, the theory of supersymmetry breaking has been developed to the point where it provides some understanding of possible supersymmetry breaking scenarios and, from there, (1) a detailed understanding of the generation of masses for the unobserved superpartners, and (2) the dynamic generation of weak and electron masses from theories formulated at the Planck mass.

So, for a second time, a solution of a technical problem invokes a hidden symmetry. The action of a symmetry on a set of fields determines orbits known as multiplets. Gauge symmetry and supersymmetry (together or separately) suggest a perspective in which the multiplets (or supermultiplets) are the fundamental entities and the component fields comprise an emergent low-energy perspective.

Extending the gauge principle to supersymmetry, it is natural to propose that if global supersymmetry is available, we should be able to make local supersymmetry transformations. The field $\psi_\mu^\alpha$ associated with these local transformations has spin 3/2. At energies where supersymmetry is unbroken, $\psi_\mu^\alpha$ is massless. The action of $Q_\alpha$ on $\psi_\mu^\alpha$ will pair it up with either a spin 1 field or a spin 2 field. Since there is not much likelihood of an undetected vector boson, since known symmetries unite the known vector bosons into multiplets, and since a multiplet of supersymmetries turns out to be too restrictive, we must assume that the supersymmetric partner of $\psi_\mu^\alpha$ has spin 2. We conclude that a quantum gravity is an essential part of a supersymmetric theory.

In field theory, quantum gravity is not renormalizable. Supersymmetric quantum gravity is a little better, but still suffers uncontrollable divergences. Nevertheless, starting from the tensor structure intrinsic to a supersymmetric Lagrangian for quantum gravity coupled to other fields, physicists have developed schemes for investigating the running of couplings, possible mechanisms for breaking supersymmetry, and deriving the Standard Model as a low-energy effective theory.

Meanwhile, in the 1960s, inspired by quark confinement, Veneziano considered the possibility that a quark pair could be modeled as ends of a string and proceeded to develop a string theory of hadron scattering. The development of QCD put hadronic strings in the shadow for a time, but in 1974, Green and Scherk were motivated by the inevitable presence of spin 2 excitations of strings to propose that string theory should be the fundamental theory of all physics. One can readily conceive that the internal dynamics of a quantum string field might give it the potential to appear as any one of the known quantum fields. This direction proved fruitful, and in 1984, Green and Schwarz constructed a string theory which contained a finite quantum gravity theory.

For a decade, string theory was extremely popular. Heterotic (mixed fermionic and bosonic) string emerged as the type of string best fitted to be the source of the universe. The 4-D construction of heterotic string theories was developed into a practical tool, and Witten's open string field theory indicated the possibility of a complete description of a unified field. However, physicists were not able to extend the principles of open string field theory to heterotic strings. Excitement subsided and attention generally reverted to fields in which one could make progress without being Ed Witten.

On the other hand, remarkable 'coincidences' had made it possible to construct heterotic string theories which were both free from divergences and possible sources of supergravity theories and the Standard Model. This facilitated a paradigm shift in which emphasis moved from the diversified laws of nature to the integrated state of natural law represented by unified field theories.

The hard-core giants of unified field theory, however, have pushed ahead and seen in the far distance a new level of supertheory, $M$ theory, which has all known versions of string theory and the remarkable 11-dimensional supergravity as low-energy manifestations.

In practice, this has meant that theories in particle physics are expected to have non-renormalizable terms which represent the scattering of massive states of the unified field. Indeed, in this perspective, the success of renormalizable theories needs to be explained. The general idea of an explanation comes from Wilson's renormalization group of statistical mechanics, which we explain briefly in the next section.

## 21.13 Renormalizability Reconsidered

The text has focused on strictly renormalizable theories, that is, theories in which quantization modifies a polynomial classical Lagrangian density by adding counterterms of the same form as terms already present in the Lagrangian. In this section, motivated by the desire to link the Standard Model to unified field theory, we shall consider a relaxation of these constraints to allow for non-polynomial Lagrangian densities.

The principle concerns, which we shall address below, are (1) is there evidence that polynomial approximations to a non-polynomial Lagrangian density provide a good approximation to the dynamics, and (2) can a counterterm renormalization procedure be found which will preserve classical symmetries of the non-polynomial Lagrangian density? Wilson's non-perturbative analysis of the generating functional provides the best answer to the first concern, and a result due to Gomis and Weinberg sheds light on the second.

Please note that Wilsonian analysis is most thoroughly developed for scalar field theories. Its application to gauge field theories is a relatively new and active area of research. The points made below indicate grounds for hope, possibly even for optimism, but the conclusions for realistic theories are definitely tentative at this time.

To get to grips with the details of a functional integral in Minkowski space, we make a Wick rotation to Euclidean space. Here we find a beautiful analysis of the functional integral pioneered by Ken Wilson in the context of statistical mechanics. Wilson's analysis shows that the process of performing a functional integral can be expressed as an exact renormalization group (ERG) equation for the coefficients in a non-polynomial action.

In detail, let $\mathcal{L}(\phi, J)$ be a Lagrangian density for a field $\phi$ coupled to an external source $J$, and let $A_\Lambda(\phi, J)$ be the action determined by $\mathcal{L}$ subject to momentum cut-off at $\Lambda$:

$$A_\Lambda(\phi, J) = \int_{p^2 \leq \Lambda^2} d^4p \, \mathcal{L}(\phi, J). \tag{21.13.1}$$

The associated generating functional $Z_\Lambda(J)$ is defined by:

$$Z_\Lambda(J) = \int [d\phi(p)] \, e^{-A_\Lambda(\phi, J)}. \tag{21.13.2}$$

To represent integration over the high-momentum interval $b\Lambda \leq |p| \leq \Lambda$ for $b = 1 - \epsilon$, we first split the function $\phi$ into low and high energy parts,

$$\left.\begin{array}{l} \phi = \phi_< + \phi_> \\ \phi_<(p) = 0 \quad \text{for} \quad b\Lambda < |p| \leq \Lambda \\ \phi_>(p) = 0 \quad \text{for} \quad |p| \leq b\Lambda \end{array}\right\} \tag{21.13.3}$$

separate the functional measure into two parts,

$$[d\phi(p)] = [d\phi_<(p)] \, [d\phi_>(p)], \tag{21.13.4}$$

define $A_\Lambda^{(2)}$ by

$$e^{-A_\Lambda^{(2)}(\phi_<,J)} \stackrel{\text{def}}{=} \int [d\phi_>(p)] \, e^{-A_\Lambda(\phi,J)}, \tag{21.13.5}$$

and then write the generating functional in terms of an integral over small momenta:

$$Z_\Lambda(J) = \int [d\phi_<(p)] \, e^{-A_\Lambda^{(2)}(\phi_<,J)}. \tag{21.13.6}$$

We now have a transformation $\mathcal{R}: A_\Lambda \to A_\Lambda^{(2)}$ on the space of actions. Iteration leads to:

$$A_\Lambda^{(n)} = \mathcal{R}(A_\Lambda^{(n-1)}), \tag{21.13.7}$$

where $A_\Lambda^{(n)}$ is associated with performing the original integration over the interval $b^n \Lambda \le |p| \le \Lambda$. If the integration as a whole is to make sense, then the sequence $A_\Lambda^{(n)}$ should converge to a fixed point $A_\Lambda^*$ of $\mathcal{R}$:

$$A_\Lambda^* = \lim_{n \to \infty} A_\Lambda^{(n)}. \tag{21.13.8}$$

**Remark** 21.13.9. Purists will note that since the range of integration shrinks at each stage, the map $\mathcal{R}$ defined by the first integration cannot be applied again. If, as happens in statistical mechanics, one is not interested in external momenta, then one can easily rescale the momentum parameter in $A_\Lambda^{(2)}$ to restore the original range of integration. If, as in our case, we are interested in the external momenta, then we must define $\mathcal{R}$ as a map on the product space actions×cut-offs.

It is possible to express the integration over large momenta in terms of Feynman integrals. In more detail, if we split $\phi$ into high and low energy parts according to 21.13.3 for any value of $b$ in $(0,1)$, then in 21.13.5 we can split $A_\Lambda$ as follows:

$$\begin{aligned} A_\Lambda(\phi, J) &= A_\Lambda(\phi_< + \phi_>, J) \\ &= A_\Lambda(\phi_<, J) + A_\Lambda'(\phi_>, \phi_<, J). \end{aligned} \tag{21.13.10}$$

The first term passes through the integral over $\phi_<$, and the second term can be interpreted as a field theory for $\phi_>$ coupled to sources $\phi_<$ and $J$. Now we can use the trick of 11.8.1 and expand the functional integral into a sum of Feynman integrals subject to appropriate upper and lower bounds on loop momenta. The result of Section 4.5 applies, enabling us to express the sum of integrals arising from all diagrams as an exponential of the sum $A_\Lambda^{\text{con}}(\phi_<, J)$ arising from connected diagrams.

Finally, since

$$(\phi, J) = (\phi_<, J_<) + (\phi_>, J_>), \tag{21.13.11}$$

integration over $\phi_>$ does not generate any new terms in $J_<$. Consequently, we find a new form for the action,

$$A_\Lambda^{(2)}(\phi_<, J) = A_\Lambda(\phi_<, J_<) + A_\Lambda^{\text{con}}(\phi_<, J_>), \tag{21.13.12}$$

and a corresponding new expression for the generating functional:

$$Z_\Lambda(J) = \int [d\phi_<(p)] \, e^{-A_\Lambda(\phi_<,J_<) - A_\Lambda^{\mathrm{con}}(\phi_<,J_>)}. \tag{21.13.13}$$

Differentiating $Z_\Lambda(J)$ with respect to $J_>$ on the one hand brings out Green functions with high external momenta, and on the other hand brings out the coefficients of $A_\Lambda^{\mathrm{con}}$. Equating the two shows that these coefficients are Green functions with UV cut-off $\Lambda$ and IR $b\Lambda$. (Note that these cut-offs constrain the momentum on every internal and external line.) From this perspective, the ERG flow gives the evolution of these Green functions as the IR cut-off is reduced.

**Remark** 21.13.14. Wilson's purpose was to investigate the behavior of macroscopic (low energy) parameters in a theory specified by microscopic (high energy) parameters. For this purpose, it is natural to set $J_> = 0$ and thereby obtain the Wilson effective action.

The exact renormalization group flow is obtained from the integration map $\mathcal{R}$ by letting $1 - b \to 0$, and the ERG equations are obtained by differentiating $\mathcal{R}$ with respect to $b$ at $b = 1$. Remarkably enough, since the Feynman integral for an $n$-loop diagram has a factor of $(1 - b)^n$ in it, the ERG equations and flow are determined exactly by the one-loop integrals.

From the perturbative approach to computing integration over a slice, it is clear that if $A_\Lambda$ contains any interaction term, then $\mathcal{R}A_\Lambda$ is not a polynomial. Therefore, if $A_\Lambda$ is perturbatively renormalizable, then $A_\Lambda$ is locally the only perturbatively renormalizable action on the ERG trajectory through $A_\Lambda$. This suggests that perturbative renormalizability of $A_\Lambda$ is not relevant to the existence of $Z_\Lambda$.

To gain information about the effect of non-renormalizable terms in the original action $A_\Lambda$ on $Z_\Lambda$, we can consider the trajectory through $A_\Lambda$. Let $g$ be the coefficient of an operator with mass dimension $4 + n$ in $\mathcal{L}$. Then we can write $g = \mu^{-n}\bar{g}$, where $\bar{g}$ is dimensionless. Much simplified, the ERG for $g$ has the form

$$\partial_t g = \beta(g; \mu), \quad \text{where} \quad t = \ln\frac{\mu}{\Lambda}. \tag{21.13.15}$$

Because we are on a fixed trajectory, the $\beta$ function is independent of $\Lambda$. This implies that $\bar{g}$ satisfies

$$\partial_t \bar{g} = \bar{\beta}(\bar{g}) + n\bar{g}, \quad \text{where} \quad \bar{\beta}(\bar{g}) = \mu^n \beta(\mu^{-n}\bar{g}; \mu). \tag{21.13.16}$$

Since $\bar{\beta}$ is dimensionless, it does not depend on $\mu$. Note also that $\bar{\beta}$ is a power series in $\bar{g}$ which satisfies $\bar{\beta}(0) = \bar{\beta}'(0) = 0$.

Take the case $n > 0$ first. The last term in the ERG dominates for sufficiently small values of $g$, resulting in exponential suppression of $g$. For this reason, operators with positive mass dimension and their coefficients are called *irrelevant*. In the real ERGs, the differential equation for each coefficient involves every other coefficient, and as the trajectory runs from high energy to low energy, the irrelevant coefficients make small contributions to the running of other coefficients. However, the phenomenon of exponential damping persists, so the effects of high-energy initial values for these coefficients makes little difference to the theory at low energy.

**Remark** 21.13.17. In statistical mechanics on a lattice, the full microscopic action should include interactions for every subset of the sites. In fact, nearest-neighbor pair interactions on the microscopic level dominate the macroscopic parameters, and higher order interactions are irrelevant operators.

It is something of a surprise to find that the operators which destroy perturbative renormalizability are classified as irrelevant in the non-perturbative view. The apparent paradox disappears when we realize that data in perturbative and non-perturbative theories flows in opposite directions along a trajectory. In the non-perturbative formalism, the bare parameters are chosen first and the ERG provides information about the resulting low energy parameters. In the perturbative formalism, the low energy parameters are fixed by comparison with experiment and the bare parameters are determined by renormalization conditions. Thus, if a running parameter is irrelevant and we try to fix its low energy value, the bare value will be seriously divergent as a function of increasing cut-off. In this case, the functional integral may not exist in the limit of infinite cut-off. For example, if we fix a positive value for the renormalized coupling in $\phi^4$-theory, then the bare coupling diverges at finite cut-off.

Returning to the qualitative behavior of the ERG trajectory, if $n < 0$ (as in the case of a mass term), then the trajectory tends to diverge as the energy scale decreases. Such terms are said to be *relevant*. Their initial values need fine-tuning at high energy if the theory is to match observations at low energy. From the perspective of perturbative field theory, this justifies the principle that it is unnatural for a fundamental theory to have mass terms; masses should arise from spontaneous symmetry breaking or somehow be generated dynamically.

When $n = 0$, the operator is called *marginal*. Even to determine the running of the coefficient of a marginal operator qualitatively, it is necessary to investigate the $\bar{\beta}$ function in some detail. In the case of four-dimensional scalar theories, there are only Gaussian fixed points, so all marginal operators are effectively irrelevant. (Note that in this context the Gaussian is defined by kinetic terms alone. Mass terms are not included.)

The classification of operators as relevant, irrelevant, and marginal is more properly derived from analysis of the eigenvalues of an ERG at a fixed point. It is possible for the trajectory of a scalar theory to pass close to more than one fixed point. In such a case, because the trajectory will run slowly past a nearby fixed point, the theory will exhibit different phases at different energies, and each phase will have its own class of relevant operators. Thus, though it would be possible to define relevance in terms of the Lyapunov exponents of any trajectory, only the fixed point trajectories are used in physics.

**Remark** 21.13.18. In statistical mechanics, the task is to predict macroscopic, measurable, low energy parameters from microscopic interactions. The terms relevant, marginal, and irrelevant directly indicate the importance of the corresponding operators in this task.

**Remark** 21.13.19. Here we summarize briefly the differences between Wilson's exact renormalization group (ERG) and the renormalization group of perturbative field theory (PRG).

a. The ERG flow corresponds to integrating out high energy slices of a functional integral. The PRG flow provides a reparameterization of the bare Lagrangian.

b. The ERG flow generates non-polynomial effective actions out of an initial bare action. The PRG flow preserves the form of a renormalizable Lagrangian.

c. The coefficients in the action generated by the ERG flow are functions of momentum variables. The running coefficients determined by the PRG flow depend only on the scale parameter.

d. The ERG flow is determined by one-loop diagrams. The PRG flow is sequentially approximated by including the contributions of higher-loop diagrams.

Polchinski applied Wilson's ERG to demonstrate the perturbative renormalizability of $\phi^4$-theory without using diagrams or the Weinberg convergence theorem. The technique of the proof is to convert the non-perturbative irrelevance of higher-order interactions into estimates for the coefficients of the running Lagrangian.

To define the perturbative theory, Polchinski specifies three low energy renormalization conditions and a cut-off. These items determine a unique bare Lagrangian with just three terms – kinetic, mass, and interaction. The bare Lagrangian and cut-off then fix an ERG trajectory, and the ERG trajectory in turn determines an effective Lagrangian at the renormalization scale. When the cut-off is changed, the whole ERG trajectory moves, imparting cut-off–dependence to the effective Lagrangian. Polchinski shows that as the cut-off is removed, the effective Lagrangian and the Green functions converge order by order in perturbation theory. The conclusion holds for all sufficiently small values of the renormalized coupling.

Our purpose is to justify the use of non-polynomial Lagrangians derived from unified field theories. The first step is to extend the Wilson/Polchinski formalism from scalar fields to Fermi and vector fields. This indeed has been accomplished. The next step is to tackle gauge theories.

In a gauge theory like QED which is not confined at low energy, only Gaussian fixed points are expected. Because of the tendency of the coefficients of relevant operators to grow exponentially in the environment of a fixed point, small physical masses can only be achieved by specifying bare masses with unnatural precision. Even so, if the ERG trajectories pass close by a Gaussian fixed point, then at least the classification of the operators as relevant, marginal, and irrelevant makes sense, and so a non-polynomial extension of QED could give rise to a consistent perturbation theory.

**Remark 21.13.20.** We saw in Chapter 19 that there are IR divergent Green functions in QED – finite predictions require the regulating influence of soft photon contributions to observed final states. Thus perturbation theory suggests that the QED trajectory should diverge to infinity as $\mu \to 0$. Because of the wide gap between perturbative and non-perturbative analysis, this suggestion has no force.

**Remark 21.13.21.** QED is commonly investigated through simulations on a lattice. These simulations indicate that the existence of a continuum limit requires the bare mass to be zero. In this case, if the bare charge $e_0$ is small, no mass is generated, and if $e_0$ is large, then the continuum limit of the lattice theory does not exist. Only at a critical value $e_0 = e_c \sim 2.3$ is there a possibility of a continuum limit with renormalized mass $m_e$, but here the renormalized charge is zero in the continuum theory. (Experimentally, the low energy value of $e$ is about 0.3). It appears that QED does not exist as the continuum limit of a lattice theory.

**Remark 21.13.22.** Clever use of topological quantum field theory also locates a critical value $e'_c \sim \pi$. For $e_0 > e'_c$ charged particles are confined and the photon becomes massive, while for $e_0 < e'_c$ the $\beta$ function vanishes identically and $0 < e < e'_c$ is a line of fixed points.

In theories like QCD which suffer low-energy confinement, if the ERG is to be pursued all the way to low energy, there must be a switch from the fundamental fields of the theory to composite fields. The ERG trajectory, however, need not signal confinement by diverging as $\mu$ decreases to the confinement scale. After all, if the sources for confined fields are given an IR cut-off, then the functional integral should exist, and the trajectory should be well defined for all $\mu$. At this time, there is agreement that the notion of irrelevance applies to the trajectory above the confinement scale, but analysis has not yet covered the transition to a low energy theory of composite fields.

Fundamental theories in particle physics necessarily introduce the extra complexity of spontaneous symmetry breaking. This kind of symmetry breaking has its analogue in scalar field theories and has been subjected to ERG analysis. The generation of small masses by spontaneous symmetry breaking will cause an ERG trajectory to miss all Gaussian fixed points. However, the masses themselves set a natural IR cut-off and provide a terminal energy for the ERG. Furthermore, for small masses, the low-energy trajectory may well end sufficiently close to a Gaussian fixed point that the classification of operators is still meaningful, and so again a non-polynomial Lagrangian could give rise to a consistent perturbation theory.

Spontaneous symmetry breaking in itself is not a problem, but the effort to connect the Standard Model to a unified field theory has modified the nature of symmetry breaking considerably. Supersymmetric grand unified theories start from Planck-scale theories which have no mass parameters and attempt to generate a symmetry breaking potential dynamically. The expansion and cooling of the universe and the RGE running of Lagrangian parameters are invoked as mechanisms. Application of non-perturbative methods to field theories in an expanding universe are not well developed and do not yet provide grounds for extending the notion of irrelevance to theories with dynamical symmetry breaking.

**Remark** 21.13.23. Lattice field theory has used simulation techniques to study the dependence of effective Lagrangians on lattice spacing in a wide variety of discrete models. The conclusions of lattice field theory may have implications for the topics of this section, but the task of assessing this possibility is outside the scope of the current book.

To sum up, a lot of work remains to be done on confinement, composite field theories, and dynamical symmetry breaking. Nevertheless, from this excursion into Wilson's ERG, we see a possibility that the notions relevant, marginal, and irrelevant may apply to operators in a field theory derived from a unified field. In particular, though operators with mass dimension greater than four will always destroy strict renormalizability of a polynomial Lagrangian, there is a chance that they are in fact irrelevant. Then they can be included in a high energy Lagrangian and yet have negligible impact on the effective low energy theory derived from it.

In more detail, the hope is that in a fundamental theory defined by a non-polynomial Lagrangian at the Planck mass, (1) a truncation to a polynomial Lagrangian provides a valid approximation to the full theory, and (2) increasing the order of a truncation leads to a better approximation.

Assuming that this hope is fulfilled, there remains the issue of preserving fundamental symmetries of the full Lagrangian when renormalizing a truncation. In the case of a linear symmetry, the truncation can be chosen to preserve the symmetry. Then the sum of all counterterms at any order will be invariant, and the quantized truncation will also preserve the symmetry.

In the case of gauge symmetry, we saw in Section 20.8 how a BRS symmetry can be used as a test for gauge invariance of a Lagrangian or a generating functional. We also noted in Remark 20.9.29 that the central problem in finding gauge-invariant counterterms was related to the solutions of an operator $\mathbf{D}$ which satisfied $\mathbf{D}^2 = 0$. Gomis and Weinberg have generalized this style of proof to show that perturbative renormalization of non-polynomial Yang–Mills theories with matter fields can preserve the gauge symmetries of the original finite-parameter theory.

There is one significant hitch: in the general case, there are too many constraints on the set of candidate counterterms. To introduce more variables, we allow the fields to be redefined not just by a scale factor, but by a canonical transformation whose generator is perturbatively determined. (This result has not yet been extended to supergravity theories.)

To summarize the salient points, unified field theories lead to quantum field theories with non-polynomial Lagrangians. Wilson's analysis of the Euclidean functional integral indicates that non-renormalizable terms generate small perturbations in the IR Green functions. An algorithm for perturbative renormalization can be found which will preserve the symmetries of a non-polynomial Lagrangian. Consequently, non-polynomial Lagrangians are necessary and have a good chance of being perturbatively consistent.

## 21.14 Summary

In this chapter, we have investigated the significance of the arbitrary parameter $\mu$ introduced by dimensional regularization. By varying $\mu$ while holding the bare Lagrangian fixed, we find the running of the coupling parameters with energy. From here we can understand the asymptotic freedom of QCD and glimpse a confinement scale far beyond the valid domain of perturbation theory.

To close this introduction to quantum field theory, we have pursued the running energy scale all the way to the Planck mass, introduced unified field theory, and discovered the perturbative viability of non-polynomial field theories. It remains only to put the parts into a holistic understanding of Natural Law.

From what we have seen of quantum field theory, it is clear that the phenomenon of confinement forces us to use different Lagrangians at different energy scales. One simply cannot compute hadronic scattering cross sections in the Standard Model, nor can one use four-Fermi theory at energies close to $M_Z$.

Consequently, the Lagrangians of QED, the Standard Model, and indeed grand unified theories are useful approximations to Natural Law only in their respective energy ranges. Because of their limited scope, such theories are said to be *effective* theories. As the energy in a scattering process rises, new fields start to make significant contributions to the amplitude and the Lagrangian must be modified to include them. As the energy level falls, heavy fields become irrelevant and fields subject to confinement must give way to families of composite fields.

Combining this observation with the running of coupling constants, we see that the RGE flow with decreasing energy must come to a point where the Lagrangian has to be changed. At this point, there is a boundary condition to match the coefficients of the higher-energy and lower-energy Lagrangians. A simple example can be found in Section 15.7 where we matched the Standard Model to QED with four-Fermi interactions, computing in particular the value of $G_F$.

Extending this perspective to the unified field, it appears that the range of Natural Law, from a unified field at the Planck scale to QED and four-Fermi theory at low energy, can be modeled by a one-parameter family of renormalizable non-polynomial field theories which enjoy a continuous evolution under the renormalization group punctuated by brief periods of symmetry breaking.

# Appendix

A compilation of units, particle masses and decay widths,
and fundamental formulae for use in computations.

## A.1 Constants of Particle Physics

This section provides the values for constants used in the text for those readers who do not have a copy of the *Particle Physics Booklet*, available online at http://pdg.lbl.gov and http://www.cern.ch.

Writing $h$ for Planck's constant, the reduced Planck's constant $\hbar$ is given by $h/2\pi$. As mentioned on the first page, it is convenient in particle physics to use units in which $\hbar$ and the speed of light $c$ are equal to one. Conversion to other units requires insertion of appropriate factors of $\hbar$ and $c$ where:

$$\hbar = 1.055 \times 10^{-34} \text{ J s} \quad \text{and} \quad c = 299,792,458 \text{ m s}^{-1}.$$

The unit used for particle masses is the *electron volt* (eV), which relates to the more common mechanical units and to $\hbar$ as follows:

$$1\text{eV} = 1.602 \times 10^{-19} \text{ J} \quad \text{and} \quad 1\text{eV}/c^2 = 1.783 \times 10^{-36} \text{ kg},$$
$$\hbar = 6.5832 \times 10^{-22} \text{ MeV s}.$$

Decay widths and scattering cross sections are measured in units of area called *barns*:
$$1 \text{ barn} = 10^{-28} \text{ m}^2.$$

### Coupling Parameters

| | |
|---|---|
| Fine structure constant | $\alpha = e^2/4\pi = 1/137.036$ |
| Fermi coupling constant | $G_F = 1.166 \times 10^{-5} \text{ GeV}^{-2}$ |
| Weak mixing angle | $\sin^2 \theta_W(M_Z) = 0.2315$ |
| Strong coupling constant | $\alpha_s(M_Z) = 0.118$ |

A selection of masses and mean lives $\tau$ or decay widths $\Gamma = \hbar/\tau$ for the more common particles follows.

### Leptons

| | Mass (MeV) | Mean Life (sec) |
|---|---|---|
| Electron $e$ | 0.5110 | stable |
| Muon $\mu$ | 105.7 | $2.197 \times 10^{-6}$ |
| Taon $\tau$ | 1777 | $2.9 \times 10^{-13}$ |

### Gauge Bosons

| | Mass (GeV) | Decay Width (GeV) |
|---|---|---|
| Photon $\gamma$ | 0 | stable |
| $W^\pm$ | $80.33 \pm 0.15$ | $2.07 \pm 0.06$ |
| $Z$ | $91.187 \pm 0.007$ | $2.490 \pm 0.007$ |

## Octet Baryons

| | Mass (MeV) | Mean Life (sec) |
|---|---|---|
| Proton p (uud) | 938.3 | $> 10^{31}$ years |
| Neutron n (udd) | 939.6 | $887 \pm 2$ |
| $\Sigma^+$ (uus) | 1189 | $0.799 \pm 0.004 \times 10^{-10}$ |
| $\Sigma^0$ (uds) | 1193 | $7.4 \pm 0.7 \times 10^{-20}$ |
| $\Sigma^-$ (dds) | 1197 | $1.479 \pm 0.011 \times 10^{-10}$ |
| $\Xi^-$ (uss) | 1315 | $2.90 \pm 0.09 \times 10^{-10}$ |
| $\Xi^0$ (dss) | 1321 | $1.639 \pm 0.015 \times 10^{-10}$ |
| $\Lambda$ (uds) | 1116 | $2.632 \pm 0.020 \times 10^{-10}$ |

## Decuplet Baryons

| | Mass (MeV) | Decay Width (MeV) |
|---|---|---|
| $\Delta^{++}$ (uuu) | 1231 | 111 |
| $\Delta^+$ (uud) | 1232 | $131 \pm 2$ |
| $\Delta^0$ (udd) | 1233 | 112–119 |
| $\Delta^-$ (ddd) | ~1236 | no data |
| $\Sigma^+$ (uus) | 1183 | $35.8 \pm 0.08$ |
| $\Sigma^0$ (uds) | 1184 | $36 \pm 5$ |
| $\Sigma^-$ (dds) | 1187 | $39.4 \pm 2.1$ |
| $\Xi^-$ (uss) | 1535 | $9.1 \pm 0.5$ |
| $\Xi^0$ (dss) | 1532 | $9.9^{+1.7}_{-1.9}$ |
| $\Omega^-$ (sss) | 1672 | $8.01 \pm 0.12 \times 10^{-12}$ |

## Pseudoscalar Mesons

| | Mass (MeV) | Mean Life (sec) |
|---|---|---|
| $K^\pm$ (u$\bar{\text{s}}$, $\bar{\text{u}}$s) | 494 | $1.24 \times 10^{-8}$ |
| $K^0_L$ (d$\bar{\text{s}}$ + $\bar{\text{d}}$s) | 498 | $5.17 \pm 0.04 \times 10^{-8}$ |
| $K^0_S$ (d$\bar{\text{s}}$ + $\bar{\text{d}}$s) | 498 | $0.89 \times 10^{-8}$ |
| $\pi^\pm$ (u$\bar{\text{d}}$, $\bar{\text{u}}$d) | 140 | $2.60 \times 10^{-8}$ |
| $\pi^0$ (u$\bar{\text{u}}$ − d$\bar{\text{d}}$) | 135 | $8.4 \pm 0.6 \times 10^{-8}$ |
| $\eta$ (u$\bar{\text{u}}$ + d$\bar{\text{d}}$) | 549 | $5.5 \pm 0.5 \times 10^{-19}$ |

## Axial Vector Mesons

| | Mass (MeV) | Decay Width (MeV) |
|---|---|---|
| $K^{*\pm}$ (u$\bar{\text{s}}$, $\bar{\text{u}}$s) | 892 | $49.8 \pm 0.8$ |
| $K^{*0}$ (d$\bar{\text{s}}$ + $\bar{\text{d}}$s) | 896 | $50.5 \pm 0.06$ |
| $\rho^\pm$ (u$\bar{\text{d}}$, $\bar{\text{u}}$d) | $767 \pm 1$ | $1.49 \pm 3$ |
| $\rho^0$ (u$\bar{\text{u}}$ − d$\bar{\text{d}}$) | $768 \pm 1$ | $1.51 \pm 3$ |
| $\omega$ (u$\bar{\text{u}}$ + d$\bar{\text{d}}$) | 782 | $8.43 \pm 0.10$ |

**Current Quark Masses**

| | | | |
|---|---|---|---|
| $u$ | 2–8 MeV | $d$ | 5–15 MeV |
| $c$ | 1.0–1.6 GeV | $s$ | 0.1–0.3 GeV |
| $t$ | 180±12 GeV | $b$ | 4.1–4.5 GeV |

# A.2 Dirac Algebra

**Pauli Matrices** (extended to a 4-vector)

$$\sigma_0 = \begin{pmatrix} 1 & 0 \\ 0 & 1 \end{pmatrix} \quad \sigma_1 = \begin{pmatrix} 0 & 1 \\ 1 & 0 \end{pmatrix} \quad \sigma_2 = \begin{pmatrix} 0 & -i \\ i & 0 \end{pmatrix} \quad \sigma_3 = \begin{pmatrix} 1 & 0 \\ 0 & -1 \end{pmatrix} \qquad (\S 6.3)$$

**Defining Relations and Special Elements**

$$\{\gamma_\mu, \gamma_\nu\} = 2g_{\mu\nu} \qquad (\S 8.1)$$

$$\gamma^5 \overset{\text{def}}{=} i\gamma^0\gamma^1\gamma^2\gamma^3 \qquad (\S 8.2)$$

$$P_L \overset{\text{def}}{=} \tfrac{1}{2}(1 - \gamma^5), \quad P_R \overset{\text{def}}{=} \tfrac{1}{2}(1 + \gamma^5) \qquad (\S 8.2)$$

$$\sigma^{\mu\nu} \overset{\text{def}}{=} \frac{i}{2}[\gamma^\mu, \gamma^\nu] \qquad (\S 8.7)$$

**Boosts and Rotations**

$$\text{Boosts:} \quad \tfrac{1}{2}\gamma_r\gamma_0; \quad \text{Rotations:} \quad \tfrac{1}{2}\gamma_r\gamma_s \qquad (\S 8.2)$$

**Dirac Adjoint**

$$\bar{\psi} = \psi^\dagger \gamma^0$$
$$\bar{M} = \gamma_0^{-1} M^\dagger \gamma_0 = \gamma_0 M^\dagger \gamma_0$$
$$\bar{\gamma}^\mu = \gamma^\mu, \quad \bar{\gamma}^5 = -\gamma^5 \qquad (\S 8.2)$$
$$\bar{D}(\Lambda) = D(\Lambda)^{-1}$$

**Bases for Dirac Algebra**

| | $\gamma^0$ | $\gamma^1$ | $\gamma^2$ | $\gamma^3$ | $\gamma^5$ | |
|---|---|---|---|---|---|---|
| Weyl | $\begin{pmatrix} 0 & 1 \\ 1 & 0 \end{pmatrix}$ | $\begin{pmatrix} 0 & \sigma^1 \\ -\sigma^1 & 0 \end{pmatrix}$ | $\begin{pmatrix} 0 & \sigma^2 \\ -\sigma^2 & 0 \end{pmatrix}$ | $\begin{pmatrix} 0 & \sigma^3 \\ -\sigma^3 & 0 \end{pmatrix}$ | $\begin{pmatrix} 1 & 0 \\ 0 & -1 \end{pmatrix}$ | |
| Dirac | $\begin{pmatrix} 1 & 0 \\ 0 & -1 \end{pmatrix}$ | $\begin{pmatrix} 0 & -\sigma^1 \\ \sigma^1 & 0 \end{pmatrix}$ | $\begin{pmatrix} 0 & -\sigma^2 \\ \sigma^2 & 0 \end{pmatrix}$ | $\begin{pmatrix} 0 & -\sigma^3 \\ \sigma^3 & 0 \end{pmatrix}$ | $\begin{pmatrix} 0 & 1 \\ 1 & 0 \end{pmatrix}$ | $(\S 8.1)$ |
| Majorana | $\begin{pmatrix} 0 & \sigma^2 \\ \sigma^2 & 0 \end{pmatrix}$ | $\begin{pmatrix} -i\sigma^3 & 0 \\ 0 & -i\sigma^3 \end{pmatrix}$ | $\begin{pmatrix} 0 & \sigma^2 \\ -\sigma^2 & 0 \end{pmatrix}$ | $\begin{pmatrix} i\sigma^1 & 0 \\ 0 & i\sigma^1 \end{pmatrix}$ | $\begin{pmatrix} -\sigma^2 & 0 \\ 0 & \sigma^2 \end{pmatrix}$ | |

**Change of Basis Matrices**

$$\psi_W = \frac{1}{\sqrt{2}}\begin{pmatrix} 1 & 1 \\ 1 & -1 \end{pmatrix}\psi_D, \quad \psi_M = \frac{1}{\sqrt{2}}\begin{pmatrix} 1 & \sigma_2 \\ \sigma_2 & -1 \end{pmatrix}\psi_D \qquad (\S 8.1)$$

**Adjoint Identities in the Weyl, Dirac, and Majorana Representations**

$$\gamma_0^\dagger = \gamma_0, \quad \gamma_r^\dagger = -\gamma_r, \quad \gamma_5^\dagger = \gamma_5 \qquad (\S 8.2)$$

**Transpose Equalities in the Weyl, Dirac, and Majorana Representations**

$$
\begin{pmatrix} \gamma_0^\mathsf{T} \\ \gamma_1^\mathsf{T} \\ \gamma_2^\mathsf{T} \\ \gamma_3^\mathsf{T} \\ \gamma_4^\mathsf{T} \\ \gamma_5^\mathsf{T} \end{pmatrix}
=
\begin{matrix} \mathbf{W} \\ \left\{ \begin{matrix} + \\ - \\ + \\ - \\ + \\ + \end{matrix} \right\} \end{matrix}
\begin{matrix} \mathbf{D} \\ \left\{ \begin{matrix} + \\ - \\ + \\ - \\ + \\ + \end{matrix} \right\} \end{matrix}
\begin{matrix} \mathbf{M} \\ \left\{ \begin{matrix} - \\ + \\ + \\ + \\ - \\ - \end{matrix} \right\} \end{matrix}
\begin{pmatrix} \gamma_0 \\ \gamma_1 \\ \gamma_2 \\ \gamma_3 \\ \gamma_4 \\ \gamma_5 \end{pmatrix}
\qquad (\S 8.6)
$$

Thus, for example, in the Dirac representation, $\gamma_3^\mathsf{T} = -\gamma_3$.

**For Fierz Transformations, see §8.7**

# A.3 Miscellaneous Conventions and Definitions

**Metric**
$$g^{\mu\nu} = \operatorname{diag}(1, -1, -, 1-, 1)$$

**Fourier Transform Conventions**

$$\hat{f}(k) \stackrel{\text{def}}{=} \int dx \, e^{-ixk} f(x) \quad \Rightarrow \quad f(x) = \int \frac{dk}{2\pi} e^{ixk} \hat{f}(k)$$

**Lorentz Invariant Measure**

$$d\lambda = \frac{d^3\bar{k}}{(2\pi)^{3/2} \big(2\omega(\bar{k})\big)^{1/2}} \qquad (\S 1.4)$$

**Poincaré Group Action**

$$U(\Delta_a) = \exp(ia \cdot P) \qquad (\S 1.3)$$

$$\alpha(k) = (2\pi)^{3/2} \big(2\omega(\bar{k})\big)^{1/2} a(\bar{k}) \quad \Rightarrow \quad U(\Lambda)\alpha(k)U(\Lambda)^\dagger = \alpha(\Lambda k) \qquad (\S 2.3)$$

$$U(\Lambda)^\dagger \Phi(x) U(\Lambda) = D(\Lambda)\Phi(\Lambda^{-1}x) \qquad (\S 6.1)$$

**Conserved Current and Quantity**

$$j^\mu(X) = \Pi^\mu \mathbf{D}\phi - f^\mu = \Pi_a^\mu X^a{}_b \phi^b - f^\mu, \qquad Q(X) = \int d^3\bar{x} \, j^0(x) \qquad (\S 3.4)$$

## Effect of Discrete Symmetries on Fermi Bilinear Terms

|  | $C$ | $P$ | $T$ | $CPT$ |
|---|---|---|---|---|
| $\bar{\eta}\zeta$ | $\bar{\zeta}\eta$ | $\bar{\eta}\zeta$ | $\bar{\eta}\zeta$ | $\bar{\zeta}\eta$ |
| $\bar{\eta}\gamma^\mu\zeta$ | $-\bar{\zeta}\gamma^\mu\eta$ | $\bar{\eta}\gamma_\mu\zeta$ | $\bar{\eta}\gamma_\mu\zeta$ | $-\bar{\zeta}\gamma^\mu\eta$ |
| $\bar{\eta}\sigma^{\mu\nu}\zeta$ | $-\bar{\zeta}\sigma^{\mu\nu}\eta$ | $\bar{\eta}\sigma_{\mu\nu}\zeta$ | $-\bar{\eta}\sigma_{\mu\nu}\zeta$ | $\bar{\zeta}\sigma^{\mu\nu}\eta$ |
| $\bar{\eta}\gamma^5\gamma^\mu\zeta$ | $\bar{\zeta}\gamma^5\gamma^\mu\eta$ | $-\bar{\eta}\gamma^5\gamma_\mu\zeta$ | $\bar{\eta}\gamma^5\gamma_\mu\zeta$ | $-\bar{\zeta}\gamma^5\gamma^\mu\eta$ |
| $\bar{\eta}i\gamma^5\zeta$ | $\bar{\zeta}i\gamma^5\eta$ | $-\bar{\eta}\gamma^5 i\zeta$ | $-\bar{\eta}i\gamma^5\zeta$ | $\bar{\zeta}i\gamma^5\eta$ |

(§8.12)

Note that the intrinsic parity of a Bose particle is equal to that of its antiparticle, while the intrinsic parity of a Fermi particle is opposite to that of its antiparticle. Note further that to make QED invariant under the discrete symmetries, we take

$$U_C^\dagger A(x)U_C \overset{\text{def}}{=} -A(x), \quad U_P^\dagger A(x)U_P \overset{\text{def}}{=} PA(\tilde{x}), \quad \Omega_T A(x)\Omega_T^{-1} \overset{\text{def}}{=} PA(-\tilde{x}) \quad (\S 9.9)$$

## For Fermionic Calculus Conventions, see §7.4

**Wick Rotation** from Minkowski space to Euclidean space changes the time/energy components of tensors and leaves their spatial components unchanged. In the following display, the equalities are the substitutions which implement the Wick rotation. The essential definitions are on the left, and their consequences are on the right.

$$k_0 \overset{\text{W}}{=} ik_{\text{E}}^4 \quad \text{and} \quad x_0 \overset{\text{W}}{=} -ix_{\text{E}}^4 \implies x{\cdot}k = x_{\text{E}}{\cdot}k_{\text{E}} \quad \text{and} \quad k^2 = -k_{\text{E}}^2;$$

$$\gamma^0 \overset{\text{W}}{=} -i\gamma_{\text{E}}^4 \implies \{\gamma_{\text{E}}^\alpha, \gamma_{\text{E}}^\beta\} = -2\delta^{\alpha\beta}, \quad \slashed{k} = \slashed{k}_{\text{E}}, \quad \text{and} \quad \slashed{D} = \slashed{D}_{\text{E}}; \quad (\S 13.3)$$

$$\gamma^5 \overset{\text{W}}{=} \gamma_{\text{E}}^5 = -\gamma_{\text{E}}^1\gamma_{\text{E}}^2\gamma_{\text{E}}^3\gamma_{\text{E}}^4, \quad \epsilon^{\mu\nu\rho\sigma} \overset{\text{W}}{=} i\epsilon_{\text{E}}^{\alpha\beta\gamma\delta} \quad \text{where } \epsilon^{1234} = 1.$$

These definitions also imply that $\gamma_{\text{E}}^{\mu\dagger} = -\gamma_{\text{E}}^\mu$, $\slashed{\partial}_{\text{E}}^2 = -\Delta$, and a Euclidean trace formula:

$$\text{tr}(\gamma_{\text{E}}^5\gamma_{\text{E}}^\alpha\gamma_{\text{E}}^\beta\gamma_{\text{E}}^\gamma\gamma_{\text{E}}^\delta) = -4\epsilon_{\text{E}}^{\alpha\beta\gamma\delta}. \quad (\S 13.3)$$

# A.4 Free Fields

## Free-Field Lagrangians

$$\mathcal{L}_{\text{Real Scalar}} = \partial_\mu\phi\,\partial^\mu\phi - \frac{\mu^2}{2}\phi^2 \qquad (\S 2.6)$$

$$\mathcal{L}_{\text{Complex Scalar}} = \partial_\mu\psi^\dagger\partial^\mu\psi - m\psi^\dagger\psi \qquad (\S 3.5)$$

$$\mathcal{L}_{\text{Weyl}} = \psi_L^\dagger[i\partial]\psi_L, \quad \text{where} \quad [\partial] = \sigma_\mu\partial^\mu \qquad (\S 7.5)$$

$$\mathcal{L}_{\text{Dirac}} = \bar{\psi}i\slashed{\partial}\psi - m\bar{\psi}\psi, \quad \text{where} \quad \slashed{\partial} = \gamma_\mu\partial^\mu \qquad (\S 8.4)$$

$$\mathcal{L}_{\text{Massive Vector}} = -\frac{1}{4}F_{\mu\nu}F^{\mu\nu} + \frac{\mu^2}{2}A_\mu A^\mu, \quad \text{where} \quad F_{\mu\nu} = \partial_\mu A_\nu - \partial_\nu A_\mu \qquad (\S 9.2)$$

$$\mathcal{L}_{\text{Gauge}} = -\frac{1}{4}G_{\mu\nu}^a G^{a\mu\nu}, \quad \text{where} \quad G_{\mu\nu}^a = \partial_\mu A_\nu^a - \partial_\nu A_\mu^a + gf^{abc}A_\mu^b A_\nu^c \qquad (\S 12.6)$$

(The gauge field, of course, is self-interacting and hence not actually free.)

**Free-Field Expansions**

$$\phi(x) = \int \frac{d^3\bar{k}}{(2\pi)^{3/2} \left(2\omega(\bar{k})\right)^{1/2}} \left(e^{ix\cdot k}a(\bar{k})^\dagger + e^{-ix\cdot k}a(\bar{k})\right)$$

$$= \int \frac{d^3\bar{k}}{(2\pi)^3 \left(2\omega(\bar{k})\right)} \left(e^{ix\cdot k}\alpha(k)^\dagger + e^{-ix\cdot k}\alpha(k)\right) \tag{§2.4}$$

$$\psi(x) = \int \frac{d^3\bar{k}}{(2\pi)^{3/2} \left(2\omega(\bar{k})\right)^{1/2}} \left(e^{ix\cdot k}c(\bar{k})^\dagger + e^{-ix\cdot k}b(\bar{k})\right) \tag{§3.5}$$

$$\left.\begin{aligned}
\psi_L(x) &= \int \frac{d^3\bar{k}}{(2\pi)^3 2\omega(\bar{k})} \left(e^{ix\cdot k}u_L(k)\beta_R(k)^\dagger + e^{-ix\cdot k}u_L(k)\alpha_L(k)\right) \\
\psi_R(x) &= \int \frac{d^3\bar{k}}{(2\pi)^3 2\omega(\bar{k})} \left(e^{ix\cdot k}u_R(k)\beta_L(k)^\dagger + e^{-ix\cdot k}u_R(k)\alpha_R(k)\right)
\end{aligned}\right\} \tag{§7.6}$$

$$\left.\begin{aligned}
\psi(x) &= \int \frac{d^3\bar{k}}{(2\pi)^3 2\omega(\bar{k})} \left(e^{ix\cdot k}v_r(k)\beta^r(k)^\dagger + e^{-ix\cdot k}u_r(k)\alpha^r(k)\right) \\
\bar{\psi}(x) &= \int \frac{d^3\bar{k}}{(2\pi)^3 2\omega(\bar{k})} \left(e^{ix\cdot k}u_r(k)\alpha^r(k)^\dagger + e^{-ix\cdot k}v_r(k)\beta^r(k)\right)
\end{aligned}\right\} \tag{§8.5}$$

$$A_\mu(x) = \int \frac{d^3\bar{k}}{(2\pi)^3 2\omega(\bar{k})} \left(e^{ix\cdot k}e^r(k)^*_\mu \alpha_r(k)^\dagger + e^{-ix\cdot k}e^r(k)_\mu \alpha_r(k)\right) \tag{§9.2}$$

**Weyl Fields and Particles**

|            | Creates                    | Destroys                   | Spinor      |           |
|------------|----------------------------|----------------------------|-------------|-----------|
| $\psi_L$   | $\bar{\nu}_R, \frac{1}{2}$  | $\nu_L, -\frac{1}{2}$       | $u_L$       |           |
| $\psi_L^*$ | $\nu_L, -\frac{1}{2}$       | $\bar{\nu}_R, \frac{1}{2}$  | $u_L^*$      | (§7.11)   |
| $\psi_R$   | $\bar{\nu}_L, -\frac{1}{2}$ | $\nu_R, \frac{1}{2}$        | $u_R$       |           |
| $\psi_R^*$ | $\nu_R, \frac{1}{2}$        | $\bar{\nu}_L, -\frac{1}{2}$ | $u_R^*$      |           |

**Weyl Polarization Spinors**

Normalization:  $u_L(p)^\dagger \sigma^\mu u_L(p) = 2p^\mu$   $u_R(p)^\dagger \sigma^\mu u_R(p) = 2(Pp)^\mu$

Completeness:  $u_L(p)u_L(p)^\dagger = [Pp]$   $u_R(p)u_R(p)^\dagger = [p]$   (§7.5)

where $P$ is the $4 \times 4$ parity matrix.

**Dirac Polarization Spinors**

Normalization:  $\bar{u}_r(p)u_s(p) = 2m\delta_{rs}$   $\bar{v}_r(p)v_s(p) = -2m\delta_{rs}$

Completeness:  $u^r(p)\bar{u}_r(p) = \not{p} + m$   $v^r(p)\bar{v}_r(p) = \not{p} - m$   (§8.3)

**Vector Field Polarization Vectors**

Normalization:    $e^r(k)^* \cdot e^s(k) = -\delta^{rs}$   and   $e^r(k) \cdot k = 0$

Completeness:    $\displaystyle\sum_{r=1}^{3} e^r(k)_\mu^* e^r(k)_\nu = -g_{\mu\nu} + \frac{k_\mu k_\nu}{k^2}$                      (§9.2)

**Canonical Commutation Relations**

$$\big[\phi(t,\bar{x}),\phi(t,\bar{y})\big] = \big[\Pi(t,\bar{x}),\Pi(t,\bar{y})\big] = 0$$
$$\big[\phi(t,\bar{x}),\Pi(t,\bar{y})\big] = i\delta^{(3)}(\bar{x} - \bar{y})$$
(§2.6)

$$\big[\psi(t,\bar{x}),\psi(t,\bar{y})\big] = \big[\psi(t,\bar{x}),\partial_0\psi(t,\bar{y})\big] = \big[\partial_0\psi(t,\bar{x}),\partial_0\psi(t,\bar{y})\big] = 0$$
$$\big[\psi(t,\bar{x})^\dagger,\psi(t,\bar{y})^\dagger\big] = \big[\psi(t,\bar{x})^\dagger,\partial_0\psi(t,\bar{y})^\dagger\big] = \big[\partial_0\psi(t,\bar{x})^\dagger,\partial_0\psi(t,\bar{y})^\dagger\big] = 0 \quad \text{(§3.5)}$$
$$\big[\psi(t,\bar{x}),\partial_0\psi(t,\bar{y})^\dagger\big] = \big[\psi(t,\bar{x})^\dagger,\partial_0\psi(t,\bar{y})\big] = i\delta^{(3)}(\bar{x} - \bar{y})$$

$$\big\{\psi_L(t,\bar{x}),\psi_L(t,\bar{y})\big\} = \big\{\psi_L(t,\bar{x})^\dagger,\psi_L(t,\bar{y})^\dagger\big\} = 0$$
$$\big\{\psi_L(t,\bar{x}),\psi_L(t,\bar{y})^\dagger\big\} = \mathbf{1}\delta^{(3)}(\bar{x} - \bar{y})$$
(§7.6)

$$\big\{\psi(t,\bar{x}),\psi(t,\bar{y})\big\} = \big\{\bar{\psi}(t,\bar{x}),\bar{\psi}(t,\bar{y})\big\} = 0$$
$$\big\{\bar{\psi}(t,\bar{x}),\psi(t,\bar{y})\big\} = \gamma_0\delta^{(3)}(\bar{x} - \bar{y})$$
(§8.4)

$$\big[A_i(t,\bar{x}),A_j(t,\bar{y})\big] = \big[F_{i0}(t,\bar{x}),F_{j0}(t,\bar{y})\big] = 0$$
$$\big[A_i(t,\bar{x}),F_{j0}(t,\bar{y})\big] = i\delta_{ij}\delta^{(3)}(\bar{x} - \bar{y})$$
(§9.2)

# A.5 Functional Integral Quantization

**Bosonic Hamiltonian Form**

$$Z(j) = \int (d\Pi)(d\phi)\,\exp\!\Big(i\int d^4x\,\Pi\partial_0\phi - \mathcal{H}(\Pi,\phi) + j\phi\Big) \qquad \text{(§11.7)}$$

**Bosonic Lagrangian Form**

$$Z(j) = N\int (d\phi)\,\exp\!\Big(i\int d^4x\,\mathcal{L}(\phi,\partial\phi) + j\phi\Big) \qquad \text{(§11.7)}$$

**Fermionic Form**

$$Z^0(\bar{\jmath},j) = \Big(\frac{\partial}{\partial\psi}\frac{\partial}{\partial\bar{\psi}}\Big)\,\exp\!\Big(i\int d^4x\,\mathcal{L}(\bar{\psi},\psi) + \bar{\jmath}\psi + \bar{\psi}j\Big) \qquad \text{(§11.12)}$$

**Evaluation of Bosonic Gaussian Functional Integrals**

$$\int (d\phi)\,P(\phi)e^{iQ(\phi)} = \frac{1}{\sqrt{\det(A)}}P\Big(-i\frac{\delta}{\delta j(x)}\Big)e^{iQ(\phi_{\rm cl})}$$

where   $Q(\phi) = -\tfrac{1}{2}(\phi,A\phi) + (j,\phi) + c$   and   $\phi_{\rm cl} = A^{-1}j,$      (§11.5)

which implies   $Q(\phi_{\rm cl}) = \dfrac{1}{2}(j,A^{-1}j) + c$

**Evaluation of Fermionic Gaussian Functional Integrals**

$$\left(\frac{\partial}{\partial\psi}\frac{\partial}{\partial\bar\psi}\right)e^{-i(\bar\psi,A\psi)} = \det(A) \qquad (\S11.11)$$

**Perturbative Expansion of Functional Integrals**

$$D\left(-i\frac{\delta}{\delta j(y)}\right)Z(j) \equiv Z\left(-i\frac{\delta}{\delta\phi(x)}\right)\left(D(\phi)\exp\left(i\int d^4z\,\phi(z)j(z)\right)\right)\Bigg|_{\phi=0} \qquad (\S11.8)$$

**Matrix of Propagators**

$$\text{Matrix of Propagators} = -i\hat A(-ik)^{-1} \qquad (\S11.9)$$

**Vertex Rules** are derived from application of

$$\mathbf{C} = \int d^4x\, e^{-ix\cdot k}\frac{\delta}{\delta\phi(x)} \qquad (\S11.8)$$

to the interactions in the action, with $k$ the incoming momentum on the $\phi$ line.

**Faddeev–Popov Quantization** for a gauge-fixing constraint $C(A) = 0$:

$$Z(J) = N\int (dA)\, e^{iS(A,J)}\delta\big(C(A)\big)\Delta_C(A),$$
$$\text{where}\quad \Delta_C(A) = \text{abs}\det\left(\frac{\delta C}{\delta\tilde g}\right) \qquad (\S12.3)$$

**Expressing the Faddeev–Popov Determinant in Terms of Ghosts**

$$\Delta_C(A) = \left(\frac{\partial}{\partial\eta}\frac{\partial}{\partial\bar\eta}\right)\exp\left(-i\int d^4x\,\bar\eta\left(\frac{\delta C}{\delta\tilde g}\right)\eta\right) \qquad (\S12.7)$$

**Elimination of** $\delta\big(C(A)\big)$ **in the case** $C(A) = \partial_\mu A^\mu - f$:

$$Z = N'\int (df)\,\exp\left(-\frac{i}{2\alpha}\int d^4x\,f^2(x)\right)Z \qquad (\S12.4)$$
$$= NN'\left(\frac{\partial}{\partial\psi}\frac{\partial}{\partial\bar\psi}\right)\int (df)(dA)\,\exp\left(-\frac{i}{2\alpha}\int d^4x\,f^2(x)\right)e^{iS(A,\bar\psi,\psi)}\delta(\partial_\mu A^\mu - f)$$
$$= NN'\left(\frac{\partial}{\partial\psi}\frac{\partial}{\partial\bar\psi}\right)\int (dA)\psi\,\exp\left(iS(A,\bar\psi,\psi) - \frac{i}{2\alpha}\int d^4x\,(\partial_\mu A^\mu)^2\right).$$

**The Faddeev–Popov Effective Lagrangian** is derived from the classical Lagrangian $\mathcal{L}$ by adding ghost and gauge-fixing terms:

$$\mathcal{L}_{\text{FP}} = \mathcal{L} + \mathcal{L}_{\text{ghost}} + \mathcal{L}_{\text{gf}} \qquad (\S12.7)$$

In the case $C(A) = \partial_\mu A^\mu - f$, this yields:

$$\mathcal{L}_{\text{eff}} = \mathcal{L} + \bar{\eta}^a \partial^\mu (\delta^{ac} \partial_\mu + g f^{abc} A_\mu^b) \eta^c - \frac{1}{2\alpha} (\partial_\mu A^{a\mu})^2 \qquad (\S12.7)$$

# A.6 Lagrangians

**QED**

$$\mathcal{L} = -\frac{1}{4} F_{\mu\nu} F^{\mu\nu} - \frac{1}{2\alpha} (\partial_\mu A^\mu)^2 + \bar{\psi}(i\slashed{\partial} - m)\psi - e\bar{\psi}\slashed{A}\psi \qquad (\S12.4)$$

**Scalar QED**

$$\mathcal{L} = -\frac{1}{4} F_{\mu\nu} F^{\mu\nu} - \frac{1}{2\alpha} (\partial_\mu A^\mu)^2$$
$$+ (\partial_\mu + ieA_\mu)\phi^\dagger (\partial^\mu - ieA^\mu)\phi - m^2 \phi^\dagger \phi - \frac{\lambda}{4}(\phi^\dagger \phi)^2 \qquad (\S9.3, \S12.4)$$

**Yang–Mills**

$$\mathcal{L}_{\text{eff}} = -\frac{1}{4} G_{\mu\nu}^a G^{a\mu\nu} + \bar{\eta}^a \partial^\mu (\delta^{ac} \partial_\mu + g f^{abc} A_\mu^b) \eta^c - \frac{1}{2\alpha} (\partial_\mu A^{a\mu})^2 \qquad (\S12.7)$$

# A.7 The Standard Model Lagrangian

The Lagrangian density for the Standard Model of the electroweak and strong interactions is

$$\mathcal{L} = -\frac{1}{4} G_{\mu\nu}^a G^{a\mu\nu} - \frac{1}{4} W_{\mu\nu}^\alpha W^{\alpha\mu\nu} - \frac{1}{4} B_{\mu\nu} B^{\mu\nu}$$
$$+ (D_\mu \phi)^\dagger (D^\mu \phi) + \bar{\psi}_L^j i \slashed{D} \psi_L^j + \bar{\psi}_R^j i \slashed{D} \psi_R^j$$
$$+ \bar{\chi}_L^j i \slashed{D} \chi_L^j + \bar{U}_R^j i \slashed{D} U_R^j + \bar{D}_R^j i \slashed{D} D_R^j$$
$$- f_{ij} \bar{\psi}_L^i \phi \psi_R^j - g_{ij} \bar{\chi}_L^i \phi D_R^j - h_{ij} \bar{\chi}_L^i \tilde{\phi} U_R^j - \text{h.c.}$$
$$- \lambda \left( \phi^\dagger \phi - \frac{v^2}{2} \right)^2, \qquad (\S15.1, \S15.9)$$

where the covariant derivative is given by

$$D_\mu = \partial_\mu - ig_s G_\mu^a \frac{\lambda^a}{2} - igW_\mu^\alpha T^\alpha - ig' B_\mu Y. \qquad (\S15.1, \S15.9)$$

The left-handed doublets and right-handed singlets (with color indices suppressed) are

$$\psi_L = \begin{pmatrix} \nu_e & \nu_\mu & \nu_\tau \\ e_L & \mu_L & \tau_L \end{pmatrix} \qquad \psi_R = (e_R \quad \mu_R \quad \tau_R)$$

$$\chi_L = \begin{pmatrix} u_L & c_L & t_L \\ d_L & s_L & b_L \end{pmatrix} \qquad \begin{matrix} U_R = (u_R \quad c_R \quad t_R) \\ D_R = (d_R \quad s_R \quad b_R) \end{matrix} \qquad (\S15.1, \S15.9)$$

$$\phi = \begin{pmatrix} \phi^+ \\ \phi^0 \end{pmatrix} \qquad \tilde{\phi} = \begin{pmatrix} \phi^0 \\ -\phi^+ \end{pmatrix}^*$$

and the associated representations of the gauge group $SU(3) \times SU(2) \times U(1)$ are

$$\left.\begin{array}{ll} \psi_L \quad (1,2)_{-1/2} \qquad \psi_R \quad (1,1)_{-1} \\[1mm] \\ \chi_L \quad (3,2)_{1/6} \qquad \begin{array}{l} U_R \quad (3,1)_{2/3} \\[1mm] D_R \quad (3,1)_{-1/3} \end{array} \\[1mm] \\ \phi \quad (1,2)_{1/2} \qquad \tilde{\phi} \quad (1,2)_{-1/2} \end{array}\right\} \qquad (\S15.1, \S15.9)$$

After symmetry breaking, the symmetry of the kinetic terms is used to simplify the mass matrices, leaving residual lepton-flavor and baryon-number symmetries according to the following schema:

$$\underbrace{U(3)_{\psi_L} \times U(3)_{\psi_R}}_{f_{ij}} \times \underbrace{U(3)_{\chi_L} \times U(3)_{U_R}}_{g_{ij}} \times \underbrace{U(3)_{D_R}}_{h_{ij}} \qquad (\S15.9, \S10)$$

$$U(1)_e \times U(1)_\mu \times U(1)_\tau \quad U(1)_u \times U(1)_c \times U(1)_t \quad \xrightarrow{\quad\mid\quad} U(1)_B$$

Diagonalization of the mass terms introduces mass-eigenstate down-type quarks written $(d, s, b)$ which differ from the original charge-eigenstate quarks now written $(d', s', b')$ by the KM matrix acting on the left-handed fields:

$$\begin{pmatrix} d' \\ s' \\ b' \end{pmatrix}_L = \begin{pmatrix} V_{ud} & V_{us} & V_{ub} \\ V_{cd} & V_{cs} & V_{cb} \\ V_{td} & V_{ts} & V_{tb} \end{pmatrix} \begin{pmatrix} d \\ s \\ b \end{pmatrix}_L. \qquad (\S15.10)$$

The fermions interact through vector-boson exchange. Each vector boson couples to the fermionic part of the Noether current. The effective four-Fermi interactions which approximate the exchange of heavy vector bosons at low energy are current-current interactions with $G_F$ for coupling constant:

$$\mathcal{L}' = -\frac{G_F}{\sqrt{2}}\left(4J_{+\mu}J_-^\mu + 4(J_3^\mu - \sin^2\theta_W J_{EM}^\mu)(J_{3\mu} - \sin^2\theta_W J_{EM\mu})\right). \qquad (\$15.7)$$

The currents here involve the charge-eigenstate quarks; when expressed in terms of the mass-eigenstate quarks, the KM matrix induces flavor-changing effects.

There are nineteen parameters in the Standard Model: (1) the electromagnetic and strong coupling constants, (2) the two Higgs-potential parameters, (3) the Weinberg angle, (4) the nine lepton and quark masses, (5) the four parameters in the KM matrix, and (6) the $\theta$ from the $SU(3)$ global anomaly (§15.13).

The weak coupling constants satisfy

$$g = \frac{e}{\sin\theta_W} \quad \text{and} \quad g' = \frac{e}{\cos\theta_W}, \qquad (\S15.3)$$

and the masses of the $W$ and $Z$ are given by

$$M_W = \frac{ev}{2\sin\theta_W} \quad \text{and} \quad \frac{M_W}{M_Z} = \cos\theta_W. \qquad (\S15.5)$$

# A.8 Feynman Rules

Note that for simplicity of presentation, the factors $\int d^4p/(2\pi)^4$ and $(2\pi)^4\delta^{(4)}(P_{\text{in}})$ in propagator and vertex rules respectively has been omitted everywhere. Note further when deriving a Feynman integral from a Feynman diagram, there are two overall factors to include:

1. A symmetry factor $1/s(D)$, where $s(D)$ is the number of symmetries of $D$ which leave the external lines fixed, and
2. A sign factor of $(-1)^n$, where $n$ is the number of Fermi loops in $D$.

## Feynman Rules for Scalar Fields

| | | | |
|---|---|---|---|
| Neutral scalar propagator: | $\dfrac{i}{p^2 - \mu^2 + i\epsilon}$ | ——— | (§4.9) |
| Charged scalar propagator: | $\dfrac{i}{p^2 - m^2 + i\epsilon}$ | | |
| $-g\psi^\dagger\psi\phi$ vertex: | $-ig$ | | |

## Feynman Rules for Weyl Fields

| | | | |
|---|---|---|---|
| Incoming $\nu_L$: | $u_L(p)$ or $u_L(p)^\top$ | | (§7.12) |
| Outgoing $\bar{\nu}_R$: | $u_L(p)$ or $u_L(p)^\top$ | | |
| Incoming $\bar{\nu}_R$: | $u_L(p)^*$ or $u_L(p)^\dagger$ | | (§7.13) |
| Outgoing $\nu_L$: | $u_L(p)^*$ or $u_L(p)^\dagger$ | | |
| $-\frac{1}{2}g\psi_L^\top\varepsilon\psi_L\phi$ vertex: | $-ig\varepsilon$ | | (§7.12) |
| $\frac{1}{2}g^*\psi_L^\dagger\varepsilon\psi_L^*\phi$ vertex: | $ig^*\varepsilon$ | | |

## Feynman Rules for Dirac Fields

| | | | |
|---|---|---|---|
| Incoming particle spin $r$: | $u^r(p)$ | | (§8.5) |
| Incoming anti-particle spin $r$: | $\bar{v}^r(p)$ | | |
| Outgoing particle spin $r$: | $\bar{u}^r(p)$ | | |
| Outgoing anti-particle spin $r$: | $v^r(p)$ | | |
| $-g\bar{\psi}_1\Gamma\psi_2\phi$ vertex: | $-ig\Gamma$ | | |

## Feynman Rules for Abelian Vector Fields

| | | | |
|---|---|---|---|
| Massive field: | $\left(-g_{\mu\nu} + \dfrac{k_\mu k_\nu}{\mu^2}\right)\dfrac{i}{k^2 - \mu^2 + i\epsilon}\dfrac{d^4k}{(2\pi)^4}$ | | (§9.3) |

Massless field:

$$\frac{i}{k^2 + i\epsilon}\left(-g_{\mu\nu} + (1-\alpha)\frac{k_\mu k_\nu}{k^2}\right) \qquad \text{〜〜}\qquad (\S12.4)$$

Incoming vector particle:

$$e_r^\mu(p) \qquad \text{●〜}$$

Outgoing vector particle:

$$e_r^\mu(p)^* \qquad \text{〜●}$$

$e^2 g^{\mu\nu} A_\mu \phi^\dagger A_\nu \phi$ vertex:

$$2ie^2 g^{\mu\nu}$$

$-e\bar{\psi}\!\!\not{A}\psi$ vertex:

$$-ie\gamma^\mu$$

$ieA_\mu(\phi^\dagger \partial_\mu \phi - \phi \partial_\mu \phi^\dagger)$ vertex:

$$(-e)(-ip_\mu - iq_\mu) \qquad \overleftarrow{q} \text{〜} \overleftarrow{p} \qquad (\S11.8)$$

## Feynman Rules for Gauge Fields

Gauge field:

$$\frac{i\delta_{ab}}{k^2 + i\epsilon}\left(-g_{\mu\nu} + (1-\alpha)\frac{k_\mu k_\nu}{k^2}\right) \qquad \text{〜〜}\qquad (\S12.7)$$

Ghost field:

$$-\frac{i\delta_{ab}}{k^2 + i\epsilon} \qquad \text{--◄--}$$

$-g(\partial_\mu \bar{\eta}^a)A^{b\mu}\eta^c$ vertex

$$gf^{abc}p_\mu \qquad a,\overleftarrow{p}\!\cdot\!\underset{b,\mu}{\overset{}{\text{〜}}}\!\cdot\, c$$

$-\frac{1}{2}gf^{abc}A^{a\mu}A^{b\nu}(\partial_\mu A_\nu^c - \partial_\nu A_\mu^c)$ vertex:

$$c,\nu,\overrightarrow{k_3}\,\text{〜〜}\,\overleftarrow{k_1},a,\lambda$$
$$b,\mu,k_2\uparrow$$

$$gf^{abc}\big((k_2 - k_3)_\lambda g_{\mu\nu} + (k_3 - k_1)_\mu g_{\nu\lambda} + (k_1 - k_2)_\nu g_{\lambda\mu}\big)$$

$-\frac{1}{4}g^2 f^{abe}f^{cde}A_\mu^a A_\nu^b A^{c\mu}A^{d\nu}$ vertex:

$$c,\rho\,\text{〜〜}\,a,\mu$$
$$d,\sigma\,\text{〜〜}\,b,\nu$$

$$-ig^2(A_{\mu\nu\rho\sigma}^{abcd} + A_{\mu\rho\sigma\nu}^{acdb} + A_{\mu\sigma\nu\rho}^{adbc}),$$

$$\text{where} \quad A_{\mu\nu\rho\sigma}^{abcd} = f^{abe}f^{cde}(g_{\mu\rho}g_{\nu\sigma} - g_{\mu\sigma}g_{\nu\rho})$$

# A.9 Feynman Integral Technique for Loop Diagrams

**1.**   A Feynman diagram will put a propagator in the integrand of the Feynman integral for each loop in the diagram. Feynman's trick generalizes to the many propagator case:

$$\prod_{r=1}^n \frac{1}{a_r + i\epsilon} = (n-1)! \int_0^\infty da_1 \cdots da_n \, \frac{\delta(1 - \sum_r \alpha_r)}{(\sum_r \alpha_r(a_r + i\epsilon))^n}. \qquad (\S18.1)$$

**2.**   Set up momentum variables $k_a$, $p_b$ and $l_c$ as follows:

$$\begin{aligned} k_a &- \text{ internal momenta,} && 1 \le a \le N; \\ p_b &- \text{ external momenta,} && 1 \le b \le E; \qquad (21.9.1) \\ l_c &- \text{ loop momenta,} && 1 \le c \le L. \end{aligned}$$

Then the $k$'s are linear functions of the $p$'s and $l$'s. Rotate to Euclidean space, so that all the momenta satisfy $q^2 = q_0^2 + \bar{q}^2$. Then, suppressing the integration over Feynman parameters, the Feynman integral is

$$I = \int d^4 l_1 \cdots l_L \left( \sum_a \alpha_a (k_a^2 + m_a^2) \right)^{-I}.$$

**3.**   Let $D = \sum \alpha(k_a^2 + m_a^2)$. Then $D$ must have a minimum at some point $l_c = \hat{l}_c$. Expand about the minimum: $l_c = \hat{l}_c + l_c^{(1)}$. Then

$$D = C(\alpha, p) + \sum_{r,s} A^{rs}(\alpha) l_r^{(1)} l_s^{(1)}.$$

**4.**   Make an orthogonal change of coordinates to diagonalize $A$:

$$D = C(\alpha, p) + \sum_r A^{rr}(\alpha) l_r^{(2)} l_r^{(2)}.$$

**5.**   Normalize the $l^{(2)}$ by $l_c^{(3)} = \sqrt{A^c}\, l^{(2)}$. Use the fact that the product of the eigenvalues is equal to the determinant to find

$$I = \left( \frac{1}{\det(A)} \right)^2 \int d^4 l_1 \cdots d^4 l_L \left( \sum_c l_c^{(3)} \cdot l_c^{(3)} + C(\alpha, p) \right)^{-I}.$$

Since $C$ does not depend on the integration parameters, this integral can be reduced using simple algorithms.

**6.**   The hard part will be doing the integration over the Feynman parameters.

## A.10  Dimensional Regularization

**Regularized values of Feynman Integrals**

$$\text{Reg}_D \int \frac{d^4 p}{(2\pi)^4} \frac{1}{(p^2 - a)^n} = \frac{i}{16\pi^2} \frac{\Gamma(n - 2 + \delta)}{\Gamma(n)} \left( -\frac{1}{a} \right)^{n-2} \left( \frac{4\pi\mu^2}{a} \right)^\delta$$

$$\text{Reg}_D \int \frac{d^4 p}{(2\pi)^4} \frac{p^\mu}{(p^2 - a)^n} = 0 \qquad\qquad\qquad (\S 18.7)$$

$$\text{Reg}_D \int \frac{d^4 p}{(2\pi)^4} \frac{p^2}{(p^2 - a)^n} = \frac{i(2 - \delta)}{16\pi^2} \frac{\Gamma(n - 3 + \delta)}{\Gamma(n)} \left( -\frac{1}{a} \right)^{n-3} \left( \frac{4\pi\mu^2}{a} \right)^\delta$$

$$\text{Reg}_D \int \frac{d^4 p}{(2\pi)^4} \frac{p^\mu p^\nu}{(p^2 - a)^n} = \frac{ig^{\mu\nu}}{32\pi^2} \frac{\Gamma(n - 3 + \delta)}{\Gamma(n)} \left( -\frac{1}{a} \right)^{n-3} \left( \frac{4\pi\mu^2}{a} \right)^\delta$$

where $D = 4 - 2\delta$. The powers of $\mu^\delta$ come from regularization of coupling factors.

## The Gamma Function

$$\Gamma(\delta) = \frac{1}{\delta} - \gamma + O(\delta)$$

$$\Gamma(\delta - n) = \frac{(-1)^n}{n!} \left( \frac{1}{\delta} + 1 + \frac{1}{2} + \cdots + \frac{1}{n} - \gamma \right) + O(\delta),$$

(§18.8)

where $\gamma$, Euler's constant, and $\gamma_n$ are given by

$$\gamma \simeq .5772 \quad \text{and} \quad \gamma_n \overset{\text{def}}{=} 1 + \frac{1}{2} + \cdots + \frac{1}{n} - \gamma$$

(§18.8)

A convenient convention defines $\gamma_0 = -\gamma$.

## Dirac Algebra in $D$ Dimensions

$$\{\gamma^\alpha, \gamma^\beta\} = 2g^{\alpha\beta} 1,$$

$$(\gamma^\alpha)^\dagger = \gamma_\alpha = \begin{cases} \gamma^\alpha, & \text{if } \alpha = 0; \\ -\gamma^\alpha, & \text{if } \alpha \geq 1. \end{cases}$$

(§18.9)

$$\gamma^\beta \gamma^{\alpha_1 \cdots \alpha_s} \gamma_\beta = 2 \sum_{r=1}^{s} (-1)^{r-1} \gamma^{\alpha_1 \cdots \widehat{\alpha_r} \cdots \alpha_s \alpha_r} + (-1)^s D \gamma^{\alpha_1 \cdots \alpha_s},$$

(§18.9)

where the hat indicates an index to omit and $\gamma^{\alpha_1 \cdots \alpha_s}$ is short for $\gamma^{\alpha_1} \ldots \gamma^{\alpha_s}$.

The usual trace operator is available, and with the standard convention that $\text{Tr}(1) = 4$, traces of products of $\gamma$ matrices are independent of the choice of matrices that represent the Dirac algebra. Thus:

$$\text{Tr}(\gamma^\alpha \gamma^\beta) = 4g^{\alpha\beta};$$

$$\text{Tr}(\gamma^\alpha \gamma^\beta \gamma^\eta \gamma^\zeta) = 4(g^{\alpha\beta} g^{\eta\zeta} - g^{\alpha\eta} g^{\beta\zeta} + g^{\alpha\zeta} g^{\beta\eta}).$$

(§18.9)

The trace of an odd number of $\gamma$ matrices is zero.

The definition of $\gamma_5$ is not modified:

$$\gamma^5 = i\gamma^0 \gamma^1 \gamma^2 \gamma^3;$$

$$\epsilon_{0123} = 1.$$

(§18.9)

A useful and fairly standard notation separates the range of indices into 0–3 and 4–$D$ using bars and hats respectively. In particular,

$$\gamma_\mu = \bar{\gamma}_\mu + \hat{\gamma}_\mu \quad \text{and} \quad g_{\mu\nu} = \bar{g}_{\mu\nu} + \hat{g}_{\mu\nu},$$

(§18.9)

where $\bar{\gamma}_m u$ is zero for $\mu > 3$, $\hat{\gamma}_\mu$ is zero for $\mu < 4$, and similarly for $g_{\mu\nu}$. (The use of the bar here conflicts with its earlier use as the Dirac adjoint operator, but in practice the context makes the intended meaning obvious.)

## A.11  Amplitudes, Decay Widths, and Cross Sections

**$S$ Matrix**

$$(2\pi)^4 \delta^{(4)}(P_f - P_i) i\mathcal{A} = \langle f | S - 1 | i \rangle \qquad (\S 4.9)$$

**Squaring Amplitudes**

$$|u_L(q)^\top \varepsilon u_L(p)|^2 = 2p{\cdot}q \qquad (\S 7.15)$$

$$i\mathcal{A}_{sr} = F\bar{u}_s(p')\slashed{q}u_r(p) \quad \text{implies}$$

$$A^2 = \frac{1}{2}|F|^2 \operatorname{tr}\big(\slashed{q}(\slashed{p}+m)\slashed{q}(\slashed{p}'+m)\big) \qquad (\S 8.5)$$

$$\sum_{r=1}^{3} |e^r(p)^*_\mu \mathcal{A}^\mu|^2 = -\mathcal{A}^*_\mu \mathcal{A}^\mu \qquad (\S 9.3)$$

**Differential Transition Probability per Unit Time**

$$\frac{\text{DTP}}{\text{Time}} = |\mathcal{A}|^2 \Big(\prod_{\text{in}} \frac{1}{2E_a}\Big)\mathcal{D}$$

$$\text{where} \quad \mathcal{D} = (2\pi)^4 \delta^{(4)}(P_{\text{in}} - P_{\text{out}}) \prod_{\text{out}} \frac{d^3\bar{p}}{(2\pi)^3 2E_b} \qquad (\S 5.1)$$

**Decay and Cross Section Formulae**

$$\Gamma(E_i) = \frac{1}{2E_i} \sum_{\substack{\text{decay} \\ \text{modes}}} \int_{\substack{\text{final} \\ \text{momenta}}} |\mathcal{A}_f|^2 \mathcal{D} \qquad (\S 5.4)$$

$$d\sigma = \frac{1}{4E_T p_{\text{in}}} |\mathcal{A}|^2 \mathcal{D} \qquad (\S 5.5)$$

$$\frac{d\sigma}{d\Omega} = \frac{1}{64\pi^2 E_T^2} \frac{p_{\text{out}}}{p_{\text{in}}} |\mathcal{A}|^2 \qquad (\S 5.5)$$

$$d\sigma = \frac{1}{256\pi^5} \frac{1}{4E_T p_{\text{in}}} |\mathcal{A}|^2 dE_1\, dE_2\, d\Omega_1\, d\phi_{12} \qquad (\S 5.5)$$

# References

The best way to find the latest work on any topic in physics is to search the internet. The references given here are therefore to older papers and outstanding textbooks that the author found useful. All the books referenced below include copious reference sections.

Itzykson & Zuber and Cheg & Li give a broad view of the physics of quantum field theory, Georgi provides a concise introduction to weak interaction theory, and Cummins & Bucksraum are particularly strong in presenting connections to experiment. Peskin & Schroeder is absolutely outstanding, surely the best physics text of its kind.

**Cheng, T.-P. and Li, L.-F.** (1984), *Gauge Theory of Elementary Particle Physics*, (Oxford University Press).

**Cummins, E.D. and Bucksraum, P.H.** (1983), *Weak Interactions of Leptons and Quarks*, (Cambridge University Press).

**Georgi, H.** (1984), *Weak Interactions and Modern Particle Physics*, (Benjamin / Cummings, Menlo Park).

**Itzykson, C. and Zuber, J.-B.** (1980), *Quantum Field Theory*, (McGraw–Hill, London).

**Peskin, M.E. and Schroeder, D.V.** (1995), *An Introduction to Quantum Field Theory*, (Addison Wesley, New York).

The background in quantum mechanics and special relativity for Chapter 1 can be found in many excellent texts. For rigged Hilbert spaces, the following monograph is handy:

**Böhm, A.** (1978), *The Rigged Hilbert Space and Quantum Mechanics*, (Springer–Verlag, Berlin).

With reference to technical remarks in Chapters 2 and 4, a useful introduction to axiomatic quantum field theory and its travails is presented in Bogoliubov et al. The Haag and Wightman Theorems provide particular points of difficulty, showing that (a) the axioms imply triviality of the theory, and (b) the evolution of states in the interaction picture cannot be unitary.

**Bogoliubov, N.N., Logunov, A.A. and Todorov, I.T.** (1975), *Introduction to Axiomatic Quantum Field Theory*, (Benjamin/Cummins, Reading).

**Wightman, A.S.** (1964), 'La théorie locale et la théorie quantique des champs', *Ann. Inst. Henri Poincaré.* **A1**, 401.

With reference to Chapter 11, a nice presentation of the relation between operator ordering and functional integral quantization is given by:

**Kashiwa, T., Sakoda, S. and Zenkin, S.V.** (1994), 'Ordering, symbols and finite-dimensional approximations of path integrals', *Progress of Theoretical Physics* **92** (3), 669–85.

With reference to Chapter 13 and the closing sections of Chapter 18, for the theory of anomalies we can start with Schwinger for an approach in operator theory, cite Adler and Bardeen for Feynman diagrams, move on to 't Hooft for dimensional regularization, add Fujikawa for functional integration, and Tsutsui for regularization of the functional derivative.

**Adler, S., Bardeen, W.A.** (1969), 'One loop calculation of anomaly is correct to all orders', *Phys. Rev.* **182**, 1517.

**Fujikawa, K.** (1979), 'Path-integral measure for gauge-invariant fermion theories', *Phys. Rev. Lett.* **42**, 1195.

**'t Hooft, G., and Veltman, M.** (1972), 'Regularization and Renormalization of Gauge Fields', *Nucl. Phys.* **B44**, 189–213.

**Schwinger, J.** (1951), 'On gauge invariance and vacuum polarization', *Phys. Rev.* **82**, 664.

**Tsutsui, J.** (1989), 'On the origin of anomalies in the path integral formalism', *Phys. Rev.* **D40**, 3543.

With reference to Chapter 18, Weinberg's theorem on the asymptotic behavior of Feynman integrals provides the basis for the Zimmermann forest formula and BPHZ renormalization theory. Breitlohner & Maison give the application to theories with dimensional regularization. Collins offers a detailed treatment of renormalization in field theory.

**Breitlohner, P., and Maison, D.** (1977), 'Dimensional renormalization and the action principle', *Comm. Math. Phys.* **52**, 11–38.

**Collins, J.** (1984), *Renormalization*, (Cambridge University Press).

**Weinberg, S.** (1960), 'High-energy behavior in quantum field theory', *Phys. Rev.* **118** (9), 939–849.

**Zimmermann, W.** (1969), 'Convergence of Bogoliubov's method of renormalization in momentum space', *Comm. Math. Phys.* **15**, 208.

With reference to Chapter 21, Polchinski started a very fruitful application of Wilson's exact non-perturbative renormalization group to generate proofs of perturbative renormalizability. Morris gives a helpful interpretation of the process. The paper by Ball & Thorne is a beautiful and careful analysis of the renormalizability of $\phi^4$-theory. Lowell Brown provides a wealth of details about $\phi^4$-theory.

**Ball, R.D. and Thorne, R.S.** (1994), 'Renormalizability of effective scalar field theory', *Annals of Physics* **236**, 117–204.

**Brown, L.S.** (1992), *Quantum Field Theory*, (Cambridge University Press).

**Morris, T.R.** (1994), 'The exact renormalization group and approximate solutions', *International Journal of Modern Physics* **9** (14), 2411–2449.

**Polchinski, J.** (1984), 'Renormalization and effective Lagrangians', *Nucl. Phys.* **B231**, 269–295.

With reference to Chapter 21, despite the disclaimers in the prologue, in his review article Georgi actually provides an excellent introduction to the principles and practice of effective field theory.

**Georgi, H.** (1993), 'Effective field theory', *Annu. Rev. Nucl. Part. Sci.* **43**, 209–252.

# Index